Advances in Bioenergy

Advances in Bioenergy

The Sustainability Challenge

Edited by

PETER D. LUND
Aalto University School of Science, Finland

JOHN BYRNE
University of Delaware, USA

GÖRAN BERNDES
Chalmers University of Technology, Sweden

IACOVOS A. VASALOS
Centre for Research & Technology Hellas, Greece

WILEY

This edition first published 2016

Registered office
John Wiley & Sons Ltd, The Atrium, Southern Gate, Chichester, West Sussex, PO19 8SQ, United Kingdom

For details of our global editorial offices, for customer services and for information about how to apply for permission to reuse the copyright material in this book please see our website at www.wiley.com.

A catalogue record for this book is available from the British Library.

ISBN: 9781118957875

Cover image: pandacube, Soft_Light, fotofjodor, jeancliclac/iStockphoto

Set in 9/11pt Times by Aptara Inc., New Delhi, India

Printed in Singapore by C.O.S. Printers Pte Ltd

1 2016

Contents

About the Editors

Peter D. Lund is Professor in Advanced Energy Systems at Aalto University (Helsinki) where he chairs the multidisciplinary Energy Science Initiative. He is also a Visiting Professor in Wuhan, China. His primary interest is in sustainable energy systems, innovations, and policies. Dr. Lund is active in senior roles with EU initiatives in energy: he chaired the Advisory Group Energy of E.C. 2002-06, he chairs the Energy Steering Panel of European Academies Science Advisory Council (EASAC) and is a steering committee member of Euro-CASE energy platform. He has served in advisory role in many energy programmes world wide, including IEA, EIT, Baltic and Russia cooperation. Dr. Lund is Editor-in-Chief for *Interdisciplinary Reviews: Energy and Environment* and European Editor for *International Journal of Energy Research* (Wiley). Peter Lund has given numerous invited talks and written over 500 research papers on energy.

John Byrne is Director of the Center for Energy and Environmental Policy (CEEP) and Distinguished Professor of Energy and Climate Policy at the University of Delaware. He has contributed to Working Group III of the United Nations-sponsored Intergovernmental Panel on Climate Change (IPCC) since 1992 and shares the 2007 Nobel Peace Prize with the Panel's authors and review editors. He holds several senior academic, advisory and policy advisor positions, among others in South Korea, China, India and the USA. His work in the US includes service as an advisor to the National Council for Science and the Environment (US) and the Interagency Working Group on Environmental Justice, coordinated by the US EPA. Dr Byrne created the first undergraduate and graduate degrees in the United States in the combined area of energy and environmental policy and a recent external review conducted by a committee of internationally respected scholars rated the CEEP graduate program among the three best in the field. He has published 17 books and over 150 research articles.

Göran Berndes is Associate Professor in Energy and Environment at Chalmers University of Technology in Sweden. Göran Berndes' research integrates land use and energy systems at scales ranging from local case studies to the global context. The research is in particular directed towards the production and use of biomass for food, energy and materials purposes. Important aspects include: (i) the effectiveness of different ways to produce and use biomass for energy, using long-term energy system modeling and Life Cycle/Well-to-Wheel analyses linked with land use change modeling and assessment of the associated flows of C; and (ii) the resource (e.g., land and water), environmental and socioeconomic implications of bioenergy strategies and also of the alternative strategy to use land for enhancement of biospheric carbon sequestration.

Iacovos A. Vasalos was a Professor at Aristotle University of Thessaloniki, Department of Chemical Engineering from 1979 to 2005. He is currently Emeritus Researcher at the Chemical Process and Energy Resources Institute (CPERI), Centre for Research and Technology Hellas (CERTH). He started his career in the Amoco Research Center in Naperville, Illinois, where he worked for 10 years from 1969 to 1979, as a research engineer, project manager, process specialist in catalytic cracking, research supervisor, and as a consultant until 1999. He is the author or coauthor of 132 scientific publications and 43 international patents many of which have been successfully applied in refineries. He was instrumental in founding and organizing CPERI and CERTH. He was also actively involved in a series of EU-funded projects in areas such as clean fossil fuels, syngas production, biofuels, environmental catalysis with emphasis on catalytic reaction engineering and modelling coupled with overall process simulation.

Preface

Bioenergy is one of mankind's most important natural energy sources. It dominated energy supply until the industrial revolution and still plays an important role in many countries. Today around 10 percent of the world's primary energy is derived from the use of biological materials. Its potential is far beyond the present use even when we restrict development to observe a "food first" principle and nature conservation objectives.

From a technology point of view, bioenergy is a versatile fuel as it can be converted to all final energy forms such as electricity, thermal energy or fuel. Contrary to many other renewable energy sources it can be stored for a long time. It is a local energy source which is available practically everywhere in different forms except for very harsh climatic conditions. In addition, when properly utilized, bioenergy is a highly sustainable energy source.

By nature biomass and bioenergy are strongly coupled to ecosystems and both are closely linked to biodiversity and development issues as well. Because of these linkages, bioenergy cannot be evaluated only from a technology or energy supply point-of-view, but requires a multidisciplinary approach to estimating its utilization. This book seeks to provide a multidisciplinary, "whole-picture" view while at the same time identifying advances in the different fields of bioenergy research and technology development.

Measures to reduce energy demand growth and the promotion of bioenergy and other renewable energy sources are nowadays cornerstones in climate and energy security agendas around the world. Support for bioenergy is also in many countries part of policy packages for promotion of rural development with intentions to improve energy access, increase employment, and stimulate positive development in agriculture and forestry. In this respect, bioenergy must be a part of any serious green energy economy agenda.

The use of biomass for cooking, space heating, and lighting in developing countries presently accounts for roughly 80 percent of global bioenergy use. However, recently a rapid increase in interest in other biomass uses for energy has emerged as countries contemplate steps to address concerns about energy insecurity and climate change. These so-called "modern" biomass uses for energy are so far mainly restricted to the burning of municipal organic waste, straw, wood and forest industry by-flows to provide heat and electricity, anaerobic digestion of organic waste to produce biogas, and the use of conventional agriculture crops such as cereals, oil seeds, and sugar crops to produce biofuels.

However, the technologies used for converting biomass to fuels, heat and electricity continue to develop and can be expected to change the way we produce and use bioenergy products. Especially, emerging options for converting lignocellulosic biomass into refined solid, liquid and gaseous fuels gives access to new feedstock resources. New production systems for lignocellulosic biomass offer a broadened resource base. Perennial grasses and trees grown in short rotations (both coppice and single-stem plantations) represent new feedstock supply options in agriculture. Similarly, technology development in planting, silvicultural treatments and biomass extraction support an increasing biomass harvest from forests. This development will not the least be important in the case of biofuels for transport, which hitherto have been produced based on either easily degradable organic waste or food/feed crops. While different types of organic waste can be relatively cheap feedstock sources, cultivated feedstocks cost more and the ones used today make up a substantial part of the total production cost of such biofuels.

In addition to technological development and new types of feedstock for biomass, there are several other factors that contribute to the utilization of biofuels such as renewable energy related standards mandating the use of certain types of biofuels as well as tax incentives and other measures associated with low carbon fuels policies, which also limit the carbon intensity of biofuels.

The society's "footprint" on Earth will inescapably expand in order to provide food, energy and materials for an increasing human population. Yet, society expects that emerging bioenergy systems should reduce impacts caused by the existing - primarily fossil - energy systems, and that policies are developed to address risks associated with bioenergy implementation. Much attention is being directed to the possible consequences of land-use change (LUC), referring to well-documented effects of forest conversion and cropland expansion into previously uncultivated areas, possibly resulting in biodiversity loss, increased greenhouse gas emissions and degradation of soils and water bodies. There are also concerns about risks for negative social and economic impacts, including land-use conflicts, human rights violations and food-security impacts.

The management of natural resources to provide needs for human society while recognizing environmental balance is the challenge facing society. As for other human activities, governance of bioenergy development is much about balancing trade-offs between partly incompatible environmental and socioeconomic objectives. There are currently several initiatives to develop sustainability certification systems. These may hedge against some of the undesired consequences of expanding bioenergy systems and promote a positive development where implemented effectively. Complementary to sustainability certification, there is a need to develop competitive business cases that are efficient along the entire bioenergy supply chain, from feedstock production to energy markets.

The policy challenge for those wishing to utilize the planet's bioenergy potential is complex. One clear principle is balance striking the proper balance between specific energy and food needs, and more broadly between socioeconomic development and environmental sustainability goals. This challenge is fraught with uncertainties: we cannot readily know if and to what extent the delicate web of biodiverse life would be disturbed under different utilization scenarios; and we do not know if social changes in diet and family formation or international efforts to significantly alleviate poverty and hunger will succeed in enabling greater use of bioenergy of certain kinds and on certain scales. Risks embedded in these uncertainties are large, leaving many researchers unsatisfied with our statistical ability to estimate them. When policy confronts "wicked problems" like these, it is common to counsel that we observe a precautionary principle in our actions and aspirations. This second principle also has a large role in shaping policy options. For example, "food first" and other scenario evaluation approaches exemplify this thinking. Several chapters in this book contribute to this understanding, indicating that research has come a long way recently in helping us to understand the bioenergy opportunity in a precautionary manner.

It is also important that policymakers consider the bioenergy opportunity in an integrated manner as part of a multiprong strategy. There have never been "perfect" or "silver bullet" solutions in the energy field and we must be careful not to examine bioenergy in a rarefied light which would undermine its consideration altogether simply because we want an outcome free of any risk. Precaution should lead us to minimize the magnitude of adverse impacts while also balancing the need for social change sufficient to meet our pressing challenges of energy security and climate change. Precaution, balance and integrated thinking are cross themes of this book and serve as guideposts for its contribution to discussions of policy design and implementation.

This book offers an authoritative overview of opportunities and challenges associated with bioenergy utilization. In addition to up-to-date and detailed information on key issues for biomass supply and conversion to energy, the book discusses conditions for the mobilization of sustainable bioenergy supply chains and outlines governance systems to support this mobilization.

The idea for this book came from the Publisher when John Wiley & Sons established a new journal, the *Wiley Interdisciplinary Reviews: Energy and Environment*, which publishes review-type articles by authoritative authors. This book is based on selected articles from this journal. The initial vision and effort of Tony Carwardine from Wiley was decisive for starting this book project. He is now retired, but he provided helpful inputs into the shaping of the book.

The Editors would also like to thank Peter Creaton, Dan Finch, Ella Mitchell, Peter Mitchell, Faith Pidduck, Kerry Powell, and Prachi Sinha Sahay from Wiley for their valuable help during the different phases of the book process.

Peter D. Lund
John Byrne
Göran Berndes
Iacovos A. Vasalos

PART I

PROMISING INNOVATION IN BIOMASS CONVERSION

1

Metabolic Engineering: Enabling Technology for Biofuels Production

Mitchell Tai and Gregory N. Stephanopoulos

Department of Chemical Engineering, Massachusetts Institute of Technology, Cambridge MA, USA

ENGINEERING THE FUTURE OF BIOFUELS

The past few years have introduced a flurry of interest over renewable energy sources. Biofuels have attracted attention as renewable alternatives to liquid transportation fuels. There are numerous potential advantages over fossil fuels: sustainable supply, diversification of energy sources, energy independence and security, rural development, and reduction in greenhouse emissions.[1] However, achieving adequate scale requires a tremendous effort in research and development beyond what has thus far been achieved. The field of metabolic engineering is well suited to develop the future technologies that will give us widespread, cost-effective, and sustainable transportation fuels.

Metabolic engineering is the improvement of cellular activities by manipulation of metabolic networks through the use of recombinant deoxyribonucleic acid technology.[2] Interdisciplinary advances in metabolic engineering have yielded powerful strategies and methods to understand and manipulate whole metabolic pathways with confidence.[3,4] To date, numerous efforts have successfully engineered and optimized metabolic networks to produce high-value targets for use in the pharmaceutical and fine chemicals industries.[5] However, attention is now being turned toward commodity-scale processes, which require both cost-efficiency and robustness.[1]

Currently, the most prevalent biofuels are ethanol produced from corn or sugarcane and biodiesel produced from vegetable oils. Under current production processes, however, neither biofuel is economically competitive or well integrable into existing petroleum-based technologies and infrastructure.[6] Two developmental challenges underpin these shortcomings: (1) the need for a better feedstock and (2) the need for a better fuel. However, these challenges also represent key opportunities to develop the next generation of biofuel technologies. A central element in these technologies will be the use of metabolic engineering to develop the biological platforms that produce these biofuels.

Engineering for Improved Feedstocks

For the past few years, production of ethanol from corn and biodiesel from vegetable oils has been increasing rapidly. Last year, the United States production capacity of corn ethanol exceeded 13 billion gallons per year (bgy), approaching 10% of the national gasoline demand.[7] Meanwhile, global biodiesel production is approaching 5.0 bgy, with a majority coming from Europe.[8]

However, production of these biofuels from plants like corn or rapeseed also competes for arable cropland needed for food. This adds undesirable price sensitivities between biofuels and food and has already shown adverse effects on food prices. Transforming forests or existing cropland can also sometimes have the effect of increasing greenhouse gas emissions, counteracting the carbon emissions benefit of biofuels.[9]

Advances in Bioenergy: The Sustainability Challenge, First Edition. Edited by Peter D. Lund, John Byrne, Göran Berndes and Iacovos A. Vasalos.

Originally published in: *WIREs Energy Environ* 2012, 1: 1–10 doi: 10.1002/wene.5

The primary cost for producing biofuels is the cost of the feedstock: 60% in the case of corn ethanol and 80% for soybean biodiesel.[10,11] Even with gains in process yield, current crop-based feedstocks will still limit the overall profitability of biofuels. Currently, upward of half of the production cost of these biofuels needs to be supported by government subsidies.[6]

The next generation of feedstocks will need to have lower land requirements and lower production cost, yet maintain high production capacity to bring biofuels closer to economic viability. Metabolic engineering allows us to bridge the feedstock gap by enabling the utilization of cheaper and more sustainable substrates by introducing catabolic pathways and optimizing metabolic networks for the conversion of feedstock to fuel. Indeed, yield optimization has been a critical aspect of virtually all biochemical engineering processes in recent history. Metabolic engineering of organisms toward this end only serves to continue this tradition, pushing yields beyond what is naturally observed. Furthermore, microbe-based biofuel production also reduces the cropland requirements compared to crop-based methods, decreasing competition with food production.

Engineering for Improved Fuels

Although new feedstocks are explored, a simultaneous search continues for the next generation of fuel types. Current biofuels have some persistent disadvantages that limit their incorporation into existing infrastructure.

Ethanol, although widely produced, has relatively poor fuel characteristics. Ethanol is hygroscopic, capable of absorbing water, which can lead to corrosion. The energy content is also low, containing only 70% of the energy per volume of gasoline. Also, as ethanol is produced by fermentation, the resulting beer is dilute, containing roughly 10% ethanol. Subsequent distillation to separate the ethanol is very energy intensive.[10]

Biodiesel is a better fuel, but also has some disadvantages. It is not well suited for use at low temperatures because of a high cloud point, and still often requires large quantities of petroleum-derived methanol as part of its production. It also has only 89% of the energy content of its analog, petrodiesel.[11]

Current biofuel characteristics limit their integration into existing infrastructure. Because of this, there is a high transition barrier to adoption of biofuels, and both ethanol and biodiesel are often blended only at low concentrations into conventional fuels.

Development of better fuels that have high energy density and can be integrated into existing pipelines and engines will be needed if biofuels are to be more widely adopted and have a reasonable hope to replace fossil fuels. Through metabolic engineering of production pathways, alternative products can be made that have characteristics closer to their petroleum equivalents, easing the barrier for adoption. These alternatives range from slight modifications to existing metabolites, to new pathways that create naturally unique compounds. These naturally rare products will also require extensive pathway engineering and optimization to achieve effective production capacities—one of the central strengths of metabolic engineering.

TOOLS OF METABOLIC ENGINEERING

To understand how metabolic engineering plays a role in biofuels development and how it takes an interdisciplinary approach to problem solving, it is important to first understand its main strategies and tools. The strategies of metabolic engineering can be compartmentalized into three steps: (1) understanding, (2) designing, and (3) engineering the metabolic network. Each of these steps uses tools and technologies adopted from a range of disciplines. An overview of the strategies of metabolic engineering can be found in Figure 1.1.

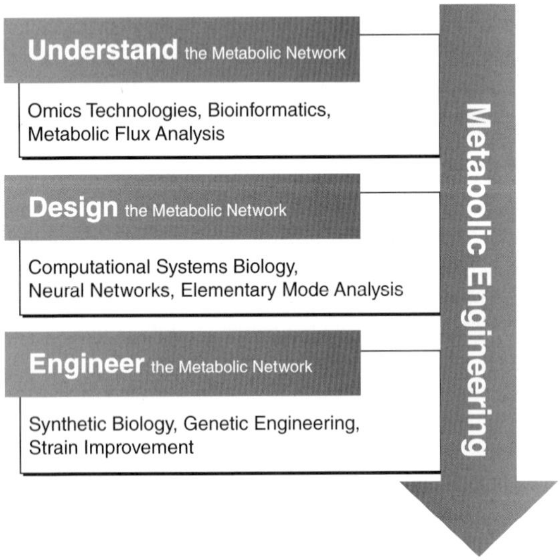

Figure 1.1 Strategies of metabolic engineering revolve around the understanding, design, and engineering of metabolic networks and pathways to produce desired molecular products from biological platforms. These strategies employ techniques and technologies from a range of disciplines, from omics technology to synthetic biology.

Understanding the Metabolic Network

The first step in metabolic engineering is to understand the complex network of enzymatic reactions that compose a cell's metabolism. In addition to the enzymology of participating enzymes, this requires information on the structure and behavior of the pathways that connect these enzymes. Knowledge of the pathway chemistry and stoichiometry allows us to calculate theoretical yields, which are often used as benchmarks for pathway engineering efficacy. Comprehensive systems-level data about these complex networks is acquired through omics technologies and bioinformatics. Omics technologies involve using genomic, transcriptomic, proteomic, or metabolomic data to quantify the system behavior of the cell along various functional axes (e.g., growth, tolerance, productivity).[12] Bioinformatics is the method of extracting biological meaning by identifying significant patterns, motifs, and connections within these large, complex data sets. These techniques enable us to develop a systems-level perspective on cellular activity and an understanding of important contributing networks.[4]

As an example, metabolic flux analysis derived from metabolomic data allows us to observe the flow of material through cellular metabolic pathways. Like a material balance, these fluxes describe the distribution of material throughout the cell's metabolic network and can help identify branch points and competing pathways relevant to our desired product. Fluxes also help to determine the degree of engagement of various enzymes in the pathway, allowing us to identify rate-limiting steps and control points.[3]

Because any biological manipulation will rarely ever produce only an isolated response, it is important to observe the system-level response of our engineering efforts. Using bioinformatics and omics technologies allows us to understand the interactions, connections, and responses between different parts of the system to predict and control the metabolic network.

Designing the Metabolic Network

Once we have sufficient understanding of the organism and its cellular activities, we are then able to develop and design specific strategies to obtain our desired product. Although we can introduce, remove, or otherwise modify pathways, identifying the most effective actions *a priori* can help save much time and effort. Modern methods to do so are found in the field of computational systems biology.

A main goal of computational systems biology is to reconstruct cellular networks *in silico*, which can model the behavior of the cell. Starting with a cellular model, one is able to simulate and characterize how possible pathway manipulations will affect the system overall. Evaluation of these changes can help identify the ideal genetic targets that will maximize our objectives.

One such method of evaluation is called elementary mode analysis, which uses a systems engineering approach to decompose metabolic networks into uniquely organized pathways that can be used to evaluate cellular phenotypes, metabolic network regulation, network robustness, and network fragility.[13] As an extension, neural networks can also be used to make sense of exceptionally difficult systems and to subsequently predict future behavior.[14]

Engineering the Metabolic Network

Once targets and pathways are identified, the next task is to implement these changes *in vivo*. This involves genetic manipulation of the host organism using molecular biology. The term synthetic biology describes the systematic approach to pathway manipulation through standardized biological components for the purpose of increasing their programmability and robustness.[15] Under this framework, genetic elements are modularized to simplify the process of genetic engineering. These elements can then be used to introduce new genes, knockout existing genes, or modify existing deoxyribonucleic acid sequences. Modules can be built up to produce whole pathways and can also be rearranged to optimize expression.

Numerous nonrational techniques are also available that extend the reach of traditional strain improvement: high throughput screening, directed evolution, gene shuffling, and combinatorial engineering. These techniques increase the efficiency of strain improvement by sampling a much larger phenotypic search space, opening up the possibility to select for changes in less-intuitive or distal targets. An interesting example is the use of transcriptional engineering to achieve phenotypic diversity.[16] This technique involves mutagenesis of transcription factors to produce global changes in the transcription and expression of genes in the cell, which would not be feasible through isolated point mutations.[17]

Finally, optimizing to maximize flux through the production pathway is often the most difficult engineering task. Tuning the expression of genes many times is necessary as intermediates and cofactors need to be balanced within the pathway. Furthermore, troubleshooting the pathway often involves alleviating any number of potential bottlenecks that may impede flux, such as competing pathways, lack of enzymatic driving force, cofactor imbalances, insufficient enzyme activity, unbalanced enzyme expression, transport issues, enzyme regulation, and toxicity.

Because of its interdisciplinary nature, metabolic engineering will be limited to the available tools and technologies of the fields from which it draws. However, as the

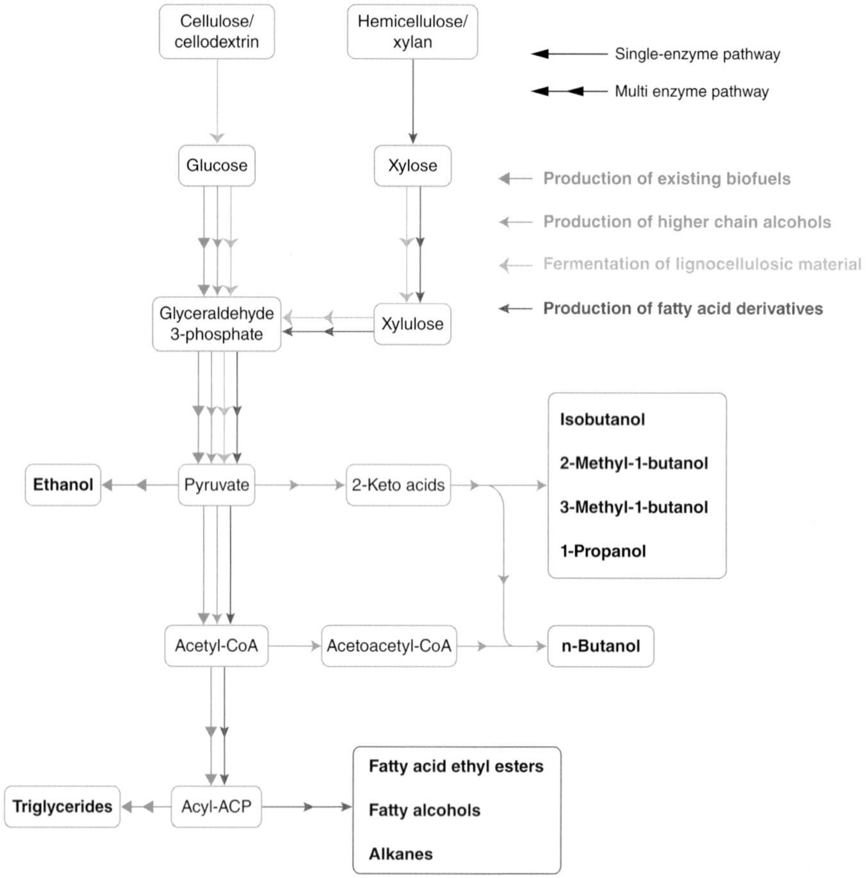

Figure 1.2 Metabolic network of biofuel production pathways and intermediates for the conversion of feedstocks to fuels (bold text): current biofuels (◄—), higher chain alcohols (◄—), lignocellulosic fermentation (◄—), and fatty acid derivatives (◄—). Engineering the desired biofuel pathway requires maximizing flux through the relevant nodes while minimizing metabolite flux to competing branches. This can involve tuning expression of intermediate reaction steps, deletion of competing pathways, or manipulation of distal enzymatic or regulatory targets.

state of the art progresses and development improves at the interface of these related disciplines, metabolic engineers will have increased power to evaluate, design, and manipulate cells to produce the desired products that will be in demand in the future. These abilities will most certainly be necessary for the development of the next generation of biofuels.

METABOLIC ENGINEERING ENABLES BIOFUELS DEVELOPMENT

In the following section, we will review various researches that attempt to address these biofuel challenges and how metabolic engineering is central to enabling these technologies. A schematic of relevant metabolic pathways for biofuels production can be found in Figure 1.2.

Production of Higher Chain Alcohols

Because ethanol has several undesirable fuel properties, higher chain alcohols have received attention as possible fuels. For example, *n*-butanol has 20% higher energy content than ethanol. In addition, it is more hydrophobic and thus less susceptible to inducing corrosion.[18] Traditionally, butanol is produced through refining petroleum or from acetone-butanol-ethanol fermentation. The ability to robustly produce higher chain alcohols represents a step toward biofuels with characteristics approaching that of gasoline.

Utilizing metabolic engineering, pathways for the production of higher chain alcohols have been introduced into organisms such as *Escherichia coli*.[18,19] The use of the model bacteria allows for the study and efficient development of novel pathways. Two pathways have recently emerged to successfully produce C4 or C5 alcohols: coenzyme A (CoA)-mediated and nonfermentative pathways.

The CoA-mediated pathway involves utilizing the native pathway of the butanol-producing organism, *Clostridium acetobutylicum*. Introduction of five *Clostridium* genes into *E. coli* is sufficient for production of *n*-butanol from acetyl-CoA. However, initial titers were at best about 1 g/L, which is much lower than the 10 g/L typically produced by native *n*-butanol producers.[19]

More recently, extensive engineering of the pathway has led to insights into maximizing the production through this pathway. Three major bottlenecks were discovered that could be alleviated by (1) balancing the expression of upstream and downstream enzymes, (2) balancing cofactor utilization and generation, and (3) engineering driving force to increase flux toward the product.[20,21] Indeed, these bottlenecks seem to be recurring obstacles in many efforts to engineer high production pathways. After addressing these bottlenecks, titers of 30 g/L could be achieved at about 70% theoretical maximum yield.[20]

The second pathway utilizes a creative nonfermentative approach, producing C4 and C5 branched chain alcohols from intermediates in the amino acid metabolic network.[22] The introduction of a 2-keto-acid decarboxylase and an alcohol dehydrogenase allows the conversion of a variety of 2-keto acid metabolites found in amino acid synthesis pathways into their analogous branched chain alcohols: 1-propanol, isobutanol, *n*-butanol, 2-methyl-1-butanol, 3-methyl-1-butanol, and 2-phenylethanol. Branched chain alcohols have higher octane numbers than their straight-chained counterparts, making them better fuels. This process has the advantage of avoiding CoA-mediated chemistry, while also leveraging the wealth of understanding from decades of research on metabolic engineering for amino acid production. As such, they were able to obtain isobutanol production of over 20 g/L at 86% of the theoretical maximum yield.[22] Optimization of cofactor imbalances produced strains that achieved 100% theoretical maximum yield, suggesting some of the same bottlenecks may exist in this alternative pathway.[23]

Fermentation of Lignocellulosic Material

In the search for improved feedstocks, the push toward cellulosic biofuels is a clear choice. Cellulosic biomass eliminates the need to compete with food crop production as an estimated 1.3+ billion dry tons per year of biomass is potentially available in the United States.[24] Two issues highlight how metabolic engineering can enable industrial utilization of this feedstock: xylose utilization and cellulose utilization.

Because hydrolysis of lignocellulosic biomass results in 20–30% carbohydrates in the form of xylose, utilization of pentose sugars is one of the first steps toward efficiently using cellulosic materials. *Saccharomyces cerevisiae*, the most productive of ethanologenic organisms, cannot ferment xylose; it lacks the ability to convert xylose into xylulose, although xylulose is metabolized within the pentose phosphate pathway (PPP). Transferring the xylose reductase (XR) and xylitol dehydrogenase (XDH) enzymes from *Scheffersomyces stipitis* (formerly *Pichia stipitis*) enables the growth of yeast on xylose and production of ethanol.[25]

However, growth and production are considerably slower than on glucose, and significant amounts of xylitol are often produced. Xylitol is the intermediate of the XR/XDH pathway, and most understand this to result from differences in cofactor specificity between reduced nicotinamide adenine dinucleotide phosphate (NADPH)-dependent XR and nicotinamide adenine dinucleotide (NAD)-dependent XDH.[26] The cofactor imbalance has been addressed in two different ways: use of (1) xylose isomerase pathway or (2) protein mutagenesis to switch cofactor specificity.

Additional factors to the limited productivity are the lack of dedicated pentose transporters, low PPP flux, and inability for the cell to identify xylose as a fermentable sugar.[25–27] However, more recently, progress has been made in these areas through additional metabolic engineering strategies: introducing heterologous xylose transporters, overexpressing PPP enzymes, engineering cofactor specificity, and evolutionary engineering (for a comprehensive review see Matsushika et al.).[26]

Cellulose on the contrary is a polysaccharide composed of $\beta(1\rightarrow4)$ linked glucose molecules. Enzymatic digestion is most commonly used to break these chains down into free glucose molecules. However, this process is somewhat inefficient, requiring a separate enzymatic unit operation. To improve the efficiency of this process, recently, cellodextrin transporters from *Neurospora crassa* were introduced into *S. cerevisiae* allowing for utilization of cellobiose, cellotriose, and cellotetraose.[28] Upon cellodextrin uptake, β-glucosidase breaks down cellodextrin into monomeric glucose, allowing for immediate catabolism. When coupled with enzymatic digestion, this process improves utilization efficiency and allows for the simultaneous saccharification and fermentation of cellulose.

Because of its preference for glucose, *S. cerevisiae* will natively repress the utilization of alternative substrates as well as down-regulate fermentative pathways after glucose is depleted.[27] This results in inefficient and delayed utilization of alternative substrates like xylose, a phenomenon known as diauxic shift. However, the cellodextrin pathway seems to avoid this by eschewing the generation of extracellular glucose. As a result, strains harboring both cellodextrin and xylose utilization pathways remarkably exhibit the ability to simultaneously coferment glucose and xylose.[29]

Metabolic engineering has shown how organisms can be adapted to utilize originally foreign substrates and feedstocks. It can also be used to circumvent regulatory bottlenecks to achieve more robust feedstock utilization. Although further regulatory and metabolic understanding is still necessary, current progress has shown great promise in enabling the production of biofuels from lignocellulosic materials.

Production of Fatty Acid Derivatives

The use of fatty acids and their derivatives as biofuels has numerous potential benefits. These long-chain hydrocarbon products have high energy density, are hydrophobic, and chemically resemble their fossil fuel analogs. Termed as drop-in biofuels, these products could be quickly integrated into existing infrastructure with little modification.

The most attractive pathway for production is the use of cytosolic thioesterase enzymes to produce free fatty acids, which can then be converted enzymatically or catalytically into a variety of derivative products: alkanes, alkenes, fatty acid esters, and alcohols. First discovered by Cho and Cronan,[30] expression of cytosolic thioesterase '*tesA* converts the lipid synthesis intermediate acyl carrier protein (ACP) into its fatty acid form. Disrupting the fatty acid synthesis pathway in this manner circumvents many of the typical fatty acid products that are often used as triggers for feedback regulation. This engineered pathway remarkably decouples production from growth by relieving feedback inhibition of a tightly regulated pathway, providing driving force for fatty acid synthesis at all stages of the cell cycle.[30] The fatty acids can then be modified into a variety of derivatives: free fatty acids, alkanes, and fatty acid ethyl esters (FAEE).[31–33] The process has even been extended into microalgae, which promises biofuels production from photosynthesis.[34] Additional engineering modifications to increase precursor availability included overexpression of acetyl-CoA carboxylase and coexpression of plant thioesterase.[33,35] The best reported efforts produced 4.5 g/L/day of fatty acids through these engineering strategies.[35]

In addition, this pathway has been used to advance the concept of a consolidated bioprocess (CBP). CBP entails having one organism, which can convert feedstock to product with little to no additional processing in between. This saves on operating costs and dramatically simplifies the production process.

Steen et al.[31] report engineering an organism that can produce FAEE from xylan. As xylan is one of the primary constituents of hemicellulose, this process demonstrates the use of a single organism to consolidate initial enzymatic degradation of xylan into xylose, the conversion of xylose into fatty acids, and the transesterification of those fatty acids into ethyl esters. The basis of this work was the discovery and utilization of a wax ester synthase that is able to perform the biodiesel transesterification reaction enzymatically.[36] Expression of xylanase and introduction of the ethanologenic pathway allowed for the *in vivo* transesterification of fatty acids to produce FAEE that can be used directly as biodiesel.

Robust control of the length and saturation of products will be a key achievement in the development of fatty acid derivatives production as a biofuel technology. Also, because fatty acid synthesis is tightly regulated and critical to many cellular functions, some creative approaches to metabolic engineering will be needed to increase yields and productivity without adversely affecting cell function. Nonetheless, a biofuel that chemically resembles petroleum fuels and can easily be integrated into existing technology has the potential to quickly become a relevant component of the biofuel industry.

CONCLUSION

A systems-level interdisciplinary approach is necessary for effective strategies to tackle today's global energy and environmental problems. The tools and strategies of metabolic engineering are well suited for addressing the persistent challenges facing a successful transition away from petroleum transportation fuels. As such, metabolic engineering will be instrumental in developing the next generation of cost-effective and robust transportation fuels, which will come from cheaper, more sustainable feedstocks and have better fuel characteristics.

It remains to be seen what processes and technologies will successfully establish sustainable alternatives to petroleum transportation fuels. However, because of obstacles in current feedstocks and fuels, it is necessary to continue research in technologies that can overcome existing limitations. Metabolic engineering is uniquely poised to develop and implement the next-generation biofuels using a systems-level approach from multiple disciplines.

Furthermore, metabolic engineering allows us to explore unconventional strategies that are naturally uncommon: nonfermentative production of branched higher alcohols, cellodextrin transport for xylose and glucose cofermentation, and derivatization of fatty acid products. Radical pathway manipulation is also on the horizon: engineering of nitrogen flux for the conversion of protein into biofuels could further improve overall yields while recycling reduced nitrogen[37]; design of artificial nonphotosynthetic carbon fixation pathways could open a new means of production processes without light (and thus land) as a limiting factor.[38] Metabolic engineering enables both promising and exciting opportunities for alternative energy, which have the potential for great societal ramifications. It will also be central in the long road ahead to develop these opportunities into robust, efficient industrial-scale technologies.

REFERENCES

1. Stephanopoulos G. Challenges in engineering microbes for biofuels production. *Science* 2007, 315:801–804.
2. Bailey JE. Toward a science of metabolic engineering. *Science* 1991, 252:1668–1675.
3. Stephanopoulos G. Metabolic fluxes and metabolic engineering. *Metab Eng* 1999, 1:1–11.
4. Tyo KE, Alper HS, Stephanopoulos GN. Expanding the metabolic engineering toolbox: more options to engineer cells. *Trends Biotechnol* 2007, 25:132–137.
5. Keasling JD. Manufacturing molecules through metabolic engineering. *Science* 2010, 330:1355–1358.
6. Hill J, Nelson E, Tilman D, Polasky S, Tiffany D. Environmental, economic, and energetic costs and benefits of biodiesel and ethanol biofuels. *Proc Natl Acad Sci* 2006, 103:11206–11210.
7. Ethanol Industry Overview. Renewable Fuels Association. Available at: http://www.ethanolrfa.org/pages/statistics. (Accessed June 14, 2011).
8. Production of biofuels in the world in 2009. Biofuels Platform. Available at: http://www.biofuels-platform.ch/en/infos/production.php. (Accessed June 14, 2011)
9. Fargione J, Hill J, Tilman D, Polasky S, Hawthorne P. Land clearing and the biofuel carbon debt. *Science* 2008, 319:1235–1238.
10. Pimentel D, Patzek TW. Ethanol production using corn, switchgrass, and wood; biodiesel production using soybean and sunflower. *Nat Resour Res* 2005, 14:65–76.
11. Balat M, Balat H. A critical review of bio-diesel as a vehicular fuel. *Energ Convers Manag* 2008, 49:2727–2741.
12. Joyce AR, Palsson BO. The model organism as a system: integrating 'omics' data sets. *Nat Rev Mol Cell Biol* 2006, 7:198–210.
13. Trinh C, Wlaschin A, Srienc F. Elementary mode analysis: a useful metabolic pathway analysis tool for characterizing cellular metabolism. *Appl Microbiol Biotechnol* 2009, 81:813–826.
14. Bosl W. Systems biology by the rules: hybrid intelligent systems for pathway modeling and discovery. *BMC Syst Biol* 2007, 1:13.
15. Leonard E, Nielsen D, Solomon K, Prather KJ. Engineering microbes with synthetic biology frameworks. *Trends Biotechnol* 2008, 26:674–681.
16. Alper H, Stephanopoulos G. Global transcription machinery engineering: A new approach for improving cellular phenotype. *Metab Eng* 2007, 9:258–267.
17. Alper H, Moxley J, Nevoigt E, Fink GR, Stephanopoulos G. Engineering yeast transcription machinery for improved ethanol tolerance and production. *Science* 2006, 314:1565–1568.
18. Atsumi S, Cann AF, Connor MR, Shen CR, Smith KM, Brynildsen MP, Chou KJY, Hanai T, Liao JC. Metabolic engineering of *Escherichia coli* for 1-butanol production. *Metab Eng* 2008, 10:305–311.
19. Inui M, Suda M, Kimura S, Yasuda K, Suzuki H, Toda H, Yamamoto S, Okino S, Suzuki N, Yukawa H. Expression of *Clostridium acetobutylicum* butanol synthetic genes in *Escherichia coli*. *Appl Microbiol Biotechnol* 2008, 77:1305–1316.
20. Shen CR, Lan EI, Dekishima Y, Baez A, Cho KM, Liao JC. Driving forces enable high-titer anaerobic 1-butanol synthesis in *Escherichia coli*. *Appl Environ Microbiol* 2011, 77:2905–2915.
21. Bond-Watts BB, Bellerose RJ, Chang MCY. Enzyme mechanism as a kinetic control element for designing synthetic biofuel pathways. *Nat Chem Biol* 2011, 7:222–227.
22. Atsumi S, Hanai T, Liao JC. Non-fermentative pathways for synthesis of branched-chain higher alcohols as biofuels. *Nature* 2008, 451:86–89.
23. Bastian S, Liu X, Meyerowitz JT, Snow CD, Chen MMY, Arnold FH. Engineered ketol-acid reductoisomerase and alcohol dehydrogenase enable anaerobic 2-methylpropan-1-ol production at theoretical yield in *Escherichia coli*. *Metab Eng* 2011, 13:345–352.
24. Perlack RD, Wright LL, Turhollow AF, Graham RL, Stokes BJ, Erbach DC. Biomass as feedstock for a bioenergy and bioproducts industry: the technical feasibility of a billion-ton annual supply. 2005.
25. Jeffries TW. Engineering yeasts for xylose metabolism. *Curr Opin Biotechnol* 2006, 17:320–326.
26. Matsushika A, Inoue H, Kodaki T, Sawayama S. Ethanol production from xylose in engineered *Saccharomyces cerevisiae* strains: current state and perspectives. *Appl Microbiol Biotechnol* 2009, 84:37–53.
27. Jin Y-S, Laplaza JM, Jeffries TW. *Saccharomyces cerevisiae* engineered for xylose metabolism exhibits a respiratory response. *Appl Environ Microbiol* 2004, 70:6816–6825.
28. Galazka JM, Tian C, Beeson WT, Martinez B, Glass NL, Cate JHD. Cellodextrin transport in yeast for improved biofuel production. *Science* 2010, 330:84–86.
29. Ha S-J, Galazka JM, Rin Kim S, Choi J-H, Yang X, Seo J-H, Louise Glass N, Cate JHD, Jin Y-S. Engineered *Saccharomyces cerevisiae* capable of simultaneous cellobiose and xylose fermentation. *Proc Natl Acad Sci* 2011, 108:504–509.
30. Cho H, Cronan JE. Defective export of a periplasmic enzyme disrupts regulation of fatty acid synthesis. *J Biol Chem* 1995, 270:4216–4219.

31. Steen EJ, Kang Y, Bokinsky G, Hu Z, Schirmer A, McClure A, del Cardayre SB, Keasling JD. Microbial production of fatty-acid-derived fuels and chemicals from plant biomass. *Nature* 2010, 463:559–562.
32. Lu X, Vora H, Khosla C. Overproduction of free fatty acids in *E. coli*: Implications for biodiesel production. *Metab Eng* 2008, 10:333–339.
33. Lennen RM, Braden DJ, West RM, Dumesic JA, Pfleger BF. A process for microbial hydrocarbon synthesis: overproduction of fatty acids in *Escherichia coli* and catalytic conversion to alkanes. *Biotechnol Bioeng* 2010, 106:193–202.
34. Liu X, Sheng J, Curtiss III R. Fatty acid production in genetically modified cyanobacteria. *Proc Natl Acad Sci* 2011, 108:6899–6904.
35. Liu T, Vora H, Khosla C. Quantitative analysis and engineering of fatty acid biosynthesis in *E. coli*. *Metab Eng* 2010, 12:378–386.
36. Kalscheuer R, Stolting T, Steinbuchel A. Microdiesel: *Escherichia coli* engineered for fuel production. *Microbiology* 2006, 152:2529–2536.
37. Huo Y-X, Cho KM, Rivera JGL, Monte E, Shen CR, Yan Y, Liao JC. Conversion of proteins into biofuels by engineering nitrogen flux. *Nat Biotechnol* 2011, 29: 346–351.
38. Bar-Even A, Noor E, Lewis NE, Milo R. Design and analysis of synthetic carbon fixation pathways. *Proc Natl Acad Sci* 2010, 107:8889–8894.

2

Hydrolysis and Fermentation for Cellulosic Ethanol Production

Charilaos Xiros,[1]* Evangelos Topakas[2]* and Paul Christakopoulos[3]*

[1]Industrial Biotechnology, Department of Chemical and Biological Engineering, Chalmers University of Technology, Gothenburg, Sweden
[2]BIOtechMASS Unit, School of Chemical Engineering, National Technical University of Athens, Zografou Campus, Athens, Greece
[3]Biochemical and Chemical Process Engineering, Division of Sustainable Process Engineering, Department of Civil, Environmental and Natural Resources Engineering, Luleå University of Technology, SE-97187, Luleå, Sweden

INTRODUCTION

The depletion of fossil fuels together with the global economic and environmental consequences encourage the search for renewable energy sources, such as biofuels. One of the most promising feedstocks for biofuels production is plant biomass that contains large amounts of sugar polymers, such as cellulose and hemicelluloses.[1] When subjected to enzymatic hydrolysis, these polysaccharides are transformed into glucose and other fermentable pentoses, which might further be converted to liquid fuels such as bioethanol. The last few years, the processes involved in the conversion of cellulose and hemicellulose have been closely related to the word 'recalcitrance', to emphasize obstacles that can impede the conversion.[1] Recalcitrance is linked to each step in biomass conversion, which can be costly, driving the production costs to exceed those of the transportation fuel competitors derived from fossil fuels or those derived from starch, sucrose, and vegetable oils (e.g., first-generation bioethanol and biodiesel). The strong glycosidic bonds of cellulose[2] together with its associated crystal structure (Figure 2.1) prescribe application of either harsh physicochemical conditions or use of a consortium of different cellulose acting enzymes. The applied harsh conditions and cellulolytic enzymes add considerable expense to the construction of saccharification reactors and downstream processing. However, the protein production efficiency of cellulases has been increased more than 10-fold nowadays resulting in a serious decrease in their price, rendering enzymatic saccharification more economical than physicochemical methods.[3]

At present, recombinant strains of *Saccharomyces cerevisiae*, and in a lesser extent of *Escherichia coli* and *Zymomonas mobilis* strains, are considered as the most successful microorganisms for biofuel production in industrial scale. Bottlenecks and obstacles, such as the narrow range of fermentable sugars for some microorganisms, the imbalanced anaerobic metabolism for some others, or the low uptake rates for some sugars, are challenges which have to be nearly solved or overpassed. Although several improvements are yet to be done, significant progress has been done on all these issues. On the other hand, consolidated bioconversion or, in other words, direct conversion of cellulosic materials into advanced biofuels is, up to now, partially successful. Cell-surface display engineering enabled the expression of cellulolytic activities and

*These authors contributed equally to this work

Advances in Bioenergy: The Sustainability Challenge, First Edition. Edited by Peter D. Lund, John Byrne, Göran Berndes and Iacovos A. Vasalos.

Originally published in: *WIREs Energy Environ* 2013, 2: 11–32 doi: 10.1002/wene.49

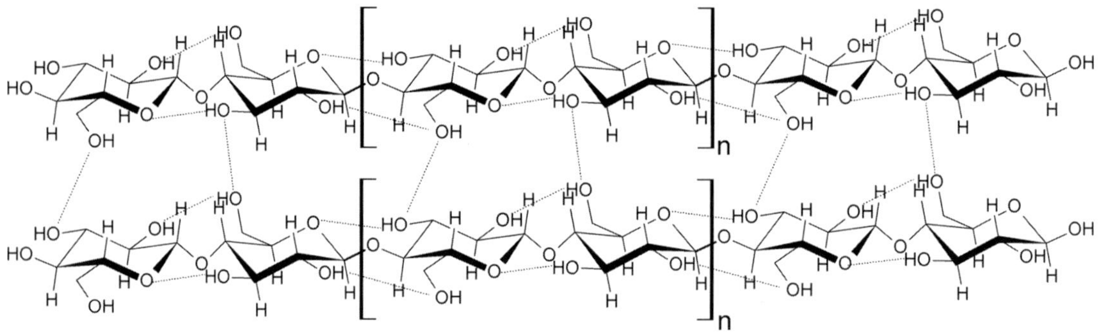

Figure 2.1 Cellulose structure is formed by β-(1,4)-linked D-glucose units, where adjacent D-glucoses are flipped making cellobiose the fundamental repeating unit. The inter- and intramolecular hydrogen bonds (shown as dots) and van der Waals interactions form recalcitrance microfiber structures.

minicellulosomes on *S. cerevisiae* cell surfaces.[4] The present review is aimed at summarizing both the hydrolysis technologies and bioconversion processes employed for cellulosic ethanol production.

HYDROLYSIS TECHNOLOGY

Hydrolysis is one of the main technologies which are in use for converting cellulosic feedstocks into ethanol. In hydrolysis, the polysaccharides (cellulose and hemicellulose) present in a feedstock are broken down to free sugar molecules (glucose, mannose, galactose, xylose, and arabinose). These free sugar molecules are then fermented to produce ethanol. As lignin cannot be used for ethanol production, it is removed during the conversion process, and is generally utilized to meet electricity or heat requirement of an ethanol mill.

In order to ensure an efficient enzymatic hydrolysis of cellulose, pretreatment (also called first-stage hydrolysis) is needed to break down the shield formed by lignin and hemicellulose, resulting in the disruption of the crystalline structure and the reduction of the degree of polymerization of cellulose. The mixture obtained at the pretreatment stage is separated into liquid (hemicelluloses hydrolysate consisted by xylose, mannose, arabinose, galactose, and small amount of hydrolyzed cellulose in the form of glucose) and solid (lignin + unhydrolyzed cellulose). Solids are sent for another round of enzymatic hydrolysis (also called second-stage hydrolysis).

Pretreatment Methods (First-Stage Hydrolysis)

Physical pretreatment, such as hammer and ball milling,[5] involve the breakdown of biomass feedstock into smaller particles to increase the surface area of cellulose and have been shown to improve hydrolysis yields by rendering them more accessible to attack by cellulases.

Thermal pretreatment, such as steam pretreatment/steam explosion, liquid hot water, ammonia fiber explosion, and CO_2 explosion, involve heating of feedstock to about 150°C–180°C to break down hemicellulose and lignin.[6] At higher temperatures (above 250°C), phenolic compounds are formed which later retard the fermentation process, so care is taken not to pretreat the feedstock in severe thermal conditions.

Acid pretreatment, such as dilute acid pretreatment, is performed by soaking the material in dilute acid solution (0.5–1.5%) and then heating to temperatures between 140°C and 200°C for a certain time (from several minutes up to an hour). Sometimes, concentrated sulfuric acid is also utilized for feedstock pretreatment. The hemicellulose is hydrolyzed and the main part is usually obtained as monomer sugars.[7]

Alkaline pretreatment uses bases such as calcium hydroxide (lime), potassium hydroxide, aqueous ammonia, ammonia hydroxide, and sodium hydroxide in combination with hydrogen peroxide or others and breaks the bonds between lignin and carbohydrates making the latter more accessible to enzymatic attack.[8]

In the *organosolv process*, an organic or aqueous–organic solvent mixture with addition of an inorganic acid catalyst (H_2SO_4 or HCl) is used to break the internal lignin and hemicellulose bonds. Solvents that are used are typically methanol, ethanol, acetone, ethylene glycol, triethylene glycol, and phenol.[9]

Biological pretreatments typically utilize wood degrading fungi (soft, brown, and white rot) to modify the chemical composition of the lignocellulosic feedstock. Generally, soft and brown-rot fungi primarily degrade the hemicellulose, while imparting minor modifications to lignin. White-rot fungi can more actively attack the lignin component.[10]

Enzymatic Hydrolysis (Second-Stage Hydrolysis)

Enzymatic hydrolysis takes place at mild conditions of temperature and pressure. As a result, glucose yields are high, chances for the production of fermentation inhibiting compounds are less, equipment requirements are not significant, and there is a reduction in the total environmental impact of the whole process. In the following sections, a detailed description of the function and structure of the cellulases used in the second-stage hydrolysis is provided, prior to the fermentation toward ethanol production.

ENZYMATIC BREAKDOWN OF CELLULOSE

Cellulose

Cellulose, that is the most abundant biopolymer on earth, is formed by β-(1,4)-linked D-glucose units, where adjacent D-glucoses are flipped making cellobiose the fundamental repeating unit (Figure 2.1).[11] Polymers of cellulose form robust microfiber structures, which are stabilized by inter- and intramolecular hydrogen bonds and van der Waals interactions between glucose residues contributing to its recalcitrance.[12] The bond network results in a mostly uniform arrangement of fibers forming crystalline cellulose that lacks enzyme-accessible surface morphologies, further enhancing resistance to hydrolysis.[13] Crystallinity varies widely among plant cell wall from approximately 40–50% in plant cellulose to 65–80% in bacterial and algal cellulose.[14] Glucan chains occur in hexagonal arrays of 36 (3 nm × 5 nm width) with exceptionally high degrees of polymerization (DPs).[15] Although cellulose is mainly present as crystalline fibers that are highly resistant to hydrolysis, its content in biomass is typically larger compared to hemicellulose and, as a result, cellulases are the key enzymes for bioethanol production.[3]

Cellulose Degrading Enzymes

Cellulases have been widely applied in industry in different sectors, such as in the textile industry for cotton softening and denim finishing, in the detergent market for color care, cleaning, and antideposition, in the food industry for mashing and in the pulp and paper industries for deinking, improvement, and fiber modification.[16] The cellulase market is expected to expand dramatically for the hydrolysis of pretreated cellulosic materials into sugars, which can be fermented to produce commodities such as bioethanol and other bioproducts on a large scale. However, the high cost of fuel ethanol produced from lignocellulosic biomass, is heavily depended on the cost of cellulases that needs to be significantly reduced for their commercial use in biorefineries.[17]

Box 2.1 Functional and Structural Diversity of Cellulases

Enzymes that modify complex carbohydrates, such as cellulases, together with their accessory noncatalytic CBMs, have been grouped into sequence-based families on the continuously updated CAZy database (http://www.cazy.org/). At the moment, CAZy database describes around 130 GH families, whereas around 16 of them are populated by many different cellulolytic enzymes. From a structural point of view, cellulases belong to at least eight unrelated protein folds that are diffentiated into even more GH families, underpinning that these enzymes have evolved convergently from diverse ancestors.

Classification and Structure of Cellulases

Enzymes that modify complex carbohydrates, such as cellulases, together with their accessory noncatalytic carbohydrate binding modules (CBMs), have been grouped into sequence-based families on the continuously updated Carbohydrate-Active EnZymes (CAZy) database (http://www.cazy.org/).[18] Cellulolytic enzymes are classified into different glycoside hydrolase (GH) families (Box 2.1). Starting with β-1,4-endoglucanases (EC 3.2.1.4), these enzymes catalyze the endohydrolysis of (1,4)-β-D-glucosidic bonds in cellulose, which are members of GH families 5–9, 12, 44, 45, 48, 51, 74, and 124. Cellobiohydrolases (EC 3.2.1.91) that are known to catalyze the release of cellobiose from either the non-reducing end or the reducing end of cellulose, belong to GH families 5, 6, 7, 9, and 48. Finally, β-glucosidases (EC 3.2.1.21) act on soluble cello-oligosaccharides, including cellobiose, toward the release of D-glucose, belong to GH families 1, 3, 5, 9, 30, and 116. Recently, members of GH family 61 that were known in literature for many years as 'weak endoglucanases', have been shown to lack GH activity by exhibiting monoxygenase profile,[19–21] displaying little or no activity on polymeric cellulose.

All cellulolytic enzymes known to date, except GH61s, are GHs that utilize the same catalytic mechanism of acid–base catalysis, with inversion or retention of D-glucose anomeric configuration. The cellulase active sites are categorized in two common types. GHs with open (groove, cleft) active sites, which typically exhibit endocellulase activity (endo-hydrolases) that bind almost anywhere along the length of the cellulose polysaccharide and hydrolyze the β-1,4 glycosidic bonds between D-glucose units, whereas there are also tunnel-like active sites that exhibit exocellulase activity (exo-hydrolases) that bind at the ends of the cellulose polysaccharide producing specific oligosaccharide products[22] (Table 2.1). Exocellulolytic enzymes are

Table 2.1 Structural Conservation and Diversity of Cellulases

Topology	Glycoside Hydrolase Family	Activity	Structure	Reference
$(\beta/\alpha)_8$ barrel	**5, 44, 51** 5 **1**, 5, **30**	*endo*-β-1,4-Glucanase Cellobiohydrolase β-Glucosidase	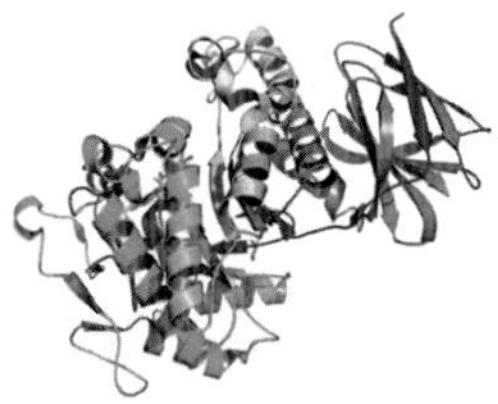 [PDB: 2E0P]	Kitago et al.[24]
Atypical $(\beta/\alpha)_8$ barrel	6	*endo*-β-1,4-Glucanase Cellobiohydrolase	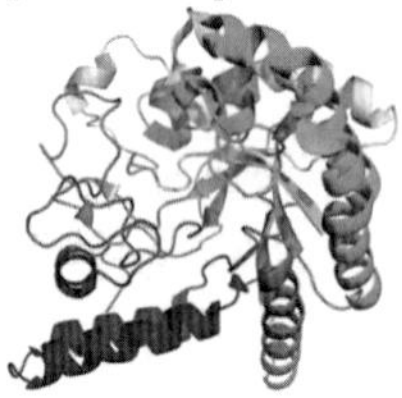 [PDB:1DYS]	Davies et al.[25]
β-jelly roll	**7, 12** **7**	*endo*-β-1,4-Glucanase Cellobiohydrolase	 [PDB:1H8V]	Sandgren et al.[26]
$(\alpha/\alpha)_6$ Thoroid	8, 9, 48 9, 48 9	*endo*-β-1,4-Glucanase Cellobiohydrolase β-Glucosidase	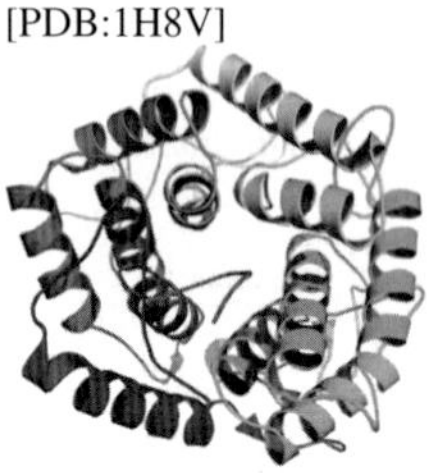 [PDB:1CEM]	Alzari et al.[27]
α_8 Super- helical	124	*endo*-β-1,4-Glucanase	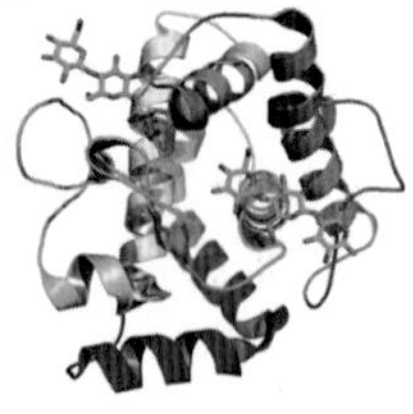 [PDB:2XQO]	Brás et al.[28]
β_6-Barrel	45	*endo*-β-1,4-Glucanase	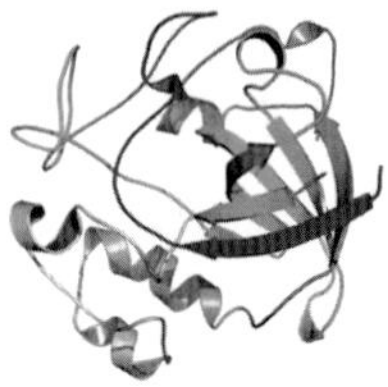 [PDB:2ENG]	Davies et al.[29]

(*Continued*)

Table 2.1 Continued

Topology	Glycoside Hydrolase Family	Activity	Structure	Reference
β-Sandwich, like Ig	61	Polysaccharide monoxygenases (former weak *endo*-β-1,4-glucanases)	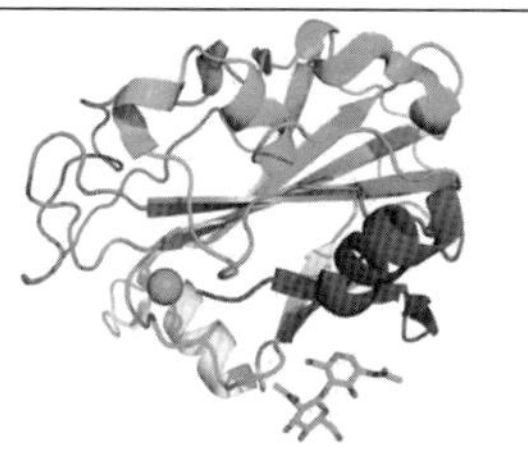 [PDB:2VTC]	Karkehabadi et al.[30]
Sevenfold β-propeller	74	*endo*-β-1,4-Glucanase	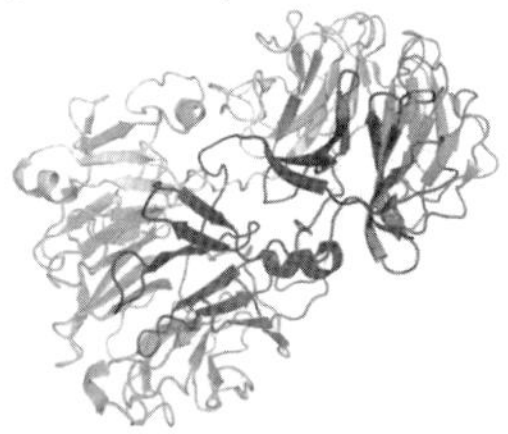 [PDB:2CN2]	Martinez-Fleites et al.[31]
Multiple structural domains: 1. $(\beta/\alpha)_8$-Barrel 2. $(\alpha/\beta)_6$ -Sandwich 3. FnIII fold	**3**	β-Glucosidase	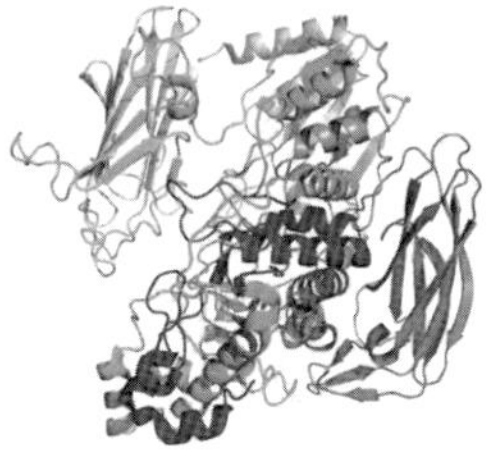 [PDB:3ABZ]	Yoshida et al.[32]
–	**116**	β-Glucosidase	No available structure	

Carbohydrate-Active EnZymes' glycoside hydrolase family numbers are shown in bold referring to retaining mechanism, otherwise the corresponding enzymes follow inverting mechanism. Images are taken from the RCSB PDB (http://www.pdb.org) and visualized using PyMOL (http://www.pymol.org/).

typically processive enzymes, which means that they are attached to the cellulose chain until it is completely hydrolyzed, whereas endocellulases can be both processive and nonprocessive enzymes.[22] The efficiency of the processive cellulases can greatly influence the rate-limiting step of cellulose hydrolysis.[23]

Structurally, β-1,4-endocellulases are represented by a larger number of protein folds compared to the exocellulases (Table 2.1). The first structure of a cellulase was determined in 1990 (PDB: 3CBH), which was an exoglucanase Cel6A from *Hypocrea jecorina* (anamorph *Trichoderma reesei*). The structure revealed an active site that was in a tunnel and there were multiple subsites containing aromatic residues such as tryptophan, for binding glucose residues. The CAZy database describes, at the moment of writing, around 130 GH families, whereas around 16 of them are populated by many different cellulolytic enzymes (Table 2.1). From a structural point of view, many of these families have completely different three-dimensional folds, underpinning that cellulases have evolved convergently from diverse ancestors.[33] As previously mentioned, β-1,4-endoglucanases have an 'open groove' topography in which the polysaccharide chain is accommodated in a canyon running across the enzyme surface, allowing the access of the internal glycosidic bonds of the substrate. On the other hand, exoglucanases or cellobiohydrolases, which are often members of the same GH family thus sequence identical to some extent, possess extended surface loops that generate a tunnel-shaped topography to facilitate a processive action on the substrate.[33] It is possible that loop breathing can impart an occasional endo-activity to exoglucanase processive enzymes and, depending on whether the initial attach occurs solely at chain ends or randomly in the main chain of cellulose, these enzymes might be subdivided into exo- and endo-processive categories.[34] In comparison to exoglucanases, β-glucosidases are considered as true exo-acting cellulases aiming solely to the end of cellooligosaccharides, having active centers that are buried in pockets or caves; therefore, these enzymes have little or no activity on polymeric cellulose.[34]

Cellulases belong to at least eight unrelated protein folds (Table 2.1), which are diffentiated into even more protein families in the CAZy database. For instance, Cel44A endoglucanase from *Clostridium thermocellum* (PDB: 2E0P) has an $(\beta/\alpha)_8$-fold belonging to family GH44, sharing the same topology with other cellulases including cellobiohydrolases and β-glucosidases, members of completely different GH families. Recently, an endoglucanase from *C. thermocellum* (CtCel124)[28] that is distantly related to lytic transglycosylases initially positioned in family GH23, introduced a novel GH family 124 that exhibits only members with endoglucanase activity. As cellulases belong to many different GH families, although having different structural characteristics, we can presume that these enzymes are representatives of a large class of nonhomologous isofunctional enzymes.[35] This means that these enzymes catalyze the same reaction, while having evolved independently and are unrelated in structure and sequence. Therefore, cellulases from each GH family must be treated as independent cases in any type of genomic analysis, resulting in a huge increase of the amount of data analysis.[36]

In literature, two are the main classical mechanisms leading to the inversion or retention of the anomeric carbon configuration. Transglycosylation events are restricted to retaining enzymes. The stereochemistry of catalysis is dictated by the spatial arrangement of the catalytic groups in the active site of the enzyme.[33] Enzymatic glycoside hydrolysis normally requires the presence of two catalytic groups, which in most cases have been observed to be either aspartate or glutamate residues. A novel cellulase fold was determined for a *C. thermocellum* cellulosomal enzyme that showed low activity by itself but acted synergistically with other cellulases.[28] It is surprising that there is no evidence for a catalytic base, although it has a catalytic acid. It seems that the limiting step for the hydrolysis of crystalline cellulose is not the bond cleavage but the way that cellulose molecule binds into the active site of the catalytic domain of cellulase, leading to the complete understanding of the important residues that participate in the binding step.[37]

Synergistic Interaction

Multiple enzymatic activities are required in cellulose hydrolysis leading to soluble sugars that can be metabolized by fermenting microorganisms. These activities include β-1,4-endoglucanases, exoglucanases including D-cellodextranases and cellobiohydrolases and, finally, β-glucosidases (Figure 2.2). Endoglucanases randomly cut internal sites on amorphous cellulose surfaces, generating new chain ends. Cellobiohydrolases act in a processive manner on the reducing or nonreducing ends of cellulose and liberate cellobiose as major product. β-Glucosidases hydrolyze soluble cellodextrins and cellobiose to D-glucose and thus relieve the system from end product inhibition.[38] Cellodextrinases hydrolyze soluble cellooligosaccharides, producing cellobiose but exhibit little or no activity against insoluble cellulose or carboxymethylcellulose (CMC). In comparison with many other GHs that act on natural substrates, cellulases are known to interact on crystalline cellulose showing low catalytic activity.[39] For example, several endoglucanases and cellobiohydrolases from *H. jecorina* display k_{cat} values of 2–20 sec^{-1} on amorphous cellulose[40] that corresponds to sixfold lower rates on crystalline cellulose.[41] On the other hand, the efficiency of the protein production has been improved approximately 10-fold, resulting in a serious decrease of the cost of cellulases from approximately \$0.20 to \$0.50 per gallon of ethanol produced.[39]

There are several types of synergies that have been observed among the aforementioned cellulolytic enzymes during the breakdown of crystalline cellulose, but convincing explanations at the molecular level have yet to be found, especially for the cooperation between cellobiohydrolases. For example, the cooperation between endo- and exo-acting cellulases is extensively described in literature, however, the degree of synergy observed is usually less than that expected for a system in which the cellobiohydrolases depend entirely on β-1,4-endoglucanases for the provision of free chain ends.[34] Cellobiohydrolases are critical to the function of fungal systems, as, for example, Cel7A from *Chrysosporium lucknowense* comprises in excess of 50% of the cellulases secreted by this organism.[42] However, the poor understanding of the role of cellobiohydrolases is indicated by a recent investigation showing that the deletion of two major cellobiohydrolases of *C. thermocellum* does not affect substantially cellulose breakdown.[43] Moreover, cellulolytic systems may lack the pair of cellobiohydrolases that act from the reducing and nonreducing ends of cellulose chains.[44,45] For example, the cellulose-degrading bacterium *Cytophaga hutchinsonii* does not contain GH6, GH48, and GH7 cellobiohydrolases, although it is possible that *C. hutchinsonii* contains novel-type cellobiohydrolases.[46] Recently, a complete cellulolytic system was reported to occur in the marine bacterium *Saccharophagus degradans* 2–40.[45] The genome sequence of this bacterium predicts that this organism does

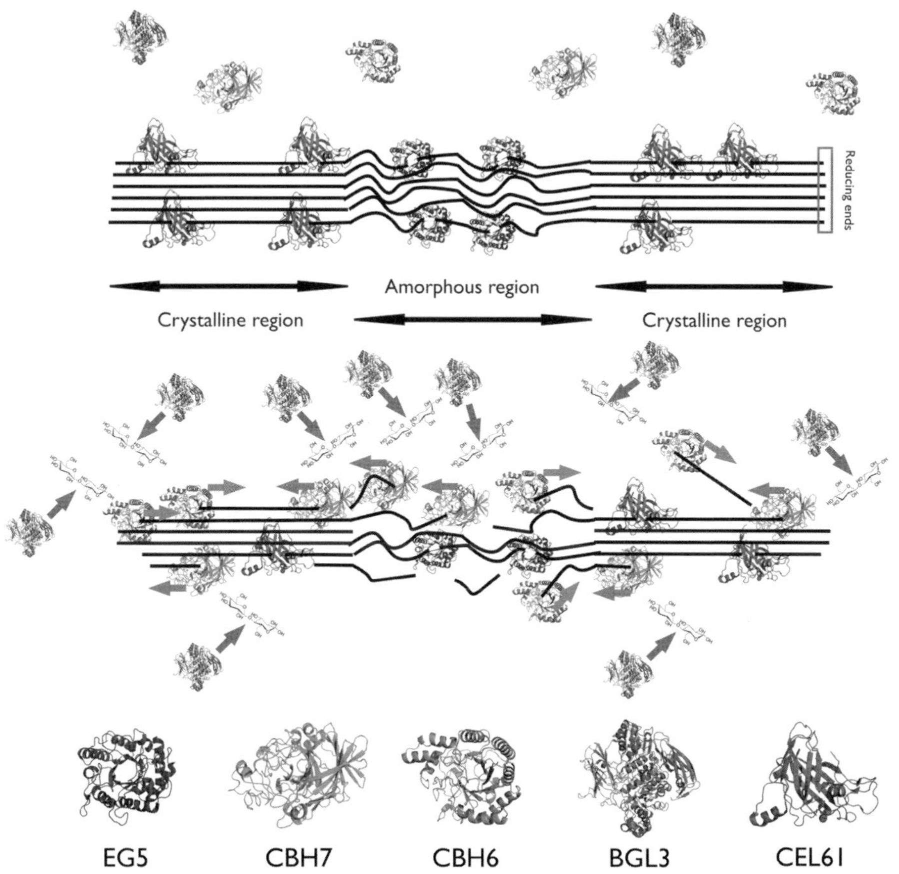

Figure 2.2 An overview of cellulose hydrolysis by the synergistic action of cellulolytic enzymes; the β-1,4-endoglucanases (EG5) catalyze the hydrolysis of the main chain of cellulose located in the amorphous region, resulting in nonoxidized chain ends, whereas polysaccharide monooxygenases of GH family 61 (CEL61) catalyze oxidatively possibly the crystalline region, resulting in oxidized chain ends. Cellobiohydrolases hydrolyze cellulose chain ends from the reducing (CBH7) or nonreducing (CBH6) end in a processive manner to produce cellobiose or oxidized cellobiose, depending on the preceded family of enzymes that made the nick on cellulose surface. The processive action of cellobiohydrolases generates a majority of cellobiose that could be further hydrolyzed to D-glucose by β-D-glucosidases (BGL3).

not contain any known processive cellobiohydrolase genes and has 13 endocellulase genes of which 9 belong to a new class of processive enzymes members of GH5 family. These processive endo-acting cellulases do not require an additional domain for processivity, such as CBMs, suggesting that their processivity may result from unusual subsite affinities that need to be determined by structural crystallography.[47]

Most of the cellulases, discovered to date, are modular proteins that in addition to their catalytic domain, exhibit accessory domains such as CBMs, which are connected to the catalytic part with a flexible linker.[48] The main role of these accessory modules is helping the binding of cellulolytic enzymes on the polymer of cellulose, although CBMs might also participate in the initial disruption of cellulose fibers.[49] It is considered that β-1,4-endocellulases which exhibit CBMs, are keen on binding on the amorphous regions of cellulose fibers creating accessible ends, whereas cellobiohydrolases continue hydrolysis by removing oligosaccharides from the ends of the disrupted polymer.[50] Moreover, the processivity of most endoglucanases is dependent upon their associated CBM modules, for example, several bacterial GH9 endoglucanases depend upon the resident CBM3 module.[51]

Anaerobic bacteria, such as *Clostridium* and *Ruminococcus* spp., or fungi such as *Chytridomycetes* have developed complexed cellulase systems called 'cellulosomes'.[52] The assembly and spatial organization of enzymes in naturally occurring cellulosomes, which are regarded as nanomachines for the efficient cellulose degradation, constitute the base of their synergistic activity. In most cases, these complexes are bound to the cell surface; however, this does not seem to be true for all cellulosome biosynthesizing microorganisms.[53] The cellulosome architecture consists of multiple enzyme subunits, with different substrate specificities and catalytic mechanisms, organized by scaffolding proteins.[52,53] Cellulosome hydrolases, apart from the catalytic module, always contain at least one supplementary domain, which is called the dockerin module involved in enzyme interaction with the scaffolding proteins. Scaffoldins are large multidomain, multifunctional proteins that recruit catalytic proteins interacting with GH dockerin domains and improve complex affinity for the substrate and catalytic efficiency via CBMs. Such a complex architecture enables the enzymatic components to act in a synergistic and coordinated manner via intra- and intermolecular interactions and makes the cellulosomes the most efficient biochemical systems for cellulose degradation.

Oxidative Enzymes for Cellulose Degradation

Recently, a revolutionary mode of action was discovered for the hydrolysis of cellulose, revealing the mechanism that the members of the known GH61 family follow (Table 2.1), using oxidative chemistry. This chemistry occurs in the cleavage of chitin (a fibrous, crystalline polymer of β-1,4-*N*-acetylglucosamine residues) by the bacterial chitin-binding protein 21 (CBP21) that belongs to CBM family 33.[54] This novel mode of action opens up the crystalline polysaccharide material that is inaccessible for hydrolysis by other GHs. A homologue to CBP21 in *Streptomyces coelicolor* was shown to act synergistically with cellulases in the digestion of cellulose.[55] The CBP21 oxidative mechanism led to the assumption that the structurally similar GH61 enzymes, which were known for many years as weak endoglucanases, might also follow the same mode of action on crystalline cellulose. The first example that supported this assumption, reported that some members of GH61 family could significantly enhance the effectiveness of common cellulases in the breakdown of complex lignocellulosic substrates.[56] Interestingly, another study revealed that the combination of a *Thermoascus aurantiacus* GH61 with a *Humicola insolens* cellobiose dehydrogenase resulted in cellulose cleavage, yielding a mixture that included reducing end-oxidized and nonreducing end-modified cellooligosaccharides.[20] In addition, the authors showed a synergistic effect of both *Thielavia terrestis* GH61 and cellobiose dehydrogenase on microcrystalline cellulose hydrolysis by *T. terrestris* canonical cellulases.[20] It was hypothesized that an oxidoreductive cellulolytic system coexists with the well-studied fungal cellulases resulting in efficient lignocellulose conversion. In GH61 enzymes, copper is the most likely redox center[19] and taking in consideration that the catalytic metal-binding site is conserved, it is highly probable that members of both the GH61 and CBP21 families are oxidative enzymes that target recalcitrant plant cell wall material such as cellulose (Figure 2.2). It has been demonstrated that the presence of lignin, which presents a significant source of reductant residues, is probably responsible for the increase in activity of GH61 from *Myceliophthora thermophila* by acting as electron donors.[57] This boosting effect was first described by the use of the reductant ascorbic acid by increasing the efficiency of CBP21.[54] These fungal derived, copper-depended oxidative enzymes lacking hydrolase activity, are recently named polysaccharide monooxygenases (PMO) that catalyze the O_2-dependent oxidative cleavage of recalcitrant polysaccharides.[58] In addition to this surprising discovery in the theory of cellulose hydrolysis, a new type of enzymes with possibly a very important role in cellulose utilization was discovered in the *M. thermophila* supernatant termed aldonolactonases.[58] These recently discovered enzymes catalyze the hydrolysis of glucono-δ-lactone and cellobiono-δ-lactone that are produced by the enzymatic oxidation of cellulose.[58] Sugar lactones have been shown to inhibit many different types of glycosyl hydrolases,[59] for example, glucono-δ-lactone is a potent inhibitor of β-glucosidases.[60]

BIOCONVERSION PROCESSES

There are three distinguished (functional) stages of cellulose bioconversion into ethanol: the pretreatment of the raw cellulosic material, the enzymatic saccharification, and the fermentation. Depending on the material used, the fermentation stage may include hexoses fermentation and/or pentoses fermentation. The stage of enzymes production needed for the enzymatic saccharification completes the list of necessary biological stages for ethanol production from lignocellulosic materials. Depending on the process design, these stages may take place in different chronic combinations, being consecutive or simultaneous. The major bioconversion processes are: The separate (or sequential) hydrolysis and fermentation (SHF), the simultaneous saccharification and fermentation (SSF) and the consolidated bioconversion process (CBP) in which, enzyme production, growth of microorganism, enzymatic hydrolysis and fermentation are performed in the same reactor. When

C5 sugars (pentoses) are involved, a fourth process, the simultaneous saccharification and cofermentation (SScF) is often considered.

SHF and SSF Processes

In SHF processes the enzymatic SSF stages are performed in different reactors. The main advantage of this process is the fact that each step can be conducted at optimal conditions of pH and temperature. However, the accumulation of glucose and cellobiose (occurring in most experimental set ups during the hydrolysis step) has inhibitory effects on the activity of the cellulases, decreasing significantly after some time the saccharification and accordingly the overall yield of the process. In a SSF process, the enzymatic hydrolysis and fermentation are not discernible stages, neither chronically nor topically. They are both performed in the same reactor and thus, glucose released by the action of cellulases is converted directly to ethanol by the fermenting microorganism. This continuous removal of glucose from the medium has a beneficial effect on the overall yield due to the minimization of the end-product inhibition on cellulolytic activities.

Most reported experimental work on SSF, as thoroughly reviewed by Olofsson et al.,[61] have focused on improving the process by increasing the substrate loading (i.e., the content of water insoluble solids: WIS), decreasing enzyme and yeast concentration, and varying temperature and pH. In most cases SSF is compared favorably to SHF. According to Wingren et al.,[62] using softwood as the raw material, the SSF process results in 10% lower ethanol production cost (~4.81 SEK/L) than the SHF process (~5.32 SEK/L), mainly due to the higher ethanol yield that in most cases is achieved from SSF processes. Increase in the initial dry matter concentration of the raw material, decrease in the microorganism dry weight used and the recycling of the stillage stream would substantially reduce the cost of SSF process (by 20–30% of the above calculated value). Continuous and fed batch processes are often adopted to cope with the inhibitory effects of compounds present in lignocellulosic hydrolysates. Continuous systems are most often used with immobilized cells because bacterial contamination with *Lactobacilli* is a very often problem in stirred tank reactors using free yeast cells.[63] The results obtained by Hoyer et al.[64] who studied the effects of different enzyme feeding strategies on ethanol yield from spruce under high gravity (HG) conditions, show that fed batch or continuous fermentation strategies not only for the enzymes addition, but also for the cells or the raw material could result in significant process improvements.

Consolidated Bioconversion Processes (CBP)

Lignocellulose bioconversion technology has exhibited a trend toward increasing consolidation over the last decades. In CBP, the saccharification and the fermentation of biomass are carried out in a single stage by the same microorganism. The major advantage of CBP technology is the avoidance of the cost of dedicated cellulase production which is associated with significant potential energy savings.[65] Therefore, the majority of scientific efforts dealing with this issue have tried to develop 'CBP microorganisms' based on two strategies: (category I) engineering a cellulase producer to be ethanologenic and (category II) engineering an ethanologen to be cellulolytic.

Significant part of the research done so far focused on *Thermocellum cellulolyticum* (category I) and *S. cerevisiae* (category II). On building *S. cerevisiae* acquiring cellulolytic activity, several approaches have been tested: expressing heterologous cellulases, mimicking the cellulosome system of complexed enzymes bound to the cell surface, or codisplaying different fungal cellulolytic activities.[66] Fujita et al.[67] described the efficient direct fermentation of amorphous cellulose to ethanol, which was achieved by developing a yeast strain codisplaying three types of cellulolytic enzymes. *H. jecorina* endoglucanase II and cellobiohydrolase II (CBH-II) and *Aspergillus aculeatus* β-glucosidase were simultaneously codisplayed as individual fusion proteins in yeast. It was shown that codisplay of CBH-II significantly affected cellulose hydrolysis, suggesting that combination of cellulases with various functions is effective in producing efficient degradation of cellulose. Cell-surface engineering systems established using the yeast *S. cerevisiae* are expected to be used as paradigm for many more applications. The surface-displayed proteins are self-immobilized on the cell surface, with this feature being passed on to daughter cells as long as the genes are retained by the cells. The ability to display enzymes should further enhance the status of *S. cerevisiae* as a novel and attractive microorganism able to act as a whole-cell biocatalyst. This is particularly important when target substrates cannot be taken up by the cells.[68]

Biorefinery

With emerging research trends in the production of bio-based products, the concept of biorefinery will continue to gain momentum. The definition of the term 'biorefinery' has been subject to debate, but the overall goal of the biorefinery production approach is the generation of a variety of goods from different biomass feedstocks, within a particular biorefinery, through a combination of technologies. Ideally, a biorefinery should integrate biomass conversion

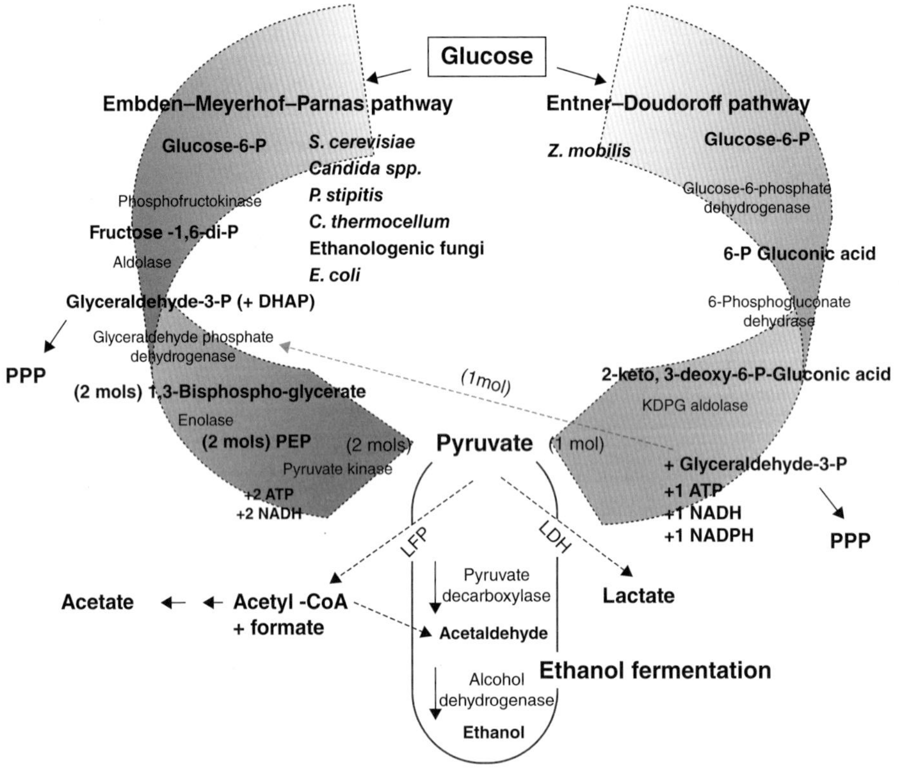

Figure 2.3 Microbial conversion of glucose to ethanol under anaerobic conditions. The enzymes catalysing the main biochemical steps are indicated. Most microorganisms catabolize glucose through the glycolytic pathway (EMP). Although there are many aerobic bacterial species that use the ED pathway, *Zymomonas* is the only known microbial genus that uses this pathway under anaerobic conditions. LFP, pyruvate formate lyase; LDH, lactate dehydrogenase; PEP: phosphoenolpyruvate; PPP, pentose phosphate pathway; DHAP, dihydroxyacetone phosphate; KDPG, 2-keto-3-deoxy-6-phosphogluconate.

processes to produce a range of fuels, power, materials, and chemicals from biomass.[69] Lately, a trend toward high value, lower volume bio-based chemicals has been exhibited, aiming at the increase of the overall cost-effectiveness of a lignocellulose-based biorefinery. However, this trend has to face two major challenges to become really applicable and efficient in large scale: (1) The bio-based chemical production is challenged by a lack of conversion technology; and (2) the bio-based chemical production is challenged by an overabundance of targets.[70]

BIODIVERSITY IN ETHANOL FERMENTATION

Significant differences are observed in the biochemical paths used by different fermenting microorganisms (Figure 2.3). The differences may refer to very important characteristics for ethanol production in industrial scale, such as the enzymes used, the metabolites produced, the energy gain for the cell, and the final fermentation products. In recent trends of biofuels production processes (Biorefinery concept), the successful fermentation of undiluted, not fractionated lignocellulosic hydrolysates is substantial not only to exploit all biomass components but also to make the process simpler. Thus, apart from ethanol yields and productivities, the 'robustness' of the fermenting microorganisms (mainly reflected to toxicity and ethanol tolerances, and to a lesser extent to thermotolerance and to pH tolerance) is also important for their success in industrial applications. The formation of by-products is considered a major disadvantage not only because of their impact on the overall ethanol yield, but also because of the difficulties in separation and down processing of end products. Although the natural fermenting microorganisms, *S. cerevisiae* and *Z. mobilis*, are the more suitable for large-scale applications, using genetic engineering and system biology tools, the fermentative performances of bacteria such as *E. coli*,

Klebsiella oxytoca, and *C. thermocellum* have also been significantly improved.

Fermentation of Biomass-Derived Sugars by *Zymomonas mobilis*

The anaerobic bacterium *Z. mobilis* ferments glucose through the Entner–Doudoroff (ED) pathway, in which one mole of glucose is converted to one mole of pyruvate, one mole of NADPH, one mole of NADH and one mole of ATP (Figure 2.3). Although many bacterial species are using the ED pathway, *Z. mobilis* is the only known natural species that uses this path under anaerobic conditions. The ED pathway yields only half as much ATP per mole of glucose as the Embden–Meyerhof (EM) pathway. As a consequence, *Zymomonas* produces less biomass than yeast, and more carbon is channelized to fermentation products. Also, as a consequence of the low ATP yield, *Zymomonas* maintains a high glucose flux through the ED pathway. All the enzymes involved in fermentation are expressed constitutively, and fermentation enzymes comprise as much as 50% of the cells' total protein.[71]

The complete genome sequence of *Z. mobilis* ZM4 has been reported at 2005,[72] revealing that 2056416 base pairs form a circular chromosome with 1998 open reading frames (ORFs) and three ribosomal RNA transcription units. Recognizable genes for 6-phosphofructokinase, and for two enzymes in the tricarboxylic acid (TCA) pathway were absent and, therefore, it was confirmed that glucose could be metabolized only by the ED pathway.

Escherichia coli

Although *E. coli* is naturally able to ferment a wide spectrum of sugars (much wider than natural *S. cerevisiae* and *Z. mobilis* strains), the ethanol yields are very low because of the formation of other fermentation products in higher rates and final concentrations. In *E. coli*, the fermentation of glucose may be regarded as a process in which the four extra reducing equivalents (or $4H^+ + 4e^-$) released, as the metabolic price for making two ATPs by substrate-level phosphorylation, are disposed of by the further metabolism of two pyruvate molecules. Glucose is converted to a mixture of metabolic products consisting primarily of acetate and formate, as well as smaller amounts of lactate, succinate, and ethanol.[73] Furthermore, the native ethanol pathway is not redox balanced when sugars, such as glucose or xylose, are fermented. Only one $NADH^+$ is formed per acetyl-CoA synthesized, but two $NADH^+$ are needed to produce one ethanol molecule via this pathway.[74]

Most scientific efforts dealing with the ethanol fermentation using *E. coli* as the fermenting microorganism have focused on the improvement of ethanol production by introducing *Z. mobilis* genes to *E. coli* strains. It was shown that the introduction of *Z. mobilis* homoethanol pathway (PET operon) increased significantly ethanol production, reaching yields similar to those achieved by some *S. cerevisiae* strains. Directed engineering and metabolic evolution of *E. coli* strains have also been able to improve the tolerance toward inhibitory compounds present in lignocellulosic hydrolysates, giving more promises for the future.[75] In 2007, Kim et al.[76] showed that *E. coli* strains could be efficient ethanol producers without adding genes from other microorganisms. In that study, the development of an ethanologenic *E. coli* mutant devoid of foreign genes was described. An essential mutation in this mutant was mapped within the *pdh* operon (*pdhR aceEF lpd*), which encodes components of the pyruvate dehydrogenase complex. Anaerobic ethanol production by this mutant is apparently the result of a novel pathway that combines the activities of pyruvate dehydrogenase (typically active during aerobic, oxidative metabolism) with the fermentative alcohol dehydrogenase.

Ethanologenic Fungi

Monilia, *Fusarium*, *Neurospora*, *Rhizopus*, *Aspergillus*, *Neocallimastix*, *Trametes*, *Phlebia*, and *Trichoderma* are some of the filamentous fungi genera that are known to produce ethanol under anaerobic and/or microaerobic conditions. The fermentation characteristics by various fungal species are shown in Table 2.2. The strategies to improve ethanol production by fungi have focused on metabolic engineering techniques aiming at (1) the introduction of heterologous genes, such as *S. cerevisiae* pyruvate decarboxylase (PDC) and alcohol dehydrogenase (ADH) to enhance the classical ethanol synthetic pathway; (2) the knockout of some fungal genes responsible for the production of byproducts; or (3) the overexpression of existing genes playing a key role in sugars metabolism.[77] The enhancement of the efficiency of sugar transporters has also been proposed as a strategy to enhance the ethanol producing rate by *H. jecorina*.[78]

ETHANOL FERMENTATION OF CELLULOSE-DERIVED SUGARS BY YEASTS

Many yeast species have been reported to convert simple sugars to ethanol under anaerobic conditions. *S. cerevisiae*, *Pichia stipitis*, *P. kudriavzevii* (*Candida krusei*), *Kluveromyces marxianus*, *C. shehatae*, *C. tropicalis*, *C. guilliermondii*, and *Pachysolen tannophilus* are among the most often used yeast species in biomass-derived sugars–to–ethanol conversion processes. In Table 2.3, the

Table 2.2 Fermentation Characteristics of Major Ethanologenic Fungi

Microorganism	Substrate	Ethanol Yield: $g\ g^{-1}$ of Sugars Consumed	Ethanol Yield: $g\ g^{-1}$ of Available Sugars	Ethanol Productivity $g\ L^{-1}\ h^{-1}$	Process, Conditions (pH, T)	Formation of By-products	References
Aspergillus oryzae	Glucose	n.d.	0.49	n.d.	30°C	n.d.	Skory et al.[79]
	Xylose		0.09				
	Crystalline cellulose		0.01				
A. niger	Glucose	n.d.	0.11	n.d.	30°C	n.d.	Skory et al.[79]
	Xylose		0.02				
	Crystalline cellulose		0.00				
A. sojae	Glucose	n.d.	0.24	n.d.	30°C	n.d.	Skory et al.[79]
	Xylose		0.11				
	Crystalline cellulose		0.01				
Fusarium oxysporum	Cellulose	n.d.	0.35	0.044	SSF pH 6, 30°C	Succinate, glycerol, acetate, pyruvate	Panagiotou et al.[80]
Neurospora crassa	Alkali-treated BSG	0.17	0.08	n.d.	CBP, pH 6, 30°C	Acetate	Xiros et al.[81]
Phlebia sp.	Glucose	n.d.	0.44	n.d.	SSF, 28°C, PH ~ 6.5	n.d.	Kamei et al.[82]
	Mannose		0.41				
	Galactose		0.40				
	Fructose		0.41				
	Xylose		0.33				
	Waste		0.20				
	Newspaper		0.42				
	Unbleached						
	Hardwood kraft						
	Pulp						
Trichoderma harzianum	Glucose	n.d.	0.1	n.d.	30°C	No organic acids	Stevenson and Weimer[83]
	Cellulose		0.04				
Rhizopus javanicus	Glucose	n.d.	0.47	n.d.	30°C, pH 4–7	Lactic acid, fumaric acid, trehalose	Skory et al.[79]
	Xylose		0.23				
	Crystalline		0.03				
	Cellulose		0.15				
	Corn fibers		0.24				
	Cellobiose						
R. oryzae	Glucose	0.43	n.d.	1	30°C, pH 4.3–5.7	Lactic acid	Taherzadeh et al.[84]
	Xylose	0.28		0.11		Glycerol	
	Dilute-acid hydrolyzate	0.34				Xylitol	
Trametes hirsuta	Glucose	0.49	n.d.	n.d.	28°C	n.d.	Okamoto et al.[85]
	Mannose	0.48					

(*Continued*)

Table 2.2 Continued

		Ethanol Yield					
Microorganism	Substrate	$g\ g^{-1}$ of Sugars Consumed	$g\ g^{-1}$ of Available Sugars	Ethanol Productivity $g\ L^{-1}\ h^{-1}$	Process, Conditions (pH, T)	Formation of By-products	References
	Cellobiose	0.47					
	Maltose	0.47					
	Xylose	0.44					
	Starch	0.45					
	Wheat bran	0.39					
	Rice straw	0.29					
Mucor indicus	Glucose	0.39	n.d.	1.41	30°C, pH 4.3–5.7	Glycerol	Taherzadeh et al.[84]
	Xylose	0.22		0.18		Xylitol	
	Dilute-acid Hydrolyzate	0.44					
M. hiemalis	Glucose	0.39	n.d.	1.44	30°C, pH 4.3–5.7	Glycerol	Taherzadeh et al.[84]
	Xylose	0.18		0.12		Xylitol	
	Dilute-acid hydrolyzate	0.44					

CBP, consolidated bioconversion process; SSF, simultaneous saccharification and fermentation.

Table 2.3 Selected Publications on Fermentation Applications by Yeast Species Other than *S. cerevisiae*

		Ethanol Yield					
Microorganism	Substrate/Pretreatment	$g\ g^{-1}$ of Sugars Consumed	$g\ g^{-1}$ of Available Sugars	Ethanol Productivity	Process, Conditions (pH, T)	Formation of By-products	References
Pichia stipitis	Corn stover/dilute H_2SO_4	0.42	0.39	n.d.	30°C SHF	n.d.	Agbogbo et al.[86]
Pichia kudriavzevii	Sugarcane juice	n.d.	0.42	$4\ g\ L^{-1}\ h^{-1}$	40°C–45°C	Arabitol, glycerol	Dhaliwal et al.[87]
Candida krusei	Glucose	n.d.	0.4	$55\ g\ L^{-1}\ day^{-1}$	30°C,	Significantly low	Nakayama et al.[88]
Kluyveromyces marxianus[1]	β-Glucan	0.47	0.42	n.d.	Direct conversion, 45–48°C	n.d.	Yanase et al.[89]
Kluyveromyces marxianus	Poplar, sweet sorghum bagasse, wheat straw/ steam explosion	n.d.	0.31–0.37	n.d.	SSF, 42°C	n.d.	Ballesteros et al.[90]
Pachysolen tannophilus	Sugarcane bagasse/acid hydrolysis, electrodialysis	n.d.	0.34	$0.57\ g\ L^{-1}\ h^{-1}$	SHF, pH 5, 30°C	n.d.	Cheng et al.[91]

[1]Genetically engineered strain to display *Trichoderma reesei* endoglucanase and *A. aculeatus* β-glucosidase on the cell surface.

fermentative performances of selected yeast strains other than *S. cerevisiae* are compared.

P. stipitis has been mainly used in applications where not only cellulose but also hemicellulose is to be fermented, as it is considered one of the most promising naturally occurring C5 fermenting organisms. *P. stipitis* has been used as an ethanologen to ferment various types of pretreated biomass such as red oaks, wheat straw, sugarcane bagasse, rice straw, corn cob, corn stover, aspen wood, pine wood, and poplar wood (Table 2.3).[86] Lately, *K. marxianus* has received much attention because of the potential cost savings, which could be obtained by continuous evaporation of ethanol from the broth under reduced pressure. Promising ethanol yields from various lignocellulosic woody and herbaceous materials, during SSF in the presence of commercial cellulases, have been achieved by Ballesteros et al.,[90] while two years ago Yanase et al.[89] have reported ethanol production by *K. marxianus* displaying cellulolytic enzymes.

Saccharomyces cerevisiae

Although *S. cerevisiae* is used in industrial fermentation applications for many years, the commercialization of cellulosic ethanol using this microorganism has been proved to be a strong challenge. The efficient cofermentation of C6 and C5 sugars (simultaneous fermentation of both pentoses and hexoses) by *S. cerevisiae* has always been (and still is) considered as a major concern. Apart from this, the inhibitory effects of (toxic) lignocellulose derivatives on the fermentative performance of *S. cerevisiae*—the stressful conditions due to high ethanol concentration or due to high sugars concentrations during the fermentation, the high cost of separation of ethanol from the fermentation broth—are major restraints which the advance of biofuels technology is trying to overcome. The majority of the scientific reports published so far could be divided in two distinguished, though supplementary categories; the process optimization strategies and the strain development strategies.

High Cell Density Processes

S. cerevisiae immobilization has been applied in industrial alcohol fermentation aiming mainly at the increment in fermentation rates. High cell density, greater volumetric productivity, increased tolerance to high concentrations of inhibitors, relative easiness of downstream processing are some of the properties that have been attributed to these processes.

Flocculating *S. cerevisiae* strains is another way to increase cell density during fermentation process. Flocculation is the asexual, reversible, and Ca^{+2}-dependent aggregation of thousands of vegetative cells into flocs. Detailed biochemical and structural studies of the responsible adhesins (flocculins) have been performed, enlightening the biological aspects of this mechanism based on a lectin–carbohydrate interaction.[92] Recently, Ruiz et al.[93] reported the successful production of ethanol from hydrothermally treated wheat straw during SSF by a thermotolerant flocculating yeast strain. As shown, the flocculating strain performed very well at 45°C. In other words, it was possible to perform the SSF at a temperature close to the optimal for the enzyme action overcoming a major restraint of SSF processes.

A novel type of cell–cell flocculation induced by trehalose-6-phosphate synthase 1 (*TPS1*) promoter was reported recently by Li et al.[94] From the experimental results, it was shown that the fermentative performance of this flocculating engineered *S. cerevisiae* strain acquired some beneficial characteristics for ethanol production processes from agro-industrial substrates under HG. Therefore, this differential induction of gene expression directed by *TPS1* promoter (under ethanol induction) added evidence that controlled gene expression can endow beneficial yeast phenotypes for industrial applications.

High Gravity Ethanol Production Processes

The main advantage of biomass bioconversion under HG (initial dry matter concentrations above 12% w/w) or very high gravity (VHG, above 30% w/w) conditions is the generation of high final ethanol concentrations (above 5% v/v), decreasing the cost of the distillation step, which is considered one of the main constraints in the bioethanol industry.[95] However, serious inconveniences come along with increase in initial dry matter: (1) The high sugar content may cause osmotic stress which leads to a number of cellular responses which in turn affect the ethanol yield and productivity. (2) Mixing of solids, enzymes, and yeast cells becomes very difficult using conventional reactors due to low water content, which increases the viscosity of the substrate material. (3) Ethanol has an inhibitory effect on cell growth and viability already at relatively low concentrations, which becomes more severe as the ethanol concentration increases. (4) The presence of toxic compounds, generated during the pretreatment step in high concentrations, makes the bioconversion almost impossible in some cases.

The design of new type of reactors was a major concern at the beginning of the research in the field. Free fall rotating systems were proven very efficient for HG processes. Jørgensen et al.[96] reported a successful free fall bioreactor design for the bioconversion of lignocellulose into ethanol using initial dry matter up to 40%. Recently, Matano et al.,[4] in an attempt to improve the mixing efficiency, used a similar rotary fermentation system, consisting of a fermentation vessel positioned horizontally in a drum-type heat block which axially rotated under controlled temperature. By using a recombinant cellulolytic *S. cerevisiae* strain and 10 FPU g^{-1} DM, 89% of the theoretical ethanol yield was achieved from 20% (w/v) rice straw.

Various process improvement approaches have been tested to cope with the stressful HG conditions imposed to yeast cells. The detoxification of the initial material has in many cases proved beneficial to the overall fermentative performance,[97] whereas, in other cases, the addition of nutrient supplementations during fermentation enhanced the ethanol yield and rate of the process.[98]

Evolutionary Engineering and Inverse Engineering

The majority of evolutionary engineering applications have focused on two areas: evolutionary engineering of stress resistance and substrate utilization. *S. cerevisiae* is probably the most robust microorganism used in biofuels applications. It has an innate tolerance to some of the inhibitory compounds in lignocellulose hydrolysates, such as hydroxymethylfurfural (HMF), furfural, and phenolics, due to its ability to convert them to less harmful molecules. However, it can only tolerate certain levels of these inhibitory compounds, whereas its fermentative performance can be completely inhibited in the presence of low concentrations of some other inhibitors. A way to exploit these intrinsic metabolic capabilities of *S. cerevisiae*, to improve its fermentative performance in the presence of toxic compounds, is to adapt it to toxic conditions by constant exposure to tolerable inhibitor concentrations in repetitive batch, fed-batch, or continuous cultivation. A truly mechanistic understanding of the action of inhibitors is not needed, but a well-designed selection procedure, on the other hand, is necessary to have a successful adaptation.[99] Cultivation in increasing concentrations of selected individual inhibitors or in increasing concentration of toxic hydrolysates will result in the increase in frequency of adapted cells in the culture and accordingly in the isolation of an adapted strain. However, the adapted properties of the strain may be inducible only under the selection pressure imposed by the experimental conditions, as in the recent study on the adaptation of a xylose-fermenting *S. cerevisiae* to acetic acid performed by Wright et al.[100]

The promise of a successful coconsumption of xylose and glucose really boosted the commercial bioethanol perspective. Kuyper et al.[101] managed to develop through a very systematic experimental setup, an adapted *S. cerevisiae* xylose fermenting strain, showing significantly improved xylose consumption rates under various conditions. More recently, combined approaches of genetic engineering, chemical mutagenesis, and evolutionary engineering were used to obtain high xylose-consuming rates by a recombinant *S. cerevisiae* strain.[102] A significantly improved strain was developed from a parental chemically mutagenized, when the latter was exposed to serial batch cultivations for growth in xylose medium, first by aerobic and then under oxygen-limiting conditions using increasing xylose concentrations for selection.

If not only improved strains, but also knowledge output, through evolutionary engineering are desirable, the genetic alterations connected with the improved performance of the strains developed have to be identified (Box 2.2). Inverse metabolic engineering (in later articles also reported as reverse metabolic engineering) is the term introduced by Bailey et al.[103] to describe biotechnological approaches which in contrast to the conventional 'forward' metabolic engineering cycle, begin from successful phenotypes toward the knowledge of the factors and mechanisms that create or affect them. Genome-wide analytical approaches are probably the key tool to a deeper understanding of them. As Oud et al.[104] recently underlined, although whole genome sequencing applied for the reverse metabolic engineering of yeasts is still small, the available information consistently indicates that this technique is a real game changer. Adopting microarray and statistical two-way ANOVA analysis for analyzing 'successful' *S. cerevisiae* phenotypes, Stanley et al.[105] identified significant transcriptional changes relevant to ethanol tolerance, revealing that improved ethanol tolerance is related to increased mitochondrial and NADH oxidation activities, which also stimulates glycolysis.

Box 2.2 Directed Evolution's Black Box

Evolutionary engineering is often considered as a nontargeted strain improvement strategy. It is though widely recognized as a very powerful tool for strain improvement. Insight on the improved cell population is a major challenge for scientists in the field. Better and deeper knowledge of the genetic modifications and the mechanisms responsible for the improvements observed will allow the transfer of relevant traits among different microorganisms. For inverse metabolic engineering, or in other words, for the elucidation of the genetic basis for improved performance, the choice of the correct tools is of major importance. *S. cerevisiae*, being the model microorganism once more, is showing the way as long as integral analysis at different information levels (genome, proteome, and metabolome) is already successful and economically reasonable for academic and industrial research. Among all available tools, wide genome techniques seem the most powerful and most promising.

Metabolic Engineering Strategies

Metabolic engineering is defined by Bailey[106] as the improvement of cellular activities by manipulation of enzymatic, transport, and regulatory functions of the cell with the use of recombinant DNA technology. Metabolic engineering efforts aim at introducing heterologous enzymes in *S. cerevisiae*, overexpressing existing genes, or introducing whole enzymatic paths in the cell. Deletion of one or more genes is also an interesting approach often supplementary to the others, which aim at pointing the cell metabolism toward the desirable direction.

Increase in yeast tolerance to lignocellulosic hydrolysates has been achieved by overexpressing homologous or heterologous genes encoding enzymes that confer resistance toward specific inhibitors such as furan derivatives and phenolic compounds.[107] Recently, rational experimental design combined a metabolomics approach with metabolic engineering, giving promising results.[108] In that study, capillary electrophoresis–mass spectrometry (CE–MS) and gas chromatography–mass spectrometry (GC–MS) were used to determine the effect of acetic acid on xylose fermentation. It was shown that accumulation of metabolites involved in the nonoxidative pentose phosphate pathway (PPP) increased with an increase in acetic acid concentration. Based on this, *S. cerevisiae* genes encoding PPP-related enzymes, transaldolase (TAL) or transketolase (TKL), were overexpressed in the strain, which achieved increased ethanol productivity in the presence of acetic and formic acids.

Although pentose metabolism does not fall in the scope of this review, some key points have to be mentioned because the cofermentation of hexoses and pentoses is a significant trend in biotechnology for biofuels. Key points of the enzymatic paths of pentoses anaerobic metabolism in *S. cerevisiae* are as follows: the xylose transport into the cell, bridging xylose and xylulose, the relief of the redox imbalance, the flux of xylulose to xylulose-5-P, the expression of key enzymes in the PPP, in many cases have been the targets of metabolic engineering approaches.

Among the strategies that have been adopted, some aim to alter the cofactor specificities of D-xylose reductase (XR) and xylitol dehydrogenase (XDH) from *P. stipitis*, whereas others adjust the redox balance through regulating the central pathways.[109] A strain developed by chromosomal integration of the genes encoding XR, XDH, and xylulokinase (XK) fermented whole pretreated slurry of corn stover at high WIS preferably in a fed-batch SSF process, showing that the simultaneous coutilization of xylose was not favorable in the presence of high amounts of glucose.[110] On the other hand, when fungal xylose isomerase (XI) genes were functionally expressed in *S. cerevisiae* at high levels, the growth on xylose was very slow. As mentioned above, a combined research approach implementing evolutionary engineering of the constructed strain, improved significantly the fermentation rate of xylose in the presence of glucose.[101] Promising results have recently been published describing the construction of engineered yeasts, which coferment mixtures of xylose and cellobiose.[111] In these yeast strains, hydrolysis of cellobiose takes place inside yeast cells through the action of an intracellular β-glucosidase following import by a high-affinity cellodextrin transporter. As shown, intracellular hydrolysis of cellobiose minimizes glucose repression of xylose fermentation allowing co-consumption of cellobiose and xylose.

CONCLUSION

After some decades of scientific efforts, the production of cellulosic ethanol in commercial scale seems feasible, as demonstration and industrial scale ethanol plants pullulate year by year in various countries. The great majority of the scientific reports on the field have so far focused on one of three main targets: (1) the amelioration of the feedstock (plant or residue) with regard to a readily convertible and efficient substrate, (2) the improvement of the bioconversion process, and (3) the development of a desirable microbial factory which efficiently carries out the bioconversion. When bioethanol feedstocks refer to residues, the improvement is attempted through the pretreatment of the raw materials by chemical, physical, and/or biological means, whereas the amelioration of the raw material refers to the genetic engineering of plant cells, targeted to biofuels production, when whole plants are cultivated on this purpose. Enzymatic hydrolysis is advanced by the discovery of a new class of oxidative enzymes that cleave cellulose (GH61s), offering an exciting possibility for the improvement of cellulose deconstruction. Metagenomic DNA and genetically engineered libraries should be explored for the discovery of novel enzymes with improved cellulolytic activity on natural substrates and increased inhibitor tolerance. The integration of the bioconversion process in a more wide production scheme, where all biomass components are exploited and, at the same time, different kind of products (fuels, chemicals, electricity, heat) are produced, is considered as the most effective way to make cellulosic ethanol production profitable.

S. cerevisiae, the microorganism most widely and longer used in fermentation processes is still a paradigm and a model for most modifications and improvements. Powerful tools as the metabolic engineering and the inverse metabolic engineering based on evolutionary engineered *S. cerevisiae* strains advance day by day. Solutions to the problem of the cofermentation of pentoses and hexoses have been proved feasible through the combination of these techniques, showing the way for the integral exploitation of plant biomass. Moreover, the robustness of microorganisms is no longer considered as a black box belonging to industrial processes. Insight on the biochemical mechanisms and the genetic individualities of the robust strains is increasing, as wide genome approaches become accessible to more and more scientists worldwide.

REFERENCES

1. Lynd LR, Laser MS, Brandsby D, Dale BE, Davison B, Hamilton R, Himmel M, Keller M, McMillan JD, Sheehan J, et al. How biotech can transform biofuels. *Nat Biotechnol* 2008, 26:169–172.
2. Popper ZA, Michel G, Herve C, Domozych DS, Willats WGT, Tuohy MG, Kloareg B, Stengel DB. Evolution and diversity of plant cell walls: from algae to flowering plants. *Annu Rev Plant Biol* 2011, 62:567–588.
3. Ragauskas AJ, Williams CK, Davison BH, Britovsek G, Cairney J, Eckert CA, Frederick WJ, Hallett JP, Leak DJ, Liotta CL, et al. The path forward for biofuels and biomaterials. *Science* 2006, 311:484–489.
4. Matano Y, Hasunuma T, Kondo A. Display of cellulases on the cell surface of *Saccharomyces cerevisiae* for high yield ethanol production from high-solid lignocellulosic biomass. *Bioresour Technol* 2012, 108:128–133.
5. Rivers DB, Emert GH. Lignocellulose pretreatment—a comparison of wet and dry ball attrition. *Biotechnol Lett* 1987, 9:365–368.
6. Ramos LP. The chemistry involved in the steam treatment of lignocellulosic materials. *Quim Nova* 2003, 26:863–871.
7. Xiao WP, Clarkson WW. Acid solubilization of lignin and bioconversion of treated newsprint to methane. *Biodegradation* 1997, 8:61–66.
8. Hamelinck CN, van Hooijdonk G, Faaij APC. Ethanol from lignocellulosic biomass: techno-economic performance in short-, middle- and long-term. *Biomass Bioenergy* 2005, 28:384–410.
9. Pan X, Xie D, Kang KY, Yoon SL, Saddler JN. Effect of organosolv ethanol pretreatment variables on physical characteristics of hybrid poplar substrates. *Appl Biochem Biotechnol* 2007, 137:367–377.
10. Lee JW, Gwak KS, Park JY, Park MJ, Choi DH, Kwon M, Choi IG. Biological pretreatment of softwood *Pinus densiflora* by three white rot fungi. *J Microbiol* 2007, 45:485–491.
11. Mansfield SD, Mooney C, Saddler JN. Substrate and enzyme characteristics that limit cellulose hydrolysis. *Biotechnol Progr* 1999, 15:804–816.
12. Cheng G, Varanasi P, Li CL, Liu HB, Menichenko YB, Simmons BA, Kent MS, Singh S. Transition of cellulose crystalline structure and surface morphology of biomass as a function of ionic liquid pretreatment and its relation to enzymatic hydrolysis. *Biomacromolecules* 2011, 12:933–941.
13. Zhou W, Schuttler HB, Hao ZQ, Xu Y. Cellulose hydrolysis in evolving substrate morphologies I: a general modeling formalism. *Biotechnol Bioeng* 2009, 104:261–274.
14. Klemm D, Heublein B, Fink HP, Bohn A. Cellulose: fascinating biopolymer and sustainable raw material. *Angew Chem Int Ed* 2005, 44:3358–3393.
15. Ding SY, Himmel ME. The maize primary cell wall microfibril: a new model derived from direct visualization. *J Agric Food Chem* 2006, 54:597–606.
16. Bhat MK. Cellulases and related enzymes in biotechnology. *Biotechnol Adv* 2000, 18:355–383.
17. Zhang YHP, Himmel ME, Mielenz JR. Outlook for cellulase improvement: screening and selection strategies. *Biotechnol Adv* 2006, 24:452–481.
18. Cantarel BL, Coutinho PM, Rancurel C, Bernard T, Lombard V, Henrissat B. The Carbohydrate-Active EnZymes database (CAZy): an expert resource for glycogenomics. *Nucleic Acids Res* 2009, 37:D233–D238.
19. Quinlan RJ, Sweeney MD, Lo Leggio L, Otten H, Poulsen JCN, Johansen KS, Krogh KBRM, Jorgensen CI, Tovborg M, Anthonsen A, et al. Insights into the oxidative degradation of cellulose by a copper metalloenzyme that exploits biomass components. *Proc Natl Acad Sci USA* 2011, 108:15079–15084.
20. Langston JA, Shaghasi T, Abbate E, Xu F, Vlasenko E, Sweeney MD. Oxidoreductive cellulose depolymerization by the enzymes cellobiose dehydrogenase and glycoside hydrolase 61. *Appl Environ Microbiol* 2011, 77:7007–7015.
21. Phillips CM, Beeson WT, Cate JH, Marletta MA. Cellobiose dehydrogenase and a copper-dependent polysaccharide monooxygenase potentiate cellulose degradation by *Neurospora crassa*. *Acs Chemical Biology* 2011, 6:1399–1406.
22. Kurasin M, Valjamae P. Processivity of cellobiohydrolases is limited by the substrate. *J Biol Chem* 2011, 286:169–177.
23. Beckham GT, Bomble YJ, Bayer EA, Himmel ME, Crowley MF. Applications of computational science for understanding enzymatic deconstruction of cellulose. *Curr Opin Biotechnol* 2011, 22:231–238.
24. Kitago Y, Karita S, Watanabe N, Kamiya M, Aizawa T, Sakka K, Tanaka I. Crystal structure of Cel44A, a glycoside hydrolase family 44 endoglucanase from *Clostridium thermocellum*. *J Biol Chem* 2007, 282:35703–35711.
25. Davies GJ, Brzozowski AM, Dauter M, Varrot A, Schülein M. Structure and function of *Humicola insolens* family 6 cellulases: structure of the endoglucanase, Cel6B, at 1.6 Å resolution. *Biochem J* 2000, 348:201–207.
26. Sandgren M, Shaw A, Ropp TH, Wu S, Bott R, Cameron AD, Ståhlberg J, Mitchinson C, Jones TA. The X-ray crystal

structure of the *Trichoderma reesei* family 12 endoglucanase 3, Cel12A, at 1.9 Å resolution. *J Mol Biol* 2001, 308:295–310.

27. Alzari PM, Souchon H, Dominguez R. The crystal structure of endoglucanase CelA, a family 8 glycosyl hydrolase from *Clostridium thermocellum*. *Structure* 1996, 4:265–75.
28. Brás JLA, Cartmell A, Carvalho ALM, Verze G, Bayer EA, Vazana Y, Correia MAS, Prates JAM, Ratnaparkhe S, Boraston AB, et al. Structural insights into a unique cellulase fold and mechanism of cellulose hydrolysis. *Proc Natl Acad Sci USA* 2011, 108:5237–5242.
29. Davies GJ, Tolley SP, Henrissat B, Hjort C, Schülein M. Structures of oligosaccharide-bound forms of the endoglucanase V from *Humicola insolens* at 1.9 Å resolution. *Biochemistry* 1995, 34:16210–16220.
30. Karkehabadi S, Hansson H, Kim S, Piens K, Mitchinson C, Sandgren M. The first structure of a glycoside hydrolase family 61 member, Cel61B from *Hypocrea jecorina*, at 1.6 Å resolution. *J Mol Biol* 2008, 383:144–54.
31. Martinez-Fleites C, Guerreiro CI, Baumann MJ, Taylor EJ, Prates JA, Ferreira LM, Fontes CM, Brumer H, Davies GJ. Crystal structures of *Clostridium thermocellum* xyloglucanase, XGH74A, reveal the structural basis for xyloglucan recognition and degradation. *J Biol Chem* 2006, 281:24922–24933.
32. Yoshida E, Hidaka M, Fushinobu S, Koyanagi T, Minami H, Tamaki H, Kitaoka M, Katayama T, Kumagai H. Role of a PA14 domain in determining substrate specificity of a glycoside hydrolase family 3 β-glucosidase from *Kluyveromyces marxianus*. *Biochem J* 2010, 431:39–49.
33. Henrissat B, Davies G. Structural and sequence-based classification of glycoside hydrolases. *Curr Opin Struct Biol* 1997, 7:637–644.
34. Medie FM, Davies GJ, Drancourt M, Henrissat B. Genome analyses highlight the different biological roles of cellulases. *Nat Rev Microbiol* 2012, 10:227-U.
35. Omelchenko MV, Galperin MY, Wolf YI, Koonin EV. Non-homologous isofunctional enzymes: a systematic analysis of alternative solutions in enzyme evolution. *Biol Direct* 2010, 5:31.
36. Sukharnikov LO, Cantwell BJ, Podar M, Zhulin IB. Cellulases: ambiguous nonhomologous enzymes in a genomic perspective. *Trends Biotechnol* 2011, 29:473–479.
37. Koivula A, Kinnari T, Harjunpaa V, Ruohonen L, Teleman A, Drakenberg T, Rouvinen J, Jones TA, Teeri TT. Tryptophan 272: an essential determinant of crystalline cellulose degradation by *Trichoderma reesei* cellobiohydrolase Cel6A. *FEBS Lett* 1998, 429:341–346.
38. Himmel ME, Ding SY, Johnson DK, Adney WS, Nimlos MR, Brady JW, Foust TD. Biomass recalcitrance: engineering plants and enzymes for biofuels production. *Science* 2007, 315:804–807.
39. Gusakov AV. Alternatives to *Trichoderma reesei* in biofuel production. *Trends Biotechnol* 2011, 29:419–425.
40. Gruno M, Valjamae P, Pettersson G, Johansson G. Inhibition of the *Trichoderma reesei* cellulases by cellobiose is strongly dependent on the nature of the substrate. *Biotechnol Bioeng* 2004, 86:503–511.
41. Zhang YHP, Cui JB, Lynd LR, Kuang LR. A transition from cellulose swelling to cellulose dissolution by o-phosphoric acid: evidence from enzymatic hydrolysis and supramolecular structure. *Biomacromolecules* 2006, 7:644–648.
42. Gusakov AV, Sinitsyn AP, Salanovich TN, Bukhtojarov FE, Markov AV, Ustinov BB, van Zeijl C, Punt P, Burlingame R. Purification, cloning and characterisation of two forms of thermostable and highly active cellobiohydrolase I (Cel7A) produced by the industrial strain of *Chrysosporium lucknowense*. *Enzyme Microb Technol* 2005, 36:57–69.
43. Olson DG, Tripathi SA, Giannone RJ, Lo J, Caiazza NC, Hogsett DA, Hettich RL, Guss AM, Dubrovsky G, Lynd LR. Deletion of the Cel48S cellulase from *Clostridium thermocellum*. *Proc Natl Acad Sci USA* 2010, 107:17727–17732.
44. Xie G, Bruce DC, Challacombe JF, Chertkov O, Detter JC, Gilna P, Han CS, Lucas S, Misra M, Myers GL, et al. Genome sequence of the cellulolytic gliding bacterium *Cytophaga hutchinsonii*. *Appl Environ Microbiol* 2007, 73:3536–3546.
45. Weiner RM, Taylor LE, Henrissat B, Hauser L, Land M, Coutinho PM, Rancurel C, Saunders EH, Longmire AG, Zhang HT, et al. Complete genome sequence of the complex carbohydrate-degrading marine bacterium, *Saccharophagus degradans* strain 2–40(T). *PLoS Genet* 2008, 4:e1000087.
46. Gilbert HJ. The biochemistry and structural biology of plant cell wall deconstruction. *Plant Physiol* 2010, 153:444–455.
47. Watson BJ, Zhang HT, Longmire AG, Moon YH, Hutcheson SW. Processive endoglucanases mediate degradation of cellulose by *Saccharophagus degradans*. *J Bacteriol* 2009, 191:5697–5705.
48. Fontes CMGA, Gilbert HJ. Cellulosomes: highly efficient nanomachines designed to designed to deconstruct plant cell wall complex carbohydrates. *Annu Rev Biochem* 2010, 79:655–681.
49. Shoseyov O, Shani Z, Levy I. Carbohydrate binding modules: biochemical properties and novel applications. *Microbiol Mol Biol Rev* 2006, 70:283–295.
50. White AR, Brown RM. Enzymatic-hydrolysis of cellulose—visual characterization of the process. *Proc Natl Acad Sci USA* 1981, 78:1047–1051.
51. Sakon J, Irwin D, Wilson DB, Karplus PA. Structure and mechanism of endo/exocellulase E4 from *Thermomonospora fusca*. *Nat Struct Biol* 1997, 4:810–818.
52. Doi RH. Cellulases of mesophilic microorganisms—cellulosome and noncellulosome producers. *Ann N Y Acad Sci* 2008, 1125:267–279.
53. Bayer EA, Lamed R, White BA, Flint HJ. From cellulosomes to cellulosomics. *Chem Record* 2008, 8:364–377.
54. Vaaje-Kolstad G, Westereng B, Horn SJ, Liu ZL, Zhai H, Sorlie M, Eijsink VGH. An oxidative enzyme boosting the enzymatic conversion of recalcitrant polysaccharides. *Science* 2010, 330:219–222.
55. Forsberg Z, Vaaje-Kolstad G, Westereng B, Bunaes AC, Stenstrom Y, MacKenzie A, Sorlie M, Horn SJ, Eijsink

VGH. Cleavage of cellulose by a CBM33 protein. *Protein Sci* 2011, 20:1479–1483.
56. Harris PV, Welner D, McFarland KC, Re E, Poulsen JCN, Brown K, Salbo R, Ding HS, Vlasenko E, Merino S, et al. Stimulation of lignocellulosic biomass hydrolysis by proteins of glycoside hydrolase family 61: structure and function of a large, enigmatic family. *Biochemistry (Mosc)* 2010, 49:3305–3316.
57. Dimarogona M, Topakas E, Olsson L, Christakopoulos P. Lignin boosts the cellulase performance of a GH-61 enzyme from *Sporotrichum thermophile*. *Bioresour Technol* 2012, 110:480–487.
58. Beeson WT, Phillips CM, Cate JH, Marletta MA. Oxidative cleavage of cellulose by fungal copper-dependent polysaccharide monooxygenases. *J Am Chem Soc* 2012, 134:890–892.
59. Conchie J, Levvy GA. Inhibition of glycosidases by aldonolactones of corresponding configuration. *Biochem J* 1957, 65:389–395.
60. Parry NJ, Beever DE, Owen E, Vandenberghe I, Van Beeumen J, Bhat MK. Biochemical characterization and mechanism of action of a thermostable beta-glucosidase purified from *Thermoascus aurantiacus*. *Biochem J* 2001, 353:117–127.
61. Olofsson K, Bertilsson M, Liden G. A short review on SSF—an interesting process option for ethanol production from lignocellulosic feedstocks. *Biotechnol Biofuels* 2008, 1:7.
62. Wingren A, Galbe M, Zacchi G. Techno-economic evaluation of producing ethanol from softwood: comparison of SSF and SHF and identification of bottlenecks. *Biotechnol Prog* 2003, 19:1109–1117.
63. Brethauer S, Wyman CE. Review: continuous hydrolysis and fermentation for cellulosic ethanol production. *Bioresour Technol* 2010, 101:4862–4874.
64. Hoyer K, Galbe M, Zacchi G. Effects of enzyme feeding strategy on ethanol yield in fed-batch simultaneous saccharification and fermentation of spruce at high dry matter. *Biotechnol Biofuels* 2010, 3:14.
65. Lynd LR, Weimer PJ, van Zyl WH, Pretorius IS. Microbial cellulose utilization: fundamentals and biotechnology *Microbiol Mol Biol Rev* 2002, 66:506–579.
66. Olson DG, McBride JE, Joe Shaw A, Lynd LR. Recent progress in consolidated bioprocessing. *Curr Opin Biotechnol.* 2012, 23:396–405.
67. Fujita Y, Ito J, Ueda M, Fukuda H, Kondo A. Synergistic saccharification, and direct fermentation to ethanol, of amorphous cellulose by use of an engineered yeast strain codisplaying three types of cellulolytic enzyme. *Appl Environ Microbiol* 2004, 70:1207–1212.
68. Kondo A, Ueda M. Yeast cell-surface display—applications of molecular display. *Appl Microbiol Biotechnol* 2004, 64:28–40.
69. FitzPatrick M, Champagne P, Cunningham MF, Whitney RA. A biorefinery processing perspective: treatment of lignocellulosic materials for the production of value-added products. *Bioresour Technol* 2010, 101:8915–8922.
70. Bozell JJ, Petersen GR. Technology development for the production of biobased products from biorefinery carbohydrates-the US Department of Energy's "Top 10" revisited. *Green Chem* 2010, 12:539–554.
71. Sprenger GA. Carbohydrate metabolism in *Zymomonas mobilis*: a catabolic highway with some scenic routes. *FEMS Microbiol Lett* 1996, 145:301–307.
72. Seo JS, Chong HY, Park HS, Yoon KO, Jung C, Kim JJ, Hong JH, Kim H, Kim JH, Kil JI, et al. The genome sequence of the ethanologenic bacterium *Zymomonas mobilis* ZM4. *Nat Biotechnol* 2005, 23:63–68.
73. Chatterjee R, Millard CS, Champion K, Clark DP, Donnelly MI. Mutation of the ptsG gene results in increased production of succinate in fermentation of glucose by *Escherichia coli*. *Appl Environ Microbiol* 2001, 67:148–154.
74. Orencio-Trejo M, Utrilla J, Fernandez-Sandoval MT, Huerta-Beristain G, Gosset G, Martinez A. Engineering the *Escherichia coli* fermentative metabolism. *Adv Biochem Eng Biotechnol* 2010, 121:71–107.
75. Miller EN, Jarboe LR, Yomano LP, York SW, Shanmugam KT, Ingram LO. Silencing of NADPH-dependent oxidoreductase genes (yqhD and dkgA) in furfural-resistant ethanologenic *Escherichia coli*. *Appl Environ Microbiol* 2009, 75:4315–4323.
76. Kim Y, Ingram LO, Shanmugam KT. Construction of an *Escherichia coli* K-12 mutant for homoethanologenic fermentation of glucose or xylose without foreign genes. *Appl Environ Microbiol* 2007, 73:1766–1771.
77. Kourtoglou E, Anasontzis GE, Mamma D, Topakas E, Hatzinikolaou DG, Christakopoulos P. Constitutive expression, purification and characterization of a phosphoglucomutase from *Fusarium oxysporum*. *Enzyme Microb Technol* 2011, 48:217–224.
78. Xu Q, Singh A, Himmel ME. Perspectives and new directions for the production of bioethanol using consolidated bioprocessing of lignocellulose. *Curr Opin Biotechnol* 2009, 20:364–371.
79. Skory CD, Freer SN, Bothast RJ. Screening for ethanol-producing filamentous fungi. *Biotechnol Lett* 1997, 19:203–206.
80. Panagiotou G, Christakopoulos P, Olsson L. Simultaneous saccharification and fermentation of cellulose by *Fusarium oxysporum* F3—growth characteristics and metabolite profiling. *Enzyme Microb Technol* 2005, 36:693–699.
81. Xiros C, Topakas E, Katapodis P, Christakopoulos P. Hydrolysis and fermentation of brewer's spent grain by *Neurospora crassa*. *Bioresour Technol* 2008, 99:5427–5435.
82. Kamei I, Hirota Y, Mori T, Hirai H, Meguro S, Kondo R. Direct ethanol production from cellulosic materials by the hypersaline-tolerant white-rot fungus *Phlebia sp.* MG-60. *Bioresour Technol* 2012, 112:137–142.
83. Stevenson DM, Weimer PJ. Isolation and characterization of a *Trichoderma* strain capable of fermenting cellulose to ethanol. *Appl Microbiol Biotechnol* 2002, 59:721–726.

84. Millati R, Edebo L, Taherzadeh MJ. Performance of *Rhizopus, Rhizomucor*, and *Mucor* in ethanol production from glucose, xylose, and wood hydrolyzates. *Enzyme Microb Technol* 2005, 36:294–300.
85. Okamoto K, Nitta Y, Maekawa N, Yanase H. Direct ethanol production from starch, wheat bran and rice straw by the white rot fungus *Trametes hirsuta. Enzyme Microb Technol* 2011, 48:273–277.
86. Agbogbo F, Coward-Kelly G. Cellulosic ethanol production using the naturally occurring xylose-fermenting yeast, *Pichia stipitis. Biotechnol Lett* 2008, 30:1515–1524.
87. Dhaliwal SS, Oberoi HS, Sandhu SK, Nanda D, Kumar D, Uppal SK. Enhanced ethanol production from sugarcane juice by galactose adaptation of a newly isolated thermotolerant strain of *Pichia kudriavzevii. Bioresour Technol* 2011, 102:5968–5975.
88. Nakayama S, Morita T, Negishi H, Ikegami T, Sakaki K, Kitamoto D. Candida krusei produces ethanol without production of succinic acid; a potential advantage for ethanol recovery by pervaporation membrane separation. *FEMS Yeast Res* 2008, 8:706–714.
89. Yanase S, Hasunuma T, Yamada R, Tanaka T, Ogino C, Fukuda H, Kondo A. Direct ethanol production from cellulosic materials at high temperature using the thermotolerant yeast *Kluyveromyces marxianus* displaying cellulolytic enzymes. *Appl Microbiol Biotechnol* 2010, 88:381–388.
90. Ballesteros M, Oliva JM, Negro MJ, Manzanares P, Ballesteros I. Ethanol from lignocellulosic materials by a simultaneous saccharification and fermentation process (SFS) with *Kluyveromyces marxianus* CECT 10875. *Process Biochem* 2004, 39:1843–1848.
91. Cheng KK, Cai BY, Zhang JA, Ling HZ, Zhou YH, Ge JP, Xu JM. Sugarcane bagasse hemicellulose hydrolysate for ethanol production by acid recovery process. *Biochem Eng J* 2008, 38:105–109.
92. Goossens K, Willaert R. Flocculation protein structure and cell-cell adhesion mechanism in *Saccharomyces cerevisiae. Biotechnol Lett* 2010, 32:1571–1585.
93. Ruiz HA, Silva DP, Ruzene DS, Lima LF, Vicente AA, Teixeira JA. Bioethanol production from hydrothermal pretreated wheat straw by a flocculating *Saccharomyces cerevisiae* strain—effect of process conditions. *Fuel* 2012, 95:528–536.
94. Li Q, Zhao XQ, Chang AK, Zhang QM, Bai FW. Ethanol-induced yeast flocculation directed by the promoter of TPS1 encoding trehalose-6-phosphate synthase 1 for efficient ethanol production. *Metab Eng* 2012, 14: 1–8.
95. Mussatto SI, Dragone G, Guimaraes PMR, Silva JPA, Carneiro LM, Roberto IC, Vicente A, Domingues L, Teixeira JA. Technological trends, global market, and challenges of bio-ethanol production. *Biotechnol Adv* 2010, 28:817–830.
96. Jorgensen H, Vibe-Pedersen J, Larsen J, Felby C. Liquefaction of lignocellulose at high-solids concentrations. *Biotechnol Bioeng* 2007, 96:862–870.
97. Alriksson B, Cavka A, Jonsson LJ. Improving the fermentability of enzymatic hydrolysates of lignocellulose through chemical in-situ detoxification with reducing agents. *Bioresour Technol* 2011, 102:1254–1263.
98. Laopaiboon L, Nuanpeng S, Srinophakun P, Klanrit P, Laopaiboon P. Ethanol production from sweet sorghum juice using very high gravity technology: effects of carbon and nitrogen supplementations. *Bioresour Technol* 2009, 100:4176–4182.
99. Almeida JRM, Modig T, Petersson A, Hahn-Hagerdal B, Liden G, Gorwa-Grauslund MF. Increased tolerance and conversion of inhibitors in lignocellulosic hydrolysates by *Saccharomyces cerevisiae. J Chem Technol Biotechnol* 2007, 82:340–349.
100. Wright J, Bellissimi E, de Hulster E, Wagner A, Pronk JT, van Maris AJA. Batch and continuous culture-based selection strategies for acetic acid tolerance in xylose-fermenting *Saccharomyces cerevisiae. FEMS Yeast Res* 2011, 11:299–306.
101. Kuyper M, Toirkens MJ, Diderich JA, Winkler AA, van Dijken JP, Pronk JT. Evolutionary engineering of mixed-sugar utilization by a xylose-fermenting *Saccharomyces cerevisiae* strain. *FEMS Yeast Res* 2005, 5:925–934.
102. Liu EK, Hu Y. Construction of a xylose-fermenting *Saccharomyces cerevisiae* strain by combined approaches of genetic engineering, chemical mutagenesis and evolutionary adaptation. *Biochem Eng J* 2010, 48:204–210.
103. Bailey JE, Shurlati A, Hatzimanikatis V, Lee K, Renner WA, Tsai PS. Inverse metabolic engineering: a strategy for directed genetic engineering of useful phenotypes. *Biotechnol Bioeng* 1996, 52:109–121.
104. Oud B, van Maris AJA, Daran JM, Pronk JT. Genome-wide analytical approaches for reverse metabolic engineering of industrially relevant phenotypes in yeast. *FEMS Yeast Res* 2012, 12:183–196.
105. Stanley D, Chambers PJ, Stanley GA, Borneman A, Fraser S. Transcriptional changes associated with ethanol tolerance in *Saccharomyces cerevisiae. Appl Microbiol Biotechnol* 2010, 88:231–239.
106. Bailey JE. Toward a science of metabolic engineering. *Science* 1991, 252:1668–1675.
107. Laadan B, Almeida JRM, Radstrom P, Hahn-Hagerdal B, Gorwa-Grauslund M. Identification of an NADH-dependent 5-hydroxymethylfurfural-reducing alcohol dehydrogenase in *Saccharomyces cerevisiae. Yeast* 2008, 25:191–198.
108. Hasunuma T, Sanda T, Yamada R, Yoshimura K, Ishii J, Kondo A. Metabolic pathway engineering based on metabolomics confers acetic and formic acid tolerance to a recombinant xylose-fermenting strain of *Saccharomyces cerevisiae. Microb Cell Fact* 2011, 10:2.
109. Cai Z, Zhang B, Li Y. Engineering *Saccharomyces cerevisiae* for efficient anaerobic xylose fermentation: reflections and perspectives. *Biotechnol J* 2012, 7:34–46.
110. Ohgren K, Bengtsson O, Gorwa-Grauslund MF, Galbe M, Hahn-Hagerdal B, Zacchi G. Simultaneous saccharification

and co-fermentation of glucose and xylose in steam-pretreated corn stover at high fiber content with *Saccharomyces cerevisiae* TMB3400. *J Biotechnol* 2006, 126:488–498.

111. Ha SJ, Galazka JM, Kim SR, Choi JH, Yang XM, Seo JH, Glass NL, Cate JHD, Jin YS. Engineered Saccharomyces cerevisiae capable of simultaneous cellobiose and xylose fermentation. *Proc Natl Acad Sci USA* 2011, 108:504–509.

FURTHER READING

Buckeridge MS, Goldman GH. *Routes to Cellulosic Ethanol.* New York: Springer Verlag; 2011, 284.

3

Lipid-Based Liquid Biofuels from Autotrophic Microalgae: Energetic and Environmental Performance

Lucas Reijnders*

Institute for Biodiversity and Ecosystem Dynamics, University of Amsterdam, Amsterdam, the Netherlands

INTRODUCTION

Growing microalgae on a commercial scale for food production has a substantial tradition, dating back to the 1950s (Refs 1–4 and references therein). For this purpose, autotrophic microalgae such as *Spirulina* and *Chlorella* are grown, usually in open ponds.[1] Omega-3 polyunsaturated fatty acids extracted from such microalgae, which are considered to be important for 'healthy diets', are commercially available.[3–6]

Autotrophic microalgae, which convert solar energy into biomass by photosynthesis, have also been proposed as a source of biofuels, especially as a source of lipids (fatty acids, oil) for biodiesel.[1] Table 3.1 outlines the main stages in an autotrophic microalgal biodiesel life cycle and the energetic inputs and co-outputs linked to this life cycle.

Proposals for lipid-based biofuels from autotrophic microalgae were firstly triggered by the oil crisis of the 1970s.[3] Substantial research and development work was done in the 1970s and Goldman[3] concluded at the end of this decade that autotrophic microalgae had little to offer as even a partial solution to national and worldwide energy needs. Since then, substantial research and development work on autotrophic microalgal biofuels has continued, but there is as yet no commercial production of lipid-based autotrophic microalgal biofuels and there is no proof that near-term commercial production would be financially competitive if compared with other sources of energy.[1,7–19] Estimated near-term costs for lipid-based autotrophic microalgal biofuels produced with currently available technologies are much, usually many times, higher than current fossil fuel prices, even if for inputs there is heavy reliance on wastes.[7,8,14,16,17,19] Estimates for near-term costs of autotrophic microalgal lipids from bioreactors exceed those of lipids from cultivation in ponds typically by a factor 4–7.[17] Also, near-term costs of autotrophic microalgal lipids originating from ponds have been estimated to be a factor 3 to 4 higher than near-term costs of competing lignocellulosic (bio)fuels, which in turn are higher than near-term costs of biofuels from edible parts of food crops including terrestrial oil crops.[19] Nor is there consensus about the future commercial viability of microalgal biofuels. Some hold that lipid-based autotrophic microalgal biofuels will become commercially viable and will outperform biofuels from terrestrial plants, whereas others feel that the prospects for commercial algal biofuels are very bleak.[13,17,20–22] In view of the many uncertainties regarding future production technologies, future costs of liquid autotrophic microalgal biofuels are highly uncertain.[22,23] It has been argued that a successful strategy to lower the prices of future microalgal biofuel is to use microalgal biomass as an input for biorefineries, which produce, apart from liquid biofuel, a variety of high-value

*Correspondence to: l.reijnders@uva.nl

Advances in Bioenergy: The Sustainability Challenge, First Edition. Edited by Peter D. Lund, John Byrne, Göran Berndes and Iacovos A. Vasalos.

Originally published in: *WIREs Energy Environ* 2013, 2: 33–44 doi: 10.1002/wene.29

Table 3.1 Main Stages in an Autotrophic Microalgal Biodiesel Life Cycle and the Energetic Aspects of Inputs and Co-Outputs

Stage of the Autotrophic Microalgal Biodiesel Life Cycle	Inputs Requiring energy	Co-Outputs Requiring Energy Inputs for Handling	Co-Outputs Which May Confer Energy Credits
(1) Inoculation and cultivation of autotrophic microalgae in open pond, bioreactor, or combination thereof	-construction, operation, and maintenance -water supply management -CO_2 -nutrients -additives for suitable conditions (where applicable)	-waste water	-treated waste water
(2) Harvesting microalgae	-construction, operation and maintenance of harvester and storage facility lipids	-waste water	-treated waste water
(3) Transport to facility for extraction of lipids	-construction, operation, and maintenance of means of transport		
(4) Extraction of lipids-wet or dry process -in case of wet process, combination with stage 5 is an option	-construction, operation, and maintenance -extractants (e.g., solvents)	-solvents, methanol, and ethanol -wastes	-algal residue (e.g., algal cake) -recycled solvents, methanol, ethanol
(5) Transesterification of algal lipids	-construction, operation, and maintenance -methanol (or ethanol)	-wastes -methanol (or ethanol)	-glycerol -treated/converted waste lipids -recycled methanol (or ethanol)
(6) Upgrading and purification of biodiesel	-construction, operation, and maintenance -substances for upgrading and purification (e.g., H_2)	-wastes	-treated/converted wastes
(7) Transport upgraded and purified biodiesel to pump	-construction, operation, and maintenance of means of transport		
(8) Storage of biodiesel and pumping to car	-construction, operation, and maintenance of storage and pump facility		
(9) Burning biodiesel in car			

Stages 1–6: cradle-to-gate; stages 1–7: cradle-to-pump; all stages: cradle-to-wheel.

substances as coproducts.[23–26] A problem with this strategy is that to make a substantial contribution to future energy supply, the production volume of biofuels should be large. This may easily lead to an oversupply of current high-value substances. This has already happened to glycerol, which is produced as a coproduct of biodiesel. The price thereof has dropped dramatically.[27]

There is also no consensus on the energetic desirability of lipid-based autotrophic microalgal biofuels. Whereas some hold that energetic inputs in commercial autotrophic microalgal biofuel production will be much smaller than the energetic output embodied in those biofuels,[12,13,28] others hold that energetic outputs of microalgal biofuel production are and/or will be lower than, or not much exceed, inputs.[22,27] There is also disagreement about comparative life cycle water consumption and greenhouse gas emissions, while comparing autotrophic microalgal lipid-based biofuels with terrestrial biofuels.[27,29]

It may be noted that there are also other ways to produce liquid biofuels from autotrophic microalgae than from lipids. The microalga *Botrycoccus braunii* generates terpenoids, which may be upgraded to liquid fuels.[30] Alternatively, microalgal biomass may be thermochemically converted to liquid biofuels.[31–35] In this context, there has been research into hydrothermal liquefaction, (preferably) catalytic pyrolysis, and gasification, followed by methanol or Fischer–Tropsch synthesis.[31–35] Research into these alternatives for autotrophic microalgal lipid-based liquid biofuels is still at an early stage, and here the focus will be on the better researched autotrophic microalgal lipid-based liquid biofuels, especially biodiesel.

In the following, first, the major ways to produce biomass from which autotrophic microalgal lipid-based biofuels can be derived and their yields will be surveyed. An energetic criterion for suitable energy sources will be discussed. Then, life cycle assessment (LCA) that can address the energetic and environmental performance of biofuels will be briefly outlined. Thereafter, life cycle energetic inputs and outputs and the life cycle emissions of greenhouse gases and pollutants will be considered.

TECHNOLOGIES FOR THE PRODUCTION OF AUTOTROPHIC MICROALGAE AND THEIR BIOMASS AND LIPID YIELDS

There are essentially two ways to grow autotrophic microalgae: in closed bioreactors and in open systems such as ponds.[11–18,36–46] Open ponds are often raceway ponds with a depth of 10–35 cm, with water mixed by paddlewheels. Three types of closed bioreactors have been found suitable for large-scale algal cultivation. These are the flat plate or flat panel type, the horizontal tubular type, and the vertical column type, whether or not with circulating loop (e.g., Refs 9, 34, 42, and 45). If compared with ponds, bioreactors have the advantage of higher yields (m^{-2}) of area used for algal cultivation and lower evaporative loss of water and the disadvantage of higher inputs of energy and materials and higher costs. There is no agreement on the best type of photobioreactor in view of the efficiency in converting solar irradiation to algal biomass.[9,41,43] Some have argued in favor of vertical column reactors,[41,43] whereas others have come out preferring flat plate reactors with short light path length and little shading.[9] In the latter case, east–west orientation and between-panel distances ranging from 0.2 to 0.4 m are favorable to relatively high yields.[44] There are also approaches to algal biomass production in which autotrophic microalgae are firstly cultivated in closed bioreactors and subsequently in open ponds.[40,47]

When microalgal biofuels are to make a large contribution to future energy supply, the use of open systems would seem to be preferable from the perspective of cost and scale.[11,14,17,37,38]

Following harvesting of algal biomass, several options are open. One option is to concentrate on biofuel production, e.g., producing liquid biofuel and using the residual biomass to generate methane by anaerobic conversion[13,39,48,49] or to be subjected to thermochemical conversion (e.g., Ref 42). Another option is to generate a variety of partially nonenergetic microalgal coproducts, using a biorefinery (e.g., Refs 23–26, 50, and 51). Still another option is the coproduction of biodiesel, glycerol, and algal cake (which may serve as feed) (e.g., Ref 27).

Estimating realistic future biomass and lipid yields from open ponds is hard. On one hand, there are efforts to increase biomass and lipid yields (m^{-2}) of algal cultivation area by selection, by genetic engineering including reconfiguration of the photosynthetic system and metabolic engineering, and by synthetic biology.[11,52–58] It has been proposed to increase biomass yields by genetic engineering for truncated antennae, and to increase biomass and lipid yield by metabolic engineering. Demonstrations of the intended increases in yield under conditions used for commercial production have however been absent so far. Modalities of cultivation, which may increase lipid yields such as additives and N-limitation are also studied.[59–61]

On the other hand, in practice, commercial production tends to generate lower yields than achievable in well controlled experimental settings.[2,61] The differences in yields can apparently be large when production is in open ponds. For the cultivation of *Spirulina*, Vonshak and Richmond,[2] e.g., found the yield from commercial production to be a factor 5 to 6 lower than from well controlled small-scale production. A major problem of growing microalgae in open systems such as ponds is the vulnerability of autotrophic microalgal cultures to contamination with competing microorganisms, grazers (consumers of algae), and infective agents, including viruses and bacteria.[37,61–64] In practice, strategies to keep out competing microorganisms and grazers in open systems have led to commercial growing of autotrophic algae suitable for lipid-based liquid biofuel production under extreme conditions (such as a high pH or a high NaCl concentration).[61,63] Such extreme conditions are not conducive to high biomass yields. And in such systems, production may still be lowered because of, e.g., heavy rainfall (leading to less extreme conditions) and infections.[61] So far, there has apparently been no practical demonstration of successful open systems continuously producing autotrophic microalgae for biofuels, while not employing extreme conditions.[61,63] Moreover, expansion of algal cultivation may well lead to an increase of infection pressure, negatively affecting future yields.[64] All in all, future lipid yields in commercial open ponds, which—as pointed out above—would seem preferable from the perspective of cost and scale, would seem as yet highly uncertain.

AN ENERGETIC CRITERION FOR ENERGY SOURCES

To make energetic sense, energy sources should show a good energetic return on energy investment.[65–68] This return, energy return on investment, abbreviated as EROI, is the ratio of energy delivered to the life cycle input of (man-made) energy. Current major transport energy sources have EROIs of more than 5 (e.g., Refs 27, 66, and 67). This holds both for mineral oil-based fuels such as gasoline and diesel and for electricity.[27,66,67] An EROI of more than 5 also holds for novel sources of electricity such as photovoltaic cells and wind power systems.[67] A meta-analysis of operational studies did, e.g., show an EROI of on average 19.8 for wind power systems.[67] It has been suggested that EROIs of more than 5 will be required to maintain quality of life in the absence of readily abundant fossil energy and this criterion has been used previously in the context of microalgal biofuels.[68,69] In line with this, an EROI of at least 5 is used here as energetic criterion for a suitable liquid algal biofuel.

LIFE CYCLE ASSESSMENT

Products, such as biofuels, have life cycles, which stretch 'from cradle to grave', or from resource extraction to emissions and waste disposal. LCAs are often considered the proper way to compare the energetic and environmental performance of specific biofuels with other fuels or types of energy supply.[27,36,38,39,70,71]

LCA has been developed for analyzing current products from resource extraction to final waste disposal or emissions. Apart from analyzing the status quo, LCAs may also deal with novel products and production processes. The latter type of assessment has been called *consequential*, as distinguished from the analysis of status quo products, which has been called *attributional*.[72–74] LCAs of microalgal biofuels are often consequential, as they refer to future commercial production systems, which are as yet not operational.

Different data and approaches may be needed in attributional and consequential LCA.[27,73–77] Regarding microalgal biofuels, knowledge often partly or fully relates to the research and development stage or to the limited production stage. Extrapolation of such knowledge to future production systems gives rise to large uncertainties. The same holds, as pointed out before, for estimation of future yields of algal biomass and lipids.

Also, future microalgal biofuels may be subject to currently unavailable environmental improvement options and may have to operate under conditions that diverge from those that are currently common. The latter conditions may, e.g., include higher resource costs, which are conducive to resource efficiency. Also, the comparison with other products may be different in consequential LCA. For instance, it is expected that life cycle emissions of fossil transport fuels will go up in the future, whereas their EROI will go down,[77–79] and this may increase the relative attractiveness of microalgal biofuels if compared with fossil fuels, although not if compared with solar and wind energy.[63]

LCA is generally divided in the following four stages[70,71]:

- goal and scope definition;
- inventory analysis;
- impact assessment; and
- interpretation.

In the *goal and scope definition* stage, the aim and the subject of LCA are chosen. This implies the establishment of 'system boundaries' and usually the definition of a 'functional unit'. A functional unit is a quantitative description of the service performance of products. It may for instance be: 1 GJ (GigaJoule) of biofuel (measured as lower heating value). This allows for comparing different products providing the same service.

In the *goal and scope definition* stage, system boundaries are drawn between technological systems and the environment, between significant and insignificant processes, and between technological systems. For instance, in dealing with microalgal biofuel production, one has to decide whether, when applicable, water management facilities serving algal cultivation and wastewater treatment of effluents from bioreactors or ponds and harvesting facilities (cf. Table 3.1) should be included. It would seem proper to do so. When algal cultivation includes the use of wastes as inputs in the production of microalgae, the extent to which the generation and treatment of such wastes should be included in LCA has to be decided upon. Also when considering the emission of greenhouse gases, in establishing new algal cultivation facilities, one should consider the inclusion of land use references, reflecting what would otherwise have happened with the land used for the biofuel production studied.[27] When algal cultivation for biofuels replaces algal cultivation for food and greenhouse gas emissions are considered, the indirect effect on land use linked to the inelasticity of demand for food should be considered for inclusion in LCA.[80–82] The choice of system boundaries may have a substantial effect on the outcomes of LCAs.[27,83,84]

The *inventory analysis* gathers the necessary data for all processes involved in the product life cycle and the *impact assessment* stage deals with estimating impacts of the life cycle.

When dealing with the environmental impacts of microalgal products in these stages of LCA, the problem arises that the conversion of microalgae into biodiesel may have more than one output (e.g., Refs 27 and 39). For instance, processing *Chlorella* biomass into biodiesel not only may lead to the output oil (lipids), which may be used for biodiesel production, but also to algal cake, which may be used as (animal) feed ingredient, and to the output glycerol (cf. Table 3.1). Also, biorefineries that may process microalgal biomass can produce a variety of product outputs.[29,50,51] Biofuel production processes may generate wastes (nonproduct outputs)[27] and nonproduct outputs of other processes (such as flue gases and waste water) may be used as inputs for the production of microalgae.

In the case of multi-output processes, extractions of resources and emissions have to be allocated to the different outputs. There are several ways to do so. Major ways to allocate are based on physical units (e.g., energy content or weight of outputs) or on monetary value (price). There may also be allocation on the basis of substitution. In the latter case, the environmental burden of a coproduct is established on the basis of another, similar product. Occasionally, there is disagreement about the question whether there should be allocation to any biofuel coproduct at all.[27] There is often, although not always, no allocation to 'wastes',[27] but this can change when the nonproduct output is found to be useful as an input in a production process. In the latter case, 'wastes' might be considered coproducts. On the other hand, coproducts may be 'overproduced' to such an extent that they are handled as 'wastes'. This has for instance happened with glycerol, a coproduct of biodiesel production.[27] Changes of coproduct into nonproduct output vice versa may lead to changes in allocation. Different kinds of allocation may lead to different outcomes of LCA.[27,81,82]

The *interpretation* stage of LCA connects the outcome of the assessment done to the real world. In this stage, uncertainties in outcome can be considered and conclusions can be drawn.

RESULTS OF LCAs RELEVANT TO THE EROI OF AUTOTROPHIC MICROALGAL LIPID-BASED BIOFUELS

A number of attributional and consequential LCAs of biomass, lipids, biodiesel, and other biofuels derived from autotrophic microalgae are available (Refs 13, 19, 22, 27, 29, 36, 38, 40, 68, 85–102, and references therein). This is a large increase if compared with an earlier review of the energetic performance of microalgal biofuels, which appeared in 2009.[61] Available LCAs tend to vary much in their quality, system boundaries, ways of allocation, assumptions about inputs and outputs and outcomes. Table 3.2 summarizes the EROIs and key assumptions of peer-reviewed LCAs of biodiesel from autotrophic microalgae, as they have appeared in the 2009–2011 period.

From the available LCAs, several generalizing conclusions may be drawn regarding their energetic implications.

It is noteworthy that several energetically relevant aspects of the life cycle tend to be outside system boundaries of LCAs as they have been done so far (see also Table 3.2).

The first aspect which is rather often absent is (waste) water treatment, although there are exceptions such as the LCAs of Clarens et al.[38,68] and Lardon et al.[90] To the extent that waste water is discharged, loading of waste water with nutrients and, in the case of ponds, with substances conducive to achieving extreme conditions would seem such that waste water treatment is warranted. To the extent that water is recycled, the removal of inhibitory waste products and pathogens is necessary, which may require substantial energy inputs, as suggested by water recycling in aquaculture and horticultural irrigation water for greenhouses.[103–105] Energy inputs into water supply management preceding algal cultivation is also often outside the system boundaries. An exception is for instance the study of Murphy and Allen.[29]

Another aspect that tends to be absent in LCA is the treatment of microalgal lipids to obtain purified and/or upgraded biodiesel, beyond the transesterification step of the extracted microalgal lipids. Purified and/or upgraded biodiesel is needed for a wide range of applications.[9,106–108] Because microalgal lipids as yet tend to be polyunsaturated,[3–6] upgrading by, e.g., hydrogenation is needed when biodiesel for the internal combustion engine is produced.[8,27,108] Both upgrading and purification require substantial energetic inputs.[19,27,30,107–109] It may be noted, although, that algal strain selection may focus on increased levels of saturated fatty acids.[30]

Available LCAs which include inputs in hardware (ponds and photobioreactors) suggest that, per functional unit, open ponds are more energy efficient than current photobioreactors in spite of the much higher autotrophic microalgal yields of the latter, although there are substantial differences between photobioreactors dependent on design and materials use (e.g., Refs 46, 64, 91, and references therein).

With currently operational technologies, the EROIs of lipid-based biofuels derived from autotrophic microalgae may well be < 1 e.g.[22,61,91,95]. None of the available LCAs performed so far for future liquid microalgal biofuels finds an EROI of more than 5, whatever the system boundaries and type of allocation chosen (see also Table 3.2). Even with optimistic assumptions about future biomass and lipid yields, estimated EROIs do not always turn out to be more than 1. A recent case in point is the paper of Brentner et al.,[93] who studied cultivation in both open ponds and bioreactors. They suggest an EROI of less than 1 while using an optimistic assumption about future lipid yield (53,200 l oil ha^{-1} $year^{-1}$).[93] Khoo et al.,[99] who consider a bioreactor/pond combination, assume a yearly

Table 3.2 Key Assumptions and Estimated EROIs in Peer-Reviewed Life Cycle Assessments of Biodiesel from Autotrophic Microalgae Published in the 2009–2011 Period

Authors and References	Type of Biofuel and Cultivation	System Boundaries	Type of LCA	Approximate Yield (in Dry weight ha^{-1} $year^{-1}$)	Allocation to Coproducts	Estimated Energetic Return on Investment (EROI)
Clarens et al.[38,68]	Biodiesel from pond	Cradle-to-wheel[1]; water supply management and biodiesel upgrading and purification not included	Consequential	91 Mg biomass, 19.6% lipids	Energy credits	0.65–4.1
Stephenson et al.[40]	Biodiesel from pond, waste CO_2	Cradle-to-wheel[1]; water supply management, waste water treatment, and biodiesel upgrading and purification not included	Consequential	20 Mg lipids	Price based	1.87
Xu et al.[88]	Biodiesel from pond	Cradle-to-wheel[1]; water supply management, waste water treatment, and biodiesel upgrading and purification not included	Consequential	38.3–50 Mg biomass, 19.7% lipids	Energy credits	< 1.5; 1.83–2.53 if waste heat is available
Lardon et al.[90]	Biodiesel from pond	Cradle-to-wheel[1]; water supply management and biodiesel upgrading and purification not included	Consequential	68–89 Mg biomass, 17.5–38.5% lipids	Energetic	<1–4.3
Sander and Murthy[91]	Biodiesel from pond	Cradle-to-pump[1]; water supply management and biodiesel upgrading and purification not included	Attributional	'Maximized growth'; base case: 30% lipids	Energy credits on the basis of substitution	0.16–0.3
Brentner et al.[93]	Biodiesel from pond or bioreactor	Cradle-to-gate[1]; water supply management and biodiesel upgrading and purification not included	Consequential	>15 Mg biomass, 'potential production' lipids: 53200 l	Energy credits	0.13–0.93
Razon and Tan[95]	Biodiesel from pond	Cradle-to-gate[1]; water supply management, waste water treatment, and biodiesel upgrading and purification not included	Consequential	58–91 Mg biomass, 29.6–35.7% lipids	Energy credits	<0.13
Khoo et al.[99]	Biodiesel from bioreactor/ pond combination	Cradle-to-gate[1]; solvent recycling, coproduct and water management and biodiesel upgrading and purification outside system boundaries	Consequential	91 Mg biomass, 25–45% lipids	None	<0.25–0.6
Batan et al.[101]	Biodiesel from bioreactor	Cradle-to-pump[1]; water supply management, waste water treatment, and biodiesel upgrading and purification not included	Consequential	91 Mg, 50% lipids	Energy credits	0.93
Lohrey and Kochergin[102]	Biodiesel from pond; waste inputs from sugar factory	Cradle-to-gate[1]; waste water treatment and biodiesel upgrading and purification not included	Consequential	73 Mg, 30% lipids	Energy credits	About 1.5

[1]Terms explained in footnote of Table 3.1.

microalgal biomass yield of approximately 91 mg ha^{-1}, a lipid content of 45% and a reduction of energy input in processing algal culture to biodiesel by 60%, but still estimate an EROI of less than 1. Moreover, it may be noted that currently available LCAs may often overestimate EROIs due to noninclusion within system boundaries of energetically important aspects of the life cycle, such as water management preceding algal cultivation (see also Table 3.2). For instance, Murphy and Allen[29] studying fresh water-fed ponds for the mass cultivation of algae in the continental United States estimated that energy inputs associated with water management would exceed the energy content of biomass produced (43.2 mg dry weight ha^{-1} year^{-1}) by a factor 7.[29]

To the extent that available LCAs deal with the aspects upstream from harvesting autotrophic microalgae in ponds, relatively positive energetic outcomes (though the EROI is still <5) seem to be strongly dependent on the assumptions that microalgal lipid yields per hectare are high (often assuming biomass yields in the order of 80–90 mg year^{-1} and/or lipid contents of 25–50%), and/or assuming that there are low or no energy inputs in water supply and/or that inputs of nutrients, CO_2, or heat are available from wastes, with no environmental burden allocated on these inputs.

Assumed yields of 70–90 mg algal dry weight ha^{-1} year^{-1} are much higher than actual microalgal yields from current commercial ponds (e.g., Refs 22, 27, 61, 92, and references therein). Whether these high yields combined with very high lipid contents can be achieved in the future is uncertain. The assumption that very high biomass yields can be combined with very high lipid contents does not take sufficient account of the increase in photon demand, which is associated with high lipid contents.[57] The proposal to achieve much higher algal biomass yields than currently achieved by genetic engineering for truncated antennae is also problematic. Algae with truncated antennae are less competitive than their counterparts with full antennae.[58] This may well have a negative impact on their yields in open ponds where contamination by airborne full-antenna organisms may occur.[58]

Relatively good energetic outcomes for liquid microalgal lipid-based biofuels have, as pointed out above, been achieved in available LCAs when wastes (e.g., waste heat, waste water, including nutrients present therein) are used as inputs and the environmental burden of these inputs is valued at 0 (e.g., Refs 68, 88, 93, and 102). The assumption that the environmental burden of 'wastes' can be valued at 0 merits further consideration. In line with the tenets of industrial ecology,[109] wastes are increasingly viewed as resources. This also holds for waste water. The nutrients present therein are increasingly considered for recovery to be applied in food production.[110] Moreover, when microalgal biofuels are to become major suppliers of energy limitations to the availability of CO_2, water and nutrient wastes may emerge.[111] In view thereof, the absence of any allocation of environmental burdens to the use of 'wastes' as inputs in microalgal production is debatable, as such inputs might also be considered coproducts. If the latter assumption is subscribed to, the environmental burden of the production process should partly be allocated to the inputs in the production of microalgae, which are currently considered to be 'wastes'. This would negatively affect the estimated EROI of microalgal biofuels.

To the extent that aspects downstream from the production of microalgae in ponds are concerned, the assumed use of algal biofuel residues to generate energy and to provide for nutrient inputs leads to relatively low life cycle energy inputs. So does the assumption that energy inputs in processing of algal cultures to biodiesel can be lowered much. The substitution of dry technologies for the isolation and processing of algae by wet technologies has been assumed to be beneficial to expected energy efficiency in producing microalgal biofuels (e.g., Ref 93). Examples of such wet technologies are treatments with supercritical methanol,[93] ethanol,[112] or CO_2 plus solvent[113] and the *in situ* transesterification with very high amounts of methanol.[114] Still, even when wet technologies are applied, the algae *Chlorella*, *Spirulina* and *Dunaliella*, which are grown at a density typically around approximately 0.1–0.6 g/L, have to be concentrated to about 10–15% solids in water, which is resource intensive.[18,115] Also, algal cell walls are sturdy, which requires relatively large life cycle energy inputs to get the lipids out.[113] Supercritical technologies are furthermore characterized by relatively large life cycle energy inputs if compared with conventional catalytic transesterification.[116] Also, supercritical transesterification of polyunsaturated fatty acids, as commonly present in autotrophic microalgae,[3–6] may lead to substantial reductions in biodiesel yield.[117] In the case of *in situ* transesterification of algal biomass, furthermore, very high inputs of methanol and methanol recovery are needed,[114] which are not conducive to life cycle energy efficiency.[116] Thus, the assumptions about the beneficial effects of wet technologies on EROI may be overly optimistic.

It has furthermore been argued that energy inputs may be reduced when benthic cyanobacteria are used for biofuel production, as these may form dense mats at the surface of the culture medium.[118] Another option which has been suggested to facilitate harvesting is culturing attached microalgal biofilms.[18] Liquid biofuel yields achievable by these approaches are as yet highly uncertain as research into these options is at an early stage.

It has also been suggested that the increase in photon demand linked to high lipid concentrations might be countered by the excretion of lipids from algal cells, which may be achieved by genetic engineering of algae or cyanobacteria.[56,119,120] This would also circumvent the need for breaching sturdy cell walls to get the lipids out.[113] What liquid biofuel yields might be achieved by using algae or cyanobacteria which excrete lipids is, however, again highly uncertain as research into this option is at an early stage.

The many uncertainties about the future development of biofuel production on the basis of autotrophic microalgal lipids do not allow for firm predictions about future EROIs. However, it seems likely that an EROI of more than 5 will remain elusive unless major breakthroughs emerge, which may, or may not, occur. This also seems to hold for alternative options to generate liquid biofuels from autotrophic algae.[121,122]

LCAs OF GREENHOUSE GAS EMISSIONS LINKED TO AUTOTROPHIC MICROALGAL LIPID-BASED BIOFUELS

Nonenergetic environmental aspects have been addressed in LCAs in a relatively limited way. Quite a number of LCAs have however addressed life cycle greenhouse gas emissions of lipids or methanol from autotrophic microalgae (e.g., Refs 19, 27, 38, 68, 86, 87, 90, 91, 93, 94, 100, 101, 121, and 122). Several types of emissions can be expected to contribute significantly to such emissions.[27] These are the CO_2 emissions linked to fossil fuel inputs, life cycle N_2O emissions largely linked to nutrient N inputs, and CO_2 emissions or sequestration linked to changes in ecosystem carbon stocks.[27] CO_2 emissions linked to fossil fuel inputs have been addressed in LCAs, but the questions previously raised regarding the system boundaries chosen, such as the exclusion of waste water treatment, also apply to estimates of these emissions. Fossil fuel inputs linked to building the infrastructure for growing algae, such as ponds, are not always included (e.g., Ref 19). And, as emerged in the previous section, future fossil fuel inputs in generating autotrophic microalgal biodiesel are highly uncertain. Life cycle N_2O emissions have been neglected or estimated using assumptions, which were not based on empirical data about the cultivation of autotrophic microalgae and the downstream fate of N-compounds in waste water of such cultivation.[123] Reference land use, which allows for estimates of changes in ecosystem carbon stocks, is excluded from currently available LCAs. This exclusion may impact estimated greenhouse gas balances. This impact would be small in settings that sequester little C, such as deserts, but may matter much in settings where forests would be natural ecosystems.[27] Indirect effects of land use change on ecosystem carbon stocks are also not included in available LCAs.

As it stands, available LCAs of greenhouse gas emissions linked to autotrophic microalgal biofuels do not seem to allow for firm estimates of future greenhouse gas emissions linked to such biofuels.

LIFE CYCLE POLLUTION LINKED TO AUTOTROPHIC MICROALGAL LIPID-BASED BIOFUELS

Life cycle pollution linked to autotrophic microalgal lipid-based biofuels has been addressed by Hou et al.,[100] assuming a yield of 109 mg dry weight biomass ha^{-1} $year^{-1}$ and a lipid content of 45%. When waste water treatment is not included, emissions of substances characterized by human toxicity and ecotoxicity and of eutrophying substances were estimated to be larger for autotrophic microalgal lipid-based biofuels than for fossil fuels. Lardon et al.,[90] who included waste water treatment in their assessment, found microalgal biodiesel and fossil diesel performing roughly similar as to eutrophication, but also found microalgal biodiesel to do worse than fossil diesel regarding aspects such as acid deposition, human and marine toxicity, and photochemical oxidation. Clarens et al.,[68] who also included waste water treatment, found autotrophic microalgal biofuels in terms of pollution to be a larger environmental burden than biofuels derived from switchgrass or canola. Thus, apart from greenhouse gases, life cycle pollutant emissions of autotrophic microalgal biodiesel may well be more of an environmental burden than the life cycle emissions linked to fossil diesel or biofuels based on canola or switchgrass.

CONCLUSIONS

Near-term liquid autotrophic microalgal lipid-based biofuel production with an EROI of more than 5 and costs which are not much higher than the prices for fossil fuels is highly unlikely. Whether the prospects for microalgal biofuel production in the more distant future will much improve is uncertain. As discussed above, in several cases, assumptions about the future impact of proposals for improving energy return on investment seem overly optimistic. In other cases, the impact of proposals for improvement of EROI is highly uncertain because research is at an early stage. So, prospects for autotrophic microalgal lipid-based biofuels appear to be largely dependent on breakthroughs in production technology which may, or may not, occur.

ACKNOWLEDGEMENTS

The comments of the anonymous reviewers are gratefully acknowledged.

REFERENCES

1. Mata T, Martins AA, Caetano NS. Microalgae for biodiesel production and other applications. *Renew Sustain Energy Rev* 2010, 14:217–232.
2. Vonshak A, Richmond A. Mass production of the blue-green alga Spirulina: an overview. *Biomass* 1988, 15:233–247.
3. Goldman JC. Outdoor algal mass cultures—applications. *Water Res* 1979, 13:1–19.
4. Spolaore P, Joannis-Cassan C, Duran E, Isambert A. Commercial applications of microalgae. *J Biosci Bioeng* 2006, 101:87–96.
5. Rubio-Rodriguez N, Beltran S, Jaime I, de Diego SM, Sanz MT, Carballido JR. Production of omega-3 poly unsaturated fatty acid concentrates: a review. *Innov Food Sci Emerg Technol* 2010, 11:1–12.
6. Mercer P, Armenta RE. Developments in oil extraction from microalgae. *Wur J Lipid Sci Technol* 2011, 113:539–547.
7. Davis R, Aden A, Pienkos PT. Techno-economic analysis of autotrophic microalgae for fuel production. *Appl Energy* 2011, 88:3524–3531.
8. Carriquiry MA, Du X, Timilsina GR. Second generation biofuels: economics and policies. *Energy Policy* 2011, 39:4222–4234.
9. Morweiser M, Kruse O, Hankamer B, Posten C. Development and perspectives of photobioreactors for biofuel production. *Appl Microbiol Biotechnol* 2010, 87:1291–1301.
10. Service RF. Algae's second try. *Science* 2011, 333:1238–1239.
11. Hall CAS, Benemann JT. Oil from algae? *BioSci* 2011, 61:741–742.
12. Chisti Y. Biodiesel from microalgae beats bioethanol. *Trends Biotechnol* 2008, 28:126–131.
13. Chisti Y. Response to Reijnders: do biofuels from microalgae beat biofuels from terrestrial plants. *Trends Biotechnol* 2008, 26:351.
14. Sun A, Davis R, Starbuck M, Ben-Amotz A, Pate R, Pienkos PT. Comparative cost analysis of algal oil production for biofuels. *Energy* 2011, 36:5169–5179.
15. Adey WH, Kangas PC, Mulbry W. Algal turf scrubbing: cleaning surface waters with solar energy while producing a biofuel. *BioSci* 2011, 61:434–441.
16. Rosenberg JN, Mathias A, Korth K, Betenbaugh MJ, Oyler GA. Micoalgal biomass production and carbon dioxide sequestration from an integrated biorefinery in Iowa: a technical appraisal and economic feasibility evaluation. *Biomass Bioenerg* 2011, 35:3865–3876.
17. Amer L, Adhikari B, Pellegrino J. Technoeconomic analysis of five microalgae-to-biofuels processing of varying complexity. *Bioresour Technol* 2011, 102:9350–9359.
18. Christenson L, Sims R. Production and harvesting of microalgae for wastewater treatment, biofuels, and byproducts. *Biotechnol Adv* 2011, 29:686–702.
19. Agusdinata DB, Zhao F, Ileleji KE, DeLaurentis D. Life cycle assessment of potential bio-jet fuel production in the United States. *Environ Sci Technol* 2011, 45:9133–9143.
20. Singh A, Olsen SI, Nigam PS. A viable technology to generate third-generation biofuel. *J Chem Technol Biotechnol* 2011, 86:1349–1353.
21. Santyanarayana KG, Mariano AB, Vargas JVC. A review on microalgae, a versatile source for sustainable energy and materials. *Int J Energy Res* 2011, 35:291–311.
22. van Beilen JB. Why microalgal biofuels won't save the internal combustion machine. *Biofuels Bioprod Biorefin* 2010, 4:41–52.
23. Wijffels R, Barbosa MJ. An outlook for microalgal biofuels. *Science* 2010, 329:796–799.
24. Zhang YHP. What is vital (and not vital) to advance economically-competitive biofuels production. *Process Biochem* 2011, 46:2091–2110.
25. Subhadra BG, Edwards M. Coproduct market analysis and water footprint of simulated algal biorefineries. *Appl Energ* 2011, 88:3515–3523.
26. Ghatak HR. Biorefineries from the perspective of sustainability. *Renew Sustain Energ Rev* 2011, 15:4042–4052.
27. Reijnders L, Huijbregts MAJ. *Biofuels for Road Transport: A Seed to Wheel Perspective*. London: Springer; 2009.
28. Brennan L, Owende P. Biofuels from microalgae—a review of technologies for production, processing, and extractions of biofuels and coproducts. *Renew Sustain Energ Rev* 2010, 14:557–577.
29. Murphy CF, Allen DT. Energy-water nexus for mass cultivation of algae. *Environ Sci Technol* 2011, 45:5861–5868.
30. Tran NH, Bartlett JR, Kannangara GSK, Milev AS, Volk H, Wilson MA. Catalytic upgrading of biorefinery oil from micro-algae. *Fuel* 2010, 89:265–274.
31. Miao X, Wu Q, Yang C. Fast pyrolyis of microalgae to produce renewable fuels. *J Anal Appl Pyrolysis* 2004, 71:855–863.
32. Vardon DR, Sharma BK, Scott J, Yu G, Wang Z, Schideman L, Zhang Y, Strathmann TJ. Chemical properties of biocrude oil from the hydrothermal liquefaction of Spirulina algae, swine manure, and digested anaerobic sludge. *Bioresour Technol* 2011, 102:8295–8303.
33. Babich IV, van der Hulst M, Lefferts L, Moulijn JA, O'Connor P, Seshan K. Catalytic pyrolysis of microalgae to high-quality liquid bio-fuels. *Biomass Bioenerg* 2011, 35:3199–3207.
34. Hirano A, Hon-Nami K, Kunito S, Hada M, Ogushi Y. Temperature effect on continuous gasification of microalgal biomass: theoretical yield of methanol production and its energy balance. *Catal Today* 1998, 45:399–404.
35. Amin S. Review of biofuel oil and gas production processes from microalgae. *Energy Conv Manage* 2009, 50:1834–1840.
36. Jorquera O, Kiperstock A, Sales EA, Embirucu M, Ghirardi ML. Comparative energy life-cycle analyses of microalgal

biomass production in open ponds and photobioreactors. *Bioresour Technol* 2010, 101:1406–1413.
37. Smith VH, Sturm BSM, de Noyelles FJ, Billings SA. The ecology of algal biodiesel production. *Trends Ecol Evol* 2009, 15:301–309.
38. Clarens AF, Resurreccion EO, White MA, Colosi LM. Environmental life cycle comparison of algae to other bioenergy feedstocks. *Environ Sci Technol* 2010, 44:1813–1819.
39. Singh A, Olsen SI. A critical review of biochemical conversion, sustainability and life cycle assessment of algal biofiuels. *Appl Energy* 2011, 88:3548–3555.
40. Stephenson AL, Kazamia E, Dennis JS, Howe CJ, Scott SA, Smith AG. Life cycle assessment of potential algal biodiesel production in the United Kingdom: a comparison of raceways and air lift tubular bioreactors. *Energy Fuels* 2010, 24:4062–4077.
41. Kunjapur AM, Eldridge RB. Photobioreactor design for commercial biofuel production from microalgae. *Ind Eng Chem Res* 2010, 49:3516–3526.
42. Hulatt CJ, Thomas DN. Productivity, carbon dioxide uptake and net energy return of microalgal bubble column photobioreactors. *Bioresour Technol* 2011, 102:5775–5787.
43. Cuaresma M, Janssen M, Vilchez C, Wijffels RH. Horiontal or vertical bioreactors? How to improve microalgae photosynthetic efficiency. *Bioresour Technol* 2011, 102:5129–5137.
44. Slegers PM, Wijffels RH, van Straten G, van Boxtel AJB. Design scenarios for flat panel bioreactors. *Appl Energ* 2011, 88:3342–3353.
45. Sasi D, Mitra P, Vigueras A, Hill GA. Growth kinetics and lipid production using *Chlorella* vulgaris in a circulating loop bioreactor. *J Chem Technol Biotechnol* 2011, 86:875–880.
46. Hundt K, Reddy BV. Algal biodiesel production from power pant exhaust and its potential to replace petrodiesel and reduce greenhouse gas emissions. *Int J Low Carbon Technol* 2011, 6:294–298.
47. Huntley ME, Redalje DG. CO_2 mitigation and renewable oil from photosynthetic microbes. A new appraisal. *Mitig Adapt Stat Global Change* 2006, 12:573–608.
48. Ehimen EA, Sun ZD, Carrington CG, Birch EJ, Eaton-Rye JJ. Anaerobic digestion of microalgae residues resulting from the biodiesel production process. *Appl Energ* 2011, 88:3454–3463.
49. Phukan MM, Chutia RS, Konwar BK, Kataki R. Microalgae *Chlorella* as a potential bio-energy feedstock. *Appl Energ* 2011, 88:3307–3312.
50. Sing J, Gu S. Commercialization potential of microalgae for biofuels production. *Renew Sustain Energy Rev* 2010, 14:2596–2610.
51. Jacob-Lopes E, Franco TT. Microalgae-based systems for carbon dioxide sequestration and industrial biorefineries. In: Momba M, Bux F, eds. *Biomass*. Croatia: Sciyo.com; 2010
52. Zeng X, Danquah MK, Chen XD, Lu Y. Microalgae engineering: from CO_2 fixation to biofuel production. *Renew Sustain Energ Rev* 2011, 15:3252–3260.
53. Day JG, Slocombe SP, Stanley MS. Overcoming biological constraints to enable the exploitation of microalgae for biofuels. *Bioresour Technol* 2012, 109:245–251.
54. Quintana N, van der Kooy F, van de Rhee MD, Voshol GD, Verpoorte R. Renewable energy from cycanobacteria: energy production optimization by metabolic pathway engineering. *Appl Microbiol Biotechnol* 2011, 91:471–490.
55. Alper H, Stephanopoulos G. Engineering for biofuels: exploiting innate microbial capacity or importing biosynthetic potential? *Nat Microbiol* 2009, 7:715–723.
56. Connor MR, Stsumi S. Synthetic biology guides biofuel production. *J Biomed Biotechnol* 2010, 541698. doi:10.1155/2010/541698
57. Wilhelm C, Jakob T. From photons to biomass and biofuels: evaluation of different strategies for the improvement of algal biotechnology based on comparative energy balances. *Appl Microbiol Biotechnol* 2011, 92:909–919.
58. Ort DR, Zhu X, Melis A. Optimizing antenna size to maximize photosynthetic efficiency. *Plant Physiol* 2011, 155: 79–85.
59. Gomaa AE, Hafez EE, Lim HS, Chung GH, Sun SM. Physiological studies on microalgal culture additives to optimize growth rate and oil content. *Bioprocess Bosyst Eng* 2012, 35:135–143.
60. Griffith MJ, van Hille RP, Harrison STL. Lipid productivity, settling potential and fatty acid profile of 11 microalgal species grown under nitrogen replete and limited conditions. *J Appl Phycol* 2011. doi:10.1007/s10811-011-9723-y
61. Reijnders L. Microalgal and terrestrial transport biofuels to displace fossil fuels. *Energies* 2009, 2:48–56.
62. Amaro HM, Guedes AC, Malcata FX. Advances and perspectives in using microalgae to produce biodiesel. *Appl Energ* 2011, 88:3402–3410.
63. Park JNK, Craggs RJ, Shilton AN. Wastewater treatment high rate algal ponds for biofuel production. *Bioresour Technol* 2011, 102:35–42.
64. Gachon CMM, Sime-Ngando T, Strittmatter M, Chambouvet A, Kim GH. Algal diseases: spotlight on a black box. *Trends Plant Sci* 2010, 15:633–640.
65. Cleveland CJ, Costanza R, Hall CAS, Kaufman R. Energy and the US economy: a biophysical perspective. *Science* 1984, 225:890–897.
66. Cleveland CJ. Energy quality and energy surplus in the extraction of fossil fuels in the US. *Ecolog Econ* 1992, 6:139–162.
67. Kubiszewski I, Cleveland CJ, Endres PK. Meta-analysis of net energy return for wind power systems. *Renew Energ* 2010, 35:218–223.
68. Clarens AF, Nassau H, Resurreccion EP, White MA, Colosi LM. Environmental impacts of algae-derived biodiesel and bioelectricity for transportation. *Environ Sci Technol* 2011, 45:7554–7560.
69. Hall C, Balogh S, Murphy D. What is the minimum EROI that a sustainable society must have? *Energies* 2009, 2:25–47.

70. Guinee JB, ed. *Handbook on Life Cycle Assessment*. Dordrecht, the Netherlands: Kluwer; 2002.
71. Rebitzer G, Ekvall T, Frischknecht R, Hunkeler D, Norris G, Rydberg T, Schmidt W, Suh S, Weidema B, Pennington DW. Life cycle assessment. Part I: framework, goal and scope definition, inventory analysis and applications. *Environ Int* 2004, 30:701–720.
72. Sanden B, Kalström M. Positive and negative feedback in consequential life cycle assessment. *J Clean Prod* 2007, 15:1469–1481.
73. Frischknecht R, Büsser S, Krewitt W. Environmental assessment of future technologies: how to trim LCA to fit this goal? *Int J Life Cycle Assess* 2009, 14:584–588.
74. Hospido A, Davis J, Berlin J, Sonesson U. A review of methodological issues affecting LCA of novel food products. *Int J Life Cycle Assess* 2010, 15:44–52.
75. Wernet G, Conradt S, Isenring HP, Jimenez-Gonzales C, Hungerbühler K. Life cycle production of fine chemical production: a case study of pharmaceutical synthesis. *Int J Life Cycle Assess* 2010, 15:294–303.
76. Mohr N, Meijer A, Huijbregts MAJ, Reijnders L. Environmental impact of thin-film GaInP/GaAs and multicrystalline silicon solar modules produces with solar energy. *Int J Life Cycle Assess* 2009, 14:225–235.
77. Reijnders L. Fuels for the future. *J Integr Environ Sci* 2009, 6:279–294.
78. Farrell A, Brandt A, Arons S. The race for 21st century auto fuels. In: Hafemeister D, Levi B, Levine M, Schwarts P, eds. *Physics of Sustainable Energy, Using Energy Efficiently and Producing it Renewably*. Melville, New York: American Institute of Physics; 2008, 235–250.
79. Hughes L, Rudolph J. Future world oil production: growth, plateau of peak? *Curr Opin Environ Sustain* 2011, 3:225–234.
80. Searchinger T, Heimlich R, Houghton RA, Dong F, Elobeid A, Fabiosa J, Tokgoz S, Hayes D, Yu T. Use of US cropland for biofuels increases greenhouse gases through emissions from land use change. *Science* 2008, 319:1238–1240.
81. Cherubini F, Stromman AH. Life cycle assessment of bioenergy systems: state of the art and future challenges. *Bioresour Technol* 2011, 102:437–451.
82. Cherubini F, Bird ND, Cowie A, Jungmeier G, Schlamadinger B, Woess-Gallasch S. Energy and greenhouse gas-based LCA of biofuel and bionergy systems. *Resour Conserv Recycl* 2009, 53:434–477.
83. Finnveden G, Hauschild MZ, Ekvall T, Guiné J, Heijungs R, Hellweg S, Koehler A, Pennington D, Suh S. Recent developments in life cycle assessment. *J Environ Manage* 2009, 91:1–21.
84. Gandreault C, Samson R, Stuart PR. Energy decision making in a pulp and paper mill: selection of LCA system boundary. *Int J Life Cycle Assess* 2010, 15:198–211.
85. Kadam KL. Environmental implications of power generation via coal-microalgae cofiring. *Energy* 2002, 27:905–722.
86. Soratana K, Landis AE. Evaluating industrial symbiosis and algae cultivation from a life cycle perspective. *Bioresour Technol* 2011, 102:6892–6901.
87. Campbell PK, Beer T, Batten D. Life cycle assessment of biodiesel production from microalgae in ponds. *Bioresour Technol* 2011, 102:50–56.
88. Xu L, Brilman SWP, Withag JAM, Brem G, Kersten S. Assessment of a dry an wet route for the production of biofuels from microalgae: energy balance analysis. *Bioresour Technol* 2011, 102:5113–5122.
89. Yang J, XU M, Zhang X, Hu Q, Sommerfeld M, Chen Y. Life-cycle analysis of biodiesel production from microalgae: water footprint and nutrients balance. *Bioresour Technol* 2011, 102:159–165.
90. Lardon L, Helia A, Sialve B, Steyer J, Bernard O. Life cycle assessment of biodiesel production from microalgae. *Environ Sci Technol* 2009, 43:6475–6481.
91. Sander K, Murthy GS. Life cycle analysis of algal biodiesel. *Int J Life Cycle Assess* 2010, 15:704–714.
92. Sturm BSM, Lamer SL. An energy evaluation of coupling nutrient removal from wastewater with algal biomass production. *Appl Energ* 2011, 88:3499–3506.
93. Brentner LB, Eckelman MJ, Zimmerman JB. Combinatorial life cycle assessment to inform process design of industrial production and algal biodiesel. *Environ Sci Technol* 2011, 45:7060–7067.
94. Collet P, Helias A, Lardon L, Ras M, Goy R, Steyer J. Life-cycle assessment of microalgae culture coupled to biogas production. *Bioresour Technol* 2011, 102:207–214.
95. Razon LF, Tan RR. Net energy analysis of the production of biodiesel and biogas form microalgae: *Haematococcus pluvialis* and *Nanochloropsis*. *Appl Energ* 2011, 88:3507–3514.
96. Pfromm PH, Amanor-Boadu V, Nelson R. Sustainability of algae derived biodiesel: a mass balance approach. *Bioresour Technol* 2011, 102:1185–1192.
97. Dinh LTT, Guo YY, Mannan MS. Sustainability evaluation of biodiesel production using multicriteria decision making. *Environ Progr Sustain Energ* 2009, 28:38–46.
98. Stephens E, Ross IL, Mussnug JH, Wagner LD, Borowitka MA, Posten C, Kruse O, Hankamer B. Future prospects in microalgal production systems. *Trends Plant Sci* 2010, 15:554–563.
99. Khoo HH, Sharatt PN, Das P, Balasubramanian RK, Naraharisetti PK, Shaik S. Life cycle energy and CO_2 analysis of microalgae-to-biodiesel: preliminary results and comparisons. *Bioresour Technol* 2011, 102:5800–5807.
100. Hou J, Zhang P, Yuan X, Zheng Y. Life cycle assessment of biodiesel from soybean, jatropha and microalgae in China conditions. *Renew Sustain Energ Rev* 2011, 15:5081–5091.
101. Batan L, Quinn J, Willson B, Bradley T. Net energy and greenhouse gas emission evaluation of biodiesel derived from microalgae. *Environ Sci Technol* 2010, 44:7975–7980.
102. Lohrey C, Kochergin V. Biodiesel production from microalgae: co-location with sugar mills. *Bioresour Technol* 2012, 108:76–82.
103. Newman SE. Disinfecting irrigation water for disease management. In: *Proceedings of 20th Annual Conference on Pest Management on Ornamentals. Society of American Florists*. San Jose: California; 2004.

104. Zaisheng Y, Changfa L, Shihe W, Jie H, Yuan L, Liong Z, Junxin Z. Water treatment and performance characteristics evaluation of a pilot-scale recirculation aquaculture system. In: *Conference on Bioinformatics and Biomedical Engineering*. Shanghai, China; 2008, 3300–3303.
105. Qin G, Liu CCK, Richman NH, Moncur JET. Aquaculure wastewater treatment and reuse by wind-driven reverse osmosis membrane technology: a pilot study on Coconut Island, Hawaii. *Aquacul Eng* 2005, 32:365–378.
106. Rasoul-Amini S, Manotazeri-Najafabady N, Mobasher MA, Hoseini-Alhashemini S, Ghasemi Y. Chlorella sp: a new strain with highly saturated fatty acids for biodiesel production in bubble-column photobioreactor. *Appl Energ* 2011, 88:3354–3356.
107. Atadashi IM, Aroua MK, Abdul Aziz AR, Sulaiman NMN. Refining technologies for the purification of crude biodiesel. *Appl Energ* 2011, 88:4239–4251.
108. Hoekman SK, Broch A, Robbins C, Ceniceros E, Natarajan M. Review of biodiesel composition, properties, and specifications. *Renew Sustain Energy Rev* 2011, 16:149–163.
109. Hardy C, Graedel TE. Industrial ecosystems as food webs. *J Ind Ecol* 2002, 6:29–38.
110. McCarthy PL, Bae J, Kim J. Domestic wastewater treatment as a net energy producer—can this be achieved? *Environ Sci Technol* 2011, 45:7100–7106.
111. Pate R, Klise G, Wu B. Resources demand implications for US algae biofuels production scale up. *Appl Energ* 2011, 88:3377–3388.
112. Levine BR, Pinnarat T, Savage PE. Biodiesel production from wet algal biomass through in situ lipid hydrolysis and supercritical transesterification. *Energ Fuels* 2010, 24:5235–5243.
113. Cheng C, Du T, Pi H, Jang S, Lin Y, Lee H. Comparative study of lipid extraction from microalgae by organic solvent and supercritical CO_2. *Bioresour Technol* 2011, 102:10151–10153.
114. Haas NJ, Wagner K. Simplifying biodiesel production: the direct or in situ transesterification of algal biomass. *Eur J Lipid Sci Technol* 2011, 113:1219–1222.
115. Clarens A, Colosi L. Putting algae's promise into perspective. *Biofuels* 2010, 1:805–808.
116. Kiwjaroun C, Tubtimdee C, Piumsomboon P. LCA studies comparing biodiesel by conventional and supercritical methanol methods. *J Clean Prod* 2009, 17:143–153.
117. Quesada-Medina J, Olivares-Carillo P. Evidence of thermal decomposition of fatty acid methyl esters during the synthesis of biodiesel with supercritical methanol. *J Supercrit Fluids* 2011, 56:56–63.
118. Esson D, Wood SA, Packer MA. Harnessing the self-harvesting capability of benthic cyanobacteria for use in benthic photobioreactors. *AMB Express* 2011, 19, 1–7.
119. Parmar A, Singh NK, Pandey A, Gnansounou E, Madamwar D. Cyanobacteria and microalgae: a positive prospect for biofuels. *Bioresour Technol* 2011, 102:10163–10172.
120. Liu A, Sheng J, Curtiss R. Fatty acid production in genetically modified cyanobacteria. *Proc Nat Acad Sci U S A* 2011, 108:6899–6904.
121. Liu J, Ma X. The analysis on energy and environmental impacts of microalgae-based fuel methanol. *Energ Policy* 2009, 37:1479–1488.
122. Luo S, Hu Z, Choi DG, Thomas VM, Realff MJ, Chance RR. Life cycle energy and greenhouse gas emissions for an ethanol production process based on blue green algae. *Environ Sci Technol* 2010, 44:8670–8677.
123. Fagerstone KD, Quinn JC, Bradley T, de Long SK, Marchese AJ. Quantitative measurement of direct nitrous oxide emissions from microalgae cultivation. *Environ Sci Technol* 2011, 45:9449–9456.

4

Catalytic Pyrolysis of Biomass for Transportation Fuels

Angelos A. Lappas,* Kostas G. Kalogiannis, Eleni F. Iliopoulou, Kostas S. Triantafyllidis and Stylianos D. Stefanidis

Chemical Process and Energy Resources Institute, Center for Research and Technology Hellas, Thermi-Thessaloniki, Greece

INTRODUCTION

Catalytic pyrolysis of biomass is a thermochemical conversion process where a heterogeneous catalyst is used as a heat carrier in the pyrolysis reactor for the *in situ* upgrading of the quality of the pyrolysis oil (bio-oil). Bio-oil upgrading refers to minimizing its well-known undesirable properties [high water (H_2O) and oxygen content, high viscosity, corrosivity, instability, low heating value, etc.). Heterogeneous catalysis is widely used in the petrochemical industry for the conversion of heavy oil fractions into lighter fuels and chemicals. This concept is transferred to the biomass pyrolysis process where, ideally, the heavy oxygenated volatiles from the decomposition of biomass are deoxygenated and converted to lighter fuels and chemicals by coming in contact with a suitable catalyst.[1] Ultimately, the goal of the process is to produce a liquid with improved properties that could be either used directly as a liquid fuel or as a feedstock (or co-feedstock) in modern refineries, much like crude oil.[2] The downstream upgrading of the conventional (thermal) bio-oil has been extensively studied in the literature. However, it was proved to be a very complicated issue, creating many operating problems.

On the contrary, use of a catalytic bio-oil is believed to overcome most of these problems. An ideal catalyst for the biomass pyrolysis process should produce high quality and yield of bio-oil with low amounts of oxygen and H_2O, minimize the undesirable compounds present in the bio-oil, and exhibit both resistance to deactivation and thermal stability. The decrease of oxygenated, bio-oil component species can increase bio-oil's heating value and result in a product with improved physical and chemical properties. Oxygen is removed from the pyrolysis vapors in the form of CO, CO_2, and H_2O. The removal of oxygen in the form of CO_2 is more preferable than as CO because only one carbon atom is required for the removal of two oxygen atoms, whereas in the case of CO formation, one carbon atom is required for each oxygen atom that is removed. Oxygen removal by CO_2 or CO is in turn more preferable than H_2O formation in order for the hydrogen to be preserved for the hydrocarbon-forming reactions.[3] Oxygen removal as CO_2 would effectively enhance the H–C ratio and thus lead to reduced coke deposition. Improving the quality of bio-oil using heterogeneous catalysis has received extreme attention from several research groups. The challenges in developing this technology involve the processing of lignocellulosic materials and the development/optimization of advanced porous materials as efficient monofunctional and bifunctional catalysts for the production of transportation fuels from biomass. In this review, the recent developments on catalysts and processes will be discussed.[4,5]

*Correspondence to: angel@cperi.certh.gr

Advances in Bioenergy: The Sustainability Challenge, First Edition. Edited by Peter D. Lund, John Byrne, Göran Berndes and Iacovos A. Vasalos.

Originally published in: *WIREs Energy Environ* 2012, 1: 45–56 doi: 10.1002/wene.16

REVIEW ON CATALYSTS FOR BIOMASS CATALYTIC PYROLYSIS

A plethora of catalytic materials such as zeolites, mesoporous materials with uniform pore size distribution (MCM-41, MSU, and SBA-15), microporous/mesoporous hybrid materials doped with noble and transition metals, and base catalysts have been investigated as candidate catalysts for biomass pyrolysis. These catalysts should be able to selectively favor the decarboxylation reactions, producing high-quality bio-oil with low amounts of oxygen and H_2O. They should also inhibit formation of undesirable oxygenated compounds, such as ketones, acids, and carbonyl compounds, which are known to be detrimental for the direct use or further coprocessing of bio-oil. The hydrothermal stability of the catalysts must also be improved; thus, resistance to deactivation and catalyst behavior upon regeneration must be investigated to optimize new catalysts. In general, catalyst development comprises of controlled formation of appropriate catalyst particles and tailoring the porosity, acidity, basicity, and metal–support interactions of the candidate catalytic materials.

Microporous Acidic Catalysts

Microporous acidic catalysts are widely used in oil refineries and are known to catalyze the scission of carbon–carbon bonds of heavier oil fractions. Biomass pyrolysis requires a similar mechanism for the conversion of heavier oxygenates to lighter ones. In this respect, several groups have studied the catalytic pyrolysis of biomass over acidic zeolite catalysts, whereas ZSM-5 has been the main zeolite studied for the upgrading of biomass pyrolysis vapors.[6–11] In a series of studies, Horne and Williams investigated the effect of the ZSM-5 zeolite on the pyrolysis of biomass. They found that after catalysis, the oxygenated species in the bio-oil were markedly reduced, whereas aromatic species increased, producing a premium grade gasoline-type fuel. Detailed analysis of the upgraded oils showed that there were high concentrations of economically valuable chemicals as well, but also revealed hazardous, biologically active polycyclic aromatic species that increased in concentration with increasing catalyst bed temperature.[12] They reported oxygen removal from the pyrolysis vapors mainly as H_2O at lower catalyst bed temperatures and as CO and CO_2 at higher bed temperatures. They observed a shift toward lower molecular weight species with increasing catalyst bed temperature as well.[13] Detailed analysis of the aromatic hydrocarbons of the bio-oil derived from catalytic pyrolysis of biomass with HZSM-5 catalyst showed that the single aromatic ring species were mainly benzene, toluene, and alkylated benzenes. The polycyclic aromatic hydrocarbons (PAHs) were mainly naphthalene, phenanthrene, fluorene, and their alkylated homologues.[3]

The complex nature of bio-oil components means that there is only limited understanding of the reactions that take place during zeolite catalysis. Horne and Williams investigated the reaction of oxygenated biomass pyrolysis model compounds over a ZSM-5 catalyst in an attempt to simplify the problem. They found that methanol could be catalytically converted to hydrocarbon products at temperatures 300–350°C, whereas furfural, anisole, and cyclopentanone required higher temperatures. With the exception of anisole, increasing catalyst bed temperature reduced coke formation for the upgrading of the oxygenated compounds. The optimum catalysis temperature was 500–550°C. At that temperature, however, anisole gave high yields of coke, suggesting that the phenolic compounds are the major coking components present in biomass-derived pyrolysis vapors. The oxygen bond in anisole appeared to be refractory to the catalytic upgrading, leading to the formation of large quantities of phenolic compounds and gave only low yields of aromatic hydrocarbons.[14]

Gayubo et al.[15,16] studied the transformation of oxygenated bio-oil model compounds over an HZSM-5 zeolite as well. They found that alcohols transformed via dehydration from olefins at lower temperatures to higher olefins at 250°C and to paraffins and aromatic hydrocarbons above 350°C. Phenol exhibited low reactivity on the HZSM-5 zeolite. 2-Methoxyphenol had low reactivity as well and furthermore generated thermal coke. These observations for the phenolic compounds were in good agreement with the observations of Horne and Williams.[14] The rate of deactivation by coke deposition was low for both alcohols and phenols. Acetaldehyde had a low reactivity to hydrocarbons with noticeable tendency to form coke depositions, which was attributed to its capacity for oligomerization. The transformation of ketones and acetic acid occurred through decarboxylation and, to a lesser extent, dehydration. Above 400°C, this transformation gave olefins and aromatics, similar to the reaction scheme of alcohols. The generation of coke was more significant than in the corresponding process for alcohols and increased the formation of olefins, which were the intermediate products of the reaction scheme.[15,16] In a later study, Gayubo et al. investigated the transformation of mixtures of bio-oil model compounds on the HZSM-5 zeolite and found that the results agree well with the results in the case of using pure compounds. However, synergistic effects were observed in the case of using model compound mixtures instead of pure compounds due to the high reactivity of some of the primary pyrolysis products. The catalyst underwent severe deactivation by coke deposition, which was enhanced by the thermal coke from the degradation of acetaldehyde. The presence of furfural enhanced the tendency of 2-methoxyphenol to degrade to coke and led to the conclusion that the feasibility of the process of bio-oil upgrading into hydrocarbons requires the previous separation of certain components of the bio-oil such as aldehydes, oxyphenols, furfural, and their derivatives in order to reduce the amount of coke formed.[17]

Vitolo et al.[18] studied the upgrading of bio-oil over HZSM-5 and tentatively attributed catalytic activity to its acidic sites that, through a carbonium ion mechanism, promote deoxygenation, decarboxylation, and decarbonylation of the oil constituents as well as cracking, oligomerization, alkylation, isomerization, cyclization, and aromatization. They also investigated the behavior of the zeolite, using it in repeated upgrading-regenerating cycles and observed that the material undergoes irreversible poisoning after the fifth upgrading-regenerating cycle. The deactivation was due to the disappearance of a significant amount of acid sites, mainly the stronger ones.[18]

The catalytic upgrading of the pyrolysis vapors is known to reduce the bio-oil yield and increase the gas, coke, and H_2O yields. Some groups have also reported a reduction in the biomass residue when the catalyst came in contact with the biomass feed. Atutxa et al.[9] studied the effect of an HZSM-5 zeolite on the *in situ* pyrolysis of sawdust at 400°C. The gas yields increased with increasing catalyst mass, whereas the liquid yield decreased notably. They also observed a slight reduction in char. The proportion of CO over CO_2 increased with increasing catalyst mass. Atutxa et al.[9] mainly related the reduction in the total liquid product with the transformation of the heavy liquid fraction into light liquid fraction and gases. The light fraction was more severely deoxygenated than the heavy fraction, which is evidence of the higher global reactivity of the compounds of the lighter fraction, especially of alcohols and acetic acid. This observation was in good agreement with the work with bio-oil model compounds of Horne and Williams[14] and Gayubo et al.[15,16] The catalytic pyrolysis oil was less oxygenated, less viscous, less corrosive, and more stable.[9] Wang et al.[19] studied the pyrolysis of three kinds of biomass impregnated with HZSM-5 zeolite and similarly found that the presence of HZSM-5 in the biomass increased the maximum weight loss rate. Zhang et al.[20] studied the pyrolysis of corncob in a fluidized bed with HZSM-5 catalyst and observed that the catalyst caused a marked decrease of heavy oil fraction and an increase of H_2O, coke, and noncondensable gas yields. The use of HZSM-5 also led to a remarkable increase of aromatic hydrocarbons in the oil fraction and to a decrease of all the other types of compounds. The oxygen content of the catalytic oil decreased to 14.69% from 40.28% in the noncatalytic oil.[20]

Other groups have demonstrated the selectivity of ZSM-5 toward aromatics and its ability to reduce oxygenates as well. Samolada et al.[21] studied the conversion of representative bio-oil model compounds over different catalytic materials and concluded that the HZSM-5 zeolite completely converted undesirable carbonyls to hydrocarbons with a simultaneous loss of the organic liquid fraction and a dramatic increase of H_2O. Pattiya et al.[22] investigated the upgrading of the vapors from the pyrolysis of cassava rhizome in a pyrolysis-gas chromatography–mass spectrometry (Py-GC–MS) system using various catalysts. They concluded that the ZSM-5 catalyst was the most active catalyst and increased the formation of aromatic hydrocarbons, phenols, and acetic acid. It also decreased oxygenated lignin-derived compounds and carbonyls.[22] Lastly, Adjaye et al.[23] studied the effect of mixtures of HZSM-5 and silica alumina on the product distribution during the conversion of maple wood to bio-oil and observed that the organic liquid that was produced with silica–alumina consisted mainly of aliphatic hydrocarbons, whereas the organic liquid product that HZSM-5 produced consisted mostly of aromatic hydrocarbons. The addition of HZSM-5 to silica–alumina reduced coke formation and increased organic liquid product and gas yields. The gradual increase of HZSM-5 in the mixture changed the hydrocarbon products from aliphatic to aromatic. These observations suggest that the HZSM-5 is a more effective hydrogen transfer catalyst than the less acidic silica–alumina and the aliphatics formed during upgrading are converted into the thermodynamically favored aromatics. As aromatics have a lower H–C ratio than aliphatic hydrocarbons, more hydrogen has to be put into the aliphatics, which reduces the overall formation of hydrocarbons from pyrolysis oil due to its low hydrogen content.[23]

Several groups have studied the catalytic upgrading of lignin pyrolysis vapors over a ZSM-5 zeolite. Lignin is a main constituent component of lignocellulosic biomass and is also a by-product of paper pulp mills. Thring et al.[8] studied the catalytic pyrolysis of lignin solubilized in acetone with HZSM-5 catalyst in a fixed bed reactor and observed high yields of gasoline range hydrocarbons such as benzene, toluene, and xylene (BTX). Mullen and Boateng[24] observed an increased production of aromatic hydrocarbons as well during the pyrolysis of lignin in the presence of HZSM-5 catalyst in a Py-GC–MS system. The increased aromatic hydrocarbon production was likely to be due to the enhanced depolymerization efficiency of the catalyst that released and converted the aliphatic linkers of lignins to olefins and then to aromatic compounds by aromatization. They also observed a catalyst deactivation by coke deposition due to the release of simple phenols from the decomposition of lignin, which degraded to coke.[24] Jackson et al.[25] also studied the pyrolysis of lignin over five catalysts and concluded that HZSM-5 was the best catalyst for producing a deoxygenated liquid fraction.

In an effort to further improve the catalytic effect of the ZSM-5 zeolite on the upgrading of pyrolysis vapors, some groups studied the incorporation of metals in the zeolite framework. According to French and Czernik,[26] the presence of transition metals would affect the mode of oxygen rejection by producing more carbon oxides and less H_2O, making in that way more hydrogen available for incorporation into hydrocarbons. Zhu et al.[27] performed pyrolysis of xylan in a Py-GC–MS system over HZSM-5 catalyst and metal-impregnated Fe/HZSM-5 and Zn/HZSM-5 catalysts and found that the metal-impregnated HZSM-5 materials were more effective in reducing oxygenated compounds and hence produced higher contents of hydrocarbons. French and Czernik[26] evaluated a set of commercial and laboratory synthesized catalysts for their hydrocarbon production performance and concluded that the best performing catalysts belonged to the ZSM-5 group, whereas the highest hydrocarbon yields were achieved with nickel, cobalt, iron, and gallium substituted ZSM-5. Zeolites with larger pores exhibited less deoxygenation activity. Park et al.[28] compared the catalytic effects of HZSM-5, Ga/HZSM-5, and H-Y zeolite and concluded that Ga/HZSM-5 produced more bio-oil than HZSM-5 and had better selectivity to aromatic hydrocarbons. H-Y was less efficient in bio-oil upgrading than HZSM-5.

Besides ZSM-5, other microporous zeolites and acidic materials have been studied in the literature as well. During the conversion of bio-oil over HZSM-5 zeolite, H-Y zeolite, H-mordenite zeolite, silicalite, and silica–alumina catalysts, Adjaye and Bakhshi[29] observed that the acidic zeolites were more effective in converting the bio-oil to hydrocarbons than the less acidic silica–alumina and nonacidic silicalite, and the highest yield of hydrocarbons was obtained with the HZSM-5 zeolite. HZSM-5 and H-mordenite zeolites produced more aromatic than aliphatic hydrocarbons, whereas H-Y, silicalite, and silica–alumina produced more aliphatic than aromatic hydrocarbons.[20,29] Aho et al.[30] pyrolyzed pinewood in a fluidized bed reactor using proton forms of beta, Y, ZSM-5, and mordenite zeolites and found that the chemical composition of the bio-oil depended on the structure of the zeolite used. The ZSM-5 zeolite gave the highest liquid product yield and exhibited lower selectivity to acids and alcohols and higher selectivity to ketones, whereas the mordenite zeolite was effective for minimizing the formation of PAHs.[30] Uzun and Sarioglu[31] also studied different zeolite structures (ZSM-5, H-Y, and USY) for the catalytic pyrolysis of corn stalks. They observed the highest oil yield with the ZSM-5 zeolite and the lowest with the USY. They also observed the highest amount of aromatics with the USY catalyst. The H-Y zeolite caused an increase of the aliphatic hydrocarbons.[31] Carlson et al.[32,33] studied ZSM-5, beta, and Y zeolites for the conversion of biomass compounds to aromatics in a Py-GC–MS system and found that high catalyst to feed ratios favor aromatic compounds production over coke formation by avoiding undesirable thermal decomposition reactions in the homogeneous phase. At lower catalyst to feed ratios, volatile oxygenates were formed including furan-type compounds, acetic acid, and hydroxyacetaldehyde. ZSM-5 had the highest aromatic yields and the least amount of coke.[32,33] Lu et al.[34] also studied HZSM-5 and H-Y zeolites for the upgrading of biomass fast pyrolysis vapors, along with three other mesoporous materials. The two microporous zeolites proved to be effective for the deoxygenation of the pyrolysis vapors and formed abundant aromatic hydrocarbons and some PAHs after catalysis.[34] Qi et al.[35] performed catalytic pyrolysis of bamboo in a fixed bed reactor in the presence of NaY zeolite and observed that the bio-oil yield increased after catalysis and consisted mainly of carboxylic and carbonylic compounds. Acetic acid was the main component of the catalytic bio-oil and its content was higher than in the thermal pyrolysis oil, whereas the content of carbonylic compounds was markedly lower.[35] Aho et al. studied the influence of beta, Y, and ferrierite zeolites and their Fe-modified counterparts on the upgrading of pinewood pyrolysis vapors. They concluded that beta zeolite was the most active in the deoxygenation reactions, followed by Y and ferrierite. The amount of methyl-substituted phenols in the bio-oil increased, whereas methoxy-substituted phenols decreased in the presence of the Fe-modified zeolites.[36] In an earlier study, Aho et al. also investigated the influence of the beta zeolite acidity on the catalytic pyrolysis of biomass. They observed that zeolites with stronger acidity tend to form less organic oil and more H_2O and polyaromatic hydrocarbons than less acidic zeolites.[6]

Mesoporous Acidic Catalysts

The significant increase in the production of H_2O and gas at the expense of the organics yield, the formation of undesirable PAHs, and the rapid catalyst deactivation by coke deposition are serious drawbacks of the zeolitic materials. To overcome these problems, catalysts with a larger pore size than that of zeolites have recently attracted much attention, as they are expected to allow the larger molecules of the pyrolysis products, particularly lignin-derived compounds, to enter, reformulate, and exit the catalyst matrix with less chance of coke deposition and blocking of pores.[22]

Among the mesoporous acidic materials, the MCM-41 catalyst has received the most focus in the literature. The MCM-41 is the main representative of the mesoporous molecular sieve of M41S family, which was discovered in 1992, and possesses a hexagonal array of uniform mesopores whose dimensions can vary from 1.4 nm to greater that 10 nm in size and exhibits high surface areas (>1000 m^2/g). SBA-15 catalysts have been studied as well. The SBA-15 has long-range order, large monodispersed mesopores (up to 50 nm), and thicker walls (typically between 3 and 9 nm), which make them more thermally and hydrothermally stable than MCM-41-type materials.[37] Samolada et al.[21] evaluated a mesoporous Al-MCM-41 material for the catalytic pyrolysis of biomass and found that it exhibited poor hydrothermal stability, suggesting that further optimization of this material is needed for the pyrolysis process. Iliopoulou et al.[38] studied the steam stability and the effect of acidity of Al-MCM-41 catalysts. They found that lower numbers of acid sites had a beneficial effect for the production of liquid organic product, whereas higher number of acid sites favored the conversion of the pyrolysis vapors toward gas and coke. Moderate steaming of the materials resulted in reduction of their surface area and number of acid sites by 40–60%. However, steamed samples were still active for the upgrading of biomass pyrolysis vapors.[38] Adam et al. studied the pyrolysis of spruce wood biomass in a Py-GC–MS system in the presence of four different Al-MCM-41-type catalysts (Si–Al ratio of 20) and modified Al-MCM-41 catalysts (by pore enlargement and introduction of copper (Cu) cations into the material structure). They found that after catalysis, the yield of acetic acid and furans increased and the yield of high molecular mass phenols decreased. The overall yield of phenols increased and there was also a slight increase of the hydrocarbon yield. The enlargement of the catalyst's pores reduced the yield of acetic acid and H_2O and in general, pore enlargement and Cu cation introduction to the catalyst led to higher molecular mass products.[39] In another study, Adam et al. investigated the effect of four Al-MCM-41 catalysts with Si–Al ratio of 20, a pure siliceous SBA-15 and an aluminum SBA-15 on the upgrading of biomass pyrolysis vapors in a fixed bed reactor. They also studied Al-MCM-41 materials modified by pore enlargement and Cu cation introduction into their structure. They found that gas and H_2O yields increased after catalysis with all materials. Hydrocarbon, phenol, and PAHs yields increased as well, whereas carbonyl and acid yields decreased. They observed that all catalysts reduced the amount of undesirable compounds present in the bio-oil. The pore enlargement of the Al-MCM-41 seemed to have a deteriorating effect on the quality of the bio-oil. Cu cation introduction on the other hand had a beneficial effect and increased the amount of desirable products. The incorporation of aluminum into the SBA-15 framework resulted in very high content of desirable products in the bio-oil.[37] Antonakou et al.[40] evaluated three Al-MCM-41 materials with different Si–Al ratios and three metal containing (Cu, Fe, and Zn) Al-MCM-41 materials for the biomass pyrolysis in a fixed bed reactor. The production of bio-oil and its organic fraction decreased in comparison to the noncatalytic runs and the production of coke increased. All catalysts were found to increase the amount of phenolic compounds. Fe-Al-MCM-41, Cu-Al-MCM-41, and the lowest Si/Al Al-MCM-41 were the best catalysts for phenols production. They also observed a decrease in the undesirable fractions of acids, carbonyls, and heavy compounds with almost all tested materials. They found that lower Si–Al ratios positively affect the product yields and composition.[40] Nilsen et al.[41] also studied metal modified and their parent Al-MCM-41 materials with Si–Al ratio of 20 for the catalytic pyrolysis of biomass. They observed a better bio-oil quality with respect to phenols yield with all materials. The Zn-Al-MCM-41 catalyst led to the lower yield of phenols but gave the best results with respect to coke produced.

MSU materials for the catalytic pyrolysis of biomass have also been tested. Triantafyllidis et al.[42] assembled, from zeolite beta seeds, two aluminosilicate MSU-S materials with different mesopore structure and compared them with conventional Al-MCM-41 in the upgrading of biomass pyrolysis vapors in a fixed bed reactor. The MSU-S materials significantly reduced the bio-oil's organic phase and led to higher coke yields as compared with noncatalytic pyrolysis and to the Al-MCM-41 material. The MSU-S materials were also very selective toward PAHs and heavy fractions, but produced negligible amounts of acids, alcohols, carbonyls, and very few phenols. The MSU-S materials possessed stronger acid sites than the Al-MCM-41, resulting in enhanced yields of aromatics, PAHs, and coke.[42] Pattiya et al. studied Al-MCM-41 and Al-MSU-F materials in a Py-GC–MS system and found them to demonstrate a potential for bio-oil improvement in terms of initial viscosity and heating value, as they tend to reduce the oxygenated lignin-derived compounds and produce hydrocarbons, although the effect was less pronounced, when compared with that of a ZSM-5 zeolite. On the downside, they observed an increase of the acetic yield.[22] Jackson et al. also compared an HZSM-5 zeolite to a mesoporous Al-MCM-41 for the upgrading of lignin pyrolysis vapors and observed that the Al-MCM-41 catalyst behaved similarly to the HZSM-5, although it was not as effective at deoxygenating the liquid phase and gave more naphthalenics than simple aromatics.[25] Lu et al. also used a Py-GC–MS system to study siliceous SBA-15 and Al-SBA-15 materials and found them to be mild catalysts as compared with acidic zeolites. They favored the formation of acetic acid, light phenols, furan, furfural, and other light furan compounds and decreased the yields of heavy phenols, heavy furans, and many light carbonyls. The Al-SBA-15 materials were more active than the siliceous SBA-15 and the catalytic effects were enhanced when decreasing the Si–Al ratio of the catalyst.[34,43]

Lately, mesoporous zeolites have been synthesized and studied for the catalytic pyrolysis of biomass. Lee et al. synthesized ordered mesoporous aluminosilicates (MMZ) using commercially available beta and ZSM-5 zeolites and applied them to the pyrolysis of woody biomass. The material had well-developed mesoporosity and exhibited excellent hydrothermal stability. Considering both their selectivity to phenolics and organic fraction yield and also the reduction of undesirable compounds such as oxygenates, the synthesized catalysts were found to be promising catalysts for the upgrading of bio-oil. Moreover, the high catalytic activities of the synthesized catalysts were maintained through the regeneration-upgrading cycles, whereas the activity of Al-MCM-41 significantly decreased.[44] Park et al. synthesized a mesoporous mordenite framework inverted (MFI) zeolite and compared its catalytic activity with those of conventional HZSM-5 and mesoporous material from HZSM-5 (MMZ_{HZSM-5}). The mesoporous MFI zeolite exhibited the best activity in terms of deoxygenation and aromatization. In particular, they observed high selectivity toward valuable BTX hydrocarbons, but organic fraction yield was decreased. Incorporation of gallium into the mesoporous MFI zeolite resulted in less cracking of the pyrolytic vapors, as well as an increase in the organic fraction of bio-oil and resistance to coke deposition. The amount of gallium incorporated into the mesoporous MFI zeolite played an important role in the selectivity of aromatics. The addition of the appropriate amount of gallium optimized the bifunctional mechanism and the consequent enhancement of the selectivity for BTX aromatics, whereas the excess gallium brought a negative effect on formation of aromatics because of more loss of protons.[45]

Basic Catalysts

Solid acid catalysts are the predominately tested materials in the effort to convert biomass to oils using pyrolysis. However, Nokkosmaki et al.[46] and Fabbri et al.[47] also applied basic oxides in catalytic pyrolysis, ZnO and MgO, respectively. The first group pyrolyzed pine sawdust with three different zinc oxide (ZnO) catalysts at 600°C, with 30 ms residence time passing the pyrolysis vapors of pine biomass through a fixed bed of catalyst to investigate catalysts' influence on the composition and stability of the bio-oil. They found that ZnO was a mild catalyst for producing bio-oils showing only a small reduction in the liquid yield with only a 2 wt% gas increase, but found that the stability of the catalytic oil was significantly increased. The formation of anhydrosugars and formic acid was reduced when increasing the temperature of the catalyst. At 500°C, the usual temperature of biomass pyrolysis, neither the former nor the latter of the products could be detected.[46] Lu et al. studied the upgrading of biomass pyrolysis vapors in a Py-GC–MS system using various metal oxides as catalytic materials and observed different catalytic capabilities. CaO significantly reduced the levels of phenols and anhydrosugars and eliminated the acids. It also increased the formation of cyclopentanones, hydrocarbons, and several light compounds. ZnO was found to be a mild catalyst and Fe_2O_3 resulted in the formation of various hydrocarbons.[48] Wang et al. also studied CaO by thermo gravimetric analysis-fourier transform infrared spectroscopy analysis and indeed found that it was very effective in the reduction of acids and promoted the formation of hydrocarbons.[49] Finally, Torri et al. investigated the catalytic pyrolysis of pine sawdust over various metal oxides in a pyrolysis-gas chromatography–microwave induced plasma–atomic emission detector (Py-GC–MIP–AED) system and observed reduction of heavy compounds after catalysis at the expense of bio-oil yields. The most interesting materials from their study were CuO, which exhibited the highest yields in semivolatile compounds, mixed metal oxide catalysts (Fe_2O_3, and mixed metal oxides containing Cu and cobalt), and ZnO, which reduced the proportion of heavy fraction in the bio-oil with a limited decrease in the bio-oil yield.[50]

Lu et al. used TiO_2 Rutile, TiO_2 Anatase, and ZrO_2/TiO_2 catalysts and their modified counterparts with incorporation of Ce, Ru, and Pd in a Py-GC–MS system for the upgrading of biomass pyrolysis vapors. TiO_2 Rutile catalysts, especially the Pd-containing counterpart, exhibited very promising effect to convert the lignin-derived oligomers to monomeric phenols and favored the reduction of aldehydes and sugars, while increasing the ketones, acids, and cyclopentanones. The ZrO_2/TiO_2 catalysts reduced the phenol and acid yields remarkably, eliminated sugars and meanwhile, increased hydrocarbons, light linear ketones, and cyclopentanones.[51]

The research group of Chemical Process and Energy Resources Institute (CPERI) has also worked with basic materials, i.e., with TiO_2, ZrO_2/TiO_2, and MgO. The evaluation results of these catalysts tested in biomass pyrolysis indicated different deoxygenation pathways compared with acidic zeolites. Specifically, high CO_2 yields were achieved, whereas the liquid products presented high yields of ketones, especially cyclopentanones and cyclopentenones. The mechanisms of the CO_2 production have been presented elsewhere.[52] In short, aldol condensation and ketonization reactions are prevalent, mainly leading to ketones that are formed from aldehydes and acids condensation and CO_2 as the main by-product.[53,54]

Moreover, these condensation reactions are essentially carbon–carbon (C-C) coupling reactions that can condensate smaller aldehyde and acid molecules toward the formation of larger molecules with molecular weights closer to those of liquid transportation fuels. In this case, the basic catalysis achieved with MgO presents a very interesting alternative to the acidic route that has been studied so far.

Chen et al. performed catalytic pyrolysis of pinewood sawdust by microwave heating, applying inorganic additives with different basicity. All of the eight additives altered yields and composition of pyrolysis products. However, they caused a significant increase in the yield of the solid product and a decrease in the yield of gases. Yields of liquid products were not subjected to dramatic change. All of the eight additives made the gaseous products evolve earlier. The gases produced from pyrolysis consist mainly of H_2, CH_4, CO, and CO_2, whereas alkaline sodium compounds NaOH, Na_2CO_3, and Na_2SiO_3 favored H_2 formation in the pyrolytic gases.[55] Babich et al.[56] also used Na_2CO_3 to catalyze the pyrolysis of chlorella algae. In the presence of Na_2CO_3, gas yield increased and bio-oil yield decreased, when compared with noncatalytic pyrolysis at the same temperatures, but the pyrolysis oil from the catalytic runs carried higher heating value and had lower acidity.[56]

Recently, strongly basic zeolites have been synthesized from reaction of zeolites with ammonia at elevated temperatures, providing unique activity and selectivity for base-catalyzed reactions.[57] Such basic zeolites, e.g., amine-substituted ZSM-5, are also suggested as promising candidate catalysts for biomass conversion to hydrocarbons via catalytic pyrolysis.[32,33] Peralta et al.[58] investigated the deoxygenation of benzaldehyde on basic CsNaX and NaX zeolites. They observed that the direct decarbonylation of benzaldehyde to benzene could be readily promoted over highly basic catalyst containing Cs in excess. In parallel, condensation of surface products also occurred, which decomposed and formed toluene.[58]

REVIEW ON CATALYTIC BIOMASS PYROLYSIS PROCESSES

Reactors for Catalytic Biomass Pyrolysis

As biomass pyrolysis reactions are endothermic, a heat carrier should be available to provide heat to the reacting biomass particles. Different types of heating have been proposed, but the most effective one is the direct heat transfer using a solid inert material (usually silica sand). The type of the heat carrier dictates also the type of the reactor and in general the process scheme of biomass pyrolysis. A review on the various process schemes for biomass pyrolysis is given in a recent publication.[59] Several laboratory catalytic pyrolysis/upgrading experiments are reported. In general, it seems that the most efficient process technology for biomass pyrolysis is based on fluid bed reactors. Biomass particles are introduced into a bed of hot sand, fluidized by a gas, which is usually a recirculated product gas. High heat transfer rates from fluidized sand result in the rapid heating of biomass particles. This process has been proved as very efficient for maximum bio-oil yield and for this reason it has been scaled up by companies such as Dynamotive.[59] This technology, although suitable for the conventional biomass technology, is difficult to be applied for biomass catalytic pyrolysis. The reason is that catalyst deactivation usually takes place during catalytic pyrolysis that makes the catalyst inactive within very short time. Thus, even by replacing the inert heat carrier by a catalytic material, the life of this material will be limited as very rapidly the catalyst will be deactivated and will behave as an inert material. This technology can be used for downstream upgrading of bio-oil vapors by putting a fixed or fluid bed of a catalyst at the reactor exit. However, again catalyst deactivation in this second bed can be a serious issue, except if the catalyst could be continuously replaced.

The most appropriate process for biomass catalytic pyrolysis should be based on the circulating fluid bed (CFB) technology that includes a regeneration step for the continuous catalyst regeneration. In this way, the catalyst is mixed with the biomass particles always in an active state. The technology is similar to fluid catalytic cracking (FCC) process that is the biggest conversion unit in a refinery for the conversion of heavy hydrocarbons to lighter fuels. Ensyn, CPERI, and KiOR have developed technology for CFB pyrolysis.[59–61] The rotating cone technology developed by BTG works also in the circulating mode.[62] Commercial CFB technology for biomass catalytic pyrolysis is not available today. However, it has been proved in pilot scale, while currently demonstration scales are investigated. KiOR develops technology on biomass catalytic pyrolysis and they are today in a demonstration phase producing 15 barrels per day bio-crude. KiOR granted funding for the construction of five commercial plants in Mississippi based on its technology.[59] The success of these commercial plants will open new horizons in the establishment of biomass catalytic pyrolysis process.

The catalytic biomass CFB technology is described in Figure 4.1. The biomass is loaded at ambient temperature into a feed hopper (D-61), and using a screw feeder (SF-61) is fed into the reactor. The catalyst is loaded into regenerator (D-101) that is a fluid bed reactor aiming to burn off the coke that is deposited on the catalyst and provide the required heat for the cracking of biomass. The reactor usually consists of a mixing zone vessel (D-201) and the riser.

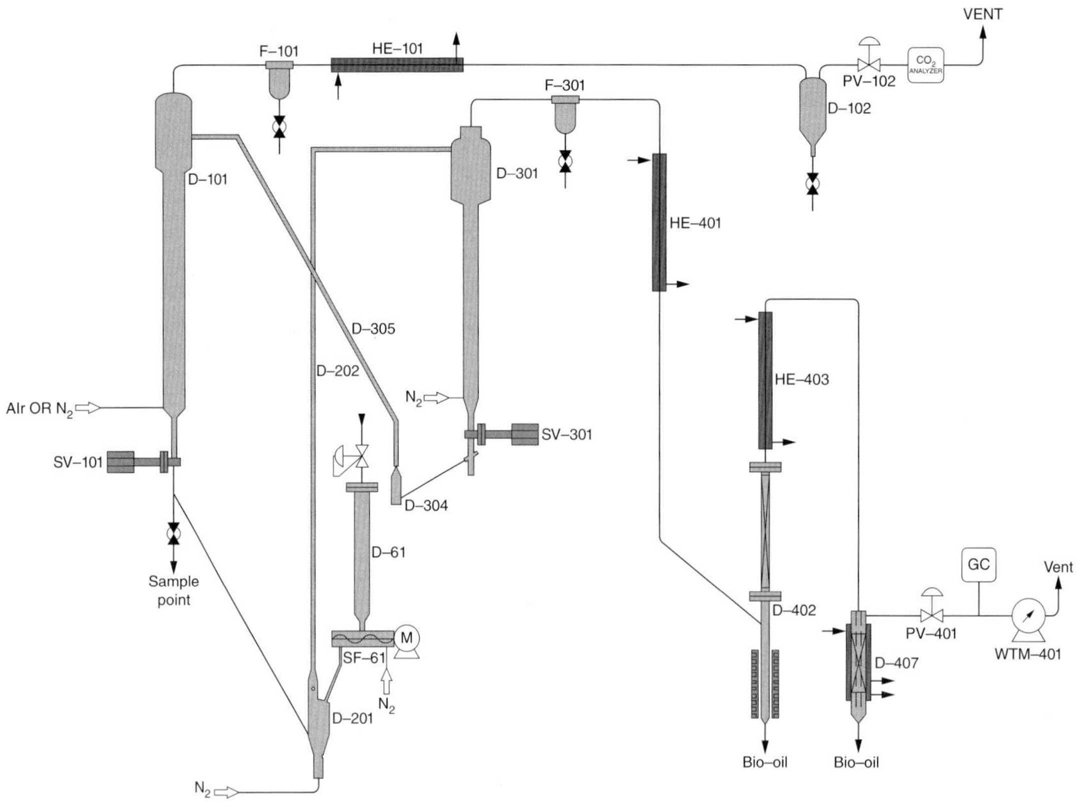

Figure 4.1 Schematic flow diagram of a biomass catalytic pyrolysis unit: regenerator (D-101), biomass feed hopper (D-61), mixing zone (D-201), reactor/riser (D-202), stripper (D-301), lift line (D-305), bio-oil recovery vessels (D-402, D-407), and heat exchangers (HE-101, HE-401, and HE-403).

The mixing zone promotes intimate contact of hot catalyst (coming from regenerator) and the biomass in order to have very high heating rates of the biomass particles. Once mixed, the combination is transported out of the mixing zone into the riser (D-202), where the pyrolysis reactions continue to take place. The mixture of the biomass pyrolysis vapors and the catalyst then enters the stripper (D-301) that allows the removal of trapped vapors within the solids and the separation of oil vapor from catalyst. Solids pass through the stripper and a lift line (D-305) to the regenerator. Downstream of the stripper and after separation of entrained solids (F-301), the vapors are cooled in heat exchangers and the liquid bio-oil is recovered, whereas lighter gases can be used for heat recovery and electricity generation. The factors of this technology leading to an optimum bio-oil yield include the type of catalyst, the reactor design, the residence time in the reactor, the fast separation of the solid, and the vapors in stripper and the fast quench of the produced vapors in the heat exchangers.

Product Yield and Quality from Catalytic Biomass Pyrolysis

Butler et al. presented a summary of the oil yield and quality from various fast pyrolysis reactors.[59] Depending on biomass type and the reactor used, the liquid yield (oil and H_2O) is between 50 and 80 wt% (on biomass). However, oxygen in the liquid is always around 50%. The literature reviews showed that, in contrast with thermal pyrolysis, lower oil yields are obtained from catalytic pyrolysis but better oil quality is achieved. More significantly, Agblevor et al.[63] charged a bubbling fluidized bed (BFB) pyrolysis reactor with 2 kg/h HZSM-5 catalyst at 450–500°C and produced 33 wt% liquid, 53 wt% gas, and 12.5 wt% char and coke yields. The oils produced were stable, exhibiting insignificant aging over 10 months, had low viscosity, contained mainly phenolics, creosols, and methyl-substituted phenols and their total acid number (TAN) number was 41 (as compared with 90 for the corresponding raw bio-oil).

In a similar work, Aho et al.[30,36] loaded a BFB pyrolysis reactor with various catalysts (beta, Y, ZSM-5, and mordenite) and observed that the catalyst structure influenced the composition of the bio-oils but not the yield. Pyrolysis in the presence of a zeolite catalyst resulted in a bio-oil yield ranging between 43.5 and 52.7 wt%. A higher selectivity for organics was observed that was directly related to more H_2O and CO production. Li et al.[64] also used a BFB pyrolysis reactor with a packed bed for the upgrading of the pyrolysis vapors with an FCC catalyst producing a bio-oil with lower amounts of acids (14.5–3.021 wt%) and increased amounts of aromatic hydrocarbons (0.488–16.795 wt%). Park et al.[28] also tested the second stage approach using a BFB pyrolysis reactor and a fixed bed for the vapor upgrading. Among the catalysts tested (HZSM-5, HY, Ga/HZSM-5, and Ga/HY), HZSM-5 was more effective at upgrading the bio-oil than HY. Deoxygenation was achieved mainly by conversion of oxygen to H_2O, CO, and CO_2.

Wang et al.[65] attempted to produce light aromatic hydrocarbons from biomass by catalytic pyrolysis using a BFB pyrolyis reactor with CoMo-S/Al_2O_3 in the catalyst bed and achieved a maximum yield of benzene, toluene, xylene, and naphthalene of about 6.3 wt% at 590°C. Zhang et al.[20] also used a BFB pyrolysis reactor with HZSM-5 catalyst achieving a 56.8 wt% liquid yield with very low oxygen content in the organic fraction (14.69 wt%). The H–C and O–C of the oil was 1.51 and 0.15 respectively, whereas its higher heating value (HHV) was 34.6 MJ/kg. The HZSM-5 catalyst reduced concentrations of ketones and phenols and caused a dramatic increase in aromatics in the bio-oil (7.62–74.22 wt%). The same group recently tested FCC in a BFB pyrolysis reactor.[66] The optimal catalyst to biomass ratio was 1:10 for fresh FCC and 1:5 for spent FCC, resulting in the first case (fresh FCC) in lower maximum oil yields (11.8 wt%) than when using the spent FCC (18 wt%). Increasing catalyst mass led to increased hydrocarbons in the oil, whereas it was concluded that less C and O were transferred to the oil in the presence of the catalyst. Finally, Hew et al.[67] performed catalytic cracking of bio-oil with HZSM-5 in a batch reactor and obtained 91.67 wt% of organic liquid product and 46.67 wt% of gasoline range product.

In the past few years, CPERI did a systematic study on catalytic pyrolysis. In particular, CPERI studied the catalytic biomass pyrolysis in small scale,[38–42] in a modified FCC unit,[11] in an once-through transported unit[61] and, recently, in a fully CFB pilot plant[68] (Figure 4.1). CPERI has tested different acidic materials in this CFB pilot plant unit in order to determine the effect of heterogeneous catalysis on product yields and quality. Table 4.1 presents the product yields obtained by catalytic pyrolysis (using a commercially diluted ZSM-5 and an FCC catalyst) and by thermal

Table 4.1 Products Yield and Oil Properties from Biomass Catalytic Pyrolysis

Catalyst	Silica Sand	ZSM-5	FCC
Product yields (wt% on dry biomass)			
Organic bio-oil	44.0	32.0	18.2
Water	19.4	30.5	30.4
Coke + Char	8.2	11.1	23.0
CO	13.3	14.6	14.7
CO_2	12.7	9.6	11.9
Other gases (HC)	2.4	2.2	1.8
Elemental analysis of organic bio-oil (wt%)			
C	60.5	68.8	64.0
H	7.0	6.7	7.5
O	32.5	24.5	28.5
Organic bio-oil Properties			
TAN	93.4	58.2	54.6
HHV (Mj/kg)	19.9	33.2	30.3

pyrolysis (using silica sand as heat carrier). Both catalysts gave lower oil yields with higher H_2O, coke, and gas yields. However, the quality of the bio-oil (Table 4.1) was better during the catalytic tests, as both catalysts decreased the oxygen content and the TAN, whereas they also increased the HHV. From Table 4.1 it is clear that H_2O and CO were the main deoxygenation products. The ZSM-5 catalysts produced the best bio-oils with significant yields (about 30 wt%) and the lowest oxygen content. The two-dimension channel-like pores of the ZSM-5 were responsible for the low coke yields, which in turn facilitated that the carbon of the biomass would end up in the liquid product, which was originally the main target. In contrast with this, the USY zeolite achieved lower deoxygenation, leading to the lowest organic oil yield. This was attributed mainly to the fact that the USY catalyst produces higher coke yields; therefore, a significant portion of the biomass' carbon is lost to solid by-products (Table 4.1).

CONCLUSIONS

Biomass catalytic pyrolysis seems to be a very promising technology that could result in a marketable bio-oil to be finally used as transportation fuel. A catalytic biomass pyrolysis system is simpler as compared with the biomass to liquid process, whereas catalytic pyrolysis oil can be produced in the region where the biomass is generated. This will stimulate the local (decentralized) use of biomass for green transportation fuels. A lot of research is now in progress for the development and evaluation of new catalysts for the catalytic biomass pyrolysis producing a bio-oil with less oxygen content, as compared with the conventional bio-oil. The research till today has shown that the presence of a catalyst in biomass pyrolysis can significantly affect the bio-oil yield and quality. Zeolites constitute a promising class of heterogeneous catalysts, and developments in recent years have demonstrated their potential to find broad use in the conversion of biomass. ZSM-5 catalysts seem to be very active for this process, as they crack the bio-oil giving lower yields (around 30 wt%) but with a better quality (oxygen content around 21 wt%). This bio-oil is very promising and could be used as bio-crude for the production of transportation fuels after a hydrotreating or a catalytic cracking step.

Research challenges on biomass catalytic pyrolysis remain focused on the understanding of the mechanism of the catalytic conversion of lignocellulosic biomass into bio-oil, including structure–property relationships and product distribution, process conditions, and large-scale production and also novel catalysts development that do not sacrifice the bio-oil yield for its quality. Optimum catalyst development includes tailoring the porosity, acidity, basicity, metal–support interactions, controlled formation of appropriate catalyst particles, improved hydrothermal stability, and resistance to catalyst deactivation. It should be underlined that even a small improvement in the quality of the bio-oil with a catalytic process would be beneficiary for any downstream process of bio-oil upgrading.

REFERENCES

1. Bridgwater AV. Catalysis in thermal biomass conversion. *Appl Catal A* 1994, 116:5–47.
2. Graça I, Fernandes A, Lopes JM, Ribeiro MF, Laforge S, Magnoux P, Ramôa Ribeiro F. Bio-oils and FCC feedstocks co-processing: impact of phenolic molecules on FCC hydrocarbons transformation over MFI. *Fuel* 2011, 90:467–476.
3. Williams P, Horne P. Analysis of aromatic hydrocarbons in pyrolytic oil derived from biomass. *J Anal Appl Pyrol* 1995, 31:15–37.
4. Bulushev DA, Ross JRH. Catalysis for conversion of biomass to fuels via pyrolysis and gasification: a review. *Catal Today* 2011, 171:1–13.
5. Stocker M. Biofuels and biomass-to-liquid fuels in the biorefinery: catalytic conversion of lignocellulosic biomass using porous materials. *Angew Chem Int Ed Engl* 2008, 47:9200–9211.
6. Aho A, Kumar N, Eränen K, Salmi T, Hupa M, Murzin DY. Catalytic pyrolysis of biomass in a fluidized bed reactor: influence of the acidity of H-beta zeolite. *Process Saf Environ Prot* 2007, 85:473–480.
7. Williams PT, Horne PA. The influence of catalyst type on the composition of upgraded biomass pyrolysis oils *J Analyt Appl Pyrolys* 1995, 31:39–61.
8. Thring RW, Katikaneni SPR, Bakhshi NN. The production of gasoline range hydrocarbons from Alcell® lignin using HZSM-5 catalyst. *Fuel Process Technol* 2000, 62: 17–30.
9. Atutxa A, Aguado R, Gayubo A, Olazar M. Kinetic description of the catalytic pyrolysis of biomass in a conical spouted bed reactor. *Energy Fuels* 2005, 19:765–744.
10. Horne PA, Williams PT. Upgrading of biomass-derived pyrolytic vapours over zeolite ZSM-5 catalyst: effect of catalyst dilution on product yields. *Fuel* 1996, 75:1043–1050.
11. Lappas AA, Samolada MC, Iatridis DK, Voutetakis SS, Vasalos IA. Biomass pyrolysis in a circulating fluid bed reactor for the production of fuels and chemicals. *Fuel* 2002, 81:2087–2095.
12. Horne P. Premium quality fuels and chemicals from the fluidised bed pyrolysis of biomass with zeolite catalyst upgrading. *Renewable Energy* 1994, 5:810–812.
13. Williams P, Horne P. Characterisation of oils from the fluidised bed pyrolysis of biomass with zeolite catalyst upgrading. *Biomass Bioenergy* 1994, 7:223–236.
14. Horne P, Williams, P. Reaction of oxygenated biomass pyrolysis model compounds over a ZSM-5 catalyst. *Renewable Energy* 1996, 7:131–144.
15. Gayubo A, Aguayo A, Atutxa A, Aguado R, Bilbao J. Transformation of oxygenate components of biomass pyrolysis oil on a HZSM-5 zeolite. I. Alcohols and phenols. *Ind Eng Chem Res* 2004, 43:2610–2618.
16. Gayubo A, Aguayo A, Atutxa A. Transformation of oxygenate components of biomass pyrolysis oil on a HZSM-5 zeolite. II. Aldehydes, ketones, and acids. *Ind Eng Chem* 2004, 43:2619–2626.
17. Gayubo A, Aguayo A, Atutxa A, Valle B, Bilbao J. Undesired components in the transformation of biomass pyrolysis oil into hydrocarbons on an HZSM-5 zeolite catalyst. *J Chem Technol Biotechnol* 2005, 80:1244–1251.
18. Vitolo S, Bresci B, Seggiani M, Gallo M. Catalytic upgrading of pyrolytic oils over HZSM-5 zeolite: behaviour of the catalyst when used in repeated upgrading-regenerating cycles. *Fuel* 2001, 80:17–26.
19. Wang J, Zhang M, Chen M, Min F, Zhang S, Ren Z, Yan Y. Catalytic effects of six inorganic compounds on pyrolysis of three kinds of biomass. *Thermochim Acta* 2006, 444:110–114.
20. Zhang H, Xiao R, Huang H, Xiao G. Comparison of non-catalytic and catalytic fast pyrolysis of corncob in a fluidized bed reactor. *Bioresour Technol* 2009, 100:1428–1434.
21. Samolada M, Papafotica A, Vasalos IA. Catalyst evaluation for catalytic biomass pyrolysis. *Energy Fuels* 2000, 14:1161–1167.
22. Pattiya A, Titiloye JO, Bridgwater AV. Fast pyrolysis of cassava rhizome in the presence of catalysts. *J Anal Appl Pyrolysis* 2008, 81:72–79.
23. Adjaye J, Katikaneni S, Bakhshi N. Catalytic conversion of a biofuel to hydrocarbons: effect of mixtures of HZSM-5 and silica-alumina catalysts on product distribution. *Fuel Process Technol* 1996, 48:115–143.
24. Mullen CA, Boateng AA. Catalytic pyrolysis-GC/MS of lignin from several sources. *Fuel Process Technol* 2010, 91:1446–1458.

25. Jackson MA, Compton DL, Boateng AA. Screening heterogeneous catalysts for the pyrolysis of lignin. *J Anal Appl Pyrolysis* 2009, 85:226–230.
26. French R, Czernik S. Catalytic pyrolysis of biomass for biofuels production. *Fuel Process Technol* 2010, 91: 25–32.
27. Zhu X, Lu Q, Li W. Fast and catalytic pyrolysis of xylan: effects of temperature and M/HZSM-5 (M=Fe, Zn) catalysts on pyrolytic products. *Front Energy Power Eng China* 2010, 4:424–429.
28. Park H, Dong J, Jeon J, Yoo K. Conversion of the pyrolytic vapor of radiata pine over zeolites. *J Ind Eng Chem* 2007, 13:182–189.
29. Adjaye J, Bakhshi N. Production of hydrocarbons by catalytic upgrading of a fast pyrolysis Bio-oil 1. Conversion over various catalysts. *Fuel Process Technol* 1995, 45:161–183.
30. Aho A, Kumar N, Eranen K, Salmi T, Hupa M, Murzin DY. Catalytic pyrolysis of woody biomass in a fluidized bed reactor: influence of the zeolite structure. *Fuel* 2008, 87:2493–2501.
31. Uzun BB, Sarioglu N. Rapid and catalytic pyrolysis of corn stalks. *Fuel Process Technol* 2009, 90:705–716.
32. Carlson T, Tompsett G, Conner W, Huber G. Aromatic production from catalytic fast pyrolysis of biomass-derived feedstocks. *Top Catal* 2009, 52:241–252.
33. Carlson TR, Vispute TP, Huber GW. Green gasoline by catalytic fast pyrolysis of solid biomass derived compounds. *Chem Sus Chem* 2008, 1:397–400.
34. Lu Q, Zhu XF, Li WZ, Zhang Y, Chen DY. On-line catalytic upgrading of biomass fast pyrolysis products. *Chin Sci Bull* 2009, 54:1941–1948.
35. Qi W, Hu C, Li G, Guo L, Yang Y, Luo J. Catalytic pyrolysis of several kinds of bamboos over zeolite NaY. *Green Chem* 2005, 8:183–190.
36. Aho A, Kumar N, Lashkul AV, Eranen K, Ziolek M, Decyk P, Salmi T, Holmbom B, Hupa M, Murzin DY. Catalytic upgrading of woody biomass derived pyrolysis vapours over iron modified zeolites in a dual-fluidized bed reactor. *Fuel* 2010, 89:1992–2000.
37. Adam J, Antonakou E, Lappas A, Stocker M, Nilsen MH, Bouzga A, Hustad JE, Oye G. *In situ* catalytic upgrading of biomass derived fast pyrolysis vapours in a fixed bed reactor using mesoporous materials. *Microporous Mesoporous Mater* 2006, 96:93–101.
38. Iliopoulou EF, Antonakou EV, Karakoulia SA, Vasalos IA, Lappas AA, Triantafyllidis KS. Catalytic conversion of biomass pyrolysis products by mesoporous materials: effect of steam stability and acidity of Al-MCM-41 catalysts. *Chem Eng J* 2007, 134:51–57.
39. Adam J, Blazso M, Meszaros E, Stocker M, Nilsen M, Bouzga A, Hustad J, Gronli M, Oye G. Pyrolysis of biomass in the presence of Al-MCM-41 type catalysts. *Fuel* 2005, 84:1494–1502.
40. Antonakou E, Lappas A, Nilsen MH, Bouzga A, Stocker M. Evaluation of various types of Al-MCM-41 materials as catalysts in biomass pyrolysis for the production of bio-fuels and chemicals. *Fuel* 2006, 85:2202–2212.
41. Nilsen MH, Antonakou E, Bouzga A, Lappas A, Mathisen K, Stocker M. Investigation of the effect of metal sites in Me-Al-MCM-41 (Me = Fe, Cu or Zn) on the catalytic behavior during the pyrolysis of wooden based biomass. *Microporous Mesoporous Mater* 2007, 105:189–203.
42. Triantafyllidis KS, Iliopoulou EF, Antonakou EV, Lappas AA, Wang H, Pinnavaia TJ. Hydrothermally stable mesoporous aluminosilicates (MSU-S) assembled from zeolite seeds as catalysts for biomass pyrolysis. *Microporous Mesoporous Mater* 2007, 99:132–139.
43. Lu Q, Li W-Z, Zhang D, Zhu X-F. Analytical pyrolysis-gas chromatography/mass spectrometry (Py-GC/MS) of sawdust with Al/SBA-15 catalysts. *J Anal Appl Pyrolysis* 2009, 84:131–138.
44. Lee HI, Park HJ, Park Y, Hur JY, Jeon J, Kim JM. Synthesis of highly stable mesoporous aluminosilicates from commercially available zeolites and their application to the pyrolysis of woody biomass. *Catal Today* 2008, 132:68–74.
45. Park H, Heo H, Jeon J, Kim J, Ryoo R. Highly valuable chemicals production from catalytic upgrading of radiata pine sawdust-derived pyrolytic vapors over mesoporous MFI zeolites. *Appl Catal B: Environ* 2010, 95:365–373.
46. Nokkosmaki M, Kuoppala E, Leppamaki E, Krause A. Catalytic conversion of biomass pyrolysis vapours with zinc oxide. *J Anal Appl Pyrolysis* 2000, 55:119–131.
47. Fabbri D, Torri C, Baravelli V. Effect of zeolites and nanopowder metal oxides on the distribution of chiral anhydrosugars evolved from pyrolysis of cellulose: an analytical study. *J Anal Appl Pyrolysis* 2007, 80:24–29.
48. Lu Q, Zhang Z, Dong C, Zhu X. Catalytic upgrading of biomass fast pyrolysis vapors with nano metal oxides: an analytical Py-GC/MS study. *Energies* 2010, 3:1805–1820.
49. Wang D, Xiao R, Zhang H, He G. Comparison of catalytic pyrolysis of biomass with MCM-41 and CaO catalysts by using TGA-FTIR analysis. *J Anal Appl Pyrolysis* 2010, 89:171–177.
50. Torri C, Reinikainen M, Lindfors C, Fabbri D, Oasmaa A, Kuoppala E. Investigation on catalytic pyrolysis of pine sawdust: catalyst screening by Py-GC-MIP-AED. *J Anal Appl Pyrolysis* 2010, 88:7–13.
51. Lu Q, Zhang Y, Tang Z, Li W, Zhu X. Catalytic upgrading of biomass fast pyrolysis vapors with titania and zirconia/titania based catalysts. *Fuel* 2010, 89:2096–2103.
52. Stefanidis SD, Kalogiannis KG, Iliopoulou EF, Lappas AA, Pilavachi PA. *In-situ* upgrading of biomass pyrolysis vapors: catalyst screening on a fixed bed reactor. *Bioresou Technol* 2011, 102:8261–8267.
53. Deng L, Fu Y, Guo QX. Upgraded acidic components of bio-oil through catalytic ketonic condensation. *Energy Fuel* 2009, 23:564–568.
54. Gaertner C, Serrano-Ruiz JC, Braden D, Dumesic J. Catalytic coupling of carboxylic acids by ketonization as a processing step in biomass conversion. *J Catal* 2009, 266:71–78.
55. Chen M, Wang J, Zhang M, Chen M, Zhu X, Min F, Tan Z. Catalytic effects of eight inorganic additives on pyrolysis of pine wood sawdust by microwave heating. *J Anal Appl Pyrolysis* 2008, 82:145–150.

56. Babich I, van der Hulst M, Lefferts L. Catalytic pyrolysis of microalgae to high-quality liquid bio-fuels. *Biomass Bioenergy* 2011, 35:3199–3207.
57. Ernst S, Hartmann M, Sauerbeck S. A novel family of solid basic catalysts obtained by nitridation of crystalline microporous aluminosilicates and aluminophosphates. *Appl Catal A* 2000, 200:117–123.
58. Peralta MA, Sooknoi T, Danuthai T, Resasco DE. Deoxygenation of benzaldehyde over CsNaX zeolites. *J Mol Catals A Chem* 2009, 312:78–86.
59. Butler E, Devlin G, Meier D, McDonnell K. A review of recent laboratory research and commercial developments in fast pyrolysis and upgrading. *Renewable Sustainable Energy Rev* 2011, 15:4171–4186.
60. Gayubo AG, Valle B, Aguayo AT, Olazar M, Bilbao J. Pyrolytic lignin removal for the valorization of biomass pyrolysis crude bio-oil by catalytic transformation. *J Chem Technol Biotechnol* 2010, 85:132–144.
61. Lappas AA, Dimitropoulos VS, Antonakou EV, Voutetakis SS, Vasalos IA. Design, construction, and operation of a transported fluid bed process development unit for biomass fast pyrolysis: effect of pyrolysis temperature. *Ind Eng Chem Res* 2008, 47:742–747.
62. Ensyn website 2009. http://www.ensyn.com/about.htm.
63. Agblevor FA, Mante O, Abdoulmoumine N, McClung R. Production of stable biomass pyrolysis oils using fractional catalytic pyrolysis. *Energy Fuels* 2010, 24:4087–4089.
64. Li H-Y, Yan Y-J, Ren Z-W. Online upgrading of organic vapors from the fast pyrolysis of biomass. *J Fuel Chem Technol* 2008, 36:666–671.
65. Wang C, Hao Q, Lu D, Jia Q, Li G, Xu B. Production of light aromatic hydrocarbons from biomass by catalytic pyrolysis. *Chin J Catal* 2008, 29:907–912.
66. Zhang H, Xiao R, Wang D, Zhong Z, Song M, Pan Q, He G. Catalytic fast pyrolysis of biomass in a fluidized bed with fresh and spent fluidized catalytic cracking (FCC) catalysts. *Energy Fuels* 2009, 23:6199–6206
67. Hew KL, Tamidi AM, Yusup S, Lee KT, Ahmad MM. Catalytic cracking of bio-oil to organic liquid product (OLP). *Bioresour Technol* 2010, 101:8855–8858.
68. Kalogiannis KG, Stephanidis SD, Voutetakis SS, Lappas AA. *In situ Catalytic Upgrading of Biooil with Novel and Commercial Catalysts, from Bench to Pilot Plant Scale*, Salt Lake City, Utah: *AICHE*; 2010.

5

Integrated Biomass Hydropyrolysis and Hydrotreating: A Brief Review

Martin Linck,* Larry Felix, Terry Marker and Michael Roberts

Department of Energy Conversion, Gas Technology Institute, Des Plaines, IL, USA

PYROLYSIS

Pyrolysis involves heating biomass in the absence of air or oxygen to drive off water and vapors evolved by the thermal breakdown of biopolymers—cellulose, hemicellulose, and lignin.[1] Biomass tends to thermally decompose above approximately 230°C, with pyrolysis occurring at temperatures nearer 500°C. The products formed during pyrolysis depend on temperature, heating rate, and product residence time at elevated temperatures.[2] Three types of products are usually obtained: species that remain gaseous under standard conditions; species that condense to form liquids under standard conditions; and solid char, which contains mostly carbon and ash. It has been observed that very rapid heating of biomass, combined with immediate quenching of the vapors obtained, produces a large proportion of liquid products. This approach is referred to as 'fast' pyrolysis.[3] Although a range of results is reported in the literature, in the case of fast pyrolysis, approximately 70% of the biomass can be recovered as liquids, 15% as solid char, and another 15% is evolved as noncondensable gases (such as CO_2).[4] If fast pyrolysis is carried out at very high temperatures, (above 650°C), the fraction of char produced can be driven well below 10%.[5]

However, in this case, a shift from condensable liquid products toward gaseous products is observed (i.e., CO, H_2, CO_2, and CH_4). The liquid product of fast pyrolysis is known as bio-oil, bio-crude, or pyrolysis oil and has been the focus of great interest because it has a much higher volumetric energy density than biomass. It has been suggested that pyrolysis oil can be upgraded to replace crude oil in certain applications.[3] Finally, another, more time-consuming biomass conversion approach is referred to as 'slow' pyrolysis, and produces more gas and more solid product, with relatively little liquid being produced. Traditional charcoal burning is an example of slow pyrolysis.[1]

Unfortunately, pyrolysis oil has a number of undesirable properties if the desired end product is a deoxygenated liquid hydrocarbon fuel. The oil consists of a great variety of chemical species, many of which remain unidentified, and many of which contain oxygen.[6] These oxygenated species are generally chemically reactive. The compounds are also hydrophilic, and they mix readily with water. Indeed, pyrolysis oil is usually approximately 25% water; this water reduces the viscosity of the oil and considerably reduces its energy density.[7] The oil is inherently acidic, which means that systems designed to handle pyrolysis oil must be made of special materials that resist corrosion. The mix of compounds found in pyrolysis oil depends to some extent on the type of biomass that is processed, but it is not possible to distill the oil in the same way traditional petroleum-derived oils have been distilled in the past.[6] When pyrolysis oil is heated, it reacts and breaks down, driving off gaseous products and forming a thick, viscous tar that may react further until only coke remains. While the volumetric energy density of pyrolysis oil is higher

*Correspondence to: martin.linck@gastechnology.org

Advances in Bioenergy: The Sustainability Challenge, First Edition. Edited by Peter D. Lund, John Byrne, Göran Berndes and Iacovos A. Vasalos.

Originally published in: *WIREs Energy Environ* 2014, 3: 57–64 doi: 10.1002/wene.119

than that of biomass, it is approximately half that of comparable petroleum-derived oils, because it contains large amounts of chemically bonded oxygen and water.[3] Pyrolysis oil obtained via fast pyrolysis is thermodynamically and chemically unstable, and the natural tendency of the oil is to continue to react to form waxes, tars, and pitch, as well as gaseous products.[7] The rates of these reactions increase rapidly with temperature, and, as a result, the viscosity of stored pyrolysis oil may increase significantly even if it is exposed only to ambient temperatures. The reactive oil also tends to create deposits in fuel lines, injectors, and burner components.[4] Microscopic particles of char entrained in the oil appear to catalyze reactions among various chemical species in the oil.[7] These unavoidable char particles are produced by the rapid abrasion and disintegration of biomass during the process of fast pyrolysis. Researchers report that the oil undergoes aging processes, even when great care has been taken to avoid conditions where aging could occur, and samples have been found to separate into distinct aqueous and hydrophobic phases as the aging process proceeds to completion. Temperatures as low as −20°C may be required to arrest the chemical mechanisms involved in the aging of the oil.[4] While technologies that make hot gas filtration possible for fast pyrolysis have been developed, and appear to make it easier to remove the char before the oil condenses to a liquid, the char particles in question can be extremely small. The smallest of these are almost impossible to remove, and will remain in the liquid product.[7] It is possible to stabilize pyrolysis oils by adding alcohols such as methanol and ethanol. As little as 5% by mass of an alcohol can reduce the viscosity of the oil, and significantly slow the aging of the oil.[4] However, it should be noted that alcohols can attack fuel lines and polymer components in fuel delivery systems, and generating these alcohols requires additional effort and cost. It may never be possible to store and transport pyrolysis oil as easily as petroleum-derived products.

Recent work has also shown that single-stage catalytic conversion of pyrolysis oil to aromatic hydrocarbons is possible.[8,9] If pyrolysis is carried out in a fluidized bed consisting of an appropriate catalyst (e.g., a zeolite-based catalyst), then oxygen-containing functional groups associated with many of the species in the pyrolysis vapors can be largely removed as H_2O, CO, and CO_2. Because much of the oxygen present in the feedstock is removed by the formation of water, a significant amount of hydrogen is also removed from the structure of the products. For this reason, aromatic hydrocarbons represent the majority of the stable products when this approach is taken, and these aromatic species may not be usable in every type of liquid transportation fuel. These aromatic products are primarily useful as gasoline blendstocks. However, while this pyrolysis oil contains less oxygen and less water than pyrolysis oil produced in the absence of a catalytically active fluidized bed, at best only about 30% of the available carbon is converted to aromatic products.[10] Rapid catalyst deactivation remains a problem, as up to approximately 30% of the carbon initially present in the feedstock has been found to form coke. The catalyst must be regenerated in order to be reused and unless the coke can be burned off in a way that produces useful heat, it represents carbon that is wasted. The liquid products of this type of catalytic pyrolysis separate into a carbon-rich phase (which usually contains about 10% oxygen) and an aqueous phase. A significant amount of carbon is usually still dissolved in the aqueous phase, and may be lost. These considerations reduce the economic viability of this approach.

It is possible to upgrade condensed liquid pyrolysis oil, eliminating oxygen from the molecular structure of the species in the oil.[11–13] It may also be possible to produce a significant yield of liquid containing far less oxygen. In practice, catalytic hydrodeoxygenation is required, utilizing pressurized hydrogen. Through hydrodeoxygenation, oxygen can be removed from the oil, which reduces the ability of compounds in the oil to react with one another and creates oil that is more similar to that derived from traditional petroleum feedstocks. However, this approach is also problematic. Work carried out with solid catalysts, gaseous hydrogen, and pyrolysis oil in pressurized reactors has demonstrated that partial pressures of hydrogen up to 2500 psig (170 barg) are required for the desired reactions to proceed.[13] Undesirable rearrangements and polymerizations of molecules in the oil occur simultaneously with the desired deoxygenation reactions, and coke formation may rapidly deactivate the catalyst. The presence of liquid water in pyrolysis oil further complicates this approach, as catalysts and their supports must be capable of tolerating exposure to water.

Recently, a major developer of petroleum-related technologies[14] worked to evaluate the potential for utilizing biomass-derived feedstocks in petroleum refineries. Two classes of materials were examined: vegetable oils and pyrolysis oils derived from forest waste. The former can be readily hydrotreated to form 'green diesel' (which consists of hydrocarbons without the oxygenates found in biodiesel). Green diesel was found to be economically attractive only if derived from a low-cost feedstock, such as brown grease, or if vegetable oil feedstocks were subsidized. This approach has been implemented on a commercial scale, but the oils required are not available from lignocellulosic feedstocks. The authors reported that economically attractive hydrocarbon fuels can be generated from the pyrolysis of a lignocellulosic feedstock only if a

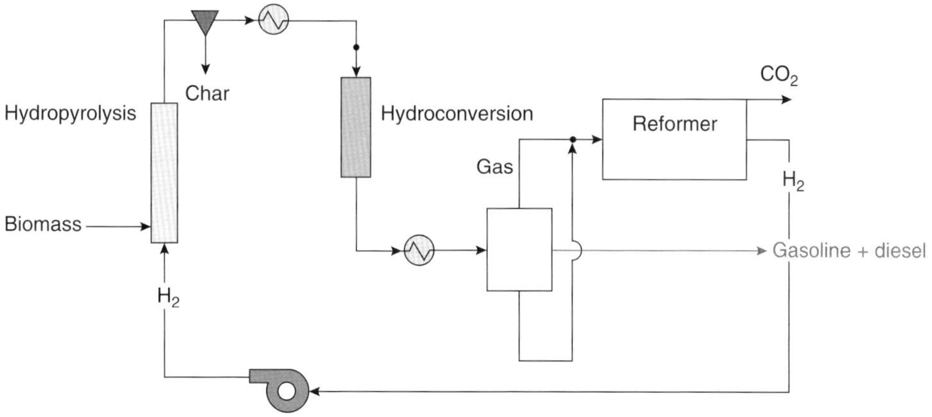

Figure 5.1 Schematic diagram of integrated hydropyrolysis and hydroconversion of biomass in the IH^2 ™ Process. (Reproduced with permission from Ref 16. Copyright 2009, John Wiley & Sons)

significant number of practical and economic hurdles are overcome, particularly the above-discussed ones.

INTEGRATED BIOMASS HYDROPYROLYSIS AND HYDROTREATING

As discussed, there are a number of considerations involved in the conversion of biomass to fungible hydrocarbon transportation fuels (i.e., gasoline, kerosene, diesel, and jet fuel). Most importantly, oxygen must be removed from the molecular structure of any species derived from biomass. Petroleum hydrotreating processes can readily remove oxygen and other heteroatoms from biomass-derived oxygenates as long as hydrogen is available in sufficient quantity and partial pressure.[14] An extensive review of recent work on pyrolysis in the presence of hydrogen (hydropyrolysis) is presented by Balagurumurthy et al.[15] However, the focus of work examined subsequently is on balancing the biomass conversion process, so that hydrogen can be produced from some portion of the process products in order to sustain the hydrotreating of molecules derived from the biomass feedstock. Given that fast pyrolysis of biomass has been shown to liberate vapors that contain much of the carbon initially present in the feedstock, a process involving rapid devolatilization of biomass that minimizes the formation of oxygenates, followed by hydrodeoxygenation of pyrolysis vapors, would appear to be a practical approach.

Such a balanced, integrated process has been developed for converting biomass to fungible liquid hydrocarbon transportation fuels.[16] A schematic of this approach, using integrated hydropyrolysis and hydroconversion reactions and referred to as the IH^2 ™ process, is shown in Figure 5.1. After proof-of-principle work at the Gas Technology Institute, funding for the concept demonstration and experimental work was provided by a grant from the United States Department of Energy (U.S. DOE Award DE-EE-0002873).

This process involves two reactors, the first of which contains a fluidized bed of proprietary catalyst, and the second containing a fixed bed of a second catalyst. Both reactors operate at pressures that are high enough to sustain exothermic deoxygenation reactions (~200–500 psig or ~13.6–34.0 barg). It should be noted that these pressures are much lower than those needed for upgrading of condensed liquid pyrolysis oil.[13]

The temperature of the fluidized bed in the first-stage reactor is sufficiently high to enable rapid devolatilization of biomass (~400–500°C). The fluidizing gas is primarily hydrogen. A stream of biomass feedstock is introduced into this reactor, where it undergoes rapid devolatilization, releasing the oxygenated vapors that are central to this process. Owing to the operating conditions, very little of the carbon present in the feedstock is converted to a solid char product (approximately 10% or less). Once oxygenated vapors have been generated from the feedstock, the vapor molecules are brought into immediate contact with the catalyst and the hydrogen atmosphere in the first-stage reactor. In this context, this process is referred to as hydropyrolysis, and it differs from more traditional pyrolysis, as the species generated during devolatilization are reacted primarily with hydrogen, not with each other. The majority of the chemically bonded oxygen associated with the product vapors is immediately converted to H_2O, CO, and CO_2. Significantly, the deoxygenation reactions occurring in the first stage are exothermic, and will generate more than enough heat to sustain the endothermic initial pyrolysis of biomass occurring in this reactor. The catalyst system also promotes and regulates polymerization and hydrogenation in the first stage of the IH^2 process, so that greater yields of desirable hydrocarbon products can be obtained.

In order for this process to operate in a balanced and economically viable manner, the proportion of each of the product vapors must be controlled, so that the amount of hydrogen needed for deoxygenation can be provided via reforming of noncondensable hydrocarbon products.

The char produced in the first stage is elutriated from the catalyst bed, and is separated from the process in a downstream cyclone, barrier filter, or gravimetric separator. The product vapor stream is then conveyed to the second stage of the process, where any remaining heteroatoms (including species such as nitrogen and sulfur) can be removed. In this integrated process, this stage is referred to as hydroconversion.

The stream of product vapor (consisting of hydrogen, condensable hydrocarbons, noncondensable hydrocarbons, and other vapors such as H_2O, CO, CO_2, H_2S, and NH_3) is then cooled and streams of condensed liquid hydrocarbon products, water with certain dissolved species, and noncondensable vapors, are obtained. The hydrocarbons produced by this IH^2 process are almost completely deoxygenated and the water produced in the process contains almost no carbon, as the hydrocarbons remaining after hydrotreating are not water soluble. As a result, a very significant proportion of the energy content of the biomass is recovered in the liquid hydrocarbon fuel.

In addition, the IH^2 process produces noncondensable hydrocarbon vapors, which are not suitable for use as liquid transportation fuel (CH_4, C_2H_6, C_3H_8). Bench-scale experiments have demonstrated that if these vapors are sent to a steam reformer, they exist in sufficient quantity to supply the hydrogen required to sustain the reactions in the first and second stages of the process. The entire stream of non-condensable vapors (comprising light hydrocarbons, as well as CO and CO_2) is sent to a steam reformer in industrial-scale facilities based on the process. The reformer produces a product vapor stream comprising H_2 and CO_2, and no CO remains. After separation, a stream of nearly pure CO_2 is vented to the atmosphere, and a stream of nearly pure H_2 is sent back into the process (see Figure 5.1). Another attractive feature of this process is that some or all of the water required for steam reforming is produced within the process.

In an initial experimental investigation,[16–18] the first and second stages of the process were demonstrated in a laboratory-scale pilot unit. The feedstock was wood, and the experiment involved a proprietary catalyst developed by CRI Catalyst Company. A condensed liquid hydrocarbon product was obtained, and analyzed in detail. The liquid hydrocarbons were found to consist of gasoline and diesel-range hydrocarbons and found to contain less than 1% by mass oxygen. The liquid hydrocarbons were nonreactive, with a total acid number (TAN) less than 2. The vapor stream from the first stage of the process was sent through a hot sintered-metal barrier filter in order to remove char. Significantly, no increase in pressure drop across the filter was observed as the experiments proceeded, and solids collected by the filter were easily removed from the filter surface. This is not possible when similar filters are used to remove solids from vapor streams produced by traditional fast pyrolysis systems.

Other aspects of the process have been investigated, in the course of subsequent work carried out using the same laboratory-scale pilot unit.[19–21] The temperature of the fluidized bed in the first stage of the process was varied to study the effect of this process variable on char production. An increase in bed temperature was found to reduce char production. However, the maximum allowable first-stage temperature is ultimately limited by the properties of the catalyst in the bed and by the observation that, as the temperature of the first stage is increased, more noncondensable vapors are formed. In consequence, the yield of condensable liquid hydrocarbons is reduced. In this study, conversion of an aquatic plant (lemna) was addressed, and promising yields of liquid hydrocarbons were documented.

Work on feedstocks has expanded to include a wide range of biomass-derived materials.[22–26] Results obtained in the course of continued laboratory-scale pilot unit testing are shown in Table 5.1.

Tests were conducted using feedstocks consisting of wood, lemna, micro-algae, bagasse, macro-algae, and corn stover. Every test demonstrated the production of commercially viable yields of liquid hydrocarbon fuels, along with the consistent deoxygenation of liquid hydrocarbon products. The effect of catalyst composition on the boiling-point range of liquid products was demonstrated, and proprietary catalysts that shift liquid hydrocarbon production into more desirable ranges have been identified. For example, in a comparison of two catalysts employed under identical experimental conditions, one was found to produce significantly more diesel and significantly less gasoline product, although the overall yield of liquid hydrocarbon product was found to remain nearly identical.

A new IH^2 pilot unit, capable of continuously converting approximately 50 kg/day of feedstock, was designed and constructed. Beginning in 2012, Marker and others presented initial results of continuous, long-term testing, carried out in the 50 kg/day, continuously operated pilot plant.[27–31] Long-term performance of catalysts was examined, and the stability of the process was demonstrated over

Table 5.1 Liquid Yields and Feedstock Flexibility Associated with the IH² Process[23]

	Wood	Lemna (Minor Duckweed)	Aquaflow Micro-Algae	Bagasse	Blue Marble Macro-Algae	Corn Stover
Feed %C	49.7	46.3	43.1	43.1	34.0	40.2
Feed %H	5.8	5.8	6.1	5.0	4.43	5.0
Feed %O	43.9	35.7	20.4	35.3	23.6	35.7
Feed %N	0.11	3.7	6.5	0.34	4.6	1.0
Feed %S	0.03	0.3	0.7	0.10	1.9	0.05
Feed %Ash	0.5	8.2	23.1	16.2	29.4	18.1
Feed %H/C	1.40	1.50	1.70	1.39	1.56	1.49
Typical %C_4 + Liquid Yield (MAF)	25–28	30	46	30	35	21
C_4 + gallon/ton MAF	82	100	157	100	119	67
%Oxygen	<1%	<1%	<1%	<1%	<1%	<1%
TAN #	<1	<1	<1	<1	<1	<1

Reproduced with permission from Terry Marker.

significant periods of time on stream. Earlier work had been carried out in the context of experiments lasting no more than a few hours, where at most a few kilograms of biomass were converted. As shown in Figure 5.2, continuous testing of the 50 kg/day pilot unit has demonstrated consistent yields of liquid hydrocarbons over almost 700 of hours of operation.

As long as feedstock composition (in this case, maple and pine), operating conditions, and catalyst composition were held constant, the liquid hydrocarbon product composition remained consistent. Considerable technical challenges were addressed; many of these were associated with the introduction of solid feedstocks into pressurized process vessels along with the recovery of a solid char product. It was shown that the pilot plant consistently operated in a way that would allow the required hydrogen for hydrotreating to be produced via reforming of the light, non-condensable hydrocarbon vapors. While a reformer was not integrated into the 50 kg/day pilot plant, the composition of the process vapor was documented, and found to be compatible with commercially available reformers.

A detailed technoeconomic analysis of the IH² process was developed by Roberts et al.[31] This analysis is based on yield data generated during operation of the 50 kg/day pilot plant, and involved a plant converting 2000 metric tons per day of wood.

The analysis was carried out using the value of a dollar in 2007 as a basis when calculating the magnitude of costs and revenue streams. Sixty percent of the required capital expenditure was assumed to be financed at 8%

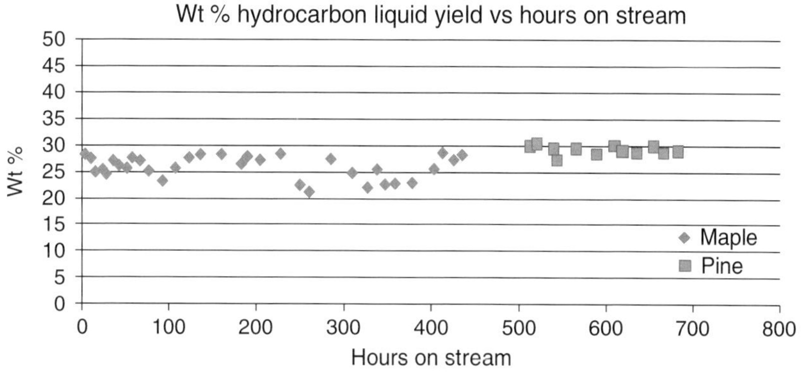

Figure 5.2 Weight % of C4+ liquid hydrocarbon yield observed during operation of 50 kg/day continuously operated IH² pilot plant.[30] (Reproduced with permission from Terry Marker.)

interest. The wood was assumed to cost \$71.97/dry US ton (\$79.31/dry metric ton), and an annual fuel production rate of 60.9 million gallons (230 million L) of liquid hydrocarbons was projected. Eight percent of the catalyst in the first stage of the process was assumed to be replaced each week, and the analysis required a 10% rate of return on capital invested. The eight percent per week first-stage catalyst replacement rate indicates the rate at which the fresh catalyst is required in the fluidized bed of the first-stage reactor. In other words, this replacement rate is anticipated to be sufficient to counteract the phenomena of catalyst deactivation and attrition. A minimum liquid hydrocarbon fuel selling price of \$1.60/gallon (\$0.42/L) was projected, with this cost including no subsidies or carbon credits of any kind. On the basis of this analysis, it appears that liquid hydrocarbon transportation fuels generated from biomass by the IH^2 process compete effectively with fuels derived from petroleum feedstocks.

CONCLUSIONS

Considerable challenges have long impeded the conversion of biomass-derived feedstocks to fungible liquid hydrocarbon blendstocks compatible with refineries and the existing global vehicle infrastructure. Fast pyrolysis of biomass, in the absence of pressurized hydrogen or catalysts, has been shown to convert much of the feedstock into pyrolysis oil. Unfortunately, pyrolysis oil is characterized by high water content, high acidity, extreme chemical instability, and high oxygen content. It can be converted into a deoxygenated hydrocarbon fuel only with considerable difficulty and expense. Char produced during pyrolysis cannot be easily separated from fast pyrolysis vapors, or from condensed liquid pyrolysis oil. If pyrolysis oil is hydrotreated, high-pressure hydrogen from nonrenewable sources may be required, and the very high pressures that are required significantly increase the cost of associated equipment.

In order to address the issues associated with pyrolysis of biomass, the feedstock can be converted to a fungible hydrocarbon blendstock fuel via integrated hydropyrolysis and hydroconversion in the IH^2 process. This process can be implemented in such a way that practical problems are addressed. These include removal of char from the hydropyrolysis vapor and the generation of process heat and hydrogen required to sustain the process. Significant yields of liquid hydrocarbons can be obtained, and, if implemented at commercial scale, with realistic feedstock costs, the cost of production of the liquid hydrocarbon product is low enough to compete effectively with petroleum-derived liquid hydrocarbon fuels.

REFERENCES

1. Bridgwater AV. Renewable fuels and chemicals by thermal processing of biomass. *Chem Eng J* 2003, 91:87–102.
2. Peacocke GVC, Russell PA, Jenkins JD, Bridgwater AV. Physical properties of flash pyrolysis liquids. *Biomass Bioenergy* 1994, 7:169–177.
3. Bridgwater AV. Principles and practice of biomass fast pyrolysis processes for liquids. *J Anal Appl Pyrolysis* 1999, 51:3–22.
4. Diebold, JP. A review of the chemical and physical mechanisms of the storage stability of fast pyrolysis bio-oils. National Renewable Energy Laboratories, NREL-SR-570-27613. Golden, CO: NREL; 2000.
5. Garcia L, Salvador ML, Arauzo J, Bilbao R. Catalytic pyrolysis of biomass: influence of the catalyst pre-treatment on gas yields. *J Anal Appl Pyrolysis* 2001, 58–59:491–501.
6. Elliott, DC. Chemical analysis of biomass fast pyrolysis oils. In *Biomass Pyrolysis Oil Properties and Combustion Conference*, Estes Park, CO, September 26–28, 1994.
7. Diebold, JP, Scahill, JW, Czernik, S, Phillips, SD, Feik, CJ. Progress in the production of hot-gas filtered biocrude oil at NREL. National Renewable Energy Laboratory technical paper NREL_TP_431_7971. Golden, CO: NREL; 1995.
8. Huber GW, Iborra S, Corma A. Synthesis of transportation fuels from biomass: chemistry, catalysts, and engineering. *Chem Rev* 2006, 106:4044–4098.
9. Carlson TR, Vispute TP, Huber GW. Green gasoline by catalytic fast pyrolysis of solid biomass-derived compounds. *ChemSusChem* 2008, 1:397–400.
10. Agblevor FA, Mante O, Abdoulmoumine N, McClung R. Production of stable biomass pyrolysis oils using fractional catalytic pyrolysis. *Energy Fuels* 2010, 24:4087–4089.
11. Elliott DC. Historical developments in hydroprocessing bio-oils. *Energy Fuels* 2007, 21:1792–1815.
12. Jones SB, Valkenburg C, Walton CW, Elliott DC, Holladay JE, Stevens DJ, Kinchin C, Czernik S. Production of gasoline and diesel from biomass via fast pyrolysis, hydrotreating and hydrocracking: a design case. Pacific Northwest National Laboratories report, PNNL-18284. Richland, WA: PNNL; 2009.
13. French RJ, Stunkel J, Baldwin RM. Mild hydrotreating of bio-oil: effect of reaction severity and fate of oxygenated species. *Energy Fuels* 2011, 25:3266–3274.
14. Marinangeli R, Marker T, Petri J, Kalnes T, McCall M, Mackowiak D, Jerosky B, Reagan B, Nemeth L, Krawczyk M. Opportunities for biorenewables in oil refineries final technical report. UOP technical report, Department of Energy Award DE-FG36-05GO15085. Des Plaines, IL: UOP; 2005.
15. Balagurumurthy B, Oza T, Bhaskar T, Adhikari DK. Renewable hydrocarbons through biomass hydropyrolysis process: challenges and opportunities. *J Mater Cycles Waste Manage* 2013, 15:9–15.
16. Marker T, Felix L, Linck M. Integrated, fast hydropyrolysis and hydroconversion process for production of gasoline and diesel fuel from biomass. In *Proceedings of the TC Biomass 2009 Conference*, Chicago, IL, USA, September 16–18, 2009a

17. Marker T, Felix L, Linck M. Integrated hydropyrolysis and hydroconversion process for production of gasoline and diesel fuel from biomass. In *Proceedings of the 2009 AIChE Midwest Regional Conference*, Chicago, IL, USA, October 5–6, 2009b.
18. Marker T, Felix L, Linck M. Integrated hydropyrolysis and hydroconversion process for production of gasoline and diesel fuel from biomass. In *Proceedings of the 2009 AIChE Annual Meeting*, Nashvill e, TN, USA, November 8–13, 2009c.
19. Marker T, Felix L, Linck M, Roberts M. Integrated hydropyrolysis and hydroconversion (IH2) for direct production of gasoline and diesel fuel from biomass. In *Proceedings of the Biomass 2010 Conference*, March 30–31, 2010. Arlington, VA: United States Department of Energy; 2010a.
20. Marker T, Felix L, Linck M, Roberts M. The production of fungible fuels from biomass via integrated hydropyrolysis and hydroconversion. In *Proceedings of the 18th European Biomass Conference and Exhibition*, Lyon, France, May 3–7, 2010b.
21. Marker T, Felix L, Linck M, Roberts M. Direct production of gasoline and diesel from biomass using integrated hydropyrolysis and hydroconversion (IH2). In *Proceedings of the TCS 2010 Symposium on Thermal and Catalytic Sciences for Biofuels and Biobased Products*, Ames, IA, USA, September 21–23, 2010c.
22. Marker T, Felix L, Linck M, Roberts M. Integrated hydropyrolysis and hydroconversion process (IH2) for production of gasoline and diesel fuel from wood, algae, cornstover, lemna, and bagasse feedstocks. In *Proceedings of the 2011 AIChE Spring Meeting*, Chicago, IL, USA, March 13–17, 2011a.
23. Marker T, Linck M, Felix L, Roberts M. Direct production of gasoline and diesel from biomass through integrated hydropyrolysis and hydroconversion (IH2). In *Proceedings of the 19th European Biomass Conference and Exhibition*, Berlin, Germany, June 6–10, 2011b.
24. Marker T, Linck M, Felix L, Roberts M. Direct production of gasoline and diesel from biomass using catalytic hydropyrolysis or integrated hydropyrolysis and hydroconversion (IH2) in a petroleum refinery. In *Proceedings of the 242nd ACS National Meeting & Exposition*, Denver, CO, USA, August 28–September 1, 2011c.
25. Marker T, Linck M, Felix L, Roberts M. The advantages of catalytic hydropyrolysis compared to pyrolysis for biomass conversion to gasoline and diesel. In *Proceedings of the TC Biomass 2011 Conference*, Chicago, IL, USA, September 28–30, 2011d.
26. Marker T, Linck M, Felix L, Roberts M. The Production of gasoline and diesel from biomass through integrated hydropyrolysis and hydroconversion (IH2). In *2011 AIChE Midwest Regional Conference*, Chicago, IL, USA, November 10–11, 2011e.
27. Marker T, Linck M, Felix L, Roberts M. Long term testing of integrated hydropyrolysis and hydroconversion (IH2) for the direct production of gasoline and diesel from biomass. In *Proceedings of the 20th European Biomass Conference and Exhibition*, Milan, Italy, June 18–22, 2012a.
28. Marker TL, Felix LG, Linck MB, Roberts MJ. Integrated hydropyrolysis and hydroconversion (IH2) for the direct production of gasoline and diesel fuels or blending components from biomass, part 1: proof of principle testing. *AIChE J* 2012b, 31:191–199.
29. Linck M, Ortiz-Toral P, Roberts M, Marker T, Wangerow J, Felix L. Direct production of gasoline and diesel from biomass using integrated hydropyrolysis and hydroconversion (IH2): continuous pilot plant testing. In *Proceedings of the 244th ACS National Meeting & Exposition*, Philadelphi a, PA, USA, August 19–23, 2012.
30. Marker T, Linck M, Roberts M, Ortiz-Toral P, Wangerow J, Felix L, Scale up of IH2 process for converting biomass to gasoline and diesel. In *Proceedings of the 2012 AIChE Annual Meeting*, Pittsburg h, PA, USA, October 28–November 2, 2012c.
31. Roberts M, Marker T, Linck M, Felix L, Ortiz-Toral P, Wangerow J. Biomass to gasoline and diesel using integrated hydropyrolysis and hydroconversion. Gas Technology Institute technical report, Department of Energy Award DE-EE-0002873. Des Plaines, IL: Gas Technology Institute; 2012.

6

Transportation Fuels from Biomass via Fast Pyrolysis and Hydroprocessing

Douglas C. Elliott*

Pacific Northwest National Laboratory, Richland, WA, USA

INTRODUCTION

Biomass is a renewable source of carbon, which could provide a means to reduce the greenhouse gas impact from fossil fuel combustion in the transportation sector and provide for an option in an increasingly petroleum-constrained world. Recycling of carbon dioxide from the atmosphere, either by direct chemical conversion or via biomass growth based on solar energy, provides the only renewable source of liquid fuels, which could displace petroleum-derived products. Using appropriate methods, such as fast pyrolysis and catalytic hydroprocessing, solid biomass can be converted to hydrocarbon liquids.[1,2]

Extensive processing is required for conversion of solid biomass, which is a collection of highly oxygenated biopolymer structural components, to liquid fuels, which are typically free of oxygen content. Fast pyrolysis is a method of direct thermochemical conversion (nonbioconversion) of biomass to a liquid product. Utilization of the thermal instability of the biostructures in biomass is a key element in the process. Although the direct conversion product, called bio-oil, is liquid, it is not compatible with the fuel handling systems currently used for transportation. As a mixture of oxygenated components, it has neither the physical nor chemical properties to compete with petroleum distillates as transportation fuel. However, its liquid form does facilitate further processing.

Upgrading the product via catalytic processing with hydrogen gas, hydroprocessing, is a means that has been demonstrated in the laboratory. A comparison of the oil products is presented in Table 6.1. Current research focuses on bringing this technology out of the laboratory. By hydroprocessing, the bio-oil can be deoxygenated to hydrocarbons, which can be renewable replacements of the refined hydrocarbon distillates from petroleum. Although the hydrocarbons from biomass tend to be more in cyclic structures, rather than the straight chains typically found in petroleum, their use should not be limited by this difference.

Commercial entities are currently bringing the fast pyrolysis of biomass into the marketplace. Although small-scale plants have been operated since the 1990s, they were and are for production of chemical products for use in the food industry. Large-scale demonstration plants of fast pyrolysis of biomass for whole bio-oil production have been built and operated, and commercial facilities are now under construction in North America and Europe. The upgrading of the liquid product by hydroprocessing remains in development, although it is moving out of the laboratory into scaled-up process demonstration systems.

BIOMASS FAST PYROLYSIS

Pyrolysis is the thermal decomposition and devolatilization by heating in the absence of oxygen. Biomass

*Correspondence to: dougc.elliott@pnnl.gov

Advances in Bioenergy: The Sustainability Challenge, First Edition. Edited by Peter D. Lund, John Byrne, Göran Berndes and Iacovos A. Vasalos.

Originally published in: *WIREs Energy Environ* 2013, 2: 65–72 doi: 10.1002/wene.74

Table 6.1 Composition of Oil Products

Characteristic	Fast Pyrolysis Bio-oil	Hydroprocessed Bio-oil	Heavy Petroleum Fuel
Water content, wt%	25	0.1	0.1
Insoluble solids, wt%	0.5	0.00	0.01
Carbon, wt%	45	87.5	85.2
Hydrogen, wt%	7.5	12	11.1
Oxygen, wt%	47	0.5	1.0
Nitrogen, wt%	0.15	<0.05	0.3
Sulfur, wt%	0.02	<0.005	2.3
Ash, wt%	0.3	0.00	<0.1
HHV, MJ/kg	17	45	43
Density, g/ml	1.2	0.84	0.94
Viscosity, cp	50 at 40°C	1 at 40°C	180 at 50°C

pyrolysis is an ancient technology, as old as fire itself. The terra preta soils of the Amazon basin have been attributed to charcoal production by pyrolysis and burial of the charcoal by prehistoric inhabitants.[3] Wood tar was produced from pyrolysis by the ancient Egyptians.[4] The biostructures in biomass begin to thermally decompose at temperatures as low as 200°C. By processing biomass in an oxygen-deficient atmosphere, a range of products is derived as a function of temperature and amount of oxygen. A second operating axis to consider is the time at temperature, which at extended times leads to the formation of secondary products from the primary pyrolysis products. In the case of biomass pyrolysis, low-temperature processing releases primary decomposition products mostly water vapor and carbon dioxide with a residual solid condensation product. With increasing temperature, the decomposition of the biopolymers increases and volatile products tend toward a higher molecular weight, which, if cooled and condensed quickly, can be recovered as liquid with minimal amounts of gas or solid products. At higher temperature, the decomposition products continue to be broken down to lower molecular weight gaseous products. The old, simple technology of pyrolysis as practiced at relatively lower temperatures and long heating and cooling times, maximizes the solid product yield. More recent interest has moved the process research to focus on liquid organic products, and, as a result, fast pyrolysis has been developed with moderately higher temperatures but very short residence times.

Biomass resources are broadly distributed and their solid nature makes their collection and transporting problematic. To take advantage of the efficiencies of scale, biomass conversion plants are designed to be fed by biomass feedstock collected from a surrounding radius inside of 100 miles.[5] Even at the scale of some hundreds of tons per day of processing, the scale of fuel production is dwarfed by typical fossil energy refineries. Fast pyrolysis provides a means to take advantage of this distributed nature of biomass feedstock and have smaller scale pyrolysis plants feeding collectively to a larger central upgrading plant to produce finished fuels. In this scheme, the bio-oil becomes the transported form of the biomass. Bio-oil is more energy dense allowing for efficient transportation. The liquid handling properties also add to the advantage over solid biomass transportation. However, bio-oil properties, such as its acidic nature and its thermal instability, must be addressed in any transport system.

Reactor Configurations

Reactor engineering for fast pyrolysis must account for the properties of biomass and the process requirements. Biomass is a solid. It can be reduced in particle size in either the wet (green) stage or the dry stage. As biomass is not a good conductor of heat, rapid heating requires that it be in a small particle size. Furthermore, a good heat transfer mechanism needs to be incorporated to accomplish the fast heat-up required. The most efficient means, with scaleable designs, have been fluidized beds using a solid particulate heat transfer medium mixed with the biomass particles. Fast heat-up of the biomass particles (<2 s) is facilitated by drying the biomass prior to the pyrolysis step. The introduction of the large heat load required by the water in green biomass (30–50 wt% moisture) has been difficult to accomplish for short residence times at reactor temperatures of 450–550°C, well above the boiling point of water. As a result, fast pyrolysis typically requires a dry biomass with a moisture content less than 10 wt%. The heat input requirement to pyrolyze biomass (and to some degree, dry the biomass) can be met by combustion of the solid and gas by-products.

Fluid Bed Systems

In the bubbling fluid bed option, the gaseous by-product is burned and the hot flue gas is injected to fluidize the bed. Char can be recovered as a by-product. In the circulating bed option, solid char is recovered by cyclones with carried-over heat transfer bed solids. The char is combusted to reheat the solids, which are returned to the bed that is fluidized by the flue gases. Transport reactors have also been used that do not include fluidized bed solids but incorporate radiation heating and partial heating by conduction from the reactor wall.

Ablative Systems

In ablative systems, the heat input comes not from small particles but by continuous contact with solid surfaces. A similar short residence time on the hot surface is accomplished by the nearly immediate volatilization of the bio-oil products from the hot surface. The surfaces are typically metal to maximize the heat transfer rate. The force to contact the biomass to the hot surface can come from centrifugal forces or mechanical pressure application, depending on the engineered design. There are even designs that combine ablative surface pyrolysis with fluidized bed operation.

Moderate Heating Rate Systems

Other engineered designs attempt to accomplish fast pyrolysis with perceived simpler designs. These designs, often using auger transport systems, typically accomplish pyrolysis more akin to slow pyrolysis used in charcoal production.[6] In these systems, the liquid yield of bio-oil is less than recovered in true fast pyrolysis, while the gas and char yields are higher. Also, the bio-oil product composition will include a lower level of residual oxygen, suggesting secondary reactions, and a higher yield of water product, which often causes the bio-oil to phase separate. An improved laboratory-scale auger system has been recently reported with a high gas sweep, which can achieve results more similar to the fluid bed systems.[7]

Alternatives: Catalytic Pyrolysis, Hydropyrolysis, Hydrothermal Processing

There are other technology options under development, which attempt to produce a more deoxygenated product bio-oil, which will require less upgrading to hydrocarbon fuels products. These other options include catalytic pyrolysis, hydropyrolysis, and hydrothermal liquefaction. Catalytic pyrolysis involves the addition of catalysts, typically similar to the acid catalyst used in petroleum cracking.[8] The point of catalyst addition remains an issue of development with some attempting direct incorporation of the catalyst into the fluidized pyrolysis bed and others working on vapor phase catalysis of the pyrolysis product following char separation. Hydropyrolysis also involves the addition of catalyst but in addition high-pressure hydrogen is injected into the reactor system.[9] Hydrothermal liquefaction is an entirely different processing concept,[10] which operates at higher pressure and lower temperature for longer residence time. The reactor typically incorporates a liquid water-processing phase, although some concepts substitute a solvent other than, or in addition to, water.

Bio-Oil Product Properties

Fast pyrolysis has been developed to maximize the liquid product yield from biomass. Examination of the liquid phase suggests that it is a single phase, which includes a wide range of oxygenated functional types with a substantial level, typically less than 30 wt%, of water.[11] The water is both dissolved in the liquid organic phase and is also found as microdroplets coating the small pieces of char that escape the cyclones and are collected in the condensate.[12]

Bio-oil from fast pyrolysis has an assigned CAS number of 1207435-39-9. The product is described as liquid condensate recovered by thermal treatment of lignocellulosic biomass at short hot vapor residence time (typically less than about 5 seconds) typically at between 450 and 600°C at near atmospheric pressure or below, in the absence of oxygen, using small (typically less than 5 mm) dry (typically less than 10% water) biomass particles.

A number of engineered systems have been used to effect high heat transfer into the biomass particle and quick quenching of the vapor product, usually after the removal of solid by-product "char," to recover a single-phase liquid product.

Bio-oil is a complex mixture of, for the most part, oxygenated hydrocarbon fragments derived from the biopolymer structures. It typically contains 15–30% water. Common organic components include acetic acid, methanol, aldehydes and ketones, cyclopentenones, furans, alkyl-phenols, alkyl-methoxy-phenols, anhydrosugars, and oligomeric sugars and water-insoluble lignin-derived compounds. Nitrogen- and sulfur-containing compounds are also sometimes found depending on the biomass source.

Chemical Composition

Bio-oil is a complex mixture of oxygenated components containing hundreds of component types and isomers. The specifics of the component structures are dependent to a significant degree on the type of biomass feedstock and the biostructures found in the feedstock. With lignocellulosic biomass the CHO portions are similar with about equal parts of carbon and oxygen and about 6% hydrogen. The amount of nitrogen and sulfur is directly attributable to the biomass feedstock with white wood containing near zero and herbaceous biomass having higher amounts of nitrogen, up to 6%, and sulfur, up to half a percent. There are also trace elements, which are also directly attributable to the elements in the biomass, particularly when significant amounts of char remain in the bio-oil. Present in some recovered bio-oils is also a portion of the fluid bed material (silica or alumina sand/dust).

Essentially all the bio-oil components contain oxygen in one of the functional group types including phenols, carboxylic acids, dehydrated carbohydrates, carbonyls (both aldehydes and ketones), ethers, and alcohols. The oxygenated components and the dissolved water component cause the bio-oil to have a low energy density, like biomass itself and about half that of pure hydrocarbon. The high level of dissolved water appears to severely complicate the measurement of flash point. Recent studies have concluded that flash point is not relevant to bio-oil, and it has been determined as an alternative that bio-oil does not support sustained combustion as determined in the ASTM method D4260. The acidic nature of some of the components causes the bio-oil to have a corrosive nature.[13] It is more of an irritant to the skin but can corrode metals in long-term exposure or at elevated temperature. The toxicity has been measured in a few cases and generally been found to be low. The carcinogenic nature of bio-oil has also been found to be low, but not entirely absent. Biodegradability measurements have found a significant level of degradability, but the time limits may not be met in all cases. These results apply broadly to woody biomass-derived bio-oil, but more assessment is required for bio-oil produced from other biomasses to validate similarities or differences.

Physical Properties

As a result of its chemical composition, bio-oil is not thermally stable. Considering that it is a nonequilibrium reaction product, recovered by rapid quenching of decomposition products, it is not surprising that the bio-oil would have a tendency to continue to react. At room temperature this reaction is barely noticeable. The viscosity of bio-oil will slowly increase, about doubling in a year's time when stored at room temperature with limited exposure to light and air. An accelerated aging test has been developed for bio-oil in which it has been found that 24 h at 80°C has about the same effect on viscosity as a year's time at room temperature.[14] This level of viscosity increase assumes that the bio-oil is kept in a sealed container without exposure to oxygen. Evaporation of water and volatiles or oxidation will lead to an increase in the viscosity of bio-oil at faster rates. The chemical reaction causing the increase in viscosity has not been entirely elucidated yet. It has been determined that the concentration of acids in the bio-oil does not change in correlation with the viscosity increase. However, the change in the concentration of carbonyls does appear to change with the change in viscosity. Other evidence suggests that the addition of light alcohols (methanol, ethanol, or isopropanol) can "stabilize" the viscosity of bio-oil.

The high oxygen content of the bio-oil has other effects on the physical properties. The density of bio-oil is relatively high, about 1.2 g/mL, compared to water at 1 and hydrocarbon liquids at <1. The oxygen content also causes the volatility of the oil to be low. As a result, higher temperatures are required to distill bio-oil. Distillation of bio-oil, at the higher temperatures required, leads to decomposition and unacceptable results. For example, in a simple batch vacuum distillation of bio-oil (ASTM D1160), less than half of the bio-oil (mostly water) was distilled before the residue solidified in the still pot at about 170°C.[15]

Reactivity

The reactivity of the oxygenates in bio-oil results in difficulties in further use and processing of the bio-oil. Bio-oil can be preheated to only a limited degree, up to about 40°C appears reasonable. Therefore, the catalytic processing required to remove the oxygen and produce a hydrocarbon fuel, which is typically done at 350–450°C, is done only with some special handling. The overall concept of biomass to liquid fuels via fast pyrolysis and hydroprocessing is shown in Figure 6.1.

Direct processing of bio-oil in conventional hydroprocessing systems has led to failure in several cases. Moderate pretreatment at lower temperatures (sometimes referred to as hydrotreating, HT) prior to the finishing hydroprocessing (sometimes referred to as hydrocracking, HC) has shown promise. The concept was first patented based on the early work in the 1980s.[16] Further developments and improvements have been more recently reported.[17] The long-term trouble-free operation has not yet been definitively reported, but there are claims in the nontechnical literature.[18]

BIO-OIL HYDROPROCESSING

Bio-oil can be deoxygenated by two theoretical routes: The oxygen can be thermally cracked from the bio-oil in the form of water and carbon oxides, or it can be removed as water by addition of hydrogen. Both routes have been tested using different catalysts. Catalytic cracking has a capital cost advantage because of a lower cost reactor, which operates at low pressure and higher throughput. Catalytic hydroprocessing uses more expensive equipment

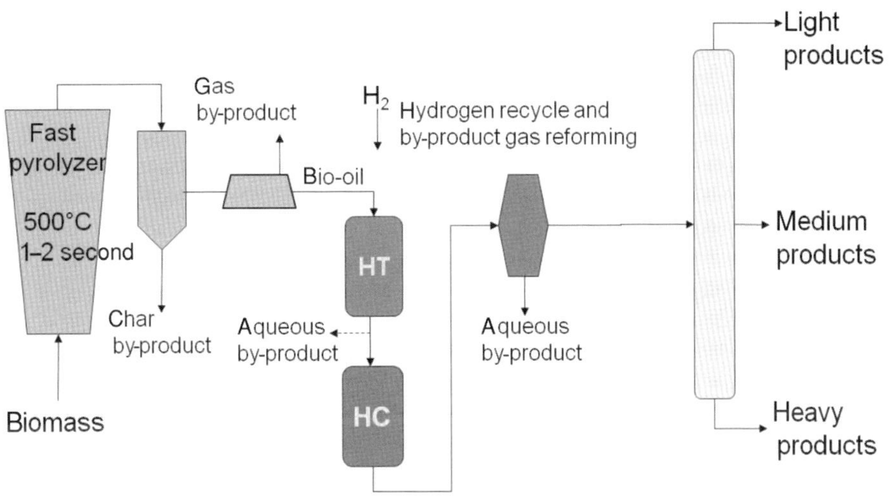

Figure 6.1 Hydrotreating of fast pyrolysis bio-oils.

due to the higher pressure and slower throughput and also has a hydrogen reagent cost; however, it typically has a higher yield overall. The current process development efforts focus for the most part on hydroprocessing. Some of the work has used model compounds and been directed at more fundamental chemistry questions.[19] Other studies have investigated traditional sulfided catalysts, whereas some have investigated nonsulfided precious metal catalysts reactors.[20] Studies have determined that much, but not all, of the hydrogen requirement can be derived within the process from reformed gas by-products.

Reactor Configurations

Conventional hydroprocessing of petroleum to finished fuels is performed almost exclusively with fixed catalyst bed operation. Similarly, most of the bio-oil processing has also been performed in fixed bed reactors. The scale of such operations has been limited to bench-scale reactors. Scale-up of the process is underway and should be online in 2014.[21]

The operation of the bio-oil catalytic hydrotreating process has similar concerns as that found with petroleum hydroprocessing. The oil feedstock needs to be relatively free of unreactive solids and preferably will be low in metals content. Distribution of the oil over the whole bed is a concern to maintain the highest efficiency of contacting the catalyst. The flow rate over the catalyst bed needs to be carefully controlled in relation to the process kinetics. The reported rates of hydrodeoxygenation of bio-oil are relatively low, typically about 0.15 volume of bio-oil per volume of catalyst bed per hour, to approach complete deoxygenation. The exothermic nature of these hydrodeoxygenation reactions, which form water, is severe. Temperature increases as much as 60°C have been reported in small fixed beds and are expected to be a significant design factor in scale-up of the system.

The catalysts used have included both conventional promoted molybdenum sulfides (CoMo and NiMo) and also precious metal catalysts. When using sulfided catalysts an important concern is maintaining the sulfided nature of the catalyst when the bio-oil may actually be deficient in sulfur. Alternatively, when using precious metal catalysts, the amount of sulfur in bio-oils, even from clean wood at 50 ppm or less, may still be sufficient to deactivate the catalyst in the long term.

The primary issue interfering with long-term operation has been the fouling of the catalyst bed by carbonaceous deposits. These deposits appear to be condensation reaction products forming at elevated temperature from the most reactive components in the bio-oil. Low-temperature hydroprocessing of the bio-oil prior to conventional hydroprocessing appears to ameliorate this problem. Another option under development is the use of an ebullated bed reactor, see Box 6.1. Refinement of the processing conditions to allow long-term operation remains at the forefront of the process R&D.

Product Properties

Dependent on the processing severity, a range of deoxygenated products can be produced from bio-oil catalytic hydroprocessing. Nearly deoxygenated hydrocarbons (<0.5 wt% oxygen) have been of most interest. These product mixtures are different from petroleum crudes in that

Box 6.1 Ebullated-Bed Hydroprocessing Reactor

An innovative reactor configuration* is under development for bio-oil hydroprocessing. As an alternative to fixed catalyst bed processing, an ebullated bed has been designed, built, and is being tested at Pacific Northwest National Laboratory.** The bench-scale ebullated bed operates effectively as a liquid phase fluidized bed. The bio-oil is processed as a dilute phase in the hydroprocessed bio-oil. The use of a dilute phase and the agitation of the catalyst particles are expected to lead to less fouling of the catalyst particles and eliminate the possibility of plugging the catalyst bed. As the process is effectively a continuous stirred tank reactor, it is not expected to produce highly deoxygenated products directly but will be operated in combination with finishing hydroprocessing in a conventional fixed bed reactor. It is likely that the product will be sufficiently deoxygenated to allow it to be fed directly as petroleum refinery feedstock.

*Rana MS, Samano V, Ancheyta J, Diaz JAI. A review of recent advances on process technologies for upgrading of heavy oils and residua. *Fuel* 2007, 86:1216–1231.

**Brown C, Beckman D, Zacher AH, Elliott, DC. Development of an ebullated bed reactor for bio-oil upgrading. In tcbiomass 2011: International Conference of Thermochemical Conversion Science, September 28–30, 2011, Chicago, IL. http://media.godashboard.com//gti/tcbiomass2011-Upgrading-Posters.pdf.

Figure Ebullated bed reactor pilot plant.

they are high in cyclic hydrocarbon structures, both saturated (naphthenes) and aromatic. These cyclic structures are clearly derived from the cyclic structures in the bio-oil itself. With less severe processing, higher space velocity or lower temperature, more oxygen is left in the product mix. The most stable oxygenates, and the last to be eliminated, have been reported to be the phenolics.[22]

Chemical Composition

The use of ^{13}C nuclear magnetic resonance spectrometry has been found useful for identifying the functional groups residing in the process bio-oils. Although this method does not specify the actual chemical compounds in the product, it is helpful to distinguish the relative amounts of functional types of carbon in the hydroprocessed bio-oil. Through quantitative methods, carbon can be categorized into saturated and unsaturated carbon, as well as different oxygen-containing functional types, such as carbonyls, phenolics, ethers, and carboxylic acids.

As for specific chemical components, alkylated benzenes and alkylated cyclohexanes are most commonly found by gas chromatography-mass spectrometry (GC-MS). Bicyclics, both saturated and aromatic, are also found by GC-MS. Through the use of other types of liquid and gas chromatography, other oxygen containing components can be identified in the products.

Physical Properties

Since oxygen plays a big role in the physical properties of the chemical components in the hydroprocessed bio-oil, physical properties such as density, viscosity and color can be correlated to the oxygen content in the oil. Density measurement is a simple way to determine the state of the catalytic processing. Effective hydrodeoxygenation to remove oxygen to levels of less than 10% will reduce the density of the product to less than 1 g/mL from 1.2 g/mL for the starting bio-oil; and the trend continues downward with density nearing 0.9 g/mL with 5% oxygen remaining and closing toward 0.82 g/mL as the oxygen is further reduced to levels approaching 0.[23]

BIO-OIL DERIVED FUELS

Only limited fuel testing has been done on these hydroprocessed bio-oil products due to lack of product because of the small-scale operations, so far. For example, research products from tests performed over a 2-year period were collected and segregated to recover sufficient material to distil a gallon of jet fuel range product.[24] In another case, a liter each of an 8.2%, 4.9%, and 0.4% oxygen-containing hydroprocessed bio-oil was produced from products collected over 4 years. The amount of these products was just sufficient to process through typically crude petroleum analytical methods first to fractionate them into representative fractions, and then perform typical fuel analyses.[21]

Use as Petroleum Refinery Feedstock

One goal of the analyses mentioned above was to determine the appropriate level of hydroprocessing required to produce products of sufficient quality to be processed within the existing petroleum refining infrastructure. A liter each of an 8.2%, 4.9%, and 0.4% oxygen-containing hydroprocessed bio-oil was produced by collection and combining of products collected from many bench-scale hydrotreating tests. These products were processed through typical petroleum analytical methods first to fractionate them into representative fractions of light naphtha, gasoline, diesel, light cycle oil, heavy cycle oil and resid, and then perform typical fuel analyses. The study concluded that the 0.4% oxygen product had several options for refinery processing and should be acceptable feedstock. The 4.9% product had some interesting possibilities that would need to be analyzed and developed. The 8.2% product was not considered a reasonable feedstock for existing petroleum refinery operations.[25]

The rationale for these analyses is the potential to produce bio-oil in small scale in a distributed fashion in proximity to the biomass resource, thereby reducing the transportation costs for raw biomass. The transporting of bio-oil to a central upgrading facility allows the scale-up efficiencies to be attained. In addition, if the upgrading facility is close to a petroleum refinery, which would be operating at even a larger scale, there is potential for additional savings for by-product gas utilization for hydrogen production and other infrastructure interactions.[26]

Direct Fuel Blending Possibilities

Hydrotreated bio-oil products from bench-scale tests were collected and segregated to recover sufficient material of appropriate quality (less than 0.5% residual oxygen) to distil a jet fuel range product. This distillate was further treated to reduce the oxygen content from 0.2% following hydroprocessing to 0.02% before blending. The upgraded fast pyrolysis product was blended with synthetic paraffinic kerosene from vegetable oil to produce the first entirely biomass-based turbine fuel. The cyclic and aromatic nature of the pyrolysis product provided the appropriate jet fuel properties in combination with the straight chain aliphatics from the lipids, and the fuel blend met the requirements for the engine manufacturer. The demonstration of the fuel was the operation of an unlimited class hydroplane for a single test of three laps around a 2.5-mile course.[23]

The corrosion issue is also addressed by the deoxygenation of the bio-oil. Corrosion tests of hydroprocessed bio-oil, as a pure component, confirm that the reduction to <0.5% residual oxygen reduces the corrosion properties to a very low level. Based on 500 h exposure at 50°C, the rate on carbon steel as well as the several stainless steels tested was negligible.[27]

CONCLUSIONS

The use of fast pyrolysis to produce liquid products from solid biomass has been demonstrated commercially. Catalytic hydroprocessing can be applied to the bio-oil product from fast pyrolysis to deoxygenate the bio-oil and produce hydrocarbon products that can serve as renewable transportation fuels. The process development of the hydroprocessing is just moving to scale up from the laboratory. Larger quantities of product should be available in the near future to allow extensive testing and validation of the fuel properties.

REFERENCES

1. Bridgwater AV. Review of fast pyrolysis of biomass and product upgrading. *Biomass Bioenergy* 2012, 38:68–94.
2. Venderbosch RH, Prins W. Fast pyrolysis technology development. *Biofuels Bioproducts Biorefining* 2010, 4:178–208.
3. Lehmann J, Gaunt J, Rondon M. Bio-char sequestration in terrestrial ecosystems—a review. *Mitigation Adaptation Strategies Global Change* 2006, 11:403–426.
4. Shreve NR. Wood chemicals. In: *Chemical Process Industries*, 3rd ed. New York: McGraw-Hill Book Company; 1967, Chapt. 32, 617.
5. Searcy E, Flynn P, Ghafoori E, Kumar A. Relative cost of biomass energy transport. *Appl. Biochem. Biotechnol.* 2007, 136–140:639–652.
6. Badger PC, Fransham P. Use of mobile fast pyrolysis plants to densify biomass and reduce biomass handling costs—a preliminary assessment. *Biomass Bioenergy* 2006, 30: 321–325.
7. Brown JN, Brown RC. Process optimization of an auger pyrolyzer with heat carrier using response surface methodology. *Bioresource Technol* 2012, 103:405–414.
8. Samolada MC, Papafotica A, Vasalos IA. Catalyst evaluation for catalytic biomass pyrolysis. *Energy Fuels* 2000, 14:1161–1167.
9. Marker TL, Felix LG, Linck MB, Roberts MJ. Integrated hydropyrolysis and hydroconversion (IH2) for the direct production of gasoline and diesel fuels or blending components from biomass, Part 1: Proof of principle testing. *Environ Progr Sustainable Energy* 2012, 31:191–199.

10. Elliott DC. Hydrothermal processing. In: Brown RC, ed. *Thermochemical Processing of Biomass: Conversion into Fuels, Chemicals and Power*, 1st ed. West Sussex, UK: John Wiley & Sons, Ltd.; 2011, 200.
11. Oasmaa A, Czernik S. Fuel oil quality of biomass pyrolysis oils—state of the art for the end users. *Energy Fuels* 1999, 13:914–921.
12. Elliott DC. Water, alkali and char in flash pyrolysis oils. *Biomass Bioenergy* 1994, 7:179–185.
13. Oasmaa A, Källi A, Lindfors C, Elliott DC, Springer D, Peacocke C, Chiraramonti D. Guidelines for transportation, handling, and use of fast pyrolysis bio-oil. 1. Flammability toxicity. *Energy Fuels* 2012, 26:3864–3873.
14. Elliott DC, Oasmaa A, Preto F, Meier D, Bridgwater AV. Results of the IEA round robin on viscosity and stability of fast pyrolysis bio-oils. *Energy Fuels* 2012, 26:3769–3776.
15. Elliott DC, Baker EG. Hydrotreating biomass liquids to produce hydrocarbon fuels. In *Energy from Biomass and Waste X*. Chicago, IL: IGT; 1987, 765–784.
16. Elliott DC, Baker. *EG process for upgrading biomass pyrolyzates*. U.S. Patent Number 4,795,841, January 3, 1989.
17. Elliott DC, Hart TR, Neuenschwander GG, Rotness LJ, Zacher AH. Catalytic hydroprocessing of biomass fast pyrolysis bio-oil to produce hydrocarbon products. *Environ Prog Sustainable Energy* 2009, 28:441–449.
18. IFPEN. Axens and Dynamotive to develop and commercialize process for upgrading bio-oil to transportation fuels. Available at: http://www.dynamotive.com/. (Accessed February 11, 2013).
19. Honkela ML, Viljava T-R, Gutierrez A, Krause AOI. Hydrotreating for bio-oil upgrading. In Crocker M, ed. *Thermochemical Conversion of Biomass to Liquid Fuels and Chemicals, RSC Energy and Environment Series, No. 1*. Cambridge, UK: Royal Society of Chemistry; 2010, 288–306.
20. Elliott, DC. Historical developments in hydroprocessing bio-oils. *Energy Fuels* 2007, 21:1792–1815.
21. Biello D. First Dedicated biorefinery could wean Hawaii off imported oil. *Scientific American* April 10, 2012. Available at: http://www.scientificamerican.com/article.cfm?id=first-biorefinery-could-wean-hawaii-military-off-imported-oil. (Accessed February 11, 2013).
22. Christensen ED, Chupka GM, Luecke J, Smurthwaite T, Alleman T, Iisa, K, Franz J, Elliott D, McCormick R. Analysis of oxygenated compounds in hydrotreated biomass fast pyrolysis oil distillate fractions. *Energy Fuels* 2011, 25:5462–5471.
23. Elliott DC, Hart TR, Neuenschwander GG, Rotness LJ, Olarte MV, Zacher AH, Solantausta Y. Catalytic hydroprocessing of fast pyrolysis bio-oil from pine sawdust. *Energy Fuel* 2012, 26: 3891–3896.
24. Bertelli C. Current status of biofuels production and use for commercial aviation. In BIO 2010-V: Seminario Latinoamericano y del Caribe de Biocombustibles, Santiago, Chile, August 17–18, 2010. Available at: http://www.olade.org/biocombustibles/Documents/Ponencias%20Chile/Sesion%204_C%20Bertelli_UOP_USA.pdf. (Accessed February 11, 2013).
25. Arbogast S, Bellman D, Paynter JD, Wykowski J. Advanced bio-fuels from pyrolysis oil: the impact of economies of scale and use of existing logistic and processing capabilities. *Fuel Process Technol*, 2012, 104:121–127. doi: 10.1016/j.fuproc.2012.04.036.
26. Holmgren J, Marinangeli R, Nair P, Elliott D, Bain R. Consider upgrading pyrolysis oils into renewable fuels. *Hydrocarbon Process* 2008, 87:95–103.
27. Keiser JR, Howell M, Connatser RM, Lewis SA Sr. Corrosivity and composition of raw and treated pyrolysis oils. In TAPPI, October 2012.

FURTHER READING

Brown RC, ed. *Thermochemical Processing of Biomass: Conversion into Fuels, Chemicals and Power*. Chichester, UK: John Wiley & Sons, Ltd.; 2011, 200–231.

7

Biomass Gasification for Synthesis Gas Production and Applications of the Syngas

Reinhard Rauch,[1]* Jitka Hrbek[2] and Hermann Hofbauer[2]

[1]Bioenergy 2020+, Güssing, Austria
[2]Institute of Chemical Engineering, Vienna University of Technology, Vienna, Austria

SPIRIT AND PURPOSE OF BIOMASS GASIFICATION FOR SYNTHESIS GAS PRODUCTION

Synthesis gas is one important intermediate to produce fuels for transportation and chemicals. Currently, synthesis gas is produced mainly from natural gas, coal or by-products from refineries. The usage of synthesis gas is about 50% to ammonia, 25% to hydrogen, and the rest is to methanol, Fischer-Tropsch (FT) products and others.[1,2]

Owing to the global warming problem caused by fossil CO_2 emissions, and also due to issues related to security of energy supply, much R&D on the production of synthesis gas from biomass was carried out in the past decade. The main investigated applications of the synthesis gas was to produce transportation fuels over FT synthesis, methanol synthesis or others, like mixed alcohol synthesis.

TECHNOLOGIES FOR PRODUCTION OF SYNTHESIS GAS FROM BIOMASS

In the past decades much R&D was done to develop special gasification systems to convert biomass to synthesis gas, or to adapt coal gasification technology for biomass. These systems can be divided into several groups according to Ref 3:

- Producing and transport of the heat into the gasification reactor: allothermal-autothermal
- Type of reactor: fixed bed-fluidized bed-entrained flow
- Type of gasification agent: oxygen-steam-mixtures of steam and oxygen

Allothermal–Autothermal

Autothermal gasifiers provide the necessary heat of reaction by means of partial oxidation within the gasification reactor. If air is used as oxidizing agent during the process, the product gas contains a high amount of nitrogen. So for synthesis gas production either pure oxygen (in entrained flow reactors) or mixtures of oxygen and steam (in fluidized bed reactors) are used as gasification agent.

The great advantage of the autothermal gasification is the direct heating of the reactants and therefore more efficient energy utilization. The process is simpler than by allothermal gasification and it is easier to operate it under pressurized conditions.

Allothermal (or indirect) gasification is characterized by the separation of the processes of heat production and heat consumption.[4] The allothermal gasification facility almost always consists of two reactors, connected by an energy flow. Biomass is gasified in the first reactor and the remaining solid residue (char) or product gas is combusted in the second reactor to produce the heat for the first process. The transport of the heat can

*Correspondence to: reinhard.rauch@tuwien.ac.at

Advances in Bioenergy: The Sustainability Challenge, First Edition. Edited by Peter D. Lund, John Byrne, Göran Berndes and Iacovos A. Vasalos.

Originally published in: *WIREs Energy Environ* 2014, 3: 73–92 doi: 10.1002/wene.97

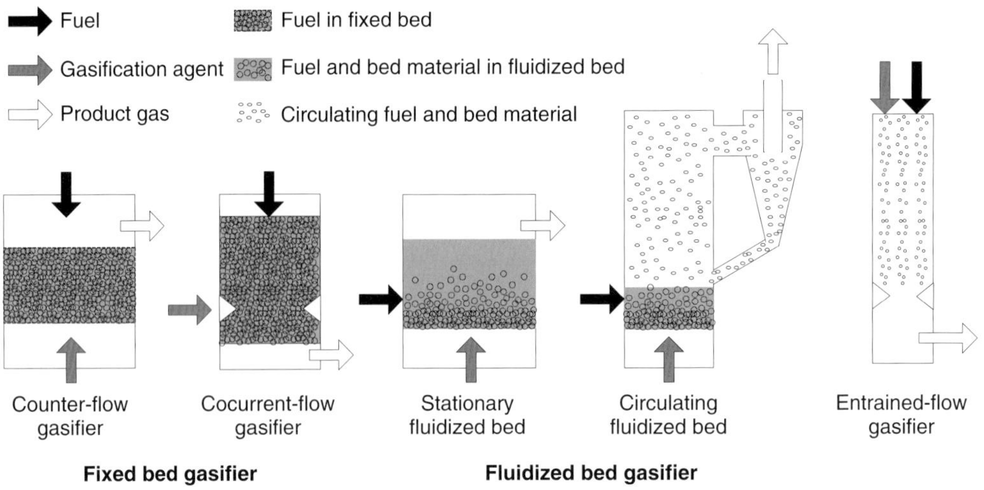

Figure 7.1 Different types of gasifiers.

be done either by circulating a bed material or by heat exchangers.

Allothermal gasifiers generally produce two gas streams: a medium calorific product gas (gasification reactor) with little or no nitrogen and a flue gas (combustion reactor). The production of a N_2-free gas without the need of pure oxygen is one of the advantages over autothermal gasification processes. Another important advantage is the complete carbon conversion and that there is no problematic waste produced. All carbon containing streams from the product gas cleaning (e.g., dust, tars) can be recycled to the combustion zone and there converted to heat, which is used for the gasification reactions.[5]

Table 7.1 Typical Composition of Dry Gas During the Oxygen and Steam Gasification of Biomass

Compound	Oxygen Gasification (Entrained Flow)	Oxygen Gasification (Fluidized Bed)	Steam Gasification
CO (vol%)	40–60	20–30	20–25
CO_2 (vol%)	10–15	25–40	20–25
H_2 (vol%)	15–20	20–30	30–45
CH_4 (vol%)	0–1	5–10	6–12
N_2 (vol%)	0–1	0–1	0–1
LHV (MJ/Nm^3)	10–12	10–12	10–14
Tar content (g/Nm^3)	<0.1	1–20	1–10

Fixed Bed–Fluidized Bed–Entrained Flow

According to the design of the fuel bed, the gasifiers can be divided into fixed bed, fluidized bed, and entrained flow. The differences in the design of the gasification reactor are shown in Figure 7.1.

The composition of the gas and the level of undesirable components (tars, dust, ash content) produced during biomass gasification process are dependent on many factors such as feedstock composition, reactor type, and operating parameters (temperature, pressure, oxygen fuel ratio).

Oxygen Blown–Steam Blown

Typical composition of a dry gas produced during the biomass gasification process is shown in Table 7.1. As can be seen, the concentration of the gas compounds during oxygen and steam gasification is completely different.

During oxygen gasification, the combustion products (CO_2, H_2O) are in the product gas and take part in the chemical reactions, mainly in water-gas shift reaction.

On the other hand, higher amount of hydrogen in the product gas can be found during the steam gasification. The hydrogen found in the product gas does not originate only from the fuel as in the case of oxygen gasification but also from the steam (H_2O).

The tar content depends not only on the type of gasifier but also mainly on the operation temperature. In entrained flow gasification, where the temperature is above 1000°C, there is no tar produced, whereas in fluidized bed gasification, where the temperature is below 1000°C, tar is produced and has to be removed from the product gas.

Requirements on the Gasification Reactor

To use the product gas from a biomass gasifier as synthesis gas, there are several properties, which have to be taken into account:

- H_2:CO ratio
- Amount of inerts, like nitrogen
- Amount of methane and higher hydrocarbons
- Catalyst poisons like sulfur-, nitrogen-, and chlorine components
- Operation pressure

Synthesis gas consists mainly of hydrogen and carbon monoxide, in some cases small amounts of carbon dioxide (methanol synthesis) or methane (BioSNG) are also present. For most syntheses a H_2:CO ratio of 2 is required. This ratio is normally adjusted in a separate catalytic reactor before the synthesis reactor, where some CO is converted to hydrogen by the water-gas shift reaction. If the gasifier produces the correct H_2:CO ratio, the exothermal water-gas shift reaction can be avoided, which reduces investment and operation costs and also increases the efficiency.

Impurities like nitrogen act as inert during the synthesis and their concentration has to be as low as possible. The inerts reduce the partial pressure and by this effect reduce the conversion. Especially for synthesis reactions, where the product is separated as a liquid and where a recycle of remaining unconverted gas is done (e.g., methanol), the inerts have to be bled off as they would otherwise be accumulated. Also for production of BioSNG the inerts have to be below 1 vol%, as otherwise the heating value of the BioSNG will not fulfill the requirements of natural gas.

Methane and higher hydrocarbons normally act as inert during the synthesis reaction, so they have to be treated similar to inerts. These components are mainly present in fluidized bed reactors and not in high temperature gasification (Table 7.1). As the heating value of the hydrocarbons is much higher than that of H_2 and CO, a small amount of hydrocarbons already present in the gas can contain most of the energy of the synthesis gas. So the overall conversion from biomass to final product (e.g., FT liquids) is reduced by the small amounts of hydrocarbons that are already produced in the gasifier. So in most cases the hydrocarbons are converted to H_2 and CO in a reformer to maximize the conversion from biomass to the final desired product. The only exemption is the production of BioSNG. Here a high content of methane and nontar hydrocarbons in the synthesis gas are favored, as then the highest conversion efficiency to BioSNG is achieved.

Catalyst poisons deactivate the synthesis catalyst and have to be removed to very low levels. The most well-known poison is sulfur, which can be in form of H_2S, COS, mercaptans, or thiophens in the synthesis gas. The organic sulfur components are mainly present in fluidized bed gasifiers, and not in high temperature gasification. So the removal technology has to be adapted to the type of the sulfur components, e.g., thiophens cannot be removed by ZnO adsorbers.

Almost all synthesis reactions are under elevated pressure and the synthesis gas has to be compressed during the gas treatment. By the choice of a pressurized gasifier, the electricity consumption of the compressor can be reduced or avoided. Until now only oxygen or oxygen-steam blown gasifiers were operated under elevated pressure, in which case the investment and operation costs for an air separation unit have to be taken into account. For this reason in small scale (<100MW fuel input), mainly indirect atmospheric gasifiers are discussed to avoid the costs of the oxygen production.[6]

The choice of the gasification reactor is not easy, as it influences the whole conversion chain, e.g., the gas treatment has to be adjusted to the type of gasifier and the synthesis should also fit to the properties of the synthesis gas (e.g., H_2:CO ratio, amount of inerts). Up to date no single 'winner' has been identified and R&D on the different types of conversion chains from biomass to transportation fuels and chemicals is going on.

APPLICATIONS FOR SYNTHESIS GAS FROM BIOMASS

R&D on applications for synthesis gas from biomass was in the past decade mainly focused on transportation fuels and less on chemicals. Most of the R&D was done on FT, hydrogen, methanol (including MtG, MtD, MtO, DME), ethanol, mixed alcohols, and SNG. Therefore, below, there are short descriptions of the different synthesis routes.

FT Synthesis

The FT synthesis is based on the conversion of a mixture of carbon monoxide and hydrogen into liquid hydrocarbons.[7]

The FT synthesis can be represented by the following chemical reactions:

$$(2n + 1)H_2 + nCO \rightarrow C_nH_{(2n+2)} + nH_2O$$

$$2nH_2 + nCO \rightarrow C_nH_{2n} + nH_2O$$

The process, a key component of gas to liquids technology, produces synthetic fuels and chemicals from mainly natural gas and coal at a large scale, and in the near future also from biomass. This process is best known for being used in South Africa by Sasol and in the gas to liquid (GTL) plant operated and built by Shell in Bintulu, Malaysia.[8]

Depending on the reaction temperature, the process can be divided into low-temperature (LTFT, 200–260°C) and high-temperature (HTFT, 300–350°C) FT synthesis.[9]

During LTFT (up to 260°C) a higher fraction of higher-boiling hydrocarbons (above 360°C) is produced.[10] Also the total distillate yield is higher than during HTFT. Higher temperature leads to faster reactions and higher conversion rates, also tends to favor shorter chains and methane, olefin, and aromatics production. Typical pressures range from one to several tens of atmospheres. Increasing the pressure leads to both higher conversion rates and formation of the much-desired long-chained alkanes. Even higher pressures would be favorable, but the benefits may not justify the additional costs of high-pressure equipment and costs for pressurizing.

A variety of catalysts can be used for the FT process, but the most common are the transition metals such as cobalt, iron, and ruthenium. Nickel could also be used, but tends to favor methane formation. In LTFT cobalt- or iron-based catalysts are mostly used, whereas in HTFT only iron-based catalysts are used.

The advantage of the cobalt-based catalysts is a much longer lifetime. Less unsaturated hydrocarbons and alcohols are produced compared with the iron-based catalysts. On the other hand, iron-based catalysts are more tolerant to sulfur and can be also used to adjust the H_2:CO ratio, if it is lower than 2 by means of water-gas shift reaction.

Hydrogen

Hydrogen can be produced from the gasification product gas through the steam reforming and water-gas shift reaction. Using a dual fluidized bed (DFB) gasification system with CO_2 adsorption along with suitable catalysts, it is possible to achieve a hydrogen yield up to 70 vol% direct in the gasifier.[11]

Furthermore, the costs of hydrogen production by biomass gasification in very large scale are competitive with natural gas reforming.[12]

Hydrogen is one of the most promising future energy carriers.[13] Therefore, it is expected that biomass gasification process will be available for large-scale hydrogen production.

Series of investigations on hydrogen production methods have been conducted over the past several decades.[14] Biomass is potentially a reliable energy source for hydrogen production. It is renewable, abundant, easy to use, and CO_2 neutral.

Methanol

Methanol, also known as methyl alcohol, wood alcohol, or wood spirits, is often abbreviated as MeOH. It can be produced from fossil or renewable resources and can be used either directly as a transportation fuel or can be converted further to hydrocarbons.[15]

Methanol is produced from synthesis gas by the hydrogenation of carbon oxides over a suitable (copper oxide, zinc oxide, or chromium oxide-based) catalyst:

$$CO + 2H_2 \leftrightarrow CH_3OH$$

$$CO_2 + 3H_2 \leftrightarrow CH_3OH + H_2O$$

The first reaction is the primary methanol synthesis reaction. A small amount of CO_2 in the feed (2–10%) acts as a promoter of this primary reaction and helps maintain catalyst activity. The stoichiometry of both reactions is satisfied when R in the following relation is 2.03 minimized.[16] H_2 builds up in the recycle loop, this leads to an actual R value of the combined synthesis feed (makeup plus recycle feed) of 3 to 4 typically.

The reactions are exothermic and give a net decrease in molar volume. Therefore, the equilibrium is favored by high pressure and low temperature. During production, heat is released and has to be removed to keep optimum catalyst life and reaction rate; 0.3% of the produced methanol reacts further to form side products such as dimethyl ether, formaldehyde, or higher alcohols.[17]

The catalyst deactivates primarily because of loss of active copper due to physical blockage of the active sites by large by-product molecules; poisoning by halogens or sulfur in the synthesis gas, which irreversibly form inactive copper salts; and sintering of the copper crystallites into larger crystals, which then have a lower surface to volume ratio.

Conventionally, methanol is produced in two-phase systems: the reactants and products forming the gas phase and the catalyst being the solid phase. The production of methanol from synthesis gas was first developed by the company BASF in Germany already in 1992. This process used a zinc oxide/chromium oxide catalyst with poor selectivity, and required extremely vigorous conditions—pressures ranging from 300 to 1000 bar, and temperatures of about 400°C. In the 1960s and 1970s, the more active Cu/Zn/Al catalyst was developed allowing more energy-efficient and cost-effective plants, and larger scales. Processes under development at present focus on shifting the equilibrium to the product side to achieve higher conversion per pass than the currently achieved 20–30% in tubular- or quench-type reactors. Examples are the gas/solid/solid trickle flow reactor, with a fine adsorbent powder flowing down a catalyst bed and picking up the produced methanol; and liquid phase methanol processes where reactants, product, and catalyst are suspended in a liquid. Fundamentally different could be the direct conversion of methane to methanol, but despite a century of research this method has not yet proved its advantages.

Methanol to Diesel

The methanol to diesel (MtD) process first converts methanol into propylene, this is followed by olefin oligomerization (conversion to distillates), then product separation-plus-hydrogenation.[18]

The process would yield mostly kerosene and diesel, along with a small yield of gasoline and light ends. The near-zero sulfur/polyaromatics diesel fuel resulting from this process would differ from more conventional FT diesel only in cetane number (>52 via 'Methanol-to-Synfuel' vs >70 cetane for FT diesel). The incidental gasoline stream not only would be near-zero sulfur but also have commercial octane ratings (92 RON, 80 MON) and maximally 11% aromatics.

Methanol to Gasoline

In the 1970s, Mobil developed and commercialized a methanol to gasoline (MtG) process. A plant was built in Montunui, New Zealand in 1985 and sold to Methanex. It produced gasoline until 1997 when the plant was permanently idled. If the gasoline is to be sold without additional blending, then further treating is necessary to reduce the amount of benzenes.[19]

Methanol to Olefins

Methanol to olefin (MtO) synthesis is a commercially attractive process because of the high demand of propylene and ethylene in the market. Nowadays, these compounds are produced mainly through noncatalytic steam cracking of fossil fuels (naphtha).[19]

The MtO process, however, uses zeolite-based catalysts that efficiently convert methanol into propylene and ethylene.

Dimethyl Ether

Dimethyl ether (DME) or (CH3OCH3) is generally produced by dehydration of methanol. In recent R&D projects, the methanol production and dehydration processes are combined in one reactor, such that DME is produced directly from synthesis gas slightly more efficient than methanol. In addition, the direct synthesis of DME allows H_2:CO ratios of about 1, which is an advantage for oxygen blown gasifiers.

Like methanol, DME has promising features as fuel candidate with both the otto and the diesel engine (www.biodme.eu). With adaptations to engine and fuel system, DME can be used in diesel engines, leading to higher fuel efficiency and lower emissions. In otto engines, DME can be used with liquefied petroleum gas (LPG). Since DME is as easily reformed as methanol, it has a big potential as fuel for fuel cell vehicles.[20] DME has similar physical properties as LPG and can be handled as a liquid, using the same infrastructure as LPG.[21]

Mixed Alcohols

The catalytic conversion of synthesis gas to mixed alcohols is nowadays of great importance because mixed alcohols are valuable additives to gasoline to increase the octane number and reduce the environmental pollution. Furthermore, a great benefit of the mixed alcohol synthesis is the high resistance of the catalysts against sulfur poisoning and the fact that the gas cleaning facilities can be simpler as in other syntheses. Mixed alcohols can also be converted to high-quality fuels via dehydration and oligomerization.

Normally, the alkali-doped oxides (zinc and chromium oxides) and alkali-doped sulfides (molybdenum sulfides) are used as catalysts for mixed alcohols synthesis (MAS).

Depending on process conditions and catalysts, the main primary products are generally methanol and water. Ethanol, propanol, butanol, etc. are gradually including methanol synthesis followed by its successive homologation.[22] The reaction mechanism of the MAS is following:[23]

$$nCO + 2nH_2 \rightarrow C_nH_{2n+1}OH + (n - 1)H_2O$$

Owing to reaction stoichiometry the proposed CO/H_2 ratio is 2, but the optimal ratio is in practice closer to 1 because of water-gas shift reaction that occurs in parallel with the alcohol formation.

BioSNG

Natural gas that is produced from coal or biomass is known as 'synthetic natural gas' or 'substitute natural gas' (SNG).[24,25]

The typical catalyst for methanation is nickel and the main reaction is

$$CO + 3H_2 \rightarrow CH_4 + H_2O$$

Ni-based catalysts are also active in water-gas shift and hydration of higher hydrocarbons, like olefins. Normally a H_2:CO ratio of 3 is necessary, which is obtained by a water-gas shift reactor before methanation. In some types of reactors, e.g., fluidized beds, the water-gas shift can be carried out also in parallel with the methanation, so no external adjustment of the H_2:CO ratio is necessary.

Methanation can be done at atmospheric pressure, although according to the thermodynamics higher pressure is preferred.

As Ni-based catalysts are sensitive to sulfur poisoning, gas treatment is quite important before the methanation, and sulfur compounds have to be removed to below 0.1 ppm.

There is one main difference in the synthesis gas treatment compared with other synthesis. Almost all gasifiers have some methane in the product gas, depending on the operation temperature. As methane has not to be converted in the methanation, high methane content in the synthesis gas is a big advantage for the production of BioSNG (if the overall efficiency is higher, the higher the methane content in the product gas is). Thus especially indirect gasifiers are advantageous for methanation, which have a syngas with a methane content of 10 vol% or more.

EXAMPLES OF SYNTHESIS GAS UTILIZATION

In the following chapter some examples of gasification systems for production of synthesis gas are described. They were selected, as there are some innovative aspects included, like:

- Innovative consortium as here forest industry works together with a refinery (Neste Oil-Stora Enso)
- Concept of decentral biomass upgrading by pyrolysis to increase the energy density for transport and central gasification in extremely large scale (Bioliq process)
- Using coal gasification technology and adjusting the biomass properties by torrefaction (BioTfueL)
- Synthesis gas platform, where R&D on different synthesis is performed, taking slip streams of synthesis gas from a commercial biomass CHP (biomass CHP Güssing)
- Dual fluidized bed gasification optimized for SNG (ECN)
- Sorption-enhanced reforming, to adjust the H_2:CO ratio direct inside the gasifier
- Microchannel reactor technology (Oxford Catalysts, Velocys)
- Decentral small scale methanation for production of BioSNG (Distributed SNG)
- Large-scale FT synthesis (Choren)
- Large scale of alcohol synthesis (Range Fuels)
- Large-scale BioSNG production (GoBiGas)

There are several other projects going on which are not described in this paper, like the UPM Andritz-Carbona project on FT synthesis in Stracel, the Ajos BTL project in Finland, the Woodspirit project (Methanol) in Oosterholm (the Netherlands), or the Ethanol demonstration project from Enerkem, the TIGAS Project from Haldor Topsoe at GTI in Chicago, or Ineos Bioenergy who are doing syngas fermentation.

The reason for selecting the above projects for this review, was on the one side, that they have an interesting concept, on the other side, that there was information available. This is one main problem in commercial and large demonstration projects, that there is almost no information available.

Innovative Consortium as Here Wood Industry Works Together with a Refinery (Neste Oil-Stora Enso)

Neste Oil Corporation is a refining and marketing company concentrating on low-emission, high-quality traffic fuels. The company produces a comprehensive range of major petroleum products and is the world's leading supplier of renewable diesel (NextBTL).[26]

Stora Enso is a forest industry and procures most of its wood from private forest owners, state forests and companies in Finland, Sweden, the Baltic states, Continental Europe, and Russia.

Stora Enso and Neste Oil have joined forces to develop the technology for, and later produce, new-generation biofuel from forest industry residues.

NSE Biofuels Oy, is owned by Neste Oil Oy and Stora Enso Oy. It combines competencies of two different industries: forest industry and oil refining industry.[27]

The goal is an implementation of the first commercial biomass-to-liquids (BtL) plant and further to develop profitable businesses based on experiences and on market demands (NOSE-Project).

The objectives of NOSE-project are to develop and verify BtL technology based on biomass gasification route, further to provide the overall commercial technology concept and evaluate the technoeconomic feasibility of the production concept.

Project partners are VTT, Foster Wheeler, NesteJacobs, Stora Enso, and Neste Oil.

The biomass gasification technology being employed was initially developed as part of the Ultra Clean Gas project, which was led by VTT Technical Research Centre of Finland, acting as the main research and testing partner for the joint venture. Gasification equipment is being

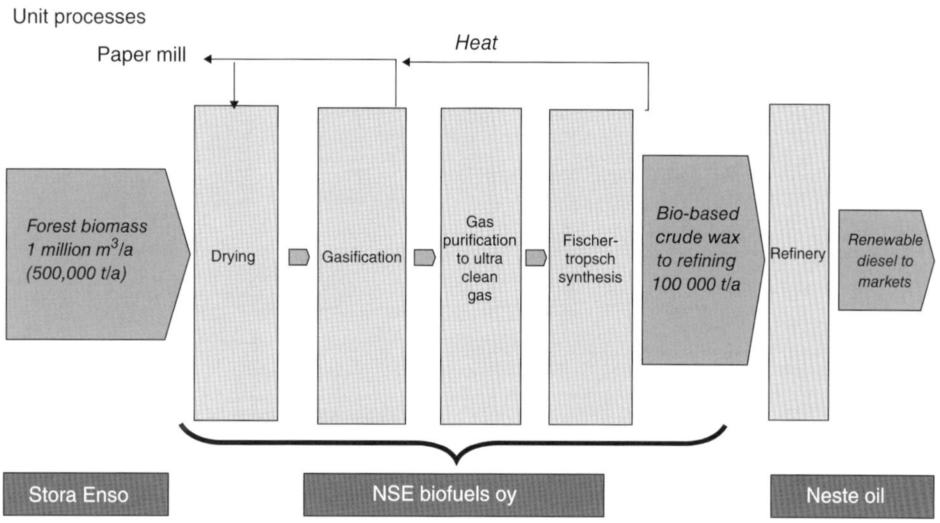

Figure 7.2 Stora Enso-Neste Oil joint venture.

supplied by Foster Wheeler, a leader in the field of fluidized bed reactors.

The demonstration process is based on the use of FT synthesis and covers the entire chain from drying wood-based biomass to test the FT catalysts needed to produce bio-based crude wax that can then be refined into premium-quality renewable diesel by Neste Oil. Much of the development effort has focused on ensuring good syngas output from an oxygen blown CFB gasifier and that it can be reformed and cleaned catalytically to a sufficiently high level of purity to yield the best results from the latest generation of FT synthesis (Figure 7.2).

High energy efficiency, the ability to handle different types of feedstock, and the potential for integration in existing pulp and paper mills have also been prioritized in the design.

The demonstration plant will start using sawmill by-products, shavings, and sawdust before testing the main feedstock intended for a commercial-scale plant: forest residues, such as logging waste, stumps, and small-dimension logs that cannot be used as pulpwood.

The tests of BtL have been completed and technology verified in Varkaus test plant at the end of August 2011. The plan was to build a full-scale commercial facility at one of Stora Enso's mills capable of producing in the order of 100,000 t/annum of bio-based crude wax for final refining into renewable diesel by Neste Oil.

Unfortunately the consortium decided not to continue at the moment with this development, as they pronounced by a press release from 17th August 2012: 'We have calculated the cost of the project very carefully and realistically. It would have represented a very significant investment and we concluded that we would not have gone ahead in any case, even if we had won public funding,' according to Neste Oil's President & CEO, Matti Lievonen, and Stora Enso's CEO, Jouko Karvinen.[28]

Concept of Decentral Biomass Upgrading by Pyrolysis to Increase the Energy Density for Transport and Central Gasification in Extremely Large Scale (Bioliq Process)

The Karlsruhe BtL concept[29] combines decentralized production of energy-rich bioSynCrude® by means of rapid pyrolysis of biomass (straw, wood) and central processing with final industrial-scale refinement.

Since the volumetric energy density of bioSynCrude® is by more than one order of magnitude higher relative to straw, it is clear that the method's efficiency is enhanced by decentralized energy densification and conversion in a central large-scale unit. The process technology can be seen in Figure 7.3.

The dry, comminuted biomass is mixed with hot sand at ambient pressure in the absence of air in a twin-screw mixing reactor. Pyrolytic conversion of the biomass particles at approximately 500°C, and condensation of the pyrolysis vapors take a few seconds only. Depending on the operating conditions and on the biomass selected, one obtains 40–70% of liquid pyrolysis oil and 15–40% of pyrolysis char as well as a fraction of noncondensable pyrolysis gas

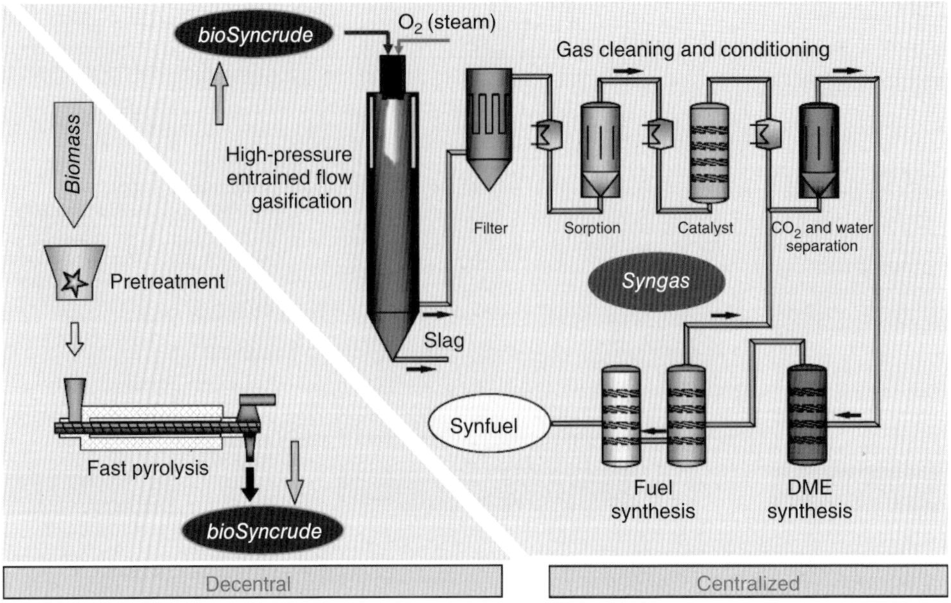

Figure 7.3 Bioliq process. (Reprinted with permission from Ref 29. Copyright 2012 Karlsruhe Institute of Technology)

whose combustion heat can be used for heating or drying. In the pilot plant, 500 kg/h (2 MW) of biomass are converted by rapid pyrolysis into bioSynCrude®.

Pyrolysis char and pyrolysis oil are mixed to obtain a slurry (bioSynCrude®). The rapid and efficient conversion of the mixture during gasification essentially depends on the size distribution of the coke particles. Rapid pyrolysis enables a pyrolysis conden-sates/pyrolysis char mixing ratio that is ideal for the slurry and contributes to optimizing product yields.

The bioSynCrude® is atomized with hot oxygen in an entrained flow gasifier and is converted above 1200°C into a tar-free, low-methane raw synthesis gas. The gasifier used is particularly suited for the high amounts of biomass ashes that are produced during gasification. The process is performed at pressures that are determined by the subsequent synthesis. There is no need for complex gas compression procedures. While FT syntheses require process pressures of up to 30 bar, methanol or DME syntheses are carried out at up to 80 bar. The bioliq pilot gasifier is designed for 5 MW (1 t/h) and two pressure stages of 40 and 80 bar.[30]

Cleaning of raw synthesis gases: particles, alkaline salts, H_2S, COS, CS_2, HCl, NH_3, and HCN are removed to avoid catalyst poisoning during fuel synthesis. The pilot plant is equipped with an innovative hot-gas cleaning system for particle filtration, pollutant decomposition, and adsorption at 500°C.[31]

The clean syngas is finally converted over methanol/DME to gasoline.

Using Coal Gasification Technology and Adjusting the Biomass Properties by Torrefaction (BioTfueL)

The aim of the BioTfueL project is to develop market and end-to-end chain of technologies to produce second-generation biodiesel and biojet fuel. The biodiesel and biojet fuel produced will be suitable for use in all types of diesel and jet engines, either pure or in blends.[32]

The BioTfueL project will develop technologies for converting lignocellulosic biomass—such as straw, forest waste, and dedicated crops—into synthetic fuel, by validating their technical and economic feasibility in demonstration units and optimizing their energy efficiency and environmental impact.

The whole process chain, which can be seen in Figure 7.4, includes R&D (CEA, IFP Energies nouvelles[33]), licensors of the technology (Axens, Uhde), and fuel producers (Sofiprotéol, Total).

BioTfueL project is planned for 7 years (2010–2016) and total funds allocated are 1127M.

The BioTfueL project uses torrefaction as first step, to convert biomass into a product similar to coal. Afterward a coal gasifier (PRENFLO) is used to produce the synthesis gas. The biomass pretreatment and torrefaction take place

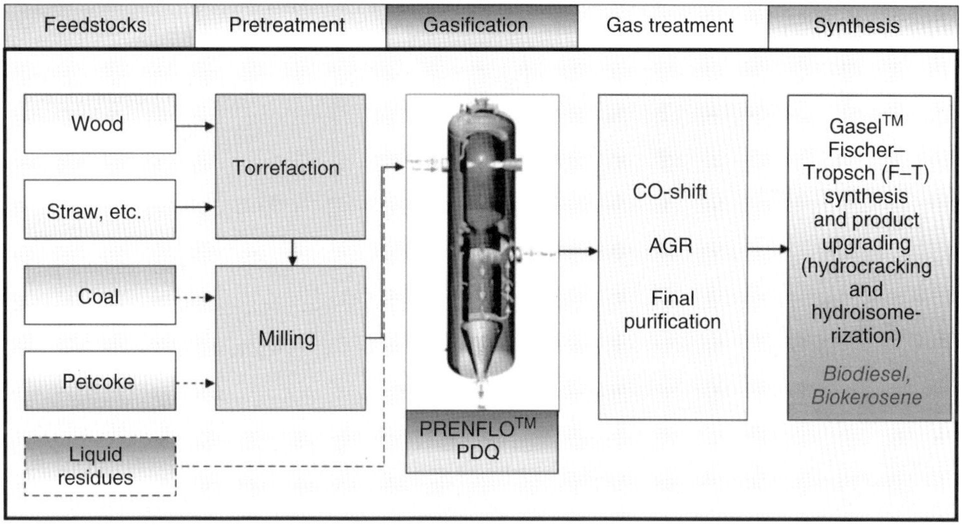

Figure 7.4 BioTfueL concept.

at Sofiprotéol site in Venette and gasification, purification, and synthesis at Total site in Dunkirk.

The biomass will be pretreated during the torrefaction process, pulverized and afterward injected under pressure into the gasifier with direct quench.

The syngas must be very pure and have the correct chemical composition to be converted into liquid fuel. Thus, the technological developments will focus mainly on the final stages of the syngas cleaning and conditioning, when the gas is pretreated for FT synthesis.

The FT process will convert purified syngas into a mixture of hydrocarbons at low temperature, using a cobalt-based catalyst. No R&D work will be carried out on the FT process because it is already available commercially.

The opinion of the BioTfueL consortium is that second-generation biofuel production requires large industrial units to be profitable and a continuous supply of feedstock. The biomass used in BioTfueL project will be forest and plant waste, straw, and other lignocellulosic materials sourced from different locations. The availability of these materials is subject to seasonal variations. Thus, a process chain will be developed that works with a wide range of biomass feedstock as well as with liquid and solid fossil fuel feedstock.

The BioTfueL objectives are to develop, demonstrate, and commercialize a full XTL chain (XTL: X to liquids, where X can be gas, coal, waste, or biomass). First industrial XTL plants will have a capacity of 5000 barrels (bbls)/day (200kt/year) in one single train. The market launch of the process chain is planned by 2020.

Synthesis Gas Platform, Where R&D on Different Synthesis Is Performed, Taking Slip Streams of Synthesis Gas from a Commercial Biomass CHP (Biomass CHP Güssing)

In 2002 an 8 MW CHP FICFB steam blown gasifier producing heat and power (4.5 MWth, 2 MWel) with a gas engine started its operation in Güssing, Austria.[34] Renet-Austria, a competence network on energy from biomass, consisting of experts from Universities and industry started to develop this process further to a commercial stage. The flow diagram of the gasification process can be seen in Figure 7.5.

Owing to the excellent performance that was reached during the last years and the high availability, several additional research projects could be started in Güssing.

The producer gas from the DFB gasifier is nearly free of nitrogen and has a high hydrogen content. For this reason, it is well suited as synthesis gas. Therefore, projects aiming at the development of processes for the production of BioSNG, FT liquids, mixed alcohols, and hydrogen are currently carried out. The picture of the synthesis platform can be seen in Figure 7.6.

BioSNG

BioSNG was the first synthesis, investigated at the biomass CHP Güssing.[35] In 2003, a first test rig was moved to Güssing in order to use a real synthesis gas from biomass. Later, a 10 kW unit was installed and successfully

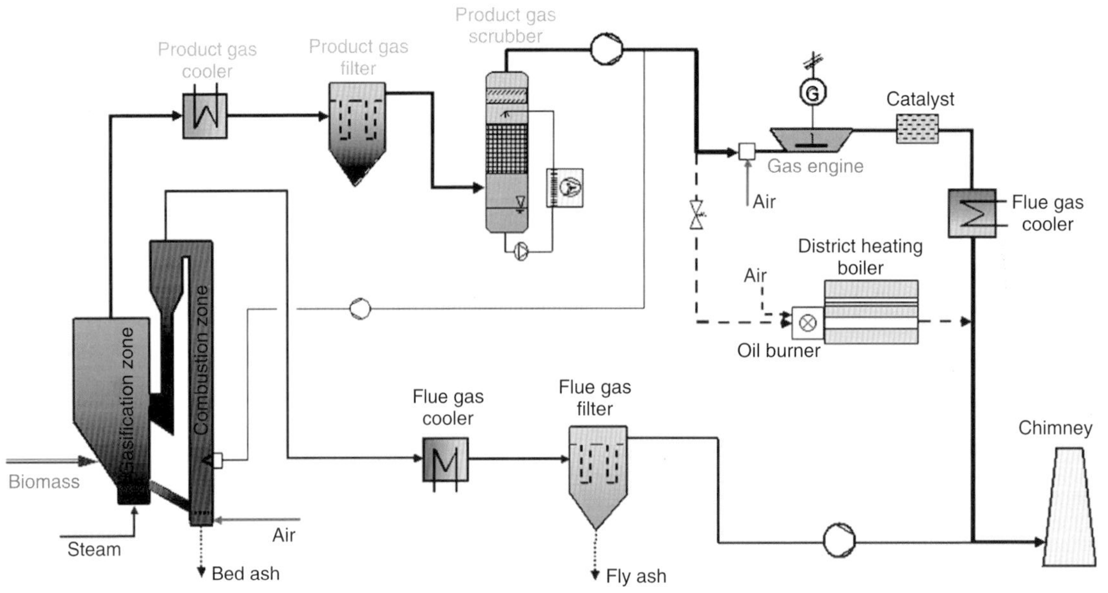

Figure 7.5 Güssing gasification plant.

operated for more than 1000 h without deactivation of catalyst. Within the BioSNG demonstration project the construction and operation of a 1 MW demo unit was realized. The purpose was to prove the concept of preprocessing of the product gas, the methanation, and upgrading of the raw BioSNG to final BioSNG ready for use (e.g., feed-in into a gas grid).

A slip stream of the product gas from the gasifier was taken before entering the gas engine and transported to the BioSNG demonstration unit. The product gas is already cleaned from particles and tars to the extent required for gas engine operation. In the BioSNG demonstration unit, additional cleaning of the gas is carried out to avoid the poisoning of the methanation catalyst. The conditioned product gas enters the fluidized bed methanation reactor, where the methanation, hydration of olefins and CO-shift take place in parallel. Then the raw SNG passes several purification steps (CO_2 removal, drying) to increase the methane content to meet the regulations for feeding into the natural gas grid.

The PDU was operated for several weeks and a high-quality gas as required by the gas grid was produced. The plant was able to operate at design process parameters. A number of parameters were tested, varying pressure, temperature and capacity of the plant and also varying the recycle flow from hydrogen separation.

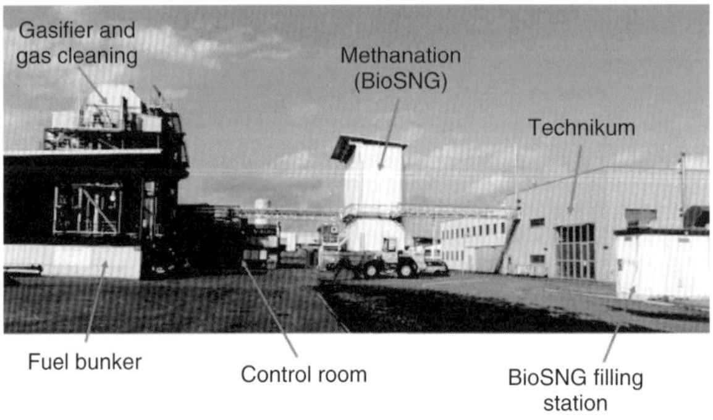

Figure 7.6 Synthesis platform Güssing. (Reprinted with permission from Ref 34. Copyright 2012 BTG Biomass Technology Group BV)

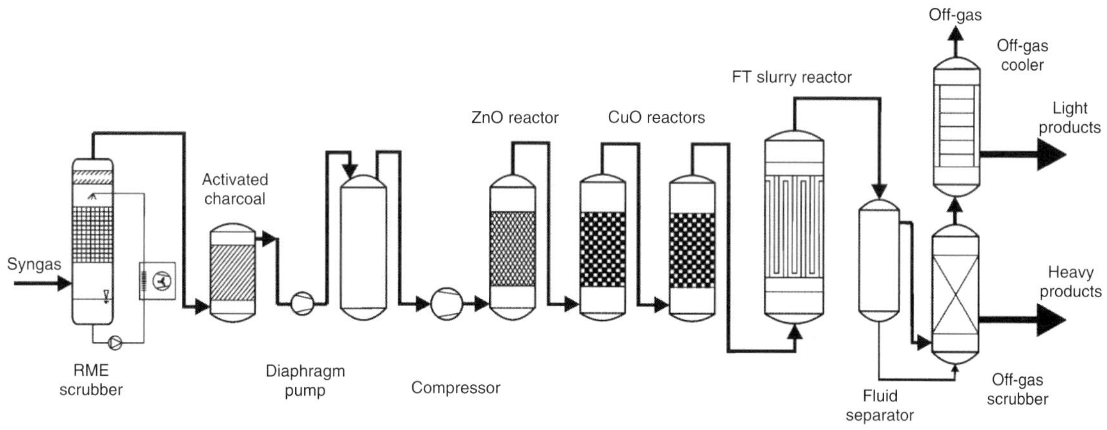

Figure 7.7 Flow sheet of the FT test rig in Güssing.

With this project it was possible to demonstrate the production of pipeline ready BioSNG from wood using the Güssing BioSNG demonstration unit. It was also shown that the chemical efficiency of more than 60% from wood chips to BioSNG could be reached in practice, which is excellent.[36]

FT Synthesis

Within the EC project RENEW the second synthesis, which was synthetic diesel production via FT synthesis was investigated.[37] A laboratory scale Fischer-Tropsch-Reactor (slurry reactor) in a side stream of the existing allothermal fluidized bed gasifier was designed and installed. During the experiments different combinations of the gas cleaning devices, different catalysts, and operation parameters were tested.

The FT liquids produced during operation were collected, analyzed, and used for further processing, like hydrotreating, or are used direct in diesel engines for testing.

The FT synthesis (Figure 7.7) consists of a gas treatment, where catalyst poisons are removed to below 10 ppb and the gas is dried and compressed. The FT reaction takes place in a slurry reactor (three phases: catalyst, gas, waxes) with a volume of 20 L. The gas and the FT product leaves the reactor over sintered metal filters. After the FT-reactor, the FT products are separated from the tailgas and collected.[38]

The collected FT products are distilled. The fraction up to 180°C is used as naphtha, from 180°C to 320°C as diesel and the fraction above 320°C are waxes.

Furthermore, investigations on hydrotreating of FT-biowaxes for generation of bio-products from lignocellulosic biomass were carried out with the industrial partner OMV.[39]

Table 7.2 Properties of FT Diesel

Properties	EN 590:2004 Min	EN 590:2004 Max	FT Diesel	Hydroprocessed FT Diesel
Cetane number (–)	51.0	–	75–85	75–80
Density at15°C (kg/m^3)	820	845	770–790	760–780
Total aromatics content% (m/m)	–	–	<1	2.6
Flash point (°C)	>55	–	87–91	75–80
Viscosity at 40°C (mm^2/second)	2.00	4.50	2.3–2.5	2.0–2.1
Cold filter plugging point (CFPP, °C)	–	–20	–5 to 0	–16 to – 58

Schablitzky et al. investigated the ability of bifunctional catalysts for cracking and hydrogenation of paraffinic feedstocks with aim to obtain second-generation biofuels in the diesel bowling range in a pilot scale hydroprocessing unit.

Analytical data from the produced diesel is given in Table 7.2.

Mixed Alcohols

Another synthesis investigated in Güssing is MAS. For that purpose a pilot plant was built to produce mixed alcohols from synthesis gas generated in commercial CHP gasification plant. The synthesis gas used for the MAS is also taken after the gas cleaning system from the biomass CHP (particle and tar removal).[40]

To adjust a H_2:CO ratio from 1.8:1 of the synthesis gas from gasification plant to a ratio of 2:1 a steam reformer is used. It consists of two electrically heated tubes. In the first one the gas is heated up to 900°C. The second tube is filled with a catalyst and acts as reforming reactor, followed by air cooling and glycol scrubber. After the drying, compression up to 300 bar is done before the MAS reactor. The condensation of the mixed alcohols takes place immediately after the reactor at the same operation pressure.

At the MAS pilot plant in Güssing a MoS_2 catalyst is used, as this type of catalyst is resistant against sulfur poisoning. So the gas treatment is much simpler which reduces the operation and investment costs. For this reason the MAS is from the economics favorable, as also work by NREL in the US showed.[41]

Hydrogen from Biomass

'Hydrogen from biomass for Industry' is the most recent project within the synthesis platform.[42]

The common method for hydrogen production is natural gas steam reforming, but if hydrogen should be produced from renewable, biomass gasification is probably the most economic option. The aim of the project was the production of hydrogen with high purity for the direct integration in a refinery. First, a simulation of the whole process was done. The hydrogen production was based on steam gasification, a CO-shift step, CO_2-separation with a pressurized water scrubber, a PSA system, a steam reformer, and advanced gas cleaning components (Figure 7.8). The results showed that 30 MW of hydrogen can be produced from 50 MW of wood chips.

Based on the simulation results, the laboratory unit for hydrogen production is now under construction.

Dual Fluidized Bed Gasification Optimized for SNG (ECN)

The Energy Research Centre of the Netherlands (ECN) has demonstrated SNG generation from biomass.[43] In this process, indirect gasification is used, and both gasification and methanation are carried out at atmospheric pressure.[44] The biomass is gasified in the riser of a DFB and the remaining char is circulated to the combustor (downcomer of DFB). In this process, the heat required for gasification is supplied by char combustion in the combustor. Steam is used for gasification and air is used for char combustion. The lab-scale gasifier, developed in 2004, has a biomass capacity of 5 kg/h and operates at temperatures of 750–900°C.

Based on the experiments, a SNG system was optimized which consists of an indirect gasifier, a solvent scrubber tar removal system which recycles tar to the gasifier, a gas cleaning reactor including CO-shift, and a methanation combined reactor. The indirect gasifier that is working at 850°C produces nearly nitrogen-free syngas and a high amount of methane. Tars are recycled to the gasifier in order to increase efficiency, whereas the tar-free syngas is cleaned from other contaminants (e.g., sulfur and chlorine). The clean syngas is fed to a combined shift and methanation process, converting the syngas into SNG. After methanation, further upgrading (e.g., CO_2 and H_2O removal) is required in order to comply with the desired SNG specifications. The overall net thermal efficiency is reported as 70% based on low heat value (LHV). Forty percent of the carbon of the biomass becomes part of the SNG and an equal amount of carbon can be captured as CO_2. The

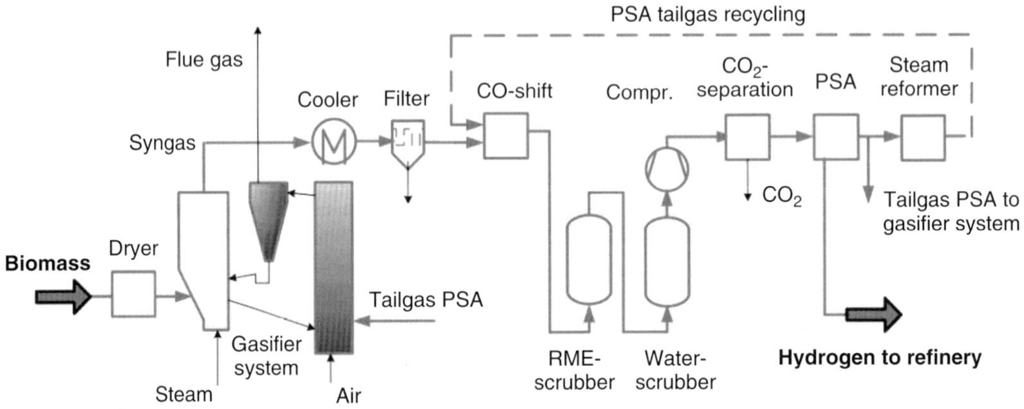

Figure 7.8 Process design for hydrogen production from biomass.

remaining 20% of the carbon is part of the flue gas from the process.

Sorption-Enhanced Reforming, to Adjust the H_2:CO Ratio Direct Inside the Gasifier

The composition of the syngas from gasification processes or syngas production technologies is generally not suited for direct use in the downstream synthesis system. In several cases, H_2/CO syngas ratio adjustment techniques such as downstream shift reactors, membrane separators or pressure swing adsorption are employed to meet the ratio requirements. This adds costs and complexity to the overall process.

In the sorption-enhanced reforming process, hydrocarbon reforming, water-gas shift, and CO_2 separation reactions occur simultaneously in a single reaction step over a reforming catalyst mixed with a CO_2 sorbent. Depending on reaction conditions, the product gas may contain as much as 98% H_2 and only ppmv levels of CO and CO_2, thereby minimizing the final H_2 purification process or even eliminating it for some applications.[45]

There is a number of CO_2 sorbents such as Ca-based oxides, HTC (hydrotreating catalysts, which are double-layered hydroxides) and mixed oxides of Li and Na. Limestone and dolomite are the precursors for Ca-based oxides. These have the advantage of good availability and are not expensive. They offer high CO_2 capacity and react rapidly over a wide range of temperatures and pressures. Their disadvantage is associated with the high temperature required for sorbent regeneration. High temperature promotes sorbent sintering that leads to rapid deactivation. Another disadvantage is the high attrition of the adsorbens, so much R&D was done to improve the mechanical properties.

The concept of sorption-enhanced hydrogen production in the presence of Ca-based sorbent is not new. The first description of this process was published already in 1867 by DuMotay and Marechal who used limestone to enhance the gasification of carbon with steam.[46] Later, Curran et al.[47] developed the 'CO_2 acceptor process', utilizing the carbonation reaction for the gasification of lignite coal to produce a H_2 rich gas.

The carbon capture reaction for the specific case of CaO sorbent is

$$CO_2(g) + CaO(S) \rightarrow CaCO_3(S)$$

HTCs are members of the family of double-layered hydroxides that, when doped with K_2CO_3, can serve as high-temperature CO_2 sorbents. They react rapidly but have much lower CO_2 capacity than Ca-based sorbents and are also considerably more expensive. But their advantage is regeneration at lower temperature so that sorbent durability can be improved. The HTC-CO_2 complex may be regenerated by pressure swing, temperature swing, or a combination of the two.

Recently, mixed metal oxide sorbents of Li and Na such as Li_2ZrO_3, Li_4SiO_4, and Na_2ZrO_3 have become subject of great interest. They have a higher CO_2 capacity than HTC and lower temperature is needed for regeneration as by the Ca-based sorbents.

In situ CO_2 capture involves suitable reaction conditions under which the whole process can be carried out. This requirement limits the available techniques for CO_2 capture, e.g., membrane and cryogenic separation processes require high pressure or low temperature for efficient CO_2 capture,[48] adsorption of CO_2 with carbon based adsorbents is limited to low temperatures (150–250°C)[49] and hydrotalcite compounds have a significantly diminished capacity for CO_2 capture beyond 300°C.[50]

Although many metal oxides react with CO_2 to form carbonates, not all metal oxides are suitable for CO_2 capture under conditions for biomass gasification.[51]

CaO is capable of scavenging CO_2 to very low concentrations at moderate temperatures (450–750°C) and atmospheric pressure. Furthermore, it is low cost and abundant because it can be derived from naturally occurring precursors including limestone, dolomite, and calcium hydroxide. The enhanced output from biomass gasification when coupled with CO_2 capture using CaO is demonstrated based on thermodynamic equilibrium modeling.[52]

The increase of H_2 concentrations corresponding to the capture of CO_2 by CaO is shown in Figure 7.9. As can be seen operating window for temperature is 460–600°C. In this temperature range the highest H_2 yield was achieved. With rising temperature H_2 content is decreasing because of deactivation of CaO sorbent. When the sorbent reaches its ultimate conversion, it can be regenerated to produce CO_2 and CaO by heating to 700–950°C, depending on the CO_2 partial pressure. Although the regeneration process requires the input of energy, the heat generated by the exothermic absorption reaction partially offsets this demand.

The application of CO_2 capture to enhance H_2 production from biomass coupled with the production of a highly concentrated stream of CO_2 represents a suitable and renewable pathway for hydrogen production. Furthermore, if the process is combined with CO_2 storage a negative balance of CO_2 in the atmosphere can be achieved.

Demonstration of this process was already done at the biomass CHP Güssing,[53] but the demonstration plant in Germany was not realized for different reasons.

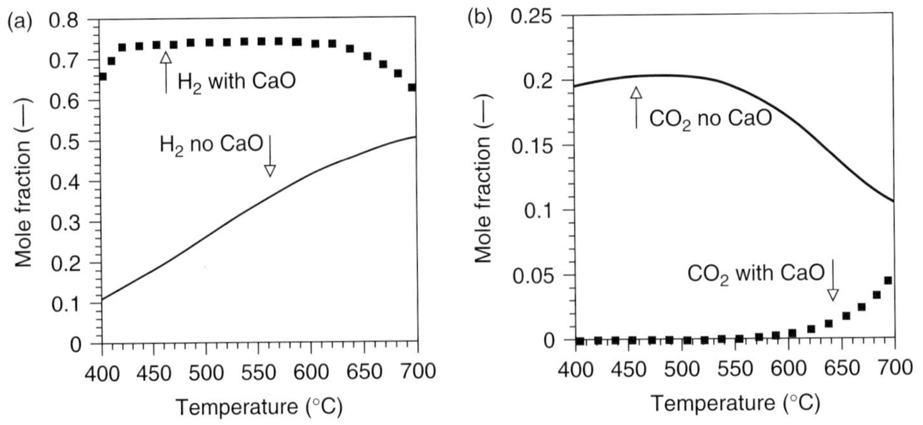

Figure 7.9 Predicted H_2 and CO_2 molar concentrations in sorption-enhanced reforming.

Microchannel Reactor Technology (Oxford Catalysts, Velocys)

Oxford Catalysts Group PLC[54] is the leading technology innovator for clean synthetic fuel production. Their two operating subsidiaries, Oxford Catalysts Ltd and Velocys, Inc. design, develop and commercialize technology for the production of synthetic oil from waste gas (which would otherwise be flared or reinjected), stranded gas, renewable sources such as waste biomass, as well as coal via FT synthesis.

Microchannel technology is suited for carrying out highly exothermic catalytic reactions, such as FT synthesis or highly endothermic, such as methane reforming.

Microchannel reactors are compact reactors that have channels with diameters in the millimeter range (Figure 7.10). These small channels dissipate heat more quickly than conventional reactors with larger channel diameters in the 2.5–10 cm range, so more active catalysts can be used. Mass and heat transfer limitations reduce the efficiency of the large conventional reactors used for FT, steam reforming reactions and hydroprocessing. The use of microchannel processing makes it possible to greatly intensify chemical reactions enabling them to occur at rates 10 to 1000 times faster than in conventional systems.

These reactors exhibit conversion efficiencies in the range of 70% per pass, and are designed for economic production in a small scale. A single microchannel reactor block produces some 30 bbls of synthetic fuel per day. In contrast, conventional FT plants are designed to work at minimum capacities of 5000 bbls/day, and function well and economically at capacities of 30,000 bbls/day or higher. They typically exhibit conversion efficiencies in the range of 50% or less per pass.

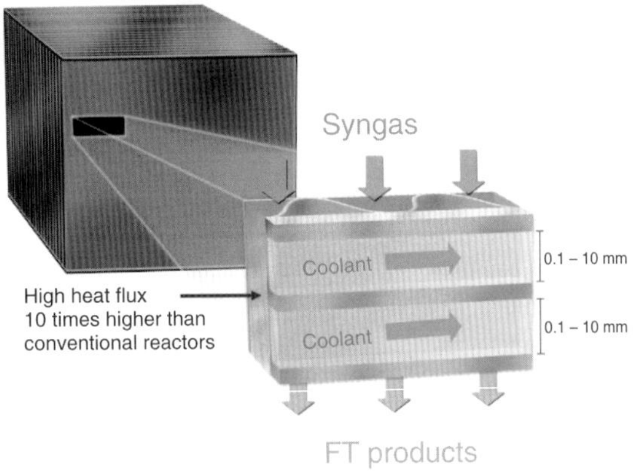

Figure 7.10 Velocys microchannel reactor. (Reprinted with permission from Ref 54. Copyright 2012)

Microchannels are claimed to have a smaller footprint, can be scaled up or down more flexibly as they can be 'numbered up', and have efficiencies superior to those achievable using conventional process technology. Microchannel reactors therefore have the potential to unlock the distributed production of fuels and other materials on a small, decentralized scale.

Velocys did already a demonstration program over 4000 h with a 0.3 bbls/day FT microchannel reactor, using the synthesis gas from the biomass CHP Güssing.[55]

Decentral Small Scale Methanation for Production of BioSNG (Distributed SNG)

A conversion of woody biomass into SNG is a promising alternative to other applications of synthesis gas. The main advantage of the production of SNG against other technologies like the conversion of biomass into power or liquid fuels is high energy efficiency in particular in case that waste heat of the process is used for combined heat and power production—that are 'polygeneration' applications. Furthermore, the necessary technologies for the upgrading of the methanized syngas require the same technologies, which are used for the feeding of biogas into the pipelines.

Steam gasification is advantageous for this process because of the high content of methane in the product gas, which gives a higher overall efficiency for the process. Agnion uses in the distributed SNG concept no cold gas cleaning, because of the heat loss. As the heat pipe reformer is operated at 5 bars, the condensation of the steam in the product gas can be done at a temperature level, where additional district heat can be provided. In this way the overall efficiency is quite high, as also the condensation heat of the steam can be used. So the chemical efficiency from biomass to SNG is up to 70% and the overall efficiency above 90%.[56]

A fixed bed methanation process, suitable for decentralized SNG production was studied by Kienberger et al. In this process,[57] catalytic methanation and catalytic tar-reforming was combined within one reactor. The influence of the gasifier operation parameters (temperature, residence time, input steam content) on syngas quality and composition was discussed. Actually the gasification technology of heat pipe reformer is demonstrated only for CHP in Germany.

Large-Scale FT Synthesis (Choren)

On January 31,1990, UET Umwelt- und Energietechnik Freiberg GmbH (later the Engineering division of CHOREN Industries GmbH) was formed. The portfolio included process technology for steam fluidized bed drying, analysis services for optimizing power plant equipment and many other energy processes.[58]

The construction and commissioning of a 1 MWth Carbo-V® pilot plant in 1998 marked an important milestone. The plant's objective was to demonstrate the effectiveness of Carbo-V® technology for the production of fuel gas and synthesis gas from carbonaceous feedstocks. The process was then thoroughly tested from 1998 to 2004. The Carbo-V® pilot plant was successfully operated with a wide variety of feedstocks (such as untreated wood, various types of waste timber, 'dry stabilate' material from waste processing, meat and bone meal, hard coal, and lignite). In 2001, a 150-kW gas engine successfully ran under load on Carbo-V® gas for 600 h, thus verifying the potential of the Carbo-V® process for electrical power production from biomass.

R&D activities for production of renewable transport fuels began in 2000. A FT test rig for fuel production was built. In September 2001, a joint research project with DaimlerChrysler was started with funding from ZIP, an innovation program from the German Federal Ministry for the Environment. Subsequently, Volkswagen joined the research activities being pursued by CHOREN and DaimlerChrysler in September 2002. One objective was the physical use of the raw gas from Carbo-V® gasification to produce renewable synthetic fuels. A synthesis pilot plant was thus built, initially for methanol synthesis and later for FT synthesis, as an extension of the already existing Carbo-V® pilot plant. The Carbo-V® pilot plant with FT synthesis was later named the a-Plant. By the end of 2004, the plant had been operating successfully for more than 22,500 h.

In November 2002, the foundation stone was laid for the construction of the first industrial-scale Carbo-V prototype plant (*f*-Plant) with a capacity of 45 MWth at the Freiberg site.

It was planned to produce 18,000 m^3 of FT fuel per year. Investment costs were more than 100 M.[59]

Choren Industries filed for insolvency in July 2011. Choren's biomass gasification technology was sold to Linde Engineering Dresden, who will further develop the Choren Carbo-V® technology used to produce syngas.[60] The EE-Plant in Freiberg is still not in operation and here also no investor was found till now, who takes over this demonstration plant for FT synthesis. Recently, it was announced by Linde that they had licensed the project to the Finish Forest Biofuels project, one of the NER 300 contractors.

Large Scale of Alcohol Synthesis (Range Fuels)

Range Fuels tried to develop a technology that converts woody biomass into ethanol over the thermochemical pathway. The technology was based on staged pressurized gasification of biomass, syngas purification, and its catalytic conversion. The aim was to produce up to 100 million gallons of biofuel per year as well as green power.[61]

The Soperton site, where the technology should be installed is located in a region rich in biomass resources in the United States. Test plots of high-yield energy crops on the plant site demonstrated the potential this feedstock has to supplement the woody biomass already available in the region.

Woody biomass was delivered by trucks to a wood yard, where the biomass was processed and stored prior to conversion. The biomass passed through a pressurized gasifier, where it was converted to a gas and residual char and minerals. The product gas from gasification passed through

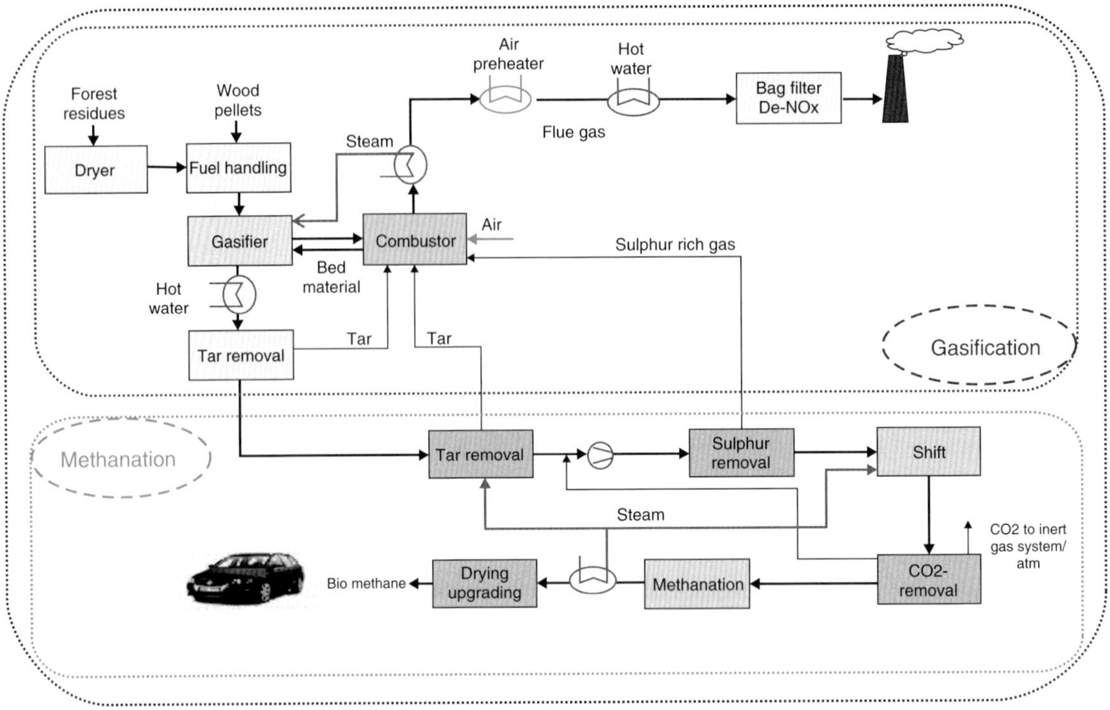

Figure 7.11 Flow chart of GoBioGas project

a reforming section where it was converted to syngas. Finally, the syngas was cleaned and compressed before injection into a catalytic converter that performed the final transformation into ethanol.

Unfortunately, Range Fuel was officially closed down in late 2011 with a foreclosure sale of its plant held on January 3, 2012. In 2012, LanzaTech NZ Ltd took over the facilities and will use it probably with their enzymatic conversion of syngas to ethanol.[62]

Large-Scale BioSNG Production (GoBiGas)

Göteborg Energi, the energy company in the western part of Sweden, has the goal to convert to renewable energy long term. One step is to replace fossil natural gas by renewable gas. Here much R&D on demonstration and also commercial plants on upgrading biogas to natural gas quality was done in the past. Also, to convert wood into methane the GoBiGas project was initiated.[63]

The goal of the GoBiGas project is to produce biomethane (BioSNG) by thermal gasification of forest residues such as branches, roots, and tops. The gasification plant is scheduled to be built in two stages, the first stage (about 20 MWgas) is planned to be operational in 2013. The second stage (about 80–100 MWgas) is scheduled to be in service in 2016. How and when to implement stage 2 will be decided after an evaluation of the stage 1 plant.

The biomass is converted to a synthesis gas in the gasification plant. This so-called synthesis gas is purified and then upgraded in a methanation plant to BioSNG with a quality comparable with natural gas to enable the two types of gases to be mixed in the gas network.

As gasifier, a dual fluidized bed system, based on the principle of the Güssing biomass CHP was selected. The methanation is supplied by Haldor Topsoe and here fixed bed methanation is used. Details about the flow chart can be seen in Figure 7.11.

The 20 MW BioSNG plant is actually under construction and start-up is planned for 2013. For the second stage grants from the NER300 program of the EC will be received with high probability. There appear to be good chances that also in near future also the second stage of 80–100 MW BioSNG production will be realized.[64]

CONCLUSION

The product gas from thermal biomass gasification processes is an outstanding energy carrier. It can be used as a stand-alone fuel (heat and power applications) or it can be further treated and transformed into another energy carrier by chemical upgrading and synthesis. The energy density of synthesis gas is only about 30% that of natural gas and is therefore not suitable for distribution as-is, unless converted to transportation fuels and other chemical products such as synthetic natural gas, methanol, and synthetic petroleum fuel. In a purified state, the hydrogen component of synthesis gas can also be used directly for industrial applications (e.g., oil refineries) or to power hydrogen fuel cells for electricity generation.

In this review, different biomass-based synthesis gas applications are described. The processes for alternative fuels production from synthesis gas as well as other applications are presented here. The heat and/power applications were developed in the last decades, but more recently the interests for transportation fuels production (trucks, jet) have come more into the focus, explaining the intensive R&D work performed and planned in this area. The reasons are both the climbing prices of fossil fuels and also the climate change issue in combination with an increasing energy demand in transportation sector. These factors are reflected in policy measures to increase the use of advanced biofuels such as from the second and third generation.

Recently, there are several projects going on to produce advanced biofuels for transportation. An overview about ongoing projects can be found in several databases like:

- IEA Bioenergy Task 33 Thermal Gasification of Biomass (http://www.ieatask33.org)
- IEA Bioenergy Task 39 Liquid Biofuels (http://demoplants.bioenergy2020.eu/projects/mapindex)
- US DOE Bioenergy Technologies Office (http://www1.eere.energy.gov/biomass/integrated_biorefineries.html)

Several different gasification technologies are used for production of synthesis gas from biomass, but there is only the DFB steam gasification technology in commercial operation (for CHP, not for synthesis gas). All others have done extensive R&D and also demonstration, but no commercial plant is at the moment in operation.

There are several demonstration plants in the planning or construction phase (GoBiGas, BioTfueL, Stracel BTL or Ajos BTL, Woodspirit, Enerkem demo on Ethanol, or TIGAS), where different gasification technologies are used and also different products are produced. It looks like, that the United States focuses more on ethanol (as blending component for gasoline) and Europe more on diesel substitutes. This is logical, looking at the market shares of the gasoline and diesel in these countries. Europe has a high share of diesel and the United States a high share of gasoline in the transportation sector, so their focus of R&D is according to the fuels used in the countries. In general, the impression is that so-called drop-in fuels are the focus of demonstration in the near future.

Some projects focus also on gaseous fuels like BioSNG or hydrogen. Here the main advantage is the high conversion efficiency of up to 70%. The main disadvantage in case of hydrogen is that the infrastructure still has to be developed.

On the long term there is no alternative to advanced biofuels. Some sectors like heavy transport or aviation will depend also in future on hydrocarbons, as they need fuels with high energy density. In the private transport sector there is a trend to electric cars and hydrogen, but a change in this sector needs decades, so also here hydrocarbons will be the main fuel for the next decades. So it is necessary to develop this technology further, although there were some drawbacks and failures of companies in the past.

REFERENCES

1. Hiller H, Reimert R, Stönner H M. Gas production. In: Ullmann's Encyclopedia of Industrial Chemistry; 2011. 9783527306732
2. Boerrigter H, Rauch R. Syngas production and utilisation. In: *Handbook Biomass Gasification. BTG* biomass technology group, Enschede, The Netherlands; 2005.
3. http://www.ieatask33.org (Accessed December 2012).
4. Pfeifer C, Puchner B, Hofbauer H. Comparison of dual fluidized bed steam gasification of biomass with and without selective transport of CO_2. *Chem Eng Sci* 2009, 64:5073–5083.
5. Tijmensen MJA, Faaij APC, Hamelinck CN, van Hardeveld MRM. Exploration of the possibilities for production of Fischer Tropsch liquids and power via biomass gasification. *Biomass Bioenergy* 2002, 23:129–152.
6. Göransson K, Söderlind U, He J, Zhang W. Review of syngas production via biomass DFBGs. *Ren and Sust En Rev* 2011, 15:482–192.
7. Fischer F, Tropsch H. Über die Herstellung syn-thetischer Ölgemische (syn-thol) durch Aufbau aus Kohlenoxyd und Wasserstoff. *Brennst. Chem* 1923, 4:276–285.
8. Klerk A, Furimsky E. *Catalysis in the Refining of Fischer-Tropsch Syncrude*. RSC Catalysis Series, No. 4. Cambridge: Royal Society of Chemistry; 2010, 1–294. ISBN: 9781849730808.
9. Dry ME. Special catalytic systems. In: Ertl G, Knözinger H, Schuth F, Weitkamp J, eds. *Handbook of Heterogeneous Catalysis*. 2nd ed. Weinheim: Wiley-VCH; 2008.
10. Klerk A. Can Fischer-Tropsch syncrude be refined to on-specification diesel fuel? *Energy Fuel* 2009, 23:4593–4604.
11. Soukup G, Pfeifer C, Kreuzeder A, Hofbauer H. In situ CO_2 capture in a dual fluidized bed biomass steam gasifier—bed material and fuel variation. *Chem Eng Technol* 2009, 32:348–354.

12. Bowen DA, Lau F, Zabransky R, Remick R, Slimane R, Doong S. Techno-economic analysis of hydrogen production by gasification of biomass. In: NREL FY 2003 Progress Report, National Renewable Energy Laboratory, 2003.
13. OECD/IEA. *World Energy Outlook 2010*. Paris: IEA; 2010, 435–437.
14. Miltner A, Wukovits W, Pröll T, Friedl A. Renewable hydrogen production—a technical evaluation based on process simulation. *J Clean Prod* 2010, 18:1–12.
15. Fiedler E, Grossmann G, Kersebohm DB, Weiss G, Witte C. Methanol. In: Ullmann's Encyclopedia of Industrial Chemistry; 2012. doi: 10.1002/14356007.a16_465.pub3.
16. Katofsky RE. *The Production of Fluid Fuel from Biomass*. Princeton, NJ: Center for Energy and Environmental Studies, Princeton University; 1993.
17. van Dijk KM, van Dijk R, van Eekhout VJL, van Hulst H, Schipper W, Stam JH. *Methanol from Natural Gas—Conceptual Design and Comparison of Processes*. Delft: Delft University of Technology; 1995.
18. Peckham J. Lurgi: cheaper gtl diesel route through methanol. *Gas-to-Liquids News*, Vol. 6; 2003.
19. Jackson PJ, White N. Technologies for the conversion of natural gas. In: Australian Institute of Energy Conference, Held at the Architecture School, University of Melbourne, 1985.
20. van Walwijk M, Bückmann M, Troelstra WP, Achten PAJ. *Automotive Fuels Survey—Part 2 Distribution and Use*. Breda: IEA/AFIS; 1996, 319 pp + annexes.
21. Ekbom T, Lindblom M, Berglin N, Ahlvik P. *Technical and Commercial Feasibility Study of Black Liquor Gasification with Methanol/DME Production as Motor Fuels for Automotive Uses—BLGMF*. EU Altener Contract 4.1030/Z/01–087/2001. Stockholm: Nykomb Synergetics AB; 2003.
22. Quarderer GJ. Mixed alcohols from synthesis gas. In: 78th National AICHE Meeting, New Orleans, LA, USA, 1986.
23. Spath PL, Dayton DC. *Preliminary Screening—Technical and Economic Assessment of Synthesis Gas to Fuels and Chemicals with Emphasis on the Potential for Biomass-Derived Syngas*. NREL/TP-51034929. Golden, CO: National Renewable Energy Laboratory; 2003.
24. van der Drift B, Biollaz S, Waldheim L, Rauch R. The status and future of bioSNG. In: IEA Bioenergy Conference 2012, Wien, Germany, November 13–15, 2012, S.13-S.20. 978–3–9502754–4–5.
25. Kopyscinski J, Schildhauer TJ, Biollaz SMA. Production of synthetic natural gas (SNG) from coal and dry biomass—a technology review from 1950 to 2009. *Fuel* 2010, 89:1763–1783.
26. http://www.nesteoil.com/ (Accessed December 2012)
27. T. Räsänen. NSE biofuels project activities. Presentation at IEA Bioenergy Task 33 Thermal Gasification of Biomass First Semi-Annual Task Meeting, Helsinki, Finland, June 1–3, 2010.
28. http://www.nesteoil.com/default.asp?path=1;41;540;1259;1260;18523;19844 (Accessed December 2012)
29. www.bioliq.de (Accessed July 2012)
30. Otto TN, Zimmermann M, Henrich T, Abeln J, Dinjus E. Synthesis of gasoline from biomass derived dimethyl ether. In: 20th European Biomass Conference and Exhibition, Milan, Italy, 2012.
31. Leibold H. Dry HTHP syngas cleaning—status and results. Presentation at IEA Bioenergy Task 33 Thermal Gasification of Biomass Task Meeting, Karlsruhe, Germany, May 13–14, 2009.
32. Radtke K. ThyssenKrupp Uhde's fluidised bed and entrained flow gasification technologies for biomass and coal. Presentation at SGC International Seminar on Gasification 2012, Stockholm, Sweden, October 18–19, 2012.
33. www.ifpenergiesnouvelles.com (accessed October 2012)
34. Rauch R. CHP-Plant Güssing, Austria. In: Knoef H, ed. *Handbook Biomass Gasification Second Edition*. Enschede: BTG Biomass Technology Group BV; 2012, S.32-S.36. ISBN: 978–90–819385–0–1.
35. Rehling B, Hofbauer H, Rauch R, Aicherning C. BioSNG-process simulation and comparison with first results from a 1-MW demonstration plant. *Biomass ConverBiorefi* 2011,1:111–119. doi: 10.1007/S13399-011-0013-3.
36. Rehling B. Development of the 1 MW Bio-SNG plant, evaluation on technological and economical aspects and upscaling considerations. PhD thesis, Vienna University of Technology, 2012.
37. Ripfel-Nitsche K, Hofbauer, H. Project status and commercialisation prospects of the Güssing BTL Project. In: BtLtec, Vienna, Austria, 2007.
38. Sauciuc A, Abosteif Z, Weber G, Potetz A, Rauch R, Hofbauer H, Schaub G, Dumitrescu L. Influence of operating conditions on the performance of biomass-based Fischer-Tropsch synthesis. *Biomass Conver Bioref.* 2012, 2:253–263. doi: 10.1007/s13399-012-0060-4.
39. Schablitzky H, Lichtscheidl J, Rauch R, Hofbauer H. Investigations on hydrotreating of Fischer Tropsch-biowaxes for generation of bio-products from lignocellulosic biomass. *Modern Appl Sci* 2012, 6.
40. Falk G, Weber G, Rauch R, Hofbauer H, Weiß C. Mixed alcohols from biomass steam gasification. In: Proceedings of the International Conference on Polygeneration Strategies (ICPS11), Wien, Austria, August 30-September 1, 2011.
41. Dutta A, Talmadge M, Hensley J, Worley M, Dudgeon D, Barton D, Groendijk P, Ferrari D, Stears B, et al. Process design and economics for conversionof lignocellulosic biomass to ethanol. Technical report NREL/TP-5100–51400, 2011.
42. Muller S, Kotik J, Proll T, Rauch R, Hofbauer H. Hydrogen from biomass for industry—biomass gasification for integration in refineries. In: Proceedings of the International Conference on Polygeneration Strategies (ICPS11), Wien, Austria, 30 August-1 September, 2011.
43. Mozaffarian M, Zwart RWR, Boerrigter H, Deur-waarder EP, Kersten SRA. ''Green gas'' as SNG a renewable fuel with conventional quality. In: Proceedings of Science in Thermal and Chemical Biomass Conversion Conference, Victoria, BC, Canada, 30 August-2 September, 2004.
44. http://www.biosng.com/ (accessed in December 2012)

45. Harrison DP. Sorption-enhanced hydrogen production: are-view. *Ind. Eng. Chem. Res.* 2008, 47:6486–6501.
46. Squires AM. Cyclic use of calcined dolomite to desulfurize fuels undergoing gasification. *Adv Chem Ser* 1967, 69:205–229.
47. Curran GP, Fink CE, Gorin E. CO_2 acceptor gasification process. In: *Fuel Gasification*. Washington, D.C.: American Chemical Society; 1966, 141.
48. Rao AB, Rubin ES. A technical, economic and environmental assessment of amine-based CO_2 capture technology for power plant greenhouse gas control. *Environ Sci Technol* 2002, 36:4467–4475.
49. Yong Z, Mata V, Rodrigues AE. Adsorption of carbon dioxide at high temperature—a review. *Sepa Purifi Technol* 2002, 26:195–205.
50. Yong Z, Mata V, Rodrigues AE. Adsorption of carbon dioxide onto hydrotalcite-like compounds (HTLcs) at high temperatures. *Ind Eng Chem Res* 2001, 40:204–209.
51. Feng B, An H, Tan E. Screening of CO_2 adsorbing materials for zero emission power generation systems. *Energy and Fuels* 2007, 21:426–434.
52. Gupta H, Fan LS. Carbonation-calcination cycle using high reactivity calcium oxide for carbon dioxide separation from flue gas. *Ind and Eng Chem Res* 2002, 41:4035–4042.
53. Zuberbuehler, U., Specht, M., Marquard-Moellenstedt, T., Sturmer, B., Sichler, P., Diurrbeck, M., Rauch, R., Pfeifer, C., Hofbauer, H., Koch, M. AER-Gas II: Poly-Generation from Solid biomass—test of the AER-Process in the 8 MW Gasification Plant in Guessing. In: 15th European Biomass Conference & Exhibition, Berlin, Germany, May 7–11, 2007, 1017–1020.
54. www.oxfordcatalyst.com (Accessed September 2012)
55. http://www.oxfordcatalysts.com/press/pr/ocgtpr20100816.php (Accessed December 2012)
56. Karl J. Distributed generation of substitute natural gas from biomass. In: Proceedings of the Biomass Conference and Exhibition, Valencia, Spain, June 2–6, 2008.
57. Kienberger T, Mair S, Karl J. Fixed bed methanation of biogeneous syngas—ideas for the decentralized approach. In: Proceedings of European Biomass Conference and Exhibition, Lyon, France, May 3–7, 2010.
58. Schulze O. Advanced gas cleaning for biomass gasification. Presentation at IEA Bioenergy Task 33 Thermal Gasification of Biomass Task Meeting 2009, Karlsruhe, Germany, May 13–14, 2009.
59. Heinl E. Optimized fuels for a sustainable transport. In: 18th European Biomass Conference, Lyon, France, May 3–7, 2010.
60. http://www.the-linde-group.com/en/news_and_media/press_releases/news_120209.html (Accessed December 2012)
61. Bacovsky D. Status of 2nd Generation Biofuels Demonstration Facilities in June 2010. Report by IEA Bioenergy Task39 Liquid Biofuels, 2010.
62. http://www.lanzatech.com/content/freedom-pines-bio-refinery (accessed in December 2012)
63. Gunnarsson I. The GoBiGas Project—efficient transfer of biomass to bio-SNG of high quality. In: SGC International Seminar on Gasification, Malmo, Sweden, October 7, 2011.
64. European Commision. NER300—Moving towards a low carbon economy and boosting innovation, growth and employment across the EU, Brussels, July 12, 2012, SWD (2012) 224 final. Downloaded from http://ec.europa.eu/clima/funding/ner300/docs/2012071201_swd_ner300.pdf (Accessed December 2012)

8

Hydrogen Generation from Biomass Materials: Challenges and Opportunities

Pravakar Mohanty,[1,2] Kamal K. Pant[1]* and Ritesh Mittal[1]

[1]Department of Chemical Engineering, Indian Institute of Technology Delhi, New Delhi, India
[2]Sardar Patel Renewable Energy Research Institute (SPRERI), V V Nagar, Gujarat, India

INTRODUCTION

There is a wide scale of increase in the demand of energy and its resources, owing to rapid increase in industrialization, urbanization, and population outgrowth. The discovery of crude oil in the 19th century has supplied an inexpensive fossil fuel source that has helped to industrialize the world and to improve the standards of living. Since then, various fossil fuels and their derivatives have remained the most favourite and widely exploited resource of energy for the world. The global use of petroleum and other liquid fossil fuels was 85.7 million barrels/day in 2008 with a projection for an increase to 97.6 million barrels/day in 2020 and 112.2 million barrels/day in 2035, respectively.[1] With these concerns for amplifying the demand of petroleum resources by emerging economies, declining reserves at rapid stage, and rising fuel prices combined with environmental apprehensions of global warming, it has become imperative to develop eco-friendly and energy-efficient processes for the sustainable production of fuels and chemicals. The total world energy consumption in 2008 was 535 quadrillion kJ; however, the consumption is expected to rise to 655 quadrillion kJ in 2020 and to 812 quadrillion kJ in 2035.[1,2]

In this scenario, second to fourth generation biomasses seem to be the most potential sources among the emerging renewable sources of organic carbon, and the fuels derived from their processing would be the only alternative liquid fuels to replace the conventional fossil fuels. Moreover, biofuels are considered to be carbon neutral as the net amount of CO_2 released from their combustion is consumed by the plants during photosynthesis to produce new biomass. These waste biomasses are lignocellulosic in composition, i.e., they contain two major structural carbohydrates, cellulose and hemicellulose, bound together by the polymeric lignin. These three biopolymers are highly energy-rich components as they can easily extract the energy from the sun.[2,3] Lignocellulosic biomass (second generation biomass) is potential feedstock for the production of biofuels and bio-H_2 as it is inexpensive, available in abundance, supplied worldwide and does not retain any competency towards food supply. For instance, the annual lignocellulosic biomass availability in the United States and Canada ranges from 6 to 256 and 425 to 1000 million dry tons, respectively.[2,3] The worldwide biomass energy potential in 2050 has been estimated to be in the range of 150–450 exajoule/year with an energy equivalence of $\sim 76\times 10^9$ barrels of oil energy equivalent.[2,4] Lignocellulosic biomasses are obtained as nonedible residues from agriculture, forestry, municipal, industrial, and urban refuse. These organic residues are suitable raw material for biofuel production through efficient conversion technologies employing physicochemical, thermochemical, or biochemical routes, which have huge potential for hydrogen production and carbon neutralization (Figure 8.1).

*Correspondence to: kkpant@chemical.iitd.ac.in

Advances in Bioenergy: The Sustainability Challenge, First Edition. Edited by Peter D. Lund, John Byrne, Göran Berndes and Iacovos A. Vasalos.

Originally published in: *WIREs Energy Environ* 2014, 4: 93–108 doi: 10.1002/wene.111

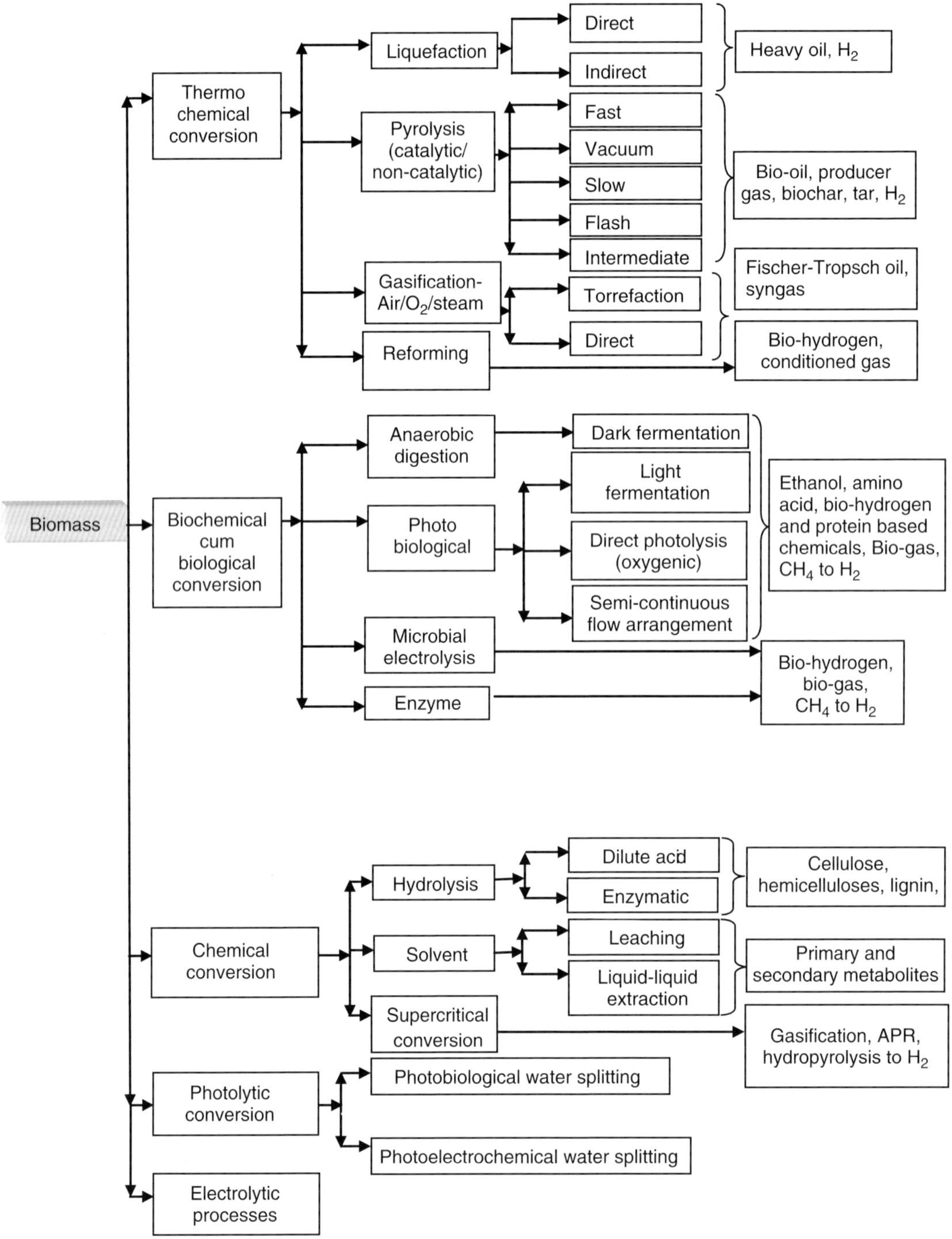

Figure 8.1 Different conversion routes of biomass to biofuels.

Inevitably hydrogen is available in each organic material. It can be produced from different sources, e.g., coal, natural gas, liquefied petroleum gas (LPG), propane, shale gases/methane (CH_4), paraffin, gasoline, light diesel, dry biomass, biomass-derived liquid fuels, as well as from water. Hence, this review has tried to focus on getting an insight about the current status of H_2 production, its usage in industry, including the chemical processes involved, suitable catalysts, supporting materials, operating conditions for maximum yield of H_2 from biomass or its derived solvents/oxygenates, with higher conversion and selectivity. The objective of this paper is to review the literature from all possible areas through graphical representation (Figure 8.1) and to make a comparative analysis.

The worldwide production of biomass from terrestrial plants is 170–200 billion tons, with an estimated 70% made of plant cell walls.[2–5] From an average estimate of 146 billion metric tons of world biomass production per annum, about 20 metric tons/acre are produced from farm crops and trees.[5–7] Lignocellulosic materials are usually categorized into agricultural and forage residues (sugarcane bagasse, corn stover, wheat straw, rice straw, rice husk etc.), dedicated energy crops (e.g., switch grass, timothy grass, hybrid poplar), forest wood residues (e.g., saw mill residues, paper mill discards), and municipal paper waste. Typically, lignocellulosic biomass contains 35–55% cellulose, 20–40% hemicellulose, and 10–25% lignin,[7,8] while the rest consists of extractives and ash.[8] With ash representing the mineral components in biomass, extractives include structural biopolymers such as terpenoids, steroids, resin acids, fats, lipids, waxes, and phenolic components.

Thermochemical conversion is the most widely known biogenic waste conversion pathway that involves pyrolysis, hydropyrolysis, reforming, gasification, and liquefaction.[8–10] Through gasification, the biomass is converted to hydrocarbons and synthesis gas (or syngas), while pyrolysis and liquefaction directly convert biomass at high temperatures to bio-oils, gases, and char. Biomass has tremendous potential to accelerate the realization and utilization of H_2 as a major fuel of the future, as biomass is the fourth largest source of energy in the global context. It accounts for 15% of world's primary energy consumption and about 38% of the primary energy consumption in developing countries such as India, China, and Brazil.

LIGNO(HEMI)CELLULOSIC BIOMASS CHEMISTRY

Cellulose is a glucose polymer consisting of β (1, 4) linked D-glucose subunits, whereas hemicellulose is a mixture of polysaccharides composed of C_5 (e.g., xylose and arabinose) and C_6 (e.g., galactose, glucose, and mannose) sugars as well as sugar acids such as methylglucuronic and galacturonic acids. In the polymeric form, cellulose consists of linear chains of 1, 4-D-glucopyranose units with an average molecular weight of around 100,000 Da.[10–12] In the monomeric form, hemicellulose consists of D-glucose, D-mannose, D-galactose, D-xylose, L-arabinose, L-rhamnose, D-glucuronic acid and D-galacturonic acid, with an average molecular weight less than 30,000 Da. On the other hand, lignin is a phenyl propane polymer linked with ester bonds that hold cellulose and hemicellulose together as a matrix. The lignin in biomass is highly branched, substituted mononuclear aromatic polymer, and consists of macromolecules containing phenolic groups. The average molecular weight of lignin is around 20,000 Da.

During the biomass pretreatment, lignin can form furan compounds such as furfural and hydroxymethyl furfural that could inhibit fermentation.[12–14] In contrast, thermochemical conversion of biomass yields various products after degradation from lignin that have high industrial values in the production of aromatic compounds, flavoring agents, food preserving agents, and other chemicals.[13] The major structural components of lignin are the monolignols (hydroxyl cinnamyl alcohols), such as coniferyl alcohols (guaiacyl propanol), coumaryl alcohols (*p*-hydroxyphenyl propanol), and sinapyl alcohols (syringyl propanol).

During thermochemical conversions, the polymeric components of biomass such as cellulose, hemicellulose, and lignin undergo a series of complex thermal degradation reactions. In most cases, cellulose and hemicellulose undergo cycloreversion and dehydration followed by transglycosylation reactions. This series of reactions results in the yield of both low- and high-molecular weight compounds in the bio-oil fractions. The typical degradation products of cellulose and hemicellulose in bio-oils include acids, esters, alcohols, ketones, aldehydes, sugars, furans, and oxygenates, while lignin derivatives include phenols, guaiacols, and syringols.[12–17] The primary products of cellulose and hemicellulose decomposition are condensable vapors (yields to liquid products) and gases. Extractives contribute to liquid and gas products either through simple volatilization or decomposition. This distribution of components into products is shown schematically in Figures 8.1 and 8.2. Vapors formed by primary decomposition of biomass components can be involved in secondary reactions in the gas phase, forming soot and/or hot char surfaces, where secondary char is formed.[5,10]

LIGNO(HEMI)CELLULOSIC BIOMASS CONVERSION

Current status defines that 90% of the H_2 is produced through the reforming of CH_4 or the light oil fraction with steam at high temperatures thereby producing CO_2, which

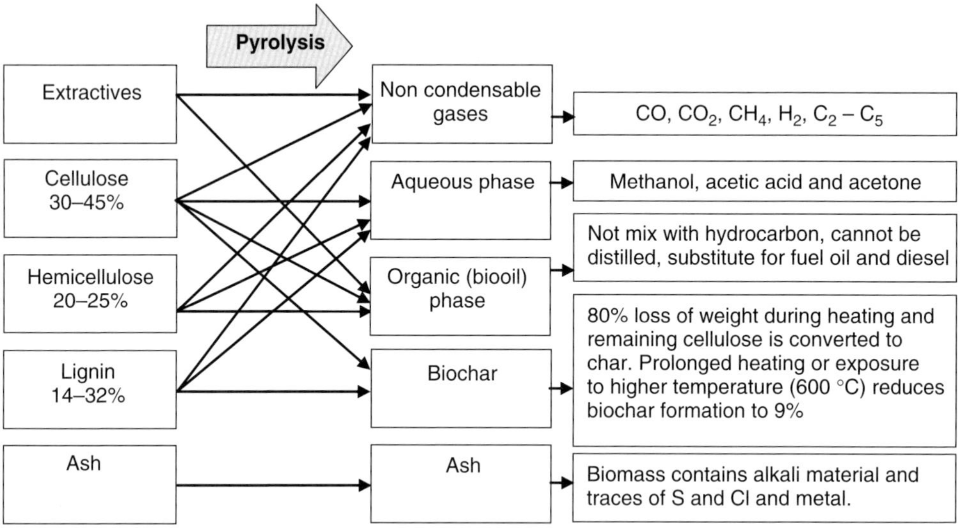

Figure 8.2 Biomass components and their pyrolytic conversions.

has forced the research community to explore new horizons through renewable sources such as biomass or its derived solvents to produce hydrogen. Hydrogen as a fuel possesses the highest energy content per unit of mass (i.e., 120.7 kJ/g), compared with any of the known fuels. The advantage involved with hydrogen production from biomass is that the CO_2 emissions reduced manifold when using biomass as the starting material. Steam reforming of methane produces approximately 13.6 kg of CO_2 per kilogram of hydrogen production. It can easily replace older, unclean fuels, and reduce greenhouse gas emissions by improving ambience air quality. H_2 is an ideal fuel source for fuel cells and many applications, as its by-products are H_2O, heat, electricity, and it contributes to country's economic growth through job development, investment opportunities, and the creation of a sustainable, secure energy supply.[14–19]

Many renewable sources like biomass and CO_2-enriched algae-derived hydrogen, most essential for fuel cell and other industrial applications, are a major influential part of a 'hydrogen society'. Varieties of aqueous-phase carbohydrates such as sugars (e.g., glucose) and polyols (e.g., methanol, ethanol, ethylene glycol, glycerol, acetic acid, phenol and sorbitol) generated from biomass can be efficiently converted in the aqueous phase over appropriate heterogeneous catalysts, at temperatures ranging from 275 to 800°C to produce H_2 and CO_2. Further, this CO_2 can be recycled back in a closed loop for algae growth and to the environment to consume and to grow more biomass. Direct storage and use of H_2 on PEM fuel cell vehicles have certain limitations; therefore, on-board hydrogen production from biomass-derived oxygenates such as methanol, ethanol, dimethyl ether, is being considered as a potential option. All such alcohol and ether groups are considered as the best 'storable' form of the hydrogen because of (1) low temperature profile that is required for maximum conversion, (2) high hydrogen-to-carbon ratio (4:1), (3) no C–C bond, thus minimizing coke formation, (4) high hydrogen content in the product stream (up to 75%), (5) no sulfur present in the fuel, and (6) current infrastructure of fossil fuels such as gasoline and biomethanol can be used for storage and distribution in decentralized manner.

To reduce the dependence on conventional oil, bio-oil is focused on as the gasoline substitute. Crude glycerol comprises a mixture of several foreign constituents, such as methanol, water, inorganic salts, free fatty acids, unreacted mono-, di-, and triglycerides, and methyl esters. It is no longer blended with fuel oil and burnt as fuel; as a result it bears low commercial value whereas further refining is expensive.[16–20] Bioethanol, glycerol, and bio-oil derived from biomass sources are increasingly available because: (1) they are easy to transport, biodegradable, and low in toxicity; (2) they can be easily decomposed in the presence of water to generate a hydrogen-rich mixture; and (3) they are free from catalyst toxins such as sulfur. This review has emphasized on hydrogen generation via pyrolysis of biomass, gasification, gasification combined with pyrolysis, from supercritical water (fluid-gas) extraction. Steam reforming (SR), autothermal reforming (ATR), dry reforming (DR), liquid phase reforming (LPR), aqueous-phase reforming (APR) and partial oxidation (POX) of renewable

Table 8.1 Typical Operating Conditions and Product Yields from Different Pyrolysis Methods[6,7,21,22,29]

Pyrolysis Process		Slow	Intermediate	Fast	Flash
Temperature (°C)	Range	250–750	320–500	550–1050	1050–1300
	Typical	350–400	350–450	550–750	1050–1150
Time	Range	2–30 min	15 min	0.5–10 s	<1 s
	Typical	2–30 min	4 min	0.5–5 s	<1 s
Heating rate (°C/s)		1–50	10–100	10–200	>1000
Biomass particle size (mm)		5–50	5–50	<1.0	<0.2
Yields (wt.% on dry basis)					
Char	Range	2–60	19–73	28–35	20–35
	Typical	25–35	30–40	29–34	20–32
Liquid	Range	0–60	18–60	20–71	20–60
	Typical	20–50	35–45	46–53	46–51
Gas	Range	0–60	9–32	10–25	11–28
	Typical	20–50	22–32	11–15	15–22

carboxylates, such as bioethanol, glycerol, carbohydrates, and bio-oil are some usual processes to be explored for their widespread awareness. Among all these processes SR is endothermic, (T > 300°C); for LPR, high pressure and T < 300°C are required; for ATR, both partial oxidation and steam reforming in the temperature range of (550°C < T < 850°C) are necessary. However, supercritical reforming always occurs at high pressure and at T > 374°C, and to carry out photocatalytic conversion $T \approx$ 25°C is appropriate.[19–23]

PYROLYSIS OF BIOMASS TO PRODUCE BIO-OILS AND HYDROGEN

Thermal or thermocatalytic decomposition of ligno(hemi)cellulosic biomass in the absence of air or oxygen delivers bio-oils, biochar, and gases. Various pyrolysis processes categorized based on different vapor residence time, feed rate, temperature range, and heating rate are summarized in Table 8.1. The process results in energy-rich bio-oils, carbon-rich biochar, condensable organic liquids, noncondensable gases, acids (e.g., acetic acid, carboxylic acid etc.), solvents (e.g., acetone), and alcohols (e.g., phenol, methanol etc.).[14] Thermochemical conversions are preferred because they convert solid biomass into energy-densified liquid products with an ease of transport, storage, combustion, and marketing.[7,10] Bio-oil is a synthetic fuel obtained from the pyrolysis of biomass, which contains 70% of the energy of the biomass feed and is a complex mixture of alcohols, acids, aldehydes, esters, ketones, and many other aromatic compounds.

It is particularly important to understand the differences between slow, intermediate, fast, and flash pyrolysis and the factors affecting bio-oil, gas, and biochar yields. In most cases, the inorganic metals in biomass, particularly the alkali metals, can have a catalytic effect on pyrolysis reactions leading to increased biochar and producer gas yields. As reported by Bridgwater and Peacocke and Mohan et al. high heating rates and fine particle size of biomass are required for fast pyrolysis leading to rapid heat transfer from the heat source, whereas slow and intermediate pyrolysis produces (> 35%) higher yield of producer gas.[21–23] Meanwhile in case of intermediate pyrolysis utilizing pyroformer, the hot volatiles may come in contact with unpyrolised solids and condense, followed by secondary reactions to form tars. Depending upon the type of reactor and feed type, the oxygen content varies typically in the range of (30–40%) in bio-oil and it carries a lower heating value (16–19 MJ/kg) compared with conventional fuel oil (40 MJ/kg). The high oxygen content of bio-oil makes it incompatible with petroleum-derived oils. Bio-oils are less volatile, have cold flow problems, are dark brown in color, chemically unstable, produce many unwanted chemical reactions with increasing time and temperature, leading to an increase in viscosity and cloud point temperature. By increasing the severity of pyrolysis conditions, feed rate, and optimized size of feed with low moisture content it is flexible to produce more gases (CO, H_2, CO_2, and CH_4). The remaining bio-oil after condensation can then be treated further to produce additional H_2 (Table 8.2).[24–28] The major challenge in this process is minimization of CO and methane formation and its easy separation up to parts per billion (ppb) level to get purified hydrogen.

Further, bio-oils can be separated into water-soluble (aqueous phase) and insoluble components (heavy organic phase). The insoluble heavy organics can be treated to produce chemicals and the soluble aqueous organic phase can

Table 8.2 Reaction Pathways of Different Oxygenates Pyrolysis, Gasification, Steam Reforming with Enthalpies of Selected C_6 Compound Reactions[6,7,15–17,19–22,34]

Reaction	Equation	(ΔH) and Remarks
Pyrolysis	$C_xH_yO_z \rightarrow (1-x)CO + (y/2)H_2 + C$	180 (KJ/gmol)[1]
	$C_xH_yO_z \rightarrow (1-x)CO + ((y-4)/2)H_2 + CH_4$	300 (KJ/gmol)[1]
Partial oxidation	$C_xH_yO_z + (1/2)O_2 \rightarrow xCO + (y/2)H_2$	Exothermic reaction difficult to continue with lower H_2 yield
	$C_xH_yO_z + O_2 \rightarrow (1-x)CO + CO_2 + (y/2)H_2$	
	$C_xH_yO_z + 2O_2 \rightarrow (x/2)CO + (x/2)CO_2 + (y/2)H_2$	
Steam reforming (SR)	$C_xH_yO_z + H_2O \rightarrow xCO + yH_2$	310 (KJ/mol), high temperature required
	$C_xH_yO_z + nH_2O \rightarrow aCO + (x-a)CO_2 + yH_2$	230 (KJ/mol)[1]
	$C_xH_yO_z + (2x-z)H_2O \rightarrow xCO_2 + (2n + (y/2) - z)H_2$	64 (KJ/mol)[1]
SR of acetic acid	$CH_3COOH + 2H_2O \rightarrow 2CO_2 + 4H_2$	131.4 (KJ/mol), suitable catalyst with high stability has not well understood.
SR of glycerol	$C_3H_8O_3 + 3H_2O \rightarrow 7H_2 + 3CO_2$	128 (KJ/mol)
Autothermal reforming (ATR)	$C_xH_yO_z + (x\text{-}z/2)2O_2 \rightarrow xCO + (y/2)H_2$	Lower H_2 yield
	$C_xH_yO_z + [(x\text{-}z + y/2)/2]O_2 \rightarrow xCO + (y/2)H_2O$	
	$C_xH_yO_z + [(2x\text{-}z + y/2)/2]O_2 \rightarrow xCO + (y/2)H_2$	
Aqueous phase reforming	$C_xH_{2x+2}O_x + xH_2O \rightarrow (2x+1)H_2 + xCO_2$	Reaction needs high 'T' and 'P' for H_2 yield
Supercritical water (SCW)	$CH_xO_y + (2-y)H_2O \rightarrow CO_2 + (2\text{-}y + x/2)H_2$	Here 'x' and 'y' are the elemental molar ratios of H/C and O/C in biomass, respectively. Reaction is endothermic. High temperature and pressure (>220 atm.) is required for better H_2 Yield
Surplus steam avail	$C_2H_6O + 3H_2O \rightarrow 2CO_2 + 6H_2$	Theoretically the highest H_2 production
Limited steam supply	$C_2H_6O + H_2O \rightarrow 2CO + 4H_2$	With low (X mole) H_2 and high (X/2 mole) CO products
Dehydrogenation	$C_2H_6O \rightarrow C_2H_4O + H_2$	H_2 production by spontaneous dehydrogenation
Acetaldehyde decomposition	$C_2H_4O \rightarrow CH_4 + CO$	Undesired reaction
SR of acetaldehyde	$C_2H_4O + H_2O \rightarrow 3H_2 + 2CO$	CO production are prominent
Dehydration	$C_2H_6O \rightarrow C_2H_4 + H_2O$	Undesired reaction
Coke formation	$C_2H_4 \rightarrow$ Plymeric deposit (coke)	Undesired pathway, main source of coke generation
Decomposition	$C_2H_6O \rightarrow CO + CH_4 + H_2$	CO and methane production are higher with undesired CO_2 and comparative lower selectivity of H_2
	$2C_2H_6O \rightarrow C_3H_6O + CO + 3H_2$	
	$C_2H_6O \rightarrow 0.5CO_2 + 1.5\ CH_4$	
Reaction of decomposition	$CO + 3H_2 \rightarrow CH_4 + H_2O$	−206 (KJ/mol) products of methanation
	$CO_2 + 4H_2 \rightarrow CH_4 + 2H_2O$	Methanation reaction needs to be suppressed
Sequential cracking	$CH_4 + \text{metal} \rightarrow C_{metal} + 2H_2$	C_{metal} utilisation and its periodic removal is essential for better H_2 Yield
	$C_{metal} + O2 \rightarrow CO2$	
Methane decomposition	$CH_4 \rightarrow 2H_2 + C$	206.11 (KJ/mol), highly endothermic where catalyst deactivation is an issue
	$CH_4 + H_2O \rightarrow CO + 3H_2$	
Boudouard reaction	$2CO \rightarrow CO_2 + C$	Leads to unwanted coke deposition on catalyst
Water gas shift reaction (WGSR)	$CO + H_2O \rightarrow CO_2 + H_2$	Reduces coke formation; enhances H_2 production with −41 (KJ/mol). Favourable for better H_2 Yield

[1]Enthalpies calculated at ref. Temperature 27°C and $x = 6$.

be further steam reformed to produce H_2. Alternatively, the bio-oil can be treated in an autothermal or steam reformer to produce H_2. Just by increasing the pyrolysis temperature range above 550–750°C, it is easy to enhance the gaseous yield to 45–50% compared with the conventional yield of 30–35% at below 500°C. Demirbas (2002) has discussed about H_2 yield which increased from 27–41% to 41–55% of the total gas yield on a volumetric basis under the same condition.[29] Catalytic pyrolysis and *in situ* pyrolysis vapor upgrading are other routes to achieve higher yield of H_2, where Vaidya and Rodrigues, reported about the usage of Ca and Cr oxides, and Garcia et al.[20,23] have discussed about the Ni/Al catalyst to enhance H_2 gas yield. Other methods to avoid the N_2 and CO_2 dilution and to achieve the energy density of gaseous effluents from pyrolysis of biomass were combined with the secondary decomposition of gaseous intermediate with the introduction of additional water (>800°C) for production of hydrogen-rich gas (65 g of H_2/kg of rice husk), known to be a high temperature pyrolysis >600°C, as discussed by Zhao et al.[30] Yield of biomass pyrolysis can be maximized and summarized as follows: (1) charcoal—a low temperature, low- and moderate heating rate process as followed in slow and intermediate pyrolysis, (2) liquid products—a medium temperature, high heating rate, short gas residence time process (as in fast, flash, and intermediate pyrolysis), and (3) fuel gas—a high temperature, low heating rate, long gas residence time process (high temperature pyrolysis as in fast and intermediate cases).

GASIFICATION OF BIOMASS FOR H_2 PRODUCTION

Coke, coal gasification at higher temperature and blended biomass co-gasification in the presence of partial O_2/air and/or steam above 700°C, used as syngas (CO, CO_2, and H_2) can be delivered as output of the process, whereas only air utilization gives rise to producer gas. In order to maximize H_2 and syngas yield with lower tar formation, catalytic cracking, gasifier type, design, heating rate, temperature, space time, space velocity in catalytic bed, and residence time can be optimized. Moreover, dolomite and CeO_2/SiO_2-supported Ni, Pt, Pd, Ru, Rh, and alkaline metal oxides can be used to catalyze the gasification process, in order to reduce tar formation, improve the product gas purity, and to enhance conversion efficiency. As ICP-MS analysis of biomass and char samples revealed about the inorganic contents of the biomass such as sodium, potassium, calcium, and other alkali contents, they sometimes act as catalysts by enhancing the decomposition rate of carbon, further converted to ash which is collected at the bottom of the gasifier or carried away with the product gas as fly ash. This ash deposited in the gasifier may cause sintering, fouling, agglomeration, and slogging, which can be removed periodically to make the process systematic and for better production of syngas.[30,24–29]

When fast or intermediate pyrolysis process is carried out around 700°C, it propagates the tendency to remove tar content from the gas and thus helps in improving the quality of the product gas. At this stage, addition of catalysts, usually dolomites and Ni, at high temperatures with steam and oxygen could be helpful to deliver higher product gases. In this context, Ramage and Scurlock have discussed about the high temperature pyrolysis (between 800 and 900°C) that delivers as low as 10% of biochar while converting more than 60% of biomass into gases rich in H_2 and CO.[28] Such high temperature pyrolysis can be a competitor to various conventional gasification methods such as (1) oxygen gasification, (2) air gasification, (3) steam gasification. But during all three kinds of gasification, the heating value varies from 3 to 15 MJ/Nm^3, while in air gasification the heating value ranges from 4 to 6 MJ/Nm^3 with different by-products such as water, CO_2, hydrocarbons, and tar etc. Steam gasification (SG) converts the ligno(hemi)cellulosic materials to H_2, CO, CO_2, CH_4, light hydrocarbons, char, and tar. This SG process can be compared with that of pyrolysis at elevated temperatures (600–1000°C). With an overlapping of some aspects of both gasification and pyrolysis process, the material decomposition, hydrogen yield, and energy yield were better with gasification in comparison with pyrolysis process. Florin and Harris have analyzed SG with the utilization of calcium oxide (CaO) as a sorbent to capture CO_2, a novel approach to manifold the H_2 concentrations ranging between 40–50 vol% in a conventional manner, to 40–80 vol% on dry basis and by utilization of CaO.[31]

Hydrogen production through SCW gasification is an emerging and promising technology for the utilization of high moisture content biomass. Critical point of H_2O ($T > 374$°C, $P > 22$ MPa) plays a significant role in chemical reactions because of their dielectric constant variation, both in the gasification method and hydropyrolysis process. When arrived at a critical point the properties of both liquid and gas become identical, making the supercritical water (SCW) completely miscible with organic substance and other gases. This SCW gasification process has the advantage of processing any kind of biomass with any moisture availability (>50%), by avoiding the unit operation such as drying and pelletization. First one is at temperatures (HTSWG) higher than 500°C and second one is at temperatures (LTSWG) lower than 500°C. In the HTSWG case, decomposition of biomass proceeds without the presence of catalyst, whereas in the LTSWG, the use of catalyst is necessary. Shimura and Yoshida reported that the Ru

Table 8.3 Comparison between experimental and thermodynamic results for ethanol steam reforming reaction using Ni-Ce-Zr catalyst (Reproduced with permission from Ref 34. Copyright 2013, American Chemical Society)

	H_2 Yield		CO_2 Selectivity		CO Selectivity		CH_4 Selectivity	
Temperature (°C)	Expt.	Thermo.	Expt.	Thermo.	Expt.	Thermo.	Expt.	Thermo.
600	4.5	5.2	23.4	20.6	5.8	5.9	0.8	1.2
650	4.8	5.3	18.8	19.3	7.6	7.5	0.3	0.3
700	5.3	5.3	15.1	18.2	10.9	8.9	0.2	0.08

catalyst has shown higher activity; where the Ni catalyst has also been explored for HTSWG processes extensively.[32]

When H_2 is produced by applying higher pressure in SCW reforming process it can be stored directly, thus avoiding the large energy expenditures associated with its compression. The cost associated with hydrogen production from SCW gasification of wet biomass is comparatively higher than the present cost of hydrogen production from steam reforming of methane.[26,27,30] With an increase in the temperature, the hydrogen and carbon dioxide yield increases, while the methane yield gradually decreases. According to IEA's long-term scenario, Balat and Kirtay reported that the cost comparison of hydrogen stands in favor of biomass gasification with \$14–25/GJ H_2 making it at third position in the list after natural gas and coal gasification along with CO_2 capture and storage (CCS) process, whereas the cost of H_2 at pump holds at \$12–18 and \$13–18/GJ H_2, respectively.[33]

CATALYTIC STEAM AND OXIDATIVE STEAM REFORMING OF BIOMASS DERIVED OXYGENATES

C–C scission and C–H scission are the indispensable functions during reforming which can be materialized using several metallic active phases mainly such as nickel, cobalt, and noble metals such as rhodium, palladium, ruthenium, platinum, and iridium. During ethanol steam reforming (ESR) ethanol interacts with the metallic surface and an ethoxide species is formed followed by an aldehyde intermediate carbon- and oxygen-bonded configuration. The addition of Na promoter to Co/ZnO catalysts inhibits the carbon formation under ethanol steam reforming conditions and enhances the stability of catalysts, while addition of Fe modifies the reduction properties of cobalt.

Many supports such as ZnO, CeO_2, and La_2O_3 along with metallic phases help to interact with ethanol by cleaving the C–C bond effectively and produce its transformation, thus influencing the reaction selectivity. These supports may also be able to promote water splitting and favor OH migration. An acid center available in the support helps the ethanol molecule for dehydration to produce ethylene, which further polymerizes and becomes a precursor for carbon deposits.[8,9,17,34] Hence, to avoid acid sites in Al_2O_3, the addition of alkaline additives is essential. Patel et al. have discussed about the comparative analysis between experimental and thermodynamic results for ethanol steam reforming reaction using Ni-Ce-Zr catalysts as shown in Table 8.3.[34] On the other hand, oxides having basic sites may lead to the formation of acetaldehyde, through an aldol condensation reaction to higher oxygenates production; as a result, the formation of dimethyl ketone was favored over ZnO during steam-reforming (SR) conditions at higher space time (h).[35] Fermoso et al. tried to achieve high yield (88%) and purity of H_2 up to (99.7 vol%) at atmospheric pressure and at 550–600°C with a $S/C = 3$, in a fixed-bed reactor by applying Ni/Co catalyst derived from hydrotalcite-like material (HT) and dolomite as CO_2 sorbent.[36] They revealed that the concentration of methane is lowest at 575°C, while the CO concentration increases concurrently with increasing temperature from 525 to 600°C. The high coking potential of glycerol and fatty acid methyl esters (C_{17}–C_{19}) resulted in the increased formation of coke, thus lowering the hydrogen yield. The reaction rates of methane reforming and WGS reactions are much higher than the steam reforming of crude glycerol on Co-Ni catalysts. The oxidative steam reforming of bio-oil under controlled O_2 environment can yield better H_2 by utilizing the exothermic heat of reactions.

The promoter effect of CeO_2 is related to the ability of (1) acting as an oxygen storage material in oxidation reactions, (2) dispersing active noble metal phases inhibiting their sintering, (3) promoting WGS, and (4) facilitating the coke gasification. Addition of ZrO_2 to CeO_2 manifolds the activity of ceria while increasing the OH mobility at the surface of catalyst with the help of added oxygen mobility. Several ESR reaction mechanisms have been proposed by different groups, whereas WGS reaction plays an important role towards the distribution of products.[33]

Bio-oil is difficult to reform completely, but steam reforming of the main components in bio-oil such as acetic acid (12–14% by wt.), ethanol, and glycerol is feasible.

Table 8.4 Comparative Analysis of Different Catalyst, (*S*/*C*) Ratio, Temperature, and Space Velocity Used for Steam Reforming of Model Compounds of Bio-Oil for H_2 Production

Catalyst (Feed)	T(°C)	(*S*/*C*) or (RM/H_2O) Ratio	%H_2 Yield	%MC	Opt. Cond (P; T = °C; SV = h^{-1})	References
Co-Fe (AAc)	350–600	9.2	96	96	1 atm, 400, 4	42
Co(R) (AAc)	350–600	9.2	100	100	1 atm, 400, 4	42
(Pd-Ag MR-Counter current) (AAc)	400	10	53.4	53.4	2.5 bar, 400, —	43
NiCoAl (AAc)	500–600	3		100	1 atm, 600, —	44
Ni/Al_2O_3 (AAc)	600	4	30	46	1 atm, 600, 5.01	45
Ni/CeO_2-ZrO_2 (AAc)	550–800	3	90	95	1 atm, 700, —	46
Co-Ce/Al_2O_3 (AAc)	350–450	7.5	91.8	91.8	1 atm, 450, 10.1	46
Co-La/Al_2O_3 (AAc)	350–450	7.5	95.5	95.5	1 atm, 450, 10.1	46
Ni/ZrO_2 (AAc)	250–600	7.5	90	100	1 atm, 400, 5.1	47
Ni-Co (AAc)	500–700	15 (wt.%)	56 (mol%)	92	1 atm, 600, —	48
$CuZnCa_{12}/Al_7$ (AAc)	600–800	10 (wt.%)	80	81	1 atm, 800, —	49
Ni-Ca-Al (AAc)	550–750	3	70	100	1 atm,750, 30,000	50
Pt/ZrO_2 (AAc)	450–600	5	87	100	1 atm, 600, 40,000	40
Ni/γ-Al_2O_3 (AAc)	600	4	30	50	1 atm, 600, 5.01	45
Ni/CeO_2 (Gly)	550–650	12	74.7	100	1 atm, 600,	51
Ni/MgO (Gly)	550–650	6	65.6	100	1 atm, 600	51
Ni/TiO_2 (Gly)	550–650	6	62.2	98.1	1 atm, 600	51
Ni/Al_2O_3 (Gly)	600–900	9	80	94	1 atm, 900, 15,300	52
Ni/CeO_2 (Gly)	450–650	9	75	100	1 atm, 450, 15,300	53
Ni/Al_2O_3/ZrO_2 (Gly)	450–650	9	71	100	—	54
Ni-Cu-Al (Gly)	450–650	3	92.9	90	1 atm, 650, —	55
NiCeZrAl (YGly)	550–650	9 (0.5-O_2/RM)	67 (69%Sel)	100	1 atm, 650, —	56
NiCeZrAl (Gly)	550–650	9 (0.5-O_2/RM)	80.9	100	1 atm, 650, —	56
Rh/Ce-Al-Si (Gly)	400–1000	—	(46%Sel)	3 (1.5-O_2/RM)	1 atm, 600, —	57
Pt/Al-Si (Gly)	400–1000	—	(14%Sel)	9 (0.5-O_2/RM)	1 atm, 700, —	57
Pd/Cu/Ni/K (Gly)	550–850	9 (0.45-O_2/RM)	(54%Sel)	100		58
Ir/La_2O_3 (Gly)	400–850	2 (1.0-O_2/RM)	(60%Sel)	—	1 atm, 650, —	59
Pt/SiO_2 (Gly)	250–450	(RM/H_2O-1:10)	(69%Sel)	100	1 atm, 350, 0.167 min^{-1}	19
Pt/CeZr (Gly)	250–450	(RM/H_2O-1:10)	(72%Sel)	78		19
Ni-Co/HT (Gly)	550–600	3	88 (99%Sel)	100	1 atm, 550, 0.917 min^{-1}	36

'AAc', acetic acid; 'Gly', glycerol, and 'YGly', denotes to yellow glycerol.

Takanabe et al. have studied the CSR of acetic acid over Pt/ZrO_2 catalyst, and Basagiannis and Verykios, and Xun and Gongxuan investigated the influence of Ni and noble metals with supports (Al_2O_3, La_2O_3/Al_2O_3, MgO/Al_2O_3) on CSR of acetic acid.[37–39] Takanabe et al. and Vagia and Lemonidou studied the thermal decomposition and its reforming tendency in the absence and presence of steam for different metals supported by calcium aluminate.[40,41] Coke deposition on the catalyst surface is a serious problem and the rate of coke formation is higher in case of nickel catalyst in comparison to noble metal catalyst; besides, noble metal catalysts are very expensive. The catalyst preparation and selection of support are the two crucial factors for higher hydrogen yield, low methane and carbon monoxide formation, and low deactivation rate of the active sites of the catalyst. This review has tried to summarize different catalysts such as Co-Fe, Co(R), Ni-Al-Ca, Ni-Al, Ni-Co-Al, Ni-Ce-Zr, Co-Al, Co-Ce-Al, Co-La-Al, Ni-Zr, Ni-Co, Cu-Zn-Ca-Al, Ni-Ca-Al, Mg-Ca, Pt-Zr and Ni-Pd-Ag in membrane reactor at counter current operation for H_2 generation, as shown in Table 8.4.[40,42–50,60–62] Some metals such as Cu-Zn-Ca-Al, Co-Fe, Co(R), and 5%Ni-Ca-Al have delivered higher %H_2 yield (>70–90%) with >80% acetic acid conversion. These selected catalysts are inexpensive and recovery of metal for recycling usage is also easy to generate hydrogen from acetic acid. Similarly,

Table 8.5 Comparative Analysis of Different Catalyst, (*S*/*C*) Ratio, Temperature, and Space Velocity Used for Steam Reforming of Different Oxygenates and Other Model Compounds of Bio-Oil for H_2 Production

Catalyst	Model compound (MC)	T (°C)	(*S*/*C*) or (RM/H_2O) Ratio	%H_2 Yield	%MC	Opt. Cond: Temp. (°C); (SV = h^{-1})	References
Ni-Al	n-butanol	550–750	5.5–14.7	76 (0.291)	90.01	650,>30,000	63
Ni-Al	Acetol	550–750	5.5–14.7	– (0.169)	97	750,>57,000	63
Pt/Al_2O_3	VFAs	300–600	13.7	70	99.1	600,>25,000	64
Mg-Ca	Acetol	650	5.58	63.5 (0.115)	99.8	650, 6800	65
Mg.Ca0	Acetol	650	5.58	67.02 (0.166)	101	650, 8247	65
MgAl	Aq. Fraction Pyro-liquid	650	7.64	67 (0.132)	85	650, 5411	65
Rh-Ca-Al	Acetone	550–750	3	80	100	750, 30,000	50
$C_{12}A_7O$-Mg	Bio-oil	400–800	> 4.0	80	96	1 atm, 750, 10,000,	66
Ag nanocrystals	Formaldehyde	25–45	Alkaline (NaOH)	99	—	1 atm, 25	67

different catalysts such as Ni-Ce, Ni-Mg, Ni-Ti, Ni-Al, Ni-Mg-Al, Ni-Cu-Al, Ni-CuMg, Ni-Mg, Ni-Ce-Zr-Al, Rh-Ce-Al-Si, Pt-Al-Si, Pd-Cu-Ni-K, and novel metals such as Ir-La, Pt-Si, Pt-Ce-Zr were also analyzed and tested for steam reforming of glycerol. However Ni-Al_2O_3 and Ni-Cu-Al are cheaper materials in comparisons with other metals available, which also delivered better %H_2 yield (>80–92%) with >90–94% glycerol conversion as tabulated in Table 8.4.[19,36,44,50–59]

Different catalysts such as Ni-Al, Mg-Ca, Mg-Al, and $Ca_{12}Al_7O$-18%Mg have been analyzed with n-butanol, acetol, aqueous fraction of pyro-liquid and bio-oil, respectively, whereas some precious metals (Pt, Rh) 5%Pt-Al_2O_3, Rh-Ca-Al and Ag nanocrystals have been tested with volatile fatty acids (VFAs), acetone, and formaldehyde, respectively. Catalysts as tabulated in Table 8.5,[63–67] reveal that the conversion of each selected model compound (MC) occurs in the range of 85–100% for %H_2 yield. Comparative analysis of different catalysts, (*S*/*C*) ratio, temperature and space velocity used for steam reforming of different oxygenates, and other model compounds of bio-oil for H_2 production are discussed.[50,68–72]

AQUEOUS-PHASE REFORMING (APR)

Aqueous-phase reforming (APR) process provides a route to generate hydrogen as a value-added chemical from aqueous-phase carbohydrates generated from fruit processing and sugar industry, as found in the wastewater to be utilized for H_2 generation. In APR, water gas shift reaction is favored by the availability of controlled pressure (5–50 bar) and desired temperature, which drives the outcome with lower amounts of CO and higher yield of hydrogen. APR has several advantages over the existing methods to produce H_2 via the steam reforming of hydrocarbons.

- Comparative energy consumption in heating and vaporising both water and oxygenated hydrocarbon is less in APR than in SR.
- Most raw materials used for APR are nonflammable and nontoxic, providing a safe operational platform for storage and handling. Glycerol, ethanol, acetic acid of low grade purity can be used as feedstock.
- Availability of adequate pressure in the process makes it possible to purify hydrogen by implementation of pressure-swing adsorption or membrane technologies and to separate the CO_2 for either sequestration or to be used as a chemical entity for industry.
- Undesirable decomposition reactions are minimized during APR process as it occurs with hydrocarbons at elevated temperatures leading to char and tar formation.
- Multistage reactor configuration can be eliminated in APR, as APR is appropriate in single stage and low temperature profile (<300°C) and the purified H_2 can be obtained by using well-known pressure-swing adsorption technology.

On the same line among different precious metal catalysts tested, the activity of the different metals for H_2 production was in the following descending order: Pt > Ru > Pd > Ni, and alumina support is most unfavourable for APR, whereas precious metal catalysts over carbon and activated carbon exhibited significantly better activity than alumina. As discussed by Meryemoglu et al., the solubilized lignocellulosic biomass of wheat straw in subcritical water and aqueous-phase reforming of solubilized organic materials in the presence of various types of reforming catalysts follows the descending order, Raney-Ni > Pt > Ru > Pd > Ni, for the production of H_2-rich gas mixture.[68] But Raney-nickel catalysts showed good performance on degradation of organic compounds solubilized from lignin

Table 8.6 Optimum Reaction Parameters and Yields, as Determined by a Thermodynamic Study of Autothermal Reforming of Some Model Bio-Oil Compounds

Reaction Parameters	Acetic Acid[50]	Acetone[50]	Ethylene Glycol[50]	Bio-oil[4] (Beech Wood)	EtOH[17]	Beef Tallow[16]
S/C ratio or (S/F)	3 (6/1)	3 (9/1)	3 (6/1)	10.8	1.5	5
O_2/feed ratio	0.33	0.62	0.26	0.54	0.4	0.423
Reaction temp (°C)	627	627	627	830	900	600
Reaction press (atm)	1	1	1	1	1	1
Conversion (%)	100	100	100	100	100	—
Yield % (Based on carbon except for H_2)						
H_2 (Stoichiometric H_2)	72.59	70.97	77.40	70	90.1	70.1
CO	19.74	27.80	24.46	25	—	2.92
CO_2	79.69	70.13	74.07	57	—	26.91
CH_4	0.58	2.07	1.47	6	—	0.07

fraction during APR process. As discussed by Davda et al., the catalysts based on Pt and Ni–Sn alloys are promising materials for hydrogen production by APR.[69]

For analysis of APR process, focus on its thermodynamic and kinetics considerations is required, whereas different factors such as nature of catalyst, nature of feed, reaction conditions, and its pathways are most influential for APR of any oxygenated compound during its reforming process. Optimum reaction parameters and yields as determined by a thermodynamic study of autothermal reforming of some model bio-oil compounds are depicted in Table 8.6. When oxygen reacts with oxygenated compounds such as glycerol and phenol to produce energy, it is *in situ* transferred to dry reforming process. Dry autothermal reforming (DATR) is a better process than dry reforming (DR), as in DR the external energy is necessary to carry out the process and more carbon is formed in the reactor while DATR does not require external energy and minimum carbon is formed at optimized conditions. Kale and Kulkarni have elaborated about dry autothermal reforming of glycerol process, where they emphasize for such operations based on four main thermodynamic parameters such as temperature, pressure, OCGR [feed O_2/C (C of glycerol only) ratio], and CGR (feed CO_2 to glycerol ratio).[70] The role of oxygen addition was to enhance the conversion of C_1–C_4 products and to promote the oxidation of coke and coke precursors from the catalyst surface by enhancing the production of hydrogen, as supported by Swami and Abraham and Wang et al.[58,71–79] On the basis of Damkohler number and Wagner criteria, Swami and Abraham concluded that the steam reforming reaction for glycerol was surface-reaction controlled at lower temperatures (550–650°C) and mass-transfer controlled at higher temperatures (700–850°C).[58,80]

AUTOTHERMAL REFORMING (ATR)

Autothermal reforming is another alternative to produce H_2 by steam reforming of oxygenates or bio-oils. ATR is a combination of both steam reforming (SR) and partial oxidation (PO) of hydrocarbons known to be single pot process to produce CO, CO_2, and H_2. The reactions involved in the ATR including both PO and SR are discussed in Table 8.2. Added advantage of the ATR process over SR is that the exothermic heat generated by oxidation/partial oxidation of bio-oil or oxygenates can compensate toward the heat requirement for endothermic reforming reaction steps. *In situ* water formation helps the SR to propagate without adding external water and energy in terms of excess steam. Controlled addition of oxygen through the feed can be well managed to match the heat required for SR or both for preheating and steam reforming as well. The O_2/feed molar ratio stands approximately >0.3–0.5 for ATR, whereas H_2 yields are high comparable to that of steam reforming (O_2/feed > 7) and are in the range of 71–80% of the stoichiometric yield with a good reproducibility.[16,50,57,70] However, the autothermal reforming reaction was mass-transfer limited at all temperatures except at 550°C, because at this temperature, Damkohler number matches that of (0.254) for SR and (0.202) for ATR, respectively. Microchannel and monolith reactor systems may be designed for these reactions. High reaction rate can be achieved by depositing the active catalyst metal in a thin layer on the channels, thus reducing the diffusion resistance.

SEQUENTIAL CRACKING METHOD

Recent development throws a stone toward sequential cracking method to produce H_2 from bio-oils. The

Table 8.7 Hydrogen Yields and Energy Ratios Obtained from Various Processes[29,73]

Processes	H_2 Yield (wt.%)	Energy Ratio[1]	
Pyrolysis + steam reforming	12.6	91	
Gasification + WGS	11.5	83	
Biomass + steam + except heat (Theoretical maximum)		17.1	124

[1]Energy ratio = heating value of product H_2/heating value of the biomass feed.

sequential cracking is a two-step process in which catalytic cracking of feed is alternated with that of catalyst regeneration step. In sequential cracking process, first step (Table 8.2) is carried out with the hydrocarbon feed that decomposes into H_2 and solid carbon on the metal site which further accumulates as coke on a metal catalyst surface.[20–22,33,69] During the second stage (regeneration step), the coke deposited on metal sites is exposed to air/O_2 or steam in order to combust or gasify to produce CO_2 and this restores the catalytic activity in the process in a recycling mode of operation to be followed sequentially. If required at least two or more reactors are placed in parallel, to produce H_2 in continuous mode by switching the hydrocarbon and air/O_2 feed between the reactors in cycles. This sequential operational arrangement affords an added advantage where it is easy to produce the H_2 and CO_2 in two different steps, thereby saving the energy required to purify H_2. Varieties of supports such as Al_2O_3, ZrO_2, and CeO_2-ZrO_2 have been used with platinum group of metals to produce purified H_2 through this sequential cracking process.[33,69,73,81–85] For comparison, energy ratio and H_2 yields from different processes and their combinations are illustrated in Table 8.7.[29,73]

BIOLOGICAL PROCESS FOR HYDROGEN PRODUCTION

Different biological methods toward hydrogen production processes are (1) direct biophotolysis, (2) indirect biophotolysis, (3) photo-fermentation, (4) dark fermentation, and (5) hybrid reactor system (combined dark and photofermentation). Comparing with thermal, thermochemical, and electrolytic processes, the photolytic process or biological hydrogen production (BHP) processes are found to be more eco-friendly and less energy intensive. The term 'Biohydrogen' has much significance and it can be produced from water, renewable organic wastes or biomass, either biologically (biophotolysis and fermentation) or photobiologically (photodecomposition) and photochemically. BHP processes are much dependent on the presence of a hydrogen-producing enzyme. Hydrogenases (two subcategories, hydrogenases and reversible hydrogenases) and nitrogenases are two known enzymes that catalyze biological hydrogen production.[20–22,33,69,73,81–85]

Biological process and photocatalytic hydrogen production from water and biomass derivatives such as different saccharides are under extended research as it may be through homogeneous or heterogeneous catalytic routes. Methane production from an anaerobic condition (fermentation) broadly proceeds through three steps. In first stage, all large molecules of organic compounds such as carbohydrates, proteins, and fatty oils are decomposed to smaller molecules such as monosaccharides, amino acids, and higher fatty acids by acid-producing bacteria and those are further decomposed to lower fatty acids (e.g., propionic acid and butyric acid), lactic acid and ethanol. Further, all these $<C_6$ to CH_4 molecules are converted into hydrogen, while acetic acid is converted to H_2 by the hydrogen-producing bacteria. Finally, they are transformed into methane and carbon dioxide by the methanogenic bacteria. The obtained methane can be further converted into hydrogen by either thermal decomposition or cracking into pure hydrogen and carbon nano-tubes as a by-product. Alternately, conventional steam reforming and the water gas shift reaction are other options with dry reforming or bireforming which utilize CO_2 to produce syngas in bulk. As this fermentation proceeds at ambient temperature, the reaction rate is low for H_2 production, but in a decentralized manner in a remote location it can be achieved to a larger extent for self-sufficiency in domestic use.[73,83–94]

CONCLUSION

Hydrogen production from biomass is a promising method. Different reaction schemes involved in cracking, reforming, pyrolysis, and co-pyrolysis processes can be thermodynamically controlled after understanding the enthalpy associated with them. Physicochemical properties of steam, supercritical water, different solvents associated with reforming, and co-gasification are largely different from those of gas and liquid at normal condition, which not only make these solvents to associate in the reaction as a reactant but also affect the reaction pathway as a catalyst. Design of suitable reactor systems such as microchannel reactors or monolith reactors might be useful for reforming reactions. High effectiveness factor can be obtained by using the thin layer of the catalyst on these reactors. This overall review depicts about the thermochemical processes for hydrogen production as well as the hydrogen generation via pyrolysis of biomass, gasification, gasification combined with pyrolysis, from supercritical water (fluid-gas) extraction, steam reforming (SR), auto thermal reforming

(ATR), dry reforming (DR), liquid phase reforming (LPR), aqueous-phase reforming (APR), and partial oxidation (POX) of renewable carboxylates such as bioethanol, acetic acid, phenol, glycerol, carbohydrates, different water-soluble sugars, bio-oil, and its aqueous phases. Few processes are referred to as emerging processes for hydrogen production such as plasma reforming, thermochemical water splitting, and biochemical/biological process, where hydrogen can be produced by varied renewable sources and helps in minimizing and eradicating global warming. A number of research groups around the world have shown much interest to make a novel approach for hydrogen production methods that are attractive yielding a higher efficiency and minimal CO formation.

ACKNOWLEDGMENTS

This work was generously supported by the Department of Science and Technology (DST), Government of India, through a grant to K.K.P. [RC-UK DST (EP/G021937/1) India funded project (C-16/2008)]. P.M. would like to thank Aston University, Birmingham, UK, for facilitating understanding of the Pyroformer facility and its operations, and Indian Institute of Technology Delhi (IRD-section) for the financial assistantship, goodwill, and generosity.

REFERENCES

1. USEIA, US. Energy Information Administration, Independent Statistics and Analysis. International Energy Outlook 2011. Available at: www.eia.gov/ieo. (Accessed July 30, 2013).
2. Huber GW, Iborra S, Corma A. Synthesis of transportation fuels from biomass: chemistry, catalysts, and engineering. *Chem Rev* 2006, 106:4044–4098.
3. Gronowska M, Joshi S, MacLean HL. A review of U.S. and Canadian biomass supply studies. *Bioresources* 2009, 4:341–369.
4. Rioche C, Kulkarni S, Meunier FC, Breen JP, Burch R. Steam reforming of model compounds and fast pyrolysis bio-oil on supported noble metal catalysts. *Appl Catal B* 2005, 61:130–139.
5. Mohanty P, Nanda S, Pant KK, Naik S, Kozinski JA, Dalai AK. Evaluation of the physiochemical development of biochars obtained from pyrolysis of wheat straw, timothy grass and pinewood: effects of heating rate. *J Anal Appl Pyrol* 2013, 104:485–493.
6. Demirbas A. Biorefineries: current activities and future developments. *Energy Convers Manage* 2009, 50:2782–2801.
7. Demirbas A. The influence of temperature on the yields of compounds existing in bio-oils obtained from biomass samples via pyrolysis. *Fuel Process Technol* 2007, 88:591–597.
8. Balat M. Production of bioethanol from lignocellulosic materials via the biochemical pathway: a review. *Energ Convers Manage* 2011, 52:858–875.
9. Sukumaran RK, Surender VJ, Sindhu R, Binod P, Janu KU, Sajna JV, Rajasree KP, Pandey A. Lignocellulosic ethanol in India: prospects, challenges and feedstock availability. *Bioresour Technol* 2010, 101:4826–4833.
10. Nanda S, Mohanty P, Pant KK, Naik S, Kozinski JA, Dalai AK. Characterization of North American lignocellulosic biomass and biochars in terms of their candidacy for alternate renewable fuels. *Bioenergy Res* 2013, 6:663–677.
11. Goyal HB, Seal D, Saxena RC. Bio-fuels from thermochemical conversion of renewable resources: a review. *Renew Sust Energ Rev* 2008, 12:504–517.
12. Zaldivar J, Martinez A, Ingram LO. Effect of selected aldehydes on the growth and fermentation of ethanologenic *Escherichia coli*. *Biotechnol Bioeng* 1999, 65:24–33.
13. Ayyachamy M, Cliffe FE, Coyne JM, Collier J, Tuohy MG. Lignin: untapped biopolymers in biomass conversion technologies. *Biomass Conv Bioref* 2013, 3:255–269.
14. Joshi J, Lawal A. Hydrodeoxygenation of pyrolysis oil in a microreactor. *Chem Eng Sci* 2012, 74:1–8.
15. Dave CD, Pant KK. Renewable hydrogen generation by steam reforming of glycerol over zirconia promoted ceria supported catalyst. *Renew Energy* 2011, 36: 3195–3202.
16. Hajjaji N, Pons M. Hydrogen production via steam and autothermal reforming of beef tallow: a thermodynamic investigation. *Int J Hydrogen Energy* 2013, 38:2199–2211.
17. Laosiripojana N, Assabumrungrat S, Charojrochkul S. Steam reforming of ethanol with co-fed oxygen and hydrogen over Ni on high surface area ceria support. *Appl Catal A* 2007, 327:180–188.
18. Dickinson RE, Errico RM, Giorgi F, Bates GT. A regional climate model for western United States. *Clim Change* 1989, 15:383–422.
19. Pompeo F, Santori G, Nichio NN. Hydrogen and/or syngas from steam reforming of glycerol Study of platinum catalysts. *Int J Hydrogen Energy* 2010, 35:8912–8920.
20. Vaidya PD, Rodrigues AE. Glycerol reforming for hydrogen production: a review. *Chem Eng Technol* 2009, 32:1463–1469.
21. Mohan D, Pittman CU Jr, Steele PH. Pyrolysis of wood/biomass for bio-oil: a critical review. *Energy Fuel* 2006, 20:848–889.
22. Bridgwater AV, Peacocke GVC. Fast pyrolysis processes for biomass. *Renew Sustain Energy Rev* 2000, 4:1–73.
23. Garcia L, Salvador ML, Bilbao R, Arauzo J. Influence of calcination and reduction conditions on the catalyst performance in the pyrolysis process of biomass. *Energy Fuel* 1998, 12:139–143.
24. Uddin MN, Daud WMAW, Abbas H. Potential hydrogen and non-condensable gases production from biomass pyrolysis: insights into the process variables. *Renew Sustain Energy Rev* 2013, 27:204–224.
25. Wang C, Dou B, Chen H, Song Y, Xu Y, Du X, Zhang L, Luo T, Tan C. Renewable hydrogen production from steam reforming of glycerol by Ni-Cu-Al, Ni-Cu-Mg, Ni-Mg catalysts. *Int J Hydrogen Energy* 2013, 38:3562–3571.
26. Guo Y, Wang SZ, Xu DH, Gong YM, Ma HH, Tang XY. Review of catalytic supercritical water gasification for hydrogen

production from biomass. *Renew Sustain Energy Rev* 2010, 14:334–343.
27. Iulianelli A, Ribeirinha P, Mendes A, Basile A. Methanol steam reforming for hydrogen generation via conventional and membrane reactors: a review. *Renew Sustain Energy Rev* 2014, 29:355–336.
28. Ramage J, Scurlock J. Biomass. In: Boyle G, ed. *Renewable Energy-Power for a Sustainable Future*. Oxford: Oxford University Press; 1996.
29. Demirbas A. Gaseous products from biomass by pyrolysis and gasification: effects of catalyst on hydrogen yield. *Energy Convers Manage* 2002, 43:897–909.
30. Zhao B, Zhang X, Sun L, Meng G, Chen L, Xiaolu Y. Hydrogen production from biomass combining pyrolysis and the secondary decomposition. *Int J Hydrogen Energy* 2010, 35:2606–2611.
31. Florin NH, Harris AT. Mechanistic study of enhanced H_2 synthesis in biomass gasifiers with in-situ CO_2 capture using CaO. *Environ Energy Eng* 2008, 54:1096–1109.
32. Shimura K, Yoshida H. Heterogeneous photocatalytic hydrogen production from water and biomass derivatives. *Energy Environ Sci* 2011, 4:2467–2481.
33. Balat H, Kirtay E. Review hydrogen from biomass present scenario and future prospects. *Int J Hydrogen Energy* 2010, 35:7416–7426.
34. Patel M, Jindal T. *Pant K*, vol. xx. Ind Eng Chem Res: Kinetic study of steam reforming of ethanol on Ni based Ceria Zirconia catalyst; 2013, 52:15763–15771.
35. de la Piscina PR, Homs N. Use of biofuels to produce hydrogen (reformation processes). *Chem Soc Rev* 2008, 37:2459–2467.
36. Fermoso J, He L, Chen D. Production of high purity hydrogen by sorption enhanced steam reforming of crude glycerol. *Int J Hydrogen Energy* 2012, 37:14047–14054.
37. Takanabe K, Aika K, Seshan K, Lefferts L. Sustainable hydrogen from bio-oil-steam reforming of acetic acid as a model oxygenate. *J Catal* 2004, 227:101–108.
38. Basagiannis AC, Verykios XE. Reforming reactions of acetic acid on nickel catalysts over a wide temperature range. *Appl Catal A* 2006, 308:182–193.
39. Xun H, Gongxuan L. Investigation of steam reforming of acetic acid to hydrogen over Ni–Co metal catalyst. *J Mol Catal A* 2007, 261:43–48.
40. Takanabe K, Aika K, Seshan K, Lefferts L. Catalyst deactivation during steam reforming of acetic acid over Pt/ZrO2. *Chem Eng J* 2006, 120:133–137.
41. Vagia EC, Lemonidou AA. Thermodynamic analysis of hydrogen production by steam reforming of bio-oil components. *Int J Hydrogen Energy* 2007, 32:212–223.
42. Wang Z, Roberts RR, Naterer GF, Gabriel KS. Comparison of thermochemical, electrolytic, photo-electrolytic and photochemical solar-to-hydrogen pro-duction technologies. *Int J Hydrogen Energy* 2012, 37:16287–16301.
43. Iulianelli A, Longo T, Basile A. CO-free hydrogen production by steam reforming of acetic acid carried out in a Pd–Ag membrane reactor: the effect of co-current and counter-current mode. *Int J Hydrogen Energy* 2008, 33:4091–4096.
44. Assaf PGM, Nogueira FGE, Assaf EM. Ni and Co catalysts supported on alumina applied to steam reforming of acetic acid: representative compound for the aqueous phase of bio-oil derived from biomass. *Catal Today* 2013, 213:2–8.
45. An L, Dong C, Yang Y, Zhang J, He L. The influence of Ni loading on coke formation in steam reforming of acetic acid. *Renew Energy* 2011, 36:930–935.
46. Hu X, Lu G. Acetic acid steam reforming to hydrogen over Co–Ce/Al_2O_3 and Co–La/Al2O3 catalysts—the promotion effect of Ce and La addition. *Catal Commun* 2010, 12: 50–53.
47. Li Z, Hua X, Zhang L, Liu S, Lu G. Steam reforming of acetic acid over Ni/ZrO2 catalysts: effects of nickel loading and particle size on product distribution and coke formation. *Appl Catal A* 2012, 417–418: 281–289.
48. Pant K, Mohanty P, Agarwal S, Dalai AK. Steam reforming of acetic acid for hydrogen production over bifunctional Ni–Co catalysts. *Catal Today* 2013, 207:36–43.
49. Mohanty P, Patel M, Pant KK. Hydrogen production from steam reforming of acetic acid over Cu–Zn supported calcium aluminate. *Bioresour Technol* 2012, 123:558–565.
50. Vagia EC, Lemonidou AA. Thermodynamic analysis of hydrogen production via autothermal steam reforming of selected components of aqueous bio-oil fraction. *Int J Hydrogen Energy* 2008, 33:2489–2500.
51. Adhikari S, Fernando SD, Haryanto A. Hydrogen production from glycerin by steam reforming over nickel catalysts. *Renew Energy* 2008, 33:1097–1100.
52. Adhikari S, Fernando S, Haryanto A. Production of hydrogen by steam reforming of glycerin over alumina-supported metal catalysts. *Catal Today* 2007, 129:355–364.
53. Zhang B, Tang X, Li Y, Xu Y, Shen W. Hydrogen production from steam reforming of ethanol and glycerol over ceria-supported metal catalysts. *Int J Hydrogen Energy* 2007, 32:2367–2373.
54. Iriondo A, Barrio VL, Cambra JF, Arias PL, Guemez MB, Navarro RM, Sanchez-Sanchez MC, Fierro JLG. Hydrogen production from glycerol over nickel catalysts supported on Al_2O_3 modified by Mg, Zr, Ce or La. *Top Catal* 2008, 49:46–58.
55. Wang W, Xie L, Luo G, Zhou Q. Enhanced fermentative hydrogen production from cassava stillage by co-digestion: the effects of different co-substrates. *Int J Hydrogen Energy* 2013, 38:6980–6988.
56. Kamonsuangkasem K, Therdthianwong S, Therdthian-wong A. Hydrogen production from yellow glycerol via catalytic oxidative steam reforming. *Fuel process Technol* 2013, 106:695–703.
57. Rennard DC, Kruger JS, Schmidt LD. Autothermal catalytic partial oxidation of glycerol to syngas and to non-equilibrium products. *ChemSusChem* 2009, 2:89–98.
58. Swami SM, Abraham MA. Integrated catalytic process for conversion of biomass to hydrogen. *Energy Fuel* 2006, 20:2616–2622.

59. Yang G, Yu H, Peng F, Wang H, Yang J, Xie D. Thermodynamic analysis of hydrogen generation via oxidative steam reforming of glycerol. *Renew Energy* 2011, 36:2120–2127.
60. Bimbela F, Oliva M, Ruiz J, Garcı;a L, Arauzo J. Hydrogen production by catalytic steam reforming of acetic acid, a model compound of biomass pyrolysis liquids. *J Anal Appl Pyrolysis* 2007, 79:112–120.
61. Galdamez JR, Garcı;a L, Bilbao R. Hydrogen production by steam reforming of bio-oil using coprecipitated Ni–Al catalysts. Acetic acid as a model compound. *Energy Fuel* 2005, 19:1133–1142.
62. Medrano JA, Oliva M, Ruiz J, Garcı;a L, Arauzo J. Catalytic steam reforming of model compounds of biomass pyrolysis liquids in fluidized bed reactor with modified Ni/Al catalysts. *J Anal Appl Pyrolysis* 2009, 85:214–225.
63. Bimbela F, Oliva M, Ruiz J, Garcı;a L, Arauzo J. Catalytic steam reforming of model compounds of biomass pyrolysis liquids in fixed bed: acetol and nbutanol. *J Anal Appl Pyrolysis* 2009, 85: 204–213.
64. Jeong CM, Park GW, Choi J, Kang JW, Kim SM, Lee W, Woo SI, Chang HN. Steam reforming of volatile fatty acids (VFAs) over supported Pt/Al2O3 catalysts. *Int J Hydrogen Energy* 2011, 36:7505–7515.
65. Medrano JA, Oliva M, Ruiz J, Garcı;a L, Arauzo J. Catalytic steam reforming of model compounds of biomass pyrolysis liquids in fluidized bed reactor with modified Ni/Al catalysts. *J Anal Appl Pyrolysis* 2009, 85:214–225.
66. Wang Z, Pan Y, Dong T, Zhu X, Kan T, Yuan L, Torimoto Y, Sadakata M, Li Q. Production of hydrogen from catalytic steam reforming of bio-oil using C12A7-O-based catalysts. *Appl Catal A* 2007, 320: 24–34.
67. Bi Y, Lu G. Nano-Cu catalyze hydrogen production from formaldehyde solution at room temperature. *Int J Hydrogen Energy* 2008, 33:2225–2232.
68. Meryemoglu B, Hesenov A, Irmak S, Atanur OM, Erbatur O. Aqueous-phase reforming of biomass using various types of supported precious metal and raney-nickel catalysts for hydrogen production. *Int J Hydrogen Energy* 2010, 35:1258–12587.
69. Davda RR, Shabaker JW, Huber GW, Cortright RD, Dumesic JA. A review of catalytic issues and process conditions for renewable hydrogen and alkanes by aqueous-phase reforming of oxygenated hydrocarbons over supported metal catalysts. *Appl Catal B* 2005, 56:171–186.
70. Kale GR, Kulkarni BD. Thermodynamic analysis of dry autothermal reforming of glycerol. *Fuel Process Tech* 2010, 91:520–530.
71. Lu YJ, Guo LJ, Ji CM, Zhang XM, Hao XH, Yan QH. Hydrogen production by biomass gasification in supercritical water: Aparametric study. *Int J Hydrogen Energy* 2006, 31:822–831.
72. Zhu LJ, Guo PJ, Chu XW, Yan SR, Qiao MH, Fan KN, Zhang XX, Zong BN. An environmentally benign and catalytically efficient non-pyrophoric Ni catalyst for aqueous-phase reforming of ethylene glycol. *Green Chem* 2008, 10:1323–1330.
73. Tanksale A, Beltramini JN, Lu GM. A review of catalytic hydrogen production processes from biomass. *Renew Sustain Energy Rev* 2010, 14:166–182.
74. Bartels JR, Pate MB, Olson NK. An economic survey of hydrogen production from conventional and alternative energy sources. *Int J Hydrogen Energy* 2010, 35:8371–8384.
75. Biagini E, Masoni L, Tognotti L. Comparative study of thermochemical processes for hydrogen production from biomass fuels. *Bioresour Technol* 2010, 101:6381–6388.
76. Zhao B, Zhang X, Sun L, Meng G, Chen L, Xiaolu Y. Hydrogen production from biomass combining pyrolysis and the secondary decomposition. *Int J Hydrogen Energy* 2010, 35:2606–2611.
77. Li Y, Wang J, Peng S, Lu G, Li S. Photocatalytic hydrogen generation in the presence of glucose over ZnS-coated ZnIn2S4 under visible light irradiation. *Int J Hydrogen Energy* 2010, 35:7116–7126.
78. Wang C, Dou B, Chen H, Song Y, Xu Y, Dua X, Luo T, Tan C. Hydrogen production from steam reforming of glycerol by Ni–Mg–Al based catalysts in a fixed-bed reactor. *Chemical Engineering J* 2013, 220: 133–142.
79. Chang ACC, Louh RF, Wong D, Tseng J, Lee YS. Hydrogen production by aqueous-phase biomass reforming over carbon textile supported PteRu bimetallic catalysts. *Int J Hydrogen Energy* 2011, 36: 8794–8799.
80. Kirtay E. Recent advances in production of hydrogen from biomass. *Energy Convers Manage* 2011, 52:1778–1789.
81. James OO, Maity S, Mesubi MA, Ogunniran KO, Siyanbola TO, Sahub S, Chaubey R. Towards reforming technologies for production of hydrogen exclusively from renewable resources. *Green Chem* 2011, 13:2272–2284.
82. Wei Z, Sun J, Li Y, Datye AK, Wang Y. Bimetallic catalysts for hydrogen generation. *Chem Soc Rev* 2012, 41:7994–8008.
83. Mohanty P, Pant KK, Naik SN, Das LM, Vasudevan P. Fuel production from biomass: Indian perspective for pyrolysis oil. *J Sci Ind Res* 2011, 70:668–674.
84. Trane R, Dahl S, Skjøth-Rasmussen MS, Jensen AD. Review catalytic steam reforming of bio-oil. *Int J Hydrogen Energy* 2012, 37:6447–6472.
85. Chaubey R, Sahu S, James OO, Maity S. A review on development of industrial processes and emerging techniques for production of hydrogen from renewable and sustainable sources. *Renew Sustain Energy Rev* 2013, 23:443–462.
86. Hu R, Yan C, Zheng X, Liu H, Zhou Z. Carbon deposition on Ni/ZrO2eCeO2 catalyst during steam reforming of acetic acid. *Int J Hydrogen Energy* 2013, 38:6033–6038.
87. Avasthi KS, Reddy RN, Patel S. Challenges in the production of hydrogen from glycerol-a biodiesel byproduct via steam reforming process. *Procedia Eng* 2013, 51:423–429.
88. Zhao B, Zhang X, Sun L, Meng G, Chen L, Xiaolu Y. Hydrogen production from biomass combining pyrolysis and the secondary decomposition. *Int J Hydrogen Energy* 2013, 35:2606–2611.

89. Pairojpiriyakul T, Croiset E, Kiatkittipong W, Kiatkittipong K, Arpornwichanop A, Assabumrungrat S. Hydrogen production from catalytic supercritical water reforming of glycerol with cobalt-based catalysts. *Int J Hydrogen Energy* 2013, 38:4368–4379.
90. Cohce MK, Dincer I, Rosen MA. Thermodynamic analysis of hydrogen production from biomass gasification. *Int J Hydrogen Energy* 2010, 35:4970–4980.
91. Agrafiotis C, Storch HV, Roeb M, Sattler C. Solar thermal reforming of methane feedstocks for hydrogen and syngas production - a review. *Renew Sustain Energy Rev* 2014, 29:656–682.
92. Demirbas A, Arin G. An overview of biomass pyrolysis. *Energy Source* 2002, 24:471–482.
93. Demirbas A. Yields of hydrogen-rich gaseous products via pyrolysis from selected biomass samples. *Fuel* 2001, 80:1885–1891.
94. Xiu S, Shahbazi A. Bio-oil production and upgrading research: a review. *Renew Sustain Energy Rev* 2012, 16:4406–4414.

9

Production of Renewable Hydrogen by Reformation of Biofuels

Paraskevi Panagiotopoulou,[1] Christina Papadopoulou,[2] Haris Matralis[2] and Xenophon Verykios[3]*

[1]Catalysis Center for Energy Innovation and Center for Catalytic Science and Technology, Department of Chemical and Biomolecular Engineering, University of Delaware, Newark, DE, USA
[2]Department of Chemistry, University of Patras, Patras, Greece
[3]Department of Chemical Engineering, University of Patras, Patras, Greece

INTRODUCTION

Considerable changes in the use of global energy and material resources have occurred during the last decades. The rising demand for energy, which is associated with economic growth, imparts a heavy impact on local and regional environment. On the other hand, the volatility of oil and natural gas prices illustrates the necessity of diversity, sustainability, and security in energy supply. Energy production is still mainly based on combustion of fossil fuels, a fact which is unlikely to change in the near future. For this reason, CO_2 emissions are expected to increase. Furthermore, deforestation and changes in land usage aggravate the problem of the greenhouse phenomenon, causing global warming and unpredictability of weather patterns. Restrictions concerning gas emissions urgently demand the development of clean energy technologies with minimum environmental impact.

The use of hydrogen as an energy vector has been proposed as a solution to the global environmental problem related to the greenhouse effect. It must be realized, however, that for an effective solution, hydrogen would have to be produced from renewable resources. Although hydrogen can be produced from excess renewable electricity via electrolysis, with zero greenhouse gas (GHG) emissions, this cannot be adapted in distributed power generation due to severe difficulties in hydrogen storage and transportation. To overcome these difficulties, the idea of hydrogen production on site and on demand has been gaining ground. This implies that an appropriate hydrogen carrier, which could be liquid or gas, can be used to extract hydrogen from. The employment of hydrogen as an energy carrier, combined with fuel cell systems, will lead to low or no carbon emissions and no emissions of atmospheric pollutants. The most appropriate renewable hydrogen carrier seems to be biomass components and derivatives, such as alcohols, bio-oil, and biogas. Reformation of these, in a distributed power generation mode, could offer significant energy and environmental advantages.

PRODUCTION OF REFORMABLE BIOFUELS

Biomass can be used directly for the production of hydrogen via gasification processes. However, these processes are hampered with difficulties and inconveniences, especially when the objective is related to distributed power production, due to the high cost of transportation of raw biomass. For such applications, the most appropriate

*Correspondence to: verykios@chemeng.upatras.gr

Advances in Bioenergy: The Sustainability Challenge, First Edition. Edited by Peter D. Lund, John Byrne, Göran Berndes and Iacovos A. Vasalos.

Originally published in: *WIREs Energy Environ* 2014, 3: 109–130 doi: 10.1002/wene.93

approach is to produce transportable biofuels, or biomass derivatives, which concentrate a large fraction of the energy of raw biomass and posses reduced mass and volume. The most convenient biofuels for reformation purposes are bioethanol, bio-oil, and biogas, whereas glycerol is produced as by-product of diesel production processes.

Bioethanol is a first generation biofuel that is already in the market for many years. The well-known conventional fermentation process is applied for its production.[1,2] Sugars and starches are the most convenient raw materials. However, their use creates limitations and controversy as these materials are also used for food.[3,4] Furthermore, only part of the plant energy is converted to biofuel, whereas fossil energy is often used in cultivations.[1] Thus, in many cases, there is only modest energy gain and GHG emission benefits. Furthermore, the production process is considered expensive due to high feedstock cost (fertilizers, harvesting, pretreatment) and energy-intensive distillation and dehydration steps.

For the reasons stated above, second generation bioethanol processes have been proposed aiming at: (1) exploitation of nonedible biomass, (2) increase of energy efficiency, reducing fossil fuel consumption and GHG emissions, and (3) cultivation of energy crops suitable to local climate. Nonedible biomass is composed of cellulose, hemicellulose, and lignin. Cellulose is a crystalline glucose polymer, strong, and resistant to acid or enzymatic hydrolysis.[1,2] Hemicellulose is a sugar polymer of five different sugars, having a random, amorphous structure and being easily hydrolyzed to its monomer sugars by dilute acid or base. Cellulose and hemicellulose make up 60–90 wt% of terrestrial biomass, they are abundant in plants and wood, and represent a large percentage of domestic and industrial wastes.[1,2] Lignin composes 10–25% of biomass and consists of highly branched, substituted, mononuclear aromatic polymer.[1] Lignin, cellulose, and hemicellulose are known as lignocellulose and, due to their abundance, are the biomass materials that can guarantee the security of raw materials supply. However, decomposing these highly complex compounds presents technical difficulties.

Production of second generation bioethanol, via the biochemical conversion of lignocellulose, is difficult because of the refractory nature of biomass and the different nature of sugar monomers, which cannot be fermented by the same microorganism. Processing includes pretreatment, hydrolysis, fermentation, and product separation/distillation. Although pretreatment increases production cost, it is essential to enhance hydrolysis yields, from typically <20 to 90% after pretreatment.[2] Hydrolysis methods convert cellulose into monomers and can be classified in two categories, acid and enzymatic hydrolysis. Fermentation that follows can be performed under the same conditions as sugar and starch fermentation. The major disadvantages of bioethanol production with the known enzymatic technology is the limited fraction of biomass converted and the larger capital intensity of the process, as compared with first generation bioethanol.[4]

Various types of thermochemical processes, such as liquefaction, pyrolysis, and gasification, can be applied to produce gaseous and liquid products.[5–7] Liquefaction and pyrolysis take place in the absence of air and produce mainly liquid products, accompanied by solid and gaseous fractions.[6] Liquefaction (50–200 atm and 250–325 °C) and pyrolysis (1–5 atm and 375–525 °C) proceed via thermal depolymerization of the lignocellulosic matter, producing bio-oils.[1] A wide range of feedstock can be used, including wood, black liquor, agricultural wastes, and forest wastes. This bio-oil production path is advantageous as it requires only a single reactor and a large fraction of the biomass energy (50–90%) can be preserved into the liquid. Slow pyrolysis produces large amounts of coke, whereas fast pyrolysis produces bio-oils in high yields, of up to 80 wt% of dry feed.

Bio-oils are complex mixtures of more than 400 highly oxygenated compounds, including acids, alcohols, aldehydes, esters, ketones, and aromatic compounds, and can contain up to 50 wt% water.[1,8] A representative distribution of components from a fast pyrolysis reactor, operated to maximize bio-oil liquid, is: 65 wt% organics, 10 wt% water, 12 wt% char, and 13 wt% gas.[1] However, the relative ratios of these fractions are highly dependent on reaction conditions, reactor design, and biomass alkali content. Liquefaction oils have higher heating content, lower oxygen content, and lower moisture content than fast pyrolysis oils.

Biomass can be decomposed via the microbiological action of bacteria under anaerobic conditions. This decomposition occurs naturally in landfills, whereas the commercially proven technology of anaerobic digestion is widely used for treating high moisture content organic wastes.[7] The main advantage of this approach is the fact that the digester's feedstock can be any biodegradable raw material, originating from sewage sludge, municipal solid wastes, animal manure, agroindustrial wastes, energy crops, and by-products from biofuels' production.[5,7,9–11] In many cases, the raw material is not only of negative cost, but it also results in a considerable reduction of disposable wastes, having thus both financial and environmental benefits. The main product is biogas, a gas mixture consisting mainly of methane and carbon dioxide in various ratios, depending on the type of feedstock, the technical design of the digester and operating conditions.[9,11] CH_4 content in biogas derived from sewage is approximately 70%, that originating from waste of food industry can be as high as

Table 9.1 Main Reactions Occurring Under Conditions of Reformation of Biofuels

Reaction Stoichiometry	ΔH° (kJ mol^{-1})	Equation #
$C_2H_5OH + 3H_2O \rightarrow 6H_2 + 2CO_2$	196.7	1
$C_2H_5OH \rightarrow CH_4 + H_2 + CO$	49.4	2
$C_2H_5OH \rightarrow CH_3CHO + H_2$	68.5	3
$C_2H_5OH \rightarrow C_2H_4 + H_2O$	−45.3	4
$C_2H_5OH + 2H_2O \rightarrow 2CO + 4H_2$	497.5	5
$CO + H_2O \leftrightarrow CO_2 + H_2$	−41.1	6
$CH_4 + CO_2 \rightarrow 2CO + 2H_2$	247	7
$CO + 3H_2 \rightarrow CH_4 + H_2O$	−206.2	8
$CH_3CHO \rightarrow CH_4 + CO$	−19	9
$2CO \leftrightarrow CO_2 + C$	−172.5	10
$CH_4 \leftrightarrow 2H_2 + C$	74.9	11
$C_2H_4 \rightarrow$ polymers $\rightarrow$ coke		12
$C_xH_yO_z + (x - z)\,H_2O = xCO + (x + 0.5y - z)\,H_2$		13
$CO + H_2 = C + H_2O$	−131.3	14
$C_3H_8O_3 + 3H_2O \rightarrow 3CO_2 + 7H_2$	128	15
$C_3H_8O_3 + 3/4O_2 + 3/2H_2O \rightarrow 3CO_2 + 11/2H_2$		16

85%, whereas CH_4 content of landfill gas can be as low as 30%.[12]

An attractive biomass derivative for renewable hydrogen production is glycerol ($C_3H_8O_3$ 1,2,3-propanetriol), which is a viscous, biodegradable, and nonflammable alcohol, with a high boiling point. Crude glycerol is the main by-product of biodiesel production, following the transesterification method. About 10 wt% of the biomass processed to produce biodiesel is converted to crude glycerol. It has low commercial value and applications. Thus, a glut of crude glycerol is accumulating in biodiesel plants and has already become a waste by-product with an attached disposal cost increasing overall biodiesel production costs.[13,14] Its exploitation as a raw material for the production of hydrogen is expected to offer economic and environmental advantages.[14,15]

Concluding, biomass can be converted to transportable energy carriers following different routes, depending on the source and the type of biomass feedstock available, the conversion process, the infrastructure, the form in which the energy is required, economic aspects, and the environmental requirements. These energy carriers, being liquid or gas, can be reformed to produce hydrogen and, via fuel cells, power.

REFORMATION OF BIOETHANOL

The use of bioethanol as renewable hydrogen carrier is attractive because of its wide availability, storage, and handling safety, and because it can be produced from several biomass sources. Reformation of ethanol by steam [Eq. (1), Table 9.1] offers significant advantages over competing processes. Steam reforming (SR) of ethanol produces high yields of hydrogen at elevated temperatures. Recently, alternative processes operating at lower temperatures are being investigated.

High Temperature Ethanol SR

SR of ethanol is a strongly endothermic reaction and, thus, it exhibits maximum performance with respect to hydrogen yield at high temperatures, typically >700°C. Depending on the catalyst used and operating conditions, several reactions may run in series and in parallel, producing acetaldehyde, hydrogen, methane, carbon oxides, and ethylene, as per Eqs (2)–(8) in Table 9.1. Equation (2) can occur either directly, or via intermediate formation of acetaldehyde as the sum of Eqs (3) and (9).

The production of hydrogen via SR of ethanol at high temperatures has been studied over a variety of metal catalysts, including Co,[16] Rh,[16,17] Pd,[16,17] and Ni.[18] Catalytic performance tests obtained over Pd, Rh, Ni, and Co supported on MgO showed that Rh exhibits the best performance in terms of ethanol conversion and stability, whereas Ni exhibits the highest hydrogen selectivity (>95%).[16] For low-loaded Al_2O_3 supported catalysts, Rh is significantly more active and selective toward hydrogen formation compared to Ru, Pt, and Pd, which show a similar behavior.[17] Catalytic activity and selectivity is strongly affected by the nature of the support. For 10 wt% Ni-based catalysts ethanol conversion reaches almost 100% at 650°C for all catalysts investigated.[19] However, selectivity toward H_2 follows the order: Ni/ZnO $\approx$ Ni/La_2O_3 > Ni/MgO > Ni/γ-Al_2O_3. Similarly, it has been found that Ni catalyst exhibits

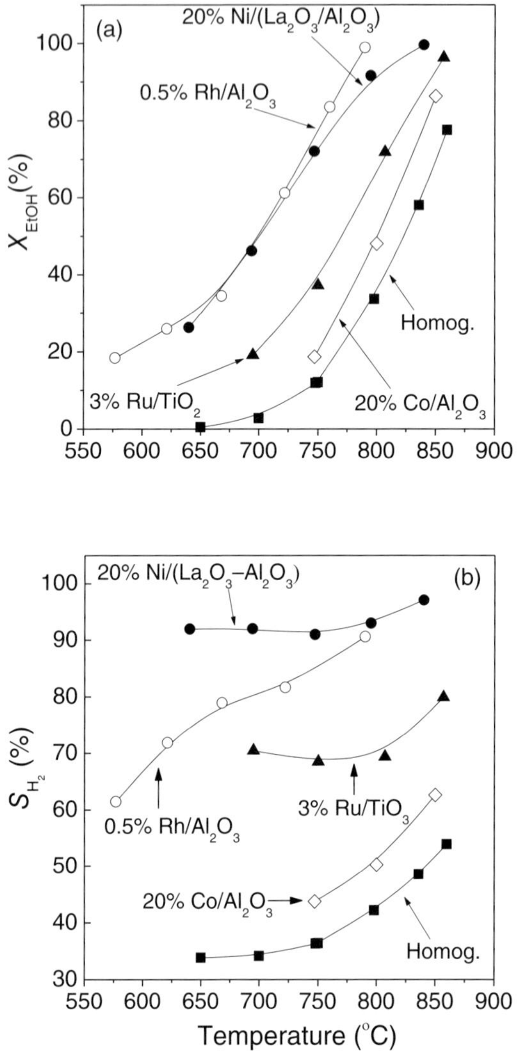

Figure 9.1 (a) Conversions of ethanol and (b) selectivities toward hydrogen obtained as functions of reaction temperature over 20% $Ni/(La_2O_3/Al_2O_3)$, 0.5% Rh/Al_2O_3, 3% Ru/TiO_2, and 20% Co/Al_2O_3 catalysts and in the absence of catalyst (homogeneous reactions). Experimental conditions: mass of catalyst: 0.1 g; particle diameter, $0.25 < d_p < 0.50$ mm; total flow: 350 cm^3/min; feed composition: 25% EtOH, 75% H_2O; P = 1 atm. (Reproduced with permission from Ref 18. Copyright 2002, Elsevier.)

higher activity and selectivity toward hydrogen production and, most important, long term stability when supported on La_2O_3 compared with Al_2O_3, YSZ, and MgO.[18]

Typical results obtained over several catalytic materials are presented in Figure 9.1, in which conversion of ethanol and selectivity toward hydrogen are plotted as functions of reaction temperature. It is observed that the homogeneous conversion of ethanol becomes significant at temperatures above 750°C and reaches 80% at 850°C (Figure 9.1a). However, selectivity toward hydrogen is very low under these conditions and increases from ca. 35 to 55% upon increasing reaction temperature from 650 to 850°C (Figure 9.1b). The $Ni/(La_2O_3/Al_2O_3)$ catalyst exhibits superior catalytic performance with X_{EtOH} reaching 100% at temperatures higher than 800°C and selectivity toward H_2 exceeding 95% at 840°C. The activity of 0.5% Rh/Al_2O_3 catalyst is comparable with that of $Ni/(La_2O_3/Al_2O_3)$. However, selectivity toward hydrogen is poor over the rhodium catalyst, especially at temperatures below 750°C. The 20% Co/Al_2O_3 catalyst is almost inactive under the experimental conditions employed. The 3%Ru/TiO_2 catalyst exhibits better performance but both X_{EtOH} and S_{H2} are significantly lower than the corresponding values obtained over $Ni/(La_2O_3/Al_2O_3)$. The best performing $Ni/(La_2O_3/Al_2O_3)$ catalyst exhibits good long-term stability for SR of ethanol to hydrogen production and is a good candidate for ethanol reforming processors for fuel cell applications.[18]

Bimetallic or alloy catalysts have also been investigated.[20–22] It has been found that the presence of Ni improves Rh dispersion on the surface of CeO_2, leading to higher catalytic activity and selectivity toward H_2. Dispersion of Rh on MgAl-based spinel oxide supports results in a significant decrease of surface acidity compared with Rh/Al_2O_3 catalysts, improving catalyst stability.[21] A PdZn alloy is formed by codeposition of Pd and Zn on ZnO support, which enhances dehydrogenation reaction and, consequently, hydrogen production.[22]

One of the major problems in ethanol SR reaction is related to coke formation and deposition onto the catalyst surface, which may lead to catalyst deactivation. Carbon formation may take place via several processes, such as the 'Boudouard' reaction[18] [Eq. (10)], methane decomposition[23] [Eq. (11)], and cracking of ethylene produced by dehydration of ethanol [Eq. (12)].[18,24]

Carbon deposition can be avoided at high temperatures, where high hydrogen yield can be also achieved.[23,25] However, high reaction temperatures enhance the formation of byproduct CO, due to the thermodynamically favored reverse water–gas shift (WGS) [Eq. (6)] and dry methane reforming [Eq. (7)]. For PEM fuel cell applications, the concentration of CO should be reduced to less than 50 ppm, as dictated by the poisoning limit of Pt electrodes.[26] This is usually accomplished by treatment of the reformate gas in a WGS unit, followed by a final purification step, which may involve preferential oxidation or methanation of residual CO,[27,28] resulting in higher hydrogen cost and larger size and weight of the fuel processor.

Low Temperature SR of Ethanol

As ethanol can be decomposed at low temperatures (300–400°C), it is of interest to develop alternative processes and catalytic materials for hydrogen and/or power production. For example, ethanol could be reformed at low temperatures (300–400°C) toward a gaseous mixture containing H_2, CH_4, and carbon oxides. Depending on operating conditions and catalyst, the amount of byproduct CO may be significant and should be reduced. This could be accomplished by the development of novel catalytic materials capable to simultaneously enhance the WGS reaction. The elimination of CO from the reformate gas can also result in a decrease of CH_4 selectivity by retarding CO methanation, which may run in parallel in the temperature range of 300–400°C, consuming significant quantities of hydrogen. The resulting gas mixture can then be used to feed fuel cells for the production of electricity and heat. Advantages of the low temperature SR (LTSR) of ethanol include: (1) the reformer requires lower quality heat to operate at low temperatures, making the thermal independence of the process easier, (2) heat can be obtained from other processes, thus increasing system efficiency, and (3) CO levels can be very low, making the final CO clean-up step easier.

The LTSR of ethanol has been investigated over a variety of metal catalysts, including Ni, Co, Cu, Ir, Pt, Rh, Pd, and Ru dispersed on different supports (e.g., Al_2O_3, CeO_2, La_2O_3, ZrO_2, Y_2O_3, ZnO, Sm_2O_3, TiO_2, V_2O_5, and MgO).[24,29–32] In particular, it has been found that activity and selectivity varies in the order of Ir > Ni > Co over CeO_2-supported catalysts, with Ir exhibiting complete conversion of ethanol to H_2, CO_2, and CH_4 and zero concentration of CO at 350°C.[32] Cobalt catalysts supported on ZnO exhibit high activity and selectivity toward H_2, resulting in the production of CO-free hydrogen at temperatures as low as 350°C. The presence of Pt has been found to promote the WGS reaction at low temperatures over alumina-supported Ru–Pt catalysts.[33]

In general, supported Pt catalysts seem to present higher activity and selectivity compared with other metals for the LTSR of ethanol and are more widely studied.[30,34] Activity at 450°C was found to vary in the order of Pt > Pd > Rh > Ru over CeO_2-supported catalysts.[34] Similarly, it has been found that the ranking of noble metal catalysts supported on Al_2O_3 is the following: 1% Pd/Al_2O_3 ≈ 0.5% Pt/Al_2O_3 > 5% Ru/Al_2O_3 > 2% Rh/Al_2O_3 (Figure 9.2),[24] with ethanol conversion reaching 100% at 360°C over Pt and Pd catalysts. Product distribution showed that over Pt, the main products are H_2, CH_4, CO, CO_2 and, to a smaller extent, CH_3CHO. A symmetric variation of CO and CO_2 selectivities has been observed, implying production of CO_2 at the expense of CO, due to occurrence of the WGS reaction. Although Pd catalyst exhibits comparable activity with Pt, it is characterized by low efficiency for the WGS reaction.[24]

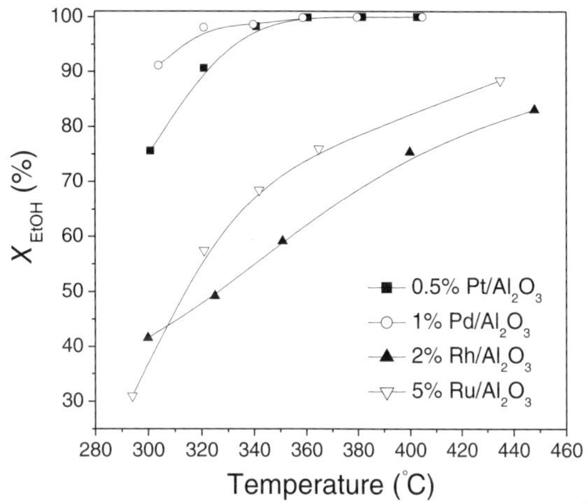

Figure 9.2 Ethanol conversion (X_{EtOH}) as a function of reaction temperature over 0.5% Pt/Al_2O_3, 1% Pd/Al_2O_3, 2% Rh/Al_2O_3, and 5% Ru/Al_2O_3 catalysts. Experimental conditions: mass of catalyst: 0.65 g; particle diameter: $0.18 < d_p < 0.25$ mm; TOTAL Flow: 120 cm^3/min; GHSV = 9350 h^{-1}; feed composition: 12.5% EtOH, 37.5% H_2O (balance He); P = 1 atm. (Reproduced with permission from Ref 24. Copyright 2008, Springer.)

The nature of the support has also been found to affect significantly catalytic performance for the LTSR reaction. In particular, investigations over Ni catalysts have shown that the activity, stability and selectivity toward hydrogen follows the order Ni/La_2O_3 > Ni/Y_2O_3 > Ni/Al_2O_3.[29] Cobalt supported on ZnO, La_2O_3, Sm_2O_3, and CeO_2 exhibit superior performance compared to MgO, γ-Al_2O_3, SiO_2, TiO_2, V_2O_5, resulting in the production of a CO-free hydrogen gas stream.[35] The beneficial effect of CeO_2 support has been also reported by Ciambelli et al.,[31] who found that hydrogen yield is higher and more stable for Pt/CeO_2 than for Pt/Al_2O_3 catalysts. Product distribution is also influenced by the nature of the support, with Pt/CeO_2 catalyst exhibiting higher selectivities toward H_2 and CO_2 and lower selectivity toward CO because of the enhancement of the WGS reaction. The improved catalytic performance of Pt/CeO_2 catalysts has been related to the greater ability of CeO_2 to release and store oxygen, resulting in higher stability compared with Pt/Al_2O_3 catalysts.[36]

Recent studies over supported Pt catalysts showed that the conversion of ethanol is higher when Pt is dispersed on ZrO_2, compared with CeO_2, whereas Al_2O_3-supported catalyst exhibits intermediate performance (Figure 9.3a).[37]

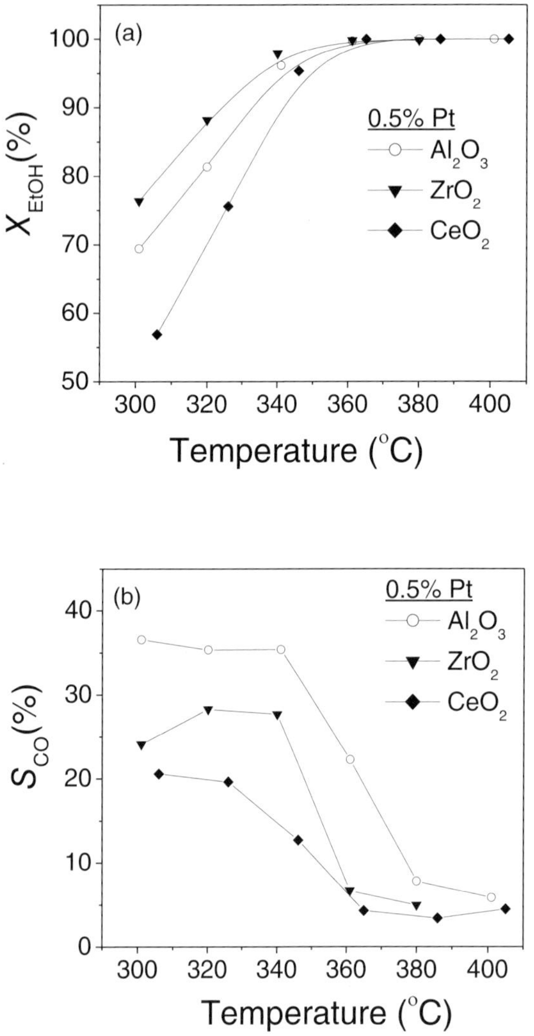

Figure 9.3 (a) Effect of the nature of the support on the catalytic performance and (b) selectivity toward CO over Pt (0.5 wt%) supported on the indicated commercial oxide carriers. Experimental conditions: same as in Figure 9.2. (Reproduced in part from Ref 37. Copyright 2012, Elsevier.)

The main difference observed between the investigated supports is related to variations of CO and CO_2 selectivities. As it can be seen in Figure 9.3(b), selectivity to CO is significantly affected by the nature of the support and is lower over Pt supported on ZrO_2 and CeO_2, which enhance the WGS reaction.

Investigations regarding the contribution of the metal phase and the support material in ethanol conversion as well as in products selectivity showed that neither Pt alone nor Al_2O_3 alone are capable of effectively catalyzing the reformation of ethanol.[24] Product distribution results demonstrated that the carrier catalyzes mostly dehydration reactions. Its specific activity is low but it provides a large surface area and its overall contribution may be significant. Platinum catalyzes the dehydrogenation reaction, and, mostly, the decomposition/reformation of dehydration products, as well as the WGS reaction. At the metal/support interface, Al_2O_3 functions to adsorb ethylene and diethyl ether, whereas Pt functions to promote their fragmentation and decomposition toward smaller molecules. The WGS reaction also takes place at the metal/support interface.

The WGS reaction is a critical step in fuel processors, which is used to improve the H_2 yield and to reduce CO levels in the reformate gas, before a final purification step, which may involve preferential oxidation or methanation of residual CO.[38,39] WGS catalysts for fuel cell applications should be sufficiently active and selective in the temperature range of 200–280°C, thermally stable, and resistant to poisoning under the reformer's conditions. Noble metal catalysts may offer significant advantages, compared with conventional WGS catalysts and therefore have been widely investigated in the last few years.

Among the various metals examined so far, including Pt,[40–46] Rh,[40–43,45,47] Ru,[40,42,43,45,48] Pd,[40,42–45,49] Au,[40,46,50–53] Re,[40] Os,[40] Ir,[40,44,54] Ag,[52] Cu,[40,52,55] Ni,[40,49,55] Fe,[40,49] and Co,[40,49] Pt catalysts seem to be promising candidates, characterized by high activity at low-to-medium reaction temperatures. In particular, investigations over Pt, Rh, Ru, or Pd catalysts supported on TiO_2,[42] CeO_2,[43] or Al_2O_3[43] showed that the turnover frequency (TOF) of CO conversion at 250°C varies in the order of Pt > Rh ≈ Ru > Pd, with Pt being 20–50 times more active than Pd, depending on the metal oxide support, whereas Rh and Ru exhibited an intermediate performance.[42,43] Similarly, Sato et al.[44] reported that Pt/TiO_2 catalyst exhibits higher activity compared to Pd and Ir, whereas Radhakrishnan et al.[45] found that the order of activity for the WGS reaction is Pt > Rh > Ru > Pd over catalysts supported on ceria–zirconia oxides.

Results of kinetic measurements obtained with the use of Pt or Ru catalysts of variable metal content (0.1–5 wt%) supported on TiO_2,[42] CeO_2,[43] or Al_2O_3[43] showed that the specific reaction rate (TOF) does not depend on metal loading (0.1–5.0 wt%) or Pt crystallite size (1.2–16.2 nm). Thus, it can be concluded that the WGS reaction over supported noble metal catalysts is structure insensitive, as far as the metallic phase is concerned, in agreement with other studies.[40,56–58]

Detailed investigations of a variety of metal–support combinations showed that the key parameters that determine the WGS activity of dispersed noble metal catalysts are related mainly to the nature and the physicochemical characteristics of the metal oxide support.[43,59] Generally, it has been shown that noble metals exhibit significantly higher activities when supported on 'reducible' (e.g., TiO_2, CeO_2, La_2O_3) compared with 'irreducible' oxides (e.g., Al_2O_3, MgO, SiO_2).[43,59,60] Titania-supported Pt catalysts exhibit sufficiently high activity, which can be further improved by addition of small amounts of alkali (Li, Na, K, Cs) or alkaline earth promoters (Ca, Ba, Sr, Mg).[61–63] Optimum results have been obtained over Na- and Ca-promoted samples, compared with samples containing Cs, Li, Sr, Ba, or Mg,[62,63] where the specific activity (TOF) is about three times higher, compared with that of the unpromoted catalyst. Qualitatively similar results were obtained for Ru and Pd catalysts.[61,62] The beneficial effect of alkalis on WGS activity has been also observed over Pt/ZrO_2[64] and Pt/CeO_2[65] catalysts doped with Na, K, Rb, or Cs. According to the authors, there is an optimum concentration of alkali where the C–H bond of the surface formate intermediate is weakened and enhances the WGS activity at low temperatures.

Mechanistic Aspects of the Ethanol Reforming Reaction

Investigation of the mechanism of ethanol SR has been the subject of several studies.[30,37,66–71] It has been proposed that adsorbed ethanol or ethoxy species dehydrogenate toward acetaldehyde, which can be either oxidized by hydroxyl groups over the alumina surface, resulting in the formation of surface acetate species, or it can be decomposed to methane and carbon monoxide.[30,37,66–71] Acetate species can be decomposed toward $CH_{3(ad)}$ species, which can further react with surface hydroxyl groups producing CO.[37,66,67,70,71] Although the overall reaction network of ethanol SR is highly complicated,[24,37] it is generally agreed that ethanol interacts strongly with the support, promoting mainly dehydration reactions. Platinum enhances ethanol decomposition and dehydrogenation reactions at low temperatures, producing CH_4, CO, and H_2, while reforming, WGS, and methanation reactions dominate at higher temperatures resulting in the formation of hydrogen and carbon oxides.[24,37]

REFORMATION OF BIOGAS

Biogas refers to a mixture primarily composed of methane and carbon dioxide, produced via anaerobic fermentation or digestion of organic matter due to microbiological action of bacteria. The chemical composition and physical properties of biogas are strongly related to the type of feedstock, the technical design of the digester–fermenter and operating conditions. The two major components are always methane and carbon dioxide, but their ratio varies significantly. Hydrogen sulfide is the impurity that can cause the most important nuisance in biogas applications, whereas higher hydrocarbons, aromatics, chroro/fluorocarbons, organic sulfur component, oxygenated compounds, chlorine, fluorine, as well as silicon compounds (siloxanes) can also be present, especially in landfill-derived biogas.[72–74] Thus, purification is necessary for almost any biogas use, especially if it is to be used as catalytic reaction feedstock.

Various physicochemical methods, adopting chemical adsorption and absorption processes, have been scientifically and technically explored for biogas purification and most of them are now commercially available.[74] The most interesting H_2S and NH_3 removal processes involve chemical absorption in aqueous solutions and can be divided into two categories, those involving oxidation of S^{2-} to S^{0}[75,76] and those based on the capture of S^{2-} through precipitation of its metallic salts owing to their very low Ksp.[74] Semi-batch processes based on selective adsorption of H_2S on solid adsorbents have been also developed. Materials based on iron oxides and activated carbons, alkali doped or undoped have been explored for this purpose.[77] Biological methods are also employed based on biofilters, bioscrubbers, and biotrickling filters.[74] Thiopaq is one of the most known technologies for large-scale industrial biogas desulfurization using microorganisms to oxidize the sulfide to elemental sulfur.[78]

Another problem that concerns only biogas from landfill facilities or waste composting is the presence of siloxanes in the landfill gas.[74] Siloxanes, silicones containing Si–O bonds with organic groups, are widely employed in industry for the production of shampoos, detergents, cosmetics, pharmaceuticals, and so on. The decomposition of these products generates siloxanes that are found in landfill gas because of their high vapor pressure and low water solubility.[79] Upon landfill gas combustion siloxanes are converted to silicon dioxide (crystalline silica), that adheres and is deposited inside the boiler. Similar problems have been reported for other engines or catalytic processes.[79,80] There are various methods for removing siloxanes through adsorption, absorption, cooling (cryogenic condensation), biological, and catalytic methods, some of them being commercialized.[74,79] A detailed description of biogas cleaning methods is beyond the scope of this review. The interested reader is referred to the relevant literature.

Among reforming processes, CO_2 reforming or dry reforming of methane (DRM) [Eq. (7), Table 9.1] is more suitable for biogas, as both major constituents are incorporated in the final product. However, the process faces difficulties due to the fact that it is energy demanding, as it is thermodynamically favored at high temperatures (above 973 K), whereas catalysts may suffer from sintering and from the formation of carbonaceous deposits. The latter constitutes the most important challenge of the process.

Reaction Mechanism

The first step in the DRM reaction sequence is the adsorption of methane. At low temperatures, the adsorption is precursor mediated, whereas at higher temperatures, it is direct.[81] It has been suggested that CH_4 adsorbs reversibly on the surface of transition metals, arriving at equilibrium. This conclusion is derived from results of steady-state isotopic transient kinetic analysis, which detected methane on the surface of various Ni- and Pd-based catalysts under reaction conditions.[82,83] There is general agreement that one of the slow reaction steps in the reaction sequence is the cracking of methane on the metal surface, as the dissociation energy of the CH_3–H(g) bond is high (439.3 kJ/mol). However, the total dissociation energy of the bond CH_x–H depends on the hosting surface and the entire catalytic system which may be controlling the surface metal work function. Consequently, lower CH_x–H bond dissociation energies are required in catalyzed decomposition. Nevertheless, for many catalytic systems, methane decomposition is considered as the rate-determining step (RDS).[81,84,85]

Attempts have been made to define the number of hydrogen atoms in CH_x species adsorbed on transition metal surfaces. Such species have been detected using both steady-state and transient isotopic tracing experiments, the latter being more representative of phenomena on working metal surfaces.[81,85] The dissociation of CH_4 is thought to yield a distribution of CH_x species with x depending on the metal and the support.[81]

Like methane, the dissociation and reduction of CO_2 are also reported to be structure sensitive, promoted at defect sites, such as corner atoms.[81] *Ab initio* valence-bond calculations and spectroscopic data indicate that CO_2 can adsorb dissociatively on various metal surfaces (Pt, Pd, Rh, Re, Ni, Fe, Cu, Ag, Al, and Mg), involving electron transfer to the CO_2 moiety, which is accompanied by an elongation of the C–O bond with respect to the free molecule.[12]

In the presence of methane, the dissociation of CO_2 is promoted, although there are no indications for the formation of any surface complexes between the two reactants.[86] The enhanced dissociation of CO_2 is attributed to the presence of surface hydrogen species and the consecutive formation of carbonyl-hydride species. The effect of the support for the dissociation of CO_2 to CO is crucial. For example, for Pd, the order of activity for the dissociation at 773 K: Pd/TiO_2 > $Pd/A1_2O_3$ > Pd/SiO_2 > Pd/MgO.[86] As the DRM activity of the catalysts, based on turnover frequencies, follows the same order. Enhanced carbon dioxide decomposition results in higher surface concentration of reactive oxygen species. As the latter are required for the activation of methane, their increase leads to enhancement of the DRM rate. As more oxygen vacancies are present on the titania surface, promoting the adsorption and the dissociation of carbon dioxide.[86]

Differences in carbon dioxide adsorption behavior related to the support have been reported for various catalytic systems. The chemisorption of carbon dioxide on the Ni/TiO_2 catalyst occurred with a heat of adsorption in the order of 1 kcal/mol, indicative of weak adsorption. On Ru/SiO_2, both CH_4 and CO_2 are activated on the metallic phase, whereas a bifunctional mechanism is proposed for Ru/Al_2O_3.[87] Although Ru is able to activate and dissociate CO_2, in the presence of a support such as alumina, a bifunctional mechanism takes place. While methane adsorbs on ruthenium, an alternative and more effective path for CO_2 activation is followed comprising the formation of HCO_2^- on the alumina surface and its decomposition to CO and hydroxyl groups on the support. The latter diffuse toward the metal particles where oxidation of the carbonaceous adspecies, located on the metallic surface, takes place.[87]

An important category of oxide materials, which are used as catalyst components for the DRM, are the oxides of rare earths, particularly ceria and lanthana. There are good reasons for that, one being the high activity of these materials for the adsorption and activation of CO_2.[82,85,88] A mechanism of interaction between M/CeO_2 (M = Rh, Ru, Pt, Pd, and Ir) and CO_2 influenced by reduction temperature.[88] By increasing reduction temperature, a progressive reduction of bulk CeO_2 takes place, which is not promoted by the presence of the metal. CO_2 adsorption and activation takes place on a surface Ce^{3+} site with formation of CO, whereas Ce^{3+} is oxidized to Ce^{4+}. Oxygen vacancies act as 'additional driving force' for the reduction of CO_2 to CO.[88] Adsorption and dissociation of the CO_2 molecule has been observed to be faster over Ni/La_2O_3 than over Ni/Al_2O_3. Over Ni/La_2O_3, the CO_2 molecule interacts with the carrier to form $La_2O_2CO_3$ species, which decompose to produce CO and oxygen species, important for the DRM reaction mechanism. The much higher affinity of Ni/La_2O_3 catalyst for CO_2 chemisorption could be attributed to the higher basicity of lanthana as compared with that of alumina.[82]

Primary elementary steps of the DRM reaction are the adsorption and decomposition of CH_4 on active sites on the metallic surface, forming hydrogen and methyl-like adsorbed species and the dissociative adsorption of CO_2 on the metal oxide surface, preferably on the metal–support interface, forming CO and oxygen adsorbed species

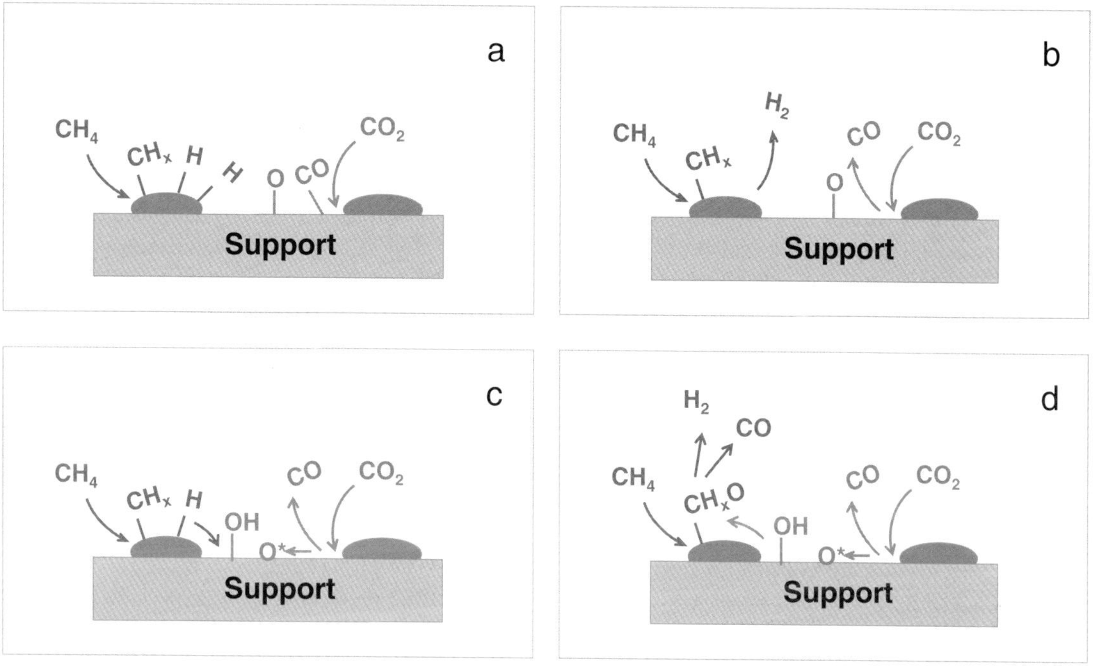

Figure 9.4 Reaction steps for the dry reforming of methane. (a) Adsorption and dissociation of CH_4 and CO_2 on the metal and the metal–support interface, respectively. (b) CO and H_2 desorption are fast steps. (c) Surface hydroxyls are formed from hydrogen and oxygen spillover. (d) Surface oxygen species or hydroxyls oxidize the hydrogen depleted surface methyl-like species ($*CH_x$), forming $*CH_xO$ species and finally CO and H_2. (Reproduced from Ref 12. Copyright 2012, Springer.)

(Figure 9.4a). Once methane and carbon dioxide are adsorbed, many surface reactions occur, leading to desired or undesired products. Most of the reaction steps are fast and arrive at equilibrium, e.g. desorption of CO from the support and of hydrogen from the metallic surface (Figure 9.4b). Kinetic investigations of DRM have shown that the WGS reaction is near equilibrium over a wide range of temperatures. The quasi-equilibrium of the WGS reaction implies that the surface reaction steps related to this reaction are fast. Most kinetic models predict hydrogen spillover from the metal surface to the support where it reacts with oxygen species forming hydroxyl groups (Figure 9.4c), while oxygen spillover from the support to the metal also occurs (Figure 9.4d). However, at temperatures higher than 1073 K, the presence of hydroxyl groups on the support is not likely. Oxygen migrating on the metal surface reacts with hydrogen depleted S_1–CH_x species ($0 \leq x \leq 3$) forming S_1–CH_xO species (Figure 9.4d). It has been suggested that H_2O, produced on the support and migrating to the metal–support interfacial region, participates in the formation of CH_xO. The formation and/or the decomposition of S_1–CH_xO species to CO and H_2 are considered as RDS.[82,84] What seems to be crucial is the relative rate of oxidation of S_1–CH_x species as compared to its dissociation: higher rates of oxidation means $x > 0$ in S_1–CH_xO species, whereas higher S_1–CH_x decomposition rate leads to the complete decomposition of the S_1–CH_x species forming surface carbon (S_1–C). If the rate of oxidation of S_1–C is not fast enough, carbonaceous species accumulate, leading to catalyst deactivation.

Deactivation Due To Coking

As in many hydrocarbon reforming processes, the most important mechanism of deactivation is the formation of carbon-rich deposits, known as carbon or coke. In general, amorphous and graphitic carbon islands are initially formed, whereas carbon accumulation results either in the encapsulation of the metallic phase or in the growth of filamentous, whisker type. The nature, structure, and rate of formation depend on various parameters such as catalyst characteristics (nature of the metal and the support, composition, texture, and structure), feed composition, reaction conditions, and environment.

The mechanism of formation of carbonaceous deposits is believed to include coke formation and coke gasification reactions proceeding through several different routes. Net accumulation of coke depends on the difference of the rates of the deposition–removal reactions. In DRM, carbon originates from methane decomposition [Eq. (11)], carbon monoxide disproportionation [Eq. (10)] and carbon condensation reactions and there is no thermodynamic hindrance for any of these reactions. CH_4 decomposition is dominating at temperatures higher than 973 K and the Boudouard reaction at temperatures lower than 873 K. Both reactions proceed on the metal surface and the extent of the contribution of each on carbon accumulation depends on operating conditions and catalyst composition. At temperatures higher than 1050 K, the amount of carbon due to the Boudouard reaction is expected to be very low as compared with that originating from methane decomposition.

Nickel catalysts are prone to carbon formation. As already described, methane is known to adsorb dissociatively on nickel nanocluster surfaces, releasing hydrogen and forming S_1–CH_x surface species. If reactive oxygen and/or hydroxyl surface species are available to react with the hydrogen-depleted species, S_1–CH_xO species are rapidly formed, eventually producing CO and H_2 (Figure 9.4d). In the opposite case, S_1–CH_x undergoes further decomposition to produce carbon atoms and/or groups of atoms. As carbon has a high solubility in the nickel lattice, carbon atoms can either remain on the surface or diffuse in the subsurface toward carbon growth centers, such as grain boundaries.[89] Probably coke accumulation proceeds through the formation of carbides as intermediates, which decompose to give free carbon, which can migrate through the nickel lattice. These carbonaceous species are incorporated into the graphene layer, which forms at the nickel–support boundaries, and may encapsulate the particles or detach them from the support and grow filamentous carbon (whiskers), nanofiber and/or nanotubes of carbon with nickel particles at their top edge.[89] In many Ni-based systems, the formation of filamentous carbon predominates.

Catalyst Studies

Composition, texture, and structure of the support, nature, dispersion, and loading of the supported metal, metal–support interactions (MSI), the presence of a second metal or an additive (modifier–promoter), the preparation method and thermal pretreatment procedures as well as reaction conditions, are important factors affecting catalytic properties.

Noble Metal Catalysts

Catalysts based on noble metals are very active for DRM, without significant coke formation.[85,90–93] Activity and resistance to carbon deposition largely depend on the metal, but also on the nature of the support.[83,85,90–96] An early study by Erdohelyi and coworkers[90] on Pt metals supported on alumina showed that the specific activities in terms of turnover numbers decreased in the order Ru, Pd, Rh, Pt, Ir. Among transition metals, Co, Ni, Ru, Rh, Ir, and Pt supported on either silica or γ-alumina, activity at 723 K, expressed as TOF, followed the order: Rh > Ni > Ir > Pt, Ru > Co for the alumina supported catalysts and: Ni > Ru > Rh, Ir for the silica supported series (Figure 9.5).[91] According to the results of Rostrup-Nielsen and Bak Hansen,[92] for magnesia supported catalysts, the sequence of DRM activity follows the order Ru > Ni, Rh $\gg$ Ir, Pt > Pd. It can be inferred that the support exerts a significant influence on TOF of a given metal (Figure 9.5), but also on the reaction mechanism and on catalyst stability under reaction conditions. The rate of carbon generation and its migration toward the support are factors affecting activity and stability of a given catalyst.

Supported Rh catalysts have been extensively investigated under DRM conditions.[90–96] Results show that the specific activity of Rh strongly depends on the carrier employed to disperse the metal. Metal particle size is important for the initial intrinsic activity as well as the rate of deactivation, both decreasing with increasing Rh particle size. However, the degree of these dependences seems to be affected by the nature of the carrier, suggesting different MSI. Mechanistic studies indicate that the origin of carbon formed on Rh/Al_2O_3 catalyst is mainly the disproportionation of CO, whereas the contribution of CH_4 is small.[93]

Ni-Based Catalysts

Ni-based catalysts are very active for reforming reactions, however, they are more prone to carbon formation. As the adsorptive dissociation of methane, occurring on the metal surface, is one of the rate-limiting steps, obviously the support must offer maximum metal dispersion, ensuring, in addition, stability under reaction conditions and avoidance of sintering.

Surface acidity–basicity is a major factor influencing activity and carbon formation. Increased Lewis basicity of the support leads to increased adsorption of CO_2, which produces surface species that react with carbon to form CO. The role of MSI on the performance and stability of Ni catalysts is crucial. Alumina is one of the most common supports. Ni/Al_2O_3 is considered as the state-of-the-art catalyst and many articles have been devoted to the elucidation of the reaction mechanism and the effective suppression of carbon formation on this catalytic system. Its acidity–basicity depends on the preparation method and the impurities it contains. Its CO_2 adsorption capacity is moderate. Crystallographic forms, preparation procedure,

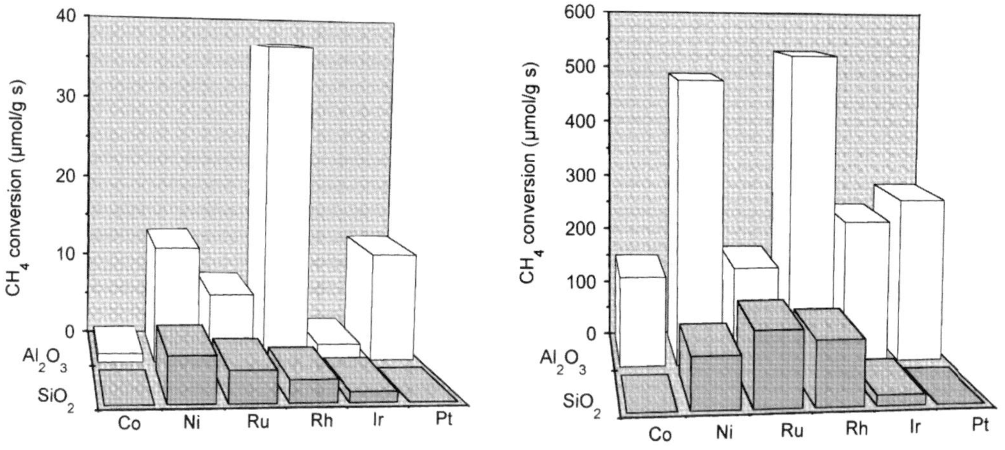

Figure 9.5 Methane conversion for transition metal catalysts supported on either silica or alumina [reacting mixture: CH_4:CO_2:He (10:10:80); total flow rate: 100 mL/min]: (a) at 723 K, TOS 45 min; (b) at 1023 K TOS 5 min. (Reproduced from Ref 50. Copyright 1996, Elsevier.)

and Ni loading play an important role on catalytic performance. High surface area aluminas with low Ni loadings (e.g., 5–10 wt% Ni) give satisfactory results, provided that good Ni dispersion is achieved by the preparation and the thermal treatment procedure.[97] The thermal treatment procedure is an important factor for the formation of $NiAl_2O_4$, the degree of reduction and thus the catalytic behavior and stability of these catalysts.[98] Reduction of Ni/Al_2O_3 is a prerequisite for an active catalyst.

MgO is a unique carrier for Ni catalysts, having multiple beneficial effects. MgO has the same crystal structure as NiO and Ni can diffuse in its lattice creating NiO–MgO solid solutions with strong interactions between the two phases.[85] As a result, NiO is well dispersed and can only be partially reduced, forming very small nickel crystallites on the surface of the solid solution.[85] Therefore, metal sintering and carbon deposition are hindered. In addition, magnesia is an alkaline earth oxide of high surface basicity and thus high affinity to adsorb CO_2, contributing further to the suppression of carbon deposition. NiO–MgO catalysts have been the subject of many studies that have shown that active and relatively stable catalysts can be based on this solid solution, provided that they are suitably prepared.[99] Impregnation of MgO with an aqueous Ni^{2+} solution seems the most appropriate preparation method, whereas coprecipitation is unsuitable.[85,100] The textural and crystallographic properties of MgO, that is, surface area, pore size distribution, and lattice parameters may affect notably the performance of the final Ni/MgO catalyst.

Materials such as rare earth oxides, which can enhance the availability of $O_{(ads)}$/$OH_{(ads)}$ species are expected to contribute significantly to the efficiency of a DRM catalyst. Ceria is known for its high oxygen storage/transport capacity, that is, its ability to release oxygen under oxygen poor environments and quickly reoxidize under oxygen rich environments. In redox reactions, which utilize oxygen, ceria is expected to play a key role as it can supply oxygen originating from its lattice and replace it through the dissociation of carbon dioxide on its surface, in other words a Mars Van Krevelen mechanism.[101] Under dry reforming conditions, the catalytic behavior of ceria is the outcome of a complicated series of surface reactions. Ceria can be reduced by H_2 and CO, the latter being even better reducing agent than H_2. CO_2 adsorption and activation takes place on Ce^{3+} surface sites with formation of CO and simultaneous oxidation of Ce^{3+} to Ce^{4+}, whereas bulk oxygen vacancies act as 'additional driving force' for the reduction of CO_2 to CO.[88] In addition, ceria can chemisorb large amounts of H_2 and CO, the uptake being also dependent on its physicochemical properties as well as thermal treatment conditions. The extent of ceria reduction is expected to affect its reactivity; CO_2 adsorption and activation depends on the electron donor ability of the surface, consequently the more reduced is ceria, the higher its electron donor ability. Lanthana has also been used as support for DRM catalysts.[82,84,87,96] Results have shown that no matter which is the active metal, Ni or Rh, lanthanum oxycarbonate ($La_2O_2CO_3$), formed by the interaction of La_2O_3 with CO_2 plays a vital role in the reaction mechanism.

Being more robust and cost effective and thus suitable for industrial applications, Ni-based catalysts have been also tested using feeds simulating biogas. One parameter that might play an important role, when using biogas feedstock, is the presence of hydrogen sulfide. There are a few studies regarding this issue that have given interesting and unexpected results. Fidalgo et al.[102] have tested the catalytic performance of a commercial activated carbon, 5% Ni/Al_2O_3 and a mixture of them for the CO_2 reforming of methane in the absence and in the presence of hydrogen sulfide. In the case of activated carbon, CH_4 and CO_2 conversions increased when H_2S (0.6 vol%) was added to the feed, about 6% higher that of methane and 20% higher that of carbon dioxide. In contrast, H_2S acts as a poison in the case of 5% Ni/Al_2O_3, decreasing CO_2 conversion from 80% to around 20%, and CH_4 conversion decreased from 70 to 8%. In the absence of H_2S, a synergistic action is demonstrated by the mechanical mixture of these two catalysts. However, no synergistic effect is observed in the presence of H_2S, the mechanical mixture exhibiting a similar to 5% Ni/Al_2O_3 behavior. This effect was also observed for the decomposition of methane. Pinilla et al.,[103] studying the catalytic decomposition of methane over a commercial carbon black, found similar results; that is, the presence of H_2S has a catalytic effect on the decomposition rate of methane, increasing considerably the hydrogen production. The effect is completely reversible upon H_2S removal from the feed. According to Muradov[104] it can be explained in terms of intermediate formation of relatively active HS• radicals, which attack methane molecules at elevated temperatures, increasing its decomposition rate, and finally, forming hydrogen and carbon. The promotional effect of H_2S for the SR of methane was also observed when using nanoscale CeO_2 and doped CeO_2 (with Gd, Y, Nb, La, and Sm). This was related to the formation of various Ce–O–S phases [i.e., $Ce(SO_4)_2$, $Ce_2(SO_4)_3$] during the reaction. It was concluded that $Ce(SO_4)_2$ can promote the OSC, the lattice oxygen mobility, and eventually the reforming activity, whereas the formation of Ce_2O_2S reduces the OSC, the lattice oxygen mobility, and the reforming rate.[105] The H_2S impact on the effectiveness and deactivation behavior of Ni-based catalysts was also investigated for the SR of biogas.[106] Ashrafi et al.[106] showed that the catalytic performance of supported nickel catalysts in the presence of H_2S strongly depends upon operating temperature, catalyst deactivation being stronger at 800°C compared with 700 or 900°C. Nevertheless, the activity of the catalyst is regained rather quickly when H_2S is removed from the feed gas at 900°C.

Bimetallic Catalysts

Beneficial effects of small amounts of precious metals such as Ru, Pt, Pd, Ir, and Rh on Ni catalysts have been reported[107] suggesting that these metals function as 'a porthole for hydrogen spillover' toward the main component catalyst, and keep the catalyst surface a reductive one. The addition of small amounts of noble metals [atomic ratio of M/(Ni+Mg) between 0.007 and 0.032%, M = Pt, Pd, or Rh] improved catalytic activity and stability. This was attributed to the formation of Pt–Ni alloy particles which favor catalyst reducibility and accelerate the dissociation of CH_4. The presence of small amount of noble metal was also proven to be beneficial for different types of Ni catalysts.[108, 109]

SR OF BIO-OIL AND GLYCEROL

Bio-oil is a free-flowing, dark red-brown, tarry liquid with an acrid and eye-irritating smell. It is a complex mixture consisting of a great number of oxygenated organic compounds, formed during pyrolysis of the three main components of lignocellulosic biomass (cellulose, hemicelluloses, and lignin), and of water. The precise chemical composition of bio-oil, as well as its physical and chemical properties, depend strongly on both the feedstock biomass and the conditions of the pyrolysis process used for its production.[110]

The major components of bio-oil and their typical content range are: water (20–30 wt%), lignin fragments (15–30 wt%), aldehydes and hydroxyaldehydes (10–20 wt%), carboxylic acids (10–15 wt%), carbohydrates (5–10 wt%), phenols (2–5 wt%), alcohols (2–5 wt%), ketones and hyroxyketones (1–5 wt%), and furfurals (1–4 wt%).[63] In this wide spectrum of oxygenated compounds acetic acid, hydroxyacetaldehyde, acetol, butanol, and levoglucosan are typically the most abundant.[111]

Bio-oil is generally considered to be a two-phase microemulsion. The continuous phase is an aqueous solution of the polar hydrophilic products of the decomposition of holocellulose (cellulose and hemicelloses) in the original biomass (such as low molar mass acids, alcohols, hydroxyaldehydes, and ketones) and of small molecules produced from the lignin decomposition, whereas the discontinuous phase is mainly formed by the water-insoluble, high molar mass, lignin fragments.[110]

SR Reactions

The SR of bio-oil is the process by which the oxygenated compounds forming bio-oil react with steam, at elevated temperatures and in the presence of a catalyst, and produce synthesis gas, according to Eq. (13) (Table 9.1). The reforming reactions of the oxygenated compounds are endothermic and their equilibrium is favored at high temperatures and low pressures. Along with these reforming reactions, the exothermic WGS [Eq. (6)] and methanation [Eq. (8)] reactions also take place at an extent

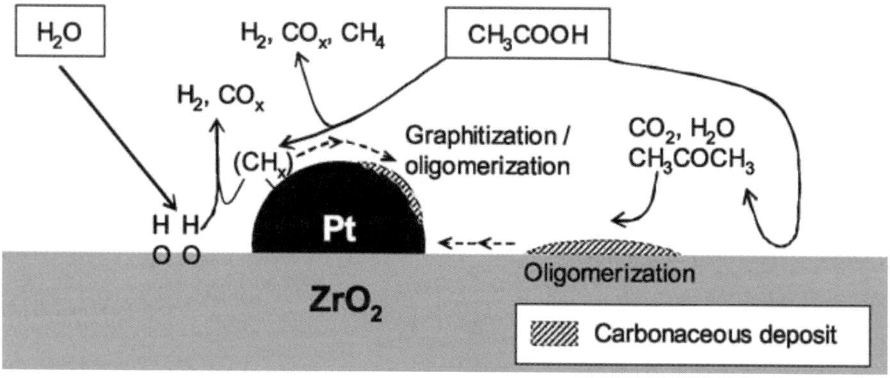

Figure 9.6 Schematic representation of the bifunctional mechanism proposed by Takanabe for the steam reforming of acetic acid over a Pt/ZrO_2 catalyst. (Reproduced from Ref 115. Copyright 2004, Elsevier.)

depending on operating conditions. At the high temperatures necessary for the reforming reaction, the thermally unstable oxygenated compounds of bio-oil also decompose forming smaller compounds and pyrolytic coke, which may encapsulate the catalyst pellets and form deposits on the reactor walls.[112,113] Besides the above reaction, coke can also be formed, as whisker carbon, on the catalyst surface by the same catalytic reactions occurring during SR of hydrocarbons, namely the CH_4 dissociation [Eq. (11)], the CO disproportionation [Boudouard reaction, Eq. (10)] and the CO reduction [Eq. (14)] reactions. The carbon whiskers are graphene tubes with high mechanical strength able, as they grow in length, to disintegrate the catalyst by destroying its pore walls and, thus, lead to increased pressure drop.

Similar to the SR of hydrocarbons, for the SR of oxygenates the steam to carbon ratio (S/C) constitutes a parameter of major importance. Increase of this ratio favors the conversion of the oxygenated compounds and hydrogen yield and, most importantly, reduces carbon deposition by favoring steam gasification of carbonaceous deposits.[114]

Mechanisms of the SR of Oxygenated Compounds

The mechanism and kinetics of the SR of simple oxygenated compounds, namely of acetic acid[112,115,116] and ethanol (Section *Reformation of Bioethanol*), have been extensively investigated. For the SR of acetic acid a bifunctional mechanism, similar to that for the reformation of methane, has been proposed (Figure 9.6) for SR over a Pt/ZrO_2 catalyst.[115] Based on this, acetic acid is dissociatively adsorbed on the metal surface forming adsorbed acetate (CH_3COO) and acyl (CH_3CO) species. These species can then decompose forming CO, CO_2, and adsorbed methyl (CH_3*) species. H_2 and CH_4 are formed by the recombination of adsorbed hydrogen atoms and the reaction of adsorbed methyl species with adsorbed hydrogen atoms, respectively. However, methyl species, or further dehydrogenated CH_x* species ($x < 3$), can also graphitize and/or oligomerize, forming deposits on the metal surface, which eventually block the active sites and inhibit subsequent adsorption and decomposition of acetic acid, unless they are removed by an oxidant. This is accomplished by the reaction of CH_x* species with OH species, formed by the dissociative adsorption of steam on the support. This reaction takes place via the metal–support interface, produces CO, CO_2, and H_2 and keeps the metal surface available for the adsorption of acetic acid.

Ethoxy, acetate, acetaldehyde, and formic acid have been proposed as intermediate species involved in the SR of ethanol (Section *Reformation of Bioethanol*). Reaction mechanisms for the SR of other molecules, such as hydroxyacetaldehyde,[112] ethylene glycol,[117] phenol,[118] and m-cresol,[119] or larger oxygenated compounds are not abundant in the literature due to the complexity induced by the large number of possible intermediate and surface species and the numerous ways by which these can react.[111]

Catalysts for SR of Bio-Oils

Both, base metals (Ni, Co, Fe, and Cu) and noble metals (Rh, Ru, Pt, and Pd) have been investigated for the SR of, mainly, model oxygenated compounds, but also of aqueous fractions of bio-oils and of whole bio-oils produced from various types of biomass. Among the noble metals tested for the SR of acetic acid,[113,115,120] the Al_2O_3-supported Rh and Ru catalysts have been reported to be much more active than similarly loaded Pd/Al_2O_3 and Pt/Al_2O_3 catalysts.[113] Noble metal catalysts, especially Rh and Pt supported on

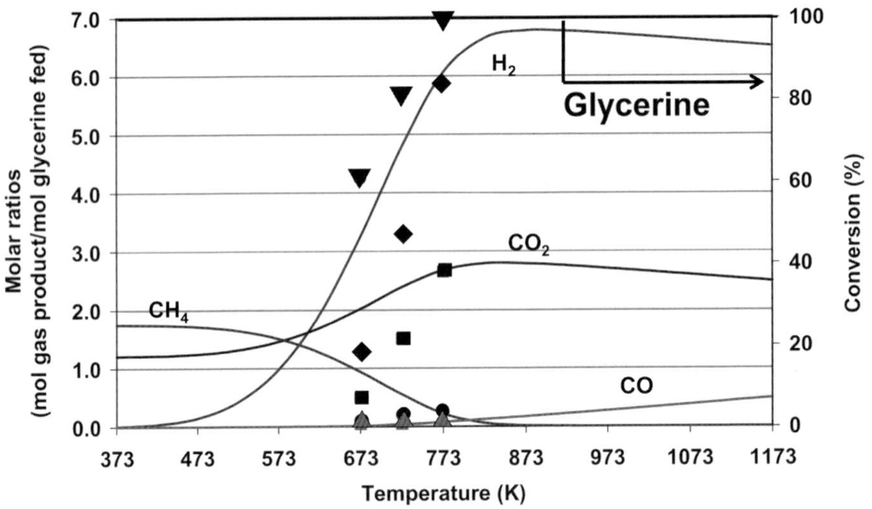

Figure 9.7 Comparison among thermodynamic data (lines) and experimental results (symbols) of molar ratios of H_2 (♦), CO_2 (■), CH_4 (•) and CO (▲), and the conversion of glycerine (▼) on 13 wt% Ni–6 wt% La_2O_3/Al_2O_3 catalyst at steady state, 0.4 MPa, different temperatures, and 1:9 glycerol/water ratio. (Reproduced from Ref 135. Copyright 2012, Elsevier.)

various single and mixed supports (Al_2O_3, MgO, CeO_2, ZrO_2, $MgAl_2O_3$, $MgCeO_2$, $MgZrCeO_2$, $MgZrO_2$, $CeZrO_2$, and $CeZrLaO_2$) have also been tested for the SR of phenol, the aqueous fractions of bio-oils and of whole bio-oils. Although high conversions and H_2 yields have been achieved, the catalysts suffer from rapid deactivation due to coke formation, despite of the high S/C ratios employed.[121,122]

Ni-based catalysts are most often used for the SR of individual oxygenated compounds (f.e. acetic acid,[113,123–125] higher oxygenates,[119,124,126] and bio-oils.[127–130] Among base metals, Ni is much more active than Fe or Cu and exhibits better catalytic performance than Co, mainly due to the higher tendency of Co-based catalysts for coke formation.[125] As compared with noble metal-based catalysts, nickel catalysts exhibit similar or even higher activity. However, they are more prone to coke formation.[113] Depending on the support, Ni loadings of 5–30 wt% have been used, with a preference for the range 10–20 wt% because of the high sintering tendency and the resulting severely enhanced coke formation observed at higher loadings.[131]

The choice of the support is of major importance regarding the activity, H_2 yield and stability for both nickel- or noble metal-based catalysts. Because of its acidic character, alumina favors the formation of carbonaceous deposits. For this reason, mixed supports composed of Al_2O_3 and basic oxides (CeO_2, La_2O_3, or MgO) are often used, and exhibit significantly improved catalytic behavior and stability for the SR of either model oxygenated compounds or bio-oils.[21,113,119,120,127] The beneficial influence of basic oxide addition to alumina, especially in the case of MgO, has been attributed to an enhanced rate of spillover of O and/or OH radicals from the magnesium aluminate spinel onto the metal surface where they participate in reforming reaction steps and facilitate coke gasification.[120] Promising results have also been obtained with mixed CeO_2–ZrO_2 supports.[121,132,133]

Carbon deposition, sintering of the metallic phase and sulfur poisoning are the main causes of catalyst deactivation in the SR of oxygenated compounds and bio-oils, as it is also the case for the reforming of hydrocarbons. Moreover, due to the thermally unstable nature of large oxygenated compounds and of bio-oils, these cannot be vaporized before the reactor without forming considerable amounts of residues, whereas rapid heating in the reactor enhances thermal decomposition and formation of pyrolytic coke. Minimization of these problems necessitates advanced feeding techniques (such as spray nozzles or the encapsulation of bio-oil in water droplets), experimental setups and reforming concepts, which have been recently reviewed.[111]

SR of Glycerol

Glycerol reforming has been intensively investigated during the last years focusing on three processes: steam, aqueous, and autothermal reforming.[1,14] Reforming of glycerol in the presence of steam/water is an endothermic reaction

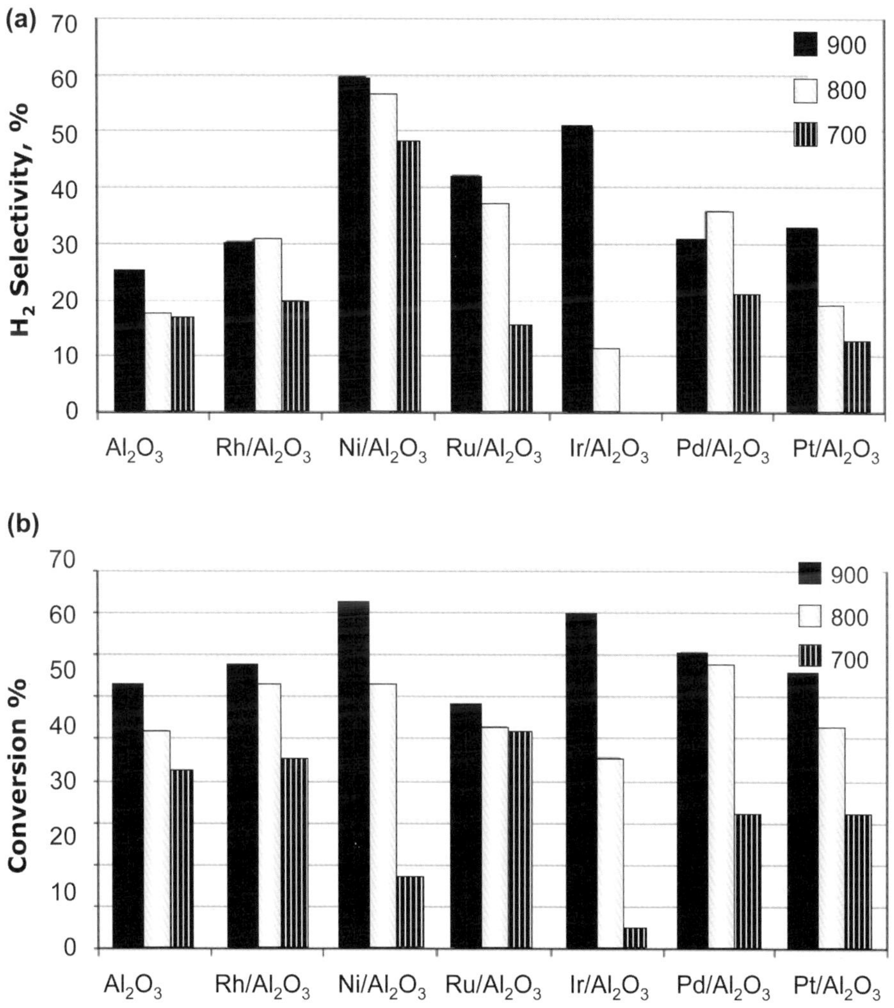

Figure 9.8 (a) H_2 selectivity and (b) glycerol conversion at selected temperatures over Al_2O_3-supported catalysts. Reaction conditions: Water/glycerol ratio = 6:1, FFR: 0.5 mL/min (GHSV = 51,000 h^{-1}), data collected after 1 h of operation. (Reproduced from Ref 139. Copyright 2004, Elsevier.)

generating CO_2 and H_2 [Eq. (15), Table 9.1] and involves complex reactions. In the absence of a catalyst, several intermediate by-products can be formed, depending on reaction conditions.[134] Thermodynamically, it is entirely feasible at temperature >800 K and a molar ratio of water to glycerol of 9:1 (Figure 9.7).[134, 135]

The SR process is a vapor phase catalytic reaction, which occurs at high temperature (e.g., 800°C) and low pressure (e.g., 0.1 MPa).[14] The main problems that the process faces are the formation of by-products, catalyst deactivation, and high energy consumption.[14] The heat required may be supplied to the system either externally, by burning some type of fuel (e.g., glycerol itself, product hydrogen, etc.), or internally. In the latter case, oxygen (air) is cofed to the reformer unit to burn a portion of the fuel and thus provide the necessary heat of reaction [oxidative SR (OSR), Eq. (16)]. OSR has certain advantages, compared with SR, including the potential use of simpler, lighter, and smaller reactor configurations, capable of rapid start-up and performance under highly variable power loads. The drawbacks of the OSR process are related to the higher volumetric flow rates and the significantly lower concentration of hydrogen in

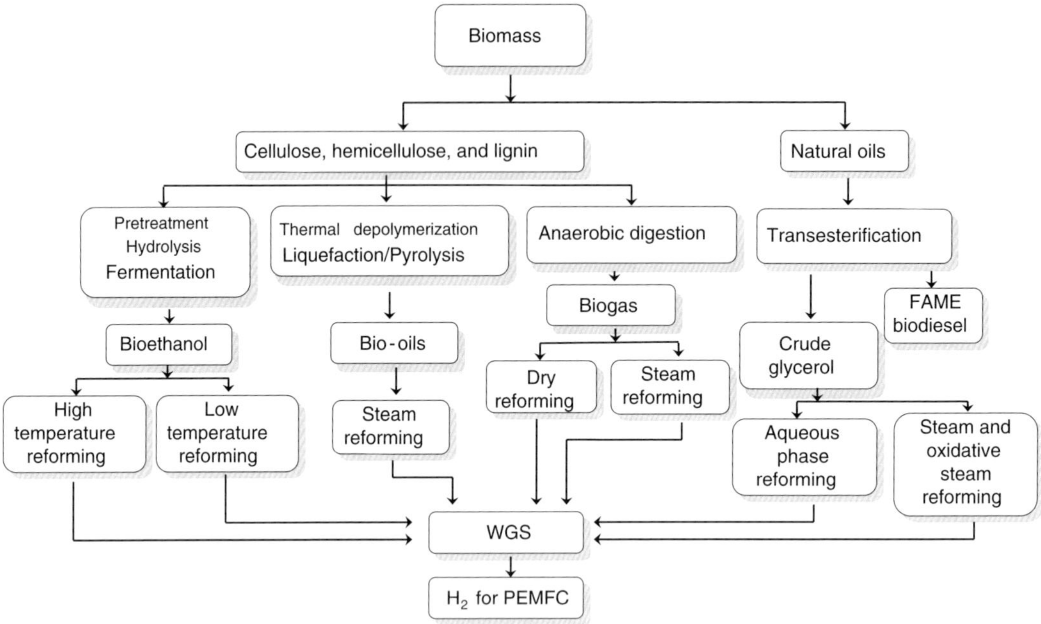

Scheme 1 Hydrogen productions paths from biomass derived energy carriers.

the reformate gas, due to the presence of nitrogen at the reactor effluent.

Reforming can also be carried out in the aqueous phase (APR) at mild temperatures (e.g., 225°C) and high pressures (e.g., 2 MPa). APR presents some advantages compared with SR as it can use a feedstock with high water content (25% glycerol in the solution), it is not so energy demanding, simpler installations are required, whereas CO amount and coke formation rate is reduced.[136] Limitations of the APR process are high pressure requirement and low H_2 selectivity due to hydrogenation reactions forming alkanes.[137]

Catalytic reaction pathways of glycerol reforming involve cleavage of C–C, C–H, O–H bonds, as well as of C–O bonds. Series and parallel reactions can lead to the formation of a variety of chemical compounds such as H_2, CO, and CO_2, alcohols and organic acids, methane, and other alkanes, and the undesired carbonaceous deposits due to the decomposition of CO, CH_4, and CO_2, or the CO and CO_2 reactions with H_2.[14,138] Depending on catalyst and operating parameters, several intermediates and byproducts may be formed. Therefore, the major challenge for efficient glycerol reforming are: (1) the kinetic control of the processes by using an effective catalyst, resulting in enhanced selectivity toward hydrogen, (2) the inhibition of catalysts coking/deactivation, and (3) reduction of energy consumption

Catalysts facilitating C–C bond cleavage are selective toward the production of synthesis gas, whereas catalysts promoting C–O bond cleavage produce various compounds including alcohols, alkenes, and carboxylic acids. Thus, it is difficult to control the formation of desired compounds from glycerol reforming and efficient catalysts must be carefully formulated in order to achieve high selectivity, stability at high temperatures and resistance to coke deposition. Catalysts investigated so far for glycerol reforming are mainly based on Ni, Co, Pt, Rh, and Ru dispersed on different supports (e.g., CeO_2, TiO_2, ZrO_2, Y_2O_3, La_2O_3, MgO, and Al_2O_3).[14,135–147] There is no common agreement regarding the most efficient catalyst, due to different operating parameters and different methods used for the calculation of glycerol conversion. Adhikari et al.[139] studying the performance of various metals supported on Al_2O_3 supports, have concluded that Ni/Al_2O_3 was the best performing in terms of hydrogen selectivity and glycerol conversion (Figure 9.8). On the other hand, Hirai et al.[147] report that the activity follows the order $Ru \approx Rh > Ni > Ir > Co > Pt > Pd > Fe$. For the APR reaction, Pt catalyst showed the best performance.[136] Nevertheless, most recent studies focus on Ni as it is active for C–C bond rupture and at the same time is more cost effective than noble metals.[14,135,140,143–146] Only few studies explored the use of crude glycerol.[142,143]

CONCLUSIONS

Biomass can be used for the production of renewable hydrogen, which is thought to be an appropriate energy carrier, addressing issues of global climate change and energy supply. In most cases, especially for distributed power generation, biomass-derived components and materials, such as biofuels, are most appropriate for this application. The most important biofuels, which can be reformed to produce hydrogen, are: bioethanol, which is produced by fermentation of sugars and starch or by enzymatic hydrolysis/fermentation of lignocellulosic materials, biogas which is produced by anaerobic digestion of mostly waste materials, bio-oil which is produced by fast pyrolysis of any type of biomass, and glycerol which is produced as by-product of biodiesel processes. These molecules can react with steam at elevated temperatures, over the surface of appropriate catalysts, to yield mostly hydrogen and carbon dioxide. Scheme 1 describes the production of hydrogen from biomass derived energy carriers through reforming processes. Knowledge of the various catalytic steps and surface transformations is essential for the development of efficient catalytic materials. The most critical problem which catalysts face is deactivation due to deposition on their surface of carbonaceous materials which originate from various reaction routes. Catalysts based on noble metals (Pt and Rh) are very active in catalyzing the reformation reactions and rather resistant to deactivation by carbon deposition. However, because of high costs associated with such materials, less rare and less expensive catalyst formulations are sought. Among many metals investigated, nickel seems to be the metal of choice, primarily due to its high reformation activity and low cost. However, Ni- based catalysts are hampered with severe carbon deposition problems and other deactivation mechanisms. Solutions to the stability issues are sought in the catalyst support which is used to disperse the metal and to its participation in reaction routes which result in combustion of the deposited carbon. For example, certain metal oxides can provide active oxygen species or other oxidizing species which enhance the rate of combustion of the deposited carbon. In general, the relative rates of carbon deposition and removal by combustion define the stability characteristics of the catalyst. Furthermore, the dispersion of the metal and the size of the metal crystallites, which can be influenced by preparation techniques, seem to be important, not only with respect to activity but also with respect to deactivation characteristics. Finally, various additives and promoters on the metal itself or on the carrier are used to enhance the ability of the catalyst to resist deactivation by carbon deposition or by sintering.

REFERENCES

1. Huber GW, Iborra S, Corma A. Synthesis of transportation fuels from biomass: chemistry, catalysts, and engineering. *Chem Rev* 2006, 106:4044–4098.
2. Balat M, Balat H. Recent trends in global production and utilization of bio-ethanol fuel. *Appl Energy* 2009, 86:2273–2282.
3. Kessler E. Our food and fuel future. Chapter 11. In: Pimentel D, ed. *Biofuels, Solar and Wind as Renewable Energy Systems-Benefits and Risks*. The Netherlands: Springer; 2008, 259–294.
4. Larson ED. *Biofuel Production Technologies: Status, Prospects and Implications for Trade and Development*. New York and Geneva: United Nations Conference on Trade and Development; 2008.
5. Anex RP, Aden A, Kazi FK, Fortman J, Swanson RM, Wright MM, Satrio JA, Brown RC, Daugaard DE, Platon A, Kothandaraman G, Hsu DD, Dutta A. Techno-economic comparison of biomass-to-transportation fuels via pyrolysis, gasification, and biochemical pathways. *Fuel* 2010, 89:S29–S35.
6. Faaij A. Modern biomass conversion technologies. *Mitig Adapt Strategies Glob Chang* 2006, 11:343–375.
7. McKendry P. Energy production from biomass (part 2): conversion technologies. *Bioresour Technol* 2002, 83:47–54.
8. Kunkes EL, Simonetti DA, West RM, Serrano-Ruiz JC, Gärtner CA, Dumesic JA. Catalytic conversion of biomass to monofunctional hydrocarbons and targeted liquid-fuel classes. *Science* 2008, 322:417–421.
9. Martins das Neves LC, Converti A, Penna TCV. Biogas production: new trends for alternative energy sources in rural and urban zones. *Chem Eng Technol* 2009, 32:1147–1153.
10. Holm-Nielsen JB, Al Seadi D, Oleskowicz-Popiel P. The future of anaerobic digestion and biogas utilization. *Bioresour Technol* 2009, 100:5478–5484.
11. Ferreira-Aparicio P, Benito MJ, Sanz JL. New trends in reforming technologies: from hydrogen industrial plants to multifuel microreformers. *Catal Rev* 2005, 47:491–588.
12. Papadopoulou C, Matralis H, Verykios XE. Utilization of biogas as a renewable carbon source: dry reforming of methane. Chapter 3. In: Guczi L, Erdohelyi A, eds. *Catalysis: Alternative Energy Generation*. New York: Springer Science+Business Media; 2012, 57–129.
13. Rapier R. Renewable diesel. Chapter 7. In: Pimentel D., ed. *Biofuels, Solar and Wind as Renewable Energy Systems-Benefits and Risks*. The Netherlands: Springer; 2008, 153–172.
14. Vaidya PD, Rodrigues AE. Glycerol reforming for hydrogen production: a review. *Chem Eng Technol* 2009, 32:1463–1469.
15. Rostrup-Nielsen JR, Bøgild Hansen J. Steam reforming for fuel cells. Chapter 4. In: Shekhawat D, Spivey JJ, Berry DA, eds. *Fuel Cells: Technologies for Fuel Processing*. Oxford: Elsevier; 2011, 62.

16. Frusteri F, Freni S, Spadaro L, Chiodo V, Bonura G, Donato S. H_2 production for MC fuel cell by steam reforming of ethanol over MgO supported Pd, Rh, Ni and Co catalysts. *Catal Commun* 2004, 5:611–615.
17. Liguras DK, Kondarides DI, Verykios XE. Production of hydrogen for fuel cells by steam reforming of ethanol over supported noble metal catalysts. *Appl Catal B* 2003, 43:345–354.
18. Fatsikostas AN, Kondarides DI, Verykios XE. Production of hydrogen for fuel cells by reformation of biomass-derived ethanol. *Catal Today* 2002, 75:145–155.
19. Yang Y, Ma JX, Wu F. Production of hydrogen by steam reforming of ethanol over a Ni/ZnO catalyst. *Int J Hydrogen Energy* 2006, 31:877–882.
20. Kugai J, Subramani V, Song C, Engelhard MH, Chin YH. Effects of nanocrystalline CeO_2 supports on the properties and performance of Ni–Rh bimetallic catalyst for oxidative steam reforming of ethanol. *J Catal* 2006, 238:430–440.
21. Aupretre F, Descorme C, Duprez D, Casanave D, Uzio D. Ethanol steam reforming over $Mg_xNi_{1-x}Al_2O_3$ spinel oxide-supported Rh catalysts. *J Catal* 2005, 233:464–477.
22. Casanovas A, Llorca J, Homs N, Fierro JL, de la Piscina PR. Ethanol reforming processes over ZnO-supported palladium catalysts: effect of alloy formation. *J Mol Catal A Chem* 2006, 250:44–9.
23. Szijjarto GP, Tompos A, Margitfavi JL. High-throughput and combinatorial development of multicomponent catalysts for ethanol steam reforming. *Appl Catal A Gen* 2011, 391:417–426.
24. Basagiannis AC, Panagiotopoulou P, Verykios XE. Low temperature steam reforming of ethanol over supported noble metal catalysts. *Top Catal* 2008, 51:2–12.
25. Klouz V, Fierro V, Denton P, Katz H, Lisse J, Bouvot-Mauduit S, Mirodatos C. Ethanol reforming for hydrogen production in a hybrid electric vehicle: process optimization. *J Power Sources* 2002, 105:26–34.
26. Panagiotopoulou P, Kondarides DI. Effect of morphological characteristics of TiO_2-supported noble metal catalysts on their activity for the water–gas shift reaction. *J Catal* 2004, 225:327–336.
27. Panagiotopoulou P, Kondarides DI, Verykios XE. Chemical reaction engineering and catalysis issues in distributed power generation systems. *Ind Eng Chem Res* 2011, 50:523–530.
28. Panagiotopoulou P, Kondarides DI, Verykios XE. Selective methanation of CO over supported Ru catalysts. *Appl Catal B* 2009, 88:470–478.
29. Sun J, Qiu XP, Wu F, Zhu WT. H_2 from steam reforming of ethanol at low temperature over Ni/Y_2O_3, Ni/La_2O_3 and Ni/Al_2O_3 catalysts for fuel-cell application. *Int J Hydrogen Energy* 2005, 30:437–445.
30. Raskó J, Dömök M, Baán K, Erdőhelyi A. FTIR and mass spectrometric study of the interaction of ethanol and ethanol–water with oxide-supported platinum catalysts. *Appl Catal A* 2006, 299:202–211.
31. Ciambelli P, Palma V, Ruggiero A. Low temperature catalytic steam reforming of ethanol. 1. The effect of the support on the activity and stability of Pt catalysts. *Appl Catal B* 2010, 96:18–27.
32. Casanovas A, Leitenburg C, Trovarelli A, Llorca J. Catalytic monoliths for ethanol steam reforming. *Catal Today* 2008, 138:187–192.
33. Koh ACW, Chen L, Leong WK, Ang TP, Johnson BFG, Khimyak T, Lin J. Ethanol steam reforming over supported ruthenium and ruthenium-platinum catalysts: comparison of organometallic clusters and inorganic salts as catalyst precursor. *Int J Hydrogen Energy* 2009, 34:5691–5703.
34. Erdőhelyi A, Raskó J, Kecskés T, Tóth M, Dömök M, Baán K. Hydrogen formation in ethanol reforming on supported noble metal catalysts. *Catal Today* 2006, 116:367–376.
35. Llorca J, Homs N, Sales J, Ramírez de la Piscina P. Efficient production of hydrogen over supported cobalt catalysts from ethanol steam reforming. *J Catal* 2002, 209:306–317.
36. Ciambelli P, Palma V, Ruggiero A, Iaquaniello G. Low temperature catalytic steam reforming of ethanol. 2. Preliminary kinetic investigation of Pt/CeO_2 catalysts. *Appl Catal B* 2010, 96:190–197.
37. Panagiotopoulou P, Verykios XE. Mechanistic aspects of the low temperature steam reforming of ethanol over supported Pt catalysts. *Int J Hydrogen Energy* 2012, 37(21):16333–16345.
38. Trimm DL, Önsan ZI. Onboard fuel conversion for hydrogen-fuel-cell-driven vehicles. *Catal Rev* 2001, 43:31–84.
39. Ghenciu AF. Review of fuel processing catalysts for hydrogen production in PEM fuel cell systems. *Curr Opin Solid State Mater Sci* 2002, 6:389–399.
40. Grenoble DC, Estadt MM, Ollis DF. The chemistry and catalysis of the water gas shift reaction: 1. The kinetics over supported metal catalysts. *J Catal* 1981, 67:90–102.
41. Barbier J Jr, Duprez D. Reactivity of steam in exhaust gas catalysis I. Steam and oxygen/steam conversions of carbon monoxide and of propane over PtRh catalysts. *Appl Catal B* 1993, 3:61–83.
42. Panagiotopoulou P, Kondarides DI. Effect of morphological characteristics of TiO_2-supported noble metal catalysts on their activity for the water–gas shift reaction. *J Catal* 2004, 225:327–336.
43. Panagiotopoulou P, Kondarides DI. Effect of the nature of the support on the catalytic performance of noble metal catalysts for the water–gas shift reaction. *Catal Today* 2006, 112:49–52.
44. Sato Y, Terada K, Soma Y, Miyao T, Naito S. Marked addition effect of Re upon the water gas shift reaction over TiO_2 supported Pt, Pd and Ir catalysts. *Catal Commun* 2006, 7:91–95
45. Radhakrishnan R, Willigan RR, Dardas Z, Vanderspurt TH. Water gas shift activity of noble metals supported on ceria-zirconia oxides. *AIChE J* 2006, 52:1888–1894.
46. Thinon O, Diehl F, Avenier P, Schuurman Y. Screening of bifunctional water–gas shift catalysts. *Catal Today* 2008, 137:29–35

47. Dictor R. A kinetic study of the water–gas shift reaction over $RhAl_2O_3$ catalysts. *J Catal* 1987, 106:458–463.
48. Basińska A, Kępiński L, Domka F. The effect of support on WGSR activity of ruthenium catalysts. *Appl Catal A* 1999, 183:143–153.
49. Hilaire S, Wang X, Luo T, Gorte RJ, Wagner J. A comparative study of water–gas-shift reaction over ceria supported metallic catalysts. *Appl Catal A* 2001, 215:271–278.
50. Andreeva D, Idakiev V, Tabakova T, Andreev A, Giovanoli R. Low-temperature water–gas shift reaction on Auα-Fe_2O_3 catalyst. *Appl Catal A* 1996, 134:275–283.
51. Andreeva D, Idakiev V, Tabakova T, Ilieva L, Falaras P, Bourlinos A, Travlos A. Low-temperature water–gas shift reaction over Au/CeO_2 catalysts. *Catal Today* 2002, 72:51–57.
52. Boccuzzi F, Chiorino A, Manzoli M, Andreeva D, Tabakova T, Ilieva L, Iadakiev V. Gold, silver and copper catalysts supported on TiO_2 for pure hydrogen production. *Catal Today* 2002, 75:169–175.
53. Fu Q, Kudriavtseva S, Saltsburg H, Flytzani-Stephanopoulos M. Gold–ceria catalysts for low-temperature water–gas shift reaction. *Chem Eng J* 2003, 93:41–53.
54. Erdőhelyi A, Fodor K, Suru G. Reaction of carbon monoxide with water on supported iridium catalysts. *Appl Catal A* 1996, 139:131–147.
55. Li Y, Fu Q, Flytzani-Stephanopoulos M. Low-temperature water–gas shift reaction over Cu- and Ni-loaded cerium oxide catalysts. *Appl Catal B* 2000, 27:179–191.
56. Wang X, Gorte RJ, Wagner JP. Deactivation mechanisms for Pd/ceria during the water–gas-shift reaction. *J Catal* 2002, 212:225–230.
57. Iida H, Kondo K, Igarashi A. Effect of Pt precursors on catalytic activity of Pt/TiO_2 (rutile) for water gas shift reaction at low-temperature. *Catal Commun* 2006, 7:240–244.
58. Iida H, Igarashi A. Characterization of a Pt/TiO_2 (rutile) catalyst for water gas shift reaction at low-temperature. *Appl Catal A* 2006, 298:152–160.
59. Panagiotopoulou P, Kondarides DI. A comparative study of the water–gas shift activity of Pt catalysts supported on single (MO_x) and composite (MO_x/Al_2O_3, MO_x/TiO_2) metal oxide carriers. *Catal Today* 2007, 127:319–329.
60. Sandoval A, Gomez-Cortes A, Zanella R, Dıaz, G, Saniger JM. Gold nanoparticles: support effects for the WGS reaction. *J Mol Catal A* 2007, 278:200–208.
61. Panagiotopoulou P, Kondarides DI. Effects of alkali additives on the physicochemical characteristics and chemisorptive properties of Pt/TiO_2 catalysts. *J Catal* 2008, 260:141–149.
62. Panagiotopoulou P, Kondarides DI. Effects of alkali-promotion of TiO_2 on the chemisorptive properties and water–gas shift activity of supported noble metal catalysts. *J Catal* 2009, 267:57–66.
63. Panagiotopoulou P, Kondarides DI. Effects of promotion of TiO_2 with alkaline earth metals on the chemisorptive properties and water–gas shift activity of supported platinum catalysts. *Appl Catal B* 2011, 101:738–746.
64. Hagemeyer A, Carhart RE, Yaccato K, Strasser P, Herrmann M, Grasselli RK, Brooks CJ, Phillips CB. Catalyst formulations containing Groups 11 metals for hydrogen generation. U.S. Patent 7,179,442; 2007.
65. Evin HN, Jacobs G, Martinez JR, Thomas GA, Davis BH. Low temperature water gas shift: alkali doping to facilitate C–H bond cleaving over Pt-Ceria catalysts—an optimization problem. *Catal Lett* 2008, 120:166–178.
66. Dömök M, Tóth M, Raskó J, Erdőhelyi A. Adsorption and reactions of ethanol and ethanol–water mixture on alumina-supported Pt catalysts. *Appl Catal B* 2007, 69:262–272.
67. Sanchez-Sanchez MC, Navarro Yerga RM, Kondarides DI, Verykios XE, Fierro JLG. Mechanistic aspects of the ethanol steam reforming reaction for hydrogen production on Pt, Ni, and PtNi catalysts supported on γ-Al_2O_3. *J Phys Chem A* 2010, 114:3873–3882.
68. Fatsikostas AN, Verykios XE. Reaction network of steam reforming of ethanol over Ni-based catalysts. *J Catal* 2004, 225:439–452.
69. Lima SM, Silva AM, Costa LOO, Graham UM, Jacobs G, Davis BH, Mattos LV, Noronha FB. Study of catalyst deactivation and reaction mechanism of steam reforming, partial oxidation, and oxidative steam reforming of ethanol over Co/CeO_2 catalyst. *J Catal* 2009, 268:268–281.
70. Mattos LV, Noronha FB. Hydrogen production for fuel cell applications by ethanol partial oxidation on Pt/CeO_2 catalysts: the effect of the reaction conditions and reaction mechanism. *J Catal* 2005, 233:453–463.
71. Mattos LV, Jacobs G, Davis BH, Noronha FB. Production of hydrogen from ethanol: review of reaction mechanism and catalyst deactivation. *Chem Rev* 2012, 112:4094–4123.
72. Rasi S, Veijanen A, Rintala J. Trace compounds of biogas from different biogas pro-duction plants. *Energy* 2007, 32:1375–1380.
73. Rasi S, Lehtinen J, Rintala J. Determination of organic silicon compounds in biogas from wastewater treatments plants, landfills, and co-digestion plants. *Renew Energy* 2010, 35:2666–2673.
74. Abatzoglou N, Boivin S. A review of biogas purification processes. *Biofuels Bioprod Bioref* 2009, 3:42–71.
75. Grierson J, Allen M. Process and apparatus for removal of H_2S from gaseous streams using membrane purification of recycle solution, Assignee: The Dow Chemical Company. PCT Patent, WO 92/17401. International Publication date: October 15, 1992.
76. The SulFerox® process:. Available at: http://www.gazintegral.com/sulferox/. (Accessed September 26, 2013).
77. Abatzoglou N, Truong LVA. Scale-up of a reactive adsorption process for the purification of biogas produced by psychrophilic anaerobic fermentation of swine manure biomass. In Bridgewater AV, Boocock DGB, eds. *Proceedings of Science in Thermal and Chemical Biomass Conversion*. Victoria, BC, Canada: CPL Press; 2006, 847–857.
78. Paqell-Thiopaq O&G the innovative desulphurisation technology. Available at: http://www.paqell.com/thiopaq/about-thiopaq-o-and-g/. (Accessed September 26, 2013).

79. Nam S, Namkoong W, Kang J-H, Park J-K, Lee N. Adsorption characteristics of siloxanes in landfill gas by the adsorption equilibrium test. *Waste Management* 2013, 33:2091–2098.
80. Matsui T, Imamura S. Removal of siloxane from digestion gas of sewage sludge. *Bioresour Technol* 2010, 101:S29–S32.
81. Bradford MCJ, Vannice MA. CO_2 reforming of CH_4. *Catal Rev Sci Eng* 1999, 41:1–42.
82. Tsipouriari VA, Verykios XE. Carbon and oxygen reaction pathways of CO_2 reforming of methane over Ni/La_2O_3 and Ni/Al_2O_3 catalysts studied by isotopic tracing techniques. *J Catal* 1999, 187:85–94.
83. Yamaguchi A, Iglesia E. Catalytic activation and reforming of methane on supported palladium clusters. *J Catal* 2010, 274:52–63.
84. Verykios XE. Catalytic dry reforming of natural gas for the production of chemicals and hydrogen. *Int J Hydrogen Energy* 2003, 28:1045–1063.
85. Hu YH, Ruckenstein E. Catalytic conversion of methane to synthesis gas by partial oxidation and CO_2 reforming. *Adv Catal* 2004, 48:297–345.
86. Erdohelyi A, Cserenyi J, Papp E, Solymosi F. Catalytic reaction of methane with carbon dioxide over supported palladium. *Appl Catal A: Gen* 1994, 108:205–219.
87. Ferreira-Aparicio P, Rodrıguez-Ramos I, Anderson JA, Guerrero-Ruiz A. Mechanistic aspects of the dry reforming of methane over ruthenium catalysts. *Appl Catal A: Gen* 2000, 202:183–196.
88. De Leitenburg C, Trovarelli A, Kaspar J. A temperature-programmed and transient kinetic study of CO_2 activation and methanation over CeO_2 supported noble metals. *J Catal* 1997, 166:98–107.
89. Abild-Pedersen F, Norskov JK, Rostrup-Nielsen JR, Sehested J, Helveg S. Mechanisms for catalytic carbon nanofiber growth studied by ab initio density functional theory calculations. *Phys Rev B* 2006, 73:115419–1–115419–13.
90. Solymosi F, Kutsfin G, Erdohelyi A. Catalytic reaction of CH_4 with CO_2 over alumina-supported Pt metals. *Catal Lett* 1991, 11:149–156.
91. Ferreira-Aparicio P, Guerrero-Ruiz A, Rodriguez-Ramos I. Comparative study at low and medium reaction temperatures of syngas production by methane reforming with carbon dioxide over silica and alumina supported catalysts. *Appl Catal A: Gen* 1998, 170:177–187.
92. Rostrup-Nielsen JR, Bak Hansen J-H. CO_2 reforming of methane over transition metals. *J Catal* 1993, 144:38–49.
93. Verykios XE. Mechanistic aspects of the reaction of CO_2 reforming of methane over Rh/Al_2O_3 catalyst. *Appl Catal A: Gen* 2003, 255:101–111.
94. Maestri M, Vlachos DG, Beretta A, Groppi G, Tronconi E. Steam and dry reforming of methane on Rh: microkinetic analysis and hierarchy of kinetic models. *J Catal* 2008, 259:211–222.
95. Donazzi A, Beretta A, Groppi G, Forzatti P. Catalytic partial oxidation of methane over a 4% Rh/α-Al_2O_3 catalyst Part II: role of CO_2 reforming. *J Catal* 2008, 255:259–268.
96. Munera JF, Irusta S, Cornaglia LM, Lombardo EA, Cesar DC, Schmal M. Kinetics and reaction pathway of the CO_2 reforming of methane on Rh supported on lanthanum-based solid. *J Catal* 2007, 245:25–34.
97. Al–Fatish ASA, Ibrahim AA, Fakeeha AH, Soliman MA, Siddiqui MRH, Abasaeed AE. Coke formation during CO_2 reforming of CH_4 over alumina-supported nickel catalysts. *Appl Catal A: Gen* 2009, 364:150–155.
98. Sahli N, Petit C, Roger CA, Kiennemann A, Libs S, Bettahar MM. Ni catalysts from $NiAl_2O_4$ spinel for CO_2 reforming of methane. *Catal Today* 2006, 113:187–193.
99. Rostrup-Nielsen JR, Sehested J, Norskov JK. Hydrogen and synthesis gas by steam- and CO_2 reforming. *Adv Catal* 2002, 47:65–139.
100. Djaidja A, Libs S, Kiennemann A, Barama A. Characterization and activity in dry reforming of methane on NiMg/Al and Ni/MgO catalysts. *Catal Today* 2006, 113:194–200.
101. Kambolis A, Matralis H, Trovarelli A, Papadopoulou C. Ni/CeO_2-ZrO_2 catalysts for the dry reforming of methane. *Appl Catal A: Gen* 2010, 377:16–26.
102. Fidalgo B, Muradov N, Menendez JA. Effect of H_2S on carbon-catalyzed methane decomposition and CO_2 reforming reactions. *Int J Hydrogen Energy* 2012, 37:14187–14194.
103. Pinilla JL, Suelves I, Lázaro MJ, Moliner R. Influence on hydrogen production of the minor components of natural gas during its decomposition using carbonaceous catalysts. *J Power Sources* 2009, 192:100–106.
104. Muradov N, Thermocatalytic CO_2-free production of hydrogen from hydrocarbon fuels. Proceedings of the 2002 U.S. DOE Hydrogen Program Review NREL/CP-610–32405; 2002. Available at: http://www1.eere.energy.gov/hydrogenandfuelcells/pdfs/32405a17.pdf. (Accessed September 26, 2013).
105. Laosiripojana N, Charojrochkul S, Kim-Lohsoontorn P, Assabumrungrat S. Role and advantages of H_2S in catalytic steam reforming over nanoscale CeO_2-based catalysts. *J Catal* 2010, 276:6–15.
106. Ashrafi M, Pfeifer C, Proll T, Hofbauer H. Experimental study of model biogas catalytic steam reforming: 2. Impact of sulfur on the deactivation and regeneration of Ni-based catalysts. *Energy Fuels* 2008, 22:4190–4195.
107. Inui T, Saigo K, Fujii Y, Fujioka K. Catalytic combustion of natural gas as the role of on-site heat supply in rapid catalytic CO_2-H_2O reforming of methane. *Catal Today* 1995, 26:295–302.
108. Nagaoka K, Jentys A, Lercher A. Methane autothermal reforming with and without ethane over mono- and bimetal catalysts prepared from hydrotalcite precursors. *J Catal* 2005, 229:185–196.
109. Arbag H, Yasyerli S, Yasyerli N, Dogu G. Activity and stability enhancement of Ni-MCM-41 catalysts by Rh incorporation for hydrogen from dry reforming of methane. *Int J Hydrogen Energy* 2010, 35:2296–2304.
110. Basu P. *Biomass Gasification and Pyrolysis: Practical Design and Theory*. Amsterdam: Elsevier; 2010.

111. Trane R, Dahl S, Skjøth-Rasmussen MS, Jensen AD. Catalytic steam reforming of bio-oil. *Int J Hydrogen Energy* 2012, 37:6447–6472.
112. Wang D, Montané D, Chornet E. Catalytic steam reforming of biomass-derived oxygenates: acetic acid and hydroxyacetaldehyde. *Appl Catal A: Gen* 1996, 143:245–270.
113. Basagiannis AC, Verykios XE. Catalytic steam reforming of acetic acid for hydrogen production. *Int J Hydrogen Energy* 2007, 32:3343–3355.
114. Wang D, Czernik S, Montané D, Mann M, Chornet E. Biomass to hydrogen via fast pyrolysis and catalytic steam reforming of the pyrolysis oil or its fractions. *Ind Eng Chem Res* 1997, 36:1507–1518.
115. Takanabe K, Aika K, Inazu K, Baba T, Seshan K, Lefferts L. Steam reforming of acetic acid as a biomass derived oxygenate: bifunctional pathway for hydrogen formation over Pt/ZrO_2 catalysts. *J Catal* 2006, 243:263–269.
116. Wang S, Li X, Guo L, Luo Z. Experimental research on acetic acid steam reforming over Co–Fe catalysts and subsequent density functional theory studies. *Int J Hydrogen Energy* 2012, 37:11122–11131.
117. Dauenhauer PJ, Salge JR, Schmidt LD. Renewable hydrogen by autothermal steam reforming of volatile carbohydrates. *J Catal* 2006, 244:238–247.
118. Constantinou DA, Efstathiou AM. Low-temperature purification of gas streams from phenol by steam reforming over novel supported-Rh catalysts. *Appl Catal B: Environ* 2010, 96:276–289.
119. Wu C, Liu R. Carbon deposition behavior in steam reforming of bio-oil model compound for hydrogen production. *Int J Hydrogen Energy* 2010, 35:7386–7398.
120. Basagiannis AC, Verykios XE. Influence of the carrier on steam reforming of acetic acid over Ru-based catalysts. *Appl Catal B: Environ* 2008, 82:77–88.
121. Rioche C, Kulkarni S, Meunier FC, Breen JP, Burch R. Steam reforming of model compounds and fast pyrolysis bio-oil on supported noble metal catalysts. *Appl Catal B: Environ* 2005, 61:130–139.
122. Basagiannis AC, Verykios XE. Steam reforming of the aqueous fraction of bio-oil over structured $Ru/MgO/Al_2O_3$ catalysts. *Catal Today* 2007, 127:256–264.
123. Basagiannis AC, Verykios XE. Reforming reactions of acetic acid on nickel catalysts over a wide temperature range. *Appl Catal A: Gen* 2006, 308:182–193.
124. Hu X, Lu G. Investigation of the steam reforming of a series of model compounds derived from bio-oil for hydrogen production. *Appl Catal B: Environ* 2009, 88:376–385.
125. Hu X, Lu G. Comparative study of alumina-supported transition metal catalysts for hydrogen generation by steam reforming of acetic acid. *Appl Catal B: Environ* 2010, 99:289–297.
126. Marquevich M, Czernik S, Chornet E, Montané D. Hydrogen from biomass: steam reforming of model compounds of fast-pyrolysis oil. *Energy Fuels* 1999, 13:1160–1166.
127. Garcia L, French R, Czernik S, Chornet E. Catalytic steam reforming of bio-oils for the production of hydrogen: effects of catalyst composition. *Appl Catal A: Gen* 2000, 201:225–239.
128. Kechagiopoulos PN, Voutetakis SS, Lemonidou AA, Vasalos IA. Hydrogen production via steam reforming of the aqueous phase of bio-oil in a fixed bed reactor. *Energy Fuels* 2006, 20:2155–2163
129. Czernik S, Evans R, French R. Hydrogen from biomass-production by steam reforming of biomass pyrolysis oil. *Catal Today* 2007, 129:265–268.
130. Medrano JA, Oliva M, Ruiz J, García L, Arauzo J. Hydrogen from aqueous fraction of biomass pyrolysis liquids by catalytic steam reforming in fluidized bed. *Energy* 2011, 36:2215–2224.
131. An L, Dong C, Yang Y, Zhang J, He L. The influence of Ni loading on coke formation in steam reforming of acetic acid. *Renew Energy* 2011, 36:930–935.
132. Yan C-F, Cheng F-F, Hu R-R. Hydrogen production from catalytic steam reforming of bio-oil aqueous fraction over $Ni/CeO_2–ZrO_2$ catalysts. *Int J Hydrogen Energy* 2010, 35:11693–11699.
133. Güell BM, Babich IV, Lefferts L, Seshan K. Steam reforming of phenol over Ni-based catalysts – a comparative study. *Appl Catal B: Environ* 2011, 106:280–286.
134. Adhikari S, Fernando S, Gwaltney SR, To SDF, Bricka RM, Steele PH, Harytano AH. A thermodynamic analysis of hydrogen production by steam reforming of glycerol. *Int J Hydrogen Energy* 2007, 32:2875.
135. Iriondo A, Barrio VL, El Doukkali M, Cambra JF, M Güemez MB, Requies J, Arias PL, Sánchez-Sánchez MC, Navarro RM, Fierro JLG. Biohydrogen production by gas phase reforming of glycerine and 140ethanol mixtures. *Int J Hydrogen Energy* 2012, 37:2028–2036.
136. Luo N, Fu X, Cao F, Xiao T, Edwards PP. Glycerol aqueous phase reforming for hydrogen generation over Pt catalyst—effect of catalyst composition and reaction conditions. *Fuel* 2008, 87:3483–3489.
137. Huber GW, Cortright RC, Dumesic JA. Renewable alkanes by aqueous-phase reforming of biomass-derived oxygenates. *Angew Chem Int Ed* 2004, 43:1549–1551.
138. Slinn M, Kendall K, Mallon C, Andrews J. Steam reforming of biodiesel by-product to make renewable hydrogen. *Bioresour Technol* 2008, 99:5851–5858.
139. Adhikari S, Fernando S, Haryanto A. Production of hydrogen by steam reforming of glycerine over alumina-supported metal catalysts. *Catal Today* 2007, 129:355–364.
140. Ebshish A, Yaakoba Z, Narayanan B, Bshish A, Daud WRW. Steam reforming of glycerol over ni supported alumina xerogel for hydrogen production. *Energy Procedia* 2012, 18:552–559.
141. El Doukkali M, Iriondo A, Arias PL, Cambra JF, Gandarias I, Barrio VL. Bioethanol/glycerol mixture steam reforming over Pt and PtNi supported on lanthana or ceria doped alumina catalysts. *Int J Hydrogen Energy* 2012, 37:829–839.
142. Czernik S, French R, Feik C, Chornet E. Hydrogen by catalytic steam reforming of liquid byproducts from biomass thermoconversion processes. *Ind Eng Chem Res* 2002, 41:4209–4215.

143. Dou B, Rickett G, Dupont V, Blakeman N, Williams PT, Chen H, Ding Y, Ghadiri M. Hydrogen production by sorption-enhanced steam reforming of glycerol. *Bioresour Technol* 2010, 101:2436–2442.
144. Profeti LPR, Ticianelli EA, Assaf EM. Production of hydrogen via steam reforming of biofuels on $Ni/CeO_2–Al_2O_3$ catalysts promoted by noble metals. *Int J Hydrogen Energy* 2009, 34:5049–5050.
145. Nichele V, Signoretto M, Menegazzo F, Gallo A, Dal Santo V, Cruciani G, Cerrato G. Glycerol steam reforming for hydrogen production: design of Ni supported catalysts. *Appl Catal B* 2012, 111–112:225–232.
146. Chiodo V, Freni S, Galvagno A, Mondello N, Frusteri F. Catalytic features of Rh and Ni supported catalysts in the steam reforming of glycerol to produce hydrogen. *Appl Catal A* 2010, 381:1–7.
147. Hirai T, Ikenaga N, Miyake T, Suzuki T. Production of hydrogen by steam reforming of glycerin on ruthenium catalyst. *Energy Fuels* 2005, 19:1761–1762.

Fischer–Tropsch Conversion of Biomass-Derived Synthesis Gas to Liquid Fuels

Andreas Helland Lillebø, Anders Holmen, Bjørn Christian Enger and Edd Anders Blekkan*

Department of Chemical Engineering, Norwegian University of Science and Technology (NTNU), Trondheim, Norway

INTRODUCTION

A possible link between carbon emissions and climate change, as well as diminishing nonrenewable energy resources, have motivated the development of carbon neutral technologies based on renewable feedstocks. The conversion of biomass to liquid fuels may in the near future become an important complimentary source of carbon-neutral transportation fuels compatible with present infrastructure. However, it is crucial that the biomass used for transportation fuels does not compete with the production of food and that the biomass production itself does not harm the environment.[1–3] The analysis of these questions is a complex issue. In a recent paper, Schaidle et al.[4] used an approach involving economic, environmental as well as societal criteria for the comparison of different routes to biofuels. They concluded that the Fischer–Tropsch (FT)-based route discussed in this paper is comparable to grain-based ethanol production in terms of overall sustainability, but is better in terms of societal factors, where the impact of fuel production on food prices is accounted for.

Lignocelluloses from wood or agricultural byproducts, is an attractive raw material, but the chemical composition is such that extensive processing is necessary in order to produce fuels. Key issues are molecular weight, composition (the biomass contains large amounts of oxygen, as well as trace amounts of other elements), and chemical structure.[5]

Several routes are possible for the conversion of lignocellulose to fuels. The building blocks of cellulose and hemicellulose are sugars, which after depolymerization (e.g., hydrolysis) can be fermented to fuels such as ethanol or butanols,[6] or chemically converted via so-called platform molecules.[7] The aromatic lignin structure is less easy to convert using these rather mild conditions.[8] Pyrolysis is a high temperature process where the lignocellulose is broken down to char, oil, and gaseous products. A large fraction of the stored energy can be in the liquid fraction, which needs upgrading in order to be used as a fuel.[9] A very promising way for producing biofuels is through gasification and conversion of the resulting synthesis gas ($CO + H_2$) to liquid fuels (diesel) by the Fischer–Tropsch synthesis (FTS). The technology for conversion of biomass to liquid fuels (BTL) is similar to the technology for coal to liquids (CTL) or gas to liquids (GTL). The objective of this paper is to focus on the differences and challenges specific to FTS using biomass-derived synthesis gas. In particular, biomass-derived synthesis gas can contain significant amounts of contaminants like alkali and alkaline earth species, sulfur (H_2S, COS), nitrogen (NH_3, HCN), dust, and tars.[10] The H_2:CO ratio in synthesis gas derived from biomass is typically lower than the ratio from obtained

*Correspondence to: edd.blekkan@chemeng.ntnu.no

Advances in Bioenergy: The Sustainability Challenge, First Edition. Edited by Peter D. Lund, John Byrne, Göran Berndes and Iacovos A. Vasalos.

Originally published in: *WIREs Energy Environ* 2013, 2: 131–148 doi: 10.1002/wene.69

natural gas reforming, but higher than the ratio from coal gasification. Due to the high oxygen content of biomass a significant concentration of CO_2 is usually present in the synthesis gas.[11]

The industrially relevant FTS catalysts are based on cobalt or iron as the active phase. Cobalt has the highest activity and is best suited for fuel production from natural gas, but is more expensive and more sensitive to certain poisons than iron.[12] It is possible with state-of-art technology to remove the contaminants down to acceptable levels, but the cost may be high.[13] Also, the economic risk related to a trip or failure, where a breakthrough of contaminants reaches the catalyst, is very high because of the high replacement cost for the catalyst. Thus, knowing the severity and managing this risk as well as designing catalysts which are more tolerant to contaminants will help to reduce the commercial risk for BTL. In this respect, iron catalysts are an attractive alternative. They are the conventional choice for CTL processes with a long history of large-scale production. In general, choosing and developing the optimal catalyst for liquid fuel production may require the consideration of numerous parameters including scale of operation, operating conditions, feedstock, desired product range, infrastructure at the production site, and proximity to market.

BRIEF HISTORICAL PERSPECTIVE AND MOTIVATION FOR RESEARCH ON FT-BTL

The FT process has an interesting and rich history spanning back to the beginning of the 20th century. In 1913, BASF patented a process for hydrogenation of CO to produce hydrocarbons.[14] BASF later abandoned this research and decided to focus their efforts on developing other processes. However, their patent sparked an interest in Professor Franz Fischer at the Kaiser-Wilhelm Institute in Mulheim, who sought to test their claims together with Hans Tropsch. The first pilot plant was constructed in 1932. By the end of 1939, nine FT plants with an annual capacity of 5.4 million barrels were in operation in Germany. A cobalt/thorium catalyst developed by Roelen was in use at that time. The main products were motor vehicle fuels while 28% were different chemicals ranging from alcohols to soft waxes that produced fatty acids when oxidized.[15]

The period after the 2nd World War (1950–1974) has been labeled the 'Iron Age' for FT synthesis. Iron catalysts were developed and commercialized, mainly in South Africa, for the conversion of coal to liquids; a process which still remains important later. Cobalt was later 'rediscovered' (1975–1990) and became the preferred catalyst for the conversion of natural GTL fuels.[14]

Commercial FT-GTL and CTL processes are in operation today and from this one can argue that the realization of FT-BTL plants is largely dependent on the availability of clean and cheap enough bio-derived synthesis gas. Kirkels et al.[16] investigated the annual number of publications on biomass and coal gasification from 1976 until 2009. Publication numbers for both coal and biomass spiked after the oil crisis and remained high until late 1980s. Publication numbers on coal gasification has remained stable from the early 1990s while there has been a remarkable increase in publications on biomass gasification in the same time-period. This lack in congruence in publication trends between these two closely related technologies demonstrates the increasing interest in green processes for production of electricity, fuels, and chemicals via synthesis gas.

An additional motivation for studying the FTS synthesis is the beneficial properties of the products, in particular the quality and environmental benefits of the fuel products. The environmental impact of a fuel is linked to local pollution [particulate emissions (PM), hydrocarbons, CO, NOx (NO and NO_2)] as well as the total (well-to-wheel) CO_2-emissions linked with production and use of the fuels. FT diesel has, independent of the feedstock for the production, very desirable combustion properties, leading to reduced particulate matter and NOx emissions, compared to conventional fuels.[17] The actual reductions depend on the engine technology, degree of blending of fuels, and the basis for comparison. Wang Hewu et al.[18] provide typical results from a heavy duty engine study: Using the ESC cycle (European stationary test cycle, a cycle focusing on high engine loads and high exhaust gas temperatures)[19] on a turbocharged and intercooled Cummins Euro III diesel engine, they report reductions amounting to approx. 34% in PM, 5% in NOx, and 20% on CO and total hydrocarbon emissions, respectively. Similar results have been reported by others.[17] The NOx and PM emissions can to a certain degree only be reduced as trade-offs between the two, since engine conditions that minimize PM emission will give less reduction in NOx, and vice versa. In terms of CO_2 emissions, BTL diesel from wood shows very low values (close to zero net CO_2 emissions) compared to fossil fuels, this is in spite of a larger amount of energy consumed in the production. This is because the conversion process is fuelled by the wood itself, thus not contributing to anthropogenic CO_2 emissions.[20]

DESCRIPTION OF THE FISCHER–TROPSCH PROCESS

In short, FTS is a process where a gas mixture of H_2 and CO (synthesis gas or syngas) is converted into hydrocarbons by using a transition metal catalyst, typically iron or cobalt. The hydrocarbon products are formed via a polymerization mechanism. The Anderson–Schulz–Flory (ASF) distribution is a statistical model that predicts the carbon number distribution for a polymerization process in terms of the chain growth probability, usually labeled α. The product distribution from most FTS catalysts resemble the ASF distribution, with some important exceptions. Methane and alkanes of different length are typically the main products, but some 1-alkenes and oxygenates also formed.

Gasification, the technology used to produce the synthesis gas, is in principle an incomplete combustion, sometimes combined with gasification with steam. The process is a high temperature reaction, with steam and a limited amount of air or oxygen giving a product gas containing CO, H_2, CO_2, and CH_4, and depending on the conditions, some heavier hydrocarbons and tar. Traditionally this is done in fixed-bed reactors with controlled inlet of air from the top (downdraft) or bottom (updraft) of the slowly moving bed, giving a gas (often called producer gas). The most relevant technology for chemical applications is entrained flow gasification, where small biomass particles are blown co-currently with oxygen into the reactor. The feed can also be coal particles or liquid fuel droplets. The temperature is very high, so that tar is not formed, but the inorganic part of the biomass will melt and form slag. In a fluid-bed gasifier the temperature is kept below the melting point of the ash. Several configurations are possible. The dual fluid-bed gasifier in Güssing in Austria uses air in a combustion bed where a part of the biomass is burnt to provide heat. The hot-bed material (sand and biomass) passes to a separate bed where the rest of the biomass is gasified with steam. This way the syngas is not diluted with nitrogen.

From the ASF distribution it can be seen that the only two products that can be produced directly from the FT reaction with high selectivity is methane (low α) and heavy paraffinic wax (high α). Instead of reporting a calculated α, the C_{5+} selectivity is often used to quantify the liquid product selectivity. Products containing four carbon atoms or less are gaseous at ambient pressure and temperature. These by-products are typically recycled or utilized for heat and/or power production. The build-up of small inert alkanes (and CO_2) in a recycle loop decreases efficiency and increases reactor size and costs. Because of this, it is normally desired to have as high selectivity toward heavy hydrocarbons as possible, corresponding to a high value of α. Heavy paraffinic wax can be used for production of chemicals or can easily be converted into liquids fuels (diesel) via hydrocracking. It has been reported that cobalt-based processes combined with hydrocracking can approach 80% selectivity toward diesel based on the total plant output.[21] The main overall reactions in FT synthesis are listed in Table 10.1.

Table 10.1 Main Reactions in Fischer–Tropsch Synthesis

Main reactions	
Paraffins	$nCO + (2n+1)H_2 \rightarrow C_nH_{2n+2} + nH_2O$
Olefins	$nCO + 2nH_2 \rightarrow C_nH_{2n} + nH_2O$
Water–gas shift	$CO + H_2O \rightarrow CO_2 + H_2$
Side reactions	
Oxygenates	$nCO + 2nH_2 \rightarrow C_nH_{2n+2}O + (n-1)H_2O$
Boudouard reaction	$2CO \rightarrow C + CO_2$

The FT synthesis is a complex reaction, providing a product spectrum with a wide range of properties, including high concentrations of either water or carbon dioxide. The hydrocarbon products range from methane to heavy wax, and thus product recovery and separation are important issues. In addition, the reaction is strongly exothermic with a typical heat of reaction of approximately −165 kJ/mol CO converted (the actual value depends on the product spectrum). The choice of reactor configuration and design is therefore the key to solving issues related to product flow and heat removal, such as controlling the catalyst surface temperature and intraparticle mixing of reactants and products.[22] Three different reactor types have been used commercially: fixed bed, fluid bed, or slurry bubble-column reactors. The fluid-bed reactors are two-phase (gas–solid) reactors operated at high temperatures where all the products are in gas phase. Both circulating fluid beds and bubbling fluid beds have been used, both have good heat-transfer properties. The low-temperature operation requires a reactor system capable of handling wax products in liquid phase. Slurry reactors are at the outset cheap to build and operate, and the slurry medium provides good heat transfer, but the catalyst-product separation can be a challenge. In multitubular fixed-bed reactors, the wax produced flows by gravity out of the catalyst bed. These reactors are tubular heat exchangers with simple temperature control, but requires a catalyst with a long lifetime since catalyst replacement requires stopping operation.[22]

CATALYSTS

According to Vannice et al.[23] ruthenium, iron, nickel, and cobalt are the most active metals for FT-synthesis, and the average weight of product molecules is decreasing in the following sequence: Ru > Fe > Co > Rh > Ni > Ir > Pt > Pd. Ruthenium is too expensive, leaving cobalt and iron as the only viable metals for industrial applications. The most

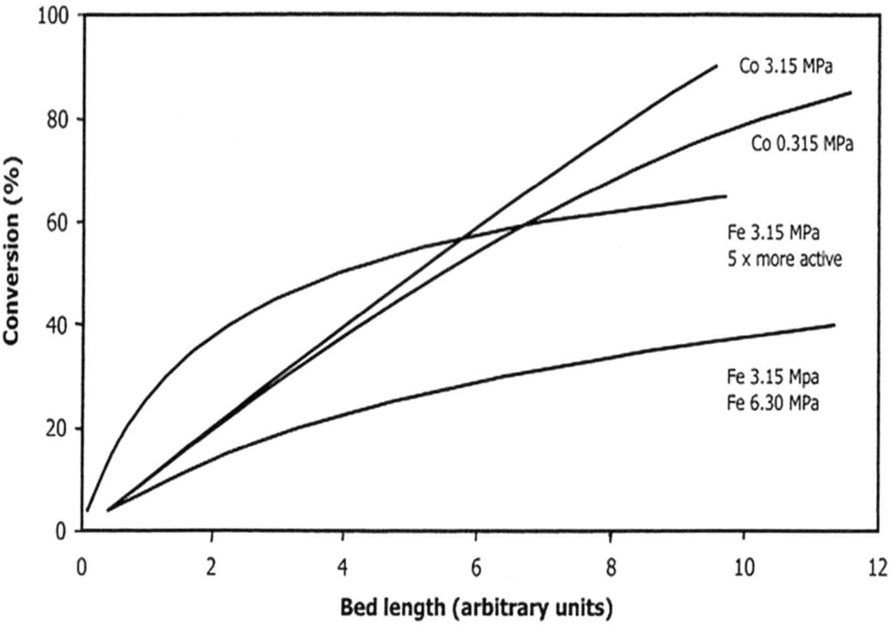

Figure 10.1 Calculated conversion profiles for low temperature Fischer–Tropsch (LTFT) operation for cobalt and iron catalysts. Reprinted with permission from Ref 27. Copyright 2002, Elsevier.

important difference between cobalt and iron catalysts is that iron has significant water gas shift activity while cobalt exhibits almost no such activity.

It is not straightforward to compare the activity of iron and cobalt catalysts. Site time yields can easily be determined for cobalt catalysts with H_2 or CO chemisorption data together with catalytic activity measurements. Active iron catalysts can contain a mixture of oxides, carbides, and metallic phases[24,25] and there is an on-going debate on the nature of the active phase and the active site in Fe-based FT catalysts.[24] Because of this it is probably more appropriate to use some other unit for comparison. According to van de Loosdrecht et al.,[26] cobalt catalysts are more active than iron on a per gram basis. In agreement with apparent activation energies on their catalysts, (~110 and ~80 kJ/mol for Co and Fe, respectively) the activity for cobalt catalysts also increases more rapidly with increasing temperatures. The catalytic activity of iron catalysts also displays a strong dependence on conversion. Dry[27] used kinetic equations for iron catalysts derived from studies at fixed and fluidized bed pilot plants at Sasol R&D[28] and the Satterfield equation[29] for cobalt catalysts to calculate conversion profiles. In Figure 10.1 the activity of the catalysts were deliberately set to be equal at 4% CO conversion except for one of the plots where the intrinsic activity of the iron catalyst was set to be five times higher. The figure shows how cobalt retains more of its activity at higher conversions and also is more dependent on pressure. The curve where the intrinsic activity of the iron catalyst was set to be five times higher illustrates how the activity would start to drop below that of cobalt catalysts at high conversions.

Cobalt Catalysts

Cobalt catalysts are usually prepared with impregnation techniques where a cobalt salt, often cobalt nitrate, is dissolved in a solvent, for example, water, and added to a highly porous high surface area SiO_2, Al_2O_3, or TiO_2 support material. Different noble metal or oxide promoters are also typically added by pre-, co- or postimpregnation in order to improve a range of properties such as reducibility, selectivity, activity, and inhibition of catalyst deactivation or to improve mechanical and attrition properties. After drying to remove water and calcination to remove residual nitrates, cobalt is present as Co_3O_4. The catalyst becomes catalytically active after reduction in pure H_2 or CO or a mixture of these gases to obtain metallic cobalt. Site time yields on cobalt catalysts have traditionally been seen as independent of support variables and dispersion.[30–32] However, it has been demonstrated how site time yield decreases dramatically for very small cobalt particles, that is, below a cobalt particle size of 6–8 nm,[33] which defines an optimum on how efficiently cobalt can be used. Readers who are interested in more information on preparation and characteristics of cobalt catalysts are referred to the review by Khodakov et al.[12]

The temperature range cobalt catalysts can be operated in is fairly narrow (200–240°C) as higher temperatures will lead to a high selectivity toward methane.[12,34] Sintering can also become a significant problem at higher temperatures as solid-state diffusion becomes faster. The Hüttig temperature of cobalt is fairly low (~250°C) and may indicate the temperature range where sintering becomes a problem.[35] The C_{5+} selectivity of cobalt catalyst (220°C, H_2/CO = ~1.9) increases significantly until ~15 bar pressure is reached and a further increase in pressure increases selectivity only moderately.[36] According to Schanke et al.,[37] catalytic activity remains more or less constant in the range of 0–60% CO conversion with varying space velocities. They also found an optimum for C_{5+} selectivity at ~70% CO conversion and that CO conversions above 80% will lead to a dramatic increase in CO_2 formation.

Cobalt only displays a negligible water gas shift (WGS) activity, thus a H_2/CO ratio slightly above 2 is required to satisfy the ratio of H_2 and CO consumed in the FT reaction. This ratio is not always attainable in biomass gasification, so in order to provide sufficient H_2 some WGS activity can be desirable. Chanenchuck et al.[38] suggested to use a mechanical mixture of a cobalt catalyst with a Cu-ZnO/Al_2O_3 WGS catalyst in a slurry FT reactor. This way it would be possible to benefit from the characteristics of cobalt catalysts in FT-synthesis while also being able to feed CO rich synthesis gas. They reported a stable FT activity for 400 h comparable to the activity with the cobalt catalyst operating alone. Lualdi et al.[39] performed a similar study but used a fixed-bed plug flow reactor in their experimental work. They reported a lower productivity of hydrocarbons per gram Co-catalyst with increasing amounts of WGS catalyst. Escalona et al.[40] added Cu and Co on the same support but found a dramatic decrease in FT activity. They speculated that formation of surface spinels or possibly blocking of surface Co sites by Cu could explain this behavior. Because of the interaction between Cu and Co, it can prove difficult to make this kind of bimetallic catalysts with both WGS and FT activity.

With increasing partial pressures of CO_2, the ratio of paraffin to olefins has been found to increase exponentially until $CO_2/(CO+CO_2) = 0.7$ where paraffins are formed almost exclusively.[41] Several researchers believe that readsorption of α-olefins to the catalyst surface for further chain growth plays an important role in the selectivity of cobalt catalysts,[42,43] and lower formation of olefins might correlate with lower C_{5+} selectivity in CO_2 rich feeds. Riedel et al.[44] found a methane selectivity at 10% with pure CO and as they increased the partial pressure of CO_2 they observed an exponential increase in methane selectivity. When pure CO_2 was fed to the reactor, the selectivity toward methane was 95%. They speculated that with lower concentrations of CO in the synthesis gas, hence also lower CO coverage on the catalyst surface, the probability of desorption of the growing chains increases. In this way CO_2 acts mainly as a diluent by replacing CO in the synthesis gas. It is well known that high H_2/CO ratios lead to a methanation regime for cobalt catalysts. This is also important for the design of the catalyst in terms of the physical parameters of supported catalyst. If diffusion limitations become significant, hydrogen diffusion in pores will be faster than CO diffusion. This could lead to CO-depleated areas in catalyst particles, with methanation, lower weight hydrocarbon products and higher paraffin/olefin ratio as the consequence.[45,46]

Iron Catalysts

Iron catalysts can be used under two different process regimes: high temperature (HTFT) or low temperature Fischer–Tropsch (LTFT). HTFT (300–350°C and 20–40 bar pressure) yields hydrocarbons in the C_1 to C_{15} range, with a large fraction of light olefins. This regime is suitable for fluid-bed reactors, which do not tolerate liquid phase products at reaction conditions. In LTFT (200–240°C and 20–45 bar pressure), the main product is heavy paraffinic wax (C_1–C_{100}), comparable to a cobalt-based process.

Conventional iron catalysts typically contain SiO_2, Cu, and potassium. The role of Cu is to improve the reducibility of iron. In hydrogen, Fe_2O_3 reduces to Fe_3O_4 and then to metallic Fe, while in CO, a mixture of $Fe_{2.5}$ and Fe_3C is formed.[47] de Smit et al.[48] reported that CuO was reduced to Cu^0 at 180°C and at this temperature reduction of Fe_3O_4 also started. They explained the promotion effect by spillover of hydrogen and/or CO from metallic Cu^0. According to Li et al.,[47] Cu increases CH_4 and paraffin selectivity while addition of K suppresses these effects. The effect of alkali promoters is complex as they increase WGS activity, FT activity, hydrocarbon selectivity, 1-alkene selectivity, and resistance against re-oxidation by water.[49,50] Iron-based FTS processes are typically operated with synthesis gas that is lean in CO_2 and with a H_2/CO ratio lower than the required stoichiometric ratio consumed by the FT reaction. The required amount of H_2 is obtained by formation of H_2 along with CO_2 at the expense of CO and H_2O through the WGS reaction. With an inlet H_2/CO ratio of about 0.7 WGS can become rate controlling at approximately 50% CO conversion, when the production of hydrogen becomes a rate limiting factor.[51,52] If the ratio of partial pressures between H_2O and CO reaches high enough levels (significantly higher than one) the catalyst can suffer severe oxidation and lose more or less all of its activity.[52] In general, the advantages of operating at conversion levels of 50% or lower is a higher activity per gram of iron catalysts (as demonstrated in Figure 10.1), avoiding excess production of H_2 and CO_2 and also a higher selectivity for CO toward hydrocarbon products.[52]

Effect of Alkali Promoters

Comparison of different alkali promoters and loadings is not straightforward as their ranking in terms of activity differs with the test conditions, especially with the CO conversion. According to Ngantsoue-Hoc et al.,[51] the activity ranking for different alkali elements is in the order Li = K > Na = un-promoted > Rb = Cs at 20% CO conversion, while at 60% CO conversion the ranking is different, K > un-promoted > Na > Rb > Cs > Li. The data were gathered with the same atomic ratio of 1.44/100 for alkali/iron on a 100Fe/4.6Si precipitated catalyst.[51] Riedel et al.[44] performed a similar series of experiments with Fe/Y-zeolite catalysts and found an activity ranking in the following order Na > K > Li = Rb > un-promoted. The catalysts were tested with the same feed flows and CO conversions were in the range of 17–22%. They also found that the heavier alkali elements increased C_{5+} selectivity the most. The high cost of Rb would most likely prohibit the use of this rare element as a promoter in industrial applications. This makes potassium the most practical choice in order to obtain both high activity and selectivity. Yang et al.[53] found 0.7 wt% potassium to be the optimal loading on an iron-manganese catalyst. At this loading, FT and WGS reaction rates as well as C_{5+} selectivity were at a maximum. It should be noted that other studies have identified different optimums for potassium loadings, most likely because of other differences in the catalyst design or in the operating conditions.

Effect of CO_2 *with Different Mechanical Promoters*

Iron-based catalysts typically contain one or several promoters for improvement of mechanical and attrition properties. However, the choice of metal oxide promoter seems to have a profound effect on the performance of the catalysts with different synthesis gas compositions. In order to study the effects of SiO_2 as a mechanical and structural promoter, Dlamini et al.[54] measured the activity with $H_2/CO = 2$. No CO_2 was present in the feed. Their SiO_2 free catalysts had higher activity than catalysts prepared by adding SiO_2 before or after precipitation and also for catalysts where SiO_2 were added after drying or calcination. Yang et al.[55] reported similar effects on iron-manganese catalysts. With increasing SiO_2 loadings, the surface area increased while catalytic activity decreased. They speculated that a lower degree of reduction and carburization with increasing SiO_2 loading was responsible for this behavior. These observations are also supported by Zhang et al.[56] who found that the degree of reduction for iron catalysts containing SiO_2 is lower than for silica-free catalysts. Wan et al.[57] found that SiO_2 suppresses H_2 adsorption and increases CO adsorption and carburization compared to Al_2O_3. Both FT and WGS activity, and also C_{5+} selectivity was higher in samples containing SiO_2 than Al_2O_3 under CO hydrogenation conditions. However, Al_2O_3 containing samples did display better stability under their experiments. From the available literature there seems to be an agreement that SiO_2 is the preferred mechanical promoter in CO_2 lean environments.

However, during CO_2 hydrogenation ($H_2/CO_2 = 3$) the activity of iron catalysts on different supports follow this order $Al_2O_3 > TiO_2 > SiO_2$.[44] In a gas composition where the volume percent of CO, CO_2, Ar, and H_2 is 11, 32, 5, and 52%, respectively, comparable to bio-derived synthesis gas,[58] it is also found that Fe/Cu/Al/K catalysts have significantly higher activity than the silica containing catalysts.[59] The difference in activity under such conditions seems to arise from a higher resistance toward oxidation and higher stability of the carbide phases with alumina as structural promoter.[44,59] One significant difference between alumina and silica supports is the ability of K containing alumina to chemisorb CO_2.[44,60] Addition of K to co-precipitated Fe-Cu-Al catalysts was also found to promote CO_2 hydrogenation activity and hydrocarbon selectivity, it also shifted the product distribution in the direction of olefins and long-chain hydrocarbons.[61] Addition of silica was reported to be unfavorable in spite of an increased surface area. This was explained by a decreased interaction between Fe and K.[62] These results indicate that Fe/K interaction could play an important role during CO_2 hydrogenation, and that conventional iron catalysts may not be appropriate for a BTL process with CO_2 rich synthesis gas.

According to Yao et al.,[41] iron catalysts follow the same trend as cobalt in regard of paraffin to olefin ratio with increasing partial pressures of CO_2. A previous study of 10 wt% Fe on TiO_2 also showed higher selectivity toward lighter hydrocarbons with increasing amounts of CO_2 in the feed.[63] Riedel et al.[44] performed FT synthesis over a 100 Fe/13 Al_2O_3/10 Cu/10 K with $H_2/CO = 2.3$ and $H_2/CO_2 = 2.3$ synthesis gas. The product distribution in terms of hydrocarbon weight was near identical at both conditions, while the olefin selectivity was slightly higher in the CO_2 containing synthesis gas. Using an intermediate $H_2/CO/CO_2$ mixture the olefin selectivity fell between the selectivity at pure CO and pure CO_2 conditions. The

Table 10.2 Analysis of Two Types of Coal and Three Types of Biomass

	Pittsburgh No. 8 coal	Eastern Kentucky coal	Red oak	Danish wheat straw	Imperial wheat straw
			Proximate (wt% as received)		
Moisture	1.14	1.63	4.76	5.41	7.99
Ash	7.90	7.43	1.15	7.33	13.49
Volatile	36.80	35.44	82.88	73.39	65.14
Fixed carbon	54.16	55.50	11.21	13.87	13.38
Heating value (HHV as received – Btu/lb)	13,691	13,351	7800	7355	6318
			Ultimate (wt% as received)		
Moisture	1.14	1.63	4.76	5.41	7.99
C	78.02	76.83	49.02	44.44	39.14
H	4.87	4.88	5.34	5.10	4.34
N	1.36	1.47	0.19	0.98	0.99
S	2.78	0.88	0.02	0.16	0.30
Ash	7.90	7.43	1.15	7.33	13.49
O	3.93	6.88	39.52	36.58	33.75
Cl	0.09	0.17	<0.01	0.60	2.00
			Elemental ash analysis (wt% of fuel as received)		
SiO_2	3.411	3.897	0.464	2.909	5.016
Al_2O_3	1.786	2.453	0.097	0.062	0.097
TiO_2	0.073	0.107	0.065	0.003	0.005
Fe_2O_3	1.318	0.340	0.092	0.058	0.063
CaO	0.385	0.080	0.155	0.586	0.495
MgO	0.066	0.030	0.012	0.136	0.177
Na_2O	0.134	0.017	0.005	0.052	2.064
K_2O	0.108	0.098	0.104	2.507	3.049
P_2O_5	0.045	0.024	0.012	0.355	0.360
SO_3	0.885	0.096	0.022	0.302	0.762

very different results obtained with different catalyst designs exemplify the importance of careful selection of an appropriate catalyst for FT-BTL processes.

IMPURITIES IN BIO-DERIVED SYNTHESIS GAS

Coal, which originates from biomass, typically contains all of the same inorganic impurities found in biomass.[64] Hence, technologies for the removal of these impurities have already been developed in connection with large-scale CTL plants. However, these plants normally operate with iron catalysts, while most BTL concepts to this date are based on cobalt catalysts. Iron and cobalt catalysts share the sensitivity toward some, but not all of the impurities commonly found in coal and biomass-derived synthesis gas. The difference between the catalysts are mainly the detrimental effect of alkali and alkaline earth elements on cobalt catalysts, while these elements often can have a beneficial or negligible effect on iron catalysts. In addition, cobalt catalysts seem to be very sensitive toward NH_3 and HCN, while iron catalysts are largely unaffected by these impurities. This raises some interesting challenges when determining sensitivity levels and synthesis gas cleaning requirements. Table 10.2 shows the content of inorganic materials in two types of coal and three types of biomass.

Biomass-derived synthesis gas can contain both organic and inorganic impurities such as tars, benzene, toluene, xylene, NH_3, HCN, H_2S, COS, HCl, volatile metals, dust, and soot.[66] Table 10.3 summarizes proposed impurity limits by different authors, though without reference to a specific catalyst or cleaning process.

Impurities in synthesis gas can be removed in wet processes such as the Rectisol process. This is essentially an absorption process for removal of acid gases (H_2S and CO_2) with methanol at low temperatures (−20 to −60°C) and high pressures (30–80 bar).[69] This process requires cooling and re-heating of the synthesis gas,[13,70] which results in exergy loss. Therefore, large efforts are now directed toward the development of high temperature synthesis gas cleaning processes based on sorbents or catalysts.[69,71]

Table 10.3 Gas cleaning requirements: target levels of major contaminants

	SASOL	Newby[67]	Vogel (Ref 6 in Ref 12)	Boerrigter[68]
CO_2	<10%	n.s.	<5 vol%	<5 vol%
Particulate	n.s.	<0.1 ppm	n.s.	<0.1 ppm
HCN	<20 ppb	<10 ppb	<20 ppb	<1 ppm
NH_3		<10 ppm		
H_2S	<10 ppb	<60 ppb	<10 ppb	<1 ppm
COS				
HCl	<10 ppb	<10 ppb	<10 ppb	<10 ppb
Br, F		n.s.	n.s.	<10 ppb
Alkalis	<10 ppb	n.s.	<10 ppb	<10 ppb
Tar	n.s.	n.s.	Below dew point	Below dew point

Note: n.s., not specified.
Reproduced from Ref 13. Copyright 2008, Elsevier.

According to a recent review paper, there are several technological challenges to overcome before hot gas-cleaning is commercially ready.[72] However, it is evident from the ratio of impurities in common feedstocks and the suggested impurity limits that extensive cleaning will be required. It is also necessary to determine tolerance levels and possible consequences in case of a process upset, which could cause impurities to enter the FT-reactor. Especially cobalt catalysts are known for high stability and Shell has indicated a lifetime of five or more years for the catalysts at their GTL plants.[73] With run times of this length it is essential to remove potential catalyst poisons down to very low concentrations.

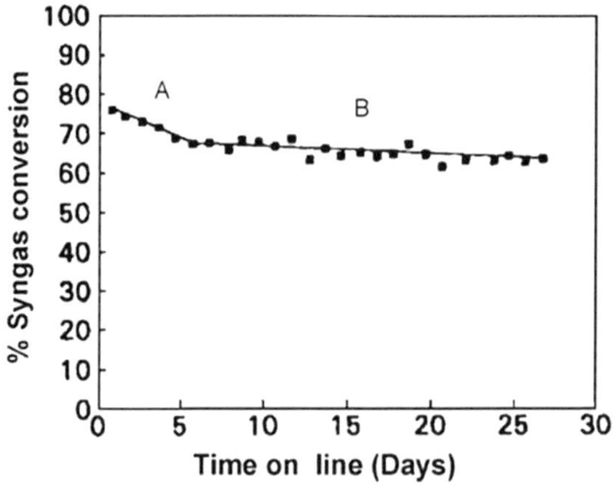

Figure 10.2 Typical deactivation profile for cobalt catalysts during Fischer–Tropsch synthesis. Reproduced with permission from Ref 36. Copyright 1997, Elsevier.

Sulfur on Cobalt

Sulfur is known as a poison to several catalysts because of its tendency to chemisorb strongly to metallic surfaces. In addition to physically blocking up to several surface atoms, it can possibly alter neighboring atoms electronically and by that affecting their ability to adsorb and/or dissociate reactant or product molecules.[74] Pansare et al.[75] investigated the effect of low concentrations of H_2S in the synthesis gas on a cobalt catalyst. During experiments lasting ~150 h they did not find any detectable deactivation if the synthesis gas contained 50 ppbV H_2S, while higher concentrations (≤300 ppbV) caused a significant and irreversible deactivation. In Figure 10.2, the deactivation of a cobalt catalyst in a slurry reactor shows two regimes.[30] The loss in activity during period A was mainly attributed to initial build-up of heavy hydrocarbon wax inside the catalyst pores, while the gradual loss of activity in period B was correlated to the sulfur content (~0.03 mgS/m^3) in the coal-derived synthesis gas. Run times exceeding 1 year with this level of sulfur in the synthesis gas may not be obtainable for cobalt catalysts unless a proper sulfur guard is used.[21]

Borg et al.[76] tested a cobalt catalyst at 20 bar pressure, 210°C, and $H_2/CO = 2.1$ in fixed-bed. After 82 h on stream, they introduced 2.5 ppmV H_2S to the synthesis gas feed. After 168 h on stream, they cut the sulfur addition and adjusted the CO conversion back to 50% before again increasing the sulfur feed rate; first to 4.8 and then to 9.5 ppmV. Figure 10.3 displays the catalytic activity during this experiment. Note that the feed flow is adjusted to obtain 50% CO conversion after 24 h on stream, which explains the coinciding shift in activity. The decline in activity when adding 2.5 ppmV sulfur was extrapolated to zero activity using a linear expression for catalyst deactivation. This was used to calculate the catalyst dispersion assuming a 1:1 stoichiometry of S on Co. This *in operando* dispersion measurement correlated well with dispersion measurements from H_2-chemisorption on a fresh catalyst. In their study, they also found an apparent decline in C_{5+}

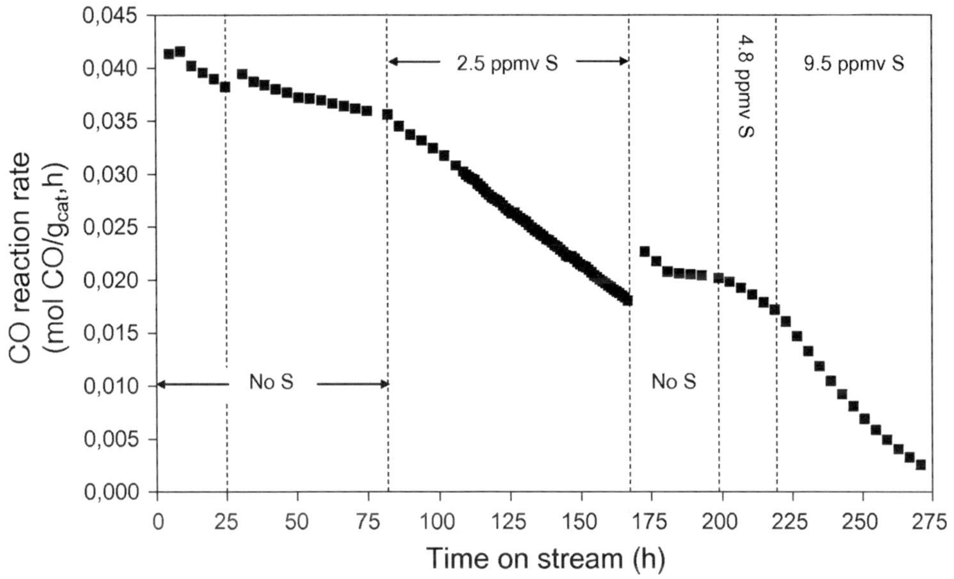

Figure 10.3 CO reaction rate versus time on stream with H_2S addition. Fischer–Tropsch synthesis conditions were 210°C, 20 bar pressure, $H_2/CO = 2.1$, and a 12 wt% Co/0.3 wt% Re/$NiAl_2O_4$ catalyst. Reproduced with permission from Ref 76. Copyright 2011, Elsevier.

selectivity with time on stream during sulfur addition. This reflects the effect of CO conversion on selectivity. When CO conversion was adjusted back to 50%, the initial selectivity was obtained. Hence, it can be concluded that sulfur itself did not influence C_{5+} selectivity to any significant degree.

Phosphorus, which is also present in biomass, seems to have a similar effect to sulfur on cobalt catalysts. Borg et al.[76] added H_2PO_4 via incipient wetness impregnation (IWI) to one of their samples and measured H_2-chemisorption and catalytic activity. The results from H_2-chemisorption and catalyst testing confirmed that the loss in catalytic activity was proportional to the loss in cobalt surface area.

Sulfur on Iron

According to Kritzinger,[77] early studies of sulfur on iron catalysts were performed at accelerated conditions, meaning high levels of sulfur. The conclusions from these studies were a strong poisoning effect and rapid loss of activity. This has led to an operating philosophy where sulfur was seen as highly undesirable and kept at as low levels as possible (lower than 5 ppb). However, new data indicate that sulfur in slightly higher concentrations can have positive effects on olefin selectivity and catalyst stability in iron HTFT processes.[77]

In a pilot plant, Duvenhage et al.[78] studied precipitated iron catalyst using fixed-bed reactors with industrially purified synthesis gas (low ppb sulfur levels). They were able to remove parts of the catalyst from different positions in the bed during the study. The samples were tested using microreactors to determine the relative activity compared to a fresh catalyst. In a plot of activity versus position in the reactor, a clear inverse V-shape appears with the maximum activity present at around a quarter of the bed height down from the top of the pilot plant reactors. Further characterization revealed high sulfur content in the top fraction of the pilot plant bed. Iron carbides were also mainly present near the top whereas iron oxide phases increased in the lower parts of the bed. The fraction of iron oxides also increased with time on stream, coinciding with loss in catalyst activity. This indicates two different deactivation regimes. Sulfur poisoning blocks active sites at the beginning of the bed, while oxidation becomes dominant further down as the H_2O/H_2 ratio increases with increasing CO conversion.

Alkali and Alkaline Earth Elements on Co and Fe

Nordheim et al.[79] calculated the thermodynamic amounts of different alkali species present in bio-derived synthesis gas after steam gasification. Their results indicate that there will be significant amounts of alkali in the synthesis gas. Potassium and sodium will mainly be present as

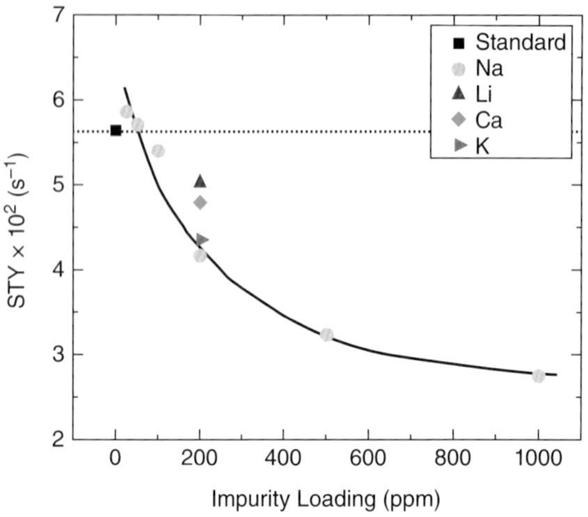

Figure 10.4 Site time yield with increasing alkali impurity loading at 50% CO conversion with a γ-alumina supported 20 wt% Co, 0.5 wt% Re catalyst. Impurity loading (ppm) actually denotes the weight fraction of impurities in the sample. Reproduced with permission from Ref 80. Copyright 2002, Springer.

hydroxides, but chlorides are favored with high Cl content in the synthesis gas. The question is then how this will affect long-term catalyst activity? Borg et al. calculated that if the synthesis gas contains 5–50 ppbw Ca and all of it sticks to the catalyst (in this case a supported Co catalyst), an average concentration of 100–1000 ppmw is obtained after 20,000 h on stream (2.5 years).[76]

Several studies show that low concentrations of alkali and alkaline earth impurities have a detrimental effect on the catalytic activity of Co-based FT catalysts.[49,76,80–82] However, C_{5+} selectivity increases moderately with increasing loadings of alkali elements and CO_2 selectivity increases slightly.[76,80] The alkaline earth elements Ca and Mg decrease C_{5+} selectivity while they have a similar effect on CO_2 selectivity as the alkali metals.[76,80]

Figure 10.4 illustrates results on a γ-alumina supported cobalt-rhenium catalyst.[80] It shows the effect on site time yield from sodium in weight fractions ranging from 0 to 1000 ppm, and 200 ppm of lithium, potassium, or calcium. Impurities in this concentration range had no measurable effect on dispersion from H_2-chemisorption and only slightly increased reduction temperatures during temperature programmed reduction. The effect on reducibility was in agreement with previous studies.[76,81] The activity data in Figure 10.4 were obtained at 50% CO conversion, 210°C, 20 bar pressure, and $H_2/CO = 2.1$.

With 200 ppm K, the number of potassium atoms per cobalt surface atom is slightly less than 2% of the number of cobalt sites. In contrast, the loss in activity was roughly 22%. Assuming that all the cobalt surface atoms contribute equally to the activity, the authors suggest that site blocking is an unlikely explanation for the loss in activity, especially when taking into consideration that some of the alkali is likely to be present on the support material. A previous study on SiO_2 supported Ru catalysts by Uner et al.[83] supports the latter conclusion.

Balonek et al.[80] attributed the effect of alkali contaminants to electronic effects, but the conclusion was not supported by direct evidence of electronic interactions. Uner[82] and Trépanier et al.[81] speculated that alkali atoms may block edge and corner sites for dissociative adsorption of hydrogen. In this way the alkali can act as modifiers on adsorption/desorption mobility of hydrogen while not interfering with the electronic structure.

The studies reported here typically added alkali *ex situ* by IWI in the form of nitrates and treated the samples in flowing air or argon at high temperature (300–350°C) to remove the nitrates before characterization and activity testing. It is unclear if addition of impurities in this manner gives a representative picture of the effect of these species under real process conditions in a FT-BTL plant. The available literature indicates that alkali and alkaline earth elements may be the impurities commonly found in biomass that have the strongest impact on catalytic activity for Co catalysts.

In contrast to the effect on Co catalysts, alkali is known to promote desirable effects in iron catalysts and potassium is commonly added as a promoter.[27,33] A recent study has also shown that *in operando* addition of 400 ppbw KCl and NaCl, and also 40 ppmw $NaHCO_3$ and $KHCO_3$ with the synthesis gas, did not have any measurable impact on catalytic activity on a precipitated iron catalyst after 400 h of operation.[84]

Ammonia and HCN

The amount of information on the effect of impurities such as NH_3 and HCN in literature is very limited. An Exxon patent[85] claims that a combined concentration of 100 ppb of NH_3 and HCN in synthesis gas will result in a catalyst half-life of only 4 days for supported cobalt catalysts in a slurry reactor. However, the patent also describes that the catalyst can be rejuvenated by hydrogen treatment to restore the initial activity. More recent work has found that the effect of NH_3 on cobalt catalysts is quantitatively similar to that of H_2S and that activity was not recovered after removing ammonia from the feed gas.[86] In the same study they found no effect of ammonia on the activity of iron catalysts. This is supported by Ma et al.,[87] who did not find any negative effect of ammonia on the catalytic activity or selectivity for iron catalysts.

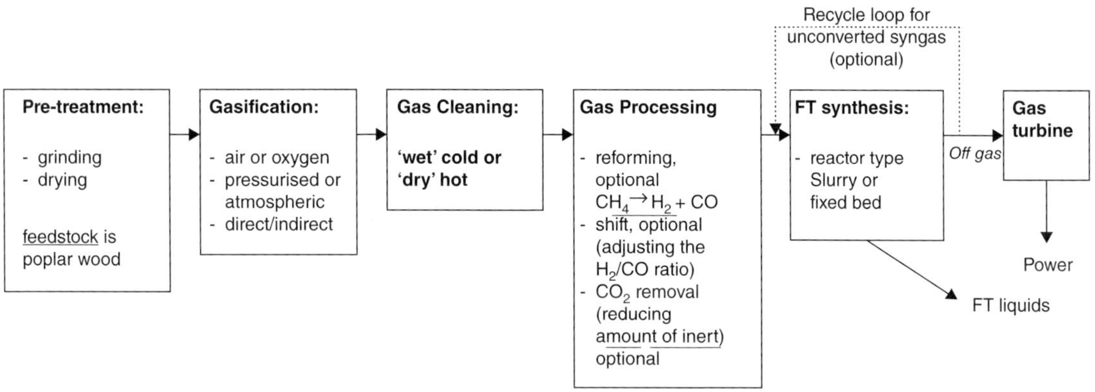

Figure 10.5 A schematic of key components in a Fischer–Tropsch-biomass to liquid fuel (FT-BTL) plant including a gas turbine (combined cycle) for power generation. Reproduced with permission from Ref 88 Copyright 2010, Elsevier.

BTL PLANT LAYOUT

There are several different design options for a FT-BTL process. The choice of gasification technology and FT-catalyst not only dictates most of the design requirements, but also opens up different possibilities for upstream process design. A schematic of key components is shown in Figure 10.5.

Production of synthesis gas from solid carbon-containing materials is performed in a gasification reactor. There are several different designs and optimum process conditions depend on the gasifier. However, the main difference is the choice of gasification medium, including direct heating by partial oxidation with air, oxygen enriched air or pure oxygen in atmospheric, or pressurized reactors, all possible in the presence of varying partial pressures of steam.[88] Another possibility for synthesis gas production is gasification by indirect heating with steam as a gasification medium. This allows production of N_2 free synthesis gas without the need of an air separation unit (ASU) for production of pure oxygen.[89]

In many cases biomass-derived synthesis gas has a low H_2/CO ratio and contains light hydrocarbons such as methane. Additional hydrogen must be introduced prior to the FT reactor unless WGS activity is present in it. Adjustments of the H_2/CO ratio can be done with addition of steam in the gasifier, with a dedicated WGS reactor and/or from steam reforming of natural gas or the C_1–C_4 by-product from the FT process. Hydrogen enrichment using the WGS reaction also increases the amount of CO_2. The effects of CO_2 on the FT-synthesis with different catalysts have already been discussed, and in many cases it may be required to remove this gas prior to the FT reactor. This can be done with amine absorption or physical absorption (i.e., Selexol process).[90] It is also possible to include a reformer to convert light hydrocarbons from the gasifier into synthesis gas. In once-through (OT) concepts, auxiliary units are typically added, such as a gas turbine for power generation or other units where the energy content of the light gases can be utilized if an additional reformer is not included. However, it is for the processes with low inert concentrations that the full potential of auxiliary reformers can be achieved, since recycling of unconverted synthesis gas is more economic in the absence of inert gases. The design of recycle loops becomes more complicated if WGS activity is present in the reactor, as CO_2 removal must be incorporated in the loop, otherwise higher total pressures are required to maintain the desired partial pressures of H_2 and CO. Water is known to inhibit the activity of iron catalysts.[25,27] Because of this it can be beneficial to use synthesis gas that is lean in hydrogen in order to consume water in the WGS reaction. This will enable higher conversions per pass.

OT designs with one reactor or several reactors in series become more relevant when the synthesis gas contains significant amounts of inert gases. Higher total pressure is probably also required in order to compensate for the dilution with inert. Using membranes to separate CO, H_2, and C_1–C_4 compounds from the tail gas may be an option, and this valuable feedstock can be injected back into the gasifier to increase the efficiency.[91] Another option for utilization of the tail gas is power generation.[92] Processes with cobalt catalysts can attain higher conversions than iron-based catalysts per pass, but require higher pressures in order to maintain high selectivity.[36] Reactors in series with water and product knock-out in between might be a more relevant option for iron-based processes,[93,94] as product selectivities are less dependent on pressure and the conversion

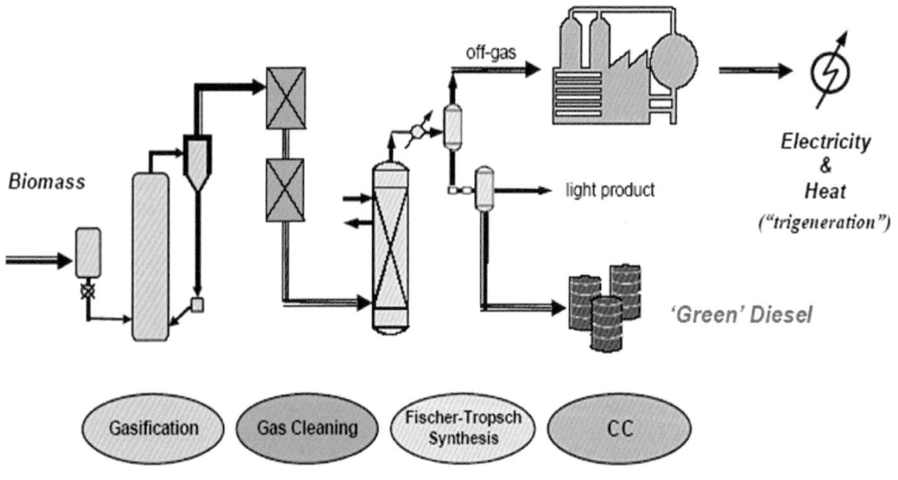

Figure 10.6 An example of a possible process layout for a Fischer–Tropsch-biomass to liquid fuel (FT-BTL) process including electricity and heat production from the tail gas. Reproduced from Ref 95 Copyright 2005, European Commission.

per pass is lower.[36] In summary, the determination of an optimal plant design is complex and depends on several factors. Most likely several different designs for commercial plants will appear as different process solutions can be optimal depending on variable factors such as the type of feedstock, plant scale, and location. Figure 10.6 shows a possible layout of an FT-BTL process including the option for production of electricity and heat from the tail gas.[95]

ECONOMY OF FT-BTL PLANTS

In general, feedstock transportation costs are not decisive for the economy for GTL and CTL plants. Compared to coal, geographical distribution of biomass is much higher and energy density in terms of both volume and weight is significantly lower. Given that wood chips are milled down to the same particle size (100 μm) as in coal gasification, it can still not be used in the same type of plants due to the fibrous nature of biomass.[96] It will not fluidize properly and 'fluffs' are formed which can plug the piping. Hence, the feedstock must go through pretreatment before it can be introduced to the gasifier, otherwise new feeding technology must be developed. Torrefaction or pyrolysis prior to gasification has been proposed. If the feedstock has to be transported over large distances, it may be feasible to perform pretreatment near the location of biomass production in order to increase the bulk density before shipment.[96] These factors complicate economic modeling of FT-BTL plants as feedstock transportation and pretreatment must be included in the evaluations. Table 10.4 shows the required land area biomass must be gathered from for sustainable biomass supply for relevant plant sizes.[97]

Table 10.4 Required minimum radius for biomass harvest for sustainable supply of biomass for FT-BTL processes

	Radius of collection (km)		
Plant size (Barrels per day)	North European forest (1 t/ha/year)	Switch grass (15 t/ha/year)	Arundo Donax (40 t/ha/year)
11–16000	109	28	17
22–32000	154	40	24

Adapted from Ref 97.

Boerrigter[98] estimated the investment costs and costs of scale using the 34,000 bbld ORYX-1 GTL plant of Sasol-QP as the basis for comparison. This study employed an oxygen-blown pressurized slagging entrained flow gasification process for synthesis gas production and a low temperature cobalt-based synthesis process. According to his estimates, a FT-BTL plant would require 60% higher capital investment costs compared to the GTL plant with similar production capacity. The largest source of added investment cost was the ASU. This must provide 50% more oxygen compared to a GTL process for the production of an equivalent amount of heating value synthesis gas. The gasifier is also more expensive than natural gas reformers and additional costs also arise from more complicated synthesis gas cleaning and conditioning processes. The sensitivity analysis showed that the decrease in specific capital investment costs with increasing plant size offsets the added transportation costs of biomass as seen from Figure 10.7. Because of this, larger plants

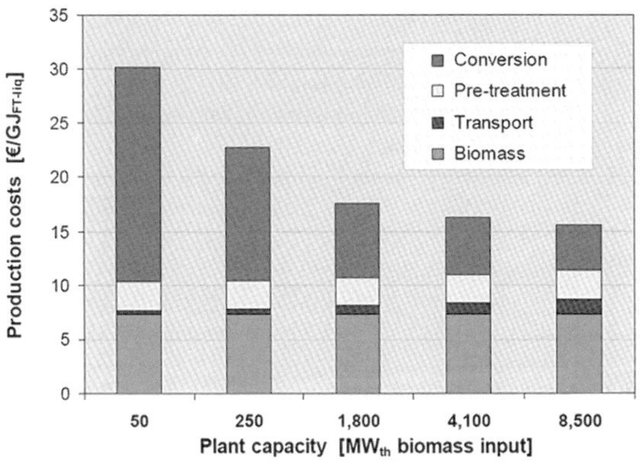

Figure 10.7 Scale dependency of Fischer–Tropsch (FT) fuel production costs. ~4000 MW biomass input equals ~34,000 bbld FT products. Reproduced from Ref 98 Copyright 2007, ECN.

(16,000–32,000 bblp capacity) are favored with this type of plant design.

From Figure 10.7 it appears that the increase in transportation costs are significantly lower than the decrease in conversion costs with increasing plant scales. The effect of scale on plant economy is strong on FTS processes, and new plants are typically of very large size. It appears that small-scale plants may require new technology, which allows lower conversion costs compared to large plants based on conventional technology. One possibility is so-called process intensification using microstructured reactor technology. This may play an important part in bringing down the cost of small-scale plants, which can benefit from close proximity to biomass production areas. Significant advances in the development of FT processes based on this technology have been made in the last few years.[99] Better heat transfer properties allow more active catalysts to be used under severe conditions without losing control of the highly exothermic FT reaction.[100] However, it is disputed whether the productivity of microreactors per reactor volume is higher due to the thin layer of catalyst that can be applied to the reactor walls.[99,101,102] Currently both Velocys[103] and CompactGTL[104] have demonstrated their proprietary microchannel reactor technology; both are reported to be commercially ready.[105] Velocys have also demonstrated their process with bio-derived synthesis gas at Güssing in Austria.[106] An additional development which may contribute to bringing down the costs of small-scale plants is the bifunctional cobalt/zeolite catalysts developed by Chevron. These catalysts are reported to yield slightly more C_1–C_4 products, but also a wax free product that does not require downstream hydrocracking or hydroisomerization.[107,108]

CONCLUSIONS

Research on catalysts for biomass conversion by FT synthesis is mostly aimed at achieving better understanding of the effects of impurities in the synthesis gas and also the effects of different synthesis gas compositions. From current research it appears that alkali, alkaline earth, ammonia, and HCN only have a negative impact on cobalt catalysts while iron catalysts are not affected by these impurities to any significant degree. The benefit of cobalt catalysts is higher CO conversion per pass and also longer catalyst lifetime. Biomass-derived synthesis gas also typically contains significant amounts of CO_2. If this gas is not removed it will mainly act as a diluent, replacing other gases in cobalt-based processes, whereas the effects are more significant for iron-based processes. Iron-based catalysts containing SiO_2 and TiO_2 appears to have low activity and selectivity toward heavier hydrocarbons in CO_2-rich feeds, while Al_2O_3 promoted catalysts seem to retain more of its activity and selectivity.

The availability of biomass feedstock and transportation costs will have an effect on plant scale and design. A consensus on the optimal approach is not yet established. Most likely several different concepts will appear, as different designs may be feasible with depending on location and feedstock. The benefit of costs of scale with increasing plant size significantly outweighs the higher transportation costs of biomass with increasing size. Breaking this relation will require new technologies allowing the constructing of smaller plants with improved economic performance. Promising developments in microstructured reactor technology and bifunctional cobalt/zeolite catalysts may help with bringing down the costs of small-scale plants. Large-scale plants will require the development of an extensive infrastructure including facilities for biomass densification in close proximity to the areas where it is harvested. This is required to increase the energy to volume ratio and the stability of the biomass before shipment. Small-scale plants do not suffer from this problem and can benefit from closer proximity to the sources of biomass.

REFERENCES

1. Bartle JR, Abadi A. Toward sustainable production of second generation bioenergy feedstocks. *Energy Fuels* 2010, 24: 2–9.
2. Sims REH, Mabee W, Saddler JN, Taylor M. An overview of second generation biofuel technologies. *Bioresource Technol* 2010, 101:1570–1580.
3. Petrou EC, Pappis CP. Biofuels: a survey on pros and cons. *Energy Fuels* 2009, 23:1055–1066.
4. Schaidle JA, Moline CJ, Savage PE. Biorefinery sustainability assessment. *Environ Prog Sustain Energy* 2011, 30:743–753.

5. Vlachos DG, Chen JGG, Gorte RJ, Huber GW, Tsapatsis M. Catalysis center for energy innovation for biomass processing: research strategies and goals. *Catal Lett* 2010, 140:77–84.
6. Thirmal C, Dahman Y. Comparisons of existing pretreatment, saccharification, and fermentation processes for butanol production from agricultural residues. *Can J Chem Eng* 2012, 90:745–761.
7. Serrano-Ruiz JC, Pineda A, Balu AM, Luque R, Campelo JM, Romero AA, Ramos-Fernández JM. Catalytic transformations of biomass-derived acids into advanced biofuels. *Catal Today* (2012, 148:115–118.)
8. Nigam PS, Singh A. Production of liquid biofuels from renewable resources. *Prog Energy Combust Sci* 2011, 37:52–68.
9. French R, Czernik S. Catalytic pyrolysis of biomass for biofuels production. *Fuel Process Technol* 2010, 91:25–32.
10. Ragauskas AJ, Williams CK, Davison BH, Britovsek G, Cairney J, Eckert CA, Frederick WJ, Hallett JP, Leak DJ, Liotta CL, et al. The path forward for biofuels and biomaterials. *Science* 2006, 311:484–489.
11. Gautam G, Adhikari S, Bhavnani S. Estimation of biomass synthesis gas composition using equilibrium modeling. *Energy Fuels* 2010, 24:2692–2698.
12. Khodakov AY, Chu W, Fongarland P. Advances in the development of novel cobalt Fischer–Tropsch catalysts for synthesis of long-chain hydrocarbons and clean fuels. *Chem Rev* 2007, 107:1692–1744.
13. Leibold H, Hornung A, Seifert H. HTHP syngas cleaning concept of two stage biomass gasification for FT synthesis. *Powder Technol* 2008, 180:265–270.
14. Bartholomew CH. History of cobalt catalyst design for FTS. *National Spring Meeting of the American Institute of Chemical Engineers*. New Orleans: American Institute of Chemical Engineers; 2003.
15. Stranges AN. Germany's synthetic fuel industry 1927–1945. *National Spring Meeting of the American Institute of Chemical Engineers*. New Orleans: American Institute of Chemical Engineers; 2003.
16. Kirkels AF, Verbong GPJ. Biomass gasification: still promising? A 30-year global overview. *Renewable Sustain Energy Rev* 2011, 15:471–481.
17. Gill SS, Tsolakis A, Dearn KD, Rodríguez-Fernández J. Combustion characteristics and emissions of Fischer–Tropsch diesel fuels in IC engines. *Prog Energy Combust Sci* 2011, 37:503–523.
18. Wang H, Hao H, Li X, Zhang K, Ouyang M. Performance of Euro III common rail heavy duty diesel engine fueled with gas to liquid. *Appl Energy* 2009, 86:2257–2261.
19. EU directive 1999/96/EC of the European Parliament and of the Council. Official Journal of the European Communities, L44, 16 February 2002, 56–67.
20. Joint report from EUCAR, CONCAWE and JRC (the Joint Research Centre of the EU Commission), with an evaluation of the Well-to-Wheels energy use and greenhouse gas (GHG) emissions for a wide range of potential future fuel and powertrain options. Version 2c, March 2007. Available at: http://ies.jrc.ec.europa.eu/uploads/mediya/WTW_Report_010307.pdf (Accessed December 23, 2012).
21. Eisenberg B, Fiato RA, Mauldin CH, Say GR, Soled SL. Exxon's advanced gas-to-liquids technology. *Surf Sci Catal* 1998, 119:943–948.
22. Dry ME. High quality diesel via the Fischer–Tropsch process—a review. *J Chem Technol Biotechnol* 2002, 77:43–50.
23. Vannice MA. Catalytic synthesis of hydrocarbons from H_2–Co mixtures over group-8 metals. 1. Specific activities and product distributions of supported metals. *J Catal* 1975, 37:449–461.
24. de Smit E, Cinquini F, Beale AM, Safonova OV, van Beek W, Sautet P, Weckhuysen BM. Stability and reactivity of ϵ-χ-θ iron carbide catalyst phases in Fischer–Tropsch synthesis: controlling μ_c. *J Am Chem Soc* 2010, 132:14928–14941.
25. Dry ME. Fischer–Tropsch synthesis over iron catalysts. *Catal Lett* 1991, 7:241–251.
26. van de Loosdrecht JFGB, Niemantsverdriet JW. A comparison of cobalt and iron based slurry phase Fischer–Tropsch synthesis. *SynFuel 2012*. Munich; 2012.
27. Dry ME. The Fischer–Tropsch process: 1950–2000. *Catal Today* 2002, 71:227–241.
28. Dry ME. Practical and theoretical aspects of the catalytic Fischer–Tropsch process. *Appl Catal A* 1996, 138:319–344.
29. Yates IC, Satterfield CN. Intrinsic kinetics of the Fischer–Tropsch synthesis on a cobalt catalyst. *Energy Fuels* 1991, 5:168–173.
30. Iglesia E, Soled SL, Fiato RA. Fischer–Tropsch synthesis on cobalt and ruthenium—metal dispersion and support effects on reaction-rate and selectivity. *J Catal* 1992, 137:212–224.
31. Iglesia E. Design, synthesis, and use of cobalt-based Fischer–Tropsch synthesis catalysts. *Appl Catal A* 1997, 161:59–78.
32. Iglesia E. Fischer–Tropsch synthesis on cobalt catalysts: structural requirements and reaction pathways. *Stud Surf Sci Catal* 1997, 107:153–162.
33. Bezemer GL, Bitter JH, Kuipers HPCE, Oosterbeek H, Holewijn JE, Xu XD, Kapteijn F, van Dillen AJ, de Jong KP. Cobalt particle size effects in the Fischer–Tropsch reaction studied with carbon nanofiber supported catalysts. *J Am Chem Soc* 2006, 128:3956–3964.
34. Schulz H. Short history and present trends of Fischer–Tropsch synthesis. *Appl Catal A* 1999, 186:3–12.
35. Moulijn JA, van Diepen AE, Kapteijn F. Catalyst deactivation: is it predictable? What to do? *Appl Catal A* 2001, 212:3–16.
36. van Berge PJRCE. Cobalt as an alternative Fischer–Tropsch catalyst to iron for the production of middle distillates. *Stud Surf Sci Catal* 1997, 107:207–212.
37. Dag Schanke PL, Sigrid Eri, Erling Rytter, Bente Helgeland Sannæs, Keijo J. Kinnari. Optimisation of Fischer–Tropsch reactor design and opearation in GTL plants. *Stud Surf Sci Catal* 2001, 136:239–243.
38. Chanenchuk CA, Yates IC, Satterfield CN. The Fischer–Tropsch synthesis with a mechanical mixture of a cobalt

catalyst and a copper-based water gas shift catalyst. *Energy Fuels* 1991, 5:847–855.
39. Lualdi M, Logdberg S, Regali F, Boutonnet M, Jaras S. Investigation of mixtures of a Co-based catalyst and a Cu-based catalyst for the Fischer–Tropsch synthesis with bio-syngas: the importance of indigenous water. *Top Catal* 2011, 54:977–985.
40. Escalona N, Medina C, Garcia R, Reyes P. Fischer–Tropsch reaction from a mixture similar to biosyngas: influence of promoters on surface and catalytic properties of Co/SiO_2 catalysts. *Catal Today* 2009, 143:76–79.
41. Yao Y, Liu X, Hildebrandt D, Glasser D. Fischer–Tropsch synthesis using $H_2/CO/CO_2$ syngas mixtures: a comparison of paraffin to olefin ratios for iron and cobalt based catalysts. *Appl Catal A* 2012, 433–434:58–68.
42. Kapteijn F, de Deugd RM, Moulijn JA. Fischer-Tropsch synthesis using monolithic catalysts. *Catal Today* 2005, 105:350–356.
43. Shi BC, Davis BH. Fischer–Tropsch synthesis: the paraffin to olefin ratio as a function of carbon number. *Catal Today* 2005, 106:129–131.
44. Riedel T, Claeys M, Schulz H, Schaub G, Nam SS, Jun KW, Choi MJ, Kishan G, Lee KW. Comparative study of Fischer–Tropsch synthesis with H_2/CO and H_2/CO_2 syngas using Fe- and Co-based catalysts. *Appl Catal A* 1999, 186:201–213.
45. Xu B, Fan Y, Zhang Y, Tsubaki N. Pore diffusion simulation model of bimodal catalyst for Fischer–Tropsch synthesis. *Am Inst Chem Eng J* 2005, 51:2068–2076.
46. Iglesia E, Reyes SC, Madon RJ, Soled SL. Selectivity control and catalyst design in the Fischer–Tropsch synthesis—sites, pellets, and reactors. *Adv Catal* 1993, 39:221–302.
47. Li S, Li A, Krishnamoorthy S, Iglesia E. Effects of Zn, Cu, and K promoters on the structure and on the reduction, carburization, and catalytic behavior of iron-based Fischer–Tropsch synthesis catalysts. *Catal Lett* 2001, 77:197–205.
48. de Smit E, de Groot FMF, Blume R, Havecker M, Knop-Gericke A, Weckhuysen BM. The role of Cu on the reduction behavior and surface properties of Fe-based Fischer–Tropsch catalysts. *Phys Chem Chem Phys* 2010, 12:667–680.
49. Gaube J, Klein HF. The promoter effect of alkali in Fischer–Tropsch iron and cobalt catalysts. *Appl Catal A* 2008, 350:126–132.
50. Pendyala VRR, Jacobs G, Mohandas JC, Luo MS, Ma WP, Gnanamani MK, Davis BH. Fischer–Tropsch synthesis: attempt to tune FTS and WGS by alkali promoting of iron catalysts. *Appl Catal A* 2010, 389:131–139.
51. Ngantsoue-Hoc W, Zhang YQ, O'Brien RJ, Luo MS, Davis BH. Fischer–Tropsch synthesis: activity and selectivity for group I alkali promoted iron-based catalysts. *Appl Catal A* 2002, 236:77–89.
52. Davis BH. Fischer–Tropsch synthesis: relationship between iron catalyst composition and process variables. *Catal Today* 2003, 84:83–98.
53. Yang Y, Xiang HW, Xu YY, Bai L, Li YW. Effect of potassium promoter on precipitated iron-manganese catalyst for Fischer–Tropsch synthesis. *Appl Catal A* 2004, 266:181–194.
54. Dlamini H, Motjope T, Joorst G, ter Stege G, Mdleleni M. Changes in physico-chemical properties of iron-based Fischer–Tropsch catalyst induced by SiO_2 addition. *Catal Lett* 2002, 78:201–207.
55. Yang Y, Xiang HW, Tian L, Wang H, Zhang CH, Tao ZC, Xu YY, Zhong B, Li YW. Structure and Fischer–Tropsch performance of iron-manganese catalyst incorporated with SiO_2. *Appl Catal A* 2005, 284:105–122.
56. Zhang CH, Wan HJ, Yang Y, Xiang HW, Li YW. Study on the iron–silica interaction of a co-precipitated Fe/SiO_2 Fischer–Tropsch synthesis catalyst. *Catal Commun* 2006, 7:733–738.
57. Wan HJ, Wu BS, Li TZ, Tao ZC, Xia AN, Xiang HW, Li YW. Effects of SiO_2 and Al_2O_3 on performances of iron-based catalysts for slurry Fischer–Tropsch synthesis. *J Fuel Chem Technol* 2007, 35:589–594.
58. Ruggiero M, Manfrida G. An equilibrium model for biomass gasification processes. *Renewable Energy* 1999, 16:1106–1109.
59. Jun KW, Roh HS, Kim KS, Ryu JS, Lee KW. Catalytic investigation for Fischer–Tropsch synthesis from bio-mass derived syngas. *Appl Catal A* 2004, 259:221–226.
60. Choi PH, Jun KW, Lee SJ, Choi MJ, Lee KW. Hydrogenation of carbon dioxide over alumina supported Fe–K catalysts. *Catal Lett* 1996, 40:115–118.
61. Yan SR, Jun KW, Hong JS, Choi MJ, Lee KW. Promotion effect of Fe–Cu catalyst for the hydrogenation of CO_2 and application to slurry reactor. *Appl Catal A* 2000, 194:63–70.
62. Jun K-W, Lee S-J, Kim H, Choi M-J, Lee K-W. Support effects of the promoted and unpromoted iron catalysts in CO_2 hydrogenation. *Stud Surf Sci Catal* 1998, 114:345–350.
63. Yao YL, Liu XY, Hildebrandt D, Glasser D. Fischer–Tropsch synthesis using $H_2/CO/CO_2$ syngas mixtures over an iron ctalyst. *Ind Eng Chem Res* 2011, 50:11002–11012.
64. Cayan FN, Zhi MJ, Pakalapati SR, Celik I, Wu NQ, Gemmen R. Effects of coal syngas impurities on anodes of solid oxide fuel cells. *J Power Sources* 2008, 185:595–602.
65. Dayton DC, Belle-Oudry D, Nordin A. Effect of coal minerals on chlorine and alkali metals released during biomass/coal cofiring. *Energy Fuels* 1999, 13:1203–1211.
66. Hamelinck CN, Faaij APC, den Uil H, Boerrigter H. Production of FT transportation fuels from biomass: technical options, process analysis and optimisation, and development potential. *Energy* 2004, 29:1743–1771.
67. Newby RA, Smeltzer EE, Lippert TE, Slimane RB, Akpolat OM, Pandya K, Lau FS, Abbasian J, Williams BE, Leppin D. Novel gas cleaning/condi-tioning for integrated gasification combined cycle. Final report. DOE Award Number: DE-AC26–99FT40674, August 2001. Available at: http://www.netl.doe.gov/technologies/coalpower/gasification/projects/gas-clean/docs/BaseProgramFinalReport.PDF. (Accessed December 23, 2012).
68. Boerrigter H, Uil Hd. Green diesel from biomass via Fischer–Tropsch synthesis: new insights in gas cleaning and

process design. *Pyrolysis and Gasification of Biomass and Waste*. Strasbourg, France; 2002.
69. Weiss H. Rectisol wash for purification of partial oxidation gases. *Gas Sep Purif* 1988, 2:171–176.
70. Meng X, de Jong W, Pal R, Verkooijen AHM. In bed and downstream hot gas desulphurization during solid fuel gasification: a review. *Fuel Process Technol* 2010, 91:964–981.
71. Yung MM, Jablonski WS, Magrini-Bair KA. Review of catalytic conditioning of biomass-derived syngas. *Energy Fuels* 2009, 23:1874–1887.
72. Sharma SD, Dolan M, Ilyushechkin AY, McLennan KG, Nguyen T, Chase D. Recent developments in dry hot syngas cleaning processes. *Fuel* 2010, 89:817–826.
73. Davis BH. Fischer–Tropsch synthesis: comparison of performances of iron and cobalt catalysts. *Ind Eng Chem Res* 2007, 46:8938–8945.
74. Bartholomew CH. Mechanisms of catalyst deactivation. *Appl Catal A* 2001, 212:17–60.
75. Pansare SS, Allison JD. An investigation of the effect of ultra-low concentrations of sulfur on a Co/gamma-Al_2O_3 Fischer–Tropsch synthesis catalyst. *Appl Catal A* 2010, 387:224–230.
76. Borg Ø, Hammer N, Enger BC, Myrstad R, Lindvåg OA, Eri S, Skagseth TH, Rytter E. Effect of biomass-derived synthesis gas impurity elements on cobalt Fischer–Tropsch catalyst performance including *in situ* sulphur and nitrogen addition. *J Catal* 2011, 279:163–173.
77. Kritzinger JA. The role of sulfur in commercial iron-based Fischer–Tropsch catalysis with focus on C_2-product selectivity and yield. *Catal Today* 2002, 71:307–318.
78. Duvenhage DJ, Coville NJ. Deactivation of a precipitated iron Fischer–Tropsch catalyst—a pilot plant study. *Appl Catal A* 2006, 298:211–216.
79. Norheim A, Lindberg D, Hustad JE, Backman R. Equilibrium calculations of the composition of trace compounds from biomass gasification in the solid oxide fuel cell operating temperature interval. *Energy Fuels* 2009, 23:920–925.
80. Balonek CM, Lillebø AH, Rane S, Rytter E, Schmidt LD, Holmen A. Effect of alkali metal impurities on Co–Re catalysts for Fischer–Tropsch synthesis from biomass-derived syngas. *Catal Lett* 2010, 138:8–13.
81. Trepanier M, Tavasoli A, Dalai AK, Abatzoglou N. Co, Ru and K loadings effects on the activity and selectivity of carbon nanotubes supported cobalt catalyst in Fischer–Tropsch synthesis. *Appl Catal A* 2009, 353:193–202.
82. Uner DO. A sensible mechanism of alkali promotion in Fischer–Tropsch synthesis: adsorbate mobilities. *Ind Eng Chem Res* 1998, 37:2239–2245.
83. Uner DO, Pruski M, King TS. The role of alkali promoters in Fischer–Tropsch synthesis on Ru/SiO_2 surfaces. *Top Catal* 1995, 2:59–69.
84. Wenping Ma, Gary Jacobs, Dennis E. Sparks, Jungshik Kang, Muthu Kumaran Gnanamani, Venkat Ramana Rao Pendyala, Wilson D. Shafer, Robert A K, Graham U, Davis BH. Effect of alkali compounds in syngas on Fischer–Tropsch synthesis behaviour of a precipitated iron catalyst in a slurry reactor. *SynFuel 2012*, Munich; 2012.
85. LeVinnes S, Mart C, Behrmann W, Hsia S, Neskora D. *Slurry hydrocarbon synthesis process with increased catalyst life*. International: Patent WO 98/50487. 1998.
86. Robota HJ, Alger J, Pretorius PJ. Impact of low levels of ammonia in syngas on the Fischer–Tropsch synthesis performance of cobalt and iron catalysts in fixed-bed operation. *National Spring Meeting of the American Institute of Chemical Engineers*. Houston, TX: American Institute of Chemical Engineers; 2012.
87. Ma W, Jacobs G, Kang J, Sparks D, Davis BH. Effect of ammonia in syngas on the Fischer–Tropsch behavior of a precipitated iron catalyst. *American Institute of Chemical Engineers Annual Meeting*. Minneapolis, MN: American Institute of Chemical Engineers; 2011.
88. Tijmensen MJA, Faaij APC, Hamelinck CN, van Hardeveld MRM. Exploration of the possibilities for production of Fischer–Tropsch liquids and power via biomass gasification. *Biomass Bioenergy* 2002, 23:129–152.
89. Smolinski A, Stanczyk K, Howaniec N. Steam gasification of selected energy crops in a fixed bed reactor. *Renewable Energy* 2010, 35:397–404.
90. Tock L, Gassner M, Marechal F. Thermochemical production of liquid fuels from biomass: thermo-economic modeling, process design and process integration analysis. *Biomass Bioenergy* 2010, 34:1838–1854.
91. Spyrakis S, Panopoulos KD, Kakaras E. Synthesis, modeling and exergy analysis of atmospheric air blown biomass gasification for Fischer–Tropsch process. *Int J Thermodyn* 2009, 12:187–192.
92. Mark J, Prins KJP, Frans JJG, Janssen. Exergetic optimisation of a production process of Fischer–Tropsch fuels from biomass. *Fuel Process Technol* 2005, 84:375–389.
93. Xu LG, Bao SQ, O'Brien RJ, Raje A, Davis BH. Don't rule out iron catalysts for Fischer–Tropsch synthesis. *CHEMTECH* 1998, 28:47–53.
94. Srinivas S, Malik RK, Mahajani SM. Fischer–Tropsch synthesis using bio-syngas and CO_2. *Energy Sustainable Dev* 2007, 11:66–71.
95. Kavalov B, Peteves SD. Status and perspectives of biomass-to-liquid fuels in the European Union. Report EUR 21745 EN. European Commission, Joint Research Centre, Petten, The Netherlands; 2005. p. 79.
96. Zwart RWR, Boerrigter H, van der Drift A. The impact of biomass pretreatment on the feasibility of overseas biomass conversion to Fischer–Tropsch products. *Energy Fuels* 2006, 20:2192–2197.
97. Perego C, Bianchi B. Today and tomorrow of diesel fuel from biomass: an ENI perspective. *CatBior 2011—1st International Congress on Catalysis for Biorefineries*. Torremolinos, Spain; 2011.
98. Boerrigter H. *Economy of biomass-to-liquids (BTL) plants*. Report ECN Number: ECN-C–06–019. Available at: ftp://question.ecn.nl/pub/www/library/report/2006/c06019.pdf. (Accessed December 23, 2012).
99. Lerou JJ, Tonkovich AL, Silva L, Perry S, McDaniel J. Microchannel reactor architecture enables greener processes. *Chem Eng Sci* 2010, 65:380–385.

100. Myrstad R, Eri S, Pfeifer P, Rytter E, Holmen A. Fischer–Tropsch synthesis in a microstructured reactor. *Catal Today* 2009, 147:S301–S304.
101. Guttel R, Turek T. Micro structured packed bed reactors for strong exothermic gas phase reactors: a feasibility study. *Chem Ing Tech* 2009, 81:495–500.
102. Jarosch KT, Tonkovich ALY, Perry ST, Kuhlmann D, Wang Y. Microchannel reactors for intensifying gas-to-liquid technology. *Microreact Technol Process Intensificat* 2005, 914:258–272.
103. Velocys. Available at: http://www.velocys.com/ (Accessed December 23, 2012).
104. Compact GTL. Available at: http://compactgtl.com/. (Accessed December 23, 2012)
105. Lerou JJ. Small scale gas-to-liquids processes today. In: *SynFuel 2012*. Munich; 2012.
106. Velocys Press statement: Gssing FT demonstration update. Available at: http://www.velocys.com/financial/fa/ocgfa20100816.php (Accessed December 23, 2012).
107. Saxton RJKJ, Das T, Lacheen H, Rea T, Kibby C. Chevron's gas conversion catalysis—hybrid catalyst for wax-free Fischer–Tropsch synthesis. *SynFuel 2012*. Munich; 2012.
108. Kibby CL, Jothimurugesan K, Das TK, Saxton RJ, Burton Jr AW. *Process of synthesis gas conversion to liquid fuels using mixture of synthesis gas conversion catalyst and dual functionality catalyst*. US Patent 79,973,087. 2011.

11

Critical Factors for High Temperature Processing of Biomass from Agriculture and Energy Crops to Biofuels and Bioenergy

Stelios Arvelakis and Emmanuel G. Koukios*

Department of Chemical Engineering, Bioresource Technology Unit, National Technical University of Athens, Zografou Campus, Athens, Greece

INTRODUCTION AND BACKGROUND

In the search for fuel sources that can be used as alternatives to fossil fuels, biomass represents one of the most attractive options. Utilization of biomass as feedstock for power production offers the advantages of a renewable and almost CO_2 neutral fuel. Biomass sources include wood, agricultural residues, and fast growing trees and crop plantations; these biomasses can be either directly burned in a combustor or first upgraded via gasification into a gaseous fuel and then used for the generation of heat and power, as well as for the production of liquid fuels. Since in most parts of the world agriculture dominates the land uses, agricultural residues such as straws, olive residues, hulls and pods, etc. are the most abundant biomass resources, which can be used for decentralized heat and power production, especially in less developed and environmentally sensitive areas. On the contrary, the cultivation of dedicated energy crops, such as *Arundo donax* reeds, miscanthus, cynara cardunculus, switchgrass, and so on, is expected to increase substantially over the coming years due to a combination of land availability and a high demand for biomass to be used as feedstock for energy and liquid fuels production.

As far as biomass conversion is concerned, the need to meet the Kyoto protocol directives and to decrease the environmental footprint of energy and fuels produced from fossil energy sources, such as coal and oil, has led to the development of advanced thermochemical conversion technologies, including supercritical combustion, pressurized fluidized bed combustion and gasification, as well as entrained flow combustion and gasification. These new thermochemical conversion processes operate at high temperatures (>1500°C) and pressures (>50bar) and can achieve very high conversion efficiencies (>50%). Thermal gasification processes can be also combined with Fischer–Tropsch modules or other catalytic processes to produce liquid fuels such as synthetic diesel, methanol, and ethanol.

As a rule, the use of such high-temperature and high-pressure environments puts additional requirements on the feedstock quality, as otherwise significant problems could be created that would limit the conversion potential of the specific biomass materials to energy and fuels in an economical and sustainable way.

In particular, the ash behavior of agricultural residues and energy crops during their thermochemical conversion is one of the most important subjects to be addressed.[1,2] With the term ash, we refer to all inorganic components of biomass materials present in a variety of forms, including organically bound cations, inorganic salts, and other minerals. Agricultural residues and energy crops contain mainly alkali metals, such as potassium (K) present in the

*Correspondence to: koukios@chemeng.ntua.gr

Advances in Bioenergy: The Sustainability Challenge, First Edition. Edited by Peter D. Lund, John Byrne, Göran Berndes and Iacovos A. Vasalos.

Originally published in: *WIREs Energy Environ* 2013, 2: 149–162 doi: 10.1002/wene.28

form of inorganic salts, e.g., KCl dissolved in the inherent moisture, or connected to carboxylic or other functional groups, or as complex ions and chemisorbed material, and silicon (Si) in the form of hydrated silica or as deposits on the cell walls, as their principal ash-forming inorganic constituents. They are also relatively richer in chlorine (Cl) and sulfur compared with wood fuels, and also contain small amounts of alkali earth materials, i.e., calcium (Ca) present either on the cell walls or as crystalline Ca oxalate in cytoplasm, magnesium, and also phosphorous, with the latter two present mainly in biological forms.[2,3]

The presence of large amounts of these inorganic constituents in some biomass feedstocks is associated with serious technical problems, such as deposition, sintering, agglomeration, fouling, corrosion, and erosion, all caused during the operation of high-temperature combustion and gasification plants. In particular, alkali metals, and especially K, which has high mobility, tend to react either with silica, even at temperatures far below 900°C by breaking the Si–O–Si bonds, and forming low-melting-point silicates, or with sulfur to produce alkali sulfates. Cl acts as a facilitator of these reactions by increasing the mobility of K since most of it is present as potassium chloride (KCl). KCl is among the most stable, high-temperature, gas-phase, alkali-containing species, whereas the amount of Cl in the fuel often dictates the amount of the alkali possible to be vaporized during combustion or gasification. Ca also appears to react with sulfur to form sulfates, but the lower mobility of Ca in combination with its limited quantity in these biofeedstocks does not result in significant problems.[1,2,4–6] The produced alkali silicates and/or sulfates have very low melting points that may reach as low as 700°C and tend to deposit on reactor walls or heat exchange surfaces in the case of the conventional grate-fired systems, whereas in the case of the fluidized bed reactors, they contribute significantly in bed sintering and defluidization of the bed inert material through the development of a sticky deposit on the surface of the bed particles.[3,7–9]

Ash-related problems in the forms of fouling, slagging, corrosion, and erosion constitute one of the biggest obstacles toward a safe, economical, and environmentally friendly utilization of a large variety of biomass and waste fuels [straws, hulls, municipal solid waste (MSW), and sewage sludge] for energy generation via thermochemical conversion routes. Furthermore, in the case of the MSW incineration plants, the deposition and the corrosion processes can also be affected by the presence of heavy metals, such as zinc (Zn) and lead (Pb), in the form of chlorides and sulfates. These compounds have very low melting points, e.g., 290°C for $ZnCl_2$, 501°C for $PbCl_2$, and 680°C for $ZnSO_4$, and also interact with compounds such as KCl and alkali metal sulfates present in large amounts in the ashes generated during MSW incineration. This leads to the formation of molten phases with even lower melting points compared with pure salts, i.e., 250°C for a $ZnCl_2$/KCl, 48/52% wt mixture, thus accelerating significantly the deposition and the corrosion processes.[10–15] The presence of large amounts of heavy metals [Zn, Pb, copper (Cu) etc.] and also alkali chlorides and sulfates in the ash formed during the MSW incineration, apart from their contribution to the ash-related problems, also constitute a significant problem regarding the safe treatment and disposal of the generated ash material in economical and environmentally friendly ways.[16–19]

In addition, the optimal use of solid agricultural biomass and energy crops as feedstocks in high-temperature and high-pressure conversion systems, either alone or in combination with coal, requires particular mixing, feeding, and sizing properties of the solid biomass particles. Biomass materials as a rule are difficult to be milled, mixed, and fed, both alone as well as in combination with other biomasses and solid fossil fuels. In particular, feeding in high-temperature and high-pressure environments requires the use of very small particle sizes ($<70\,\mu$), as well as of storage, mixing, and feeding systems that have been developed for use with coal feedstock. However, biomass cannot be milled efficiently at the specific particle size or handled efficiently through any of the specified infrastructure that has been developed for use with the coal materials.

All these lead to the need for the development of technologies and processes that will minimize the effects of the inorganic components on the formation of ash-related problems, and help with the various, sizing, mixing, feeding, and storage problems during thermochemical conversion of biomass materials to energy and fuels. This need becomes especially imperative in today's world, where the demand for deployment in the utilization of low cost biomass and waste feedstocks in order to mitigate rising environmental problems, such as the greenhouse effects, is growing fast.

CRITICAL FACTORS FOR THE USE OF AGRICULTURAL AND ENERGY CROP BIOMASS

Biomass Feedstocks: Agricultural Residues

Agricultural residues are a potentially huge source of biomass. Crop and animal wastes already provide significant amounts of energy, coming second only to wood as the dominant biomass fuel worldwide. Waste streams from agriculture include the portions of crop plants discarded such as straw, weather-damaged or surplus supplies, and animal wastes (dung). It was estimated, e.g., that 110 Mt of dung and crop residues are used each year as fuel in

India, compared with 133 Mt per year of wood; in China, the annual potential mass of available agricultural residues has been estimated at 2.2 times the mass of fuelwood.

Every year, millions of tons of straw are produced worldwide with usually half of it being surplus to local and market needs. In many countries, this surplus is still burned on the field or ploughed back into the soil, but in some developed countries, environmental legislation that restricts field burning has drawn attention to the potential use of straw as an energy resource. Efforts to remove crop residues from soils and use them for energy purposes lead to a central question: how much residue should be left back and recycled into the soil in order to sustain agricultural and biomass production? According to the experiences from developed countries, on the average, around 35% of the total crop residues can be sustainably removed from the soil without adverse effects on future plant productivity.

Organic industrial wastes from the processing of biomass may also be used to produce energy. For example, the residue from alcohol production (known as vinasse) can be utilized to produce flammable gas. Other useful biowaste feedstocks include wastes from food processing, and fluff from the cotton and textiles industries.[20]

Biomass Feedstocks: Energy Crops

Biomass can be also produced from short-rotation plantations of trees and other plants, e.g., grasses (sorghum, sugarcane, and switchgrass). Biomass from all these plants can be used as fuel such as wood, with the main advantage of the short time span between plantation and harvesting—typically between 3 and 8 years. For some grasses, harvesting can take place every 6–12 months.

Parameters that are important in evaluating species as short-rotation plants include availability of planting stock, ease of propagation, survival ability under adverse conditions, and yield potential, measured as dry matter production per hectare per year (t/ha/y). Yield is a good measure of a plant's ability to utilize the site resources. It is the most important factor when considering biomass utilization, due to the need to optimize/maximize yield from a given area of land within a given time frame at the least possible cost. As a rule, high yielding species are therefore preferred for biomass energy systems.

Some plant communities have shown superiority in dry matter production compared with others grown under similar conditions. Although reported dry matter production of different tree species varies over a wide range depending on soil types and climate, certain species stand out. For *Eucalyptus* species, yields of up to 65 dry t/ha/y have been reported, compared with 30 and 43 t/ha/y from *Salix* and *Populus* species, respectively.

Table 11.1 shows some examples of the analysis and characterization of various biomass feedstocks, whereas Figures 11.1 and 11.2 illustrate the biomass-to-energy cycle for energy crops and agricultural residues, respectively.

Despite the fact that biomass plantations can be of great importance for most developed countries, experience has shown that they are unlikely to be established on a large scale in many developing countries, especially in low income, rural areas, so long as other types of biofuels, particularly wood, can be obtained at zero or near zero costs.[20]

Thermochemical Conversion Technologies

Low-quality biomass and waste materials may be used today in a variety of applications for the production of energy and fuels using thermochemical conversion processes. These processes include biomass and waste combustion and gasification in fluidized bed boilers and gasifiers, in stokers and fixed bed systems, such as updraft and downdraft gasifiers, pressurized fluidized bed boilers and gasifiers, as well as entrained flow reactors.

There is also a growing interest for co-combustion and co-gasification applications where biomass and waste could be utilized together with fossil fuels in conventional pulverized fuel (pf) boilers for the production of energy, as well as in more advanced systems such as integrated gasification combined cycle, supercritical combustion boilers, and entrained flow reactors where the percentages of the biomass and waste materials could be as high as 40% of the fuel mixture in energy basis.

There is also a strong interest for fast and slow pyrolysis applications for the production of biooils as well as solid char that could be used as an advanced energy carrier.

Table 11.2 presents the operating characteristics regarding temperature and pressure of the main reactors that are used today for the production of energy and fuels.

Atmospheric combustion and gasification can be divided in three main categories depending on the operating temperature. From these three categories, the first one, where the operating temperature is limited below 900°C, is mainly used for small-scale energy and fuels production; the second category, where the temperature is limited below 1300°C, is mainly used today for large-scale energy production, whereas the third category, where the operating temperature is limited below 1600+°C, constitutes the future option for large-scale, high-conversion efficiency energy production.

Pressurized combustion and gasification can be divided into two main categories where the first category, where the operating temperature is limited below 900°C, is

Table 11.1 Analysis and Characterization of Various Biomass Feedstocks

	Switchgrass (Columbus, OH)		Almond Shells		Wood Hybrid Poplar		Straws Alfalfa Stems		Straws Wheat (CA)		Straws Rice	
Fuel Type	As Received	Dry	As Received	Dry	As Received	Dry	As Received	Dry	As Received	Dry	As Received	Dry
Proximate Analysis												
Fixed carbon	12.93	14.34	19.28	20.71	11.63	12.49	14.34	15.81	16.47	17.71	14.60	15.86
Volatile matter	69.14	76.69	70.73	76.00	78.97	84.81	71.59	78.92	69.97	75.27	60.28	65.47
Ash	8.09	8.97	3.06	3.29	2.51	2.70	4.78	5.27	6.52	7.02	17.19	18.67
Moisture	9.84	–	6.93	–	6.89	–	9.29	–	7.04	–	7.93	–
Total	100.00	100.00	100.00	100.00	100.00	100.00	100.00	100.00	100.00	100.00	100.00	100.00
Ultimate Analysis												
Carbon	42.09	46.68	45.88	49.30	46.72	50.18	42.79	47.17	41.75	44.92	35.20	38.24
Hydrogen	5.25	5.82	5.55	5.97	5.64	6.06	5.44	5.99	5.07	5.46	4.79	5.20
Oxygen	33.87	37.57	37.83	40.64	37.66	40.44	35.09	38.69	39.08	42.00	33.92	36.84
Nitrogen	0.69	0.77	0.71	0.76	0.56	0.60	2.43	2.68	0.40	0.44	0.80	0.87
Sulfur	0.17	0.19	0.04	0.04	0.02	0.02	0.18	0.20	0.14	0.16	0.17	0.18
Ash	8.09	8.97	3.06	3.29	2.51	2.70	4.78	5.27	6.52	7.02	17.19	18.67
Moisture	9.84	–	6.93	–	6.89	–	9.29	–	7.04	–	7.93	–
Total	100.0	100.00	100.00	100.00	100.00	100.00	100.00	100.00	100.00	100.00	100.00	100.00
HHV, Btu/lb	7002	7766	7797	8378	7615	8178	7280	8025	7171	7714	7330	8123
Chlorine%	0.17	0.19	<0.01	<0.01	0.01	0.01	.045	0.50	0.21	0.23	.054	0.58
Water-Soluble Alkalis (%)												
Na_2O	–	–	–	–	–	–	0.109	0.120	–	–	–	–
K_2O	–	–	–	–	–	–	1.050	1.158	–	–	–	–
CaO	–	–	–	–	–	–	–	–	–	–	–	–
Elemental Composition												
SiO_2	–	65.19	–	8.71	–	5.90	–	5.79	–	55.32	–	74.67
Al_2O_3	–	4.51	–	2.72	–	0.84	–	0.07	–	1.88	–	1.04
TiO_2	–	0.24	–	0.09	–	0.30	–	0.02	–	0.08	–	0.09
Fe_2O_3	–	2.03	–	2.30	–	1.40	–	0.30	–	0.73	–	0.85
CaO	–	5.60	–	10.50	–	49.92	–	18.32	–	6.14	–	3.01
MgO	–	3.00	–	3.19	–	18.40	–	10.38	–	1.06	–	1.75
Na_2O	–	0.58	–	1.60	–	0.13	–	1.10	–	1.71	–	0.96
K_2O	–	11.60	–	48.70	–	9.64	–	28.10	–	25.60	–	12.30
SO_3	–	0.44	–	0.88	–	2.04	–	1.93	–	4.40	–	1.24
P_2O_5	–	4.50	–	4.46	–	1.34	–	7.64	–	1.26	–	1.41
CO_2/Other	–	–	–	17.38	–	8.18	–	14.80	–	–	–	–
Undetermined	–	2.32	–	-0.53	–	1.91	–	11.55	–	1.82	–	2.68
Total	–	100.00	–	100.00	–	100.00	–	100.00	–	100.00	–	100.00
Alkali, Lb/MMBtu	–	1.41	–	1.98	–	0.32	–	1.92	–	2.49	–	3.05

SOURCE: Ref 21.

Figure 11.1 Cycle for energy crops biomass-to-energy cycle.

commercial today regarding combustion and in the demonstration phase regarding gasification, and is used mainly for energy and in a smaller degree for fuels production, whereas the second category, where the operating temperature is limited below 1600°C, is in the demonstration phase today, with a few commercial plants in the case of combustion and some large demos in the case of gasification and is considered to be the technology of tomorrow for energy and fuels production.

All these technologies regardless of their commercial status are currently using very low amounts of low-quality biomass and waste materials, either as firing or cofiring feedstock, due to the substantial ash-related problems associated with the use of these low-quality materials in high volumes.[2,4,6,8,21]

Scope for Biomass Pretreatment: Leaching

Figures 11.3 and 11.4 present the recommended minimum removal rate targets for the leaching process, when using low cost biomass and waste materials as feedstock for the production of energy and fuels in large scale for the different conversion technologies. These data are based on our research that includes extensive laboratory-, and pilot-scale testing performed during the last 15 years with various combustors and gasifiers operating at temperatures and pressures as high as 1300°C and 40 bars, respectively, and after a thorough study and understanding of the composition and properties of a large number of low cost biomass and waste material.

As it is seen from these Figures, the first and most important condition to target is the almost complete removal of Cl to substantially reduce/eliminate the ash-related

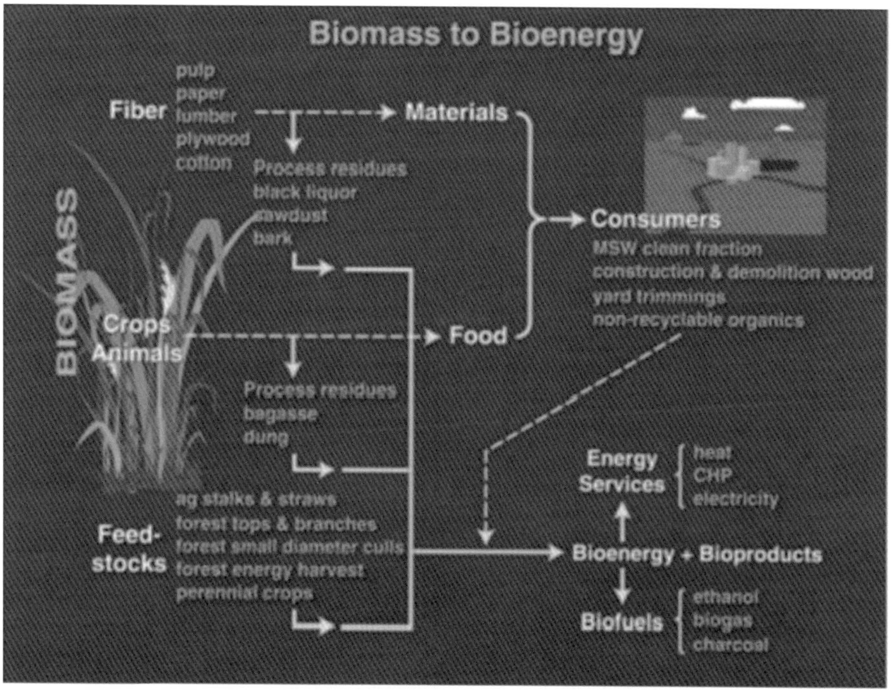

Figure 11.2 Biomass to bioenergy route for agricultural residues.

Table 11.2 Operating Characteristics of Boilers and Gasifiers for Energy and Fuels Production

Thermochemical Conversion Processes	Operating Pressure (bar)	Operating Temperature (°C)
Atmospheric combustion/gasification	1	900
Atmospheric combustion/gasification	1	1300
Super critical combustion/OxyFuel combustion	1	1600+
Pressurized combustion/gasification	2–50	1000
High-temperature-pressurized gasification	2–50	1600
Pyrolysis production of biooil–char	1	600

SOURCE: Ref 22.
As it is seen from this Table, pyrolytic applications have the lowest temperature and pressure requirements, but the produced biooil–char have to be almost free from alkali metals, chlorine, and sulfur in order to ensure their stability and thermal behavior during their later utilization in combustion and gasification applications.

problems from the use of biomass and waste materials. The role of Cl in the inorganic transformations during thermochemical conversion was explained in the *Introduction and Background* section of this paper.

The second condition is the removal of the reactive forms of alkali metals, such as those being present in water-soluble and ion-exchangeable forms, in percentages higher than 99% for combustion and gasification applications and at least higher than 90% for biooil production.

Finally, the reactive forms of phosphorus and sulfur in low-quality biomass and waste materials should also be removed to below 50% of the respective amount present in the raw feedstock.

Although heavy metals are not included in the above targets, our recommendation is that for metals such as zinc and lead, the removal rates have to be similar to those proposed for alkali metals, whereas for metals such as copper, cadmium, nickel, and so on, the removal rates have to be similar to those proposed for phosphorus and sulfur.

Ash Chemistry

Biomass ash chemistry plays a very significant role regarding the ash-related problems appearing during thermochemical conversion of biomass at high-temperature and high-pressure conversion environments. In specific, the alkali metals, sodium and K, as well as Cl have long been associated with the fouling tendencies of biomass ash. Volatile forms of these elements are vaporized in the furnace/gasifier at combustion/gasification temperatures,

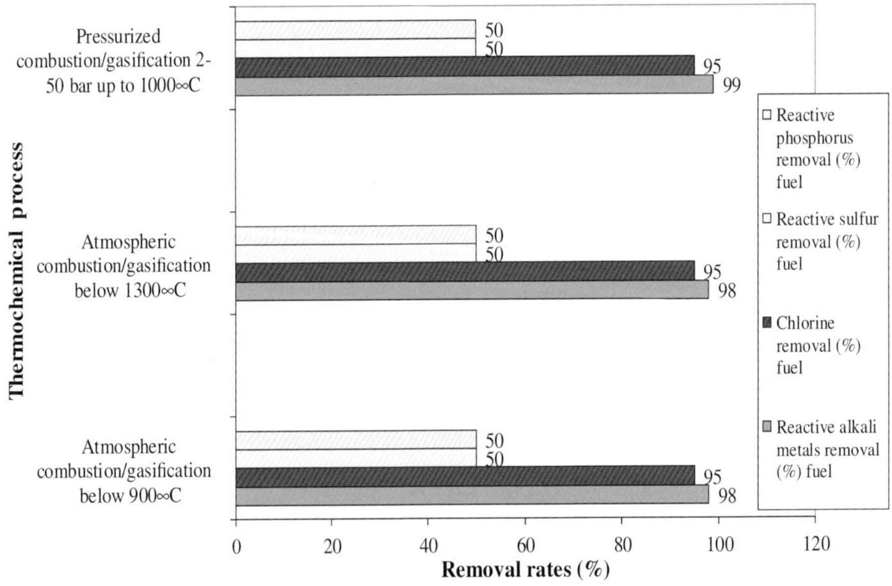

Figure 11.3 Targets for minimum removal rates of inorganics after the leaching pretreatment—Part A.

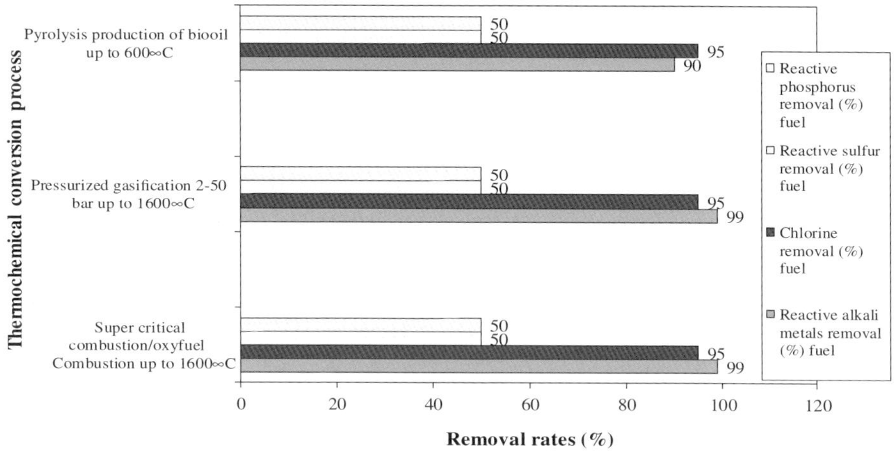

Figure 11.4 Targets for minimum removal rates of inorganics after the leaching pretreatment—Part B.

and react with sulfur in the flue gas and other elements in the ash, thus forming compounds that deposit on convection surfaces.

The base-to-acid ratio of ash also affects the slagging potential. The basic constituents are iron, the alkaline earth metals Ca and magnesium, and the alkali metals sodium and K. Acidic constituents are Si, aluminum, and titanium. Ash that has either high or low base-to-acid ratio generally has high ash fusion and melting temperatures. However, ash with moderate values of base-to-acid ratio tends to have low fusion and melting temperatures. Agricultural residues and energy crops appear to be rich in alkali metals especially K and Cl and contain also low to average amounts of sulfur and in most cases high amounts of Si and Ca. As a result, significant ash-related problems are caused during the thermochemical conversion of these materials that inhibit substantially the use of these biomass materials in high-efficiency thermochemical conversion systems for the production of energy and fuels.[21,23–25]

Ash Melting Point

Apart from the ash chemistry, another parameter that plays a significant role in the ash-related problems caused when agricultural and energy crops biomass is used in thermochemical conversion systems is the ash melting point.

Ash melting temperatures provide an indication of the temperature range over which portions of the ash will be a molten fluid or semi-molten, plastic state. High melting temperatures indicate that ash released in the furnace will cool quickly to a nonsticky state, resulting in minimal potential for ash-related problems. Conversely, low melting temperatures indicate that ash will remain in a molten or plastic state longer, exposing more of the furnace to potential deposition, agglomeration, and corrosion.

In the case of the agricultural residues and the energy crops, the ash melting point might not be a very accurate method to predict the ash-related problems in a few cases. This is due to the ash chemistry of the specific materials. A combination of both the ash melting point and the ash chemistry is required in most cases to be able to make a better estimation regarding the ash melting behavior and the ash-related problems during the combustion/gasification of these materials.[21,26,27]

Ash-Related Problems

Ash-related problems constitute a major factor affecting the use of low-quality biomass such as agricultural residues and energy crops in thermochemical conversion systems. The ash content of these biomass materials is rich in alkali metals, Cl and contains also significant amounts of phosphorus and sulfur in reactive forms. High temperature conversion of these biomass materials results in severe ash-related problems when temperature exceeds 800°C that makes impossible the use of these materials in high-temperature/pressure thermochemical environments.

Corrosion

Corrosion is the deterioration of materials by chemical interaction with their environment. The consequences of corrosion are many and varied and the effects of these on the safe, reliable, and efficient operation of equipment or structures are often more serious than the simple loss of a mass of metal. Failures of various kinds and the need for expensive replacements may occur even though the amount of metal destroyed is quite small.

In the case of combustion/gasification processes, corrosion is defined as the reaction of the metal surface of a tube, wall, and so on with chemical compounds such as alkali metals, chlorides, sulfur, and oxygen to form a layer of oxides that gradually grows until it results in the wastage and mechanical failure of the specific surface. In the case of biomass firing/cofiring, corrosion is caused due to the reaction of alkali, sulfur, and Cl with the metal surfaces of the reactors as well as the tubes used downstream for heat exchange and gas transfer. Corrosion can be accelerated when deposition of biomass ash is a serious problem during combustion and gasification processes.[21,27]

Deposition

Deposition is a process that includes two different processes that are defined as slagging and fouling. Slagging is the formation of molten or partially fused ash solid phases that are called deposits on the reactor walls or convection surfaces exposed to radiant heat during combustion and gasification. Fouling is defined as the formation of ash deposits on convection heat surfaces such as superheaters and reheaters.

Slag is formed when melt or soften ash particles are not cooled down to solid state when they reach the heat surface. Typical initial deformation temperatures could vary depending on the feedstock and its ash composition. In case of biomass such as agricultural residues and energy crops with high alkali, phosphorus, and Cl content, these temperatures could be as low as 800°C. Most ash tends to be resolidified due to the relatively lower temperature at the tube surface and the particle factures on impact and partially bounce back into the gas stream. However, if the reactor is too small, the exit gas temperature is too high or the melting point of ash is relatively low, molten ash may not have enough time to be resolidified when they hit the heat surface; therefore, they are easy to stick to the surface and cause the accumulation of deposits leading to slagging.

Fouling is generally caused by the vaporization of volatile inorganic elements in the feedstock (biomass and coal) during combustion/gasification. When heat is absorbed and temperatures decrease in the convection area of the boiler, compounds formed by these elements condense on ash particles and heating surface, forming a glue, which initiates deposition.

Power plant deposits occur on the grate, in the bed or on refractory as hard fused glassy deposits, agglomeration of grate ash or bed media, and as accumulations of fly ash in hot gas ducts. Fouling of the boiler convection passes includes buildup on screen tubes or superheater tubes, bridging between tubes, and hard deposits on tubes. Deposits on convection surfaces develop quickly when furnace exit temperatures exceed 750°C.

Corrosion is found under many deposits. Deposits collected from similar temperature zones in different boilers have similar compositions. Grate and wall deposits are seen to be similar in composition to the fuel ash. K and Ca silicates and sulfates deposit on screen tubes and upstream superheaters. Chlorides and carbonates appear in the cooler convection passes.

There are three principal undesirable effects of deposits: (1) deposits retard the heat transfer and lead to an eventual decline in boiler efficiency and capacity if they cannot be removed according to the design assumptions for the boiler; (2) deposits can grow to the extent that flow through the boiler is restricted, often bridging across tubes and tube bundles, and causing mechanical damage; and (3) deposits are associated with corrosion.

Alkali and alkaline earth metals as well as sulfur, phosphorus, and Cl in the biomass fuel ash are important to the formation of fireside deposits. Furthermore, the presence of silica in large amounts in the case of many agricultural residues and energy crops biomass in combination with high amounts of alkali metals and Cl leads to increased deposition problems when these biomass materials are used in combustion and gasification environments.

Especially, Cl can be an important facilitator in fouling, leading to the condensation of alkali chlorides on heat transfer surfaces in the boiler, and promoting the development of alkali sulfates. Cl may be an important element in the vaporization of alkali species, leading to the formation of more severe deposits.[21,27–31]

Agglomeration–Sintering

Agglomeration occurs in fluidized bed combustion and gasification reactors and is caused because of the reaction of the bed material such as silica with the reactive elements of the biomass ash such as alkali metals, phosphorus, Cl, and sulfur. In a second stage, alkali earth metals such as Ca and magnesium are also added into the agglomerate. In case the biomass material contains large amounts of Si in its ash, then agglomeration occurs, regardless of the type of the bed material.

In most fluidized bed combustion/gasification systems, the tendency of the bed particles to sinter and form agglomerates is a key process, the avoidance of which is an important reactor design and operational issue. Sintering is defined as the process through which the ash particles of the feedstock (biomass and coal) tend to coalesce when heated at high temperatures and form an initially solid phase that with time and temperature and depending on the ash composition could be transformed to a viscous liquid that could attract and capture more ash particles while be used as a bridge to connect bed particles together. The presence of large amounts of alkali metals, phosphorus, sulfur, and Cl in the ash of the biomass materials as well as of Si favors the ash sintering and the formation of a viscous semiliquid phase that could be used to bond the bed particles together and cause severe agglomeration problems during high-temperature and high-pressure thermochemical conversion processes (>800°C).

In bubbling fluidized bed systems, in particular, excessive agglomeration of the bed particles can lead to poor air distribution and eventually defluidization of the bed. In extreme cases, it may be necessary for the operators to reduce the bed temperatures or to bring the reactor offline to remove and replace the bed material.

In operating fluidized bed combustion/gasification plants, the key issue is the control over the bed temperatures and the fuel distribution. The formation of agglomerates is often associated with the formation of fuel-rich hot spots in the bed. These can also be formed as a result of slag falls from the upper furnace surfaces or the detachment of pieces of furnace wall refractory material.[21,26,27]

Tar Content in the Producer Gas

The product gas formed from biomass gasification contains the major components CO, H_2, CO_2, CH_4, H_2O, and N_2, in addition to organic (tars) and inorganic (H_2S, HCl, NH_3, alkali metals) impurities and particulates. The organic impurities range from low-molecular-weight hydrocarbons to high-molecular-weight polynuclear aromatic hydrocarbons. The lower molecular weight hydrocarbons can be used as fuel in gas turbine or engine applications but are undesirable products in fuel cell applications and methanol synthesis. The higher molecular weight hydrocarbons are known as 'tar'.

A definition of tars is that 'tars' are considered to be the condensable fraction of the organic gasification products and are largely aromatic hydrocarbons, including benzene. The diversity in the operational definitions of 'tars' usually comes from the variable product gas compositions required for a particular end-use application and how the 'tars' are collected and analyzed.

Tars are problematic in integrated biomass gasification systems for a number of reasons. Tars can condense in exit pipes and on particulate filters leading to blockages and clogged filters. Tars also have varied impacts on other downstream processes. Tars can clog fuel lines and injectors in internal combustion engines. Luminous combustion and erosion from soot formation can occur in pressurized combined cycle systems where the product gases are burned in a gas turbine. The product gas from an atmospheric pressure gasification process needs to be compressed before it is burned in a gas turbine and tars can condense in the compressor or in the transfer lines as the product gas cools.

The removal of tars from the producer gas using catalysts as well as physical processes has been seen to be problematic and to cause large drops in efficiency, while it has high capital cost that could reach 50% of the total gasification capital cost. Thermal tar cracking could be a more efficient and less costly tar elimination process but requires high temperatures (>1000°C) that in the case of agricultural residues and energy crops biomass could cause significant ash-related problems. As a result, a pretreatment technology that eliminates these problems could also assist significantly with the efficient removal of tars from the producer gas.[32–35]

Milling–Feeding Behavior

Biomass has been proved to be a difficult-to-handle material. Pulverization (<200 μ) of 1 metric ton of biomass such as wheat straw or olive residue to be used in a pulverized coal combustion boiler requires more than 11 times the time for pulverization of 1 metric ton of coal. In addition, the pulverization of biomass using existing coal pulverization mills causes significant problems to them because of the high temperatures developed during the milling process and as a result of high wear and tear, resulting in substantially higher milling costs when co-combustion and or co-gasification is attempted. Furthermore, the pulverized biomass still contains large percentage of particles above the particle size required by the pf boiler feeding system for an optimum feeding and efficient combustion.

The use of screw-type feeders has shown to have the lowest problems when biomass is fed into boilers as well as gasifiers. However, problems are caused mainly in case pneumatic-type feeders are used for feeding the biomass as well as when valves or other kind of intermediate mechanisms are used during the moving of the biomass from the storage bins to the reactor. Biomass tends to bridge and the smaller the particle size, the higher the bridging problems observed. Bridging is a very significant problem when agricultural as well as energy crops biomass is used, regardless of the biomass particle size.

Particle Size

As mentioned before, biomass is very difficult to be milled in small particle size (<200 μ) required normally for feeding into pf boilers or entrained flow gasifiers, where the

Table 11.3 Energy Requirements During Milling of Various Biomass Materials

Material	Initial/Final Size (mm)	Moisture Content (% wet basis)	Equipment	Specific Energy Requirement (kWh $t-1$]
Switchgrass	NA/0.8–3.2	8	Hammer mill	62.62 –23.5
Corn stover	07.15/0.8–3.2	12	Hammer mill	34.3 –11.04
Corn silage	19.05/3–12.5	7	Knife mill	20 –1
Wheat straw	07.67/0.8–3.2	8	Hammer mill	51.55 –10.77
Wheat straw	22.4/1.6–6.3	7	Knife mill	07.50 –05.50
Rye straw	20.52/0.8–3.2	6.9	Hammer mill	53.00 –13.56
Hardwood	19.05/1.6–9.5	6	Hammer mill	130 –15
Alfalfa	18 –12/1.6	9	Hammer mill	1.6

optimum particle size is even lower (<60 μ). However, even milling at particle sizes below 6 mm requires a significant amount of energy and could result in high costs.

The energy requirement to reduce the particle size of biomass to below 2 cm is estimated to be around 10–20 kWh/metric ton. As a result, the milling energy requirements for a pretreatment plant that will process 120 ktons per year will be around 1200–2400 MWh. As a result, depending on the cost of electricity per MWh that could range from $60–120, the energy cost of milling will be $0.6–2.4 per dry metric ton. A value closer to $0.6–1.8/metric ton is considered to be most likely today in the United States as well as in most parts of the world, taking into consideration the current prices of electricity for industrial use.

Table 11.3 presents experimental data from various groups regarding the energy required for milling various biomass materials in different particle sizes.[36]

As it is seen from Table 11.3, milling below 3 mm is very costly and decreases significantly the viability of the milling process and each relative process.

Mixing Behavior

Biomass materials are seen to be difficult to be mixed with either other biomasses and/or coal. This is mainly because of the different bulk density, particle size, and flowing characteristics that are very different in each biomass material. In general, pits and woody biomass are seen to have the best mixing and flowing properties, compared with the other biomass materials and they also have average to high bulk density. Straw and grasses are seen to be among the worst regarding the mixing and flowing properties. They also have very low bulk densities that make very difficult their mixing with other biomass and/or coal materials in rather large ratios. As a result, the application of a technology/process to improve the mixing as well as flowing properties and bulk density is required in order these biomass materials to be able to be used in thermochemical conversion systems in rather high ratios with other biomasses and/or coal materials.

Remediation Measures and Techniques

Biomass pretreatment technologies have gained a lot of interest in the last few years as a way to overcome the problems associated with the use of large amounts of agricultural residues and energy crops in high-temperature and high-pressure thermochemical conversion systems for the production of energy and liquid fuels.

These pretreatment technologies comprise both physicochemical as well as thermal processes and have been seen to include a number of advantages such as:

- they are simple and easy to be applied;
- no sophisticated equipment is required;
- they are cost-effective; and
- they transform the pretreated biomass materials into a clean and easy to handle, mix and feed feedstock that can be used in modern thermochemical conversion systems with high efficiency and without ash-related problems.

Among the most important of these pretreatment technologies is leaching and prepyrolysis/low temperature gasification as well as their combination.

Leaching Pretreatment

Leaching biomass via solid–liquid extraction with water as the extraction mean has been shown to improve biomass fuel properties for high temperature processes and may be a useful process for pretreatment of food processing residues. Water leaching effectively removes water-soluble alkali metals, Cl as well as phosphorus and sulfur from biomass, and upgrades crude biomass to better quality feedstock for biofuel and bioenergy production.[23,24] Leaching biomass with water has demonstrated improvements for reducing deposit formation during thermal conversion.[24–26,28–30,37] Leaching generally improves the properties of biomass by lowering the ash concentration, increasing the heating value, and altering slag-forming ash compositions.[38] Removing Cl from biomass reduces acid gas formation, corrosion, and toxic species production, such as dioxins and furans, during thermal processing.[39–41]

In general, large fractions of K, sodium, Cl, sulfur, and phosphorous in biomass fuels can be removed by leaching with water. Several biomass leaching techniques have been studied, such as repeated soaking and draining.[27,42] Another method is to rinse biomass with water, drain, and mechanically dewater the wet biomass.[37] Biomass leaching can also be done naturally by rain washing in the field.[24–31,37,42,43] As it has been shown, leaching can be performed using any kind of water.[27] Combustion and gasification tests with the water-leached biomass proved to expand the operation of the reactors by more than 10 times compared with the original untreated biomass materials.

Depending on the leaching method, and the structure of biomass, significant amounts of alkali metals, Cl, sulfur, and phosphorous can be removed within a few minutes although the fraction removed varies depending on ion type.[23] Water leaching also removes organic compounds such as organic acids, water-soluble carbohydrates, and ethanol from biomass.[25]

Although the possibility of use of water-leached biomass in some combustion and gasification applications such as atmospheric fluidized bed systems has been successfully demonstrated, the limitations of water only leaching and its limited success in combustion and gasification applications, where higher temperatures and pressures are utilized, in combination with the higher cost of the leached biomass and the low prices of electricity prevented the commercial application of such solutions until today.[25,30,31,37]

Table 11.4 presents the best estimates on the main properties range of the leached biomass versus the original untreated biomass for the main contaminants of interest.

These properties may vary depending on the type of the biomass materials, solvents used for the leaching pretreatment, and any additional special properties and characteristics we would like to give them to maximize and/or minimize some of the impacts their utilization may have in the energy conversion systems.

Table 11.4 Main Properties of Raw Versus Leached Biomass

Thermochemical Conversion Process	Original Biomass	Leached Biomass
Ash content	100%	−80%/+10%
HHV	100%	+5% – +20%
Chlorine content	100%	−99%
Reactive alkali metals content	100%	−99%
Reactive sulfur content	100%	−50%
Reactive phosphorus content	100%	−50%

Prepyrolysis, Low Temperature Gasification

Prepyrolysis, also known as torrefaction, includes the thermal treatment of the biomass materials in the temperature range 200–300°C under inert atmosphere aiming at the partial decomposition of the biomass and change of its chemical and physical structure. Figures 11.5 presents the main steps of the prepyrolysis process. Prepyrolysis has been seen to decrease the volume of biomass, increasing its energy density, reduce the amount of volatiles, and increasing fixed carbon as well as the ash content of the prepyrolyzed biomass. It also increases considerably the grindability of the biomass and retains most of its energy value.

Serious consideration of development of this process as a source of fuels, chemicals, and materials began with the work of Bourgeois in the 1980s. Pechiney used torrefied wood as a replacement for charcoal in production of metals from metal oxides. Work on related processes that are not directed to fuel and chemical products continues, especially in the Netherlands, France, and Finland.

Prepyrolysis achieves equilibrium moisture content of 3%, reduction of mass by 20–70% (primarily by release of water, carbon oxides, and volatiles) while retaining 50–90% of the biomass's original energy content in the solid phase. The highest mass loss as well as energy loss is observed in the case of the agricultural residues and waste biomass due to its higher reactivity and the lower lignine content compared with the woody biomass. Prepyrolysis provides fuels that are rich in energy, low in moisture content, resistant to moisture, and easily applied via direct cofiring with coal at conventional pf-fired power plants.

The prepyrolyzed biomass can be easily ground, mixed, and fed into commercial boilers and gasifiers using grinding, mixing, and feeding systems having developed for coal materials. As a result, substantial savings are realized when prepyrolyzed biomass is used as feedstock in direct firing as well as cofiring applications.[44,45]

In case that the thermal treatment of the biomass material is performed in a low oxygen (<30% of the stoichiometric) environment, then the process is called low temperature gasification and it results in similar carbonized material as in the case of prepyrolysis.

It has to be mentioned that in both cases, apart from the energy that remains bound in the solid phase after the end of the pretreatment process, a substantial part of the initial biomass energy content that could vary from 10–40% is available as a low Btu gas stream that could be used for the production of energy simultaneously with the solid prepyrolyzed/carbonized material. This improves substantially the economics of the process and increases its efficiency. A major challenge for both prepyrolyzed and

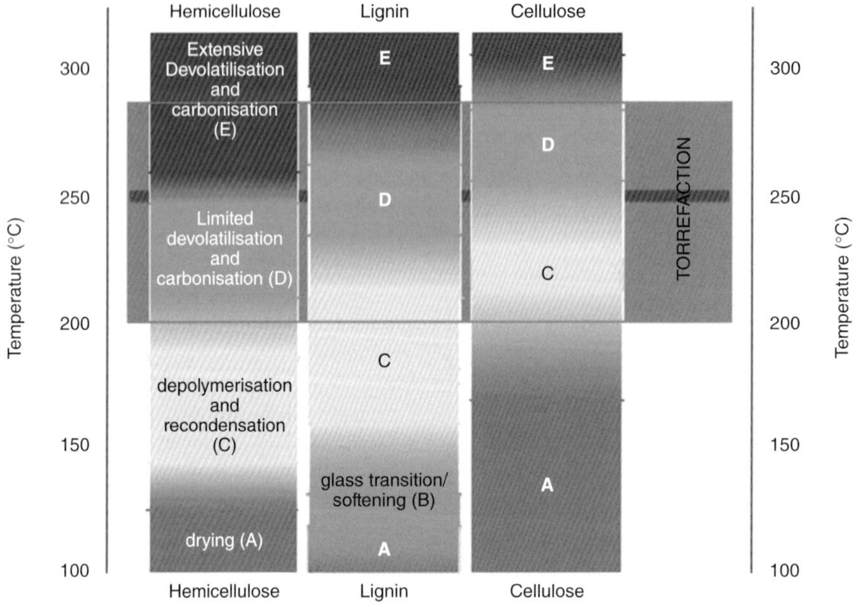

Figure 11.5 Main steps of the pre-pyrolysis/low temperature gasification rretreatment process.[46]

carbonized biomass plant developers is the availability of commercial equipment that will reduce the processing time, energy inputs, and production costs for manufacturing bulk prepyrolyzed/carbonized materials to acceptable levels.

The combination of both, pretreatment technologies, prepyrolysis/low temperature gasification and leaching leads to the production of a very clean coal substitute material that contains trace amounts of alkali metals and Cl and very low amounts of sulfur and phosphorus and has grinding, mixing, and feeding characteristics similar to coal. As a result, the derived material could be used for energy and fuels production in high-temperature and high-pressure thermochemical conversion systems without ash-related problems, catalyst deactivation problems, and with high efficiencies and improved economics.

CONCLUSIONS

On the basis of a detailed survey of the literature and long years of original research,[4,25–27,44,46] we have identified a number of critical factors affecting the successful conversion of biomass feedstocks from agricultural and energy crop sources to energy and fuels by thermochemical methods. The list includes the following:

- the type, composition, availability, and sustainability of the bioresource;
- the type and operating conditions of the conversion technology;
- the composition and chemistry of the biomass ash, and particularly
- the melting point of the ash, plus the main
- ash-related problems, such as corrosion, deposition, agglomeration, and sintering;
- the tar content of the producer gas;
- the milling and feeding behavior of the biomass;
- the particle size of the feedstock; and
- the mixing behavior of biomass.

To meet the needs of the conversion and optimize the feedstock for one of more of the above factors, a number of biomass pretreatments appear to show promise for large-scale application, including

- biomass leaching;
- biomass prepyrolysis/low temperature gasification; and
- their combinations.

REFERENCES

1. Baxter LL. Ash deposition during biomass and coal combustion: a mechanistic approach. *Biomass Bioenerg* 1993, 4:85–105.
2. Miles TR PE, Miles TR Jr., Baxter LL, Bryers RW, Jenkins BM, Oden LL. Alkali deposits found in biomass power plants.

Vol. I and II. Summary Report for the National Renewable Energy Laboratory. NREL Subcontract TZ-2- 11226–1; 1995.
3. Sander B. Properties of Danish biofuels and the requirements for power production. *Biomass Bioenerg* 1997, 12:177–183.
4. Arvelakis S. *PhD Dissertation*. National Technical University of Athens. Athens, Greece; 2001.
5. Baxter LL, Jenkins BM, Miles TR, Miles TR Jr., Milne T, Dayton D, Bryers RW, Olden LL. Inorganic material deposition in biomass boilers. In: *Proceedings of the 9th European Bioenergy Conference*. Copenhagen, Denmark; 1996, 1114.
6. Benson SA. *Ash Formation and Behavior in Utility Boilers*. Newsletter published by Microbeam Technologies Inc.; 1998.
7. Folkedahl BC, Steadman EN, Brekke DW, Zygarlicke CJ. Inorganic phase characterization of coal combustion products using advanced SEM techniques. In: *Proceedings of the Engineering Foundation Conference*. Solihull, England; 1993, 399.
8. Andersen KH, Hansen PFB, Hansen KW, Frandsen F, Dam-Johansen K. Co-firing coal and straw in a 150MWe utility boiler. In: *Proceedings of the 9th European Bioenergy Conference*. Copenhagen, Denmark; 1996, 1102.
9. Mann MD, Galbreath KC. The role of ash chemistry and operating parameters on ash agglomeration and deposition in FBC systems. *5th Engineering Foundation Conference on Inorganic Transformations and Ash Deposition During Combustion*. Palm Coast; 1991, 773.
10. Lide DR, ed. *CRC Handbook of Chemistry and Physics*. 77th Ed; 1996–1997.
11. Spiegel M. The role of molten salts in the corrosion of metals in waste incineration plants. *Molten Salt Forum* 2003, 7:253–268.
12. Stringer J, Wright IG. The Future for High Temperature Materials and Coatings in the New Millennium. *Mater Sci Forum* 2001, 369–372:1–22.
13. Otero E, Pardo A, Perez FJ, Utrilla MV, Levi T. Corrosion behavior of 12CrMoV steel in waste incineration environments: hot corrosion by molten chlorides. *Oxidation of Metals*. 1998, 49:467–484.
14. Otero E, Pardo A, Merino MC, Utrilla MV, Lopez MD, del Peso JL. Corrosion behavior of IN-800 superalloy in waste-incineration environments: hot corrosion by molten chlorides. *Oxidation of Metals*. 1999, 51:507–525.
15. Krause HH. Corrosion by chlorine in waste-fueled boilers. In: *Proceedings of the International Conference on Fireside Problems While Incinerating Municipal and Industrial Waste*, Engineering Foundation Conference, Florida 1989.
16. Khandra S, Mallick D, Dutta SN, Chaudhuri SK. Studies on the phase mineralogy and leaching characteristics of coal fly ash. *Water, Air and Soil Pollution*, 1998, 107:251–275.
17. Steenari BM, Schelander S, Lindqvist O. Chemical and leaching characteristics of ash from combustion of coal, peat and wood in a 12 MW CFB–a comparative study. *Fuel* 1999, 78:249–258.
18. Abbas Z, Steenari BM, Lindqvist O. A Study of Cr(VI) in ashes from fluidized bed combustion of municipal solid waste: leaching, secondary reactions and the applicability of some speciation methods. *Waste Manage* 2001, 21:725–739.
19. Wiles CC. Municipal solid waste combustion ash: state of the knowledge. *J Hazardous Mater* 1996, 47:325–344.
20. http://www.inforse.dk/europe/dieret/Biomass/biomass.html
21. Baxter L, Miles T, Miles T Jr., Jenkins B, Dayton D, Milne T, Bryers R, Oden L. Alkali deposits found in biomass boilers: the behavior of inorganic material in biomass-fired power boilers—field and laboratory experiences; *Report for US Department of Energy*. Contract no. AC36–83CH10093; 1996.
22. Program on Technology Innovation: Biomass Leaching Pretreatment Technology Bench Testing, *EPRI Report ID#1022720*; 2011
23. Bakker RR, 2000. *Biomass Fuel Leaching for the Control of Fouling, Slagging, and Agglomeration in Biomass Power Generation. PhD Thesis Biological System Engineering*. Davis: University of California.
24. Jenkins BM, Bakker RR, Wei JB. On the properties of washed straw. *Biomass Bioenerg* 1996, 10:177–200.
25. Arvelakis S, Gehrmann H, Beckmann M, Koukios EG. Preliminary results on the ash behavior of peach stones during fluidized bed gasification: evaluation of fractionation and leaching as pre-treatments. *Biomass Bioenerg* 2005, 28:331–338.
26. Arvelakis S, Vourliotis P, Kakaras E, Koukios EG. Effect of leaching on the ash behavior of wheat straw and olive residue during fluidized bed combustion. *Biomass Bioenerg* 2001, 20:459–470.
27. Arvelakis S, Koukios EG. Physicochemical upgrading of agroresidues as feedstocks for energy production via thermochemical conversion methods. *Biomass Bioenerg* 2002, 22:331–348.
28. Davidsson KO, Korsgren JG, Pettersson JBC, Jaglid U. The effects of fuel washing techniques on alkali release from biomass. *Fuel* 2002, 81:137–142.
29. Dayton DC, Jenkins BM, Turn SQ, Bakker RR, Williams RB, Belle-Oudry D, Hill LM. Release of inorganic constituents from leached biomass during thermal conversion. *Energy Fuels* 1999, 13:860–870.
30. Jenkins BM, Bakker RR, Baxter LL, Gilmer JH, Wei JB. Combustion characteristics of leached biomass. In: *Developments in Thermochemical Biomass Conversion*; 1997.
31. Turn SQ, Kinoshita CM, Ishimura DM, Zhou J. The fate of inorganic constituents of biomass in fluidized bed gasification, *Fuel* 1998, 77:135–146.
32. Abatzoglou N, Bangala D, Chornet E. An integrated modular hot gas conditioning technology." from Biomass: a growth opportunity in green energy and value-added products. Vol. 2. In: Overend RP, Chornet E, eds. *Proceedings of the 4th Biomass Conference of the Americas, Oakland, California* Oxford, UK: Elsevier Science; 1999, 953–959.
33. Aldén H, Hagstrom P, Hallgren A, Waldheim L. High temperature catalytic gas cleaning for pressurized gasification processes. In: Chartier P, ed. *Biomass for Energy and the Environment. Vol. 2. Proceedings of the 9th European Bioenergy Conference, Copenhagen, Denmark*. Oxford, UK: Elsevier Science; 1996, 1410–1415.

34. Dayton D. A review of the literature on catalytic biomass tar destruction, milestone completion report. National Renewable Energy Laboratory. Colorado.
35. Milne TA, Evans RJ, Abatzoglou N. Biomass gasifier "tars": their nature, formation, and conversion. National Renewable Energy Laboratory. Colorado; 1998.
36. Kratky L, Jirout T. Biomass size reduction machines for enhancing biogas production. *Chem Eng Technol* 2011, 34:391–399.
37. Turn SQ, Kinoshita CM, Ishimura DM, Removal of inorganic constituents of biomass feedstocks by mechanical dewatering and leaching. *Biomass Bioenerg* 1997, 12:241–252.
38. Jenkins BM, Baxter LL, Miles TR Jr, Miles TR. Combustion properties of biomass. *Fuel Process Technol* 1998, 54:17–46.
39. Demirbas MF. Emissions of polychlorinated dibenzo-p-dioxins and dibenzofurans from biomass combustion and solid waste incineration. *Energy Sources Part A—Recovery Utilization and Environmental Effects* 2007, 29:1041–1047.
40. Lavric ED, Konnov AA, De Ruyck J. Dioxin levels in wood combustion—a review. *Biomass Bioenerg* 2004, 26:115–145.
41. Lavric ED, Konnov AA, De Ruyck J. Modeling the formation of precursors of dioxins during combustion of woody fuel volatiles. *Fuel* 2005, 84:323–334.
42. Jenkins BM, Mannapperuma JD, Bakker RR. Biomass leachate treatment by reverse osmosis. *Fuel Process Technol* 2003, 81:223–246.
43. Bakker RR, Jenkins BM. Feasibility of collecting naturally leached rice straw for thermal conversion. *Biomass Bioenerg* 2003, 25:597–614.
44. Arvelakis S, Spliethoff H, Koukios EG.Torrefaction and leaching: a novel pre-treatment for the effective thermal conversion of agricultural residues. In: *15th European Biomass Conference*. Berlin, Germany; 2007.
45. Pach M, Zanzi R, Bjornbom E. Torrefied biomass a substitute for wood and charcoal. In: *6th Asia-Pacific international symposium on combustion and energy utilization*. Kuala Lumpur; 2002.
46. Koukios EG, Maurokoukoulakis J, Abatzoglou N. Energy densification of biomass. In: *Proceedings of 1st National Conference on Soft Energy Forms*. Thessaloniki, Greece; 1982.

12

Second-Generation Biofuels: Why They are Taking so Long

Daniel J. M. Hayes*

Carbolea Biomass Research Group, Department of Chemical and Environmental Sciences, University of Limerick, Limerick, Ireland

BACKGROUND

Second-Generation Biofuels

This paper considers second-generation biofuels (2GBs) to be those that have been derived from lignocellulosic feedstocks. Such biomass is primarily composed of the biopolymers of cellulose, hemicellulose, and lignin. Cellulose is a homopolysaccharide consisting of D-anhydroglucopyranose units linked through β-(1→4)-glycosidic bonds and contains both crystalline and amorphous regions. Hemicelluloses are amorphous heteropolysacchrides, of shorter chain lengths than cellulose, that can contain the pentoses D-xylose and L-arabinose and the hexoses D-glucose, D-mannose, and D-galactose, as well as some uronic acids and acetylated derivatives.[1] Lignin is a complex three-dimensional polymer of phenylpropane units. The lignocellulose macrostructure consists of complex arrangements and conformations of these three polymers. For more information regarding the chemistry of lignocellulosics, the reader is referred to other sources.[2–4] Examples of lignocellulosic biomass feedstocks include municipal/industrial wastes (e.g., household and green wastes, paper wastes), agricultural residues (e.g., corn stover, sugarcane bagasse, straws), and dedicated energy crops (e.g., *Miscanthus*, switchgrass).

In contrast, first-generation biofuels (1GB) are obtained from oil or sugar/starch yielding biomass. For example, sucrose, from crops such as sugarcane, and starch, from crops such as wheat and corn, can be relatively easily hydrolyzed to their constituent monomer sugars, which can then be fermented to products such as ethanol. Oils and fats can be converted to fatty acid methyl esters (biodiesels) via transesterification processes.[5] However, the supply side costs of 1GB feedstocks tend to be significantly higher than those of 2GB feedstocks, both in energetic and environmental/energy terms.[6,7] Furthermore, there are food versus fuel conflicts associated with the use of 1GB feedstocks for biofuel production whereas these do not exist, or are much reduced, for most 2GB feedstocks.

Hence, there has been a focus on developing commercially viable technologies for the production of biofuels, compatible with existing vehicle engines and fuel infrastructure, that do not require food/oil-based feedstocks.[8] The strategies tend to focus on the production of either 2GBs or biodiesels from algae. The key to exploiting the chemical value of lignocellulosics is to depolymerize the lignocellulosic matrix to obtain smaller molecules that can be utilized as (or further converted to) platform chemicals and biofuels. There are two main pathways for achieving this, through hydrolytic mechanisms that aim to liberate free monosaccharides from the lignocellulosic polysaccharides, and through thermochemical processes (e.g., gasification and pyrolysis) that degrade more extensively the

*Correspondence to: danieljohnhayes@gmail.com

Advances in Bioenergy: The Sustainability Challenge, First Edition. Edited by Peter D. Lund, John Byrne, Göran Berndes and Iacovos A. Vasalos.

Originally published in: *WIREs Energy Environ* 2013, 2: 163–192 doi: 10.1002/wene.59

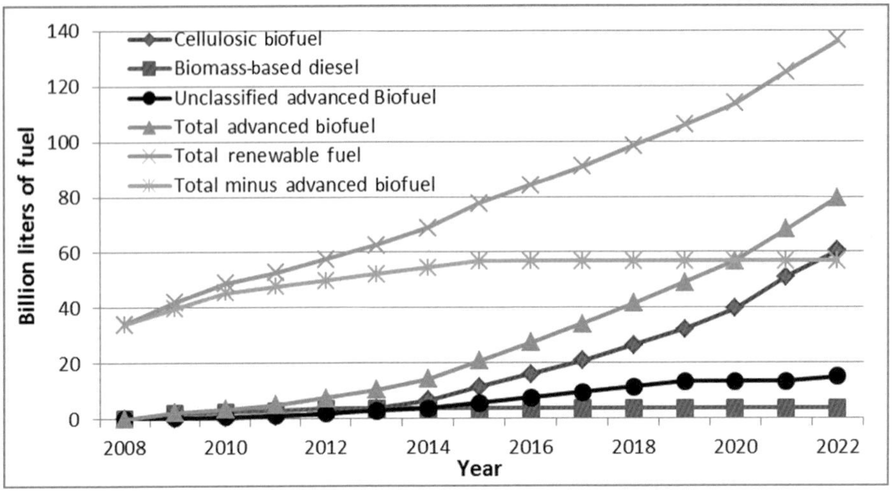

Figure 12.1 Volumes for the Renewable Fuels Standards of the US Energy Independence and Security Act, 2007.[17] Cellulosic biofuel = renewable fuels produced from cellulose, hemicellulose, or lignin (must meet a 60% + greenhouse gas (GG) reduction threshold compared to fossil fuels); Biomass-based diesel = biodiesel or renewable diesel (must achieve 50% + GG reduction); total advanced biofuel = anything but corn-based ethanol (must achieve 50% + GG reduction threshold); unclassified advanced biofuel = total advanced biofuel minus cellulosic and biomass-based diesel fuel; total renewable fuel = ethanol from corn starch or any other qualifying renewable fuel (must achieve 20% + GG reduction); total minus advanced biofuel = expected volumes of corn-starch ethanol.

components of both polysaccharides and lignin. Such technologies can be termed biorefineries, an analogy to oil refineries because, like these, biorefineries obtain a variety of end products according to the chemical components of the starting materials. This paper is not meant to serve as a detailed review regarding the concepts of these technologies, for that the reader should refer to a previous article by the author[2] and the numerous other articles cited. The reader is also referred to other review papers for discussions on the production of valuable fine chemicals from lignocellulosic biomass[9–12] and on the production of algal biofuels.[13–15]

Governmental Support for Second-Generation Biofuels

Because of the deleterious environmental effects associated with the combustion of fossil fuels,[16] as well as uncertainties regarding their future supply and price, governments around the world have been promoting the increased utilization of biofuels and, particularly, the commercial development of 2GBs.[8]

In his 2006 State of the Union Address, the US President George W. Bush famously stated that America was addicted to oil. He also launched the Advanced Energy Initiative, a program that targeted the development of advanced energy technologies to reduce this dependence, and stated that 2GBs 'would be practical and competitive within 6 years'. A specific Renewable Fuels Standard (RFS) was provided by the US Energy Independence and Security Act of 2007, which mandated for 136 billion liters of renewable fuel by 2022 with 80 billion liters coming from 'advanced biofuels' (with a cellulosic mandate of 57 billion liters for that year). There were also cellulosic ethanol targets of 378 million liters for 2010 and 946 million liters for 2011, as well as for subsequent years.[17] These targets are illustrated in Figure 12.1.

To facilitate the achievement of these targets, the US Department of Energy announced in 2007 a loan guarantee scheme for the construction of six commercial-scale biorefineries covering a variety of process technologies. These facilities, their outputs, conversion mechanisms, and level of funding are outlined in Table 12.1.

At a similar time, the EU was also promoting the development of biofuels. First, a 2003 EU Directive set nonbinding indicative targets of 2% biofuels in petrol and diesel for 2005 and 5.75% by 2010.[18] This was followed by EU Directive 2009/28/EC which mandated each Member State to provide 10% of the energy of transport fuels (land based) by renewable resources by 2020.[19] This Directive also stated eligibility criteria that must be met including, by 2017, greenhouse gas savings of 50% compared with fossil fuels (rising to 60% by 2018 for facilities constructed after January 1, 2017), and that biofuels from high biodiversity lands and peat lands would not be eligible.[19] Furthermore, biofuels derived from wastes and lignocellulosic feedstocks are afforded double weighting (compared to 1GBs) when

Table 12.1 Details and Eventual Fates of the Six Commercial-Scale Biorefineries that Were Considered Suitable for Loan Guarantees from the US Department of Energy (DoE) in 2007

Company	Output (Per Year)	Cost (Loan)	Location	Feedstock(s)	Process	Fate
Abengoa Bioenergy	Ethanol: 56 million liters	\$300 million (\$76 million)	Kansas	Agricultural residues	Enzymatic hydrolysis	A \$132 million loan guarantee was finally secured in Sept 2011.
Alico	Ethanol: 53 million liters Hydrogen Ammonia	\$120 million (\$33 million)	Florida	Waste wood	Syngas fermentation	Alico sold the license to the BRI syngas fermentation technology to INEOS BIO in 2008, see Table 12.2. Loan guarantee for the new facility will be provided by the US Department of Agriculture, not DoE.
BlueFire Ethanol	Ethanol: 72 million liters	\$200 million (\$49 million)	California	Waste wood	Concentrated acid hydrolysis	Has yet to secure a loan guarantee from the DoE. Planned location has changed to Mississippi, see Table 12.2.
Broin (now known as Poet)	Ethanol: 113 million liters	\$220 million (\$80 million)	Iowa	Corn stover	Enzymatic hydrolysis	Poet has commenced construction of a smaller scale facility, see Table 12.2, but has declined the DoE guarantee.
Iogen	Ethanol: 68 million liters	\$200 million (\$80 million)	Idaho	Straws	Enzymatic hydrolysis	Iogen canceled the plant in Idaho in 2008 and has canceled a further 2 planned commercial facilities since then, the most recent in April 2012.
Range Fuels	Ethanol: 150 million liters Methanol: 34 m liters	\$150 million (\$76 million)	Georgia	Forestry residues	Gasification and catalytic upgrading	The facility started operations in 2010. However, there were technological issues with the process; the facility was unable to produce ethanol in high quantities, instead only producing 15 million liters of methanol in 2010. The plant shut down in 2011 and Range Fuels is no longer operational. The site was sold at a heavily discounted price to LanzaTech, who have a syngas fermentation process.

contributing to the 10% mandate. All of these criteria favored the development of 2GBs compared with 1GBs.

China has a production target of 10 million tonnes of ethanol and 2 million tonnes of biodiesel for 2020.[20] Importantly, there is a requirement that future expansion of biofuels should not be based on cereal crops and that there should be a focus on the use of marginal land for the production of biomass, favoring lignocellulosic feedstocks. Japan has directives to allow for 20% of the total oil demand to be met with biofuels by 2030.[21]

Where are the Commercial Facilities?

On the basis of these biofuel targets and the high levels of financial support that was being offered, particularly by the US government, the commercial prospects for 2GBs looked extremely promising in 2007/2008. However, as of May 2012 there are no commercial-scale 2GB facilities currently operational. Commercial scale can be considered to be an output of 25 million liters or more of biofuel per year.[22] Indeed, the US Environmental Protection Agency had to scale back the 2GB targets outlined in the US Energy Independence and Security Act to 24 million liters (from 378 million liters) in 2010 and to 25 million liters (from 946 million liters) in 2011. However, these revised targets were not met. The failures of the deployment of commercial biorefineries over the last 5 years are illustrated in Table 12.1, which outlines what happened to the six facilities offered loan guarantees by the US Department of Energy in 2007.

The simplest answer as to why no commercial-scale biorefineries are operational is that they are not currently cost competitive with 1GBs and fossil fuels, even given the extra financial incentives (grants, excise exemptions) that may be in place. For example, Stephen et al.[23] developed a model to compare the costs, both current and those projected for 2020, of producing ethanol from softwoods against that of 1GB ethanol from corn. The 2GB route involved the pretreatment, enzymatic hydrolysis, and fermentation stages and assumed a facility size of 50 million liters of ethanol per year in the base case. Under the assumption of an ethanol yield of 250 L tonne^{-1} (with a feedstock price of \$50 tonne^{-1}), the estimated current cost of ethanol was \$0.90 L^{-1}, 250 to 300% higher than the cost for ethanol from corn. It was found that amortization of the capital cost contributed most to this cost, followed by feedstock costs and the cost of enzymes. Wright and Brown[24] also calculated significant capital costs, for both hydrolysis and thermochemical 2GB facilities, that were over five times those of corn ethanol plants of a similar capacity (ranging from \$1.06–1.48 L^{-1} of installed capacity). The reader is also referred to a paper by Carriquiry et al.[8] that reviews numerous articles in the literature concerning the cost of 2GBs.

The study by Stephen et al.[23] predicted that the cost of corn ethanol would decrease to \$0.22–0.25 L^{-1} by 2020 and that it could be possible for cellulosic ethanol to be competitive with this but that significant efficiency and cost savings would need to be achieved at all stages of the production process. A reliance on such savings being achieved was the primary basis for the governmental 2GB targets outlined by several countries approximately 5 years ago, as, at the time, the estimated cost of producing 2GBs was highly uncompetitive. However, it is now clear that these savings have not been met in the desired timeframe.

The remainder of this paper will examine the problems that have been encountered in achieving commercially viable 2GBs and will, where such examples exist, outline recent breakthroughs in research that may be able to address these. To date there has been no clear consensus on whether the hydrolysis or thermochemical route can provide the most cost-effective 2GBs.[25] However, each route possesses its own unique set of problems and will be examined separately.

HYDROLYSIS TECHNOLOGIES

Most hydrolysis processes use either enzymes or acids (concentrated or dilute) to catalyze the degradations of the polysaccharides.[2] Subsequently, the liberated monosaccharides are typically fermented to biofuels (e.g., ethanol) although abiotic conversion mechanisms can be employed (see *Abiotic Processing of Sugars*). The process of releasing the sugars economically and in high yields has been more difficult than initially expected. For example, it has been estimated that this stage can contribute 40–45% of the total biofuel cost.[26] It has been suggested that once fermentable sugars can be isolated from lignocellulosic material at a cost of approximately \$0.22 kg^{-1} then biorefineries could be commercially viable.[27]

Recalcitrance of the Lignocellulosic Matrix

Many of the problems associated with achieving high sugar yields in a cost-effective manner are, particularly in enzymatic hydrolysis, related to the complex macrostructure of the lignocellulosic matrix, and in particular to the inhibitory effect of lignin. Lignin can act as a physical barrier to enzymes and the polysaccharide–lignin bonds can be particularly resistant to hydrolysis.[28,29] The extent of this recalcitrance not only depends on the type of linkage involved but also on the types of sugars associated with the linkage, and on the chemical structure of the lignin unit attached to the sugar.[30]

Indeed, it has been considered that the lignin content is the primary property limiting enzymatic hydrolysis[31] and that lignin contents need to be less than 12% for efficient enzymatic hydrolysis.[32] However, an important recent study[33] found that the negative correlation between lignin and enzymatic hydrolysis rates could not be explained simply by lignin content. The authors carried out organosolv pretreatments that led to improved cellulose enzymatic hydrolysis but these pretreatments did not fully remove the lignin polymer. It was considered that the structural changes occurring in the pretreatment were more important than the absolute lignin content. These changes led to the formation of lignin droplets that increased the pore volume of the biomass, thereby improving the enzymatic hydrolysis. Hence, targets for future pretreatments may not necessarily be for full lignin removal (which may be more costly in chemical, capital, and energy terms) but instead aim for appropriate modifications of the structure of the lignocellulosic matrix to allow for improved rates of enzymatic hydrolysis.

It is important also to consider that hemicellulose and lignin are covalently linked, and so hydrolysis of hemicellulose might facilitate enzymatic hydrolysis of cellulose; it has been suggested that removal of over 50% of the hemicellulose would be necessary to significantly increase cellulose enzymatic hydrolysis rates.[32] The effect of cellulose crystallinity on the rates of enzymatic hydrolysis is unclear. There have been studies indicating a proportional negative correlation between the two[34] and some that do not.[35] It is considered that the means by which the crystallinity is determined can account for these discrepancies. There can be a relative increase in crystallinity associated with a pretreatment that removes the amorphous polymers hemicellulose and lignin even if there is no direct effect on the crystallinity of the cellulose polymer itself. In such an instance, a rise in crystallinity could be associated indirectly with an improvement in enzymatic hydrolysis even though this improvement is most likely directly attributable directly to the removal of lignin and hemicellulose.[33] More specific methods of analyzing the crystallinity of the cellulose polymer are needed to ascertain the relevance of crystallinity to enzymatic hydrolysis. In a recent study,[36] it was found that there was a linear relationship between the crystallinity of pure cellulose and the initial rate of enzymatic hydrolysis.

Pretreatment for Hydrolysis

Given the recalcitrance of lignocellulose, it is very important that pretreatments are carried out prior to hydrolysis. It was found, for example, that yields of sugars from enzymatic hydrolysis were typically less than 20% where there was no pretreatment, and could be over 90% when pretreatment was employed.[37]

Types of pretreatment include particle size reduction, steam explosion,[38] ammonia fiber explosion (AFEX),[39] liquid hot water (LHW) treatment,[37,40,41] dilute acids, alkalis, wet oxidation,[42] organosolv, formic acid (FA), hydrogen peroxide,[43] supercritical CO_2,[44] ionic liquids (IL),[45–47] ozonolysis,[48] and biological pretreatments. Reviews of pretreatment processes are available in the literature.[49–53]

The pretreatment stage is one of the current limiting factors in allowing for cost-effective 2GBs. The Biomass Program at the US Department of Energy calculated that pretreatment was the second largest production cost, after the cost of the feedstock, and predicted that this would still be the case in future commercial-scale facilities.[54] These high costs are said to be the result of the need for reactors capable of operating under high temperatures and pressures, often with corrosive catalysts, and the need to employ mechanisms for the efficient recovery of these catalysts.[23,55] Banjeree et al.[26] estimated that approximately 18–20% of the total cost for the production of lignocellulosic ethanol (via the biochemical hydrolysis and fermentation route) could be attributed to pretreatment.

The target for an attractive pretreatment process would be a low cost for reagent/chemical/water use, a high process rate, a high extent of delignification, a minimal of enzymatic-hydrolysis/fermentation inhibitors formed, a low production of wastes, low capital and energy costs, and a process suitable for a wide range of lignocellulosic materials. Furthermore, if delignification, or hemicellulose depolymerization, takes place in the pretreatment process it is important that these biopolymers can ultimately be utilized for the economical production of fuels, chemicals, or energy. Unfortunately, no current technology excels in all of these requirements.

The most appropriate pretreatment method can vary according to the conversion process (and mechanisms involved) to be used, after pretreatment, and according to the feedstock used. Feedstock sensitivity in pretreatment can be a problem when designing plants to accommodate multiple biomass types. For instance, AFEX has been shown to result in a significant improvement in the fermentation rate of herbaceous materials.[56] However, this method is less effective when used on feedstocks with higher lignin contents, for example, woods.[57] The same relationship with lignin content has been shown for standard alkali pretreatment.[53] Steam (or steam-explosion) treatments have been shown to be less effective for softwoods (whose xylan polymers do not contain acetyl groups) than for hardwoods and herbaceous species, which means that an acid catalyst is usually required.[53]

Dilute acid hydrolysis pretreatments are the most established but can result in the production of fermentation inhibitors (see *Fermentation Inhibitors*) and the capital costs of these processes can be more expensive than other methods, particularly if a two-stage process, as described by Nguyen et al.,[58] is employed. The use of organic acids (e.g., maleic and fumaric acids) has recently been compared with mineral acids, however, and it was found that these can allow for hemicellulose hydrolysis in pretreatment with furfural production levels being three to five times less than when dilute sulfuric acid was used.[59] These organic acids also allowed solids loadings of up to 30% by weight.[59]

An organosolv pretreatment can yield a high-quality dry lignin product, hemicellulose in the aqueous phase, and relatively pure cellulose. This has been achieved in the Carbolea process.[43] In this the biomass of Miscanthus × giganteus was treated with FA, hydrogen peroxide, and catalytic amounts of ferric sulfate and sodium hydroxide. In an autothermal reaction (in the presence of 5% or 7.5% hydrogen peroxide), the lignin and hemicellulose dissolved and cellulose was recovered. The lignin precipitated when the solution was poured into water, and hemicellulosic sugars may be recovered from the aqueous solution. Unlike many other pretreatment methods, the organosolv processes are suitable for high lignin content feedstocks. There are, however, significant costs associated with the use of chemicals for the processes and in some cases products may be formed that can inhibit fermentation. In contrast, supercritical CO_2 pretreatment has low chemical costs but high capital costs due to the high pressures and temperatures required. Furthermore, this technology has not yet been fully demonstrated beyond the laboratory scale. Ozonolysis, in which ozone gas is used to breakdown the lignin and hemicellulose and increase the biodegradability of the cellulose, is an interesting pretreatment given that, unlike supercritical CO_2, the conditions are mild (room temperature). Also, the formation of toxic by-products is minimal.[60] However, the large amounts of ozone that are needed may hinder the commercial viability of the process.

Ionic liquids (IL), see Ref 61 for a comprehensive review, have been considered to show promise for the pretreatment of biomass for some time because these can be less energy intensive and more environmentally friendly than other pretreatment methods.[62] However, this promise has yet to be fulfilled for several reasons. For example, IL may not always be compatible with downstream biological processes (enzymatic hydrolysis, fermentation). Research to address this problem is taking place, however.[63] For example, researchers at the Joint BioEnergy Institute of the US Department of Energy have found, in a tropical rainforest, a bacterium, *Enterobacter lignolyticus*, that can tolerate IL in concentrations seen in the cellulose residue after IL pretreatment.[64] They have studied the mechanisms for this tolerance, opening up the possibility that it can be engineered in other microorganisms.

IL are also currently highly expensive, because they are manufactured on a small scale, meaning that all of the solvent would need to be recovered for economic utilization, and that has not been satisfactorily achieved on pilot scales to date.[65] Furthermore, it has been shown that the efficiency of the enzymatic hydrolysis of wood decreased with an increase in the number of IL reuse cycles, a phenomenon that may be due to the presence of lignin in the recycled IL.[66] There is still belief, however, that IL could ultimately achieve their potential. A recent paper[67] undertook a techno-economic evaluation of a biorefinery incorporating an IL pretreatment step and identified the main areas that need to be addressed so that IL could be competitive with other pretreatment methods. They considered that, by addressing these challenges, IL could be commercial in the future.

LHW pretreatment has also been mentioned as an attractive process for the future.[37] It has a similar mode of action to steam pretreatments but water is kept in the liquid state and lower temperatures tend to be used, minimizing the formation of fermentation inhibitors. However, more of the biomass tends to be solubilized in a dilute solvent stream, which means that energy intensive concentration steps will be required for recovery.[53] Furthermore, it is generally considered that this pretreatment process is still far from commercial deployment, as most of the work has been carried out on the laboratory scale.

Biological pretreatments that use fungi and bacteria to make the lignocellulosic matrix more amenable to enzymes have also received little commercial attention to date. The focus has been on basic research in identifying suitable microorganisms. The process requires low energy inputs but is slow and needs careful engineering so that only the lignin, not the polysaccharides, is consumed. An economic analysis of pretreatment technologies found this approach to be commercially unattractive on the basis of the current state of the art.[68]

Combined pretreatments offer the potential to exploit the particular benefits of each process. For example, Chen and Qui[69] suggest various pretreatments aimed at hemicellulose and lignin removal (e.g., alkaline peroxide, IL, organosolv) to precede a steam explosion pretreatment targeted toward breaking apart crystalline cellulose. Clearly, however, the capital and operational costs of combined pretreatments will need to be evaluated to determine whether such an approach is economical.

It is also important to consider the fate of the noncellulosic fractions of the biomass in pretreatment. For example, hemicellulose occupies a significant mass proportion of biomass (approximately a quarter for many energy crops)[70] and biofuel production is unlikely to be commercial unless this polysaccharide is also utilized. In pretreatments that solubilize the hemicellulose leaving a cellulosic pulp, such as alkali extraction, it is important that the hemicellulosic sugars can be recovered and utilized. The concentration of hemicellulose in the filtered hydrolysates of a pretreatment process can be low (2–3% depending on the amount of solvent used).[71] Thus concentration will be necessary for the production of ethanol, or other products, to be economical. This may require acids and/or ethanol or ultrafiltration, adding to the chemical/capital costs of the process. Research involving nanofiltration[72] has been able to achieve higher hemicellulose recovery rates at lower capital costs. The need to effectively utilize lignin is discussed in *Processes Need to Use the Lignin Residue Effectively*.

Enzymatic Hydrolysis

Refer to useful articles giving background information with regard to enzymatic hydrolysis and the modes of action.[34,49,73,74] The costs involved in purchasing/producing the enzymes, and the excessive amount of enzymes needed for effective hydrolysis of cellulose are the major problems to date regarding the production of commercial 2GBs via the enzymatic hydrolysis route. For example, it has been estimated that it takes 40–100 times more enzymes to break down cellulose than starch.[73] The total cost of enzymes can be reduced by improvements in the efficiency of their production, by increasing their specific activity (thereby allowing for reductions in the required enzyme loadings for hydrolysis), and by the recycling of the enzymes so that they may be used again for hydrolysis.

As discussed in *Recalcitrance of the Lignocellulosic Matrix*, lignin is a strong inhibitor of enzymatic hydrolysis. However, surfactants can be used to adsorb onto the lignin thereby preventing the enzymes from binding to it.[75] This allowed for the cellulase loading in one study to be reduced by up to 50% while maintaining ethanol yields and also reducing the time required to achieve the maximum ethanol concentration.[76] Ultrasonic treatment of biomass has also been shown to reduce enzyme loadings (by 50%) while also increasing ethanol yields (by 20%),[77] because the swelling of the biomass improves cellulase accessibility.

The recycling of enzymes needs to take account both of those that are free in solution as well as those that are bound to the residual substrate (cellulose and, predominately, lignin). In several studies, the free enzymes have been shown to be recoverable.[78,79] However, it has been more difficult to recover the bound cellulases.[80] Tu et al.[81] undertook experiments to optimize the conditions used (temperature, pH, ionic strength, surfactant concentration) for the desorption of bound cellulases in the hydrolysis of mixed softwoods. They found that the optimum conditions used resulted in a 25% greater hydrolysis yield, after three rounds of hydrolysis, than were obtained when the residues were reused (without desorption) in the hydrolysis process. An approach for the recycling of enzymes that is being investigated at South Dakota State University, with funding from enzymes company Novozymes, involves immobilizing the enzymes by attachment to beads.[82] That means that the enzymes are retained rather than being removed with the hydrolysate. With this approach, the enzymes could be used for five cycles, although their activity had decreased to 40% by the fifth cycle.

The products of hydrolysis (glucose and cellobiose) can inhibit cellulase activity, thereby increasing the required enzyme loadings. Methods to overcome this problem include utilizing a simultaneous saccharification and fermentation (SSF), as opposed to a separate hydrolysis and fermentation (SHF), setup so that these products are fermented to ethanol as they are produced.[49] However, the optimum conditions for hydrolysis tend to be different from those for fermentation. Products of the pretreatment process, for example, acetate,[83] may also be inhibitory to enzyme activity as may contaminants such as silica.[84]

Enzymatic hydrolysis can also be a lengthy process and does not follow a linear progression.[85] Instead, the initial release of sugars is rapid (associated with the hydrolysis of the amorphous portion of cellulose) with the release of the remaining (mostly crystalline) cellulosic sugars taking longer. An alternative to full hydrolysis might be to employ a reduced hydrolysis time, resulting in an increased average rate of hydrolysis and using the remaining unhydrolyzed material for power generation or for the production of other products. Such an approach would reduce capital and process costs but would also reduce the yields of sugars, and hence the biofuel yield. Thus an effective means of processing the residue (e.g., pyrolysis; see *Pyrolysis*) would be required.

To date most research has focused on improving the efficacy and commercial prospects of cellulases and there has been less activity regarding the development of hemicellulases. This is partly because of the complexity and diversity of the hemicellulose polymers and also because many of the current chemical pretreatment methods degrade hemicellulose as well as lignin. For many herbaceous feedstocks (e.g., *Miscanthus*) arabinoxylan is the principal hemicellulose. This consists of a xylan backbone with β-(1 $\rightarrow$ 4) linkages that are branched mainly with arabinofuranosyl and 4-*O*-methyl-glucopyranosyl units, and the degree of branching varies according to the feedstock.[86,87] The complexity of this polymer, and of other hemicelluloses, means that a variety of enzymes are required to work on the different glycosidic linkages. Each of these hemicellulases may have different substrate specificities and optimum conditions with regards to pH, temperature, and resistance to contaminants. There also can be conflicts between the activities of hemicellulases and cellulases; for example, the xylo-oligosaccharides produced by xylanases can be strong inhibitors of cellulases.[88] A thorough review of the types of different hemicellulases and recent developments in the field was provided by Dumon et al.[89]

There has been significant progress in recent years, despite all the problems encountered, toward achieving commercial 2GBs from enzymatic hydrolysis. For example, in March 2012 Novozymes introduced the Cellic CTec3 cellulase and hemicellulase complex. The company claimed that the efficiency of this product in biomass to sugar conversion was 50% greater than that of the previous iteration (CTec2) and that a lower enzyme loading was required; 50 kg would allow for a tonne of ethanol to be produced from biomass.[90] These developments are said to have reduced the cost of enzymes significantly from over $5 per US gallon of ethanol in 2001[91] to a cost point that can now allow for total cellulosic ethanol production costs at $2–2.5 per US gallon, similar to that for corn-based ethanol at $2.2 per US gallon.[90] Other companies, such as Dyadic (which has developed the C1 enzyme platform), Genencor (which sells Accellerase TRIO), and DSM, are working specifically at developing advanced enzyme formulations. Some biorefining companies (e.g., Sud-Chemie) are also developing proprietary enzyme formulations while many others (e.g., Beta Renewables, Poet) collaborate with the specialist enzyme developers in developing formulations optimized for their feedstocks and technologies.

Acid Hydrolysis

The acid hydrolysis of biomass is not a new concept, the process was commercialized in the late nineteenth century,[37] several dilute-acid facilities existed in the USA, Germany, Japan, and Russia by World War I, while concentrated acid hydrolysis facilities were being built between 1937 and the late 1960s.[92] However, these were uneconomic where fossil fuels were available and no commercial-scale facilities currently exist. There are three main problems associated with dilute acid hydrolysis processes: (1) low sugar yields; (2) the production of fermentation inhibitors (see *Fermentation Inhibitors*); and (3) the capital and chemical costs associated with the use of acids. A major challenge for commercially-viable dilute acid hydrolysis technologies is to raise glucose yields above 70% while maintaining a high rate of hydrolysis.[49] In contrast, enzymes are much more selective and can achieve high sugar yields, estimated at 75–85% and expected to rise to 85–95% in the future.[37] Concentrated acid hydrolysis processes can achieve significantly higher sugar yields (than dilute acid technologies) and the formation of chemicals inhibitory to fermentation is less of a problem. The capital and chemical costs become the primary causes for concern. Commercial concentrated acid hydrolysis processes will require that the acid is recovered in high yields, and that has not been possible in commercial-scale facilities to date. However, some modern companies, such as BlueFire Renewables and Virdia, claim to have achieved recovery through the development of advanced product separation technologies. BlueFire Renewables is an integrated biorefining company in the USA that has produced ethanol on a pilot scale using concentrated sulfuric acid to hydrolyze the biomass (the Arkenol process),[93] and recently has spun-off a separate company, SucreSource, which plans to market the front end hydrolysis part of this process. Virdia is a US company that is solely focused on the production of sugars from biomass and has developed a hydrolysis technology called CASE (Cold Acid Solvent Extraction) that uses concentrated hydrochloric acid.

FERMENTATION OF LIBERATED SUGARS

Fermentation Inhibitors

The presence of fermentation inhibitors derived from pretreatment/hydrolysis can be a significant obstacle in achieving high product yields. These inhibitors can be classified into three groups: aliphatic acids (e.g., formic, levulinic), furan derivatives (e.g., furfural and hydroxymethylfurfural), and phenolic compounds (e.g., phenol, vanillin). Their mechanisms of inhibition have been reviewed.[94] The types and concentrations of inhibitors will be dependent on the feedstock, pretreatment, and hydrolysis methods used, whereas the problem of inhibitors is increased if process water is recycled (something that is necessary to improve the commercial viability of biorefining).

Inhibitors can be removed by evaporation, solvent extraction, over-liming with calcium hydroxide, activated charcoal, ion exchange resins, enzymatic detoxification, and electrodialysis.[95] Many of these processes have been reviewed by Huang et al.[71] However, the cost of detoxification of the process hydrolysates to a level suitable for fermentation can be significant; Sivers et al.[96] calculated that it amounted to 22% of the total cost of ethanol production. There have been promising developments in this area, however. A recent paper by Alriksson et al.[97] outlined a procedure for the treatment of lignocellulosic hydrolysates (from enzymatic hydrolysis) using a reducing agent (dithionite). This addition was carried out *in situ* in the fermentation vessel, in contrast to many previous studies on the neutralization of inhibitors that require separate (and often costly) steps. The treatment also did not require any pH or temperature adjustments. The authors found that the treatment raised the ethanol productivities of a spruce hydrolysate from 0.2 to 2.5 g L^{-1} h^{-1} and of a bagasse hydrolysate from 0.9 to 3.9 g L^{-1} h^{-1}. These values were even higher than those for fermentations with reference sugar solutions without inhibitors.

An alternative to the detoxification of hydrolysates is to develop microorganisms that are more tolerant of inhibitors. There has been progress in this area; researchers at the US Department of Energy's Oak Ridge National Laboratory have identified a gene in a mutant strain of *Zymomonas mobilis* that allows the microorganism to tolerate acetate.[98] This gene has been inserted into nonmutated strains of *Z. mobilis*, as well as into other microorganisms such as yeasts, and the tolerance was maintained. There exists the possibility for development of tolerances to other inhibitors through a similar approach.

Limitations of Conventional Microorganisms

Historically, the yeast *Saccharomyces cerevisiae* has been used for the fermentation of ethanol from glucose. However, there are problems associated with utilizing this microorganism for the fermentation of hydrolysates from lignocellulosics. For instance it cannot ferment pentoses. This is a significant limitation given that these can account for 25% or more of the mass balance of many feedstocks.[70] One means of addressing this problem involves the fermentation of pentoses, using capable microorganisms, in a separate reactor to that used for hexose fermentation, although this will incur additional capital costs. Pentose fermentation can take place after pretreatment (if pentoses are liberated in that stage) or after the distillation of the resulting beer from hexose fermentation. However, the yeasts that can ferment pentoses tend to be sensitive to ethanol concentrations > 4%, can be highly sensitive to inhibitors, and cannot ferment xylose at low pH values.[26] Enteric bacteria (e.g., *Escherichia coli*) can ferment pentoses as well as a hexoses and the soil bacterium *Klebsiella oxytoca* can also ferment cellobiose; however, ethanol is not the main product of the fermentation pathways of these microorganisms.[99] Researchers have, however, been able to modify these pathways for ethanol production by introducing genes of the ethanol-producing *Zymomonas* into these bacteria.[100]

While there has been a significant amount of laboratory research on the development of recombinant strains of conventional fermentative micro-organisms[101] it has often been found that these strains are not sufficiently robust to function in large-scale industrial environments.[102] Recently, however, a genetically engineered strain of the industrial-scale-capable yeast *S. cerevisiae* capable of fermenting pentose sugars has been developed.[103,104] This was achieved by introducing a gene, xylose isomerase, which allows for the conversion of xylose to xylulose, a form which the microorganism can naturally utilize. Similarly, the Danish biorefining company BioGasol will use a *Saccharomyces* yeast that has had incorporated the xylose-fermentation pathway gene from the *Piromyces* fungus.[105]

The development of a robust microorganism capable of fermenting pentoses and hexoses would allow for the deployment of simultaneous saccharification and cofermentation (SSCF) processes where both sugar types are fermented together in the same reactor that is used for enzymatic hydrolysis.[106] This simplified process would be significantly less capital intensive than process schemes requiring separate reactors for each of these stages. However, such processes have been slow to develop commercially because the optimal conditions required for hydrolysis are different from those for fermentation and improved micro-organisms are necessary, such as thermophiles for the (co)fermentation of sugars.

SSF processes still require the yeast to be produced in a separate reactor, however. In pilot biorefining schemes approximately 9% of the carbohydrate released from hydrolysis is used to grow yeast cellular mass.[107] The ultimate simplification of the enzymatic hydrolysis process is presented by the concept of consolidated bioprocessing (CBP) whereby a single microorganism, capable of both hydrolysing polysaccharides and fermenting the products of hydrolysis, produces a 2GB from (pretreated) biomass in a single reactor. The CBP process is considered to have the potential to offer the lowest costs for biofuels and chemicals.[108] Unfortunately, CBP has been found in the past to result in low yields of ethanol and to require long periods for fermentation (up to 12 days).[109] However, there has been a significant amount of research in recent years toward developing capable microorganisms. For example, researchers affiliated with Oak Ridge National Laboratory in the US developed a genetically engineered strain of the microbe *Clostridium celluloyticum* so that it could produce isobutanol directly from cellulose in a single step.[110] Its activity has only been demonstrated on the laboratory scale but there are plans to scale up the process. Most importantly in the near term, the US biorefining company Mascoma is planning to start construction of a commercial-scale CBP facility in 2012 (see Table 12.2).

Extraction of Chemicals discusses the problems associated with the recovery of the products of biorefining. The energy and capital costs associated with the concentration of the products of fermentation would be reduced if the fermentative microorganisms could tolerate

Table 12.2 A Summary of the Commercial-Scale Biorefineries Currently in Construction and Those That Are Planned to be under Construction before the End of 2012

Company	Output (Per Year)	Cost	Location	Construction Period	Feedstock(s)	Process	Notes
Under Construction							
Abengoa Bioenergy	Ethanol: 87 million liters	$400 million	Kansas, USA	Sep. 2011 → 2013	Corn stover, wheat straw, switchgrass	Enzymatic hydrolysis	Has operated a pilot plant in Spain since 2006; another commercial facility is planned for Nebraska
Beta Renewables	Ethanol: 50 million liters	$350 million	Crescentino, Italy	Apr. 2011 → 2012	Wheat straw and (later) Arundo donax	Enzymatic hydrolysis	Also plans for commercial facilities in Colombia and Brazil
Enerkem	Ethanol: 38 million liters	$75 million	Alberta, Canada	2010 → 2013	Municipal solid waste	Gasification and catalytic upgrading	Plans for another two facilities of the same scale by 2015
INEOS Bio	Ethanol: 30 million liters Electricity: 6 MW	$130 million	Florida, USA	Feb. 2011 → 2012	Municipal solid and other wastes	Syngas fermentation	Another facility of a similar scale is planned in the UK (Teeside)
KiOR	Biocrude: 42 million liters	$222 million	Mississippi, USA	2011 → 2012	Wood	Catalytic pyrolysis, hydrodeoxy-genation	Facility will house an HDO unit; three more, larger scale, pyrolysis facilities are planned; their bio-oil will be sent to this HDO unit
Poet	Ethanol: 76 million liters	$250 million	Iowa, USA	Mar. 2012 → 2013	Corn stover	Enzymatic hydrolysis	Will be adjacent to a 1GB facility
Planned in Near Term							
BlueFire Renewables	Ethanol: 72 million liters	$300 million	Mississippi, USA	2012 → 2013	Waste wood, municipal solid waste	Concentrated acid hydrolysis	Site clearance began in 2010; second phase needs additional financing
BP Biofuels	Ethanol: 136 million liters	$400 million	Florida, USA	2012 → 2014	Energy cane/grasses, *Miscanthus*	Enzymatic hydrolysis	BP bought the Verenium technology; a 5 million L yr^{-1} demo facility has operated since 2008
China New Energy	Acetone, butanol, ethanol (50k tonnes)	$32 million	Jilin, China	2012 → 2013	Straws, corn stover	Enzymatic hydrolysis	Uses the butanol technology of Green Biologics

(Continued)

Table 12.2 Continued.

Company	Output (Per Year)	Cost	Location	Construction Period	Feedstock(s)	Process	Notes
COFCO-Sinopec	Ethanol: 64 million liters	\$93 million	Heilongjiang Province, China	2012 → 2013	Corn stover	Enzymatic hydrolysis	A pilot facility has operated since 2007; enzymes from Novozymes
Coskata	Ethanol: 60 million liters	\$192 million	Alabama, USA	2012 → 2013	Wood	Syngas fermentation	Phase II (2015), 295 million liters target
DuPont Cellulosic Ethanol	Ethanol: 106 million liters	\$300 million	Iowa, USA	2012 → 2014	Corn stover	Enzymatic hydrolysis	Will be colocated with an existing corn to ethanol facility
Fiberight	Ethanol: 22 million liters	\$60 million	Iowa, USA	2012 → 2013	Municipal solid waste	Enzymatic hydrolysis	Is working with Novozymes
Fulcrum Bioenergy	Ethanol: 40 million liters Electricity: 16 MW	\$120 million	Nevada, USA	2012 → 2013	Municipal solid waste	Gasification and catalytic upgrading	Requires a loan guarantee from the US Department of Energy to begin
Mascoma	Ethanol: 76 million liters	\$232 million	Michigan, USA	2012 → 2013	Hardwoods	CBP	150 million liters output target for 2014

CBP, consolidated bioprocessing.

higher ethanol concentrations (the maximum current level is about 10% at 30°C and less at higher temperatures).[37] Researchers at the BioEnergy Science Center of the US Department of Energy recently found the gene that controls the ethanol production capacity of the anaerobic bacterium, *Clostridium thermocellum*, a microorganism that can tolerate higher ethanol concentrations.[111] It is possible that this discovery can facilitate the development of tailor-made microorganisms capable of breaching current ethanol concentration limits.

Alternative Fermentation Pathways

The conventional fermentation of sugars to ethanol may not be the most efficient fermentative approach. Even assuming that the microorganism worked at 100% efficiency and required none of the sugar for the growth of cellular mass, the weight of ethanol would be 51.11% of that of glucose. That means that 48.89% of the sugar mass from glucose will be converted to carbon dioxide, a product of little or no value on mass scales of production. The US company ZeaChem utilizes acetogens to ferment the hydrolyzed (C_6 and C_5) sugars to ethanoic acid. This is a pathway that does not produce carbon dioxide as a by-product allowing for a carbon efficiency, according to their Web site, of close to 100% for this stage. This ethanoic acid is esterified and then hydrogenated, using hydrogen produced from the gasification of the lignin residue from hydrolysis, to produce ethanol. ZeaChem claim significantly greater ethanol yields through this approach than through conventional fermentation. In January 2012, the company began operations at a 1 million liter per year demonstration facility in Oregon although the initial products will only be ethanoic acid and ethyl acetate; the production of ethanol is planned for later in 2012. The company also plans for a commercial-scale facility producing approximately 100 million liters of product per year to be operational in the same state by 2014 and has received a \$232.5 million governmental loan guarantee to facilitate this.

LS9 and Synthetic Genomics have been developing tailor-made microorganisms for the production of drop in biofuels (biofuels fully compatible with existing fossil-based fuels) from free sugars. The complex engineered pathways allow for the synthesis of hydrocarbons of a longer carbon-chain length than the original carbohydrate.[112] The commercial progress of these companies has been relatively slow as they emerged from stealth mode. However, LS9 plans to open a demonstration facility in Florida in 2012 although this will initially produce high-value chemicals rather than biofuels.[113] The company has also, in conjunction with researchers from the University of California at Berkeley and the Joint BioEnergy Institute of the Department of Energy, recently claimed to have engineered *Escherichia coli* to produce biodiesels, fatty alcohols, and waxes from free sugars and also from the sugars that are liberated from wood chips by hemicellulases expressed by the microbe.[114]

Abiotic Processing of Sugars

There are ways, other than through the use of microorganisms, in which biofuels can be obtained from the sugars liberated in hydrolysis. For instance, the acid degradation of biomass could be targeted so that the sugars themselves are degraded into products such as levulinic acid (LvA), hydroxymethylfurfural, and FA (from the hexoses) and furfural (from the pentoses). These are problematic fermentation inhibitors in fermentation processes but are also high value platform chemicals that can be used to synthesize a range of biofuels.[115] Lange et al.[116] proposed a new platform of LvA derivatives, termed valeric biofuels, that can provide gasoline and diesel miscible biofuels that have several advantages over the conventional LvA-derived biofuels such as methyltetrahydrofuran, ethyl levulinate, and gamma valerolactone (GVL). The researchers outlined an integrated catalytic process incorporating the acid hydrolysis of lignocellulose to LvA, the hydrogenation of LvA to GVL and then valeric acid and its esterification to alkyl (mon/di)valerate esters. Another recent paper[117] demonstrated that the production of gasoline-range fuels was possible via the catalytic fast pyrolysis of furfural. Catalytic hydrogenation of furfural can yield the gasoline-type fuel 2-methylfuran or, if the aldehyde group is hydrogenated, furfural alcohol, a valuable platform chemical that can be etherified with alcohols to produce biodiesel components.[32]

The commercial production of such biofuels will depend to a large extent on whether high yields of LvA and furfural can be achieved in a cost-effective manner. The Biofine technology, which employs the dilute acid hydrolysis of biomass in two reactors, was initially developed in Dartmouth College in 1988, and has been considered to show promise.[115] However, while the technology has been deployed in two pilot plants over some periods during the last 15 years, no commercial facilities using this technology are operational. Mercurius biofuels have developed an integrated process for the production of biofuels via the ethanolysis of biomass to produce ethyl levulinate along with furfural.[118] These products are then combined to increase the chain length of the molecule which is then hydrotreated to produce a range of chemicals, including biofuels. This method should allow for the products to be concentrated in a more cost-effective manner than would be possible for the low concentrations of LvA produced in dilute acid hydrolysis. However, the Mercurius technology has not yet been demonstrated on the pilot scale.

Murat Sen et al.[119] present a techno economic evaluation for an integrated biofuel production process involving (1) the conversion of cellulose to LvA and FA; (2) the conversion of this LvA + FA solution to GVL; (3) the separation of GVL from the sulfuric acid solution using liquid–liquid extraction with butyl acetate; (4) the catalytic decarboxylation of GVL to butene and carbon dioxide followed by; (5) the oligomerization of butene to higher molecular weight alkanes (C_8 to C_{20}). Unfortunately, the minimum selling price calculated for this mix of alkenes ($4.31 per US gallon of gasoline equivalent) is prohibitively high, based on the costs of petroleum-based fuels at this time. Also, these calculations were derived from LvA production experiments that took place using cellulose and not lignocellulosic biomass.

'Aqueous phase reforming' is the name often given to the liquid-phase catalytic processing of sugars to produce drop-in biofuels.[120] The process involves the acid catalyzed dehydration of sugars which is then followed by aldol condensation over solid base catalysts to form large organic compounds that are converted to alkanes by dehydration/dehydrogenation over bifunctional catalysts. These alkanes (C_7 to C_{15} chain length) are readily separated from the reaction mixture and are said to contain 90% of the energy of the carbohydrate and hydrogen feeds.[120] However, this process is focused only on the conversion of sugars, not on the hydrolysis of polysaccharides. That means that a stream of hydrolyzed lignocellulosic sugars will be required. The work of Huber and others at the University of Wisconsin led to the formation of the 'Virent Bioforming' technology that Virent (funded by Shell and Cargill, among others) is currently looking to commercialize. The company is collaborating with Virdia to develop gasoline and jet fuels, and the latter are being tested by the US Air Force Research Laboratory.[121] If the Virent Bioforming and Virdia CASE technologies can be integrated in a cost-effective manner then this combined approach may have significant potential for commercial 2GB production.

Processes Need to Use the Lignin Residue Effectively

It has been estimated[122] that approximately 40% of the lignin product from a hydrolysis biorefinery would be necessary for the energy requirements of the process. That means that 60% of the lignin residue could be available for other uses. The end uses for the lignin product will depend on its purity. If it is extracted prior to hydrolysis (e.g., by an organosolv process as with the Lignol technology)[123] then the production of high-value biomaterials may be possible.

If the lignin is a residue of the hydrolysis process, it will be less pure as it will also contain unconverted sugars, unhydrolyzed cellulose, bound cellulases, and ash and hence may be less suitable for the production of biomaterials but still a viable feedstock for liquid fuel synthesis. Sannigrahi et al.[122] outline some of the research undertaken to synthesize gasoline/diesel miscible biofuels from lignin. Such approaches can involve the hydroprocessing of lignin that has been depolymerized by an alkaline solvent[124] or the catalytic pyrolysis of lignin (see *Pyrolysis of Lignin*) and subsequent hydrodeoxygenation (HDO) of the bio-oil.[125]

Thesis Chemistry, a company based in Iowa, is planning to develop facilities that will utilize the lignin residue from 2GB (biological hydrolysis) biorefineries for the production of chemicals and fuels. The company plans to develop and sell lignin derivatives in four phases: (1) the production of vanillin (of use in flavoring); (2) the production of cresols (of use in disinfectants and deodorisers); (3) the production of phenols; and (4) the production of butanes and other fuel additives.[126] The company plans for the first facility to be operational in 2014.

THERMOCHEMICAL TECHNOLOGIES

Gasification

Gasification (see Refs 127–129 involves the thermal decomposition of biomass (lignin and the polysaccharides) under a limited supply of oxygen. This results in the production of a syngas, which can contain carbon monoxide, carbon dioxide, hydrogen, water, and short-chain hydrocarbons, among other products. Chemicals and biofuels can be synthesized from this gas by utilizing either metal catalysts or fermentative microorganisms.

There have been some high profile technological failures in the gasification of biomass (for power production). The ARBRE plant in Yorkshire, England, is one example. However, there are facilities that have demonstrated that gasification of biomass is feasible over prolonged periods; a 2 MW (electric) and 4.5 MW (thermal) fluidized bed gasification facility in Austria has been running at 90% availability since 2001.[22] However, cheap, large-scale biomass gasifiers have not, as yet, been demonstrated nor has the successful commercial-scale production of fuels from the syngas.

Fuel Synthesis Using Metal Catalysts

There are several articles that focus on the synthesis of chemicals and fuels (e.g., methanol, ethanol, higher alcohols, Fischer–Tropsch (FT) diesel, gasoline, dimethyl ether) via metal catalysts.[130–133] The synthesis of these products depends on the ratio of carbon monoxide to hydrogen in the syngas. The amount of hydrogen (and hence carbon dioxide) produced can be increased via the water gas shift reaction, which involves mixing steam with the syngas, as insufficient water vapor is usually released from the biomass.

The production of fuels from fossil-fuel-derived syngas is well established; however, there have been several problems associated with catalytic fuel synthesis when dealing with biomass-derived syngas. For instance, the production of tars is a significant problem because these need to be cleaned before chemical synthesis can occur. The purity of syngas required for the production of synthetic fuels is approximately one order of magnitude greater than that for power generation.[134] The recent failure of the commercial-scale biorefinery built by Range Fuels (see Table 12.1) has been attributed to the build-up of tars during the process.[135] Catalysts have a vital role in the reforming of these tars and nonmetallic calcined dolomites and nickel-based catalysts have been the focus of much research. However, there are problems with both of these, and further research is needed.[136] Less tars will be formed if higher steam/fuel ratios and process temperatures are used but these will decrease the efficiency of the process.[137] Alternatively, catalysts that are less sensitive to tars/impurities could be developed, allowing them to be functional for longer periods.

The ash compositions and contents of biomass can also cause problems; for instance, many catalysts can be deactivated by ash.[138] A study[132] estimated that, for mixed alcohol synthesis from forest residues, the minimum selling price for ethanol rose by approximately 50% when the ash content was increased from 1–15%. The ash compositions of some feedstocks may lower ash melting points meaning that the ash will melt during thermochemical conversion, resulting in the blockage, erosion, and corrosion of equipment.[139] Such reductions in ash melting points can be attributable to the presence of high concentrations of alkali and alkaline earth metals, silica, phosphorus, sulfur, and chlorine in the biomass.[140]

Fluidized bed gasifiers, by operating at less than 1000°C, can reduce these ash slagging problems, but such conditions produce more tars and methane in the syngas. Furthermore, it is considered that such reactors may be inappropriate at the scales needed for the economical synthesis of fuels via the conventional FT mechanism.[136] Entrained flow gasifiers are suitable for larger scales[141] and are engineered to minimize slagging but they require the biomass particles to be very fine, which may incur a significant energy expense.[142] Torrefying the biomass prior to gasification may facilitate biomass grinding at reduced energy costs, however, allowing biomass to be utilized effectively in an entrained flow gasifier.[143] The Choren process[141] consists of a slow-pyrolysis stage which produces a gas, which goes to a gasifier, and a char that is ground down and injected in the lower part of the gasification chamber. This facilitates a reaction with the hot gases. The syngas then feeds an FT chamber, based on a shell gas-to-liquids process,[144] for the synthesis of a diesel-type fuel, termed SunDiesel.

These ash problems can be tackled on the supply side, harvesting the biomass when the concentrations of the problematic constituents are lower. For example, harvesting *Miscanthus* in the spring rather than at senescence (October) is associated with a much lower mineral content in the crop, an affect mostly attributable to the loss of harvestable leaf matter over this period.[145] As a complement, the delayed harvest of *Miscanthus* also provides lower moisture contents. This is a bonus for thermochemical processing because higher moisture contents will lower the calorific value of the biomass and decrease the temperature in the oxidation zone of the gasifier, resulting in more tars. It is considered that the moisture content of feedstocks should be less than 30%.[140]

The fossil fuel-to-liquids facilities that currently exist are very large; for example, a (natural) gas-to-liquids facility in Qatar produces about 34,000 barrels of FT products a day.[146] Such economies of scale are necessary for many gasification technologies, particularly those employing the FT mechanism; it has been estimated that a minimum gasifier rating of 500 MW is necessary for FT biomass-to-liquids to be viable.[147] This may lead to prohibitive transport costs associated with sourcing the biomass. Recently, however, there has been work on the development of small, yet efficient, catalytic gasifiers that combine thermal gasification, catalytic tar cleanup, and catalytic water–gas shift in a single reactor.[32] It is possible that such gasifiers, when integrated with advancements in microreactor technologies (e.g., FT synthesis)[148] could allow for smaller-sized thermochemical plants to be viable.[32]

There have also been interesting recent developments in improving the potential yields from the gasification of biomass. A flaw in current gasification schemes is that a significant amount (approximately half) of the carbon from the biomass ends up as carbon dioxide, rather than the carbon monoxide suitable for biofuel synthesis. However, researchers at the University of Massachusetts at Amherst and the University of Minnesota have claimed to convert all of the carbon dioxide produced from the biomass into carbon monoxide.[149] This is achieved in one step through the use of a catalytic reactor and by gasifying the biomass in specific quantities of carbon dioxide and methane (obtained from natural gas, for example). The carbon dioxide combines with hydrogen to produce water and carbon monoxide, whereas the addition of methane allows for the full conversion of the carbon dioxide. The carbon contained in the methane is also converted to carbon monoxide in this process. This approach has been demonstrated to date only on cellulose, and not on virgin biomass samples.

Fermentation of Syngas

An alternative to the utilization of metal catalysis is for the syngas to be fermented. Autotrophs can use C1 compounds, such as CO, CO_2, and methanol, for their carbon source and hydrogen as their energy source while unicarbonotrophs use C1 compounds alone for both these purposes.[150] These microorganisms also utilize metals such as cobalt and nickel (which are contained in their enzymes) for the conversion of C1 compounds into value added products such as ethanol. Numerous end products other than ethanol are also possible, including hydrogen, butanol, ethanoic acid, and methane; a list of the microbes used and their fermentation products are provided by Mohammadi et al.[151]

There are several advantages that the biological route has over the use of metal catalysts. These include that it involves much lower temperatures and pressures, thereby reducing the energy and capital costs of conversion. The biological route also tends to be more specific in its conversion, resulting in higher yields. For example, depending on the microorganism used, ethanol can be the only biofuel product, whereas the use of metal catalysts produces a range of alcohols that need to be separated at significant economic and energy costs. Furthermore, the adjustment of the H_2:CO ratio via the water gas shift reaction is not necessary where biological catalysts are employed[150] and the microorganisms can be less sensitive to many of the gas contaminants, such a sulfur, that poison many metal-based catalysts.[152]

There are, however, problems associated with this biological approach to utilizing syngas. First, the right environment (levels of nutrients and impurities) needs to be maintained in the reactor so that the microorganisms can survive and be active. As with the microorganisms associated with the hydrolysis pathway (see *Fermentation Inhibitors*), the syngas fermenting biota can be subject to fermentation inhibitors. It is considered that further work may be necessary to improve the efficiency of the microorganisms used for fermentation along with their sensitivity to high concentrations of end products.[151] Second, there are issues relating to the poor mass transfer of the syngas into the liquid media,[153] which results in low reaction rates. Techniques for improving this mass transfer include reducing the diameter of the gas bubbles through the use of (impeller-based) stirred tank reactors or sparging the gas into the reactor using a microsparger.[151] Soluble chemicals[151] and nanoparticles[154] have recently been shown to reduce the surface tension and increase mass-transfer.

Despite these issues, several companies are pressing ahead with plans for commercial-scale biorefineries employing the fermentation of syngas. These include Coskata, INEOS Bio (see Table 12.2), and the New Zealand company LanzaTech. The LanzaTech technology is said to be able to convert gases containing carbon monoxide into fuels and chemicals and to be flexible to both the hydrogen and impurities content of the syngas. The company has plans to apply the technology to the flue gases from the steel manufacturing industry as well as to biomass-derived gases.[155]

PYROLYSIS

Optimization of the Yield and Quality of Bio-Oil

Pyrolysis can be considered to be the thermal treatment of a material in the absence of oxygen. It results in the evolution of noncondensable gases, a liquid fraction (bio-oil) obtained from the condensable vapors generated, and a residual solid component (biochar) that does not decompose under the conditions employed. Pyrolysis conditions can be modified to alter the relative proportions and compositions of these three products. The production of biofuels via the pyrolysis route focuses on the production of bio-oil in high yields via fast pyrolysis.[156–158] The bio-oil that is produced is a complex mixture of water, carboxylic acids and lignin- and carbohydrate-derived products[159] that can deteriorate rapidly over time, with polymerization reactions forming additional water and causing phase separations.[160] It has a low heating value, a high acidity, and high oxygen and water contents[161] and is primarily used for power generation using modified diesel engines (e.g., in the commercial facilities operated by Dynamotive Energy Systems in Canada).

The type of feedstock can have a significant impact upon the quality of the bio-oil produced. For example, woody feedstocks tend to produce bio-oils with higher C and H contents and lower water contents than many herbaceous feedstocks.[158] A particular advantage of woody materials is that their ash contents tend to be low. High levels of ash can cause many problems in fast pyrolysis and result in an increased char yield and a decreased bio-oil organics yield (with an increased water content in the bio-oil).[162] An ash content of over 3% can also lead to the phase separation of bio-oil over time.[158] The high levels of chlorine and alkali metals such as potassium present in this ash may also cause problems if bio-oil is to be combusted.[163] However, Javaid et al.[164] recently demonstrated that the removal of char particles from the bio-oil by microfiltration could significantly reduce its ash content, improving its stability and performance.

These issues will prohibit the economical use of high ash feedstocks (such as agricultural and municipal wastes) in fast pyrolysis technologies unless mechanisms to limit their impact can be implemented. One option may be to wash such materials with acid or water to reduce the ash content.[165] Washing biomass will increase its moisture content, however, and an energy expense may be incurred to reduce this to an acceptable level prior to pyrolysis. With some feedstocks, for example, agricultural residues, it may be possible to leave the harvested biomass on the field so that rain may leach some of the inorganic material.

The highly complicated processes needed to develop a pyrolysis procedure can potentially lead to the production of a range of desirable chemicals in good yields. However, these processes are not fully understood in terms of reaction pathways for polymers such as cellulose and lignin despite the significant amount of work that has been carried out in this area.[166–168] The rapidity, variety, and complexity of the reactions taking place are considered to be too great for current models to handle. However, recent research at the University of Massachusetts, Amherst, has shown that α-cyclodextrin acts in a similar way to cellulose during conversion.[169] As it is much smaller than cellulose, computer modeling of the conversion mechanisms of this ‘mini-cellulose’ can take place. This discovery opens up the possibility for optimized pyrolysis processes to be developed according to the desired end-products.

Catalysts can be used during the pyrolysis process to improve the quality of the bio-oil, for instance Al-MCM-41 type mesoporous catalysts with a Si/Al ratio of 20 have been tested at various pore sizes and have allowed the elimination of levoglucosan from pyrolysis gas and a preference for ethanoic acid, furfural, and furans from cellulose.[170] Solvents can also be used to improve bio-oil quality and yield. Sulfolane has been shown to inhibit the polymerization reactions that may decrease yields of such anhydromonosaccharides as levoglucosan.[171,172] It has, in fact, allowed for the complete pyrolysis of cellulose to soluble products, with no char formation,

Bio-Oil Upgrading

Even the most optimiszd pyrolysis process will not produce a bio-oil suitable as a drop-in biofuel. Indeed, due to its high moisture content bio-oil is polar and so is immiscible with crude oil.[173] To obtain a viable biofuel it is necessary to upgrade the bio-oil.[133,158,161] The main methods employed are HDO and zeolite cracking.

HDO involves the addition of hydrogen, under high pressure and at temperatures between 250 and 450°C, to remove oxygen from the bio-oil, forming water and a hydrocarbon-like product that allows for phase separation between the aqueous and organic layers. In the case of extensive deoxygenation there will be two organic layers, with a light fraction lying on top of the water and a heavy fraction below.[173] The use of hydrogen also saturates double bonds, increasing the H/C ratio and hence the heating value, of the final fuel.[174] However, the HDO route requires hydrogen, with increasing amounts necessary to remove greater amounts of oxygen. Furthermore, there is a trade-off between low oxygen content in the final product and the ultimate yield of that product; increasing deoxygenation leads to more bio-oil being converted to gas and water phases. Samolada et al.[175] found that increasing the degree of deoxygenation from 78 to 100% resulted in a decrease in the oil yield from 55 to 30%, while the water yield increased from 40 to 55%. Another problem currently associated with HDO is the deactivation of the catalysts, primarily as a result of coking.[176] It is considered that there needs to be significant improvements in the use of catalysts before HDO can be considered a commercially viable route.

The zeolite cracking of bio-oil is conceptually similar to fluid catalytic cracking (FCC) and involves the removal of oxygen from the bio-oil as carbon dioxide and water. The process typically takes place at 300°C–600°C but, unlike HDO, atmospheric pressure can be used as no hydrogen is required. However, the same degree of deoxygenation with zeolite cracking will result in a lower oil yield and H/C ratio than with HDO. Indeed, very low oil yields (14–23%) have been seen for zeolite cracking, with carbon formation being significant (26–39%).[157] This carbon formation severely diminishes the lifespan of the catalysts used, even in comparison to HDO catalysts. Regeneration of these catalysts has been attempted, for example, Vitolo et al.[177] washed the catalyst with acetone and heated it at 500°C over 12 h, but it was found that the regenerated catalyst had a shorter lifespan and was unable to deoxygenate the bio-oil to the same extent. This phenomenon increased with the number of regeneration cycles. Typically, zeolite cracking will result in an oil that has a higher oxygen content (13–24%) than HDO (<5%). That means that the heating value of the fuel will also be lower.[173] The oil produced by zeolite cracking will also have a higher viscosity and lower pH. It is considered by many[173, 178] that the current state of the art of zeolite processing does not produce oil of sufficient quality or yield to allow biofuel production to be commercial and that the focus should be on the HDO route. However, the hydrogen requirement for HDO needs considering; it could come from the steam reforming of a portion of the bio-oil (approximately one third)[179] or from other sources (e.g., other biomass, natural gas).

Importantly, the oils produced via these two upgrading routes will not be fully deoxygenated, and are also unlikely to be suitable for current fuel infrastructure. That means that further processing is required to produce drop-in biofuels.[180] This can be through FCC, with or without the addition of crude oil as a cofeed, or through conventional hydrocracking. However, both approaches will result in the ultimate biofuel yield being decreased further.

There has been recent interest in a relatively mild semi-upgrading process that lowers the acidity and water content of bio-oils while increasing their stability. It involves the esterification of many of the chemicals present by reaction with alcohols.[181] The efficacy of this process will be increased if the reactions are catalyzed by acids and if supercritical alcohols are used.[182] However, this approach does not upgrade bio-oil to the same extent as HDO does and the upgraded bio-oils will not be of a sufficient quality to allow for processing in an oil refinery. Thus, further upgrading steps will be necessary.[158]

In summary, a significant amount of work will be required to improve the commercial prospects of drop-in biofuels from pyrolysis. The lifespans of the catalysts need to be improved significantly and the relative proportions of the products of upgrading needs to be shifted more toward bio-oil and less toward water, gas, and coke. Furthermore, a sustainable and economical source for the hydrogen requirement for HDO needs to be put forward. There has been some interesting recent research regarding the development of superior catalysts, however. For example, Crossley et al.[183] have developed a recoverable catalyst, made from magnesium oxide with carbon nanotubes (incorporating palladium nanoparticles) grown between them, that can allow for carbon chain lengthening to take place in the water phase and for deoxygenation in the oil phase of the bio-oil. This is possible because the metal oxide nanoparticles are hydrophilic and the carbon nanotubes are hydrophobic. The catalytic process allows for the migration of the carbon chains from the water phase to the oil phase when they reach a sufficient chain length, allowing for the stabilization of water–oil emulsions. This catalyst has yet to be demonstrated on a large scale on bio-oils.

An alternative to producing an upgraded bio-oil compatible with existing engines is to produce a bio-oil that is a viable substrate for biochemical processing. This can occur, for example, via acid hydrolysis of levoglucosan to glucose,[184] although there are also microorganisms that can directly ferment levoglucosan.[185]

Pyrolysis of Lignin

As discussed in *Processes Need to Use the Lignin Residue Effectively*, the pyrolysis of the ligneous residue from hydrolysis biorefining process may offer the potential for increased total biofuel yields. There are issues associated with the pyrolysis of lignin and hydrolysis residues,[186] however. Lignin tends to produce a greater quantity of biochar than bio-oil, as found by the pyrolysis network (PyNe) of IEA Bioenergy Task 34.[187] This group found that the lignin was difficult to process in the conventional fast-pyrolysis reactors and concluded that, for highly concentrated lignins, new reactor designs would be necessary. They did find, however, that a feedstock of 50% lignin and 50% cellulose behaved like typical biomass and produced bio-oils of typical qualities with reasonable yields. This would imply that, given a feedstock of 20% lignin and 40% cellulose, the residue after hydrolysis would need to contain 20% lignin and 20% cellulose, that is, the hydrolysis would liberate only one half of the cellulose polymer. There could be a commercial synergy between the need for cellulosic material to be present during the pyrolysis of lignin and the higher rates of enzymatic hydrolysis seen for noncrystalline cellulose. That suggests that decreasing the amount of time for hydrolysis could allow for improved pyrolysis bio-oil yields of the remaining residue.

However, such an approach may not be necessary if bio-oil yields from lignin can be improved. A study in the pyrolysis of lignin utilized FA and ethanol at a temperature of 380°C to depolymerize and deoxygenate the lignin.[188] This resulted in little biochar (<5%) and more bio-oil of a higher quality (5–10% oxygen, 76–83% carbon, and 9–14% hydrogen) that phase separates from the water layer. Another means for improving lignin bio-oil yields was put forward by researchers at the Energy Research Centre of the Netherlands, as a response to the observations of the PyNe Network. They developed a lignin-pyrolysis process called LIBRA (LIgnin BioRefinery Approach), based on bubbling fluidized bed pyrolysis technology.[125] The process has recently been tested on the organosolv lignins derived from wheat straws.[189] The researchers achieved bio-oil yields of 40–60% and biochar yields of 30–40%. This bio-oil contained 25–40% (based on the dry lignin weight) of a phenolic fraction constituting of monomeric (7–11%) and oligomeric (14–24%) components

Hydrothermal Liquefaction

Hydrothermal upgrading (HTU) involves reacting biomass with liquid water at high temperatures and pressures (subcritical) and results in the production of a biocrude (bio-oil). This biocrude is separate from the water phase, can contain as little as 10% oxygen, and can be upgraded using similar means as for bio-oils (e.g., HDO).[190] It has been indicated that this approach may be more attractive for the production of liquid fuels than pyrolysis or gasification.[190] However, to date, most of the research has taken place on the pilot scale and there are significant obstacles in the way before commercial scale-ups can be achieved (such as the corrosion of equipment and the capital costs of the high pressure reactors).[191] Importantly, however, this process, unlike conventional gasification and pyrolysis, allows for the processing of biomass that can contain up to 80% water.[140] There is currently a lack of understanding with respect to the reaction mechanisms taking place in this process. However, it is generally accepted that, initially, cellulose and hemicellulose degrade to simple sugars and furans and that lignin degrades to phenolic and methoxy phenyl residues.[191] There then follows dehydrogenation and decarboxylation reactions that reduce the oxygen content of the biocrude. This could be a promising technology for the future as it may allow for the thermochemical processing of high moisture content waste materials.

Extraction of Chemicals

The separation and concentration of the products (and solvents) of biorefining processes are important considerations in determining their commercial viability. In the case of ethanol production, the liquid product stream after fermentation, the 'beer', only contains between 6% and 12% ethanol. Economical techniques for ordinary distillation in a single column will not achieve fuel-grade ethanol concentrations because ethanol forms an azeotrope with water at 95.6% by weight. Obtaining fuel-grade ethanol is typically a multistage process that can be energy intensive and incur significant capital costs. Huang et al.[71] review the methods available for increasing the concentrations of ethanol. These include azeotropic distillation, extractive distillation, liquid–liquid (solvent) extraction, adsorption, and membrane separation. The authors considered that cost and energy savings could be achieved in the future by employing promising technologies such as: extractive distillation with IL or hyperbranched polymers; extractive fermentation (a solvent extraction method carried out in-situ in the fermentation reactor that removes ethanol and fermentation inhibitors); and adsorption with molecular sieve and bio-based adsorbents.

Products such as *n*-butanol and iso-butanol have higher boiling points than water meaning that their recovery from the fermenter stream can be energy intensive, particularly given their low concentrations (15–20 g L^{-1}).[192] In such instances extractive fermentation can be a superior option to distillation. If the concentration levels of these alcohols could be increased sufficiently (to 63 and 76 g L^{-1} for *n*-butanol and iso-butanol, respectively) then recovery could be possible through phase separation.[192] There is on-going research aimed at developing microorganisms that can tolerate higher butanol concentrations although it is questioned as to whether these threshold levels can be reached.[192]

Financing Problems

While technological issues have been important in preventing commercial biorefineries from being deployed to date, it is also likely that financing issues have played a role. Securing governmental support for first-of-a-kind facilities can often be crucial to offset the perceived risks that commercial banks may have regarding the provision of finance. In this context there has been much press regarding the announcements of loan guarantees, particularly in the US; however, many of these have failed to materialize (see Table 12.1). Companies have complained that the regulations and procedures that must be adhered to secure them are too stringent and complex.[193] It appears that these problems are starting to be addressed, however. For example, the US Department of Agriculture (USDA) in 2011 proposed to allow loan guarantees to apply to the bond market, not just commercial banks.[194] The USDA also proposed to remove the requirement that the commercial facility be majority owned by a US citizen and to also remove the 'rural' requirement. That would mean that areas with a population over 50,000 would be eligible. The European Commission has also recently demonstrated a strong focus on funding advanced biofuel projects with the announcement of candidates to receive the proceeds from the sale of 300 million carbon credits via the NER-300 programme of the EU Emissions Trading System.[195]

There is also the issue of the uncertainty of tax credits in some countries. In the US, tax incentives for ethanol and biodiesel need to be renewed each year.[21] There needs to be a greater confidence in the longer-term stability of these support mechanisms so that sound financial projections can be made for potential facilities. Such current financial uncertainties, coupled with the technological uncertainties of 2GB facilities may lead those banks that may consider providing finance for commercial facilities to expect high returns on their investment to cover the perceived risks.[196] This will put an added strain on the already high capital costs and long payback periods.

OTHER WAYS TO ALLOW FOR ECONOMICAL BIOFUELS

Produce High Value Chemicals in Addition to Biofuels

The ability to modify the relative amounts/types of products produced in a biorefinery would allow the operator to hedge against price fluctuations that may occur for any one particular product (e.g., ethanol) or its competitors (e.g., gasoline). The range of products available will depend on the composition of the feedstock and on the main processing technologies and pretreatment technologies employed. For instance, the use of an organosolv pretreatment can allow for a high quality lignin product to be isolated but a steam pretreatment method would not.[23] Arato et al.[123] put forward an economic analysis of the Lignol organosolv process and found that the revenues from the coproducts (e.g., lignin) could allow for much smaller biorefineries to be economically viable; for example, a facility processing 100 dry tonnes of wood per day (equivalent to the residue production from a single sawmill).

Instead of targeting the, historically difficult, production of ethanol from pentoses, alternative means of utilizing hemicellulosic sugars could be explored, for example, the production of chemicals (furfural, xylitol, ethanoic acid, etc.) from the soluble hemicellulose stream liberated in some pretreatments.[89] Luo et al.[197] undertook an economic analysis of a biorefinery processing corn-stover and producing ethanol, succinic acid, ethanoic acid, and electricity. They found that such a biorefinery would be significantly more profitable than a facility only producing ethanol and that the profits would be less sensitive to fluctuations in ethanol prices and stover costs.

Some companies originally designed for the biofuels market plan instead to target first the specialty chemicals market. This will allow them to take advantage of the increased revenues these products can generate enabling them to develop profits and infrastructure early. The companies can then diversify into the biofuels market when their technologies have been sufficiently refined at the commercial scale for this to be economical. An example is Gevo, which plans to sell the isobutanol it produces to the chemical companies Lanxess and Sasol. Other companies using this approach include Amyris, Codexis, LS9, and Virent.

Integration with Existing Facilities

The capital costs associated with a biorefinery can be significantly reduced if it is colocated with an existing facility, particularly if the waste resources of that facility are feedstock. An example would be the colocation of a biorefinery with a sugar mill. Dias et al.[198] carried out an economic analysis to compare a standalone 2GB (enzymatic hydrolysis) facility producing ethanol from sugarcane bagasse and field-residues with an integrated facility (first and second generation). They found that ethanol production costs were lower and yields higher in scenarios where both facilities were integrated and that the total ethanol and electricity production costs in such an integrated facility would be less than that associated with a 1GB sugarcane to ethanol facility (assuming 2015 technologies).

Another example would be the use of hemicellulosic sugars, present in the black liquor that is generated in traditional pulping processes, for the production of biofuels or platform chemicals rather than for combustion. The integration of biorefineries with municipal waste facilities is another opportunity; the US waste services company Waste Management is collaborating with Canadian biorefining company Enerkem on such an approach. Poet is currently constructing a commercial biorefinery in Iowa that will process corn stover on a site adjacent to an existing 1GB corn mill (see Table 12.2). Several other companies are planning a similar approach.

Reduce Feedstock Costs

The total cost of the feedstock is very important; it can be up to 30–50% of the total production cost of the biofuel.[71] Ideally, feedstock costs could be zero or even negative, as may be the case where waste resources (e.g., municipal solid waste) are used at their point of source. However, there is unlikely to be sufficient extant waste biomass to allow for mass substitution of petroleum by 2GBs[70] and not all biorefining technologies may be capable of handling negative value feedstocks. Hence, the utilization of dedicatedly grown energy crops is likely to be necessary.

The prices that the biorefinery would need to pay for these crops could be reduced if their yields were increased and costs of production decreased. Other desirable traits for energy crops would include increased cold-temperature photosynthesis, abiotic and biotic stress tolerance, delayed flowering, and rapid early season growth.[8] 'Energy cane' is a recent example of a crop developed for 2GBs. It is a type of sugarcane that has a lower sucrose content, and instead is more focused on the production of fiber allowing for higher yields and a wider range of possible climatic conditions for its growth.[199] There have also been recommendations[179] for breeding programs that increase the stem to leaf ratio of energy crops so that biomass partitioning favors the fraction most likely to be harvested.

There is a significant amount of research underway to increase the photosynthetic efficiencies of energy crops toward the theoretical maximum, as described by Heaton et al.[200] who also suggest that there may be means in which these theoretical limits could be raised (e.g., by limiting the degree of plant respiration). It has been predicted that the crop development activities currently underway could allow for an average energy crop productivity of 37 tonnes per hectare per year in the next decade.[201]

Modify Feedstocks

There has been work on genetically modifying the lignin contents of feedstocks.[202] For instance, researchers at Purdue University recently identified a gene responsible for the production in plants of phenylalanine, a building block for lignin.[203] Suppression of the gene expression reduced the phenylanine content in petunias by 80%. While lignin reduction could be of value for hydrolysis processes, increased lignin contents could be advantageous for thermochemical processes due to the greater heating value of this polymer (compared with cellulose and hemicellulose). There could also be advantages in modifying the compositions of lignin and the structural polysaccharides. For instance, softwoods only contain coumaryl and guaiacyl units, whereas hardwoods and grassy plants also contain syringyl units. The simpler nature of the softwood lignin has been shown to result in delignification being more difficult as the polymer is more stable upon acid treatment.[204] There can also be efficiency savings associated with lower moisture contents; a 20% increase in dry matter content has been shown to reduce the energy costs for a SO_2-catalyzed pretreatment by nearly 50%.[26]

Companies such as Agrivida and Leaf Energy have also genetically modified plants so that they can produce the enzymes needed to hydrolyze their own polysaccharides. These enzymes would be inactive until subjected to the conditions of the biorefinery.

Use Multiple Feedstocks

The seasonality of the feedstock and the predictabilities of its supply and yields are important factors in determining the commercial viability and risks of a biorefining venture. The storage of large quantities of feedstock can be associated with dry matter losses,[205] compositional changes,[163] and risks of fire.[206] A biorefinery that can source multiple feedstocks may reduce the need for excessive amounts of biomass in storage and will be less affected by supply fluctuations of any one given feedstock. Kou and Zhao[207] undertook an economic analysis to compare the costs of producing ethanol (via the thermochemical platform) under one feedstock and three-feedstocks scenarios. The multiple feedstock strategy improved the net present value by 18% when compared to the single feedstock strategy. When extreme weather effects were considered, this margin was increased to 57%. Furthermore, the three-feedstocks facility had no risk of bankruptcy during the payback period, while the one-feedstock plant had a 75% chance. The option of multiple and varied feedstocks may be more applicable to thermochemical facilities due to their increased feedstock flexibility, whereas enzymatic processes can, in some cases, be more targeted to a specific feedstock type.

Share Progress Between Companies

Both the hydrolysis and thermochemical pathways incorporate numerous distinct and important stages to produce 2GBs. In the (biological) hydrolysis route (where CBP is not employed) the stages include pretreatment, enzymatic hydrolysis, and fermentation. While some companies (e.g., Sud-Chemie) try to develop their own proprietary technologies for each step in the chain, others focus on a specific link. For example, there are companies (e.g., SucreSource, Virdia, Renmatrix, and Comet Biorefining) that focus solely on the extraction of fermentable sugars from lignocellulose. Then there are companies (e.g., LS9, Amyris, Virent) that develop processes to utilize these sugars. There are also companies that work in the field of enzyme production (e.g., Novozymes, Dyadic, and Genencor). The strategy of focusing on optimizing a specific link in the biofuel production chain and then using the experience of other companies in other areas may be more practical for most companies than attempting to be competitive in all stages with proprietary technologies, although such an approach will incur licensing fees.

Develop Biorefinery Networks

The transport costs associated with large scale thermochemical biorefineries could be reduced by increasing the density of the energy carrier through the production of bio-oil via small scale distributed facilities that supply a larger scale upgrading (or gasification) plant. Envergent (USA) plans such an approach, involving four smaller scale pyrolysis facilities, utilizing the flash pyrolysis technology currently employed in Ensyn's plants, supplying a centralized facility using UOP's HDO technology. The company is currently not ready for commercial deployment. It is focused instead on a pilot facility in Hawaii that is expected to be fully operational in 2014.[208] KiOR (USA) plans a similar approach (see Table 12.2). Dynamotive Energy Systems claims to have tested syngas produced from a bio-oil/biochar slurry.[209,210]

Increase the Size of the Biorefinery

Huang et al.[211] investigated the effect of plant size (up to 5000 dry tonnes of feedstock per day) on feedstock costs and total ethanol production costs (with a technology incorporating dilute acid hydrolysis pretreatment, enzymatic hydrolysis, and SSCF). They found that the total ethanol production cost decreased in a nonlinear fashion (largest decreases seen earlier in the curve) with increased plant size up to 4000 tonnes per day, due to economies of scale. Cost reductions were significantly less for further increases in plant sizes (and there was a cost increase in the case of using hybrid poplars). The authors concluded that the most suitable plant sizes were between 2000 and 4000 tonnes per day. Stephen et al.[23] also concluded that an increase in scale would be the best way in which to achieve capital cost reductions, citing the example of the sugarcane industry where this method, rather than technology improvements, provided the majority of capital cost reductions between 1975 and 2004.[212]

Ng and Sadhukan[213] carried out a techno-economic analysis of bio-oil processing gasification to FT-liquids facilities of various sizes (employing the biorefinery network concept outlined in *Develop Biorefinery Networks*). They also found that economic competitiveness increased with scale but that the shipping of bio-oil to the United Kingdom from other countries (principally bio-oil obtained from palm oil empty fruit bunches in Malaysia) amounted to 2–4 times more than the operating cost and was not considered feasible.

Current Commercial Activities

Table 12.2 details the commercial-scale biorefineries that, as of May 2012, were under construction. This classification requires that the main biorefinery is being built; site clearances and other preparatory activities are not considered eligible. The total expected capacity of these six biorefineries is 323 million liters of biofuel (281 million liters of ethanol and 42 million liters of biocrude, a HDO-upgraded bio-oil) at a total capital cost of \$1.4 billion. Importantly, these technologies cover a range of different conversion mechanisms and feedstocks, as illustrated in Figure 12.2, which may be less risky than focusing on a single process that may encounter technical/financial difficulties.

Three of these biorefineries (Abengoa, INEOS Bio, Poet) are based on technologies that were selected for the US Department of Energy loan guarantees in 2007 (Table 12.1), although only one of these companies (Abengoa) is taking assistance from this source. Most of the companies constructing these biorefineries are also developing plans for other commercial facilities. Indeed, the KiOR facility being constructed in Mississippi is meant to represent the central HDO processing unit for a further three pyrolysis facilities (the company expects the total capital cost for the planned infrastructure to be over \$1 billion).[214]

Enerkem also has ambitious plans for two further commercial facilities, of the same scale as the biorefinery being constructed in Alberta, Canada, to be operational before 2015. The Enerkem process involves the gasification of municipal waste and the subsequent catalytic synthesis of methanol. To produce ethanol the methanol will be reacted with the carbon dioxide from the syngas to produce methyl acetate which is then hydrogenated to ethanol.[215] However, an important caveat stressed by the company in its filing for an initial public offering was that the production of ethanol has not yet been demonstrated at their demonstration scale

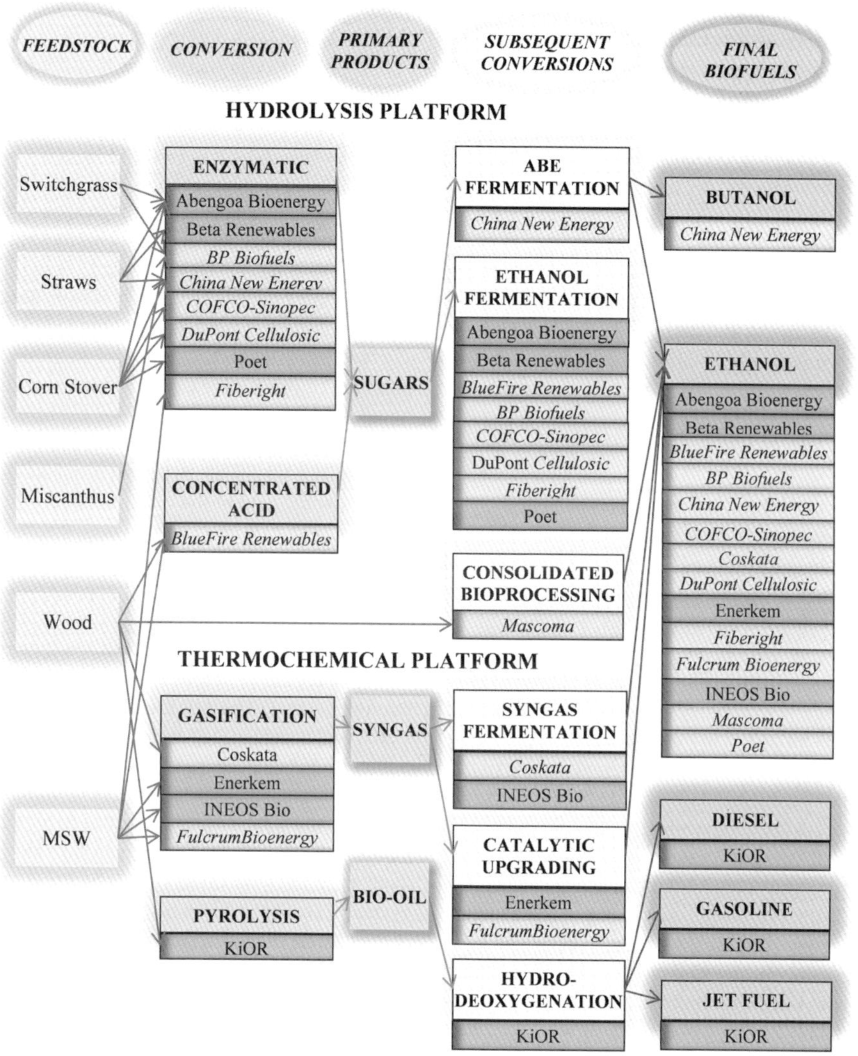

Figure 12.2 Graphical representation of the different feedstocks and conversion technologies used, and the different primary and final biofuel products obtained, from the commercial biorefineries listed in Table 12.2. The companies written in italics are those that plan to start construction by the end of 2012, whereas those in normal typeface are currently (May, 2012) constructing commercial-scale facilities. MSW = municipal solid waste; ABE = acetone, butanol, ethanol.

facility in Quebec, with only methanol having being produced to date since June 2011.[215] The filing claims that the capacity for ethanol production will be added to that facility during 2012. The Enerkem process is conceptually very similar to that of the failed Range Fuels process (see Table 12.1). That company also did not demonstrate the production of ethanol prior to the construction of its commercial-scale facility and, indeed, never achieved this in a continuous manner before going bankrupt.[216]

Table 12.2 also provides details for some important commercial facilities that are planned to be under construction by the end of 2012. In some cases (e.g., BP Biofuels, DuPont Danisco Cellulosic Ethanol, COFCO-Sinopec), the Author considers it highly likely that construction will occur as the majority of the funding is in place. However, construction by other companies (e.g., BlueFire Renewables, Fulcrum Bioenergy) is dependent on loan-guarantees/grants being received. The planned biorefineries in Table 12.2 cover two conversion mechanisms not represented in the facilities under construction (concentrated acid hydrolysis and CBP). The fate of the Mascoma facility will be of particular interest because the CBP technology it will look to employ represents a significant advance, theoretically, to conventional enzymatic hydrolysis (see *Limitations of Conventional Microorganisms*).

There are numerous other biorefining companies planning to develop commercial-scale facilities in the longer term. A more detailed database of these companies, as well as interactive maps illustrating the locations of the planned facilities, is available on the Carbolea Web site, http://www.carbolea.ul.ie.

CONCLUSION

There are important obstacles in each conversion pathway, and at every stage, which have prevented commercial 2GBs from being produced to date. The problem is complicated further as each process modification has implications further down the process flow. For example, a specific pretreatment may present specific obstacles, and opportunities, in the hydrolysis and fermentation stages. A holistic approach in examining process refinements will be very important in achieving cost-effective 2GBs. The compositions and levels of impurities can also vary significantly between different biomass feedstocks and these are likely to influence which conversion technology would be most suitable for them. Hence, it is of benefit that there are a large number of companies working to advance the art in various mechanisms of biomass conversion using a variety of biomass feedstocks (Figure 12.2).

There has been significant progress in research and commercial activities to address many of the technical obstacles. Various means of improving the supply side economics of 2GB production and the financial viability of biorefineries have also been outlined. The current commercial state of play in the biorefining sector, as covered in Table 12.2 and Figure 12.2, appears to be much more promising than at any previous time. There are six commercial-scale facilities currently (May, 2012) under construction, whereas to date only one such facility has been built (the now defunct Range Fuels plant). Of course, the failure of the Range Fuels technology illustrates that the construction of commercial facilities is not in itself enough to guarantee their survival or competiveness. Indeed, it may be expected that not all of these first-of-a-kind facilities will be successful, particularly if they are dependent on technologies not even fully proven on the demonstration scale. However, numerous different conversion processes and intellectual properties are represented in the facilities that are planned and under construction (Figure 12.2), lessening the risk associated with any one approach.

It should also be considered that the estimates of biofuel production costs presented in scientific journals, such as those referenced in this paper, typically use publically available data and are unlikely to be privy to industrial developments and trade secrets that may lead to improved economics. The real competitiveness of near-term commercial 2GBs will only be demonstrated when some of the facilities in Table 12.2 and Figure 12.2 are operational. Although some of these may fail, it may be necessary only for a small number of these to be commercially viable. These can then provide models for future facilities that should be expected to offer increased efficiencies with each generational increment.

Whether this will be enough to allow for the biofuel targets for the next decade to be met is questionable; however, given how far behind the original targets the industry currently is. The US Renewable Fuels Standard targeted 1.9 billion liters of cellulosic ethanol to be produced in 2012; however, the combined output capacity of the US-based facilities in Table 12.2 that are expected to be operational by the end of 2012 is only 72 million liters of biofuel and their actual biofuel output for that year will only be a small fraction of that. The US Department of Agriculture estimated that a total of 527 biorefineries, each producing 150 million liters of 2GBs a year, would be required to achieve the RFS target of 80 billion liters of 'advanced biofuels' by 2022, at a total estimated cost of $168 billion.[217] Given that the construction process may take up to three years, there may only be three further generations of facilities from 2013 to 2022. Even if some commercially viable processes can be demonstrated from the near-term facilities in Table 12.2, there will need to be a massive large-scale global deployment of these in order for the biofuel targets to be achieved in the next decade. Such levels of expansion look unlikely.

In summary, 2GBs are taking so long because the assumptions of future efficiency improvements that were made around 2007 failed to manifest in the expected timeframe. The state of the art has advanced in this time, however, and it now appears likely that commercial 2GBs will be produced within the next year. However, the ambitious targets of 2007 are unlikely to be met in the short to medium term.

ACKNOWLEDGEMENTS

The author acknowledges financial support via research grants from DIBANET, funded by the European Community's Seventh Framework Programme (FP7/2007–2013) under grant agreement #227248-2; the Research Stimulus

Fund Programme of the Irish Department of Agriculture, Fisheries and Food; and the STRIVE Programme of the Irish EPA.

REFERENCES

1. Bungay HR. *Energy, the Biomass Options*. New York: John Wiley and Sons; 1981.
2. Hayes DJ. An examination of biorefining processes, catalysts and challenges. *Catal Today* 2008, 145:138–151.
3. Jung GA, Buxton DR, Hatfield RD, Ralph J. *Forage Cell Wall Structure and Digestibility*. Madison, WI: ASA-CSSA-SSSA; 1993.
4. Rees DA. *The Shapes of Molecules: Carbohydrate Polymers Contemporary Science Paperbacks 14*. Edinburgh, UK: Oliver and Boyd; 1967.
5. Leung DYC, Wu X, Leung MKH. A review on biodiesel production using catalyzed transesterification. *Appl Energy* 2010, 87:1083–1095.
6. Venturi P, Venturi G. Analysis of energy comparison in European agricultural systems. *Biomass Bioenergy* 2003, 25:235–255.
7. Hill J, Nelson E, Tilman D, Polasky S, Tiffany D. Environmental, economic and energetic costs and benefits of biodiesel and ethanol biofuels. *Proc Natl Acad Sci USA* 2006, 103:11206–11210.
8. Carriquiry MA, Du X, Timilsina GR. Second generation biofuels: economics and policies. *Energy Policy* 2011, 39:4222–4234.
9. Aden A, Bozell JJ, Holladay J, White J, Manheim A. *Top Value Added Chemicals From Biomass*. Joint NREL/PNNL Report, August 2004: Golden, CO: National Renewable Energy Laboratory/Pacific Northwest National Laboratory; 2004.
10. Bozell JJ. *Chemicals and Materials from Renewable Resources*. Washington, D.C.: American Chemical Society; 2001.
11. Gallezot P. Catalytic routes from renewables to fine chemicals. *Catal Today* 2007, 121:76–91.
12. Koutinas AA, Du C, Wang RH, Webb C. Production of chemicals from biomass. In: *Introduction to Chemicals from Biomass*: John Wiley & Sons; 2008, 77–101.
13. Huang G, Chen F, Wei D, Zhang X, Chen G. Biodiesel production by microalgal biotechnology. *Appl Energy* 2010, 87:38–46.
14. Singh A, Olsen SI. A critical review of biochemical conversion, sustainability and life cycle assessment of algal biofuels. *Appl Energy* 2011, 88:3548–3555.
15. Singh A, Nigam PS, Murphy JD. Mechanism and challenges in commercialisation of algal biofuels. *Bioresource Technol* 2011, 102:26–34.
16. IPCC. *Climate Change 2001: The Scientific Basis*. Cambridge, UK: Cambridge University Press; 2001.
17. EPA. *National Renewable Fuel Standard Program—2010 and Beyond*. Available at: http://www.usbiomassboard.gov/pdfs/epa_rfs_feb2011.pdf. (Accessed May 28, 2012).
18. European Commission. Directive 2003/30/EC of the European Parliament and of the Council of 8 May 2003 on the promotion of the use of biofuels or other renewable fuels for transport. *Official J Eur Union* 2003, 123:42–46.
19. Smyth BM, Ó Gallachóir BP, Korres NE, Murphy JD. Can we meet targets for biofuels and renewable energy in transport given the constraints imposed by policy in agriculture and energy? *J Cleaner Prod* 2010, 18:1671–1685.
20. Qiu H, Sun L, Huang J, Rozelle S. Liquid biofuels in China: current status, government policies, and future opportunities and challenges. *Renew Sustain Energy Rev* 2012, 16:3095–3104.
21. Timilsina GR, Shrestha A. How much hope should we have for biofuels? *Energy* 2011, 36:2055–2069.
22. Sims REH, Mabee W, Saddler JN, Taylor M. An overview of second generation biofuel technologies. *Bioresource Technol* 2010, 101:1570–1580.
23. Stephen JD, Mabee WE, Saddler JN. Will second-generation ethanol be able to compete with first-generation ethanol? Opportunities for cost reduction. *Biofuels Bioprod Biorefin* 2012, 6:159–176.
24. Wright MM, Brown RC. Comparative economics of biorefineries based on the biochemical and thermochemical platforms. *Biofuels Bioprod Biorefin* 2007, 1:49–56.
25. Foust T, Aden A, Dutta A, Phillips S. An economic and environmental comparison of a biochemical and a thermochemical lignocellulosic ethanol conversion processes. *Cellulose* 2009, 16:547–565.
26. Banerjee S, Mudliar S, Sen R, Giri B, Satpute D, Chakrabarti T, Pandey RA. Commercializing lignocellulosic bioethanol: technology bottlenecks and possible remedies. *Biofuels Bioprod Biorefin* 2010, 4:77–93.
27. Zhang YHP. What is vital (and not vital) to advance economically-competitive biofuels production. *Process Biochem* 2011, 46:2091–2110.
28. Palonen H. Role of lignin in the enzymatic hydrolysis of lignocellulose. *Doctor Technol* 2004.
29. Jung HG, Deetz DA. Cell wall lignification and degradability. In: Jung HG, Buxton DR, Hatfield RD, Ralph J, eds. *Forage Cell Wall Structure and Digestibility*. Madison, WI: ASA-CSSA-SSSA; 1993, 315–346.
30. Joniak D, Kosikova B. Stability of acetal type lignin carbohydrate bonds against acid and alkali catalysed hydrolysis. *Cellulose Chem Technol* 1976, 10:691–694.
31. Yoshida M, Liu Y, Uchida S, Kawarada K, Ukagami Y, Ichinose H, Kaneko S, Fukuda K. Effects of cellulose crystallinity, hemicellulose and lignin on the enzymatic hydrolysis of *Miscanthus sinensis* to monosaccharides. *Biosci Biotech Bioch* 2008, 72:805–810.
32. Centi G, Lanzafame P, Perathoner S. Analysis of the alternative routes in the catalytic transformation of lignocellulosic materials. *Catal Today* 2011, 167:14–30.
33. Koo B-W, Min B-C, Gwak K-S, Lee S-M, Choi J-W, Yeo H, Choi I-G. Structural changes in lignin during organosolv pretreatment of *Liriodendron tulipifera* and the effect on enzymatic hydrolysis. *Biomass Bioenergy* 2012, 42: 24–32.

34. Laureano-Perez L, Teymouri F, Alizadeh H, Dale BE. Understanding factors that limit enzymatic hydrolysis of biomass. In: Davison BH, Evans BR, Finkelstein M, McMillan JD, eds. *Twenty-Sixth Symposium on Biotechnology for Fuels and Chemicals*. New York: Humana Press; 2005, 1081–1099.
35. Yu Z, Jameel H, Chang H-m, Park S. The effect of delignification of forest biomass on enzymatic hydrolysis. *Bioresource Technol* 2011, 102:9083–9089.
36. Hall M, Bansal P, Lee JH, Realff MJ, Bommarius AS. Cellulose crystallinity—a key predictor of the enzymatic hydrolysis rate. *FEBS J* 2010, 277:1571–1582.
37. Hamelinck CN, van Hooijdonk G, Faaij A. Ethanol from lignocellulosic biomass: techno-economic performance in short-, middle- and long-term. *Biomass and Bioenergy* 2005, 28:384–410.
38. Toussaint B, Excoffier G, Vignon MR. Effect of steam explosion treatment on the physico-chemical characteristics and enzymatic hydrolysis of poplar cell wall components. *Anim Feed Sci Technol* 1991, 32:235–242.
39. Teymouri F, Laureano-Perez L, Alizadeh H, Dale BE. Optimization of the ammonia fiber explosion (AFEX) treatment parameters for enzymatic hydrolysis of corn stover. *Bioresource Technol* 2005, 96:2014–2018.
40. Mosier NS, Hendrickson RRD, Welch G, Dien BS, Bothast R, Ladisch MR. Principles and economics of pretreating cellulose in water for ethanol production. In: *225th American Chemical Society Meeting, 26 March 2003*. Paper 103. New Orleans, LA; 2003.
41. Weil JR, Sarikaya A, Rau S-L, Goetz J, Ladisch CM, Brewer M, Hendrickson R, Ladisch MR. Pretreatment of yellow poplar sawdust by pressure cooking in water. *Appl Biochem Biotechnol* 1997, 68:21–40.
42. Schmidt AS, Thomsen AB. Optimization of wet oxidation pretreatment of wheat straw. *Bioresource Technol* 1998, 64:139–151.
43. Haverty D, Dussan K, Piterina AV, Leahy JJ, Hayes MHB. Autothermal, single-stage, performic acid pretreatment of Miscanthus × giganteus for the rapid fractionation of its biomass components into a lignin/hemicellulose-rich liquor and a cellulase-digestible pulp. *Bioresource Technol* 2012, 109:173–177.
44. Kim HK, Hong J. Supercritical CO_2 pretreatment of lignocellulose enhances enzymatic cellulose hydrolysis. *Bioresource Technol* 2001, 77:139–144.
45. Fort DA, Remsing RC, Swatloski RP, Movna P, Movna G, Rogers RD. Can ionic liquids dissolve wood? Processing and analysis of lignocellulosic materials with 1-*n*-butyl-3-methylimidazolium chloride. *Green Chem* 2007, 9:63–69.
46. Zhu S. Use of ionic liquids for the efficient utilization of lignocellulosic materials. *J Chem Technol Biotechnol* 2008, 83:777–779.
47. Murugesan S, Linhardt RJ. Ionic liquids in carbohydrate chemistry—current trends and future directions. *Curr Organ Synth* 2005, 2:437–451.
48. García-Cubero MT, González-Benito G, Indacoechea I, Coca M, Bolado S. Effect of ozonolysis pretreatment on enzymatic digestibility of wheat and rye straw. *Bioresource Technol* 2009, 100:1608–1613.
49. Balat M. Production of bioethanol from lignocellulosic materials via the biochemical pathway: a review. *Energy Convers Manag* 2011, 52:858–875.
50. Alvira P, Tomás-Pejó E, Ballesteros M, Negro MJ. Pretreatment technologies for an efficient bioethanol production process based on enzymatic hydrolysis: a review. *Bioresource Technol* 2010, 101:4851–4861.
51. Zhu JY, Pan XJ. Woody biomass pretreatment for cellulosic ethanol production: technology and energy consumption evaluation. *Bioresource Technol* 2010, 101:4992–5002.
52. Mosier NS, Wyman C, Dale B, Elander R, Lee YY, Holtzapple M, Ladisch M. Features of promising technologies for pretreatment of lignocellulosic biomass. *Bioresource Technol* 2005, 96:673–686.
53. Agbor VB, Cicek N, Sparling R, Berlin A, Levin DB. Biomass pretreatment: fundamentals toward application. *Biotechnol Adv* 2011, 29:675–685.
54. Humbird D, Aden A. *Biochemical Production of Ethanol from Corn Stover: 2008 State of Technology Model*. NREL/TP-510-46214. Golden, CO: National Renewable Energy Laboratory; 2008
55. Eggeman T, Elander R. Process and economic analysis of pretreatment technologies. *Biores Technol* 2005, 96:2019–2025.
56. Zheng Y, Pan Z, Zhang R. Overview of biomass pretreatment for cellulosic ethanol production. *Int J Agric Biol Eng* 2009, 2:51–68.
57. Taherzadeh MJ, Karimi K. Pretreatment of lignocellulosic wastes to improve ethanol and biogas production: a review. *Int J Mol Sci* 2008, 9:1621–1651.
58. Nguyen QA, Tucker MP, Keller FA, Eddy FP. Two-stage dilute-acid pretreatment of softwoods. *Appl Biochem Biotechnol* 2000, 84–86:561–575.
59. Kootstra AMJ, Beeftink HH, Scott EL, Sanders JPM. Comparison of dilute mineral and organic acid pretreatment for enzymatic hydrolysis of wheat straw. *Biochemical Engineering Journal* 2009, 46:126–131.
60. Vidal PF, Molinier J. Ozonolysis of lignin — Improvement of in vitro digestibility of poplar sawdust. *Biomass* 1988, 16:1–17.
61. Liu C-Z, Wang F, Stiles AR, Guo C. Ionic liquids for biofuel production: opportunities and challenges. *Appl Energy* 2012, 92:406–414.
62. Rogers RD, Seddon KR. Ionic liquids-solvents of the future? *Science* 2003, 302:792–793.
63. Roosen C, Müller P, Greiner L. Ionic liquids in biotechnology: applications and perspectives for biotransformations. *Appl Microbiol Biotechnol* 2008, 81:607–614.
64. Khudyakov JI, D'haeseleer P, Borglin SE, DeAngelis KM, Woo H, Lindquist EA, Hazen TC, Simmons BA, Thelen MP. Global transcriptome response to ionic liquid by a tropical rain forest soil bacterium, *Enterobacter lignolyticus*. *Proc Natl Acad Sci* 2012.

65. Menon V, Rao M. Trends in bioconversion of lignocellulose: biofuels, platform chemicals and biorefinery concept. *Prog Energy Combust Sci* 2012, 38:522–550.
66. Li B, Asikkala J, Filpponen I, Argyropoulos DS. Factors affecting wood dissolution and regeneration of ionic liquids. *Ind Eng Chem Res* 2010, 49:2477–2484.
67. Klein-Marcuschamer D, Simmons BA, Blanch HW. Techno-economic analysis of a lignocellulosic ethanol biorefinery with ionic liquid pre-treatment. *Biofuels Bioprod Biorefin* 2011, 5:562–569.
68. Sendich ED, Dale BE. Environmental and economic analysis of the fully integrated biorefinery. *GCB Bioenergy* 2009, 1:331–345.
69. Chen H, Qiu W. Key technologies for bioethanol production from lignocellulose. *Biotechnol Adv* 2010, 28:556–562.
70. Hayes DJ, Hayes MHB. The role that lignocellulosic feedstocks and various biorefining technologies can play in meeting Ireland's biofuel targets. *Biofuels Bioprod Biorefin* 2009, 3:500–520.
71. Huang H-J, Ramaswamy S, Tschirner UW, Ramarao BV. A review of separation technologies in current and future biorefineries. *Separat Purif Technol* 2008, 62:1–21.
72. Schlesinger R, Götzinger G, Sixta H, Friedl A, Harasek M. Evaluation of alkali resistant nanofiltration membranes for the separation of hemicellulose from concentrated alkaline process liquors. *Desalination* 2006, 192:303–314.
73. Merino ST, Cherry J. Progress and challenges in enzyme development for biomass utilization. *Adv Biochem Eng Biotechnol* 2007, 108 95–120.
74. Teter SA, Xu F, Nedwin GE, Cherry JR. Enzymes for biorefineries. In: Kamm B, Gruber VR, Kamm M, eds. *Biorefineries: Industrial Processes and Products*. Vol. 1. Weinheim, Germany: John Wiley & Sons; 2005, 357–384.
75. Eriksson T, Börjesson J, Tjerneld F. Mechanism of surfactant effect in enzymatic hydrolysis of lignocellulose. *Enzyme Microb Technol* 2002, 31:353–364.
76. Alkasrawi M, Eriksson T, Börjesson J, Wingren A, Galbe M, Tjerneld F, Zacchi G. The effect of Tween-20 on simultaneous saccharification and fermentation of softwood to ethanol. *Enzyme Microb Technol* 2003, 33:71–78.
77. Wood BE, Aldrich HC, Ingram LO. Ultrasound stimulates ethanol production during the simultaneous saccharification and fermentation of mixed waste office paper. *Biotechnol Prog* 1997, 13:232–237.
78. Tjerneld F. Enzyme-catalyzed hydrolysis and recycling in cellulose bioconversion. In: Harry Walter GJ, ed. *Methods in Enzymology*. Vol. 228. Waltham, MA: Academic Press; 1994, 549–558
79. Ramos LP, Breuil C, Saddler JN. The use of enzyme recycling and the influence of sugar accumulation on cellulose hydrolysis by Trichoderma cellulases. *Enzyme Microb Technol* 1993, 15:19–25.
80. Deshpande MV, Eriksson KE. Reutilization of enzymes for saccharification of lignocellulosic materials. *Enzyme Microb Technol* 1984, 6:338–340.
81. Tu M, Zhang X, Paice M, MacFarlane P, Saddler JN. The potential of enzyme recycling during the hydrolysis of a mixed softwood feedstock. *Bioresource Technol* 2009, 100:6407–6415.
82. Biomass Magazine. SDSU research: re-using enzymes in industry. Available at: http://www.biomassmagazine.com/articles/3672/sdsu-research-re-using-enzymes-in-industry. (Accessed May 9, 2012).
83. Pan X, Gilkes N, Saddler JN. Effect of acetyl groups on enzymatic hydrolysis of cellulosic substrates. *Holzforschung* 2006, 60:398–401.
84. Tolan JS. Iogen's demonstration process for producing ethanol fromc ellulosic biomass. In: Kamm B, Gruber VR, Kamm M, eds. *Biorefineries: Industrial Processes and Products*. Vol. 1. Weinheim, Germany: John Wiley & Sons; 2005, 193–208.
85. Gan Q, Allen SJ, Taylor G. Kinetic dynamics in heterogeneous enzymatic hydrolysis of cellulose: an overview, an experimental study and mathematical modelling. *Process Biochem* 2003, 38:1003–1018.
86. Sun J-X, Sun R, Sun X-F, Su Y. Fractional and physico-chemical characterization of hemicelluloses from ultrasonic irradiated sugarcane bagasse. *Carbohydr Res* 2004, 339:291–300.
87. Sun R, Lawther JM, Banks WB. Fractional and structural characterization of wheat straw hemicelluloses. *Carbohydr Polym* 1996, 29:325–331.
88. Kumar R, Wyman CE. Effect of enzyme supplementation at moderate cellulase loadings on initial glucose and xylose release from corn stover solids pretreated by leading technologies. *Biotechnol Bioeng* 2009, 102:457–467.
89. Dumon C, Song L, Bozonnet S, Fauré R, O'Donohue MJ. Progress and future prospects for pentose-specific biocatalysts in biorefining. *Process Biochem* 2012, 47:346–357.
90. Nichols W. Novozymes: new enzyme brings biofuels into petrol cost bracket. Available at: http://www.businessgreen.com/bg/news/2154011/novozyme-enzyme-brings-biofuels-petrol-cost-bracket. (Accessed May 27, 2012).
91. Green Car Congress. Novozymes and NREL reduce cost of enzymes for biomass-to-ethanol production 30-fold. Available at: http://www.greencarcongress.com/2005/04/novozymes_and_n.html. (Accessed May 27, 2012).
92. Wenzl HFJ. Chapter IV: the acid hydrolysis of wood. In: *The Chemical Technology of Wood*. New York: Academic Press; 1970, 157–252.
93. Farone WA, Cuzens JE. Method of producing sugars using strong acid hydrolysis of cellulosic and hemicellulosic materials. *Biotechnol Adv* 1997, 15:548–549.
94. Palmqvist E, Hahn-Hägerdal B. Fermentation of lignocellulosic hydrolyzates II: inhibitors and mechanisms of inhibition. *Bioresource Technol* 2000, 74:25–33.
95. Cardona CA, Quintero JA, Paz IC. Production of bioethanol from sugarcane bagasse: status and perspectives. *Bioresource Technol* 2010, 101:4754–4766.
96. Sivers MV, Zacchi G, Olsson L, Hahn-Hügerdal B. Cost analysis of ethanol production from willow using recombinant *Escherichia coli*. *Biotechnol Prog* 1994, 10:555–560.

97. Alriksson B, Cavka A, Jönsson LJ. Improving the fermentability of enzymatic hydrolysates of lignocellulose through chemical in-situ detoxification with reducing agents. *Bioresource Technol* 2011, 102:1254–1263.
98. Yang S, Land ML, Klingeman DM, Pelletier DA, Lu T-YS, Martin SL, Guo H-B, Smith JC, Brown SD. Paradigm for industrial strain improvement identifies sodium acetate tolerance loci in *Zymomonas mobilis* and *Saccharomyces cerevisiae*. *Proc Natl Acad Sci USA* 2010, 107:10395–10400.
99. Cheng JJ, Timilsina GR. Status and barriers of advanced biofuel technologies: a review. *Renew Energy* 2011, 36:3541–3549.
100. Lawford HG, Rousseau JD. Fuel ethanol from hardwood hemicellulose hydrolysate by genetically engineered *Escherichia coli* B carrying genes from *Zymomonas mobilis*. *Biotechnol Lett* 1991, 13:191–196.
101. Meijnen JP, Winde JH, Ruijssenaars HJ. Engineering Pseudomonas putida S12 for efficient utilization of D-xylose and L-arabinose. *Appl Env Microbiol* 2008, 74:5031–5037.
102. Hahn-Hägerdal B, Karhumaa K, Fonseca C, Spencer-Martins I, Gorwa-Grauslund M. Towards industrial pentose-fermenting yeast strains. *Appl Microbiol Biotechnol* 2007, 74:937–953.
103. Hughes SR, Hector RE, Rich JO, Qureshi N, Bischoff KM, Dien BS, Saha BC, Liu S, Cox EJ, Jackson JS, et al. Automated yeast mating protocol using open reading frames from *Saccharomyces cerevisiae* genome to improve yeast strains for cellulosic ethanol production. *J Assoc Lab Autom* 2009, 14:190–199.
104. Hughes SR, Rich JO, Bischoff KM, Hector RE, Qureshi N, Saha BC, Dien BS, Liu S, Jackson JS, Sterner DE, et al. Automated yeast transformation protocol to engineer *Saccharomyces cerevisiae* strains for cellulosic ethanol production with open reading frames that express proteins binding to xylose isomerase identified using a robotic two-hybrid screen. *J Assoc Lab Autom* 2009, 14:200–212.
105. Kuyper M, Hartog MMP, Toirkens MJ, Almering MJH, Winkler AA, van Dijken JP, Pronk JT. Metabolic engineering of a xylose-isomerase-expressing *Saccharomyces cerevisiae* strain for rapid anaerobic xylose fermentation. *FEMS Yeast Res* 2005, 5:399–409.
106. Lynd LR. Overview and evaluation of fuel ethanol from cellulosic biomass: technology, economics, the environment, and policy. *Annu Rev Energy Environ* 1996, 21:403–465.
107. Wooley R, Ruth M, Sheehan J, Ibsen K. Lignocellulosic biomass to ethanol process design and economics utilizing co-current dilute acid prehydrolysis and enzymatic hydrolysis current and futuristic scenarios. Technical Report NREL/TP-580-26157. National Renewable Energy Laboratory: US Department of Energy; 1999.
108. Lynd LR, Elander RT, Wyman CE. Likely features and costs of mature biomass ethanol technology. *Appl Biochem Biotechnol* 1996, 57/58:741–761.
109. Szczodrak J, Fiedurek J. Technology for conversion of lignocellulosic biomass to ethanol. *Biomass Bioenergy* 1996, 10:367–375.
110. Higashide W, Li Y, Yang Y, Liao JC. Metabolic engineering of Clostridium cellulolyticum for isobutanol production from cellulose. *Appl Environ Microbiol* 2011, 77:2727–2733.
111. Brown SD, Guss AM, Karpinets TV, Parks JM, Smolin N, Yang S, Land ML, Klingeman DM, Bhandiwad A, Rodriguez M, et al. Mutant alcohol dehydrogenase leads to improved ethanol tolerance in *Clostridium thermocellum*. *Proc Natl Acad Sci USA* 2011, 108:13752–13757.
112. Patrinos A. The promises of biotechnology, Synthetic Genomics Inc. USA paper. In: *OECD Conference: Scientific Challenges for Energy Research Paris, 17-18 May*. Paris: OECD; 2006.
113. Voegele E. LS9 scales up fermentation platform for biobased fuels, chemicals. Available at: http://biorefiningmagazine.com/articles/5934/ls9-scales-up-fermentation-platform-for-biobased-fuels-chemicals. (Accessed May 26, 2012).
114. Steen EJ, Kang Y, Bokinsky G, Hu Z, Schirmer A, McClure A, del Cardayre SB, Keasling JD. Microbial production of fatty-acid-derived fuels and chemicals from plant biomass. *Nature* 2010, 463:559–562.
115. Hayes DJ, Fitzpatrick SW, Hayes MHB, Ross JRH. The biofine process: production of levulinic acid. furfural and formic acid from lignocellulosic feedstocks. In: Kamm B, Gruber PR, Kamm M, eds. *Biorefineries: Industrial Processes and Products*. Vol. 1. Weinheim, Germany: John Wiley & Sons; 2005, 139–164.
116. Lange J-P, Price R, Ayoub PM, Louis J, Petrus L, Clarke L, Gosselink H. Valeric biofuels: a platform of cellulosic transportation fuels. *Angew Chem Int Ed* 2010, 49:4479–4483.
117. Fanchiang W-L, Lin Y-C. Catalytic fast pyrolysis of furfural over H-ZSM-5 and Zn/H-ZSM-5 catalysts. *Appl Catal A: Gen* 2012, 419–420:102–110.
118. Olsen E. Cellulosic Fuel Production: One Step Closer to Standard Refining. Available at: http://www.biorefiningmagazine.com/articles/5536/cellulosicfuel-production-one-step-closer-to-standard-refining. (Accessed May 26, 2012).
119. Murat Sen S, Henao CA, Braden DJ, Dumesic JA, Maravelias CT. Catalytic conversion of lignocellulosic biomass to fuels: Process development and technoeconomic evaluation. *Chem Eng Sci* 2012, 67:57–67.
120. Huber GW, Chheda JN, Barrett CJ, Dumesic JA. Production of liquid alkanes by aqueous-phase processing of biomass-derived carbohydrates. *Science* 2005, 308:1446–1450.
121. Lane J. Virdia, Virent pioneer new super-performing, parity-cost renewable jet fuel pathway. Available at: http://www.biofuelsdigest.com/bdigest/2012/03/27/double-drop-in-aviation-biofuels-virdia-virent-pioneer-new-super-perfoming-same-cost-jet-fuel/. (Accessed May 27, 2012).
122. Sannigrahi P, Pu Y, Ragauskas A. Cellulosic biorefineries—unleashing lignin opportunities. *Curr Opin Environ Sustain* 2010, 2:383–393.
123. Arato C, Pye EK, Gjennestad G. The lignol approach to biorefining of woody biomass to produce ethanol and

chemicals. *Appl Biochem Biotechnol* 2005, 121/124:871–882.
124. Shabtai JS, Zmierczak WW, Chornet E, Johnson D. Process for converting lignins into a high-octane blending component for gasoline. *US Pat Appl* US 2003/0115792 A1 2003.
125. de Wild P, Van der Laan R, Kloekhorst A, Heeres E. Lignin valorisation for chemicals and (transportation) fuels via (catalytic) pyrolysis and hydrodeoxygenation. *Environ Prog Sustain Energy* 2009, 28:461–469.
126. Sims B. Thesis Chemistry to build Iowa biorefinery using stover lignin. Available at: http://biorefiningmagazine.com/articles/5976/thesis-chemistry-to-build-iowa-biorefinery-using-stover-lignin. (Accessed May 16, 2012).
127. Lange J-P. Lignocellulose conversion: an introduction to chemistry, process and economics. *Biofuels Bioproducts Biorefin* 2007, 1:39–48.
128. Overend RP. Thermochemical biomass gasification technologies and products. In: *Global Climate and Energy Project (GCEP)*. Stanford, CA: Stanford University; 2004.
129. Sutton D, Kelleher BP, Ross JRH. Review of literature on catalysts for biomass gasification. *Fuel Process Technol* 2001, 73:155–173.
130. Stöcker M. Methanol-to-hydrocarbons: catalytic materials and their behavior. *Micropor Mesopor Mater* 1999, 29:3–48.
131. Dry ME. The Fischer–Tropsch process: 1950–2000. *Catal Today* 2002, 71:227–241.
132. Phillips S, Aden A, Jechura J, Dayton D, Eggeman T. *Thermochemical Ethanol via Indirect Gasification and Mixed Alcohol Synthesis of Lignocellulosic Biomass*. Technical Report NREL/TP-510-41168. Golden, CO: National Renewable Energy Laboratory; 2007.
133. Bulushev DA, Ross JRH. Catalysis for conversion of biomass to fuels via pyrolysis and gasification: a review. *Catal Today* 2011, 171:1–13.
134. Bridgwater AV. The technical and economic feasibility of biomass gasification for power generation. *Fuel* 1995, 74:631–653.
135. Bullis K. The death of range fuels shouldn't doom all biofuels. Available at: http://www.technologyreview.com/printer_friendly_article.aspx?id=39327. (Accessed May 27, 2012).
136. Smith W. *Literature Review: State of the Art in Biorefinery Development*. York, UK: NNFCC; 2007.
137. Pfeifer C, Koppatz S, Hofbauer H. Steam gasification of various feedstocks at a dual fluidised bed gasifier: impacts of operation conditions and bed materials. *Biomass Convers Biorefin* 2011, 1:39–53.
138. Ahlgren S, Baky A, Bernesson A, Nordberg Å, Norén O, Hansson P-A. Future fuel supply systems for organic production based on Fischer–Tropsch diesel and dimethyl ether from on-farm-grown biomass. *Biosyst Eng* 2008, 99:145–155.
139. Jenkins BM, Baxter LL, Miles TR Jr, Miles TR. Combustion properties of biomass. *Fuel Process Technol* 1998, 54:17–46.
140. Robbins MP, Evans G, Valentine J, Donnison IS, Allison GG. New opportunities for the exploitation of energy crops by thermochemical conversion in Northern Europe and the UK. *Prog Energy Combust Sci* 2012, 38:138–155.
141. Rudloff M. Biomass-to-liquid fuels (BtL)—made by CHOREN process, environmental impact and latest developments. In: *Automobile and Environment at Belgrade EAEC Congress*. Belgrade, Serbia: Yugoslav Society of Automotive Engineers; 2005.
142. Wooley R, Ruth M, Glassner D, Sheehan J. Process design and costing of bioethanol technology: a tool for determining the status and direction of research and development. *Biotechnol Prog* 1999, 15:794–803.
143. Bergman PCA, Boersma AR, Kiel JHA. Torrefaction for entrained flow gasification of biomass. In: *Second World Conference and Technology Exhibition on Biomass for Energy, Industry and Climate Protection, 10 May 2004*. Rome, Italy; 2004.
144. Sie ST, Senden MMG, van Wechem HMH. Conversion of natural gas to transportation fuels via the Shell Middle Distillate Synthesis process (SMDS). *Catal Today* 1991, 8:371–394.
145. Lewandowski I, Kicherer A. Combustion quality of biomass: practical relevance and experiments to modify the biomass quality of Miscanthus × giganteus. *Eur J Agronom* 1997, 6:163–177.
146. Qatar Petroleum. Oryx GTL inauguration—a key step in Qatar's bid to become the GTL capital of the world. Available at: http://www.qp.com.qa/qp.nsf/8c264276b952633c432571290026c60e/4cebdaa21d0dc426432571860018 0400?OpenDocument. (Accessed December 8, 2008).
147. Smith W. *Mapping the Development of UK Biorefinery Complexes*. York, UK: NNFCC; 2007.
148. Lerou JJ, Tonkovich AL, Silva L, Perry S, McDaniel J. Microchannel reactor architecture enables greener processes. *Chem Eng Sci* 2010, 65:380–385.
149. Bullis K. Increasing yield from gasification. Available at: http://www.technologyreview.com/energy/24838./ (Accessed May 16, 2012).
150. Brown RC. Biomass refineries based on hybrid themochemical-biological processing—an overview. In: Kamm B, Gruber VR, Kamm M, eds. *Biorefineries—Industrial Processes and Products*. Vol. 1. Weinheim, Germany: John Wiley & Sons; 2005, 227–252.
151. Mohammadi M, Najafpour GD, Younesi H, Lahijani P, Uzir MH, Mohamed AR. Bioconversion of synthesis gas to second generation biofuels: a review. *Renew Sustain Energy Rev* 2011, 15:4255–4273.
152. Grethlein AJ, Jain MK. Bio-processing of coal-derived synthesis gases by anaerobic bacteria. *Trends Biotechnol* 1993, 10:418–423.
153. Klasson KT, Ackerson MD, Clausen EC, Gaddy JL. Bioreactor design for synthesis gas fermentations. *Fuel* 1991, 70:605–614.
154. Zhu H, Shanks BH, Choi DW, Heindel TJ. Effect of functionalized MCM41 nanoparticles on syngas fermentation. *Biomass Bioenergy* 2010, 34:1624–1627.

155. Jessen H. LanzaTech to use funding to further develop core technology. Available at: http://www.ethanolproducer.com/articles/8522/lanzatech-to-use-funding-to-further-develop-core-technology. (Accessed May 27, 2012).
156. Bridgwater AV, Meier D, Radlein D. An overview of fast pyrolysis of biomass. *Org Geochem* 1999, 30:1479–1493.
157. Balat M, Balat M, Kırtay E, Balat H. Main routes for the thermo-conversion of biomass into fuels and chemicals. Part 1: pyrolysis systems. *Energy Convers Manag* 2009, 50:3147–3157.
158. Butler E, Devlin G, Meier D, McDonnell K. A review of recent laboratory research and commercial developments in fast pyrolysis and upgrading. *Renew Sustain Energy Rev* 2011, 15:4171–4186.
159. Huffman DR, Vogiatzis AJ, Bridgwater AV. The characterization of fast pyrolysis bio-oils. In: Bridgwater AV, ed. *Advances in Thermochemical Biomass Conversion*. Vol. 2. London: Blackie; 1994, 1095.
160. James P, Diebold PE. A review of the chemical and physical mechanisms of the storage stability of fast pyrolysis bio-oils. In: Bridgewater AV, ed. *Fast Pyrolysis of Biomass—A Handbook*. Vol. 2. Newbury, UK: CPL Press; 2002, 243.
161. Czernik S, Maggi R, Peacocke GVC. Review of methods for upgrading biomass-derived fast pyrolysis oils. In: Bridgwater AV, ed. *Fast pyrolysis of biomass-a handbook*. Vol. 2. Newbury, UK: CPL Press; 2002, 141.
162. Agblevor FA, Besler S, Evans RJ. Inorganic compounds in biomass feedstocks: their role in char formation and effect on the quality of fast pyrolysis oils. In: *Specialists Workshop, Biomass Pyrolysis Oil Propoerties and Combustion*. Estes Park, CO; 1994.
163. Wiselogel AE, Agblevor FA, Johnson DK, Deutch S, Fennell JA, Sanderson MA. Compositional changes during storage of large round switchgrass bales. *Bioresource Technol* 1996, 56:103–109.
164. Javaid A, Ryan T, Berg G, Pan X, Vispute T, Bhatia SR, Huber GW, Ford DM. Removal of char particles from fast pyrolysis bio-oil by microfiltration. *J Membrane Sci* 2010, 363:120–127.
165. Eom I-Y, Kim K-H, Kim J-Y, Lee S-M, Yeo H-M, Choi I-G, Choi J-W. Characterization of primary thermal degradation features of lignocellulosic biomass after removal of inorganic metals by diverse solvents. *Bioresource Technol* 2011, 102:3437–3444.
166. Demirbas A. Mechanisms of liquefaction and pyrolysis mechanisms of biomass. *Energy Convers Manag* 2000, 41:633–646.
167. Fisher T, Hajaligol M, Waymack B, Kellogg D. Pyrolysis behavior and kinetics of biomass derived materials. *J Anal Appl Pyrolysis* 2002, 62:331–349.
168. Yang H, Yan R, Chen H, Zheng C, Ho Lee D, Liang T. In-depth investigation of biomass pyrolysis based on three major components: hemicellulose, cellulose and lignin. *Energy Fuels* 2006, 20:388–393.
169. Mettler MS, Mushrif SH, Paulsen AD, Javadekar AD, Vlachos DG, Dauenhauer PJ. Revealing pyrolysis chemistry for biofuels production: Conversion of cellulose to furans and small oxygenates. *Energy Environ Sci* 2012, 5:5414–5424.
170. Adam J, Blazso M, Meszaros E, Stocker M, Nilsen MH, Bouzga A, Hustad JE, Grønli M, Øye G. Pyrolysis of biomass in the presence of Al-MCM-41 type catalysts. *Fuel* 2005, 84:1494–1502.
171. Kawamoto H, Hatanaka W, Saka S. Thermochemical conversion of cellulose in polar solvent (sulfolane) into levoglucosan and other low molecular-weight substances. *J Anal Appl Pyrolysis* 2003, 70:303–313.
172. Kawamoto H, Saka S. Heterogeneity in cellulose pyrolysis indicated from the pyrolysis in sulfolane. *J Anal Appl Pyrolysis* 2006, 76:280–284.
173. Mortensen PM, Grunwaldt JD, Jensen PA, Knudsen KG, Jensen AD. A review of catalytic upgrading of bio-oil to engine fuels. *Appl Catal A: Gen* 2011, 407: 1–19.
174. de Miguel Mercader F, Groeneveld MJ, Kersten SRA, Way NWJ, Schaverien CJ, Hogendoorn JA. Production of advanced biofuels: co-processing of upgraded pyrolysis oil in standard refinery units. *Appl Catal B: Environ* 2010, 96:57–66.
175. Samolada MC, Baldauf W, Vasalos IA. Production of a bio-gasoline by upgrading biomass flash pyrolysis liquids via hydrogen processing and catalytic cracking. *Fuel* 1998, 77:1667–1675.
176. Furimsky E, Massoth FE. Deactivation of hydroprocessing catalysts. *Catal Today* 1999, 52:381–495.
177. Vitolo S, Bresci B, Seggiani M, Gallo MG. Catalytic upgrading of pyrolytic oils over HZSM-5 zeolite: behaviour of the catalyst when used in repeated upgrading—regenerating cycles. *Fuel* 2001, 80:17–26.
178. Huber GW, Iborra S, Corma A. Synthesis of transportation fuels from biomass: chemistry, catalysts, and engineering. *Chem Rev* 2006, 106:4044–4098.
179. Jones SB, Valkenburg C, Walton CW, Elliott DC, Holladay JE, Stevens DJ, Kinchin C, Czernik S. *Production of Gasoline and Diesel from Biomass via Fast Pyrolysis, Hydrotreating and Hydrocracking: A Design Case*. PNNL-18284. U.S. Department of Energy; 2009.
180. Baldauf W, Balfanz U, Rupp M. Upgrading of flash pyrolysis oil and utilization in refineries. *Biomass Bioenergy* 1994, 7:237–244.
181. Mahfud FH, Melián-Cabrera I, Manurung R, Heeres HJ. Biomass to fuels: upgrading of flash pyrolysis oils by reactive distillation using a high boiling alcohol and acid catalysts. *Process Safety Environ Protect* 2007, 85:466–472.
182. Peng J, Chen P, Lou H, Zheng X. Catalytic upgrading of bio-oil by HZSM-5 in sub- and super-critical ethanol. *Bioresource Technol* 2009, 100:3415–3418.
183. Crossley S, Faria J, Shen M, Resasco DE. Solid Nanoparticles that Catalyze Biofuel Upgrade Reactions at the Water/Oil Interface. *Science* 2010, 327:68–72.
184. Yu Z, Zhang H. Ethanol fermentation of acid-hydrolyzed cellulosic pyrolysate with Saccharomyces cerevisiae. *Biores Technol* 2004, 93:199–204.
185. Zhuang XL, Zhang HX, Yang JZ, Qi HY. Preparation of levoglucosan by pyrolysis of cellulose and its citric acid fermentation. *Bioresource Technol* 2001, 79:63–66.

186. Melligan F, Dussan K, Auccaise R, Novotny EH, Leahy JJ, Hayes MHB, Kwapinski W. Characterisation of the products from pyrolysis of residues after acid hydrolysis of Miscanthus. *Bioresource Technol* 2012, 108:258–263.
187. Nowakowski DJ, Bridgwater AV, Elliott DC, Meier D, de Wild P. Lignin fast pyrolysis: results from an international collaboration. *J Anal Appl Pyrolysis* 2010, 88:53–72.
188. Gellerstedt Gr, Li J, Eide I, Kleinert M, Barth T. Chemical structures present in biofuel obtained from lignin. *Energy Fuels* 2008, 22:4240–4244.
189. de Wild PJ, Huijgen WJJ, Heeres HJ. Pyrolysis of wheat straw-derived organosolv lignin. *J Anal Appl Pyrolysis* 2012, 93:95–103.
190. Demirbas A. Biorefineries: Current activities and future developments. *Energy Convers Manag* 2009, 50:2782–2801.
191. Toor SS, Rosendahl L, Rudolf A. Hydrothermal liquefaction of biomass: a review of subcritical water technologies. *Energy* 2011, 36:2328–2342.
192. Blanch HW. Bioprocessing for biofuels. *Curr Opin Biotechnol* 2012.
193. Fehrenbacher K. No DOE loan guarantee for KiOR, at least not this year. Available at: http://gigaom.com/cleantech/no-doe-loan-guarantee-for-kior-at-least-not-this-year/. (Accessed May 27, 2012).
194. Ethanol Producer Magazine. USDA proposes reform to loan guarantees for advanced biofuels. Available at: http://www.ethanolproducer.com/articles/7526/usda-proposes-reform-to-loan-guarantees-for-advanced-biofuels. (Accessed May 10, 2012).
195. European Commission. *Commission Staff Working Document: NER300—Moving Towards a Low Carbon Economy and Boosting Innovation, Growth and Employment Across the EU*. Brussels: European Commission; 2012.
196. Keller JB, Plath PB. Financing biotechnology projects: lender due diligence requirements and the role of independent technical consultants. *Appl Biochem Biotechnol* 1999, 77/79:641–648.
197. Luo L, van der Voet E, Huppes G. Biorefining of lignocellulosic feedstock—technical, economic and environmental considerations. *Bioresource Technol* 2010, 101:5023–5032.
198. Dias MOS, Junqueira TL, Cavalett O, Cunha MP, Jesus CDF, Rossell CEV, Maciel Filho R, Bonomi A. Integrated versus stand-alone second generation ethanol production from sugarcane bagasse and trash. *Bioresource Technol* 2012, 103:152–161.
199. Mark T, Darby P, Salassi M. Energy cane usage for cellulosic ethanol: estimation of feedstock costs. In: *Southern Agricultural Economics Association—2009 Annual Meeting, January 31-February 3*. Atlanta, GA. College Station, TX: SAEA; 2009.
200. Heaton EA, Flavell RB, Mascia PN, Thomas SR, Dohleman FG, Long SP. Herbaceous energy crop development: recent progress and future prospects. *Curr Opin Biotechnol* 2008, 19:202–209.
201. Lynd LR, Laser MS, Bransby D, Dale BE, Davison B, Hamilton R, Himmel M, Keller M, McMillan JD, Sheehan J, et al. How biotech can transform biofuels. *Nat Biotech* 2008, 26:169–172.
202. Chen F, Dixon RA. Lignin modification improves fermentable sugar yields for biofuel production. *Nat Biotech* 2007, 25:759–761.
203. Maeda H, Shasany AK, Schnepp J, Orlova I, Taguchi G, Cooper BR, Rhodes D, Pichersky E, Dudareva N. RNAi suppression of arogenate dehydratase1 reveals that phenylalanine is synthesized predominantly via the arogenate pathway in petunia petals. *Plant Cell Online* 2010, 22:832–849.
204. Shimada K, Hosoya S, Ikeda T. Condensation reactions of softwood and hardwood lignin model compounds under organic acid cooking conditions. *J Wood Chem Technol* 1997, 17:57–72.
205. Sanderson M, Egg RP, Wiselogel AE. Biomass losses during harvest and storage of switchgrass. *Biomass Bioenergy* 1997, 12:107–114.
206. Buggeln R, Rynk R. Self-heating in yard trimmings: Conditions leading to spontaneous combustion. *Compost Sci Utili* 2002, 10:162–182.
207. Kou N, Zhao F. Effect of multiple-feedstock strategy on the economic and environmental performance of thermochemical ethanol production under extreme weather conditions. *Biomass Bioenergy* 2011, 35:608–616.
208. Green Car Congress. Honeywell's UOP breaks ground on integrated biorefinery project in Hawaii for drop-in gasoline, jet and diesel from fast pyrolysis of biomass and catalytic upgrading. Available at: http://www.greencarcongress.com/2011/08/kapolei-20110830.html. (Accessed May 27, 2012).
209. Nexant Ltd. *The Feasibility of Second Generation Biodiesel Production in the UK*. London: Nexant Chem Systems; 2007.
210. Henrich E. Clean Syngas form Biomass by Pressurised Entrained Flow Gasification of Slurries from fast Pyrolysis. In: *Second Generation Automotive Biofuel Conference*. Stockholm, Sweden; 2005.
211. Huang H-J, Ramaswamy S, Al-Dajani W, Tschirner U, Cairncross RA. Effect of biomass species and plant size on cellulosic ethanol: A comparative process and economic analysis. *Biomass Bioenergy* 2009, 33:234–246.
212. Van den Wall Bake JD, Junginger M, Faaij A, Poot T, Walter A. Explaining the experience curve: cost reductions of Brazilian ethanol from sugarcane. *Biomass Bioenergy* 2009, 33:644–658.
213. Ng KS, Sadhukhan J. Techno-economic performance analysis of bio-oil based Fischer–Tropsch and CHP synthesis platform. *Biomass Bioenergy* 2011, 35:3218–3234.
214. Biofuels Digest. KiOR files $100M IPO—the complete digest analysis. Available at: http://www.biofuelsdigest.com/bdigest/2011/04/13/kior-files-100m-ipo-the-complete-digest-analysis./ (Accessed May 27, 2012).
215. Enerkem. *United States Securities and Exchange Commission Form F-1*. Montreal, Quebec: Enerkem, Inc.; 2012.
216. Moore P. Range fuels at 'end of road'. *Boulder County Business Report* 2012. Available at: http://www.bcbr.com/article.asp?id = 61980. (Accessed May 5, 2012)
217. USDA. *A USDA Regional Roadmap to Meeting the Biofuels Goals of the Renewable Fuels Standard by 2022 – USDA Biofuels Strategic Production Report*. Washington, DC: US Department of Agriculture; 2010.

13

Separation Technologies for Current and Future Biorefineries—Status and Potential of Membrane-Based Separation

Lan Ying Jiang* and Jia Ming Zhu

Central South University, Changsha, Hunan, China

INTRODUCTION

Biorefinery refers to the integrated processes that convert or fractionate biomass feedstocks to produce a broad range of value-added chemicals, or sometimes direct electricity and heat. As far as the basic principle is concerned, the biorefinery concept is not contemporary. For example, using sugarcane to produce bioethanol dates back to as early as 6000 BC.[1] For modern society, biorefinery is foremost poised as a strategic industry that provides energy sources to crack the problems associated with fossil fuels.[2] Being renewable, bioenergy related with biofuels in forms of solid, liquid, or gas and biomass-derived electricity power and heat as well has eased the worldwide concern caused by energy consumption mode heavily relying on traditional fossil fuels, of which the depletion cannot be avoided. Geologically, the major reservoirs of fossil fuels are located at certain areas with political instability. The evenly distributed biomass resources and hence the supply of renewable energy in a similar vein are expected to break the energy bottleneck restricting the development of many countries by import uncertainty and price fluctuation. More than these, the reasons underlying the shift toward biofuels are also related to their properties and performance in applications. The balance in carbon footprint by using biofuels undoubtedly can curb the greenhouse effect due to the fossil fuel combustion.[3] Compared with the other types of renewable energy sources, biofuels are more appreciated in transportation sector because of its cutting edge of energy density and its similarity to gasoline and diesel, the two global transportation fuels.

With the above considerations, biorefinery is regarded as a drop-in replacement of crude oil refinery, and biofuels produced are expected to supplement or even take the place of fossil fuel in some fields. The fact is that biofuels have already acquired a long list of market orders. In 2001, the policy of using fuels with gasoline and ethanol blend was actualized in China. Concerning motor vehicles, it has already been put to effectuation for many years in several provinces of China that the ethanol content in the fuels must be at least 10 wt%. Biomass-derived energy sources furnish industrial activity and social life with 50 EJ (exajoules) units of energy, as reported in the statistics conducted by some related organizations in 2011. This amount takes up 10% of global annual consumption in primary energy (renewable and nonrenewable energy). The energy of 50 EJ from biofuels accounts for 75% of the total amount of renewable energy (e.g., solar, wind, and geothermal energy) consumed annually. Nowadays, Brazil producing bioethanol from sugarcanes, and the United States deriving bioethanol from corn rank first and second, respectively, in biofuel production worldwide. The European Union

*Correspondence to: lanyingjiang99@hotmail.com

Advances in Bioenergy: The Sustainability Challenge, First Edition. Edited by Peter D. Lund, John Byrne, Göran Berndes and Iacovos A. Vasalos.

Originally published in: *WIREs Energy Environ* 2013, 2: 193–208 doi: 10.1002/wene.73

ranking third catches up with the trend in biofuel production in recent years. Among the European countries, the two leading producers of biofuels are Germany and France.[5] Their effort is on the production of biodiesel with oilseeds.

Other than the role of renewable energy supplier, biorefinery industry also functions as an essential provider of renewable raw materials or additives for many commercialized products such as food, cosmetics, medicines, package, and so on.[5] These materials or additives include hemicelluloses, antioxidants (phenolics or polyphenolics), glucose, citric acid, lactic acid, starch carboxylate, lysine, vitamins, biocolorant, enzymes, glycerol, and the list could still be extended.

In all, biorefinery associates with many aspects of social development and its potential in establishing a sustainable future cannot be denied. It has been accepted as the most strategically important industry, especially in the scenario of foreseeable future, when technologies of transforming and utilizing other alternative renewable energy resources are still immature. Notwithstanding, there exist several significant problems that cause its actual effect below the expectation. The high costs of biomass feedstock and their transportation and processing are among the top list.[6] With more and more resources being identified as substitutive feedstocks that are sufficient in quantity and do not compete with human in consuming edible food and arable land, the focus of research and development about biorefinery is how to improve the transformation processes. Just as the word of 'refinery' indicates, separation is an intrinsic constituent throughout the biorefinery process. With multiple products being delivered in a complicated way, actually, advanced separation technologies have been always pursued so as to maximize the biorefinery efficacy and biomass potential. In this review paper, the application status and potentials of membrane-based separation in biorefinery will be reviewed on the basis of the introduction about feedstocks' classification and major biorefinery routes. Membrane separation is biofriendly and energy efficient. Additionally, many researches and applications have shown its advantages in lowering operation cost of separation.[7,8] With regard to these edges, it is also an interest of the review to tentatively identify some potential applications regarding specific processes.

BACKGROUND

Biomass Feedstocks

Biomass that can be used for biorefinery feedstocks exist adequately on the earth. Forestry, agricultural plants, microbial cells, and the biological wastes and residues associated with human life can all be counted. The energy existing in various carbonaceous compounds inside biomass is solar energy in nature. Plants and microbial cells synthesize hydrocarbon with atmospheric CO_2, water, and sunlight. This process is the well-known photosynthesis.

First-generation feedstock is mainly derived from agricultural products, with starch crops, sugar crops, and oilseed being the most popularly utilized ones.[9] They are abundant and characterized of high carbohydrate or energy content. The conventional biochemical technologies are normally employed to generate biofuels from the first-generation feedstock. Nevertheless, biorefinery relying on first-generation feedstock is not a panacea for renewable energy. The tremendously increased demand of energy in the recent years puts the strategy under heavy pressure. Firstly, more arable croplands in tropical regions are needed for their cultivation so as to meet the extra energy requirement. Secondly, the competition for crops supply between bioenergy production and food and animal feed drives the price of crops up. There also exists an environmental worry with the first generation of feedstock because of the nitrogen oxides (NO_x) emission during the crops cultivation.[10] NO_x are classified as gaseous chemicals causing greenhouse effect.

Lignocellulosic biomass and organic waste residues from various agricultural and forestry processes are the major resources for second-generation feedstock.[11] The schematic representation of woody mass biorefinery is demonstrated in Figure 13.1. They circumvent the disadvantages associated with first-generation feedstocks. These raw materials are not food/animal feed choices. They can grow on land with a wide range of soil qualities and have no strict requirements on environment conditions. More than these, utilizing the wasted organics help reduce the dissipation of natural resources. Currently, the superior challenge with biorefinery using second-generation feedstocks is the complex and sophisticated conversion processes. The major biorefinery routes for lignocellulosic biomass are thermal and biochemical conversion that will be introduced in section *Conversion Routes in Biorefinery*.

Because of the high oil–lipid, carbohydrate, or protein contents and the high yield (up to 75% dry weight), a host of bacteria and algae are being utilized as biofuel feedstocks.[12] They are generally known as 'third-generation feedstock'. The oil stored within algae is accrued by phototrophic, heterotrophic, and mixtropic approaches. As far as biodiesel production is concerned, estimation shows that in the near future, microalgae available may offer 10–300 times more oil than the conventional oil crops now cultivated.[13] The major problem hampering the immediate scale-up of algae or bacteria biorefinery is the immature conversion technology.

In recent years, many other types of resources have been proposed as future generation of biorefinery feedstocks.

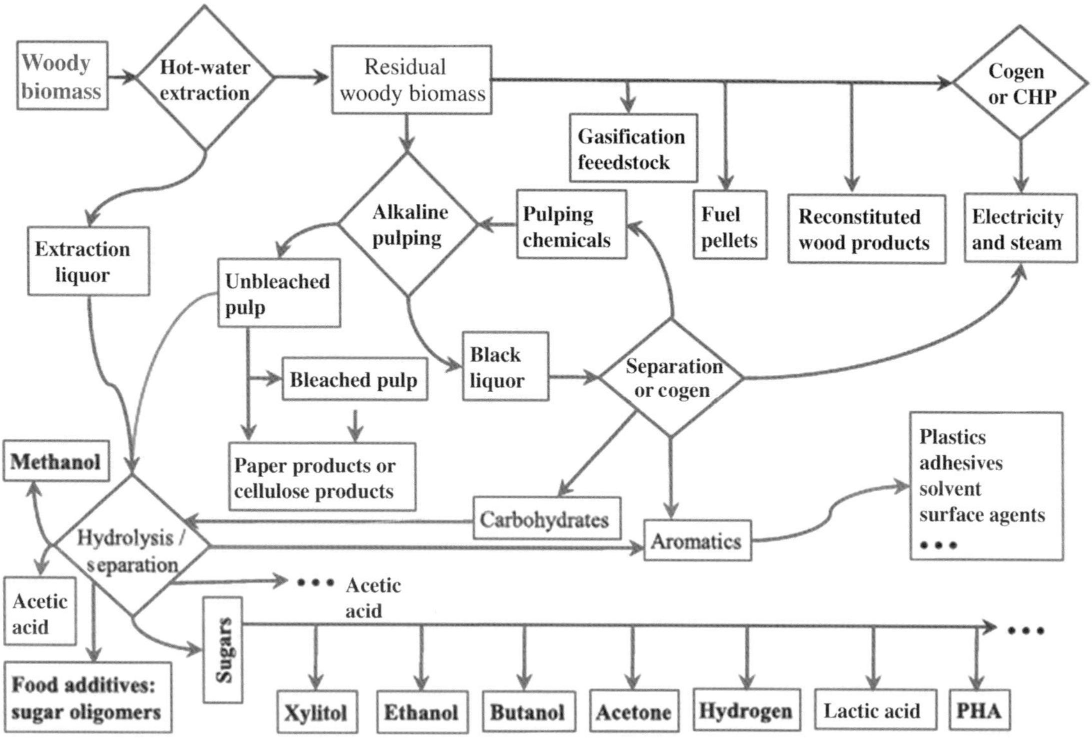

Figure 13.1 Schematic representation of woody biomass biorefinery. (Reproduced from Ref 27. Copyright 2012, Elsevier.)

For example, Lammens et al.[14] reviewed the feasibility of amino acids for production of bulk chemical via biorefinery. The investigated sources are maize and wheat dried distillers grains with solubles, sugarcane vinasse and its leaves, sugar beet vinasse and its leaves, cassava leaves, press cakes of rapeseed, sunflower, soybean, palm oil and *Jatropha*, animal slaughter waste, grass, and alfalfa. The progress of biorefinery using glycerol for producing biofuels and other chemicals can be found in the article of Rahmat et al.[15]

Conversion Routes in Biorefinery

Depending on the mechanisms for getting the final product, the conversion routes that are widely employed in biorefinery process are broadly classified into three major types: (1) thermochemical and mechanical conversions, (2) biochemical and biological conversions, and (3) physiochemical conversions.[2] One common feature of these conversion routes is the simultaneous deoxygenization and depolymerization of the carbonaceous compounds with high molecular weight to release mono molecules of sugar for subsequent processing.

In thermal–chemical conversion, the biomass is treated at high temperature so as to initiate the fraction of hydrocarbon structure.[2] In biochemical conversions, the mono sugars released in preceding pretreatment and hydrolysis are changed into fuels or other chemicals via a series of chemical reactions that are biologically catalyzed inside microorganisms.[2] Physiochemical conversion commonly refers to the process where oils are extracted (e.g., pressurized solvent extraction) from oilseed feedstocks.[6] Thereafter, the oils are transformed in biodiesels by catalytically mediated transesterification with alcohols, which is a biochemical process.[1] The chemical structure of biodiesel is fatty acid alkyl esters. The oils extractives can also be used directly in transportation fuel system. The subcategories of these routes are summarized in Table 13.1 and compared in terms of their advantages and disadvantages.[1,6]

STATUS AND POTENTIAL OF APPLYING MEMBRANE-BASED SEPARATION IN BIOREFINERY

Whichever the conversion process is, separation is an indispensible and integral part of it. Those separation

Table 13.1 Analysis of Different Biorefinery Processes

Category	Subcategory	Feedstock	Major Product	Advantage	Disadvantage
Thermal–chemical	Direct combustion (oxygen rich)	Fuel wood, bagasse, municipal solid waste, residue from pulp and paper	Electricity, heat	Power output is the same as that of traditional carbonaceous combustion	High cost in network distribution and biomass; processing;Toxic emission
	Gasification (800–900°C, partial oxidation)		Gases of CO, H_2, CH_4	The syngas can be used to produce many useful products	Formation of tar and other undesired by-product
	Pyrolysis (350–more than 800°C; anoxygenic)	Wood, wood charcoal	Bio-oils (organic acid, esters, alcohols, ketones, phenols, aldehydes, alkenes, furfurals, and sugars)	The product is easy for transport and can be converted into valuable chemicals, and so on	The oil is inconsistent in composition, stability
	Liquefaction (250–350°C, high pressure; reducing environment)		Bio-oil (heavy oily compounds) similar to pyrolysis products	Product has appropriate molecular weight	High cost due to complex reactor and feed system
Biochemical	Anaerobic digestion	Agricultural, municipal and industrial waste, manure, sludge, leftover food, and so on	Biogas (CH_4, CO_2), digestate	It can reduce waste disposal	The sustainability and reliability depends on transportation and accessibility of biomass
	Microbial/enzymatic process	Fermentable feedstock	Various biofuels (liquid, gas), by-product of food additives, animal feedstuff, heat, electricity	Less by-products and pollutants, broader range of feedstocks	Complex pretreatment
Physiochemical		oilseeds, algae	Various seed oil for biodiesel production	Environmentally amicable and clean	Technology is relatively immature and cost of raw material is high

technologies that have been investigated for or applied in biorefinery include distillation, gas stripping, adsorption, membrane-based separation, and chemical separation (i.e., chemical reactions bonding certain compounds to other molecules that make the separation easier). The advantages of membrane-based separation have been highlighted in previous section. A membrane is a semi-permeable barrier that permits the preferential transport of one or more component(s) of a feed mixture, thereby enabling the separation.[16] In the process, membrane hinders the gross mass movement between the two phases, while permits preferred passage of certain species from one phase (upstream) to the other (downstream);[17] as a result, the bulk phase on one side of the membrane will be enriched of certain components, whereas the other side almost depleted of them.

For most of the membrane separations, the transport is mainly determined by its separation layer structure in terms of pore size, whereby the particles larger than the pore size are generally rejected by the membranes. In industrial application, pore size is often indicated by corresponding molecular-weight cut-off (MWCO) at certain retention, facilitating customers' selection. The major types of membranes are summarized in Figure 13.2 in the order of increasing pore size: gas separation (GS), pervaporation (PV), dialysis, reverse osmosis (RO), nanofiltration (NF), ultrafiltration (UF), and microfiltration (MF).[18] Membrane with smaller pore size provides more delicate discrimination on size and shape and removes finer particles. For example, GS membrane can separate O_2 (3.46 Å) and N_2 (3.64 Å),[19] whereas difference of 1000 times in sizes of two particles is the basis for membrane MF giving reasonable separation.[20] In some cases, the other physiochemical properties of the membrane materials are also involved or even become the decisive factors. Even though membrane separation sometimes is simply referred to as 'filtration', its mechanism is more complicated than the superficial meaning of the word. Because of the utilization of the various influential factors, so-called 'reverse' membranes are invented, such as hydrophobic PV.[21] In these separations, the smaller particles are delayed more than the bigger ones. Yet another category is the ion-charged membrane that selectively transports particles with charge opposite the one grafted on the membrane. The transport of species across the membrane is caused by chemical potential gradient due to concentration difference, pressure difference, or electrical potential difference between the upstream and downstream.

In the biorefinery process, the particle sizes of various products range from a few angstroms for gaseous and liquid species to larger than macro size for the germs, fibers, and so on. Most of the membranes can find a separation step with the product size matching its functional pore size. At first, therefore, we plan to introduce and analyze membrane-based separation applied in biorefinery according to the membrane classifications aforementioned. Further consideration makes it clear that hybrid processes combining different separation technologies are generally installed for achieving specific targets and the role of membrane is more than just separation. Therefore, the above design might not be helpful in understanding the scenario of biorefinery as a whole, which gives rise to the forthcoming section based on the outline constructed by the broadly classified separation segment (pretreatment, product recovery, wastewater treatment, etc.) in biorefinery with the background of each task and the characteristics of the membrane technology employed being briefed along.

Pretreatment

Pretreatment is an indispensible procedure in biochemical biorefinery. It is usually employed to crack or fraction biomass structure and reduce the complex hydrocarbons to simple ones suitable for hydrolysis. The coproducts released in these pretreatment include germ, fibers, and gluten for corn-to-ethanol biorefinery. Regarding the biorefinery using second-generation lignocellulosic feedstocks to produce ethanol, extraction of hemicelluloses and separation of short and long fibers in biorefinery liquid are the major separation tasks in the pretreatment. Removal of these coproducts from the main line of the biorefinery will allow following advantages: (1) more efficient fermentation, (2) the coproducts can be used as additive for food and feedstock for chemicals production (i.e., polymer), and (3) incremental usage of biomass. The major chemicals of second-generation biomass are cellulose, hemicelluloses, and lignin, with hemicelluloses being the easiest to be degraded and extracted. Therefore, separation of hemicelluloses from biomass first followed by the transformation of cellulose in another step will realize more thorough decomposition and utilization of woody biomass.

In the work of Gabrielii et al.,[22] extraction of hemicelluloses from aspen (*Populus tremula*) was performed by alkali solution. Thereafter, hydrogen (H_2) peroxide treatment, membrane UF, and spray drying were employed to get solid product.[5,22] Membrane UF is a separation process using membranes with pore sizes in the range of 0.1–0.001 μm. Typically, UF will remove high-molecular-weight substances, colloidal materials, and organic and inorganic polymeric molecules with molecular weight 10^3–10^6 Da. Hemicelluloses are composed of five- and six-carbon sugars such as xylose, glucose, rhamnose, mannose, and arabinose. Therefore, hemicelluloses actually are a collection of different polymers heterogeneous in

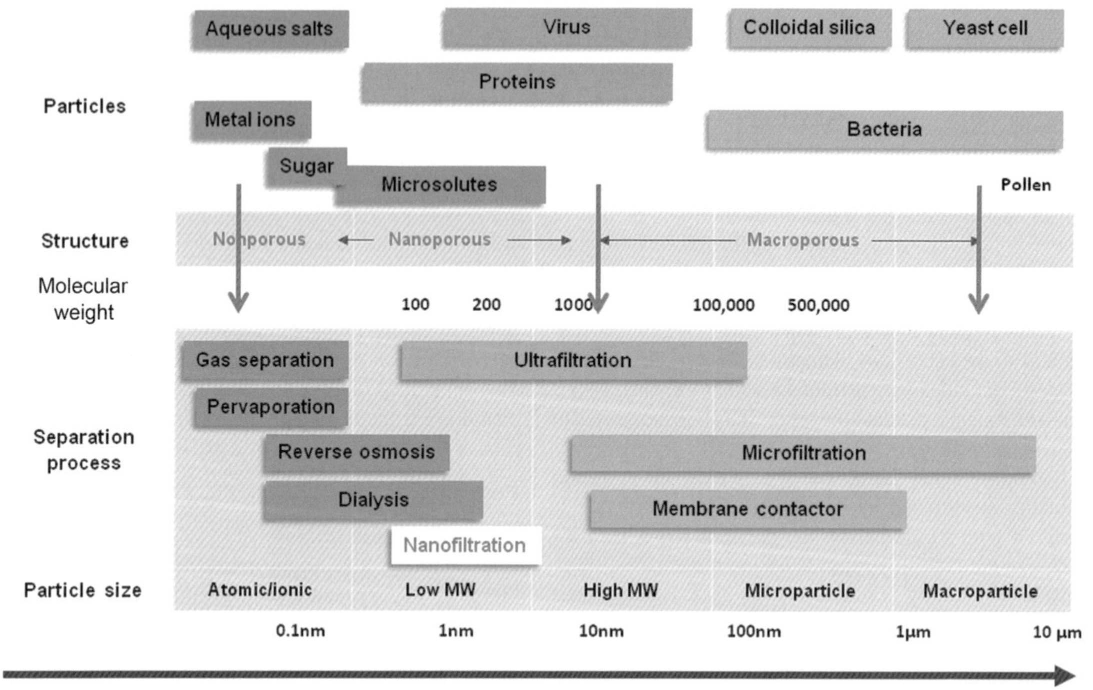

Figure 13.2 Classifications of conventional membrane separations.

nature.[23] In the process,[22] the pectins, starch, and fat are extracted and removed by ammonium hydroxide and the lignin was solubilized by 1% NaOH in 70% ethanol solution. After solid–liquid separation by centrifugation, hemicellulose left in the solid residue was dissolved with 4% NaOH.[22] The extractant was peroxide bleached to further reduce the lignin. Finally, the dissolved hemicellulose was concentrated by 3000 MWCO UF and dried to obtain solid hemicelluloses.[22]

In the twin-screw extrusion method for extracting hemicelluloses,[24,25] filtration (not necessarily membrane filtration) was employed to remove the fine solids and concentrate the hemicelluloses precipitated by ethanol at low temperature with 90% of the initial hemicelluloses (mainly pentosans) being recovered, followed by filtration, washing, and drying.[25] In the twin-screw process, the biomass is subjected to mechanical force originated from two conveying screws or counter-screws turning within a fixed barrel.[5,24,25] This technology integrates solid–liquid extraction and filtration (solid–liquid separation) in one step with extremely thorough processing of the plant matter.

In viscose fiber production, hemicelluloses extracted into soda liquor are detrimental to fiber quality. Removal of hemicelluloses with least loss of soda sometimes is necessary.[26] In the extraction operated under severe pretreatment conditions (e.g., alkaline extraction), low-molecular-weight hemicelluloses are produced. Therefore, UF process might not provide enough retention of the hemicellulose in the extractant liquid. In the work of Schlesinger et al.,[26] NF using membranes with different MWCO was investigated to separate hemicellulose from the caustic soda liquor (alkaline extracting solution containing 200 g/L NaOH). NF with pore size between 8 and 50 Å is most suitable for processing component dissolved in liquid with molecular weight of 200–1000 Da and molecular size around 1 nm. The hemicelluloses with molecular weight higher than 1000 g/mol are almost all retained. The membranes with the MWCO of 200–300 (SelRO MPF-34; Koch Membrane Systems, Inc., Wilmington, MA) and 200–250 (NTR-7470; Nitto-Denko Corporation, Osaka, Japan) g/mol achieved retention of around 90%. In addition to low retention efficiency, application of UF is also limited by more severe membrane fouling due to the deposition onto and intrusion of macromolecules into the microporous membrane.

In biorefinery using different feedstocks, the extractives from various pretreatment procedures such as hot-water extraction, steam explosion contains not only hemicelluloses, but also polysaccharides, monosaccharides, acetic acid, degraded lignin, and many other substances such as

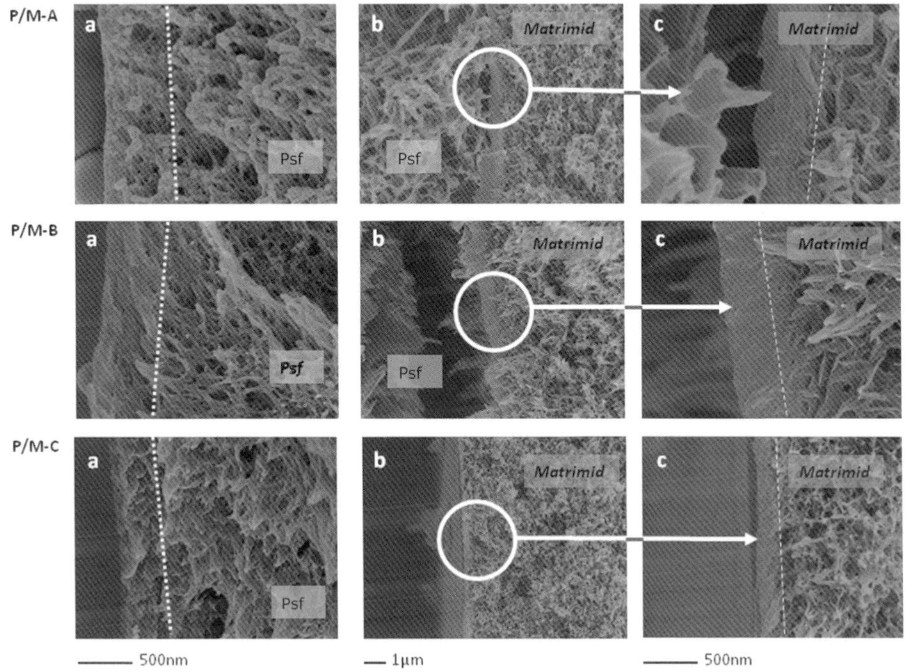

Figure 13.3 Partial cross-section morphology of different layers of polysulfone/Matrimid dual-layer hollow fibers. (Reproduced from Ref 49. Copyright 2009, John Wiley & Sons, Ltd.)

aliphatic and alicyclic, phenolics, inorganics, ash, and so on.[27] The presence of several nonsugar compounds in the sugar stream will hamper the conversion efficiency. Yet phenolics are antioxidants that could be used as cheap and renewable food additive.

In the works of Vazquez et al.[28] and Duarte et al.,[29] the extractive liquor from hot-water wood extraction is concentrated by NF membrane, resulting in lowering pH conditions in retentate. As higher acid concentration is favorable for hydrolysis, the follow-up hydrolysis produced more monosugars (i.e., xylose, arabinose, glucose, rhamnose, mannose, and galactose) in a relatively shorter period of reaction time. The extractant generated in the pretreatment procedure were generally concentrated by a factor of about 10. Another result of membrane concentration is that acidic conditions facilitate aromatic inhibitors' removal as phenolics precipitates easily in acidic conditions. The precipitant is removed from the liquor phase via membrane UF.

Zein existing in corn is a prolamine-rich protein (gluten). It is widely used as environmental-benign feedstock in producing adhesive, fiber, cosmetic, textile, and plastics.[5] In the total protein of corn, zein represents 45–50%. With its high price (US$ 10–40/kg), it could help reduce the cost of operating a corn-to-ethanol biorefinery plant if being extracted from the feedstocks. The general procedure for zein extraction in the corn-to-ethanol biorefinery is described below: (1) zein and/or oil from milled maize or corn are extracted by using ethanol and (2) the liquid phase of ethanol containing dissolved oil and zein is isolated from the residual solids of corn. The molecular weight of α-zein and β-zein are 21–25 and 17–18 kDa, respectively. Therefore, membrane UF could be a good choice to retain zein so as to separate it from oil and ethanol. Cheryan[30] reported this novel utilization in one of his patents.

Hydrolysis

Hydrolysis usually refers to the breakage of chemical bonds by water, whereby the complex hydrocarbon produced in pretreatment is reduced to mono sugar suitable for fermentation. Generally, it is necessary to catalyze hydrolysis by using acidic (either strong or weak) conditions, enzymes, and/or hydrothermal environment. Because of the complex components in the extractives and catalytic mechanism of breakdown, the products of hydrolysis are complicated with pentose and hexose sugars and a series of nonsugar constituents. In the hardwood extract hydrolysate samples, the nonsugar components having high concentration include acetic acid and furfural, and minor components include formic acid, hydroxymethylfurfural (HMF), and so on.[27] Some of the nonsugar products will impede the fermentation as inhibitors. The major inhibitors include furan derivatives furfural and 5-HMF, aliphatic acids (i.e., acetic, formic, and levulinic acid), and phenolic compounds (i.e., phenol, vanillin, and *p*-hydroxybenzoic acid). Therefore, as demonstrated in Figure 13.1 using woody biomass refinery as an example, the separation tasks in hydrolysis step include (1) the removal/recovery of inhibitors, so that they can find alternative usages; chemical feedstock, direct energy production through combustion, for example; (2) purification and concentration of sugar chemicals.

Liu et al.[27] proposed a series of separation using membrane technology to achieve above targets according to the molecular of the chemical present. The molecular weights of hexoses, pentoses, and acetic acid are 180.2, 150.1, and 60.1 g/mol, respectively; furfural and HMF have molecular weights of 96.09 and 110.11 g/mol, respectively. Firstly, chemicals existing in the hydrolysate with high molecular weight (e.g., aromatics and polysaccharides) are retained by NF membrane with MWCO of 200. Thereafter, furfural and acetic acid in permeate will pass through the membrane with MWCO of 100, while leaving the concentrated sugar solution as retentate.

Conversion of hemicelluloses extracted by hot-water extraction from *Scots Pine* to monosaccharides and other chemicals was studied by Sainio et al.[31] A process comprising membrane UF, acid hydrolysis, and chromatographic separation was proposed and evaluated. Polyphenols and high molar mass hemicelluloses exist in the retentate of UF stage one (UFX5 MWCO 20,000); the permeate of UF stage two is further treated by chromatography with acetic acid and a mixture of monosaccharides (glucose, xylose, galactose, and mannose) being concentrated. The hydrolysis acid released from chromatography can be reused.

The literature review shows that membrane separation is seldom directly used in inhibitor removal; instead, solvent extraction, overliming, adsorption, and enzymatic methods are widely used. Minor components separation is a strenuous task by requiring high sensitivity of the separation media. However, the trivial difference in molecular size for these chemicals may not be sufficiently discriminated by membrane structure with relatively low specification. Furthermore, only a limited concentration coefficient can be achieved by the traditional membrane separations (i.e., NF and UF) subjected to osmotic pressure. Another reason might be that the irreversible sorption inside the membrane porous structure will significantly influence the recovery of the value-added component with low quantity.

In this regard, supported liquid membrane that combines the solvent extraction with membrane separation may provide an alternative separation technology.[32] By employing a highly specific extractant, the minor components can be preferentially transported through liquid phase of the membrane; the solid membrane structure makes the space needed for the extraction separation more compact.

Fuel Product Recovery and Purification

Biorefinery releases a host of by-products that are useful in different fields in aforementioned pretreatment and hydrolysis procedures. As the major target of biorefinery, nevertheless, the generation of clean and efficient energy sources is the focus of biorefinery. As indicated in Table 13.1, the major products for energy provision in biorefinery are biofuels in gas, liquid, heat, and electricity. They represent a big portion of biomass yield. Because of the special existing status, heat and electricity can be straightforwardly isolated and exported from its environment without any 'contamination'. Separation technologies adopted in biorefinery are all concerning gaseous and liquefied biofuels.

Bioalcohol Products

The alcoholic biofuels include bioethanol and biobutanol. A very important motivation for bioethanol industry development is oil embargo of 1970s. Currently, two major fermentation platforms for ethanol production exist, that is, the corn–ethanol program in the United States and the sugarcane–ethanol program in Brazil, with annual production of ~13 and ~7 billion gallons, respectively. Besides greenhouse gas benefit, bioethanol has high production efficiency and high octane rating (108).

In the bioethanol production, the fermentation liquid contains only 5–10 wt% of ethanol. In achieving removal or recovery of ethanol from the fermentation broth, several hybrid processes with hydrophobic PV or membrane distillation as the key separation unit have been employed. PV is membrane-based molecular-level liquid–liquid separation based on the solution–diffusion mechanism.[7] According to the downstream conditions, two different PV processes are classified; namely, vacuum and sweep gas PV. The preferential passage of ethanol bigger than water in principle requires the membrane made of hydrophobic materials that assure the sorption of ethanol molecule weighs down that of water, so as to counterbalance the effects of its slower diffusion. There exist three major categories of polymer materials with hydrophobicity; namely, silicon-containing polymers, fluorine-containing polymers, and poly[1-(trimethylsilyl)-1-propyne] (PTMSP).[33,34] The other hydrophobic polymers investigated include

Table 13.2 Commercialized Hydrophobic and Hydrophilic Pervaporation Membranes

Hydrophobic	Structure
PERVAP 1060	PDMS cross-linked/PAN
PERVAP 1070	PDMS cross-linked + silicalite/PAN
MTR 100	PDMS cross-linked/porous support
MTR 200	EPDM/PDMS cross-linked/porous support
GKSS PEBA	PEBA/porous support
GKSS PDMS	PDMS cross-linked/porous support
GKSS PMOS	PMOS cross-linked/porous support
Hydrophilic	**Structure**
PERVAP 2200	PVA cross-linked/PAN support
PERVAP 2201, flux	PVA highly cross-linked/PAN
PERVAP 2202	PVA specially cross-linked/PAN
PERVAP 2205	PVA specially cross-linked/PAN
PERVAP 2210	PVA lightly cross-linked/PAN
PERVAP 2510	PVA specially cross-linked/PAN
CM-Celfa	PVA cross-linked/PAN
GKSS simplex	Complex polyelectrolytes/PAN

Reproduced from Ref 35. Copyright 2002, Elsevier.

heptadecafluorodecyl acrylate, styrene–fluoroalkyl acrylate graft copolymer, copolymer of polysiloxane and phosphate ester, plasma polymerized silane, poly-*N*-isopropylacrylamide-*co*-*N*-isopropylmethacrylamide, polysiloxaneimide, and poly ether block amide (PEBA). Review of the PV performance shows that PTMSP presents the best separation factor. Nevertheless, PTMSP is a material of which the performance in terms of flux and selectivity gradually degrades. As shown in Table 13.2, the commercialized hydrophobic membranes are composed of multilayer structures with polydimethylsiloxane (PDMS), ethylene–propylene–diene monomer (EPDM), PEBA, and so on as selective material.[35]

Another membrane separation that have been investigated is vacuum membrane distillation (VMD) using porous membrane (0.1 μm diameter pores).[36] The mechanism for VMD is combined vapor–liquid equilibrium and Knudsen flow/Poisseuille flow (viscous).[37] The fully porous structure on the selective layer undoubtedly indicates higher flux in VMD than that in PV using dense membrane. Regardless, the risk of the hydrophobic membranes being wetted by organic ethanol makes it not so attractive. Wetting signifies loss of membrane's separation function.[38] Additionally, fouling by microorganism, enzymes, proteins, and so on will decrease productivity and increase maintenance and operation cost.[38] To solve the fouling in VMD, prefiltration using MF will be useful.

Prefiltration is also necessary for separation using PV which is generally conducted at temperatures higher than ambient ones. Therefore, temperature-sensitive substances must be taken out from the fermentation broth before being exposed to PV.

In most biochemical platform, the major fouling particles as well as the temperature-sensitive materials are microorganisms, enzymes, and so on. Therefore, membrane MF or centrifugation is viable separation approaches for deleting the above solid from the fermentation liquid, with MF being theoretically able to remove all the solids larger than the selected MWCO. UF or NF may replace MF if different particles' retention efficiency is required.[33]

In the competing dehydration technologies for recovery of bioethanol, distillation is a well-established technology that requires much lower energy input than PV in recovering ethanol. The ethanol to water selectivity of currently available hydrophobic membranes cannot insure energy efficiency of PV to be higher than that of distillation,[33,39] which is a very important motivation for the intense research in hydrophobic material design and membrane formation.

Butanol production by acetone–butanol–ethanol (ABE) fermentation is one of the oldest processes providing chemical for human kind. Nowadays, it arouses great interest as a biofuel production pathway. There exist several advantages of applying butanol in the liquid fuel system. Butanol will reduce carbon emissions, it has **higher energy content** than ethanol, almost 20% more by density, it is able to blend much better than ethanol with gasoline because of its similarities to conventional gasoline, to name just a few.

Liu et al.[40] investigated the PV using PEBA membranes to recover ABE from dilute aqueous solutions. It was shown that the PEBA membrane was more selective to butanol than the other organics coproduced in the fermentation process. The efficiency of several commercialized membrane in recovery of butanol from modeled ABE solution are summarized by Qureshi and Blaschek[41] in a review.

Unlike bioethanol production with highly specific product of ethanol, butanol production by fermentation cogenerates acetone and ethanol. As far as butanol is concerned, hydrophobic PV is more profitable than distillation in recovering it from the ABE solution. This is because the energy needed by former is used for evaporation minor component of butanol, whereas latter will vaporize the bulk water. As for the other two chemicals, they are more volatile than water. Therefore, a hybrid separation system comprising *in situ* stripping of ethanol and acetone by gas and butanol recovery by PV will be more reasonable than the separation mode only adopting membrane.

Another aspect of bioalcohols treatment is dehydration. The materials suitable for dehydration are abundant. Poly(vinyl alcohol) (PVA) is the major choice for commercial dehydration membrane. The majority of the composite hydrophilic membranes commercialized by Gesellschaft fur Trenntechnik Co. (GFT) use PVA with different cross-linking degrees as the selective material. GKSS (Geesthacht, Germany) Simplex has developed polyelectrolyte–polyacrylonitrile (PAN) composite membranes.[35] Other selective materials include PAN, sodium alginate, chitosan, cellulose acetate, and many glassy polymers such as polyimide (PI), polysulfone (Psf). In recent years, the PIs become the popular membrane materials for PV application.[42–45]

Table 13.2 listed out the commercialized hydrophilic membranes with multilayer composite structures formed by step by step casting method.[35] Recently, several studies focused on thin-film composite[46–48] and novel dual-layer coextrusion technologies for fabricating composite PV membranes.[49,50] In the work of Chao et al.,[48] novel Thin film composite membranes were formed by interfacial polymerization with 5-nitrobenzene-1,3-dioyl dichloride or 5-*tert*-butylbenzene-1,3-dioyl dichloride reacting with triethylenetetraamine on the surface of a PAN porous membrane. Xu et al.[47] presented a novel thin-film-composite membrane based on layer-by-layer self-assembly of polyelectrolytes onto an interfacially polymerized polyamide membrane for dehydration of ethylene glycol by PV.

Regarding the dual-layer coextrusion technology in just one step through phase inversion, Jiang et al.[49] reported Psf/Matrimid dual-layer hollow fiber membranes that exhibited extremely high separation performance far surpassing the prior art of polymeric membrane and approaching ceramic membranes in *t*-butanol dehydration.[49] The overall performance, especially the flux of the polymeric hollow fibers in this study, was significantly higher than those of commercially available polymeric membranes. Central to this achievement is the synergy of unique pair of miscible polymeric blends with the aid of controlled intermolecular diffusion by means of molecular engineering and membrane fabrication made feasible by the coextrusion concept. Figure 13.3 shows the morphologies of these dual-layer hollow fibers having a interfacial dense selective layer.

In a similar vein to recovery using hydrophobic membranes, higher selectivity of dehydration membrane implies less energy consumed.[51] Some of the polymeric membranes including the commercial ones present very high selectivity and reasonable flux, which makes the dehydration by PV very promising.

In both hydrophobic recovery and hydrophilic dehydration, inorganic zeolite and ceramic membrane are considered as more efficient than polymeric membrane.[33,52–54] Research showed that inorganic hydrophobic membranes by two types of hydrophobic zeolites (i.e., silicalite-1 and ZSM-5) delivered average ethanol–water separation factors of around 40. This value is three more times than that of PDMS membranes. The tabular zeolite NaA membrane module has the PV flux of about 2.35 kg/m^2 h and the separation factor of above 5000 for dehydrating the aqueous solution of 95 wt% ethanol at 95°C.[55] It is generally accepted that the rigid nature and narrow pore size distribution are the major contributors to higher performance observed with hydrophilic zeolite membranes. Nevertheless, some of the lab-made prototype polymeric membrane exhibited higher performance in terms of both separation factor and flux than some of the zeolite membranes.[42]

For both the recovery and dehydration of bioalcohol, a new generation of material is the mixed matrix material, which combines the advantages of inorganic and polymeric membrane for obtaining high ratio of membrane performance/cost. The inorganic materials that favor hydrophobic PV comprise silicalite and a series of activated carbon (e.g., CP, SX, W20, NY, 4S), whereas those facilitating dehydration include zeolite 3A, 4A, 5A, NaX, NaY, beta and metal oxides (e.g., magnesium oxide, titanium oxide).[7,56,57]

Biodiesel Products

Biodiesel is a fuel commonly prepared through alkali-catalyzed transesterification with vegetable oil, animal fats, or grease as the raw material. The typical biodiesel production process is shown in Figure 13.4. In addition to the land oil crops, algae represented a nascent platform to be actively exploited for biodiesel production as its harvested oils can be extracted for conversion into biodiesels. The use of algal oils as a feedstock appears to be more effective in biodiesel production than land oil crops.[58]

Separation of glycerol from biodiesel is the major separation problem encountered in biodiesel production. The mechanism for the separation based on droplet size is dominantly utilized. Gomes et al.[59] investigated the efficiency of MF with ceramic membranes in the separation of biodiesel and glycerol. Simulated mixtures of biodiesel, ethanol, and glycerol were used as feed. Tubular Al_2O_3/TiO_2 ceramic membranes with average pore size of 0.2, 0.4, and 0.8 μm were employed. With increasing pore size, the permeate flux range from around 12 to around 40 kg/m^2 h with corresponding glycerol retention ranging from 99.6% to 99.0%. Higher ethanol concentrations give rise to smaller glycerol droplet diameters dispersed in biodiesel, leading to lower separation factor.

In the work of Wang et al.,[60] ceramic UF membrane with average pore size of 0.02 μm was used along with treatment of acidified water to achieve biodiesel product meeting both ASTMD6751 and EN14241 standards. The pretreatment step for UF is rotary evaporation of residual methanol. Afterward, acidified water was added to the biodiesel samples, which led to the formation of aqueous phase containing

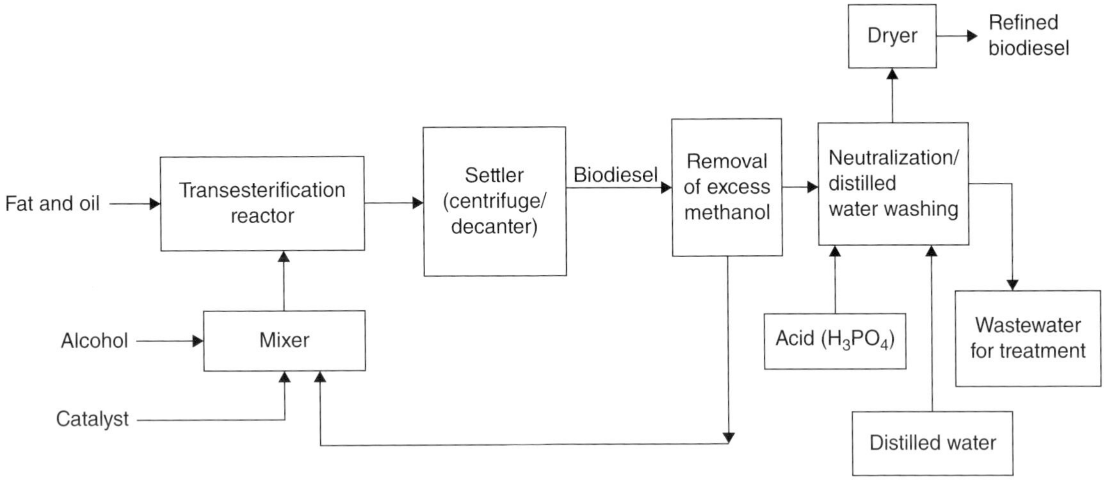

Figure 13.4 Process flow schematic diagram for biodiesel production and purification process. (Reproduced from Ref 59. Copyright 2010, Elsevier.)

catalyst, salt, glycerol, and other related water-soluble substances. The resultant larger droplets of glycerol are too big for the membrane pores. Besides, acidified water helps in breaking down the soap into water-soluble salt and free fatty acids and neutralizing the catalyst as well, forming salt and water. Under conditions of transmembrane pressure of 0.15 MPa and temperature of 60°C, the transmembrane was 300 L/m^2 h. The residual-free glycerol in the permeate estimated by water extraction was 0.0108 wt%. Therefore, the novel process reported in this work generates biodiesel with less impurity.

The potential of polymeric solvent-resistant NF membranes was investigated by Othman et al.[61] for biodiesel separation. Eight commercial polymeric NF membranes were examined for their abilities to separate the desired biodiesel products from its mixture of glycerin, excess methanol, and catalyst after the transesterification process. Even though the polymeric membrane (Solsep 030306F) has a high retention to glycerol, its flux is much lower than inorganic membranes as introduced above, probably due to the much smaller pores of NF membrane. Another problem arise with using polymeric membrane in biodiesel purification is the membranes' damage by the feed solution containing various organic components mentioned above.

As have been mentioned in previous section, fouling is a challenge for separation using porous membrane. Choi et al.[62] attributed the continuous flux reduction of UF membrane to concentration polarization, pore blocking, or molecule adsorption followed by the formation of fouling layer. However, back flushing of the membrane module could lead to high permeate flux.[63] More works on using membrane in biodiesel production and purification have been reviewed by Atadashi et al.[64]

Gaseous Products

Gasification at high temperature is the principle route for gaseous product. Besides, a minor route similar to the gasification exists in the lignocelluloses biorefinery, where the lignin inside black liquor from the pulping is used to produce syngas via gasification. Anaerobic digestion is well established and widely practiced in China and some European countries such as Germany.[65,66] The composition of the gas product from the two biorefinery routes differs a lot. The gasification process delivers product with H_2, carbon monoxide (CO), and carbon dioxide (CO_2) as major components, whereas the gas product from anaerobic process mainly has methane (CH_4) and CO_2. The gaseous products are important energy carriers. They can also be used as feedstock to generate transport fuel or other chemicals.

Whatever the gaseous products' final applications, the removal of contaminants such as H_2 sulfide (H_2S), ammonia (NH_3), and siloxane must be conducted to prevent the detrimental effect to biogas thermal or thermocatalytic conversion device (e.g., corrosion, erosion, fouling) and the harmful environmental emissions.[67] The only example using membrane is regarding the removal of siloxane.[68] In this work, the membrane of rubbery polymethylsiloxane was employed to remove the condensable siloxane. Chemical absorption and cryogenic condensation are the more popular methods for removing H_2S and NH_3.

If intended for fuel application, additionally, H_2 and CH_4 must be separated from the other bulk components (mainly CO_2). Either a molecular-sieving membrane or a reverse selective membrane can be utilized to preferentially remove H_2 or CO_2, respectively, depending on their relative concentration.[69–71]

H_2-selective membrane can be fabricated from following polymeric materials: ethyl cellulose, polyetherimide (PEI), polyphenyleneoxide, Psf, polymethylpentene, PI, polyethersulfone, polystyrene, poly(vinylidene fluoride), and poly(methyl methacrylate). Detailed information about these materials and the corresponding references have been given in the work of Shao et al.[72] Even though the two gas molecules have a big difference in their size, the competition of CO_2 in sorption due to its condensability makes the separation of H_2 from CO_2 by membrane a difficult task. The highest selectivity presented by PEI is only around 6.

The materials available for CO_2-selective membrane mainly include PTMSP, poly(4-methyl-2-pentyne), polyphosphazene, poly(*tert*-butylacetylene), PEBA (trade name PEBAx 1657), poly(ethylene glycol) diacrylate, cross-linked PEG copolymer, polyether, poly(styrene-co-butadiene), and PDMS.[73–77] It was found that cross-linked PEG exhibited the highest CO_2/H_2 selectivity of 30.

Concerning CH_4 and CO_2 separation, a number of research works examining the various materials for GS membranes have been published over the last two decades.[16] Among these materials, intensive research has been performed to investigate and improve the separation properties of PI.[7,16,78–81]

Two serious problems involved in the GS tasks using glassy polymeric membrane are: (1) the plasticization of the polymer highly condensable gases such as CO_2, H_2S, and H_2O[82] and (2) the aging of the polymeric materials.[83] To prevent the negative effect of plasticization and aging in actual application, several solutions have been proposed such as cross-linking via ultraviolet and ion beam radiations, thermal treatment, and chemical modification.[83,84] Other than suppressing the plasticization and aging, the side-effect of cross-linking is higher size- and shape-discrimination property. A series of diamino-modified PIs were examined by Tin et al.[85] It was found that the ideal H_2/CO_2 selectivity in pure gas tests is 101, which is well above the Robeson's upper-bound curve. Another way turns to employing more stable inorganic membranes such as carbon membrane, zeolite membrane, and metal membrane.[86–88] The GS efficiency of membrane separation relies also on the process operation condition. Take CH_4/CO_2 separation, the impurity of CO_2 within the feed may not be adequately removed, as its partial pressure at permeation side will accumulate and increase to be the same as that of feed side.

The separation of H_2 from CO is used in syngas ratio adjustment. Syngas is commonly produced with a H_2:CO ratio between 1 and 3. In order to use a syngas mixture to produce acetic acid, methanol, plasticizers, and other valuable petrochemical products, this ratio must be reduced. The separation of H_2 from CO in syngas ratio adjustment with PI membranes was investigated by Peer et al.[89] This polymer has shown large permeability and selectivity for H_2 separation (selectivity of ca. 30). Measured membrane permeances for H_2 and CO were about 70–100 GPU (10^{-6} cm^3 (STP) cm^{-2} $second^{-1}$ $cmHg^{-1}$) and 3–5.5 GPU, respectively.

Although polymeric membranes have been used for H_2 separation in industries, it is mainly applied in occasions with low temperature. Their general instability at high temperatures restricts its suitability for membrane reactors generating highly purified H_2 either for direct usage as energy or for supplying the membrane fuel cell. One of the polymeric materials that have been investigated at elevated temperature is polybenzimidazole, which exhibited excellent thermal stability.[77] In most cases, nevertheless, zeolite membranes are preferred because of their intrinsic thermal resistance as well as capability of integrating catalysts facilitating deep purification of H_2 by membrane reactor.[90]

Membrane Reactor

According to International Union of Pure and Applied Chemistry, a membrane reactor is defined as a device that combines reaction and separation in a single unit.[91] Membrane reactors can have several advantages over conventional reactors: (1) it integrates reaction and separation in a single system, hence mitigating operational costs and recycling requirements, (2) higher conversions per cycle can be achieved as the prohibitive effect imposed by product on the reaction equilibrium is removed or reduced. In other words, the function a membrane-based separation in a membrane reactor is more than separation of two or more components as shown in the many examples aforementioned, which makes it reasonable to summarize the cases regarding membrane reactors utilized in biorefinery in a separate section.

H_2 purification is very critical for power generation by proton-exchange membrane fuel cells. CO can lead to cell performance degradation; therefore, its concentration must be below 10 ppm in the H_2 streams.[92] Among various alternative approaches for achieving this target, CO selective oxidation (CO Selox) is the most effective one that can achieve H_2 deep purification.[93] With this method, membrane reactor–separator system simultaneously performing oxidation and separation of H_2 and other gas components is applied. Bernardo et al.[90] explored CO Selox using catalytic zeolite (Pt/Na-Y) membranes in a continuous flow membrane reactor. Depending on the operating conditions, the catalytic membranes effectively reduced the CO content from 10,000 ppm down to 10–50 ppm.[90]

Biodiesel production using a membrane reactor was studied by Dube et al.[94] Because of the immiscibility of canola oil and methanol as well as various surface forces, the canola oil will exist in the form of an emulsion; that is, droplets suspended in methanol. Therefore, transesterification occurs at the surface of the canola oil droplets. The canola oil droplets could not pass through the pores of the membrane, whereas the biodiesel product will pass through the membrane pores along with the methanol, glycerol, and catalyst. As a result, the reaction equilibrium is shifted to the product side. In the research, experiment was performed in the membrane reactor in semi-batch mode and at different temperatures, catalyst concentrations, and feed flow rates. The observation shows that increasing temperature, catalyst concentration, and feedstock (methanol/oil) flow rate significantly promote the conversion of canola oil to biodiesel.

O'Brien and Craig[95] utilized the commercially available PDMS membrane module in a continuous fermentation/membrane PV system to produce ethanol, resulting in permeate of 20–23 wt% ethanol, whereas 4–6 wt% ethanol level was retained in stirred tank fermentor. The selectivity of ethanol over water ranged from 1.8 to 4.1.

In the work of Qureshi et al.,[96] ABE were produced in a membrane reactor integrating fermentation and PV recovery. The bacterium used was *Clostridium acetobutylicum* and the PV membrane is lab-synthesized silicalite–silicone mixed matrix membrane.[96] In the experiment, cells of *C. acetobutylicum* were removed from the cell culture using a UF membrane of MWCO 500,000 and fed to the fermentor. Subsequently, the ABE components were extracted from the UF permeate by PV using the silicalite–silicone membrane. The fed-batch reactor was operated for 870 h. Actually, the removal of the products from the fermentation solution can reduce their toxic effect. Groot et al.[97] have proved that membrane reactor can increase the conversion and yield of biomass in fermentation. Membrane reactor is not a new topic in bioalcohol biorefinery. More literatures or patents are introduced in Refs 33 and 98.

CONCLUSION

Biorefinery is an integrated process to produce biofuels and many other value-added chemicals useful for human being. It takes a very important position in modern industry. Separation technology is indispensable for biorefinery with products being generated in a complicated manner.

In this article, academic investigations and industrial applications concerning membrane-based separation in biorefinery have been reviewed. Obviously, most of the conventional membrane separation technologies can identify some biorefinery segments where they have the potential to solve the existing separation problems. In the thermal conversion and anaerobic conversion, GS membrane is helpful in recovering and purifying biogases including H_2 and CH_4. In the biorefinery depending on biochemical conversions, a spectrum of membrane separations is employed; a series of filtration relying on MF, UF, and NF play very important roles in retrieving value-added by-product in fermentation and realizing incremental usage of biomass; PV and membrane distillation find application in recovery and purification of bioalcohols. UF is highly evaluated in physiochemical process of producing biodiesel; it is utilized to remove impurities such as glycerol, diglycerides, and so on. Membrane-based separation is not the sole technique employed in biorefinery. Sometimes, hybrid system synergizing membrane and other separation practices is more effective.

Membrane separation efficiency is always a key aspect of membrane-based separation. For GS and PV, membrane intrinsic performance is closely related with material intrinsic physiochemical properties. With advent of a large number of materials, membrane with recommendable performance has already been commercialized. Industrialization of dehydration is realized worldwide. With respect to the performance of hydrophobic membranes, inorganic membrane is much superior to polymeric membranes. Nevertheless, their high fabrication cost hinders its immediate application. Another factor that needs attention in hydrophobic PV is the complexity of the feed system. Generally, the fermentation liquor subject to recovery by hydrophobic PV contains several organics (i.e., ABE in butanol biorefinery). Fundamental research on improving the discriminating ability of membrane among these organic chemicals so as to make the overall separation more economical and energy efficient should be the focus. As for the porous membranes employed in different procedures, current works are more emphasizing on application engineering. Within this scope, membrane service life threatened by fouling should be addressed more intensively.

There exist many publications about using membrane separation in biorefinery. More importantly, the biorefinery industry is progressing continuously, which might create new separation problems. Within this single article, it is impossible to review all the cases regarding membrane separation in biorefinery. In addition, some aspects such as economic analysis and process design are of great interest to membranologists. More effort will be needed to fully understand this important member of modern industry.

ACKNOWLEDGEMENTS

The authors would like to thank National Natural Science Foundation of China (Project number 21176265) and Planned Science and Technology Project of Hunan Province, China (Project Number 2010FJ2006) for funding this work.

REFERENCES

1. Fatih Demirbas M. Biorefineries for biofuel upgrading: a critical review. *Appl Energy* 2009, 86:151–161.
2. Demirbas A. Biorefineries: current activities and future developments. *Eneryg Convers Manag* 2009, 50:2782–2801.
3. Centi G, Lanzafame P, Perathoner S. Analysis of the alternative routes in the catalytic transformation of lignocellulosic materials. *Catal Today* 2011, 167:14–30.
4. Kamm B, Kamm M. Principles of biorefineries. *Appl Microbiol Biotechnol* 2004, 64:137–145.
5. Huang H-J, Ramaswamy S, Tschirner UW, Ramarao BV. A review of separation technologies in current and future biorefineries. *Sep Purif Technol* 2008, 62:1–21.
6. Srirangan K, Akawi L, Young MM, Perry Chou C. Towards sustainable production of clean energy carriers from biomass resources. *Appl Energy* 2012, 100:172–186.
7. Jiang LY, Wang Y, Chung TS, Qiao XY, Lai J-Y. Polyimides membranes for pervaporation and biofuels separation. *Prog Polym Sci* 2009, 34:1135–1160.
8. Chung T-S, Jiang LY, Li Y, Kulprathipanja S. Mixed matrix membranes (MMMs) comprising organic polymers with dispersed inorganic fillers for gas separation. *Prog Polym Sci* 2007, 32:483–507.
9. Yu J, Zhang T, Zhong J, Zhang X, Tan T. Biorefinery of sweet sorghum stem. *Biotechnol Adv* 2012, 30:811–816.
10. Swaina PK, Dasa LM, Naikb SN. Biomass to liquid: a prospective challenge to research and development in 21st century. *Renew Sustain Energy Rev* 2011, 15:4917– 4933.
11. Margeot A, Hahn-Hagerdal B, Edlund M, Slade R, Monot F. New improvements for lignocellulosic ethanol. *Curr Opin Biotechnol* 2009, 20:372–380.
12. Khan SA, Rashmi A, Hussain MZ, Prasad S, Banerjee UC. Prospects of biodiesel production from microalgae in India. *Renew Sustain Energy Rev* 2009, 13:2361–2372.
13. Nigam PS, Singh A. Production of liquid biofuels from renewable resources. *Prog Energy Combust Sci* 2011, 37:52–68.
14. Lammens TM, Franssen MCR, Scott EL, Sanders JPM. Review: availability of protein-derived amino acids as feedstock for the production of bio-based chemicals. *Biomass Bioenergy* 2012, 44:168–181.
15. Rahmat N, Zuhairi Abdullah A, Rahman Mohamed A. Recent progress on innovative and potential technologies for glycerol transformation into fuel additives: a critical review. *Renew Sustain Energy Rev* 2010, 14:987–1000.
16. Paul DR, Yampolskii YP. *Polymeric Gas Separation Membranes*. Boca Raton, FL: CRC Press; 1994.
17. Koros W J, Flemming GK. Membrane-based gas separation. *J Membr Sci* 1993, 83:1–80.
18. Mulder M. *Basic Principle of Membrane Technology*. Dordrecht, The Netherlands: Kluwer Academic Publisher; 1996.
19. Jiang LY. Fabrication and characterization of composite membranes for gas separation. PHD thesis; 2006.
20. Zeman LJ, Zydney AL. *Microfiltration and Ultrafiltration: Principles and Applications*. New York: Marcel Dekker; 1996.
21. Freeman BD, Pinnau I. Separation of gases using solubility-selective polymer. *Trends Polym Sci* 1997, 5:167–173.
22. Gabrielii I, Gatenholm P, Glasser WG, Jain RK, Kenne L. Separation, characterization and hydrogel-formation of hemicellulose from aspen wood. *Carbohydr Polym* 2000, 43:367–374.
23. Fang X, Shen Y, Zhao J, Bao X, Qu Y. Review: status and prospect of lignocellulosic bioethanol production in China. *Bioresour Technol* 2010, 101:4814–4819.
24. Diaye SN, Rigal L, Larocque P, Vidal PF. Extraction of hemicelluloses from poplar, *Populus tremuloides*, using an extruder-type twin-screw reactor: a feasibility study. *Bioresour Technol* 1996, 57:61–67.
25. Sun RC, Fang JM, Tomkinson J, Geng ZC, Liu JC. Fractional isolation, physico-chemical characterization and homogeneous esterification of hemicelluloses from fast-growing poplar wood. *Carbohydr Polym* 2001, 44:29–39.
26. Schlesinger R, Gotzinger G, Sixta H, Friedl A, Harasek M. Evaluation of alkali resistant nanofiltration membranes for the separation of hemicellulose from concentrated alkaline process liquors. *Desalination* 2006, 192:303–314.
27. Liu S, Lu H, Hu R, Shupe A, Lin L, Liang B. Research review paper: a sustainable woody biomass biorefinery. *Biotechnol Adv* 2012, 30:785–810.
28. Vazquez MJ, Alonso JL, Dominguez H, Parajo JC. Production of xylose-containing fermentation media by enzymatic post-hydrolysis of oligomers produced by corn cob autohydrolysis. *World J Microbiol Biotechnol* 2001, 17:817–822.
29. Duarte LC, Silva-Fernandes T, Carvalheiro F, Girio FM. Dilute acid hydrolysis of wheat straw oligosaccharides. *Appl Biochem Biotechnol* 2009, 153:116–126.
30. Cheryan M. Method and system for extraction of zein from corn. Patent US7,045,607; 2006.
31. Sainio T, Kallioinen M, Nakari O, Mänttäri M. Production and recovery of monosaccharides from lignocellulose hot water extracts in a pulp mill biorefinery. *Bioresour Technol.* DOI: 10.1016/j.biortech.2012.08.126.
32. Kocherginsky NM, Yang Q, Seelam L. Recent advances in supported liquid membrane technology. *Sep Purif Technol* 2007, 53:171–177.
33. Vane LM. A review of pervaporation for product recovery from biomass fermentation processes. *J Chem Technol Biotechnol* 2005, 80:603–629.
34. Kaseno A, Miyazawa I, Kokugan T. Effect of product removal by a pervaporation on ethanol fermentation. *J Ferment Bioeng* 1998, 86:488–493.

35. Jonquières A, Clément R, Lochon P, Néel J, Dresch M, Chrétien B. Industrial state-of-the-art of pervaporation and vapour permeation in the western countries. *J Membr Sci* 2002, 206:87–117.
36. Lawson KW, Lloyd DR. Membrane distillation. I. Module design and performance evaluation using vacuum membrane distillation. *J Membr Sci* 1996, 120:111–121.
37. Izquierdo-Gil MA, Jonsson G. Factors affecting flux and ethanol separation performance in vacuum membrane distillation (VMD). *J Membr Sci* 2003, 214:113–130.
38. Alkhudhiri A, Darwish N, Hilal N. Membrane distillation: a comprehensive review. *Desalination* 2012, 287:2–18.
39. Zhu JY, Zhuang XS. Conceptual net energy output for biofuel production from lignocellulosic biomass through biorefining. *Prog Energy Combust Sci* 2012, 38:583–598.
40. Liu F, Liu L, Feng X. Separation of acetone–butanol–ethanol (ABE) from dilute aqueous solutions by pervaporation. *Sep Purif Technol* 2005, 42:273–282.
41. Qureshi N, Blaschek HP. Butanol recovery from model solution/fermentation broth by pervaporation: evaluation of membrane performance. *Biomass Bioenergy* 1999, 17:175–184.
42. Jiang LY, Chung T-S. β-Cyclodextrin containing Matrimid sub-nanocomposite membranes for pervaporation application. *J Membr Sci* 2009. 327:216–225.
43. Qiao XY, Chung T-S. Fundamental characteristics of sorption, swelling, and permeation of P84 co-polyimide membranes for pervaporation dehydration of alcohols. *Ind Eng Chem Res* 2005, 44:8938–8943.
44. Guo WF, Chung T-S. Study and characterization of the hysteresis behavior of polyimide membranes in the thermal cycle process of pervaporation separation. *J Membr Sci* 2005, 253:13–22.
45. Wang Y, Jiang LY, Matsuura T, Chung T-S, Goh SH. Investigation of the fundamental differences between polyamide-imide (PAI) and polyetherimide (PEI) membranes for isopropanol dehydration via pervaporation. *J Membr Sci* 2008, 318:217–226.
46. Huang S-H, Li C-L, Hu C-C, Tsai HA, Lee K-R, Lai J-Y. Polyamide thin-film composite membranes prepared by interfacial polymerization for pervaporation separation. *Desalination* 2006, 200:387–389.
47. Xu J, Gao C, Feng X. Thin-film-composite membranes comprising of self-assembled polyelectrolytes for separation of water from ethylene glycol by pervaporation. *J Membr Sci* 2010, 352:197–204.
48. Chao W-C, Huang S-H, An Q, Lia D-J, Huang Y-C, Lee K-R, Lai J-Y. Novel interfacially-polymerized polyamide thin-film composite membranes: studies on characterization, pervaporation, and positron annihilation spectroscopy. *Polymer* 2011, 52:2414–2421.
49. Jiang LY, Chen H, Jean YC, Chung T-S. Ultra-thin polymeric interpenetration separation performance approaching ceramic membranes for biofuel. *AIChE J* 2009, 55:75–86.
50. Wang Y, Goh SH, Chung T-S, Peng N. Polyamide-imide/polyetherimide dual-layer hollow fiber membranes for pervaporation dehydration of C1–C4 alcohols. *J Membr Sci* 2009, 326:222–233.
51. Shao P, Huang RYM. Polymeric membrane pervaporation. *J Membr Sci* 2007, 287:162–179.
52. Bowen TC, Noble RD, Falconer JL. Fundamentals and applications of pervaporation through zeolite membranes. *J Membr Sci* 2004, 245:1–33.
53. Peters TA, Fontalvo J, Vorstman MAG, Benes NE, van Dam RA, Vroon ZAEP, van Soest-Vereammen ELJ, Keurentjes JTF. Hollow fiber microporous silica membranes for gas separation and pervaporation. *J Membr Sci* 2005, 248:73–80.
54. Gallego-Lizon T, Edwards E, Lobiundo G, Freitas dos Santos L. Dehydration of water/*tert*-butanol mixtures by pervaporation: comparative study of commercially available polymeric, microporous silica and zeolite membranes. *J Membr Sci* 2002, 197:309–319.
55. Kondo M, Komori M, Kita H, Okamoto K. Tubular type pervaporation module with zeolite NaA membrane. *J Membr Sci* 1997, 133:133–141.
56. Ji W, Sikdar SK. Pervaporation using adsorbent-filled membranes. *Ind Eng Chem Res* 1996, 35:1124–1132.
57. Smitha B, Suhany D, Sridhar S, Ramakrishna M. Separation of organic–organic mixtures by pervaporation—a review. *J Membr Sci* 2004, 241:1–21.
58. Nigam PS, Singh A. Production of liquid biofuels from renewable resources. *Prog Energy Combust Sci* 2011, 37:52–68.
59. Gomes MCS, Pereira NC, Barros STD. Separation of biodiesel and glycerol using ceramic membranes. *J Membr Sci*, 2010, 352:271–276.
60. Wang Y, Wang X, Liu Y, Ou S, Tan Y, Tang S. Refining of biodiesel by ceramic membrane separation. *Fuel Process Technol* 2009, 90:422–427.
61. Othmana R, Wahab Mohammad A, Ismail M, Salimon J. Application of polymeric solvent resistant nanofiltration membranes for biodiesel production. *J Membr Sci* 2010, 348:287–297.
62. Choi H, Kai Z, Dionysios DD, Daniel BO, George AS. Influence of cross-flow velocity on membrane performance during filtration of biological suspension. *J Membr Sci* 2005, 248:189–199.
63. Almandoz C, Pagliero C, Ochoa A. Marchese J. Corn syrup clarification by microfiltration with ceramic membranes. *J Membr Sci* 2010, 363:87–95.
64. Atadashi IM, Aroua MK, Abdul Aziz AR, Sulaiman NMN. Membrane biodiesel production and refining technology: a critical review. *Renew Sustain Energy Rev* 2011, 15:5051–5062.
65. Chen Y, Yang G, Sweeney S, Feng Y. Household biogas use in rural China: a study of opportunities and constraints. *Renew Sustain Energy Rev* 2010, 14:545–549.
66. Zeng X, Ma Y, Ma L. Utilization of straw in biomass energy in China. *Renew Sustain Energy Rev* 2007, 11:976–987.
67. Abatzoglou N, Boivin S. A review of biogas purification processes. *Biofuels Bioprod Biorefing* 2009, 3:42–71.
68. Abatzoglou N, Ajhar M, Melin T. Siloxane removal with gas permeation membranes. *Desalination* 2006, 200:234–235.
69. Abetz V, Brinkmann T, Dijkstra M, Ebert K, Fritsch D, Ohlrogge K. Developments in membrane research: from

material via process design to industrial application. *Adv Eng Mater* 2006, 8:328–358.
70. Li Y, Chung T-S. Highly selective sulfonated polyethersulfone (SPES)-based membranes with transition metal counterions for hydrogen recovery and natural gas separation. *J Membr Sci* 2008, 308:128–135.
71. Orme CJ, Stone, ML, Benson MT, Peterson ES. Testing of polymer membranes for the selective permeability of hydrogen. *Sep Sci Technol* 2002, 38:3225–3238.
72. Shao L, Low BT, Chung T-S, Greenberg AR. Polymeric membranes for the hydrogen economy: contemporary approaches and prospects for the future. *J Membr Sci* 2009, 327:18–31.
73. Orme CJ, Harrup MK, Luther TA, Lash RP, Houston KS, Weinkauf DH. Characterization of gas transport in selected rubbery amorphous polyphosphazene membranes. *J Membr Sci* 2001, 186:249–257.
74. Patel NP, Miller AC, Spontak RJ. Highly CO_2-permeable and selective polymer nanocomposite membranes. *Adv Mater* 2003, 15:729–733.
75. Lin HQ, Van Wagner E, Freeman BD, Toy LG, Gupta RP. Plasticization enhanced hydrogen purification using polymeric membranes. *Science* 2006, 311:639–642.
76. Simmons JW. Block polyurethane-ether and polyurea-ester gas separation membranes. Patent US6,843,829; 2005.
77. Orme CJ, Klaehn JR, Harrup MK, Luther TA, Peterson ES, Stewart FF. Gas permeability in rubbery polyphosphazene membranes. *J Membr Sci* 2006, 280:175–184.
78. Wang R, Cao C, Chung T-S. A critical review on diffusivity and characterization of 6FDA-6FpDA polyimide membranes for gas separation. *J Membr Sci* 2002, 198:259–267.
79. Cheng SX, Chung T-S, Wang R, Vora RH. Gas-sorption properties of 6FDA–durene/1,4-phenylenediamine (pPDA) and 6FDA–durene/1,3-phenylenediamine (mPDA) copolyimides. *J Appl Polym Sci* 2003, 90:2187–2193.
80. Liu SL, Wang R, Chung T-S, Chng ML, Liu Y, Vora RH. Effect of diamine composition on the gas transport properties in 6FDA-durene/3,3′-diaminodiphenyl sulfone copolyimides. *J Membr Sci* 2002, 202:165–173.
81. Hosseini SS, Teoh MM, Chung T-S. Hydrogen separation and purification in membranes of miscible polymer blends with interpenetration networks. *Polymer* 2008, 49:1594–1603.
82. Bos I, Punt H, Strathmann M, Wessling M. Suppression of gas separation membrane plasticization by homogeneous polymer blending. *AIChE* 2001, 47:1088–1093.
83. Zhou C, Chung T-S, Wang R, Liu Y, Goh SH. The accelerated CO_2 plasticization of ultra-thin polyimide films and the effect of surface chemical cross-linking on plasticization and physical aging. *J Membr Sci* 2003, 225:125–134.
84. Xiao Y, Dai Y, Chung T-S, Guiver MD. Effects of brominating matrimid polyimide on the physical and gas transport properties of derived carbon membranes. *Macromolecules* 2005, 38:10042–10049.
85. Tin PS, Chung T-S, Shao L. Surface modification of polyimide membranes by diamines for H_2 and CO_2 separation. *Macromol Rapid Commun* 2006, 27:998–1003.
86. Caro J, Noack M, Kolsch P, Schafer R. Zeolite membranes—state of their development and perspective. *Micropor Mesopor Mater* 2000, 38:3–24.
87. Uemiya S. State-of-the-art of supported metal membranes for gas separation. *Sep Purif Methods* 1999, 28:51–85.
88. Tin PS, Xiao Y, Chung T-S. Polyimide-carbonized membranes for gas separation: structural, composition, and morphological control of precursors. *Sep Purif Rev* 2006, 35:285–318.
89. Peer M, Kamali SM, Mahdeyarfar M, Mohammadi T. Separation of hydrogen from carbon monoxide using a hollow fiber polyimide membrane: experimental and simulation. *Chem Eng Technol* 2007, 30:1418–1425.
90. Bernardo P, Algieri C, Barbieria G, Drioli E. Hydrogen purification from carbon monoxide by means of selective oxidation using zeolite catalytic membranes. *Sep Purif Technol* 2008, 62:629–635.
91. Caro J. Catalysis in micro-structured membrane reactors with nano-designed membranes. *Chin J Catal* 2008, 29:1169–1177.
92. Rohland B, Plzak V. The PEMFC-integrated CO oxidation—a novel method of simplifying the fuel cell plant. *J Power Sources* 1999, 84:183–186.
93. Avgouropoulos G, Ioannides T. Selective CO oxidation over CuO-CeO_2 catalysts prepared via the urea-nitrate combustion method. *Appl Catal A* 2003, 224:155–167.
94. Dube MA, Tremblay AY, Liu J. Biodiesel production using a membrane reactor. *Bioresour Technol* 2007, 98:639–647.
95. O'Brien DJ, Craig JC. Ethanol production in a continuous fermentation/ membrane pervaporation system. *Appl Microbiol Biotechnol* 1996, 44:699–704.
96. Qureshi N, Meagher MM, Huang J, Hutkins RW. Acetone–butanol–ethanol (ABE) recovery by pervaporation using silicalite–silicone composite membrane from fed-batch reactor of *Clostridium acetobutylicum*. *J Membr Sci* 2001, 187:93–102.
97. Groot WJ, Kraayenbrink MR, Waldram RH, van der Lans RGJM. Ethanol production in an integrated process of fermentation and ethanol recovery by pervaporation. *Bioprocess Biosyst Eng* 1992, 8:99–111.
98. Qureshi N, Ezeji C. Butanol, 'a superior biofuel' production from agricultural residues (renewable biomass): recent progress in technology. *Biomass Bioprod Biorefining* 2008, 2:319–330.

14

Catalysis at Room Temperature: Perspectives for Future Green Chemical Processes

Frank Leung-Yuk Lam,[1]* Michael C.L. Li,[1] Rock S.L. Chau,[1] Rick A.D. Arancon,[2] Xijun Hu[1] and Rafael Luque[1,2]

[1]Department of Chemical and Biomolecular Engineering (CBME), Hong Kong University of Science and Technology (HKUST), Kowloon, Hong Kong
[2]Departamento de Quimica Organica, Universidad de Cordoba, Edificio Marie Curie (C-3), Campus de Excelencia Agroalimentario CeiA3, Cordoba, Spain

INTRODUCTION

Future society will experience a number of daunting challenges related to water, resource allocation, and energy demand. Facing these challenges is not a simple task and will require multidisciplinary efforts to improve efficiencies, energy savings, and the use of renewable resources and more sustainable practices.[1] For example, the stoichiometric use of reactants to obtain high product yields and avoid excessive waste byproducts (atom economy) was proposed as a key green metric for green chemical protocols.[2]

Catalysis has played a pivotal role in the chemical industry for several decades where significant efforts have been made to develop innovative strategies for the production of chemicals, materials, and fuels. In this regard, recent advances on the preparation of *benign by design* materials and nanocatalysis are among the most promising approaches with remarkable potential for future development. However, catalytic protocols have been the subject of intensive investigation in recent years in order to switch to a bio-based economy able to maximize the value of renewable resources including biomass and waste. Aqueous chemistries, biomass deconstruction strategies, and waste valorization to chemicals, fuels, and materials using environmentally sound methodologies can be considered as important steps forward for a more sustainable chemical industry.

The development of room temperature catalytic protocols recently emerged as a promising alternative to minimize energy consumption in catalytic reactions, at the same time, obtaining high product yields and/or better atom economies (lowering byproduct generation). In principle, both high yield and satisfactory atom economy can be achieved at low temperatures as side reactions are often minimized with less collision among reacting molecules. A catalyst, either homogeneous or heterogeneous, is typically involved in a reaction to lower the activation energy of the reaction by providing a unique mechanism. Nevertheless, the development of advanced room temperature methodologies requires careful catalyst and process design to provide a sufficiently active system able to promote reactions at room temperature. Many of these systems involve the utilization of noble metal and metal complexes, ionic liquids (ILs), heterogeneous catalysts, supported

*Correspondence to: kefrank@ust.hk

Advances in Bioenergy: The Sustainability Challenge, First Edition. Edited by Peter D. Lund, John Byrne, Göran Berndes and Iacovos A. Vasalos.

Originally published in: *WIREs Energy Environ* 2014, 4: 209–232 doi: 10.1002/wene.141

nanoparticles, and related nano-materials. This contribution has been aimed to provide an overview of key catalytic systems to perform chemistries at room temperature, different types of catalysts (e.g., solid acid/base catalysts, nanocatalysts, ILs, etc.) as well as a number of synthetically relevant catalytic processes (e.g., reduction/oxidation and coupling reactions) will be discussed in this work.

HOMOGENEOUS/HETEROGENEOUS CATALYSIS

Ionic Liquids

ILs have been widely used in recent years as an alternative to organic solvents due to their unique properties including low vapor pressure, solubility with organic reactants, and catalyst recoverability and reusability.[3,4] They have found particular applications in analytical methods,[5–8] biocatalysis,[9,10] electrochemical devices,[11] and as a reaction medium.[12] Common cations include 1-alkyl-3-methylimidazolium, 1-alkyl pyridinium, 1-alkyl-1-methylpyrrolidinium, tetraalkyl ammonium, tetraalkylphosphonium, and trialkyl sulfonium anions, however, include bis(trifluoro methylsulfonyl)imide, trifluoromethanesulfonate triflate, alkyl sulfate, dicyanamide, hexafluorophosphate, and tetrafluoroborate. Depending on their specific applications, various combinations of ILs have been used in many researchers citing the determining properties such as solubility and reactivity in the selection.

With a wide range of composition, ILs have been extensively utilized as a solvent in different chemical reactions such as acid-base reactions,[13] substitutions,[14,15] eliminations,[16] additions (e.g., electrophilic,[17] nucleophilic,[18] Diels-Alder reaction,[19] metal alkyl and aryl additions[20]), electrophilic aromatic substitution, esterification, Beckmann rearrangement, and glycosylation), and many others.

Greener electrophilic aromatic substitution processes have been a long-time goal of the synthetic chemical industry. Friedel-Crafts alkylations of benzene, conventionally catalyzed by either mineral acid (HF) or homogeneous Lewis acids ($AlCl_3$ or $FeCl_3$), can also be catalyzed using the chloroaluminate ILs at room temperature.[21] The reaction proceeds via arene alkylation immediately after dissociation of carbonium ions. Such reaction hardly goes to completion because the alkylated product is comparatively more reactive to the parent arene (generally leading to over-alkylation and a mixture of product isomers). Apart from IL alkylation catalysts, sulfonated [$-SO_3$] functionalized ILs have also been shown to catalyze efficiently the synthesis of α-aminophosphonates,[22] which is an essential com-

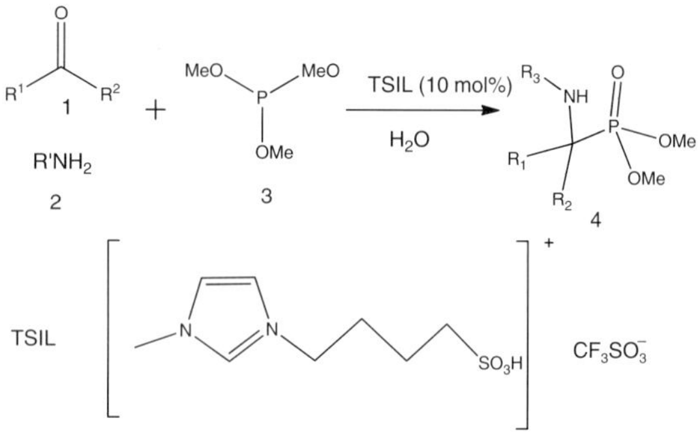

Figure 14.1 Three-component reaction catalyzed by a task-specific ionic liquid (TSIL).[22]

pound for various biological applications such as peptide mimicry,[23] enzyme inhibition,[24] antibacterial agents,[25,26] herbicides, fungicides, insecticides,[27] and plant growth regulation.[28] Typically, synthesis of α-aminophosphonates can be accomplished using Lewis acid catalyst such as $SnCl_4$, $ZnCl_2$, BF_3OEt_2, $MgBr_2$, and $InCl_3$.[29,30] However, the reaction cannot be accomplished effectively in one step with carbonyl, amine, and phosphite functionalities because the products of imine formation (amine and water) can deactivate the Lewis acid catalyst. In the work of Akabari and Heydari, the reaction of trimethyl phosphite with the imine generated from the reaction of benzalde-hyde and aniline at room temperature was studied.[22] (Figure 14.1) Using the sulfonic acid-functionalized IL as the Brφnsted catalyst, the α-aminophosphonates were obtained in 98% yield within 10 min. Different conditions were also studied and shown in Table 14.1.

It was found that aromatic aldehydes possessing either electron-donating or electron-withdrawing substituents were effectively tolerated giving excellent yields to the expected α-aminophosphonates (Table 14.1, entries A, B, E, H, and I). Further work was conducted for a ketone under similar experimental conditions, in which, 4-methoxyacetophenone with aniline and trimethyl phosphite were used as reactants (Table 14.1, entry L). This gave an excellent product yield. The durability of the catalysts was also evaluated, and no significant catalyst deactivation was observed for six repeated experimental runs.

Esterification is not only one of the most important industrial processes but is also one of the more challenging reactions because of the unfavorable equilibrium of ester formation which always requires reactive acid or coupling agents to achieve high product yields. Acidic chloroaluminate ILs were first reported as a trial homogeneous catalyst for Fischer esterification in 2001.[31] However, moisture sensitivity is a problem of the ILs, making it impractical

Table 14.1 Synthesis of α-Aminophosphonates from Aldehydes or a Ketone, Amines, and Trimethylphosphite[22]

	R_1	R_2	R_3	Time (h)	Yield (%)
A	$4\text{-ClC}_6\text{H}_4$	H	Ph	15 min	96
B	$4\text{-O}_2\text{NC}_6\text{H}_4$	H	Ph	15 min	98
C	2-Furyl	H	Ph	1	95
D	4-Pyridyl	H	Ph	1	94
E	$4\text{-MeOC}_6\text{H}_4$	H	Ph	2	96
F	Benzyl	H	Ph	0.5	98
G	Phenyl	H	Ph	10 min	98
H	$4\text{-NCC}_6\text{H}_4$	H	Ph	0.5	96
I	$4\text{-MeC}_6\text{H}_4$	H	Ph	0.5	95
J	*i*-Propyl	H	Ph	1	97
K	Cinnamyl	H	Ph	1	94
L	$4\text{-MeOC}_6\text{H}_4$	Me	Ph	1	92
M	Phenyl	H	Ph	15 min	98
N	Phenyl	H	Ph	2	95
O	Phenyl	H	Ph	1	92
P	Phenyl	H	Ph	45 min	96

for the esterification. Another trial of the Fischer esterification between ethanol and acetic acid has been conducted using acid-functionalized ILs.[32] A bi-phase IL, $[(HO_3\ S)^3\ C_3\text{-}(Ph)_3P][pTSA]$ (which is liquid at the tem perature of the reaction but solid at room temperature) has also been used for esterification reactions due to its efficiency demonstrated by the high product yield obtained and its ease of separation from the reaction mixture. The yields of the ethyl acetate product were improved by the presence of water because of the ability of water to act as proton bridge in reactions involving proton transfers.

Another kind of IL, e-caprolactamium tetrafluorobarate, was also found to be effective as a catalyst for Beckmann rearrangement without the need to add aqueous acid.[33] The Beckman rearrangement of a number of ketoximes into amides, catalyzed by Lewis acids ($AlCl_3$, $TiCl_4$, $SnCl_4$, or BF_3) has been tried with different ILs.[34] Changing the catalyst of the system into ILs had significant effects on the reactions, such as a small decrease in the reactivity when alkyl chain lengths of the cation were increased, and reaction rates with C^2 substitution of the imidazolium ring were also found to have increased because of the change. However, to date, the impact of the IL anion on the rates of the reaction is not clear, as well as their ability to interact with various Lewis acids and hydrate themselves during the catalysis.

Being considered as green solvents, ILs are also not without disadvantages. It should be clarified that the term 'green solvent' refers to the property of ILs to be separated from the reacting mixture; however, they are also chemicals with varying degrees of toxicity depending on the level of exposure. This makes these substances not totally fit for industrial application. Also, ILs might have low vapor pressures, but they can also become flammable when exposed to air. With these two disadvantages, it is necessary to seek another solution for the catalysis at room temperature condition.

Heterogeneous Catalysts

Heterogeneous catalysis has become the one of the core categories of catalysis, with the main advantage of providing extra surface area for enhancing collisions of reactant molecules to obtain high yields of reaction product. Moreover, only simple techniques such as filtration and sedimentation are required for catalyst recycling and reuse. In a conventional heterogeneous catalytic system, porous solid materials are used as a carrier of embedded catalytic metal elements. As the porous material has a high specific surface area, it allows the metal elements to be situated on the surface and become available for catalysis.

In this review, the catalytic performances achieved in various chemical reactions by different metals and their effects to surface morphologies (metal nanoparticles, grafting metal complexes, etc.) will be discussed extensively.

ZSM-5-based Catalysts

As mentioned in homogeneous catalysis, using Lewis acids such as $AlCl_3$, $ZnCl_2$, and BF_3 with ILs improve the efficiency of the whole catalytic system. However, it is still a question whether to apply such dangerous ILs in large-scale industrial applications. A siliceous-based metal catalyst, ZSM-5, which was discovered by Landolt and Argauer in 1965,[35] was first used in 1983 for the first commercial-scale trial cracking unit by Mobil Technology Company.[36] It is a kind of acidic zeolite and is found to be an effective catalyst for catalytic cracking of the additive for propylene production and gasoline octane improvement.[37–39] To date, ZSM-5 is still a good catalyst for cracking processes at high reaction temperatures to produce light olefins (ethylene, propylene, and butylenes) from various sources, including ketones, naphtha feedstocks, bio-oil, and alcohols. The catalytic activity of ZSM-5 does not only come from its surface acidity but also from its shape selectivity,[44–49] special porous structure, and ion-exchanging properties. Scientists have performed some work on ZSM-5 catalyst at room temperature to find its 'own' catalytic activity without any harsh reaction condition provided.

According to the work of Stepnanov and his coworkers in 1996, on the formation of carboxylic acids from alcohols and olefins in H-ZSM-5 zeolitic materials,[50] hydrocarbon

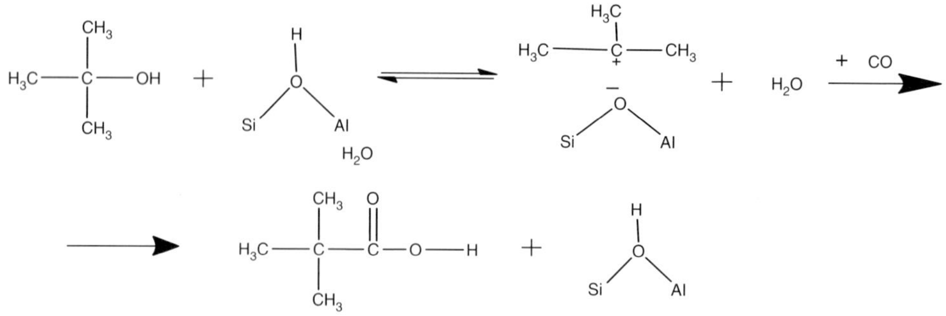

Scheme 14.1 Production of carboxylicacids using zeolites.[50]

conversion via solid acid catalyst (zeolite in this case) typically involves the car-benium ion formation.[51,52] The *in situ* ^{13}C solid-state NMR study was used to characterize entrapment of the alkyl carbenium ion by means of carbon monoxide (CO). They found that butyl alcohols (*t*-BuOH, *i*-BuOH, and *n*-BuOH) and olefins (ethane, *iso*-butene, and octene-1) undergo carbonylation to form carboxylic acids (known as the Koch reaction) with high conversion on H-ZSM-5 zeolite at 296–373 K. Under the same reaction conditions, *t*-BuOH is carbonylated to form trimethylacetic acid with full conversion within a few days at ambient conditions. In the presence of water and CO, *iso*-butene at room temperature can be carbonylated and hydrated with full conversion to form more than 80% trimethylacetic acid and less than 20% of *t*-BuOH. The CO in this case selectively reacts with the tertiary cations to produce tertiary carboxylic acids, as shown in Scheme 14.1. This successful attempt has led some scientists to examine the feasibility of H-ZSM-5 to be a room temperature catalyst in many other chemical reactions.

The same group by Stephanov and his coworkers published findings on the room temperature oligomer-ization using H-ZSM-5 as acidic catalysts in 1998.[53] It is interesting to note that the transformation of the oligomeric alkoxides to the oligomeric carboxylic acids is highly dependent on the oxy-functional group of the zeolite, as shown in Scheme 14.2.

Apart from its application as a catalyst in organic synthesis, H-ZSM-5 doped with metal elements has also been attempted for methane activation and benzene oxidation processes. Xu et al. did some work on the fabrication of Zn-modified H-ZSM-5 for methane activation.[54] The theoretical calculation of methane activation shown by Santen et al. indicates that the polarized $C^{\delta-}$–$H^{\delta+}$ bond undergoes heterolytic cleavage and forms CH_3^- ions which readily react with Zn^{2+} ions producing the zinc methyl species (Zn-CH_3).[55] Experimentally, it has been proven by Ivanova and his coworkers that the $ZnCH_3$ species at room temperature do not proceed further even with heating up to 773 K.[56] However, in the presence of a coreactant, the coconversion of zinc methyl species and CO_2 into acetic acid at 523 K was achieved.[57,58] In addition, Zn^{2+} ions acting as hydride acceptors to promote inverse polarization of $C^{\delta+}$–$H^{\delta-}$; bond were theoretically achieved as well.[59] Apart from this study, dimerization and oxidation of methane through homolytic cleavage mechanism have been observed in the Zn-modified zeolite-catalyzed system.[60]

Another example of a use of a metal doped zeolite is on the synthesis of phenol from the oxidation of benzene which is a challenging reaction in catalysis.[61–66] Phenol is one of the most important industrial feedstocks that serve as a precursor for phenol resins, fibers, dyestuffs, and medicine. Attempts have been undertaken to study the reaction of benzene with oxygen species on Fe/

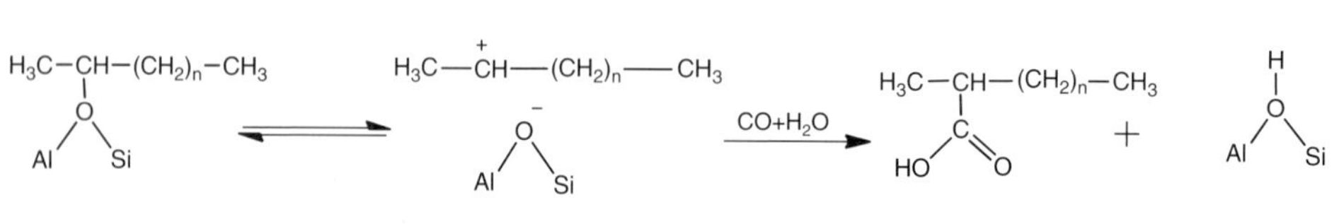

Scheme 14.2 Transformation of the oligomeric alkoxides into oligomeric carboxylic acids via interaction of equilibrated oligomeric carbenium ions with CO and H_2O.[53]

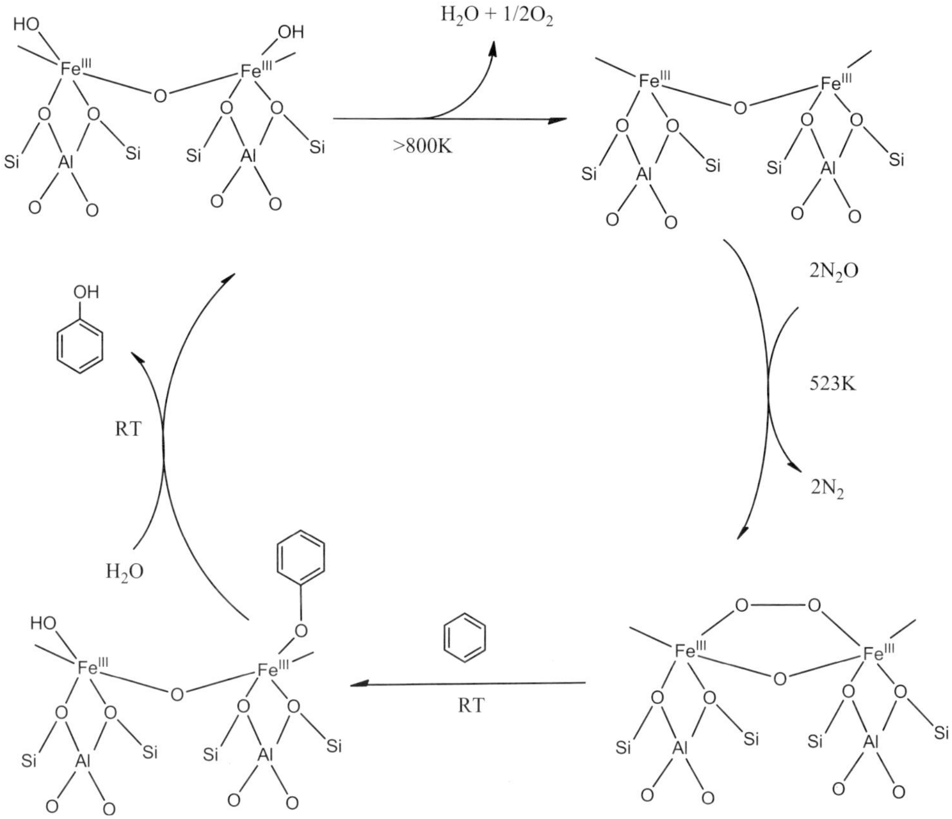

Figure 14.2 Schematic description of the reaction pathway of the active oxygen species with benzene on Fe/ZSM-5 at room temperature.[61]

ZSM-5; however, the involved active sites and mechanisms are still unclear. Xia et al. investigated the reaction of benzene with peroxide species using a direct spectroscopic approach. They found that a complex, Fe(III)-phenolate complex, was formed during the reaction as can be seen in the proposed mechanism in Figure 14.2.

The formation of the Fe(III)-phenolate complex is due to the reaction between benzene and the peroxide that was bridged on the binuclear iron sites of the ZSM-5. The complex bound to the binuclear iron sites releases the phenol molecules after hydrolysis.[65,67] The hydroxylated binuclear iron sites could be converted into the dehydroxylated binuclear iron sites by releasing H_2O and O_2 on high temperature treatment. It is important to note that the reaction of benzene with the oxygen species at room temperature is a stoichiometric reaction, which is different from a common catalytic thermal oxidation.

Apart from the catalytic systems of iron- and zinc-based zeolites, gold-, palladium-, iron-, ruthenium-, and cobalt-based supported catalysts also demonstrate good catalytic performances for certain chemical reactions.

Gold-based Catalysts

Gold, having physical interactions rather than chemical to the oxygen molecules in the presence of air and water, offers an advantage as an effective oxidative catalyst. Liu et al. did some work in the selective oxidation of hydrocarbons at a temperature of around 120°C[68] where gold and palladium nanoparticles have been doped onto the metal organic frameworks (MOFs). It was found that the Lewis acidity of the MOFs significantly promoted the activities of the catalysts. Table 14.2 shows the performance of various catalytic systems (Entry 1–6) on the oxidation of toluene in the absence of solvent.

Using the optimized Au-Pd catalyst, a maximum conversion of 98.6% was achieved with 93% selectivity to benzyl benzoate at 120°C in a short reaction time of 36

Table 14.2 Results of the Oxidation of Toluene in the Absence of Solvent[68]

Entry[1]	Pd:Au (mol:mol)	Conversion (mol%)	TON	Selectivity (mol%) BZ	BZBZ	PhBZ	C–C prod.
1[2]		1.0		61			
2[3]		1.2		65			
3	1.5:1	75.6	1663	4	89	4	3
4[3]	1.5:1	98.6	986	2	93	3	2
5[4]	1.5:1	32.0	2080	13	81	3	3
6[5]	1.5:1	73.9	1626	5	89	4	2

[1]Toluene (5 mL), 1 wt% Au-Pd/MIL-101, substrate/metal = 2200, 120°C, 1MPa O_2,48h.
[2]No catalyst used. Other products: benzyl alcohol (39.0%) and benzoic acid (0.5%).
Catalyst: MIL-101. The other product is benzyl alcohol (35%).
[3]Substrate/metal = 1000, 36h.
[4]Substrate/metal = 6500.
[5]Third run of Au-Pd/MIL-101 under the conditions of entry 3. Products: benzaldehyde (BZ); benzylbenzoate (BZBZ); phenyl benzoate (PhBZ); C–C coupling products from 2 toluene molecules (C–C prod).

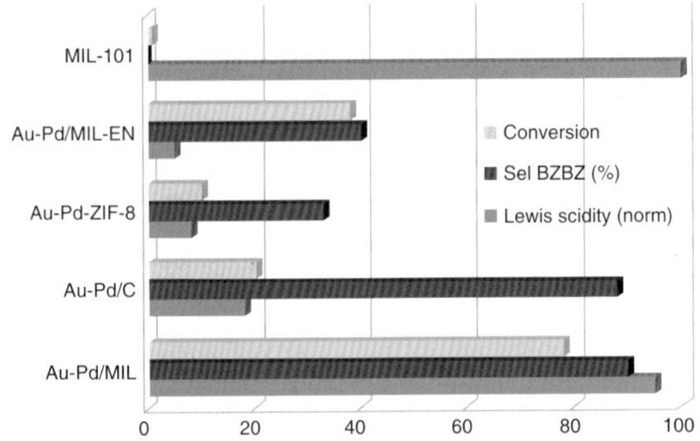

Figure 14.3 Influence of the Lewis acidity (normalized values with respect to MIL-101) in the conversion and selectivity toward benzyl benzoate (mol%) for different catalysts.[68]

h. The excellent catalytic activity is attributed to the presence of Lewis acid sites in the Au-Pd/MIL-101 catalysts, which potentially influence the activity of the proposed system in aromatic oxidations. Different kinds of substrates, including carbon, ZIF-8, MIL-EN (MIL-101 grafted with ethylenediamine), and MIL-101, have been examined.

The results show that the presence of gold and palladium nanoparticles demonstrates a significant promoting effect in the presence of a Lewis acid substrate, as illustrated in Figure 14.3.

Lewis acids have certain affinities to aromatic rings.[69,70] Toluene may interact with the Lewis acid site of the MIL-101 (Cr sites) and affect the electron distribution of the methyl group. Consequently, the carbon on the methyl group becomes electron-deficient making it more susceptible for attack by the activated oxygen species adsorbed on the Au-Pd nanoparticles.

Recently, gold has also been found to be catalytically active for amidation processes. Kobayashi's research team has worked for amide synthesis from alcohols and amines under aerobic conditions using gold nanoparticle catalysts.[71] Amide is an essential functional group for drug development with an increasing demand in the pharmaceutical industry. It is a useful compound for the synthesis of efaproxiral intermediate, which is manufactured for metastatic cancer therapy.[72] Typically, it is synthesized by the reactions of carboxylic acids with amines (using coupling agents) or through the direct catalytic amidation of aldehydes and amines with a catalyst. For the former approach, Kobayashi's group developed gold nanoparticles supported on polymer-incarcerated carbon blacks (PICB), denoted as Au/PICB. Their work mainly explores the catalytic mechanism of a tandem oxidative process. The main advantage of a tandem oxidative process is that the only byproduct is water and that high product yields are usually obtained. Their results show that a conversion of 86% was obtained at a temperature of 40°C using Au/PICB. A better yield was achieved by using bimetallic (gold and cobalt) catalyst, resulting in a full conversion and a good selectivity of around 96%. The incorporation of cobalt catalysts produces two effects: (1) decreases the reactivity of gold through electronic interactions, and (2) stabilizes the cabinolamine during the oxidation step.

Besides mounting gold nanoparticles on carbon substrate, silica-based gold catalysts for direct amidation between aldehyde and amines were also recently studied.[73] In his work, laser has been introduced as the activation source of gold nanoparticles as they have a sharp plasmon absorption band at 532 nm. Photonic nanoparticles have a strong surface plasmon resonance (SPR) absorption in the visible region due to the collective oscillation of free electrons in the conduction band.[74] Scheme 14.3 shows the laser-driven amidation between benzaldehyde and morpholine.

Figure 14.4 shows that with the laser-assisted catalytic system, the desired product, 4-benzoyl morpholine could be efficiently produced at room temperature within 4–5 h of reaction. Without the aid of laser, the gold/silica catalysts poorly performed with a conversion of 30% after 6 h of reaction. A plausible reason for the poor conversion is due to the inactivation of gold nanoparticles by the laser, consequently decreasing the reaction rate and the product yield. This work has been further confirmed through a laser-driven tandem oxidation process for amide formation from benzyl alcohol and morpholine as shown in Scheme 14.4.

Benzyl alcohol was first oxidized to benzalde-hyde by means of O_2 bubbling or H_2O_2 addition. The results were found to give 38% conversion with >99% selectivity for a

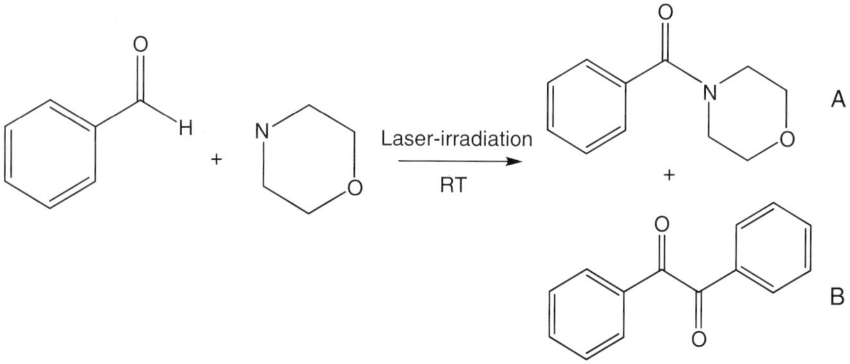

Scheme 14.3 Laser-driven amide formation between benzaldehyde and morpholine at room temperature.[73]

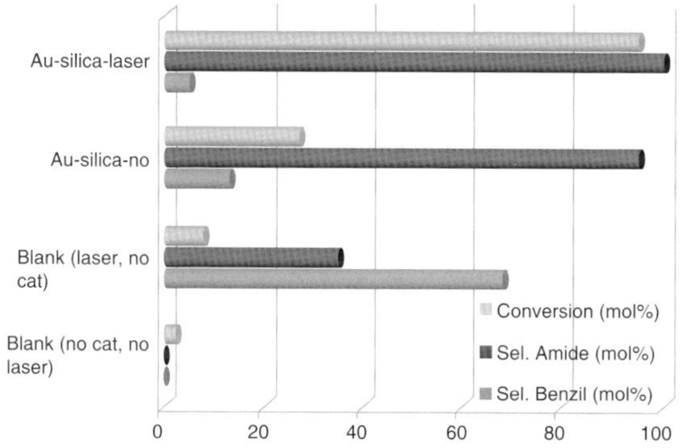

Figure 14.4 Comparative experiments in the amidation of benzaldehyde and morpholine at room temperature conducted under identical conditions. Reaction conditions: 0.2 mL benzaldehyde, 0.2 mL morpholine, 19 mL THF, 1 mL H_2O_2, 0.01 g KOH, and 0.025 g Au-silica catalyst, 4 h of reaction, 25°C.[73]

4-h reaction, as shown in Figure 14.5. It is interesting to note that by providing extra excitation source such as laser, UV irradiation, ultrasonic, or microwave, the reaction conversion as well as the product yield was improved. It is an important aspect for room temperature catalysis particularly to some thermodynamically unfavorable processes.

Imines, which are highly desirable electrophilic reagents in various reactions (reductions, additions, condensations, and cycloadditions[75–78]), can also be synthesized from the condensation of amines and carbonyl compounds in the presence of an acid catalyst.[79,80] A number of transition-metal-catalyzed carbonylative reductions of nitroarenes and aldehydes with CO and water to produce imines have been reported since the 1970s.[81–83] However, the main problem of applying such catalysts is their recovery and recyclability because they are usually converted to metal complexes after the reaction. In recent years, heterogeneous catalysts such as Au/TiO_2[84] and Ni/SiO_2[85] have also been applied in such systems with hydrogen as a reducing agent. The imperfection of this approach is that because of its high reaction temperature (>110°C), elevated pressures, and long reaction time, it is difficult to apply in an upscale process. Huang et al. recently reported gold nanoparticles supported on titania for the direct imination of nitroarenes using aldehydes and carbon monoxide (CO) as reducing agents, as shown in Scheme 14.5.[86]

The breakthrough of this work is that the catalytic reaction is held at room temperature and the choice of gold nanoparticles is mainly due to its ability to allow the rapid and chemoselective reduction of polysubstituted nitro and carbonyl compounds under relatively mild conditions.[87,88] Synthesis of *N*-benzylidenebenzenamine from nitrobenzene and benzaldehyde in different solvents was studied and the results are shown in Table 14.3.

Direct imination of nitrobenzene is also demonstrated in Entry 1. In this work, the formation of byproducts, which are intermediates used in the synthesis of biologically active compounds[89,90] was observed. According to many studies, the imination process is solvent-dependent. A 100% conversion with 95% selectivity for imination can be obtained using triethylamine as solvent (Entry 11). Other studies did not just present the catalytic activities of various catalysts, but they also explored on the most plausible mechanism of the process. It was found that the Au/TiO_2 in the final imine formation step plays an important role in the reaction due to the mild acidity of the TiO_2 support.

The catalyst Au/TiO_2, is also applicable for the one-pot reductive imination reaction which is an aza Diels-Alder reaction of the imine, as illustrated in Scheme 14.6.

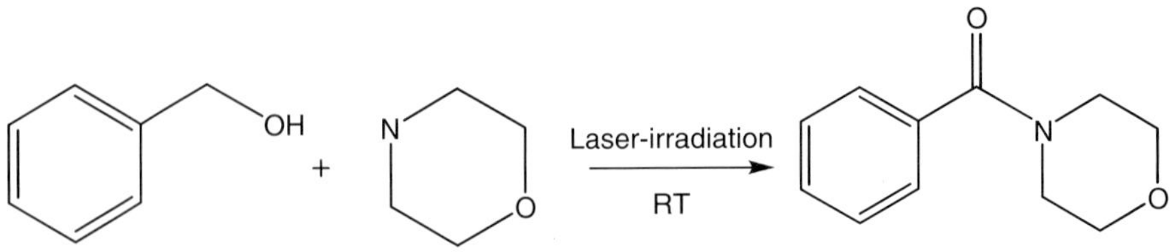

Scheme 14.4 Laser-driven tandem oxidation/amidation reaction. Benzyl alcohol and benzyl amine to N-benzylbenzamide. Reaction conditions: 0.2 mL benzyl alcohol, 0.2 mL morpholine, 19 mL THF, 1 m LH_2O_2,0.01 g KOH, 2 m LH_2O_2 (added dropwise), and 0.025 g Au-SiO_2 catalyst, 4 h of reaction, 25°C.[73]

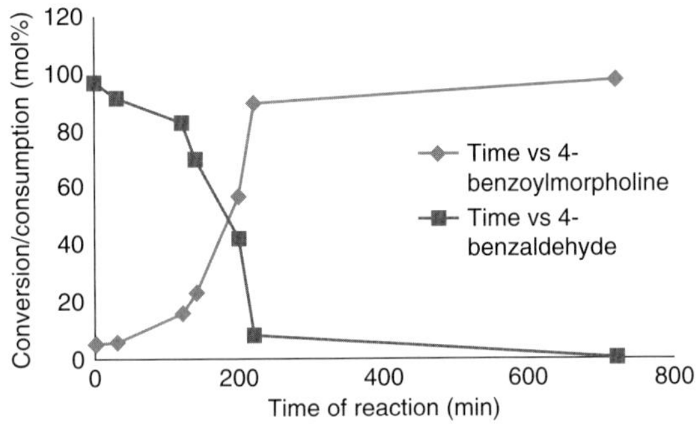

Figure 14.5 Reaction profiles of benzaldehyde consumption and product generation in the laser-assisted amidation of benzaldehyde and morpholine. Reaction conditions: 0.2 mL benzaldehyde, 0.2 mL morpholine, 19 mL THF, 1 mL H_2O_2, 0.01 g KOH, and 0.025 g Au-silica catalyst, 25°C. The proposed induction period can be clearly observed at the beginning of the reaction (the first 30–60 min, see the line of 4-benzoylmorpholine formation).[73]

After this one-pot synthesis, the desired product, 1,2-diphenyl-2,3-dihydropyridin-4-one, was obtained with a yield up to 85%. This demonstrates a promising approach for the application of ligand-free supported gold catalysts to the green synthesis of pharmaceutically important compounds.

In addition to the usage of gold nanoparticles for the synthesis of pharmaceutical compound intermediates, reduction of aldehydes is a very useful reaction in many industries. Among the numerous reduction methodologies for hydroxyl compound synthesis,[91,92] the catalytic transfer hydrogenation is environmentally benign because it does not use both flammable hydrogen and metal hydrides.[93,94] Gold-supported catalysts have an advantage because they can perform the chemoselective reduction of unsaturated carbonyl or nitro compounds using molecular hydrogen. However, gold catalysts tend to have low hydrogen delivery rates compared with other noble metals.[95] He et al. has developed gold-supported on ceria catalysts for transfer hydrogenation (Table 14.4).[96]

The improvement in synthesis condition of aldehyde hydrogenation was achieved by using increasing the loading of gold (3 mol%) and prolonging the reaction duration (3 h), to achieve a yield of around 99%. Most importantly, the Au/meso-CeO_2 catalyst kept its catalytic activity even after five consecutive runs. The transfer reduction of benzaldehyde (10 mmol) in the presence of 1 mol% Au/meso-CeO_2 and 5 equivalents of HCOOK in aqueous media attained 96% isolated benzyl alcohol yield.

The catalytic role of Au/meso-CeO_2 in this work has been investigated and hypothesized.[96]

It is proposed that the dehydrogenation of formate salt occurs specifically on the ceria surface, which acts as a sink for the reactive formate ions comparable with phenomenon of water gas shift.[97] Ceria sites (Ce^{3+} and Ce^{4+}) were involved in the rate-determining step (dehydrogenation of formate to bicarbonate species, step 1 in scheme). The hydrogen species formed could then transfer to vicinal gold clusters via reverse hydrogen spillover to form Au-H complexes. The formed Au-H complexes were suggested

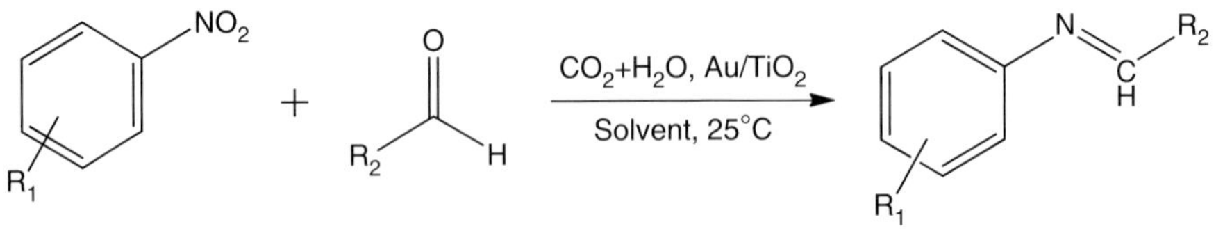

Scheme 14.5 Imination of nitroarenes using aldehydes and carbon monoxide by supported gold nanoparticles at room temperature.[86]

Table 14.3 Synthesis of *N*-benzylidenebenzenamine from Nitrobenzene and Benzaldehyde in Different Solvents (reprinted from Ref 86)

Entry[1]	Solvent	Water (equiv.)[2]	Conv.(%)[3]	Selectivities(%)[3] Imine	Aniline	Nitrone[4]
1	Ethanol/Water (v/v = 2:1)	33	100	27	65	8
2	Ethanol	0	25	78	2	20
3	Ethanol	0.5	71	74	8	18
4	Ethanol	1	90	72	13	15
5	Ethanol	2	100	64	23	13
6	Ethanol	10	100	43	47	10
7	Acetonitrile	1	43	53	28	19
8	DMF	1	34	50	35	15
9	Toluene	1	78	79	10	11
10	Tetrahydrofuran	1	100	85	10	5
11	Triethylamine	1	100	95	5	0
12[5]	Triethylamine	1	97	93	7	0
13[6]	Triethylamine	1	*n.r.*	–	–	–

[1]Reaction conditions: 1 mmol nitrobenzene, 1 mmol benzaldehyde, Au/TiO2-VS (1 mol% Au), 2 mL solvent, 5 atm CO, 25°C, 1 h.
[2]Relative to that of nitrobenzene.
[3]GC conversion and selectivity obtained by using anisole as an internal reference.
[4]Confirmed by GC/MS.
[5]1 atm CO, 5h.
[6]H_2 (5 atm) as the hydrogen source, n.r. = no reaction.

to be key active intermediates for transfer hydrogenation as shown in step 2 of the scheme. This work brings an insight on the role of the gold catalyst and how it significantly improves the reaction temperature of traditional organic synthesis. This not only provides an alternative route to organic synthesis but it also encourages an energy-efficient process for a sustainable environment.

Karimi and Esfahani have developed gold-supported periodic mesoporous organosilica (PMO) with a large surface area, tunable pore structure, and tailored compositions for oxidation of alcohols.[98] They claim that the size of gold nanoparticles is a critical factor for its catalytic activity. As gold has high surface tension and easily agglomerates to become a large gold particle when sufficient thermal energy is provided, the control of nanoparticle synthesis is crucial for catalyst development. The catalytic performance of the synthesized Au@PMO catalysts have been evaluated for oxidation of different alcohols and the results are shown in Table 14.5.

The developed Au/PMO catalysts are able to oxidize toluene in the presence of oxygen with a low gold loading of 0.4%, achieving 99% product yield within a very

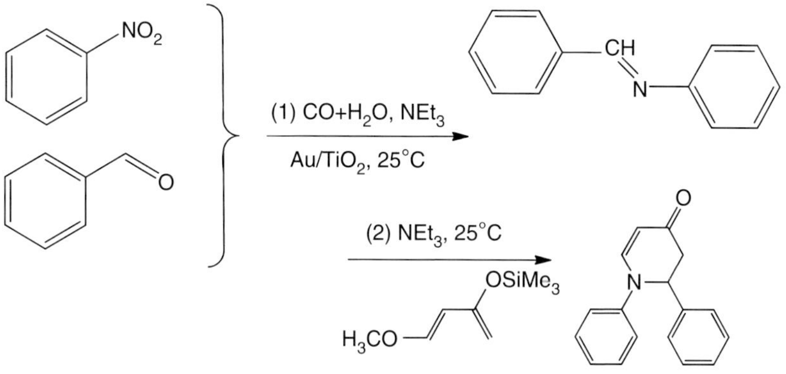

Scheme 14.6 One-pot two-step aza Diels-Alder reaction.[86]

Table 14.4 Catalytic Results of Transfer Hydrogenation of Benzaldehyde Under Various Conditions

Entry[1]	Catalyst	%Metal (mol)	T (°C)	Time (h)	Yield (%)[5]
1[2]	Au/meso-CeO_2	1	80	3	>99 (97)
2[3]	Au/CeO2 (Adnano 90)	1	80	3	21
3	Au/TiO_2 (WGC)	1	80	3	13
4	Au/Fe_2O_3 (WGC)	1	80	3	8
5	Au/C (WGC)	1	80	3	1
6	Au/Al_2O_3 (Mintek)	1	80	3	12
7	Pd/C (Alfa Aesar)	1	80	3	20
8	Pd/C (Alfa Aesar)	1	80	3	17
9	Meso-CeO_2	1	80	3	—
10	Au/meso-CeO_2	3	25	6	>99
11[4]	Au/meso-CeO_2	3	25	6	94

[1]Reaction conditions: 1 mmol substrate, HCOOK (5 equiv), water (15 mL), in air.
[2]BET surface area of meso-CeO_2 is 263 m^2/g.
[3]BET surface area of CeO_2 (Adnano 90) is 90 m^2/g.
[4]Fifth run.
[5]Yield was determined by GC. Numbers in parenthesis refer to yields of isolated products.

Table 14.5 Effects of Various Solvents and Bases on the Aerobic Oxidation of Benzyl Alcohol with Au@PMO Catalyst

Entry[1]	n (mol%)[3]	Solvent	Base	Time (h)	Yield (%)[2]
1	1	Toluene	K_2CO_3	24	10
2	0.4	Toluene	Cs_2CO_3	0.75	99
3	0.4	Toluene	Cs_2CO_3	1	99
4	0.4	Toluene	Et_3N	24	14
5	0.4	Toluene	AcO^-Na+	24	9
6	0.4	Toluene	K_3PO_4	24	11
7	0.4	Toluene	–	24	<5
8	0.4	TFT	Cs_2CO_3	24	78
9	0.4	CH_2Cl_2	Cs_2CO_3	24	65
10	0.4	H_2O	Cs_2CO_3	24	12
11	0.4	TFT:H_2O	Cs_2CO_3	24	25

[1]Reaction conditions: benzyl alcohol (1 mmol), Au/PMO (0.4 mol%), toluene, Cs_2CO_3 (3 mmol), O_2 (1 atm), room temperature.
[2]GC yield.
[3]n refers to the gold loading (mol%).

short time (0.75–1 h) in the presence of alkaline Cs_2CO_3 (Entries 2 and 3). Changing or not adding the base reagent would significantly reduce the yield to the range of 9–14% (Entries 1 and 4–7). Such finding shows that the catalytic activity of gold is also dependent on the experimental condition. Unfortunately, no proposed mechanism has been proposed for Au/PMO. Benzyl alcohols examined for the aerobic oxidation processes using Au/PMO, are illustrated in Table 14.6.

Studied alcohols as shown in Table 14.6 were successfully oxidized to form the products of ketones or aldehydes with yields ranging between 72 and 100%. It was suggested that the excellent catalytic activity of Au/PMO is driven by the nanosized gold particles as well as the special porous support. The PMO having aromatic bridges in their framework is an excellent alternative matrix to stabilize the gold nanoclusters mainly acting as an electron sink. This protects the gold nanoclusters in the solution phase and prolongs the catalyst durability throughout the reaction process.

In the application of different catalytic processes, catalyst sintering is always a concern to the scientist, particularly, in a large-scale industrial operation. Of course, sintering is mainly due to the deactivation of the catalyst caused by certain reagents such as reaction substrates, intermediates, or even products made from the chemical processes. Organsulfur compounds are well-known poisons to metallic nanoparticles. They will bind to the catalyst active sites and hence deactivate by forming sulfide (or sulfate) compounds. Rodriguez et al. demonstrated their work on the hydrogenolysis of thiophene (C-S bond) using gold nanoparticles (AuNP) supported on titanium carbide. However, only a little catalytic effect was observed.[99] Recently, Ansar et al. presented their work on the desulfurization of mercaptobenzimidazole and thioguanine using gold nanoparticles at room temperature.[100] They illustrates the scheme on the desulfurization of two compounds: (1) 2-mercaptobenzimidazole (2-MBI) and (2) 6-thioguanine (6-TG) by gold nanoparticles in the presence of $NaBH_4$. With the gold catalyst, the sulfide group can be isolated from the organic compounds. They optimized the catalytic system with parameters of AuNP such as size, loading, concentrations of $NaBH_4$, and 6-TG as well. Figure 14.6 shows the effect of the AuNP size on the product formation of 2-AP.

6-TG desulfurization on the 10 and 15 nm diameter of AuNPs occurs 8.7 and 3.7 times faster than that on the 50 nm AuNPs. The complete 6-TG desulfurization can be achieved even if the 6-TG concentration is higher than the full packing density of the AuNPs. The results confirm that the 6-TG does not have to be pre-adsorbed onto the AuNP to undergo desulfurization. The dependence of the catalytic activity on the size of AuNP is similarly found in other catalytic systems.[101,102] The role of AuNP is to serve as catalytic sites for the activation of BH_4^-during desulfurization. The smaller AuNPs exhibit higher activity per unit surface area which has a higher density of surface defects. This in turn provides more catalytic sites for the $NaBH_4$ activation. They have optimized the reaction system and investigated the reaction mechanism by means of isotope labeling technique and found that the gold nanoparticles play several roles in this catalytic system: (1) the AuNP captures and concentrates 2-MBI and 6-TG from the

Table 14.6 Aerobic Oxidation of Alcohols Using Au@PMO Catalyst at Room Temperature[98]

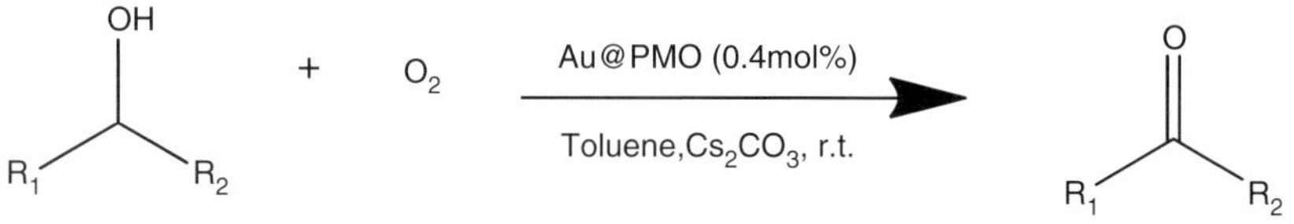

Entry[1]	Substrate	Time (h)	Yield (%)[2]	Entry	Substrate	Time (h)	Yield (%)[2]
1	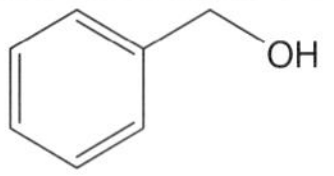	1	97	15	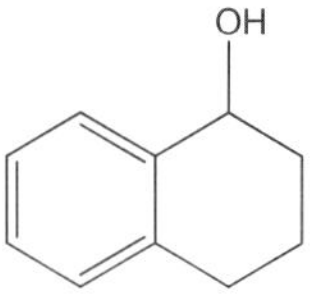	1.5	100
2	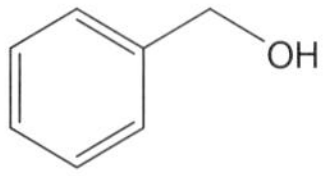	1.75	88[3]	16	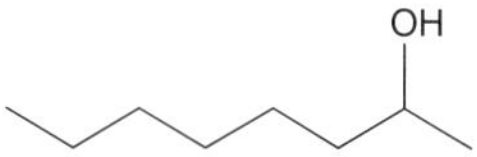	9	95
3	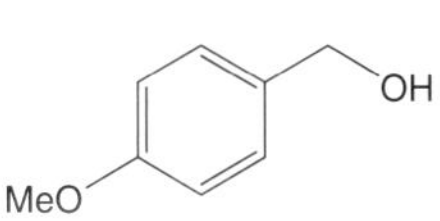	0.75	100	17	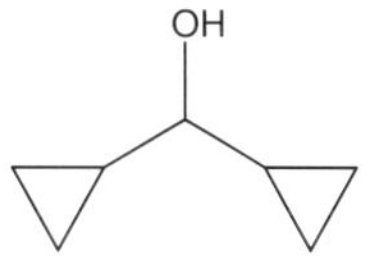	10	88
4	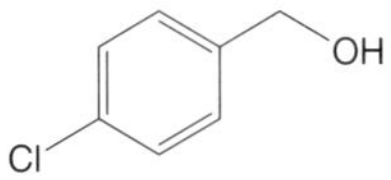	1.25	100	18	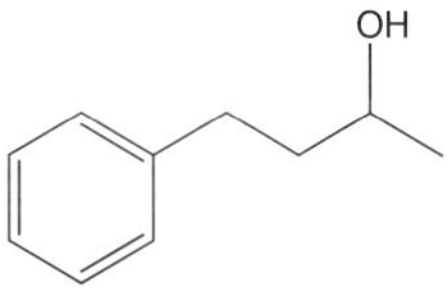	5	95
5	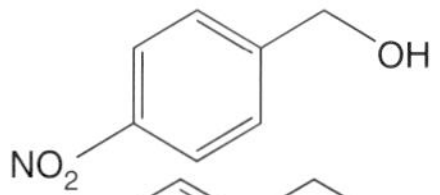	15	99	19	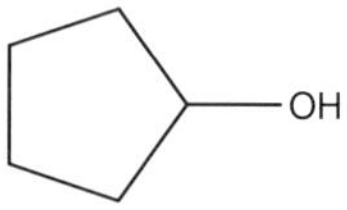	24	95
6	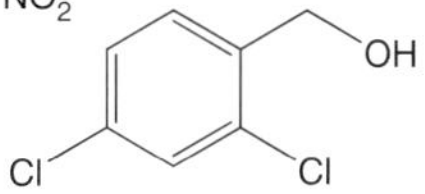	8.5	100	20	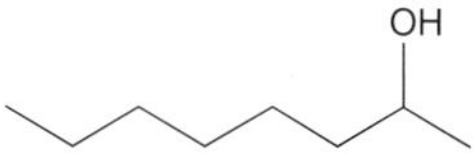	24	15[4]
7	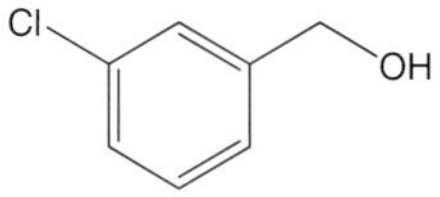	2.75	100	21	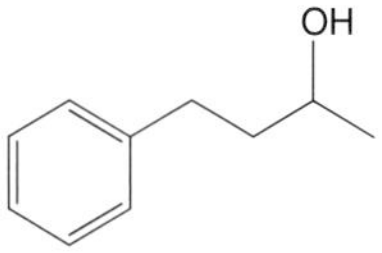	24	10[4]
8	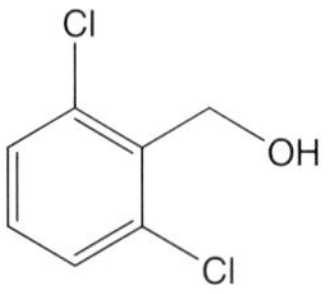	24	–	22	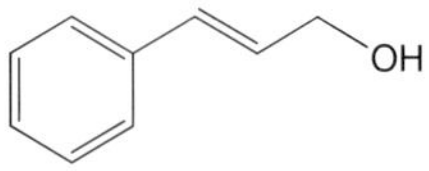	8	52
9	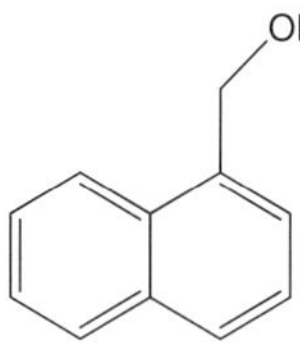	6	75	23	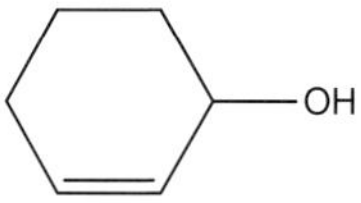	4	80

Table 14.6 Aerobic Oxidation of Alcohols Using Au@PMO Catalyst at Room Temperature[98] (*Continued*)

Entry[1]	Substrate	Time (h)	Yield (%)[2]	Entry	Substrate	Time (h)	Yield (%)[2]
10	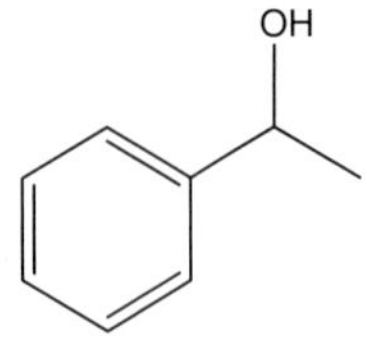	2.5	100	24	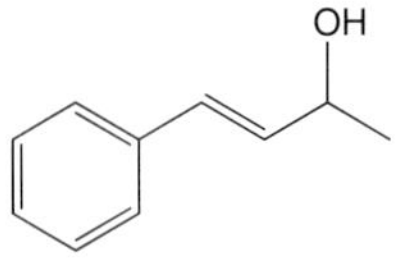	30	98
11	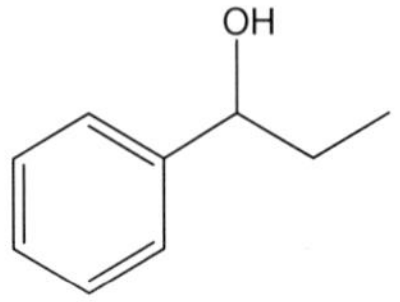	2.5	100	25	1-octanol	14	98[5]
12	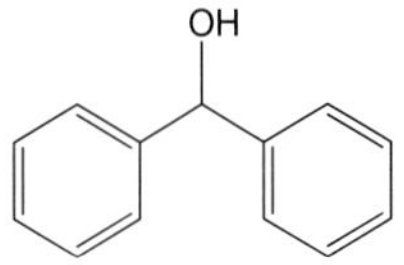	2.5	72	26	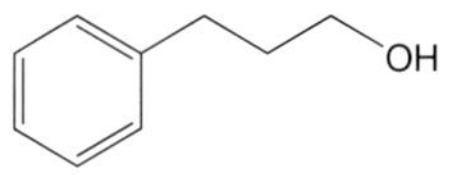	15	99[5]
13	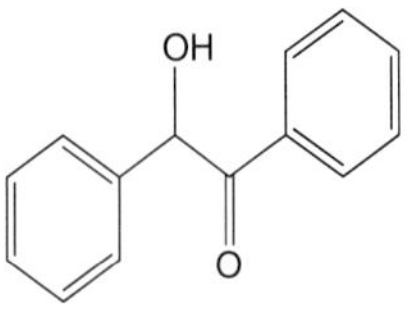	8.5	95	27	1-octanol	24	—
14	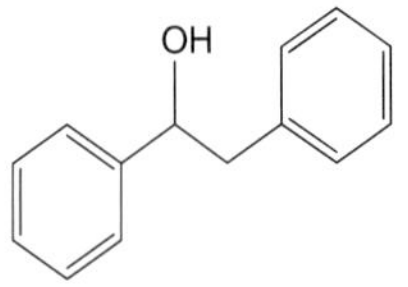	9	90	28	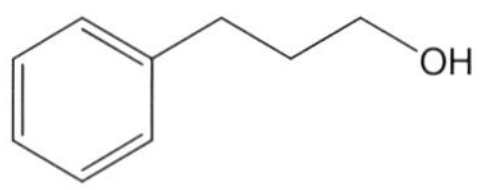	24	—

[1]Reaction conditions: alcohols (1 mmol), Au@PMO (0.4 mol%), toluene, Cs_2CO_3 (3 mmol), O_2 (1 atm), room temperature.
[2]GC yield.
[3]Reaction conditions: benzyl alcohol (1 mmol), Au@PMO (0.4 mol%), toluene, Cs_2CO_3 (3 mmol), O_2 (1 atm), room temperature.
[4]Reactionconditions:alcohols(1mmol), $NaAuCl_4$(0.4mol%), toluene, Cs_2CO_3 (3mmol), O_2 (1atm), roomtemperature.
[4]Reaction conditions: alcohols (1 mmol), Au@PMO (2 mol%), toluene, Cs_2CO_3 (3 mmol), O_2 (1 atm), room temperature.

solution, (2) they provide a pathway for hydrogen chemisorption on the gold surface, and (3) they capture the sulfides cleaved from 2-MBI and 6-TG. The mechanism involves the H-transfer from the AuNP surfaces to 2-MBI (or 6-TG) leading to the cleavage of the carbon–sulfur bond in 2-MBI and 6-TG, thus providing more insights on the role of gold in the cleavage of C–S bonds.

In the production of organic compounds, alkenes play an important role as an intermediate compound for organic synthesis. Typically, the alkene production requires high reaction temperature during the cracking process which makes a room temperature reaction quite impossible. Gold recently has also been found to catalyze the synthesis of alkenes from epoxides.[103] Deoxygenation of epoxides to alkenes is a useful step in organic synthesis because the oxirane ring can serve as a protective group for carbon–carbon double bonds.[104–106] Mitsudome et al. also investigated the catalytic activity of gold-supported catalysts in the liquid phase through deoxygenation of epoxides into alkenes with alcohols as reducing agents.[107] However, the required reaction temperature is comparatively higher (110°C) than the ambient chemical processes, causing incompatibility with functionalized epoxides. In this work, the researchers developed a gold catalyst supported on hydrotalcite (HT), which is a kind of clay with an anion exchange capacity. Using this material, the catalyst was able to deoxygenate a phenyl epoxide at room temperature, moving away from the usually high temperature reaction. Table 14.7 shows different kinds of catalysts that

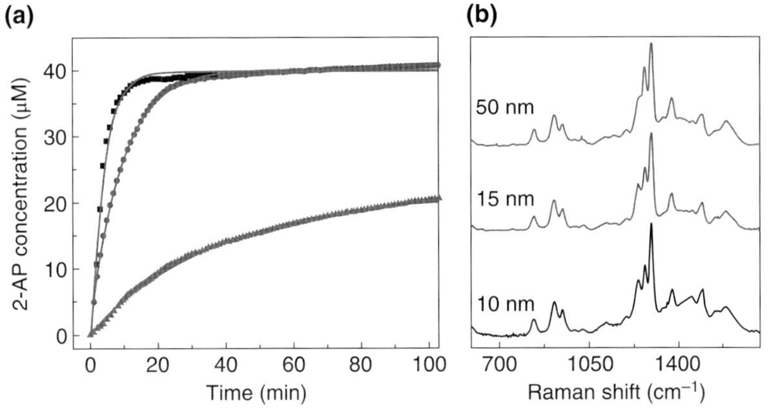

Figure 14.6 (a) The rate of 2-AP formation with AuNPs of different sizes. (b) SERS spectra of 6-TG and $NaBH_4$ are 40 μM and 33 mM, respectively. The nominal particle concentrations of the 10,15, and 50 nm-sized AuNPs are 4.20,1.87, and 0.17 nM, respectively. The volume of the reaction solution is 3 mL. The dots represent the experimental data, and the solid line is the fitting curve.[100]

Table 14.7 Catalytic Deoxygenation of Styrene Oxide[103]

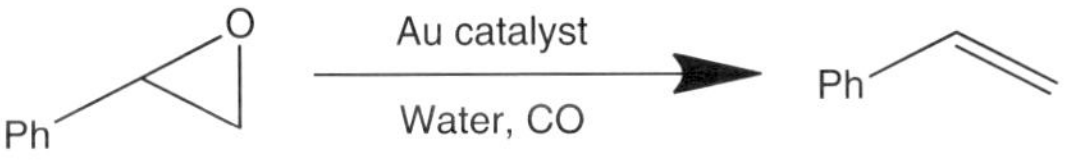

Entry[1]	Catalyst	Yield (%)[2]	Selectivity(%)[2]
1	Au/HT	>99	>99
2[3]	Au/HT	>99	>99
3[4]	Au/HT	97	>99
4[5]	Au/HT	73	>99
5[6]	Au/HT	60	>99
6[7]	Au/HT	6	>99
7	Au_2O_3	<1	–
8	$HAuCl_4$	0	–
9	BulkAu metal	0	–
10	Au/Al_2O_3	79	>99
11	Au/MgO	36	>99
12	Au/TiO_2	18	>99
13[8]	$Au/TiO_2 + Na_2CO_3$	57	>99
14	Au/SiO_2	3	>99
15	Au/BN	0	–
16	Ag/HT	0	–
17	Pd/HT	0	–
18	Pt/HT	0	–
19	Rh/HT	0	–
20	Ru/HT	0	–

[1]Reaction conditions: 1a (0.5 mmol), Au catalyst (Au: 0.9 mol%), water (5 mL), CO (1 atm), 27°C, 6h.
[2]Determined by GC by using an internal standard technique.
[3]Reuse 1.
[4]Reuse 2.
[5]Toluene (5 mL), water (0.1 mL).
[6]THF (5 mL), water (0.1 mL).
[7]DMA (5 mL), water (0.1 mL).
[8]Na_2CO_3 (1.5 mmol) was added.

have been evaluated for deoxygenation process of styrene oxide.

Table 14.7 shows the catalytic performance of different developed metal-supported catalysts. The synthesized Au/HT provided an excellent yield (>99%) and selectivity (>99%) toward styrene (Entry 1). Repeated runs were performed and it showed that the catalytic activity is constant after 2 runs (Entries 2 and 3). It was also found that organic solvents can reduce the yield and selectivity of the reaction (Entry 1 vs. entries 4–6). Other supporting materials such as alumina, magnesium oxides, titania, silica, and even with the aid of alkaline medium have also been attempted; however, most of them showed no significant catalytic activity for the deoxygenation process (Entry 1 vs. entries 10–14). Surprisingly they found that the Au/HT is the best catalyst for the deoxygenation of styrene oxide as compared with other metal-supported catalysts (Entry 1 vs. entries 16–20). Besides the detailed optimization of catalyst synthesis, other epoxides were also tested using the Au/HT catalytic system, as shown in Table 14.8.

Table 14.8 shows almost all epoxides were converted to alkenes with selectivity up to >99% with water as the only solvent of the catalytic system. Based on the proposed mechanism, a gold-hydride species is involved during in the catalysis. The species is generated *in situ* from the reaction between H_2O and CO, and performs the deoxygenation through a series of reduction. In addition, the hydrotalcite also has basic sites that can explain the formation of the gold-hydride species, which are formed through the nucleophilic attack of OH^- on a gold-CO species followed by decarboxylation. The heterolytic hydrogen species of H^+ and H^- generated on Au/HT would

Table 14.8 Deoxygenation of Various Epoxides by Using Au/HT[103]

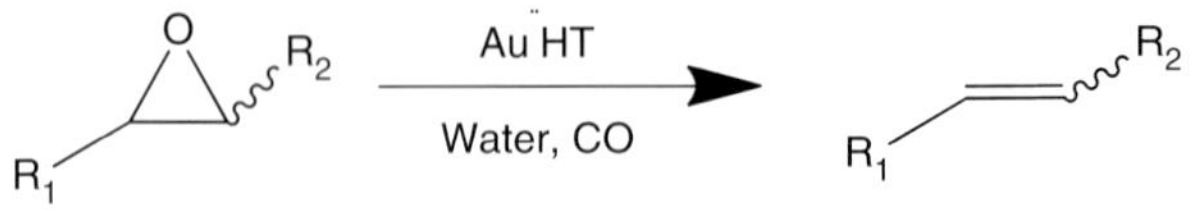

Entry[1]	Substrate	Product	Time (h)	Yield[2](%)	SeL[2](%)
1	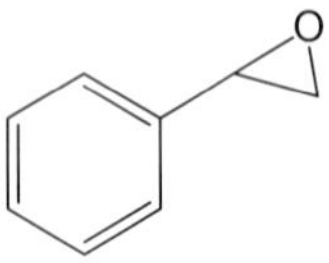	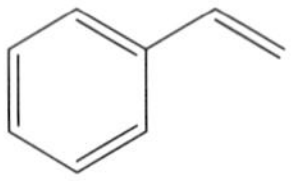	6	>99(94)	>99
2	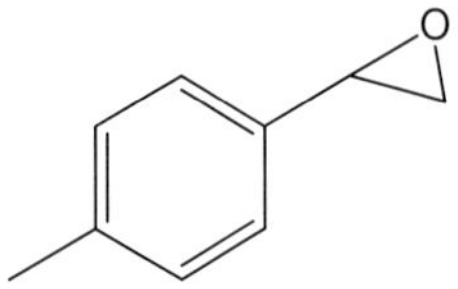	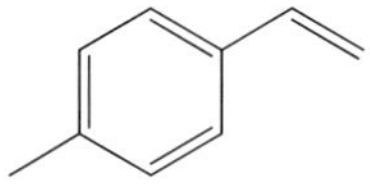	12	90(85)	>99
3	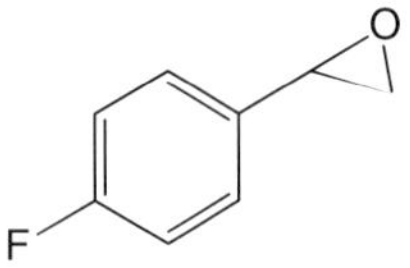	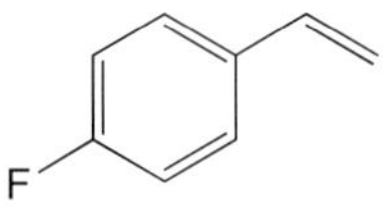	12	99(94)	>99
4	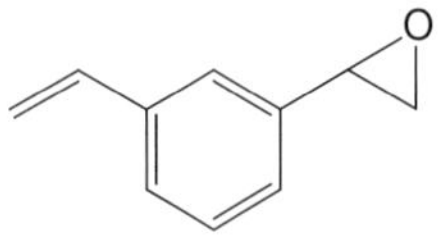	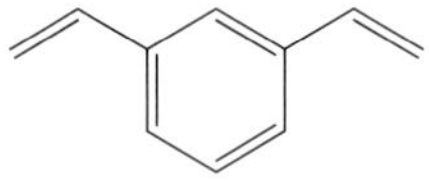	16	90(85)	>99
5	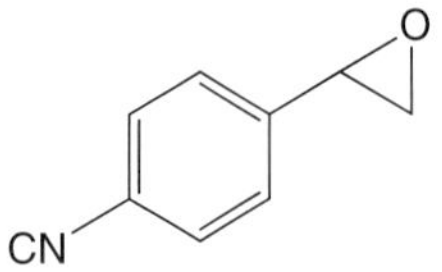	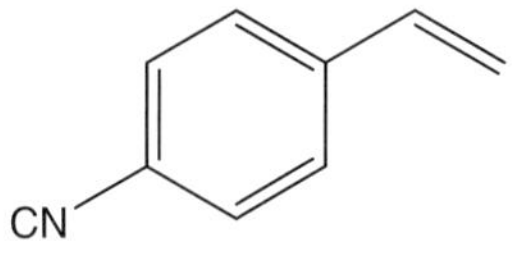	24	Trace	—
6[3]			12	61	>99
7[3]	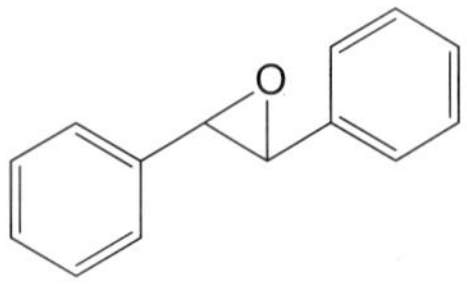	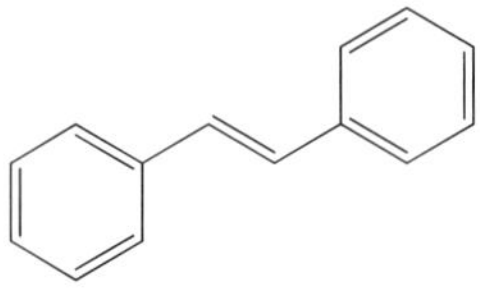	3	>99	>99
8[3]	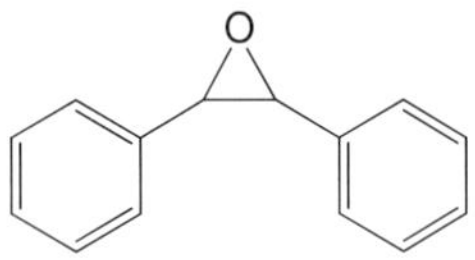	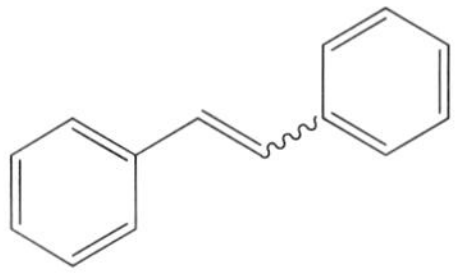	3	>99	>99 (E/Z-38:62)
9[3]	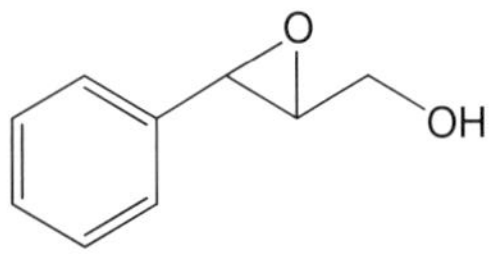	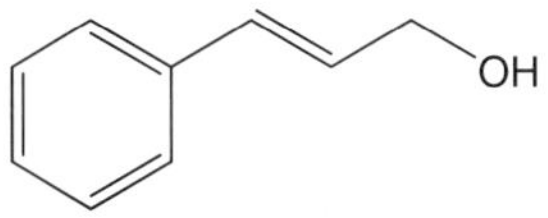	16	96(90)	96
10[3,4]			16	82	82
11[5,6]			16	36	36

Table 14.8 Deoxygenation of Various Epoxides by Using Au/HT[103]

Entry[1]	Substrate	Product	Time (h)	Yield[2](%)	SeL[2](%)
12	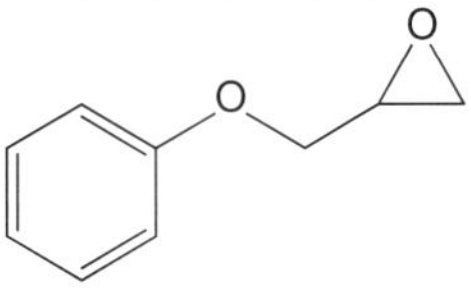	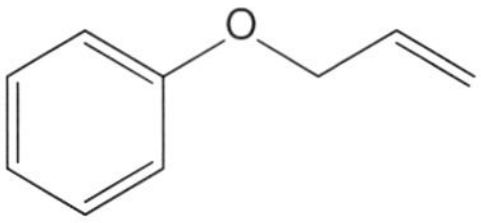	24	90(83)	>99
13	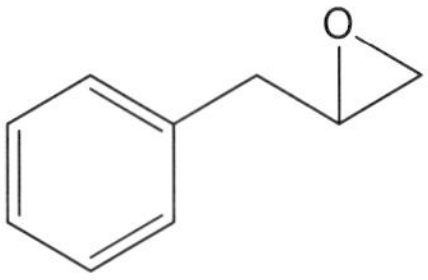	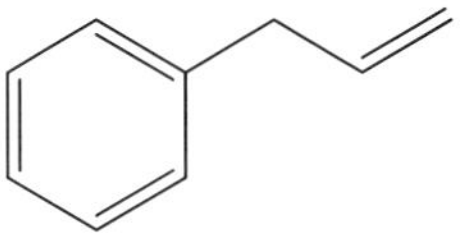	24	75(69)	>99
14	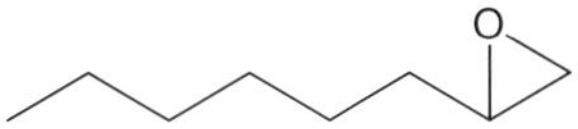		24	12	>99
15[4]	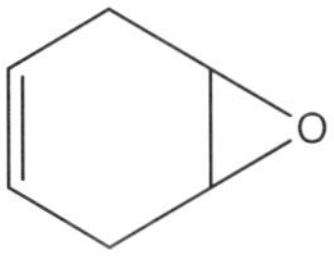	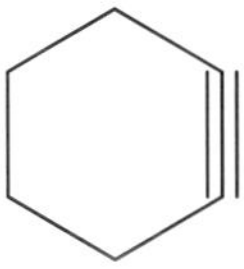	12	90	>99
16			24	Trace	–
17[4]	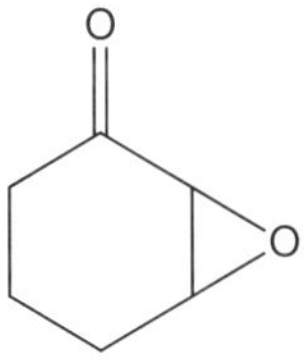	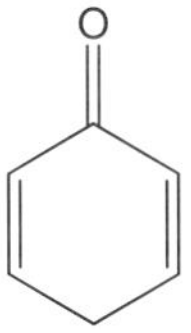	6	89	>99
18			24	Trace	–
19[4]			12	99	>99

[1]Reactions conditions: Substrate (0.5 mmol), Au/HT (0.1g Au: 0.9 mol%), water (5mL), CO (1atm), 27°C.
[2]Determined by GC by using an internal standard technique. Values in parentheses are the balanced yield.
[3]Cinnamaldehyde was formed as a byproduct.
[4]Toluene (5 mL) was used as solvent.
[5]THF (5 mL), water (0.1 mL), CO (5atm), 110°C.
[6]THF(5 mL) wasusedasasolvent.

deoxygenate the epoxide to form the corresponding alkene and water.[106, 107]

The application of gold nanoparticles is not limited to the area of organic synthesis but it is also applied for years in environmental remediation. Studies of ceria-supported gold catalysts have been previously presented for the oxidation of carbon monoxide.[108–112] Being a very common toxic pollutant, CO is a major cause of concern as they are usually formed in many of our everyday devices.[113–115] Precious metal catalysts are also considered promising catalysts for low-temperature CO oxidation.[116–119] This work further illustrates the development of an Au/CeO_2 catalyst using a deposition–precipitation approach, showing high efficiencies in the oxidation of carbon monoxide to carbon dioxide, as illustrated in Figure 14.7.

Figure 14.7 shows that the prepared 1 wt% Au/CeO_2 promotes the oxidation of carbon monoxide at room temperature. It should also be noted that the disappearance of CO is not due to the adsorption of CO on the catalyst but due to its conversion to CO_2. The estimated specific activity for the highest active catalyst is approximately 490×10^{-4} mol CO s^{-1} $g_{Au}{}^{-1}$ at 30°C. The catalytic activity is attributed to the presence of highly dispersed gold clusters and to the availability of active gold sites in close contact with the surface defects on ceria.

Apart from the CO pollutant, volatile organic compounds (VOCs) are also points of concern in recent years due to the increasing use of volatile solvents among common items. An example of a common VOC is formaldehyde, an indoor air pollutant that is emitted from solvents used in household decorative items.

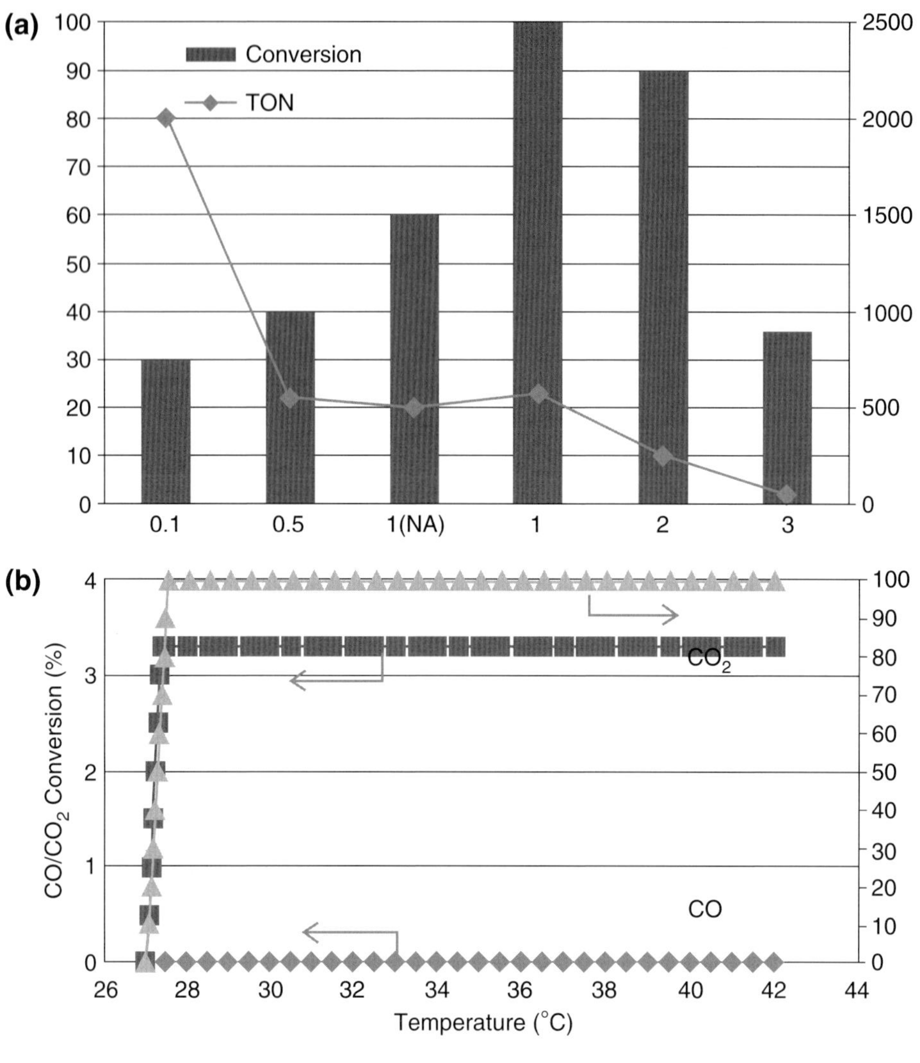

Figure 14.7 (a) CO conversion and turn over number (TON) as a function of Au loading (1-NA = 1% Au/CeO_2 without aging) (experimental conditions: 3.6% CO-21% O_2-Ar = 1,0 L/min, catalyst = 50 mg, room temperature; TON, number of molecules of product formed per gold atom), (b) CO conversion over 1% Au/CeO_2 (dp, us, and age). Experimental conditions same as Figure 14.9(a).[108]

Different kinds of catalysts have also been attempted for the aerobic oxidation of aldehydes requiring high reaction temperatures with catalysts MoO_3-SnO_2, MnO_x-CeO_2, and Ag/CeO_2.[120–125] In contrast, noble metals (Pt, Ru, and Pd) as catalysts are favorable at low reaction temperatures.[126–131] Ceria-supported gold catalysts prepared by deposition–precipitation using urea (DPU) or NaOH (DPN) have been reported as effective catalysts for formaldehyde oxidation at room temperature.[132]

Chen et al. performed a catalyst evaluation of the developed catalysts of Au/CeO_2. Two kinds of catalysts were synthesized: (1) 1 wt% Au/CeO_2 (DPU) and (2) 1wt% Au/CeO2 (DPN), where 1 wt% of gold was coated on the surface of the ceria catalyst through deposition–precipitation approach in the presence of urea (DPU) and sodium hydroxide (DPN), respectively. Figure 14.8 shows the catalytic oxidation of formaldehyde over the prepared catalysts. The Au/CeO_2 (DPU) catalyst showed excellent performance at a wide range of examined temperature. Even under a humid environment as illustrated in Figure 14.9, formaldehyde was oxidized within 3 h of reaction time.

Durability test conducted, as illustrated in Figure 14.10, shows that the catalytic activity is kept constant throughout the 60-h period. Incorporating some findings from *in situ* DRIFTS measurement, it is found that the interaction of formaldehyde with surface oxygen results in the generation of $[HCOO^-]_s$, which is the rate-limiting step for catalytic oxidation of formaldehyde over Au/CeO_2. Water also has a promoting effect on both formation and consumption of the formate reaction intermediates, resulting in a constructive effect on removal of formaldehyde under wet air environment.

In summary, gold nanoparticles supported on porous carriers have shown their unique catalytic activities on organic synthesis. It should be noted that the activity of the gold nanoparticles is not solely from its particle size, morphology, or the loading but it is also dependent on its synergistic interaction with the support. Catalyst recovery, reusability, and catalytic activity are still important factors that need to be evaluated for room temperature catalysis reactions. With these, we move a small step forward toward achieving our goal of a more sustainable chemical process.

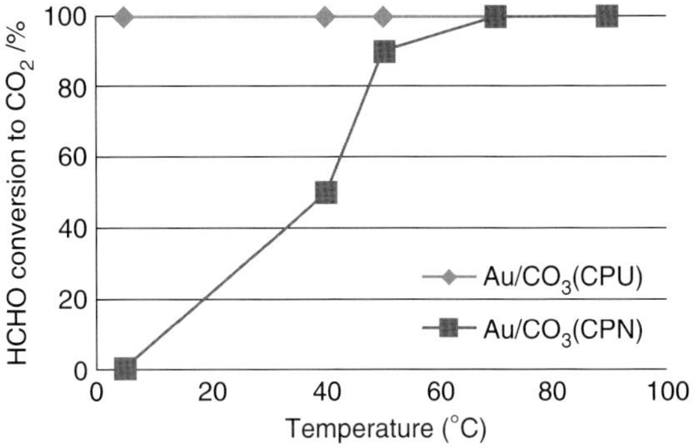

Figure 14.8 Catalytic oxidation of HCHO over 1 wt% Au/CeO2(DPU) and 1 wt% Au/CeO_2 (DPN) catalysts. Reaction conditions: 80 ppm HCHO/21% O2/N2, RH=50% (25°C); temperature: 25°C; GHSV = 34,000 h^{-1}.[132]

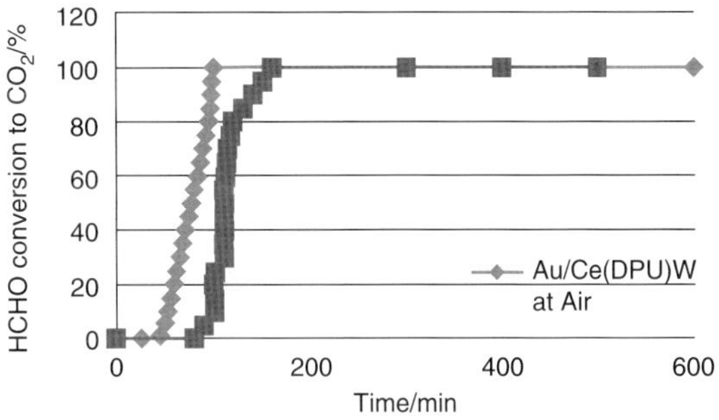

Figure 14.9 Effect of water on HCHO conversion over the 1 wt% Au/CeO_2 (DPU) catalyst. Reaction conditions: wet air: 80 ppm HCHO/21% O_2/N_2, RH = 50% (25°C); temperature: 25°C; GHSV = 34,000 h^{-1}; dry air: 80 ppm HCHO/21% O_2/N_2; temperature: 25°C; GHSV = 34,000 h^{-1}.[132]

Palladium-based Catalysts

Prior to the discovery of palladium-catalyzed cross-coupling reactions, copper was a popular choice for carbon–carbon bond formation. Palladium (Pd) has gained popularity in the 21st century when the role of Pd on C–C bond coupling was discovered. This catalysis was mainly developed by three Nobel-winning scientists in Heck, Suzuki, and Negishi. Since then, there have been plenty of scientific publications about metal-catalyzed couplings.[133–137] Old et al. reported the room temperature Suzuki couplings of aryl chlorides/bromides with Pd-based catalysts.[138,139] The general idea is to construct a rapid and high-yielding method for the production of acetylenic ketones by means of Pd-based catalysts.

A new type of C–C bond formation emerged recently about the Pd-catalyzed decarboxylative cross-coupling reaction with simple carboxylic acids as coupling reagents. The advantage of using carboxylic acids instead of traditional organometallic reagents is on its good stability, selectivity, and functional group tolerance during the reaction. Myers et al. discovered a decarboxylative Heck coupling of olefin with arene carboxylic acids catalyzed by palladium.[140–142] However, this idea has some drawbacks such as a wild reaction condition and limitations in the type of carboxylic acids that can be used. With current methods, it is still challenging to perform C–H activation at room temperature.

The work on C–H activation by Pd-catalyzed material at room temperature can be demonstrated by *ortho*-carbonylation of aniline derivatives.[143] Synthesis of functionalized aromatic rings and the development of transition metal catalysts for achieving the transformation is a challenging area to organic chemists due to the requirement of high reaction temperature and powerful oxidants. Houlden et al. reported the development of the Pd-based catalyst, known as $Pd(OTs)_2(MeCN)_2$ for the *ortho*-selective carbonylation sequence for aniline derivatives.[143] They conducted a series of studies with aryl urea derivatives, as shown in the following scheme and Table 14.9.

Based on the optimization, a range of aryl urea derivatives proceed through rapid carbonylation in the presence of Pd-based catalyst at room temperature to produce cyclic imidates with 67% yield. Changing the solvent from CH_2Cl_2 to THF/MeOH significantly increases the product yield up to 88%. The developed $Pd(OTs)_2(MeCN)_2$ in

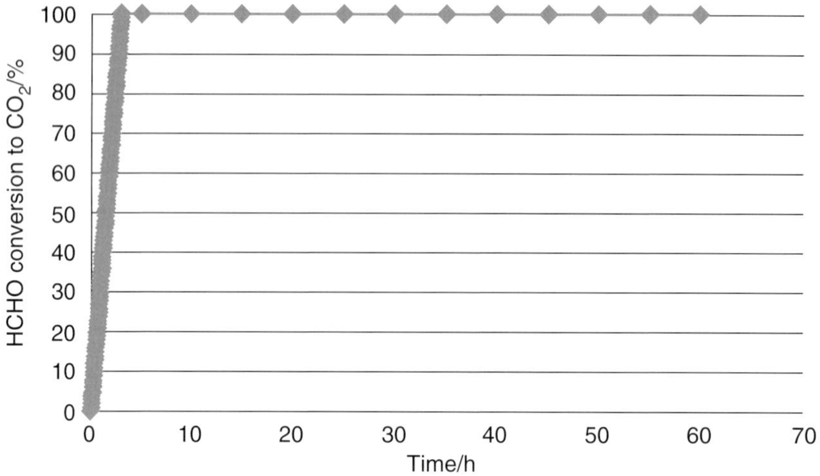

Figure 14.10 Durability test using 1 wt% Au/CeO_2 (DPU) catalyst. Reaction conditions: 80 ppm HCHO/21% O_2/N_2, RH = 50% (25°C); temperature: 25°C; GHSV = 34,000 h^{-1}.[132]

Table 14.9 Room-Temperature Palladium (II)-Catalyzed Ortho-carbonylation of Aryl Urea Derivatives[143]

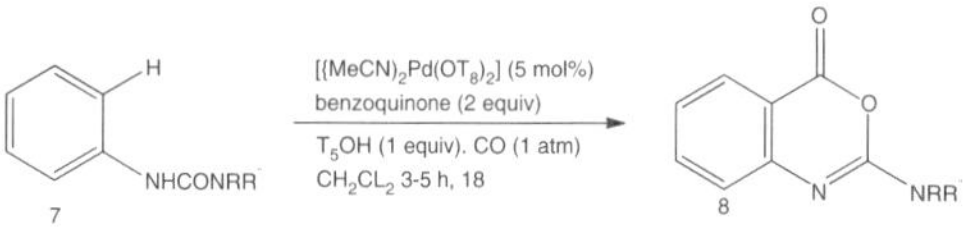

Entry[1]	7	R	R′	8	Yield (%)
1	7a	Me	Me	8a	67 (75)[14]
2	7b	Et	Et	8b	50
3	7c	*i*Pr	*i*Pr	8c	40
4	7d	Morpholine		8d	49
5	7e	Me	H	8e	34
6	7f	Et	H	8f	40
7	7g	*i*Pr	H	8g	66
8	7h	tBu	H	8h	85

[1]Reaction conducted with 10 mol% pre-catalyst.

Table 14.10 Optimization of the Reaction Conditions for the Suzuki Reaction of 4-Bromoanisole with Phenylboronic acid[144]

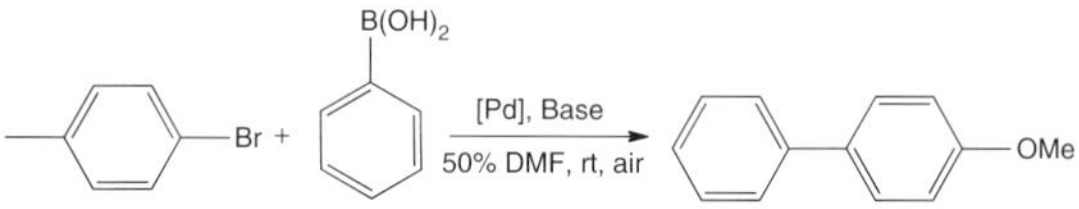

Entry[1]	[Pd]	Base	Time/min	Yield(%)[2]
1	$PdCl_2$	K_2CO_3	5	99
2[3]	$PdCl_2$	K_2CO_3	120	84
3	$Pd(OAc)_2$	K_2CO_3	5	94
4	Pd/C	K_2CO_3	5	11
5	$Pd_2(dpa)_3$	K_2CO_3	5	Trace
6	$PdCl_2$	$K_2PO_4 . 7H_2O$	5	18
7	$PdCl_2$	NaOH	5	17
8	$PdCl_2$	CH_3ONa	5	16
9	$PdCl_2$	$KF \cdot 2H_2O$	5	20

[1]Reaction conditions: 4-bromoanisole (0.5 mmol), phenylboronic acid (0.75 mmol),[Pd](0.0025 mmol), base(1 mmol), DMF-H_2O (2 mL/2mL), room temperature, in air. The reaction was monitored by TLC.
[2]Isolated yield.
[3]$PdCl_2$ (0.0005mmol,0.088mg).

this study is a pre-catalyst which activates the aniline-urea moiety.

The previous example of palladium catalysts are mediated with organic ligands. Palladium (II) chloride-based catalysts have been attempted by Liu et al. using DMF as solvent for the Suzuki reaction between arylbromides and arylboronic acids at room temperature. Suzuki reaction between 4-bromoanisole and phenylboronic acids has firstly been tried to confirm the activity of the developed catalysts, showing that 99% of product yield is obtained with K_2CO_3 as a base in 5 min of reaction duration, as shown in Table 14.10.

Different bases attempted resulted in poor product yields. Besides, only the performance of $Pd(OAc)_2$ catalyst is comparable to the designed catalysts (Entries 1 and 3). Further work was performed by testing a series of aryl bromides and arylboronic acids, and the results are shown in Table 14.11, achieving the corresponding product yields of 70–99% in general.

This work offers a vision of the utilization of ligand-free Pd-based catalysts, solving the conventional problem

Table 14.11 Suzuki Reactions of Aryl Bromides with Arylboronic Acids[144]

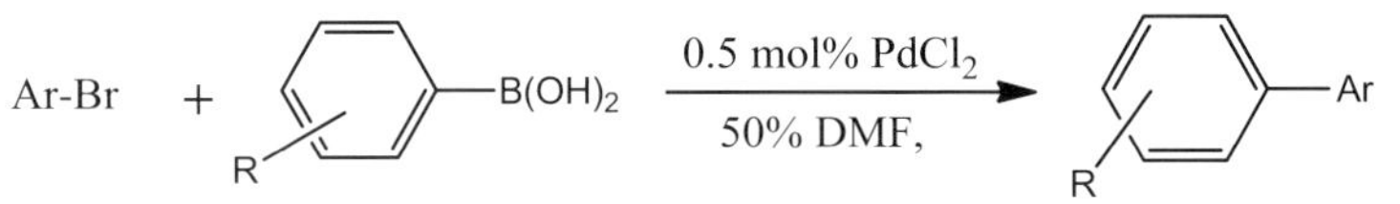

Entry[1]	Ar-Br	R	Product	Time/min	Yield (%)
1	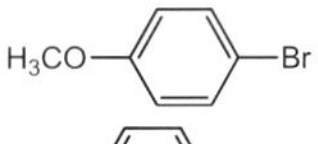	H	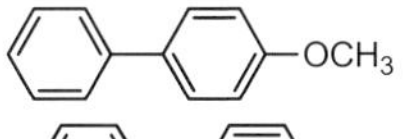	3	99
2	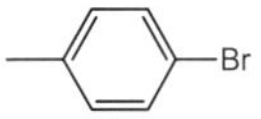	H	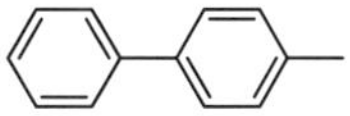	15	85
3	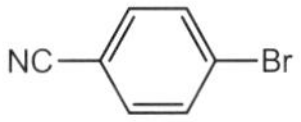	H	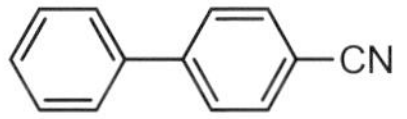	3	99
4	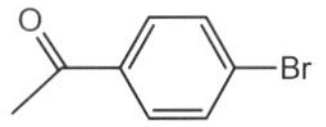	H	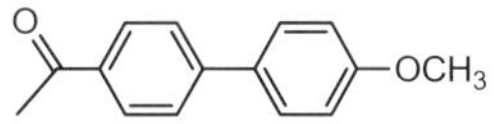	10	99
5	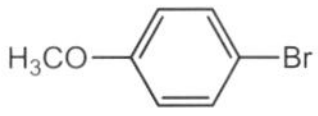	4-CH_3	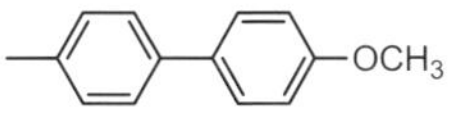	6	94
6	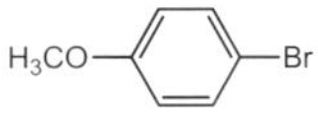	3-CH_3	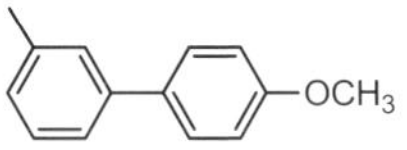	6	99
7	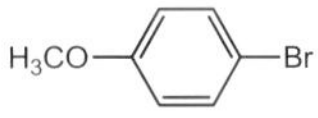	4-OCH_3	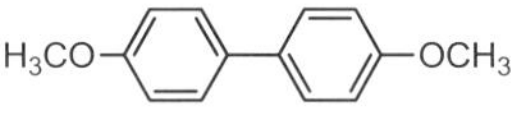	15	97
8	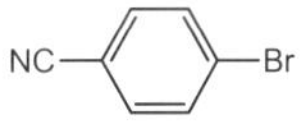	4-OCH_3	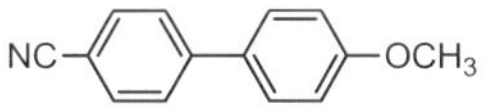	25	99
9	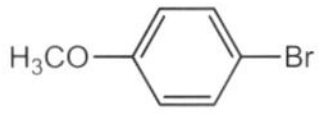	4-F	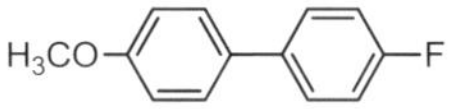	25	99
10	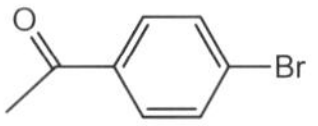	4-F	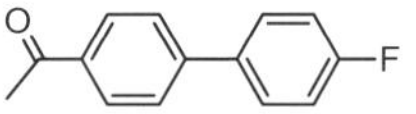	14	76
11	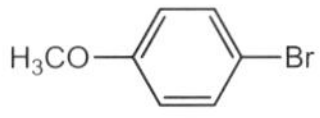	4-CHO	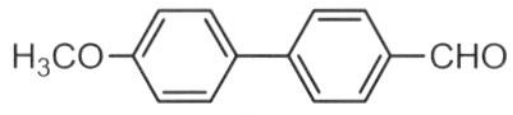	120	72
12	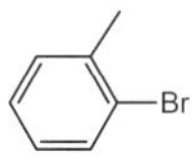	H	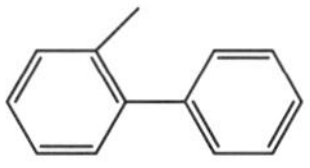	15	97
13	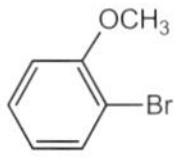	H	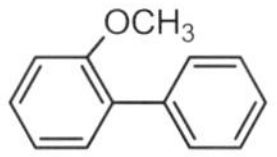	6	99
14	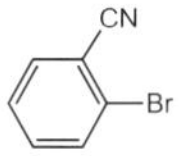	H	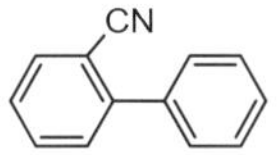	6	99
15	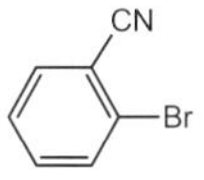	4-CH_3	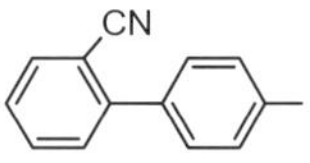	5	99
16	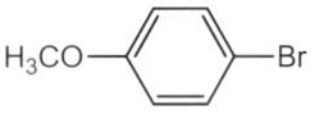	2-CH_3	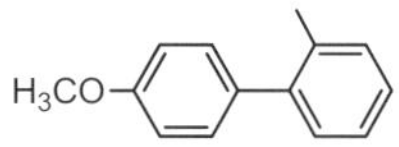	12	99
17	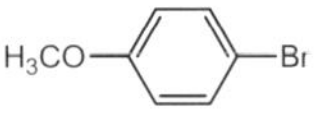	2-OCH_3	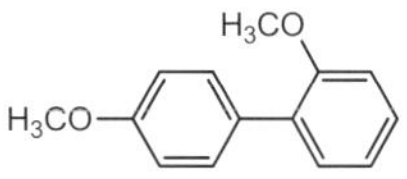	5	99
18	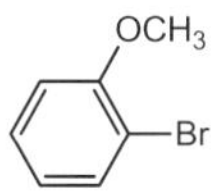	2-OCH_3	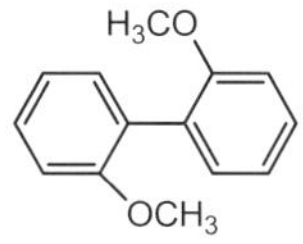	10	99

[1]Reaction conditions: aryl bromides (0.5 mmol), arylboronic acids (0.75 mmol), K_2CO_3 (1 mmol), $PdCl_2$ (0.0025 mmol, 0.44 mg), DMF H_2O (2 mL/2 mL), room temperature, in air. The reaction was monitored by TLC.

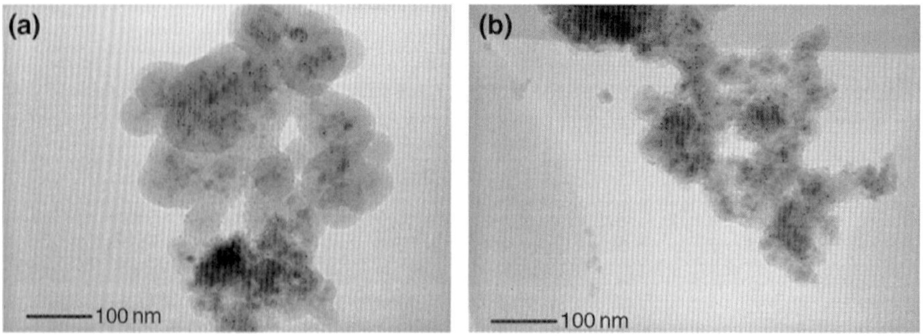

Figure 14.11 TEM images of freshly prepared Pd/IL-NH_2/SiO_2/Fe_3O_4 (a) and after being reused three times (b).[146]

to get rid of involving complicated procedures for synthesis of ligand-mediated Pd-based catalyst, and also enhancing their stability in the air/water environment.

Aside from using ligand-based or ligand-free palladium salt catalysts, palladium nanoparticles (PdNP) alone or supported on porous carrier are the considered alternatives for the room temperature Suzuki-Miyaura coupling reaction. PdNPs have been reported as catalysts for some carbon–carbon bond formation in organic synthesis; however, they mostly involved heating or reflux to obtain high product yields. Sawai et al. used palladium nanoparticles stabilized by protective ligands such as phosphine.[145] The developed chiral PdNP, having particles diameter of 1.6 nm, have been shown to efficiently catalyze asymmetric Suzuki–Miyaura cross-coupling reactions of naphthyl halides with naphthyl boronic acid at room temperature within short reaction time.

Interestingly, a catalytic system utilizing IL-modified magnetic carrier as a substrate with the deposited palladium nanoparticles has also been demonstrated.[146] Wang et al. developed the Pd catalyst by immobilizing the PdNP on amine functionalized IL on ferric oxide. The prepared catalysts were characterized by TEM, showing that the PdNPs were finely dispersed over the ferric oxide particles, as shown in Figure 14.11.

The activities of the developed catalysts have been evaluated and shown in Table 14.12.

A high product yield of around 99% was achieved within 6 h reaction time with a solvent mixture of water and ethanol using NaOH as a base (Entry 10). It is also found that the smaller the catalyst loading to the system, the longer the reaction time required for obtaining the high product yield (Entries 10–12). The optimized catalysts have been evaluated for the Suzuki coupling reaction of aryl halides and arylboronic acids as shown in Table 14.13.

Table 14.12 Optimization of the Reaction Conditions[146]

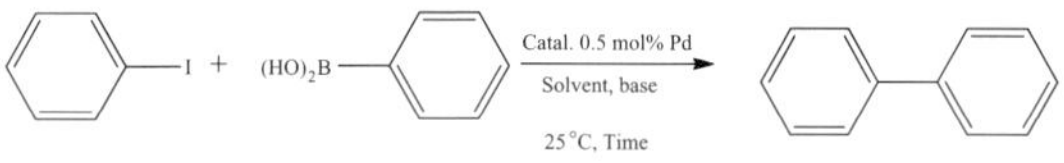

Entry[1]	Solvent	Base	Time (h)	Yield (%)[2]
1	EtOH	K_2CO_3	3	56
2	EtOH	NaOH	3	71
3	Acetone	NaOH	3	4
4	DMF	NaOH	3	Trace
5	H_2O	NaOH	3	21
6	Acetone:H_2O (1:1)	NaOH	3	34
7	EtOH:H_2O (1:1)	NaOH	3	99
8	EtOH:H_2O (1:1)	K_2CO_3	3	94
9	EtOH:H_2O (1:1)	Net_3	3	66
10	EtOH:H_2O (1:1)	NaOH	6	99[3]
11	EtOH:H_2O (1:1)	NaOH	10	99[4]
12	EtOH:H_2O (1:1)	NaOH	14	99[5]

[1]For each reaction, 0.5 mmol iodobenzene, 0.75 mmol phenylboronic acid, 1 mmol base, 19 mg Pd/IL-NH_2/SiO_2/Fe_3O_4 (0.5mol%Pd), and 3 mL solvent were used. The amount of Pd was based on iodobenzene.
[2]GC yields (internal standard: dodecane).
[3]0.2 mol% Pd used.
[4]0.1 mol% Pd used.
[5]0.05 mol% Pd used.

Both electron-rich and electron-deficient aryl iodides produced products with good yields within 4h of reaction time (Entries 1–10). Prolonged reaction time was required for the less active aryl bromides to finish the reaction (Entries 12–17). The coupling of 4-bromochlorobenzene and 4-iodochlorobenzene with arylboronic acids delivered the products with high yields, showing good selectivity (entries 4, 8, 15, and 16). Favorably, the catalysts in the reaction can be magnetically recovered from the mixture and reused without significant loss of catalytic activity.

In the case of zeolite-based Pd catalyst, the work is similar to previous examples wherein a mixture of DMF

Table 14.13 Suzuki Coupling Reactions of Aryl Halides and Arylboronic Acids[146]

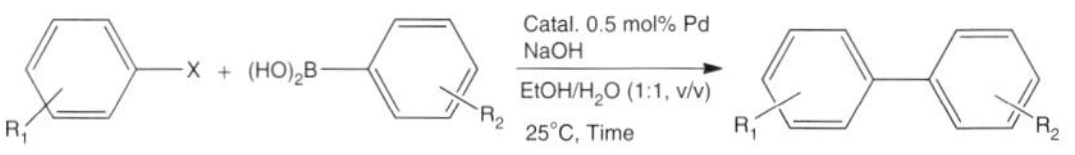

Entry[1]	R_1	X	R_2	Time (h)	Yield (%)[2]
1	H	I	H	3	96
2	*p*-Me	I	H	3.5	90
3	*p*-OMe	I	H	3.5	88
4	*p*-Cl	I	H	3	96
5	H	I	*p*-Me	3	98
6	*p*-Me	I	*p*-Me	3	98
7	*p*-OMe	I	*p*-Me	4	91
8	*p*-Cl	I	*p*-Me	3	94
9	H	I	*m*-NO_2	3	90
10	*p*-Cl	I	*m*-NO_2	3	92
11	*o*-CF_3	I	H	18	60
12	H	Br	H	5	87
13	*p*-Me	Br	H	6	82
14	*p*-Me	Br	*p*-Me	4.5	92
15	*p*-Cl	Br	H	6	86
16	*p*-Cl	Br	*p*-Me	4.5	98
17	*m*-NO_2	Br	H	10	88
18	H	Cl	H	24	3[3,4]

[1]For each resction, 0.5 mmol aryl halide, 0.75 mmol phenylboronic acid, 1 mmol NaOH, 19 mg Pd/IL-NH2/SiO_2/Fe_3O_4 (0.5 mol% Pd), and 3 mL EtOH-H_2O (1:1) were used. The amount of Pd was based on aryl halides.
[2]Isolated yields.
[3]The reaction temperature is 80°C.
[4]GC yield (internal standard dodecane).

and water was used as a solvent. While in the case of SBA15-based Pd catalysts, highly dispersed Pd catalysts were synthesized by incorporating Pd salt into pore channels of SBA15 through electrostatic interaction. The developed catalysts have been evaluated for a Suzuki coupling reaction at room temperature, achieving high product yield in results. It should be noted that controlling the supported Pd salts located in the interior of SBA15 can effectively restrain Pd agglomeration because of the pore containment effect of silica and stabilization of imidazolium salts on the inner walls.

To date, room temperature catalysis using palladium-based catalysts is mainly for Suzuki coupling, decarbonylation, C–H activation-related reactions, and Heck coupling. Palladium has a unique property in mediating with carbon–carbon bond. Some works are promising for Pd application particularly in areas such as environmental control and biofuel generation, which are the most challenges in the 21st century.

FUTURE PROSPECTS AND CONCLUSIONS

Living in the energy-demanding 21st century, energy generation, storage, as well as utilization have become the key issues to the society and also the challenges to the scientists. Conventional chemical processes, consuming large amounts of thermal energy and generating lots of byproducts and wastes, will gradually become our energy burdens in the near future. Digging a 'greener' chemical process particularly discovering new kinds of catalysts is a right track for our sustainable environment. Room temperature catalysis is critically important for the organic chemists as well as environmentalists. This short review has summarized several existing room temperature catalysts, including homogeneous and heterogeneous systems concentrating on precious metals (gold, palladium, and ruthenium) and also transition metals (iron), accordingly. The mentioned catalysts have shown their potentials and merits in specific organic synthesis by means of their unique chemical properties. Although the room temperature catalysis is still at an early stage, it will be a new chapter to the catalysis science in the 21st century.

REFERENCES

1. Arno B, Neubert P. *Applied Homogeneous Catalysis*. Weinheim: Wiley-VCH; 2012, 1.
2. Iwata T, Miki H, Fujita Y. *Ullmann's Encyclopedia of Industrial Chemistry*, vol. A19. Weinheim: VCH; 1991, 347.
3. Welton T. *Chem Rev* 1999, 99:2071.
4. Earle MJ, Seddon KR. *Pure Appl Chem* 2000, 72:1391.
5. Shamsi SA, Danielson ND. *J Sep Sci* 2007, 30:1729.
6. Bankmann D, Giernoth R. *Prog Nucl Magn Reson Spectrosc* 2007, 51:63.
7. Pandey S. *Anal Chim Acta* 2006, 556:38.
8. Koel M. *Crit Rev Anal Chem* 2005, 35:177.
9. van Rantwijk F, Sheldon RA. *Chem Rev* 2007, 107:2757.
10. Kragl U, Eckstein M, Kaftzik N. *Curr OpinBiotechnol* 2002, 13:565.
11. Hagiwara R, Lee JS. *Electrochemistry* 2007, 75:23.
12. Zhao H. *Chem Eng Commun* 2006, 193:1660.
13. Angelini G, Chiappe C, De Maria P, Fontana A, Gasparrini F, Pieraccini D, Pierini M, Siani G. *J Org Chem* 2005, 70:8193.
14. Reichardt C. *Solvents and Solvent Effects in Organic Chemistry*. 3rd ed. Weinheim: Wiley-VCH; 2003.
15. Ingold CK. *Structure and Mechanism in Organic Chemistry*. 2nd ed. London: Bell; 1969.
16. D'Anna F, Frenna V, Pace V, Noto R. *Tetrahedron* 2006, 62:1690.
17. Chiappe C, Conte V, Pieraccini D. *Eur J Org Chem* 2002:2831.
18. Byrne R, Fraser KJ, Izgorodina E, MacFarlane DR, Forsyth M, Diamond D. *Phys Chem Chem Phys* 2008, 10:5919.
19. Pegot B, van Buu ON, Gori D, Vo-Thanh G. *Beilstein J Org Chem* 2006, 2:18.

20. Gordon CM, McClusky A. *Chem Commun* 1999:1431.
21. Boon JA, Levisky JA, Pflug JL, Wilkes JS. *J Org Chem* 1986, 51:480.
22. Akbari J, Heydari A. *Tetrahedron Lett* 2009, 50:4236.
23. Kafarski P, Lejczak B. *Relat Elem* 1991, 63:1993.
24. Allen MC, Fuhrer W, Tuck B, Wade R, Wood JM. *J Med Chem* 1989, 32:1652.
25. Baylis EK, Campbell CD, Dingwall JG. *J Chem Soc PerkinTrans* 1984, 1:2845.
26. Atherton FR, Hassal CH, Lambert RW. *J Med Chem* 1986, 29:29.
27. Maier L. *Relat Elem* 1990, 53:43.
28. Emsley J, Hall D. *The Chemistry of Phosphorous*. London: Harper and Row; 1976, 494.
29. Petrov KA, Chauzov VA, Erokhina TS. *Usp Khim* 1974, 43:2045.
30. Atmani A, Combret JC, Malhiac C, Mulengi K. *J Tetrahedron Lett* 2000, 41:6045.
31. Deng Y, Shi F, Beng J, Qiao K. *J Mol Catal A* 2001, 165:33.
32. Cole AC, Jensen JL, Ntai I, Tran KLT, Weaver KJ, Forbes DC, Davis JH. *J Am Chem Soc* 2002, 124:5962.
33. Guo S, Du Z, Zhang S, Deng Y. *J Mol Catal A* 2005, 237:80.
34. Zicmanis A, Katevica S, Mekss P. *Catal Commun* 2009, 10:614.
35. Argauer RJ, Landolt GR. *US Patent 3702886*, November 14, 1972.
36. Degnan TF, Chitnis GK, Schipper PH. *Microporous Mesoporous Mater* 2000, 35–36:245.
37. Yanik SJ, Demmel EJ, Humphries AP, Campagna RJ. *Oil Gas J* 1985, 83:108.
38. Buchanan JS. *Appl Catal* 1991, 74:83.
39. Buchanan JS. *Appl Catal A* 1998, 171:57.
40. Tago T, Konno H, Sakamoto M, Nakasaka Y, Masuda T. *Appl Catal A:General* 2011, 403:183.
41. Rahimi N, Karimazdeh R. *Appl Catal A: General* 2011, 398:1.
42. Gong F, Yang Z, Hong C, Huang W, Ning S, Zhang Z, Xu Y, Li Q. *Bioresour Technol* 2011, 102:9247.
43. Furumoto Y, Harada Y, Tsunoji N, Takahashi A, Fujitani T, Ide Y, Sadakane M, Sano T. *Appl Catal A: General* 2011, 399:262.
44. Bellussi G, Pollesel P. *Stud Surf Sci Catal* 2005, 158:1201.
45. Thomas J, Degnan F. *J Catal* 2003, 216:32.
46. Marcilly CR. *Top Catal* 2000, 13:357.
47. Creyghton EJ, Downing RS. *J Mol Catal A: Chem* 1998, 134:47.
48. Rigutto MS, van Veen R, Huve L, Cejka J, van Bekkum H, Corma A, Schueth F. *Introduction to Zeolite Science and Practice*. Amsterdam: Elsevier; 2007, 855–913.
49. PEP Report 153A, Zeolite Catalysts, 1996.
50. Stepanov AG, Luzgin MV, Romannikov VN, Sidelnikov VN, Zamaraev KI. *J Catal* 1996, 164:411.
51. Germains JE. *Catalytic Conversion of Hydrocarbons*. New York: Academic Press; 1969.
52. Jacobs PA. *Carboniogenic Activity of Zeolites*. Amsterdam: Elsevier; 1977, 86.
53. Stepanov AG, Luzgin MV, Romannikov VN, Sidelnikov VN, Paukshtis EA. *J Catal* 1998, 178:466.
54. Xu J, Zheng A, Wang X, Zi G, Su J, Du J, Gan Z, Wu J, Wang W, Deng F. *Chem Sci* 2012, 3:2932.
55. Pidko EA, van Santen RA. *J Phys Chem C* 2007, 111:2643.
56. Kolyagin YG, Ivanova II, Pirogov YA. *Solid State Nucl Magn Reson* 2009, 35:104.
57. Wang XM, Qi GD, Xu J, Li BJ, Wang C, Deng F. *Angew Chem Int Ed* 2012, 51:3850.
58. Wu JF, Wang WD, Xu J, Deng F, Wang W. *Chem Eur J* 2010, 16:14016.
59. Kazansky VB, Subbotina IR, Rane N, van Santen RA, Hensen EJM. *Phys Chem Chem Phys* 2005, 7:3088.
60. Li L, Li GD, Yan C, Mu XY, Pan XL, Zou XX, Wang KX, Chen JS. *Angew Chem Int Ed* 2011, 50:8299.
61. Xia H, Sun K, Sun K, Feng Z, Li WX, Li CJ. *Phys Chem C* 2008, 112:9001.
62. Jia JF, Pillai KS, Sachtler WMH. *J Catal* 2004, 221:119.
63. Hensen EJM, Zhu Q, Janssen RAJ, Magusin P, Kooyman PJ, van Santen RA. *J Catal* 2005, 233:123.
64. Hensen EJM, Zhu Q, van Santen RA. *J Catal* 2005, 233: 136.
65. Panov GI, Uriarte AK, Rodkin MA, Sobolev VI. *Catal Today* 1998, 41:365.
66. Chemyavsky VS, Pirutko LV, Uriate AK, Kharitonov AS, Panov GI. *J Catal* 2007, 245:466.
67. Wood BR, Reimer JA, Bell AT, Janicke MT, Ott KC. *J Catal* 2004, 225:300.
68. Liu H, Li Y, Jiang H, Vargas C, Luque R. *Chem Commun* 2012, 48:8431.
69. Liu HL, Li YW, Luque R, Jiang HF. *Adv Synth Catal* 2011, 353:3107.
70. Tarakeshwar P, Lee JY, Kim KS. *J Phys Chem A*1998, 102:2253.
71. Soulé J-F, Miyamura H, Kobayashi S. *J Am Chem Soc* 2011, 133:18550.
72. Anderson JE, Davis R, Fitzgerald RN, Haberman JM. *Synth Commun* 2006, 36:2129.
73. Pineda A, Gomez L, Balu AM, Sebastian V, Ojeda M, Arruebo M, Romero AA, Santamaria J, Luque R. *Green Chem* 2013, 15:2043.
74. Mulvaney P. *Langmuir* 1996, 12:788.
75. Uematsu N, Fujii A, Hashiguchi S, Ikariya T, Noyori R. *J Am Chem So c* 1996, 118:4916.
76. Adams JP. *J Chem Soc Perkin Trans* 2000, 1:125.
77. Hoepping A, Johnson KM, George C, Flippen-Anderson J, Kozikowski AP. *J Med Chem* 2000, 43:2064.
78. Kosower EM, Radkowsky AE, Fairlamb AH, Croft SL, Nea RA. *Eur J Med Chem* 1995, 30:659.
79. Layer RW. *Chem Rev* 1963, 63:489.
80. Das B, Ravikanth B, Laxminarayana K, Rao BV. *J Mol Catal A:ChemA:Chem*2006, 253:92.
81. Iqbal AFM. *J Org Chem* 1972, 37:2791.
82. Cenini S, Ragaini F. *Catalytic Reductive Carbonylation oforganic nitro Compounds*. Dordrencht: Kluwer Academic Publishers; 1997.

83. Ragaini F, Cognolato C, Gasperini M, Cenini S. *Angew Chem Int Ed* 2003, 42:2886.
84. Santos LL, Serna P, Corma A. *Chem Eur J* 2009, 15:8196.
85. Zheng YF, Ma K, Li H, Li J, He JG, Sun X, Li R, Ma JT. *Catal Lett* 2009, 128:465.
86. Huang J, Yu L, He L, Liu Y-M, Cao Y, Fan K-N. *Green Chem* 2011, 13:2672.
87. He L, Wang LC, Sun H, Ni J, Cao Y, He HY, Fan KN. *Angew Chem Int Ed* 2009, 48:9538.
88. He L, Yu FJ, Lou XB, Cao Y, He HY, Fan KN. *Chem Commun* 2010, 46:1553.
89. Murray RW, Iyana K. *J Org Chem* 1996, 61:8099.
90. Chatterjee A, Maiti DK, Bhattacharya PK. *Org Lett* 2003, 5:3967.
91. Larock RC. *Comprehensive Organic Transformations: A Guide to Functional Group Preparation*. Weinheim: Wiley-VCH; 1999, 823.
92. Everaere K, Mortrex A, Capentier J-F. *Adv Synth Catal* 2003, 345:67.
93. Johnstone R, Wilby AH, Entwistle ID. *Chem Rev* 1985, 85:129.
94. Ikariya T, Blacker AJ. *Acc Chem Res* 2007, 40:1300.
95. Mohr C, Hofmeister H, Radnik J, Claus P. *J Am Chem Soc* 2003, 125:1905.
96. Jacobs G, Ricote S, Graham UM, Patterson PM, Davis BH. *Catal Today* 2005, 106:259.
97. He L, Ni J, Wang L-C, Yu F-J, Cao Y, He H-Y, Fan K-N. *Chem Eur J* 2009, 15:11833.
98. Karimi B, Esfahani FK. *Adv Synth Catal* 2012, 354:1319.
99. Rodriguez JA, Liu P, Takahashi Y, Nakamura K, Vines F, Illas F. *J Am Chem Soc* 2009, 131:8595.
100. Ansar S, Perera GS, Ameer FS, Zou S, Pittman CU Jr, Zhang D. *J Phys ChemC* 2013, 117:13722.
101. Ketchies WC, Fang YL, Wong MS, Murayama M, Davis RJ. *J Catal* 2007, 250:94.
102. Tsunoyama H, Sakurai H, Negishi Y, Tsukada T. *J Am Chem Soc* 2005, 127:9374.
103. Mitsudome T, Noujima A, Mikami Y, Mizugaki T, Jitsukawa K, Kaneda K. *Chem Eur J* 2010, 16:11818.
104. Corey EJ, Su WG. *J Am Chem Soc* 1987, 109:7534.
105. Kraus GA, Thomas TJ. *J Org Chem* 1988, 53:1395.
106. Johnson WS, Plummer MS, Reddy SP, Barlett WR. *J Am Chem Soc* 1993, 115:515.
107. Mitsudome T, Noujima A, Mikami Y, Mizugaki T, Jitsukawa K, Kaneda K. *Angew Chem Int Ed* 2010, 122: 5677.
108. Pillai UR, Deevi S. *Appl Catal A: General* 2006, 299:266.
109. Liu W, Flytzani-Stephanopoulos M. *J Catal* 1995, 153:304.
110. Epling WS, Hoflund GB, Weaver JF, Tsubota S, Haruta M. *J Phys Chem* 1996, 100:9929.
111. Bera P, Hegde MS. *Catal Lett* 2002, 79:75.
112. Venezia AM, Pantaleo G, Longo A, Carlo GD, Casaletto MP, Liotta FL, Deganello G. *J Phys Chem B* 2005, 109:2821.
113. Choi Y, Stenger J. *Power Sources* 2004, 129:246.
114. Whittle DM, Mirzaei AA, Hargreaves JSJ, Joyner RW, Kiely CJ, Taylor SH, Hutchings GJ. *Phys Chem Chem Phys* 2002, 4:5915.
115. Taylor SH, Hutchings GJ, Mirzaei AA. *Chem Commun* 1999:1373.
116. Wolf A, Schuth F. *Appl Catal A: General* 2002, 226:1.
117. Qiao B, Deng Y. *Chem Commun* 2003:2192.
118. Moreau F, Bond GC, Taylor AO. *Chem Commun* 2004: 1642.
119. Yan W, Chen B, Mahurin SM, Dai S, Overbury SH. *Chem Commun* 2004:1918.
120. Wen Y, Tang X, Li J, Hao J, Wei L, Tang X. *Catal Commun* 2009, 10:1157.
121. Shen WJ, Tang XF, Xu YD, Zhu HQ, Wang JG. *Appl Catal B* 2006, 62:265.
122. Imamura S, Uchihori D, Utani K, Ito T. *Catal Lett* 1994, 24:377.
123. Lochař V. *Appl Catal A* 2006, 309:33.
124. Shi C, Wang Y, Zhu A, Chen B, Au C. *Catal Commun* 2012, 28:18.
125. Shi C, Chen B-B, Li X-S, Crocker M, Wang Y, Zhu A-M. *Chem Eng J* 2012, 200–202:729.
126. Peng J, Wang S. *Appl Catal B* 2007, 73:282.
127. Leung DYC, Huang HB. *ACS Catal* 2011, 1:348.
128. He H, Zhang CB, Tanaka K. *Appl Catal B* 2006, 65:37.
129. Leung DYC, Huang HB, Ye DQ. *J Mater Chem* 2011, 21:9647.
130. Jia MJ, An NH, Yu QS, Liu G, Li SY, Zhang WX. *J. Hazard. Mater* 2011, 186:1392.
131. Shen WJ, Tang XF, Chen JL, Huang XM, Xu Y. *Appl Catal B* 2008, 81:115.
132. Chen B-B, Shi C, Crocker M, Wang Y, Zhu A-M. *Appl Catal B: Environ* 2013, 132–133:245.
133. Scientific Background on the Nobel Prize in Chemistry in 2010: Palladium-catalyzed cross couplings in organic synthesis, The Royal Swedish Academy of Sciences, Stockholm, Sweden, 6 October; 2010.
134. Shi B-S, Zhang Y-H, Lam JK, Wang D-H, Yu J-Q. *J Am Chem Soc* 2010, 132:460.
135. Chu J-H, Tsai S-L, Wu M-J. *Synthesis* 2009:3757.
136. Shi B-F, Maugel N, Zhang Y-H, Yu JQ. *Angew Chem Int Ed* 2008, 47:4882.
137. Wang D-H, Wasa M, Giri R, Yu JQ. *J Am Chem Soc* 2008, 130:7190.
138. Old DW, Wolfe JP, Buchwald SL. *J Am Chem Soc* 1998, 120:9722.
139. Cox RJ, Ritson DJ, Dane TA, Berge J, Charmant JPH, Kantacha A. *Chem Commun* 2005:1037.
140. Myers AG, Tanaka D, Mannion MR. *J Am Chem Soc* 2002, 124:11250.
141. Tanaka D, Myers AG. *Org Lett* 2004, 6:433.
142. Tanaka D, Romeril SP, Myers AG. *J Am Chem Soc* 2005, 127:10323.
143. Houlden CE, Hutchby M, Bailey CD, Gair Ford J, Tyler SNG, Gagne MR, Lloyd-Jones GC, Booker-Milburn KI. *Angew Chem Int Ed* 2009, 48:1830.
144. Liu C, Ni Q, Bao F, Qiu J. *Green Chem* 2011,13:1260.
145. Sawai K, Tatumi R, Nakahodo T, Fujihara H. *Angew Chem Int Ed* 2008, 47:6917.
146. Wang J, Xu B, Sun H, Song G. *Tetrahedron Lett* 2013:238.

15

Co-Firing of Biomass with Coal in Thermal Power Plants: Technology Schemes, Impacts, and Future Perspectives

Emmanouil Karampinis,[1,2]* Panagiotis Grammelis,[1] Michalis Agraniotis,[1] Ioannis Violidakis[1] and Emmanuel Kakaras[1,2]

[1]Centre for Research and Technology Hellas, Chemical Process and Energy Resources Institute, Athens, Greece
[2]Laboratory of Steam Boilers and Thermal Plants, Department of Mechanical Engineering, National Technical University of Athens, Athens, Greece

INTRODUCTION

Co-firing is defined as the simultaneous combustion of two or more fuels in the same energy plant in order to produce one or more energy carriers.[1] The most common application of co-firing involves the partial replacement of coal in large-scale, pulverized fuel (PF) power plants by solid biomass fuels. Recent reviews of co-firing experiences identify more than 100 successful field demonstrations in 16 countries that use essentially every major type of biomass (herbaceous, woody, animal wastes, and wastes) combined with essentially every rank of coal and combusted in essentially every major type of PF boiler (tangential, wall, and cyclone fired)[2,3] (Table 15.1).

The main advantage of co-firing is its potential to mitigate the CO_2 emissions of the coal sector at a very low cost and short implementation time[4,5] compared with other technologies. The reported investment costs for a co-firing retrofit range[6,7] from 50 to 300 USD/kW$_e$ to less than 1000 USD/kW$_e$ of biomass capacity, compared with 2500–5000 USD/kW$_e$ for dedicated biomass power plants.[8] This is comparable to other lower cost RES-e, such as 2500 USD/kW$_e$ for hydropower and 1000 USD/kW$_e$ for onshore wind power. However, being a thermal process, co-firing, as well as dedicated biomass combustion, can produce power on demand, unlike solar or wind energy. This contributes to the stability of electric grids and accelerates the capital investment payoff rate by utilizing higher capacity factors.[1] Overall, the total electricity cost for co-firing is comparable to that of coal, with any higher prices being attributed mostly to the higher cost of the biomass fuel. Apart from the above, co-firing is currently the biomass conversion technology with the highest electrical efficiency. Dedicated biomass power plants have a typical report efficiency[9] of only 25%, whereas conventional, subcritical coal-fired power plants in OECD countries operate at efficiencies[10] around 36%, with state of the art units[11] reaching or exceeding 43%. Given the negligible impact of co-firing on the generating efficiency of a coal plant, these are essentially the same efficiencies in which co-firing power plants operate.[6,7]

Despite these and other advantages, co-firing is not universally adopted. Issues related to policies, biomass markets, and sustainability schemes are the first consideration for the conversion of a coal-fired power plant to co-firing; variations in the support schemes that are required for

*Correspondence to: karampinis@certh.gr

Advances in Bioenergy: The Sustainability Challenge, First Edition. Edited by Peter D. Lund, John Byrne, Göran Berndes and Iacovos A. Vasalos.

Originally published in: *WIREs Energy Environ* 2014, 3: 233–248 doi: 10.1002/wene.100

Table 15.1 Number of Coal-Fired Power Plants That Have Co-Firing Experience per Country and Combustion Technology, as of 2009

	Combustion Technology				
	Fluidized Beds (BFB, CFB)	Grate	Pulverized Fuel	Not Reported	Total
Austria	3	1	1		5
Belgium			3		3
Denmark	1	4	4		9
Finland	58	4	10	6	78
Germany		1	4	22	27
Italy			6	1	7
Netherlands			6		6
Poland	1		9		10
Spain	1			1	2
Sweden	10	2	3		15
UK	2		19		21
EU-27	76	12	65	30	183
Australia			8		8
Canada			7		7
Indonesia	2				2
Norway	1				1
Taiwan	1				1
Thailand	1				1
USA	6	5	29		40
Total	87	17	109	30	243

CFB, circulating fluidized bed.
Data from Ref 3.

large-scale biomass co-firing in different EU countries have led or are expected to lead to different market developments and co-firing deployments.[12] Boxes 15.1 and 15.2 provide a further discussion on some of these issues.

Once the decision for the adoption of co-firing has been taken and the boundary conditions resulting from policies and markets have been set, it is mostly technical and environment issues that affect the retrofit decisions, such as the type of biomass fuel to be used, the co-firing share, and the co-firing scheme. Several technical and environmental challenges have been identified,[2,9] which affect different aspects of the operation of a coal-fired power plant and involve impacts on the fuel handling, fuel conversion, slagging/fouling and corrosion, emissions formation and gas cleaning equipment, and, finally, ash utilization. This article focuses on the current experiences and recent progress relevant to the technical and environmental issues. First, an overview of the available technical solutions in implementing co-firing schemes to a coal-fired boiler will be considered. Then, each technical and environmental impact category will be examined separately. Finally, conclusions and future outlooks based on the technical direction of the co-firing sector will be presented.

CO-FIRING TECHNOLOGY OPTIONS

In all co-firing cases, the idea is that biomass provides a thermal input that replaces an equal amount of thermal input from the main fuel, coal. The technology solution adopted for this replacement depends on several parameters, such as the power plant type, the fuels employed, the biomass thermal share, the desired complexity, and economics. Generally, as the process complexity and investment cost rise, the biomass thermal share can be increased and more 'problematic' biomass fuels can be co-fired. The co-firing technology options can be classified into three broad categories[13]:

- Direct co-firing.
- Parallel co-firing.
- Indirect/Gasification co-firing.

Direct co-firing is the most common and economic solution; however, it poses several limitations on the range of fuels and thermal shares. Parallel and indirect co-firing schemes are more suitable for biomass fuels containing problematic compounds or when the ash quality is of importance for subsequent sale or disposal.[14,15] The different co-firing technology options are presented schematically in Figure 15.1.[16]

Direct Co-Firing

Direct co-firing is the co-combustion of coal and biomass in the same boiler. This is the simplest solution to adopt co-firing, because the required additional installations are kept to a minimum. Different technology schemes are available for the direct co-firing option, which differ mostly on the conditioning of biomass prior to combustion and the entry point of the biomass fuel in the boiler. The basic options[17,18] for coal-fired power plants are presented in Figure 15.2.[19]

The first option (route 1 in Figure 15.2) involves the treatment of biomass in the same milling equipment used for coal and its injection in the furnace through the existing coal burners. This approach is called the co-milling option and is the simplest method for co-firing, involving the lowest capital costs. However, as the combustion conditions are not optimized, the flame stability and heat transfer characteristics,[20] along with the possibility for increased slagging/fouling and corrosion propensity, are highest. In addition, there is a higher risk of biomass interfering with the coal feeding and milling system, e.g., by causing blockages or by increasing the particle size distribution. Thus, co-milling is applied only when the biomass thermal share is low and the biomass fuel is not particularly challenging, e.g., clean wood pellets.

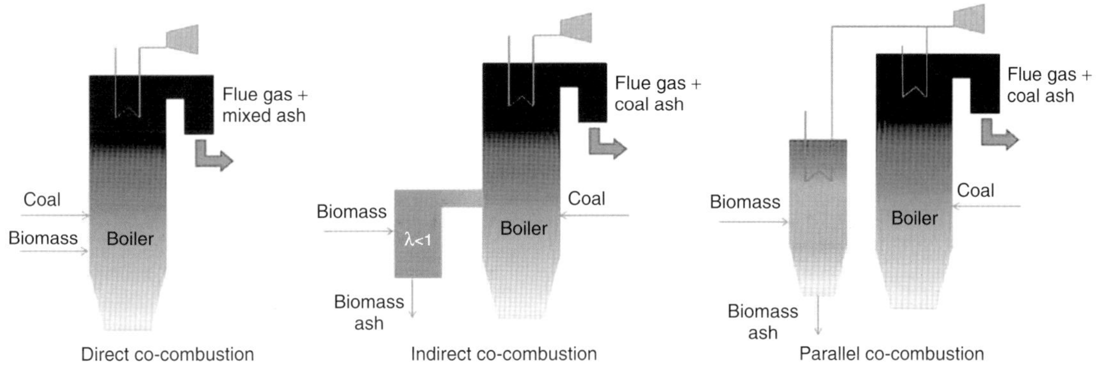

Figure 15.1 Schematic presentation of co-firing technology options.

In the second option (route 2 in Figure 15.2), biomass is pretreated through a separate system, which includes new, dedicated mills or modified coal mills. The entry of the biomass fuel in the coal furnace is still achieved through the existing coal burners. This setup requires modifications that are external to the boiler, while still allowing for a better control of the biomass and coal particle sizes. On the other hand, the installation of biofuel transport pipes requires space that may be limited around the boiler block, while the control of the burner operating characteristics is still not optimal. The Studstrupværket power plant in Aarhus, Denmark co-fires straw with coal using this feeding concept; its layout is presented in Figure 15.3.[21]

The third option (route 3 in Figure 15.2) expands upon the previous one by installing not only a dedicated biomass feeding and milling system but also dedicated biomass burners. This solution requires higher capital investment and involves a more significant technological risk, because the modifications also affect the boiler itself. However, the degree of control that can be achieved on combustion conditions is greater and the risk to boiler operation through the impact of biomass is minimized. Unit 8 of the Amer power plant of Essen is an example of this co-firing option.

The final option is the use of the biomass fuel in a specially designed reburn system located in the upper furnace with the aim of controlling NO_x emissions. The system is in the development phase and only results from small-case tests are available.[22] It is expected that the capital cost of this system will be higher than other co-firing options.

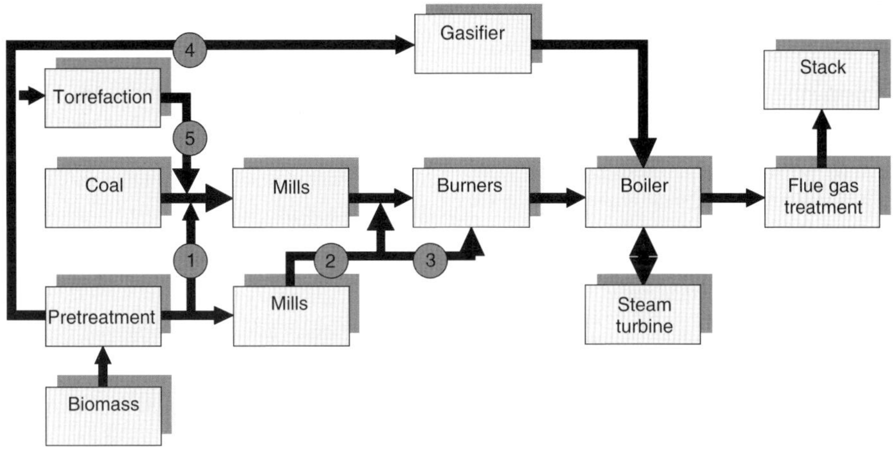

Figure 15.2 Direct and indirect/gasification co-firing technology options, including the possibility of using torrefied biomass. (Reproduced with permission from Ref 19. Copyright 2009, DNV KEMA)

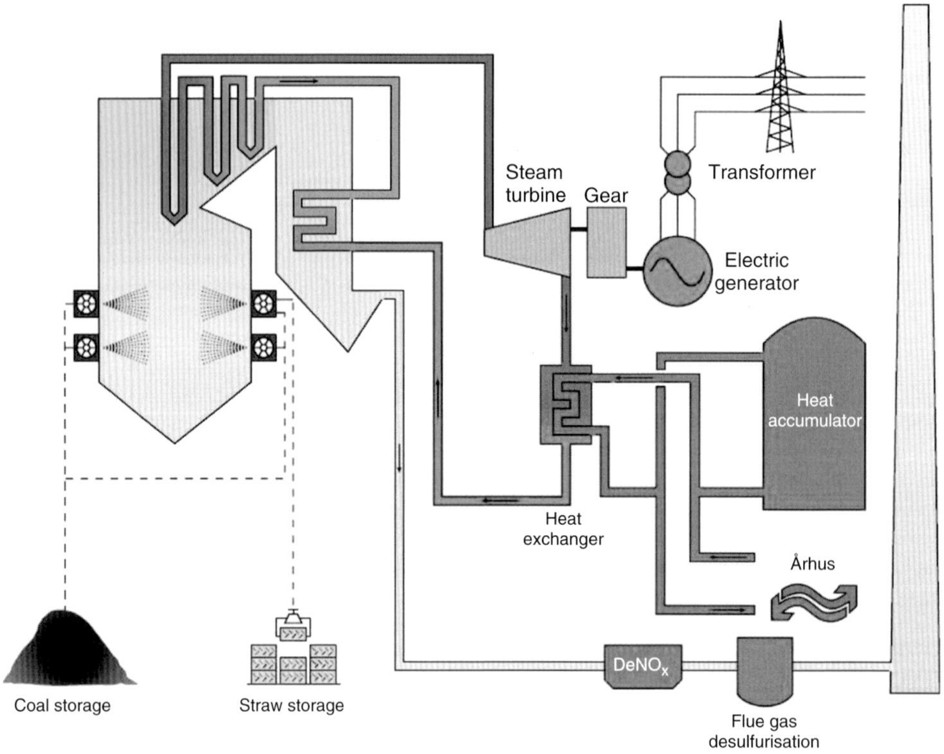

Figure 15.3 The Studstrupværket power plant, including separate handling lines for coal and straw and one type of burner. (Reproduced with permission from Ref 21. Copyright 2007, DONG Energy)

In most cases, the biomass thermal input in direct co-firing schemes is around 10% owing to technical and economic restraints.

Parallel Co-Firing

Parallel co-firing involves the combustion of biomass in a separate, distinct biomass boiler. The integration of the coal and biomass systems takes place in the steam cycle. Typically, the biomass boiler produces low-grade steam, which is upgraded in the coal boiler; thus, the efficiencies that can be achieved are higher than a stand-alone biomass power plant, while the operation of the coal boiler is not affected by the addition of a biomass fuel. Combustion residues are also produced separately and can be handled as per the residues of single fuel combustion. Furthermore, although the investment cost is higher than that of the direct co-firing option, it is still lower than a dedicated biomass power plant, because the system design is developed to utilize already existing components of the coal power plant.

The Avedøre Unit 2 (Figure 15.4), which provides district heating for the Copenhagen metropolitan area, is an example of a power plant utilizing parallel co-firing. Avedøre 2 uses a new multifuel concept; while the main, ultra supercritical 800 MW boiler is operating using a mixture of wood pellets, heavy fuel oil, and natural gas, a supplementary steam is also produced by a 105 MW straw boiler. The high sophistication of the Avedøre 2 design makes it one of the most efficient plants in the world that utilize biomass, with a reported total efficiency of 92% and electrical efficiency of 42%, rising to 49% if no district heating is provided.[23]

Indirect/Gasification Co-Firing

The indirect, also known as gasification, co-firing option involves gasification of the biomass fuel in a separate gasifier; the combustion of the product gas takes place in the furnace of the coal boiler (route 4 in Figure 15.2). As in the case of parallel co-firing, indirect co-firing reduces the impact of the biomass fuel in the coal furnace and allows

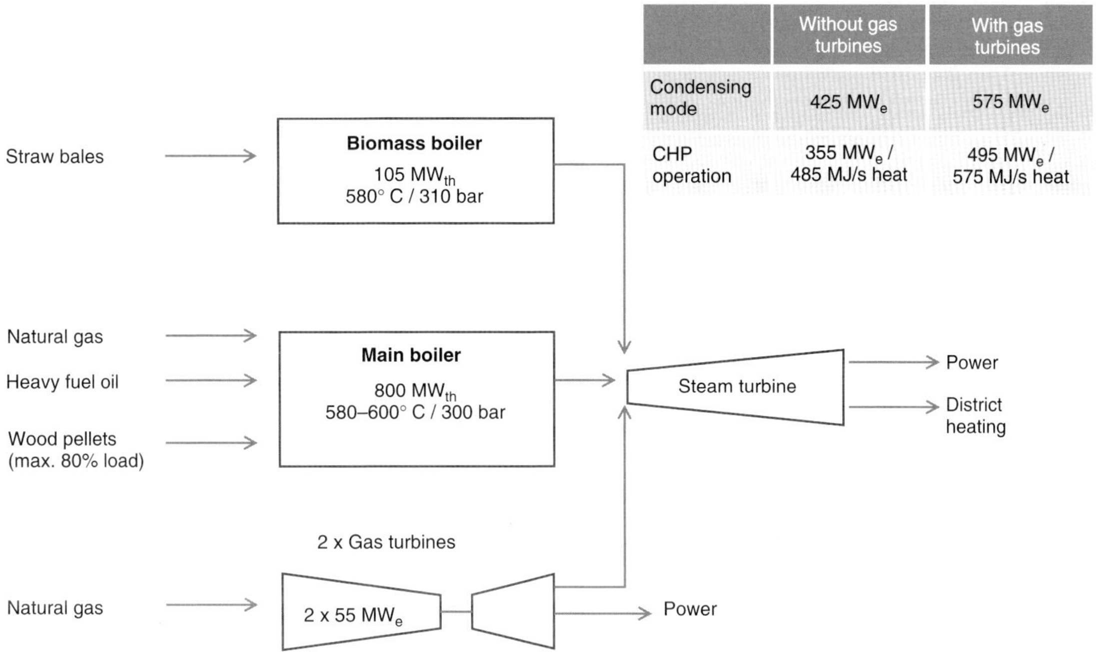

Figure 15.4 The Avedore Unit 2 concept of parallel co-firing.

for a separate collection of residues. In addition, the inject of the gasification gas in the furnace can be used as a gas over-firing for NO_x minimization. Finally, as the gas can be injected in many cases directly in the furnace, the plant can avoid the energy conversion losses and expensive flue gas cleaning, which are common issues in biomass gasification. However, depending on the biomass fuel quality and the presence of elements such as chlorine and alkalis, flue gas cleaning may be required.[13] The capital cost expenditure is likewise high, owing to the need for installing a new gasifier.

Foster Wheeler has developed an atmospheric circulating fluidized bed (CFB) gasifier suitable for indirect co-firing that has been installed in the Kymijärvi power plant of Lahden Lämpövoima Oy at Lahti, Finland and in the Ruien power plant of GDF Suez, Belgium.[24] In both cases, the raw, hot gas is fed directly in the main coal furnace; the Lahti gasifier has a successful record of operation with increasing quantities of 'difficult' fuels, such as in-origin classified waste fuel (REF), railway sleepers, and shredded tyres. Other indirect co-firing installations include the Enstedværket power plant in Aabenraa, Denmark; the Amergas power station in Geertruidenberg, The Netherlands; and the Mälarenergi CHP plant in Västerås City, Sweden. The Pyroneer gasification concept of DONG Energy has also been successfully tested on a 6-MW demonstration unit.[25]

BIOMASS HANDLING, STORAGE, AND PRETREATMENT

Large-scale combustion plants require significant bulk storage, handling, and pretreatment of fuels. For solid biomass co-firing power plants, these processes are severely affected by the low energy density of biomass and its physical and biochemical properties, such as its fibrous nature and potential to decompose. Such issues typically do not occur with coals. As a result, the reliable and trouble-free operation of the biomass handling system is one of the prime objectives for co-firing retrofits.

The setup of the biomass handling and pretreatment system depends heavily on the physical properties of the fuel; generally, two types of systems can be distinguished: those that handle pellets or granular materials and systems developed for baled biomass.

Straw and other herbaceous biomass materials are usually delivered at the co-firing plant in the form of large square bales with typical dimensions of 1.2 m × 1.3 m × 2.4 m and a weight of 500 kg. Such bales are generally quite dry, with less than 20% moisture content, and are less susceptible to biological degradation. In addition, they can be easily stacked and stored indoors for long time periods. Such systems can be highly automated; straw bales are handled by overhead cranes and fed automatically to conveyor belts, where they are debaled, cleaned of impurities, and then fed to a hammer mill for the required size reduction before they are conveyed in the burner.

Granular, such as wood chips and chunks, and pelletized materials are the most common form of co-fired biomass fuels. Wood chips and pellets are both free flowing, although the former may exhibit a wider range of particle sizes, moisture contents, and elasticities, which can cause blockages issues in the feeding system.

Storage for such materials is usually done in piles. For wood chips, materials with moisture content higher than 20% may result in increased decomposition, loss of dry matter, and heating of the pile to the point of self-ignition. Wood pellets exhibit very low moisture contents as a result of their production process, but require roofed storage in order to avoid physical and biological decomposition. A combination of temperature and gas measurements can offer the proper control of the biomass storage.[26]

The generation and accumulation of dust is major issue, especially when handling pellets in bulk.[27] The proper equipment for dust suppression includes explosion vents and fire suppression systems. The removal of oversize, ferrous and other impurities is another important step before feeding the biomass. A reception hopper grid can help with the removal of large materials, while a moving screen or a cyclone can remove the smaller ones[28]; ferrous materials are separated by magnets.

For all types of biomass materials, granular, pellets, or baled, the milling of the fuel is considered as the most important pretreatment step for co-firing applications. The reason is that PF units, which are the most common type of co-firing plant, require a particle size distribution of 50–90 μm for coal particles. Because, as will be discussed in the next section, PF furnaces can achieve sufficient conversion of larger biomass particles, it is unfeasible and unnecessary to reduce biomass to the same size or shape as coal. Still, the size requirements for biomass particles are in excess of the typical requirements of the technologies most often used for dedicated biomass combustion, namely grate firing and fluidized beds.

The 'co-milling' option, where biomass is mixed with the coal stream prior to the mill system, is as discussed before the simplest approach as the only retrofits necessary for co-firing are the biomass handling and storage system. The operation of most coal mills depends on the coal particles being subject to brittle fracture, which does not apply to fibrous biomass materials. Moreover, there is a tendency for the biomass particles to accumulate in the mill, during normal operation, and to take longer to clear from the mill during shutdown. In addition, the top size product tends to increase, while milling of wet biomass is expected to impact the mill heat balance. The increase of the mill power consumption by 10–15% has also been reported in older plant modification projects,[29] where no appropriate handling of the supplementary fuel was assured. Finally, as biomass tends to release volatile combustibles at low temperatures of 180°C, the contact of dry biomass with hot primary air in the mill should be avoided, especially during intermittent mill operation.[30]

Because of the impact of biomass on the performance of the coal mill, co-milling is limited to low biomass shares. For an increase in the biomass thermal share, the use of dedicated milling equipment is recommended. In several cases, retrofits of coal mills for the milling of wood pellets have been performed with overall satisfactory results. However, experience shows that the heat input from the converted mills is significantly derated in the range[30] of 50–70%. In addition, retrofitted coal mills tend to break the pellets back to their original size distribution, while similar safety precautions as in the case of co-milling have to be adopted. For cases where a smaller particle size is required, it is recommended to install hammer mills specifically developed for biomass fuels.

FUEL CONVERSION

Because of the high energy consumption and mill wear and their fibrous nature, the milling of biomass particles typically results in large, nonspherical, elongated particles with typical lengths of 1–3 mm and aspect ratios exceeding 6.[31] PF boilers are typically designed to operate with very small, nearly spherical, coal particle sizes, in the range of 50–100 μm, with aspect ratios of no more than 2. However, biomass typically contains a much higher volatile content compared with coal and yields 90–95% of its inorganic-free mass during devolatilization, compared with 55–60% for most coals.[31] Biomass volatiles are released much more rapidly compared with coal and at lower temperatures, as demonstrated by thermogravimetric analyses experiments,[32] thus reducing the ignition temperature compared with coal and promoting flame stability. In addition, biomass chars tend to be highly porous and very reactive under the conditions found in PF furnaces, while the shape of biomass particles promotes more rapid combustion than the typically spherical shape of coal particles.[33–35] Overall, despite the nonfavorable sizing, fuel conversion can be maintained in high levels, especially when the biomass thermal share is low.

Still, biomass particles that are too large or too dense may be found in the bottom ash with minimal conversion.[2,36] This can be the case when the combustion of a big biomass particle is delayed and the aerodynamic forces on the particle may not be sufficient to sustain its weight.[34] In addition, the conversion efficiency can be affected by the biomass moisture content, which may be higher than that of coal and thus affect the maximum achievable boiler load and efficiency.[7]

Computational fluid dynamics (CFD) is becoming an increasingly important tool in accessing the impact of biomass co-firing in the operation of burners and boilers. To achieve good modeling results, the importance of using appropriate submodels for the description of the behavior of biomass particles has been stressed out by several authors. The submodels that need to be especially considered include:

- The biomass devolatilization kinetics, which are of increased importance owing to the high volatile content of biomass fuels. Typically, CFD submodels implement single pyrolysis reactions, which do not take into account variable feedstocks and variations in temperatures and heating ranges. As devolatilization gains in importance with the increase of the biomass share, the use of more sophisticated models such as multicomponent, competitive schemes can provide deeper insights.[37]
- The isothermal behavior typically used for coal particles is no longer valid and the internal heat gradient needs to be considered. The use of an isothermal model for a large particle has been shown to lead to a delayed and shortened gas release to the surrounding gas phase,[38] while experimental results and modeling evidence suggest that the surface of the particle reaches the temperature at which devolatilization starts before its core.[31]
- Finally, the effect of the nonsphericity of the biomass particles has to be considered in the evaluation of the drag force and heat transfer equations, as these affect the residence time and burnout of the biomass particles.[34]

Generally though, the impact of co-firing on the conversion efficiency is minimal and, if negative, is manageable,[7] given a proper biomass particle size and dryness. According to a scenario taken from the literature,[35] assuming a woody biomass fuel with moisture content less than 10% co-fired at a 5% share on a mass basis with a high volatile bituminous coal with 12% moisture and 15.7% ash content, the calculated loss of efficiency was in the range of 0.1%. Other publications consider a 1% loss of electrical efficiency for direct co-firing,[13] taking also into account the impact of slagging/fouling on the heat transfer.

DEPOSITS AND CORROSION

A common problem associated with solid fuels energy applications is the gradual reduction of the heat transfer rates owing to the deposits accumulated on the heat exchanging equipment. Deposits are typically classified using the terms slagging and fouling. Slagging refers to deposits formed on sections of the boiler exposed mainly to radiant heat, such as the furnace walls, whereas fouling refers to deposits formed by condensation on the convective pass, such as the heat exchanger tubes. Increases in slagging and fouling reduce heat transfer and facilitate the initiation of corrosive reactions. Corrosion refers to the deterioration of intrinsic properties of the wall material caused by complex reactions involving gas-phase species, deposits, or their interaction. Although slagging and fouling problems affect the overall efficiency and the availability of the equipment, soot blowing (for superheater tubes) and cleaning during plant shutdowns can remove deposits. Corrosion, on the other hand, is permanent and severely affects the lifetime of the equipment.

Slagging, fouling, and corrosion are complex phenomena depending on the chemical and mineralogical composition of the fuel ash as well as on conditions (temperature and velocity distribution, reducing or oxidizing atmosphere, and many others) in the furnace.[36] For PF boilers, slagging propensity is typically evaluated through a series of empirical indices, which are calculated either through the measured ash fusion temperatures (slagging index) or through the chemical composition of the main ash elements (the base-to-acid ratio). Fouling indices for coal depend mostly on the sodium content.[9] The applicability of these indices to biomass fuels has been verified only for specific cases.[37,38] In co-firing cases, the ash composition of the mixture depends mainly on the fossil fuel part, owing to its larger ash content compared with biomass fuels. Therefore, the indices can still be considered useful for the estimation of the deposition tendency of such mixed fuels when the biomass thermal share is low,[23] typically[38] below 20%.

Biomass ashes, especially those of herbaceous materials or fast-growing tree species, are characterized by higher levels of alkalis and chlorine compared with most coal types. Part of the alkali content of the biomass ash is released in the gas phase through the formation of chlorides or hydroxides. Chlorine in particular facilitates the vaporization of the alkali metals more strongly than the alkali concentration itself.[39] Thus, during combustion, part of the alkalis remain in the fuel ash in the form of silicates or sulfates, and end up in the coarser fly ash particles, which are mainly constituted of refractory species, such as Ca, Mg, and Si. The alkali silicates in ash have low melting temperatures, sometimes less than[39] 700°C, which increase the slagging propensity. The gas-phase alkalis, mostly alkali chlorides, either escape aerosols with the stack gas or condense on the superheater tubes or on coarser ash particles and increase the fouling propensity.

Corrosion phenomena occur when the protective oxide layer that is formed on tube walls is attacked by chlorine- or sulfur-containing compounds. The sulfidation and chloridation of the tube surfaces result in the formation of an outer layer that does not have the protective properties of the oxidized one. Its defective structure means that it can be scaled off easily owing to erosion and thus become subject to further corrosion.[40] Corrosion can take place either through gas-phase reactions of compounds such as Cl_2 and $NaCl_{(g)}$ with the metal surface or through solid- and molten-phase reactions with sulfates and chlorides. For coal combustion, corrosion is usually caused by the molten-phase reactions[41,42] with iron alkali trisulfates, $(Na,K)_3Fe(SO_4)_3$, whereas for biomass combustion the effect of chlorides is domineering.[41]

Experience in Denmark with dedicated combustion of straw in stoker boilers and co-firing of coal and straw in CFB boilers resulted in severe fouling and accelerated superheater corrosion (increased by a factor of 5–40), predominantly as selective chlorine corrosion. Co-firing of coal and straw in a converted PF boiler gave no significant problems with fouling and only a moderate increased rate of superheater corrosion (50%), although it should be noted that the straw thermal share was less than 10%.[42]

Corrosion is mainly associated with not only[43] alkali (Na and K) and Cl but also Pb and Zn. Fouling and corrosion phenomena during biomass combustion or co-firing can be mitigated by capturing alkalis from their chlorides before deposition. This can be achieved through the reaction with sulfur or aluminum silicate compounds,[44] as follows:

$$2MCl + SO_2 + \tfrac{1}{2}O_2 + H_2O \rightarrow MK_2SO_4 + 2HCl$$

$$Al_2O_3 \cdot 2SiO_2 + 2MCl + H_2O \rightarrow M_2O \cdot Al_2O_3 \cdot 2SiO_2 + 2HCl,$$

where M can be potassium (K) or sodium (Na). Thus, mitigation of the fouling and corrosion propensity of problematic biomass fuels can be achieved through co-firing with coal,[39,45–51] sewage sludge,[44,52–54] or even by the addition of coal ash.[52] Figure 15.5 illustrates the protective role of sulfur and aluminum silicates during the co-firing of coal with a high alkali biomass.[55] It should be noted that corrosion may still take place through gas-phase reactions with HCl[39] or with sulfur deposits, although these are considered less severe than corrosion from deposition of alkali chlorides. The use of chemical additives with Al-Si-, Ca-, S-, and P-containing substances has also been widely studied[50,56–61]; however, this practice is more expensive and the certain results have been reported to be site-specific.[62]

The S/Cl or the Al-Si/Cl molar ratios are widely used as an indication of the chlorine-influenced deposition and corrosion propensity of a fuel or fuel mixture. S/Cl ratios of less than 2 are typically indicative of high corrosion risk,[63,64] whereas noncorrosive behavior is expected when the S/Cl ratio exceeds 4. It should be noted that, contrary to the use of additives, the complex chemical composition of fuel mixtures may affect the action of the corrosion-fighting elements. For example, the retention of sulfur by calcium might reduce the availability of sulfur for the creation of alkali sulfates.[62]

Generally, deposition and corrosion phenomena in solid fuels combustion are very complex phenomena, not yet fully understood. Based on the current operational experience, co-firing at low thermal shares of even 'problematic' fuels, e.g., containing high concentrations of alkalis and chloride, results in manageable deposits or corrosion and may actually be considered an improvement over dedicated biomass combustion. Tools are currently being developed for the prediction of the deposition propensity and corrosive behavior for fuel mixtures, while significant progress has been made on monitoring and on-line cleaning systems for PF boilers.[65]

EMISSIONS

One of the main reasons for the implementation of co-firing in coal-fired power plants is the reduction of CO_2 emissions from the combustion of carbon neutral biomass. Pollutant formation and other gaseous emissions during biomass co-firing exhibit all the complexities as do the same issues for coal combustion[66–69]; major pollutants include SO_x, NO_x, CO, organic compounds (including dioxins), and dust particles. This section examines the impact of biomass co-firing on the environmental impact of power plants, as well as the performance of standard flue gas cleaning equipment, compared to that of dedicated coal combustion.

The potential for reduction of SO_x emissions is widely considered as a benefit of biomass co-firing. Generally, the decrease is higher as the biomass co-firing share increases, as the sulfur content of most biomass fuels is lower compared to that of most coal types.[70] An additional incremental reduction beyond the amount anticipated on the basis of fuel sulfur content is sometimes observed, owing to retention of sulfur in coal by alkali-based compounds in biomass ashes. Lower SO_x emissions in co-firing also affect the operation of FGD units and reduce the limestone requirements and the gypsum production rate.[9]

The NO_x emissions during co-firing present a more complicated picture and reports indicate the potential for emissions both less and higher than those from dedicated coal combustion.[9,31] Generally though, the case is that

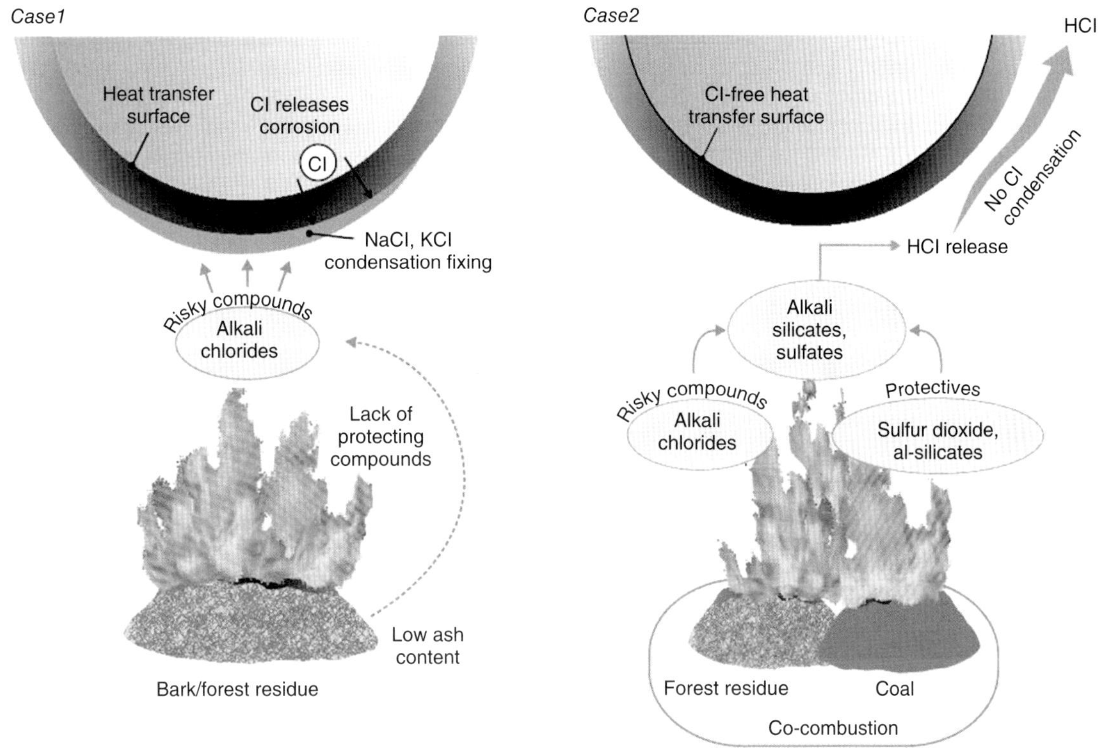

Figure 15.5 Avoidance of formation of alkali chlorides from dedicated biomass combustion through co-firing with coal containing sulfur and aluminum silicates. (Reproduced with permission from Ref 22. Copyright 2003, VTT)

co-firing reduces the level of NO_x emissions over those of coal combustion[34,71–78] because of the low fuel nitrogen content of many biomass types, especially wood, as well as because fuel nitrogen in biomass is mostly transformed to NH_3, which has a lower conversion to NO compared to HCN, the main product of coal fuel nitrogen.[79] In some cases, the NO_x reduction in co-firing has been reported as higher than the one theoretically calculated by the amount of total fuel nitrogen that enters the furnace.[18,34]

As previously mentioned, biomass fuels typically produce higher volatile yields than coals, thus creating larger fuel-rich regions compared with coal in the near-burner regions, which promote NO_x reduction reactions.[32] However, special care must be provided in the design of the air supply of the burner system such that sufficient oxygen is provided for the combustion of the fast-releasing volatile species and delays are avoided in the combustion.[80] This is especially true for the case of both tangentially fired and wall-fired boilers equipped with low NO_x burners, which are already initially designed to operate at stoichiometric or substoichiometric conditions at the primary combustion zone, while part of the secondary air is diverted to an overfire air system.[32]

There is also potential of utilizing biomass as a reburn fuel for additional NO_x reduction.[9] Biomass fuels also commonly contain more moisture than coal, decreasing peak temperatures and leading to commensurate decreases in NO_x.[2]

However, there are cases of biomass fuels, such as straw, that have quite high fuel nitrogen content. In these cases, it is suggested to evaluate the impact of co-firing on NO_x emissions through CFD modeling and results from pilot and large-scale demonstrations.

An increasing number of power plants are installing (SCR) $DeNO_x$ systems in order to comply with increasing stricter emissions limits. Contrary to (SNCR) systems, the catalyst of SCR systems is susceptible to poisoning from volatile inorganic compounds, such as alkalis and phosphates, which are expected to be present in increased amounts in the flue gases from co-firing plants, especially in the cases where herbaceous biomass species are used.[80–82] Apart from the biomass type, the impact on the catalyst depends on the location of the SCR unit in the flue gas treatment system. If the unit is placed between the FGD unit and the chimney, the flue gas contains lower amounts of damaging components, which could damage the catalyst. However, most SCR units are installed in the high-dust environment before the air preheater, because the higher temperatures in this region allow the use of more efficient zeolite and vanadium or titanium dioxide catalysts[81]; as a result, the potential impact of inorganic volatiles in the flue gas is higher. Overall, the issue of the co-firing impact on SCR operation is not yet fully resolved and requires further research.

The emissions of CO and organic pollutants depend mostly on the quality of the combustion process, which is discussed in a previous section. Given an adequate design of the combustion system, the emission levels of these species should be in the same level as that of coal firing. The higher chlorine content of some, mostly herbaceous, biomass may however lead to higher levels of HCl emissions.[9] On the other hand, the high temperature levels of the typical PF boilers indicate that the dioxins are thermally degraded in the furnace and are not emitted by the power plant.[82]

The performance of the electrostatic precipitators and the particulate emissions level depends strongly on the chemical properties of the fly ash. As a result, a major concern in co-firing applications is that the capture efficiency of the biomass ash particles will be lower than that of the coal particles, whereas the emissions level of submicron particles will increase. This has been confirmed by the operational experience of a number of plants, which report a slight increase in particulate emissions when co-firing with wood.[83] In some cases, the effectiveness of the flue gas treatment systems, including the (ESPs), might be affected owing to the increase in flue gas volume during co-firing. On the other hand, the low ash levels of solid biofuels relieve the ESP performance. Other plants, however, report no effect of co-firing on the ESP performance, although most operate with lower co-firing shares. As with other impacts of co-firing, higher thermal shares are expected to increase the impacts on the ESP performance.

Overall, co-firing can be said to have a positive or neutral impact on the environment compared with coal firing. There are still areas of uncertainties mostly related to the issue of SCR catalyst deactivation and the impact on ESP performance, which require further research and feedback from operational experience.

ASH UTILIZATION FROM CO-FIRING APPLICATIONS

Ash is the main by-product of thermal power plants and its utilization is an important part of their environmental performance. However, the issue of ash utilization is often postponed after the start-up of an installation owing to the difficulties in predicting the exact ash properties as well as the ever-present option of ashes landfilling.[84]

Utilizing routes for ashes for coal-fired power plants is well established; for example, the pozzolanic behavior of most coal fly ashes renders it a valuable additive for concrete production, while another important application is in cement and cement products. Biomass ash utilization is not as advanced, although, depending on the fuel properties, there are several options available, such as recycling in the ground as a natural fertilizer.[85,86] Co-firing ashes pose a unique challenge because the chemical properties of the combustion by-product fall in between those of the pure ashes and may prove to be out of the specifications for their utilization routes. As the amount of ash in coal is higher than that of most typical biomass types, the goal is to attain co-firing ash properties that make it possible to maintain the existing coal ash utilization routes. This poses an important limitation to the maximum co-firing ratio, along with the factors discussed in the previous paragraphs.[84]

Bottom ash forms typically 10–15% of the total ash produced in pulverized coal furnaces; its most common form of utilization for bottom ashes is in road construction and other infrastructural works, where it is used in the embankment and foundations. The key requirement in these applications is whether contamination through leaching of unwanted elements stays below the legal limits. In practice, most bottom ashes from coal-fired plants, with or without co-firing of biomass, can be used with little or no restrictions. Sometimes the ashes need aging (exposure to ambient air) so that they adsorb carbon dioxide and become less basic, which diminishes leaching of certain elements.[84]

Fly ashes form the bulk of the total ash. As previously mentioned, the main utilization route for coal fly ashes is as an additive for concrete production. Standard EN-450 sets requirements that coal fly ashes have to meet for this application; for co-firing fly ashes the standard imposes a series of direct and indirect limitations, which relate to the list of allowed biomass fuels, the maximum co-firing share and the content of chemical compounds, such as free CaO, phosphorus, and alkalis, as well as the unburnt matter.[87] For example, according to the current draft of the new version of EN-450, the mass-based co-firing share should be below 40 or 50% if the biomass fuel is green wood, whereas the share of biomass ash in the total ash should be below 30%. This is an increase over the limiting values in the 2005 version of the standard, which allows for power plants to combust higher shares of biomass without resorting to the more expensive landfilling option for noncompliant ashes; however, the limits cannot be changed beyond the point where concrete quality is jeopardized.[84]

Future expectations for increased biomass thermal shares or retrofits to 100% biomass combustion will increase the quantities of noncompliant ashes produced and will also raise the issue of identifying novel utilization options. Several routes have been suggested, including trace metal recovery, filler in metals and polymers, and phosphor production[87]; however, most are not economically attractive. A recent overview[29] indicates that production of light-weight aggregates is closest to economic success.

FUTURE PERSPECTIVE OF CO-FIRING

In recent years, two main trends related to direct biomass co-firing have been identified. The first one involves the utilization of torrefied biomass as a co-firing fuel (route 5 in Figure 15.2). Torrefaction is a thermochemical process conducted at 200–300°C and with a typical residence time of 1 h, during which biomass partially decomposes giving off volatiles and producing the remaining solid as a final product.[75,88] The breakdown of the hemicellulose matrix during torrefaction leads to significant benefits, such as increases of energy density, lower energy consumption during grinding, development of hydrophobic properties, resistance to biological degradation, and more uniform fuel properties.[88–94] Apart from economic benefits to the supply chain, torrefied biomass can be easily used in the co-milling option, thus decreasing the investment costs for co-firing, while its thermal properties are more accommodating to the conditions prevalent in coal-fired boilers. Thus, higher biomass thermal shares can be achieved with a minimum of investment costs and the concept of the retrofit of a pulverized coal boiler to 100% torrefied biomass combustion has been formulated.[95] However, large-scale production of torrefied biomass has not yet been implemented, while there are still uncertainties regarding the emissions and quality of by-products.

The second trend is an expansion of the use of dedicated biomass feeding and milling up to the point where a pulverized coal power plant is converted to 100% biomass combustion. The rationale behind this trend is tied to RES support policies that disregard co-firing in favor of dedicated biomass combustion.[12] An example of a unit retrofitted to 100% biomass combustion is Rodenhuize Unit 4 of GDF Suez. Co-firing at the Rodenhuize power plant has been implemented at a series of successive steps, starting from installation of transport, storage, handling, and milling infrastructure and the conversion to co-firing of a single burner row in 2005, to the Advanced Green Project, where two coal burner rows were switched to wood pellet firing and finally to the Max Green project, which included a complete switch from coal combustion to wood pellet, burner replacement, as well as other retrofitting options such as the installation of a catalytic DeNOx unit.[96–101] Although some issues of unburnt particles remain owing to the relatively small boiler size, the emissions of the unit are able to meet the strict requirements of local legislation.

Box 15.1 NONTECHNICAL ISSUES OF BIOMASS CO-FIRING: MARKETS, SUSTAINABILITY SCHEMES, AND POLICIES

The decision of a coal plant operator to switch to co-firing is affected by several nontechnical parameters. For example, on the positive side and unlike biomass power plants, most co-firing plants have the option to revert to dedicated coal combustion and thus mitigate the effect of biomass fuel shortages. Alternatively, if the biomass fuel utilized is low cost, the co-firing plant can be shielded against price increases and shortages of coal. In addition, co-firing can bring about a number of economic and social benefits, on different levels, such as company, e.g., higher revenue through RES support schemes, regional, e.g., job and income creation through a local biomass supply chain, or country, e.g., reduction of energy imports. On the other hand, the large scale of most coal-fired power plants means that a switch to co-firing, even with low biomass thermal shares, requires very large quantities of biomass; this raises several issues, such as competition for raw materials between the energy sector and the wood or food industry to concerns on the sustainability of long-distance biomass supply chains. Widespread adoption of co-firing might also be considered as prolonging the lifetime of coal-fired power plants, strengthening the dependence on the coal sector and reducing the market share of other bioenergy options. As a result, the level of support that co-firing receives by policy makers varies widely from country to country. For a recent discussion on these issues relevant to the EU, see Ref 12.

Box 15.2 CO-FIRING BIOMASS WITH BROWN COAL

Brown coal, or lignite, contributes to 4% of the global electricity production compared with 37% for hard coal and is the major fossil fuel in several EU countries such as Germany and Greece. However, co-firing of brown coal with lignite is not as widespread as in the case of hard coal. Biomass is typically a more expensive fuel compared with hard coal and for the economic operation of a co-firing power plant, the difference is usually bridged through the CO_2 savings and the financial support received for the production of electricity from biomass. Although brown coal is producing more CO_2 per unit of energy compared with hard coal, the low cost of brown coal is placing a significant restraint and is a major reason why demonstrations of co-firing with lignite have been performed mostly with alternative fuels with very low or even negative cost, such as RDF and sewage sludge. On the other hand, while most biomass types have a lower heating value than hard coal, the co-firing of lignite with biomass has the potential to improve combustion conditions in the furnace. In addition, the large size of brown coal furnaces provides sufficient residence times for the combustion of even large biomass particles; both of these trends can have a positive impact on fuel conversion. The normal potential benefits and negative impacts of biomass co-firing are also relevant for lignite-fired power plants; in particular, because of the high ash content of lignite compared with most types of biomass, issues of slagging/fouling and ash utilization are expected to be dominated by the lignite ash composition and the impact of biomass addition is less pronounced.

CONCLUSION

Co-firing biomass with coal is considered the most attractive short- and long-term solution to greenhouse gas emissions from power generation. The effectiveness of the co-firing application, combined with its low cost and low technical risk, classifies it as high on a priority list of the large-scale power plant operators and institutions considering mitigation of CO_2 emissions and saving of diminishing fossil fuels. Different co-firing schemes have been developed for coal-fired power plants. Direct co-firing is the most common approach owing to its low investment cost; however it is reserved mostly for nonproblematic, woody biomass fuels and is typically applied for biomass thermal shares in the range of 10%. Examples of indirect and parallel co-firing plants exist and have been demonstrated to reach high efficiencies and good operating conditions even with more problematic biomass fuels.

Biomass fuel handling poses challenges owing to the biological and fibrous nature of biomass; solutions can be drawn by the operating experiences of large-scale dedicated biomass power plants, while milling requirements are somewhat relaxed by the relatively high conversion efficiencies of biomass particles in PF boilers. Slagging/fouling and corrosion, especially in the case of herbaceous biomass fuels, is a major issue of biomass combustion. Co-firing such fuels at low thermal shares with coals can mitigate some of the effects of chlorine-induced corrosion and deposition through the retention of alkali by sulfates or Al-Si compounds. Co-firing has been demonstrated to have mostly a positive effect on the pollutant emissions, especially in the case of SOx and NOx. Uncertainties still exist regarding the issue of DeNOx catalyst deactivation and ESP performance impact. Because of the lower biomass ash content, co-firing fly ash can in most cases follow the normal coal ash utilization routes, although challenges can be expected when biomass fuels high in alkalis or phosphorus are combusted and when the co-firing share is increased. The need for identifying novel and economic utilization schemes is apparent.

A general guideline would be that co-firing is easily implemented and poses few technical issues when woody fuels are combusted at thermal shares up to 20%; for straw and herbaceous biomass fuels the typical thermal share for which no problems are observed is 10%.[16] Increasing competition with other bioenergy systems, as well as nonenergy uses of wood, is expected to push the co-firing sector in the direction of increased utilization of herbaceous and other 'problematic' biomass sources. Along with the developing trends of co-firing torrefied biomass and conversion of coal-fired power plants to increasingly higher biomass thermal shares the need for further (RTD) in biomass co-firing is apparent.

ACKNOWLEDGMENTS

The authors gratefully acknowledge the financial support of the National Strategic Reference Framework (NSRF) 2007–2013 through the National Cooperation Program 2009/ENER-BIO project (Energetic utilization of solid and liquid biofuels for the power sector)/Grant Nr. 09-32-596.

REFERENCES

1. Hansson J, Berndes G, Johnsson F, Kjärstad J. Co-firing biomass with coal for electricity generation—an assessment of the potential in EU 27. *Energy Policy* 2009, 37:1444–1455.
2. Baxter L. Biomass cofiring overview. In: 2nd World Conference and Exhibition on Biomass for Energy, Industry and Climate Protection, Rome, Italy, 2004.
3. International Intelligence Agency (IEA). Database of biomass cofiring initiatives, 2009. Available at: http://www.ieabcc.nl/database/cofiring.php. (Accessed September 14, 2012).
4. Baxter L. Biomass-coal co-combustion: opportunity for affordable renewable energy. *Fuel* 2005, 84:1295–1302.
5. Johnsson F, Berndes G, Berggren M. Cost competitive bioenergy: linking lignocellulosic biomass supply with co-firing for electricity in Poland. In: World Bioenergy

Conference and Exhibition, Jönköping, Sweden, 30 May–1 June, 2006.

6. Baxter LL, Rumminger M, Lind T, Tillman D, Hughes E. Cofiring biomass in coal boilers: pilot- and utility-scale experiences. In: Biomass for Energy and Industry: 1st World Conference and Technology Exhibition, Seville, Spain, 2000.
7. Tillman DA, Hughes E, Gold BA. Cofiring of biofuels in coal fired boilers: results of case study analysis. In: 1st Biomass Conference of the Americas, Vermont, USA, 1994.
8. Organisation for Economic Co-operation and Development/International Energy Agency (OECD/IEA). Energy Technology Perspectives—Scenarios & Strategies to 2050, Paris, France, 2006.
9. Van Loo S, Koppejan J. *Handbook of Biomass Combustion and Co-firing*. London: Earthscan; 2008.
10. Wicks R, Keay M. Can coal contribute to sustainable development? *Energy Environ* 2005, 16:767–779.
11. Beér J. High efficiency electric power generation: the environmental role. *Prog Energy Combust Sci* 2007, 33:107–134.
12. Karampinis E, Grammelis P, Kakaras E. Current status and future of co-firing in five EU countries: effect of support schemes, sustainability criteria and biomass markets. In: Proceedings of the 19th European Biomass Conference, Berlin, Germany, 2011, 2517–2522.
13. Basu P, Butler J, Leon MA. Biomass co-firing options on the emission reduction and electricity generation costs in coal-fired power plants. *Renew Energy* 2011, 36:282–288.
14. Rohan F. Experience of indirect co-firing of biomass and coal. In: CCC64—IEA Coal Research, October 2002, pp. 39.
15. Grammelis P, Agraniotis M, Kakaras E. *Co-Utilization of Biomass Based Fuels in Pulverized Coal Power Plants in Europe*. Weinheim: Wiley-VCH Verlag GmbH & Co. KGaA; 2010.
16. Spliethoff H. Biomass co-firing applications: state of the art and research requirements. In: Proceedings: TOTeM 35, Co-Firing Secondary Fuels in Power Generation, September 2010.
17. Livingston WR. Development of low cost systems for co-utilisation of biomass in large power plants, Mid-term review report, 2003.
18. Tillman DA. Biomass cofiring: the technology, the experience, the combustion consequences. *Biomass Bioenergy* 2000, 19:365–384.
19. Cremers MFG. IEA Bioenergy Task 32—Technical Status of Biomass Co-Firing, KEMA, Arnhem, the Netherlands, 11 August, 2009.
20. Dai J, Sokhansanj S, Grace JR, Bi X, Jim Lim C, Melin S. Overview and some issues related to co-firing biomass and coal. *Can J Chem Eng* 2008, 86:367e86.
21. Sander B, Skøtt T. Bioenergy for electricity and heat, DONG Energy, 2007.
22. Veijonen K, Vainikka P, Järvinen T, Alakangas E. Biomass co-firing—an efficient way to reduce greenhouse gas emissions. In: EUBIONET—European Bioenergy Networks, Finland, 2003.
23. Sørensen HD. Avedøre unit 2—the world's largest biomass fuelled CHP plant. In: IEA Clean Coal Centre Workshop on Cofiring Biomass with Coal, Drax Power Station, UK, 25–26 January, 2011.
24. Palonen J, Anttikoski T, Eriksson T. The Foster Wheeler gasification technology for bio-fuels: refuse-derived fuel (RDF) power generation, Foster Wheeler, 2010.
25. Moeller M. Demonstration of a new gasification technology for indirect co-firing of difficult biomass with coal. In: IEA Clean Coal Centre, 2nd Workshop on Cofiring Biomass with Coal, Copenhagen, Denmark, 2012. Available at: http://www.iea-coal.org.uk/site/2010/conferences/2nd-workshop-on-cofiring-biomass-with-coal? (Accessed September 14, 2012).
26. Meijer R, Gast CH. Spontaneous combustion of biomass: experimental study into guidelines to avoid and control this phenomena. In: Proceedings of the 2nd World Conference on Biomass for Energy, Industry and Climate Protection, 10–14 May 2004, Volume II, Rome, Italy, pp 1231–1233.
27. Livingston WR. A review of the recent experience in British with the co-firing of biomass with coal in large pulverized coal-fired boilers. In: IEA Exco Workshop on Biomass Co-firing, Copenhagen 25 May 2005. Available at: http://www.ieabioenergy.com/publications/ws02-co-utilisation-of-biomasswith-fossil-fuels-exco55/. (Accessed September 14, 2012).
28. DTI. Co-firing of biomass at UK Power Plant, Carbon Abatement Technologies, Programme BPB009, August 2005.
29. European Commission Directorate-General for Energy and Transport. Opportunities and markets for co-utilisation of biomass and waste with fossil fuels for power generation, 2010. Available at: http://ec.europa.eu/energy/renewables/studies/doc/bioenergy/2000_opportunities_and_markets.pdf. (Accessed September 14, 2012).
30. Livingston WR. Recent developments in biomass co-firing in large pulverised coal boilers. In: IEA Clean Coal Centre, 2nd Workshop on Cofiring Biomass with Coal, Copenhagen, Denmark, 2012. Available at: http://www.iea-coal.org.uk/site/2010/conferences/2nd-workshop-on-cofiring-biomass-with-coal? (Accessed September 14, 2012).
31. Lu H, Baxter L. Biomass combustion characteristics and implications for renewable energy. In: Grammelis P, ed. *Solid Biofuels for Energy—A Lower Greenhouse Gas Alternative*. London: Springer-Verlag; 2011, 95–121.
32. Laux S, Grusha J, Tillman D. Co-firing of biomass and opportunity fuels in low NOx burners. Available at: http://www.fwc.com/publications/tech_papers/powgen/pdfs/clrw_bio.pdf. (Accessed May 21, 2013).
33. Sjostrom K, Chen G, Yu Q, Brage C, Rosen C. Promoted reactivity of char in co-gasification of biomass and coal: synergies in the thermochemical process. *Fuel* 1999, 78:1189–1194.
34. Gera D, Mathur MP, Freeman MC, Robinson A. Effect of large aspect ratio of biomass particles on carbon burnout in a utility boiler. *Energy Fuels* 2002, 16:1523–1532.

35. Lu H, Scott J, Echols K, Foster P, Ripa B, Farr R, Baxter LL. Effects of particle shape and size on black liquor and biomass reactivity. In: Science in Thermal and Chemical Biomass Conversion, Vol. 1, CPL Press, Newbury, Berkshire, UK, 2004.
36. Karampinis E, Nikolopoulos N, Nikolopoulos A, Grammelis P, Kakaras E. Numerical investigation Greek lignite/cardoon co-firing in a tangentially fired furnace. *Appl Energy* 2012, 97:514–524.
37. Blondeau J, Jeanmart H. From coal to biomass combustion: how to adapt CFD simulations? *VGB Powertech* 2012, 4:81–85.
38. Blondeau J, Jeanmart H. Biomass pyrolysis in pulverized-fuel boilers: derivation of apparent kinetic parameters for inclusion in CFD codes. *Proc Combust Inst* 2011, 33:1787–1794.
39. Mitsui Babcock. IPA Energy consulting, The economics of co-firing, DTI, Final report URN 06/1959, July 2006.
40. Żelkowski J. *Kohleverbrennung—Brennstoff, Physik und Theorie, Technik (Coal Combustion—Fuel, Physics and Theory)*, vol. 3. Essen: VGB Kraftwerkstechnik; 1986, 320–364.
41. Pronobis M. Evaluation of the influence of biomass co-combustion on boiler furnace slagging by means of fusibility correlations. *Biomass Bioenergy* 2005, 28:375–383.
42. Carpenter AM. *Switching to Cheaper Coals for Power Generation*. London: IEA Coal Research; 1998.
43. Khan AA, de Jong W, Jansens PJ, Spliethoff H. Biomass combustion in fluidized bed boilers: potential problems and remedies. *Fuel Process Technol* 2009, 90:21–50.
44. Prabir B. *Combustion and Gasification in Fluidized Beds*. CRC Press, Taylor & Francis Group; 2006.
45. Harb NJ, Smith EE. Fireside corrosion in PC-fired boilers. *Prog Energy Combust Sci* 1990, 16:169–190.
46. Hansen PFB, Andersen KH, Wieck-Hansen K, Overgaard P, Rasmussen I, Frandsen FJ, Hansen LA, Dam-Johansen K. Co-firing straw and coal in a 150-MW e utility boiler: in situ measurements. *Fuel Process Technol* 1998, 54:207–225.
47. Nielsen HP, Frandsen FJ, Dam-Johansen K, Baxter LL. The implications of chlorine-associated corrosion on the operation of biomass-fired boilers. *Prog Energy Combust Sci* 2000, 26:283–298.
48. Aho M, Yrjas P, Taipale R, Hupa M, Silvennoinen J. Reduction of superheater corrosion by co-firing risky biomass with sewage sludge. *Fuel* 2010, 89:2376–2386.
49. Dayton DC, Belle-Oudry D, Nordin A. Effect of coal minerals on chlorine and alkali metals released during biomass/coal cofiring. *Energy Fuels* 1999, 13:1203–1211.
50. Simms NJ, Kilgallon PJ, Oakey JE. Degradation of heat exchanger materials under biomass co-firing conditions. *Mater High Temp* 2007, 24:333–342.
51. Montgomery M, Vilhelmsen T, Jensen SA. Potential high temperature corrosion problems due to co-firing of biomass and fossil fuels. *Mater Corros* 2008, 59:783–793.
52. Henderson P, Szakalos P, Pettersson R, Anderson C, Hogberg J. Reducing superheater corrosion in wood-fired boilers. *Mater Corros* 2006, 57:128–134.
53. Simms NJ, Kilgallon PJ, Oakey JE. Fireside issues in advanced power generation systems. *Energy Mater: Mater Sci Eng Energy Syst* 2007, 2:154–160.
54. Aho M, Ferrer E. Importance of coal ash composition in protecting the boiler against chlorine deposition during combustion of chlorine-rich biomass. *Fuel* 2005, 84:201–212.
55. Ferrer E, Aho M, Silvennoinen J, Nurminen RV. Fluidized bed combustion of refuse-derived fuel in presence of protective coal ash. *Fuel Process Technol* 2005, 87:33–44.
56. Coda B, Aho M, Berger R, Hein KRG. Behavior of chlorine and enrichment of risky elements in bubbling fluidized bed combustion of biomass and waste assisted by additives. *Energy Fuels* 2001, 15:680–690.
57. Aho M, Silvennoinen J. Preventing chlorine deposition on heat transfer surfaces with aluminium–silicon rich biomass residue and additive. *Fuel* 2004, 83:1299–1305.
58. Johansson L, Leckner B, Tullin C, Åmand L, Davidsson K. Properties of particles in the fly ash of a biofuel-fired circulating fluidized bed (CFB) boiler. *Energy Fuels* 2008, 22:3005–3015.
59. Veijonen K, Vainikka P, Järvinen T, Alakangas E. Biomass co-firing—an efficient way to reduce greenhouse gas emissions. VTT Processes, Jyväskylä, 2003.
60. Punjak WA, Uberoi M, Shadman F. High-temperature adsorption of alkali vapors on solid sorbents. *AIChE J* 1989, 35:1186–1194.
61. Turn SQ, Kinoshita CM, Ishimura DM, Zhou J, Hiraki TT, Masutani SM. An experimental investigation of alkali removal from biomass producer gas using a fixed bed of solid sorbent. *Ind Eng Chem Res* 2001, 40:1960–1967.
62. Aho M. Reduction of chlorine deposition in FB boilers with aluminium-containing additives. *Fuel* 2001, 80:1943–1951.
63. Iisa K, Salmenoja YL. Sulfation of potassium chloride at combustion conditions. *Energy Fuels* 1999, 13:1184–1190.
64. Punjak WA, Shadman F. Aluminosilicate sorbents for control of alkali vapors during coal combustion and gasification. *Energy Fuel* 1988, 2:702–708.
65. Kyi S, Chadwick BL. Screening of potential mineral additives for use as fouling preventatives in Victorian brown coal combustion. *Fuel* 1999, 78:845–855.
66. Becidan M, Todorovic D, Skreiberg Ø, Khalil RA, Backman R, Goile F, Skreiberg A, Jovovic A, Sørum L. Ash related behaviour in staged and non-staged combustion of biomass fuels and fuel mixtures. *Biomass Bioenergy* 2012, 41:86–93.
67. Ishikawa R, Buekens A, Huang H, Watanabe K. Influence of combustion conditions on dioxin in an industrial-scale fluidized-bed incinerator: experimental study and statistical modelling. *Chemosphere* 1997, 35:465–477.
68. Zhong ZP, Jin BS, Huang YJ, Zhou HC, Lan E. Experimental research on emission and removal of dioxins in flue gas from a co-combustion of MSW and coal incinerator. *Waste Manage* 2006, 26:580–586.

69. Stam A F, Livingston W R, Cremers M F G, Brem G. Review of models and tools for slagging and fouling prediction for biomass co-combustion. Task 32, 2009. Available at: www.ieabcc.nl/publications/slagging_fouling_in_cofiring.pdf. (Accessed September 14, 2012).
70. Surmen Y, Demirbas A. Cofiring of biomass and lignite blends: resource facilities; technological and environmental issues. *Energy Sources* 2003, 25:175–187.
71. Wu C, Damstedt B, Burt S, Tree D, Baxter L. Fuel-nitrogen chemistry during combustion of low-grade fuels in a low-NOx burner. In: Science in Thermal and Chemical Biomass Conversion, 2004.
72. Robinson AL, Junker H, Baxter LL. Pollutant formation, ash deposition, and fly ash properties when cofiring biomass and coal. In: Engineering Foundation Conference on the Economic and Environmental Aspects of Coal Utilization, 1997.
73. Dayton DC, Belle-Oudry DA. Bench-scale biomass/coal cofiring studies. In: Engineering Foundation Conference on the Impact of Mineral Impurities on Solid Fuel Combustion, 1997.
74. Wood J. Feel the burn—the truth behind co-firing biomass. In: IEE Power Engineer, 2004, 18–21.
75. Mandø M, Rosendahl L, Yin C, Sørensen H. Pulverized straw combustion in a low-NOx multifuel burner: modeling the transition from coal to straw. *Fuel* 2010, 99:3051–3062.
76. Savolainen K. Co-firing of biomass in coal-fired utility boilers. *Appl Energy* 2003, 74:369–381.
77. Blondeau J, Jeanmart H, Ryckmans Y, Allard P, Rochaya D. Biomass combustion in pulverized-fuel boilers: use of apparent pyrolysis parameters in CFD codes. In: 9th European Conference on Industrial Furnaces and Boilers, Estoril, Portugal, April 26–29, 2011.
78. Spliethoff H. Effect of co-combustion of biomass on emissions in pulverized fuel furnaces. *Fuel Process Technol* 1998, 54:189–205.
79. Maciejewska A, Veringa H, Sanders J, Peteves SD. Co-firing of biomass with coal: constraints and role of biomass pretreatment. DG JRC Institute for Energy, European Commission, Joint Research Centre, 2006.
80. Lawrence B, Annamalai K, Sweeten JM, Heflin K. Cofiring coal and dairy biomass in a 29 kWt furnace. *Appl Energy* 2009, 86:2359–2372.
81. Desroches-Ducarne E, Dolinger JC, Marty E, Martin G, Delfose L. Modelling of gaseous pollutants emissions in circulating fluidized bed combustion of municipal refuse. *Fuel* 1998, 77:1399–1410.
82. Yin CG, Rosendahl L, Kaer SK, Condra TJ. Use of numerical modeling in design for co-firing biomass in wall-fired burners. *Chem Eng Sci* 2004, 59:3281–3292.
83. Strege JR, Zygarlicke CJ, Folkedahl BC, McCollor DP. Bench-scale study of interactions between flue gas and cofired ash in an SCR. *Fuel* 2006, 85:2439–2444.
84. Wieck-Hansen K, Overgaard P, Larsen OH. Cofiring coal and straw in a 150 MWe power boiler experiences. *Biomass Bioenergy* 2000, 19:395–409.
85. Zheng Y, Jensen AD, Johnsson JE. Deactivation of V2O5-WO3-TiO2 SCR catalyst at a biomass-fired combined heat and power plant. *Appl Catal B: Environ* 2005, 60:253–264.
86. Buecker B. SCR design. In: Power Engineering, 2002, 24–28.
87. McKay G. Dioxin characterisation, formation and minimisation during municipal solid waste (MSW) incineration: review. *Chem Eng J* 2002, 86:343–368.
88. Pels JR, Saraber AJ. Utilization of biomass ashes. In: *Solid Biofuels for Energy*. A Lower Greenhouse Gas Alternative. London, 2011, 219–235.
89. Mahmoudkhani M, Richards T, Theliander H. Sustainable use of biofuel by recycling ash to forests: treatment of biofuel ash. *Environ Sci Technol* 2007, 41:4118–4123.
90. Nurmesniemi H, Mäkelä M, Pöykiö R, Manskinen K, Dahl O. Comparison of the forest fertilizer properties of ash fractions from two power plants of pulp and paper mills incinerating biomass-based fuels. *Fuel Process Technol* 2012, 104:1–6.
91. Van Eijk R, Arthur Stam A. Co-firing and ash quality. In: IEA Workshop on Cofiring Biomass with Coal, Copenhagen, March 27–28, 2012.
92. Saraber AJ, Overhof LFAG. Identification of bulk and niche applications of ashes derived from co-combustion, co-gasification and biomass combustion, Report 50780586.610-TOS/ECC 08-9280, KEMA, Arnhem, the Netherlands, 2009.
93. Arias B et al. Influence of torrefaction on the grindability and reactivity of woody biomass. *Fuel Process Technol* 2008, 89:169–175.
94. Bergman PCA, Boersma AR, Zwart RWR, Kiel JHA. Torrefaction for biomass co-firing in existing coal-fired power stations, ECN report ECN-C-05-013, 2005.
95. Kongkeaw N, Patumsawad S. Thermal upgrading of biomass as a fuel by torrefaction. In: *Second International Conference on Environmental Engineering and Applications*. Singapore: IACSIT Press; 2011, 38–42.
96. Hakkou M, Pétrissans M, Gérardin P, Zoulalian A. Investigations of the reasons for fungal durability of heat treated beech wood. *Polym Degrad Stab* 2006, 91:393–397.
97. Prins MJ, Ptasinski KJ, Janssen FJJG. Torrefaction of wood: Part 1. Weight loss kinetics. *J Anal Appl Pyrol* 2006, 77:28–34.
98. Prins MJ, Ptasinski KJ, Janssen FJJG. Torrefaction of wood: Part 2. Analysis of products. *J Anal Appl Pyrol* 2006, 77:35–40.
99. Svoboda K et al. Pretreatment and feeding of biomass for pressurized entrained flow gasification. *Fuel Process Technol* 2009, 90:629–635.
100. Li J, Brzdekiewicz A, Yang W, Blasiak W. Co-firing based on biomass torrefaction in a pulverized coal boiler with aim of 100% fuel switching. *Appl Energy* 2012, 99:344–354.
101. Savat P. Activities at Rodenhuize power plant. In: Advanced and Max Green Projects, IFRF - TOTeM 35, Pisa, Italy, 23 and 24 September, 2010.

PART II

CHALLENGES AND SOLUTIONS FOR BIOMASS SUPPLY

16

Bioenergy and Land Use Change—State of the Art

Göran Berndes,[1]* Serina Ahlgren,[2] Pål Börjesson[1] and Annette L. Cowie[3]

[1]Department of Energy and Environment, Physical Resource Theory, Chalmers University of Technology, Gothenburg, Sweden (E-mail: goran.berndes@chalmers.se)
[2]Department of Technology and Society, Environmental and Energy Systems Studies, Lund University, Lund, Sweden
[3]Rural Climate Solutions—A partnership between University of New England and New South Wales Department of Primary Industries, University of New England, New South Wales, Australia

Throughout history, human societies have depended on the management and harvest of biological resources to obtain food and materials including solid, liquid, and gaseous fuels. Thus, human beings have always influenced their habitats and the conversion of ecosystems to land for biomass production is perhaps the most evident alteration of the earth.[1] Human societies have put almost half of the world's land surface to their service, and human land use has caused extensive land degradation and biodiversity loss, and also emissions to air and water, leading to impacts such as eutrophication, acidification, stratospheric ozone depletion, and climate change—although a large portion of anthropogenic emissions come from other than land use activities.[2–4] The scale of human influence on the global biogeochemical cycles has led geologists to define the present era in earth's history as the *Anthroposcene*, noting that human actions have become the main driver of global environmental change.[5,6]

Biomass is still a very important source of energy in some regions, e.g., Sub-Saharan Africa. But the importance of biomass as an energy source has steadily declined since humans learned to exploit fossil resources, and the substitution of biomass with fossil resources has—together with the intensification of agriculture—saved large areas from deforestation. At the same time, the intensified land management and the use of oil, coal, and natural gas cause many of the environmental impacts we see today. In response to these impacts, societies take measures to reduce our dependence on fossil resources and return to relying more on biomass and other renewable resources. Of the renewable energy sources, bioenergy is unique in that it is a part of the terrestrial carbon (C) cycle. The carbon dioxide (CO_2) emitted due to bioenergy use was earlier sequestered from the atmosphere and will be sequestered again if the biomass production system is managed sustainably, although there may be a significant temporal imbalance between C emissions and sequestration, such as in long-rotation forestry.

It is not axiomatic that the use of biomass is environmentally superior to the use of the fossil resources. Both the dedicated production of biomass crops and the collection of residues in agriculture and forestry can lead to undesirable negative impacts and it is crucial that practices are found that ensure that these impacts are avoided or mitigated as far as possible. This review concerns the use of biomass for energy and the connection between increased bioenergy use and land use change (LUC). Land use and LUC associated with bioenergy can lead to a multitude of—positive and negative—environmental and socioeconomic consequences. The focus here is placed on the

*Correspondence to: goran.berndes@chalmers.se

Advances in Bioenergy: The Sustainability Challenge, First Edition. Edited by Peter D. Lund, John Byrne, Göran Berndes and Iacovos A. Vasalos.

Originally published in: *WIREs Energy Environ* 2013, 2: 249–272 doi: 10.1002/wene.41

question whether greenhouse gas (GHG) emissions associated with LUC could undermine the climate change mitigation benefits of bioenergy.

LUC can influence the climate also through other mechanisms besides GHG emissions, where changes in surface albedo might be the most important factor,[7–16] but this is outside the scope of this review. Traditional use of biomass for cooking, space heating, and lighting presently accounts for roughly 80% of global bioenergy use.[17] However, such uses will not be considered in this review. We concentrate on so-called 'modern bioenergy', which is the production and use of refined solid, liquid, and gaseous biofuels for electricity, heat, and transport. There exist many options for avoiding or mitigating negative impacts associated with LUC—and for optimizing the climate benefits of bioenergy. Because of space limitations, this review is limited to a brief overview of such options, providing selected references for further reading. Neither do we discuss in detail how LUC is influenced by development of demand for food, feed, and various materials from agriculture and forestry, but it should be noted that this demand has a major influence. For instance, many studies have shown the importance of diets and the food sector's biomass use efficiency in determining land requirements (both cropland and grazing land) for food and thus implicitly the LUC consequences of bioenergy.[18–23]

LAND USE, LUC, AND GHG EMISSIONS

GHG Emissions Over Time

Emissions from changes in land use, principally those associated with deforestation and expansion of agricultural production for food, have contributed roughly one-third of the accumulated anthropogenic C emissions to the atmosphere since 1850.[24] Most of the remaining accumulated emissions have been caused by fossil fuel use (Figure 16.1, left *y*-axis) and the relative importance of fossil fuel use has increased over time. The annual fossil C emissions are presently (2010) more than five times larger than the LUC emissions, which contribute about 15% of the annual global C emissions (Figure 16.1, red line and right *y*-axis). Two-thirds of the accumulated C emissions associated with fossil fuel use have taken place since 1970.

South and Central America have had the largest LUC emissions since 1960 and the emissions in this region were more than 0.9 Pg C per year in 1986–1991.[24] Since then, the emissions have declined and they are now about 0.6 Pg C per year, similar to South and Southeast Asia.[24] Together, these two regions presently make up about 80% of total LUC emissions. In tropical Africa, LUC emissions reached 0.2 Pg C per year during the 1980s and are now well above this level.[24] It is estimated that more than

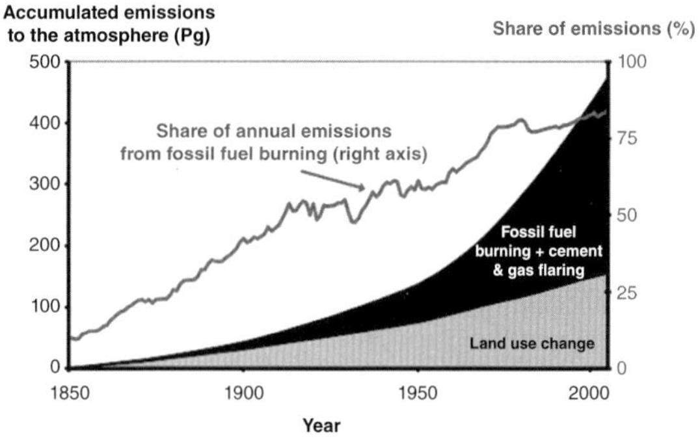

Figure 16.1 Accumulated anthropogenic C emissions to the atmosphere since 1850 (left *y*-axis). The red line (right *y*-axis) shows the share of annual GHG emissions that comes from fossil fuel burning. Most of the remaining part of the annual GHG emissions is associated with LUC. A small part comes from cement manufacturing and gas flaring, which has contributed some 1–2% of the total accumulated emissions. Source: Ref 35. (Reprinted with permission from Ref 35. Copyright 2011, IEA Bioenergy.)

80% of new tropical croplands in the 1990s replaced mature or degraded forests.[25] North America and Europe have contributed large LUC emissions in past centuries but have long been net sinks of atmospheric C because of forest regrowth.[24] The former USSR region and China have had substantial reductions in LUC emissions since 1960.[24] It is estimated that the terrestrial biosphere is presently a net sink of C on a global level, due primarily to the CO_2 fertilization effect, although uncertainty exists regarding sink strength and permanence.[26–34]

LUC Associated with Different Energy Options

Fossil Fuels

The dominance of fossil fuels also makes them a large contributor to LUC associated with energy supply; surface mining of coal, onshore oil and gas projects, and also exploitation of unconventional fossil resources such as oil shale, can cause deforestation or other land conversion for access roads, drilling platforms, and pipelines. Direct LUC emissions are estimated to contribute a small fraction of total lifecycle GHG emissions from conventional and unconventional oil, although estimates indicate up to about 10% for surface mining of oil sands.[36] However, the easier access to previously remote primary forest provided by new roads and pipeline routes can lead to increased logging, hunting, and deforestation for human settlement, causing indirect emissions.[37] In addition, besides the GHG and other emissions associated with the extraction of fossil resources, habitat loss and ecosystem fragmentation have biodiversity implications and many of the fossil reserves are located in fragile or biodiverse areas.[38,39] The 1989 Exxon Valdez oil spill and the 2010 Deepwater Horizon oil spill represent well-known examples showing that ecosystem impacts can be large also in the absence of land appropriation.

Nuclear and Other Renewable Options Other than Bioenergy

As for fossil fuels, the land requirements for other energy options can be of very different character and displacement effects vary. For instance, just a few percent of the indicated area for wind power is needed for turbine foundations, roads, or other infrastructure, and other land uses such as agriculture can often coexist with wind power. Solar thermal and PV systems can be integrated into buildings and other structures, but some solar power installations lock away land areas from other uses. However, deserts or other land areas with few alternative human uses are among the best locations for solar power plants.[17] Hydropower projects on the other hand may affect large areas of land, and entail the relocation of local communities living within or nearby the reservoir or the construction sites, and can also affect downstream communities. Displacement as well as resettlement schemes can have both socioeconomic and environmental (positive and negative) consequences including methane emissions from dams and effects associated with LUC for establishment of new agriculture land.[17] Infrastructure and employment generation associated with hydropower projects also stimulates immigration to the affected region, which can lead to deforestation and other LUC from expansion of agricultural land uses.[40] Nuclear power has land impacts associated with mining operations and waste deposits, but the major issue is associated with the risk that a nuclear accident leads to land contamination due to release of radioactive material (see Sidebar 1).

Box 16.1 SIDEBAR 1: LAND IMPLICATIONS OF ENERGY SUPPLY

Indicative quantifications of energy output per unit land area for various renewables suggest an electricity output at about 50 MWh ha^{-1} for the entire hydroelectric potential of the Amazon Region (many hydropower stations have much higher electricity output), and 250–800 MWh ha^{-1} for wind under a range of conditions. In comparison, a bioenergy plantation yielding 15 dry Mg ha^{-1} yr^{-1} (18 GJ Mg^{-1}) could provide about 23 MWh ha^{-1} yr^{-1} if the harvested biomass was used in a power plant with 30% conversion efficiency. Emissions released from land disturbed by fossil fuel extraction can be comparable or higher than biofuels, but the higher energy yield of oil production leads to lower GHG emissions from LUC per MJ fuel output. Photovoltaic and solar thermal systems for power use less land per unit electricity output than most other options. The LUC consequences of nuclear power are difficult to assess. There are land use impacts associated with mining operations, but the major issue is the risk that a nuclear accident leads to land contamination through release of radioactive material. The 1986 Chernobyl disaster in Ukraine resulted in large amounts of radioactive contamination across Europe. Most of the fallout concentrated near Belarus, Ukraine, and Russia, where some 125,000 km^2 of land (more than a third of which was in agricultural use) was contaminated. The Fukushima accident is another example where a nuclear accident causes land use impacts, currently of unknown magnitude. GHG emissions associated with LUC can take place when activities move from contaminated regions to elsewhere (source: Ref 35).

Bioenergy

Currently, less than 1% of global agricultural land (croplands and grazing land) is used for cultivating biofuel crops[17] (primarily cereals, sugar cane, oil seed crops, and oil palm), and LUC associated with biofuels and other modern bioenergy represents a small percentage of overall changes in land use. Perhaps the most significant influence of bioenergy on land use so far is associated with how forest management has changed in countries having large forest industries and where the use of biomass for energy has increased substantially, such as in Finland and Sweden. Given that biomass harvested for nonenergy purposes (e.g., paper and construction wood) often ultimately is used for energy, there is also an indirect link between bioenergy and land use. But the primary driver of LUC in such cases is the demand for wood-based products other than bioenergy.

Substantially larger areas than at present will have to be dedicated to the production of biomass for energy in a scenario where bioenergy contributes significantly to the future energy supply. Illustrative of this, a recent review by the Intergovernmental Panel on Climate Change[17] of 164 long-term energy scenarios showed bioenergy deployment levels in year 2050 ranging from 80 to 150 EJ per year for 440–600 ppm CO_2^{eq} concentration targets and from 118 to 190 EJ per year for less than 440 ppm CO_2^{eq} concentration targets (25th and 75th percentiles). In comparison, the energy content of the present global industrial roundwood production is around 15–20 EJ per year and the energy content in the global harvest of major crops (cereals, oil crops, sugar crops, roots, tubers, and pulses) corresponds to about 60 EJ per year.[17] Organic post-consumer waste and residues and by-products from the agricultural and forest industries, which contribute a major part of biomass for energy today, will not suffice to meet the anticipated levels of longer term biomass demand. Thus, much of the bioenergy feedstock would consequently have to come from dedicated production.

Given that reducing GHG emissions is one important driver for bioenergy, it is logical that emissions associated with LUC are properly taken into account when bioenergy projects are being contemplated or incentivized. Nevertheless, recent years' almost exclusive attention to how bioenergy-associated LUC impacts on bioenergy's contribution to near-term GHG targets can be argued as too narrowly focusing on marginal effects and thereby missing considerations relating to the longer term where bioenergy may expand to become '. . . a human use of photosynthesis that is comparable in scale to that for agriculture or forestry'.[41]

Bioenergy, LUC, and GHG Emissions

LUC associated with bioenergy projects can result in both positive and negative effects on biospheric C stocks and GHG emissions. The influence of land use and LUC on biospheric C stocks has earlier been noted in the context of United Nations Framework Convention on Climate Change—in particular CDM projects concerned with land use, LUC, and forestry—and it has been proposed that crediting for C sequestration could stimulate bioenergy systems that in addition function as C sinks.[42] The significance of LUC in bioenergy systems' GHG balances was demonstrated in the 1990s[43,44] when LUC effects were also considered in Life Cycle Assessment (LCA) studies.[45,46]

Direct and Indirect LUC

LUC involves either the conversion of natural ecosystems into managed lands, or changes in the management of already appropriated land, such as a shift from cultivating food crops or grazing animals to instead producing bioenergy feedstocks (Figure 16.2). Depending on definition used, LUC can include changes in agronomic practice including crop rotation patterns, altered stocking rates on pastures, changes in harvest residue management, and changes in forest management and wood extraction practices.

The LUC that is directly associated with the bioenergy project is referred to as dLUC. Indirect LUC (iLUC) refers to the changes in land use that take place elsewhere as a consequence of the bioenergy project. For example, displaced food producers may re-establish their operations elsewhere by converting natural ecosystems to agriculture land, or because of macroeconomic factors, the agriculture area may expand to compensate for the losses in food/feed/fiber production caused by the bioenergy project. The importance of considering indirect effects was stressed already in the 1990s,[45] but explicit consideration of iLUC has not been common until the recent years.

It should be noted that any measure to change a certain activity/effect could cause different types of indirect effects. For instance, when increased production of biofuels leads to lower demand for fossil fuels, this in turn could lead to lower fossil fuel prices, triggering increased fossil fuel demand, and also lower incentive to invest in some of the more expensive and more polluting sources of fuel such as fuel from deep ocean wells or tar sands.[47,48] Another example, increased availability of biofuels may reduce the interest in energy conservation and efficiency improvements as a climate change mitigation option, and owners of biofuel cars may travel more since they consider their traveling as being more environmentally acceptable (as they might do if buying more fuel efficient cars). However, iLUC is the indirect effect of bioenergy projects that has been in focus in recent years.

Present Biomass Use for Energy and the Associated LUC

Much of the biomass use for energy today utilizes post-consumer organic residues and by-products from the agricultural and forest industries that can be used for heat and power and are also viable as feedstock for biofuel production.[49–53] The present use may avoid iLUC if these biomass sources were previously not utilized for alternative purposes. However, their use may still cause iLUC if earlier users switch to using primary biomass.[54] Biomass that is burned on site—such as straw in fields or natural vegetation during forest clearing—are obvious examples of biomass that could be utilized for bioenergy without concern for LUC. The use of biomass that would otherwise be landfilled, or decompose in wet conditions, can besides avoiding iLUC also lead to additional benefits through reduced methane emissions.[55]

Removal of harvest residues that would otherwise decay on site may seem to have no risk of LUC, but may deplete soil C and fertility.[56] Overexploitation of harvest residues is one important cause of soil degradation in many places of the world and residue extraction rates must consider constraints to avoid decreasing the long-term productivity of the soil.[57–61] Fertilizer inputs can compensate for nutrient removals connected to harvest and residue extraction, but maintenance or improvement of soil fertility, structural stability, and water-holding capacity requires recirculation of

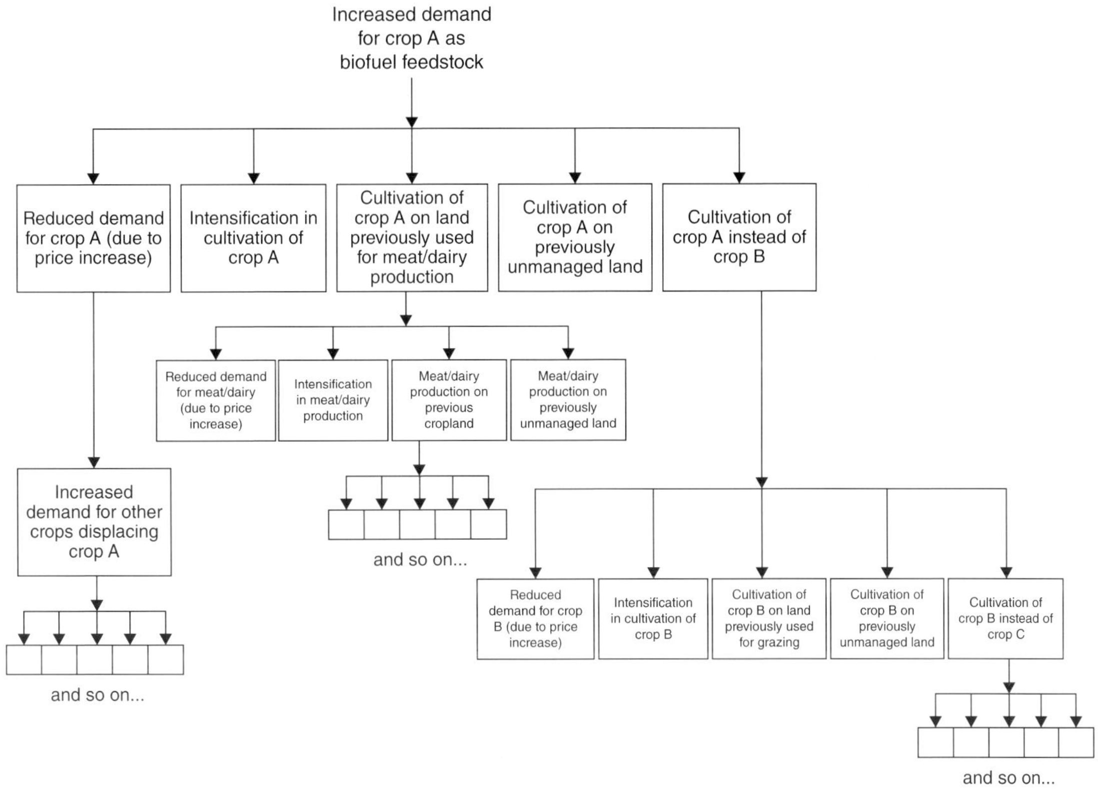

Figure 16.2 Illustration of direct and indirect land use changes arising as a consequence of a bioenergy project using crop A as feedstock. Note that in the case of iLUC, there need not be a one-to-one relationship between the land area claimed for bioenergy and the area converted to new pasture/cropland. This relationship depends on the relative productivity of the land claimed for bioenergy versus the new pasture/cropland. The iLUC taking place will also be determined by possible changes in meat/milk consumption and land use intensity (due to changes in food prices), and the displaced actors may turn to other activities than those connected to land use. The biofuel production in itself often generates protein rich by-products that are suitable for animal feeding, displacing other animal feed production and thereby reducing the net LUC effect of the bioenergy project. An example of when meat/dairy production displaces crop production is when crop rotation patterns are adjusted to include one or several years under temporary grasslands for animal feed.

organic matter to the soil.[57,62,63] Thus, the use of residues as bioenergy feedstock needs to carefully consider site-specific constraints on extraction rates. Otherwise, more rather than less land may be required on the longer term, as cultivation needs to expand to compensate for soil productivity losses.

If not utilized for bioenergy, some presently used biomass sources (e.g., felling residues left in the forest) would retain organic C for a longer time than if used for energy. This difference in timing of emissions can be considered a disbenefit for bioenergy in evaluations that only use a short time horizon, and also a relevant factor in longer term accounting in regions where biomass degradation is slow. As when considering use of land for bioenergy versus for C sinks, there is a need to consider the long-term fate of biomass that is not harvested for bioenergy; fires, insect outbreaks, and other disturbances can quickly convert a forest from net sink to emitter[64–69] (see section *The Case of Bioenergy in Long-Rotation Forestry*).

Options for Avoiding or Reducing LUC and Associated Emissions

There are several options for avoiding, or reducing, negative LUC effects associated with bioenergy. Land use restrictions and the targeting of unused marginal and degraded lands[17,70] can also reduce the risk of LUC emissions associated with bioenergy expansion. In many countries, there exist substantial areas of agricultural land that are abandoned or not actively used.[71] Using such lands for

Figure 16.3 Integrated production of biomass and milk/meat. Photo Credit: Laercio Cuoto. (Photo taken by Laercio Cuoto.)

bioenergy feedstock production can help avoid or reduce LUC and associated emissions. For instance, cereal cultivation for ethanol on excess grassland in Sweden, not needed for feed production at the current level of meat and dairy production, has low iLUC risks but may cause significant soil C emissions. If instead perennial lignocellulosic crops are cultivated, these soil C emissions may be reduced and for some feedstock systems soil C may instead increase.[72]

Productivity improvements in general reduce land requirements for both food, fiber, and bioenergy production, thus reducing the LUC pressure.[73] Integration of bioenergy systems with the existing land use can be one way to improve the land use productivity and also to reduce displacement effects. For example, double cropping and biomass processing technologies that produce leaf protein concentrates and pretreated forages can provide both animal feed and feedstock for biofuels production, while limiting nutrient leaching from cropland and enhancing soil fertility (and sequestering C) by building soil organic matter.[74] Figure 16.3 shows another example where wood production is integrated with cattle grazing. As already mentioned, the use of biofuel processing coproducts as animal feed can displace other feed production and in this way counteract negative LUC effects. A promising but presently little used option is to produce animal feed based on bagasse from sugarcane ethanol production. This integration of ethanol production with meat and dairy production releases grazing land for sugarcane plantations and also promotes intensification of animal production.[75,76]

Besides improving the productive use of land and providing CO_2 sequestration in soils and standing biomass, wisely designed, located, and managed biomass plantations can also provide benefits such as reduced erosion, reduced eutrophication, improved biodiversity, and improved socioeconomic conditions in the areas where bioenergy production expands.[17] It is however beyond the scope of this paper to provide a comprehensive review of these benefits.

Quantification of LUC Emissions

LUC can affect GHG balances in a number of ways, including: (1) when biomass is burned in the field during land clearing; (2) when the land management practice changes so that the C stocks in soils and vegetation change, and/or non-CO_2 emissions (N_2O, CH_4) change; and (3) when LUC results in changes in rates of C sequestration, i.e., the CO_2 assimilation of the land may increase or decrease relative to the case in which LUC is absent. The quantification of GHG emissions (or CO_2 assimilation) associated with LUC is subject to many uncertainties. There is variation in the amount of C stored in different ecosystem types, and the C consequences of converting natural ecosystems to production systems—or of changing land management practices—depends on the land use history and on the character of the new land use. When the amount of C in soils and aboveground biomass is well known for both the preconversion and postconversion states, it can be straightforward to calculate the GHG effects of dLUC for specific bioenergy projects. But in many instances, lack of empirical data and insufficient knowledge about how different land management practices affect the C dynamics can make quantification of dLUC emissions uncertain.

Quantification of GHG emissions associated with dLUC also requires specification of the biospheric C dynamics in the absence of the bioenergy project. One example, the cultivation of biofuel crops on cropland previously taken out of production can be considered as causing foregone C sequestration if the alternative would be natural or assisted conversion to grasslands or forests. But the future CO_2 assimilation can also be increased, for instance, when little managed and sparsely vegetated land is forested. Forests that are in stages of net C accumulation naturally lose this sink capacity if it is converted to another land cover type. Observations indicate that even very old forests can be net C sinks[77,78] but, as noted above, natural disturbances can strongly affect the C dynamics. Forests and other natural ecosystems may in addition be converted to other land use in the future, e.g., to provide food. It is therefore necessary to consider the likelihood of such events in determining the counterfactual ‘without project’ baseline against which the benefit of the bioenergy project is calculated,[79] which makes quantification a complex task. In forested lands susceptible to periodic fires, silviculture practices can lead to less frequent, lower intensity fires and can improve site conditions for replanting, leading to higher productivity,

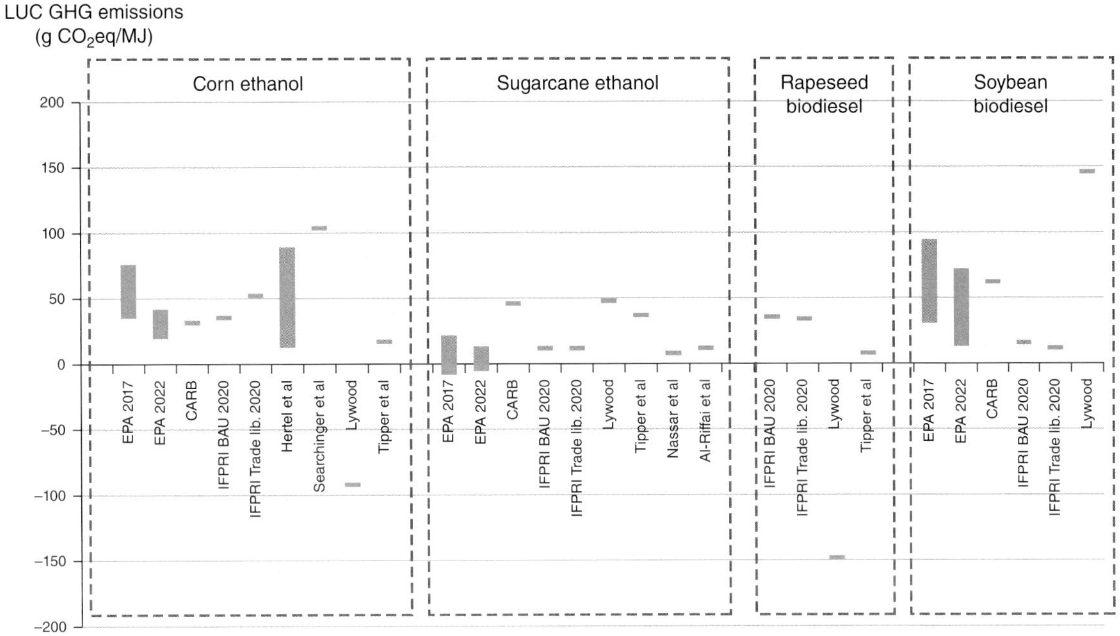

Figure 16.4 Ranges of model-based quantifications of LUC (dLUC + iLUC) emissions associated with the expansion of selected biofuel/crop combinations. The studies are reported with LUC emissions amortized over 30 years of production for comparison. (Reprinted with permission from Ref 35. Copyright 2011, IEA Bioenergy.)

i.e., accelerated forest growth rates and soil C storage. Using biomass removed in such practices for bioenergy can provide both GHG and particulate emission reductions by utilizing biomass that might otherwise burn in open-air forest fires.[80,81]

The inclusion of iLUC in quantifications of LUC effects adds greatly to the uncertainty. Agriculture expansion takes place in response to a complex set of interconnected biophysical, socioeconomic, and institutional factors and these vary in importance from one country to the other—and also within countries. The manifold causes of deforestation and other LUCs make quantification and establishment of causal chains difficult and uncertain. The modeling of such complex phenomena, involving multiple, interlinked, and time variable drivers, is a challenge. Existing methods employ either simplified approaches, where global LUC is allocated to specific biofuels/feedstocks grown on specified land types,[82–84] or economic equilibrium modeling integrating biophysical information and/or biophysical models (discussed further below).

Different indicators have been used to present the results of LUC quantification attempts. Indicators such as *carbon debt*[85] and *ecosystem carbon payback time*[86] have been used to assess the importance of upfront LUC emissions arising from the conversion of land to bioenergy production. The basic idea of these indicators is to express the upfront LUC emissions in terms of years required until these initial GHG emissions have been fully compensated for by the annual GHG savings obtained when the biofuels are produced and used instead of fossil fuels. Other indicators, such as *Cumulative warming impacts* or *global warming potential*, have been used to describe the dynamic effect of GHG emissions and discuss emissions timing and balancing between short- and long-term climate benefits of bioenergy projects. These indicators have been used, to a limited extent, to describe dynamic climate effects of bioenergy.[87–94]

Figure 16.4 shows the results from selected LUC quantifications, expressing the results in terms of g CO_2^{eq} MJ^{-1} over 30 years to facilitate comparison with other types of GHG emissions associated with the biomass production and conversion to biofuels. The focus is placed on LUC associated with so-called first generation biofuels that are produced from conventional food/feed crops. The variations for the same biofuel shown in Figure 16.4 are illustrative of the complexities and uncertainties inherent in LUC analyses. For example, six equilibrium models quantified LUC in different world regions associated with a standardized marginal increase in demand for first generation ethanol or biodiesel. All models showed large variations in both size and distribution of LUC over regions and crops (EU ethanol:

223–743; EU biodiesel: 242–1928; US ethanol: 107–863 Unit: 1000 ha $Mtoe^{-1}$).[95]

The assumed/modeled displacement effects of process coproducts used as feed can have strong influence on LUC values. Illustrative of this, the contrasting results for corn ethanol and rapeseed biodiesel (negative LUC emissions) in Figure 16.4 were obtained because of the assumption that biofuel processing coproducts would displace imported Brazilian soy as animal feed, indirectly reducing deforestation emissions (see also Refs 96 and 97). The marginal displacement effects of coproducts may have a saturation level, although new uses may be developed, e.g., to produce more biofuels. In this context, it is important to note that trade assumptions are critical and different for the various models.

Varying data resolution and differences in methodological approaches partly explain why studies report different LUC emissions. The combination of marginal and average values in LCA and other impact assessments that consider LUC effects may in addition generate results that are not sound from a methodological point of view.[84] The relative contributions of changes in yield and land area to increased crop output are, together with assumptions about trade, found to be critical factors in model-based LUC estimates.[98,99] There is a weak empirical basis for deriving price–yield relationships[100] and the yield levels will be determined by a complex set of more or less interrelated factors rather than by basic crop prices, which makes model calibration difficult. Approaches to determining yield levels on new cropland versus existing cropland vary and there are diverging views about how average yield levels change as the feedstock production area increases. Edwards et al.[95] state that the marginal area requirement per extra unit output of a particular biofuel should increase as total biofuel production increases, due to decreasing productivity of additional land converted to biofuel feedstock production. Lywood et al.,[98] however, state that the extent to which output change is met by yield or land area change varies considerably between crops and regions, and that there is no evidence that average yields decline as more land is used for the cases of EU cereals and US maize. They estimate that yield growth contributed 78% and 58% of incremental output growth for EU cereals and US maize, respectively, during the period 1961–2007. Conversely, area expansion contributed to more than 60% of output growth for EU rape seed, Brazilian sugarcane, S. American soy, and oil palm in South East Asia. The diverging views illustrate that yield levels are not solely the result of natural conditions but represent the outcome of farmers' economic allocation of resources (fertilizers, irrigation, etc.), which is influenced by expected prices of both inputs and the harvest.

The conditions for international bioenergy trade will influence the relative contribution of yield growth versus land expansion to increased bioenergy output since trade prospects influence where in the world the additional biomass fuel will be produced when demand increases in a given country, and yield levels vary strongly between countries. The link between location of bioenergy demand and biomass production varies with feedstock: cereals can be transported long distances to ethanol production facilities, while increased production of sugarcane ethanol would have to occur through construction of greenfield ethanol mills and establishment of sugarcane plantations in the vicinity of these new mills.

Illustrative of differences between models, and of how changes shaping land use and the world economy at large influence results, Tyner et al.[101] estimated LUC emissions associated with US corn ethanol expansion to be 14–20% of the value (104 g CO_2^{eq} MJ^{-1}) estimated by Searchinger et al.[102] Hertel et al.[103] estimated that market-mediated responses and use of coproducts moderated increases in cultivated land to be two-fifths of the amount estimated by Searchinger et al.[102] and iLUC emissions to be about 75% lower (see Figure 16.4). Other studies using probability modeling and other approaches to considering uncertainties report ranges of 50–250 and 21–142 (95% central interval) g CO_2^{eq} MJ^{-1} for US corn ethanol.[104,105] A study that used a bottom-up statistical approach found no correlation between biofuel production in the United States and changes in croplands for relevant crops in countries that are trading partners with the United States.[106] The approach has been debated,[107] but the conclusion that the past decade's ethanol production in the United States has not caused any large LUC (or diversion of corn exports) has also been supported.[110]

New approaches to land use—such as integrated food–bioenergy feedstock production[74–76,111] and establishment of policies, legal systems, and other mechanisms to address concerns about LUC consequences—will influence future land use and LUC emissions. This presents significant challenges for economic equilibrium modeling since incorporating effects of innovation in land use and policy can be difficult due to the lack of empirical historic data. Adding to the uncertainties, agriculture and forestry are likely to be significantly affected by climate change and this will shift the pattern of global comparative advantage in land use. The possible consequences of climate change for crop yields are not firmly established but indicate net global negative outcome, with negative impacts concentrated in tropical developing countries that will lose agriculture production potential while the northern industrialized countries might gain.[112–117]

Bioenergy options that use lignocellulosic feedstocks are projected to have lower LUC emission values than those of first generation biofuels.[17] Besides that use of lignocellulosic residues can reduce LUC effects, lower LUC emissions per unit bioenergy output can be reached by using higher yielding lignocellulosic plantations or by targeting degraded or marginal lands to reduce competition for prime croplands—although the lower productivity on such lands results in higher land requirement per given biomass output. The establishment of higher yielding biomass plantations on previously sparsely vegetated lands may in addition reduce downstream water availability and there can also be biodiversity impacts if marginal land with high biodiversity value is targeted.[17,118,119] As many lignocellulosic plants are grown under longer rotations, growers cannot readily respond to price increases since the yield can only be influenced during the plantation cycle through silvicultural means (Replace with e.g., increased fertilizer input) and by variety selection at the time of replanting. Thus, output growth in response to increasing demand is more readily obtained by area expansion for these multi-year rotation plantations.

To summarize, LUC associated with bioenergy can be large enough to significantly influence the GHG performance—in positive or negative direction. The causes behind LUC are multiple, complex, interlinked, and change over time. This makes quantification of LUC—and especially of iLUC—uncertain and reported results vary substantially. The results are sensitive to many factors that can develop in different directions—including land use productivity, trade patterns, prices and elasticities, and use of coproducts associated with biofuels production. Achieving a consensus on the extent of the LUC effects is unlikely in the near future and results reported so far do not support the ranking of biofuels with regard to LUC and associated emissions. There are however several options for avoiding or mitigating LUC emissions that can be implemented even if LUC effects are not yet described precisely.

Climate Change Mitigation Effectiveness of Bioenergy Expansion Over Time

As illustrated by the discussion above, much of recent years' attention to bioenergy and LUC has primarily been concerned with the question whether bioenergy expansion will contribute to near-term GHG targets. One objective has been to provide quantitative information to support governments that are seeking ways to address the LUC issue in their policies. Many studies have focused on the bioenergy project level and have used information from equilibrium models about marginal changes in land use systems to quantify LUC and associated GHG emissions—with uncertainties as described above.

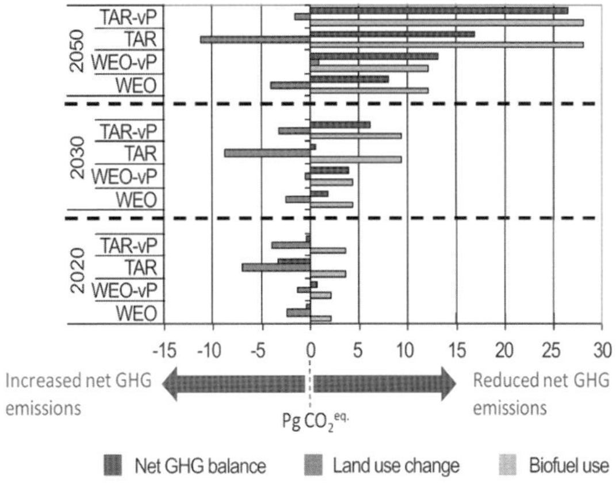

Figure 16.5 Accumulated net GHG savings in four biofuel scenarios. The green 'Biofuel use' bars show GHG savings (positive) from biofuel replacement of gasoline and diesel; the red 'Land use change' bars show GHG emissions (negative) caused by dLUC and iLUC; and the blue 'Net GHG balance' bars show the result of subtracting the LUC emissions from 'Biofuel use' savings. WEO has regional biofuel use up to 2030 as projected by the IEA World Energy Outlook 2008 reference scenario and 2nd generation biofuels are gradually deployed after 2015. TAR has roughly twice as high biofuel use as WEO and faster deployment of 2nd generation biofuels. The vP scenarios have higher agricultural productivity growth in developing countries leading to lower LUC. (Reprinted with permission from Refs 35 and 117. Copyright 2011 and 2009, IEA Bioenergy.)

Studies that focus on the near term essentially investigate to what extent there is a 'buffering capacity' in the present food system that reduces the LUC effect of near-term biofuel demand growth; increasing biomass demand (and price) stimulates increases in yields and efficiency in the use of harvested biomass, and food demand changes in response to changing prices. Conclusions from such studies should not be generalized as valid also for the case of more substantial transformation of energy and land use systems toward a longer term situation where bioenergy contributes a more significant part of the energy supply. Equilibrium models constructed to study near-term marginal changes are less suited for producing longer term outlooks and in particular for exploring innovation and paradigm shifts in land use. It is also worth noting that the use of marginal LUC values is not advisable when mean emission factor values are reported, such as in the EU RED system.

Figure 16.5 illustrates consequences of LUC on a more strategic level than Figure 16.4, which focused on the individual project level. Figure 16.5 shows how LUC emissions

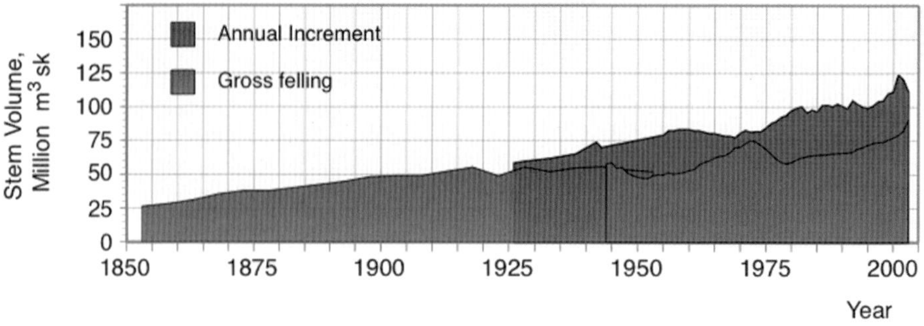

Figure 16.6 Historic overview of gross felling (1853–2003) and—placed behind the area showing gross felling—annual increment (1926–2003) in the Swedish forest. The method of estimating felling changed between 1945 and 1955, resulting in two overlapping curves (Ref 125). (Reprinted with permission from Ref 126. Copyright 2010, Peter Eliasson.)

can influence the net GHG reductions obtained by expanding the production of biofuels for transport. The quantifications shown are based on a general equilibrium approach, modeling responses of consumers and producers to price changes induced by the competition of biofuel feedstock production with conventional uses (food, feed, and fiber) of available resources. In addition to modeling LUC, this approach considers production intensification on existing agricultural land, use of biofuel coproducts such as animal feed as an additional input into feed markets, as well as consumer responses to changing availability and prices of food commodities. The LUC emissions shown are due to the *additional* need for cultivated land compared with a baseline projection without any increase in the production of crop-based biofuels (it is kept constant at the 2008 level)—thus, it is the sum of dLUC and iLUC emissions that is shown.

Figure 16.5 shows the accumulated GHG gains and losses for two baseline biofuel scenarios and their variants with higher crop productivity growth (see scenario description in the figure caption). Cumulative net GHG savings are closely linked to the effects of arable land expansion and subsequent land use conversions. The negative impact of LUC emissions is greatest in the near term and the relative importance of LUC emissions for the cumulative net GHG savings decreases over time. As a consequence, GHG savings resulting from the replacement of fossil fuels with biofuels accumulate only gradually over time and net GHG balances do not become positive until after 2020 in scenarios with baseline crop productivity growth. Therefore, one commonly used argument for promoting biofuels for transport—that it is one of few near-term options for climate change mitigation in the transport sector—may not hold unless promotion of biofuels goes hand in hand with strong efforts to improve land use productivity, especially in developing countries. However, some alternatives such as sugarcane ethanol are able to achieve a significant positive GHG balance in short-term periods. One may also take the position that too strong focus on near-term GHG saving risks missing important longer term considerations and that proper evaluation of bioenergy (or any other mitigation strategy) needs to maintain a balance between near-term targets and long-term objectives such as the 2°C target (further discussed below).

The Case of Bioenergy in Long-Rotation Forestry

The size and timing of forest C flows and C stock changes has during recent years become a topic for debate as governments and other authorities want to ensure that forest bioenergy initiatives contribute positively to climate change mitigation.[120–123] It is not questioned that long-rotation forest management with some level of biomass extraction for energy can be maintained on sustainable basis. Neither is there any disagreement on a conceptual level—i.e., that the CO_2 emitted when forest biomass is used for energy was earlier sequestered from the atmosphere and will be sequestered again if the forest bioenergy system is managed sustainably. The case of Swedish forestry illustrates that active forest management can ensure that increased biomass output need not take place at the cost of reduced forest stocks. Sweden has a considerable forest industry based on long-rotation forestry. The annual increment in Swedish forests has roughly doubled during the period 1926–2003 (Figure 16.6). Forest woodfuel production has increased considerably during the recent decades, while forest C stocks have increased.

It is rather the timing of net GHG savings that is debated and diverging standpoints can to a significant degree be explained by the different perspectives on bioenergy and climate change mitigation taken by the opponents. The disagreements arise because emissions and sequestration are

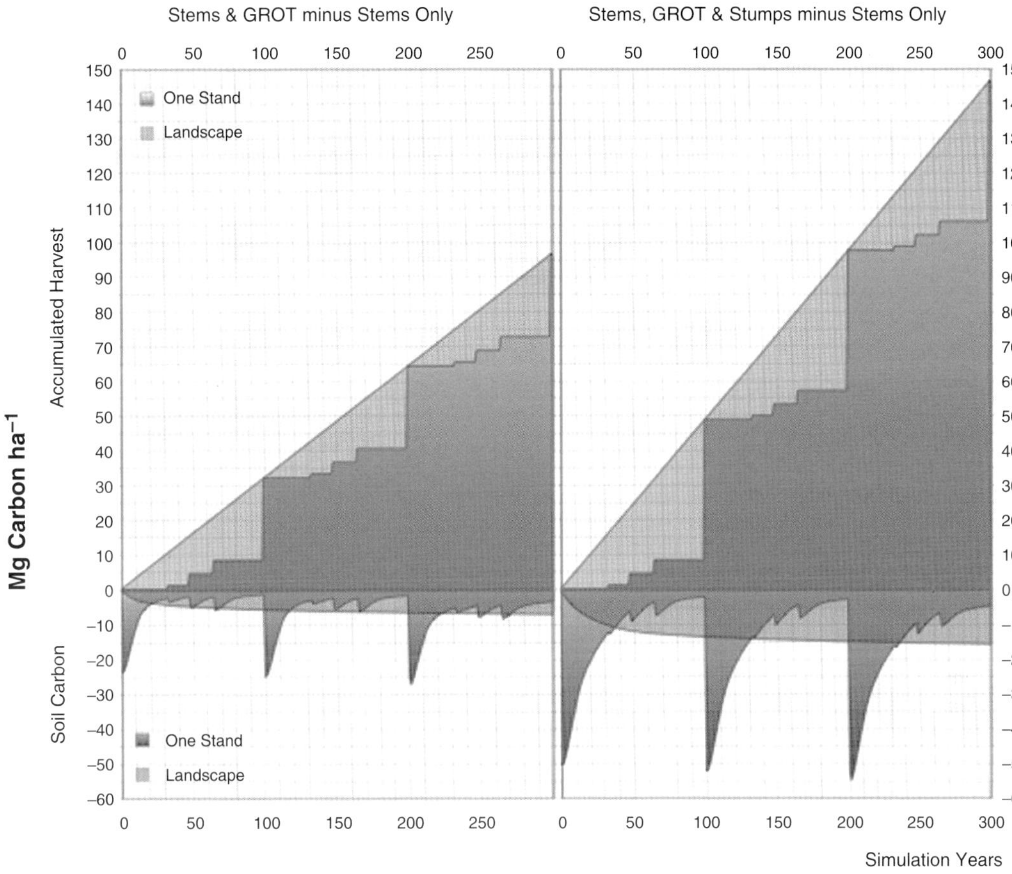

Figure 16.7 Effects on the C balance of increased removal of felling residues in a Norway spruce forest in south Sweden. In the 'Stems Only' scenario, harvest residues are left on the ground both after thinning and final felling. The 'Stems & GROT' scenario involves extraction of 80% of the logging residue after thinning and final felling (GROT is the Swedish acronym for branches and tops—GRenar Och Toppar in Swedish), and the 'Stems, GROT & Stumps' scenario includes in addition the removal of 50% of stumps-coarse root systems at final felling. The increased residue removal continues over the whole 300-year period. Upper panes show the amount of removal in comparison to the 'Stems Only' scenario and lower panes the corresponding variation in soil C. Single stands are plotted behind the landscape averages in the foreground. The sharp declines in stand level soil carbon shown at each harvest occasion are caused by the removal of residues, reducing litter addition to soil C. (Reprinted with permission from Ref 127. Copyright 2011, Peter Eliasson.)

not necessarily in temporal balance with each other; the dynamics of terrestrial C stocks in long-rotation forestry—where the temporal imbalance of C dynamics differ substantially for the bioenergy use on the one hand and decomposition/regrowth processes in the forest ecosystem on the other hand—lead to GHG mitigation trade-offs between biomass extraction for energy use and the alternative to leave the biomass in the forest, either as forest residues decaying over time or (in the case of harvesting living tress for bioenergy) as a C store that could further sequester more C over time.[43,124,125] Depending on perspective taken—e.g., spatial scale (stand level vs. landscape level) and time scale (short vs. long term)—different answers are obtained regarding the contribution of forest bioenergy to climate change mitigation.

Figure 16.7 shows the modeled *differences* in accumulated harvested biomass and soil C change between one scenario where only stems are harvested and two scenarios where also tops and branches [GRenar Och Toppar (GROT)] and stumps are extracted. In other words, the figure shows the net C effect of changing the biomass extraction to include also felling residues that can be used for

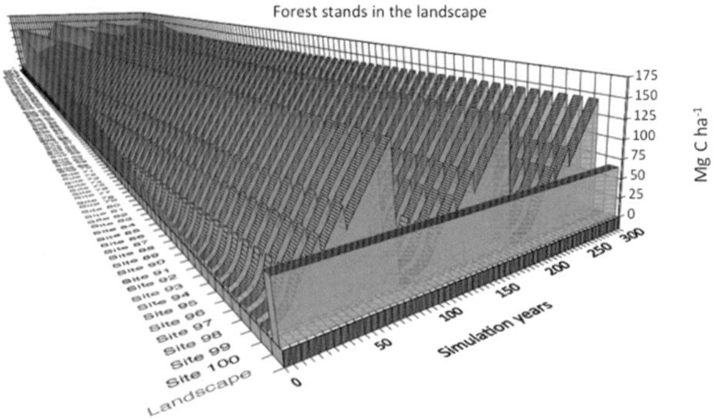

Figure 16.8 Tree biomass in a chronosequence of 100 identical simulations of Norway spruce stands, illustrating the difference between stand level and landscape level dynamics in a forest in southern Sweden. The assumed management resembles the dominant management regime during the previous decades, i.e., only stem wood has been removed at harvests, and thinning has been done at intervals prescribed by the Swedish Forest Agency. The stand is thinned three times (year 33, 48, and 65, with biomass harvest corresponding to about one-fourth of the basal area) and final harvest takes place after 100 years where only stems are removed. It is assumed that 10% of the stem biomass is left as harvest residue (tops). Each stand is planted 1 year after the other starting from year 0. The average C content of the trees included each year is shown in the front most bar representing the landscape level. The losses and gains of C from individual forest stands in the landscape counterbalance and the average C content in the forest is stable over time. (Reprinted with permission from Ref 127. Copyright 2011, Peter Eliasson.)

energy. The figure shows the C dynamics on stand level and also shows how the C dynamics for individual stands are averaged to create steady trends on the landscape level (i.e., a forest estate with equal areas of each age class).

As can be seen, the landscape level soil C stock initially declines in response to intensified harvests and stabilizes over time. But compared with the C in the additional accumulated harvest, there are no major changes in soil C stocks at the landscape level. The losses in ecosystem C are considerably less than the C in corresponding withdrawals. If the additional biomass that is extracted in the scenarios with intensified harvests displaces fossil fuels, the benefit of increased harvest occurs immediately and grows linearly over time, whereas the associated loss of soil C has a declining response at the landscape level. The relative attractiveness of increased biomass extraction, from a climate change mitigation perspective, depends on the fossil fuel displacement efficiency associated with the use of the harvested biomass.

As shown in Figure 16.7, increasing extraction from forests of biomass for any purpose (not just bioenergy) may cause a short-term decrease in C stocks over the business as usual situation, unless it is accompanied by activities that promote regeneration, enhance growth, or decrease decomposition rates. Thus, despite that C stock losses are small compared with the increase in accumulated harvest in the longer term, evaluations that assess the contribution of individual forest bioenergy projects to near-term GHG reduction targets may well come to the conclusion that it would be better to leave the biomass to decay in the forest and continue using fossil fuels. This is illustrative of how the project level approach to evaluating different options has limitations and needs complementary consideration of forest bioenergy and associated C flows on a landscape level and also consideration of how forest management is affected by the promotion and growth of bioenergy.

Forest management will likely change in response to changing forest biomass demand (including for bioenergy) and this influences forest C flows and can lead to increased or decreased forest C stocks. As shown in Figure 16.6, active forest management can ensure that increased biomass output need not take place at the cost of reduced forest stocks. Figure 16.8 illustrates how the forest C stock at the landscape level may be influenced by changes in the conditions for forest management (policy, regulation, etc.). If, for instance, a new policy is implemented at a given year—and this policy changes practices for forests planting, thinning, and final felling—then one stand is planted, one is harvested, and three stands are thinned according to the new policy this specific year. Thus, not all stands are immediately influenced by a new policy; a given stand remains under the 'old' management regime until the next thinning, or until final felling with subsequent planting occurs.

Shortened forest rotation length to obtain increased output of timber and biomass fuels leads to decreased C stock in living biomass (ceteris paribus). Intensified biomass extraction in forests, e.g., for bioenergy, can lead to a decrease in soil C or dead wood C pool compared with existing practice. Conversely, if changed forest management employing intensified extraction also involves growth-enhancing measures, forest C stocks may increase.[128] As an example of measures, site preparation increases forest productivity by drastically shortening the regeneration phase, but may on the other hand also lead to faster decomposition of soil organic matter.[129] Another example, fertilization is a conventional management option to increase forest production and experiments show that stem volume production can, even in already highly productive forests, be more than doubled with optimal fertilization and, when needed, irrigation.[130] This increased stem volume production should result in an approximately equal increase in litter production (although fine roots may not respond as much) and a similar long-term increase in soil C. It has also been shown that fertilization slows down decomposition in boreal forests contributing further to increasing soil C stocks.[131,132] Other studies of the possibilities to intensify forest management (particularly in the boreal forests) confirm that stand management can increase both the C stocks and the biomass and timber production on the same piece of land.[133–135]

To the extent that increased demand for forest bioenergy makes such measures feasible (i.e., they would not have taken place in a scenario without bioenergy demand), effects of changed forest management should be considered when evaluating the climate change mitigation benefit of forest bioenergy. This has implications for how evaluation frameworks should be designed. It is common that the time period for the evaluation starts at the point of biomass extraction, with the consequence that forest management prior to the extraction is not considered. The two cases of forest C accounting in Figure 16.9 illustrate how this accounting design prevents a holistic evaluation of forest bioenergy operations and how management of long-rotation forests contributes to climate change mitigation. As can be seen in Figure 16.9, the two cases of accounting are indicated to start at different years and both use a 20-year time frame:

- In A, the accounting starts so as to allow consideration of the higher growth rate—and consequently higher C sequestration—achieved from fertilization (the unfertilized forest might be used as a reference case). The biomass extraction during thinning operations is quite soon compensated for due to the rapid growth rate.
- In B, the accounting starts at the time when final harvest takes place. If a sufficiently large part of the extracted biomass is used for bioenergy or in short-lived products, the accounting will in most cases (depends on energy system configuration) show as a large net C emission to the atmosphere, since there is little time for forest regrowth before the accounting period comes to an end.

The size of the net C emissions will depend on fossil C displacement efficiency and the time profile for the forest regrowth that compensates for the biomass extraction—or in the case of forest residue extraction where the alternative (reference) situation is to leave the residues in the forest: the residue decay rate.[136,137] However, 'project A' in Figure 16.9 would clearly appear to be much more favorable for the climate than 'project B' in an evaluation that narrowly considers a distinct forest bioenergy project (either A or B) and that uses a relatively short time horizon. Yet, both 'project A' and 'project B' are components of the same forest management regime that have undisputable net substitution benefits (lower diagram in Figure 16.9).

NEED FOR FURTHER RESEARCH AND COMPLEMENTARY PERSPECTIVES

So far, studies of bioenergy and LUC have primarily been concerned with the contribution of specific bioenergy projects/products (primarily biofuels for transport) to near-term GHG targets. LCA and economic equilibrium modeling are commonly used tools. LCA studies are useful for comparing products and for identifying improvement options in production processes. Economic equilibrium modeling advances our understanding of complex system dynamics by indicating how small changes in conditions influence a system in a state of equilibrium (e.g., the effect of higher fertilizer prices on agriculture activities).

However, the contribution of bioenergy to climate change mitigation should be evaluated from many points of view reflecting a balance between near-term targets and the long-term objective to hold the increase in global temperature below 2°C (Copenhagen Accord). There is otherwise a risk that the development of policies and incentive structures creates a situation where the most economically rational way of managing land is very different from how land management is best developed in response to prospective demand for food, bioenergy, and other biomass products considering the ultimate requirements of a far reaching energy system transformation.

Adding landscape level considerations to complement project level indicators and metrics is one important step (see Figure 16.8), but also additional perspectives are needed. Near-term GHG targets are primarily intended to induce energy system transformation toward meeting longer term stabilization objectives. It is clear that mitigation efforts over the next two to three decades will have a large impact on opportunities to achieve lower stabilization levels. But the effectiveness of near-term mitigation efforts cannot be evaluated only in terms of their immediate GHG savings.

The concept of 'GHG emissions space' (Figure 16.10), which focuses on accumulated emissions up to a given

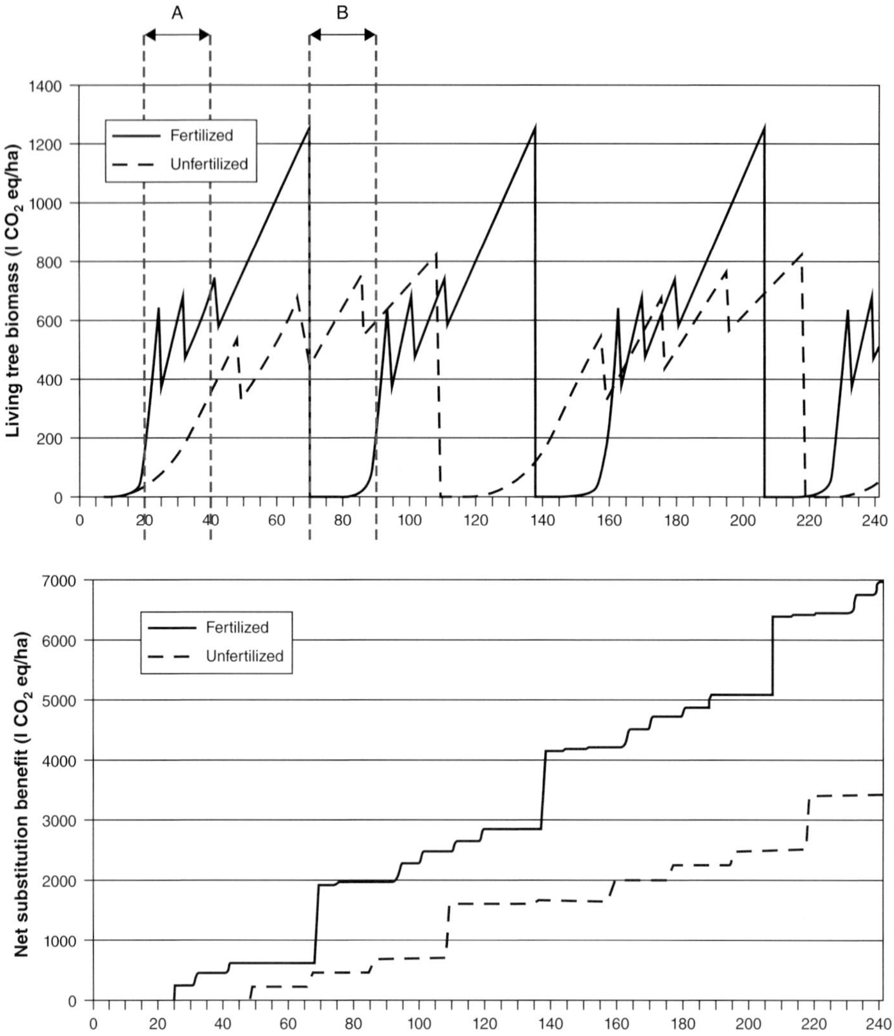

Figure 16.9 Development of C stocks and GHG flows over a 240-year period for typical fertilized and unfertilized stands in northern Sweden. The top diagram shows living tree biomass and the bottom diagram shows net substitution benefits of wood product use assuming coal reference fuel, with deductions made for N_2O, CH_4, and fossil CO_2 emissions. The dynamics of C in soils and dead biomass (not shown) is highly influenced by the forest management but occurs at a smaller scale (fluctuations are within 250 ton CO_2 ha^{-1}). A and B denote two possible cases of forest bioenergy accounting (see text). (Reprinted with permission from Ref 128. Copyright 2010, IEA Bioenergy.)

year, is relevant in relation to temperature targets since the peak warming appears to be insensitive to the CO_2 emissions pathway, i.e., timing of emissions or peak emission rate.[138,139] Depending on the atmospheric lifetime of specific GHGs, the trade-off between emitting more now and less in the future is not one to one in general. But the relationship for CO_2—the major component in LUC emissions—is practically one to one, so that one additional (less) ton CO_2 emitted today requires that one has to reduce (can increase) future CO_2 emissions by one ton. The reason for this is the close to irreversible climate effect of CO_2 emissions.[139–143]

Thus, from the perspective of temperature targets and GHG emission space, the exact shape of C fluctuations associated with bioenergy systems is not relevant. For example, it does not matter whether C in forest residues is

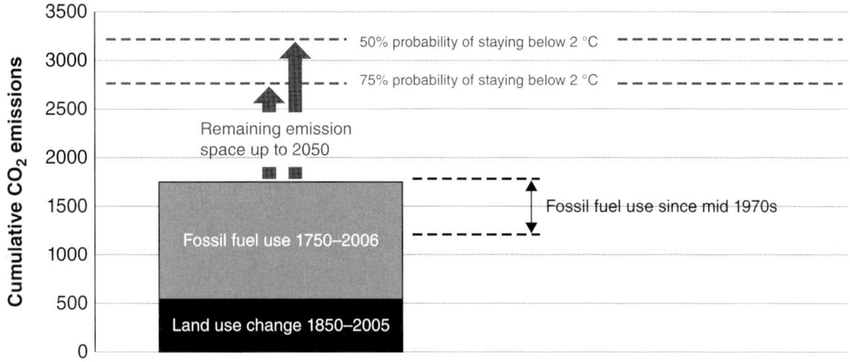

Figure 16.10 Cumulative CO_2 emissions and indicative remaining emission space in relation to 2°C target. (Reprinted with permission from Refs 35 and 138. Copyright 2011 and 2009, IEA Bioenergy.)

emitted to the atmosphere early after the forestry operations take place (such as when used for energy) or is emitted some decades after the forestry operations (such as when the residues are left in the forest to decay). What matters is whether forest bioenergy systems are part of a changed forest management paradigm that results in systematic decreases or increases in the forest C stocks. If such changes in C stocks would occur, they need to be considered in evaluations of forest management strategies that are developed in response to climate and energy policies.

At present, fossil energy infrastructure is expanding rapidly around the world and due to the long lifetime of energy infrastructure—so that the present day investments into the energy systems have implications for GHG emissions several decades into the future[144,145]—this expansion implies considerable claims for future GHG emission space. Too little attention has been paid to the question of how promotion of bioenergy affects the development of energy and associated systems, i.e., directions and size of investments (the rate at which coal power plants are built to meet the growing electricity demand, R&D into coal-based liquid fuels and unconventional oil vs. bioenergy vs. other renewables), and innovation and learning in forestry and agriculture (e.g., residue management, allocation of land to different types of crops and forest management regimes, nature protection).

It is proposed that strategic considerations and the design of policies and regulations to guide bioenergy development use complementary insights from studies employing integrated energy/industry/land use/land cover models and other methods to analyze bioenergy and land use in the context of technology evolution and longer term development of land use and energy systems—including how this development is influenced by policies and regulations put in place (see, e.g., Refs 146–154).

CONCLUSIONS AND IMPLICATIONS FOR POLICY

Bioenergy projects can lead to both direct and indirect LUC, which can affect GHG balances with both beneficial and adverse outcomes for bioenergy's contribution to climate change mitigation. Long-rotation forest management is associated with C emissions and sequestration that are not in temporal balance with each other and this leads to mitigation trade-offs between biomass extraction for energy use and the alternative to leave the biomass in the forest. The causes behind LUC are multiple, complex, interlinked, and change over time. This makes quantification of LUC effects difficult, especially when they are associated with iLUC, and quantifications reported so far vary substantially and do not support the ranking of bioenergy options with regard to LUC and associated emissions. There are however several options for mitigating these emissions that can be implemented despite the uncertainties, e.g., development of bioenergy feedstock production systems that integrate with rather than displace existing agriculture and forestry production, enhancement of land use productivity in agriculture and forestry in general, and legal protection of natural ecosystems with associated build-up of enforcement capacity—although leakage impacts effectiveness of protection as long as activities can migrate to other areas where protection is less stringent.

Project level evaluations that use a relatively short time horizon and narrow spatial perspective need to be complemented with additional analyses that balance near-term targets and the long-term objective to hold the increase in global temperature below 2°C (Copenhagen Accord). Bioenergy and the associated C flows need to be considered within the larger context of energy system transformation and adaptation in agriculture and forestry to a prospective situation where bioenergy and other renewable energy sources contribute a substantial part of the primary energy supply.

One critical strategic question is how society should make use of the remaining allowable 'space' for GHG in the atmosphere. To some extent, the establishment of new energy technologies and associated infrastructure will in itself occupy part of the remaining space for GHG emissions. For example, electric vehicle fleets will contribute to increasing atmospheric CO_2 levels as long as electricity is mainly generated from fossil fuels (although they may cause lower GHG emissions than present gasoline and diesel vehicles). Yet, promotion of electric vehicles can be justified because they can provide efficient transport services that cause low GHG emissions in a future situation when electricity is less reliant on fossil fuels.

Similarly—in view of the long-term benefit of bioenergy—it may be found attractive to use part of the GHG emission space for developing a bioenergy industry capable of providing renewable and climate friendly energy services for the world in the long term. Furthermore, possible LUC emissions associated with bioenergy expansion will decrease over time as above ground biomass and soil C stabilize at new equilibrium levels, and also other GHG emissions decrease as conversion technologies improve and use renewable process fuel, and feedstock production systems develop into less GHG intensive systems. Should Carbon Capture and Storage (CCS) technologies become available for large-scale implementation, bioenergy is currently the only energy technology that, combined with CCS, allows net removal of CO_2 from the atmosphere, making it pivotal for achieving ambitious climate protection targets if GHG emissions are not soon curbed.

There are certainly many other reasons for protecting natural vegetation than the desire to avoid GHG emissions, and these may have become too little considered when LUC impacts are considered. Future LUC rates will depend on the willingness of national governments to protect forests and other natural ecosystems—and the effectiveness of legislation and other measures to reduce deforestation. Strict focus on the climate benefits of ecosystem preservation may put undue pressure on valuable ecosystems that have a relatively low C density. Although this may have a small impact in terms of climate change mitigation, it may impact negatively on, e.g., biodiversity and water tables. In fact, research results indicate that the pricing of C from LUC emissions may not even be sufficient to protect forests with relatively high C density, since forest conversion to bioenergy plantations may remain profitable because the price of bioenergy increases along with the C prices.[155]

Policy measures implemented to minimize negative impacts of LUC should be based on a holistic perspective recognizing bioenergy's strong interconnectedness with food, feed, and various biomaterials production, and the multiple drivers and consequences of LUC. They also need to acknowledge that the conversion of native vegetation land into biomass production systems may in some places reflect the outcome of sensible land use planning, promoting more sustainable land uses in multifunctional landscapes. Sensible land development programmes can have better prospects for achieving sustainable development than the top-down establishment of global sustainability criteria using strong and inflexible measures.

REFERENCES

1. Hurtt GC, Frolking S, Fearon MG, Moore B, Shevliakova E, Malyshev S, Pacala SW, Houghton RA. The underpinnings of land-use history: three centuries of global gridded land-use transitions, wood-harvest activity, and resulting secondary lands. *Global Change Biol* 2006, 12:1208–1229.
2. Ellis EC. Anthropogenic transformation of the terrestrial biosphere. *Philos Trans R Soc A* 2011, 369:1010–1035.
3. Haberl H, Erb KH, Krausmann F, Gaube V, Bondeau A, Plutzer A, Gringrish S, Lucht W, Fischer-Kowalski M. Quantifying and mapping the human appropriation of net primary production in earth's terrestrial ecosystems. *Proc Natl Acad Sci U S A* 2007, 104:12942–12947.
4. Turner BL II, Lambin EF, Reenberg A. The emergence of land change science for global environmental change and sustainability. *Proc Natl Acad Sci USA* 2007, 104:20666–20671.
5. Steffen W, Crutzen PJ, McNeill JR. The Anthropocene: are humans now overwhelming the great forces of nature? *Ambio* 2007, 36:614–621.
6. Williams M, Zalasiewicz J, Haywood A, Ellis M. The Anthropocene: a new epoch of geological time? *Philos Trans R Soc A* 2011, 369:835–841.
7. Gibbard S, Caldeira K, Bala G, Phillips TJ, Wickett M. Climate effects of global land cover change. *Geophys Res Lett* 2005, 32:L23705.
8. Betts RA. Offset of the potential carbon sink from boreal forestation by decreases in surface albedo. *Nature* 2000, 408:187–190.
9. Bonan GB. Forests and climate change: forcings, feedbacks, and the climate benefits of forests. *Science* 2008, 320:1444–1449.
10. Bright RM, Stromman AH, Peters GP. Radiative forcing impacts of boreal forest biofuels: a scenario study for Norway in light of albedo. *Environ Sci Technol* 2011, 45:7570–7580.
11. Bala G, Caldeira K, Wickett M, Phillips TJ, Lobell DB, Delire C, Mirin A. Combined climate and carbon-cycle effects of large-scale deforestation. *Proc Natl Acad Sci U S A* 2007, 104:6550–6555.
12. Loarie SR, Lobell DB, Asner GP, Mu Q, Field CB. Direct impacts on local climate of sugar-cane expansion in Brazil. *Nat Climate Change* 2011, 1:105–109.

13. Georgescu M, Lobell DB, Field CB. Direct climate effects of perennial bioenergy crops in the United States. *Proc Natl Acad Sci U S A* 2011, 108:4307–4312.
14. Schweiger HP, Bird N. Integration of albedo effects caused by land use change into the climate balance: should we still account in greenhouse gas units? *Forest Ecol Manage* 2010, 260:278–286.
15. Boucher O, Myhre G, Myhre A. Direct human influence of irrigation on atmospheric water vapour and climate. *Climate Dyn* 2004, 22:597–603.
16. Lenton TM, Vaughan NE. The radiative forcing potential of different climate geoengineering options. *Atmos Chem Phys* 2009, 9:5539–5561.
17. Edenhofer O, Pichs-Madruga R, Sokona Y, Seyboth K, Matschoss P, Kadner S, Zwickel T, Eickemeier P, Hansen G, Schlömer S, von Stechow C [eds]. *IPCC Special Report on Renewable Energy Sources and Climate Change Mitigation*. Cambridge University Press, Cambridge, United Kingdom and New York, NY, USA; 2011.
18. Gerbens-Leenes P, Nonhebel S. Consumption patterns and their effects on land required for food. *Ecol Econ* 2002, 42:185–199.
19. Smil V. Worldwide transformation of diets, burdens of meat production and opportunities for novel food proteins. *Enzyme Microbial Tech* 2002, 30:305–311.
20. de Boer J, Helms M, Aiking H. Protein consumption and sustainability: diet diversity in EU-15. *Ecol Econom* 2006, 59:267–274.
21. Elferink EV, Nonhebel S. Variations in land requirements for meat production. *J Cleaner Prod* 2007, 15:1778–1786.
22. Stehfest E, Bouwman L, van Vuuren DP, den Elzen MGJ, Eickhout B, Kabat P. Climate benefits of changing diet. *Climatic Change* 2009, 95:83–102.
23. Wirsenius S, Azar C, Berndes G. How much land is needed for global food production under scenarios of dietary changes and livestock productivity increases in 2030? *Agric Syst* 2010, 103:621–638.
24. Houghton RA. Carbon Flux to the Atmosphere from Land-Use Changes: 1850–2005. In *TRENDS: A Compendium of Data on Global Change*. Carbon Dioxide Information Analysis Center, Oak Ridge National Laboratory, U.S. Department of Energy, Oak Ridge, Tenn., USA; 2008.
25. Gibbs HK, Ruesch AS, Achard F, Clayton MK, Holmgren P, Ramankutty N, Foley JA. Tropical forests were the primary sources of new agricultural land in the 1980s and 1990s. *Proc Natl Acad Sci USA* 2010, 107:16732–16737.
26. Fung IY, Doney SC, Lindsay K, John J. Evolution of carbon sinks in a changing climate. *Proc Natl Acad Sci USA* 2005, 102:11201–11206.
27. Canadell JG, Le Quéré C, Raupach MR, Field CB, Buitenhuis ET, Ciais P, Conway TJ, Gillett NP, Houghton RA, Marland G. Contributions to accelerating atmospheric CO_2 growth from economic activity, carbon intensity, and efficiency of natural sinks. *Proc Natl Acad Sci USA* 2007, 104:18866–18870.
28. Kurz WA, Stinson G, Rampley GJ, Dymond CC, Neilson ET. Risk of natural disturbances makes future contribution of Canada's forests to the global carbon cycle highly uncertain. *Proc Natl Acad Sci USA* 2008, 105:1551–1555.
29. Magnani F, Mencuccini M, Borghetti M, Berbigier P, Berninger F, Delzon S, Grelle A, Hari P, Jarvis PG, Kolari P, et al. The human footprint in the carbon cycle of temperate and boreal forests. *Nature* 2007, 447:849–851.
30. Goodale CL, Apps MJ, Birdsey RA, Field CA, Heath LS, Houghton RA, Jenkins JC, Kohlmaier GH, Kurz W, Liu S, et al. Forest carbon sinks in the northern hemisphere. *Ecol Appl* 2002, 12:891–899.
31. Raupach MR. Carbon cycle: Pinning down the land carbon sink. *Nat Climate Change* 2011, 1:148–149.
32. Sitch S, Cox PM, Collins WJ, Huntingford C. Indirect radiative forcing of climate change through ozone effects on the land-carbon sink. *Nature* 2007, 448:791–794.
33. Mercado LM, Bellouin N, Sitch S, Boucher O, Huntingford C, Wild M, Cox PM. Impact of changes in diffuse radiation on the global land carbon sink. *Nature* 2009, 458:1014–1017.
34. Pan Y, Birdsey RA, Fang J, Houghton R, Kauppi PE, Kurz WA, Phillips OL, Shvidenko A, Lewis SL, Canadell JG, et al. A large and persistent carbon sink in the world's forests. *Science* 2011, 333:988–993.
35. Berndes G, Bird N, Cowie A. *Bioenergy, Land Use Change and Climate Change Mitigation*. Background Technical Report. IEA Bioenergy, ExCo: 2011:04.
36. Yeh S, Jordaan SM, Brandt AR, Turetsky MR, Spatari S, Keith DW. Land use greenhouse gas emissions from conventional oil production and oil sands. *Environ Sci Technol* 2010, 44:8766–8772.
37. Southgate D, Wasserstrom R, Reider S. *Oil Development, Deforestation, and Indigenous Populations in the Ecuadorian Amazon*. Report presented to the Latin American Studies Association in Rio de Janeiro, Brazil; 2009.
38. Jordaan SM, Keith DW, Stelfox B. Quantifying land use of oil sands production: a life cycle perspective. *Environ Res Lett* 2009, 4:024004.
39. Gorissen L, Buytaert V, Cuypers D, Dauwe T, Pelkmans L. Why the debate about land use change should not only focus on biofuels. *Environ Sci Technol* 2010, 44:4046–49.
40. World Commission on Dams. The future of large dams: dealing with social, environmental, institutional and political costs. In: Scudder T, ed. *Dams and Development—A New Framework for Decision-Making*. London: Earthscan; 2000.
41. Williams R. Role for biomass energy in sustainable development. In: Socolow R, Andrews C, Berkhout F, Thomas V, eds. *Industrial Ecology and Global Change*. Cambridge: Cambridge University Press; 1994, 199–225.
42. Schlamadinger B, Grubb M, Azar C, Bauen A, Berndes G. Carbon sinks and the CDM: could a bioenergy linkage offer a constructive compromise? *Climate Policy* 2001, 1:411–417.
43. Marland G, Schlamadinger B. Forests for carbon sequestration or fossil fuel substitution? A sensitivity analysis. *Biomass Bioenergy* 1997, 13:389–397.
44. Leemans R, van Amstel A, Battjes C, Kreileman E, Toet S. The land cover and carbon cycle consequences of large-scale utilizations of biomass as an energy source. *Global Environ Change* 1996, 6:335–357.

45. DeLuchi MA. Greenhouse-gas emissions from the use of new fuels for transportation and electricity. *Transport Res Part A* 1993, 27A:187–191.
46. Reinhardt G. *Biofuels: Energy and GHG balances: Methodology and Case Study Rape Seed Biodiesel*. Heidelberg: Institut für Energie- und Umweltforschung; 1991.
47. Rajagopal D, Hochman G, Zilberman D. Indirect fuel use change (IFUC) and the lifecycle environmental impact of biofuel policies. *Energy Policy* 2011, 39:228–233.
48. Stoft S. *Renewable Fuel and the Global Rebound Effect*. Berkeley: Global Energy policy Center; 2010.
49. Cherubini F, Ulgiati S. Crop residues as raw material for biorefinery systems—a LCA case study. *Appl Energy* 2010, 87:47–57.
50. Kim S, Dale B. Global potential bioethanol production from wasted crops and crop residues. *Biomass Bioenergy* 2004, 26:361–375.
51. Uihlein A, Schebek L. Environmental impacts of a lignocellulose feedstock biorefinery system: an assessment. *Biomass Bioenergy* 2009, 33:793–802.
52. Van Loo S, Koppejan J, eds. *The Handbook of Biomass Combustion and Co-firing*. London: IEA Bioenergy Task 32/Earthscan; 2008.
53. Yin C, Rosendahl LA, Kær SK. Grate-firing of biomass for heat and power production. *Prog Energy Combust Sci* 2008, 34:725–754.
54. Cowie AL, Gardner WD. Competition for the biomass resource: Greenhouse impacts and implications for renewable energy incentive schemes. *Biomass Bioenergy* 2007, 31:601–607.
55. Börjesson P, Berglund M. Environmental systems analysis of biogas systems—part II: environmental impact of replacing various reference systems. *Biomass Bioenergy* 2007, 31:326–344.
56. Cowie AL, Smith P, Johnson D. Does soil carbon loss in biomass production systems negate the greenhouse benefits of bioenergy? *Mitig Adapt Strategies Global Change* 2006, 11:979–1002.
57. Lal R. Crop residues as soil amendments and feedstock for bioethanol production. *Waste Manage* 2008, 28:747–758.
58. Wilhelm WW, Johnson JMF, Karlen DL, Lightle DT. Corn stover to sustain soil organic carbon further constrains biomass supply. *Agron J* 2007, 99:1665–1667.
59. Wilhelm WW, Johnson JMF, Hatfield JL, Voorhees WB, Linden DR. Crop and soil productivity response to corn residue removal: a literature review. *Agron J* 2004, 96:1–17.
60. Ball BC, Bingham I, Rees RM, Watson CA, Litterick A. The role of crop rotations in determining soil structure and crop growth conditions. *Can J Soil Sci* 2005, 85:557–577.
61. Blanco-Canqui H, Lal R, Post WM, Izaurralde RC Owens LB. Rapid changes in soil carbon and structural properties due to stover removal from no-till corn plots. *Soil Sci* 2006, 171:468–482.
62. Lal R, Pimentel D. Bio-fuels from crop residues. *Soil Tillage Res* 2007, 93:237–238.
63. Blanco-Canqui H, Lal R. Corn stover removal for expanded uses reduces soil fertility and structural stability. *Soil Sci Soc Am J* 2009, 73:418–426.
64. Li Z, Apps MJ, Kurz WA, Banfield E. Temporal changes of forest net primary production and net ecosystem production in west central Canada associated with natural and anthropogenic disturbances. *Can J For Res* 2003, 33:2340–2351.
65. Kurz WA, Stinson G, Rampley G. Could increased boreal forest ecosystem productivity offset carbon losses from increased disturbances? *Philos Trans R Soc B: Biol Sci* 2008, 363:2259–2268.
66. Bond-Lamberty B, Peckham SD, Ahl DE, Gower ST. Fire as the dominant driver of central Canadian boreal forest carbon balance. *Nature* 2007, 450:89–92.
67. van der Werf GR, Dempewolf J, Trigg SN, Randerson JT, Kasibhatla PS, Giglio L, Murdiyarso D, Peters W, Morton DC, Collatz GJ, et al. Climate regulation of fire emissions and deforestation in equatorial Asia. *Proc Natl Acad Sci USA* 2008, 105:20350–20355.
68. Kurz WA, Dymond CC, Stinson G, Rampley GJ, Neilson ET, Carroll AL, Ebata T, Safranyik L. Mountain pine beetle and forest carbon feedback to climate change. *Nature* 2008, 452:987–90.
69. Lindner M, Maroschek M, Netherer S, Kremer A, Barbati A, Garcia-Gonzalo J, Seidl R, Delzon S, Corona P, Kolström M, et al. Climate change impacts, adaptive capacity, and vulnerability of European forest ecosystems. *Forest Ecol Manage* 2010, 259:698–709.
70. Wicke B, Sikkema R, Dornburg V, Faaij A. Exploring land use changes and the role of palm oil production in Indonesia and Malaysia. *Land Use Policy* 2011, 28:193–206.
71. Campbell JFM, Lobell DB, Genova RC, Field C.B. The global potential of bioenergy on abandoned agriculture lands. *Environ Sci Technol* 2008, 42:5791.
72. Börjesson P, Ahlgren S, Berndes G. The climate benefit of Swedish ethanol—present and prospective performance. *WIREs: Energy Environ*. In press.
73. Food and Agriculture Organization. *The State of the World's Land and Water Resources for Food and Agriculture (SOLAW)—Managing Systems at Risk. Earthscan*; 2011.
74. Dale BE, Bals BD, Kim S, Eranki P. Biofuels done right: land efficient animal feeds enable large environmental and energy benefits. *Environ Sci Technol* 2010, 44:8385–8389.
75. Egeskog A, Berndes G, Freitas F, Gustafsson S, Sparovek G. Integrating bioenergy and food production—a case study of combined ethanol and dairy production in Pontal, Brazil. *Energy Sustainable Dev* 2011, 15:8–16.
76. Sparovek G, Berndes G, Egeskog A, Freitas F, Gustafsson S, Hansson J. Sugarcane ethanol production in Brazil: an expansion model sensitive to socioeconomic and environmental concerns. Biofuels. *Bioprod Biorefining* 2007, 1:270–282.
77. Luyssaert S, Schulze ED, Boerner A, Knohl A, Hessenmöller D, Law BE, Ciais P, Grace J. Old-growth forests as global carbons sinks. *Nature* 2008, 455:213–215.
78. Lewis SL, Lopez-Gonzalez G, Sonké B, Affum-Baffoe K, Baker TR, Ojo LO, Phillips OL, Reitsma JM, White L,

Comiskey JA, et al. Increasing carbon storage in intact African tropical forests. *Nature* 2009, 457:1003–1006.
79. Gustavsson L, Karjalainen T, Marland G, Savolainen I, Schlamadinger B, Apps M. Project-based greenhouse-gas accounting: guiding principles with a focus on baselines and additionality. *Energy Policy* 2000, 28:935–946.
80. Ffolliott PF, Stropki CL, Chen H, Neary DG. *Impacts of a historic wildfire in southwestern ponderosa pine forests*. Fort Collins, Colorado: Rocky Mountain Research Station; 2011.
81. Neary DG, Zieroth E. Forest bioenergy system to reduce the hazard of wildfires: White Mountains, Arizona. *Biomass Bioenergy* 2007, 31:638–645.
82. Bauen A, Chudziak C, Vad K, Watson P. A causal descriptive approach to modelling the GHG emissions associated with the indirect land use impacts of biofuels. *A study for the UK Department for Transport*. E4tech; 2010.
83. Fritsche U, Hennenberg K, Hünecke K. *The 'iLUC Factor' as a Means to Hedge Risks of GHG Emissions from Indirect Land Use Change*. Darmstadt: Oeko Institute; 2010.
84. Tipper R, Hutchison C, Brander M. A Practical Approach for Policies to Address GHG Emissions from Indirect Land Use Change Associated with Biofuels. Technical Paper - TP-080212-A. Ecometrica and Greenenergy; 2009.
85. Fargione J, Hill J, Tilman D, Polasky S, Hawthorne P. Land clearing and the biofuel carbon debt. *Science* 2008, 319:1235–1238.
86. Gibbs HK, Johnston M, Foley JA, Holloway T, Monfreda C, Ramankutty N, Zaks D. Carbon payback times for crop-based biofuel expansion in the tropics: the effects of changing yield and technology. *Environ Res Lett* 2008, 3:034001.
87. Cherubini F, Peters GP, Berntsen T, Stomman AH, Hertwich E. CO2 emissions from biomass combustion for bioenergy: atmospheric decay and contribution to global warming. *GCB Bioenergy* 2011, 3:413–426.
88. Dornburg V, Marland G. Temporary storage of carbon in the biosphere does have value for climate change mitigation: a response to the paper by Miko Kirschbaum. *Mitig Adapt Strategies Global Change* 2008, 13:211–217.
89. Kendall A, Chang B, Sharpe B. Accounting for time-dependent effects in biofuel life cycle greenhouse gas emissions calculations. *Environ Sci Technol* 2009, 43:7142–7147.
90. Kirschbaum MUF. To sink or burn? A discussion of the potential contributions of forests to greenhouse gas balances through storing carbon or providing biofuels. *Biomass Bioenergy* 2003, 24:297–310.
91. Kirschbaum MUF. Temporary carbon sequestration cannot prevent climate change. *Mitig Adapt Strategies Global Change* 2006, 11:1151–1164.
92. Kirkinen J, Sahay A, Savolainen I. Greenhouse impact of fossil, forest residues and Jatropha diesel: a static and dynamic assessment. *Prog Ind Ecol, An Int J* 2009, 6:185–206.
93. Levasseur A, Lesage P, Margni M, Deschenes L, Samson R. Considering time in LCA: Dynamic LCA and its application to global warming impact assessments. *Environ Sci Technol* 2010, 44:3169–3174.
94. O'Hare M, Plevin RJ, Martin JI, Jones AD, Kendall A, Hopson E. Proper accounting for time increases crop-based biofuels' greenhouse gas deficit versus petroleum. *Environ Res Lett* 2009, 4:024001.
95. Edwards R, Mulligan D, Marelli L. Indirect land use change from increased biofuel demand. *Comparison of models and results for marginal biofuels production from different feedstocks*. Ispra: European Commission, Joint Research Centre, Institute for Energy; 2010.
96. Lywood W, Pinkney J, Cockerill SAM. Impact of protein concentrate coproducts on net land requirement for European biofuel production. *Global Change Biol Bioenergy* 2009, 1:346–359.
97. Weightman RM, Cottrill BR, Wiltshire JJJ, Kindred DR, Sylvester-Bradley R. Opportunities for avoidance of land-use change through substitution of soya bean meal and cereals in European livestock diets with bioethanol coproducts. *Global Change Biol Bioenergy* 2011, 3:158–170.
98. Lywood W, Pinkney J, Cockerill SAM. The relative contributions of changes in yield and land area to increasing crop output. *Global Change Biol Bioenergy* 2009, 1:360–369.
99. Keeney R, Hertel TW. The indirect land use impacts of United States biofuel policies: the importance of acreage, yield, and bilateral trade responses. *Am J Agric Econ* 2008, 91:895–909.
100. Keeney R, Hertel TW. *Yield Response to Prices: Implications for Policy Modeling*. Department of Agricultural Economics, Purdue University; 2008.
101. Tyner W, Taheripour F, Zhuang Q, Birur D, Baldos U. Land use changes and consequent CO2 emissions due to U.S. corn ethanol production: a comprehensive analysis. *GTAP Resource 3288, Department of Agricultural Economics*, Purdue UniversitY; 2010.
102. Searchinger T, Heimlich R, Houghton RA, Dong F, Elobeid A, Fabiosa J, Tokgoz S, Hayes D, Yu TH. Use of U.S. croplands for biofuels increases greenhouse gases through emissions from land-use change. *Science* 2008, 319:1238–1240.
103. Hertel TW, Golub AA, Jones AD, O'Hare M, Plevin JR, Kammen DM. Effects of US maize ethanol on global land use and greenhouse gas emissions: estimating market-mediated responses. *BioScience* 2010, 60:223–231.
104. Plevin RJ, O'Hare M, Jones AD, Torn MS, Gibbs HK. Greenhouse gas emissions from biofuels indirect land use change are uncertain but may be much greater than previously estimated. *Environ Sci Technol* 2010, 44:8015–8021.
105. Mullins KA, Griffin WM, Matthews HS. Policy implications of uncertainty in modeled life-cycle greenhouse gas emissions of biofuels. *Environ Sci Technol* 2011, 45((1):):132–38.
106. Kim S, Dale BE. Indirect land use change for biofuels: testing predictions and improving analytical methodologies. *Biomass Bioenergy* 2011, 35:3235–3240.
107. O'Hare M, Delucchi M, Edwards R, Fritsche U, Gibbs H, Hertel T, Hill J, Kammen D, Laborde D, Marelli L, et al. Comment on "Indirect land use change for biofuels: testing predictions and improving analytical methodologies" by Kim and Dale: statistical reliability and the definition of the

indirect land use change (iLUC) issue. *Biomass Bioenergy* 2011, 35:4485–4487.

108. Kline K, Oladosu GA, Dale VH, McBride AC. Scientific analysis is essential to assess biofuel policy effects: in response to the paper by Kim and Dale on "Indirect land-use change for biofuels: testing predictions and improving analytical methodologies". *Biomass Bioenergy* 2011, 35:4488–4491.
109. Dale BE, Kim S. Response to comments by O'Hare et al., on the paper indirect land use change for biofuels: testing predictions and improving analytical methodologies. *Biomass Bioenergy* 2011, 35:4492–4493.
110. Oladosu G, Kline K, Uria-Martinez R, Eaton L. Sources of corn for ethanol production in the United States: a decomposition analysis of the empirical data. *Biofuels Bioproducts Biorefining* 5:640–653.
111. George B, Nicholas ID. *Developing Options for Integrated Food-Energy Systems—Volume 1. IEA Bioenergy Task43: PR*; 2011.
112. Reilly J, Paltsev S, Felzer B, Wang X, Kicklighter D, Melillo J, Prinn R, Sarofim M, Sokolov A, Wang C. Global economic effects of changes in crops, pasture, and forests due to changing climate, carbon dioxide, and ozone. *Energy Policy* 2007, 35:5370–5383.
113. Lobell DB, Burke MB. Why are agricultural impacts of climate change so uncertain? The importance of temperature relative to precipitation. *Environ Res Lett* 2008, 3:034007.
114. Lobell DB, Burke MB, Tebaldi C, Mastrandrea MD, Falcon WP, Naylor RL. Prioritizing climate change adaptation needs for food security in 2030. *Science* 2008, 319:607–610.
115. Easterling WE, Aggarwal PK, Batima P, Brander KM, Erda L, Howden SM, Kirilenko A, Morton J, Soussana J-F, Schmidhuber J, et al. Food, fibre and forest products. In: Parry ML, Canziani OF, Palutikof JP, van der Linden PJ, Hanson CE, eds. *Climate Change 2007: Impacts, Adaptation and Vulnerability*. Cambridge, United Kingdom: Cambridge University Press 2007, 273–313.
116. Fischer G, van Velthuizen H, Shah M, Nachtergaele F. *Global Agro-ecological Assessment for Agriculture in the 21st Century: Methodology and Results*. Laxenburg: International Institute for Applied Systems Analysis (IIASA); 2002.
117. Fischer G, Hizsnyik E, Prieler S, Shah M, van Velthuizen H. *Biofuels and Food Security*. Laxenburg: The OPEC Fund for International Development (OFID) and International Institute of Applied Systems Analysis (IIASA); 2009.
118. Berndes G. Bioenergy and water—the implications of large-scale bioenergy production for water use and supply. *Global Environmental Change* 2002, 12:7–25.
119. Garg K, Karlberg L, Wani S, Berndes G. Biofuel production on wastelands in India: opportunities and trade-offs for soil and water management at the watershed scale. *Biofuels Bioproducts Biorefining* 2011, 5:410–430.
120. Walker T, ed. Massachusetts Biomass Sustainability and Carbon Policy Study: Report to the Commonwealth of Massachusetts Department of Energy Resources. Natural Capital Initiative Report NCI-2010- 03. Manomet Center for Conservation Sciences; 2010. (critique against the study and Manomet's responses are available at www.manomet.org/node/322.
121. Greenpeace. *Fuelling a biomess: Why Burning Trees for Energy Will Harm People, the Climate and Forests*. Canada: Greenpeace:, 2011. Available at: www.greenpeace.org/canada/en/recent/Burning-trees-for-energy-puts-Canadian-forests-and-climate-at-risk-Greenpeace/.
122. Repo A, Tuomi M, Liski J. Indirect carbon dioxide emissions from producing bioenergy from forest harvest residues. *Global Change Biol: Bioenergy* 2011, 3:107–115.
123. Law BE, Harmon ME. Forest sector carbon management, measurement and verification, and discussion of policy related to climate change. *Carbon Manage* 2011, 2:73–84.
124. Marland G, Obersteiner M, Schlamadinger B, Righelato R, Spracklen DV. The carbon benefits of fuels and forests. *Science* 2007, 318:1066–1068.
125. Righelato R, Spracklen DV. Carbon mitigation by biofuels or by saving and restoring forests? *Science* 2007, 317:902.
126. Skogsstyrelsen. *Swedish Statistical Yearbook of Forestry*. Sweden: Swedish Forest Agency Jönköping; 2010.
127. Eliasson P, Svensson M, Olsson M, Ågren G. *Forest Carbon Balances at the Landscape Scale and Responses to Intensified Harvest Investigated with the Q and COUP Models*. Eskilstuna, Sweden: Swedish Energy Agency project 34884-1; 2011.
128. Sathre R, Gustavsson L, Bergh J. Primary energy and greenhouse gas implications of increasing biomass production through forest fertilization. *Biomass Bioenergy* 2010, 34:572–581.
129. Johansson MB. The influence of soil scarification on the turn-over rate of slash needles and nutrient release. *Scand J Forest Res* 1994, 9:170–179.
130. Bergh J, Linder S, Lundmark T, Elfving B. The effect of water and nutrient availability on the productivity of Norway spruce in northern and southern Sweden. *Forest Ecol Manage* 1999, 119:51–62.
131. Ågren GI, Bosatta E, Magill AH. Combining theory and experiment to understand effects of inorganic nitrogen on litter decomposition. *Oecologia* 2001, 128:94–98.
132. Franklin O, Högberg P, Ekblad A, Ågren GI. Pine forest floor carbon accumulation in response to N and PK additions: Bomb C-14 modelling and respiration studies. *Ecosystems* 2003, 6:644–658.
133. Alam A, Kilpeläinen A, Kellomäki S. Potential energy wood production with implications to timber recovery and carbon stocks under varying thinning and climate scenarios in Finland. *BioEnergy Res* 2010, 3:362–372.
134. Routa J, Kellomäki S, Peltola H. Impacts of intensive management and landscape structure on timber and energy wood production and net CO_2 emissions from energy wood use of Norway spruce. *BioEnergy Res* 2011:1–18.
135. Routa J, Kellomäki S, Peltola H, Asikainen A. Impacts of thinning and fertilization on timber and energy wood production in Norway spruce and Scots pine: scenario analyses based on ecosystem model simulations. *Forestry* 2011, 84:159–175.

136. Melin Y, Petersson H, Nordfjell T. Decomposition of stump and root systems of Norway spruce in Sweden—a modelling approach. *Forest Ecol Manage* 2009, 257:1445–1451.
137. Palviainen M, Finér L, Laiho R, Shorohova E, Kapitsa E, Vanha-Majamaa I. Carbon and nitrogen release from decomposing Scots pine, Norway spruce and silver birch stumps. *Forest Ecol Manage* 2010, 259:390–398.
138. Meinshausen M, Meinshausen N, Hare W, Raper SC, Frieler K, Knutti R, Frame DJ, Allen MR. Greenhouse-gas emission targets for limiting global warming to 2 degrees C. *Nature* 2009, 458:1158–1162.
139. Allen MR, Frame DJ, Huntingford C, Jones CD, Lowe JA, Meinshausen M, Meinshausen N. Warming caused by cumulative carbon emissions towards the trillionth tonne. *Nature* 2009, 458:1163–1166.
140. Archer D, Brovkin V. Millennial atmospheric lifetime of fossil fuel CO_2. *Climatic Change* 2008, 90:283–297.
141. Matthews HD, Caldeira K. Stabilizing climate requires near-zero emissions. *Geophys Res Lett* 2008, 35:L04705.
142. Matthews HD, Gillett NP, Stott PA, Zickfeld K. The proportionality of global warming to cumulative carbon emissions. *Nature* 2009, 459:829–832.
143. Solomon S, Plattner GK, Knutti R, Friedlingstein P. Irreversible climate change due to carbon dioxide emissions. *Proc Nat Acad Sci USA* 2009,106:1704–1709.
144. Mignone BK, Socolow RH, Sarmiento JL, Oppenheimer M. Atmospheric stabilization and the timing of carbon mitigation. *Climatic Change* 2008, 88:251–265.
145. Vaughan NE, Lenton TM, Shepherd JG. Climate change mitigation: trade-offs between delay and strength of action required. *Climatic Change* 2009, 96:29–43.
146. Azar C. Sandén B. The elusive quest for technology-neutral policies. *Environ Innov Soc Trans* 2011, 1:135–139.
147. DeFries R, Rosenzweig C. Toward a whole-landscape approach for sustainable land use in the tropics. *Proc Nat Acad Sci USA* 2010, 107:19627–19632.
148. Di Lucia L, Ahlgren S, Ericsson K. The dilemma of indirect land-use changes in EU biofuel policy—an empirical study of policy-making in the context of scientific uncertainty. *Environ Sci Policy* 2012, 16:9–19.
149. Melillo JM, Reilly JM, Kicklighter DW, Gurgel AC, Cronin TW, Paltsev S, Felzer BS, Wang X, Sokolov AP, Schlosser CA. Indirect emissions from biofuels: how important? *Science* 2009, 326:1397–1399.
150. Palm CA, Smukler SM, Sullivan CC, Mutuo PK, Nyadzi GI, Walsh MG. Identifying potential synergies and trade-offs for meeting food security and climate change objectives in sub-Saharan Africa. *Proc Nat Acad Sci USA* 2010, 107:19661–19666.
151. Strengers B, Leemans R, Eickhout B, De Vries B, Bouwman L. The land-use projections and resulting emissions in the IPCC SRES scenarios as simulated by the IMAGE 2.2 model. *GeoJ* 2004, 61:381–393.
152. Thomson AM, Calvin KV, Chini LP, Hurtt G, Edmonds JA, Bond-Lamberty B, Frolking S, Wise MA, Janetos AC. Climate mitigation and the future of tropical landscapes. *Proc Nat Acad Sci USA* 2010, 107:19633–19638.
153. van Vuuren D, den Elzen M, Lucas P, Eickhout B, Strengers B, van Ruijven B, Wonink S, van Houdt R. Stabilizing greenhouse gas concentrations at low levels: an assessment of reduction strategies and costs. *Climatic Change* 2007, 81:119–159.
154. Wise M, Calvin K, Thomson A, Clarke L, Bond-Lamberty B, Sands R, Smith SJ, Janetos A, Edmonds J. Implications of limiting CO2 concentrations for land use and energy. *Science* 2009, 324:1183–1186.
155. Persson UM, Azar C. Preserving the world's tropical forests—a price on carbon may not do. *Environ Sci Technol* 2010, 44:210–215.

FURTHER READING

Berndes G, Bird N, Cowie A. Bioenergy, land use change and climate change mitigation. Report for Policy Advisors and Policy Makers. IEA Bioenergy: ExCo:2010:03. Available at: www.ieabioenergy.com/LibItem.aspx?id=6770.

Bird N, Cowie A, Cherubini F, Jungmeier G. Using a life cycle assessment approach to estimate the net greenhouse gas emissions of bioenergy. IEA Bioenergy:ExCo:2011:03.Available at: www.ieabioenergy.com/LibItem.aspx?id=7096.

Edenhofer O, Pichs-Madruga R, Sokona Y, Seyboth K, Matschoss P, Kadner S, Zwickel T, Eickemeier P, Hansen G, Schlömer S, von Stechow C [eds]. *IPCC Special Report on Renewable Energy Sources and Climate Change Mitigation.* Cambridge University Press, Cambridge, United Kingdom and New York, NY, USA; 2011.

17

Forest Energy Procurement: State of The Art In Finland And Sweden

Johanna Routa,* Antti Asikainen, Rolf Björheden, Juha Laitila and Dominik Röser

Finnish Forest Research Institute, Yliopistokatu 6 FI-80101, Joensuu, Finland

INTRODUCTION

The European Commission as well as countries outside the EU have set targets to increase the share of renewable energy sources, the so called 20/20/20 targets[1] with the goal to mitigate climate change. The "Directive for renewable energy" promotes the use of forest biomass for energy purposes. To meet these objectives, a large share of this increase has to be based on forest biomass. The energy sector, particularly district heating and combined heat and power production, is increasing the use of forest biomass as a fuel in Finland and Sweden. Recent studies indicate that EU's forests could supply approximately 200–400 million m^3 (1500–2900 PJ, less than 5% of total energy consumption) woody biomass for energy annually by 2020.[2,3] To reduce greenhouse gas emissions, the Council of Europe accepted the proposal of the European Commission that EU countries should produce 20% of their energy by using renewable sources, including bioenergy, by 2020. Each member state has their own target, for example, Finland should produce 38% and Sweden 49% of their energy from renewable sources by 2020.

The absolute bulk (in Sweden around 85%) of renewable fuels from forests are consequential by-products such as black liqueurs, bark saw dust, and shavings from the conventional forest industries. But since all such by-products are already utilized, Finland and Sweden are now developing supply systems for such fractions that have previously been left in the forest after logging; small trees, logging residues and stumps. The use of these fractions referred to as primary forest fuels or forest chips, has been increasing steadily over the last decade and they may contribute significantly higher and important volumes in the future. Under Nordic conditions, the biomass of tops, branches, stumps and coarse roots correspond to about 50% of the merchantable bole volume from a tree in final fellings.[4]

The annual consumption of forest chips in heating and power plants has already increased over fivefold since the year 2000, up to 6.2 million m^3 solid (m^3 refers to solid m^3 throughout this paper) per year by 2010 in Finland (Figure 17.1).[5] The target set by Ministerial Working Group of the Finnish Government for climate and energy policy is 13.5 million $m^3\ a^{-1}$ (90 PJ) by 2020. The technical harvesting potential of forest chips in Finland has been estimated to be 14–20 million m^3 per year.[6] About 45% of this amount could be obtained as small-dimension wood from early thinnings.[10] The use of small diameter stem wood in heat and power plants increased remarkably in recent years and was the most important raw material for forest chips (2.5 million m^3) in the year 2010. In Finland, the subsidies for small-diameter wood harvesting and comminution are essential that the harvesting is economical if compared to procurement of forest chips from logging residues.[11] The use of logging residues (2.2 million m^3) and stump and root wood (1 million m^3) also increased in recent years. The use of robust stem wood in 2010 was 0.5 million m^3. In 2010, forest chips were used by almost 1000 power and heating plants in Finland.[12] In Sweden, the consumption of

*Correspondence to: johanna.routa@metla.fi

Advances in Bioenergy: The Sustainability Challenge, First Edition. Edited by Peter D. Lund, John Byrne, Göran Berndes and Iacovos A. Vasalos.

Originally published in: *WIREs Energy Environ* 2013, 2: 273–284 doi: 10.1002/wene.24

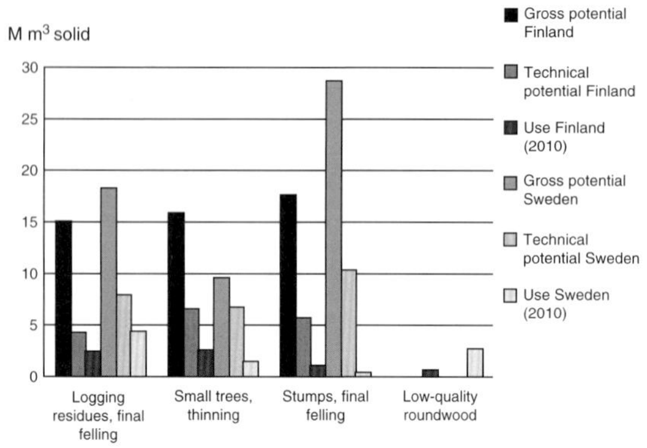

Figure 17.1 Raw material sources, technical harvesting potentials, and gross potentials of forest chips in Finland and Sweden in 2010.

forest chips was 8.4 million m^3 in 2010 (Figure 17.1).[13] The technical harvesting potential of forest chips in Sweden is approximated to be 20–25 million m^3 per year.[14] Half the gross supply (56 million m^3 per year) is exempt for ecological or technoeconomic reasons. Logging residues have been the most important raw material for forest chips in Sweden (4.3 million m^3, i.e., roughly half the potential) in the year 2010. The use of small diameter wood was 1.3 million m^3 (20% of the potential), stump and root wood 0.3 million m^3 (2–3% of the potential) and roundwood 2.5 million m^3 in 2010.[13] Although the full primary fuel potential of Swedish forests cannot be economically harvested, it is possible to more than double the annual utilization.

RAW MATERIALS USED IN FOREST ENERGY PRODUCTION IN FINLAND AND SWEDEN

General Problems

The available primary forest fuel feedstock combines difficult properties such as and great diversity in shape, bulkiness, small-piece size, and scattered occurrence with low relative value, heterogeneity and, often, contaminations increasing ash content, tear, wear, and damage to equipment and machines.[4] The demand for forest fuel varies throughout the year, while supply is generally stable. Consequently, it is necessary to store the biomass over a certain time.[15] Further, the operations must be performed under the constraints of a sensitive environment with no or low impact on soil, water, and remaining growing forest stand.[16,17] Thus, the nature of forest energy operations is challenging. This description of current systems should be seen in the light of these diverse problems.

The development of forest energy operations in Finland and Sweden reflects the need to tackle these general problems. Exploitation of existing economies of scale is facilitated through concentration of the scattered material and is a very important prerequisite for the employment of large-scale operations yielding low costs but also demanding large volumes. This is done in many steps of the supply chain to allow for efficient bulk handling. Bundling and baling may be used to compact the material increasing the density and improving the handling properties. Efficient long-distance transportation is currently the most important bottleneck for increased use of forest biomass for energy in Finland and Sweden[18] and can only be solved by even more efficient concentration, densification and improvement of material value to allow large-scale transports. The low quality of the material may be improved by drying, storing, screening, and sorting. This, combined with comminution (comminution in the context of this article refers to either chipping or crushing) will increase homogeneity, handling properties, and density of the material. Since the availability of primary forest fuels is largely dependent on conventional logging, integration with the main operations of forestry may be used as a means to cut costs and increase efficiency.

In recent years, a lot of attention has been put on improving the eco-friendliness of forest operations, including forest energy harvesting. Examples are identifying types of operations with the lowest impact under varying conditions, purposeful planning procedures as well as follow-up and training programs for machine operators and contractors. Consequently, techniques and methods for restorative and compensatory actions have been developed and put into action. Nevertheless, much remains to be done.

Logging Residues

The most productive forest sites in Finland and Sweden are dominated by Norway spruce *(Picea abies)* and recovery of logging residues is mainly carried out after the final felling of such areas. In Finland approximately 700 million m^3 of the growing stock is made up of spruce and 1100 million m^3 of pine. In Sweden, these figures are 1285 million m^3 and 1231 million m^3, respectively. Logging residues consist of tree tops, branches, needles/leaves, and unmerchantable stemwood.[18] Typical logging residue removal was approximately 35–50 m^3 ha^{-1}. In 2010, logging residues were removed from more than 50,000 ha in Finland and 70–80,000 ha in Sweden.[19,20]

In Finland, most of the logging residues are comminuted at the roadside using, for the most part, truck-mounted chippers. Furthermore, a small share of logging residues is also transported as loose to the CHP plant where it is then processed using stationary, electrically powered crushers. The importance of bundling has decreased in recent years and currently approximately 15 bundling units are operational in Finland.[21] In Sweden, the system layout is different. Almost 90% of the residues are comminuted at the roadside, two thirds by forwarder-mounted chippers. Their main advantage is that residue piles do not need to be placed within boom reach of the landing, simplifying planning and also allowing placement at optimal drying conditions. Recently, chipper trucks, that is, chip trucks also including a chipper unit, are quickly gaining in popularity, performing one third of roadside comminution. They can work independently, and since they do not have additional costs for changing between objects they can operate on small sites. Further, with only one machine involved the system becomes independent of delay times (i.e., time that machines in the same system have to wait on others). The remaining 10% of forest residues are comminuted at terminal or at end consumer by stationary, electrically powered crushers. Bundling has now practically disappeared in Sweden.

Small-Diameter Thinning Wood

Chips from small-diameter thinning wood are produced from whole trees ($d_{1.3}$ < 10 cm), delimbed trees and pulpwood-sized trees harvested in young stands. A typical small-sized energy wood harvesting site in Finland is an overstocked and unproductive hardwood or Scots pine stand, where good tending practices have been neglected.[22] Typical removal is 40–60 m^3 ha^{-1}. In 2010, thinning wood for energy use was harvested in about 40,000 ha in Finland and 20,000 ha in Sweden.[19,13] The development of machinery for the harvesting of small diameter trees has been fast in the last decade. The harvesting can be integrated to timber harvesting or it can be totally separated.[12] Most of the thinning is made as whole tree harvesting, but the harvesting as delimbed stems is increasing due to better chip quality and transport economy.[23] In Sweden, the predominating system is a small one-grip harvester with accumulation equipment producing roughly delimbed tree sections, while terrain transport is carried out by a conventional forwarder. Simpler felling heads are also used, but mainly on infrastructural objects such as roadsides and powerlines and volumes are comparatively small.[20,13]

Stump and Root Wood

In Finland, typical stump wood removal is 40–60 m^3 ha^{-1}. In 2010, stumps were removed from a total area of around 20,000 ha or roughly 10% of the annual regeneration cutting area.[19] In Sweden, the stump harvesting is approximately 2500–3500 $ha^{-1}a^{-1}$ or roughly 1% of the annual regeneration cutting area.[20] Stumps are harvested mainly from spruce regeneration areas since spruce stumps have a high wood content as well as being superficially and, thus, more loosely anchored in the ground.[24] Stump and root wood are dry and homogenous raw materials and proper storage improves their quality even further. During storage, stump wood does not absorb water as rapidly as, for example, logging residues. Therefore, the consumption of stump wood chips is concentrated in the cold winter period when the need for energy is at its highest. Stones, sand, and other impurities might cause problems during comminution and combustion, if the harvesting work has not been performed properly.[24]

Roundwood

In general, roundwood will be used for energy generation only if the value of it is higher for that purpose than it is in using it for sawlogs or pulpwood or there is no demand. Chips from roundwood are produced from stems, which are not acceptable for timber, for example, rotten wood or spoilt pulp wood, which has not been delivered to further processing in time. Furthermore, it includes wood, which does not fulfill the tight quality requirements for timber in terms of color, rotten content. Most of the roundwood for energy use is imported in Finland. The use of robust stem wood in 2010 was 0.5 million m^3 and its utilization decreased remarkably from the previous year.[5] In Sweden, the use of low quality roundwood in 2010 was 2.5 million m^3.

FOREST ENERGY SUPPLY CHAINS IN FINLAND AND SWEDEN

Forest energy supply chains are built around the comminution phase. The position of the chipper or crusher in the procurement chain determines the state of biomass during transportation and whether subsequent machines are dependent on each other, that is, *whether the system is hot or cool*. In "hot system" subsequent machines are dependent on each other. In a "cool system" the machines involved operate independently of each other which eliminates delay times between machines.[10] Comminution may take place on the logging site, at the road side landing, at a terminal, or at the plant. By concentrating the comminution to terminals or plants it is possible to work effectively and get rid of the problems of "hot systems" such as waiting and queuing at the landing. In general, forest energy supply chains can be divided to chains based on *Roadside comminution* (Figure 17.2), *Terminal comminution* (Figure 17.3) or *Comminution at the plant* (Figure 17.4). Comminution on the cutover has ceased in Sweden and Finland due to logistical challenges and the fact that it is not a cost-effective system.[12,20]

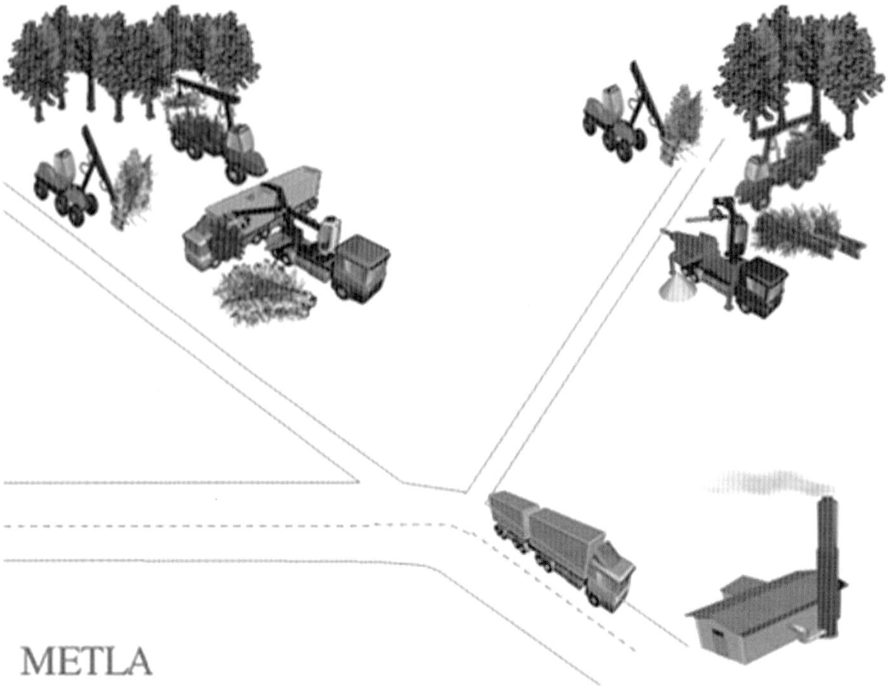

Figure 17.2 Forest fuel supply chain based on comminution at the landing. On the left logging residues from final harvest, truck-mounted chipper. On the right small diameter trees from early thinning, truck-mounted chipper.

Roadside comminution is the predominant option of forest chip production. In Finland, about 70% and in Sweden 90% of the logging residues are comminuted at the roadside landing close to the logging site. In Finland, also about 83% of the small sized wood and some 10% of the large-sized roundwood for energy was comminuted at roadside in 2010.[25] In Sweden less of the roundwood is comminuted at roadside.[15,26] The biomass is forwarded to the landing and piled there. Comminution is performed at the landing using farm tractor-driven chippers in smaller operations and heavy truck-mounted chippers or crushers in Finnish large-scale operations. Chips are blown directly into a chip truck having 100–140 m^3 bulk load space, a process that makes the system "hot" and vulnerable, that is, subsequent machines are dependent on each other.[10] Chippers or chip trucks may waste a remarkable part of the time by waiting and other stoppages consequently reducing their operational efficiency. Furthermore, large biomass storage piles and the space requirements of chipper and chip truck bring on large space requirements. Swedish operations rely on forwarder-mounted (65%) or chipper-truck (33%) comminution.[13] In the case that interchangeable containers are used, waiting and queuing can also occur, however, the associated problems are normally smaller. In the case that chips are blown directly onto the ground or snow, which is a commonly used practice in Sweden, a separate loader or a bucket crane on the chip truck is used to load the chips,[20] interference between the different units within the supply chain is insignificant. Also when a chipper-truck system is used, the system waiting and queuing is eliminated. These advantages are gained at the expense of increased capital commitment and/or lower payload.[20] The chipper truck blows the chips directly into containers or a conventional cargo hull and then hauls the load to the plant. When containers are used, the system also requires space for handling the containers. As only a single unit is needed, the chipper truck is suitable for small work sites and for delivering chips to small heating plants.[10] Roadside comminution is a flexible and well proven production chain. The availability of harvesting machinery in the Nordic countries is very good. With a separate chipper and chip truck, the chain becomes hot and utilization rate of chipper may be low and waiting times long, leading to low operational efficiency. The roadside storage space has to be large, and in practice the storage areas are often too small and muddy. For the chipper-truck system, the truck has a small payload

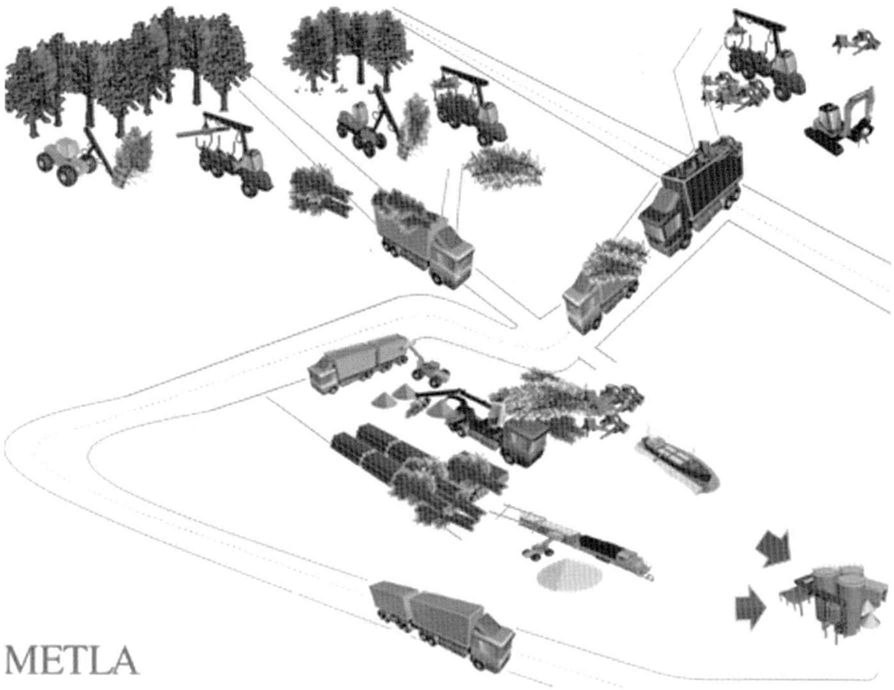

Figure 17.3 Forest fuel supply chain based on comminution at terminal.

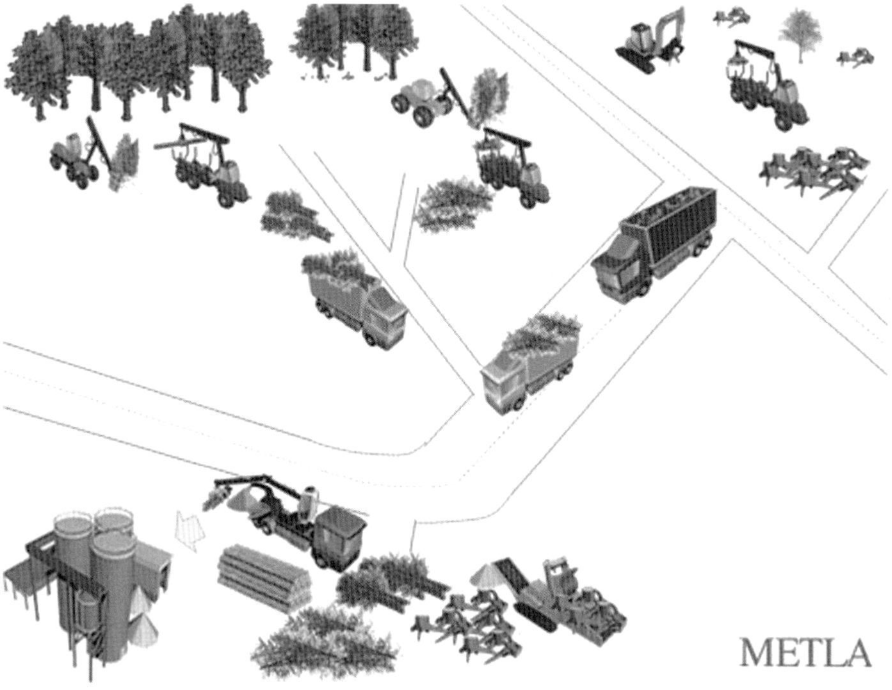

Figure 17.4 Forest fuel supply chain based on comminution at power plant.

and if transportation distances are very long the efficiency suffers, but in such cases, a container chipper-truck may act as a roadside chipper for a group of chip trucks.

Terminal comminution means that the forest biomass is transported to the terminal for comminution, and then optionally stored, mixed, and transported by truck, train or barge to the plant. About 7% of logging residues, 10% of all the chips from small-sized wood and about 41% of stump and root wood were comminuted at terminals in Finland in 2010.[25] About 23% of large-sized roundwood was comminuted at terminals. Due to high building costs, terminals require large volume flows to be competitive. Terminal comminution chains diminish the interactions between comminution and transport and the quality and the management of chip supply is high. Furthermore, the safety of chip deliveries can be guaranteed in all seasons. The comminution process is effective and can handle most types of biomass. A weakness is the low bulk density of the biomass in the transportation to the terminal which often takes place in an unprocessed form as loose residues, whole trees or pieces of stump wood. Also longer transport distances, high investment costs and small value added to the chain are weaknesses of the terminal comminution system. In Finland, the size of the load is usually limited by the bulk volume rather than legal mass capacity.[27] Comminution and long-distance transportation are independent of each other, which results in a high degree of capacity utilization and thus relatively low comminution costs. Loading of chip trucks with a front end loader, however, has interactions with the chip transportation. In addition, extensive investment in the centralized comminution system presupposes full employment and large annual comminution volumes.[28] Identifying ideal terminal areas is challenging and the total costs of supply chain can be relatively high.

Comminution at plant makes the chipper and chip truck independent of each other. About 23% of the logging residues, about 7% of all the chips from small-sized wood and about 51% of stump and root wood were comminuted at power plants in Finland in 2010.[25] In addition, in 2009, 70% of the large-sized roundwood for energy was comminuted at power plants. The technical and operative availability of the equipment increases, control of the procurement process is facilitated, demand for labor is decreased, and the control of fuel quality is improved.[10] Heavy stationary equipment may be used, chippers or crushers, which are suitable for comminution all kinds of biomass, including stumps and recycled wood. In general, fuel flow should be as high as possible in order to ensure the largest benefits. Because the investment cost is high, only large plants can afford a stationary crusher. The system can reduce interactions between transport and comminution. To be economic, the supply must be large-scale and produced more than 100,000 m^3 annually. If the transportation distances are short, comminution at plant is the most cost-efficient supply chain. The weakness of this system if the material is not compacted through, for example, bundling or precomminuting, is the low bulk density, leading to high transport costs.

Harvesting of Logging Residues

There are two alternative systems for the procurement of logging residues: (1) loose residues and (2) bundled residues (about 10% and 1% of the logging residues were bundled in 2009 in Finland and in Sweden, respectively.[20,26] The harvesting method is adapted so that logging residues are piled during the timber processing and only the residues needed to improve the soil bearing capacity are left on the strip roads.[28] After drying, the logging residues are forwarded to roadside, normally by a forwarder with an enlarged load space and a residue grapple. Forwarding may be combined with soil preparation. Combined forwarding and soil preparation has been tested in Finland and Sweden and it seems to increase productivity.[29,30]

Bundling may be a feasible alternative when long forwarding and road transportation distances are needed to ensure availability of forest chips.[31] The currently used bundling process is continuous and automatic. The bundles are hauled by a standard forwarder to the roadside storage area, from where they are transported to the power plant by a standard timber truck. Bundling aims at improving efficiency and flexibility in supply chains especially when transportation distances are long. This system is cost efficient only if the logistics and cost savings in the other work phases of the supply chain covers the costs of bundling, or if the fuel customers are willing to pay extra for the improved storing and handling properties of the bundled material.[36]

Harvesting of Small-Diameter Thinning Wood

In Finland and Sweden, logging systems in small diameter wood thinning wood can be classified into those, which are based on the *motor-manual or mechanized* cutting of trees. Manual cutting is carried out using a chainsaw, which is equipped with a felling frame. This working technique is called “lift-felling”.[37–39] Two alternative systems are used in mechanized whole tree logging: (1) *the traditional two-machine system* (a harvester and a forwarder) and (2) *the harwarder system* (i.e., the same machine performs both cutting and forest haulage to the roadside).[40,41] Both the harvester and the harwarder utilize the multitree processing technique. When using harvesters as a base machine, the cutting of whole trees can be done with purpose built

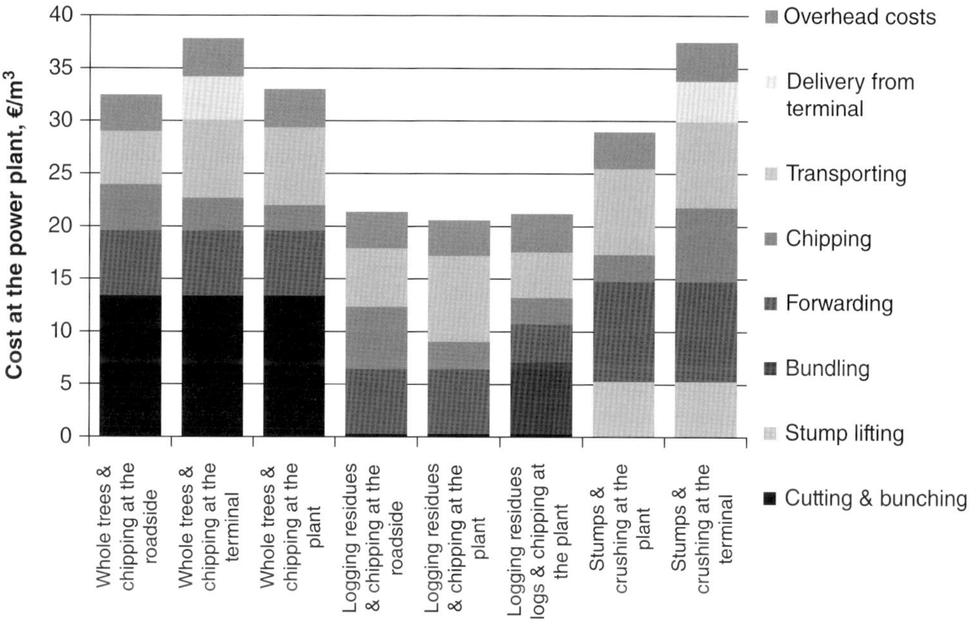

Figure 17.5 Procurement costs (»/m3 solid) of forest chips with different harvesting systems from different raw materials.[12] The forwarding distance was 250 m and transport distance was 45 km.

accumulating felling heads or by normal harvester heads equipped with multitree handling accessories.[42] It has been estimated that currently more than 300 harvesters equipped with either a felling head or energy wood harvester head with accumulating and feeding properties operate in the cutting of small-sized thinning wood in young stands in Finland.[43] Currently, there are also more than 50 energy wood harwarders in use in Finland either seasonally or throughout the year. In Sweden, practically all large-scale fuel procurement from young stands is carried out by mechanized harvesting; 2/3 by accumulating harvesting heads who normally compact, part-delimb, and buck the small trees; and 1/3 by accumulating felling heads with shears which cannot delimb nor compact the trees but may buck them into transport lengths.[13]

Multiple tree-handling is achieved by felling more than one tree in one cut or by felling one tree per cut and accumulating trees on the felling head, thereby handling several trees during one crane cycle.[44] According to Bergkvist,[45] productivity rose by 18% from 7.9 m^3 per productive hour (E15) in single-tree handling to 9.3 m^3 per hour in multi-tree handling in roundwood harvesting. The felling phase is still the most demanding and costly part of the production chain in energy thinning (Figure 17.5).[12,20,22] The costs can be reduced by increased utilization of the machines' capacity for multiple tree handling,[44,46] although the benefits of accumulation seem to abate after 4–5 trees per felling cycle.[47] Also, choosing corridor felling instead of discrete point felling reduces the costs.[44,48,49]

The harwarder is a machine for both cutting and forwarding.[50] Compared to the normal harvester–forwarder chain, translocation of machinery is cheaper when only one machine is moved between sites. A full crane cycle may be omitted if the wood is directly loaded instead of sorting it on the ground. The harwarder method provides more diverse work for the operator, since the cutting and forwarding tasks are integrated.[51–53] A dual purpose machine diminishes organizational costs and need for management, when two tasks can be accomplished with a machine and a single pass at the site.[54] It also enables better employment of machinery and a higher degree of capacity utilization, when two tasks are done with a single machine.[30] However, the harwarder is a compromise machine; it is designed both for harvesting and loading, and it is thus a little bit clumsy compared to special machines.[51,55] Finally, it is also an expensive machine, because it is a combination of a forwarder and harvester.

Integrated Timber and Energy Wood Harvesting

Integrated timber and energy wood harvesting has become more common as logistical costs can be reduced considerably. The most common method is the two-pile method where roundwood is piled in one pile and tops and small trees for energy in another. Small-diameter stems can be also delimbed. Multistem harvester heads are necessary for efficient combined harvesting of delimbed or undelimbed pulpwood and energy wood. With the two-pile system the productivity is about 10% lower than in roundwood harvesting in first thinnings.[56]

Another method for integrated harvesting is a whole tree bundler-system, which has been tested in Finland. Whole-tree bundling in early thinnings can generate significant benefits for the procurement system, by increasing the productivity of forwarding and making long-distance transport economically feasible. In the deepest mode of integration, the pulp and energy fractions are separated in the debarking drum of the pulp mill.[22,57] The main weaknesses of the bundling system are the high cost of bundle harvesting, caused by the high capital cost in combination with low productivity,[58] and the poor quality of both fuel and pulp leading to low revenues.[57]

Based on the literature the two machine system has been found to be the most competitive logging system in early thinning due to the efficient forwarding work.[7,52,53,59] High logging costs, especially felling and piling costs, are the main problems in early thinnings.[20,22] To increase the procurement volumes of energy wood in young stands, which, together with stumpwood, is the largest unutilized source of forest fuels in Finland[6] and a rapidly growing source of energy in Sweden,[13,20] logging costs have to be significantly reduced. This calls for improved methods and equipment, learning-by-doing[60] or development of novel machines, technology and systems.[20,44,61] In the transportation phase, improvements target maximizing of the payload within legal weight and dimensional limits and the minimization of idle time at the terminal points.[62,63] Optionally, logging conditions may be improved by considering new silvicultural alternatives for young forests[64] and by identifying such stands where the conditions are favorable for energy wood recovery. With current technologies, cutting of very small-sized trees (DBH 1–6 cm) should be avoided to improve the profitability of energy wood logging.[20,65] There is also a need for improvements and rationalization in the management of procurement activities.[66]

Harvesting of stumps. The stumps are uprooted using an excavator that is equipped with a stump lifting-splitting device.[20,67] Processed stump and root wood is piled into small piles in the stand. Piles are located so that the rain and weathering can rinse soil and dry the material. On smaller sites, soil preparation for forest regeneration can be combined with stump extraction using a mounding blade, which is a part of the extraction device.[67] After initial storage in the stand, the piled stump wood is forwarded to the roadside for further storage. Forwarders are similarly equipped as logging residue forwarders. After seasoning at the roadside, the stumps are transported to the terminal or end-use facility for comminution by a truck specifically developed for transporting uncomminuted biomass. If the stump parts are precrushed and screened at the landing, standard chip transportation vehicles may be used, the payload is increased and the proportion of soil contaminants is reduced.[20] Stump uprooting is possible also with separate splitting cranes for large harvesters and forwarders, but heavy-duty tracked excavators, weighing approximately 20 tonnes, are the most cost-efficient machines.[68]

THE NEXT DECADE AND BEYOND

The mobilization of additional forest biomass resources calls for new technological solutions.[49,69,70] Foresight studies show that in the developed biomass markets such as Finland and Sweden and partially Central Europe, the growth of biomass demand will be from the large combined heat and power (CHP) plants and from possibly emerging large biorefineries.[71,72] This calls for the development of supply chains that are cost efficient and sustainable also when supply stretches over very long distances.[20] The manpower requirements is estimated at over 40,000 machine operators in the EU by 2030.[73] In the Nordic countries, entrepreneurs running roundwood harvesting for industrial purposes have often extended their operation to energy biomass supply. Although forest energy use has grown fast in several countries, the low demand of wood products and paper has released manpower from roundwood harvest to energy wood harvesting. As forest industries are now recovering, the availability of skilled labor may become a serious bottleneck. Consequently, the training and education of forest machine operators has to be the focus of further developments to reach the ambitious EU targets in regards to climate change.

The productivity of machinery ultimately depends on the operator's skills and thus new technological solutions should be developed to improve the ease of use and versatility of the expensive machinery. Current forestry practices support the production and harvesting of industrial roundwood. Several studies show that by integrating the procurement of industrial roundwood and energy wood there is a possibility to improve the economy of operations.[23,64] Logistical solutions and service and management innovations for integrated supply have to be developed further. Solutions that support the mobility of machines, optimal work patterns, and economical driving are under development in the most advanced roundwood harvesting systems. This approach can be transferred to energy wood harvesting and particularly processing of the material in the terminals and other material handling units can be largely automated. The current supply chains can be improved to innovative new supply chains in cooperation between research, technology providers, fuel suppliers, and users.

The product quality and fuel efficiency of operations may also be improved. The most important factor affecting energy balance is the low net calorific value due to the high moisture content of forest fuel. By reducing the amount of foliage and moisture content of wood fuel through improved production practices, the calorific value and thus the quality of forest fuel may be improved. Also, the evaluation and feasibility of automated loading (and discharging) facilities, economics and separate testing "full load" measuring systems will be important. In many areas in Europe, procurement of biomass needs to stretch over longer distances due to the long distance between resources and end users. Supply chains for mobilizing distant forest fuel sources with large trucks and/or intermodal transportation (truck + railway/boat) have to be tested and developed.[20]

CONCLUSION: LESSONS LEARNED IN FINLAND AND SWEDEN

In the market penetration phase, the biomass supply can be subsidized. But to achieve an economically sustainable long-term supply, large investments are needed in the development of technology for forest fuel procurement. Investments and strong commitment of research organizations as well as forest and energy industries are necessary for successful operations. The end users also need to modify the raw material reception, handling, and combustion technologies to be able to use forest fuel more efficiently. Involvement of forest owners and development of forest energy trade and measurement practices are prerequisites for the market entrance of forest energy.

Forest fuel supply has to be integrated with industrial wood harvesting, because both industrial roundwood and residual forest biomass for energy are often extracted from the same stands. Using the same base machinery as in the roundwood harvesting is a reasonable strategy to start the operations. As the business grows, the capacity for investments in purpose built machinery and equipment will increase.

The experience in Finland and Sweden has also shown the importance of relying on proven and established technology and integration of operations to establish systems based on forest biomass for energy. This has helped to build trust among all stakeholders in the supply chain and in the public domain.

Because of the high costs of establishing the infrastructure for biomass-based energy production, long-term raw material availability has to be ensured. This is particularly true for larger biomass users, which must draw from a very large the procurement area.

Use of forest biomass for heating is a simple, reliable, and efficient way to use biomass. Any production of higher forms of energy includes conversion losses in the form of heat. If high-end conversion technologies (i.e., combined heat and power generation, and liquid biofuels) are used, it is essential for the overall economics that the excess heat can be used. Thus, large enough customers for the excess heat play an important role in bioenergy systems.

REFERENCES

1. European Commission. Combating climate change the EU leads the way. *Eur Move Ser* 2008, 24.
2. Asikainen A, Liiri H, Peltola S, Karjalainen T, Laitila J. Forest energy potential in Europe (EU27). In: *Working Papers Finnish Forest Research Institute, Joensuu*. Joensuu, Finland; 2008, 69.
3. Verkerk PJ, Anttila P, Eggers J, Lindner M, Asikainen A. The reliable potential supply of woody biomass from forests in the European Union. *Forest Ecol Manag* 2011, 261:2007–2015.
4. Hakkila P. *Utilization of Residual Forest Biomass*. New York: Springer; 2011, 568.
5. Ylitalo E, ed. *Wood Fuel Used for Energy Generation 2010*. Finnish Forest Research Institute, Forest Statistical Bulletin ISSN 1797-3074; 2011, 7. [In Finnish].
6. Helynen S, Flyktman, M, Asikainen, A, Laitila J. Scenarios on wood-based bioenergy business in Finland for the forestry and forest industry. VTT Research Notes 2397, 2007, 66. [In Finnish].
7. Laitila J, Asikainen A, Anttila P. Energiapuuvarat. In: Kuusinen M, Ilvesniemi H, eds. *Energiapuun korjuun ympäristövaikutukset, tutkimusraportti. Tapion ja Metlan julkaisuja*, 2008. Available at: www.metsavastaa.net/energiapuu/raportti. (Accessed January 12, 2012).
8. Kärhä K, Elo J, Lahtinen P, Räsänen T, Pajuoja H. Kiinteiden puupolttoaineiden saatavuus ja käyttö Suomessa 2020. In: *TTS tutkimuksen tiedote*. Rajamäki, Metsä: Luonnonvara-ala; 736, 2009, 8.
9. Salminen O. Energiapuun ja ainespuun hakkuumahdollisuudet. Valtakunnan metsien inventoinnin tiedotustilaisuus 22.6.2010. Available at: www.metla.fi/tiedotteet/2010/VMI/salminen-esitys.pdf. (Accessed January 12, 2012).
10. Hakkila P. Developing technology for large-scale production of forest chips. In: *Wood Energy Technology Programme 1999-2003*, Final report. Helsinki: Tekes; 2004, 99.
11. Petty, A, Kärhä K. Effects of subsidies on the profitability of energy wood production of wood chips from early thinnings in Finland. *Forest Policy Econ* 2011, 13:575–581.
12. Laitila J, Leinonen A, Flyktman M, Virkkunen M, Asikainen A. Metsähakkeen hankinta- ja toimituslogistiikan haasteet ja kehittämistarpeet. In: *Procurement and logistics of forest*

biomass- challenges and development needs. VTT-Research Notes. Vol. 2564. Helsinki: VTT; 2010, 143.
13. Brunberg T. [Skogsbränsle: sortiment, metoder och kostnader 2010.] Forest fuel: assortments, methods and costs 2010. Skogforsk Resultat nr 8, 2011. Uppsala: Skogforsk; 2011.
14. The Swedish Forest Agency. [Skogliga konsekvensanalyser 2008.] Forest sector consequential analyses 2008. *Report* 2008:25. [In Swedish].
15. Björheden R. Making bioenergy systems work. In: *Proceedings of 4th Forest Engineering Conference: Innovation in Forest Engineering—Adapting to Structural Change.* Stellenbosch, South Africa Stellenbosch University; 2011.
16. Hall DO, Scrase JJ. Will biomass be the environmentally friendly fuel for the future? *Biomass Bioenergy* 1998, 15:357–367.
17. Angelstam P, Pettersson B. Principles of present Swedish biodiversity management. *Ecol Bull* 1997, 46:191–203.
18. Björheden R, Eriksson LO. Optimal storing, transport and processing for a forest-fuel supplier. *Eur J Oper Res* 1989, 43: 26–33.
19. Ylitalo E, ed. Metsätilastollinen vuosikirja 2011-Skogsstatistisk årsbok - Finnish Statistical Yearbook of Forestry. SVT Maa-, metsä- ja kalatalous. Sastamala: Metsäntutkimuslaitos; 2011, 472 p. ISBN 978-951-40-2330-9, ISBN 978-951-40-2329-3.
20. Thorsén Å, Björheden R, Eliasson L, eds. Efficient forest fuel supply systems, composite report from a four year. R&D programme 2007–2010. Uppsala: Skogforsk; 2011. 113 p. ISBN 978-91-977649-4-0.
21. Kärhä K. Supply chains of forest chip production in Finland. *Bioenergy* 2009 Conference proceedings. Jyväskylä, 2009.
22. Laitila J. Harvesting technology and the cost of fuel chips from early thinnings. *Silva Fennica* 2008, 42:267–283.
23. Laitila J, Väätäinen K. Kokopuun ja rangan autokuljetus ja haketustuottavuus. *Metsätieteen aikakauskirja* 2011, 2: 107–126.
24. Hakkila P. Juurakot polttoainelähteenä (Stumps as a source of fuel wood). *BioEnergia* 2003, 4:32–35. [In Finnish].
25. Kärhä K. Metsähakkeen tuotantoketjut Suomessa vuonna 2010. Metsätehon tuloskalvosarja 6/2011. *Metsäteho* 2011. [In Finnish]. Available at: http://www.metsateho.fi/files/metsateho/Tuloskalvosarja/Tuloskalvosarja_2011_06_Metsayhakkeen_tuotantoketjut_2010_kk.pdf. (Accessed January 15, 2012).
26. Kärhä K. [Metsähakkeen tuotantoketjut Suomessa vuonna 2009.] Industrial supply chains of forest chips in Finland in 2009. Metsäteho 2010. [In Finnish]. Available at: http://www.metsateho.fi/files/metsateho/Tuloskalvosarja/Tuloskalvosarja_2010_09_Metsahakkeen_tuotantoketjut_Suomessa_2009_kk.pdf. (Accessed January 15, 2012).
27. Ranta T, Rinne S. The profitability of transporting uncomminuted raw materials in Finland. *Biomass Bioenergy* 2006, 30:231–237.
28. Asikainen A, Ranta T, Laitila J, Hämäläinen J. [Hakkuutähdehakkeen kustannustekijät ja suurimittakaavaisen hankinnan logistiikka.] Cost factors and large scale procurement of logging residue chips. In: Research notes 131. Faculty of Forestry, University of Joensuu; 2001. [In Finnish].
29. Laitila J, Asikainen A, Hotari S. Residue recovery and site preparation in a single operation in regeneration areas. *Biomass Bioenergy* 2005, 28:161–169.
30. Von Hofsten H, Norden B. Nytt koncept: Kombinerad risskotare och markberedare, SkogForsk Resultat nr 11 Uppsala, 2002. 4 pp. [In Swedish].
31. Andersson G. Technology of fuel chip production in Sweden. In: Alakangas E, ed. *Nordic Treasure Hunt.* Extracting energy from forest residues. VTTsymposium, vol. 208; Espoo, 2000. p. 113.
32. Laitila J. [Puupolttoaineiden hankinta Oy Alholmens Kraft Ab:n voimalaitokselle.] Wood fuel procurement for Oy Alholmens Kraft Ab power plant. M.Sc. thesis, Faculty of Forestry, University of Joensuu; 2000. [In Finnish].
33. Björheden R, Gullberg T, Johansson J. System analyses for harvesting small trees for forest fuel in urban forestry. *Biomass Bioenergy* 2003, 24:389–400.
34. Kärhä K, Vartiamäki T, Liikkanen R, Keskinen S, Lindroos J. [Hakkuutähteen paalauksen ja paalien metsäkuljetuksen tuottavuus ja kustannukset.] Productivity and costs of slash bundling and bundle forwarding. Report of Metsäteho, Helsinki; 2004, 179 [In Finnish].
35. Kärhä K, Vartiamäki T. Productivity and costs of slash bundling in Nordic conditions. *Biomass Bioenergy* 2006, 30:1043–52.
36. Korpilahti A, Poikela A. [Hakkuutähdeketjujen vertailu UPM-Kymmene OY:lle.] Comparison of different logging residues supply chains to UPM-Kymmene. *Report of Metsäteho, Helsinki*; 1997, 33, 45. [In Finnish].
37. Harstela, P, Tervo L. [Männyn taimikon ja riukuasteen metsikön korjuun tuotos ja ergonomia.] Work output and ergonomical aspects in harvesting of sapling and pole-stage stands (Scots pine). *Folia Forestalia* 1977, 294:23. [In Finnish].
38. Hakkila P, Kalaja H, Salakari M, Valonen P. Whole tree harvesting in the early thinnings of pine. *Folia Forestalia* 1978, 333:58.
39. Ihonen M. [Pienpuun siirtelykaato kahvoilla varustetulla moottorisahalla.] Felling-piling of small-diameter wood using a chainsaw equipped with a felling frame. *Työtehoseuran metsätiedote Rajamäki,* 1997, 581, 4. [In Finnish].
40. Laitila J, Asikainen A. Energy wood logging from early thinnings by harwarder method. *Baltic Forestry* 2006, 12: 94–102.
41. Kärhä K. Whole-tree harvesting in young stands in Finland. *Forestry Stud* 2006 45:118–134.
42. Laitila J, Heikkilä J, Anttila P. Harvesting alternatives, accumulation and procurement cost of small-diameter thinning wood for fuel in central Finland. *Silva Fennica* 2010, 44:465–480
43. Kärhä K. Integrated harvesting of energy wood and pulpwood in first thinnings using the two-pile cutting method. *Biomass Bioenergy* 2011, 35:3397–3403.

44. Johansson J, Gullberg T. Multiple tree handling in the selective felling and bunching of small trees in dense stands. *Int J Forest Eng* 2002, 13:25–34.
45. Bergkvist I. [Flerträdshantering höjer prestationen och ökar nettot i klen gallring.] Multi-tree handling increases both productivity and profitability in smallwood thinnings. *Resultat från Skogforsk* 2003, 5. 4. [In Swedish].
46. Brunberg B. [Flerträdsteknik- en litteraturstudie.] Multitree-Handling Technology—A Literature Study. Kista, Sweden: Forskningsstiftelsen Skogsarbeten; 1989. [In Swedish].
47. Belbo H. Efficiency of accumulating felling heads and harvesting heads in mechanized thinning of small diameter trees. Dissertations No 66/2011. School of Engineering, Linnaeus University, Växjö; 2011. 42 pp.
48. Bergström D, Bergsten U, Nordfjell T, Lundmark T. Simulation of geometric thinning systems and their time requirements for young forests. *Silva Fennica* 2007, 41:137–147.
49. Bergström D. Techniques and systems for boom-corridor thinning in young dense forests. Umeå. *Sveriges lantbruksuniv, Acta Universitatis Agriculturae Sueciae*, 2009:87, 53 pp.
50. Asikainen A. Integration of work tasks and supply chains in wood harvesting—cost savings or complex solutions? *Int J Forest Eng* 2004, 15:11–17.
51. Rieppo K, Pekkola P. [Korjureiden käyttömahdollisuuksista.] Harwarders usability in logging operations. *Metsätehon raportti*, Helsinki; 2001, 121, 51 pp.
52. Sirén M. Economic and silvicultural performance of different thinning machinery. 2nd Forest Engineering Conference 2003 (Techniques and Methods): pages 3–12.
53. Sirén, M, Aaltio H. Productivity and cost of thinning harvesters and harvester-forwarders. *Int J Forest Eng* 2003, 14:39–48.
54. Talbot B, Nordfjell T, Suadicani K. Assessing the utility of two integrated harvester-forwarder machine concepts through stand-level simulation. *Int J Forest Eng* 2003, 14:31–43.
55. Björheden R, Dahlin B. The complete logging machine. In: Kellog L, ed. *The feasibility study of the harvester forwarder in Swedish forestry*. Proceedings of the IUFRO 3.10 Conference on .Timber Harvesting and Transport Technologies for Forestry in the New Millennium, South Africa; 1999.
56. Kärhä K, Mutikainen A. Moipu 400 ES ensiharvennuspuun integroidussa hakkuussa. *TTS research note* Rajamäki: TTS; 2008, 726:6. [In Finnish].
57. Jylhä P, Dahl O, Laitila, J, Kärhä K. The effect of supply system on the wood paying capability of a kraft pulp mill using Scots pine harvested from first thinnings. *Silva Fennica* 2010, 44:695–714.
58. Kärhä K, Jylhä, P, Laitila J. Integrated procurement of pulpwood and energy wood from early thinnings using whole-tree bundling. *Biomass Bioenergy* 2011, 35:3389–3396.
59. Wester, F, Eliasson L. Productivity in final felling and thinning for a combined harvester-forwardes (harwarder). *Int J Forest Eng* 2003, 14:45–51.
60. Junginger M, Faaij A, Björheden, R, Turkenburg WC. Technological learning and cost reductions in wood fuel supply chains in Sweden. *Biomass Bioenergy* 2005, 29:399–418.
61. Jylhä P, Laitila J. Energy wood and pulpwood harvesting from young stands using a prototype whole-tree bundler. *Silva Fennica* 2007, 41:763–779.
62. Angus-Hankin C, Stokes, B, Twaddle A. The transportation of fuel wood from forest to facility. *Biomass Bioenergy* 1995, 9:191–203.
63. Sikanen L, Asikainen, A, Lehikoinen M. Transport control of forest fuels by fleet manager, mobile terminals and GPS. *Biomass Bioenergy* 2005, 28:183–191.
64. Sirén M, Heikkilä, J, Sauvula T. Combined production of industrial and energy wood in Scots pine stands. *Metsanduslikud Uurimused -Forestry Studies* 2006, 45:150–163.
65. Kärhä K, Jouhiaho A, Mutikainen, A, Mattila S. Mechanized energy wood harvesting from early thinnings. *Int J Forest Eng* 2005, 16:15–26.
66. Asikainen, A, Kuitto P.-J. 2000. Cost factors in wood fuel procurement. *New Zealand J Forestry Sci* 2000, 30:79–87.
67. Laitila J, Ranta T, Asikainen A. Productivity of Stump Harvesting for Fuel. *Int J Forest Eng* 2008, 19:37–47.
68. Laitila J. [Kantojen korjuun tuottavuus.] Productivity of stump harvesting. In: *Working Papers of the Finnish Forest Research Institute*. Vol. 150. 2010, 29. [In Finnish].
69. Anerud, E, Jirjis R. Fuel quality of Norway spruce stumps—influence of harvesting technique and storage method. *Scand J Forest Res* 2011, 26:257–266.
70. Walmsley JD, Godbold DL. Stump harvesting for Bioenergy—a review of the environmental impacts. *Forestry* 2010, 83:17–38.
71. Eriksson KP. *Börjesson*. Potential utilization of biomass in production of electricity, heat and transportation fuels including energy combines—Regional analyses and examples. Report 64. Lund University, Department of Technology and Society, Environmental and Energy Systems Studies, Lund, Sweden; 2008.
72. Ree R, Annevelink B. Status report biorefinery 2007. Report 847. Agrotechnology and Food Sciences Group, Wageningen University, The Netherlands.
73. Asikainen A, Anttila P, Verkerk H, Diaz O, Röser D. Development of forest machinery and labour in the EU in 2010-2030. Formec Austria 2011, Pushing the Boundaries with Research and Innovation in Forest Engineering. Conference proceedings in cd, 2011.

18

Options for Increasing Biomass Output from Long-Rotation Forestry

Gustaf Egnell[1]* and Rolf Björheden[2]

[1]Department of Forest Ecology and Management, Swedish University of Agricultural Sciences, Umeå, Sweden
[2]The Forestry Research Institute of Sweden, Uppsala Science Park, Uppsala, Sweden

INTRODUCTION

Today, forest biomass plays an important role, supplying renewable energy. Globally, roughly half the volume of wood harvested is used as firewood[1] and approximately 50% of the gross inland energy consumption of renewable energy within EU 27 comes from forest biomass. Legally binding targets for the proportion of renewable energy sources in 2020 as set by the EU RES directive[2] will put additional strain on biomass resources in the future. The energy sector is currently looking for biomass resources within and outside the union as potential feedstock for existing and planned production capacity. Is it possible to increase the output of forest biomass for energy? The options are scrutinized from a long-rotation forestry perspective, dividing the possible actions by the three production stages of the forest sector[3]: (1) primary production—the net production of tree photosynthesis; (2) secondary production—activities connected with harvesting, intermediate processing, and transports from the forest stand to users of forest biomass; and (3) tertiary production—the (industrial) conversion of forest products into useful commodities. Primarily, experiences from Sweden, a country with relatively large forest resources, a developed forest industry and an emerging renewable energy industry based on biomass, will be used. Swedish forests are found in the boreal and temperate zone and are managed primarily by a clearcut system, with long-rotation periods (60–100 years) and a total dominance of only two major tree species, Scots pine (*Pinus sylvestris* L.) and Norway spruce [*Picea abies* (L.) H. Karst]. Thus, the identified options to increase biomass output mainly apply for long-rotation forestry in the boreal and temperate zone.

OPTIONS IN PRIMARY PRODUCTION

An obvious way to increase biomass output from long-rotation forestry in the short term is to harvest more. But under the restrictions of sustainable forest management, annual harvest levels cannot exceed annual growth over time. In the short term, the focus therefore has to be put on harvesting more of the tree biomass that is available, but not utilized by the traditional forest industry, i.e., excess growth, small and unmerchantable trees and tops, foliage, branches, and stumps from harvested sites. As a substantial proportion (30–50%) of the total biomass of a tree is to be found in these tree fractions,[4] this adds a large physical biomass potential. The realizable potential is set by environmental constraints[5] and techno-economic constraints.

When the immediately available resources have been utilized, the only option to allow increased future harvest levels is by increased forest growth. A general idea about this potential can be found in the history behind the development of the growing stock in Swedish forests. During

*Correspondence to: gustaf.egnell@slu.se

Advances in Bioenergy: The Sustainability Challenge, First Edition. Edited by Peter D. Lund, John Byrne, Göran Berndes and Iacovos A. Vasalos.

Originally published in: *WIREs Energy Environ* 2013, 2: 285–292 doi: 10.1002/wene.25

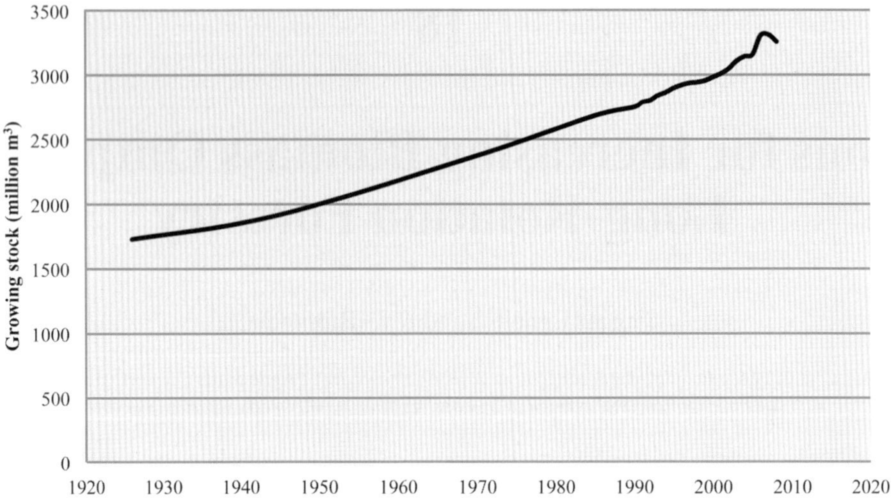

Figure 18.1 Development of the growing stock in Swedish forests from the 1920s up until today. Source: The Swedish National Forest Inventory, Swedish University of Agricultural Sciences, Umeå, Sweden.

the 18th and 19th century, Swedish forests were heavily exploited for timber, fuelwood, charcoal, tar, potash, and so on, and the forest land was also extensively used for grazing. To secure forest resources for the future, efforts to increase the growing stock were initiated in the early 20th century. The results of these efforts are reflected in data from the Swedish national forest inventory that includes data on forest resources in Sweden from the 1920s and onward (Figure 18.1). The growing stock has almost doubled, from 1760 in the mid 1920s to over 3000 million cubic meters (corresponding to approximately 1350 M ton dry matter) of stemwood today. The annual growth rate—establishing the sustainable harvest level over time—has developed from 57 to 117 M m^3 stemwood per year in the same period. There is still a potential to further increase biomass output from the forests in Sweden and more so in some other parts of the world by means of more intense or appropriate silvicultural practices and to use set aside and degraded land more wisely.[6]

STAND ESTABLISHMENT

The historical increase of the growing stock in Sweden was primarily the result of two things—namely, obligation to re-establish a new forest after harvest and regulations that secured increased future harvest levels by not allowing annual harvest levels far beyond annual growth. This shows the potential in the stand establishment phase during a rotation period in a clearcut system. But it is equally important to secure recruitment of new seedlings of preferred tree species in shelterwood and selection systems where the recruitment is based on natural regeneration.

For future forest growth, prompt re-establishment of the next tree generation is essential; therefore, treatments to establish the next tree crop should be started as soon as possible after harvest. Site preparation makes a difference when it comes to seedling survival and growth by altering a number of critical factors for the establishment of seedlings, e.g., soil moisture and temperature, nutrient availability, air temperature, amount of competing vegetation, and insect damages.[7] The choice of site preparation method is therefore crucial for future growth on many sites and the preferred site preparation method depends on site conditions, regeneration method, and tree species. Typically young seedlings suffer from competition for nutrients, water, and light from other vegetation—particularly on more fertile soils. Site preparation, together with mechanical or chemical vegetation control, therefore has a positive effect on the survival and growth of the new tree crop.[8]

The plant genotype is important for future growth and there is a high potential to increase future forest growth by using suitable and improved plants and seeds. Plants and seeds have to be produced, stored, and handled properly to make sure they retain health and vigor before they are planted or sown in the field. The clearcut system also allows for selecting tree species with the highest production potential on a particular site, including fast growing exotic species, in Sweden, e.g., lodgepole pine (*Pinus contorta* Dougl. var. *latifolia* Engelm.).[9] On top of that, genetic

Figure 18.2 Quick establishment of the new tree crop, proper site preparation, and selection of genetically improved and fast growing tree species are examples of treatments with the potential to reach large areas fast. Thus, they provide the most effective measures to substantially increase future forest production of the slow-moving long-rotation forestry.

selection of fast growing tree individuals, preferably resistant against major forest pests, will add additional growth.[10] New technology such as somatic embryogenesis and gene technology may help in speeding up this selection process in the near future (Figure 18.2).[11]

FERTILIZATION AND IRRIGATION

In many forests, growth is limited by nutrient availability, with nitrogen as the most common limiting nutrient on mineral soils—particularly in the temperate and boreal zone.[12] On drained forested peatlands, potassium and phosphorous are usually the most limiting nutrients, making wood ash, a residue from the energy industry, a suitable fertilizer with the potential to increase forest growth on such lands.[13] Fertilization is therefore one of the most common measures to increase forest growth in conventional forestry. More intense fertilization regimes where fertilizers are added earlier during the rotation period and more frequently, based on the nutritional needs of the tree crop, can increase this potential even further.[14] Forest growth is sometimes also limited by water availability where irrigation—preferably in combination with fertilization—can increase forest growth.[14]

OTHER OPTIONS

Other options to increase future forest growth include afforestation of set-aside land, ditching and maintenance of ditches, fire protection, forest growth-oriented wildlife management (reduced browser populations), and protection against major diseases. Furthermore, current silvicultural practices favor stemwood production rather than biomass production. Thinning regimes typically increase the radial growth of the remaining stems at the expense of total biomass production.[15] Thus, silvicultural systems favoring a high stem density (leaf area) will increase total biomass production. A simulation study indicated that forest production in Sweden could be almost doubled at the site level if more intense forest management regimes were applied. Thus, total forest production could increase by 1% for each percent of forest land that is used for intensive forest management.[6]

The conclusion is that there are many silvicultural practices to increase biomass output from long-rotation forestry. But even more important is the message that it takes time to achieve a significant effect on total growth rate in long-rotation forestry (cf. Figure 18.1). This is due to the fact that only a minute fraction of the forests is actively managed annually and any new treatment to improve forest growth will only reach a fraction of the forest land each year. Long rotations also mean that the profitability of large investments often is low. Typically, straightforward, uncontroversial, inexpensive treatments, that can be applied to large areas quickly, such as improved seedling stock or site preparation methods, have the highest potential to make a difference in total forest production. More intensive, environmentally questionable, and expensive treatments, that do not reach large areas fast, such as intense fertilization programmes, have a lower potential. Implementation of new silvicultural practices is also easier if the forest ownership is concentrated to few owners. In reality, it is often the opposite, with a large number of private forest owners that have to be convinced that treatments to increase forest production are a profitable investment.

OPTIONS IN SECONDARY PRODUCTION

Secondary production in forestry includes the activities connected with harvesting, intermediate processing, and transports from the forest stand to users of forest biomass.[3] Traditionally, these activities have been aimed at high diameter stem segments from preferred species, but a trend, observed already by Keays,[16] is the shift toward a more complete utilization of the forest. Much of this has already been achieved in Sweden and Finland, but even there, most of the potential remains untapped.

In principle, all forest biomass is instantly available for bioenergy purposes, but in practice, the utilization of forest biomass is regulated and delimited by a number of legal rules, economic and practical considerations, and policies. When describing the options for increased output from long-rotation forestry, exemplified by the case of Sweden, only two such delimiting elements are considered: (1) continued supply of the traditional forest industry and (2) sustained forest inventory.

The Swedish Forestry Agency[17] estimated that the harvest of nontraditional forest products (tops, foliage, branches, and stumps) could be increased from present 2.6 M tDM per year (approximately 5.8 million m^3 solid) to 28.7 M tDM per year at the current harvesting levels. In practice, however, the options are more restricted; not all available biomass may be harvested due to technical, economic, and silvicultural/ecological delimitations. On the contrary, in Sweden, there is an option to increase the harvesting level substantially without decreasing the forest inventory,[18] since annual felling falls below the net annual increment; approximately 18.9 M tDM per year is added to the standing inventory in Sweden.[19]

INCREASED OUTPUT OF BIOMASS FROM REGENERATION FELLING

Regeneration felling, representing 70–75% of the annual logging in Sweden, constitute the most sizeable potential source of additional biomass. After conventional logging for delimbed sawlogs and pulpwood, logging residues (branches, tops, and small and unmerchantable trees) can theoretically increase the harvest of biomass by approximately 25%.[20] The additional biomass is left on the logging site after the main operation and must be recovered by a second-pass system. In practice, the degree of logging residue recovery is lower. In well-planned practical operations, recovery rates of 17–20% additional biomass have been recorded after such two-pass harvesting.[21] An option for increasing the degree of utilization is to develop harvesting systems that will process the whole tree above the felling cut in a one-pass operation. Current trials in Sweden show that full-tree systems may be designed not only as longwood or tree-length operations but also as cut-to-length and chipping systems.[22] The increase of output compared with the current two-pass operation will be moderate, however, since there will always be losses due to breakage and waste. But if the cost of recovery is decreased by one-pass operations, the effects on output is greater since it counteracts the diseconomies of scale caused by large catchment areas, i.e., rising transportation costs. For every percent, the production cost may be decreased; the economically available output of biomass is in principle increased by 2%. Eriksson and Björheden[23] noted that *optimizing forest fuel production essentially means minimizing transportation costs.* The same statement seems valid for maximizing output. The lack of cost-efficient transportation systems has been identified as the single most delimiting factor for an increased use of forest biomass for energy in Sweden.[24]

Another sizeable option for increased output of biomass from regeneration felling is harvesting of the remaining stumps and coarse roots (Figure 18.3). The gross mass of stumps after regeneration felling in Sweden amounts to 11.7 M tDM per year.[17] Large-scale tests are being performed in both Sweden and Finland, mainly based on technology developed in the 1970s: 18–25 ton excavators are fitted with tong–nippers or rakes to extract and split the stumps.[25] Presently, the stumps from smaller trees and from deciduous trees are excluded from harvesting, often representing around half the number of stumps of a felling site but normally less than 25% of the volume. The need to revisit the felling area with a unique system in order to harvest the stumps leads to high costs and longer lead times for the felling–regeneration cycle. Alternative stump harvest systems have been proposed capable of recovering

Figure 18.3 Current large-scale whole-tree harvesting in Sweden is built on multiple systems; conventional logging, followed by a slash harvesting system and then a separate stump lifting operation as shown. The systems are highly productive but require extensive system resiting and high capital commitment, thus presupposing large logging sites.

the central part of the stump when harvesting the trees[25] or a novel felling technique, enabling the trees to be felled at soil surface level.[20] With the present small extent of the field trials in Sweden, technical development is slow but the effect of rationalization through practice demonstrated in Sweden is significant. For logging residue, a progress ratio of 87% has been shown, meaning that for each doubling of accumulated production, real costs decrease by 13%.[26] Unpublished data from the stump lifting trials indicate similar effects at the level 5–9% lower costs with each doubling. Stumps and split stump sections are bulky and irregularly shaped and up to one quarter of the dry weight may consist of contaminations in the form of mineral soil and stones. A rough pregrinding and screening at the landing is a promising technique for reducing these problems and improving handling and transportation characteristic. A pregrinding system has been shown to cut direct costs by 15–20% compared with transports of the unrefined stump sections to a terminal for grinding.[27]

Yet another way of increasing the output of biomass for energy is through intensified logging. If, e.g., the minimum top diameter of pulpwood is doubled from current 5–10 cm, the available logging residues from final felling would increase by 1.2–1.5 M tDM per year and an additional 1.1 M tDM year volume of *tree sections* as long tops would become available also from thinning. The reduction in pulpwood supply may be compensated for by increasing the annual cut correspondingly (5–6 M m^3s in total). An effect of such a strategy would be higher average diameter of pulpwood and more stemwood in the fuel chip fraction.

Danielsson and Liss[28] studied a more extreme option, producing only sawlogs and *long tops*, denoting the total undelimbed sub-sawlog diameter stem. They concluded that the method is feasible and that it will yield 2–3.6 times more biomass for energy in the stands where it is used, thus compensatory volumes of pulpwood must be procured from other stands. The economy of the method is depending on the price relation between pulpwood and fuel chips, the properties of the stand in question, and also of the perspective of the calculation, e.g., forest owner or pulp industry perspective. The competitiveness of this method improves with increased heterogeneity and decreased average tree diameter, wood quality, and size of the stand.

INCREASED OUTPUT OF BIOMASS FROM THINNING

The fuel fractions available from silvicultural thinning activities of young stands are principally the same as from mature stands after regeneration felling; undersized stemwood, foliage, and stumps. According to Swedish forest sector analyses,[17] a gross mass of 5.7 M tDM year of stumps and 3.9 M tDM year is annually available from thinning. But in thinning, a relatively larger fraction of the volume will consist of low diameter stems. In precommercial thinning, practically no industrial wood is harvestable. Thus, thinning for energy may be a way of (partly) financing the stand treatment. Also in later development stages, pure fuel harvest may be considered if the stand is very heterogeneous and the portion of undersized stems is large. Forest fuel from young stands is easy to handle and transport and with good fuel properties due to the high stemwood content.

The combination of high costs and no or low revenues has led to a situation where especially small, private woodlot owners refrain from early stand thinning. The accumulated Swedish stock of wood in such neglected early thinnings is considerable, in total the harvestable mass is in excess of 40 M tDM. The annual replenishment of small tree biomass in young stands is lower, the practically available volumes have been calculated to be around 2.5–5 M tDM year, of which around half is stemwood.[29]

Thus, the output of biomass can be increased through recovery of the stumps and residues currently available or through intensified thinning activities, partly as *energy thinnings,* including all or most of the stemwood of the harvested trees. In *energy thinnings*, the trees are felled and transported with the conventional Nordic two-machine system: a harvester and a forwarder. In the early thinnings and in maintenance of roadsides and powerlines, fellers sometimes substitute the harvester or a one-machine system such as a harvester–forwarder or a feller–forwarder is used. The rationale for using fellers instead of harvesters is that it is a less costly machine and that there is no real need for delimbing the trees if only fuel is to be produced. This approach has led to the definition of a new assortment in forestry *tree sections*, denoting undelimbed trees cut into transport lengths, normally 5–5.5 m. Undelimbed small trees and tree sections are unwieldy to handle and very bulky, leading to high costs of transportation. In recent years, it has become more common to produce roughly delimbed and compacted tree sections by running them through a harvesters processing unit. This practice greatly improves the handling properties of the tree sections and a load of roughly delimbed tree sections may contain 85–95% of that of conventional pulpwood load.[30]

OPTIONS IN TERTIARY PRODUCTION

Tertiary production is the term used to describe the industrial conversion of forest feedstock into useful commodities. It is a part of the forest sector but not of forestry and will be dealt with superficially.

The reshaping of the Swedish energy sector has been rapid and extensive, from 80% dependence on fossil fuels in 1970[31] to the situation in 2010 when less than 39% of the energy consumption is based on fossil sources and 48% is renewable with biomass (32.0%) as the dominating fuel.[32] This has been possible without destructive competition with Sweden's large and important forest industry. Actually, the Swedish forest industries have played an important role in this shift, based on their ability to substitute fossil fuels for biomass in their own processes (the forest industry is a major user of forest biomass for production of heat and electricity) and also on their ability to supply other sectors through their control of wood procurement and material flows. Through this development, the forest industries have converted former residues and by-products into useful resources. In recent years, deliveries of forest fuels, heat, and 'green electricity' have developed into a notable and vital business for Swedish large-scale forestry. The industrial by-products used for energy correspond to around 40–45% of the volume annually procured by the industries.[33] A remaining, substantial increase of forest biomass for energy would be available if waste management was organized so that larger portions of the wood used in products were used for energy after their useful product life. It is not clear to what extent this takes place today.[33]

Even though the Swedish forest industry itself is a forerunner, a vast expansion of the 'new' energy source has also taken place outside the sector. Most notable is the district heating sector. From 1980, the demand for wood fuel has increased steadily by 4–11 PJ per year. In 2010, the use was around 126 PJ corresponding to a volume of 17.5 M m^3 solid.[33] Initially, the core business was production of heat but in recent years, heat is seen as a by-product from combined heat and power production where the heat sink represented by premises in the residential and service sectors is of vital importance for the overall profitability. In principle, the generation of 'green electricity' creates 20–30% extra demand for solid biofuels compared with heat production only.

Current tests to convert biomass into gaseous or liquid fuels for the transport sector, if successful, will increase the demand for biomass further. Currently, around 360 PJ per year, 88% of which is diesel or petrol, are used for domestic transport. This is the last remaining large sector where fossil fuel is dominating in Sweden and close to the total 410 PJ of energy currently supplied by forest biomass.[33] Even if suitable processes are developed, it seems unlikely that all of this demand may be covered by forest biomass due to the lack of heat sinks currently needed for acceptable economy, current lack of biomass, and losses in conversion. But a significant portion may be covered.

Energy efficiency and conservation of energy are eagerly pursued by Swedish industries, and supported by various government initiatives. Although this does not increase the potentially available volumes, it is important to note that such measures will make biomass relatively more important.

CONCLUSIONS

Short-term options for increasing biomass output from long-rotation forestry include increased harvest levels up to current annual growth and increased harvest intensity by harvesting more of the available tree biomass than the stem itself. This potential is large, since the top, branches,

Figure 18.4 Environmentally friendly and economic one-pass harvesting systems, flexible enough to supply the wood for the traditional forest industry and the biomass for the energy and the biorefinery sectors can be developed. Lower production costs is an efficient way to increase the immediate availability of biomass.

stump, and coarse root systems contain around 50–60% additional biomass as compared with the traditionally merchantable roundwood assortments. But the possibilities to realize this potential are limited by both environmental and techno-economic constraints. Thus, the biomass output can be increased if more efficient and environmentally friendly harvesting and transport systems are developed. Long-term options include silvicultural treatments to increase forest growth. Because of long-rotation periods, it takes decades to substantially increase primary forest production in long-rotation forestry. Low cost silvicultural treatments with the potential to reach large areas fast have the highest potential to increase total forest yield. More readily available is a more efficient use of biomass in both the forest and energy industry and in biorefineries of the future, including both a more purposeful use of by-products and waste management aiming at extracting the energy from biomass in various products after their useful life.

A general problem when trying to increase the use of biomass for energy is that the operational setup with areas (forests) serving points (heating plants, industries) will eventually lead to diseconomies of scale due to rapidly increasing transport costs. Therefore, one of the most effective ways of increasing biomass availability is through rationalization of production, leading to lower direct production costs, not least aimed at improved infrastructure and transportation technology. In a steady-state solution, the economically available supply of biomass is doubled if production costs are cut by 41%. One-pass systems, flexible enough to retrieve both the wood in demand by traditional forest industry and biomass suited for conversion to energy, and efficient distance transportation technologies must thus be developed (Figure 18.4).

ACKNOWLEDGEMENTS

This work was funded by the Swedish Energy Agency and by Future Forests, a multidisciplinary research programme supported by the Foundation for Strategic Environmental Research (MISTRA), the Swedish Forestry Industry, the Swedish University of Agricultural Sciences (SLU), Umeå University, by the R&D programme Efficient Forest Fuel Supply Systems, and Skogforsk, the Forestry Research Institute of Sweden.

REFERENCES

FAO. *Global Forest Resources Assessment 2010—Main Report.* FAO Forestry Paper 163. Rome, Italy; 2010.

European Parliament. Directive 2009/28/EC of the European Parliament and of the Council of 23 April 2009 on the promotion of the use of energy from renewable sources; 2009.

Richardson J, Björheden R, Hakkila P, Lowe AT, Smith CT, eds. *Bioenergy from Sustainable Forestry—Guiding Principles and Practice (Forestry Sciences).* Dordrecht/Boston/London: Kluwer Academic Publishers; 2002.

Teobaldelli M, Somogyi Z, Migliavacca M, Usoltsev VA. Generalized functions of biomass expansion factors for conifers and broadleaved by stand age, growing stock and site index. *Forest Ecol Manage* 2009 257:1004–1013.

Lattimore B, Smith CT, Titus BD, Stupak I, Egnell G. Environmental factors in wood fuel production: opportunities, risks, and criteria and indicators for sustainable practices. *Biomass Bioenergy* 2009 33:1321–1342.

Nilsson U, Fahlvik N, Johansson U, Lundstrom A, Rosvall O. Simulation of the effect of intensive forest management on forest production in Sweden. *Forests* 2011, 2:373–393.

Nilsson U, Luoranen J, Kolstrom T, Orlander G, Puttonen P. Reforestation with planting in northern Europe. *Scandinavian J Forest Res* 2010, 25:283–294.

Boateng JO, Heineman J, Bedford L, Harper G, Nemec AFL. Long-term effects of site preparation and postplanting vegetation control on *Picea glauca* survival, growth and predicted yield in boreal British Columbia. *Scandinavian J Forest Res* 2009, 24:111–129.

Elfving B, Ericsson T, Rosvall O. The introduction of lodgepole pine for wood production in Sweden—a review. *Forest Ecol Manage* 2001, 141:15–29.

Li B, McKeand S, Weir R. Tree improvement and sustainable forestry—impact of two cycles of loblolly pine breeding in the U.S.A. *Forest Genet* 1999, 6:229–234.

Farnum P, Timmis R, Kulp JL. Biotechnology of forest yield. *Science* 1983, 219:694–702.

Tamm CO. Nitrogen in terrestrial ecosystems—questions of productivity, vegetational changes, and ecosystem stability. In: Billings WD, Golley F, Lange OL, Olson JS, Remmert H, eds. *Ecological Studies*. Vol. 81. Berlin, Heidelberg: Springer Verlag; 1991, 116.

Moilanen M, Silfverberg K, Hokkanen TJ. Effects of wood-ash on the tree, growth, vegetation and substrate quality of a drained mire: a case study. *Forest Ecol Manage* 2002, 171:321–338.

Bergh J, Linder S, Bergstrom J. Potential production of Norway spruce in Sweden. *Forest Ecol Manage* 2005, 204:1–10.

Simard SW, Blenner-Hassett T, Cameron IR. Pre-commercial thinning effects on growth, yield and mortality in even-aged paper birch stands in British Columbia. *Forest Ecol Manage* 2004, 190:163–168.

Keays JL. Forest harvesting of the future. *AICLE Symposium Series* 1976, 72:4–12.

The Swedish Forest Agency. Skogliga konsekvensanalyser 2008 (Forest sector consequential analyses); 2008.

Lohmander P. Guidelines for economically rational and coordinated dynamic development of the forest and bioenergy sectors with CO_2 constraints. *Proceedings from the 16th European Biomass Conference and Exhibition*. Valencia, Spain; 2008.

The Swedish Forest Agency. Swedish Statistical Yearbook of Forestry 2010. *Official Statistics of Sweden*; 2011.

Hakkila P. *Utilization of Residual Forest Biomass*. Heidelberg, New York: Springer-Verlag; 1989, 568.

Möller JJ, Hannrup B, Larsson W, Arlinger J, Barth A, Wilhelmsson L. *Ett System för Beräkning Och Geografisk Visualisering Av Avverkade Kvantiteter Skogsbränsle Baserat På Skördardata*. Uppsala, Sweden: Skogforsk; 2009.

Björheden R, Furness-Lindén A. The development of forest fuel in Sweden. In: Savolainen M, ed. *Bioenergy 2007; Proceedings of 3rd International Bioenergy Conference and Exhibition*. Jyväskylä, Finland; 2007.

Eriksson LO, Björheden R. Optimal storing, transports and processing for a forest fuel supplier. *Eur J Operational Res* 1989, 43:26–33.

Enström J. Greater efficiency in railway transport of forest fuel. In: Thorsén Å, Björheden R eds. *Efficient Forest Fuel Supply Systems—Composite Report from a Four Year R&D Programme 2007–2010*. Sweden: Skogforsk; 2010.

Nordfjell T, Athanassiades D, Lindroos O. Force requirements in stump extraction. In: Thorsén Å, Björheden R, eds. *Efficient Forest Fuel Supply Systems—Composite Report from a Four Year R&D Programme 2007–2010*. Sweden: Skogforsk; 2010.

Junginger M, Faijj A, Björheden R. Technological learning and cost reductions in wood fuel supply chains in Sweden. *Biomass Bioenergy* 2005, 29:399–418.

von Hofsten H. Pre-grinding can reduce costs. In: Thorsén Å, Björheden R, eds. *Efficient Forest Fuel Supply Systems—Composite Report from a Four Year R&D Programme 2007–2010*. Sweden: Skogforsk; 2010.

Danielsson BO, Liss JE. *Långa Toppar (Long Tops). Projekt SWX-Energi*; 2010.

Nordfjell T, Nilsson P, Henningsson M, Wästerlund I. Unutilized biomass resources in Swedish young dense forests. *Proc World Bioenergy* 2008, Jönköping; 2008.

Iwarsson-Wide M. Technology and methods for logging in young stands. In: Thorsén Å, Björheden R, eds. *Efficient Forest Fuel Supply Systems—Composite Report from a Four Year R&D Programme 2007–2010* Sweden: Skogforsk; 2010.

Hillring B, Ling E, Blad B. The potential and utilisation of biomass, pp 335–401. In: Silveira S, ed. *Building Sustainable Energy Systems—Swedish Experiences* AB Svensk Byggtjänst and Swedish National Energy Administration; 2001.

The Swedish Energy Agency Energy Indicators 2009—Follow-up of Sweden's energy policy objectives; 2009.

The Swedish Energy Agency Energy in Sweden 2010; 2010.

19

Recovery Rate of Harvest Residues for Bioenergy in Boreal and Temperate Forests: A Review

Evelyne Thiffault,[†,*] Ariane Béchard, David Paré and Darren Allen

Canadian Forest Service, Natural Resources Canada, Quebec City, Quebec, Canada

INTRODUCTION

The past few years have seen an increasing use of forest biomass for bioenergy, spurred by policy incentives for reduction of fossil fuel use and greenhouse gas emissions[1] and by efforts from governments and the forestry sector to diversify products and increase value from the forests. Harvest residues, consisting mainly of tree tops, branches, and bucking and trimming material, are byproducts of harvesting operations for traditional wood products. Given the proper economic and policy context, they are considered an attractive biomass feedstock source because their use for bioenergy does not compete with other wood uses, and their potential availability is high.[2] Moreover, it is considered as a source of biomass with short-term atmospheric benefits.[3] A large body of research has been dedicated to help increase the economic viability of biomass supply chains based on harvest residues, notably through the improvement and adaptation of existing procurement systems and new innovative technologies such as residue bundling systems.[4] Technological development and learning have been rapid over the past decades, and more cost-effective solutions are becoming more readily available.[5]

However, there has also been increasing concern about the ecological sustainability of harvest residue removal, because of the potential impacts on soils and biodiversity caused by the increased export of organic material relative to conventional stem-only harvesting, where only merchantable boles are removed from the forest site.[6] In response to this concern, many jurisdictions are developing policies and guidelines for recovery of harvest residues to ensure the ecological sustainability of this practice.[7–10] These guidelines often rely on expert knowledge to define sustainable thresholds of residue removal either through defining suitable and unsuitable sites, or defining a proportion or an amount of harvest residues to leave on site.

Notwithstanding constraints related to environmental regulations, it is generally understood that a portion of the total harvest residues is left on the cutover due to technical and profitability constraints.[11] A better knowledge of these constraints and how they influence the actual harvest residue recovery rate would help us provide realistic estimates of current and future technical potential from this feedstock source.[12] A recent study for Canada applied a net-down of 50% on the theoretical potential of harvest residues to account for ecological constraints[13]; it was assumed that this value would also roughly encompass technical net-downs. In the United States, another study[14]

*Correspondence to: evelyne.thiffault@sbf.ulaval.ca
†Present address: Department of Wood and Forest Science, Université Laval, 2405 rue de la Terrasse, Pavillon Abitibi-Price, Universitépgbrk Laval, Québec (QC), G1V 0A6, Canada

Advances in Bioenergy: The Sustainability Challenge, First Edition. Edited by Peter D. Lund, John Byrne, Göran Berndes and Iacovos A. Vasalos.

Originally published in: *WIREs Energy Environ* 2014, 4: 293–316 doi: 10.1002/wene.157

uses a 65% harvest residue removal rate in their estimates of biomass potential. However, it is also recognized that residue recovery rates will vary with equipment, operator skill, season, and stand conditions.[15] Moreover, it is possible that, in the foreseeable future, a change in fossil fuel prices and economic drivers of bioenergy and technological and logistical advances may reduce economic and operational constraints. It is therefore important to refine our understanding of the factors influencing the variability of recovery rate, so that we may better inform policy development on sustainable forest biomass procurement practices, as well as on the setting of ecological thresholds on residue removal.

The goal of the study was to characterize the variability of operational recovery rate of harvest residues in boreal and temperate forests. The objectives were to quantify the range of variation of harvest residue recovery rates based on a review of scientific studies and technical reports, and to identify the main factors that explain this variability. For the purpose of this review, the rate of residue recovery is defined as the ratio of the quantity of residues (mainly tree tops and branches) removed from the forest cutover versus the original total quantity of residues found on the ground of the cutover and/or left after conventional harvesting of the merchantable bole only.

MATERIAL AND METHODS

The geographic focus of this review was on forests from the boreal and temperate biomes, as defined by the World Wildlife Fund classification. An investigation of peer-reviewed literature was conducted using online search tools including Scopus, Web of Science, and Google Scholar based on different combinations of the following keywords: biomass harvesting; harvest residues; recovery; removal; efficiency; retention, stem-only harvesting; whole-tree harvesting; bundling; and operations. We targeted studies that used field trials focusing on harvest residue operations and reported empirical data on (1) mass or volume of residues before and after recovery operations, (2) comparison of sites with and without residue removal, or (3) reported data on recovery rates (provided the definition of recovery was consistent with our definition). The primary focus of this study was on final fellings, which are meant to encompass both clearcutting and partial cutting of mature trees/stands. We also broadened our search to include internal reports of trials conducted for the purpose of evaluating residue recovery rate. We retained in this review only studies that reported required information about harvesting methods, and trial location and description.

In order to enable a standardized measure of rate of recovery, in some instances, we recalculated the harvest residue recovery rate using data from the individual studies. Only residues present on the ground in the cutblock were considered as retained on the forest site. Residues that were left at the landing, or trees or snags left standing in the cutblock were excluded. Conversely, residue piles at the roadside or central landing, either to be left there permanently, further processed or transported to a processing plant, were calculated as removed from the harvest block. Classification information about country/region of operation, origin of stand (plantation/natural forest), type of operations, and season of operations was extracted for each study.

Data on recovery rate were analyzed for standard measures of distribution and statistical dispersal. Also, a classification tree analysis was performed using the *Rpart* package in R[16] to explore the structure of the dataset and identify decision rules for predicting recovery rate of harvest residues, based on classification information extracted from relevant studies. This method involves the segregation of different values of classification variables through a decision tree composed of progressive binary splits. Every value for each predictor variable is considered as a potential split, and the optimal split is selected based on how well the split reduces the heterogeneity within the resulting subsets. Each parent node in the decision tree produces two child nodes, which in turn can become parent nodes producing additional child nodes. This process continues with both tree building and pruning until statistical analysis indicates that the tree fits without overfitting the information contained in the dataset.

RESULTS AND DISCUSSION

Our search found studies and trials mainly from the Nordic countries, and to a lesser degree from other European countries, the United States, and Canada (Figure 19.1).

Most of the studies reviewed were conducted in thelast-decade. We were able to calculate 68 values representing recovery rates; values were either extracted directly from the study or calculated using published data therein. Tables 19.1 and 19.2 summarize the compiled studies and the details of calculations or transformations of the data. Most studies made no mention of regulations or guidelines for harvest residue or debris retention that would have been intentionally applied during the operations; it was therefore assumed that the reported data mostly or entirely reflected operational constraints to residue removal. One exception is data from Briedis et al.,[12] which explicitly noted that harvest operations followed Maine's water quality Best Management Practices for erosion control.[17]

The forest harvesting operations and related residue supply chains described in studies were mostly based on the following three systems:

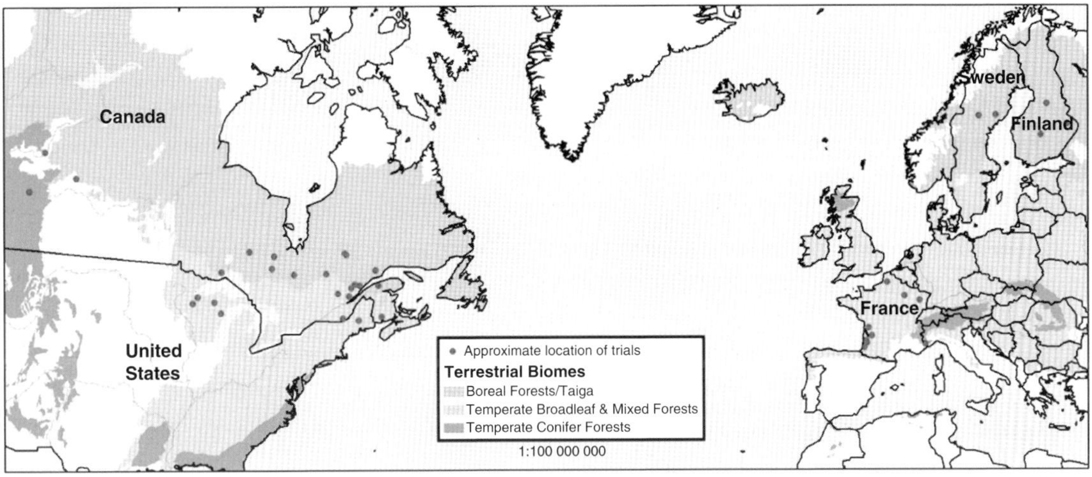

Figure 19.1 Approximate location of study sites included in the review. One point may represent several studies. When precise location of trials was not available, the middle of the region/country was used.

- Stem-only (also called cut-to-length) harvesting, in which trees are cut, delimbed, topped, and bucked in the harvested block, and the stem forwarded to the roadside; a portion of the remaining residues (mainly tree tops, branches, and bucking and trimming material) are then picked up from the forest floor as loose slash and brought to the roadside or central landing, usually with the same forwarder used for the merchantable stemwood.
- Bundling, a variant of the stem-only harvesting system, in which loose residues are picked up and compacted by a bundler, and bundles are then forwarded to the roadside.
- Whole-tree harvesting, in which trees are cut and skidded in one piece to the roadside or central landing, where they are delimbed and/or bucked to specifications; most of the residues are therefore produced directly at the roadside or landing.

The average recovery rate among the reviewed studies was 52.2%, with a standard error of 18.1% (Figure 19.2). The median value (54.1%) was very close to the average. The distribution of values was close to a normal distribution, with 70.6% of values within one standard deviation of the mean. The minimum and maximum values observed were 4.0 and 89.1%, respectively.

Along with the harvest method described above, the compiled studies were also classified by: country; harvesting intensity (clearcut or partial harvest); stand origin (plantation or natural, the latter broadly defined as stands of native species established through natural regeneration); stand species composition (hardwood, softwood, or mixedwood); and season of residue recovery operations (winter, spring, summer, or fall). A regression tree analysis was performed on 44 observations of recovery rate from studies for which all classification information was available. It was found that a pruned tree with three splits made it possible to minimize the relative error and optimize the prediction of recovery rate (Figure 19.3). Although the small number of observations impacted the power of the analysis, the regression tree nevertheless demonstrated that the country of operations was the most important factor driving recovery rate (Figure 19.4). Trials conducted in Nordic countries, i.e., Sweden and Finland, had on average higher recovery rates (71.6%) than other countries.

The main method of biomass recovery operations in the Nordic countries is to recover loose slash from the harvested block, which is then forwarded to the roadside where it is chipped directly into a truck container. The high recovery rate in Nordic countries encompasses many defining factors typical of that region.[18] First, since the oil crisis in the early 1970s there has been a strong policy backing the use of forest biomass for energy production as an alternative to fossil fuels. This backing is expressed via bioenergy R&D programs, investment subsidies in forest biomass supply chains, and policy measures that promote the increased use of forest biomass.[19–21] This has translated into a strong market for forest biomass and a high economic value for harvest residues, and therefore an incentive to increase their removal. This in turn has led to several decades of technological learning that reduces production costs through learning-by-doing,

Table 19.1 Compilation of studies on forest harvest residue recovery rate in boreal and temperate forests

No.	Reference	Country	Region	Site/Trial name	Principal Forest Operation	Stand Origin	Type of Stand	Harvest Season
1	Cormier et al.[35]	Canada	Ontario. Kapuskasing		Clearcut	Natural	Softwood	Winter (assumed)
2	Cormier et al.[35]	Canada	Ontario. Thunder Bay		Clearcut	Natural	Hardwood	Winter (assumed)
3	Cormier et al.[35]	Canada	Québec. Saguenay		Clearcut	Natural	Mixedwood	Summer (assumed)
4	Cormier et al.[35]	Canada	Québec. Matapédia		Clearcut	Natural	Mixedwood	Winter (assumed)
5	Cormier et al.[35]	Canada	Ontario. Great Lakes—St. Lawrence		Clearcut	Natural	Hardwood	Summer (assumed)
6	Cormier et al.[35]	Canada	Ontario. Great Lakes—St. Lawrence		Clearcut	Natural	Hardwood	Summer (assumed)
7	Cormier et al.[35]	Canada	Alberta. Hinton		Clearcut	Natural	Softwood	n.a
8	Cormier et al.[35]	Canada	Alberta. Lac La Biche		Clearcut	Natural	Hardwood	n.a
9	Cormier et al.[35]	Canada	Alberta. Peace River		Clearcut	Natural	Hardwood	n.a
10	Ralevic et al.[34]	Canada	Ontario. Northeastern Ontario Claybelt	Block A	Clearcut	Natural	Mixedwood	Summer
11	Ralevic et al.[34]	Canada	Ontario. Northeastern Ontario Claybelt	Block B	Clearcut	Natural	Mixedwood	Winter
12	Ralevic et al.[34]	Canada	Ontario. Northeastern Ontario Claybelt	Block C	Clearcut	Natural	Softwood	Winter
13	Ryans and Desrochers[30]	Canada	Southern New Brunswick	Site 2	Clearcut	Natural	Mixedwood	Fall
14	Ryans and Desrochers[30]	Canada	Southern New Brunswick	Site 3	Clearcut	Natural	Mixedwood	Fall
15	Volpé and Desrochers[36]	Canada	Québec. Matapédia		Clearcut	Natural	Hardwood	Fall
16	Volpé and Desrochers[36]	Canada	Québec. Ferland-et-Boilleau		Clearcut	Natural	Mixedwood	Summer
17	Thiffault et al. (unpublished data)	Canada	Québec. Nord-du-Québec	Cawcot	Clearcut	Natural	Softwood	Summer
18	Thiffault et al. (unpublished data)	Canada	Québec. Estrie	Chartierville	Partial cut	Natural	Hardwood	Summer
19	Thiffault et al. (unpublished data)	Canada	Québec. Abitibi	Disson	Clearcut	Natural	Softwood	Winter
20	Thiffault et al. (unpublished data)	Canada	Québec. Réserve faunique des Laurentides	Forêt Montmorency	Clearcut	Natural	Softwood	Fall

(Continued)

Table 19.1 Compilation of studies on forest harvest residue recovery rate in boreal and temperate forests (*Continued*)

No.	Reference	Country	Region	Site/Trial name	Principal Forest Operation	Stand Origin	Type of Stand	Harvest Season
21	Thiffault et al. (unpublished data)	Canada	Québec. Côte-Nord	Grosses-Roches	Clearcut	Natural	Mixedwood	Summer
22	Thiffault et al. (unpublished data)	Canada	Québec. Côte-Nord	Stephane	Clearcut	Natural	Mixedwood	Summer
23	Thiffault et al. (unpublished data)	Canada	Québec. Nord-du-Québec	Tournemine A	Clearcut	Natural	Softwood	Summer
24	Thiffault et al. (unpublished data)	Canada	Québec. Nord-du-Québec	Tournemine B	Clearcut	Natural	Softwood	Summer
25	Thiffault et al. (unpublished data)	Canada	Québec. Côte-Nord	Toulnustouc	Clearcut	Natural	Mixedwood	Summer
26	Thiffault et al. (unpublished data)	Canada	Québec. Mauricie	La Tuque	Clearcut	Natural	Mixedwood	Summer
27	Eräjää et al.[37]	Finland	Central Finland		Clearcut	Plantation	Softwood	n.a
28	Nurmi[25]	Finland		M1	Clearcut	Plantation	Softwood	n.a
29	Nurmi[25]	Finland		M2	Clearcut	Plantation	Softwood	n.a
30	Nurmi [25]	Finland		MC	Clearcut	Plantation	Softwood	n.a
31	Peltola et al. [24]	Finland	North Karelia	NK1	Clearcut	Plantation	Softwood	Summer
32	Peltola et al. [24]	Finland	North Karelia	NK2	Clearcut	Plantation	Softwood	Summer
33	Peltola et al. [24]	Finland	North Karelia	NK3	Clearcut	Plantation	Softwood	Summer
34	Peltola et al. [24]	Finland	Central Finland	CF4_1	Clearcut	Plantation	Softwood	Summer
35	Peltola et al. [24]	Finland	Central Finland	CF4_2	Clearcut	Plantation	Softwood	Summer
36	Peltola et al. [24]	Finland	Central Finland	CF5	Clearcut	Plantation	Softwood	Summer
37	Peltola et al. [24]	Finland	Central Finland	CF6	Clearcut	Plantation	Softwood	Summer
38	Peltola et al. [24]	Finland	Central Finland	CF7	Clearcut	Plantation	Softwood	Summer
39	Cuchet et al.[30]	France	Dogneville	1	Clearcut	Plantation	Softwood	n.a
40	Cuchet et al.[29]	France	Pérignac	2	Clearcut	Plantation	Hardwood	n.a
41	Cuchet et al.[29]	France	Pérignac	3	Clearcut	Plantation	Hardwood	n.a
42	Cuchet et al.[29]	France	Montpon	4	Clearcut	Plantation	Hardwood	n.a
43	Cuchet et al.[29]	France	Lugos	5	Clearcut	Plantation	Softwood	n.a
44	Cuchet et al.[29]	France	Soulaines	6	Clearcut	Plantation	Hardwood	n.a
45	Cuchet et al.[29]	France	Willencourt	7	Clearcut	Plantation	Hardwood	n.a
46	Cuchet et al.[29]	France	Azur	8	Clearcut	Plantation	Softwood	n.a
47	Cuchet et al.[29]	France	Taller	9	Clearcut	Plantation	Softwood	n.a
48	Cuchet et al.[29]	France	Morcenx	10	Clearcut	Plantation	Softwood	n.a
49	Cuchet et al.[29]	France	Sabres	11	Clearcut	Plantation	Softwood	n.a
50	Gullberg and Johansson[38]	Sweden	Central Sweden		Clearcut	Plantation	Softwood	Summer (assumed)
51	Rudolphi and Gustafsson[39]	Sweden	Eastern central Sweden		Clearcut	Plantation	Softwood	n.a
52	Benjamin et al.[40]	United States	Northern Maine		Partial	Natural	Softwood	n.a
53	Briedis et al.[12]	United States	Central Maine	Site 1 Location A	Clearcut	Natural	Softwood	Spring
54	Briedis et al.[12]	United States	Central Maine	Site 2 Location B	Partial harvest	Natural	Mixedwood	Winter
55	Briedis et al.[12]	United States	Central Maine	Site 3 Location C	Partial harvest	Natural	Hardwood	Winter
56	Briedis et al.[12]	United States	Central Maine	Site 4 Location C	Partial harvest	Natural	Hardwood	Winter
57	Briedis et al.[12]	United States	Central Maine	Site 5 Location C	Partial harvest	Natural	Hardwood	Winter

(*Continued*)

Table 19.1 Compilation of studies on forest harvest residue recovery rate in boreal and temperate forests (*Continued*)

No.	Reference	Country	Region	Site/Trial name	Principal Forest Operation	Stand Origin	Type of Stand	Harvest Season
58	Briedis et al.[12]	United States	Central Maine	Site 6 Location D	Clearcut	Natural	Softwood	Summer
59	Briedis et al.[12]	United States	Central Maine	Site 7 Location D	Clearcut	Natural	Mixedwood	Summer
60	Briedis et al.[12]	United States	Central Maine	Site 8 Location E	Clearcut	Natural	Softwood	Summer
61	Briedis et al.[12]	United States	Central Maine	Site 9 Location E	Clearcut	Natural	Hardwood	Summer
62	Briedis et al.[12]	United States	Central Maine	Site 10 Location F	Clearcut	Natural	Softwood	Winter
63	Briedis et al.[12]	United States	Central Maine	Site 11 Location F	Clearcut	Natural	Mixedwood	Winter
64	Briedis et al.[12]	United States	Central Maine	Site 12 Location G	Partial harvest	Natural	Mixedwood	Winter
65	Rittenhouse et al.[41]	United States	Northwestern Wisconsin. Burnett county		Clearcut	Natural	Hardwood	n.a
66	Rittenhouse et al.[41]	United States	Northwestern Wisconsin. Douglas county		Clearcut	Natural	Hardwood	n.a
67	Rittenhouse et al.[41]	United States	North-central Wisconsin. Oneida county		Clearcut	Natural	Hardwood	n.a
68	Rittenhouse et al.[41]	United States	Northeastern Wisconsin. Marinette county		Clearcut	Natural	Hardwood	n.a

equipment improvements[19] (notably forwarders) and better integration of operations related to traditional forest products and biomass.

Access to pools of skilled workers with extensive operational experience with the tasks and equipment is also another important driver of recovery rate, and one that likely distinguished Nordic countries (in the data reported here) from others with a shorter history of forest-based biomass supply chains, a smaller local market for forest biomass or a limited access to bioenergy markets. Several authors have documented the higher level of productivity related to operator experience.[22–24] For example, when the harvester operator creates high and well-positioned piles of residues along strip roads as trees are delimbed, recovery of residues by the forwarder is more efficient, notably because they have not been trampled over and are less contaminated.[25] Although it is not possible to extract this type of information from the compiled studies, it can be hypothesized that Sweden and Finland achieve, relative to other countries, a high degree of residue recovery rate due to policy support and technological learning. As illustrated by McCormick and Kåberger,[20] a combination of know-how and institutional capacity is needed to increase the mobilization of bioenergy systems. Interestingly, another international study also identified country-specific characteristics (forest policies, regulations and guidelines, historical forestry practices, differences in forest ownership tenure, etc.) as important determinants of forest harvest intensity at the country level.[26] However, the small sample size used in this analysis allowed only a crude partitioning; this probably prevented the finer delineation of areas of the world (such as Northeastern United States), which also have a long history of forest biomass supply chains, and active bioenergy markets.

Table 19.2 Calculations of Forest Harvest Residue Recovery Rate

No.	Reference	Original Total Residues on Cutblock	Recovered Residues	Leftover Residues	Recovery Rate	Notes
1	Cormier et al.[35]	47.0 tma/ha. Author's calculation. Sum of Recovered and Leftover residues (= Roadside + Cutover)	32.0 tma/ha. Roadside. ('Bord de route')	15.0 tma/ha. Cutover ('Parterre')	32.0 / 47.0 × 100 = 68.1%	
2	Cormier et al.[35]	40.0 tma/ha. Author's calculation. Sum of Recovered and Leftover residues (= Roadside + Cutover)	24.0 tma/ha Roadside. ('Bord de route')	16.0 tma/ha. Cutover ('Parterre')	24.0 / 40.0 × 100 = 60.0%	
3	Cormier et al.[35]	51.0 tma/ha. Author's calculation. Sum of Recovered and Leftover residues (= Roadside + Cutover)	19.0 tma/ha. Roadside. ('Bord de route')	32.0 tma/ha. Cutover ('Parterre')	19.0 / 51.0 × 0100 = 37.3%	
4	Cormier et al.[35]	39.0 tma/ha. Author's calculation. Sum of Recovered and Leftover residues (= Roadside + Cutover)	13.0 tma/ha. Roadside. ('Bord de route')	26.0 tma/ha. Cutover ('Parterre')	13.0 / 39.0 × 100 = 33.3%	
5	Cormier et al.[35]	62.0 tma/ha. Author's calculation. Sum of Recovered and Leftover residues (= Roadside + Cutover)	31.0 tma/ha. Roadside. ('Bord de route')	31.0 tma/ha. Cutover ('Parterre')	31.0 / 62.0 × 100 = 50.0%	
6	Cormier et al.[35]	33.0 tma/ha. Author's calculation. Sum of Recovered and Leftover residues (= Roadside + Cutover)	17.0 tma/ha. Roadside. ('Bord de route')	16.0 tma/ha. Cutover ('Parterre')	17.0 / 33.0 × 100 = 51.5%	
7	Cormier et al.[35]	63.0 tma/ha. Author's calculation. Sum of Recovered and Leftover residues (= Roadside + Cutover)	45.0 tma/ha. Roadside. ('Bord de route')	18.0 tma/ha. Cutover ('Parterre')	45.0 / 63.0 × 100 = 71.4%	
8	Cormier et al.[35]	29.0 tma/ha. Author's calculation. Sum of Recovered and Leftover residues (= Roadside + Cutover)	14.0 tma/ha. Roadside. ('Bord de route')	15.0 tma/ha. Cutover ('Parterre')	14.0 / 29.0 × 100 = 48.3%	

(*Continued*)

Table 19.2 Calculations of Forest Harvest Residue Recovery Rate (*Continued*)

No.	Reference	Original Total Residues on Cutblock	Recovered Residues	Leftover Residues	Recovery Rate	Notes
9	Cormier et al.[35]	140 tma/ha. Author's calculation. Sum of Recovered and Leftover residues (= Roadside + Cutover)	92.0 tma/ha. Roadside. ('Bord de route')	48.0 tma/ha. Cutover ('Parterre')	92.0 / 140.0 × 100 = 65.7%	
10	Ralevic et al.[34]	30.3 odt/ha Table 19.3. Method: sampling. Difference between Potentially available and Standing	19.0 odt/ha Table 19.3. Method: sampling. Sum of Roadside technically usable and post-recovery residues	11.3 odt/ha Table 19.3. Method: sampling. Cutover biomass on ground	19.0 / 30.3 × 100 = 62.71%	Two estimates of amounts of biomass were provided
11	Ralevic et al.[34]	83.2 odt/ha (Sampling) Table 19.3. Method: sampling. Difference between Potentially available and Standing	40.5 odt/ha Method: sampling. Sum of Roadside technically usable and post-recovery residues	42.7 odt/ha Table 19.3. Method: sampling. Cutover biomass on ground	40.5 / 83.2 × 100 = 48.68%	Two estimates of amounts of biomass were provided
12	Ralevic et al.[34]	48 odt/ha (Sampling) Table 19.3. Method: sampling. Difference between Potentially available and Standing	32.6 odt/ha Method: sampling. Sum of Roadside technically usable and post-recovery residues	15.4 odt/ha Table 19.3. Method: sampling. Cutover biomass on ground	32.6 /48.0 × 100 = 67.92%	Two estimates of amounts of biomass were provided
13	Ryans and Desrochers[30]	n.a	n.a	n.a	78.0% Taken from the report (Table 19.1. Debris recovery). Insufficient data to recalculate	
14	Ryans and Desrochers[30]	n.a	n.a	n.a	83.0 % Taken from the report (Table 19.1. Debris recovery). Insufficient data to recalculate	
15	Volpé and Desrochers[36]	13.4 odt/ha. Amount of residues on control stem-only harvested plots	6.0 odt/ha. Author's calculation. Difference between Original total and Leftover residues	7.4 odt/ha. Amount of residues on whole-tree harvested plots	6.0 / 13.4 × 100 = 44.8%	Comparison of stem-only harvested vs. whole-tree harvested sites

Table 19.2 Calculations of Forest Harvest Residue Recovery Rate (*Continued*)

No.	Reference	Original Total Residues on Cutblock	Recovered Residues	Leftover Residues	Recovery Rate	Notes
16	Volpé and Desrochers[36]	17.4 odt/ha Amount of residues on control stem-only harvested plots	2.2 odt/ha. Author's calculation. Difference between Original total and Leftover residues	15.2 odt/ha. Amount of residues on whole-tree harvested plots	2.2 / 17.4 × 100 = 12.6%	Comparison of stem-only harvested vs. whole-tree harvested sites
17	Thiffault et al. (unpublished data)	35.7 odt/ha Amount of residues on control stem-only harvested plots	18.2 odt/ha. Author's calculation. Difference between Original total and Leftover residues	17.5 odt/ha. Amount of residues on whole-tree harvested plots	18.2 / 35.7 × 100 = 51.0%	Comparison of stem-only harvested vs. whole-tree harvested sites
18	Thiffault et al. (unpublished data)	36.1 odt/ha Amount of residues on control stem-only harvested plots	12.6 odt/ha. Author's calculation. Difference between Original total and Leftover residues	23.5 odt/ha. Amount of residues on whole-tree harvested plots	12.6. / 36.1 × 100 = 34.9%	Comparison of stem-only harvested vs. whole-tree harvested sites
19	Thiffault et al. (unpublished data)	77.7 odt/ha Amount of residues on control stem-only harvested plots	31.9 odt/ha. Author's calculation. Difference between Original total and Leftover residues	45.8 odt/ha. Amount of residues on whole-tree harvested plots	31.9 / 77.7 × 100 = 41.1%	Comparison of stem-only harvested vs. whole-tree harvested sites
20	Thiffault et al. (unpublished data)	71.5 odt/ha. Amount of residues on control stem-only harvested plots	14.7 odt/ha. Author's calculation. Difference between Original total and Leftover residues	56.8 odt/ha. Amount of residues on whole-tree harvested plots	14.7 / 71.5 × 100 = 20.6%	Comparison of stem-only harvested vs. whole-tree harvested sites
21	Thiffault et al. (unpublished data)	15.4 odt/ha. Amount of residues on control stem-only harvested plots	8.6 odt/ha. Author's calculation. Difference between Original total and Leftover residues	6.8 odt/ha. Amount of residues on whole-tree harvested plots	8.6 / 15.4 × 100 = 55.8%	Comparison of stem-only harvested vs. whole-tree harvested sites
22	Thiffault et al. (unpublished data)	17.9 odt/ha. Amount of residues on control stem-only harvested plots	1.8 odt/ha. Author's calculation. Difference between Original total and Leftover residues	16.1 odt/ha. Amount of residues on whole-tree harvested plots	1.8 / 17.9 × 100 = 10.6%	Comparison of stem-only harvested vs. whole-tree harvested sites
23	Thiffault et al. (unpublished data)	61.2 odt/ha. Amount of residues on control stem-only harvested plots	25.1 odt/ha. Author's calculation. Difference between Original total and Leftover residues	36.1 odt/ha. Amount of residues on whole-tree harvested plots	25.1 / 61.2 × 100 = 41.0%	Comparison of stem-only harvested vs. whole-tree harvested sites

(*Continued*)

Table 19.2 Calculations of Forest Harvest Residue Recovery Rate (*Continued*)

No.	Reference	Original Total Residues on Cutblock	Recovered Residues	Leftover Residues	Recovery Rate	Notes
24	Thiffault et al. (unpublished data)	52.8 odt/ha. Amount of residues on control stem-only harvested plots	17.7 odt/ha. Author's calculation. Difference between Original total and Leftover residues	35.1 odt/ha. Amount of residues on whole-tree harvested plots	17.7 / 52.8 × 100 = 33.5%	Comparison of stem-only harvested vs. whole-tree harvested sites
25	Thiffault et al. (unpublished data)	37.4 odt/ha. Author's calculation. Sum of recovered and Leftover residues	12.1 odt/ha. Page 3	25.3 odt/ha. Page 3	12.1 / 37.4 × 100 = 32.4%	
26	Thiffault et al. (unpublished data)	55.8 odt/ha. Author's calculation. Sum of recovered and Leftover residues	24.0 odt/ha. Page 4	31.8 odt/ha. Page 4	24.0 / 55.8 × 100 = 43%	
27	Eräjää et al.[37]	24.8 m3/ha. Table 19.1. Sum of Logs, pieces of logs and branches on control traditional clear-cuts	5.6 m3/ha. Author's calculation. Difference between Original total and Recovered residues	19.2 m3/ha. Table 19.1. Sum of Logs. Pieces of logs and branches on forest fuel harvested clear-cuts	5.6 / 24.8 × 100 = 22.6%	We assumed Residues were composed of logs, pieces of logs and branches; other types of dead wood (stumps, stump pieces, roots and high stumps) were ignored
28	Nurmi[25]	113.5 green tonnes/ha. Author's calculations. Sum of Recovered residues and Leftover residues	75.9 green tonnes/ha. Table 19.1. Quotient of Recovered residues and Total stand area	37.9 green tonnes/ha. Table p. 377. Average value. Units converted from kg/m2	75.9 / 113.5 × 100 = 66.9%	
29	Nurmi[25]	90.7 green tonnes/ha. Author's calculations. Sum of Recovered residues and Leftover residues	71.3 green tonnes/ha. Table 19.1. Quotient of Recovered residues and Total stand area	19.4 green tonnes/ha. Table p. 377. Average value. Units converted from kg/m2	71.3 / 90.7 × 100 = 78.6%	
30	Nurmi [25]	119.0 green tonnes/ha. Author's calculations. Sum of Recovered residues and Leftover residues	69.1 green tonnes/ha. Table 19.1. Quotient of Recovered residues and Total stand area	49.9 green tonnes/ha. Table p. 377. Average value. Units converted from kg/m2	69.1 / 119.0 × 100 = 58.1%	

Table 19.2 Calculations of Forest Harvest Residue Recovery Rate (*Continued*)

No.	Reference	Original Total Residues on Cutblock	Recovered Residues	Leftover Residues	Recovery Rate	Notes
31	Peltola et al. [24]	39.6 odt/ha. Table 19.1. Estimated amount of residues based on inventory data. Repola et al. 3 var	22.5 odt/ha. Table 19.1. Recovered residues	17.1 odt/ha. Author's calculation. Difference between available and recovered residues	22.5 / 39.6 × 100 = 56.8%	Several estimates of amounts of residues were provided
32	Peltola et al. [24]	33.9 odt/ha. Table 19.1. Estimated amount of residues based on inventory data. Repola et al. 3 var	20.5 odt/ha. Table 19.1. Recovered residues	13.4 odt/ha. Author's calculation. Difference between available and recovered residues	20.5 / 33.9 × 100 = 60.5%	Several estimates of amounts of residues were provided
33	Peltola et al. [24]	35.2 odt/ha. Table 19.1. Estimated amount of residues based on inventory data. Repola et al. 3 var	17.0 odt/ha. Table 19.1. Recovered residues	18.2 odt/ha. Author's calculation. Difference between available and recovered residues	17.0 / 35.2 × 100 = 48.3%	Several estimates of amounts of residues were provided
34	Peltola et al. [24]	44.6 odt/ha. Table 19.1. Estimated amount of residues based on inventory data. Repola et al. 3 var	34.5 odt/ha. Table 19.1. Recovered residues	10.1 odt/ha. Author's calculation. Difference between available and recovered residues	34.5 / 44.6 × 100 = 77.4%	Several estimates of amounts of residues were provided
35	Peltola et al. [24]	44.3 odt/ha. Table 19.1. Estimated amount of residues based on inventory data. Repola et al. 3 var	39.4 odt/ha. Table 19.1. Recovered residues	4.9 odt/ha. Author's calculation. Difference between available and recovered residues	39.4 / 44.3 × 100 = 88.9%	Several estimates of amounts of residues were provided
36	Peltola et al. [24]	38.9 odt/ha. Table 19.1. Estimated amount of residues based on inventory data. Repola et al. 3 var	30.1 odt/ha. Table 19.1. Recovered residues	8.8 odt/ha. Author's calculation. Difference between available and recovered residues	30.1 / 38.9 × 100 = 77.4%	Several estimates of amounts of residues were provided

(*Continued*)

Table 19.2 Calculations of Forest Harvest Residue Recovery Rate (*Continued*)

No.	Reference	Original Total Residues on Cutblock	Recovered Residues	Leftover Residues	Recovery Rate	Notes
37	Peltola et al. [24]	43.6 odt/ha. Table 19.1. Estimated amount of residues based on inventory data. Repola et al. 3 var	26.3 odt/ha. Table 19.1. Recovered residues	17.3 odt/ha. Author's calculation. Difference between available and recovered residues	26.3 / 43.6 × 100 = 60.3%	Several estimates of amounts of residues were provided
38	Peltola et al. [24]	39.4 odt/ha. Table 19.1. Estimated amount of residues based on inventory data. Repola et al. 3 var	35.1 odt/ha. Table 19.1. Recovered residues	4.3 odt/ha. Author's calculation. Difference between available and recovered residues	35.1 / 39.4 × 100 = 89.1%	Several estimates of amounts of residues were provided
39	Cuchet et al.[30]	178.1 green tonnes/ha. Table 19.1. Initial amount	96.6 green tonnes/ha. Author's calculation. Difference between available residues and Leftover residues	81.5 green tonnes/ha. Table 19.1. Left on ground	96.6 / 178.1 × 100 = 54.2%	Assumption that amounts reported in Table 19.1 were in green tonnes
40	Cuchet et al.[29]	72.0 green tonnes/ha. Table 19.1. Initial amount	38.8 green tonnes/ha. Author's calculation. Difference between available residues and Leftover residues	33.2 green tonnes/ha. Table 19.1. Left on ground	38.8 / 72.0 × 100 = 53.9%	Assumption that amounts reported in Table 19.1 were in green tonnes
41	Cuchet et al.[29]	110.7 green tonnes/ha. Table 19.1. Initial amount	72.8 green tonnes/ha. Author's calculation. Difference between available residues and Leftover residues	37.9 green tonnes/ha. Table 19.1. Left on ground	72.8 / 110.7 × 100 = 65.8%	Assumption that amounts reported in Table 19.1 were in green tonnes
42	Cuchet et al.[29]	73.4 green tonnes/ha. Table 19.1. Initial amount	22.6 green tonnes/ha. Author's calculation. Difference between available residues and Leftover residues	50.8 green tonnes/ha. Table 19.1. Left on ground	22.6 / 73.4 × 100 = 30.8%	Assumption that amounts reported in Table 19.1 were in green tonnes
43	Cuchet et al.[29]	22.1 green tonnes/ha. Table 19.1. Initial amount	9.8 green tonnes/ha. Author's calculation. Difference between available residues and Leftover residues	12.3 green tonnes/ha. Table 19.1. Left on ground	9.8 / 22.1 × 100 = 44.3%	Assumption that amounts reported in Table 19.1 were in green tonnes

Table 19.2 Calculations of Forest Harvest Residue Recovery Rate (*Continued*)

No.	Reference	Original Total Residues on Cutblock	Recovered Residues	Leftover Residues	Recovery Rate	Notes
44	Cuchet et al.[29]	101.0 green tonnes/ha. Table 19.1. Initial amount	48.0 green tonnes/ha. Author's calculation. Difference between available residues and Leftover residues	53.0 green tonnes/ha. Table 19.1. Left on ground	48.0 / 101.1 × 100 = 47.5%	Assumption that amounts reported in Table 19.1 were in green tonnes
45	Cuchet et al.[29]	96.0 t/ha. Table 19.1. Initial amount	54.0 green tonnes/ha. Author's calculation. Difference between available residues and Leftover residues	42.0 green tonnes/ha. Table 19.1. Left on ground	54.0 / 96.0 × 100 = 56.3%	Assumption that amounts reported in Table 19.1 were in green tonnes
46	Cuchet et al.[29]	51.0 green tonnes/ha. Table 19.1. Initial amount	21.0 green tonnes /ha. Author's calculation. Difference between available residues and Leftover residues	30.0 green tonnes/ha. Table 19.1. Left on ground	21.0 / 51.0 × 100 = 41.2%	Assumption that amounts reported in Table 19.1 were in green tonnes
47	Cuchet et al.[29]	50.0 green tonnes/ha. Table 19.1. Initial amount	23.0 green tonnes/ha. Author's calculation. Difference between available residues and Leftover residues	27.0 green tonnes/ha. Table 19.1. Left on ground	23.0 / 50.0 × 100 = 46.0%	Assumption that amounts reported in Table 19.1 were in green tonnes
48	Cuchet et al.[29]	34.0 green tonnes/ha. Table 19.1. Initial amount	11.0 green tonnes/ha. Author's calculation. Difference between available residues and Leftover residues	23.0 green tonnes/ha. Table 19.1. Left on ground	11.0 / 34.0 × 100 = 32.4%	Assumption that amounts reported in Table 19.1 were in green tonnes
49	Cuchet et al.[29]	26.0 green tonnes/ha. Table 19.1. Initial amount	12.0 green tonnes/ha. Author's calculation. Difference between available residues and Leftover residues	14.0 green tonnes/ha. Table 19.1. Left on ground	12.0 / 16.0 × 100 = 46.2%	Assumption that amounts reported in Table 19.1 were in green tonnes
50	Gullberg and Johansson[38]	26.9 odt/ha. Author's calculation. Quotient of Recovered residues and Recovery rate	22.5 odt/ha. Page 1038. Removal of logging residues	4.4 odt/ha. Author's calculation. Difference between available and recovered residues	85.6%. Taken from the article (Degree of slash recovery. page 1038). Insufficient data to recalculate	

(*Continued*)

Table 19.2 Calculations of Forest Harvest Residue Recovery Rate (*Continued*)

No.	Reference	Original Total Residues on Cutblock	Recovered Residues	Leftover Residues	Recovery Rate	Notes
51	Rudolphi and Gustafsson[39]	15 m3/ha. Author's calculation. Sum of recovered and Leftover residues	9.8 m3/ha. Table 19.1. Total volume in piles	5.2 m3/ha. Table 19.1. Total volume outside piles	9.8 / 15 × 100 = 65.3%	The authors mentioned that only the piles are extracted as biofuel. We assumed that all residues in piles were recovered, and all residues outside piles were leftover
52	Benjamin et al.[40]	4.8 tons/acre. Weight of downed woody debris of cut-to-length treatment	3.3 tons/acre. Author's calculation. Difference of Original total and Leftover residues	1.5 tons/acre. Weight of downed woody debris of whole-tree treatment	3.3 / 4.8 × 100 = 68.8%	Comparison of stem-only harvested vs. whole-tree harvested sites. Calculations using debris volume as opposed to debris weight yields a recovery rate of: 306.3 / 511.9 × 100 = 59.8%
53	Briedis et al.[12]	23.6 m^3/ha. Table 19.1. Debris volume	13.1 m3/ha. Author's calculation. Difference of Original total and Leftover residues	10.5 m3/ha. Author's calculation. Product of Available residues and Recovery rate	55.5% Taken from the article (Table 19.3. 100% - %Volume remaining of total residue). Insufficient data to recalculate	
54	Briedis et al.[12]	19.7 m^3/ha. Table 19.1. Debris volume	9.6 m3/ha. Author's calculation. Difference of Original total and Leftover residues	10.1 m3/ha. m3/ha. Author's calculation. Product of Available residues and Recovery rate	48.7% Taken from the article (Table 19.3. 100% - %Volume remaining of total residue). Insufficient data to recalculate	

Table 19.2 Calculations of Forest Harvest Residue Recovery Rate (*Continued*)

No.	Reference	Original Total Residues on Cutblock	Recovered Residues	Leftover Residues	Recovery Rate	Notes
55	Briedis et al.[12]	25.4 m^3/ha. Table 19.1. Debris volume	14.2 m3/ha. Author's calculation. Difference of Original total and Leftover residues	11.2 m3/ha. Author's calculation. Product of Available residues and Recovery rate	56% Taken from the article (Table 19.3. 100% - %Volume remaining of total residue). Insufficient data to recalculate	
56	Briedis et al.[12]	17.7 m^3/ha. Table 19.1. Debris volume	9.9 m3/ha. Author's calculation. Difference of Original total and Leftover residues	7.8 m3/ha. Author's calculation. Product of Available residues and Recovery rate	56% Taken from the article (Table 19.3. 100% - %Volume remaining of total residue). Insufficient data to recalculate	
57	Briedis et al.[12]	19.4 m^3/ha. Table 19.1. Debris volume	10.9 m3/ha. Author's calculation. Difference of Original total and Leftover residues	8.5 m3/ha. Author's calculation. Product of Available residues and Recovery rate	56% Taken from the article (Table 19.3. 100% - %Volume remaining of total residue). Insufficient data to recalculate	
58	Briedis et al.[12]	28.4 m3/ha. Table 19.1. Debris volume	12.1 m3/ha. Author's calculation. Difference of Original total and Leftover residues	16.3 m3/ha. Author's calculation. Product of Available residues and Recovery rate	42.6% Taken from the article (Table 19.3. 100% - %Volume remaining of total residue). Insufficient data to recalculate	

(*Continued*)

Table 19.2 Calculations of Forest Harvest Residue Recovery Rate (*Continued*)

No.	Reference	Original Total Residues on Cutblock	Recovered Residues	Leftover Residues	Recovery Rate	Notes
59	Briedis et al.[12]	28.1 m^3/ha. Table 19.1. Debris volume	12.0 m3/ha. Author's calculation. Difference of Original total and Leftover residues	16.1 m3/ha. Author's calculation. Product of Available residues and Recovery rate	42.6% Taken from the article (Table 19.3. 100% - %Volume remaining of total residue). Insufficient data to recalculate	
60	Briedis et al.[12]	15.8 m^3/ha. Table 19.1. Debris volume	10.4 m3/ha. Author's calculation. Difference of Original total and Leftover residues	5.4 m3/ha. Author's calculation. Product of Available residues and Recovery rate	65.8% Taken from the article (Table 19.3. 100% - %Volume remaining of total residue). Insufficient data to recalculate	
61	Briedis et al.[12]	12.0 m^3/ha. Table 19.1. Debris volume	7.9 m3/ha. Author's calculation. Difference of Original total and Leftover residues	4.1 m3/ha. Author's calculation. Product of Available residues and Recovery rate	65.8% Taken from the article (Table 19.3. 100% - %Volume remaining of total residue). Insufficient data to recalculate	
62	Briedis et al.[12]	53.2 m^3/ha. Table 19.1. Debris volume	30.2 m3/ha. Author's calculation. Difference of Original total and Leftover residues	23.0 m3/ha. Author's calculation. Product of Available residues and Recovery rate	56.8% Taken from the article (Table 19.3. 100% - %Volume remaining of total residue). Insufficient data to recalculate	
63	Briedis et al.[12]	22.4 m^3/ha. Table 19.1. Debris volume	12.7 m3/ha. Author's calculation. Difference of Original total and Leftover residues	9.7 m3/ha. Author's calculation. Product of Available residues and Recovery rate	56.8% Taken from the article (Table 19.3. 100% - %Volume remaining of total residue). Insufficient data to recalculate	

Table 19.2 Calculations of Forest Harvest Residue Recovery Rate (*Continued*)

No.	Reference	Original Total Residues on Cutblock	Recovered Residues	Leftover Residues	Recovery Rate	Notes
64	Briedis et al.[12]	18.0 m^3/ha. Table 19.1. Debris volume	10.6 m3/ha. Author's calculation. Difference of Original total and Leftover residues	7.4 m3/ha. Author's calculation. Product of Original total residues and Recovery rate	59.1% Taken from the article (Table 19.3. 100% - %Volume remaining of total residue). Insufficient data to recalculate	
65	Rittenhouse et al.[41]	121.3 m3/ha. Figure 19.2. Sum of coarse and fine woody debris in roundwood treatment	62.9 m3/ha. Author's calculation. Difference between Original total and Leftover residues	58.4 m3/ha. Figure 19.2. Sum of coarse and fine woody debris in Whole-tree treatment	62.9 / 121.3 × 100 = 51.8 %	Comparison of stem-only harvested vs. whole-tree harvested sites
66	Rittenhouse et al.[41]	144.7 m3/ha. Figure 19.2. Sum of coarse and fine woody debris in roundwood treatment	50.3 m3/ha. Author's calculation. Difference between Original total and Leftover residues	94.4 m3/ha. Figure 19.2. Sum of coarse and fine woody debris in Whole-tree treatment	50.3 / 144.7 × 100 = 34.8%	Comparison of stem-only harvested vs. whole-tree harvested sites
67	Rittenhouse et al.[41]	130.3 m3/ha. Figure 19.2. Sum of coarse and fine woody debris in roundwood treatment	37.8 m3/ha. Author's calculation. Difference between Original total and Leftover residues	92.5 m3/ha. Figure 19.2. Sum of coarse and fine woody debris in Whole-tree treatment	37.8 / 130.3 × 100 = 29.0 %	Comparison of stem-only harvested vs. whole-tree harvested sites
68	Rittenhouse et al.[41]	107.7 m3/ha. Figure 19.2. Sum of coarse and fine woody debris in roundwood treatment	4.3 m3/ha. Author's calculation. Difference between Original total and Leftover residues	103.4 m3/ha. Figure 19.2. Sum of coarse and fine woody debris in Whole-tree treatment	4.3 / 107.7 × 100 = 4.0 %	Comparison of stem-only harvested vs. whole-tree harvested sites

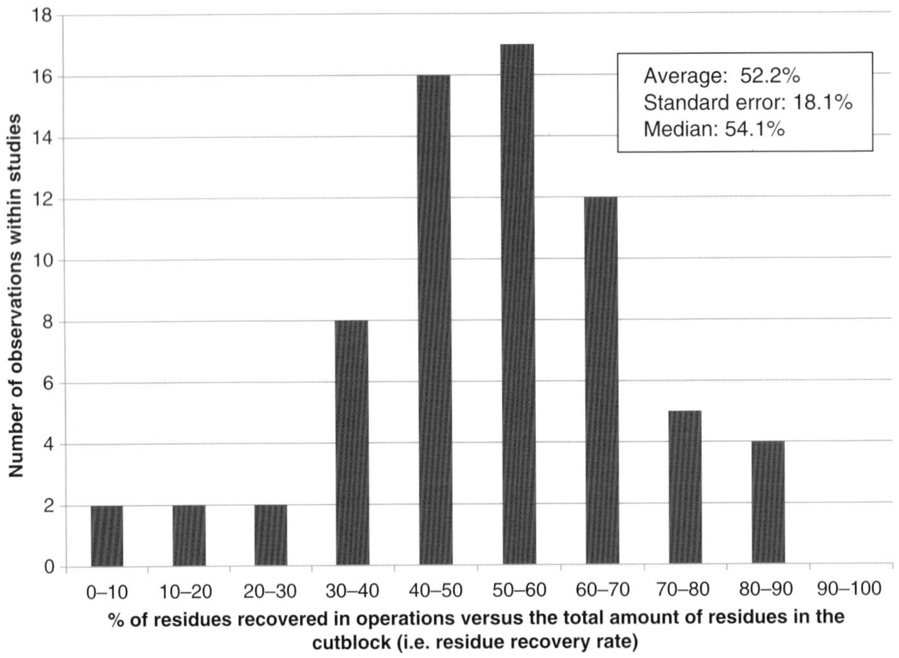

Figure 19.2 Distribution of recovery rates observed within studies ($n = 68$).

Another aspect typical of the Nordic forestry context (and embedded in operational conditions of trials reported in this study) is that operations all occur in softwood plantations. In such a context, uniformity of tree species, size and spacing, and high stand productivity facilitate machinery tasks and contribute to attaining a high level of recovery rate[15,24] relative to conditions found in natural stands (more typical of trials from the United States and Canada). For example, in Canada, the bulk of forest operations occur in naturally originated stands, which implies higher variability within and between stands than plantations, and therefore higher operational constraints. It is possible that with a shift in policy context and/or with increased technological learning (learning-by-doing and through sharing of knowledge[19]), residue recovery rate in other countries may increase, approaching the higher levels of residue removal common to Nordic countries. However, once maximum operator efficiency is attained, local stand conditions will most likely continue to set upper limits for residue removal operations.

For field trials occurring in non-Nordic countries, the most important driver of residue recovery rate was the residue recovery system: either loose slash from cutovers forwarded to the roadside (stem-only harvesting), residues produced by roadside delimbing of whole trees (whole-tree harvesting), or bundles of loose slash gathered in the harvested area and forwarded to the roadside or central landing (bundling) (Figure 19.4). Trials gathering and forwarding loose slash had a lower recovery rate (i.e., a larger proportion of residues were left on the ground) than trials using the other two residue recovery systems.

Whole-tree harvesting is widespread in many North American jurisdictions and is seen as an efficient and economical method for extracting round-wood for traditional forest products and biomass for bioenergy.[12] Several reasons may explain the lower rate of residues retained in the cutblock (i.e., the higher residue recovery rate from the cutblock) based on whole-tree harvesting, forwarding, and delimbing at the roadside. First, the main source of residues left in the cutblock in this type of operation is usually derived from the breakage of parts of the tops and branches as the whole trees are being cut and yarded from the harvest block. By contrast, in recovery of loose slash or when using slash bundling techniques, residues might be left in the cutblock due to their high level of contamination, or due to the incapacity of the forwarder to recover them (e.g., if the boom of the grapple cannot reach them, if they fall off the forwarder, etc.). Also, residue recovery in whole-tree harvesting operations is not as dependent on the skills and coordination of operators working on the harvest block and

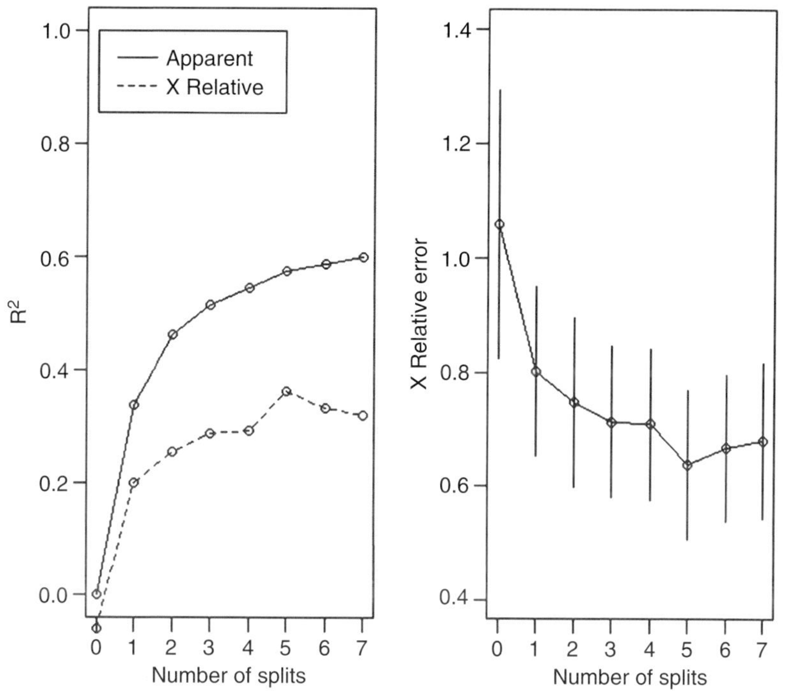

Figure 19.3 R^2 value and relative error as a function of the number of splits for the classification tree based on the harvest residue recovery rate dataset ($n = 48$).

their experience in forest bioenergy and residue recovery systems; harvest operations are more or less business-as-usual for harvesting traditional forest products only, and recoverable residues are produced at the roadside or at the landing. However, operators might be required as part of the stand prescription to leave a portion of the harvested trees on site, or to transport residues back to the site after roadside delimbing operations (e.g., as suggested in Maine's Best Management Practices for water quality protection).

Along with whole-tree operations, residue bundling was also found to be a more efficient method relative to recovery of loose slash (Figure 19.4). In Nordic countries, and especially Finland, bundling has been developed into a commercial technology.[23] Elsewhere, however, this method is rarely used in regular residue recovery operations and for the most part is still at an experimental/developmental stage.[27] The main advantages of bundling is that it increases the bulk density and homogeneity of residues, and bundles can be transported with standard log trucks to processing facilities and stored for extensive periods without loss in quality and quantity.[28] In terms of residue recovery rate from cutblocks, they should generally present similar issues to those of loose residue recovery, e.g., a reliance on well-delineated rows or piles of residues along strip roads.[22,27,29] However, the boom and grapple mounted on the bundler are likely to be more effective at collecting residues than the equipment used in field trials on loose slash recovery-forwarding; the former are specifically mounted for this task, whereas the latter also serves to collect roundwood. Operator skills may also play a role, with operators of bundlers having more specific experience in residue recovery. Consequently, bundling could also be seen as another example of how technological learning can increase residue recovery rates.

The last factor identified in the regression tree analysis as a driver of recovery rate in whole-tree and bundling operations in non-Nordic countries was the season of operations, with higher residue recovery rates in fall, winter, and spring (60.7%) than in summer (48.1%), although this segregation was based on a very small number of studies and should therefore be interpreted with caution. This could be due to the fact that it is easier to collect residues lying on a smooth snow surface than on a heterogeneous forest floor. Also, woody material may have higher moisture content in winter and spring and therefore be less brittle and prone to

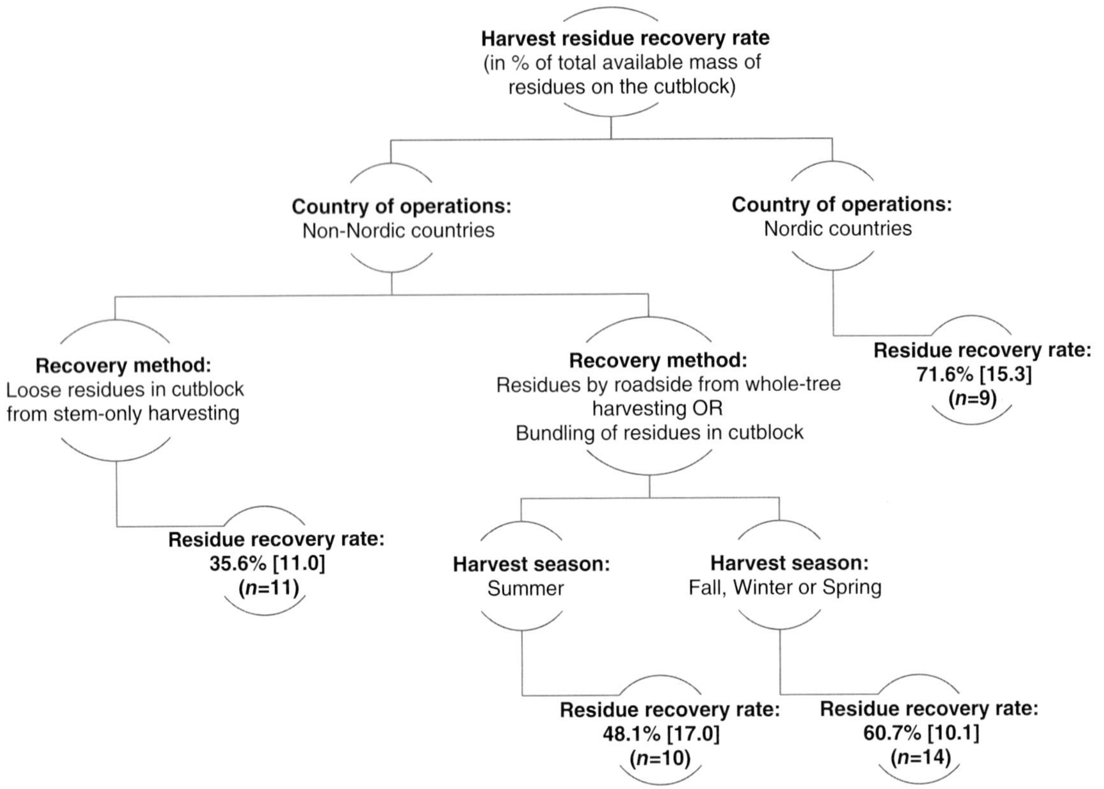

Figure 19.4 Simplified classification tree of harvest residue recovery rates ($n = 44$), with averages, standard errors [in brackets] and number of observations (in parentheses).

breakage than woody material in the summer, which may be drier as a result of warmer temperatures. Research on bundling methodology has shown that fresher material is easier to manipulate.[23,29,30] However, the overall productivity of residue supply chains in terms of the economic value of the delivered fuel is usually highest when the feedstock is drier, notably because the energy density of the material increases and transport and conditioning costs are lower per energy unit.[31,32] Moreover, allowing residues to dry on the cutover before recovery allows the shedding of nutrient-rich needles and small twigs, reducing nutrient exports from the forest site.[33] We used a small subset of trials [from Thiffault et al. (Thiffault and Paré, unpublished data) and Ralevic et al.[34] in Canada; Nurmi,[25] and Peltola et al.[24] in Finland] to plot reported values for the quantity of leftover/recovered residues in comparable units (i.e., oven-dry tonnes/ha) (Figure 19.5). In Canada, the absolute mass (in oven-dry tonnes/ha) of residues left in the cutblock after recovery did not appear to be related to the quantity of residues recovered. Instead, low postre-covery residue levels were associated with sites where the prerecovery mass of residues was low (Figure 19.5). By contrast, in Finland, the absolute mass of residues left was negatively related to the mass of residues recovered, with more efficient operations leaving less residue on site. Although based on a small number of observations, these differences may suggest that likely more homogenous forest conditions found in Finland make residue recovery intensity/rate the major driver of the absolute quantity of biomass that will remain onsite to support ecosystem functions at the site and landscape scale, such as nutrient cycling, soil productivity, biodiversity, and water quality. However, the absolute quantity of residues in Canada is mostly driven by natural characteristics and variability between stands. Consequently, predicting the amount of biomass left onsite in Canadian forests will require a better understanding of the role of factors such as site conditions (e.g., terrain conditions, stand origin, tree species composition, and stand structural characteristics), operator experience, and more importantly, the interactions between these factors. Without this knowledge, defining a

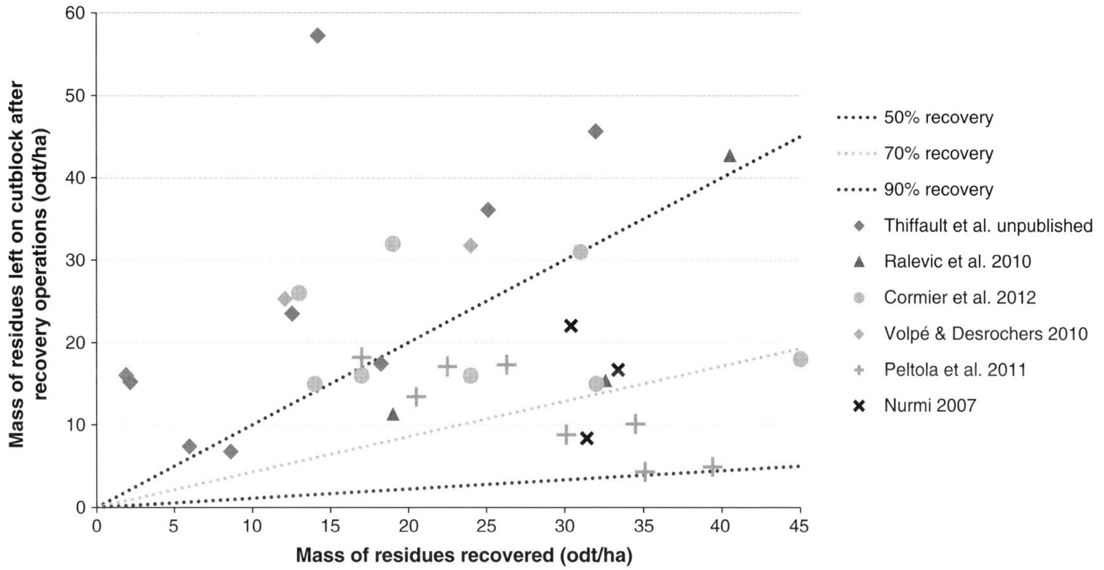

Figure 19.5 Relationship between mass of residues left in the cutblock (*y*-axis) and mass of residues recovered (*x*-axis) for a subset of field trials. See Tables 19.1 and 19.2 for trial details. Dotted lines represent values for 50, 70, and 90% harvest residue recovery rates.

threshold for harvest residue recovery would likely not have an important impact on the absolute quantity of biomass present at the site and landscape level; the goal of such a threshold could therefore be missed.

Our synthesis involved the compilation of studies encompassing a large range of methods and systems regarding forest operations, residue inventories, and estimation. This potentially introduces an unknown amount of uncertainty/variability in recovery rate estimates and comparability. For example, it is possible that studies use different definitions for harvest residues; in trials from Thiffault et al. (unpublished data) in Quebec (Canada) (Tables 19.1 and 19.2), residues were tallied in the field for material with diameters of 1 cm and greater. Moreover, due to historical and current forest management practices, cutblocks in Canadian forests often contain an amount of woody debris inherited from the previous stand rotation, and hence not produced by the current harvesting. Distinguishing fresh harvest residues from legacy debris may present a challenge, which can influence the estimated recovery rates. For example, calculating recovery rates using only the mass of fresh residues, i.e., debris with a low level of decomposition (as visually assessed in the field), and therefore presumably created by the recent harvest, yields slightly different recovery rates than calculations based on the total mass of debris (Table 19.3).

Also, when possible, we recalculated harvest residue recovery rate in order to have a standardized measure between studies. However, in some instances, recovery rate as provided by the authors had to be taken directly at face value because the raw data for determining a standardized calculation were not provided. Moreover, some studies estimated the original amount of residues potentially available for recovery (i.e., the denominator in the ratio calculating the residue recovery rate) using predictive allometric models, whereas other studies used direct field sampling to provide empirical data. Peltola et al.[24] report on significant differences between various models for estimating the potential of obtainable residues from a cutover. According to their study, for a given site, recovery rate can vary by as

Table 19.3 Recovery Rate for Trials of Thiffault et al. (Unpublished Data) (See Tables 19.1 and 19.2 for Details) Based on Total Mass of Debris and Mass of Fresh Residues Only

Site	All Debris (%)	Fresh Debris Only(%)
Cawcot	44.7	42.3
Chartierville	12.4	29.9
Disson	51.1	51.7
Foret Montmorency	34.8	40.5
Grosses-Roches	41.2	41.7
Stephane	19.9	24.8
Toulnustouc	41.0	42.4
Tournemine A	56.0	58.0
Tournemine B	10.5	17.8

much as 30% depending on the model used for estimating residue potential. In Ralevic et al.,[34] differences of 3–21% in recovery rates were demonstrated when estimates of both residue potential and remaining volumes of residues on the forest floor after recovery (both the numerator and denominator of the recovery ratio) were based on field sampling vs modeling.

Results from this study therefore raise the more fundamental question of whether and how policies and guidelines should operationally define and monitor practices and sustainability thresholds of residue removal. Statements that residue removal for bioenergy production may be unsustainable[42] clearly point out the need to define the quantity and type of biomass that may be extracted from forests. However, our analysis shows that explicit methods (e.g., protocols, guidelines, and standard operating procedures) for inventory and calculations need to be provided in order for forest residue recovery rates to be consistently implemented, monitored, and audited.

CONCLUSION

The synthesis of recovery rates of harvest residues as measured in field trials from boreal and temperate forests indicated that the average recovery rate was 52.2%, with observations more or less normally distributed around the average. The main factor explaining variation in recovery rates was country of operations, which encompasses multiple and intertwined aspects of bioenergy policy, access to bioen-ergy markets, level of technological learning, and forestry context. The small sample size did not allow a refined analysis of those aspects for different regions of the world. Nevertheless, we suggested that a shift in bioenergy policy, a growth in (and a change in access to) bioenergy markets and upward movements along the technological learning curve (e.g., improvements in machinery, better training of operators, and recovery techniques) could increase residue recovery rates approaching the highest values observed in this study, e.g., such as those routinely observed currently in Nordic countries, or perhaps even higher if economic and technological conditions keep improving for biomass extraction. However, local stand conditions may continue to constrain operational residue removal rate.

Existing studies have not yet quantified the minimum amount of organic material that should be left on site to ensure sustainable forest ecosystem functions. However, as operational recovery rates *per se* can increase when the proper conditions are met, operational considerations alone may not be sufficient to constrain residue recovery rates to levels that ensure that ecological sustainability metrics such as biodiversity and soil fertility are respected in forest biomass supply chains. Residue removal levels currently observed in countries like Canada in which forest residue-based feedstock for bioenergy production has only recently started are bound to change somewhat over time. This therefore calls for the development of policies and guidelines that define practices and thresholds based on the ecological suitability of various ecosystems to sustain harvest residue removal[43,44] and not simply rely on the assumption that operational limitations of machinery and operators will by default keep recovery rates to low levels. However, clear definitions and explicit methods for inventory and estimations would also need to be developed in order for practices to be consistently implemented, monitored, and audited.

ACKNOWLEDGMENTS

This project was made possible through funding from the EcoEnergy Innovation Initiative of Natural Resources Canada. We are grateful to Pierre Bernier and Pamela Cheers, of the Canadian Forest Service - Natural Resources Canada, for providing a scientific and linguistic review of the manuscript.

REFERENCES

1. REN21. *Renewables Global Status Report*. Paris, France: Renewable Policy Network for the 21st Century; 2012. Available at http://www.ren21.net/REN21Activities/GlobalStatusReport.aspx. (Accessed November, 2013).
2. Smeets EMW, Faaij APC. Bioenergy potential from forestry in 2050: an assessment of the drivers that determine the potentials. *Clim Change* 2007, 81:353.30. doi:10.1007/s10584-006-9163-x.
3. Haberl H, Sprinz D, Bonazountas M, Cocco P, Desaubies Y, Henze M, Hertel O, Johnson RK, Kastrup U, Laconte P, et al. Correcting a fundamental error in greenhouse gas accounting related to bioenergy. *Energy Policy* 2012, 45:18–23. doi:10.1016/j.enpol.2012.02.051.
4. Windisch J, Röser D, Mola-Yudego B, Sikanen L, Asikainen A. Business process mapping and discrete-event simulation of two forest biomass supply chains. *Biomass Bioenergy* 2013, 56:370–381. doi:10.1016/j.biombioe.2013.05.022.
5. Röser D, Asikainen A, Stupak I, Pasanen K. Chapter 2: forest energy resources and potentials. In: Röser D, Asikainen A, Raulund-Rasmussen K, Stupak I, eds. *Sustainable Use of Forest Biomass for Energy*. Dordrecht: Springer; 2008, 9–28.
6. Lamers P, Thiffault E, Paré D, Junginger M. Feedstock specific environmental risk levels related to biomass extraction for energy from boreal and temperate forests. *Biomass Bioenergy* 2013, 55:212–226. doi:10.1016/j.biombioe.2013.02.002.
7. Stupak I, Asikainen A, Jonsell M, Karltun E, Lunnan A, Mizaraite D, Pasanen K, Parn H, Raulund-Rasmussen K,

Röser D, et al. Sustainable utilisation of forest biomass for energy—possibilities and problems: policy, legislation, certification, and recommendations and guidelines in the Nordic, Baltic, and other European countries. *Biomass Bioenergy* 2007, 31:666–684. doi:10.1016/j.biombioe.2007.06.012.

8. Abbas D, Current D, Phillips M, Rossman R, Hoganson H, Brooks KN. Guidelines for harvesting biomass for energy: A synthesis of environmental considerations. *Biomass Bioenergy* 2011, 35:4538–4546. doi:10.1016/j.biombioe.2011.06.029.
9. Evans AM, Perschel RT, Kittler BA. Overview of forest biomass harvesting guidelines. *J Sustain For* 2013, 32:89–107. doi:10.1080/10549811.2011.651786.
10. Roach J, Berch, SM. A compilation of forest biomass harvesting and related policy in Canada. Prov. B.C., Victoria, B.C. technical report 081, 2014.
11. Egnell G, Björheden R. Options for increasing biomass output from long-rotation forestry. *WIREs Energy Environ* 2013, 2:465–472. doi:10.1002/wene.25.
12. Briedis JI, Wilson JS, Benjamin JG, Wagner RG. Biomass retention following whole-tree, energy wood harvests in central Maine: adherence to five state guidelines. *Biomass Bioenergy* 2011, 35:3552–3560. doi:10.1016/j.biombioe.2011.05.018.
13. Dymond CC, Titus BD, Stinson G, Kurz WA. Future quantities and spatial distribution of harvesting residue and dead wood from natural disturbances in Canada. *For Ecol Manage* 2010, 260:181–192. doi:10.1016/j.foreco.2010.04.015.
14. Perlack RD, Wright LL, Turhollow AF, Graham RL, Stokes BJ, Erbach DC. Biomass as feedstock for a bioenergy and bioproducts industry: the technical feasibility of a billion-ton annual supply. Oak Ridge National Laboratory, TN (US), technical report ORNL/TM-2005/66, 2005, 61.
15. Andersson G, Asikainen A, Björheden R, Hall PW, Hudson JB, Jirjis R, Mead DJ, Nurmi J, Weetman GF. Chapter 3: production of forest energy. In: Richardson J, Björheden R, Hakkila P, Lowe AT, Smith CT, eds. *Bioenergy from Sustainable Forestry: Guiding Principles and Practice*. Dordrecht: Kluwer Academic Publishers; 2002, 49–124.
16. R Development Core Team. R: a language and environment for statistical computing; 2004. Available at: http://www.R-project.org/. (Accessed January 2013).
17. Maine Forest Service. Best management practices for forestry: protecting Maine's water quality; 2004. Available at: http://www.maine.gov/doc/mfs/pubs.htm. (Accessed May, 2014).
18. Routa J, Asikainen A, Bjordehen R, Laitila J, Röser D. Forest energy procurement: state of the art in Finland and Sweden. *WIREs Energy Environ* 2013, 2:602–613. doi:10.1002/wene.24.
19. Junginger M, Faaij A, Björheden R, Turkenburg WC. Technological learning and cost reductions in wood fuel supply chains in Sweden. *Biomass Bioenergy* 2005, 29:399–418. doi:10.1016/j.biombioe.2005.06.006.
20. McCormick K, Kåberger T. Key barriers for bioen-ergy in Europe: economic conditions, know-how and institutional capacity, and supply chain co-ordination. *Biomass Bioenergy* 2007, 31:443–452. doi:10.1016/j.biombioe.2007.01.008.
21. Thrän D, Hennig C, Thiffault E, Heinimö J, Andrade O. Chapter 4: development of bioenergy trade in four different settings—the role of potential and policies. In: Junginger M, Goh CS, Faaij A, eds. *International Bioen-ergy Trade—History, Status and Outlook on Securing Sustainable Bioenergy Supply, Demand and Markets*. Dordrecht: Springer; 2014, 65–102.
22. Ovaskainen H, Uusitalo J, Väätäinen K. Characteristics and significance of a harvester operators' working technique in thinnings. *Int J For Eng* 2004, 15:67–77. doi:10.1080/14942119.2004.10702498.
23. Kärhä K, Vartiamäki T. Productivity and costs of slash bundling in Nordic conditions. *Biomass Bioen-ergy* 2006, 30:1043–1052. doi:10.1016/j.biombioe. 2005.12.020.
24. Peltola S, Kilpeläinen H, Asikainen A. Recovery rates of logging residue harvesting in Norway spruce *(Picea abies* (L.) Karsten) dominated stands. *Biomass Bioenergy* 2011, 35:1545–1551. doi:10.1016/j.biombioe. 2010.12.032.
25. Nurmi J. Recovery of logging residues for energy from spruce *(Picea abies)* dominated stands. *Biomass Bioenergy* 2007, 31:375–380. doi:10.1016/j.biombioe.2007.01.011.
26. Levers C, Verkerk PJ, Müller D, Verburg PH, But-sic V, Leitão PJ, Lindner M, Kuemmerle T. Drivers of forest harvesting intensity patterns in Europe. *Forest Ecol Manag* 2014, 315:160–172. doi:10.1016/j.foreco.2013.12.030.
27. Spinelli R, Magagnotti N. Logging residue bundling at the roadside in mountain operations. *Scand J Forest Res* 2009, 24:173–181. doi:10.1080/ 02827580902806593.
28. Pettersson M, Nordfjell T. Fuel quality changes during seasonal storage of compacted logging residues and young trees. *Biomass Bioenergy* 2007, 31:782–792. doi:10.1016/j.biombioe.2007.01.009.
29. Cuchet E, Roux P, Spinelli R. Performance of a logging residue bundler in the temperate forests of France. *Biomass Bioenergy* 2004, 27:31–39. doi:10.1016/j.biombioe.2003.10.006.
30. Ryans M, Desrochers L. Initial observations of the John Deere 1490D energy wood harvester. FERIC Advantage report, vol. 7 no. 7, 2006, 6.
31. Röser D, Mola-Yudego B, Sikanen L, Prinz R, Gritten D, Emer B, Väätäinen K, Erkkila A. Natural drying treatments during seasonal storage of wood for bioenergy in different European locations. *Biomass Bioenergy* 2011, 35:42 3 8–4247. doi:10.1016/j.biombioe.2011.07.011.
32. Ghaffariyan MR, Acuna M, Brown M. Analysing the effect of five operational factors on forest residue supply chain costs: a case study in Western Australia. *Biomass Bioenergy* 2013, 59:486–493. doi:10.1016/j.biombioe.2013.08.029.
33. Nurmi J. The storage of logging residue for fuel. *Biomass Bioenergy* 1999, 17:41-47. doi:10.1016/ S0961-9534(99)00023-9.
34. Ralevic P, Ryans M, Cormier D. Assessing forest biomass for bioenergy: operational challenges and cost considerations. *Forest Chron* 2010, 86:43–50. doi:10.5558/tfc86043-1.
35. Cormier D, Volpé S, Spencer S. Biomass availability related to operational constraints (In French). In: *Presentation at the Demo International Conference*, Quebec City,

Canada, September, 2012. Available at: http://www.oifq.com/pdf/congres/congres-2012/presentations/cormier.pdf.
36. Volpé S, Desrochers L. Integration of biomass forwarding in harvesting operations (in French). FPInnovations Avantage report, vol. 12 no. 12, 2010, 8.
37. Eräjää S, Halme P, Kotiaho JS, Markkanen A, Toivanen T. The volume and composition of dead wood on traditional and forest fuel harvested clear-cuts. *Silva Fenn* 2010, 44:203–211.
38. Gullberg T, Johansson J. A method for integrated extraction of logging residues and soil scarification on asmall scale. *Biomass Bioenergy* 2006, 30:1035–1042. doi:10.1016/j.biombioe.2005.12.011.
39. Rudolphi J, Gustafsson L. Effects of forest-fuel harvesting on the amount of deadwood on clear-cuts. *Scand J Forest Res* 2005, 20:235–242. doi:10.1080/02827580510036201.
40. Benjamin JG, Seymour RS, Meacham E, Jeremy W. Impact of whole-tree and cut-to-length harvesting on postharvest condition and logging costs for early commercial thinning in Maine. *North J Appl For* 2013, 30:149–155. doi:10.5849/njaf.13-016.
41. Rittenhouse TA, MacFarland DM, Martin KJ, Deelen TR. Downed wood associated with roundwood harvest, whole-tree harvest, and unharvested stands of aspen in Wisconsin. *Forest Ecol Manag* 2012, 266:239–245. doi:10.1016/j.foreco.2011.11.029.
42. Schulze ED, Körner C, Law BE, Haberl H, Luyssaert S. Large-scale bioenergy from additional harvest of forest biomass is neither sustainable nor greenhouse gas neutral. *Glob Change BiolBioenergy* 2012, 4:611–616. doi:10.1111/j.1757-1707.2012.01169.x.
43. Hazlett PW, Morris DM, Fleming RL. Effects of biomass removals on site carbon and nutrients and jack pine growth in boreal forests. *Soil Sci Soc Am J* 2014, 78:183–195. doi:10.2136/sssaj2013.08.0372nafsc.
44. Thiffault E, Barrette J, Paré D, Titus BD, Keys K, Morris DM, Hope G. Developing and validating indicators of site suitability for forest harvesting residue removal. *Ecol Indic* 2014, 43:1–18. doi:10.1016/j.ecolind.2014.02.005.

20

Forest Bioenergy Feedstock Harvesting Effects on Water Supply

Daniel G. Neary* and Karen A. Koestner

USDA Forest Service, Rocky Mountain Research Station, Flagstaff AZ, USA

INTRODUCTION

In 1990, most European countries began the process of developing management guidelines and criteria to ensure the conservation and sustainable management of forests.[1] Criterion Five of the Helsinki Process is to, 'maintain and develop the role of forests in water supply and protection against erosion'. A parallel, but independent, effort was initiated by Canada and joined by other countries with temperate or boreal forests.[2] Eight out of 67 indicators selected in the Montreal Process and endorsed by the 10 nations that drafted the Santiago Declaration in 1995 pertain to soil, watershed condition, and the quantity and quality of water resources. Although the Kyoto Protocol of 1997 did not directly address water issues, most countries that were involved in drafting the protocol now recognize that climate change is having a major impact on water supplies worldwide.[3] Thus, there is considerable sensitivity and concern as to how mitigation responses to climate change, such as the use of forest bioenergy, could potentially affect already stressed water resources.

Wood has always been used by humans as an energy source and once was the chief energy source for much of the world, although used in a dispersed manner.[4] In the coal, gas, and oil energy period that characterized the twentieth century, its consumption for energy production did not reach a large industrial scale until the search for renewable energy resulted in the development of bioenergy programs in the latter three decades.[5,6] The environmental impact of intensive forestry practices for bioenergy feedstock (woody plant material for combustion and transformation into energy) has been an issue of concern and investigation for many decades because of the growing extent and intensity of forest harvesting and the potential to impact scarce and increasingly valuable water resources.[7–13] Recent analyses have focused specifically on the potential of forest bioenergy programs to affect water supply.[14,15] Within the life cycle of forest bioenergy, operational activities during stand establishment, midrotation silviculture, harvesting, product transportation, wood storage, energy production, and ash recycling have the potential to produce disturbances that might affect water resources. There are a number of management practices that are accepted as means of reducing or eliminating the environmental effects of forestry operations and energy production. These are collectively known as best management practices (BMPs). The objective of this paper is to develop a focused analysis of the impacts of forest harvesting on water supply. Although other parts of the forest bioenergy life cycle can affect water resource quantity and quality, harvesting produces the largest changes and is the greatest focus of environmental concern. Bioenergy feedstock production adds some new twists to the forest harvesting story with short-rotation woody crops, slash harvesting, more fertilizer inputs, different plantation age classes, and different evapotranspiration (ET) scenarios. However, most of our understanding of harvesting effects come from past paired catchment studies.

*Correspondence to: dneary@fs.fed.us

Advances in Bioenergy: The Sustainability Challenge, First Edition. Edited by Peter D. Lund, John Byrne, Göran Berndes and Iacovos A. Vasalos.

Originally published in: *WIREs Energy Environ* 2012, 1: 317–332 doi: 10.1002/wene.26

WATER SUPPLY FROM FORESTS

Forests and woodlands have been relied upon as sources of water since the development of civilizations throughout the world.[16] Indian, Minoan, Mycenaean, and Egyptian civilizations recognized the need for adequate supplies of good-quality water and the importance of watershed management.[17–19] By 97 AD, Rome was a city of over 1 million people with nine aqueducts 400 km in length bringing in 450 L person^{-1} day^{-1} of fresh water from forested catchments outside the city. The Swiss were leaders in the resurrection of watershed management in Europe after the Dark Ages. Their first watershed protection forests were set aside in 1342 and, by 1777, Switzerland set had aside 322 forests as watershed reserves.[20] One of the primary objectives of the Forest Reserve Act of 1891 in the United States that created forest reserves that were to become the core of the National Forest System was protection of water supplies.

The reliance on forests for producing sustainable supplies of water is based on two fundamental reasons.[21] First, forests grow under climate conditions that produce fairly reliable amounts of streamflow. Although streams emanating from forests are affected by drought conditions, they are still more reliable than those originating in other ecosystems. The precipitation needed to sustain forest growth is also adequate to sustain stream base flows or aquifer recharge. This is more true of deep forest soils than shallow soils.[22,23] Catchment studies in Australia and South Africa have documented greater water yields off of grasslands or forests converted to grasslands[24] However, these types of catchments often have lesser water quality than forests. Second, forests generally have lower levels of disturbance so they are able to yield higher quality water than agricultural or urban catchments. In the United States alone, over 21,000 public water supplies originate on public and private forest lands.[21] In addition, there are countless private water supplies that depend on water provided by forest catchments.

Wildfires are the largest threat to forest-based water supplies.[25] Although water yields after wildfires are usually greater, the quality can be degraded.[26,27] A number of water supply catchments for major cities in eastern Australia have been affected by wildfires.[28] The forested Cotter River catchment that is one of two major water supply catchments for Canberra, Australia, was burned by a wildfire in 2003.[29] Water quality immediately after the fire was degraded by ash, sediment, and nutrient inputs, but by 2009 water quality recovered to the point where raw water input from the reforested Cotter River catchment was of a higher quality than raw water from the nearby but mainly agricultural Queanbeyan River catchment (see Box 20.1). Water supply catchments in foothills forests southwest of Denver, Colorado, were adversely affected in a similar manner by wildfires in 1996 and 2002.[30]

Another disturbance that has a potential to affect the water supply function of forests is tree harvesting.[14] The main disturbance features of forest harvesting are the change in vegetation characteristics that reduces transpiration, and the extraction and transport of biomass materials. Forest bioenergy harvesting is sometimes viewed with more concern because of the intensive nature of the removals. However, forest bioenergy harvesting is no different than intermediate thinning or final stemwood harvests, except for the intensive utilization of slash materials. Properly managed bioenergy forests are fully capable of achieving the goals of providing sustainable and high-quality water yields for beneficial natural resource, municipal, agricultural, and industrial uses.[31]

Box 20.1 CANBERRA AUSTRALIA WATER SUPPLY

The Australian Capital Territory (ACT) and the City of Canberrra draw water for municipal use from three local catchments.[29] The first developed (1912) was the Cotter River catchment, which includes part of the unmanaged Namadgi National Park, is mainly native and plantation forest. This system of three dams can store over 86.3 million m^3. The second added (1979) was the Googong system (121 million m^3 of storage) on the Queanbeyan River in New South Wales that is in a mostly agricultural landscape. The third part of the Canberra system is the Murrumbidgee River which was added as an additional source of water in 2007. The Murrumbidgee River catchment is a mixed forest, urban, agricultural watershed. Water flows by gravity to the Mount Stromolo Water Treatment Plant and then to the ACT distribution system. Daily water use ranges between 108,000 and 124,000 m^3 day^{-1}. The Cotter River catchment was the primary water supply source because it provided the most pristine water input into the Mount Strombolo Plant even though it was a managed forest. A wildfire in 2003 led to degraded water quality in the Cotter River so the Googong system became the primary supply. By 2009, the reforested Cotter River catchment's water quality had improved to the point where it was less costly to treat the Cotter River supply than the Googong water originating from a mostly agricultural watershed.

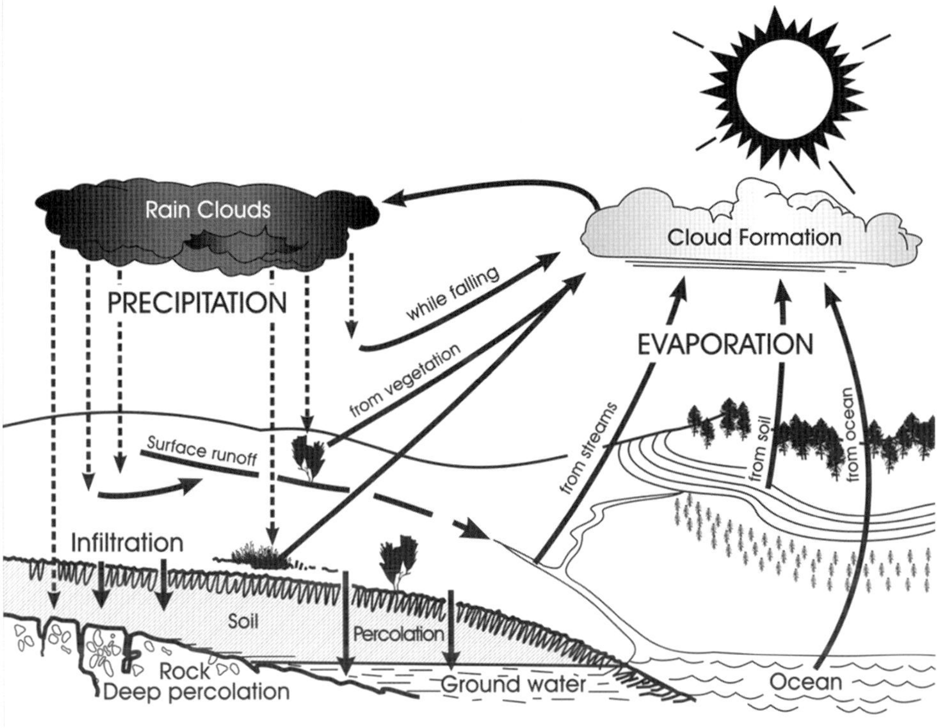

Figure 20.1 Components of the hydrologic cycle.[33]

HYDROLOGIC CYCLE

Hydrologic Processes

The hydrologic cycle quantifies the movement of water from the atmosphere to the land surface and its vegetation, movement within soils and geologic formations, distribution among land and oceanic components of the hydrosphere, and evaporation back into the hydrosphere (Figure 20.1). Water is a primary driving force in ecosystem physical, biological, and chemical processes and fluxes. The quality and quantity of water in a given ecosystem reflects the net effects of the processes, conditions, and disturbances occurring on component watersheds. Thus the quantity and quality of water emanating from watersheds are indices of ecosystem condition and frequently used as a measure of sustainability.[15] The hydrologic cycle is discussed in more detail in other publications and is represented by the simple hydrologic equation[32]:

$$I = O + dS, \quad (20.1)$$

where I is input, O is output, and dS is change in storage.

A more expanded version of this equation highlights the individual components of input, output, and storage and factors in water supply as an impact on streamflow[34]:

$$Q = P - I - T - E - L - W + R - A + dS, \quad (20.2)$$

where Q is streamflow, P is precipitation, I is interception, T is transpiration, E is evaporation from soil and plants, L is infiltration into the soil, W is water withdrawals for consumptive use (water supply for multiple uses), R is return flow from water withdrawals, A is aquifer recharge (groundwater in Figure 20.1), and dS = change in storage.

The principal inputs of water that produce *streamflow* are termed *precipitation* and consist mainly of *rainfall* and *snowfall*. In some coastal regions such as the USA Pacific Northwest, Mexico, and Chile, fog is a major contributor form of precipitation. Water outputs from catchments consist of streamflow, *evaporation*, and *transpiration* (often discussed together as *evapotranspiration* or ET). *Water withdrawals* for consumptive use consist of water uses for crop irrigation, manufacturing, mining, and municipal water supply. The storage function of the hydrologic

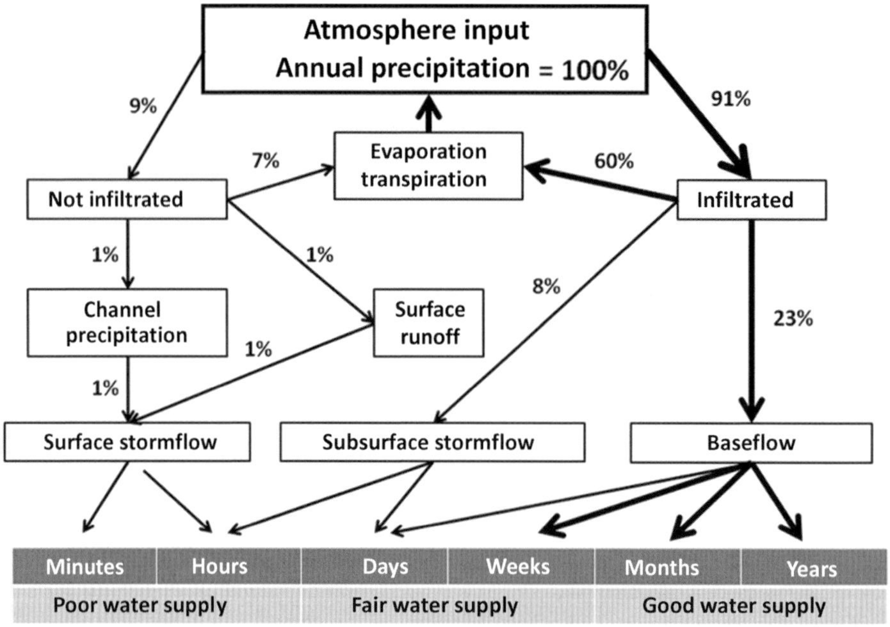

Figure 20.2 Partitioning of precipitation into output components in undisturbed forest catchments in humid regions and the effect on water supply. (Reprinted with permission from Ref 35. Copyright 2002, Elsevier.)

equation is composed of surface water detention, soil moisture, and *aquifer recharge*. Surface water detained in depressions or bodies of water such as lakes or ponds is transient in nature since it eventually is evaporated back into the atmosphere or infiltrates into the soil or geologic formations. Water incorporated into soil profiles has a range of residence times depending on its storage in quickly draining macropores, recalcitrant micropores, or clay mineral interstices. Long-term storage of water in unconsolidated or consolidated geologic formations known as aquifers can occur over periods of a few to millions of years. Water moving into aquifers can originate directly from soil *infiltration* or channel inflow into unconsolidated sediments or porous rocks. In many locations water supply comes from aquifers and not surface streamflow. Where ET is high due to aridity, latitude, altitude, clear air, and wind, streamflow is minimal, episodic, or nonexistent. In most ecosystems with precipitation <480 mm, streamflow occurs only after large storm events, during periods of the year like winter when ET is low, and as a result of snow melt.[35] Water can be temporarily stored for both short and long periods of time in both organic and mineral horizons of the soil, weathered bedrock directly beneath the soil, in porous bedrock formations, and in unconsolidated alluviual and channel sediments.[36] *Water withdrawals* are the supply function of the expanded hydrologic equation. Part of the water supply removed for consumptive use is lost back to the atmosphere as ET, and some returns to the streamflow component as *return flow* from irrigation usage and wastewater treatment.

A generalized breakdown of the inputs, fluxes, and outputs in undisturbed forested watersheds in humid regions is shown in Figure 20.2.[15] These percentage breakdown of water movement changes somewhat in arid shrub, grassland, and woodland ecosystems, and can vary considerably in watersheds disturbed by climate change, harvesting, burning, insect defoliation, windthrow, land-use conversions, mining, and agriculture. Precipitation inputs consist of rain, snow, sleet, and so on (100%). In undisturbed or well-functioning watersheds, the movement pathways for water are dominated by ET and infiltration. These processes produce the *baseflow* (streamflow between storm events) that exits the catchment area on a time frame of weeks to years and provides for a good water supply (Figure 20.2, right side). Management activities or disturbances that shift water fluxes to the left of Figure 20.2, toward surface runoff, ultimately degrade the value and usefulness of the water resource.[37]

Plant leaf, branch, stem, and woody debris surfaces intercept precipitation, causing most of it to be evaporated back into the atmosphere. This process catches and retains 7% of the initial precipitation from a storm event, depending on the degree of surface cover. Small, low-intensity storms have higher amounts of *interception* than larger, high-intensity and windier events. The reduction of intercepting surfaces after harvesting increases the amount of rainfall reaching the forest floor or mineral soil surface. Other amounts of intercepted rainfall flow down plant stems, logging slash into the soil as *stemflow* or are lost back to the atmosphere as evaporation. Snow hung up in a forest canopy eventually falls through to the forest floor or is directly evaporated back into the atmosphere through *sublimation* (water phase change from solid to gas without the intermediate liquid state.[32]

Precipitation that penetrates through a forest canopy and lands on litter, slash, or bare soil is called *throughfall*. In undisturbed forests, about 91% of precipitation infiltrates directly into the litter and mineral soil (*infiltration*), with only 1% flowing over the soil surface as surface runoff (Figure 20.2). When rainfall intensity exceeds infiltration capacity, *surface runoff* occurs. Litter cover, soil texture, vegetation, and soil porosity affect infiltration rates. With litter present as a cover on mineral soil in forests, hydraulic conductivities[38] often exceed peak rainfall intensities (>160 mm h^{-1}). Infiltrating rainfall moves laterally and vertically in the soil as *interflow*. The movement can be fast in large pore spaces in the soil (macropores) or slow in small pores (micropores).

Interflow results in temporary storage of water and ultimately produces the baseflow that is the source of the perennial flow of streams. Streamflow consists of *baseflow* between and during storm events and *stormflow* only during and immediately after rainfall. Baseflows are the consistent flows characteristic of forested watersheds that provide sustainable water supplies. Stormflows are much higher than baseflows because they are produced by processes such as in-channel precipitation and surface runoff that are minimally affected by evaporation, transpiration, and soil moisture storage capacity. The stormflow component of water yield is highest in semiarid environments and the least in higher rainfall climates. The magnitude and duration of *stormflow* is a function of climate characteristics, as well as the watershed condition after harvesting. As the period of record for watersheds increases, the probability of observing stormflows outside their known 'range of natural variability' logically increases due to the chance occurrence of extreme climatic events. Evidence is accumulating that climate change is producing more weather events outside the previously understood 'range of natural variability'.[3] The response of forested catchments to these unusual events is a function of precipitation characteristics (amount, intensity, and duration), season, topography, vegetation cover, litter, soils, and geology. Forest harvesting may have a significant effect on stormflows or none at all.

Forest harvesting *per se* does not affect the magnitude of the flow pathways shown in Figure 20.2. Cutting trees shifts some of the flow from ET to baseflow. However, high levels of soil disturbance (e.g. physical exposure, removal, or tilling, and development of chemical water repellency) can shift more water originating from precipitation to the 'water not infiltrated' side of the flow diagram. The result is higher overland flows and surface stormflows that produce damaging peakflows or floods. A concomitant impact of increased surface stormflows is a reduction in long-term soil moisture and baseflows. Soil moisture is critical for regenerating vegetation, and baseflows are important for maintaining aquatic habitat and water supplies.

Watershed (catchment) condition is a more logical term that describes the ability of a watershed system to receive, route, store, and transport precipitation. With good conditions, rainfall infiltrates into the soil, and baseflows are sustained between storms. In this situation, rainfall does not contribute to increased surface runoff and water quality degradation since most of the precipitation does not flow over the surface where it can detach and transport sediments (Figure 20.2). Severe fires, poor harvesting practices, overgrazing, conversion to agriculture, mining, and urban uses degrade watershed condition, causing the percentage of infiltrated rainfall to be reduced significantly. Rainfall then runs over the soil surface, resulting in little or no baseflow between storms (Figure 20.2), and erosion is accelerated during storm runoff events. The result is a watershed whose water supply value is poor. The surface conditions that promote good watershed condition include: (1) the presence of a surface organic horizon (O horizon) <5 mm to >20 cm thick; (2) herbaceous, shrub, and woody vegetation (variable cover); and (3) well-developed and intact soil horizons with high surface roughness. For the most part, forest bioenergy systems are able to provide these conditions.

Harvesting Effects

Forest harvesting and associated site preparation activities are the largest management disturbances in the life cycle of a forest from which fire has been excluded. They affect processes in the hydrologic cycle that determine the sustainability and quality of water supply. The specific hydrologic effects depend upon intensity, frequency, duration, timing, cutting area of the harvest; the type of harvesting equipment and extraction systems used; and the level of BMPs employed for mitigation of soil disturbances. Water quantity changes are determined by the degree and patterns of overstory and understory canopy removal while water quality changes are mostly the result of forest biomass reduction and the degree of soil

Table 20.1 General Effects of Forest Harvesting on Specific Hydrologic Processes[15]

Hydrologic Process	Type of Change	Specific Effect
1. Interception	Reduced	Moisture storage smaller; greater runoff in small storms; increased water yield
2. Throughfall	Increased	Baseflow increased; soil moisture increased
3. Evaporation	Increased	Baseflow decreased; soil moisture decreased
	Decreased	Baseflow increased; soil moisture increased
4. Litter storage	Litter reduced	Less water stored (0.5 mm cm^{-1})
	Litter not affected	No change
	Litter increased	Storage increased
5. Transpiration	Temporary elimination	Baseflow increased; soil moisture increased
6. Infiltration	Reduced	Overland flow increased; stormflow increased
	Increased	Overland flow decreased; baseflow increased
7. Streamflow	Changed	Increase in most ecosystems; decrease in snow systems; decrease with type conversion
		Decrease in fog-drip systems
8. Baseflow	Changed	Decrease with less infiltration; increase with less transpiration; summer low flows (+ and–)
9. Stormflow	Increased	Greater volume; larger peakflows; time to peakflow shorter
10. Snowpack	Charged	Cuts < 4 ha, increase snowpack cuts > 4 ha, decrease in snowpack; snowmelt rate increased; evaporation/sublimation greater

disturbances. Changes in baseflow, streamflow, and stormflow definitely affect both the quantity and quality of water delivered from forested catchments (Table 20.1). The following discussion is a general summary of the effects of forest harvesting in a bioenergy setting, not a site-specific analysis. The occurrence and magnitude of these effects is a function of the general climate, precipitation, aspect, latitude, severity of disturbance, and the percentage of a catchment harvested.

MEASURED WATER QUANTITY CHANGES

Water quantity increases due to forest harvesting for bioenergy or other products as well as intermediate thinning are very ecosystem specific. The flow of water from forested catchments is usually greatest after the first year following tree cutting.[32] Thereafter, the increased quantity of water yield associated with tree cutting or vegetation control declines as vegetation recovers and leaf area index returns to preharvesting conditions. This recovery period has been measured to be as short as 1–4 years in forests with high growth rates (e.g. *Eucalyptus* spp. in Brazil), high ET rates[11,12] and low precipitation, and over 10 years in ecosystems with high rainfall and low ET.[10,22] Water quantity increases of 80–100+% have been measured in high precipitation regions (1500–2600 mm $year^{-1}$) to <20% in low precipitation areas (450–600 mm $year^{-1}$). In some long-rotation native forest regeneration systems, harvesting results in an initial water yield increase followed by declines that have been reported to last up to 100 years.[39] There are a number of site and climate factors, which affect the hydrologic equation and determine the degree of catchment response measured in numerous studies around the world.

Annual Yield Increases

The largest effect of forest harvesting for bioenergy feedstocks is to immediately increase the water yield (Figure 20.3). The only situation where this statement is not true is in coastal forests that derive much of their precipitation input from fog drip. The amount of water yield increase after forest harvesting is a function of the proportion of a watershed that is cut, the amount of precipitation, and site factors such as aspect, soils, and vegetation. Aspect, which is a good local indicator of relative potential ET, has a strong influence on water quantity responses to forest harvesting. Slopes oriented normal to solar radiation receive the highest loadings and therefore have the highest ET. In general, mean annual streamflow increases as the percentage harvest of a forest stand or watershed increases and precipitation increases (Figure 20.3). Streamflow is usually minimal at the low end of the precipitation range for forest ecosystems (450–1000 mm $year^{-1}$) and increases due to forest harvesting are often undetectable. Evaporation is such a powerful factor in low precipitation climates, that basin-wide vegetation management or species conversion does not have much effect on streamflow.[32] Water quantity increases can be substantial (200–760 mm $year^{-1}$) in forests that occur in high precipitation zones, but some anomalies exist (e.g., rainforests).

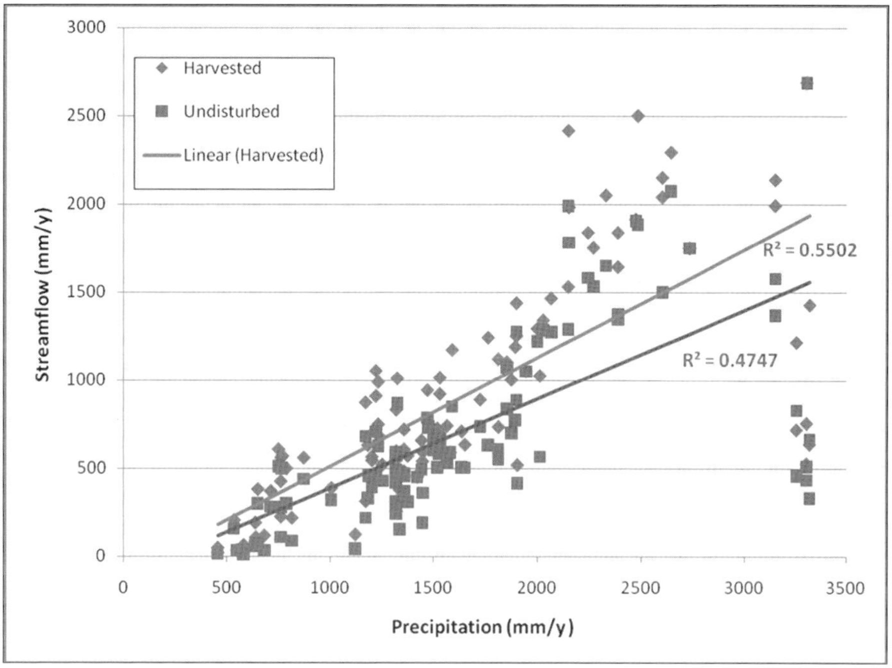

Figure 20.3 Effect of precipitation on streamflow in undisturbed and harvested forest ecosystems.[9,10,14,15]

A considerable amount of research has been conducted in the past on the hydrologic effects of forest harvesting on over 105+ individual catchments. The results have been summarized in a number of syntheses.[9,14,15,40] These studies have been very costly to install and monitor, and have required considerable dedication to their continuity by hydrologic scientists. The earliest watershed experiments were installed in Switzerland, Japan, and the United States in the first ten years of the twentieth century.[16] Some have been in existence since the 1930s. Researchers have examined various harvesting intensities, configurations, and timing.

With a 100% clearcut harvest, first-year water yield increases reported in the literature generally range from 0% to 225% (Figure 20.4). The absolute amount of water yield is strongly related to a number of factors such as the annual rainfall, vegetation type, ET regime, aspect, leaf area reduction, and soil depth.[22,32,39–41] Although the absolute water yields increase the first year after harvesting and increase with total precipitation, the percentage increase is poorly correlated to precipitation amount (Figure 20.4). Indeed, the greatest variation occurs at 100% harvest because other factors in Eq. (20.1) override transpiration reduction. Vegetation type is strongly correlated to streamflow increases after forest harvesting (Figure 20.5). Broadleaved forests have the highest mean increase in water quantity after harvesting (237 mm) compared to coniferous forests (161 mm) or mixed conifer-broadleaved forests (170 mm). They also produce the most predictable responses.

Harvesting of forests has been used to augment municipal water supplies because of the resulting increases in water yield.[10,32] The duration of the response depends on a number of factors. Generally, the increase in total water yield after harvesting is considered to be a benefit, and not of sufficient magnitude to produce adverse hydrologic or ecosystem effects (e.g., flooding). However, vigorously growing young forest stands can cause subsequent water yield declines. Short-rotation *Eucalyptus* spp. plantations are an example.[11,24,42] In a landscape managed for forest bioenergy landscape, the increases due to harvesting could offset the declines, but this is a topic for further investigation.

Annual Yield Decreases

Water yield declines after reforestation or conversion of native forests or agriculture to coniferous forests have been addressed by a number of researchers.[7,23,43–45] Water quantity increases decline to baseline levels over a period of 7 to 25 years depending on ecosystem, climate, and tree species. That may be followed by a period of three to ten decades

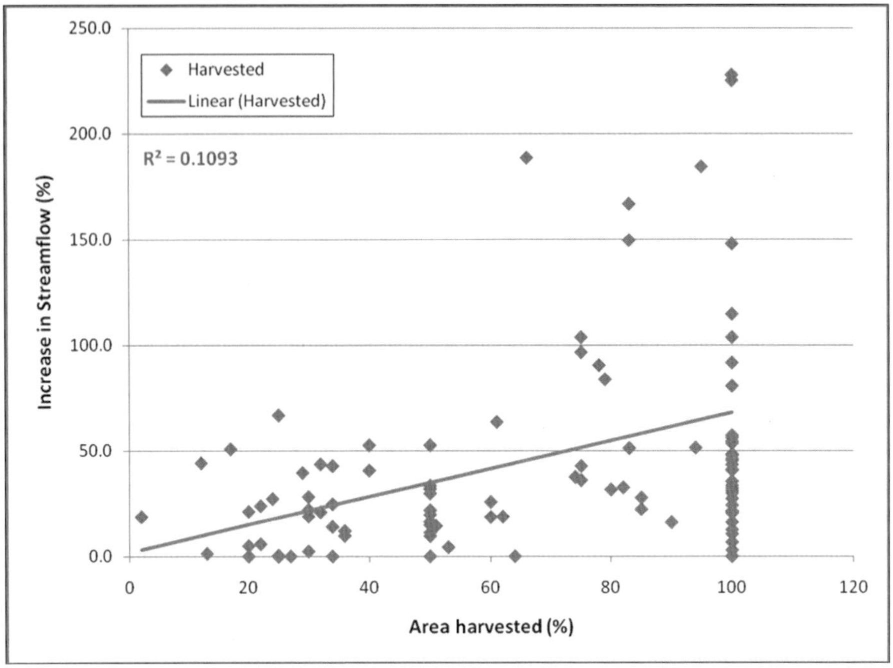

Figure 20.4 Relationship between forest area harvested and streamflow increase.[9,10,14,15]

where water yields fall below baseline levels.[39,46] This is especially the case where broadleaved trees are converted to coniferous forests, altering ET dynamics. Declines in annual flows from pretreatment conditions are typically due to increases in interception, transpiration, and evaporation [see Eq. (20.2)].

A case of water quantity decline after forest harvesting was documented in coastal conifers in Oregon that scavenged water from fog.[47] Removal of the forest cover drastically reduced oceanic fog interception, which was a major component of precipitation input, and thus produced a decline in annual water yields. In most cases, water quantity declines after forest harvesting are the result of type conversions or reversions such as agriculture to forest[48] or grassland to *Pinus* spp. or *Eucalyptus* spp.[42] This is an area requiring additional research to determine the cumulative effect on a landscape basis and for large water supply catchments. Factors causing water yield declines below base levels after harvesting are discussed elsewhere.[39] There is a scale factor in that effects observed at a small catchment scale may not hold up when analyzed at the larger landscape level.[8] Reforestation to the same tree species typical in a bioenergy forest management scheme is not likely to produce declines in the annual output of streamflow. Year-to-year climatic variation and aspect effects are constant sources of water yield increases and decreases.

The H.J. Andrews Experimental Forest is situated in the western Cascade Range of Oregon in a 6400 ha catchment of Lookout Creek, a tributary of Blue River and the McKenzie River.[49] It is representative of the rugged mountainous landscape of the United States's Pacific Northwest and it contains conifer species typical of many Cascade Range forests of the region. Lower elevation forests are dominated by Douglas-fir (*Pseudotsuga menziesii*), western hemlock (*Tsuga heterophylla*), and western redcedar (*Thuja plicata*). Upper elevation forests contain noble fir (*Abies procera*), Pacific silver fir (*Abies amabilis*), Douglas-fir, and western hemlock. The northwest-facing Watershed 1, 96 ha in size, was 100% harvested in 1962.[49] It produced a first-year increased total annual water yield of 462 mm that declined progressively to 330 mm five years later. The south-facing Watershed 6, 13 ha in area, was also 100% clear felled 12 years later in 1974 and it produced a first-year total annual yield increase of 425 mm that declined to 180 mm by year five.[41] Another south-facing catchment, Watershed 10 with an area of 9 ha, produced a first-year flow increase of 195 mm. However, water yields increased the next two years to 310 and 400 mm. The fourth year they dropped to 65 mm. The precipitation variations over those years demonstrated that water quantity changes after forest harvesting can be highly variable, particularly in forest ecosystems that are warming and drying.

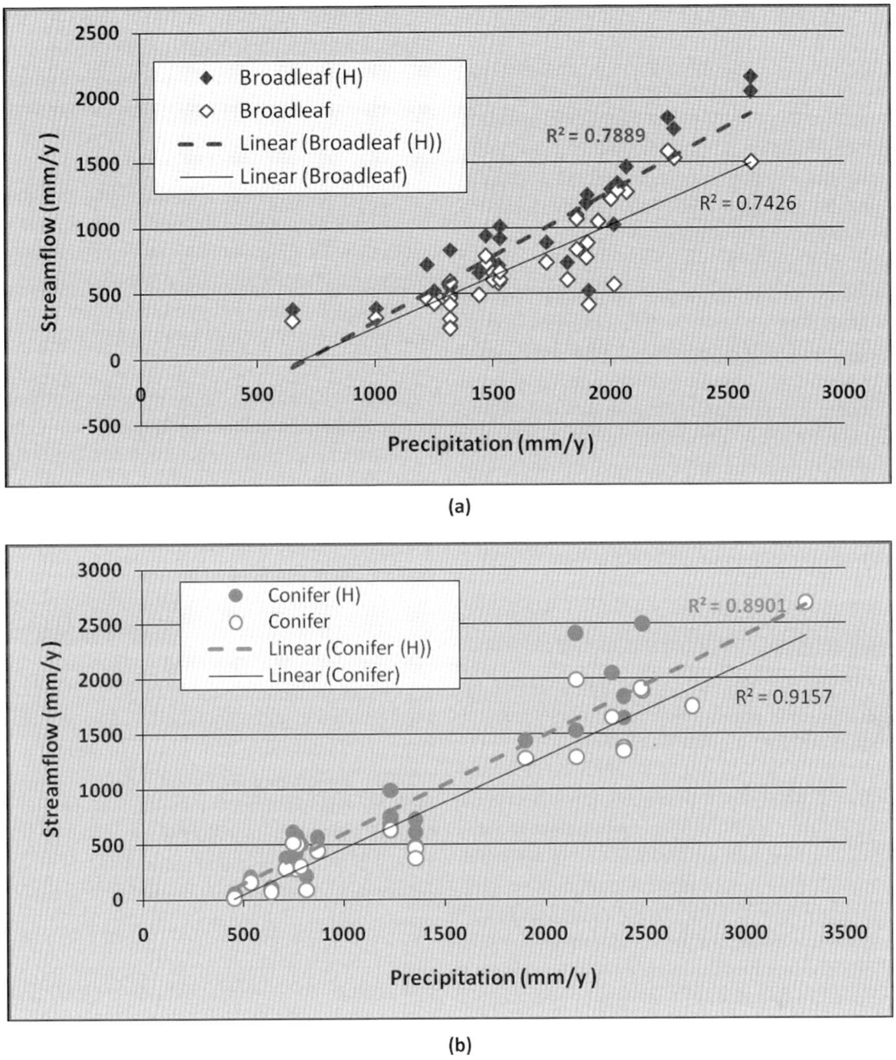

Figure 20.5 Relationship between precipitation and undisturbed and harvested forests (H): (a) broadleaved forests, and (b) conifer forests.[9,10,14,15]

Peakflows

A more important parameter of concern is flood peak flows. It has been known for centuries that forested lands are less prone to high peakflows than those converted to agricultural purposes. Study of the literature indicates that harvesting forests produces a mixed peakflow response.[7,15] Where snowmelt runoff is an important component of annual hydrographs, declines in peakflows up to 35% have been reported after forest harvesting.[50,51] Some investigators have reported no peakflow response to harvesting.[44,52] In other locations, watershed, vegetation, and climatic characteristics have produced peakflow increases of up to 1400%. However, such large-magnitude responses are rare, and more often produced by concomitant disturbances like wildfire. Several investigators have concluded that except for unusual storm events, harvesting and reforestation do not result in significant increases in peakflows.[15,46,48] Some combinations of terrain, geology, soils, vegetation, and climate may combine to create localized hazards.[14,15]

MEASURED WATER QUALITY CHANGES

Because of the recognition of forests as being major sources of water supply, a considerable number of research studies have investigated the impacts of forest harvesting on water quality in the past 50 years. The water quality parameters commonly examined by these studies are nitrate–nitrogen (NO_3–N), ammonia nitrogen (NH_4–N), phosphorus (P), major cations (sodium, Na; potassium, K; calcium, Ca; and magnesium, Mg), sediment, temperature, pH, and microorganisms. These parameters are important for streams being used as water supplies, and for their potential impact on aquatic biota, particularly threatened and endangered species. Heavy metals are a water quality concern in association with mining activities and some biosolids or waste reuse, and are not normally associated with forest bioenergy harvesting. Syntheses of the water quality studies relating to forest harvesting have been provided by several investigators.[9,14,15,40]

Nutrients

Nitrate–nitrogen is often used as an indicator of watershed health and good quality water. It is a good integrator of watershed processes and an indicator of disturbance.[13] It is used throughout the world as a key water quality standard for human health. Large increases in NO_3–N levels in streams draining harvested catchments have not been measured except where herbicides have been used to suppress postharvesting vegetation regrowth,[15,50] or nitrogen saturation of ecosystems has reached a critical level due to atmospheric deposition.[53] There is no indication in the scientific literature that water quality standards (10 mg L^{-1} NO_3–N) are threatened by postharvesting NO_3–N increases since most measured increases are <0.5 mg L^{-1} NO_3–N.

Increases of other nutrients after forest harvesting have been examined by many investigators and summarized by major syntheses.[9,14,15,54] Although the conservative nutrient cycles of undisturbed forests are temporarily interrupted by harvesting, the short-term nature of harvesting disturbances and usually vigorous regrowth of vegetation limit water quality impact. Several scientists[14,54] have concluded that there is no evidence of forest harvesting producing a significant adverse impact on the quality of water supplies from additional dissolved inorganic ions in streamflow.

Sediment

Sediment is an important water quality parameter for water supplies as it can harm aquatic organisms and habitats, render water unacceptable for drinking water supplies or recreation purposes, or result in large expenditures by municipal water treatment facilities to bring input waters up to standards. Although increased sediment movement into streams is a major environmental concern in managed forest catchments, it also occurs naturally without human-associated disturbances. Watersheds vary greatly in their natural sediment load characteristics ranging from <0.01 to 140 Mg ha^{-1} $year^{-1}$ depending on climate, geology, and soils.[14] Sediment deposits can be re-entrained after initial deposition in ephemeral or perennial stream channels, and move downstream with streamflow for long time periods and distances. The cumulative effects of erosion and sedimentation that occurred centuries ago from agriculture, mining, urbanization, or forestry present watershed managers with many current challenges.

Sediment yield increases and water turbidity produced by forest harvesting are the largest and most visible water quality impacts on water supply.[15] The degree to which sediment impairs the quality of water supplies after forest harvesting is highly variable and depends on such factors as soils, climate, topography, ground cover, watershed condition, the type of harvesting equipment used, and the transportation system. Physical disturbance of soil followed by high rainfall intensity or amounts is the main reason for sediment yield and turbidity increases after harvesting, but they are usually transient due to understory vegetation recovery. Soil disturbance and sediment movement can be significantly reduced by use of low ground pressure and wide-tracked harvesters (Figures 20.6 and 20.7). Recent research has documented the success of BMPs in limiting postharvest water quality declines.[27] (see Box 20.2).[55]

Box 20.2 BEST MANAGEMENT PRACTICES IN AUSTRALIA PROTECT WATER QUALITY

In Australia, natural resource managers are striving to enhance environmental outcomes at farm and catchment scales by planting streamside management zones (SMZs) with trees to produce a variety of ecological services including bioenergy feedstocks. A study was conducted in Tasmania, Australia, to evaluate the water quality benefits of an SMZ and the effects of tree harvesting.[55] This case study consisted of a 20-year-old *Eucalyptus nitens* pulpwood/bioenergy feedstock plantation in an SMZ of an intermittent stream that was harvested according to Code of Forest Practice of the state. A machinery exclusion zone adjacent to the stream limited machinery traffic. Tree harvesting in the Tasmanian study resulted in minimal mineral soil exposure (<11%) and increased surface roughness. Postharvesting turbidity levels in streamflow were similar to preharvest levels and an unharvested reference catchment [<2.5 nephelometric turbidity units (NTUs)] in streamflow exiting the harvested catchment. Much more significant sources of sediment were a road, a stream section that was accessible to cattle, and a cultivated paddock. These sources led to turbidities of about 300 NTUs in a dam immediately below these points and above the harvested stream reach during a postharvest storm. In-stream dams, installed many years earlier to store water for stock and irrigation, also acted as very effective sediment traps. The SMZs and other BMPs used in the agroforestry landscape were effective at protecting water quality. Forest harvesting for bioenergy feedstocks can be conducted in SMZs without increasing stream turbidity, if existing BMPs and the appropriate harvesting equipment are used.

The major sediment problems documented in the literature have been associated with log extraction methods,[55] postharvest mechanical site preparation,[56] slope instability,[57] road construction,[58] or highly erosive soils.[14] BMPs are effective in reducing on-site sediment movement when they are properly planned and implemented prior to, during, and after harvesting. Most BMPs relate to designing, constructing, and maintaining major access roads, logging roads, skid trails, and landings. These areas are the primary sources for 90% of the sediment generated by harvesting.[59] The underlying principles of BMPs are aimed at minimizing disturbances in streamside zones, reducing the erosive power of runoff on bare soils and road surfaces, and maintaining the normally high infiltration capacity of forest soils.

Temperature

Forest vegetation maintains stream temperatures that are cooler and less variable than unshaded stream reaches by blocking solar radiation.[14,15] Increases in temperature that result from forest harvesting affect physical, chemical, and biological processes. The impact on aquatic biota varies considerably, depending on whether or not individual species are adapted to wide temperature ranges, and the degree to which stream temperature is controlled by solar heating or stream baseflow. A more complete discussion of temperature impacts is summarized elsewhere.[14,32] Unacceptable increases in stream temperature can be mitigated by using buffer strips as a BMP. The cumulative effects

Figure 20.6 Harvesting of slash materials in Finland using a low ground pressure and tracked Deere bundling system (photo by Daniel G. Neary).

Figure 20.7 Stacking of a *Eucalyptus nitens* stems in a SMZ in northern Tasmania, Australia, by a Tigercat tracked harvester (photo by Daniel G. Neary).

of stream temperature increases are often moderated as streamflow from harvested areas with higher temperatures merge with those from larger, uncut areas.

CONCLUSION

Water Quantity

There is a large body of science across the world on the water yield effects of forest harvesting (>105 individually studied watersheds). The overwhelming majority of these were studies of traditional forest harvesting. In many locations, silviculture operations in bioenergy forests do not differ substantially from traditional fiber- and timber-producing ones. Forest harvesting to provide feedstocks for bioenergy developments is not likely to adversely affect the quantity of water supplies. In most instances, water yields increase after harvesting and then return to baseline levels 7–25 years later, depending on climate and forest ecosystem. Declines in water yield have been measured only in special instances such as fog-drip forests, type conversions from grasslands or agriculture to coniferous forests, and restoration of riparian and plantation forests in landscapes converted to grasslands.[24] Peakflows are probably the biggest concern relative to forest harvesting. Research has shown increases, decreases, and no change in peakflows after harvesting. The consensus of scientific opinion is that, except for unusual storm events, harvesting, and reforestation do not result in significant increases in peakflows. Future research needs to concentrate on analyzing water balances across large catchments that include a range of harvest units spanning entire rotations, water supply reservoirs, agricultural areas, and urban settings to determine the large-scale impact of forest harvesting for bioenergy feedstocks. Paired catchment research is also needed to evaluate the water yield effects of intensive short-rotation plantation forests. In most instances, there are sufficient BMPs available to mitigate potentially adverse impacts.

Water Quality

Several syntheses have concluded that there is no evidence of forest harvesting producing a significant, long-term adverse impact on the quality of water supplies from dissolved inorganic ions in streamflow such as nitrates, phosphates, and cations. Short-term increases do occur in forest catchments, but with normal harvesting and reforestation operations there is no long-term effect. Forest affected by large-scale disturbances such as wildfire and salvage logging usually recover quickly and produce higher quality water than catchments where agriculture is the dominant land use. Sediment is the main water quality problem associated with forest harvesting operations. The main sources of water quality deterioration are associated with log extraction methods, postharvest mechanical site preparation, slope instability, road construction, or highly erosive soils. Water temperature can be a concern where cold-water fisheries are important resources. BMPs are often effective in reducing sediment movement and water temperature changes when they are properly planned and implemented prior to, during, and after harvesting. Future research relative to forest bioenergy harvesting should concentrate on the cumulative effects of changes in the physical, chemical, and biological quality of water in intensively managed short-rotation plantations.

REFERENCES

1. Helsinki Process. *Proceedings of the Ministerial Conferences and Expert Meetings. Liaison Office of the Ministerial Conference on the Protection of Forests in Europe*. FIN-00171, Helsinki, Finland; 1994.
2. Montreal Process. *Criteria and Indicators for the Conservation and Sustainable Management of Temperate and Boreal Forests*. Fo42-238/1995E. Canadian Forest Service. Natural Resources Canada, Hull, Quebec, Canada; 1995.
3. Bates B, Kundzewicz ZW, Wu S, Palutikof J. *Climate Change and Water* 2008. IPCC Technical Paper VI, Intergovernmental Panel on Climate Change. Geneva, Switzerland: IPCC Secretariat; 2008.
4. Pyne SJ. *Fire: A brief History*. Seattle, WA: University of Washington Press; 2001, 204.
5. Ericksson HM, Hall JP, Helynen S. Rationale for forest energy production. In: Richardson J, Bjorheden R, Hakkila P, Lowe AT, Smith CT, eds. *Bioenergy from Sustainable Forestry: Guiding Principles and Practice*. Dordrecht, The Netherlands: Kluwer; 2002 1–17, Ch. 1.
6. Hakkila P, Parikka M. Fuel resources from the forest. In: Richardson J, Bjorheden R, Hakkila P, Lowe AT, Smith CT, eds. *Bioenergy from Sustainable Forestry: Guiding Principles and Practice*. Dordrecht, The Netherlands: Kluwer; 2002 19–48, Ch. 2.
7. Andreassian V. Water and forests: from historical controversy to scientific debate. *J Hydrol* 2004, 291:1–27.
8. Bell MA, Beckett JM, Hubabrd WF. Impact of harvesting on forest environments and resources. Ottawa, Canada, Canadian Forest Service, *Environ Canada* 1974, 39.
9. Binkley D, Brown T. Forest practices as nonpoint sources of pollution in North America. *Water Resour Bull* 1993, 29: 729–740.
10. Bosch JM, Hewlett JD. A review of catchment experiments to determine the effect of vegetation changes on water yield and evapotranspiration. *J Hydrol* 1982, 55:3–23.
11. Brown AE, Zhang L, McMahon TA, Western AW, Vertessey RA. A review of paired catchment studies for determining changes in water yield resulting from alterations in vegetation. *J Hydrol* 2005, 310:28–61.
12. Brown HE, Baker MB Jr., Rogers JJ, Clary WC, Kovner JL, Larson FR, Avery CC, Campbell RE. Opportunities for increasing water yields and other multiple use values on ponderosa pine forest lands. USDA Forest Service, Research Paper RM-129, Fort Collins, CO: Rocky Mountain Forest and Range Experiment Station; 1974, 40.
13. Stone E. The impact of timber harvest on soils and water. In: *Report of the President's Advisory Panel on Timber and the Environment*. Washington, DC: U.S. Government printing Office; 1973, 427–467.
14. Neary DG, Hornbeck JW. Impacts of harvesting practices on off-site environmental quality. In: Dyck WJ, Cole DW, Comerford NB, eds. *International Energy Agency Project A6 Book, Impacts of Harvesting on Long-Term Site Productivity*. London: Chapman and Hall; 1994, 368, Ch. 4.
15. Neary DG. Environmental sustainability of forest energy production. In: Richardson J, Smith T, Hakkila P, eds. *Bioenergy from Sustainable Forestry: Guiding Principles and Practices*. Amsterdam: Elsevier; 2002. 344, Ch. 6.
16. Neary DG. Changing perceptions of watershed management from a retrospective viewpoint. In Ffolliott PF, Baker MB Jr., Edminster CB, Dillon MC, Mora KL. tech. cords. Land Stewardship in the 21st Century: The Contributions of Watershed Management. USDA Forest Service, Rocky Mountain Research Station, Proceedings RMRS-P-13: Fort Collins, CO; 2000 167–176.
17. Illich I. *Water and the Waters of Forgetfulness*. London: Boyars Press; 1985.
18. Tainter JA. *The Collapse of Complex Societies*. New York: Cambridge University Press; 1988, 250.
19. Chandra S. *Hydrology in Ancient India*. Rookee, India: National Institute of Hydrology; 1990, 106.
20. Kittredge J. *Forest Influences*. New York: McGraw Hill Book Company; 1948, 394.
21. Ryan DF, Glasser, SP. Goals of this report. In: Dissmeyer GE, ed. *Drinking water from forests and grasslands: a synthesis of the scientific literature. General Technical Report SRS-39*. Asheville, NC: Southern Research Station; 2000, 246, Ch. 1.
22. Hewlett JD, Hibbert AR. Increases in water yield after several types of forest cutting. *Int Assoc Sci Hydrol* 1961, 6:5–17.
23. Hornbeck JW, Adams M, Corbett ES, Verry ES, Lynch JA. Long-term impacts of forest treatments on water yield: a summary for northeastern USA. *J Hydrol* 1993, 150:2043–2052.
24. Zhang L, Dawes WR, Walker GR. Response of mean annual evapotranspiration to vegetation changes at catchment scale. *Water Resour Res* 2001, 37:701–708.
25. Ice GG, Neary DG, Adams PW. Effects of wildfire on soils and watershed processes. *J Forest* 2004, 102:16–20.
26. DeBano LF, Neary DG, Ffolliott PF. *Fire's Effects on Ecosystems*. New York: John Wiley & Sons; 1998, 333.
27. Neary DG, Ryan KC, DeBano LF, EDS. Fire effects on soil and water. USDA Forest Service, General Technical Report RMRS-GTR-42, Vol. 4. Fort Collins, CO: Rocky Mountain Research Station; 2005, 250.

28. Wade A, White I, Worthy M, Gill AM, Muellet N, Wasson RJ. Natural and human induced fire impacts on water quality in water supply catchments. In: Lambert M, Daniell TM, Leonard M, eds. *Proceedings of Water Down Under 2008*. Modbury, SA: Engineers Australia, Casual Productions; 2008, 1201–1212.
29. ACTEW. 2011. Water networks. Available at: http://www.actewagl.com.au/water/networks/default.aspx. (Accessed April 20, 2011).
30. Graham RT, ed. Interim Hayman fire case study analysis. USDA Forest Service, General Technical Report, RMRS-GTR-115. Fort Collins, CO: Rocky Mountain Research Station; 2003, 281.
31. Dissmeyer GE. ed. Drinking water from forests and grasslands: a synthesis of the scientific literature. USDA Forest Service, General Technical Report SRS-39. Asheville, NC; Southern Research Station; 2000, 246.
32. Brooks KN, Ffolliott PF, Gregerson HM, DeBano LF. *Hydrology and the management of watersheds*. Iowa State University Press, Ames, IA, 2003, 502.
33. Ackerman WC, Colman, EA, Ogrosky, HO. *Where we get our water. In: The Yearbook of Agriculture*. Washington, DC: *U.S. Department of Agriculture*; 1955, 41.
34. Swanson FJ, Scatena FN, Dissmeyer GE, Fenn ME, Verry ES, Lynch JA. Watershed processes—fluxes of water, dissolved constituents, and sediment. In: Dissmeyer GE, ed. *Drinking Water from Forests and Grasslands: a Synthesis of the Scientific Literature*. USDA Forest Service General Technical Report SRS-39. Asheville, NC: Southern Research Station; 2000. 246.
35. Clary WP, Baker MB Jr., O'Connell PF, Johnsen TN Jr., Campbell RE. Effects of pinyon-juniper removal on natural resource products and uses in Arizona. USDA Forest Service, Research Paper RM-128. Fort Collins, CO: Rocky Mountain Forest and Range Experiment Station; 1974. 28.
36. Velbel MA. Weathering and soil-forming processes. In: Swank WT, Crossley DA, eds. *Forest Hydrology and Ecology at Coweeta*. New York: Springer-Verlag; 1988, 469.
37. Neary DG, Ice GG, Jackson CR. Linkages between forest soils and water quantity and quality. *Forest Ecol Manag* 2009, 258:2269–2281.
38. Neary DG. Impacts of wildfire severity on hydraulic conductivity in forest, woodland, and grassland soils. In: Elango L, ed. *Hydraulic Conductivity - Issues, Determination and Application*. Rijeka, Croatia: INTECH; 2011, 123–142. ISBN 978-953-307-288-3, 434 p.
39. Watson FGR, Vertessy RA, Grayson RB. Large-scale modelling of forest hydrological processes and their long-term effect on water yield. *Hydrol Process* 1999, 13: 689–700.
40. Stednick JD. Monitoring the effects of timber harvest on annual water yield. *J Hydrol* 2000, 176:79–95.
41. Harr RD. Effects of timber harvest on streamflow in the rain-dominated portion of the Pacific Northwest. In: *Proceedings of the Workshop on Scheduling Timber harvest for Hydrological Concerns*. Portland, OR; 1979, 45.
42. Scott DF, Lesch W. Streamflow responses to afforestation with *Eucalyptus grandis* and *Pinus patula* and to felling in the Mokobulaan experimental catchments. *J Hydrol* 1997, 199:360–377.
43. Kuczera G. Prediction of water yield reductions following a bushfire in ash-mixed species eucalypt forest. *J Hydrol* 1987, 94:215–236.
44. Bren LJ, Papworth M. Water yield effects of conversion of slopes of a eucalypt forest catchment to radiata pine plantation. *Water Resour Res* 1991, 27: 2421–2428.
45. Swank WT, Vose JM, Elliott KJ. Long-term hydrologic and water quality responses following commercial clearcutting of mixed hardwoods on a southern Appalachian catchment. *Forest Ecol Manag* 2001, 143:63–178.
46. Beschta RL, Pyles MR, Skaugset AE, Surfleet CG. Peakflow responses to forest practices in the western Cascades of Oregon. *J Hydrol* 2000, 233:102–120.
47. Ingwerson JB. Fog drip, water yield, and timber harvesting in the Bull Run municipal watershed, *Oregon Water Resour Bull* 1985, 21:469–473.
48. McGuiness JL, Harrold L. Reforestation influences on small watershed streamflow. *Water Resour Res* 1971, 7: 845–852.
49. Rothacher J. Increases in water yield following clear-cut logging in the Pacific Northwest. *Water Resour Res* 1970, 6: 653–658.
50. Pierce RS, Hornbeck JW, Likens GE, Bormann FH. Effect of elimination of vegetation on stream water quality and quantity. *Int Assoc Hydrol Sci* 1970, 96:311–328.
51. Verry ES. Effect of aspen clearcutting on water yield and quality in northern Minnesota. In: Csallany SC, McLaughlin TG, Striffler WD, eds. *Watersheds in Transition. Proceedings of a Symposium*. Fort Collins, CO: American Water Resources Association and Colorado State University; 1972, 405.
52. Kochenderfer JN, Edwards PJ, Wood F. Hydrologic impacts of logging an Appalachian watershed using West Virginia's Best Management Practices. *North J Appl Forest* 1997, 14:207–218.
53. Aber JD, Nadelhoffer KJ, Steudler P, Melillo JM. Nitrogen saturation in northern forest ecosystems. *Bioscience* 1989, 39:378–386.
54. Stednick JD. Timber management. In: Dissmeyer GE, ed. *Drinking Water from Forests and Grasslands: A Synthesis of the Scientific Literature*. General Technical Report SRS-39. Asheville, NC: Southern Research Station; 2000. 246, Ch. 10.
55. Neary DG, Smethurst PJ, Baillie B, Petrone KC, Cotching W, Baillie CC. Does tree harvesting in riparian areas adversely affect stream turbidity—preliminary observations from an Australian case study. *J Soils Sediments* 2010, 10:652–670.
56. Beasley RS, Granillo AB. Sediment and water yields from managed forests on flat coastal plain sites. *Water Resour Bull* 1988, 24:361–366.
57. O'Loughlin CL, Pearce AJ. Influence of Cenozoic geology on mass movement and sediment yield response to forest removal. *Bull Int Assoc Eng Geol* 1976, 14:41–46.
58. Swank WT. Stream chemistry responses to disturbance. In: Swank WT, Crossley DA, eds. *Forest Hydrology and Ecology at Coweeta*. New York: Springer-Verlag; 1988, 339–357, Ch. 25.

59. Swift LW Jr. Forest access roads: design, maintenance, and soil loss. In: Swank WT, Crossley DA, eds. *Forest Hydrology and Ecology at Coweeta*. Springer-Verlag, New York, 1988. 469, Ch. 23.

FURTHER READING

Buol SA, Southard RJ, Graham RJ, McDaniel PA. *Soil Genesis and Classification*. Ames, IA: Iowa State University Press; 2003. 460.

Collier KJ, Rutherford JC, Quinn JM, Davies-Colley RJ. Forecasting rehabilitation outcomes for degraded New Zealand pastoral streams. *Water Sci Technol* 2001, 43:175–184.

O'Loughlin CL, Rowe LK, Pearce AJ. Sediment yield and water quality responses to clearfelling of evergreen mixed forests in western New Zealand. In: *Proceedings of the Helsinki Symposium on the Influence of Man on the Hydrological Regime with Special Reference to Representative and Experimental Basins*. Wallingford, Oxfordshire: International Association of Hydrological Sciences, PIAHS-AISH Publication No. 130; 1980.

Swanson RH, Golding DL, Rothwell RL, Bernier PY. Hydrologic effects of clear-cutting at Marmot Creek and Streeter watersheds, Alberta. In: *Information Report NOR-X-278. Northern Forestry Centre, Canadian Forestry Service*, Edmonton, Canada: Northern Forestry Centre, Canadian Forestry Service; 1986, 27.

21

Best Management Practices for Forest Bioenergy Programs

Daniel G. Neary*

USDA Forest Service, Rocky Mountain Research Station, Flagstaff, AZ, USA

INTRODUCTION

In the search for renewable energy sources, woody plants are being viewed as an important source of environmentally friendly feedstocks for electrical generation and biofuels.[1–3] Forest bioenergy operational activities encompass activities of a continuing and cyclical nature such as stand establishment, mid-rotation silviculture, harvesting, product transportation, wood storage, energy production, ash recycling, and then back to stand establishment. All of these have the potential to produce varying levels of disturbance that might affect site quality and water resources.[4–7] There are a number of management practices that are accepted as means of reducing or eliminating the environmental effects of forestry operations and energy production.[8,9] These are collectively known as best management practices (BMPs).[10,11] This term is used in many domains from accounting and tourism to forestry. It implies that there is a widely acceptable combination of management actions that under most conditions ensure desirable outcomes.

In forestry and farming, the term BMP usually refers to practical and economic operational procedures and practices that eliminate or keep risks to environmental quality at an acceptably low level.[12,13] In most instances, the key environmental parameter is water quality, and the focus of BMPs in forest management is the streamside management zone (SMZ) (Figure 21.1).[14] However, as discussed in this paper, BMPs exist and can be used for all life cycle phases of forest products and bioenergy feedstock production. For example, one BMP for forestry operations is to keep machinery out of waterways.[15] Another set minimizes road stream crossings by efficient design of main roads and skidder tracks.[16] Still others establish sediment control treatments such as gabions, sediment fences, straw bales, or wattles, and ditch-line diversions in order to trap sediment onsite and minimize sediment runoff into streams at road crossings.[9,17,18] Still other BMPs exist for wood processing facilities like sawmills and pulp and paper manufacturing plants as well as power transmission lines pipelines associated with bioenergy production facilities. Not all BMPs are necessarily accepted by all stakeholder groups or land managers as providing the desired environmental outcome for all sites.

The term BMP can be misleading if 'Best' is understood to imply that better practices do not exist. There is always the possibility that new scientific knowledge and practical experience can be used to improve a currently accepted BMP or create new ones.[19,20] BMPs are effective, practical, structural, or nonstructural methods, which prevent or reduce environmental degradation. In the forestry and bioenergy context, they are used most commonly to protect water resources, and are usually developed to achieve a balance between environmental protection and the production of woody and herbaceous crops within natural and economic limitations.[21]

Codes of practice are collections of BMPs that are, if compulsory, prescribed in regulations and guidelines, and therefore require compliance.[14] The BMPs embodied in

*Correspondence to: dneary@fs.fed.us

Advances in Bioenergy: The Sustainability Challenge, First Edition. Edited by Peter D. Lund, John Byrne, Göran Berndes and Iacovos A. Vasalos.

Originally published in: *WIREs Energy Environ* 2013, 2: 333–350 doi: 10.1002/wene.77

Figure 21.1 A Streamside Management Zone along the Black River, Arizona, USA. Reproduced with permission of the USDA Forest Service and Daniel G. Neary.

codes of practice may be applicable to all or any combination of target groups, for example, forestry operations on public and private land, and in small or large areas. Forest practices in many developed countries tend to be regulated in this manner. However, BMPs and codes of practice can also be voluntarily developed and adopted, which is more common in the agricultural sector.[22] BMPs can be general in nature or tailored to specific activities.[8] General BMP guidelines are designed to sustain forest or agricultural resources such as cultural resources, soil productivity, riparian areas, visual quality, water quality, wetlands, and wildlife habitat. They are applicable to activities such as road construction and maintenance, harvesting, site preparation, pesticide use, reforestation, stand tending and thinning, fire management, and recreation management. Specific BMPs are activity-specific guidelines that are unique to an activity and designed to work with general guidelines to provide an integrated framework needed to ensure forest or agricultural resource sustainability.

For forest bioenergy programs, BMPs are essential to ensure long-term productivity and sustainability because management of forests for bioenergy objectives often involves intensification of forest access, harvesting, and disturbance.[23–25] Because many forest bioenergy producers seek certification of sustainability through the Forest Stewardship Council (FSC), the Programme for the Endorsement of Forest Certification, or other certification systems, adoption and use of BMPs is a necessity.[26–29] This paper provides an overview of BMPs used in forestry and forest bioenergy programs.[30] It discusses development of BMPs,[31] types of BMPs,[32] and examples of their implementation.[33] Although most forestry BMPs are directly applicable to forest bioenergy programs, there are some aspects of the forest bioenergy life cycle that are different from production forestry and require unique BMPs. These include slash harvesting, woody biomass storage, power generation, powerline right-of-way maintenance, and ash recycling. This paper examines forestry BMPs in the context of forest bioenergy programs.

FOREST BIOENERGY LIFE CYCLE

Wood, as a bioenergy feedstock, is viewed as presenting opportunities and risks to the environment throughout its life cycle.[34] Forest bioenergy has a number of components where BMPs can be employed because of the nature of disturbances associated with production components.[35–37] A key part of any bioenergy system is the periodic assessment of life cycle components to determine the net environmental benefits and costs.[38] These evaluations also measure the benefits and costs of BMPs, and identify areas

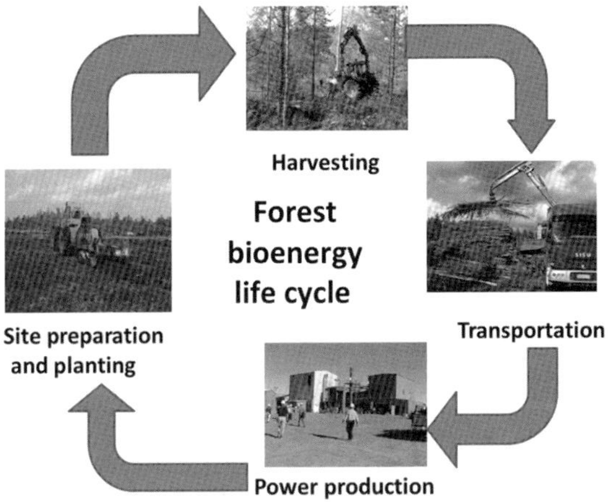

Figure 21.2 Schematic of a lifecycle for forest bioenergy. Reproduced with permission of Daniel G. Neary.

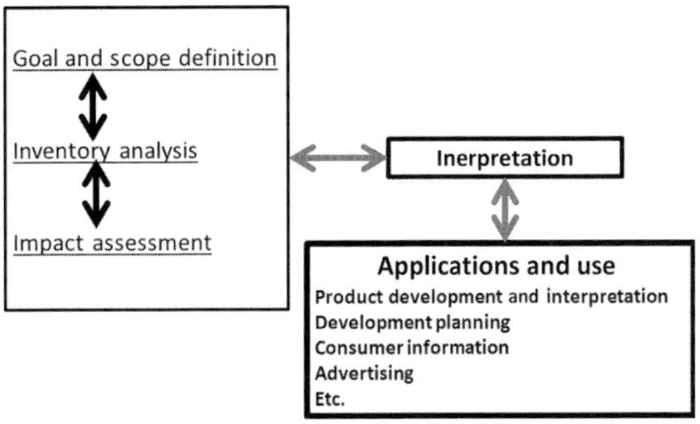

Figure 21.3 Life cycle assessment phases according to ISO 14040 (1997). Reproduced from Ref 41. Copyright 2004, Elsevier.

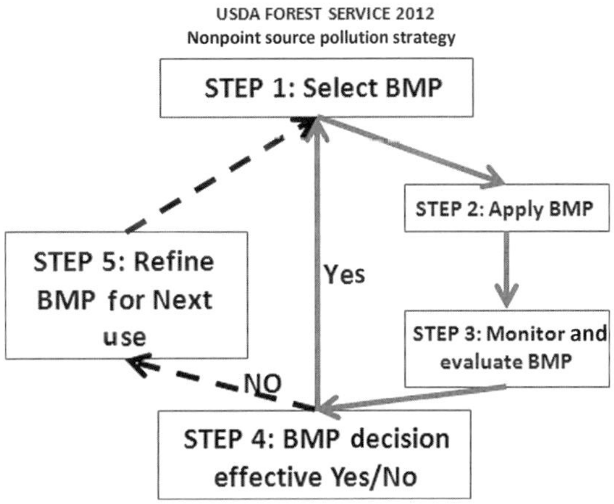

Figure 21.4 USDA Forest Service nonpoint source strategy. Reproduced from Ref 9. Copyright 2012, United States Department of Agriculture.

needing refinement, BMP improvements, or replacement BMPs.[39–41]

The forest bioenergy life cycle starts with stand establishment (Figure 21.2). In some instances, there are few impacts of this phase because naturally regenerated forests can be placed under forest bioenergy objectives and management. Operations during the forest or bioenergy plantation reestablishment phase can include site preparation, tree planting, weed control, and fertilization. The next step involves intermediate stand treatments that may or may not include harvesting of thinned stems and slash depending on the particular silvicultural prescription, economics of the management plan, and demands for feedstocks. Weed control and fertilization might be included if warranted. The final harvest usually causes the greatest site disturbances. Temporary roads used to haul felled trees or slash to landings, and permanent, but unpaved, roads are often the main sources of sediment that enters water bodies as nonpoint source pollution. Wood processing facilities and power generation stations have a minimal environmental footprint and their pollution generation activities can be more readily controlled and mitigated. Power lines used to distribute energy generated at bioenergy power plants normally require BMPs only during construction and routine weed control. However, permanent powerline access roads require ongoing BMPs for the life of the powerline. The handling of ash wastes is the final step in the cycle where BMPs can be employed.

Life cycle assessment (LCA) is a tool used to assess the environmental impacts of a product, a process or an action.[38,39] The LCA methodology consists of four stages: (1) goal and scope definition, (2) inventory analysis (life cycle inventory), (3) impact assessment, and (4) interpretation and improvement evaluation (Figure 21.3).[41,42] An integrated LCA must include all of the above four stages according to ISO 14040.[43] The last stage of an LCA is to interpret and check the results of the study and not to assess improvements. This last stage can then lead to continued use of current methods, product or operation improvement, strategic planning, public policy making, marketing, technology transfer actions, media contact planning, and so on. The LCA process can be used at large scales or focused down on specific activities as is done with a nonpoint source pollution control strategy and BMP implementation for a specific forest harvesting operation (Figure 21.4).[9] The BMPs selected by the LCA are then implemented and evaluated in a similar process to the initial LCA process.

BMP HISTORY

In the early 20th century, engineers designing industrial and municipal sewage pollution controls typically utilized physical systems (e.g., filters, clarifiers, biological reactors) to provide the core components of pollution control systems. They used the term 'BMPs' to describe the supporting functions for engineered pollution control systems.[44] The BMPs they used included support items such as operator training and equipment maintenance. Later in the century, storm water managers used the term BMP to describe both structural control devices systems as well as operational or procedural approaches to reduce pollutant loadings (e.g., reducing the use of fertilizers and pesticides).[12]

The use of BMPs is a 'best management' approach based on known science that, if followed, should allow a manager or group of managers to comply with regulatory standards or achieve desired management objectives. The term was associated with water pollution since it was specifically mentioned in the USA Clean Water Act of 1977, as amended in 1987. Functionally, BMPs are much broader in scope and application and are now associated with a diverse suite of actions aimed at protecting the environment. They are practical control measures with technological, economic, and institutional considerations that have been demonstrated to effectively minimize environmental impacts. The use of BMPs is widely accepted as the most appropriate method of controlling nonpoint sources of pollution because BMPs prevent or minimize pollution rather than retrospectively respond to it.[20,32,45] Extensive testing of BMPs and nonpoint source control programs has demonstrated that these approaches are not a lesser companion to engineered effluent treatment and point source control efforts. Assessments have demonstrated that BMPs are able to effectively address extremely complex and variable watershed conditions. BMPs continue to evolve as research identifies new challenges and practices. As the primary tool for controlling nonpoint source pollution, BMPs have the most important role in addressing water quality problems for threatened water resources.[20]

BMPs usage for complex problems is context specific and often occurs in a context of imperfect, but constantly improving knowledge. Thus, it is more appropriate to think of BMP application to land management activities as an adaptive learning process rather than a rigid set of rules or guidelines. This approach to best practice focuses on fostering continuous improvements in quality and promoting continuous learning.[13,46,47] Codes of practice that do not incorporate the adaptive learning concept embodied in BMPs revert to a regulatory approach that is reactive rather than proactive.

BMP DEVELOPMENT

BMPs development is an iterative process that begins with selection of a land management objective or activity that requires a BMP. A BMP can be selected from known, existing BMPs, or specifically designed (Figure 21.3). Once the BMP is applied for the management activity of concern, it is then evaluated through monitoring to determine if it is effective.[48] If the BMP is effective, it then becomes part of the management activity process and is applied again. If the decision is that the BMP was not effective, then it is adjusted or refined for another application and evaluation or it is abandoned. At that point, the development cycle goes back to the beginning for design or selection of a new BMP. Monitoring and evaluation are constant factors in the BMP application and use cycle, as environmental or operational conditions are subject to change, and improvement is a constant factor in BMP use.[49]

LIFE CYCLE COMPONENTS REQUIRING BMPS

Stand Establishment

Stand establishment might be done by natural regeneration processes or intensive practices like mechanical site preparation, prescribed fire, chemical weed control, mechanical planting, or fertilization. These activities are conducted as a normal part of the forestry life cycle, but are more critical where intensity forest bioenergy is the driving management consideration. The BMPs for these activities are aimed at reducing soil disturbance and damage, minimizing erosion, preventing off-site movement of pesticides and fertilizers, minimizing temporary road construction, controlling prescribed fire spread, burning in weather conditions that limit fire severity, and so on. Next to the final harvest of forest stands, this part of the forestry life cycle produces the most environmental disturbance and requires implementation of BMPs. The BMPs are often combinations of engineering controls, operations planning, and timing.

Mid-Rotation Silviculture

In the middle of or at intervals within the life cycle of a forest stand, intermediate silviculture treatments might be applied to increase tree growth. The major activities include weed control, fertilization, and thinning. Their use is a function of growth response potential and economics because of the typical costs involved. Mid-rotation thinning of noncommercial tree stems and their associated slash debris is commonly used as a feedstock source for bioenergy. Again, the BMPs are often combinations of engineering controls, operations planning, and timing.

Harvesting

Tree harvesting is the most visible and potentially site-disturbing activity in the forest bioenergy life cycle, but not necessarily the greatest impact on water quality. It involves tree cutting, on-site processing, skidding, loading onto trucks or rail cars, or floating in rivers and lakes for transportation to mills for further processing or use in energy production. Harvesting is done by ground-based equipment (skidders, forwarders, etc.) or aerial systems (cable, high lead, skyline, jammers, balloon, helicopter, etc.).[50–52] The BMPs for harvesting begin with planning and focus on engineering and operations solutions for minimizing ground disturbance, water resources impacts, and residual stand damage.

Product Transportation

Roads are most commonly used as the access mechanism to transport forest bioenergy feedstocks from forest stands to energy production facilities. The term encompasses paved highways and side roads, gravel secondary roads, temporary access roads, and skidder/hauler tracks. These areas are the primary sources for 90% of the sediment generated by harvesting.[40,53] The underlying principles of BMP guidelines in this context are aimed at minimizing disturbances in SMZs, to reduce the erosive power of runoff from bare road surfaces, and to maintain the normally high infiltration capacity of forest soils. Recent regulatory and judicial decisions in the USA have brought the role of forest roads and their drainage systems into closer scrutiny and potential change in classification from 'nonpoint source pollution' to 'point source pollution'.[54] The USA Environmental Protection Agency (USEPA) recently published a proposed rule for public comment in the Federal Register concerning stormwater discharges from logging roads (and forest roads in general) under the Clean Water Act. The proposal would place road runoff subject to National Pollution Elimination Discharge Systems permit requirements. Two recent judicial decisions have reached conflicting opinions.[54] These developments highlight the importance of using road BMPs for forest management and bioenergy harvesting programs.

Wood Storage

Log or bundled slash storage areas can become point sources of water quality degradation either at off-site storage or at power generation facilities.[48,55] Locations of concern are the actual storage sites, loading and unloading areas, chemical and liquid fuel storage facilities, and equipment and vehicle maintenance, storage, and repair areas. Pollutants include soil from storage decks and vehicle movement areas, and oils, grease, and hydraulic fluids from feedstock handling equipment. The BMPs for these areas are aimed at minimizing soil and vehicle petrochemical contact with storm runoff, dust control, or detaining surface runoff from storage sites so that it cannot mix with nearby surface waters.[56]

Energy Production

The main environmental considerations relating to power generation with biomass are feedstock storage and handling, air emissions, noise, water reuse, water treatment, and ash handling.[57] Despite these concerns, biomass is considered to be a preferred renewable energy production method. Facilities for generating electricity fall into the category of point-source risks to water quality so that BMPs can be more specifically targeted to individual site concerns. The BMPs for power plants are often combinations of engineering controls, operations planning, and timing.

Ash Recycling

Ash produced by feedstock combustion can be returned to harvesting sites as a BMP to counter nutrient loss and soil acidification.[58] Applications of ash are done by ground or air and are similar to fertilizer applications. The BMP technology used for fertilizer applications are applicable to ash spreading and are aimed at minimizing or avoiding direct inputs into streams.[59,60]

Stand Reestablishment

Reforestation of stands harvested for forest bioenergy feedstocks completes the life cycle. Reestablishment operations can be low intensity if natural regeneration is used, but are likely to be intensive.[61] Thus, BMPs become an important part of reestablishment planning and execution.

EXAMPLES OF BMPS

Forest Bioenergy BMPs

The conventional forestry BMPs that were developed in the USA and other countries since the 1970s have been generally accepted as applicable to forest bioenergy systems.[5] However, some special considerations specific to the use of harvested woody material for bioenergy feedstocks need to be considered in order to achieve the goal of sustainable biofuels from forests.[61] Some of these considerations are accelerated removal of site biomass and nutrients, more common and frequent use of fertilizers, more frequent entry into stands with associated road improvements and better harvesting equipment, better stream crossing design and construction, greater site preparation, and intensified soil disturbance. Presenting all of the known BMPs for forest bioenergy harvesting cannot be accommodated in this paper, so selected examples of the scope of forestry BMPs are presented here.

Australia

Forest practices in Australia are regulated by state governments that require that forestry operations to be conducted according to a state-based code of forest practice.[14] Federal (Commonwelath) government oversight and involvement occurs through the National Forest Policy Statement (NFPS) of 1992.[62,63] The main operating mechanism of NFPS is Regional Forest Agreements between the Commonwealth and State governments that have been negotiated by some, but not all, Australian states.

Forestry codes describing BMPs were first implemented in Australia in 1978 with most states following the lead of the Australian Capitol Territory during the 1980s along with several subsequent revisions.[64] The prime motivation for implementing a BMP-based code of forest practice was to protect environmental values during forestry operations, particularly as native (natural) forests were harvested and regenerated. Current codes include provisions for plantations that were first developed with the assumption that plantations were replacing native forest on a broad scale. The same BMP provisions have generally been applied where broad-scale plantations replace pasture or other farming enterprises. Since the earliest use of these codes, streamside reserves and machinery exclusion zones have been key BMPs used to protect soil, water, and biodiversity values. Recommended streamside reserves extend 5–100 m out from each stream bank, depending on stream size, and preclude most harvesting operations. Machinery exclusion zones are used to protect intermittent streams and drainage ways, but in some areas trees can be carefully harvested within them. Australia's BMPs are discussed elsewhere in more detail.[14]

New South Wales

Forestry harvesting operations in the State of New South Wales are more strictly regulated than in other Australian states under a licensing system known as Integrated Forestry Operations Approvals (IFOAs).[63] The BMPs required under the IFOAs center on SMZs called Drainage Feature Protection Areas (DFPAs). The DFPAs are required next to wetlands, streams, drainage lines, and drainage depressions and are based on water resource type and perennial stream order.[14] Within the DFPA concept is a whole suite of BMPs such as bridge, culvert, and road design; drainage spacing; skid trails; seasonal harvesting restrictions; weather operational restrictions; slope restrictions; soil limitations; tree felling limitations; and machinery operation guidelines. Widths of SMZs vary from 5 to 10 m but can be wider depending on stream order and harvest plans.[14]

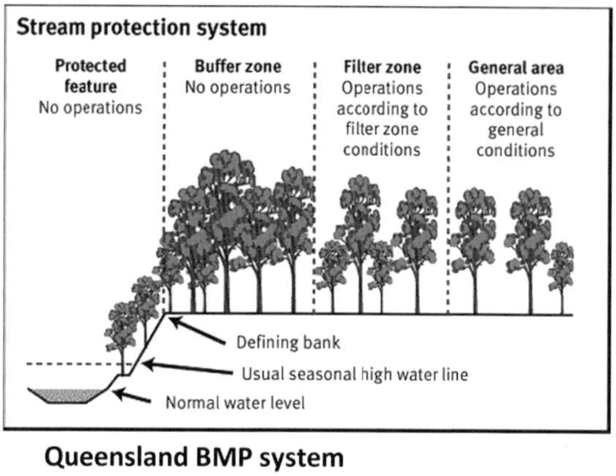

Figure 21.5 Stream protection BMP system for Queensland, Australia, based on distances from waterways and harvesting operations guidelines. Reproduced from Ref 66 with permission from the Department of Natural Resources and Mines, Queensland, Australia, 2013.

Queensland

Queensland's Code of Forest Practices provides an excellent example of BMPs for protecting streams using a three-zone SMZ.[14,66] The water resource is designated as the *protected feature*, and includes stream, river, lake, and wetland areas from normal long-term water levels to the seasonal high water mark and geomorphically defining bank (Figure 21.5). The area adjacent to the protected feature is a 2–10 m wide *buffer zone* either side of or surrounding the water resource (Table 21.1). If the protected water resource has unstable banks, then the buffer zone is expanded a distance of three times the height of the unstable banks. Tree felling is not permitted by the code into streams, into unstable drainages, within buffer zones, or into buffer zones from adjacent SMZ areas that make up the *filter zone*. This last special zone is variable in width (10–50 m) depending on stream order (Table 21.1).

Table 21.1 Queensland, Australia, Minimum Buffer and Filter Zone Widths for the Stream Protection BMP Based on Stream Order*

Stream Order	BMP: Minimum Buffer Zone	BMP: Minimum Filter Zone
	..................m..................	
1	2	10
2	2	15
3	5	20
4	5	25
5	5	40
6	10	50

*Streams are defined as drainages, creeks, or rivers that are represented on a 1:250,000 topographic map.
Reproduced from Ref 66 with permission from the Department of Natural Resources and Mines, Queensland, Australia, 2013.

Conditions under which trees may be harvested within the filter zone are clearly elucidated by Queensland's BMP guidelines. The final zone away from the protected water feature is the *general area* where harvesting operations may be conducted according to directions in the forest harvesting field guide.[66] Two other categories of protected waterways have separate BMP guidelines—drainages and wetlands. These areas carry surface runoff or store water in landscape depressions.[66]

Tasmania

The Tasmania Code of Forest Practice has an extensive set of BMPs for forest harvesting.[14, 17] Tasmania's BMP focus is on watercourses and the SMZs developed for these areas. The BMP directives in the Tasmania Code for water quality protection have separate sections for native forest reserves, plantation forests, swampy ground and surface seepage areas, and water supply catchments. The BMPs for plantation forests are further segregated by soil erodibility class (low to moderately high and high to very high) (Table 21.2).[67] A recent review of harvesting and SMZs concluded that BMPs and SMZs can be successfully established in most landscapes for a combination of environmental, agricultural, and timber production benefits if the regulatory framework allows.[14] The review recommended that regulations should be dynamic and reviewed periodically to incorporate the best science. The up-front SMZ and BMP process is viewed as the best approach to achieving environmental sustainability of forest management for bioenergy or other purposes rather than an after-the-fact regulatory approaches (Box 21.1).

Box 21.1 TASMANIAN TEST OF BMP EFFECTIVENESS

An example of forest harvesting and the effective use of BMPs in an SMZ for the protection of water quality was demonstrated in recent research in Tasmania, Australia.[33, 67] A study was conducted in northwest Tasmania to evaluate the water quality benefits of BMPs during tree harvesting in a SMZ typical of a farm forestry setting. A 20-year-old *Eucalyptus nitens* pulpwood plantation along an intermittent stream was harvested following the Tasmania code of forest practice.[17] Two provisions of the code regulated activities within a 10 m zone of the stream. A 5 m machinery exclusion zone immediately adjacent to the stream required tracked harvesters to reach into that zone to harvest trees. An additional 5 m zone, that is, 5–10 m from the stream, limited machinery traffic to tracked harvesters that were not allowed to turn in this zone. Trees were cut and extracted with almost no soil disturbance in the 0–10 m zone. Two temporary crossings of the class 4 stream were designated in the harvesting plan and used. Debarking and further processing of stems occurred outside the 0–10 m zone, with retention of woody debris as soil protection were additional BMPs. Tree harvesting produced minimal mineral soil exposure across the harvested site (<2%) and increased surface roughness needed for sediment trapping. Postharvesting turbidity in streamflow was similar to preharvest levels and an adjacent untreated stream. This study demonstrated that forest harvesting operations can be conducted without necessarily affecting water quality if existing BMPs are followed.

CANADA

Forest practices codes and BMP guidelines in Canada are regulated by each province. The general country-wide standards come from the Forest Stewardship Council of Canada, a nongovernment body that offers environmental certification for land use practices.[68] Because most of the land ownership and harvesting of Canada's 397 million ha of forest land is provincial (80%), harvesting done by the forest industry is on a permit system guided by four regionally specific forest management standards: National Boreal, British Columbia, Great Lakes St. Lawrence, and Maritimes. A couple of examples are provided here to give an indication of provincial government direction to the forest industry on BMPs.

The province of Alberta implemented forest harvesting planning and BMPS for operations to protect watersheds.

Table 21.2 Forest Code 2000 BMPs for Water Quality and Watercourse Protection Related to Tree Harvesting in and Alongside SMZs in Plantation Forests

Soil Erodibility Class	Specific BMPs
	Plantations may be harvested in streamside reserves and within 10 m of class 4 watercourses subject to the following provisions:
Low to moderate	No trees are to be harvested within 10 m of a class 1, 2, or 3 watercourse for plantations established after the commencement of this edition of the code.
	Excavator type feller bunchers may enter to within 5 m of a streambank provided slopes are less than 20°.
	Harvesting will only be carried out when soils are dry, or provided measures are taken to minimize soil disturbance.
	Trees will, wherever practicable, be felled away from watercourses.
	Machinery will move in and out of the machinery exclusion zone by the same path without slewing the machine's tracks.
	Remnant native vegetation will be retained.
	Stems will be removed for processing to a site at least 10 m from the streambank.
	Other harvesting machinery will not enter within 10 m of the streambank except at designated crossing points or to remove substantial harvesting debris.
	Where this approach is used, the 10 m machinery exclusion zone will be marked, and the marking responsibility stated in the plan.
High to very high	Selective harvesting or the felling of the planted streamside reserve at a different time to the remainder of the coupe should be considered on sensitive sites.
	Harvesting debris which results in altered watercourse conditions should be kept out of watercourses.
	Substantial harvesting debris accidentally lodged in permanently flowing watercourses should be removed where authorized by a Forest Practices Officer while causing as little damage as possible to the streambank and reserved vegetation. Machinery is excluded from watercourses.

Water is the main component of the forest environment that can be protected during harvesting operations through use of BMPS.[69] Annual operating plans and specific BMPs are intended to carry out provincial policy and legal directives. The Alberta standards for watershed protection and harvest planning begin with water body classification.[69] The layout of roads, landings, and stream protection SMZs, as well as tree felling activities, and equipment operation buffers are outlined in the Alberta forest harvesting guidelines. The BMPs for SMZs are not slope adjusted but harvesting is typically not allowed above slopes of 45%. In the Alberta system, tree harvesting is usually not allowed within SMZ buffers unless specifically permitted and mitigated by BMPs.

The Ontario code of practice for forest harvesting operations was designed to protect provincial water quality.[70] It was specifically developed for the protection of fish habitat as well as water quality. The code is a BMP approach based on common sense and the application of on-the-ground of professional experience. In order to maintain water quality, the code has a focus on minimizing or mitigation soil erosion and compaction. The Ontario guidelines are designed for flexibility of application and are not meant to be rigid. Where better local information is available, SMZ widths and other BMPs may be modified in a manner consistent with the protection of fish habitat. Harvesting practices are scheduled for routine monitoring in order to determine both compliance and effectiveness of BMP and operations. Enforcement of water quality standards violations occurs through a variety of relevant legislation.[70] Some of the features of the Ontario BMP system include both operational and equipment operation guidelines for areas adjacent to and within SMZs. The BMP goal for harvesting is to minimize the removal of residual biomass that protects the soil. Other BMPs match yarding equipment to slopes and cover the location and use of skid trails. Other operational BMP techniques include avoiding the felling trees into water bodies, designating 3 m SMZs buffers of undisturbed forest floor and vegetation, careful planning and use of designated stream crossings, and keeping lubrication and fuelling of motorized equipment in designated areas distant from waterways.

USA: USEPA

The USEPA provides detailed guidance on BMPs to control nonpoint source pollution from most land use activities including forestry.[71,72] The BMP guidance document was developed to provide state water quality and forestry managers, private forest owners, industrial forest owners, and others with interests in forest management the most current, best available, economical, and most achievable methods for mitigating water quality impacts and minimizing impacts on surface and groundwater quality.[21,73] These measures are then used, adapted, augmented, and regulated by state and local jurisdictions, and land management agencies and private companies to achieve water and environmental quality objectives specified in laws, regulations, and land management plans. The USEPA approach deals with site specific forestry activities but is designed to evaluate water quality problems on a watershed level. This approach recognizes that water quality is an integration of the impacts of multiple resource management activities and land uses.[61] Because the BMP guidance document is national in scope, it cannot address the complete matrix of forestry practices occurring on different ecosystems, soils, climates, and forest types. Ongoing research and development at state and local scales is expected to provide the refinements needed to apply the national BMP guidance at local levels. Specific BMPs for forest bioenergy operations are not spelled out in the USEPA document, so they fall under the general BMPs for tree harvesting and other silvicultural operations.

USA: USDA Forest Service

The USDA Forest Service recently published its national BMP guide.[9] The guide's objectives are to provide uniform direction to field units, improve water quality to restore impaired water resources, improve relationships with National and State regulatory agencies and the general public, improve the Forest Service's ability to clearly demonstrate progress in watershed management, improve the use of adaptive land management in forest plan implementation, establish a consistent process to monitor and evaluate the effectiveness of the BMPs, and improve compliance with federal environmental laws. There are 11 National Core BMP categories, but only six directly relate to forest bioenergy operations. These are planning, chemical use, wildland fire, roads, and mechanical vegetation management. Each BMP is organized by title, reference, objective, explanation, and recommended practices.

Planning

Planning is an important BMP in itself as the process identifies potential impacts to water quality and other resources such as soils, biodiversity (flora and fauna), and riparian areas.[9] A critical part of planning BMPs is to include trained and qualified watershed specialists (hydrologists, soil scientists, geologists, fish biologists, and aquatic ecologists) to make up an interdisciplinary team for general and project planning. Planning is a critical part of forest bioenergy programs, so these BMPs are very relevant. There are three National Core BMPs in the general planning category: Forests and Grasslands Planning, Project Planning and Analysis, and Aquatic Management Zone Planning. Examples of specific BMPs within the Project Planning and Analysis core BMP are listed in Table 21.3.

Aquatic Ecosystems Management Activities

This set of BMPs addresses the primary reason for BMP development—the protection of water quality and water resources. The objective of this group of BMPs is to avoid, minimize, or mitigate adverse impacts to water quality, riparian ecosystem components, or soils that might result from construction or maintenance operations in or around streams, lakes, and wetlands.[9] These BMPs relate primarily to access road construction in forest bioenergy operations. There are four National Core BMPs in the aquatic ecosystems management activities category: Aquatic Ecosystem Improvement and Restoration Planning, Operations in Aquatic Ecosystems, Ponds and Wetlands, and Stream Channels. Examples of specific BMPs within the Operations BMP are listed in Table 21.4.

Chemical Use

Chemical applications in forest bioenergy are used to mitigate biological or physical site problems. The most commonly used are fertilizers for nutrient deficiencies and herbicides for weed control. Other, but rarely used, chemicals include insecticides and fire retardants. Again, the objective of this group of BMPs is to avoid or minimize adverse impacts to water quality.[9] There are six National Core BMPs in the chemical use category: Chemical Use Planning, Follow Label Directions, Chemical Use Near Waterbodies, Chemical Use in Waterbodies, Chemical Handling and Disposal, and Chemical Application Monitoring and Evaluation. These BMPs are designed to minimize the contamination risk to water resources due to spray drift, transport of chemical residues in surface runoff, or leaching into groundwater and ultimately streams, lakes, ponds, and wetlands. Selected core BMPs within the Chemical Use Near Waterbodies Core are listed in Table 21.5.

Wildland Fire

Prescribed fire use is very variable in forest bioenergy stands. In some locations, it is never used at all, and in

Table 21.3 Selected USDA Forest Service Planning BMPs

General Considerations	Specific BMPs
Water Quality (WQ) Management	Identify WQ management desired conditions and objectives
	Identify and evaluate the current condition of water resources
	Identify designated beneficial uses and critical WQ parameters
	Identify dams, water supplies, diversions, and other special uses
	Identify impaired water resources
	Identify aquatic threatened, endangered, and sensitive species
Potential Impacts to WQ	Assume hydrological connections exist surface to groundwater
	Consider impacts of current use
	Consider impacts of expected in climate
	Evaluate sources of water impairment—WQ, yield, floods, and so on
	Identify unstable areas and sensitive soils
	Identify soil limitations and productivity impacts
	Develop site-specific BMP prescriptions
	Consider enhanced BMPs
	Verify preliminary findings with field inspections
	Identify federal, state, and local permit requirements
	Plan to limit surface disturbances
	Design specific Streamside Management Zones
	Plan minimal management impacts on unstable/sensitive soils
	Use suitable tools to analyze cumulative watershed effects
	Integrate restoration and rehabilitation needs into project plans
	Identify monitoring needs
	Clearly delineate protected or excluded areas on project maps

Created using data from Ref 9. Copyright 2012, United States Department of Agriculture.

Table 21.4 Selected USDA Forest Service Aquatic Ecosystem BMPs

General Considerations	Specific BMPs
Plans to protect waterways	Clearly delineate work areas
	Locate staging areas close to but outside of work areas, SMZs, and so on
	Refuel and service vehicles only in staging areas
	Develop and implement erosion control plans
	Develop plans to deal with erosion control device failures
	Consider needs for solid waste and sanitary waste disposal
	Use only small, low ground pressure equipment and manual labor
	Ensure that all equipment is free of aquatic invasive species
	Use biodegradable hydraulic fluid for heavy equipment
Minimize operations impacts	Minimize heavy equipment entry into water
	Work only during dry periods
	Stage construction to limit the extent of disturbed areas
	Promptly install and maintain erosion control devices
	Stockpile topsoil for use in restoration
	Rehabilitate disturbed areas after construction is completed
	Minimize stream bank and riparian area excavation
	Identify and use suitable off-site areas for disposal of debris
	Contour the site to disperse runoff, minimize erosion, and so on

Created using data from Ref 9. Copyright 2012, United States Department of Agriculture.

others is a common treatment to prevent catastrophic, stand replacing wildfires. Fire is used as a tool mainly in fire-evolved ecosystems such as ponderosa pine (*Pinus ponderosa*) in the western USA. Prescribed fire is a common companion to thinning in bioenergy fuels treatments in these forests. The combination of thinning and prescribed fire improves tree growth, reduces fire risk, and provides feedstock for bioenergy facilities.[74] Fire does pose a threat

Table 21.5 Selected USDA Forest Service Chemical Use Near Water BMPs

General Considerations	Specific BMPs
Project planning	Use field visits to identify surface waters, wetlands, and groundwater recharge areas
Establishing a SMZ	Identify designated water uses (e.g., irrigation, water supply, etc.)
	Determine usual climatic conditions for the planned application
	Assess terrain characteristics (e.g., slope, soils, geology, vegetation, etc.)
	Evaluate application equipment and chemical formulation alternatives
	Consider ground versus aerial application
	Flag or mark SMZs on maps and application GPS units
Application	Locate operation bases in upland areas
	Use clean and functional equipment
	Calibrate application equipment
	Apply chemicals only under favorable weather conditions
	Establish acceptable weather parameters prior to application
	Monitor weather during application
	Apply fertilizers during periods when plant nutrient uptake is high
	Follow label instructions on application rates for herbicides and insecticides
	Develop a chemical spill contingency plan
	Locate spill containment equipment and supplies near to loading areas
	Apply excess tank and hopper loads to the treatment area
	Clean application equipment away from water resources
	Monitor waters during application to determine residue contamination

Created using data from Ref 9. Copyright 2012, United States Department of Agriculture.

to watershed condition and hydrologic response by consuming vegetation, forest floor fuels, coarse woody debris, and creating some degree of water repellent conditions.[19,75] In most uses, prescribed fires are low in severity and do not pose a great risk to altered hydrologic response. However, fire BMPs are designed to avoid or minimize adverse impacts to water quality, riparian resources, and soils. There are four National Core BMPs in the Wildland Fire category, but only one addresses prescribed fire. Selected BMPs within the prescribed fire category are listed in Table 21.6.[9]

Box 21.2 ROAD IMPACTS ON WATER QUALITY WITH AND WITHOUT EFFECTIVE BMPs

The impacts of roads on water quality and the effectiveness of BMPs in protecting water quality can be seen in a study from Malaysia.[76] They investigated the effects of logging a lowland tropical rainforest in Butik Tarek Experimental Watershed in Malaysia that included 20 m SMZs. Four catchments (14–38 ha) included a control no harvest (C1), a 20 m SMZ (C2) and a low road density, a 20 m SMZ with a high road density and close proximity to the SMZ (C2T), and an SMZ partial harvest (C3). Catchment areas outside the SMZs were considered to be general logging operations areas that were clearcut. The logs were extracted by skidders and log trucks. There was no significant difference in sediment delivery to the Butik Tarek stream between the unlogged control catchment (C1) and the one with a 20 m SMZ and no significant roading. Catchments C2T and C3, characterized by a high road density or SMZ logging produced six and five times the sediment delivery into channels, respectively, than the uncut control catchment (C1). Roads and skid trails in C2T and C3 with steep gradients (>20%) and a high degree of connectivity to the catchment channels were responsible for the high sediment delivery to the Butik Tarek stream, not the harvesting operation *per se*.[77] Better planning and use of road and SMZ harvesting BMPs would have reduced the water quality impact in catchments C2T and C3.

Roads

The construction and maintenance of roads are important issues on National Forest lands in the USA and elsewhere in the world because of their costs and potential or impacting water resources (Box 21.2).[16,53] The purpose of the U.S. Forest Service BMPs for roads is to avoid, minimize, or mitigate adverse impacts on soils, water, and riparian resources. The U.S. Forest Service BMP document lists 11 National Core BMPs in the Roads category, and nine pertain to forest bioenergy activities.[9] Over 200 specific BMP standards for roads as listed within the National Core

Table 21.6 Selected USDA Forest Service Prescribed Fire BMPs

General Considerations	Specific BMPs
Site prescriptions	Keep staging areas as small as possible to minimize soil disturbance
	Store fuel for ignition devices away from water resources
	Install erosion controls to minimize water flow off of staging areas
	Collect and properly dispose of trash and wastes
	Restore staging areas after use
Fireline construction	Locate and construct fireline to minimize erosion
	Locate and construct fireline to minimize runoff into waterbodies
	Consider alternatives to ground disturbing firelines
	Maintain firebreaks that minimize soil exposure
	Rehabilitate and stabilize firelines that pose risks to water quality
	Manage fire intensity to maintain target levels of cover reduction
Riparian areas	Pretreat SMZs to reduce excessive fuel loadings
	Construct fireline to minimize area of soil disturbance
	Manage burn conditions and timing to keep fire intensities low
	Minimize complete removal of organic layer soil covering
	Rehabilitate firelines immediately after burning
	Remove debris from channels unless it improves habitat
Fire operations	Pile and burn slash only to achieve treatment objectives
	Locate slash piles where soil effects and likely to be minimal
	Construct slash piles to promote efficient burning
	Avoid burning when conditions will produce hot, damaging conditions
	Plan burn prescriptions to minimize soil and water impacts
	Avoid burning residual heavy fuels

Created using data from Ref 9. Copyright 2012, United States Department of Agriculture.

BMPs. The first two BMPs address an important aspect of road BMPs, planning and design. The remainder deals with construction, maintenance, and use. Some highlights are presented in Table 21.7. The large number of BMP standards relating to roads recognizes the potential impacts on environmental quality if roads are not properly planned, constructed, and maintained.

Mechanical Vegetation Management

Management of vegetation often requires the use of heavy machinery to accomplish project objectives. Examples of these mechanical treatments include timber harvesting, site preparation, tree planting, fuels management, prescribed fire operations, and habitat improvements.[61,78] There are eight National Core BMPs listed in the U.S. Forest Service's list for mechanical vegetation treatment.[9] The 123+ specific BMPs with the National Core BMPs pertain to operations that are conducted by forest bioenergy programs. Some BMP examples are provided in Table 21.8.

USA: State of Minnesota

The state of Minnesota has a comprehensive set of BMPs for forest management, and SMZs in general.[79] The SMZ is the core of the BMP approach in Minnesota (Figure 21.6). The guideline BMPs are worthy of a separate examination because of their completeness and packaging in an easy-to-use and easy-to-update book. The Minnesota guidebook uses the terminology 'Riparian Management Zone' (RMZ), which is congruent with SMZ terminology. An RMZ/SMZ can include areas that are not strictly 'riparian' as indicated in Figure 21.6. The Minnesota guidelines are voluntary and recognize that site variability and landowner objectives require a great deal of flexibility in developing forest management plans for conventional harvesting as well as biomass harvesting for energy production.

A number of BMP recommendations are presented by the Minnesota guidebook.[79] Because of high levels of interest and implementations of forest biomass harvesting for energy production, the Minnesota Forest Resources Council developed a separate chapter for their BMP guidebook that deals with operations that remove forest biomass of all sizes as bioenergy feedstock.[25] The environmental concerns are centered on water quality because of more intensive site access and disturbance associated with energy biomass harvesting, and the removal of biomass in excess of normal roundwood or timber extractions. But other issues also include site productivity, biodiversity, and wildlife population impacts.[80] The forest bioenergy

Table 21.7 Selected USDA Forest Service Road BMPs

General Considerations	Specific BMPs
Planning	Limit roads to the minimum practical number, width, and length
	Plan road networks to have the minimum number of water crossings
	Identify road segments with a potential to cause soil and water problems
Location and design	Locate roads to fit the terrain
	Design roads for location on stable soils and geology
	Plan road locations as far from surface water as possible
Road construction	Use suitable construction techniques to create stable fills
	Do not permit sidecasting within SMZs
	Construct pioneer roads to avoid adverse impacts to soils and waterways
Temporary roads	Install and routinely inspect sediment and runoff controls
	Avoid construction during wet periods
	Rehabilitate temporary roads after use
Road storage	Close and block roads not needed for over a year
	Establish ground cover on bare road surfaces
	Remove culverts that have a risk of failure
Stream crossings	Locate crossings where channels are narrow, straight, and uniform
	Construct bridges and culverts to prevent flow constriction
	Design crossings to allow normal passage of fish and other organisms
Parking and staging	Design parking/staging areas to minimize impacts on water
	Harden surfaces to minimize erosion
	Design detention basins to contain oil and grease discharges

Created using data from Ref 9. Copyright 2012, United States Department of Agriculture.

harvesting guidelines in the Minnesota BMP plan begin with planning and the basic BMPs outlined in the *General Guidelines* and *Timber Harvesting* chapters and then add additional BMP guidelines. For instance, there are only two additional BMPs for water quality that include avoiding harvesting additional biomass beyond tops and limbs of trees that would be removed in normal roundwood harvest, and avoiding the harvest of additional biomass within 7.5 m of ephemeral channels. There are eleven additional BMPs for managing site productivity, and nine for managing wildlife habitat and structural diversity. Reentry into stands that have already been harvested is a separate category because of the risks of soil compaction and erosion .

Box 21.3 FOREST CATCHMENTS, LAND USE, BMPs, AND WATER

New Zealand recognizes both the economic importance of forestry and the importance of forested watersheds for municipal, industrial, recreational, and wildlife habitat use.[7] A number of studies have evaluated the impact plantation forestry on water resources. One study gathered hydrologic information for 12 years from adjacent plantation forest and pasture catchments on the east coast of the North Island of New Zealand managed using BMPs that existed at the time.[81] They found that annual water yields preharvest were 6% lower in the plantation forest relative to the adjacent pasture. After a *Pinus radiata* harvest, water yields were 22% higher in the plantation catchment. This increase declined to 5% after seven years. The pasture catchment yielded three to four times more suspended sediment prior to tree harvesting when sediment yield in the forest catchment increased to two to three times greater than the pasture.[82] However, sediment yields from the forested catchment declined to preharvest levels within 3 years due to rapid reestablishment of cover. Tree harvesting did not produce any statistically significant increases in water turbidity, and nitrogen and phosphorus levels in streamflow.

NEW ZEALAND

New Zealand has an extensive set of forestry BMPs codified in its ‘Environmental Code of Practice for Plantation Forestry’ because of the need to protect its high quality freshwater and saltwater resources and the intensive forest management practiced in the country (Box 21.3).[18] The code consists of five parts: Best Environmental Management Practices or BEPs (these are the same as BMPs in other codes), Planning, Resources and References, and

Table 21.8 Examples of USDA Forest Service Mechanical Vegetation Management BMPs

General Considerations	Specific BMPs
Planning	Design treatment prescriptions to limit site disturbance, soil exposure Plan the harvesting system that best fits the topography, soils, season Evaluate the preharvest condition of system and unused roads
Erosion prevention	Apply a protective cover on exposed soil after operations are done Collect and disperse runoff from disturbed areas Avoid equipment operations on unstable, wet, or compactable soils
Aquatic areas	Clearly delineate SMZ locations and boundaries within project areas Using yarding systems or ground equipment that minimizes disturbance Directionally fell trees away from waterbody edges
Ground skidding/yarding	Designate skid trails to limit soil disturbance Limit skid trail grades on unstable geology or fragile soils Cable yard logs to skid trails to minimize soil disturbance
Cable and aerial yarding	Adjust operations in SMZs to avoid or minimize soil disturbance Use suitable mitigation/restoration techniques on erosion hazard areas Remove debris from stream channels, minimize bank disturbance
Landings	Locate landings outside of SMZs and as far from waterways as possible Locate landings to minimize skid trails Reuse existing or previously used landings
Winter logging	Use snow roads and winter harvesting where water tables are high Ensure all culverts and ditches are open and functional Plow roads to facilitate deep freezing of road grades before hauling
Site preparation	Use low ground-pressure equipment to minimize soil disturbance Use site preparation equipment that creates irregular surfaces Avoid displacing topsoil into slash piles or windrows

Created using data from Ref 9. Copyright 2012, United States Department of Agriculture.

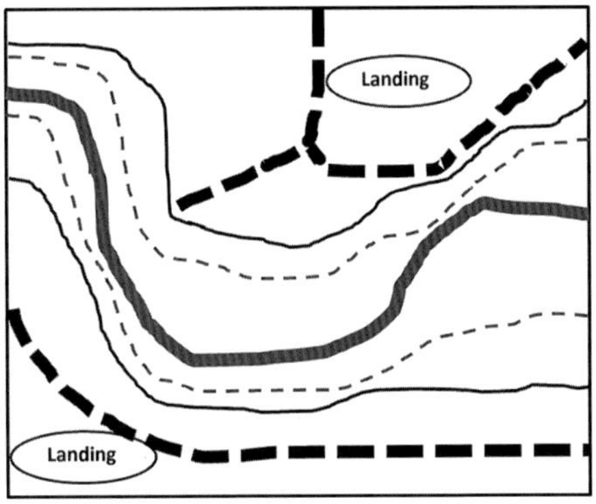

Figure 21.6 Hypothertical SMZ design typical of the flexible boundary RMZ forest harvesting designs used by the Minnesota Forest Resources Council to protect water quality. Reproduced with permission from Ref 33. Copyright 2010, ECOMED.

Training. The BEPs cover the planning of all operations and specific components such as roads, harvesting, site preparation, burning, planting, silvicultural treatments, agrichemical and petrochemical use, and forest protection. These BEPs cover many of the activities in the forest bioenergy lifecycle, but no bioenergy specific operations have been identified as requiring separate BEPs. The BEPs are organized into compulsory rules and recommended guidelines to be applied where it is safe and applicable to do so. New Zealand uses a supporting document for its BMPs, the 'Forest Road Engineering Manual', because of the recognition that roads are a major contributing factor to water quality problems in streams and lakes.[83] Road BMPs are considered to be critical for achieving goals of environmentally sound forest management.

New Zealand is in the process of going further in the use of BMPs by articulating legally enforceable and consistent forestry regulations called National Environmental Standards (NESs).[84] Plantation forestry is a significant land-use activity and a major contributor to the country's gross domestic product. There are a number of aspects of the industry that require careful management to avoid adverse environmental impacts. Plantation forestry is currently regulated by regional council and territorial authority plans required by the Resource Management Act of 1991. Because of the different approaches in local authority plans, there is a great deal of inconsistencies across jurisdictions. This new effort is aimed at bringing consistency for the New Zealand forestry industry while maintaining the high quality and consistent supply of water resources. Although many of the NESs relate to the forest bioenergy lifecycle, there are no specific provisions that focus on forest bioenergy production.

IRELAND

The Irish National Standard for sustainable forestry contains the Ireland Code of Best Forest Practice.[85] The code contains an extensive series of guideline BMPs related to water quality, archeology, landscape, tree harvesting, and biodiversity developed to ensure sustainable forestry in the country.[85,86] Forest operations and environmental protection are regulated by Irish legislation as well as European Union Directives. BMP implementation is expected to be monitored by before, during, and after forest silviculture and harvesting operations to ensure adequate BMP planning, compliance, and feedback. In addition to self-monitoring, external assessments are conducted by the Forest Service, Department of Marine and Natural Resources.

BMPs in Ireland's Code cover forest reproductive material, nursery practice, new planting sites, species selection, site preparation, planting, vegetation management, forest pests and diseases, forest maintenance, tending, silviculture and thinning, final harvesting, transport, forest roads, reforestation, biodiversity and specialized woodlands, forest planning, and training.[85] Forest bioenergy harvesting is not specifically mentioned within the context of general conventional forestry operations so BMPs for forest bioenergy lifecycle activities mostly fall within the general forestry BMPs. However, there is recognition of short rotation biomass forestry for the production of fuel feedstocks in Ireland's Code of Best Forest Practice.[85] The Irish Forest Service recognizes that this is a specialized activity requiring specialized harvesting systems, and intensified plant stocking levels, fertilizer usage, and weed and pest control. The code notes that regular interval woody biomass cropping places short-rotation forest biomass management under environmental controls specified for farming operations.[85]

CONCLUSION

Collectively, a large number of BMPs for forestry have been developed throughout the world because of national regulatory demands and the international development of 'Codes of Forest Practice'. The BMPs in the Codes and regulations cover traditional forestry activities such as planning, establishment, mid-rotation silviculture, timber harvest, roads, site preparation, fertilizer and pesticide application, product transportation, wood storage, and stand reestablishment. New BMPs have been developed for forest bioenergy applications such as energy production facilities, ash recycling, and short-rotation cropping. BMPs were originally developed in the 1970s for water quality protection but now extend to other environmental concerns. Most countries with substantial forestry sectors have BMPs for forestry. These include Australia, Brazil, Canada, Chile, Finland, Germany, Ireland, Norway, South Africa, Sweden, Russia, and the USA. Relatively few have articulated separate BMPs for forest bioenergy programs, however this is likely to change as wood becomes a larger feedstock source for renewable energy production. An important part of BMP utilization is the cycle of application, monitoring, evaluation, refinement, and reapplication. Research and development studies play a key part in the refinement and communication of improved BMPs. Existing studies of BMP effectiveness have demonstrated that most BMPs, if applied correctly, are very effective in mitigating or preventing adverse water quality impacts. Some jurisdictions have mandatory BMPs but others operate completely under voluntary systems.

The key components of a successful BMP-based code of practice for forest bioenergy systems, whether voluntary or mandatory, revolve around the cyclical strategy outlined in Figure 21.4. The initial step is planning and selection of BMPs. The minimum number of BMPs needed should come out of the planning process and is dependent on resources to be protected, site physical characteristics, regulatory requirements, and overall organization and operation goals. These will obviously vary from site to site, region to region, country to country, and organization to organization. The key BMPs revolve around SMZ establishment and management, harvesting operations and equipment, and road establishment and maintenance. Life cycle analysis should always be included in order to identify all water and ecological impacts. The next step is crucial. Monitoring and evaluation should be conducted routinely in order to decide if selected BMPs are effective and can be reapplied, or if they need to be modified, researched further, or discarded.

REFERENCES

1. Hall DG, Scrase JL. Will biomass be the environmentally friendly fuel of the future. *Biomass Bioenergy* 1998, 15:357–367.
2. Eriksson H, Hall J, Helynen S. Rationale for forest energy production. *Forestry Sci* 2002, 71:1–17.
3. Somerville C, Youngs H, Taylor C, Davis SC, Long SP. Feedstocks for lignocellulosic biofuels. *Science* 2010, 329:790–792.
4. Berndes G. Bioenergy and water—the implications of large-scale bioenergy production for water use and supply. *Global Environ Change* 2002, 12:253–271.

5. Shepard JP. Water quality protection in bioenergy production: the U.S. system of forestry best management practices. *Biomass Bioenergy* 2006, 30:378–384.
6. Fargione JE, Plevin RJ, Hill JD. The ecological impact of biofuels. *Ann Rev Ecol Evol Syst* 2010, 41:351–377.
7. Neary DG, Koestner KA. 2012. *Forest bioenergy feedstock harvesting effects on water supply*. WIRES Energy Environ 2012. doi:10.1002/wene.26.
8. Minnesota Forest Resources Council. *Sustaining Minnesota Forest Resources: Voluntary Site Level Forest Management Guidelines for Landowners, Loggers, and Resource Managers*. St Paul, MN: Minnesota Forest Resources Council; 2005, 480.
9. USDA Forest Service. *National Best Management Practices for Water Quality Management on National Forest Lands*. Vol. 1: National Core BMP Technical Guide. USDA Forest Service Technical Guide FS-990a; 2012, 165.
10. Loehr RC, Haith DA, Walter MF, Martin CS. Best Management Practices for Agriculture and Silviculture. Ann Arbor, MI: Ann Arbor Science Publishers Inc.; 1979, 740.
11. Lynch JA, Corbett ES, Mussaliem K. Best management practices for controlling non-point source pollution on forested watersheds. *J Soil Water Conserv* 1985, 40: 164–167.
12. D'Arcy B, Frost A. The role of best management practices in alleviating water quality problems associated with diffuse pollution. *Sci Total Environ* 2001, 265:359–367.
13. Broadmeadow S, Nisbet TR. The effects of riparian forest management on the freshwater environment: a literature review of best management practice. *Hydrology Earth Syst Sci* 2004, 8:286–305.
14. Neary DG, Smethurst PJ, Baillie B, Petrone KC. Water quality, biodiversity, and codes of practice in relation to harvesting forest plantations in Streamside Management Zones. Canberra, Australia: CSIRO Special Report. 2011, 99. http://www.csiro.au/Organisation-Structure/Flagships/Sustainable-Agriculture-Flagship/Plantation-forests-SMZ.aspx.
15. Phillips MJ, Swift LW Jr, Blinn CR. Chapter 16. Best management practices for riparian areas. In: Verry E, Hornbeck J, Dolloff A, eds. *Riparian Management in Forests of the Continental Eastern United States*. Washington, DC: Lewis Publishers; 2000, 273–286.
16. Forman RTT, Alexander LE. Roads and their major ecological effects. *Annu Rev Ecol Syst* 1998, 29:207–231.
17. Forest Practices Board. *Forest Practices Code*. Tasmania, Australia: Forest Practices Board; 2000, 120. Available at: http://www.fpa.tas.gov.au/_data/assets/pdf_file/0020/58115/Forest_Practices_Code_2000.pdf. (Accessed June 27, 2012).
18. New Zealand Forest Owners Association. *New Zealand Environmental Code of Practice for Plantation Forestry*. Wellington, New Zealand: New Zealand Forest Owners Association; 2007, 168.
19. Neary DG, Ryan KC, DeBano LF, eds. *Fire Effects on Soil and Water*. USDA Forest Service, Rocky Mountain Research Station, General Technical Report RMRS-GTR-42, Vol. 4. Fort Collins, CO: USDA Forest Service; 2005, 250.
20. Ice G. History of innovative best management practice development and its role in addressing water quality limited waterbodies. *J Environ Eng* 2004, 130:684–689.
21. Aust WM, Shaffer RM, Burger JA. Benefits and costs of forestry best management practices in Virginia. *South J Appl Forestry* 1996, 20:23–29.
22. Logan TJ. Agricultural best management practices for water pollution control: current issues. *Agric Ecosyst Environ* 1993, 46:223–231.
23. Dyck WJ, Bow CA. Environmental impacts of harvesting. *Biomass Bioenergy* 1992, 2:1–6.
24. Dyck WJ, Cole DW, Comerford NB, eds. *International Energy Agency Project A6 Book, Impacts of Harvesting on Long-Term Site Productivity*. London: Chapman and Hall; 1994, 368.
25. Richardson J, Smith T, Hakkila P. *Bioenergy from Sustainable Forestry: Guiding Principles and Practices*. Amsterdam, the Netherlands: Elsevier; 2002, 344.
26. Lewandowski, I, Faaij APC. Steps towards the development of a certification system for sustainable bio-energy trade. *Biomass Bioenergy* 2006, 30:83–104.
27. van Dam J, Junginger M, Faaij A, Jurgens I, Best G, Fritsche U. Overview of recent developments in sustainable biomass certification. *Biomass Bioenergy* 2008, 32:749–780.
28. Janowiak MK, Webster CR. Promoting ecological sustainability in woody biomass harvesting. *J Forestry* 2010, 108:16–23.
29. Scarlat N, Dallemand JF. Recent developments of biofuels/bioenergy sustainability certification: a global overview. *Energy Policy* 2011, 39:1630–1646.
30. Buford M, DuPont A, Widman N, Ferrell J, Miller MB, Bastian H, Steiner J, Lowenfish M, Wright B. *Bioenergy feedstock best management practices: Summary and research needs*. Biomass Research and Development Board, Report by the Feedstock Production Interagency Working Group; 2011, 26.
31. Trombulak SC, Frissell CA. Review of ecological effects of roads on terrestrial and aquatic communities. *Conserv Biol* 2001, 14:18–30.
32. Aust WM, Blinn CR. Forestry best management practices for timber harvesting and site preparation in the eastern United States: an overview of water quality and productivity research during the past 20 years. *Water Air Soil Pollut* 2004, 4:5–36.
33. Neary DG, Smethurst PJ, Baillie B, Petrone KC, Cotching W, Baillie CC. Does tree harvesting in riparian areas adversely affect stream turbidity—preliminary observations from an Australian case study. *J Soils Sediments* 2010, 10:652–670.
34. Lattimore B, Smith CT, Titus BD, Stupak I, Egnell G. Environmental factors in woodfuel production: opportunities, risks, and criteria and indicators for sustainable practices. *Biomass Bioenergy* 2009, 33:1321–1342.
35. Fantozzi F, Buratti C. Life cycle assessment of biomass chains: wood pellet from short rotation coppice using data measured on a real plant. *Biomass Bioenergy* 2010, 34:1796–1804.
36. Fazio S, Monti A. Life cycle assessment of different bioenergy production systems including perennial and annual crops. *Biomass Bioenergy* 2011, 35:4868–4878.

37. González-García S, Moreira MT, Feijoo G, Murphy RJ. Comparative life cycle assessment of ethanol production from fast-growing crops (black locust, eucalyptus, and poplar). *Biomass Bioenergy* 2012, 39:378–388.
38. Cherubini F, Strømman AH. Life cycle assessment of bioenergy systems: state of the art and future challenges. *Bioresour Technol* 2011, 102:427–451.
39. Heller MC, Keoleian GA, Volk TA. Life cycle assessment of a willow bioenergy cropping system. *Biomass Bioenergy* 2003, 25:147–165.
40. Reid LM, Dunne T. Sediment production from forest road surfaces. *Water Resour Res* 1984, 20:1753–1761.
41. Pennington DW, Potting J, Finnveden G, Lindeijer E, Jolliet O, Rydberg T, Rebitzer G. Life cycle assessment Part 2: current impact assessment practice. *Environ Int* 2004, 30:721–739.
42. Rebitzer G, Ekvall T, Frischknecht R, Hunkeler D, Norris G, Rydberg T, Schmidt WP, Suh S, Weidema BP, Pennington DW. Life cycle assessment Part 1: framework, goal and scope definition, inventory analysis, and applications. *Environ Int* 2004, 30:701–720.
43. ISO 14040. *Environmental Management—Life Cycle Assessment—Principles and Framework*. Geneva, Switzerland: International Standards Organisation (ISO); 1997.
44. Weightman D. Controlling diffuse pollution by best management practices. In: Petchey AM, D'Arcy BJ, Frost CA, eds. *Diffuse Pollution and Agriculture*. Aberdeen: The Scottish Agricultural College; 1996, 17–22.
45. Heninger RL, Terry TA, Dobkowski A, Scott W. Managing for sustainable site productivity: Weyerhaeuser's forestry perspective. *Biomass Bioenergy* 1997, 13:255–267.
46. Gitau MW, Gburek WJ, Jarrett AR. A tool for estimating best management practice effectiveness for phosphorus pollution control. *J Soil Water Conserv* 2005, 60:1–10.
47. Measham TG, Kelly GJ, Smith FP. Best Management Practice for complex problems: a case study of defining BMP for dryland salinity. *Geogr Res* 2007, 45:262–272.
48. Briggs RD, Cornier J, Kimball A. Compliance with forestry Best Management Practices in Maine. *North J Appl Forestry* 1998, 15:57–68.
49. Neary DG, Swank WT, Riekerk H. An overview of nonpoint source pollution in the Southern United States. *In Proceedings of the Symposium: Forested Wetlands of the Southern United States, July 1214, 1988*, Orlando, FL: USDA Forest Service; General Technical Report SE-50; 1989, 1–7.
50. Krag R, Higgenbotham K, Rothwell R. Logging and soil disturbance in southeast British Columbia. *Can J Forest Res* 1986, 16:1345–1354.
51. Kellogg LD, Bettinger P, Edwards RM. A comparison of logging planning, felling, and skyline yarding costs between clearcutting and five group-selection harvesting methods. *West J Appl Forestry* 1996, 11:90–96.
52. Stampfer K, Visser R, Kanzian C. Cable yarder installation times for European yarders. *Int J Forest Eng* 2006, 17:71–77.
53. Swift LW Jr. Filter strip widths for forest roads in the southern Appalachians. *South J Appl Forestry* 1986, 10:27–34.
54. Boston K, Thompson M. An argument for placing logging roads under the NPDES Program. *Ecol Law Curr* 2009, 36:169–176.
55. USEPA, Office of Water. *Industrial Storm Water Management Fact Sheet—Sector A: Timber Products Facilities*. EPA-833-F-06–016. Washington, DC: USEPA; 2006, 8.
56. Hedmark A, Jonsson M. Treatment of log yard runoff in a couch grass infiltration wetland in Sweden. *Int J Environ Stud* 2008, 65:267–272.
57. Turnbull JH. Use of biomass in electric power generation: the California experience. *Biomass Bioenergy* 1993, 4:75–84.
58. Asikainen A, Bjorheden R, Nousianen I. Chapter 4: cost of wood energy. In: Richardson J, Smith T, Hakkila P, eds. *Bioenergy from Sustainable Forestry: Guiding Principles and Practices*. Amsterdam, the Netherlands: Elsevier; 2002, 125–157, 344.
59. Neary DG, Leonard JH. Effects of forest fertilisation on nutrient losses in streamflow in New Zealand. *N Z J Forest Sci* 1978, 8:189–205.
60. Binkley D, Burnham H, Allen HL. Water quality impacts of forest fertilization with nitrogen and phosphorus. *Forest Ecol Manage* 1999, 121:191–213.
61. Neary DG, Morris LA, Swindel BF. Site preparation and nutrient management in southern pine forests. In: Stone, Earl L, eds. *Forest Soils and Treatment Impacts: Proceedings of the 6th North American Forest Soils Conference*. Knoxville, TN: University of Tennessee; June 1983. 1984, 121–144.
62. Buford MA, Neary DG. Sustainable biofuels from forests: meeting the challenge. *Biofuels and Sustainability Reports*. Washington, DC: The Ecological Society of America; 2010, 9.
63. Webb AA. Can timber and water resources be sustainably co-developed in south-eastern New South Wales, Australia? *Environ Dev Sustain* 2012, 14:233–252.
64. McIntosh P, Laffan M. Soil erodibility and erosion hazard: extending these cornerstone soil conservation concepts to headwater streams in the forestry estate in Tasmania. *Forest Ecol Manage* 2005, 220:128–139.
65. Webb AA, Dragovich D, Jamshidi R. Temporary increases in suspended sediment yields following selective eucalypt forest harvesting. *Forest Ecol Manage* 2012, 283:96–105.
66. Department of Natural Resources and Water. *Field Guide: Code Applying to a Native Forest Practice on Freehold Land Version 2*. Brisbane, Australia: Department of Natural Resources and Water, State of Queensland; 2007, 126. Available at: http://www.nrw.qld.gov.au/vegetation/clearing/pdf/forest_field_gide.pdf. (Accessed January 21, 2011).
67. McCormack RJ. A review of forest practices codes in Australia. In: Dykstra DP, Heinrich R, eds. *Forest Codes of Practice Contributing to Environmentally Sound Forest Operations*. Feldafing, Germany: Proceedings of an Food and Agriculture Organization/IUFRO Meeting of Experts on Forest Practices, 11–14 December 1994, Food and Agriculture Organization Forestry Paper No. 133, Rome, 1996, 105–116. http://www.fao.org/docrep/W3646E/w3646e0f.htm.

68. Forest Stewardship Council. *Forest Stewardship Council Principles and Criteria for Forest Stewardship*. Forest Stewardship Council-STD-01–001 (April 2004), Toronto, Ontario, Canada: Forest Stewardship Council; 2004, 11.
69. Province of Alberta. Alberta Timber Harvest Planning and Operating Ground Rules. Edmonton, Alberta, Canada: Alberta Department of Environmental Protection, Publication Ref No. 71; 1994, 57.
70. Ontario Ministry of Natural Resources. *Code of Practice for Timber Management Operations in Riparian Areas*. Toronto, Ontario, Canada: Ministry of Natural Resources. Report Number 4623; 1991.
71. USEPA. *National Management Measures to Control Nonpoint Source Pollution from Forestry*. EPA-841-B-05–001. Washington, DC: United States Environmental Protection Agency, Office of Water Report; 2005, 276.
72. USEPA. *Guidance Manual for Developing Best Management Practices*. Washington, DC: U.S. Environmental Protection Agency; 1993, 204.
73. Dissmeyer GE, ed. Drinking water from forests and grasslands. *A Synthesis of the Scientific Literature*. General Technical Report SRS-39. Asheville, NC: U.S. Department of Agriculture, Forest Service, Southern Research Station; 2000.
74. Neary DG, Zieroth E. Forest bioenergy system to reduce the hazard of wildfires: white Mountains, Arizona. *Biomass Bioenergy* 2007, 31:638–645.
75. DeBano LF, Neary DG, Ffolliott PF. *Fire's Effects on Ecosystems*. New York: John Wiley & Sons; 1998, 333.
76. Gomi T, Sidle RC, Noguchi S, Negishi JN, Abdul Rahim N, Sasaki S. Sediment and wood accumulations in humid tropical headwater streams: effects of logging and riparian buffers. *Forest Ecol Manage* 2006, 224:166–175.
77. Sidle RC, Sasaki S, Otsuki M, Noguchi S, Abdul Rahim N. Sediment pathways in a tropical forest: effects of logging roads and skid trails. *Hydrolog Process* 2004, 18:703–720.
78. Smith DM, Larson BC, Kelty MJ, Ashton PMS. *The Practice of Silviculture*. New York: John Wiley & Sons; 1997, 537.
79. Minnesota Forest Resources Council. *Biomass Harvesting Guidelines for Forestlands, Brushlands and Open Lands*. St. Paul, MN: Minnesota Forest Resources Council; 2007. Available at: http://www.frc.state.mn.us/documents/council/site-level/MFRC_FMG&Biomass_2007–12-17.pdf. (Accessed July 30, 2012).
80. Foster DR, Aber JD, Melillo JM, Bowden RD, Bazzaz FA. Temperate forest response to natural catastrophic disturbance and chronic anthropogenic stress. *BioScience* 1997, 47:437–445.
81. Eyles G, Fahey B. *The Pakuratahi Land Use Study*. New Zealand: Hawke's Bay Regional Council, Napier; 2006, 128.
82. Fahey BD, Marden M. Sediment yields from a forested and a pasture catchment, coastal Hawke's Bay, North Island, New Zealand. *N Z J Hydrol* 2000, 39:49–63.
83. New Zealand Forest Owners Association. *New Zealand Forest Road Engineering Manual*. Wellington, New Zealand: New Zealand Forest Owners Association; 2011, 150.
84. New Zealand Ministry for the Environment. *Proposed National Environmental Standard for Plantation Forestry*. Wellington, New Zealand: New Zealand Ministry for the Environment; 2010, 138.
85. Irish Forest Service. *Code of Best Forest Practice—Ireland*. Dublin, Ireland: Irish Forest Service; 2000, 101.
86. Ryan MG. *Draft Revision of Forestry and Water Quality Guidelines. Report Prepared for the Irish Forest Service, Department of Marine and Natural Resources*, Dublin, Belfield, Dublin, Ireland: COFORD, National Council for Forest Research and Development, University College; 1999, 12.

22

Principles of Nutrient Management for Sustainable Forest Bioenergy Production

Donald Mead[1] and Charles Smith[2]*

[1]Silviculture adviser, Collingwood, New Zealand
[2]Faculty of Forestry, University of Toronto, Toronto ON, Canada

INTRODUCTION

Wood for energy or increased carbon sequestration in trees are both potential products from natural and/or plantation forests that can assist in meeting local and global demands for energy and in slowing the increase in atmospheric carbon dioxide (CO_2). Similarly, using wood in construction, rather than concrete or metal, reduces energy demand. However, for increased forest bioenergy to gain wide-scale acceptance, it must be shown to be sustainable,[1] and forest owners should manage their forests in a scientific and transparent manner that ensures sustainability. Sustainable management of forests includes environmental, social, and economic dimensions, with the overall objective of ensuring that forests will be available to supply the various demands placed on them by future generations.

This paper only focuses on the nutritional aspects of forest sustainability. The environmental components of forest sustainability include productive capacity (including aspects of the extent of the forest estate as well as site and stand-level factors affecting tree growth), soil and water, biodiversity, forest health, and carbon budgets.[2,3] Although all components are affected by the utilization of forests for bioenergy, or the planting of carbon-offset forests, for this paper, we are focusing on the guiding principles for nutrient management.

This paper builds on the considerable knowledge about landscapes and soil—that is, soil types, inherent fertility, soil nutrient processes, and resilience to change—tree and forest nutrient demands, silvicultural, harvesting and other management impacts, and forest management, including adaptive forest management. The objective of this paper is to use the current state of knowledge to give forest managers harvesting tree biomass for energy a set of guiding principles by which to ensure nutrient sustainability.

PRINCIPLES

On the basis of the current state of knowledge and understanding of sustainability of forest ecosystems, it is possible to define 10 principles that can be used to sustainably manage bioenergy and CO_2-offset forests. We have taken a pragmatic approach, building on where we are now while pointing out recent research trends. Adaptive forest management is an essential component because this concept recognizes the rapid development of ideas in the area of nutrient sustainability, the huge variety of situations worldwide, the uncertainty associated with ameliorative actions, and changes that occur in ecosystems overtime.

The 10 principles we propose for developing sustainable nutrient management practices are ordered as a professional forest manager might approach the task of

The authors thank for the support of IEA Bioenergy Task 31 'Biomass Production for Energy from Sustainable Forestry' and Task 43 'Biomass Feedstocks for Energy Markets.'
*Correspondence to: tat.smith@utoronto.ca

Advances in Bioenergy: The Sustainability Challenge, First Edition. Edited by Peter D. Lund, John Byrne, Göran Berndes and Iacovos A. Vasalos.

Originally published in: *WIREs Energy Environ* 2012, 1: 351–362 doi: 10.1002/wene.3

managing nutrients to ensure sustainability and are listed as follows.

1. Define management objectives.
2. Quantify management risks for soil fertility and sustaining long-term soil quality.
3. Map and evaluate sites.
4. Decide on monitoring procedures and level.
5. Include independent auditing.
6. Do not waste resources.
7. Develop site-specific nutritional management plans.
8. Consider off-site impacts.
9. Use decision-making aids, including models, economic and energy analyses, and life-cycle assessment.
10. Apply adaptive management.

Define Management Objectives

The first step in developing a nutrient management plan is to define the forest management objectives.

With respect to forest bioenergy, there are four major production systems, which are characterized by increasing nutrient removal from the forest (Figure 22.1). Currently, the two most common situations in industrial countries are where biomass for energy is an *incidental* objective to the dominant production systems, such as the provision of industrial wood or management primarily for conservation, esthetics, soil and water, or biodiversity. In these 'conventional' situations, the provision of bioenergy either makes use of industrial wastes or is undertaken sporadically and with low intensity. In the 'conventional plus' production system, bioenergy is a *major secondary* objective. In this situation, there is likely to be more nutrient removal from a site than conventional forest harvesting. Biomass-harvesting operations usually include taking more nutrient-rich material from the forest, such as smaller diameter stem material, branch wood, foliage, and perhaps roots or understory plants. Silvicultural schedules may be altered and different harvesting techniques may be used to facilitate this.[4] The most intensive option is to grow dedicated bioenergy crops where bioenergy is the *prime* forest management objective. This system tends to place even higher nutrient demand on the site because of relatively high stocking rates and short, often coppice-regenerated, rotations.[4]

In many developing countries, natural and plantation forests and trees outside forests are utilized for fuel by local people, and they often collect nutrient-rich material such as shrubs, branches and small trees, and litter.[5,6] Where this occurs, the often unstated bioenergy management objective is often at the *major secondary* level and nutrient removals can be relatively high. Deliberate planting of fuelwood trees or woodlots as a *primary* objective is also common in many of these countries, although wood is sometimes sold for more valuable products.[4,6]

In many developing and some developed economies, agroforestry systems are common.[4,7] There are a wide range of alternative strategies for combining trees and agricultural practices. Using trees for fuel, mulch, animal bedding, or animal feed are all tree-related practices that affect soil nutrient status. Growing special trees for fruit, nuts, or other food crops will also have their own nutrient requirements that may need to be carefully managed. Because of the wide variety of practices, nutrient management of agroforestry systems is not covered in detail in this paper.

With CO_2-offset planting, the objective is to grow forests to absorb CO_2 from the atmosphere. This is the *primary* management objective, and frequently the unstated plan is that once a forest is planted, there will be no harvesting. However, there can be secondary objectives with these plantings such as increasing biodiversity, controlling erosion or other soil degradation, or providing amenity values. The nutrient demands from such plantings are likely to be modest, provided biomass for energy is not an objective, although fertilizer amendments may be required for success on infertile sites. Most afforestation can also provide a one-time gain through carbon accumulation, provided the forest is managed in the long term on a sustainable basis. Furthermore, careful management of natural forests for forest products can lead to greater sequestration of carbon.[8]

Clearly defining the primary management objective for a forest is therefore a critical first step, which helps define the likely nutrient risks that should be considered.

Define Management Risks for Soil Fertility and Sustaining Long-Term Soil Quality

The goal here is to estimate the major and likely risks for nutrient sustainability from proposed management regimes. Published studies can often assist this process. The impacts are site and species specific and may be strongly influenced by management and harvesting practices.[9,10]

If, over the crop life, nutrient losses from harvesting and establishment are greater than inputs from mineral weathering, deposition from the atmosphere or flooding, and, in the case of nitrogen, from nitrogen fixation, there will be a reduction in nutrients in the soil, with direct consequences for nutrient availability. The harvesting of biomass for energy increases nutrient removal compared to conventional harvesting of logs as more nutrient-rich materials are taken from the site (Figure 22.1). However, if soil is reasonably fertile, this increased removal may not lead to reduced tree growth of the subsequent forest because the soil may still

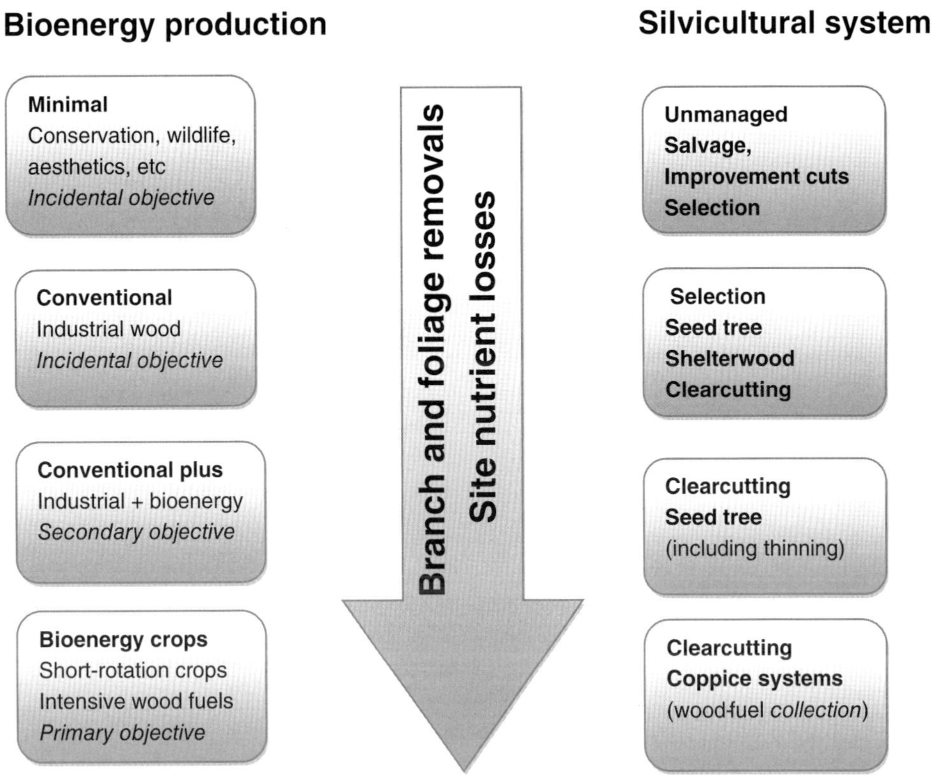

Figure 22.1 Relationship of management objectives to removal of nutrient-rich plant material and associated nutrient stress.

be able to supply adequate nutrients.[11] Determination of a site nutrient budget, or input-output balance, is a complex task and requires considerable data and expertise but the information base is steadily improving. On less fertile soil, the trees may come under nutrient stress, in which case amelioration may be required to sustain long-term soil quality.[10,11]

Poor harvesting and establishment practices and erosion, aggravated by tree removal, may also result in increased losses of nutrients or their redistribution on the site compared to careful management practices. Erosion of topsoil can be very damaging to the nutrient capital of a site, although in forestry this usually comes from using poor practices. Establishment practices that include soil cultivation or burning often result in greater mineralization of organic matter and, depending on the site, this may enhance nutrient losses.[9] However, losses from establishment practices are often short lived and are sometimes reversed with time.[9]

Where harvesting and some establishment practices redistribute nutrients on a site, they can create areas where nutrients are depleted and other areas where they may be increased. For example, this can occur when biomass is accumulated in piles for transpiration drying and needle shedding or, in the case of establishment, when slash is windrowed.[9]

Short-rotation coppice energy crops will generally have higher nutrient removals than where longer rotations are being used (Figure 22.1). However, with coppice crops, the continual presence of the root web will reduce the likelihood of erosion. Nutrient losses through harvesting and by leaching are highest when the forest is clear-cut or the seed-tree system is used for regeneration. These systems are particularly amenable to biomass harvesting and may also include thinning during the rotation. Partial cutting as with the shelterwood system or with improvement and salvage cutting will often have lower impacts as the collection of tree crown material for bioenergy is not as practical, although it is sometimes carried out. The selection system is likely to be the least amenable to intensive biomass harvesting. Some agroforestry practices can also lead to increased nutrient loss, although in some practices the presence of trees will greatly ameliorate these losses.[12]

Soil compaction is another problem to consider due to the impact on root growth, soil physical properties, water infiltration, and reduced growth rates. Compaction can occur as a result of interactions among machine and soil-specific factors.[11,13] Another potential effect of clear-cut harvesting is to increase stream flow and nutrient losses in the short term.[14]

Some forest regulatory systems place an emphasis on soil conservation objectives by prescribing the minimal amount of organic matter to retain or a minimal pH to maintain. These restrictions may need to be considered by managers.

Evaluate and Classify Sites

Because of site-specific properties affecting nutrient sustainability,[11] it is important to identify those sites with the greatest risks or sensitivities to adverse impacts. In Figure 22.2, five broad groupings are identified, ranging from very fertile sites with no limitations to very infertile sites, which often have such major limitations that they should be avoided for production forestry. The types of limitations that should be considered during a site evaluation are rooting depth, erosion risk, geology and mineralogy, soil texture, known or likely nutrient deficiencies, soil bulk density, and porosity. All these can have major influences on how the soil will react to forest management practices, and on the monitoring that will be required to ensure sustainability.

Foresters have often relied on measuring forest productivity trends to determine whether forests are being managed sustainably. Although this is very useful and may pick up trends overtime, the method is compromised if there are differences in species or genetic composition, silviculture, weed competition, disease, or climate change.[15] Unless these factors are accounted for, assessments designed to evaluate the effects of management on nutrient sustainability should concentrate on measuring changes in nutrient status overtime. These might be supplemented by other measures, such as leaf area index, which has been found to be a useful indicator of the ability of a site to sustain growth potential.[16,17]

The following procedure is suggested:

(a) Subdivide the forest into uniform units on the basis of landscape, topography, soil types, and vegetation. There is a fundamental need for identification and mapping of types of soil 9,10,18,19 because they form the basis for monitoring and nutrient management. The type of forest, its management, harvesting, roading, and streamside protection areas should be included in spatial databases. The harvesting cost of low-value bioenergy material is very dependent on ease of access, roading, and haulage distance, all of which may need to be considered.[4] The units need to be sufficiently distinct and of a size that can be practically managed. Detailed mapping at a scale of 1:24000 or greater has been recommended.[18]

New methods based on geostatistics and geographic information systems are being developed that could replace traditional methods of soil and geomorphological mapping.[20] Some of these include digital terrain modeling based on Lidar or other remote sensing which can utilize information such as altitude, slope, and aspect. An advantage of these methods is that they can incorporate previous and future data collected on the delineated management units.

(b) Allocate a management regime to each site unit. This follows on from step 2 but with a more detailed consideration of site-specific attributes. Avoid management practices that will adversely affect the nutrient status of the site, particularly on low fertility sites. Key factors to consider are the potential for excessive nutrient removal, site disturbance, compaction, and interference with site hydrology. Forest management agencies of numerous countries or their political subjurisdictions have produced management guidelines that can be helpful references.[24]

(c) Make an initial broad classification of the sites for monitoring. The purpose of classifying sites into five major groupings (Figure 22.2) is to assist in deciding on the degree and frequency of fertility monitoring that should be implemented. Five broad groupings are suggested; and depending on the proposed intensity of forest management, three levels of monitoring (plus no monitoring) are proposed (Figure 22.2). Curran et al.[18] have recommended stratification of sites for the purpose of prescribing mitigation measures because of adverse soil disturbance. Similarly, in Tasmania, Adams et al.[19] stratified sites for nutritional assessment and fertilizer applications.

On very fertile sites, such as some river bottom sites with no fertility limitations, limited or no monitoring would be necessary to ensure nutrient sustainability. At the other extreme are sites too poor to be managed for production purposes and so can be excluded from the monitoring program. Between these two extremes are sites, which could be described as moderately fertile with only a few limitations, and other sites, where limitations and fertility are known to be problematic. This grouping is subjective but is helpful for deciding on the degree of monitoring and likely fertility management regime. The allocation of sites to particular groups can be modified if experience suggests that this is needed.

In addition to these five groups, there will likely be areas that will be set aside for riparian buffer strips or for other purposes, where no harvesting will take place and nutrient monitoring may not be required.

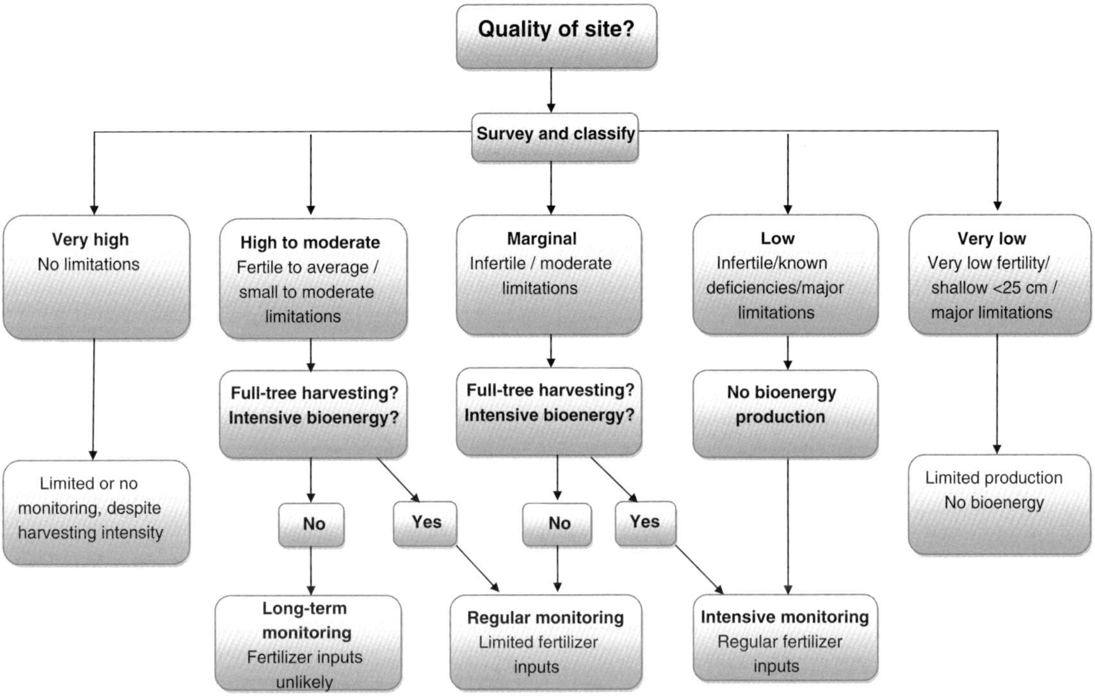

Figure 22.2 Overview of site assessment, level of monitoring, and need for nutrient amendments in relation to nutrient stresses imposed by bioenergy production (see text for details).

Assess the Nutrient Status of the Soil and/or Trees for Each Delineated Unit

Four levels of assessment and monitoring intensity are suggested on the basis of initial fertility and degree of harvesting intensity (Figure 22.2): long-term monitoring, regular monitoring, intensive monitoring, and no monitoring required.

With *long-term monitoring*, the key measures are likely to be those identified in response to the Montreal Process for ensuring that the forest soil quality or net primary productivity will not be adversely affected by normal harvesting.[2,3] Many of these measures were developed for monitoring at national and subnational scales. For example, in the United States, a nationwide assessment of a wide range of indicators has been proposed for forest inventory and analysis.[25] Other forest management agencies are currently defining appropriate indicators. For example, in New Zealand the current recommendation for *Pinus radiata* plantations involves measuring total phosphorus, carbon-to-nitrogen ratio, and soil porosity.[26] These proposed approaches to monitoring are examples of systems where substantial validation work still needs to be conducted. Choice of suitable indicators can be complex as they will vary with site, forest type, species involved, and management objectives.[9,27] For example, often a key component of biogeochemical cycling for nitrogen is organic matter decomposition and related mineralization and immobilization processes, which can change over both short- and long-time spans.[10] Measurements of litter mass, soil carbon, total nitrogen, and carbon-to-nitrogen ratio are often considered, but there may not be any one single measure that will be appropriate for all situations. On some sites, soil hydrology and rooting depth may be major factors to consider and in such case other indicators may be needed.[28] On sites where nutrient stresses are unlikely to develop (Figure 22.2), monitoring nutrient status might be undertaken only once in the rotation, unless accepted protocols for a country dictate otherwise. These examples suggest that the need for monitoring should be informed by and designed with a comprehensive conceptual model defining long-term site quality that is relevant for each type of site to be managed.

With *regular monitoring*, in addition to the long-term monitoring discussed earlier, more intensive monitoring procedures would be implemented and would focus at the forest management unit. Sensitive sites, types of soil known to be poor in some nutrients, or moderately fertile sites with intensive nutrient extraction, require additional base measurements to be made. Nitrogen and phosphorus are the most common limiting nutrients but sometimes base cations (calcium, magnesium, and potassium) or micronutrients (boron, manganese, copper, and zinc) can be important.[10] For some sites and nutrients, there may not be enough research to accurately define the best soil measurements to use. In these situations, it would be a good precaution to initially undertake a wider range of analyses on soil samples and to ensure that the samples are stored so that they can be further analyzed if required. Where trees are present, foliar analysis or using vector analysis with short-term fertilizer experiments can be helpful in determining limiting nutrients.[11,29,30] With a regular monitoring program, soil samples should be collected at planting or regeneration, and if there is any cause for concern, these should be repeated at canopy closure. Foliage samples should be collected if nutrient deficiency symptoms become apparent and at canopy closure. Sampling at canopy closure corresponds to the period of maximum nutrient demand from the soil.[31,32] With time, experience and research may suggest that the type and level of monitoring should be modified (see *Principle 10*).

A more *intensive monitoring* program is required where known low nutrient status requires more active management. Soil sampling would occur as described for regular monitoring but foliar sampling would be more frequent. Foliar analysis has a role in monitoring changes with time and managing fertilizer applications.[29,33,34] Measurement of leaf area index could also be a supplementary measure. In pine plantations in the southern United States and eucalypt plantations in Tasmania, low leaf area can indicate nutrient limitations that could be ameliorated by applying fertilizers.[16,17,19]

Efficacious soil sampling strategies need to be used. This will require consideration of spatial variability and use of geographic information systems and special analysis techniques to accurately map where the samples are taken and to design optimum sampling schemes.[34,35] Agricultural studies suggest that on virgin sites where fertilizer has not been applied, 20–30 cores per bulked sample are suitable for total nitrogen; carbon; pH; exchange capacity; and exchangeable and extractable potassium, magnesium, phosphorus, and sulfur.[36] Brown[36] provides detailed discussion of soil sampling methods and variability as related to agriculture; Mead[29] has summarized research in forestry situations; and Homann et al.[37] discuss minimum sampling needed to detect differences in highly variable types of forest soil. Similarly, foliage sampling needs to take into account between-tree variability and needs to follow the standardized procedures for the species.[29,30] To improve the accuracy of monitoring, it is advisable to collect foliage and soil samples from specific plots or specific trees that are marked for repeated sampling.

Independent Auditing

Independent, third party auditing of classification and monitoring procedures is important to ensure that they are robust. An independent audit should also check on chemical analysis reliability, possible errors in estimations of nutrients on the site, and methods of reporting. Independent auditing is typically required by forest certification schemes.

Conserve Rather than Waste Resources

Sustainability depends, in part, on the wise use of resources. Therefore managers need to be frugal with the use of inputs used to maintain or improve forest growth and yield.

In the development of nutritional management programs, wasteful use of resources should be avoided. Just because nutrients are removed from sites through biomass harvesting does not necessarily mean they need to be applied to maintain growth or the sustainability of the subsequent forest. Many types of soil have more than adequate reserves or have minerals that are weathering at a sufficient rate to supply some or all essential nutrients or have other natural inputs, as observed on sites with ‘satisfactory’ levels of foliar nutrition, as described for *Pinus radiata* in New Zealand by Will.[38] Determination of weathering rates of soil minerals is complex,[39] and the resulting estimates must be applied with caution. In other situations, nutrients deposited from the atmosphere may adequately replenish supplies. Adequate knowledge of such replacement processes may indicate that there may be no immediate need to apply certain nutrients, even though it might seem they could become limiting in the future.

For example, high calcium removals have often been flagged as a possible problem, but responses to calcium fertilizers are very rare in forestry and only implicated on extremely impoverished sites.[33] We recommend that the goal should be to maintain tree growth potential and not to maintain specific soil properties, such as calcium soil concentrations or base saturation, unless these are clearly connected with known problems. Such problems may include watershed acidification and aquatic ecosystem health, as well as terrestrial ecosystem considerations.

The application of wood ash, which usually contains high concentrations of calcium, potassium, magnesium, and boron has seldom improved growth on mineral soil in Nordic countries,[40,41] probably because of the lack of nitrogen in the wood ash. But wood ash has often increased growth on organic soil or in warm temperate climates, and it can also have other benefits.[42] Furthermore, it is a waste product of combustion that needs to be disposed of and so there is an argument that it should be used in suitable forest areas rather than disposed in landfills. Reviews by Aronsson and Ekelund[40] and Pitman[41] highlight the role of wood ash applications to replace nutrients on poor types of soil and those removed in whole-tree harvesting operations. Similarly, the use of biosolids or wastewater could be used in bioenergy plantations to recycle nutrients contained in environmentally acceptable materials.

The unwise use of energy and scarce resources is another aspect to consider. Certain fertilizers, such as those containing nitrogen, require considerable energy for their manufacture and application compared to the return of energy associated with increased biomass yield.[43] And the global scarcity of the supply of phosphorus is considered a serious threat to the sustainability of crop production in the coming decades, with demand predicted to exceed supply after 2035.[44]

Finally, if the implementation of bioenergy harvesting or short-rotation bioenergy plantations is being considered, it would be wise to avoid sites that require excessive use of external inputs for commercially acceptable yields to be realized.

Nutritional Management

To ensure sustainability of nutrient supplies, three management actions can be undertaken: replace limiting nutrients, lengthen rotations to permit natural replacement, or reduce nutrient removals from the site. Undertaking these actions will depend on four basic situations: (1) correcting gross nutrient deficiencies, particularly when establishing plantations on infertile sites for the first time; (2) managing sites where there are known problems; (3) ensuring sustainability when risk assessment and monitoring indicate possible problems are developing; and (4) no problems.

The rates and types of fertilizers to apply to forests and their application in relation to stand development have been well researched.[31,32,33,45] In general, nutrient demand increases rapidly as a stand develops and levels off after canopy closure, after which nutrient cycling through litter fall dominates. Following thinning, there can be an additional demand on the soil because the remaining trees expand their crowns and nitrogen fertilizers are sometimes used to increase productivity over a 3- to 10-year period. Utilization of applied fertilizer nitrogen by trees is typically low, and a large part of applied nitrogen is incorporated into soil organic matter or lost to the ecosystem.[32,46,47] The development of root systems in relation to soil properties and the roles of mycorrhizae also need to be considered.[48] Fertilizer practice needs to be related to these basic stand development processes.

When planting new sites either for bioenergy or for CO_2 offset, the trees need to be supplied with nutrients if the types of soil are so poor that a reasonable stand will not develop, as observed on highly weathered soil in warm temperate regions, such as the highly weathered clays of northern New Zealand,[49] southeast coastal plains of the United States,[50] and other parts of the world with highly weathered, nutrient-deficient soil. Application of fertilizer is not optional in these situations to achieve commercially acceptable crop yields. Phosphorus is often deficient in these regions, but sometimes other nutrients, for example, nitrogen on sand dunes, or micronutrients, such as boron, may be essential to achieve management objectives on some sites.[38]

Two different approaches have been taken to the use of fertilizer at planting. One approach has been to use small, individual-tree starter doses placed close to the seedlings, with the anticipation that heavier dressings would be applied as demand increased, or as needed when the trees were well established. The second approach has been to apply heavier broadcast or banded applications during the site-preparation phase, with the objective of applying sufficient nutrients until canopy closure or for most of the rotation. In both cases, it is important to control weeds to avoid increasing weed competition. Choice of fertilizer type needs to consider the interactions between nutrients, fertilizer–soil reactions, and possible leaching losses. Typical rates to apply and likely responses can be found in several publications.[33,44,51] Where phosphorus fertilizer is applied, sustained responses have often been obtained but this is less likely with nitrogen; and on some sites where both nutrients are limiting, it is essential to apply both nitrogen and phosphorus to get a response.[52]

To manage sites, where nutrient problems are known to be likely, a combination of two or three techniques are commonly used.[51] Soil analyses have been particularly suitable for phosphate deficiency, particularly for prescribing fertilizers at planting. Therefore, on these sites, sampling soil at planting is important. A second sampling at canopy closure, when the major demand by the stand on the soil has taken place, would confirm that nutrients are still at adequate levels. However, once the trees are established, foliar sampling and analysis have proved to be a good technique to monitor the nutrient status of many tree crops and to prescribe additional fertilizer.[33,51] If only a small starter fertilizer dressing has been given at planting, it is important to sample foliage at any sign of nutrient stress (deficiency symptoms) or when the stand is halfway to canopy closure. On some sites, foliage samples could be collected at regular intervals during the rotation to determine whether stands have become nutrient stressed. Where fertilizer applications cause substantial increases in soil nutrient availability, for example, as with phosphorus and boron applications, it is also possible to monitor the decrease in nutrient availability overtime using foliage analysis.[53] Foliage analysis is also useful to check on the nitrogen status of the stands where nitrogen is applied following thinning. Leaf area index measurements have also been recommended to determine which stands should be fertilized.[16,17,19] This index value could be more efficiently measured and widely adopted if aerial spectral reflectance data were collected.

Where monitoring procedures suggest that nutrient stress may be developing on reasonably fertile sites, two complementary approaches are recommended. First, check that poor management practices are not responsible and where appropriate, modify management. Often, these poor practices, such as those aggravating erosion, are obvious. Also, it may be economically and ecologically better to consider reducing nutrient removal from the site rather than replacing lost nutrients.[13] For example, crown biomass harvesting may be reduced to retain more nutrients on site. Paré et al.[54] have suggested that whole-tree harvesting of some species in the Canadian spruce-fir–northern hardwood forests, such as balsam fir (*Abies balsamea*) and trembling aspen (*Populus tremuliodes*), should not take place on soil types with low nutrient reserves because these species have greater nutrient contents than other species. Second, before applying nutrient amendments, undertake further investigations using foliage or soil analysis and perhaps fertilizer screening trials using vector analysis.[30] This will ensure that wasteful management decisions are not made.

Nitrogen-fixing plants are important in maintaining the nitrogen status of many forest ecosystems and can be incorporated into management, particularly in the establishment phase.[55,56] Nitrogen-fixing trees are sometimes successfully mixed with other tree species.[57] Pasture legumes are often part of silvopastoral agroforestry practices, although because of forage harvesting they can sometimes result in increased nitrogen stress.[58] Nitrogen-fixing associations generally are seen as having good potential and have been particularly successful on low-nitrogen sites.

Applying wood ash to the forest after combustion of forest biomass has often been recommended as a way to ensure sustainability criteria.[10,13] On peat soil, there can be a response due to increased mineralization. There is no strong experimental evidence that applying wood ash ensures sustainability of growth, although it should maintain soil base cation status, correct potassium, calcium and magnesium (perhaps phosphorus and boron) deficiencies, and may have other beneficial effects.[40–42,59] Augusto et al.[42] have also given recommendations regarding types of ash, rates to apply, and the choice of stands in which ash might be applied.

Off-Site Impacts

Off-site impacts, such as contamination of ground water or streams, have to be considered when designing and managing nutrient-amendment practices. Best management practices often stress the need for riparian buffers along streams to control the impact of logging, establishment, and fertilizers.[60] Similarly, partial cutting such as with the shelterwood system and well-designed strip cutting will often have lower off-site impacts than clear-cut systems.[61–63] Thinning is also unlikely to increase leaching losses of nutrients from the site.[64] In deciduous forests or during biomass harvesting, retention of foliage on site has been shown to reduce watershed nutrient losses.[62]

Leaching of fertilizer-applied nutrients is dependent on the fertilizer type, humus and soil properties, vegetation present, as well as rainfall. Problems are most likely to occur on sandy soil, soil with low clay content, and on peats.[46,65] The retention of nitrogen in the soil is directly associated with microbial uptake during the decomposition of residual organic matter but is less related to large slash on the forest floor.[66] Nitrate leaching and denitrification, and associated cation leaching, are related to the amount that is not retained by organic matter and taken up by developing vegetation.[64] For fertilizer-applied nitrogen, retention in organic matter also dominates but losses from the ecosystem, including gaseous and leaching outputs, can be up to half of that applied.[46,47,67] However, most catchment studies suggest that these impacts are small and short-lived because the amounts supplied in fertilizer are small compared to that in the ecosystem.[32,68]

Occasionally, the off-site movement of nutrients can be beneficial. For example, one of the major benefits of applying ash to acid forest soil could include a 'liming' effect that may reduce acidification of surface waters.[40–42]

Role of Economics, Energy Analysis, and Other Decision Aids

Forestry agencies in many countries are currently developing protocols and decision support systems to ensure that production forestry, including biomass extraction for bioenergy, is sustainable.[26,27] Akselsson et al.[59] have recently produced a nutrient balance model to assist in policy decision making in Sweden. In Canada, a regional nutrient balance model has been suggested; it looks at the impact of whole-tree harvesting of various species so as to rate the risk of degradation on various sites.[54] Stem-only harvesting is suggested for thin types of soil and glacio-fluvial sands. However, progress has been slow in developing these models for individual management units, partly because many proposed measures are weakly related to productivity.[69]

Also, the research that has suggested that there could be problems with nutrient removals has not been confirmed in practice. Thus, although various models can be useful in decision making,[70] their results need to be initially treated with caution until they have shown to be better than the more direct approach recommended earlier. Nambiar[71] has stressed that it is unwise to impose uses of poorly validated soil indicators on forest managers.

The use of nutrient cycling models in pasture management illustrates some of the advantages and limitations of modeling. For pasture agriculture, nutrient cycling models are more reliable than soil analyses for recommending fertilizer applications to maintain phosphorus nutrition after a steady state has been achieved.[72] In these developed pastures, native phosphate plays a small role in their nutrition, and fertilizer is the main source of available phosphorus. Soil heterogeneity, because of uneven fertilizer applications and dung and urine patches and seasonal variations, is seen as a major limitation with soil and foliage analyses once phosphorus deficiency has been corrected. Similar spatial variability of types of soil exists in many forest sites. Despite this, occasional soil phosphate tests are seen as having an important role in monitoring long-term trends. With potassium, modeling was not useful because potassium was largely derived from soil reserves that cannot be accurately predicted by models. Soil tests were, therefore, required to control potassium fertilizer applications.[72] Schoenholtz et al.[27] have suggested that current nutrient cycling models for forests are best used for determining research directions.

Stand growth models are usually not useful for management of fertilizers, although there are a few specific models that allow the user to specify nutrient amendments.[19,73] For Tasmanian eucalypt plantations, a fertilizer decision guide has been produced, which includes such factors as soil and site factors, stand age and growth, competing vegetation, crown form, and leaf area index.[19] Such guides are very helpful to managers.

Management practices are invariably undertaken after careful scrutiny of costs. Where nutrients are to be applied to maintain sustainability, they need to be included in the overall costs when making a decision to proceed with a particular option or not. They should not be seen as optional costs. This, however, does not preclude consideration of alternatives that may be more cost-effective, such as reducing nutrient removals by leaving crown material behind. Nor does it preclude the consideration of being able to offset part of these costs because the whole management system is more profitable. For example, whole-tree harvesting may reduce reestablishment costs and improve stand establishment results.

Similarly, the cost of fertilizer required to grow trees on severely deficient sites should not be considered optional–such costs are to be expected, if managing such sites, and should be included in economic analysis of silvicultural schedule rather than in isolation. However, where fertilizers are being considered to increase productivity to a higher level, and not just maintain productivity at the current level, then discounted cash flow analysis is appropriate for making this decision.

Mead and Pimentel[43] have argued that energy analysis also needs to be incorporated in decision-making processes. Their examples show that it was easier to justify the application of phosphate than of nitrogen fertilizers. Others have argued that other types of life-cycle assessment need to be undertaken, by estimating, for example, greenhouse gas balances.[74]

Apply Adaptive Management

The final principle requires a commitment to adaptive forest management.[2,75] The objective is to use continuing monitoring programs to evaluate the effectiveness of nutrient management plans and evaluate progress with the goals of sustainable management. If necessary, management would be altered either by reducing the impact of biomass removal or by other ameliorative techniques. Adaptive forest management is a continuous, reiterative, evolving process seeking continual improvement. Sustainable forest management is a journey rather than a destination as ecosystems, society values, and knowledge change overtime.[64,75] It is important that adaptive management be independently audited and reported to stakeholders.[2] The accuracy of risk assessment and monitoring procedures also need to be validated and updated as required.[18]

CONCLUSION

Nutrient management is a central element of ecological sustainability. If nutrient reserves and supplies become degraded to the point that forest composition and productivity are altered, ecological sustainability will be compromised. Similarly, ecological sustainability is vital to ensure social and economic sustainability. Forest owners, therefore, have strong imperatives to manage nutrients sustainably. This is even more important where forest bioenergy plans are being implemented, as they will often place additional stresses on site nutrient supplies. The wider society, particularly in developed countries, is also becoming increasingly aware of the need for sustainable land uses and insists on being involved in ensuring wise management.

The 10 principles discussed earlier are ordered as a professional forest manager might approach the task of managing nutrients to ensure sustainability. This is why, for example, defining management objectives and collecting the best available information on impacts are listed at the beginning. Site-specific mapping, coupled with evaluation and the choice of monitoring is central and will probably involve considerable costs. Although we have suggested five broad site classes in Figure 22.2, there is in fact a continuum, and in some regions not all classes may apply. The choice of methods and evaluation techniques need to be robust as well as acceptable to stakeholders—hence the recommendation for independent evaluation (Principle 5). The principle of not wasting resources is listed before the development of the site-specific management plans so that this is considered as plans are developed and written. Off-site impacts and the use of decision support systems are an integral part of developing nutritional management plans although they are listed separately. Finally, it is strongly recommended that the process of adaptive management be included as an on-going process.

These principles are built on the current state of knowledge and practice; and by adopting adaptive management, they allow for ongoing modifications due to experience and research. Furthermore, adaptive forest management can assist researchers by making them more informed about knowledge gaps. These management principles and practices are not discrete from one another but often overlap. For example, independent auditing is important in two principles whereas using the current knowledge of management impacts on nutrients runs through most principles.

Application of the 10 principles for nutrient management discussed in this paper will allow forest managers anywhere in the world to sustain the health and productivity of their forests and conform to the nutrient-related requirements of any current forest certification system. These principles are relatively easy to understand, even if requiring substantial technical expertise and site-specific data to develop nutrient budgets and nutritional management plans. Application by professional foresters will contribute to their balancing multiple demands for production, ecological services, including CO_2 offsets, and sustainability criteria. If these guidelines are followed from the outset of a new project, they could be incorporated into the budget underpinning a project.

ACKNOWLEDGEMENTS

The authors would like to express their appreciation for the very comprehensive and helpful suggestions of two anonymous reviewers.

REFERENCES

1. Powers RF. On the sustainable productivity of planted forests. *New Forest* 1999, 17:263–306.
2. Raison RJ. Environmental sustainability of forest energy production. In: Richardson J, Björheden R, Hakkila P, Lowe AT, Smith CT, eds. *Bioenergy from Sustainable Forestry: Guiding Principles and Practice*. Kluwer Academic Publishers, Dordrecht/Boston/London; 2002, 159–164.
3. MPCI. The Montreal Process. 2005. The Montreal Process. Available at: http://www.mpci.org/. (Accessed October 20, 2011).
4. Andersson G, Asikainen A, Björheden R, Hall PW, Hudson JB, Jirjis R, Mead DJ, Nurmi J, Weetman GF. Production of forest energy. In: Richardson J, Björheden R, Hakkila P, Lowe AT, Smith CT, eds. *Bioenergy from Sustainable Forestry: Guiding Principles and Practice*. Kluwer Academic Publishers, Dordrecht/Boston/London; 2002, 49–123.
5. Koopmans A. Biomass energy demand and supply for South and South-East Asia—assessing the resource base. *Biomass Bioenergy* 2005, 28:133–150.
6. Mead DJ. Forests for energy and the role of planted trees. *Crit Rev Plant Sci* 2005, 24:407–421.
7. Nair PKR. *An Introduction to Agroforestry*. Kluwer Academic Publishers, Dordrecht; 1993.
8. Ter-Mikaelian MT, Colombo SJ, Chen J. Fact and fantasy about forest carbon. *Forest Chron* 2008, 84:166–171.
9. Fox TR. Sustained productivity of intensively managed forest plantations. *Forest Ecol Manag* 2000, 138:187–202.
10. Burger JA. Soil and long-term site productivity values. In: Richardson J, Björheden R, Hakkila P, Lowe AT, Smith CT, eds. *Bioenergy from Sustainable Forestry: Guiding Principles and Practice*. Kluwer Academic Publishers, Dordrecht/Boston/London; 2002, 165–189.
11. Smith CT, Lowe AT, Skinner MF, Beets PN, Schoenholtz SH, Shengzuo Fang. Response of radiata pine forests to residue management and fertilisation across a fertility gradient in New Zealand. *Forest Ecol Manag* 2000, 138:203–223.
12. Young A. *Agroforestry for Soil Management*. CAB International, Wallingford, CT; 1997.
13. Hakkila P. Operations with reduced environmental impact. In: Richardson J, Björheden R, Hakkila P, Lowe AT, Smith CT, eds. *Bioenergy from Sustainable Forestry: Guiding Principles and Practice*. Kluwer Academic Publishers, Dordrecht/Boston/London; 2002, 244–261.
14. Neary DG. Hydrologic values. In: Richardson J, Björheden R, Hakkila P, Lowe AT, Smith CT, eds. *Bioenergy from Sustainable Forestry: Guiding Principles and Practice*. Kluwer Academic Publishers, Dordrecht/Boston/London; 2002, 190–215.
15. Burger JA, Kelting DL. Using soil quality indicators to assess forest stand management. *Forest Ecol Manag* 1999, 122:155–166.
16. Vose JM, Allen HL. Leaf area, stemwood growth, and nutritional relationships in loblolly pine. *Forest Sci* 1988, 34:547–563.
17. Allen HL, Dougherty PM, Campbell RG. Manipulation of water and nutrients—practice and opportunity in southern U.S. pine forests. *Forest Sci Ecol Manag* 1990, 30:437–453.

18. Curran MP, Miller RE, Howes SW, Maynard DG, Terry TA, Heninger RL, Niemann T, van Rees K, Powers RF, Schoenholtz, SH. Progress towards more uniform assessment and reporting of soil disturbance for operations. Research and sustainability protocols. *Forest Ecol Manag* 2005, 220:17–30.
19. Adams P, Osborn T, Volker P. EucFERT—a fertiliser decision tool for plantation eucalypts in Tasmania. Available at: http://www.forestry.org.au/pdf/pdf-public/conference2007/papers/Adams%20et%20al%20april%209.pdf (Accessed October 21, 2011).
20. Goovaerts P. Geostatistics in soil science: state-of-the-art and perspectives. *Geoderma* 1999, 89:1–45.
21. Thwaites RN, Slater BK. Soil-landscape resource assessment for plantations—a conceptual framework towards an explicit multi-scale approach. *Forest Ecol Manag* 2000, 138:123–138.
22. Scull P, Franklin J, Chadwick OA, McArthur D. Predictive soil mapping: a review. *Prog Phys Geog* 2003, 27:171–197.
23. Collerton H, Hock B, Hawke R, Payn T. Digital terrain modeling for site productivity assessment and stand management in plantation forestry. *N Z J Forest* 2008, 53:33–41.
24. NCASI. Technical Bulletin No. 0966: Compendium of Forestry Best Management Practices for Controlling Nonpoint Source Pollution in North America. 2009. Available at: http://www.ncasi.org/Publications/Detail.aspx?id=3204. (Accessed October 21, 2011).
25. O'Neill KP, Amacher MC, Perry CH. *Soils as an Indicator of Forest Health: A Guide to the Collection, Analysis, and Interpretation of Soil Indicator Data in the Forest Inventory and Analysis Program.* Gen Tech Rep NC-258. Forest Service, North Central Research Station, US Department of Agriculture, St. Paul, MN; 2005.
26. Watt MS, Davis MR, Clinton PW, Coker G, Ross C, Dando J, Parfitt RL, Simcock R. Identification of key soil indicators influencing plantation productivity and sustainability across a national trial series in New Zealand. *Forest Ecol Mang* 2008, 256:180–190.
27. Schoenholtz SH, Van Miegroet H, Burger JA. A review of chemical and physical properties as indicators of forest soil quality: challenges and opportunities. *Forest Ecol Manag* 2000, 138:335–356.
28. Kelting DL, Burger JA, Patterson SC, Aust WM, Miwa M, Trettin CC. Soil quality assessment in domesticated forests—a southern pine example. *Forest Ecol Manag* 1999, 122:167–185.
29. Mead DJ. Diagnosis of nutrient deficiencies in plantations. In: Bowen GD, Nambiar EKS, eds. *Nutrition of Plantation Forests*. Academic Press, London; 1984, 259–291.
30. Weetman GF, Wells CG. Plant analyses, an aid in fertilizing forests. In: *Soil Testing and Plant Analysis*. 3rd ed. SSSA Book Series No 3. Madison, WI: Soil Science Society of America; 1990, 659–690.
31. Miller HG. Dynamics of nutrient cycling in plantation ecosystems. In: Bowen GD, Nambiar EKS, eds. *Nutrition of Plantation Forests*. Academic Press, London; 1984, 52–78.
32. Binkley D, Carter R, Allen HL. Nitrogen fertilization practices in forestry. In: Bacon PE, ed. *Nitrogen Fertilization in the Environment*. Marcel Dekker, New York, NY; 1995, 421–441.
33. Birk EM. Fertilizer use in the management of pine and eucalypt plantations in Australia: a review of past and current practices. *N Z J Forest Sci* 1994, 24:289–320.
34. Payn TW, Hill RB, Höck BK, Skinner MF, Thorn AJ, Rijkse WC. Potential for the use of GIS and spatial analysis techniques as tools for monitoring change of forest productivity and nutrition, a New Zealand example. *Forest Ecol Manag* 1999, 122:187–196.
35. Brus DJ, Spätjens LEEM, de Gruijter JJ. A sampling scheme for estimating the mean extractable phosphorus concentrations of fields for environmental regulation. *Geoderma* 1999, 89:129–148.
36. Brown AJ. Soil sampling and sample handling for chemical analysis. In: Peverill KL, Sparrow LA, Reuter DJ, eds. *Soil Analysis: An Interpretation Manual*. Commonwealth Scientific and Industrial Research Organisation, Australia; 1999, 35–53.
37. Homann PS, Bormann BT, Boyle JR. Detecting treatment difference in soil carbon and nitrogen resulting from forest manipulations. *Soil Sci Soc Am J* 2001, 65:463–469.
38. Will GM. Nutrient deficiencies and fertiliser use in New Zealand exotic forests. FRI Bull. No. 97, Forest Research Institute, Rotorua, New Zealand; 1985, 53.
39. Zabowski D, Skinner MF, Payn TW. Nutrient release by weathering: implications for sustainable harvesting of *Pinus radiata* in New Zealand soils. *N Z J Forest Sci* 2007, 37:336–354.
40. Aronsson KA, Ekelund NGA. Biological effects of wood ash application to forest and aquatic ecosystems. *J Environ Qual* 2004, 33:1595–1605.
41. Pitman RM. Wood ash use on forestry—a review of the environmental impacts. *Forestry* 2006, 79:563–588. doi:10.1093/forestry/cpl041.
42. Augusto L, Bakker MR, Meridiue C. Wood ash applications to temperate forest ecosystems-potential benefits and drawbacks. *Plant Soil* 2008, 306:181–198.
43. Mead DJ, Pimentel D. Use of energy analyses in silvicultural decision making. *Biomass Bioenergy* 2006, 30:357–362.
44. Cordell D, Drangert J-O, White S. The story of phosphorus: global food security and food for thought. *Global Environ Chang* 2009, 19:292–305. doi: 10.1016/j.gloenvcha.2008.10.009.
45. Ballard R. Fertilization of plantations. In: Bowen GD, Nambiar EKS, eds. *Nutrition of Plantation Forests*. Academic Press, London; 1984, 327–360.
46. Carlyle JC. Opportunities for managing nitrogen uptake in established *Pinus radiata* plantations. *N Z J Forest Sci* 1994, 24:344–361.
47. Mead DJ, Chang SX, Preston CM. Recovery of ^{15}N-urea 10 years after application to a Douglas-fir pole stand in coastal British Columbia. *Forest Ecol Manag* 2008, 256:694–701.

48. Bowen GD. Tree roots and the use of nutrients. In: Bowen GD, Nambiar EKS, eds. *Nutrition of Plantation Forests*. Academic Press, London; 1984, 147–179.
49. Weston GC. Fertiliser trials in unthrifty pine plantations at Riverhead Forest. *N Z J Forest* 1956, 7:35–46.
50. Pritchett WL, Comerford NB. Long-term response to phosphorus fertilization on selected southeastern Coastal Plain soils. *Soil Sci Soc Am J* 1982, 46:640–644.
51. Mead DJ. Fertilisers. In: Colley M, ed. *Forestry Handbook*. New Zealand Institute of Forestry, Christchurch; 2005, 110–112.
52. May B, Smethurst P, Carlyle C, Mendham D, Bruce J, Baillie C. 2009. Review of fertiliser use in Australian forestry. Forest & Wood Products Australia, Melbourne. Available at: http://www.fwpa.com.au/sites/default/files/PRC072-0708_Fertiliser_Review_Research_Report.pdf (Accessed 23 May 2012).
53. Mead DJ, Gadgil RL. Fertiliser use in established radiata pine stands in New Zealand. *N Z J Forest Sci* 1978, 8:105–134.
54. Paré D, Rochon P, Brais S. Assessing the geochemical balance of managed boreal forests. *Ecol Indicators* 2002, 1:293–311.
55. Bormann BT, Gordon JC. Can intensively managed forest ecosystems be self-sufficient in nitrogen? *Forest Ecol Manag* 1989, 29:95–103.
56. Davey CB, Wollum II AG. Nitrogen fixation systems in forest plantations. In: Bowen GD, Nambiar EKS, eds. *Nutrition of Plantation Forests*. Academic Press, London; 1984, 361–377.
57. Forrester DI, Bauhus J, Cowie AL, Vanclay JK. Mixed-species plantations of *Eucalyptus* with N-fixing trees: a review. *Forest Ecol Manag* 2006, 233:211–230.
58. Goh KM, Mansur I, Mead DJ, Sweet GB. Biological nitrogen fixing capacity and biomass production of different understorey pastures in a *Pinus radiata*-pasture agroforestry system in New Zealand. *Agroforest Syst* 1996, 34:33–49.
59. Akselsson C, Westling O, Sverdrup H, Gundersen P. Nutrient and carbon budgets in forest soils as decision support in sustainable forest management. *Forest Ecol Manag* 2007, 238:167–174.
60. Shepard JP. Water quality protection in bioenergy production: the US system of forestry Best Management Practices. *Biomass Bioenergy* 2006, 30:378–384.
61. Hornbeck JW, Likens GE, Pierce RS, Bormann FH. Strip cutting as a means of protecting site and streamflow quality when clearcutting northern hardwoods. In: Bernier B, Winget CH, eds. *Forest Soils and Forest Land Management: Proceedings of the Fourth North American Forest Soils Conference*. University of Laval Press, Quebec; 1975, 209–229.
62. Horbeck JW, Smith CT, Martin QW, Tritton LM, Pierce RS. Effects of intensive harvesting on nutrient capitals of three forest types in New England. *Forest Ecol Manag* 1990, 30:55–64.
63. Wang X, Burns DA, Yanai RD, Briggs RD, Germain RH. Changes in stream chemistry and nutrient export following a partial harvest in the Catskill Mountains, New York, USA. *Forest Ecol Manag* 2006, 223:103–112.
64. Briggs RD, Hornbeck JW, Smith CT, Lemin Jr RC, McCormack Jr ML. Long-term effects of forest management on nutrient cycling in spruce-fir forests. *Forest Ecol Manag* 2000, 138:285–299.
65. Harriman R. Nutrient leaching from fertilized forest watersheds in Scotland. *J Appl Ecol* 1975, 15:933–942.
66. Vitousek PM, Matson PA. Mechanisms of nitrogen retention in forest ecosystems: a field experiment. *Science* 1984, 225:51–52.
67. Preston CM, Marshall VG, McCullough K, Mead DJ. Fate of ^{15}N-labelled fertilizer applied on snow at two forest sites in British Columbia. *Can J Forest Res* 1990, 20:1583–1592.
68. Neary DG, Hornbeck JW. Impacts of harvesting and associated practices on off-site environmental quality. In: Dyck WJ, Cole DW, Comerford NB, eds. *Impacts of Forest Harvesting on Long-Term Site Productivity*. Chapman & Hall, London; 1994, 81–118.
69. Weetman GF. A forest management perspective on sustained site productivity. *Forest Chron* 1998, 74:75–76.
70. Blanco JA, Seely B, Welham C, Kimmins JP, Seebacher TM. Testing the performance of FORECAST, a forest ecosystem model, against 29 years of field data in a *Pseudotsuga menziesii* plantation. *Can J Forest Res* 2007, 37:1808–1820.
71. Nambiar EKS. Sustained productivity of forests is a continuing challenge to soil science. *Soil Sci Soc Am J* 1996, 60:1629–1642.
72. Cornforth IS. Maintenance fertilisers for grazed pastures in New Zealand: an agriculture perspective of applying theory to management. *N Z J Forest Sci* 1994, 24:279–288.
73. Burkhart H, Amateis R, Westfall J, Daniels R. *PTAEDA4.0: Simulation of Individual Tree Growth, Stand Development and Economic Evaluation in Loblolly Pine Plantations*. Virginia Tech Department of Forest Resources and Environmental Conservation, Blacksburg, VA; 2008.
74. Schlamadinger B, Apps M, Bohilin F, Gustavsson L, Jungmeier G, Marland G, Pingoud K, Savolainen I. Towards a standard methodology for greenhouse gas balances of bioenergy systems in comparison to fossil energy systems. *Biomass Bioenergy* 1997, 13:359–375.
75. Lattimore B, Smith T, Richardson J. Coping with complexity: designing low-impact forest bioenergy systems using an adaptive forest management framework and other sustainable forest management tools. *Forest Chron* 2010, 86:20–27.

23

Crop Coefficients of *Jatropha* (*Jatropha Curcas*) and *Pongamia* (*Pongamia pinnata*) Using Water Balance Approach

Kaushal K. Garg, Suhas P. Wani* and A.V.R. Kesava Rao

International Crops Research Institute for the Semi Arid Tropics, Patancheru, Andhra Pradesh, India

INTRODUCTION

Food and energy security are becoming a major issue both for developing and developed countries all over the world. Biofuels such as ethanol and biodiesel are considered as one of the potential alternatives to reduce fossil fuel dependency,[1,2] also to reduce greenhouse gas emission, and to sequester carbon.[2–4] In recent years, governments of many countries have targeted use of biofuel by blending it with conventional petroleum fuel and are promoted for wide industrial adaptability.[1,5] In contrary, it is also hypothesized that biofuel production at large scale may threaten the food security by competing with food crops and thus affecting the food supply and prices.[6,7] On the other hand, production of biodiesel from plants grown in degraded lands offers opportunities to address the issues of environmental protection and provide options for improving rural livelihood especially in developing countries.[3,8,9] *Jatropha curcas* (*Jatropha*) and *Pongamia pinnata* (*Pongamia*) are identified as potential biodiesel crops and may be grown in marginal and degraded lands with a wide range of rainfall and ecological conditions.[4,10–12]

Although the demand for biodiesel is increasing, available data and knowledge on yield potential and water requirement of *Jatropha* and *Pongamia* are very limited and thus research efforts are required to address such issues.[13] However, few estimates on water requirements of *Jatropha* have been reported in literatures. For example, Rao et al.[14] reported that evapotranspiration (ET) from *Jatropha* plantation ranges from 1410 to 1538 mm in optimal conditions and 614 to 930 mm in field conditions in a semi-arid tropical location in India. They observed moisture extraction from soil layers 150 cm below from surface by 5-year-old plants. Transpiration requirement of *Jatropha* was estimated as 300 and 500 mm at 4th and 12th year of growth stage at Owen Sithole College of Agriculture near Empangeni on the KZN north coast and Makhathini, Southern Africa, respectively.[15] Gush and Moodley[16] estimated total transpiration of 1983 ℓ (147 mm) for a 4-year-old *J. curcas* tree grown at a relatively wetter location, and 4884 ℓ (362 mm) for a 12-year-old *J. curcas* tree grown at a dry site of South Africa. Transpiration rate was found high during the warm wet summer months and negligible during winter due to the deciduous nature of the *Jatropha* species.

Water requirement of a plant may be estimated by multiplying crop coefficients (K_c) with atmospheric water demand (ET_0). In this method, effect of crop transpiration and evaporation is integrated by a single coefficient, which is known as crop coefficient.[17] The K_c–ET_0 approach provides a simple, convenient, and reproducible way to estimate water requirement of crops in different climatic conditions.[17,18] Crop coefficient values have been reported for several crops[17] and tree plantations[19,20] and

*Correspondence to: s.wani@cgiar.org

Advances in Bioenergy: The Sustainability Challenge, First Edition. Edited by Peter D. Lund, John Byrne, Göran Berndes and Iacovos A. Vasalos.

Originally published in: *WIREs Energy Environ* 2014, 3: 363–372 doi: 10.1002/wene.88

have been widely used for irrigation scheduling,[17,21] simulation studies,[22] and decision making process.[17]

Several methodologies were developed for estimating K_c, for example, weighing lysimeter-based techniques,[23] agronomic measurements,[18] and remote sensing-based techniques.[24] Irrespective to methodologies, basis for estimating K_c in all such techniques is common, that is, computation of water and energy balance components. In the present study, we used soil moisture data collected at International Crops Research Institute for the Semi-Arid Tropics (ICRISAT) research station for analyzing water balance of *Jatropha* and *Pongamia* crops using one dimensional hydrological model. Specific objectives of the present study are: (1) to estimate K_c for *Jatropha* and *Pongamia* crops and (2) to assess water requirements and analyzing field water balance of *Jatropha* and *Pongamia* crops in semi-arid tropics.

MATERIALS AND METHODS

Experimental Setup and Data Monitoring

ICRISAT is located at 17° 53'N, 78° 27'E, and about 545 m above mean sea level in the Medak district of Andhra Pradesh, India. Climate of the study area is tropical monsoon. Long-term (1974–2010) average annual rainfall is 898 mm (standard deviation, $\sigma = 225$ mm), of which about 80–85% is received during June–October. Rainfall is highly erratic, both in terms of total amount and distribution over time. Monsoon is preceded by hot summers (minimum air temperature between 16 and 29°C and maximum air temperature between 30 and 43°C in May) and followed by cool winters (minimum air temperature in between 6 and 20°C and maximum air temperature between 23 and 32°C in December). Estimated potential evapotranspiration (PET) of the study area is 1650 mm annually.

Two microwatersheds each of 2 ha area were selected for cultivating *Jatropha* and *Pongamia* crops at ICRISAT research station. Soil of experimental site is classified as Vertisols having maximum depth up to 3 m. Topography of microwatersheds is relatively flat with average slope between 1.0 and 2.0%. Physical and chemical properties of the soil of experimental site, such as bulk density, texture, organic carbon, and soil water retention parameters were measured. Mean values of soil organic carbon was 0.65% (standard deviation, $\sigma = 0.2$) in 0–15 cm depth and 0.50% ($\sigma = 0.10$) in 15–30 cm soil depth. Soil bulk density of upper soil layers is relatively higher (1.4 g cm^{-3} up to 75 cm layer) compared with lower layers (1.2 g cm^{-3} between 120 and 150 cm). Bulk density of surface soil varied between 1.30 and 1.55 g cm^{-3} with average of 1.35 g cm^{-3} across the watersheds. Subsurface soils are high in clay content and hold more amount (soil moisture availability = 0.25 g/g) of soil moisture compared with shallow soils (soil moisture availability = 0.14 g/g).

Jatropha (3 × 2 m^2 spacing) and *Pongamia* (5 × 5 m^2 spacing) seedlings were planted in year 2004. Crops were grown under rainfed conditions, and chemical fertilizers were applied every year according to the need of plants at their different growth stages. Nitrogen and phosphorus were applied at 76 and 10 kg ha^{-1}, respectively, during 2006 and 2007. During subsequent growth stages, fertilizers dose was increased. Nitrogen and phosphorus were applied at 92 and 50 kg ha^{-1}, respectively, during 2008 and 2009. Source of N and P in above-mentioned fertilization was urea and single super phosphate, respectively.[14] Total 12 neutron probe tubes in each of *Jatropha* and *Pongamia* microwatersheds were installed for monitoring soil moisture content. Schematic layout of the experimental setup is shown in Figure 23.1 describing spacing and location of neutron probe tubes between the plants rows. Soil moisture content of each 15-cm-thick layer was monitored from surface to 225 cm depth using calibrated neutron probe (Troxler 4302) at 15 days interval between 2007 and 2010. In addition, moisture in surface soils (0–15 and 15–30 cm) was also measured by gravimetric method. Daily weather data (rainfall, maximum temperature, minimum temperature, wind speed, solar radiation, relative humidity, and pan evaporation) of experimental site were collected from meteorological station located at 200 m distance from the microwatersheds. Data on agronomic growth, seed yield, and oil content were measured both from the *Jatropha* and *Pongamia* fields.

Description of Simulation Model

A simple and one-dimensional hydrological model (Water Impact Calculator, WIC developed in ICRISAT) was used to analyze field water balance in the present study. WIC requires data on soil (soil depth and water retention), weather (reference ET and rainfall), crop growth (biomass and root growth function), topography (land slope, land form conditions), *ex situ* interventions (water harvesting capacity), and crop management (date of crop sowing and harvesting irrigation method) as an input. WIC analyzes water balance on daily time scale for each 1 cm soil layer (generate simulation matrix: time vs soil depth). Rainfall, which is the only source of water for rainfed situation, was partitioned into several hydrological components as defined by following mass balance equation:

$$\begin{aligned}\text{Rainfall} = {} & \text{Surface runoff} + \text{Deep percolation} \\ & + \text{Evapotranspiration (Evaporation} \\ & + \text{Transpiration)} + \text{Change in} \\ & \text{soil moisture storage}\end{aligned} \tag{23.1}$$

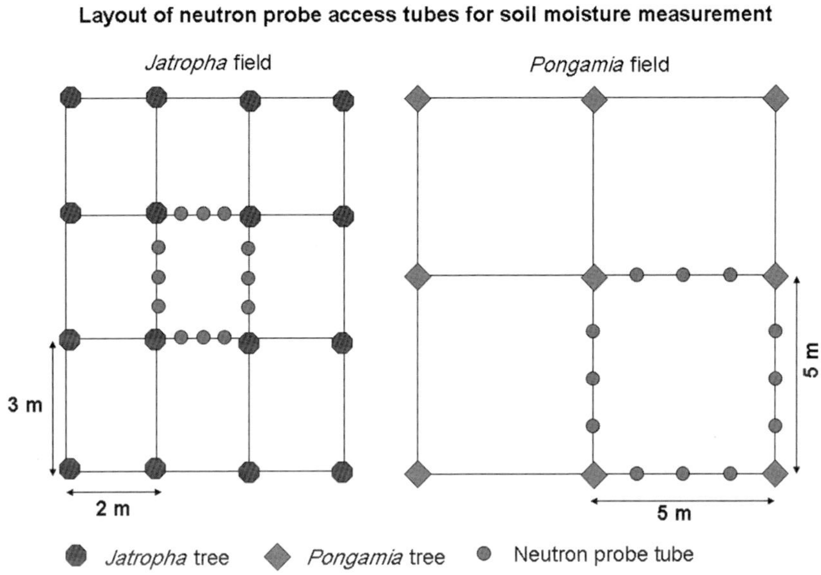

Figure 23.1 Layout of neutron probe access tubes for soil moisture measurement in *Jatropha* and *Pongamia* field at ICRISAT, Patancheru.

Surface runoff is the flow of excess rainwater after soil becomes saturated and generally occurs either during or after the rainfall event. The SCS curve number technique[25] was used to estimate surface runoff in WIC. WIC calculates daily curves number value based on antecedent moisture content of surface soil layer (assumed for 15 cm depth in current analysis) and partition rainfall into surface runoff. An excess amount of infiltrated water after satisfying soil storage capacity is percolated below the crop root zone and is defined as deep percolation. ET comprises two basic hydrological processes: evaporation and transpiration. WIC calculates evaporation and transpiration based on imposed surface boundary conditions and moisture accessibility from surface soil layer and root zone. Water available in top 10 cm soil layer is contributed toward the evaporation demand, whereas moisture available in root zone is used to meet crop water demand as defined by Allen et al.[17] Crop water requirement (CWR) for a given crop is calculated as:

$$\text{CWR} = K_c \times \text{ET}_0 \tag{23.2}$$

$$\begin{aligned} &\text{if}: \sum_{i=1}^{\text{rootzone}} \text{AWC} > \text{CWR then ET} = \text{CWR} \\ &\text{otherwise ET} = \sum_{i=1}^{\text{rootzone}} \text{AWC} \end{aligned} \tag{23.3}$$

where AWC is available water and *i* denotes each cm increment in soil layer reaching up to root zone. It was further assumed that evaporation from soil surface is inversely proportional to vegetative growth/stage. Thus, after achieving full vegetative crop growth ($K_c \geq 1.0$), evaporation from the soil surface is considered almost negligible. Detailed description of WIC, model development, testing, and validation procedure is shown by Garg et al.[26]

Estimation of Crop Coefficients

According to *FAO-56* nomenclature, K_c calculates water requirements under growing conditions having a high level of management and with little or no water stress.[17] Appropriate corrections need to be incorporated while estimating crop coefficients in case of water stress situation due to less water uptake. Allen et al.[17] described the effects of soil water stress on crop evapotranspiration (ETc) by multiplying the crop coefficient (K_c) and the water stress coefficient (K_s) with ET_0 such as:

$$\text{ET} = K_s \times K_c \times \text{ET}_0 \tag{23.4}$$

Soils of the experimental watershed are characterized by Vertisols, which has medium-to-high water holding capacity and average depth of 2.5–3.0 m. In the current study, K_c is estimated under the fully developed stage (between 2007 and 2011) and is expected to experience with minimal water stress situation. Despite this, whenever available soil moisture depleted below the threshold limit, stress adjustment in ETc was made as defined by Allen et al.,[17] and K_c was estimated using Eq. (23.4). Seventy percent of total available water (soil moisture) is considered as readily available for crop uptake and further it declined linearly. Water stress coefficient (K_s in Eq. (23.4)) is assumed 0.7 for *Jatropha* and *Pongamia* as both are stress-tolerant and deep-rooted trees.[17,27–29] Crop coefficient of *Jatropha* and *Pongamia* crops was estimated using the inverse optimization technique. Solver is an optimization program built in Microsoft excel (developed for mixed integer nonlinear parameterization) was used for the current study. Based on field measurements, field capacity and permanent wilting point are defined for every 15 cm soil layer up to 300 cm depth. Maximum root zone depth for *Jatropha* and *Pongamia* plant is considered as 150 and 250 cm, respectively.[27–29] Model is set by assigning atmospheric boundary condition (rainfall and PET) on the soil surface as upper boundary and deep percolation from the bottom layer as the lower boundary. Simulation period is started after the 4 years of plant establishment. Total soil moisture status in root zone was used as the auxiliary variable in the objective function, and crop coefficient at different crop growth stages were estimated using inverse optimization technique. WIC performance was assessed by comparing the simulated and measured soil moisture content using the root-mean-squared error (RMSE) of prediction:

$$\text{RMSE} = \sqrt{\frac{1}{n}\sum_{i=1}^{n}(x_i - y_i)^2} \tag{23.5}$$

where n is the number of data points of measured (x_i) and simulated (y_i) profile soil moisture.

Estimation of Crop Water Balance

The calibrated model setup was used to simulate water balance of 10-year time period between 2001 and 2010 capturing large variability of rainfall. Water balance components (ET, runoff, and soil moisture) are derived for dry, normal, and wet years according to following classification (India Meteorological Department, Pune, India; http://www.imdpune.gov.in): rainfall less than 20% of the long-term average = dry, rainfall between −20% and +20% of the long-term average = normal, and rainfall greater than 20% of long-term average = wet.

Model further was used to estimate ET under nonlimiting water stress situation both for *Jatropha* and *Pongamia* crops by applying auto irrigation setting in WIC between 2001 and 2010. Crop water stress was analyzed by considering difference between actual ET and ET under nonlimiting water condition such as:

$$\text{Water stress factor } (-) = 1 - \frac{\text{ET actual}}{\text{ET under nonlimiting water situtation}} \tag{23.6}$$

RESULTS AND DISCUSSION

Variation of Soil Moisture in *Jatropha* and *Pongamia* Watersheds

Soil moisture variation in block plantations of *Jatropha* and *Pongamia* is depicted in Figure 23.2a and 23.2b, respectively. Figures are divided into four panels and each one shows moisture profile of consecutive 3 months. In *Jatropha* field, new leaf flushes and biomass growth started by beginning of April, flowering initiated by May/June, and pod formation and harvesting stage occurred between September and December.[14] Previous stored soil moisture into subsurface layers was used by *Jatropha* plants to fulfill water requirements during leaf sprouting. Thus, moisture level depleted significantly between April and June (Figure 23.2a). Large variation in moisture profile is observed between July and September, as this period coincided with peak vegetative growth and occurrence of monsoonal rainfall. Total soil moisture was found highest during September. Moisture in soil profile is depleted slightly during pod formation and harvesting stage during November/December. Variation in profile moisture was found minimal between January and March because crop entered into dormancy phase and leaf started falling down (Figure 23.2a). Generally, soil moisture fluctuation was found up to 150 cm, indicating maximum effective rooting depth for *Jatropha* crop.[27]

In *Pongamia* field, low variation in moisture profile was observed during April and June, indicating lesser water uptake (Figure 23.2b). Soil moisture content was increased during monsoon season and depleted subsequently in nonmonsoonal period as expected. Data showed that *Pongamia* being a perennial crop could extract water from deeper and drier soil conditions. Several other studies indicated that *Pongamia* has an extensive lateral root system and thick-long tap root,[28] which could be developed maximum up to 5–10 m length at fully matured stage.[29] Soil moisture level in *Pongamia* cultivated field is depleted much higher compared with *Jatropha* cultivated fields. For example, soil moisture content for *Jatropha* and *Pongamia* cultivated field in months of May and June was recorded as 0.2–0.3 and 0.03–0.15 $cm^3\ cm^{-3}$, respectively.

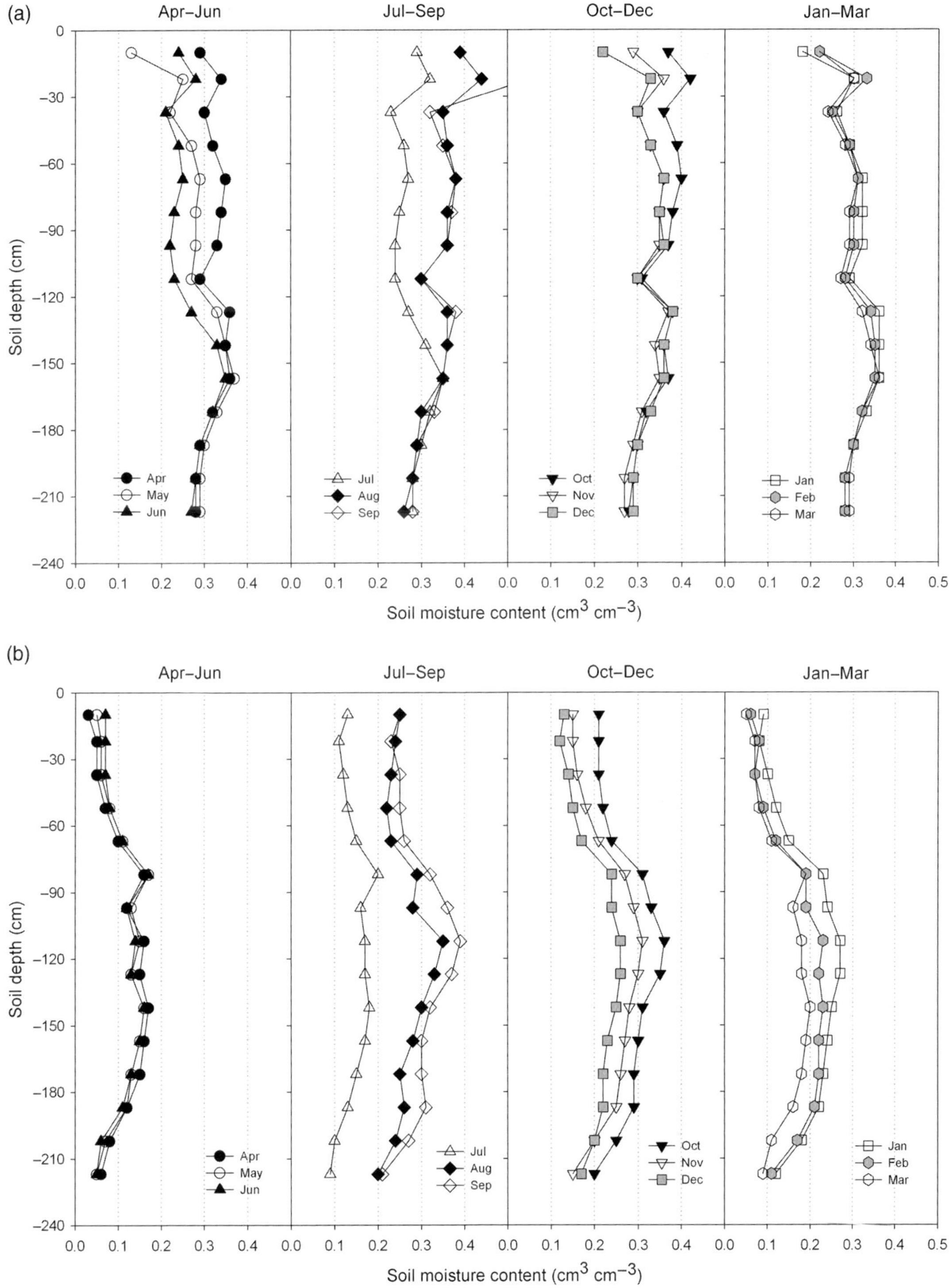

Figure 23.2 (a) Depth wise soil moisture content during 12 months in *Jatropha* experimental field; (b) depth wise soil moisture content during 12 months in *Pongamia* experimental field.

Table 23.1 Crop Coefficient for *Jatropha* and *Pongamia* Tree Crops on Monthly Time Scale

Month	K_c of *Jatropha*	K_c of *Pongamia*
April	0.20	0.30
May	0.40	0.40
June	0.60	0.40
July	0.85	0.70
August	0.90	0.90
September	0.95	1.10
October	0.80	1.10
November	0.70	1.10
December	0.50	0.75
January	0.30	0.65
February	0.10	0.60
March	0.10	0.35

Model Performance

Soil moisture measured in *Jatropha* and *Pongamia* fields is compared with model simulated moisture level between 2007 and 2011. Simulated soil moisture fluctuations are found comparable to observed numbers at most of the times both for *Jatropha* and *Pongamia* cultivated fields. RMSE in estimating profile moisture is 6.0 and 5.0 cm, which is equivalent to less than 10% of total moisture in given soil profiles, thus model performance is considered acceptable.

Crop Coefficients and CWRs

Crop coefficients estimated for *Jatropha* and *Pongamia* crops are summarized in Table 23.1. K_c for *Jatropha* and *Pongamia* are found in range between 0.10–0.95 and 0.30–1.10, respectively. Maximum K_c was found during August and September for both the crops.

Water balance components of *Jatropha* and *Pongamia* fields are summarized in Table 23.2. During normal years, rainfall recorded was 911 mm (820 mm in monsoon and 91 mm in nonmonsoon) at experimental site. In these years, ET from *Jatropha* and *Pongamia* crop was estimated to be 685 and 830 mm year^{-1}, respectively. Surface runoff generated from both the crop lands is found almost similar, that is, 130–135 mm (15% of total rainfall). In *Jatropha* field, nearly 100 mm rainfall was harvested in terms of green water. In *Pongamia* fields, 35 mm of net deficit in green water storage was observed by the end of the crop season in normal years.

ET from *Jatropha* and *Pongamia* field was found larger than the rainfall received during dry years. About 130–150 mm harvested green water in previous years was utilized by crops during dry years as shown by negative balance (Table 23.2). Surface runoff and deep percolation volume were estimated to be 90–100 mm, that is, 14% of rainfall received. Results showed that poor rainfall in dry years led to develop water shortage by 21% (in terms of ET) than the required quantity for *Pongamia* crop (shown by water stress factor in Table 23.2). On the other hand, crop water demand and availability for *Jatropha* crop were found in comparable range.

To show a detail description of mass balance, the monthly water balance of *Jatropha* and *Pongamia* field is presented for one of the selected normal year (Figure 23.3). Total 850 mm of rainfall was received during monsoon in year 2003. Thirty-four percent of monsoonal rain was captured as soil moisture, 52% (440 mm) contributed to ET, and 14% (120 mm) was exported as surface runoff. In nonmonsoon period (November–May), harvested green water acted as a source of water as shown in Figure 23.3. Results showed that water requirements of both the crops

Table 23.2 Water Balance of *Jatropha* and *Pongamia* Cultivated Fields at ICRISAT for 10-Year Simulation Period

			Water Balance Components						
			Rainfall (mm)						
Crop Cultivated	Year	No. of Year	June–October	October–May	Annual Rainfall	Surface Runoff + Deep Percolation (mm)	ET (mm)	Change in SMC (mm)	Water Stress Factor (−)
Jatropha	Dry	2	612	40	652	90	690	− 128	0.01
Jatropha	Normal	6	820	91	911	130	700	80	0.01
Jatropha	Wet	2	1179	71	1250	322	808	118	0.00
Pongamia	Dry	2	612	40	652	95	710	− 153	0.21
Pongamia	Normal	6	820	91	911	115	830	− 35	0.10
Pongamia	Wet	2	1179	71	1250	249	953	48	0.00

The calibrated model setup is used to simulate water balance for a 10-year time period between 2001 and 2010. Water balance results are presented for dry, normal, and wet years according to following classification (Indian Meteorological Department, Pune, India; http://www.imdpune.gov.in): rainfall less than 20% of the long-term average = dry, rainfall between −20% and +20% of the long-term average = normal, rainfall greater than 20% of long-term average = wet.

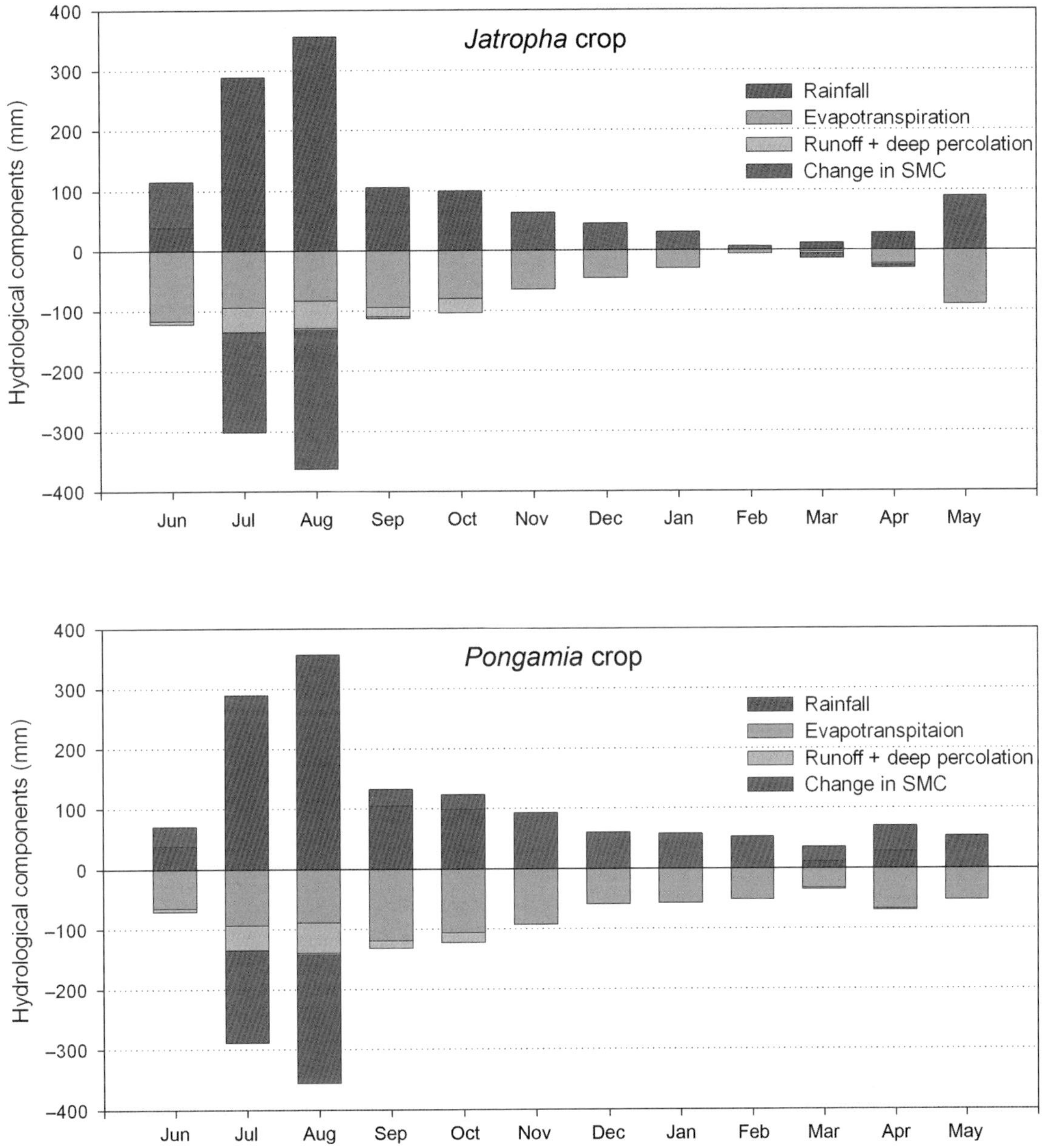

Figure 23.3 Monthly simulated water balance of *Jatropha* and *Pongamia* fields at ICRISAT for a selected normal year (2003–2004); the upper part of the graphs show the source of the water (rainfall and change in soil moisture content), and the lower part presents various sink components: ET, surface runoff, deep percolation, change in soil moisture content.

were nearly same during monsoon period (460 mm) but *Pongamia* required additional 200 mm water during non-monsoon season compared with *Jatropha* (240 mm). June is found to be the critical period for *Jatropha* growth and supplemental irrigation could be helpful in dry years.

CONCLUSIONS

Jatropha and *Pongamia* are potential biodiesel crops and could be grown in wide range of agroclimatic and soil conditions. Water requirements of these two crops are estimated in the current study. Key findings of this study are:

1. Water requirements of *Jatropha* and *Pongamia* are estimated to be 750 and 950 mm in semi-arid tropics, respectively. Water balance showed that at ICRISAT out of 820 mm rainfall received during monsoon period, 52% was contributed to ET, 34% stored as soil moisture, and 14% exported outside from watershed as surface runoff. Consumptive water use by *Jatropha* and *Pongamia* is found to be nearly the same in monsoon, that is, 460–500 mm. Water requirement (ET) of *Jatropha* during nonmonsoon period is found as 265 mm. *Pongamia* required 200 mm additional water than *Jatropha* during nonmonsoonal period.
2. Crop coefficients of *Jatropha* and *Pongamia* ranged from 0.10 to 0.95 and from 0.30 to 1.1 depending on growth stage, respectively. *Pongamia* is found to utilize green water from deeper and drier soil layers compared with *Jatropha* crop.
3. Water availability for *Pongamia* crops was found adequate in wet years but slightly less (10% shortage) in normal year. Poor rainfall in dry years led to develop water shortage by 20% than the required quantity in *Pongamia* field, whereas *Jatropha* experienced negligible water stress in all the years. *In situ* moisture availability played an important role in supplying water for both *Jatropha* and *Pongamia* crops, especially in nonmonsoon period.

Crop coefficient estimated in current study could be useful to assess technical feasibility to grow *Jatropha* and *Pongamia* crops in different agroecological regions and soil conditions.

REFERENCES

1. Government of India. National Policy on Biofuels. Ministry of New & Renewable Energy, New Delhi; 2009.
2. Achten WMJ, Maes WH, Aerts R, Verchot L, Trabucco A, Mathijs E, Singh VP, Muys B. Jatropha: from global hype to local opportunity. *J Arid Environ* 2010, 74:164–165.
3. Wani SP, Sreedevi TK, Marimuthu S, Rao AVRK, Vineela C. Harnessing the potential of *Jatropha* and *Pongamia* plantations for improving livelihoods and rehabilitating degraded lands. 6th International Biofuels Conference 2009, March 4–5, New Delhi, India.
4. Wani SP, Chander G, Sahrawat KL, Rao CS, Raghvendra G, Susanna P, Pavani M. Carbon sequestration and land rehabilitation through Jatropha *curcas* (L.) plantation in degraded lands. *Agric Ecosyst Environ* 2012, 161:112–120.
5. EC. Communication from the European Commission on An EU Strategy for Biofuels, COM 2006, 34 final. European Commission.
6. Doornbosch R, Steenblik R. *Round Table on Sustainable Development Biofuels: Is the Cure Worse than the Disease?* Paris, France: Organization for Economic Cooperation and Development (OECD); 2007, 11–12.
7. Naylor R, Steinfield H, Falcon W, Galloway J, Smil V, Bradford E, Alder J, Mooney H. Losing the links between livestock and land. *Science* 2007, 310:1621–1622.
8. Wani SP, Sreedevi TK. *Pongamia's* journey from forest to microenterprise for improving livelihood. International Crop Research Institute for the Semi-Arid Tropics. Research Report 2005 Global Theme of Agroecosystems, Patancheru, India.
9. Sreedevi TK, Wani SP, Srinivasa Rao CH, Chaliganti R, Reddy RL. *Jatropha* and *Pongamia* rain-fed plantations on wastelands in India for improved livelihoods and protecting environment. 6th International Biofuels Conference 2009, March 4–5 New Delhi, India.
10. Wani SP, Osman M, D'Silva E, Sreedevi TK. Improved livelihoods and environmental protection through biodiesel plantations in Asia. *Asian Biotechnol Dev Rev* 2006, 8:11–29.
11. Achten WMJ, Verchot L, Franken YJ, Mathijs E, Singh VP, Aerts R. Review: *Jatropha* bio-diesel production and use. *Biomass Bioenergy* 2008, 32:1063–1084.
12. Garg KK, Karlberg L, Wani SP, Berndes G. *Jatropha* production on wastelands in India: opportunities and trade-offs for soil and water management at the watershed scale. *Biofuels Bioprod Bioref* 2011, 5:410–430.
13. Raju KV. Biofuels in South Asia: an overview, editorial introduction. *Asian Biotechnol Dev Rev* 2006, 8:1–9.
14. Rao AVRK, Wani SP, Singh P, Srinivas K, Srinivasa Rao Ch. Water requirement and use by *Jatropha curcas* in a semi-arid tropical location. *Biomass Bioenergy* 2012, 39:175–181.
15. Holl MA, Gush MB, Hallowes J, Versfeld DB. *Jatropha curcas* in South Africa: An Assessment of its Water Use and Bio-Physical Potential. Report submitted to the Water Research Commission on the project "Investigation into the Impacts of Large-Scale Planting of *Jatropha curcas* on Water Resources, through Process-Based Research and Modelling". WRC Report No 1497/1/07, November 2007.
16. Gush MB, Moodley M. Water use assessment of *Jatropha curcas*. In: Holl M, Gush MB, Hallowes J, Versfeld DB, eds. *Jatropha curcas in South Africa: An Assessment of its Water Use and Bio-Physical Potential*. WRC Report No. 1497/1/07 (Chapter 4). Water Research Commission, Pretoria, RSA.
17. Allen RG, Pereira LS, Raes D, Smith M. Crop Evapotranspiration: Guidelines for Computing Crop Water Requirements, Irrigation and Drainage Paper 56. United Nations FAO, Rome; 1998, 300 p.
18. Allen RG, Pereira LS. Estimating crop coefficients from fraction of ground cover and height. *Irrig Sci* 2009, 28:17–34.
19. Teixeira AH, de C, Bastiaanssen GM, Moura MSB, Soares JM, Ahmad MD, Bos MG. Energy and water balance measurements for water productivity analysis in irrigated mango trees, Northeast Brazil. *Agric Forest Meteorol* 2008, 148:1524–1537.

20. Paco TA, Ferreira MI, Rosa RD, Paredes P, Rodrigues GC, Conceicao N, Pacheco CA, Pereira LS. The dual crop coefficient approach using a density factor to simulate the evapotranspiration of a peach orchard: IMDualKc model versus eddy covariance measurements. *Irrig Sci* 2012, 30:115–126.
21. Chen F, Robinson PJ. Estimating reference crop evapotranspiration with ET gauges. *J Irrig Drain Eng* 2009, 135:335–342.
22. Cancela JJ, Cuesta TS, Neira XX, Pereira LS. Modelling for improved irrigation water management in a temperate region of Northern Spain. *Biosyst Eng* 2006, 94:151–163.
23. Evett SR, Mazahrih NT, Jitan MA, Sawalha MH, Colaizzi PD, Ayars JE. A weighing lysimeter for crop water use determination in the Jordan Valley, Jordan. *Trans ASABE* 2009, 52:155–169.
24. Singh RK, Irmak A. Estimation of crop coefficients using satellite remote sensing. *J Irrig Drain Eng* 2009, 135:597–608.
25. USDA-SCS. *National Engineering Handbook, Section 4—Hydrology*. Washington, D.C.: USDA-SCS; 1985.
26. Garg KK, Wani SP, Patil MD. Simple and farmer-friendly decision support system for enhancing water use efficiency in agriculture: tool development, testing and validation. *Submitted to Agric Water Manag*; 2013.
27. Krishnamurthy L, Zaman-Allah M, Marimuthu S, Wani SP, Rao AVRK. Root growth in *Jatropha* and its implications for drought adaptation. *Biomass Bioenergy* 2012, 39:247–252.
28. Scott PT, Pregelj L, Chen N, Hadler JS, Djordjevic MA, Gresshoff PM. *Pongamia pinnata*: an untapped resource for the biofuels industry of the future. *Bioenerg Res* 2008, 1:2–11
29. Rangan L. *Pongamia*—a multipurpose versatile legume. *Res J Biotechnol* 2013, 8:1–3.

24

Brazilian Sugarcane Ethanol: Developments so far and Challenges for the Future

Arnaldo Walter,[1]* Marcelo Valadares Galdos,[2] Fabio Vale Scarpare,[2] Manoel Regis Lima Verde Leal,[2] Joaquim Eugênio Abel Seabra,[1] Marcelo Pereira da Cunha,[2] Michelle Cristina Araujo Picoli[2] and Camila Ortolan Fernandes de Oliveira[1]

[1]Department of Energy, Faculty of Mechanical Energy, University of Campinas, Campinas, Brazil
[2]CTBE/CNPEM – Brazilian Bioethanol Science and Technology Laboratory/National Research Center on Energy and Materials, Campinas, Brazil

INTRODUCTION

The production of liquid biofuels is rapidly increasing, as governments are setting targets to enlarge the share of biofuels in the energy matrix for the purposes of climate change mitigation, improving energy security, and fostering rural development. It was estimated that the production of fuel ethanol was nearly 86 billion liters (BL) in 2011 (it was 39.2 BL in 2006 and 17 BL in 2000), whereas in the same year the production of biodiesel was estimated at 21.4 BL (6.5 BL in 2006 and only 0.8 BL in 2000).[1] The use of fuel bioethanol in 2011 was estimated to be the equivalent of 3% (energy basis) of the gasoline consumption, considering the consumption of light distillates to be approximately 1640 BL in the same year (mainly motor gasoline).[2] World production and consumption of fuel ethanol is dominated by the United States and Brazil, with more than 75 BL of 86 BL produced in 2011 (approximately 88% of the world production). The United States is by far the main producer (54.2 BL in 2011).[1]

Despite this long-term experience, since 2008–2009 Brazil has had difficulties increasing the production of fuel ethanol (in fact, in 2011 there was a reduction of 18% compared to 2010). In spite of this, Brazil still has the best results in terms of production costs, energy balance, and avoided greenhouse gas (GHG) emissions. Except in 2010–2011, when the United States was a net exporter (surpassing 2 BL in both years), Brazil has been the main exporter of fuel ethanol—in 2011, it exported approximately 2 BL but surpassed 5 BL in 2008—and is one of the few countries capable of being a large exporter in the short to mid term.[3]

Based on recent trends (2000–2009), it was predicted that the production of ethanol should reach 64 BL in 2019, with 52.4 BL for the domestic fuel market, almost 10 BL for export, and less than 2 BL for other uses.[4] However, considering the drawbacks of ethanol production in recent years, but assuming a market recovery after 2012 (22.9 BL was produced in 2011), a production of 50 BL is predicted for 2019 (roughly 53.7 BL in 2020). The International Energy Agency predicts a world market of approximately 200 BL of fuel ethanol in 2020, which highlights the importance of Brazil in this context.[5]

However, the recovery of ethanol production in Brazil poses many challenges, as there are new players in the business, the expansion will mostly occur in less traditional areas, improving sustainability is mandatory and the industry needs to move toward production diversification and the development of second-generation biofuels. The

*Correspondence to: awalter@fem.unicamp.br

Advances in Bioenergy: The Sustainability Challenge, First Edition. Edited by Peter D. Lund, John Byrne, Göran Berndes and Iacovos A. Vasalos.

Originally published in: *WIREs Energy Environ* 2014, 3: 373–394 doi: 10.1002/wene.87

purpose of this paper is to analyze the developments so far achieved and the challenges for keeping the competitiveness of this industry. In the following section, a brief history of ethanol production and the profile of this industry in Brazil are presented. The main developments of agricultural and industrial subsectors and their challenges are presented in the subsequent sections. A further section is devoted to assessing what is already known and the required actions regarding selected sustainability aspects, such as socioeconomic impacts, land use, GHG emissions, impacts on water resources and biodiversity. Finally, the conclusions are presented.

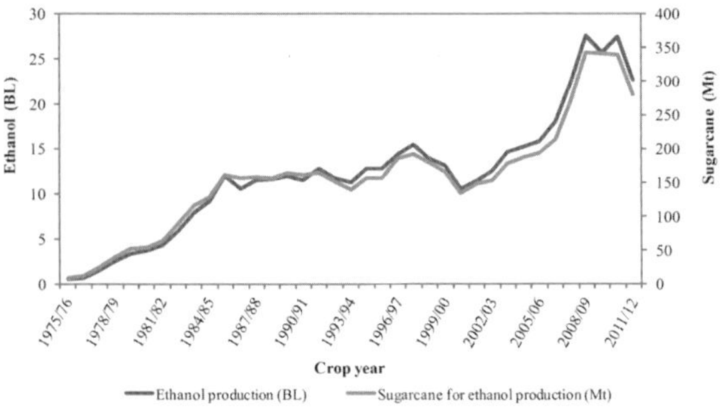

Figure 24.1 Ethanol and sugarcane production (just for ethanol) in Brazil from 1975 to 2011. Created using data from Ref 10.

ETHANOL PRODUCTION IN BRAZIL

Historical Perspective

In the early 20th century, there were several attempts by the Brazilian sugar industry to produce fuel ethanol due to the surplus of sugar and to the heavy burden of gasoline imports.[6] In 1931, a 5% blending of ethanol in all imported gasoline was mandated and as consequence, in only 4 years the production increased from almost zero to 51 million liters[7] in 1937.[6–8] The blending amount varied during World War II, after which low oil prices diminished the interest in fuel ethanol, which continued to be blended with the gasoline on an as-available basis.

The impacts of the 1973 oil shock on the balance of payments and on inflation motivated the launch in November 1975 of the National Alcohol Program (Programa Nacional do Álcool—Proalcool). The main events in ethanol production in Brazil since then are presented below.

- The first target set by the government was the substitution of 20% of the gasoline consumed in the country (this target was reached in early 1980s).[6] Production targets were 3.0 BL of anhydrous ethanol in 1980 and 10.7 BL in 1985, which were easily met. The technology of ethanol engines had been developed by Brazilian research centers due to efforts begun in the 1950s.
- In 1979, as a result of the second oil shock, the government pressured the auto industry to produce on a commercial basis vehicles able to run on neat ethanol; shortly after the introduction to the market of the first model made to run on neat hydrated ethanol (by FIAT), the remaining three other automakers operating in the country at that time closely followed the move.[6,9] At the same time, the producers received support for increasing the production of hydrated ethanol.
- In 1985, the oil prices came down causing an increase in the subsidies burden and initiating the process of stagnation of the ethanol program. During the 1990s, partly due to less support from the government, the ethanol market faced difficulties. A shortage of ethanol in 1989/1990 led to a strong drop in sales of neat ethanol cars, and this significantly impacted the consumption of hydrated ethanol during the decade.
- The deregulation of the ethanol market began in 1991, with the government slowly relinquishing control, and the deregulation process was completed in 1999.[10] Currently, the government control of the fuel ethanol market is limited to setting the ethanol blending rate (in the 18–25% range) based on an appraisal of the supply/demand balance. With the deregulation process, the Proalcool program, as initially conceived, was terminated.
- Since 2003, flex-fuel vehicles (FFVs) have been the main driving force of the domestic consumption. In Brazil, FFVs can run with any fuel mix between gasohol (18–25% ethanol, volume basis) and pure hydrated ethanol (100%). Gasohol is a blend of anhydrous ethanol and gasoline, and in Brazil all gasoline is in fact a blend.

Figure 24.1 shows the evolution of ethanol production in Brazil (hydrated and anhydrous) as well as of sugarcane production (only for ethanol; estimated on average figures of the sucrose content and on the production of sugar and ethanol) from 1975 to 2011. A deep reduction in hydrated ethanol production in the second half of the 1990s can be seen, due to the reduced fleet of ethanol-fueled vehicles. In contrast, the growth rates of ethanol production from 2000 to 2008 were even higher than the rates verified in the early years of the ethanol program; the hydrated ethanol market consolidated primarily because of the success of FFVs. Currently, all car manufacturers in Brazil have at least one FFV model and there are more than 100 models available; FFVs represent more than 90% of the new cars sold. Figure 24.2 shows the evolution of registered new light-duty vehicles in Brazil from 1975 to 2011, according to the fuel option. It is clear that the sales of neat ethanol

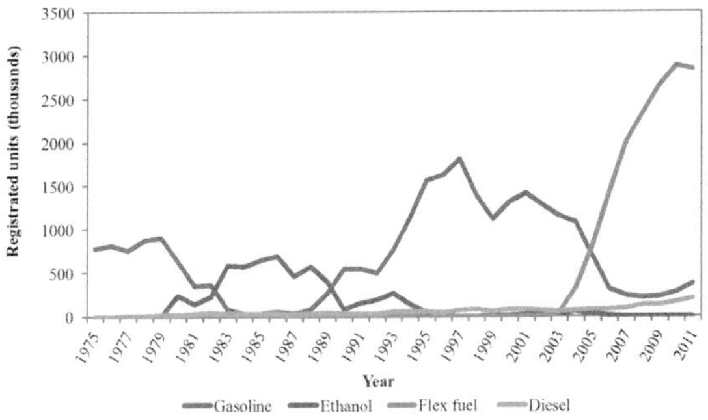

Figure 24.2 Registration of new light-duty vehicles by fuel type—two-wheel vehicles not included—1975/2011. Created using data from Ref 12.

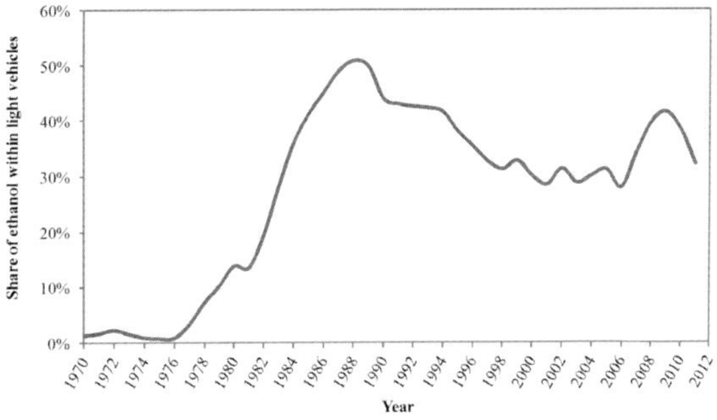

Figure 24.3 Share of fuel ethanol in light-duty vehicles in Brazil. Diesel oil is not consumed in light-duty vehicles in Brazil. All natural gas used in the transport sector is consumed in light-duty vehicles. Created using data from Ref 13.

cars were deeply reduced after the shortage of fuel ethanol in late 1980s, and they were almost nil during the 1990s. The great success of FFVs after their launch can also be observed in Figure 24.2.

However, since 2008, the production of ethanol has faced difficulties and it was drastically reduced in 2011 (see Figure 24.1). The reasons are various, and so far it is not possible to evaluate the impact of a single factor, but there are several important factors:

- The financial crisis after 2008 negatively impacted the traditional entrepreneurs of the ethanol sector that used to finance their activities with short-term loans. Owing to financial constraints, fewer investments were made along the supply chain and sugarcane yields declined.
- As a consequence of the poor financial situation of traditional producers, new players in the sector bought existing assets, who then postponed investments on new industrial plants and also on enlarging the planted area.
- Adverse weather conditions also negatively impacted sugarcane production, first with unusual rains during the harvest season of 2009–2010 and later with a long drought (2010–2011).
- The high sugar prices in the international market (from the second half of 2009 until July 2011) motivated a shift from ethanol to sugar production, further impacting the ethanol supply.
- Simultaneously, since 2008, gasoline prices to the consumers have been nearly constant (i.e., controlled to avoid inflation), reducing the competitiveness of fuel ethanol and the appeal of ethanol production.

Altogether, the supply of ethanol did not grow according to the demand and thus ethanol prices rose. On average, the price ratio between hydrated ethanol and gasoline to the final consumers (volume basis; in the city of São Paulo) that was 53% in 2007 rose to 71% in 2011. In a vicious circle, the investments for increasing sugarcane supply and producing more ethanol were reduced. Only recently (from 2011 onward), some actions were taken, such as financial credits for the sugarcane suppliers, but the recovery of production will take some years as the production of sugarcane is still inadequate.

Since 2009, the share of fuel ethanol in the energy consumption of light-duty vehicles has been reduced, as seen in Figure 24.3. The maximum share was in the late 1980s, just before the ethanol shortage crisis in the early 1990s, and the result in 2011 was almost the worst result in the last 20 years.

The negative impacts have also been noticed in the fuel ethanol trade as Brazil lost its leading position as an exporter in 2010. The exports were primarily destined for the United States (a large share via the Caribbean) and the European Union, and they reached an all-time high in 2008 (more than 5 BL); this was primarily supported by high international crude oil prices despite European Union and United States tariffs.[14] In 2010, the United States was the main exporter to Europe (with a large share of ethanol exported as gasoline blends) and this was not only because of the production difficulties in Brazil but also due to the United States surplus and to the financial incentives given in the United States to the producers. Recent ethanol trade figures are very peculiar: in 2011 Brazil imported approximately 1.2 BL (mostly from the United States), due to its lower production and to the lower prices in United States, but exported in the same year almost 2.0 BL, also mostly to the United States. Fuel suppliers in the United States needed ethanol that is classified as advanced from the point of view of the GHG emissions (at least 50% avoided GHG emissions regarding gasoline), and domestic

Table 24.1 Production and Number of Mills in Brazil, in 2010–2011[1]

Production and Number of Mills	Center-South	North-Northeast	Brazil
Sugarcane crushed (Mt)	560.56	63.27	623.83
Ethanol produced (BL)	25.62	1.99	27.61
Sugar produced (Mt)	33.57	4.61	38.18
Number of mills	349	86	435

[1]Created using data from Ref 15.

production (from corn) is not classified as such. The market for advanced biofuels was defined as 7.6 BL in 2012 and is predicted to grow to 79.5 BL in 2022.

Sugarcane Sector

The sugarcane industry in Brazil is divided into two regions based on the harvesting periods: north-northeast, where the cane harvest extends from September to March, and the center-south, which harvests cane from April to November. The center-south region is the most important and was responsible for almost 90% of the cane crushed in the 2010–2011 season; in addition, their processing units are significantly larger than the ones in the north-northeast region. Table 24.1 shows data from the industrial units under operation in the harvest season 2010–2011. The geographic location of the mills is indicated in Figure 24.4.

São Paulo, in southeast Brazil, is by far the main producer of sugarcane and ethanol. In 2011, there were 185 mills under operation that crushed 361.7 million tonnes (Mt) of sugarcane (58% of the total), produced 15.5 BL of ethanol (56%) and 23.5 Mt of sugar (62%).

In early 2011, 40 new industrial projects were identified, varying from being almost ready for operation to units in the design stage. All these new mills, except one, would be installed in the center-south region, with half of them predicted to be built in the center region (CW in Figures 24.4 and 24.5). Figure 24.5 shows the distribution of new areas cropped with sugarcane in the center-south region from 2003 to 2012 (the accumulated new area in the period was 4.85 million ha (Mha), to be compared with 5.4 Mha cropped with sugarcane in 2003); 96% of the growth of sugarcane production between 2003 and 2011 was in the center-south region, and for this reason the tiny amount of growth in the northeast is not shown in Figure 24.5.

According to the production model, there are three different types of sugarcane mills existing in Brazil[18,19]:

- A sugar mill with an annexed ethanol distillery, where both sugar and ethanol are produced in a fully integrated process. The main advantages are the flexibility in production (more sugar or ethanol depending on the market demands), the gains in economies of scale in the common systems (cane preparation, juice extraction, utilities), and the synergisms with the integrated operation of the two facilities. In 2011, there were 253 mills with annexed distilleries in Brazil.
- An autonomous distillery, which only produces ethanol. Many of the new plants (green field) being built in Brazil since 2004–2005 are of this type, and it also predominated in the second phase of the ethanol program, from 1979 to 1985. In 2011, there were 168 mills with autonomous distilleries.
- A sugar mill only produces sugar. They are less important due to the small number (14 in 2011) and to the reduced production capacity. Nevertheless, this is the model prevailing in the sugarcane mills throughout the world.

In the 2010–2011 harvest, the average output of the mills in Brazil, center-south and north-northeast were 1.4, 1.6, and 0.7 Mt of cane crushed per year, respectively; the average output of the mills in the state of São Paulo was 2 Mt $year^{-1}$. Of the 52 new units built in the 2006–2010 period, based on the projects analyzed by National Bank for Economic and Social Development (BNDES), the majority (62%) had a crushing capacity in the 1.5–2.9 Mt $year^{-1}$ range; mills larger than 3 Mt $year^{-1}$ accounted for 15% of the new projects.[20]

Electricity is rapidly becoming an important third product of the mills, and the projections are that they will tend even more to energy (i.e., ethanol and electricity); the expectation is that the proportion of electricity, ethanol, and sugar in the average gross revenues of the plants will evolve from 2%, 42%, and 56%, respectively, in the 2006–2007 season to 16%, 51%, and 32%, respectively, in the 2015–2016 season. In summary, the energy products of sugarcane in the medium term will amount to two thirds of the average gross revenues of the sector.[21] This tendency regarding surplus electricity production is due to a better regulatory environment and the vision of the new investors in the sugarcane sector. Specific regional policies have had a small impact; rather, the costs of interconnecting to the grid have been the primary determinant for the feasibility of such projects.

Despite the difficulties since 2008, it is clear that the sugarcane sector has been reorganized, motivated by the entrance of new stakeholders, many from abroad, including several oil companies.[22] One consequence of that is the increase in the size of the groups. The share of the five largest groups in the total cane production was 5% in 2005, it increased to 27% in 2010, and is expected to reach 40%

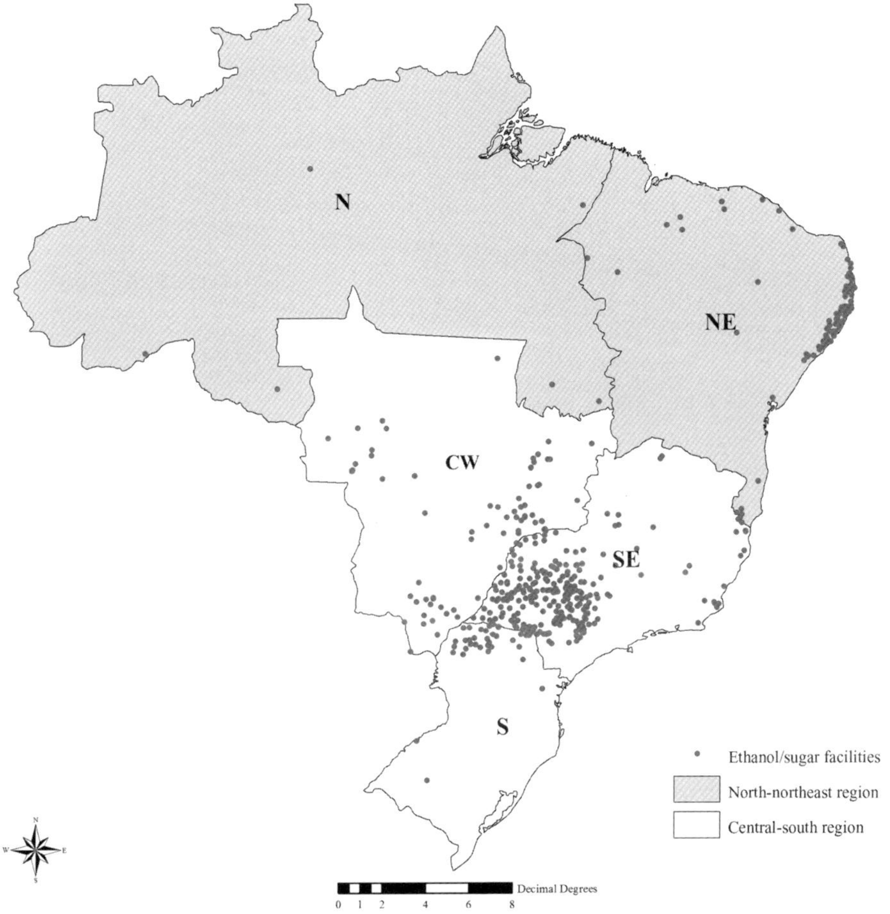

Figure 24.4 Location of sugarcane processing units in Brazil. Created using data from Ref 16.

in 2015. In 2007, there were 22 plants with a total milling capacity of 36 Mt of cane (7% of the total cane) controlled by foreign investors, and this figure is expected to increase to 31 plants with a milling capacity of 83 Mt of cane (12% of total cane) by 2012.[23]

AGRICULTURAL ISSUES

Sugarcane Productivity and Breeding Programs

Over the past 35 years, the average sugarcane yield has increased approximately 68%, from 47 t ha^{-1} in 1975 to 79 t ha^{-1} in 2010; however, owing to the reasons previously mentioned, the average yield dropped to 68 t ha^{-1} in 2011.[24] Technological advances, which include development of new varieties, genetic improvement, and best management practices played an important role in the evolution of the sugarcane crop. The yield increase was approximately 0.91 t ha^{-1} year^{-1}.[25] Figure 24.6 shows the evolution of sugarcane yields and of the total harvested area from 1975 to 2011.

The development of high-quality, high-yielding cultivars was critical for the results achieved with ethanol production in Brazil. Breeding success depends on using the genetic diversity conserved in germplasm collections. A new paradigm is emerging with the integration of biotechnology and genomic sciences for the conservation and use of genetic resources.[26]

Sugarcane varieties need to be periodically replaced as they decay after many years of cultivation. The breeding program that was established intends to adapt sugarcane cultivars for agronomic and industrial interests and to allocate those varieties to the right environment.[27] More than 550 varieties of sugarcane have been developed in the past

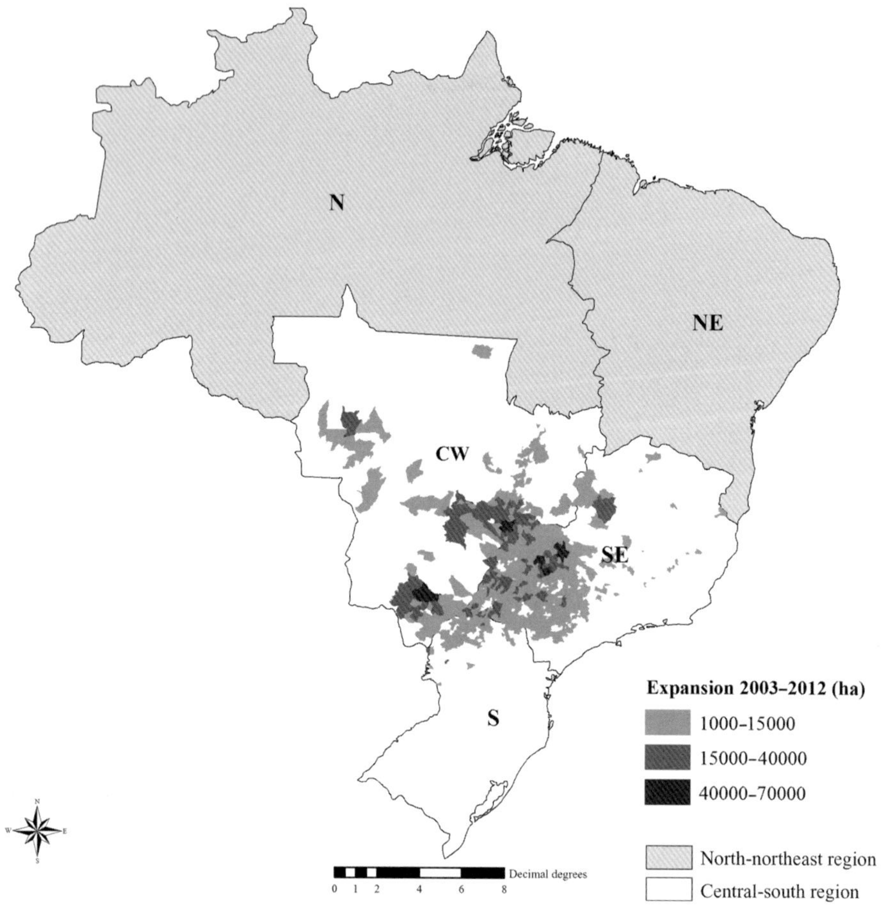

Figure 24.5 Expansion of sugarcane in center-south Brazil from 2003–2012. Created using data from Ref 17.

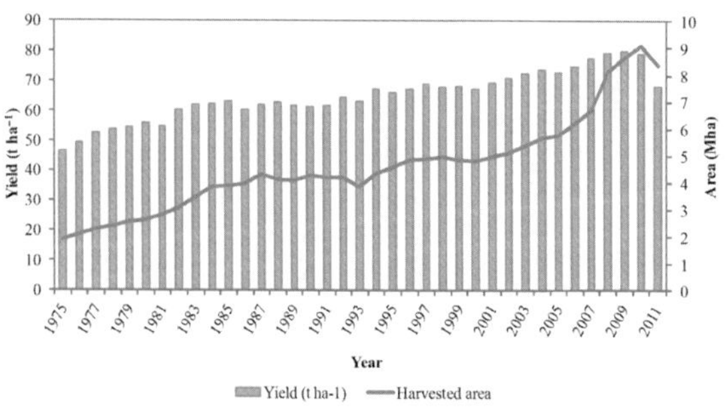

Figure 24.6 Sugarcane yield (t ha^{-1}) and harvested area (Mha) in Brazil in the 1975–2011 period. Created using data from Ref 24.

80 years, mostly after 1970. At present, there are four Brazilian breeding programs:

- IAC: The Campinas Agronomic Institute (publicly funded) has had a breeding program since 1930.
- CTC: The Sugarcane Technology Centre, formerly Copersucar Technology Centre, started its program in 1969. It is a private company that maintains the world's largest and most complete germplasm bank, consisting of more than 5000 clones of commercial and wild species of sugarcane.
- RIDESA: The Inter University Network of Sugarcane Sector Development, formerly Planalsucar (National Sugarcane Improvement Program), started in 1971 by a government decision. It was initially composed of seven Federal Universities; later, two more universities joined the group. It is the largest genetic improvement program in Brazil.
- CANAVIALIS: It is a private company established in 2003 that has developed a large breeding program dedicated to genetically modified crops.

In the Brazilian southeast region, irrigation in commercial sugarcane fields is generally not used, contributing to low production costs. However, new expansion areas in the central region, predominantly in drier areas with inadequate rainfall, have been used for sugarcane cropping. Therefore, drought tolerance is seen as an increasingly important trait for new sugarcane varieties.[28] Furthermore, sugarcane breeding programs also aim to develop pest and disease resistance and varieties better adapted to the increasingly common mechanical harvest.

The Brazilian sugarcane genome sequence has been in progress since 2000. At the University of São Paulo, a group of scientists linked to FAPESP's (São Paulo Research Foundation) Program for Research on Bioenergy obtained its first reads with a pyrosequencer—a cutting-edge technology used for genetic sequencing. The objective of this experiment is to identify the promoter regions and variants that control gene expression and to recognize patterns of genetic diversity to support the breeding programs of sugarcane.[29]

Regarding genetically modified (GM) sugarcane, the National Biosafety Technical Commission has approved field trials in Brazil, incorporating traits such as increased yield, drought tolerance, insect resistance, and herbicide tolerance. Although there are no sugarcane GM varieties in commercial use today, their introduction is expected to take place in the second half of this decade.[30]

Agricultural Management and Fertilizer Use

In Brazil, sugarcane crops use lower levels of fertilizers than in other countries. The Brazilian technical recommendation for nitrogen (N) application in planted cane is approximately 30 kg ha^{-1} in the furrow at planting with a supplementation of 30–60 kg ha^{-1} after 30–60 days[31], whereas for the ratoon cycle this level increases to 90–120 kg ha^{-1}.[32] For instance, the Australian N application levels for plant and ratoon are 30% and 54% higher, respectively, than in Brazil.[33] Intrinsic aspects, such as the lack of response of the plant cycle of sugarcane and the cost, are responsible for the application of less N in Brazil.[34] Moreover, overapplication of N can also decrease the sugar content of the cane and substantially increase N leaching from the root zone.[35–37] Furthermore, an important specific factor in Brazil's sugarcane crops is the recycling of nutrients by the application of two industrial residues, vinasse and filter cake. Vinasse, or stillage, a by-product of ethanol production, is now managed as a nutrient source (rather than residue), and its application has been optimized within the topographic, soil, and environmental control limits; filter cake is the remaining product of sugar and ethanol production that contains large amounts of nutrients and is filtered out of juice in the sedimentation process.[38]

Considering the relatively large amount of N that is exported with the stalks at harvest (80–100 kg N ha^{-1}), a depletion in soil N would be expected in the long term, particularly in regions where N fertilizer application is low.[39] This depletion has not been common, however, which has motivated research on other inputs of N into the sugarcane production system, mainly through biological N fixation.[40] Using natural abundance techniques, Boddey et al.[41] concluded that 60% of the N absorbed by a crop without mineral N fertilizer application was provided by biological fixation, mainly from endophytic and soil associative diazotrophic bacteria. However, more research is needed to understand the role of biological fixation in sugarcane and the role of N cycling in soil organic matter.

Challenges

Historically, sugarcane has been burnt prior to harvest to reduce the costs of harvesting and cane hauling, whether harvested by hand or by machine. However, in the process of burning, carbon dioxide and other GHG along with soot or ash are released. Several studies have identified a strong correlation between particle emissions from sugarcane burning and hospital admissions for asthma and other respiratory problems in children and the elderly and hypertension.[42–44]

Owing to environmental, agronomic, and economical reasons, the manual harvest of sugarcane with burning (Figure 24.7(a)) has been gradually replaced by mechanical harvest with maintenance of the dry leaves and tops (straw) on the field (Figure 24.7(b)) in a system called "green cane management." In the state of São Paulo, preharvest burning was expected to cease in all areas suitable for mechanical harvest by 2021. However, a voluntary sugarcane industry protocol, the "Environmental Protocol," has set 2014 as a target year for phasing out preharvest burning in those areas (Figure 24.8). Because the slope of the terrain is one of the limiting factors for harvesting sugarcane mechanically, different targets for nonmechanisable areas, that is, with slopes of 12% and higher, have been set for both the São Paulo State Law and the Environmental Protocol. The green management of sugarcane includes the deposition of large amounts of plant litter on the soil (Figure 24.9) after each harvest, ranging from 10 to 20 t of dry matter per hectare. This crop residue left on the soil surface can cause changes in shoot emergence and growth, soil nitrogen dynamics, soil erosion, soil water content and temperature, soil aggregate stability, and soil carbon sequestration.[45–51]

Therefore, the large areas currently being converted to mechanical harvesting in Brazil will demand significant adjustments in agronomic management, ranging from fertilizer application to cultivar selection. With increased mechanization, one of the potential problems is soil compaction due to the increased traffic of heavy machinery on the

Figure 24.7 Burnt (a) and unburnt (b) sugarcane areas with similar topography, soils, and climate conditions.

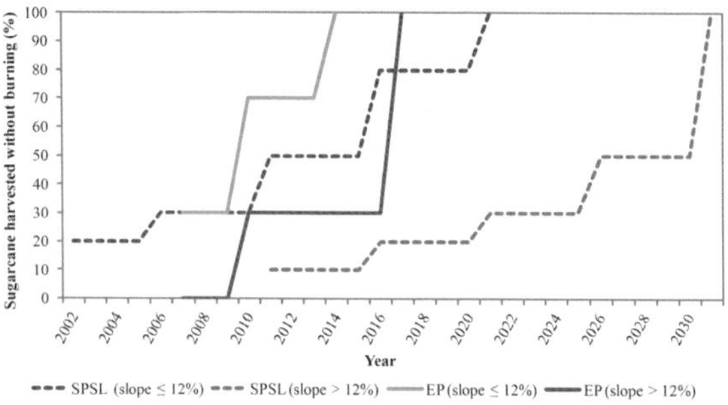

Figure 24.8 Schedule for phasing our preharvest burning in the São Paulo state. SPSL (São Paulo State Law); EP (Environmental Protocol) in areas with slope up to 12% (currently mechanizable) and above 12% (currently nonmechanizable).

field.[52] Compaction can affect soil gas exchanges, soil water dynamics, and ultimately root development and crop productivity. Currently, there are initiatives to decrease traffic from farm machinery.

In this context, the Brazilian Bioethanol Science and Technology Laboratory (CTBE) is implementing a low impact mechanization project, which is based on adopting the no-till farming system in the cane fields, precision agriculture, and the development of controlled traffic machinery. Some of the expected benefits of this project are lower machinery use, a reduction in the cost of the tillage operation, maintenance of straw on the surface, a reduction or elimination of terraces, reduced soil compaction, the possibility of planting during the rainy season, increased productivity, and an increased longevity of ratoons.

Straw represents approximately one third of the total primary energy of sugarcane and its characteristics, similar to those of the widely used bagasse, make it a promising fuel to supplement bagasse for surplus power generation at the mill. In addition, the development of commercially feasible lignocellulosic ethanol, also referred as second-generation ethanol, will generate a demand for sugarcane straw as feedstock. In Brazil, the rapid increase in the implementation of green cane management system will make large amounts of straw available in sugarcane growing regions. Variables such as soil characteristics, climate, local topography, and agricultural practices, as well as sugarcane varieties, will need to be taken into account to define the optimum amount of straw that can be removed from the field in a sustainable manner.

INDUSTRIAL ISSUES

Industrial Conversion Efficiency

During the first phase of the Proalcool program, the priority was increasing ethanol production without due consideration for process efficiencies.[53] However, after 1986 it was necessary to compete with gasoline under adverse conditions (low oil prices and fewer subsidies), and several less efficient mills were pushed out of the market.[54] This led to important improvements in sugar production, and in the 1990s the industry began down the path of becoming the world's primary sugar exporter.

Within the distillery, the largest area of sugar losses was (and still is) the fermentation, where the use of cane juice in the autonomous distilleries was a novelty for most plants. Improvements in juice treatment, beer centrifugation, microbiological control, yeast treatment and recycling, and the use of selected yeasts were important milestones in the improvement of fermentation.[53,55] Even though some

Figure 24.9 Straw accumulated on the soil surface under unburnt sugarcane in the southeastern Brazil.

continuous fermentation systems are in use, the prevailing system is still the Melle–Boinot fed-batch fermentation with yeast recycling. The distillation and dehydration technologies were imported as integrated systems and adapted to the Brazilian conditions of beer quality and fuel ethanol specifications.

The energy sector also experienced considerable improvement with mills evolving from purchasing electricity from the public grid and using wood to supplement bagasse in the boilers to the present situation with all mills being energy self-sufficient. In fact, on average, the electricity currently produced by sugarcane mills is twofold higher than the consumption. Steam conditions have been consistently increasing from the 10–15 bar range prior to 1975 to the current 65–140 bar range. The boiler technology employed at the mills is also progressing, with the first bubbling fluidized bed boiler due to start operation soon.[56] Table 24.2 summarizes the results of this technology development.

The strong modernization process of the mill power section has stood out since the late 1990s. All green field (and several brown field) units are already equipped with high-pressure cogeneration systems (e.g., 68 bar/480°C boilers with condensing extraction steam turbines) and feature better process designs to reduce energy consumption. Such configuration enables the generation of more than 60 kWh t^{-1} of cane of surplus electricity using only bagasse as fuel, whereas the implementation and evolution in cane straw recovery will eventually lead to much higher levels.[58,59]

Table 24.2 Some Figures of the Technological Evolution of Brazilian Sugar Mills and Distilleries since 1975[1]

Technological Evolution	1975	2005
Milling capacity of a 6 milling units of 78" width (tc h^{-1})	5,500	15,000
Sugar extraction efficiency (%)	93	97
Fermentation time (h)	16	8
Fermentation efficiency (%)	82	91
Distillation efficiency (%)	98	99.5
Global distillery efficiency (%)	66	86
Boiler efficiency, lower heating value (%)	66	89
Turbine generator efficiency (%)	50	75

[1]Created using data from Refs 55 and 57.

Other Products from Sugarcane

Sucrose is a relatively low-cost, renewable feedstock, which has attracted growing interest as raw material in the synthesis of different chemicals.[60–62] Several other co-products can be associated with the cane industry using different streams as feedstock, such as vinasse and bagasse. An extensive study published in Brazil in 2005 identified more than 60 technologies in several industries that use sugarcane as a raw material.[62]

In Brazil, molasses is essentially used for ethanol production in distilleries adjacent to sugar mills but it can also be used as animal feedstuff or culture media for bacteria and fungi in other fermentation processes employed in the production of chemical and pharmaceutical products, as well as for the production of baker's yeast. More recently, synthetic biology platforms have been proposed employing genetically modified yeasts, bacteria, and microalgae to produce value-added chemicals and high-energy content hydrocarbon fuels from sugars.[63–65] A pathway based on a fermentation process using genetically engineered yeast strains has been tested in pilot scale in Brazil for some years, and according to the developer production on a commercial scale is expected to start soon.[63]

In addition, in the past few years, the biofuel industry has become interested in alcohol chemistry. In 2010, Braskem inaugurated the world's largest plant producing ethylene from sugarcane ethanol and announced the construction of another unit to produce "green" propylene with a minimum production capacity of 30 kt year^{-1}.[66] In addition, the direct production of biodegradable polymers has also been pursued.[67] In 1995, a polyhydroxybutyrate pilot plant was installed attached to a sugarcane mill in the state of São Paulo.[68] Five years later, the plant was remodeled and restarted to operate with a production capacity of 60 t year^{-1}, and in 2004 a 2000 t year^{-1} commercial plant was installed.[69,70]

The solid and liquid residues of the cane industry can constitute relevant coproducts, too. As mentioned, vinasse and filter cake are recycled through field application as fertilizers whereas energy applications are expected to be in place through vinasse biodigestion.[69] Bagasse boiler ashes are recycled as a fertilizer, but studies have shown an interesting use as a partial substitute for sand in concrete.[71]

From bagasse, technologies have been developed in Brazil for producing commercial products, such as natural fiber to be used in sanitation tubes and swimming pools. It is also a source of cellulose for the paper and cardboard industries and represents an inexpensive source of raw material that can be converted into several chemical feedstocks through biotechnology.[69] One of the significant applications of bagasse has been the production of protein-enriched cattle feed and enzymes.[72]

Despite the numerous alternatives, practically all the bagasse is combusted in cogeneration systems. Some bagasse surplus is sold to other industries (e.g., ceramic and orange juice industries) to serve as energy sources, though with declining trends due to the progressive adoption of more advanced cogeneration systems at the mills. Alternatively, in the future, more advanced technologies, such as thermochemical or biochemical routes, might be employed to produce chemicals and fuels or even more power from lignocellulosic material from cane.[73–75] Owing to its relatively low cost and high availability, sugarcane residues are likely to be one of the most important feedstocks for conversion as the technologies become commercially available.

Second-Generation Ethanol

Among the diversified advanced options for biomass utilization, the biochemical conversion of cane residues into ethanol may become the principal alternative to the conventional production of bagasse-based electricity in the near term. Although significant progress continues to be made to overcome the technical and economic challenges, second-generation biofuels still face major constraints on their commercial deployment.[76] However, the deployment of bagasse-based technologies in Brazil would be favored because the production process could be attached to the sugar/ethanol units already in place, requiring lower investments, infrastructure, logistics, and energy supply. Furthermore, bagasse is available at the industrial facility and thus is free of transportation costs.[77]

In spite of the relatively limited allotted resources (especially when compared to the investments made in Europe and United States), funding bodies, government, and private companies are financing research on second-generation ethanol from sugarcane biomass in Brazil with special efforts toward the integration of first- and second-generation processes and construction of pilot plants.[78] An enzymatic hydrolysis pilot plant has been operated by Petrobras since 2008, and another plant is expected to start operation at CTBE in 2012.[79] An acid hydrolysis process was also tested in the past by CTC in pilot scale, but since 2007, a new process concept has been developed and the construction of a demonstration plant is planned.

Different levels of process integration, with different impacts on the technoeconomic performance, can be conceived. Seabra et al.[80] calculated the technoeconomic performance for the biochemical conversion of sugarcane residues, considering future stand-alone conversion plants adjacent to sugarcane mills in Brazil. Assuming a complementary use of 40% of the cane straw available in the field, the authors estimate that the adjacent conversion plant would enable an additional ethanol production of 33 L t^{-1} of cane, plus an overall electricity surplus of approximately 50 kWh t^{-1} of cane (mill and conversion plant). For the conditions assumed in the study, the minimum selling price of the second-generation ethanol was evaluated at 318 US\$ m^{-3}. Simulation results indicate that the integrated first- and second-generation processes would lead to better economic results than a stand-alone plant, especially when advanced hydrolysis technologies and pentose fermentation are included.[81] For the advanced scenario, the authors project that the integrated process would produce 116 L t^{-1} of cane (considering a complementary use of 50% of the cane straw), for which an internal rate of return of 16.8% would be verified.

Challenges

Ethanol distilleries have experienced a considerable improvement, which has led to an overall sugar stoichiometric conversion efficiency of 86%.[55,57] Juice extraction and fermentation are the two process steps responsible for most of the sugar losses, but there is very little room for improvement, which could possibly lead to 89–90% overall efficiency in the future. Still, the cogeneration and fermentation sections can present better performances in terms of costs and yields.[82]

For the near term, mills are expected to expand the diversification of sugar-based products, including some production of value-added chemicals and hydrocarbon fuels. Synthetic-biology platforms will start to operate on a commercial scale soon, but further developments are still needed to improve process yields while reducing the energy requirements. These are critical aspects in the pursuit of competitive fuel costs.

Electricity exports are projected to increase progressively, with significant contributions from the sugarcane straw collection. Major technological hurdles are not expected with respect to electricity expansion, except for the implementation of a cost-effective route for cane straw collection. However, alternative bagasse uses should also be in place in the future, especially those aimed at the production of second-generation biofuels. Despite the encouraging projections, to play a significant role in the sugarcane sector, the biochemical conversion must be not only a cost-effective alternative but also competitive with the current commercial steam cogeneration systems.[59] Results from Dias et al.[74] show that second-generation ethanol may compete favorably with electricity when sugarcane straw is used and when low-cost enzyme and improved technologies become commercially available (a major challenge today). However, Seabra and Macedo[59] indicate that the profitability of a biochemical conversion plant would not be favorable in comparison to a power plant for the reference values adopted in the study. Given all the associated uncertainties, at this point it is not clear which option will prevail and a single answer will most likely not exist.

SUSTAINABILITY ASPECTS

Production Cost and Competitiveness

The increase in sugarcane ethanol production and the development of its supply chain resulted in a continuous growth of productivity in the agricultural and industrial phases, as seen in Figure 24.6 and Table 24.2, bringing, as a consequence, continuous reductions in the cost of biofuel production, as illustrated in Figure 24.10.

The reduction of production costs with respect to the cumulative production is known as the experience curve approach. The experience curve of Brazilian sugarcane ethanol was presented by Goldemberg et al.[84] and subsequently evaluated in more detail (considering the agricultural and industrial phases) by van den Wall Bake et al.[83] At the time of this evaluation, the break-even point of Brazilian

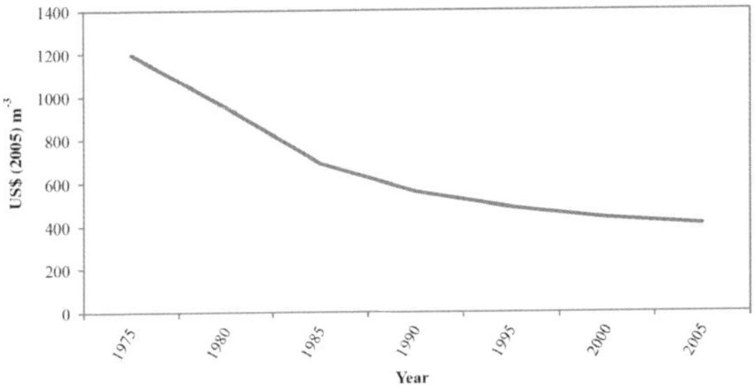

Figure 24.10 Adjusted Brazilian ethanol production costs. Created using data from Ref 83.

ethanol was estimated at US$ 38 per barrel of crude oil. The authors argued that the main driving forces to reduce sugarcane production costs were the continuous increase in sugarcane yields and, in the industrial phase, the increasing scale of the mills.[83]

First-generation sugarcane ethanol has been recognized as the only cost-competitive biofuel, and worldwide Brazil has the cheapest fuel ethanol.[85] For biofuels, raw material represents the bulk of production costs and in the case of ethanol from sugarcane in Brazil the average cost of sugarcane is approximately 60–65% of the final cost of ethanol.[1] Some of the main aspects that contribute to the low cost of Brazilian sugarcane ethanol are the joint production of sugar (currently, 85% of the cane crushed is processed in annexed ethanol distilleries), bringing the benefits of common industrial processes for both products; and the use of bagasse already available at the mill (as a result of crushing sugarcane) as a primary energy source to meet the plant's full energy demand and, more recently, to generate surplus electricity. However, this trend of reducing costs was interrupted between 2008 and 2011 due to lower yields and higher costs of inputs such as fertilizers.[86] The recovery of earlier productivities and the expected cost reduction will depend on retaining investments both in the agricultural (mainly) and the industrial phases. If these goals are realized, positive results could be noticed in just a few years.

Social and Economic Impacts

Sugarcane production grew sevenfold between 1975 and 2010.[24] Even considering all the incentives and subsidies for ethanol production in Brazil—which lasted until 1999—it is estimated that production and use of sugarcane ethanol in the country avoided US$ 100 billion (in 2000 U.S. dollars) of external debt.[87] Since 1975, the sugarcane supply chain has experienced intense development, resulting in positive side effects for the capital goods industry.

The sugar and ethanol sector directly employed 1 million workers in 2009.[88] Sugarcane production in the center-south region especially in state of São Paulo (the most developed state of the country) has been characterized as using unskilled migrant workers from less developed regions of the country, primarily for manual harvesting. This aspect is very contentious, due to the tough working conditions, but the tendency of mechanized harvesting to drastically reduce job opportunities for these workers has also been presented as a negative effect. It will be difficult to allocate so many unskilled workers (it is estimated that approximately 250,000 sugarcane cutters will lose their jobs) to other economic activities.[89]

Some authors emphasize the inadequate conditions of the sugarcane cutters, including labor exploitation;

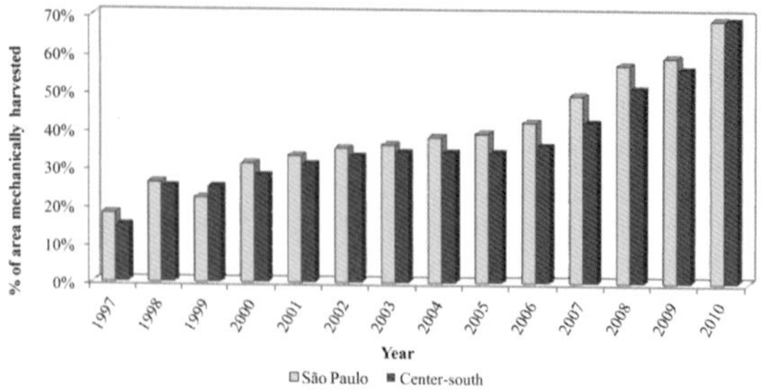

Figure 24.11 Evolution of the mechanical harvest area in São Paulo and the center-south. Created using data from Refs 93 and 94.

however, it is important to emphasize that sugarcane workers earn the second best wages in Brazilian agriculture.[90–92] Even considering the harsh conditions faced by manual sugarcane cutters, some evolution in labor aspects—such as wages and educational level—have been occurring since 1975, which is explained mainly by the increase in mechanical harvesting during recent years, as shown in Figure 24.11.

The increase in mechanically harvested sugarcane has reduced the need for sugarcane cutters, requiring, at the same time, better qualified and schooled workers.[95] There has also been an increase in job formalization in sugarcane production; in 1981, for the entire country, only 37.2% of sugarcane jobs were formal, whereas in 2008 this share was 81.4%. In the state of São Paulo, in 2008, the share of formal jobs reached 94.5% and it is currently even higher.[96]

This change in the profile of agricultural workers is a clear trend in the Brazilian sugar and ethanol industry, explained by the higher cost of manual harvesting and by the environmental protocols requiring the suppression of pre-harvest sugarcane burning.[89]

Land Use, Land Use Change, and Impacts on Food Supply

The debate regarding the advantages of biofuels replacing fossil fuels in terms of reducing GHG emissions went along another dimension from the studies conducted by Searchinger et al.[97] and Fargione et al.,[98] which tried to take into account land use changes (LUC) induced by the expansion of agricultural-based biofuels, causing the so-called indirect land use change (ILUC). Since then, the discussion over environmental and socioeconomic impacts due to the rise of using biofuels have included the need to take into consideration the direct LUC for this purpose, as well as the consequences of possible induced expansion of nonbiofuel crops elsewhere.[99]

The area cropped with sugarcane in 2011 accounted for 1.1% of the total area of Brazil, or 3.3% of the total current agricultural area. Sugarcane dedicated to ethanol production corresponds to approximately half of all sugarcane harvested.[24] Studies based on satellite images covering the center-south region (close to 90% of the sugarcane production) show that between 2000 and 2010 sugarcane expansion occurred mostly over pastures (69.7%), followed by annual crops (25%), citrus (1.3%), forest (0.6%), and sugarcane land under crop rotation (3.4%).[100]

The production of ethanol from sugarcane in Brazil was included in the debate about the indirect effects of land use change (ILUC), and the hypothesis most frequently mentioned is that the expansion of sugarcane could indirectly cause deforestation in the Amazon forest. This contention is based on the fact that sugarcane has displaced pastures and annual crops such as soybean, and the first economic activity after deforestation is livestock production, followed by soy cropping after a few years. The dynamics of deforestation is very complex, and the hypothesis that the agronomic activities displaced by sugarcane are causing deforestation is clearly a simplification.[24, 101, 102]

Global economic models have been used to evaluate and to predict LUC in response to price changes, primarily taking into consideration different crop yields and price elasticities. Typically, general equilibrium models and partial equilibrium models are used for such purposes.[103] These models are more or less simplified and have significant differences in their theoretical frameworks and in their databases, leading to different results. As a consequence, the results reached with different models that consider the same scenarios can be significantly different.[104]

Supported by the use of satellite images and by the development of a partial equilibrium model builtup specifically to evaluate LUC in Brazil (Brazilian land use model), Moreira et al.[105] estimated that the deforestation caused by indirect effects due to the sugarcane expansion from 2005 to 2008 (2.395 Mha) was 7.6%, that is, slightly more than 180,000 ha. This relatively small indirect impact due to the increase in sugarcane for ethanol production in Brazil can be explained by a variety of reasons, such as the expansion over pasture areas, considerable livestock intensification, the large amount of land available, and the continuous improvement of yields of different crops.

According to the Brazilian sugarcane Agroecological Zoning[102], there are 64 Mha suitable for sugarcane production, of which 34 Mha correspond to abandoned areas or degraded pastures. In this sense, it is not expected that the forecasted increase of ethanol production in the country in the coming 10–15 years, which will require no more than 4–5 Mha,[4] will cause considerable indirect effects of deforestation on native vegetation.

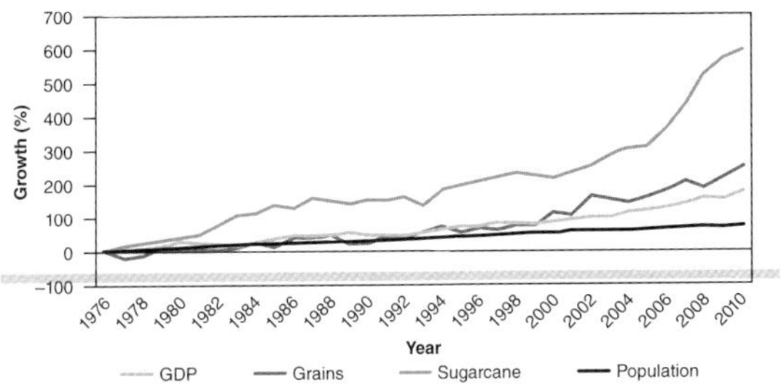

Figure 24.12 Growth of sugarcane, grains, GDP (in real terms), and population in Brazil from 1976 to 2010. Created using data from Refs 24 and 106.

Over the past three decades, the increase in sugarcane production in Brazil has occurred in parallel with a large growth in food production. Figure 24.12 shows the growth of sugarcane production, grain production, gross domestic product (GDP) (in real terms), and population from 1976 to 2010. It can be seen that the increase in grain production is greater than the increases in GDP and population. The large amount of land available for agricultural expansion, as well as the trend in livestock intensification, indicates that sugarcane ethanol will most likely not cause negative impacts on food production in Brazil, even considering the large expansion predicted.[108] In Brazil, the cattle herd is about 200 million head, occupying 200 million hectares—a population density of just one head per hectare; a 10% improvement in livestock intensification would potentially make 20 million hectares available, more than twice the area cropped with sugarcane in 2011.[108,109] In contrast, some authors argue that the expansion of crops dedicated to biofuels may displace other crops dedicated to food production at the local level.[110] These undesirable effects should be avoided as many benefits to society can accrue if bioenergy is produced in a sustainable way.[111]

Energy Balance and GHG Emissions

The environmental advantages of sugarcane ethanol for gasoline substitution and the mitigation of GHG emissions have been known since the first comprehensive analyses of energy balance and GHG emissions in the ethanol life cycle.[112–114] Updated studies have been published subsequently following the evolution of agricultural practices in the sugarcane sector and the scientific advances relating to environmental concerns.[115,116]

In the past decade, a number of studies have estimated the energy use and GHG emissions from sugarcane ethanol in Brazil using life cycle assessment (LCA) methods.[117–121] In general, these studies indicate that ethanol leads to net

Table 24.3 Fossil Energy Use and GHG Emissions in the Anhydrous Ethanol Life Cycle in 2008[1]

Emissions and Credits	Fossil Energy Use (kJ MJ^{-1})	GHG Emission (g CO_2eq MJ^{-1})
Sugarcane farming	88	6.8
Straw burning		3.8
Field emissions[2]		6.7
Agricultural inputs production	40	3.8
Sugarcane transportation	19	1.4
Sugarcane processing	4	2.6
Ethanol transport and distribution[3]	22	1.8
Tailpipe emissions		0.8
Electricity[4] (credits)	− 60	− 3.7
Bagasse[5] (credits)	− 33	− 2.7
Total	80	21.3

[1]Created using data from Ref 75.
[2]Includes emissions from the soil due to fertilizers, residues, and limestone application.
[3]It was considered road transportation using heavy-duty trucks and a total transportation distance (including distribution) of approximately 340 km.
[4]10.7 kWh t^{-1} cane displacing natural gas thermoelectricity generation.
[5]3.3% of surplus bagasse displacing fuel oil fired boilers. It was assumed 10% bagasse losses in handling and storage.

energy savings and emission reductions compared to gasoline but the magnitude of the benefits varies considerably among the analyses. However, direct comparisons are difficult to make on a consistent basis. Triana[122] standardized the results of four different studies on the ethanol energy balance and found recalculated energy ratios varying from 2.63 to 8.84. In addition to the methodological differences, the database used in each study contributes to the discrepancies between the results. Significant differences exist, for instance, between the Brazilian center-south and northeast regions in terms of the technology level employed in the sugarcane sector, which can lead to very different environmental performances.[123]

In a recent study covering the current conditions of the center-south region, ethanol life cycle emissions were estimated at 21.3 g CO_2eq MJ^{-1} (Table 24.3), with the 90% confidence interval lying between 12 g CO_2eq MJ^{-1} and 35 g CO_2eq MJ^{-1}.[75] Despite the relatively high uncertainty, the authors emphasize that there is a clear trend for the next decade, with a significant reduction in GHG emissions.

However, nitrous oxide emissions from soils and LUC emissions are key GHG emission sources and could potentially negate the environmental benefits of substituting biofuels for fossil fuels.[97,98,124] The magnitude of the flux between soil and atmosphere depends largely on soil temperature, soil water content, oxygen availability, N substrate (nitrate and ammonium) availability, and organic carbon. In addition, these regulators are strongly influenced by weather, vegetation, soil properties (bulk density, organic matter, pH and clay content), and land management.[125]

Experiments in Australia comparing burnt and unburnt harvesting systems indicate that the maintenance of sugarcane straw on the field increases soil N_2O. These results have been recently corroborated by field experiments conducted in Brazil, but with an even more marked increase when vinasse is applied. Because the soil–atmosphere exchange of N_2O depends on complex interactions, more regional and site-specific data are needed to evaluate the impact of this source on the overall GHG balance of biofuels.[126–130]

When other land uses are converted to biofuel crops, there can be either GHG emissions or removals, considering biomass burning and decomposition and soil carbon losses and sequestration processes. For the expansion that occurred in the 2002–2008 period, direct LUC emissions have been considered a carbon sink in Brazil, which could possibly be sustained in the future as the growth scenarios for 2020, indicating a progressive use of pasture lands (many with relatively low levels of carbon stocks) in the expansion.[131]

Regarding the indirect LUC effects, previous studies for specific Brazilian conditions indicate that they may be very small.[132] The Californian Air Resources Board[133] initially estimated LUC emissions for ethanol as 46 g CO_2eq MJ^{-1} (including direct and indirect effects), which is now under revision, and significantly lower values have already been proposed. Small contributions from LUC were estimated by the U.S. Environmental Protection Agency[134], which concluded that sugarcane ethanol would be able to mitigate from 59% to 91% of gasoline emissions. As a consequence, ethanol from Brazil was classified as advanced biofuel according to the RFS2 (Renewable Fuel Standard 2).

Relevant impacts may also be expected from the soil carbon stock change due to alterations in cane cultivation management. Studies have indicated the trend for carbon sequestration under green cane management, though it is conditioned by factors such as climate, soil texture, nitrogen fertilizer management, time since the adoption of the unburnt harvest, initial carbon stocks, and the level of soil disturbance during the replanting operation.[135–144] Estimates from 12 sites indicated a mean annual C accumulation rate of 1.5 Mg ha^{-1} $year^{-1}$.[145] Soil carbon can represent a significant sink of carbon dioxide at the national scale in sugarcane production if best practices are adopted in land use and management.[146]

Despite the enormous attention that has been given to the energy balance and GHG emissions, other environmental categories can also play important roles in an ethanol–gasoline trade-off analysis. However, comprehensive LCA studies of cane ethanol are relatively sparse. In addition to global warming, the life cycle impact assessment performed by Ometto et al.[147] covered ozone formation, acidification, nutrient enrichment, ecotoxicity, and human toxicity, but comparisons with counterpart fuels were not part of the study. A direct comparison between gasoline, ethanol, and blends is presented by Luo et al.[148] focused on seven impact categories. The authors concluded that in terms of abiotic depletion, GHG emissions, ozone layer depletion, and photochemical oxidation ethanol fuels are better options than gasoline, whereas gasoline is a better fuel where human toxicity, ecotoxicity, acidification, and eutrophication are concerned.

Impacts on Water Resources

Considering the availability of water resources, Brazil's location is privileged both on the surface and in water tables. The surface fresh water (rivers and lakes) covers 50,000 km^2, whereas the main water table, the Guarani Aquifer, extends for 1.2 million km^2, 70% of which is in center-west and south of Brazil. This aquifer stores approximately 40,000 km^3 of water (which is equivalent to the world's total annual runoff).[29]

For agricultural use, water consumption may vary depending on the leaf area and root density, phenological stage, cycle (plant or ratoon cane), climatic condition, and soil water content.[149] Water deficit is one of the major factors limiting the production of sugarcane, especially in areas where there is a prolonged dry period, such as in the northeast and center-west Brazilian regions.[150]

In Brazil, sugarcane is widely grown under rainfed conditions. In the state of São Paulo, sugarcane growers consider water availability to be the major cause of interannual yield variation and yield differences among soil types. Farmers apply modern management, but irrigation is not normally used because the average annual rainfall exceeds 1300 mm. In this region, the major yield reduction is blamed on dry spells of more than 2 weeks during the hot rainy season, November–March.[151]

In some sugarcane expansion regions in the Cerrado (the Brazilian Savannah) "salvage irrigation" is normally required. In this kind of irrigation, which is normally performed after planting or harvesting time, waste water and vinasse from the ethanol industrial process is normally used.[150] In contrast, in the northeast region, supplementary irrigation with different amounts of water at the most critical development stages is required to mitigate water shortages.

In addition to yield increase, irrigation provides some other benefits, such as crop longevity; however, some environmental problems, such as fresh water resource depletion, may emerge. In some countries, including Bangladesh, India, and Pakistan, the massive expansion of private sector tube well irrigation schemes has led to the rapid depletion of groundwater.[152] In Brazil, the ratio of total water supply to water use (withdrawal of water for irrigation, industry and households) was calculated to be 1% in 1995, and this figure is projected to increase from 3% to 5% in 2075, depending on the irrigation scenario.[153] A ratio of 25% or higher is generally an indication of water stress.

The vinasse produced in ethanol production is rich in organic matter and nutrients, and if released into streams it can increase the biochemical oxygen demand and promote eutrophication of surface waters.[154] However, as already mentioned, vinasse is currently recycled back to sugarcane fields as organic fertilizer.

In the industrial phase, approximately 21 m^3 of water t^{-1} of cane are used during the conversion of cane to ethanol, of which 87% is used in four processes: cane washing, juice evaporation, fermentation cooling, and ethanol condenser cooling. Catchment water in cane processing has been reduced over the past 20 years to approximately 1 m^3 t^{-1} of cane by optimization, water reuse and recycling.[29, 155]

Biodiversity Impacts

LUC is a key driver of changes in biodiversity due to its effects on habitat.[156] The magnitude of the ecological impacts of habitat loss can be exacerbated by the spatial fragmentation of remaining habitat as the surviving species are confronted with a modified environment and reduced area, increased isolation, and novel ecological boundaries.[157]

The production of biofuels can cause direct LUC and, possibly, indirect impacts. In Brazil, the impact of large-scale ethanol (and sugarcane) production on biodiversity is a relatively new issue and this feature is reflected in the limited number of detailed studies so far available. However, a common argument is that the direct impacts of sugarcane cropping on biodiversity have been limited because most plantations are in areas previously used for extensive cattle raising and annual crops. These croplands are also far away from important biomes such as the Amazon rain forest and the Pantanal.[158]

Sugarcane cropping in Brazil occurs primarily in regions where the predominant biome in the past was the Atlantic Forest and tends to expand in areas where the main biome is the Cerrado. Logging and the introduction of crops and cattle raising in the former Atlantic Forest area was disastrous, and currently less than 7% of the original area remains. Worldwide the Atlantic Forest is one of the most threatened tropical biomes with much of the standing forest in small, disturbed, and isolated patches.[159]

In most of the areas of consolidated production of sugarcane, the current forest code requires farmers to keep 20% of the agricultural land as biodiversity reserve, as well as to preserve natural vegetation to protect riparian areas. Unfortunately, law enforcement has not been effective throughout the country despite a reasonable effort (at least in some states). A good example of these efforts is described by Rodrigues et al.[160] who for many years have developed large-scale ecological restoration projects of riparian areas on private lands in the state of São Paulo. The authors describe 32 projects covering almost 530,000 ha, a reasonable proportion in large sugarcane farms. These restoration projects were developed using different methods. In the case of large sugarcane farms, the principal action was planting native tree species because of the low resilience of the target sites, reduced forest cover, and high fragmentation. The authors mention that, in general, large sugarcane farms had high portions of riparian areas occupied by highly mechanized agriculture, abandoned fields, and anthropogenic wet fields created by siltation in water courses. These projects have been developed with a close follow-up of natural vegetation restoration but without a proper study of the impacts on biodiversity and ecosystem services. It is well known that forested riparian zones may regulate energy and material transfer between terrestrial and aquatic ecosystems (e.g., providing shade and organic matter, retaining sediments, and filtering chemicals) and they are also important for maintaining habitat structure, water clarity, and food web structure.[161] In addition, well-preserved riparian zones serve as wildlife habitats and ecological corridors and help to stabilize stream banks.[162]

The impact of sugarcane cropping on fauna biodiversity has also been examined. Chiarello[163] reports a survey of mammals and birds in a forest fragment close to Ribeirão Preto, in the state of São Paulo, where there is a large concentration of sugarcane cropping. Twenty mammal species were confirmed in the area, and 49 bird species were recorded. At that time, few fragments remained and, besides being isolated from each other, they were disturbed by selective logging, hunting, and intrusion of fires. The results demonstrate that forest fragments are refuges for native fauna and the restoration of natural vegetation would be important for preserving species even in areas close to extensive sugarcane cropping. In contrast, Lyra-Jorge et al.[164] report a survey of existing species in fragments at the boarder of the Cerrado biome in the central-north portion of the state of São Paulo surrounded by eucalyptus plantations and sugarcane. The authors report that the remaining natural vegetation still shelters large mammals, including species locally threatened by extinction. However, some carnivore species expected in the region could not be detected. The authors concluded that species more vulnerable to habitat fragmentation may not have been able to adapt.

Challenges

Taking into account the international sustainability agenda of biofuels, the production of ethanol from sugarcane in Brazil has presented reasonable to good results, especially with regard to the concerns previously mentioned in this section. These results were achieved while ethanol competitiveness with gasoline was enhanced, even without subsidies. The existing disadvantages have been reduced due to factors that include strict legislation and adequate control, more consciousness and higher pressure from consumers, and concerns about corporate image.

However, a significant increase in ethanol production will require expansion of sugarcane cropping to new regions and this will bring challenges for sustainability. First, because water availability in the center-west region is less than in the southeast, local resources can be potentially impacted. Second, because the bulk of the expansion will occur in a biome (the Cerrado) with limited knowledge available and where the anthropogenic impacts have so far been minor. Third, because agricultural productivity tends to be low, at least initially, due to the availability of few varieties appropriate for the local conditions and, in general, lower soil quality and poor rain distribution. In addition, new technologies, such as intensive mechanization and new agronomic practices, including straw disposal in the soil and no-tillage practices, will be in use in the future, and their impacts are not well known.

From a socioeconomic perspective, it is necessary to investigate more thoroughly the impacts on the people directly involved in ethanol production, especially at a time that combines profound changes in the governance, intensive mechanization, introduction of new technologies and agricultural practices, and diversification of the production.

Overall, the challenge in the coming years will be reviving production and increasing it in new regions while retaining the targets of reduced costs and improved sustainability.

CONCLUSIONS

Brazil has produced fuel ethanol on a large scale for more than three decades. Despite such experience and the results of the ethanol share in the transport sector, relatively low production costs, good energy balance, and lower GHG emissions, production has been stable during the past few years and even declined in 2011. The challenge of reviving the ethanol production in Brazil is even more complex, as there are new players in the business, the expansion will mostly occur in new production areas, costs need to be reduced, and sustainability must be improved.

In the agricultural stage, the main challenge is the transition to the mechanized harvest without burning and, as a consequence, the disposal of crop litter on the soil surface and/or its use in industry. The large areas currently being converted to mechanical harvesting in Brazil will demand significant adjustments in agronomic management ranging from cultivar selection to fertilizer application. On the other hand, straw is a promising fuel to supplement bagasse for surplus power generation at the mill or even as raw material for lignocellulosic ethanol production.

There is a clear trend for product diversification in the sugarcane industry, including value-added chemicals and high-energy content hydrocarbon fuels. Electricity exports are projected to rise, possibly with significant contributions from sugarcane straw collection. Preliminary results show that second-generation ethanol may favorably compete with electricity when sugarcane straw is used and when low-cost enzyme and improved technologies become commercially available.

From the point of view of sustainability, ethanol from sugarcane in Brazil has delivered reasonable to good results despite persistent difficulties. Generally, the difficulties have been reduced steadily due to factors that vary from more pressure from consumers to stricter regulation and control. A major challenge in the coming years is that a significant increase in ethanol production will occur in new regions, and issues such as water availability, impacts on biodiversity and agricultural productivity need to be faced.

Despite the long experience and the relative success achieved, much effort will be necessary to expand the sugarcane industry in Brazil, intensifying the activities in new production areas, changing the agronomical practices and diversifying the production methods.

REFERENCES

1. Renewable energy policy network for the 21st century. *Renewables 2012 Global Status Report.* Paris: REN21 Secretariat; 2012, 172 p. Available at: http://www.map.ren21.net/GSR/GSR2012.pdf?bcsi_scan_9688b637a46568db=0&bcsi_scan_filename=GSR2012.pdf. (Accessed August 22, 2012).
2. Statistical review of world energy. Available at: bp.com/statisticalreview. (Accessed April 17, 2010).
3. Licht FO. Feedstock use for biofuels – the outlook for 2011. *World Ethanol & Biofuels Report* 2011, 9:1.

4. Empresa de Pesquisa Energética, Ministério de Minas e Energia. *Plano Decenal de Expansão de Energia 2019*. Brasília: Ministério de Minas e Energia, Empresa de Pesquisa Energética; 2010, 77.
5. International Energy Agency. *Technology Roadmap—Biofuels for Transport*. Paris: IEA; 2011, 52.
6. Gatti Junior W. 35 years of Proalcool creation: from ethanol-fuel to flex fuel vehicle. In: *Proceedings XIII SemeAd*, São Paulo, Brazil, September 8, 2010.
7. Moreira JR, Goldemberg J. The alcohol program. *Energy Policy* 1999, 24:229–245.
8. Kovalik W. Ethanol's first century: fuel blending and substitution programs in Europe, Asia, Africa and Latin America. In: *Proceedings XVI International Symposium on Alcohol Fuels*, Rio de Janeiro, Brazil, November 2006.
9. Rosillo-Calle F, Cortez LAB. Towards ProAlcool II—a review of the Brazilian bioethanol programme. *Biomass Bioenergy* 1998, 14:115–124.
10. Licht FO, World ethanol and biodiesel markets—a global industry outlook. London: Informa Agra & Commercial; 2011, 100.
11. Companhia Nacional de Abastecimento (CONAB). Acompanhamento de safra brasileira: cana-de-açúcar, segundo levantamento—agosto/2012. Brasília: Companhia Nacional de Abastecimento; 2012.
12. Brazilian Automotive Industry Association. Brazilian Automotive Industry Yearbook; 2012. Available at: www.anfavea.com.br/anuario.html. (Accessed March 2, 2012).
13. Empresa de Pesquisa Energética, Ministério de Minas e Energia. Relatório Final do Balanço Energético Nacional. Available at: https://ben.epe.gov.br/. (Accessed November 14, 2011) [in Portuguese].
14. Lamers P, Junginger M, Hamelinck C, Faaij A. Developments in international solid biofuel trade– an analysis of volumes, policies, and market factors. *Renew Sustain Energy Rev* 2012, 16:3176–3199.
15. Brazilian sugar and ethanol guide 2011. PROCANA, Ribeirão Preto; 2011, 432.
16. Companhia Nacional de Abastecimento. Mapas temáticos. Available at: http://www.conab.gov.br/conteudos.php?a=1102&t=2. (Accessed February 3, 2010) [in Portuguese].
17. CANASAT. Monitoramento da Cana-de-açúcar via imagens de satélite. Available at: http://www.dsr.inpe.br/laf/canasat/. (Accessed February 1, 2010) [in Portuguese].
18. Leite RCC. Bioetanol combustível: uma oportunidade para o Brazil. Brasília: CGEE; 2009, 536 [in Portuguese].
19. Companhia Nacional de Abastecimento. Perfil do Setor do Açúcar e do Álcool no Brasil. Brasília: Companhia Nacional de Abastecimento; 2010.
20. BNDES and CGEE. Sugarcane-based bioethanol : energy for sustainable development. Rio de Janeiro: BNDES; 2008, 304. Available at: http://www.sugarcanebioethanol.org/en/download/bioetanol.pdf. (Accessed November 9, 2010).
21. União da Indústria de Cana-de-açúcar (UNICA). Outlook for Brazilian biofuels. In: *Canadian Renewable Fuels Summit* (presentation), Canada, December 1–3, 2008.
22. Figliolino A. Visão e tendências do setor de açúcar e álcool no Brasil. In: *Canaplan Annual Meeting*, Ribeirão Preto, Brazil, April 15, 2010.
23. União da Indústria de Cana-de-açúcar (UNICA). An overview of the Brazilian sugarcane industry. In: *Better Sugarcane Initiative General Assembly*, São Paulo, Brazil, November 13, 2008.
24. Instituto Brasileiro de Geografia e Estatística. Municipal agricultural production. Available at: http://www.sidra.ibge.gov.br/. (Accessed May 14, 2012).
25. Gouvêa JRF, Sentelhas PC, Gazzola ST, Santos MC. Climate changes and technological advances: impacts on sugarcane productivity in tropical southern Brazil. *Sci Agric* 2009, 66:593–605.
26. Nass LL, Pereira PAA, Ellis D. Biofuels in Brazil: an overview. *Crop Sci* 2007, 47:2228–2237.
27. Rosseto R, Santiago AD. Programas: melhoramento genético. Agencia de informação EMBRAPA. Available at: http://www.agencia.cnptia.embrapa.br/gestor/cana-de-acucar/arvore/CONTAG01_141_22122006154842.html. (Accessed April 20, 2012).
28. Pires RCM, Arruda FB, Sakai E. Irrigação e drenagem. In: Dinardo-Miranda LL, Vasconcelos ACM, Landell MGA, eds. *Cana-de-açúcar*. Campinas, Brazil: Instituto Agronômico de Campinas; 2008, 631–670.
29. Macedo IC. A energia da cana-de-açúcar—doze estudos sobre a agroindústria da cana-de-açúcar no Brasil e sua sustentabilidade. São Paulo: Editora Berlendis & Vertecchia; 2005, 245.
30. Azevedo JL, Araujo WL. Genetically modified crops: environmental and human health. *Mutation Res* 2003, 544:223–233.
31. Espironelo A, Raij BV, Penatti CP, Cantarella H, Morelli JL, Orlando Filho J, Landell MGA, Rossetto R. Cana-De-Açúcar. In: Van Raij B, Cantarella H, Quaggio JA, Furlani AMC, eds. *Recomendações de adubação e calagem para o Estado de São Paulo*. Campinas, Brazil: Instituto Agronômico de Campinas; 1996, 237–239.
32. Trivelin PCO, Rodrigues JCS, Victoria RL. Utilização por soqueira de cana-deaçúcar de início de safra do nitrogênio da aquamônia-15N e uréia-15N aplicado ao solo em complemento à vinhaça. *Pesqui Agropecu Bras* 1996, 31:89–99.
33. Moreira JR. Water use and impacts due ethanol production in Brazil. In: *Proceedings Linkages between Energy and Water Management for Agriculture in Developing Countries*. Hyderabad, India, January 29–30, 2007. Available at: http://www.iwmi.cgiar.org/EWMA/files/papers/Jose_Moreira.pdf. (Accessed June 23, 2011).
34. Trivelin PCO, Vitti AC, Oliveira MW, Gava GJC, Sarriés GA. Utilização de nitrogênio e produtividade da cana-de-açúcar (cana-planta) em solo arenoso com incorporação de resíduos da cultura. *Rev Bras Ciênc Solo* 2002, 26:637–646.
35. Wiedenfeld RP. Effects of irrigation and N fertilizer application on sugarcane yield and quality. *Field Crops Res* 1995, 43:101–115.

36. Muchow RC, Robertson MJ, Wood AW, Keating BA. Effect of nitrogen on the time-course of sucrose accumulation in sugarcane. *Field Crops Res* 1996, 47:143–153.
37. Verburg K, Keating BA, Probert ME, Bristow KL, Huth NI. Nitrate leaching under sugarcane: interactions between crop yield, soil type and management strategies. In Michalk DL, Pratley JE. eds. In: *Agronomy: Growing Greener Future. Proceedings of the 9th Australian Society of Agronomy Conference*, July 20–23, 1998. Wagga Wagga, Australia: Charles Sturt University; 1998, 717.
38. Ribeiro BT, Lima JM, Guilherme LRG, Julião LGF. Lead absortion and leaching from an inceptisol sample amended with sugarcane vinasse. *Sci Agric* 2010, 67:441–447.
39. Boddey RM, Urquiaga S, Alves BJR, Reis V. Endophytic nitrogen fixation in sugarcane: present knowledge and future applications. *Plant Soil* 2003, 252:139–149.
40. Purchase BS. Nitrogen fixation associated with sugarcane. *Proc South African Sugar Technol Asso* 1980, 54:173–176.
41. Boddey RM, Polidoro JC, Resende AS, Alves BJR, Urquiaga S. Use of the 15 N natural abundance technique for the quantification of the contribution of N_2 fixation to grasses and cereals. *Aust J Plant Physiol* 2001, 28:889–895.
42. Arbex MA, Martins LC, Oliveira RC, Pereira LAA, Arbex FF, Cançado JED, Saldiva PHN, Braga ALF. Outdoor biomass burning and asthma hospital admissions in a sugar cane plantation area in Brazil. *J Epidemiol Community Health* 2007, 61:395–400.
43. Cançado JED, Saldiva PHN, Pereira LAA, Lara LBLS, Artaxo P, Martinelli LA, Arbex MA, Zanobetti A, Braga ALF. The impact of sugar cane—burning emissions on the respiratory system of children and the elderly. *Environ Health Perspect* 2006, 114:725–729.
44. Arbex AA, Saldiva PHN, Pereira LAA, Braga ALF. Impact of outdoor biomass air pollution on hypertension hospital admissions. *J Epidemiol Community Health* 2010, 64:573–579.
45. Alvarez IA, Castro PRC. Crescimento da parte aérea de cana crua e queimada. *Sci Agric* 1999, 56:1069–1079.
46. Gava GJC, Trivelin PCO, Vitti AC, Oliveira MW. Urea and sugarcane straw nitrogen balance in a soil-sugarcane crop system. *Pesqu Agropecu Bras* 2005, 40:689–695.
47. Sparovek G, Schnug E. Temporal erosion-induced soil degradation and yield loss. *Soil Sci Soc Am J* 2001, 65:1479–1486.
48. Tominaga TT, Cássaro FAM, Bacchi OOS, Reichardt K, Oliveira JCM, Timm LC. Variability of soil water content and bulk density in a sugarcane field. *Aust J Soil Res* 2002, 40:605–614.
49. Graham MH, Haynes RJ, Meyer JH. Changes in soil chemistry and aggregate stability induced by fertilizer applications, burning and trash retention on a long-term sugarcane experiment in South Africa. *Eur J Soil Sci* 2002, 53:589–598.
50. Galdos MV, Cerri CC, Cerri CEP. Soil carbon stocks under burned and unburned sugarcane in Brazil. *Geoderma* 2009, 153:347–352.
51. Cerri CC, Galdos MV, Maia SMF, Bernoux M, Feigl BJ, Powlson D, Cerri CEP. Effect of sugarcane harvesting systems on soil carbon stocks in Brazil: an examination of existing data. *Eur J Soil Sci* 2011, 62:23–28. doi:10.1111/j.1365-2389.2010.01315.x
52. Braunack M, Arvidsson J, Hakansson I. Effect of harvest traffic position on soil conditions and sugarcane (*Saccharum officinarum*) response to environmental conditions in Queensland, Australia. *Soil Tillage Res* 2006, 89:103–121. doi: 10.1016/j.still.2005.07.004.
53. Braunbeck O, Macedo IC, Cortez LAB. Modernizing cane production to enhance the biomass base in Brazil. In Silveira, S, ed. *Bioenergy—Realizing the Potential*. Oxford: Elsevier; 2005, 75–94.
54. Annicchino W. A resposta do empresário aos desafios do Proálcool: o ponto de vista dos produtores. In: *Proceedings Simpósio Internacional Copersucar*, São Paulo, Brazil, 1985.
55. Finguerut J. Simultaneous production of sugar and alcohol from sugarcane. In: *Proceedings XXV International Society of Sugar Cane Technologists* (ISSCT), Guatema City, Guatemala, January 30 to February 4; 2005, 315–318.
56. Machado FP. Bubbling fluidized bed at São Martinho should start operating in 2012; 2012. Available at: http://www.jornalcana.com.br/noticia/Bio-Sugar-Magazine/46940+Bubbling-Fluidized-Bed-at-Sao-Martinho-should-start-operating-in-2012. (Accessed May 2, 2012).
57. Olivério JL, Cogeração: Uma nova fonte de renda para as usinas de açúcar e etanol. In: *Proceedings Simpósio Internacional e Mostra de Tecnologia e Energia Canavieira* (SIMTEC), Piracicaba, Brazil, July 4, 2008.
58. Dias MOS, Modesto M, Ensinas AV, Nebra AS, Maciel Filho R, Rossell CEV. Improving bioethanol production from sugarcane: evaluation of distillation, thermal integration and cogeneration systems. *Energy* 2011, 36:3691–3703.
59. Seabra JEA, Macedo IC. Comparative analysis for power generation and ethanol production from sugarcane residual biomass in Brazil. *Energy Policy* 2011, 39:421–428.
60. Boscolo M. Sucrochemistry: synthesis and potentialities for applications of some sucrose chemical derivatives. *Quím Nova* 2003, 26:906–912.
61. Côté GL. Flavorings and other value-added products from sucrose. In: Rastall R, ed. *Novel Enzyme Technology for Food Applications*. Cambridge: Woodhead Publishing; 2007, 243–269.
62. BNDES and CGEE. Sugarcane-based bioethanol: energy for sustainable development. Rio de Janeiro, Brazil: BNDES; 2008, 304. Available at: http://www.sugarcanebioethanol.org/en/download/bioetanol.pdf. (Accessed November 9, 2010).
63. Amyris. Industrial Production. Available at: http://www.amyris.com/Innovation/156/IndustrialProduction. (Accessed March 29, 2012).
64. LS9. Technology Overview. Available at: http://www.ls9.com/technology/technology-overview. (Accessed March 29, 2012).
65. Solazyme. Company/Overview. Available at: http://solazyme.com/company-overview. (Accessed March 29, 2012).

66. Braskem. Chemicals and Technology. Available at: http://www.chemicals-technology.com/projects/braskem-ethanol/. (Accessed March 29, 2012).
67. Koller M, Hesse P, Kutschera C, Bona R, Nascimento J, Ortega S, Agnelli JA, Braunegg G. Sustainable embedding of the bioplastic poly(3-hydroxybutyrate) into the sugarcane industry: principles of a future-oriented technology in Brazil. In Eyerer P, Weller M, Hübner C, eds. *Polymers—Opportunities and Risks II: Sustainability, Product Design and Processing* (The Handbook of Environmental Chemistry). Berlin: Springer; 2009, 81–96.
68. Nonato RV, Mantelatto PE, Rossell CE. Integrated production of biodegradable plastic, sugar and ethanol. *Appl Microbiol Biotechnol* 2001, 57:1–5.
69. Pessoa-Jr A, Roberto IC, Menossi M, Santos RR, Ortega Filho S, Penna TCV. Perspectives on bioenergy and biotechnology in Brazil. *Appl Microbiol Biotechnol* 2005, 121–124:59–70.
70. Biocycle. Biodegradable. Available at: http://www.biocycle.com.br/imprensa_ing_01.htm. (Accessed March 29, 2012).
71. Sales A, Lima SA. Use of Brazilian sugarcane bagasse ash in concrete as sand replacement. *Waste Manage* 2010, 30:1114–1122.
72. Pandey A, Soccol CR, Nigam P, Soccol VT. Biotechnological potential of agro-industrial residues. I: sugarcane bagasse. *Bioresource Technol* 2000, 74:69–80.
73. Walter A, Ensinas AV. Combined production of second-generation biofuels and electricity from sugarcane residues. *Energy* 2010, 35:874–879.
74. Dias MOS, Cunha MP, Jesus CDF, Rocha GJM, Pradella JGC, Rossell CEV, Filho Maciel R, Bonomi A. Second generation ethanol in Brazil: can it compete with electricity production? *Bioresource Technol* 2011, 102:8964–8971.
75. Seabra JEA, Macedo IC, Chum HL, Faroni CE, Sarto CA. Life cycle assessment of Brazilian sugarcane products: GHG emissions and energy use. *Biofuels Bioproducts Biorefining* 2011, 5:519–532.
76. Sims REH, Mabee W, Saddler JN, Taylor M. An overview of second generation biofuel technologies. *Bioresource Technol* 2010, 101:1570–1580.
77. Soccol CR, Vandenberghe LPS, Medeiros ABP, Karp SG, Buckeridge M, Ramos LP, Pitarelo AP, Ferreira-Leitão V, Gottschalk LMF, Ferrara MA, et al. Bioethanol from lignocelluloses: status and perspectives in Brazil. *Bioresource Technol* 2010, 101:4820–4825.
78. Furtado AT, Scandiffio MIG, Cortez LAB. The Brazilian sugarcane innovation system. *Energy Policy* 2011, 39:156–166.
79. Noschang Neto JN. *Etanol de segunda geração*. In: *Ethanol Summit*, São Paulo, Brazil, June 6–7, 2011.
80. Seabra JEA, Tao L, Chum HL, Macedo IC. A techno-economic evaluation of the effects of centralized cellulosic ethanol and co-products refinery options with sugarcane mill clustering. *Biomass Bioenergy* 2010, 34:1065–1078.
81. Dias MOS, Junqueira TL, Cavalett O, Cunha MP, Jesus CDF, Rossell CEV, Maciel Filho R, Bonomi A. Integrated versus stand-alone second generation ethanol production from sugarcane bagasse and trash. *Bioresource Technol* 2012, 103:152–161.
82. Leal MRLV, Walter AC, Seabra JEA. Sugarcane as an energy source. *Biomass Conversion Biorefinery* 2012, 2:1–10.
83. van den Wall Bake JD, Junginger M, Faaij A, Poot T, Walter A. Explaining the experience curve: cost reductions of Brazilian ethanol from sugarcane. *Biomass Bioenergy* 2009, 33:644–658.
84. Goldemberg J, Coelho ST, Nastari PM, Lucon O. Ethanol learning curve – the Brazilian experience. *Biomass Bioenergy* 2004, 26:301–304.
85. International Energy Agency. *World Energy Outlook 2010*. Paris: IEA; 2010.
86. Duff A. Finance for the sugar & ethanol industry. F.O. In: *Licht 8th Annual Sugar & Ethanol Brazil*, São Paulo, Brazil, March 26–28, 2012.
87. International Atomic Energy Agency. *Brazil: A Country Profile on Sustainable Energy Development*. Vienna: IAEA; 2006.
88. Instituto Brasileiro de Geografia e Estatística. Pesquisa Nacional por Amostra de Domicílios (PNAD) 2009. Rio de Janeiro, Brazil, 2010. Available at: http://www.ibge.gov.br/home/estatistica/populacao/trabalhoerendimento/pnad2009/. (Accessed July 20, 2010).
89. Guilhoto JJM, Maistro MCM, Barros AMB, Istake M. Emprego e mecanização na colheita da cana-de-açúcar: diferenças regionais. In: *XLII Congresso da Sociedade Brasileira de Economia e Sociologia Rural*, Cuiabá, Brazil, 2004.
90. Silva MAM, Martins RC. A degradação social do trabalho e da natureza no contexto da monocultura canaviera paulista. *Sociologias* 2010, 24:196–241.
91. Alves F. Por que Morrem os Cortadores de Cana? *Saúde Soc* 2006, 15:90–98.
92. Oliveira FCR, Hoffmann R. Determinantes dos salários dos empregados na lavoura de cana-de-açúcar e em outras atividades agropecuárias no Brasil. *Rev Econ Agríc* 2011, 58:41–56, 2011.
93. Donzelli JL. Cana-de-açúcar no Brasil—pesquisa, desenvolvimento, produção e sustentabilidade. In: *Workshop on Impact of New Technologies on the Sustainability of the Sugarcane/Bioethanol Production Cycle*, Campinas, Brazil, May14–15, 2009. Available at: http://www.bioetanol.org.br/workshop3. (Accessed September 12, 2011).
94. Filho OF. Os desafios da mecanização em cana-de-açúcar. In: *II Simpósio Paulista de Mecanização em cana-de-açúcar* (SPMEC 2011), Jaboticabal, Brazil, February 22, 2011.
95. Moraes MAFD. O mercado de trabalho da agroindústria canavieira: desafios e oportunidades. *Econ Apli* 2007, 11:605–619.
96. Moraes MAFD. Biofuels for social inclusion. In: *Global Sustainable Bioenergy (GSB) – Latin American Convention*, São Paulo, Brazil, March 24–26, 2010.
97. Searchinger T, Heimlich R, Houghton RA, Dong F, Elobeid A, Fabiosa J, Tokgoz S, Hayes D, Yu TH. Use of U.S. croplands for biofuels increases greenhouse gases through

emissions from land-use change. *Science* 2008, 319:1238–1240.

98. Fargione J, Hill J, Tilman D, Polasky S, Hawthorne P. Land clearing and the biofuel carbon debt. *Science* 2008, 319:1235–1238.
99. Rosillo-Callé F, Tschirley J. Food versus fuel: setting the scene. In Rosillo-Calle F, Johnson FX, eds. *Food versus Fuel—An Informed Introduction to Biofuels*. London: Zed Books; 2010.
100. Adami M, Rudorff BFT, Freitas RM, Aguiar DA, Sugawara LM, Mello MP. Remote sensing time series to evaluate direct land use change of recent expanded sugarcane crop in Brazil. *Sustainability* 2012, 4:574–585.
101. Caviglia-Harris JL. Cattle accumulation and land use intensification by households in the Brazilian Amazon. *Agric Resource Econ Rev* 2005, 34:145–162.
102. Manzatto CV, Assad ED, Bacca JFM, Zaroni MJ, Pereira SEM. (org.) *Zoneamento agroecológico da cana-de-açúcar*. Rio de Janeiro, Brazil: Embrapa Solos; 2009, 55.
103. Nassar AM, Harfuch L, Bachion LC, Moreira MR. Biofuels and land-use changes: searching for the top model. *Interface Focus* 2011, 1:187–188.
104. Marelli L, Mulligan D, Edwards R. Critical issues in estimating ILUC emissions. *JRC Scientific and Technical Reports*; 2010. Available at: http://publications.jrc.ec.europa.eu/repository/handle/111111111/22908. (Accessed March 21, 2012).
105. Moreira M, Nassar A, Antoniazzi L, Bachion LC, Harfuch L. Direct and indirect land use change assessment. In Poppe MK, Cortez LAB, eds. *Sustainability of Sugarcane Bioenergy*. Brasília, Brazil: Center for Strategic Studies and Management; 2012, 183–213.
106. Instituto Brasileiro de Geografia e Estatística. Sistema de Contas Nacionais. Available at: http://www.ibge.gov.br/home/estatistica/economia/contasnacionais/2009/default.shtm. (Accessed April 5, 2012).
107. Martha GB, Alves E, Contini E. Land-saving approaches and beef production growth in Brazil. *Agric Syst* 2012, 110:173–177.
109. Cortez LAB, Leal MRLV, Sinkala T. Why biofuels are important. In Rosillo-Calle F, Johnson FX, eds. *Food versus Fuel—An Informed Introduction to Biofuels*. London: Zed Books; 2010.
108. Instituto Brasileiro de Geografia e Estatística. Censo Agropecuário 2006. Available at: http://www.ibge.gov.br/home/estatistica/economia/agropecuaria/censoagro/brasil_2006/Brasil_censoagro2006.pdf. (Accessed Oct 18, 2012).
110. Doornbosch R, Steenblik R. Biofuels: Is the cure worse than the disease? In: *Round Table on Sustainable Development*. OECD. Report SG/SD/RT, Paris; 2007.
111. Chum H, Faaij A, Moreira J, Berndes G, Dhamija P, Dong H, Gabrielle B, Goss Eng A, Lucht W, Mapako M, et al. In: Edenhofer O, Pichs-Madruga R, Sokona Y, Seyboth K, Matschoss P, Kadner S, Zwickel T, Eickemeier P, Hansen G, Schlömer S, et al. eds. IPCC *Special Report on Renewable Energy Sources and Climate Change Mitigation*. Cambridge: Cambridge University Press; 2011.
112. Silva JG, Serra GE, Moreira JR, Gonçalves JC, Goldemberg J. Energy balance for ethyl alcohol production from crops. *Science* 1978, 201:903–906.
113. Nogueira LAH. Análise da utilização de energia na produção de álcool de cana-de-açúcar. PhD Thesis, University of Campinas, Campinas, Brazil; 1987.
114. Macedo IC. The sugar cane agro-industry and its contribution to reducing CO_2 emissions in Brazil. *Biomass Bioenergy* 1992, 3:77–80.
115. Macedo IC. Greenhouse gas emission and energy balances in bio-ethanol production and utilization in Brazil (1996). *Biomass Bioenergy* 1998, 14:77–81.
116. Macedo IC, Leal MRLV, Silva JEAR. Balanço das emissões de gases de efeito estufa na produção e no uso do etanol no Brasil. *Secretaria do Meio Ambiente, Governo de São Paulo:* Produção técnica; April 2004.
117. Pimentel D, Patzek T. Ethanol production: energy and economic issues related to U.S. and Brazilian sugarcane. *Natural Resource Res* 2007, 16:235–242.
118. Wang M, Wu M, Huo H, Liu J. Life-cycle energy use and greenhouse gas emission implications of Brazilian sugarcane ethanol simulated with the GREET model. *Int Sugar J* 2008, 110:527–545.
119. Macedo IC, Seabra JEA, Silva JEAR. Green house gases emissions in the production and use of ethanol from sugarcane in Brazil: the 2005/2006 averages and a prediction for 2020. *Biomass Bioenergy* 2008, 32:582–595.
120. Soares LHB, Alves BJR, Urquiaga S, Boddey RM. Mitigação das emissões de gases efeito estufa pelo uso de etanol da cana-de-açúcar produzido no Brasil. Circular Técnica 27, Embrapa. Seropédica, RJ; 2009.
121. Oliveira MED, Vaughan BE, Rykiel EJ Jr. Ethanol as fuel: energy, carbon dioxide balances, and ecological footprint. *Bioscience* 2005, 55:593–602.
122. Triana CAR. Energetics of Brazilian ethanol: comparison between assessment approaches. *Energy Policy* 2011, 39:4605–4613.
123. Compeán G, Polenske KR. Antagonistic bioenergies: Technological divergence of the ethanol industry in Brazil. *Energy Policy* 2011, 39:6951–6961.
124. Crutzen PJ, Mosier AR, Smith KA, Winiwarter W. N2O release from agro-biofuel production negates global warming reduction by replacing fossil fuels. *Atmos Chem Phys Discus* 2007, 7:11191–11205.
125. Dobbie KE, Smith KA. Impact of different forms of N fertilizer on N_2O emissions from intensive grassland. *Nutr Cycling Agroecosyst* 2003, 67:37–46.
126. Weier Kl, Rolston DE, Thorburn PJ. The potential for N losses via denitrification beneath a greencane trash blanket. In: *20th Proceedings of the Australian Society of Sugar Cane Technologists*, Australia; 1998, 169–175.
127. Denmead OT, MacDonald BCT, Naylor T, Wang W, Salter B, White I, Wilson S, Griffith DWT, Moody PW. Whole-of-season greenhouse gas emissions from Australian sugarcane soils. In: *30th Proceedings of the Australian Society of Sugar Cane Technologists*, Australia; 2008, 105–114.

128. MacDonald BCT, Denmead OT, White I, Naylor T, Salter B, Wilson S, Griffith DWT. Emissions of nitrogen gases from sugarcane soils. In *31th Proceedings of the Australian Society of Sugar Cane Technologists*, Australia; 2009, 85–92.
129. Carmo JBD, Filoso S, Zotelli LC, Sousa Neto ER, Pitombo LM, Vargas VP, Duarte-Neto PJ, Andrade CA, Gava G JC, Rossetto R, et al. Infield greenhouse gas emissions from sugarcane soils in Brazil: effects from synthetic and organic fertilizer application and crop trash accumulation. *GCB Bioenergy* 2013, 5:267–280.
130. Saggar S, Luo J, Giltrap DL, Maddena M. Nitrous oxide emissions from temperate grasslands: processes, measurements, modeling and mitigation. In Sheldon AI, Barnhart EP, eds. *Nitrous Oxide Emissions Research Progress*. New York: Nova Science; 2009, 1–66.
131. Macedo IC, Seabra JEA. Mitigation of GHG emissions using sugarcane bioethanol. In Zuurbier P, van de Vooren J, eds. *Sugarcane Ethanol: Contributions to Climate Change Mitigation and the Environment*. Wageningen: Wageningen Academic; 2008, 95–111.
132. Nassar AM, Harfuch L, Moreira MMR, Bachion LC, Antoniazzi LB. Impacts on land use and GHG emissions from a shock on Brazilian sugarcane ethanol exports to the United States using the Brazilian land use model (BLUM). Report to the U.S. Environmental Protection Agency regarding the Proposed Changes to the Renewable Fuel Standard Program. Institute for International Trade Negotiations (ICONE); September 2009.
133. CARB (Californian Air Resources Board). Detailed California-modified GREET pathways for Brazilian sugarcane ethanol: average Brazilian ethanol, with mechanized harvesting and electricity co-product credit, with electricity co-product credit. Stationary Source Division. California Environmental Protection Agency. Version 2.2; 2009.
134. EPA (Environmental Protection Agency). Renewable Fuel Standard Program (RFS2) regulatory impact analysis. Assessment and Standards Division, Office of Transportation and Air Quality. EPA-420-R-10-006; February 2010.
135. Figueiredo EB, La Scala N. Greenhouse gas balance due to the conversion of sugarcane areas from burned to green harvest in Brazil. *Agric Ecosyst Environ* 2011, 141:77–85.
136. Robertson F. *Sugarcane trash management: consequences for soil carbon and nitrogen—final report of the project Nutrient Cycling in Relation to Trash Management*. CRC Sugar Technical Publication. Townsville: CRC for Sustainable Sugar Production; 2003, 39.
137. Silver WL, Neff J, McGroddy M, Veldkamp E, Keller M, Cosme R. Effects of soil texture on belowground carbon and nutrient storage in a lowland Amazonian Forest ecosystem. *Ecosystems* 2000, 3:193–209.
138. Hao X, Kravchenko AN. Management practice effects on surface soil total carbon: differences along a textural gradient. *Agronomy J* 2007, 99:18–26.
139. Graham MH, Haynes RJ, Meyer JH. Changes in soil chemistry and aggregate stability induced by fertilizer applications, burning and trash retention on a long-term sugarcane experiment in South Africa. *Eur J Soil Sci* 2002, 53:589–598.
140. Wood AW. Management of crop residues following green harvesting of sugarcane in north Queensland. *Soil Tillage Res* 1991, 20:69–85.
141. Vallis I, Parton WJ, Keating BA, Wood AW. Simulation of the effects of trash and N fertilizer management on soil organic matter levels and yields of sugarcane. *Soil Tillage Res* 1996, 38:115–132.
142. Graham MH, Haynes RJ, Zelles LE, Meyer JH. Long-term effects of green cane harvesting versus burning on the size and diversity of the soil microbial community. *Proc South African Sugar Technol Asso* 2001, 75:228–234.
143. Six J, Conant RT, Paul EA, Paustian K. Stabilization mechanisms of soil organic matter: Implications for C-saturation of soils. *Plant Soil* 2002, 241:155–176.
144. Resende AS, Xavier RP, Oliveira OC, Urquiaga S, Alves B Jr, Boddey RM. Long-term effects of pre-harvest burning and nitrogen and vinasse applications on yield of sugar cane and soil carbon and nitrogen stocks on a plantation in Pernambuco, N.E. Brazil. *Plant Soil* 2006, 281:339–351.
145. Cerri CC, Galdos MV, Maia SMF, Bernoux M, Feigl BJ, Powlson D, Cerri CEP. Effect of sugarcane harvesting systems on soil carbon stocks in Brazil: an examination of existing data. *Eur J Soil Sci* 2011, 62:23–28. doi:10.1111/j.1365-2389.2010.01315.x
146. Pacca S, Moreira JR. Historical carbon budget of the brazilian ethanol program. *Energy Policy* 2009, 37(11):4863–4873.
147. Ometto AR, Hauschild MZ, Roma WNL. Lifecycle assessment of fuel ethanol from sugarcane in Brazil. *Int J Life Cycle Assess* 2009, 14:236–247.
148. Luo L, van der Voet E, Huppes G. Life cycle assessment and life cycle costing of bioethanol from sugarcane in Brazil. *Renew Sustain Energy Rev* 2009, 13:1613–1619.
149. Robertson MJ, Donaldson RA. Changes in the components of cane and sucrose yield in response to drying-off before harvest. *Field Crops Res* 1998, 55:201–208.
150. Silva MA, Soares RAB, Landell MGA, Campana MP. Agronomic performance of sugarcane families in response to water stress. *Bragantia* 2008, 67(3):655–661.
151. van den Berg M, Burrough PA, Driessen PM. Uncertainties in the appraisal of water availability and consequences for simulated sugarcane yield potentials in São Paulo state, Brazil. *Agric Ecosyst Environ* 2000, 81:43–55.
152. Shrivastava AK, Shrivastava AK, Solomon S. Sustaining sugarcane productivity under depleting water resources. *Curr Sci* 2011, 101:748–754.
153. Berndes G. Bioenergy and water. The implications of largescale bioenergy production for water use and supply. *Global Environ Change* 2002, 12:253–271.
154. Martinelli LA, Filoso S. Expansion of sugarcane ethanol production in Brazil: environmental and social challenges. *Ecol Appl* 2008, 18:885–898.
155. Agência Nacional de Águas, Federação das Indústrias do Estado de São Paulo, União da Indústria de Cana-de-açúcar, Centro de Tecnologia Canavieira. *Manual de conservação*

e reuso de água na agroindústria sucroenergética. Brasília, Brazil: Agência Nacional de Águas; 2009, 288.
156. Hellmann F, Verburg PH. Impact assessment of the European biofuel directive on land use and biodiversity. *J Environ Manage* 2010, 91:1389–1396.
157. Ewers RM, Didham RK. Confounding factors in the detection of species responses to habitat fragmentation. *Biol Rev Cambridge Phil Soc* 2006, 81:117–142.
158. Smeets E, Junginger M, Faaij A, Walter AC, Dolzan P, *Sustainability of Brazilian bio-ethanol*. Universiteit Utrecht Copernicus Institute, Department of Science, Technology and Society, the Netherlands and University of Campinas, Brazil, 2006, 136.
159. Lira-Jorge PK, Tambosi LR, Ewers RM, Metzger JP. Land-use and land-cover change in Atlantic Forest landscapes. *Forest Ecol Manage* 2012, 278:80–89.
160. Rodrigues RR, Gandolfi S, Nave AG, Aronson J, Barreto TE, Vidal CY, Brancalion PHS. Large-scale ecological restoration of high-diversity tropical forests in SE Brazil. *Forest Ecol Manage* 2011, 261:1605–1613.
161. Pusey BJ, Arthington AH. Importance of the riparian zone to the conservation and management of freshwater fish: a review. *Marine Freshwater Res* 2003, 54:1–16.
162. Salemi LF, Groppo JD, Trevisan R, Moraes JM, Lima WP, Martinelli LA. Riparian vegetation and water yield: a synthesis. *J Hydrol* 2012, 454–455:195–202.
163. Chiarello AG. Conservation value of a native forest fragment in a region of extensive agriculture. *Rev Bras Biol* 2000, 60:237–247.
164. Lyra-Jorge MC, Ciocheti G, Pivello VR. Carnivore mammals in a fragmented landscape in northeast of São Paulo state, Brazil. *Biodivers Conserv* 2008, 17:1573–1580.

25

The Climate Benefit of Swedish Ethanol: Present and Prospective Performance

Pål Börjesson,[1]* Serina Ahlgren[1] and Göran Berndes[2]

[1]Environmental and Energy Systems Studies, Department of Technology and Society, Lund University, Lund, Sweden (E-mail:)
[2]Physical Resource Theory, Department of Energy and Environment, Chalmers University of Technology, Gothenburg, Sweden

INTRODUCTION

There has been considerable debate in recent years concerning the sustainability of ethanol and other biofuels, and it has been questioned whether displacing petrol and diesel with biofuels reduces the emissions of greenhouse gases (GHG) from the transportation sector. The GHG savings potential of biofuels has been debated for several years (selected studies include Refs 1–10), but increasing attention has been given to the issue as governments started to include GHG considerations as part of their policies. The connection between biofuel expansion and land use change (LUC) has caused concern, and studies have pointed to the risk that GHG emissions associated with LUC undermine the climate change mitigation benefits of biofuels (see, e.g., Refs 11–13). For example, the issue of indirect land use changes (iLUC), is currently a political topic in many countries and regions, with governments seeking ways to address the issue in their policies. Notable examples include the processes associated with the EU Renewable Energy Directive[14,15] and the implementation of the low carbon fuel standard in California.[a]

Research on biofuels, GHG balances, and LUC is widespread today, and a conclusion synthesizing the research done so far is that there is no single answer to the question whether biofuels are good or bad from a GHG perspective (for the case of ethanol, see, e.g., Refs 1, 6, 16–19). The answer to the question depends on several critical factors that are often specific to each specific biofuel production system and is to a significant degree influenced by spatial and temporal aspects—including how these are considered in the calculation methodology and in the definition of system boundaries. For instance, critical factors for grain-based ethanol production are (i) type of land used for the cultivation and the alternative land use; (ii) the efficiency in nitrogen fertilization and how the fertilizers are produced; (iii) whether the biofuel plant uses fossil fuels or biomass as process fuel; and (iv) how efficiently by-products are utilized and how their uses are accounted for in the calculations (i.e., by allocation or system expansion).[6] A constructive output from the vast research is therefore the identification of the most critical parameters concerning the GHG performance of biofuels, and solutions how to minimize negative effects and maximize positive effects (see, e.g., Ref 11). This knowledge is essential in the advancing development of national and international standards and certification schemes regarding sustainability criteria of biofuels.

This paper presents an assessment of the GHG performance of current and prospective Swedish ethanol production and outlines a strategy for improving the GHG performance on an expanding ethanol market. Calculations are presented that consider (i) process integration and establishment of biocombines producing multiple products; (ii)

*Correspondence to: pal.borjesson@miljo.lth.se

Advances in Bioenergy: The Sustainability Challenge, First Edition. Edited by Peter D. Lund, John Byrne, Göran Berndes and Iacovos A. Vasalos.

Originally published in: *WIREs Energy Environ* 2012, 1: 395–410 doi: 10.1002/wene.17

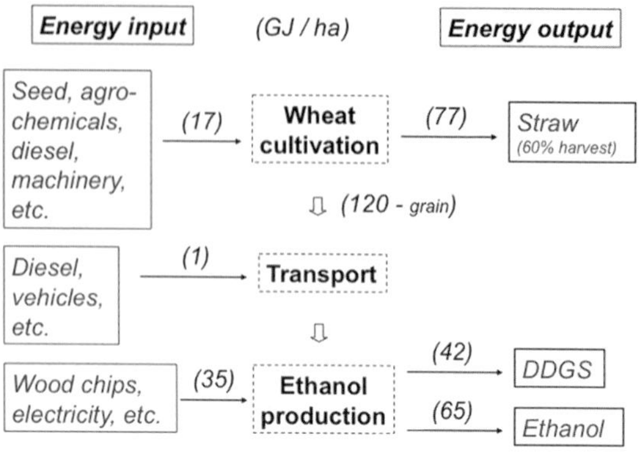

Figure 25.1 Energy flows in current production systems for wheat-based ethanol in Sweden. Straw harvest includes transport to farm-gate, and DDGS and ethanol include energy inputs to plant gate. (Data adapted from Refs 6 and 16.)

the use of a broader resource base including lignocellulosic feedstocks; (iii) and expansion options that minimize GHG emissions from LUC associated with the ethanol production. Based on this, a scenario for ethanol expansion in Sweden is presented that shows how the production could increase substantially while maintaining a very favorable GHG profile. The paper applies a broad multidisciplinary systems approach that combines an assessment of critical issues, such as (i) the technical design of ethanol systems, (ii) the category of feedstock, (iii) the direct land use changes (dLUC), (iv) iLUC, (v) the treatment of coproducts, and (vi) temporal boundaries.

CROP-BASED ETHANOL PRODUCTION IN SWEDEN

The use of biofuels in the Swedish road transport sector amounts to approximately 20 PJ today.[20] This is equivalent to 6.1% of the total vehicle fuel use, and includes ethanol (2.9%), biodiesel from rapeseed (2.4%), and biogas from organic waste material (0.8%). The ethanol is both imported, mainly from Brazil, and domestically produced from grain (mainly wheat) in a plant located in southern Sweden. The total capacity of this plant is equivalent to some 200 million liters (4.2 PJ) of ethanol per year. About 550,000 tonnes (15% water content) of grain, mainly wheat, is used as feedstock and requires approximately 100,000 ha of arable land,[21] which corresponds to about 3.8% of the total arable land in Sweden.[22]

The structure and the energy flows in the current Swedish ethanol production system are shown in Figure 25.1. The total energy output from the system is, on average, 184 GJ per hectare, including ethanol, dried distillers grains and solubles (DDGS), and straw.[6,16] The higher heating value for DDGS and straw is 17.3 and 17.9 GJ per dry tonne, respectively. The grain and straw harvests are 7.5 and 5 tonnes per hectare, respectively (15% water content, average yields in southern Sweden). The straw harvest rates are based on ecological considerations (maintaining the soil productivity and soil carbon content) as well as practical aspects (harvest losses). On average, 60% of the total straw produced is harvested. The total energy input to the wheat ethanol system is around 53 GJ, giving an energy output: input ratio of 3.5 when all energy outputs (straw, ethanol, and DDGS) are included. When only ethanol is considered as the energy output from the system, the corresponding energy balance will be approximately 1.2.

The calculations of energy inputs are based on primary energy inputs; that is, all energy flows are calculated as unconverted and untransformed natural resources.[6,16] The energy input in cultivation includes diesel fuel, commercial fertilizers, seed, pesticides, the manufacture and maintenance of field machinery, and field transport. In the conversion process, the energy inputs to produce heat and steam (forest biomass) as well as electricity are included. The energy inputs required for the transport of feedstock were calculated on the basis of considering both the diesel fuel consumption by the vehicles used and the energy needed for the manufacture and maintenance of these. The ethanol plant is integrated with a biomass-based combined heat and power plant, which delivers electricity, steam, and heat to a district-heating grid. This represents the current standard practice in Sweden where the dominating fuel in district heating plants is solid forest fuel in the form of wood chips from logging residues. The energy input in the ethanol plant consists of heat and steam (77%) and electricity (23%). Almost half of the energy input in the ethanol plant is used for drying the distillers waste.

The GHG performance of the current wheat ethanol system is summarized in Table 25.1. The calculations are based on life cycle assessment (LCA) methodology (see ISO 14044)[23] using different calculation methods to consider coproducts, that is, expansion of system boundaries, allocation based on the product's energy content and allocation based on its economic value. The GHG figures presented in Table 25.1 exclude potential GHG emissions from LUC. The coproduct in the form of DDGS is today used as protein feed in animal production (mainly dairy production) replacing soy meal imported from Brazil, and locally produced barley (1 kg dry DDGS is estimated to be equivalent to 0.6 kg dry soy bean meal and 0.4 kg dry barley).[16] Thus, when system expansion is used, life cycle inventory data for Brazilian soy meal and Swedish barley are adopted from Ref 24, which represent average cultivation systems under current conditions (excluding potential

Table 25.1 Estimated GHG emissions, expressed as g CO_2-equivalents per MJ, of present and prospective Swedish ethanol production based on wheat or sugar beet, excluding land use change[1]

	Cultivation				Total			
System	CO_2-Fossil Fuels	N_2O–Fertilizer Production	N_2O–Biogenic	EtOH Plant	No Allocation	Energy Allocation	Economic Allocation	System Expansion.[2]
Wheat–ethanol and DDGS (excl. straw)	17.6	8.0	14.0	6.2	45.8	27.9	37.0	16.5
Wheat–ethanol and DDGS (incl. straw)	20.0	8.0	11.3	6.2	45.5	15.9	33.2	12.1
Wheat–ethanol and biogas (excl. straw)	17.6	8.0	14.0	7.4	47.0	33.1	35.9	21.8
Sugar beet–ethanol and pulp (excl. tops and leaves)	14.9	3.8	6.7	4.8	30.2	19.6	25.4	15.8
Sugar beet–ethanol and biogas (incl. tops and leaves)	16.0	3.8	4.5	5.6	29.9	21.1	22.8	4.7

[1]Data adapted from Refs 16 and 36.
[2]DDGS replacing soy meal and barley, pulp replacing barley, straw replacing wood chips, and biogas replacing natural gas.

land use changes). The soy meal replacement is responsible for approximately 77% of the GHG credit, whereas the barley replacement is responsible for approximately 23% (see Table 25.1). The domestic market for DDGS as protein feed is estimated to be equivalent to an ethanol production volume of approximately 7 PJ (corresponding to 5% of current use of petrol). This is roughly twice the current ethanol production volume in Sweden.[21]

The harvest of straw for energy purposes, which to some extent already occur today, will improve the GHG performance of wheat ethanol. When system expansion is applied (see Table 25.1), it is assumed that straw replaces forest fuels, which are estimated to be the most realistic alternative fuel in Sweden today.[16] However, if the straw is assumed to replace fuel oil instead of forest fuels, the GHG credit will increase approximately 2.8 times, leading to "negative" GHG emissions, equivalent to roughly–60 g CO_2-eq./MJ ethanol.

An alternative option is to utilize distillers waste (wet distillers grains, WDG) as feedstock for biogas production by integrating ethanol fermentation with anaerobic digestion. Such a shift would lead to less displacement credits since other feed production would not be displaced but would, on the other hand, reduce heat input otherwise used for drying the distillers waste and would also generate additional energy output in the form of biogas. Upgraded and pressurized biogas can then be used as vehicle fuel, and this is a growing market in Sweden today. Plants for producing ethanol and biogas based on grain are currently under development, and the existing ethanol plant already produces a small fraction of biogas from WDG today. The biogas is assumed to replace natural gas in the system expansion presented in Table 25.1. If upgraded biogas replaces petrol instead of natural gas, then the GHG credit will increase by approximately 35%, leading to a GHG performance of about 13 g CO_2-eq./MJ ethanol.

Another option is to use sugar beet as feedstock instead of grain, and the by-product (beet pulp) can also in this case be used as animal feed or for biogas production. Table 25.1 includes all these additional options for ethanol production, which may become commercially viable within a few years. An overall conclusion is that—in the absence of LUC—all these options will reduce the GHG emissions by approximately 75%, or more, when displacing petroleum-based fuels (here assumed to emit 83.8 g CO_2-eq./MJ) and the uses described for the by-product are credited (systems expansion).

Cultivation of the feedstock crop is the dominating GHG contributor in the fuel cycle, and N_2O emissions play a significant role for most present day biofuels.[16,25] Harvest of crop residues, such as straw and tops and leaves from sugar beet, will reduce the risk of biogenic N_2O emissions from the soil due to the increased output of nitrogen from the field. The level of biogenic N_2O emissions from the soil is, however, inherently uncertain, since these levels are influenced by a large number of processes that respond differently to various environmental and soil factors (see, e.g., Refs 26–34). The role of biological N fixation is especially difficult to quantify,[29] and there is also the question of allocation of N_2O emissions as reactive N is recycled with the involvement of several actors and agricultural activities. Illustrative of this, the conclusion stated by Ref 35 that N_2O emissions from biofuels are underestimated by a factor of two to three when quantified using the Intergovernmental Panel on Climate Change (IPCC) tier 1 method—was due to the fact that IPCC tier 1 and Ref 35 estimates used different accounting approaches. Roughly two thirds of agricultural N_2O emissions occur as nitrogen is recycled internally in animal production and in plant residues used as fertilizers. The N_2O emission factors proposed in Ref 35 also make the biofuel feedstock production responsible for the part of the applied nitrogen that is recirculated into other agricultural systems and substituted for other nitrogen input.

LIGNOCELLULOSIC-BASED ETHANOL PRODUCTION SYSTEMS

The possibility to produce the so-called second-generation ethanol (and other biofuels) from lignocellulosic feedstocks creates an opening up for the exploitation of a larger resource base, including forest biomass, and it also provides options for mitigating some of the drawbacks associated with the ethanol feedstocks used at present.

Crop residues, such as straw, are viable feedstocks for ethanol production throughout the world.[37] The environmental performance of straw-based ethanol production, for example, in various biorefinery concepts (i.e., precommercial units), is in general estimated to be favorable compared with other ethanol production systems.[38,39] One important prerequisite is, however, that ecological constraints are considered regarding the maximum level of straw harvest at which a decrease in the long-term productivity of the soil is avoided.[40–43] Fertilizer inputs can compensate for nutrient removal connected to harvest and residue extraction, but maintenance or improvement of soil fertility, structural stability, and water-holding capacity require recirculation of organic matter to the soil.[44–46]

Similarly, sustainable biomass extraction in conventional forestry represents a viable resource base for second-generation ethanol. Postconsumer organic waste and by-products from the agricultural and forestry industries can also be used. The use of biomass that would otherwise be used as landfill, or decompose in wet conditions, can lead to additional benefits through reduced methane emissions.[47]

As the production of lignocellulosic feedstocks commonly requires less fuel, fertilizer, and other inputs, there is also scope for higher GHG savings than when ethanol is produced from conventional crops such as cereals and sugar beet.[16] However, if bioenergy is to provide energy for both transport and for heat and electricity production, a mix of lignocellulosic material and conventional food/feed crops is likely to be used as bioenergy feedstock over the coming decades. Thus, some of the current first-generation ethanol plant is estimated to be gradually redeveloped to coproduce first- and second-generation ethanol by implementing a variety of process integration concepts. In recent years, several pilot and demo-scale, second-generation ethanol plants have been constructed around the world. In Sweden, the demo plant in northern Sweden (Örnsköldsvik) has been in operation since 2005 and has an effective operation time of more than 30,000 h. This is one of the few plants to date that can use wood materials as feedstock. Most of the other plants are designed for the use of agricultural residues, such as corn stover and various types of straw.

The composition of lignocellulosic materials differs from one species to another. However, the main constituents are of the same type: about 50%–60% carbohydrates in the form of cellulose (made up of glucose) and hemicellulose (various pentose and hexose sugars), which can also be fermented to ethanol and some 20%–35% lignin.[48] Agricultural crops and hardwood contain more pentose sugars than does softwood. Lignin is a very complex molecule that contains numerous aromatic compounds of which some have interesting properties, and in a longer perspective lignin could be used for the production of various chemicals. Presently, lignin is used as a biomass fuel in the forest industry. In the context of cellulosic ethanol production, it can be used to meet internal process energy requirements or to produce lignin pellets as an additional fuel output.

An integrated production of ethanol and lignin pellets is estimated to give an overall conversion efficiency of approximately 70%, where the output of ethanol and solid fuel will be almost equal from an energy point of view.[49] Process integration and the production of several energy carriers, such as electricity, district heat, and biogas, can improve the overall energy efficiency and process economics. Different prospective process designs, for wood-based ethanol plants, have an expected overall conversion efficiency at 50%–90%, of which 35%–45% is ethanol.[49,50] Similarly, a Swedish case study of straw-based ethanol production found an overall energy efficiency of 57%–70%, depending on process configuration. The ethanol yield was 30%–48% with the remaining energy output being heat, electricity, and biogas.[51]

The very high conversion efficiency for both wood- and straw-based ethanol systems involves significant amounts of low energy output in the form of excess heat at a relatively low temperature and requiring sizeable heat sinks (e.g., district-heating systems) to be made useful. Since such heat sinks may always not be available, there are no credits for the use of excess heat in the assessment of second-generation ethanol production systems reported below. Thus, ethanol combines that also deliver excess heat to district-heating systems (or to other heat sinks) may perform better than is reported below.

The estimated GHG performance of future, Swedish, second-generation ethanol, excluding LUC effects, is

Table 25.2 GHG performance, expressed as g CO_2-equivalents per MJ, of Swedish second-generation ethanol production systems, excluding land use changes[1]

	Cultivation				Total			
System	CO_2-Fossil Fuels	N_2O Fertilizer Production	N_2O Biogenic	EtOH Plant	No allocation	Energy allocation	Economic Allocation	System expansion[2]
Straw–ethanol, electricity and biogas	3.6	0	–3.1	1.6	2.1	1.8	1.8	–0.6
Willow–ethanol and pellets	8.9	3.3	7.2	1.8	21.2	11.0	16.7	18.8
Poplar–ethanol and pellets	7.4	0	0	1.8	9.2	4.8	7.3	6.8
Forest residues–ethanol and pellets	5.3	0	0	1.8	7.1	3.7	5.6	4.7

[1]Data adapted from Refs 16, 36, 49, 51, and 52. The total energy output from wood-based ethanol systems is assumed to be 70%, divided between ethanol, 36%, and lignin pellets, 34%. The corresponding energy output from straw-based ethanol systems is assumed to be 57%, divided between ethanol, 48%, electricity, 6%, and biogas, 3%.
[2]Electricity is assumed to replace average Swedish electricity, biogas replaces natural gas, and pellets (lignin) replace pellets from fresh woody biomass.

shown in Table 25.2. The calculations are based on data from demo plants and modeling of ethanol plants that are considered to be technically and economically viable.[49,51] When system expansion is applied, the net emissions range from roughly zero to slightly below 20 g CO_2-equivalents per MJ ethanol, equivalent to more than a 75% GHG reduction compared with fossil vehicle fuels. The low (even negative) GHG emissions for straw-based ethanol is due to an estimated reduction of biogenic N_2O emissions from the soil, as a consequence of the straw being harvested that leads to a reduction of the N content in the soil (see, e.g., Ref 16). The recovery of logging residues in coniferous forests is estimated to have an insignificant impact on the biogenic N_2O emissions from forest soils.[52] Cultivation of willow includes fertilization, which explains the higher GHG emissions compared with poplar plantations that do not include fertilization.

The use of harvest residues, postconsumer organic waste, and by-products from the agricultural and forestry industries do not cause any indirect displacement emissions if these biomass sources were not utilized for alternative purposes. However, there is a timing issue associated with the use of residues and waste as bioenergy feedstock since some biomass sources (e.g., felling residues left in the forest) would retain organic carbon for a significantly longer time if not used for energy. Such delayed GHG emissions can be considered a benefit in short-term GHG mitigation and is also a relevant factor in longer-term accounting in regions where biomass degradation is slow, for instance in boreal forests.[53] Yet, it is commonly agreed that harvested forest systems have the best climate change mitigation capacity in the longer term.[11]

Illustrative of this, Figure 25.2 shows the C dynamics for a Norway spruce forest in south Sweden subject to three different management practices. In the "Stems Only" scenario harvest residues are left on the ground both after thinning and final felling. The "Stems & GROT" scenario involves extraction of 80% of the logging residue after thinning and final felling (GROT is the Swedish acronym for branches and tops—GRenar Och Toppar in Swedish), and the "Stems, GROT, & Stumps" scenario includes in addition the removal of 50% of stumps and coarse root systems at final felling. As can be seen, the cumulative harvest is substantially larger for the cases where residues are extracted but this larger extraction has a relatively small influence on the C dynamics of the forest.

Lignocellulosic plants can be produced on marginal lands and thereby reduce the pressure on prime cropland.[11] In cases where such plants are established on croplands, they may still reach lower LUC emissions if yielding favorably (i.e., higher biofuel output per hectare) because, first, any iLUC arising due to the displacement effect would be lower per unit biofuel produced and, second, dLUC emissions would in many instances be lower since the plants are commonly expected to be perennial plants, including woody species grown in multiyear rotations, and these cause lower soil C losses compared to conventional agricultural crops. Depending on land use history, changing to such plantations can even induce net C sequestration into soils and standing biomass, thus, improving GHG performance and also soil productivity.[54–62]

LAND USE AND LAND USE CHANGES IN SWEDEN

During the period between the First and Second World Wars, Swedish agriculture expanded to about 4.8 Mha, of which about 3.7 Mha was cropland.[63] The area under cultivation has decreased since then, and the most recent statistics show that slightly less than 3.1 Mha is classified as agricultural land, about 2.6 Mha being cropland.[22] However, the agricultural area reported in statistics does not represent the area actually used for the production of food and other products. Illustrative of this, a survey showed

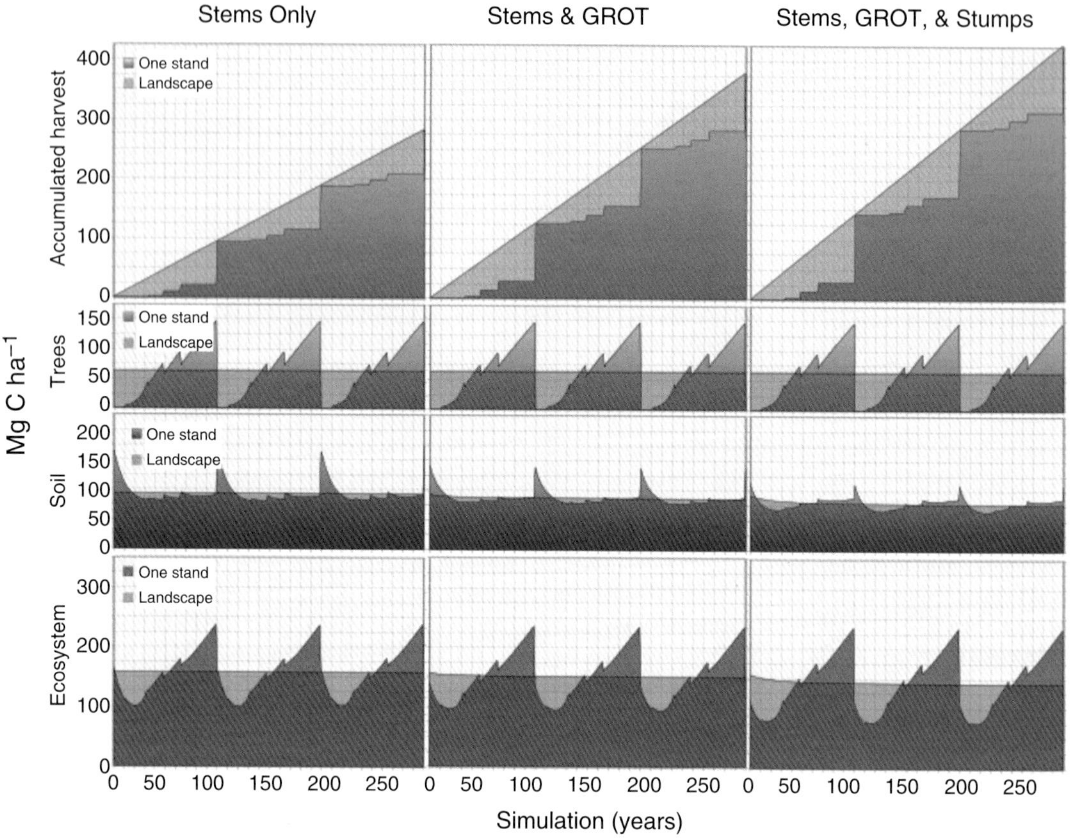

Figure 25.2 Modeled development of C stocks in a Norway spruce forest in south Sweden subject to three different management practices (described in the main text). The single stands are plotted behind the landscape averages in the foreground. (Reprinted with permission from Ref 53. Copyright 2011 Swedish Energy Agency.)

that 0.5 Mha of Swedish agricultural land was not actively being used.[64]

Based on official statistics of land use for agriculture in Sweden from 2005 and onward, covering the period when cereal-based ethanol has been produced, in Ref 36 it is estimated that a maximum of 25% of ethanol feedstock production has taken place on previous (partly unused) grassland. The remaining cereal cultivation for ethanol also took place on land previously used for annual crop production.[36] In other words, the "dLUC cost" of Swedish ethanol establishment corresponds to some level of biogenic soil C emissions from the cultivation of cereals on former grasslands.

The area used for cereal production decreased by roughly 130 kha during the period 2003–2005, more than twice as much as during the subsequent, longer period 2005–2010 (a decrease by 61 kha), when the area under wheat also increased at the expense of other cereals. The total Swedish cropland area decreased by almost 70 kha during the period 2005–2010.[22] Thus, the establishment of ethanol production in Sweden has most likely contributed to an increased share of wheat in cereal production and thereby potentially reduced the rate of cereal land conversion to other uses, for example, fallow land or grasslands that could be expected to have resulted in some CO_2 sequestration in soils.

Besides the agricultural land that is at present not actively used (and not yet converted to forests, urban infrastructure, etc.), yield increases and changes in the land allocation to different crops can release additional agricultural land for bioenergy production. Assessing these options, Ref 65 reports that 300,000–650,000 ha, equivalent to about 10%–20% of current total agricultural land, could become available for the cultivation of energy crops (see

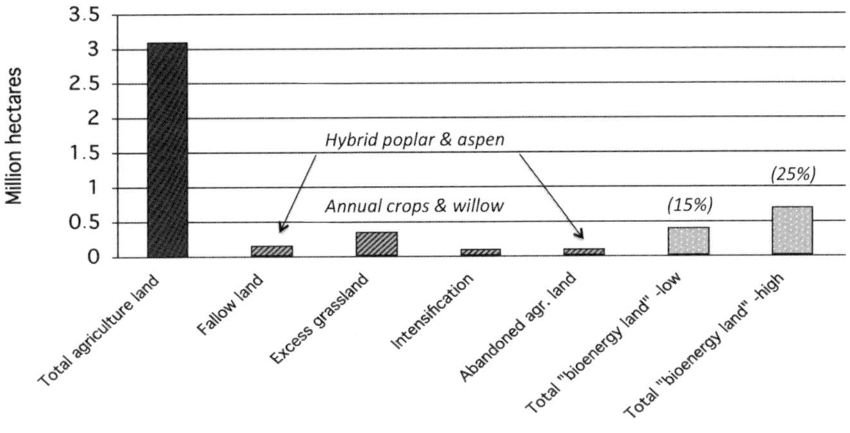

Figure 25.3 Estimated amount of agricultural land available for expanding bioenergy production in Sweden without compromising current food and feed output. The total amount of agricultural land is currently 3.1 Mha. (Data adapted from Refs 22 and 65.)

Figure 25.3). This surplus land would consist of both croplands and grasslands not needed for feed production to support the existing meat and dairy production, and abandoned, former agricultural land (marginal land) is not included in current agriculture or forestry land statistics.

INDIRECT LAND USE CHANGE

Apart from the fact that ethanol feedstock production directly influences the associated C flows—and consequently GHG balances—indirect effects may also arise. iLUC can occur when users of displaced land reestablish elsewhere, and/or because of macroeconomic factors: The agricultural area may change (or the production may change, by becoming more/less intensive) in response to the changes in food/fiber production caused by the ethanol project. The iLUC may occur in the country where the feedstock is produced, or abroad as a consequence of international trade. An example of iLUC causing increased GHG emissions is forest conversion to establish food crop cultivation that has become displaced by a bioenergy project. An example of iLUC causing decreased GHG emissions is when reduced demand for soy, because it is being displaced by DDGS as animal feed, moderates the forest conversion to croplands for soy cultivation.

The causes behind iLUC are often multiple, complex, interlinked, and change over time, and this makes it difficult to quantify iLUC effects in relation to a specific bioenergy project. Attempts at quantification commonly use general or partial equilibrium models (see, e.g., Ref 13 for an overview) of varying scope (e.g., geography and spatial resolution, sectors covered, and detail in their characterization), but there are also statistical approaches to detect iLUC[66] and approaches that make quantifications based on predefined, causal-effect chains.[1,11,12,66–71] There are many reasons for the large divergence in results, such as assumptions made in models about prices, elasticities, trade patterns, agricultural productivity, and its development, and the use of by-products within the agricultural sector. Achieving a consensus on the extent of iLUC effects in relation to specific types of bioenergy projects is unlikely in the near future.

Furthermore, the economic models used to assess iLUC have methodological features differing from LCAs, making the coupling of the two types of models complicated. There is also a risk of double counting by-products and dLUC.[1,12] A review of iLUC studies found estimates of iLUC emissions for wheat ethanol (including various production regions in the world) ranging from–53 to 176 CO_2-eq./MJ ethanol.[12] Fewer studies have investigated LUC associated with second-generation biofuels (see, e.g., Ref 72) but, generally, bioenergy options that use lignocellulosic feedstocks are projected to have lower LUC than those of first-generation biofuels (see, e.g., Refs 9, 73, 74). However, the LUC may vary significantly between different types of lignocellulosic feedstocks and depend on the local context where the production take place.

Policy makers are understandably concerned that the impacts of LUC are properly taken into account when the planting of more energy crops is being contemplated or is being considered for financial incentives, but the design of the overall mitigation regimes may be the most important in influencing the nature of bioenergy development and associated environmental consequences, including iLUC and the net GHG savings from bioenergy.[11,75] Future LUC rates will depend on the willingness of national governments to

Table 25.3 GHG performance, expressed as g CO_2-equivalents per MJ, of Swedish first-generation ethanol production systems, including direct land use change (dLUC) and a 30-year perspective

System	Total excluding dLUC[1]	dLUC[2]		Total including dLUC (Open farmland / Grassland)			
		Open Farmland[3]	Grassland	No allocation	Energy allocation	Economic allocation	System expansion
Wheat–ethanol and DDGS (excl. straw)	45.8	0	20.0	45.8/65.8	27.9/40.1	37.0/53.2	16.5/36.5
Wheat–ethanol and DDGS (incl. straw)	45.5	5.1	25.1	51.6/70.6	18.1/24.7	37.7/51.5	18.2/37.2
Wheat–ethanol and biogas (excl. straw)	47.0	0	20.0	47.0/67.0	33.1/47.2	35.9/51.2	21.8/41.8
Sugar beet–ethanol and pulp (excl. tops and leaves)	30.2	0	11.8	30.2/42.0	19.6/27.3	25.4/35.3	15.8/27.6
Sugar beet–ethanol and biogas (incl. tops and leaves)	29.9	1.5	13.3	31.4/43.2	22.2/30.5	23.9/32.7	6.2/18.0

[1]From 25.1.
[2]Data adapted from Refs 16, 36, 79, 80, and 81.
[3]Cropland previously used for the cultivation of annual crops.

protect forests and other natural ecosystems and the effectiveness of legislation and other policy measures to reduce the conversion of forests and other natural ecosystems. But they also will depend on whether sustainable land use practices become established in regions where agriculture continues to expand into new areas.[11]

A SCENARIO FOR ETHANOL EXPANSION IN SWEDEN

The above data on GHG performance for ethanol systems can be combined with the estimate of potential land availability for the production of bioenergy to produce illustrative scenarios for Swedish ethanol expansion. These scenarios combine the implementation of process integration and biorefineries producing multiple products, the use of a broader resource base including lignocellulosic feedstocks, and expansion options that minimize GHG emissions from LUC associated with the ethanol production.

Both grasslands and croplands are here estimated to be suitable for the cultivation of both conventional crops and short rotation coppice (willow). These production systems require good soil quality and relatively large fields to be profitable.[76] Part of current fallow land and former agricultural land, which have a lower soil quality, is estimated to be more suitable for the production of fast-growing, broad-leaved deciduous trees, such as hybrid poplar and aspen. These production systems are also judged to be profitable in small-size plantations since conventional forestry machines are used.[77]

In Tables 25.3 and 25.4, the GHG performance of first- and second-generation ethanol is shown when dLUC emissions are also included, applying a 30-year perspective. If, for example, wheat and sugar beet are cultivated on excess grassland, the GHG emissions will be roughly twice as high as when the cultivation takes place on cropland previously used for the cultivation of annual crops (system expansion case). As a consequence, the GHG reduction will be 50%–55% lower when wheat-based ethanol displaces fossil vehicle fuels. It is important to take the time aspect into consideration when assessing changes in the soil carbon level.[78] An estimate is that a change from annual to perennial crops on mineral soils with a long history of annual crop cultivation will lead to soil C gains over a period of about 30–50 years before a new soil C equilibrium is reached.[79] Conversely, the C losses occurring when annual crops are also cultivated on former grassland gradually decrease as a new equilibrium is reached. Similar to biogenic N_2O emissions, changes in soil carbon levels are associated with significant inherent uncertainties since these levels are influenced by a large number of local parameters (see, e.g., Refs 80, 81).

Straw harvest reduces the input of organic carbon to the soil (see Table 25.4), and this counteracts the positive effect of lower N_2O emissions (see Table 25.2). As can be seen, straw-based ethanol reduces the GHG emissions by roughly 95% when displacing fossil vehicle fuels (system expansion case). Both willow- and poplar-based ethanol could reach "negative" GHG emissions if the plantations are established on croplands with a long history of annual crop cultivation, due to the resulting C sequestration into the soil—again this C sequestration gradually decreases as a new soil C equilibrium is reached. When willow or poplar plantations are sited on excess grassland, the soil C

Table 25.4 GHG performance, expressed as g CO_2-equivalents per MJ, of Swedish second-generation ethanol production systems, including direct land use change (dLUC) and a 30-year perspective

		dLUC[2]		Total including dLUC			
System	Total excluding dLUC[1]		Reference Landuse[3]	No Allocation	Energy Allocation	Economic Allocation	System Expansion
Straw–ethanol, electricity and biogas	2.1	5.8	No straw harvest	7.9	6.6	6.7	5.2
Willow–ethanol and pellets	21.2	−19.4 0	Open farmland/ grassland	1.8/21.2	0.9/11.0	1.4/16.7	−0.6/18.8
Poplar–ethanol and pellets	9.2	−13.6 0	Open farmland/ grassland	−4.4/9.2	−2.3/4.8	−3.5/7.3	-6.8/6.8
Forest residues–ethanol and pellets	7.1	5.3	No forest residue harvest	12.4	6.5	9.7	10.0

[1]From 25.2.
[2]Adapted data from Refs 16, 36, 52, and 79–83.
[3]Open farmland is equivalent to cropland previously used for the cultivation of annual crops.

changes are estimated to be insignificant. However, more C is, on average, stored in the standing biomass stock, but such changes in carbon storage in above ground biomass are not included in the calculations here.

As discussed earlier, harvesting of logging residues (tops, branches, and thin stem wood) in long rotation forestry represents a type of bioenergy system where accounting has to consider that the temporal imbalance of C dynamics differs substantially for the bioenergy use on the one hand and decomposition/regrowth processes in the forest ecosystem on the other hand. Essentially, when harvest residues are extracted and used for bioenergy the C in this biomass is released to the atmosphere faster than when the biomass remains in the forest where the C is released over a longer time period as the residues decay.

In the case analyzed here, the harvesting of logging residues is, as in the case of straw, estimated to cause soil C losses over the mid- and long-term perspective (>30 years). This give rise to almost a doubling of the GHG emissions from logging residue-based ethanol, albeit that these are almost 90% lower than for fossil vehicle fuels (Table 25.4; system expansion using a 30-year time perspective). However, in a short-term perspective (less than 10 years), the GHG savings from displacing fossil vehicle fuels with ethanol produced from logging residues will be significantly lower since much of the C that is released to the atmosphere as the ethanol is used would in the reference situation still remain in forest residues that are left in the forest.

Improvement Options

The GHG emissions from current systems of biomass feedstock cultivation and ethanol production could be reduced by various improvements. The N_2O emissions from N fertilizer plants can be reduced by about 80% if the plants use catalytic N_2O cleaning.[16] Today, approximately half of the N fertilizer plants in Western Europe have installed catalytic cleaning equipment (which is considered in the calculations here) and in a few years all plants are expected to have this technology.[84] Ethanol from wheat, sugar beet, and willow that in the future use N fertilizers only from plants using catalytic N_2O cleaning would have roughly 10% lower fuel cycle GHG emissions (excluding dLUC; see Tables 25.1 and 25.2).

The level of biogenic N_2O emissions from the soil are, as previously discussed, uncertain since these levels are influenced by several local parameters. However, one crucial parameter is the amount of nitrogen available in the soil. Thus, an improved efficiency in the uptake of nitrogen by the crop and more efficient fertilization strategies will decrease the N_2O emissions and also reduce N leaching.[16,78] One option for improving the efficiency of N utilization is to employ the so-called precision fertilization that uses geographically explicit field data, weather forecasts, etc, to tailor the N input based on the time-specific N requirements of the crop. This technology is now commercially available and partly utilized in Swedish agriculture. A reduction of biogenic N_2O emissions by, for example, 20% will reduce the fuel cycle emissions of GHGs from crop-based ethanol by about 5% (see Tables 25.1 and 25.2).

The emissions of fossil CO_2 associated with feedstock production come mainly from the use of diesel for tractors, forestry machines, etc., and the use of N fertilizers, since their production commonly requires fossil resources (mainly natural gas). Biofuels can be used in tractors and other machinery and also in crop drying.[85,86] The fossil resources used in N fertilizer production can also be replaced by renewable alternatives.[87,88] In fact, implementation of catalytic N_2O cleaning and substitution of fossil resources in the production could make future N fertilizers almost GHG neutral. It is estimated here that the use of biofuels and other renewable energy carriers in feedstock production could reduce the fuel cycle GHG emissions by 30%–40% for crop-based ethanol and by more than 50% for ethanol produced from other sources such as straw and felling residues (see Tables 25.1 and 25.2).

The GHG performance is also related to the yield levels and the relation between yields and inputs in the production. Several studies point out that there are still yield gaps to exploit, primarily in developing countries but also in Europe.[89,90] Yields might increase faster as newer hybrid plants are adopted more widely, and theoretical yield limits also appear to provide scope to further increase the genetic yield potential.[91,92] Important agronomic measures to improve yields include irrigation and fertilization, but substantial yield gains can also be achieved with better knowledge and by better management practices. Especially dedicated biofuel crops, which have not been subject to the same breeding procedures as the major food crops, have a great potential for improvement. Selection and development of suitable plant species and genotypes for given locations to match specific soil types, climate, and conversion technologies are possible, but some dedicated energy crops are as yet at an early stage.[93–97] Traditional plant breeding, selection, and hybridization techniques are time consuming, particularly for woody plants but also for grasses. However, new biotechnological routes to produce both genetically modified (GM) and non-GM plants are possible.[98]

Investment in yield improvements has been shown to be an efficient GHG mitigation strategy.[99] But improvements in GHG performance does not necessarily require yield increases. Ref 100 reports that cultivation of wheat varieties dedicated for ethanol production, employing lower N fertilization levels and reaching somewhat lower grain yields, could reduce the GHG emissions from ethanol production by, on average, 15%. New, dedicated wheat varieties can have a starch content of about 75%, compared to 70% for conventional varieties, and this improves the ethanol yield from 55% to about 58%.[6] It is estimated here that higher yields and use of such dedicated crops could reduce the fuel cycle emissions of GHGs from crop-based ethanol by about 10%–15% in a relatively short-term perspective (see Tables 25.1 and 25.2). However, these varieties are not currently available in the Swedish market.

Process design and choice of process fuel are two important determinants of both the overall energy conversion efficiency and the GHG performance. Current ethanol production in Sweden uses biomass-based heat and steam, and the Swedish electricity mix is dominated by hydro and nuclear power. A shift to fossil-based process heat and electricity would increase the life cycle GHG emissions drastically and could, in the worst case, negate the GHG benefit of displacing fossil vehicle fuels with ethanol (see, e.g., Refs[6] and[16]). Ethanol production from lingocellulose normally requires little external energy, so that the GHG performance of such ethanol production is not much affected by changes in the primary energy use to provide the external energy.[16]

Practical measurements in an existing ethanol plant indicate that better process integration in current wheat-based ethanol production in Sweden can improve the total energy conversion efficiency by approximately 15% (net energy basis), which also leads to GHG savings.[6] This improvement includes more optimal steam pressure and integration with electricity production, better heat exchange, and recovery of waste heat, along with integration of drying processes, etc. Processes for second-generation ethanol production commonly include various biorefinery concepts with multiproduct output, including both additional energy carriers and other bioproducts. Apart from direct improvements in the GHG performance through higher energy conversion efficiency (see, e.g., Refs 49 and 51), such multiproduct output can also indirectly improve the GHG performance by replacing additional fossil-based alternatives (see, e.g., Ref 101).

Estimations of the GHG performance of future biorefineries involve methodological challenges. Consideration of certain process integration options may require that system boundaries be expanded to include the local infrastructure in energy, forestry, agriculture, and chemical industries. Supplementary functional units may be needed when ethanol is coproduced with platform chemicals and other bioproducts and also system expansion to include the indirect GHG consequences of all product outputs. Despite these uncertainties, a cautious estimation shows that improved process integration in ethanol production plants may reduce their GHG emissions by at least 15%. However, this will lead to only minor reductions (2%–4%) of the fuel cycle GHG emissions, since the ethanol plants are using mainly biomass-based energy with low GHG emissions (see Tables 25.1 and 25.2).

To sum up, if all potential technical and other improvements discussed here were implemented, then the fuel cycle GHG emissions may theoretically be reduced by 50%–60% compared with the calculated emissions in Tables 25.1 and 25.2 (excluding dLUC), both for crop-based and for residue-based ethanol. Thus, there exists a significant potential to decrease the GHG emissions along the ethanol production chain. However, all these improvements may not take place due to various limitations, for example, costs.

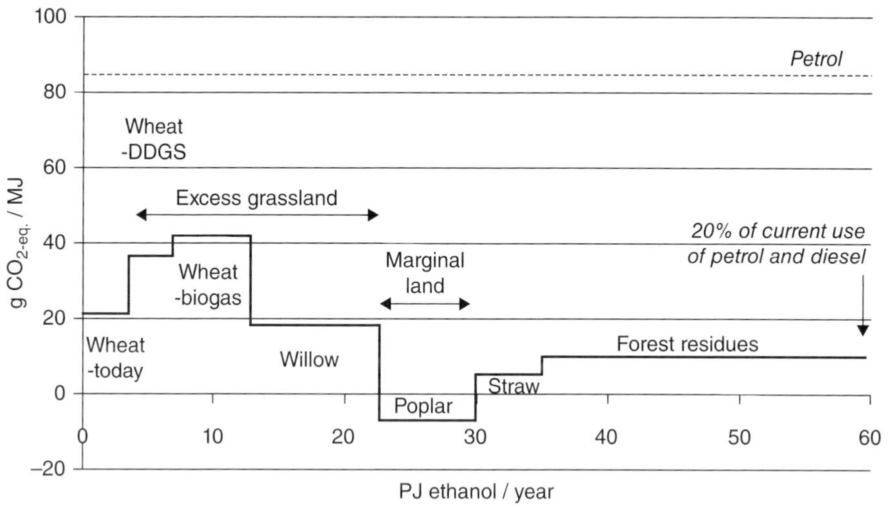

Figure 25.4 A theoretical potential of future expansion of Swedish low, indirect impact ethanol production and related GHG performance (based on system expansion), including dLUC and currently available technology.

Future Ethanol Potentials and GHG Performance

Figure 25.4 presents an illustrative scenario for ethanol expansion in Sweden, including also the estimated GHG emissions (a 30-year time frame) associated with each ethanol expansion option. Note that this scenario should not be understood as a projection of what is likely to happen in the coming years, but as a way to illustrate the relative importance of the different ethanol expansion options and also their aggregate size. Up to 7 PJ/year of wheat-based ethanol could be produced without saturating the protein feed market for DDGS. A previous estimation, based on today's feeding recommendations in Swedish animal husbandry, shows that up to 15% of the total feed intake in dairy and beef production could consist of DDGS.[21] It is estimated here that another 5–6 PJ of wheat-based ethanol could be produced annually based on using WDG produced in excess of protein feed market demand for biogas production. The required increase in feedstock cultivation could be managed by targeting excess grasslands equivalent to 7.5% of the current arable land (approximately 200,000 ha) (see Figure 25.4). The GHG emissions from this ethanol production (using current technology) is estimated to correspond to 20 g CO_2-eq./MJ, representing cultivation on open cropland, and up to 40 g CO_2-eq./MJ, representing cultivation on former grassland, of which roughly 30% would come from C losses from the grassland soils. Possible iLUC effects in this case would arise as a consequence of the increased supply of DDGS displacing other feed, possibly reducing the pressure on land expansion elsewhere. If the increased supply of wheat for ethanol production is instead based on cultivating more wheat on croplands currently used for other food and feed crops, there will be no soil C losses but a greater likelihood of GHG emissions associated with iLUC. If these GHG emissions are above 20–25 g CO_2-eq./MJ ethanol (equivalent to the calculated dLUC emissions before allocation; see Table 25.3), then the climate benefit would be lower than as shown in Figure 25.4.

Roughly 10 PJ/year of ethanol could be produced from willow cultivated on the remaining part of the excess grassland (approximately 150,000 ha; see Figure 25.3). The GHG emissions are estimated at about 20 g CO_2-eq./MJ. If, however, willow plantations are established on open cropland with a long history of annual crops cultivation, C will be sequestered into the soils, roughly canceling out the GHG emissions, making this ethanol production climate neutral (see Table 25.4). On the other hand, to the extent that the willow establishment displaces food and feed production, iLUC emissions may arise. The size of these iLUC emissions will determine whether the net outcome of LUC is positive or negative.

An additional 7–8 PJ ethanol per year could be produced based on fast-growing deciduous trees, such as hybrid poplar, grown on roughly 150,000 ha of abandoned agricultural land (marginal land). Soil C sequestration would more than outweigh the GHG emissions from the production, making this option "GHG negative" (roughly–5 to–10 g CO_2-eq./MJ). Indirect effects might occur to the extent that the farmer investing in the tree production does not invest in some other cultivation, or other farm activity. Whether this indirect effect is negative or positive for the GHG performance depends on what other activity was being considered.

The practical potential for using straw as bioenergy feedstock has been estimated to be 20–25 PJ/year in Sweden.[21] If half of this potential is used for the production of ethanol, coproduced with electricity and biogas, then 5 PJ could be produced, giving GHG emissions at about 5 CO_2-eq./MJ. The practical potential of logging residues (e.g., thin stem wood from forest thinning and stumps) in previous assessments has been estimated to be roughly 150 PJ per year.[102] Assuming that half of this biomass feedstock is used for ethanol production in cogeneration systems, about 25 PJ/year could be produced, giving GHG emissions at about 10 CO_2-eq./MJ. Both straw- and logging-residue-based ethanol production is estimated to have minor LUC effects (see Table 25.4), although they would transfer biogenic C to the atmosphere faster than if the residues were left to decay on the ground. The amount of ethanol that could be produced from straw and logging residues presented in Figure 25.4 is within the technical potential for integration with the existing district heating system and forest industry, where ethanol can be coproduced together with electricity, heat, pellets, etc.[51,103]

Taken together, these options for increased ethanol production correspond to about 60 PJ, or roughly 20% of the current use of petrol and diesel for road transport in Sweden. The average GHG emissions would be about 15 g CO_2-eq./MJ, or 83% lower than the average for the present fossil vehicle fuels. The illustration in Figure 25.4 is based on ethanol production systems using current commercial technology to produce first-generation ethanol and currently available technology to produce second-generation ethanol. However, as discussed above, there exists a significant potential for future improvements along the entire ethanol production chain, which could theoretically reduce the fuel cycle GHG emissions by up to 50%–60%, bringing the average future GHG emissions below 10 g CO_2-eq./MJ.

CONCLUSIONS

Important and realistic options for ethanol expansion exist in Sweden, and these can deliver sizeable reductions in GHG emissions. Ethanol production systems can be located, designed, and managed so as to further reduce the GHG emissions from the production and also to avoid or mitigate negative direct and indirect effects associated with LUC. Current Swedish wheat ethanol systems have already relatively low GHG emissions, and this type of production could expand somewhat further with acceptable GHG performance. The coproduct, distillers waste, could be used both as a protein feed and for biogas production. If the expanded ethanol production is based on wheat cultivation that is established on excess grassland not needed for feed production at the current level of meat and dairy production, iLUC risks will be low, albeit significant soil C emissions may result. If, on the other hand, the increased requirement of wheat for ethanol production is based on the cultivation of this wheat on cropland currently used for other food and feed crops, there will be no soil C losses, but a higher likelihood of GHG emissions associated with iLUC. If the GHG emissions from iLUC are more than 20–25 g CO_2-eq./MJ ethanol, it would be advantageous from the perspective of climate benefit, to expand the cultivation on grassland.

While there is extensive knowledge about critical aspects of GHG calculations based on the LCA methodology, including how various assumptions and methodology choices influence the resulting GHG performance of various biofuel systems, quantification of iLUC, and its effects is more complicated. In the short-term, there is little probability that a commonly agreed quantification of iLUC emissions associated with specific biofuel options will be reached. There are also methodological difficulties associated with the use of iLUC quantifications in the framework of LCA, arising from the fact that the economic models commonly used to assess iLUC have features differing from LCA models. The coupling of these two types of models is therefore complicated and may not be a realistic option.

Ethanol production based on lignocellulosic feedstocks (e.g., straw, short rotation forestry and perennial grasses, silvicultural, and felling residues) can attain a greater GHG reduction from fossil fuel displacement and also reduce LUC risks. The use of straw and logging residues as ethanol feedstock will result in minor LUC effects, and the GHG reduction will be large. Specific locations for lignocellulosic cultivation can result in C sequestration into soils and standing biomass, that is the LUC effect would improve the GHG balances. However, a shift to relying on lignocellulosic feedstocks is not a guarantee against GHG emissions associated with LUC. There is also a timing issue associated with the use of residues and waste as bioenergy feedstock since some biomass sources (e.g., felling residues left in the forest) would retain organic carbon for a significantly longer time if not used for energy.

Knowledge about how biofuel production systems can be shaped to achieve low GHG emissions—including when LUC effects are considered—is crucial in the development of governmental policy and also of other instruments such as sustainability standards and certification systems. Much of the recent years' debate on biofuels and GHG has been dominated by a narrow consideration of marginal effects. This may partly be due to the fact that policy makers tend to judge biofuels on the basis of their contribution to short-term GHG targets and partly due to a significant influence of individuals and organizations whose assessments of biofuels rely primarily on LCA and similar, rather limited tools.

It is here proposed that more attention be paid to more strategic considerations related to the much-needed transformation of the global energy system. Bioenergy's contribution to climate change mitigation needs to reflect a balance between short-term targets and the long-term objective to contain the increase in global temperature to below 2°C (the Copenhagen Accord). Clearly, bioenergy projects will not be relevant as mitigation options even on the longer term if they are associated with very large LUC emissions and avoiding LUC emissions can in many instances be advisable. But bioenergy systems causing some level of LUC emissions may still represent attractive options. From this perspective, it would be advisable to reserve some part of the remaining space for GHG emissions for the development of a biofuels industry capable of providing long-term, renewable, and climate-efficient energy services for the world. At least, a simplistic view of such emissions should not be sufficient reason to exclude bioenergy from the list of worthwhile technologies for climate change mitigation.

NOTE

[a]www.arb.ca.gov/fuels/lcfs/lcfs.htm. 2011-11-01

ACKNOWLEDGEMENTS

We gratefully acknowledge the economic support received from The Swedish Energy Agency, and two anonymous reviewers for their valuable comments.

REFERENCES

1. Kim S, Dale B. Regional variations in greenhouse gas emissions of biobased products in the United States—corn-based ethanol and soybean oil. *Int J Life Cycle Assess* 2009, 14:540–546.
2. Farrell AE, Plevin RJ, Turner BT, Jones AD, O'Hare M, Kammen DA. Ethanol can contribute to energy and environmental goals. *Science* 2006, 311;506–508.
3. Fleming JS, Habibi S, MacLean HL. Investigating the sustainability of lignocellulose-derived fuels for light-duty vehicles. *Transp Res Part D: Transp Environ* 2006, 11;146–159.
4. Larson E. A review of life-cycle analysis studies on liquid biofuel systems for the transport sector. *Energy Sustain Dev* 2006, 10;109–126.
5. von Blottnitz H, Curran MA. A review of assessments conducted on bio-ethanol as a transportation fuel from a net energy, greenhouse gas, and environmental life cycle perspective. *J Cleaner Prod* 2007, 15;607–619.
6. Börjesson P. Good or bad ethanol from a greenhouse gas perspective—what determines this? *Appl Energy* 2009, 86:589–594.
7. Cherubini F, Bird ND, Cowie A, Jungmeier G, Schlamadinger B, Woess-Gallasch S. Energy- and greenhouse gas-based LCA of biofuel and bioenergy systems: key issues, ranges and recommendations. *Resour Conserv Recycling* 2009, 53:434–447.
8. Menichetti E, Otto M. Energy balance and greenhouse gas emissions of biofuels from a life-cycle perspective. In: Howarth RW, Bringezu S, eds. *Biofuels: Environmental Consequences and Interactions with Changing Land Use.* Proceedings of the Scientific Committee on Problems of the Environment (SCOPE) International Biofuels Project Rapid Assessment, September 22–25, 2008, Gummersbach Germany. Ithaca NY: Cornell University Press. 2009, 81–109.
9. Hoefnagels R, Smeets E, Faaij A. Greenhouse gas footprints of different biofuel production systems. *Renewable Sustainable Energy Rev* 2010, 14;1661–1694.
10. Wang MQ, Han J, Haq Z, Tyner WE, Wu M, Elgowainy A. Energy and greenhouse gas emission effects of corn and cellulosic ethanol with technology improvements and land use changes. *Biomass Bioenergy* 2011, 35:1885–1896.
11. Berndes G, Bird N, Cowie A. Bioenergy, land use change and climate change mitigation. Background Technical Report, ExCo:2011:01. *New Zealand: International Energy Agency Bioenergy*; 2011.
12. Ahlgren S, Börjesson P. Indirect land use change and biofuels—a review. *Report No 73. Lund, Sweden: Environmental and Energy Systems Studies*, Lund University; 2011 (in Swedish).
13. IPCC. *Special Report: Renewable Energy Sources and Climate Change Mitigation. Intergovernmental Panel on Climate Change.* Cambridge, UK: Cambridge University Press; 2011.
14. EC. Directive 2009/28/EC of The European Parliament and of The Council of 23 April 2009 on the promotion of the use of energy from renewable sources and amending and subsequently repealing Directives 2001/77/EC and 2003/30/EC; 2009.
15. EC. Report from the Commission on indirect land-use change related to biofuels and bioliquids. COM(2010)0811, 22/12/2010; 2010.
16. Börjesson P, Tufvesson LM. Agricultural crop-based biofuels—resource efficiency and environmental performance including direct land use changes. *J Cleaner Prod* 2011, 19:108–120.
17. Kendall A, Chang B. Estimating life cycle greenhouse gas emissions from corn-ethanol: a critical review of current U.S. practices. *J Cleaner Prod* 2009, 17:1175–1182.
18. Kim H, Kim S, Dale B. Biofuels, land use change, and greenhouse gas emissions: some unexplored variables. *Environ Sci Technol* 2009, 43:961–967.

19. Gallagher E. *The Gallagher Review of the Indirect Effects of Biofuel Production*. London: Renewable Fuels Agency; 2008.
20. Swedish Energy Agency. *Energy in Sweden—Facts and Figures*. Eskilstuna, Sweden: Swedish Energy Agency; 2011.
21. Börjesson P. Bioenergy from Swedish agriculture—a growing resource. Appendix. *Official Report from the Swedish Government, SOU 2007:36*. Stockholm, Sweden: Swedish Government; 2007 (in Swedish).
22. Swedish Board of Agriculture. Use of agricultural land 2010. *Statistical Report JO 10 SM 1101*. Jönköping, Sweden: Swedish Board of Agriculture; 2011 (in Swedish).
23. International Standardization Organization. ISO 14044 Environmental management—Life cycle assessment—Requirements and guidelines. Geneva, Switzerland: International Standardization Organization; 2006.
24. Flysjö A, Cederberg C, Strid I. LCA Database for conventional feed products. *SIK Report no. 772*. Göteborg, Sweden: The Swedish Institute for Food and Biotechnology; 2008 (in Swedish).
25. Smeets EMW, Bouwman LF, Stehfest E, van Vuuren DP, Posthuma A. Contribution of N_2O to the greenhouse gas balance of first-generation biofuels. *Global Change Biol* 2009, 15;1–23.
26. Bessou C, Ferchaud F, Gabrielle B, Mary B. Biofuels, greenhouse gases and climate change. A review. *Agronomy Sustainable Dev* 2011, 31:1–79.
27. Farquharson R, Baldock J. Concepts in modeling N_2O emissions from land use, *Plant Soil* 2008, 309:147–167.
28. Mosier AR, Duxbury JM, Freney JR, Heinemeyer O, Minami K. Nitrous oxide emissions from agricultural fields: Assessment, measurement and mitigation, *Plant Soil* 1996, 181:95–108.
29. Mosier AR, Kroeze C, Nevison C, Oenema O, Seitzinger S, van Cleemput O. Closing the global N_2O budget: nitrous oxide emissions through the agricultural nitrogen cycle, *Nutr Cycling Agroecosyst* 1998, 52:225–248.
30. IPCC. Agriculture, forestry and other land use. N_2O emissions from managed soils, and CO_2 emissions from lime and urea application. In: *Guidelines for National GHG Inventories*. Geneva, Switzerland: Intergovernmental Panel on Climate Change; 2006. Vol. 4, Chap. 11.
31. Bouwman AF, Boumans LJM, Batjes NH. Modelling global annual N_2O and NO emissions from fertilized fields. *Global Biogeochem Cycles* 2002, 16: 1080.
32. Stehfest E, Bouwman L. N_2O and NO emission from agricultural fields and soils under natural vegetation: summarizing available measurement data and modeling of global annual emissions. *Nutr Cycling Agroecosyst* 2006, 74:207–228.
33. Wrage N, Velthof GL, van Beusichem ML, Oenema O. Role of nitrifier denitrification in the production of nitrous oxide, *Soil Biol Biochem* 2001, 33:1723–1732
34. Nevison CD, Esser G, Holland EA. A global model for changing N_2O emissions from natural and perturbed soils. *Climatic Change* 1996, 32:327–378.
35. Crutzen P, Moiser A, Smith K, Winiwarter W. N_2O release from agro-biofuel production negates global warming reduction by replacing fossil fuels. *Atmos Chem Phys* 2007, 7:11191–11205.
36. Börjesson P, Tufvesson L, Lantz M. Life cycle assessment of Swedish biofuels. Report No 70. Lund, Sweden: Environmental and Energy Systems Studies, Lund University; 2010.
37. Kim S, Dale B. Global potential bioethanol production from wasted crops and crop residues. *Biomass Bioenergy* 2004, 26:361–375.
38. Cherubini F, Ulgiati S. Crop residues as raw material for biorefinery systems—A LCA case study. *Appl Energy* 2010, 87:47–57.
39. Uihlein A, Schebek L. Environmental impacts of a lignocellulose feedstock biorefinery system: an assessment. *Biomass Bioenergy* 2009, 33:793–802.
40. Wilhelm WW, Johnson JMF, Hatfield JL, Voorhees WB, Linden DR. Crop and soil productivity response to corn residue removal: a literature review. *Agronomy J* 2004, 96:1–17.
41. Ball BC, Bingham I, Rees RM, Watson CA, Litterick A. The role of crop rotations in determining soil structure and crop growth conditions. *Can J Soil Sci* 2005, 85:557–577.
42. Blanco-Canqui H, Lal R, Post WM, Izaurralde RC Owens LB. Rapid changes in soil carbon and structural properties due to stover removal from no-till corn plots. *Soil Sci* 2006, 171:468–482.
43. Lal R. Crop residues as soil amendments and feedstock for bioethanol production. *Waste Manag* 2008, 28:747–758.
44. Lal R, Pimentel D. Bio-fuels from crop residues. *Soil Tillage Res* 2007, 93:237–238.
45. Wilhelm WW, Johnson JMF, Karlen DL, Lightle DT. Corn stover to sustain soil organic carbon further constrains biomass supply. *Agronomy J* 2007, 99:1665–1667.
46. Blanco-Canqui H, Lal R. Corn stover removal for expanded uses reduces soil fertility and structural stability. *Soil Sci Soc Am J* 2009, 73:418–426.
47. Börjesson P, Berglund M. Environmental systems analysis of biogas systems—Part II: The environmental impact of replacing various reference systems. *Biomass Bioenergy* 2007, 31:326–344.
48. Olofsson K, Rudolf A, Lidén G. Designing simultaneous saccharification and fermentation for improved xylose conversion by a recombinant strain of *Saccharomyces cerevisiae*. *J Biotechnol* 2008, 134:112–120.
49. Sassner P, Zacchi G. Integration options for high energy efficiency and improved economics in a wood-to-ethanol process. *Biotechnol Biofuels* 2008, 1:4. doi:10.1186/1754-6834-1-4.
50. Sassner P, Galbe M, Zacchi G. Techno-economic evaluation of bioethanol production from three different lignocellulosic materials. *Biomass Bioenergy* 2008, 32:422–430.
51. Ekman A, Wallberg O, Joelsson E, Börjesson P. Sustainable biorefineries based on agricultural residues—A case study of potential straw-based ethanol production in Sweden. *Submitted to Applied Energy*; 2012.

52. Lindholm EL, Stendahl J, Berg S, Hansson PA. Greenhouse gas balance of harvesting stumps and logging residues in Sweden. *Scand J Forest Res* 2011, 26:586–594.
53. Eliasson P, Svensson M, Olsson M, Ågren G. Forest carbon balances at the landscape scale and responses to intensified harvest investigated with the Q and COUP models. *Swedish Energy Agency project 34884-1*. Eskilstuna, Sweden: Swedish Energy Agency; 2011.
54. Anderson-Teixeira KJ, Davis SC, Masters MD, DeLucia EH. Changes in soil organic carbon under biofuel crops. *Global Change Biol Bioenergy* 2009, 1:75–96
55. Dimitriou I, Baum C, Baum S, Busch G, Schulz U, Köhn J, Lamersdorf N, Leinweber P, Aronsson P, Weih M, Berndes G, Bolte A. Quantifying environmental effects of short rotation coppice (SRC) on biodiversity, soil and water. *IEA Bioenergy Task 43,* 2011:01; 2011.
56. Lal R. Carbon sequestration. *Phil Trans R Soc B* 2008, 363:815–830
57. Tilman D, Hill J, Lehman C. Carbon-negative biofuels from low-input high-diversity grassland biomass. *Science* 2006, 314:1598–1600.
58. Liebig M, Schmer M, Vogel K, Mitchell R. Soil carbon storage by switchgrass grown for bioenergy. *BioEnergy Res* 2008, 1:215–222.
59. Robertson GP, Dale VH, Doering OC, Hamburg SP, Melillo JM, Wander MM, Parton WJ, Adler PR, Barney JN, Cruse RM, et al. Sustainable biofuels redux. *Science* 2008, 322;49–50.
60. Dondini M, Hastings A, Saiz G, Jones MB, Smith P. The potential of Miscanthus to sequester carbon in soils: comparing field measurements in Carlow, Ireland to model predictions. *Global Change Biol Bioenergy* 2009, 1:413–425.
61. Hillier J, Whittaker C, Dailey G, Aylott M, Casella E, Richter GM, Riche A, Murphy R, Taylor G, Smith P. Greenhouse gas emissions from four bioenergy crops in England and Wales: Integrating spatial estimates of yield and soil carbon balance in life cycle analyses. *Global Change Biol Bioenergy* 2009, 1:267–281.
62. Galdos MV, Cerri CC, Lal R, Bernoux M, Feigl B, Cerri CEP (2010). Net greenhouse gas fluxes in Brazilian ethanol production systems. *Global Change Biol Bioenergy* 2010, 2:37–44.
63. Swedish Board of Agriculture. Swedish agriculture in numbers 1800–2004. *Report 2005:6*. Jönköping, Sweden: Swedish Board of Agriculture; 2005 (in Swedish).
64. Swedish Board of Agriculture. Monitoring unused agricultural land. *Report 2008:6*. Jönköping, Sweden: Swedish Board of Agriculture; 2008 (in Swedish).
65. Swedish Board of Agriculture. Agriculture, bioenergy and the environment. *Report 2009:22*, Jönköping, Sweden: Swedish Board of Agriculture; 2009 (in Swedish).
66. Arima EY, Richards P, Walker R, Caldas MM. Statistical confirmation of indirect land use change in the Brazilian Amazon. *Environ Res Lett* 2011, 6, 024010.
67. E4Tech. A causal descriptive approach to modelling the GHG emissions associated with the indirect land use impacts of biofuels. *Final Report, November 2010*. London: E4Tech/U.K. Department for Transport; 2010.
68. Fritsche U, Hennenberg K, Hünecke K. *The "iLUC factor" as a means to hedge risks of GHG emissions from indirect land use change*. Darmstadt, Germany: Öko Institute; 2010.
69. Cornelissen S, Dehue B. *Summary of approaches to accounting for indirect impacts of biofuel production*. Utrecht, The Netherlands: Ecofys International BV; 2009.
70. DG Energy. The impact of land use changes on greenhouse gas emissions from biofuels and bioliquids. Litterature review. *An in-house review conducted for DG Energy*. Brussels, Belgium: DG Energy; 2010.
71. Nassar AM, Harfuch L, Bachion LC, Moreira MR. Biofuels and land use changes: searching for the top model. *Interface Focus* 2011, 1:224–232.
72. Melillo JM, Reilly JM, Kicklighter DW, Gurgel AC, Cronin TW, Paltsev S, Felzer BS, Wang XD, Sokolov AP, Schlosser CA. Indirect emissions from biofuels: how important? *Science* 2009, 326:1397–1399.
73. EPA. *Renewable fuel standard program (RFS2) regulatory impact analysis*. U.S. Environmental Protection Agency, EPA-420-R-10-006. Washington, DC: EPA; 2010.
74. Havlík P, Schneider UA, Schmid E, Böttcher H, Fritz S, Skalský R, Aoki K, Cara SD, Kindermann G, Kraxner F, et al. Global land-use implications of first and second generation biofuel targets. *Energy Policy* 2011, 39:5690–5702.
75. Wise M, Calvin K, Thomson A, Clarke L. Bond-Lamberty B, Sands R, Smith SJ, Janetos A, Edmonds J. Implications of limiting CO_2 concentrations for land use and energy. *Science* 2009, 324:1183–1186.
76. Rosenqvist H. Crops from field to energy: Calculation methodology for economic comparison between different land uses. *Report No. 1128*. Stockholm, Sweden: Thermal Engineering Research Association; 2010 (in Swedish).
77. Rytter L, Johansson T, Karacic A, Weih M. Investigation for a Swedish research program on the genus *Populus*. *SkogForsk Report No 733*. Uppsala, Sweden: SkogForsk; 2011 (in Swedish).
78. Reijnders L, Huijbregts M. (2008). Biogenic GHG emissions linked to the life cycles of biodiesel derived from European rapeseed and Brazilian soybeans. *J Cleaner Prod* 2008, 16:1943–1948.
79. Börjesson P. Environmental effects of energy crop cultivation in Sweden—Part I: identification and quantification. *Biomass Bioenergy* 1999, 16:137–154.
80. Röing K, Andrén O, Mattsson L. Long-term management effects on plant N uptake and topsoil carbon levels in Swedish long-term field experiments: cereals and ley, crop residue treatment and fertiliser N application. *Acta Agri Scand B* 2005, 55:16–22.
81. Kätterer T, Andrén O, Persson J. The impact of altered management on long-term agricultural soil carbon stocks—a Swedish case study. *Nutr Cycling Agroecosyst* 2004, 70:179–187.
82. Ågren G, Hyvönen R, Nilsson T. Are Swedish forest soils sinks or sources for CO_2—model analyses based on forest inventory data. *Biogeochemistry* 2007, 82:217–227.

83. Ågren G, Hyvönen R. Changes in carbon stores in Swedish forest soils due to increased biomass harvest and increased temperatures analysed with a semi-empirical model. *Forest Ecol Manag* 2003, 174:25–37.
84. Jenssen TK, Kongshaug G. Energy consumption and greenhouse gas emissions in fertilizer production. The International Fertilizer Society. *Proceedings No 509*; 2003.
85. Hansson PA, Baky A, Ahlgren S, Bernesson S, Nordberg A, Norén O, Pettersson O. Self-sufficiency of motor fuels on organic farms—evaluation of systems based on fuels produced in industrial-scale plants. *Agric Syst* 2007, 94:704–714.
86. Ahlgren S, Baky A, Bernesson S, Nordberg A, Norén O, Hansson PA. Future fuel supply systems for organic production based on Fischer–Tropsch diesel and dimethyl ether from on-farm-grown biomass. *Biosyst Eng* 2008, 99:145–155.
87. Ahlgren S, Baky A, Bernesson S, Nordberg A, Norén O, Hansson PA. Ammonium nitrate fertiliser production based on biomass—environmental effects from a life cycle perspective. *Bioresour Technol* 2008b, 99:8034–8041.
88. Ahlgren S, Bernesson S, Nordberg A, Hansson PA. Nitrogen fertiliser production based on biogas—energy input, environmental impact and land use. *Bioresour Technol* 2010, 101:7192–7195.
89. Liu J, Williams JR, Zehnder AJB, Yang H. GEPIC-modelling wheat yield and crop water productivity with high resolution on a global scale. *Agric Syst* 2007, 94:478–493.
90. Hengsdijk, H, Langeveld J. Yield trends and yield gap analysis of major crops in the world. *Werkdocument 170. Wageningen University*, The Netherlands: Wettelijke Onderzoekstaken Natuur & Milieu; 2009.
91. Edgerton MD. Increasing crop productivity to meet global needs for feed, food, and fuel. *Plant Physiol* 2009, 149:7–13.
92. Fischer G, Hizsnyik E, Prieler S, Shah M, van Velthuizen H. Biofuels and food security. The OPEC Fund for International Development (OFID) and International Institute for Applied Systems Analysis (IIASA) Land Use Change and Agriculture Program, Vienna, Austria: OFID and IIASA; 2009.
93. Bush DR, Leach JE. Translational genomics for bioenergy production: There's room for more than one model. *Plant Cell* 2007, 19;2971–2973.
94. Chapple C, Ladisch M, Meilan R. Loosening lignin's grip on biofuel production. *Nat Biotechnol* 2007, 25;746–748.
95. Lawrence CJ, Walbot V. Translational genomics for bioenergy production from fuelstock grasses: Maize as the model species. *Plant Cell* 2007, 19;2091–2094.
96. Carpita NC, McCann MC. Maize and sorghum: genetic resources for bioenergy grasses. *Trends Plant Sci* 2008, 13;415–420.
97. Karp A, Shield I. Bioenergy from plants and the sustainable yield challenge. *New Phytol* 2008, 179;15–32.
98. Brunner AM, DiFazio SP, Groover AT. Forest genomics grows up and branches out. *New Phytol* 2007, 174;710–713.
99. Burney J, Davis S, Lobell D. Greenhouse gas mitigation by agricultural intensification. *PNAS* 2010, 107:12052–12057.
100. Ahlgren S, Hansson PA, Kimming M, Aronsson P, Lundkvist H. *Greenhouse gas emissions from cultivation of agricultural crops for biofuels and production of biogas from manure—implementation of the Directive of the European Parliament and of the Council on the promotion of the use of energy from renewable sources*. Revised edition. Uppsala: Swedish University of Agricultural Sciences; 2011.
101. Ekman A, Börjesson P. Environmental assessment of propionic acid produced in agricultural biomass-based biorefinery system. *J Cleaner Prod* 2011, 19:1257–1265.
102. Swedish Forest Agency. Forestry consequential analyses. *Report 2008:25*. Jönköping, Sweden: Swedish Forest Agency; 2008 (in Swedish).
103. Ericsson K, Börjesson P. Potential supply of biomass for production of electricity, heat and biofuels including biorefineries. *Report ER 2008:04*. Eskilstuna, Sweden: Swedish Energy Agency; 2008 (in Swedish).

26

Performance of Small-Scale Straw-to-Heat Supply Chains in Norway

Helmer Belbo* and Bruce Talbot

Department of Forest Technology, Norwegian Forest and Landscape Institute, Steinkjer, Norway

INTRODUCTION

The European Directive on promotion of the use of energy from renewable resources[1] accentuated the urgency of accessing considerable biomass feedstocks, primarily from agricultural- and forest-based residues. An EU25 assessment mapping sustainably produced feedstocks[2] concluded a 2020 potential of 1105 TWh (95 MtOE) from agriculture (including short rotation forestry) and approximately 465 TWh (40 MtOE) from forestry. Reconsidering this potential from a resource-efficiency perspective, the European Environment Agency (EEA)[3] lists efficiency in collection, processing, and transport of feedstocks, i.e., supply logistics, as key elements determining their ultimate respective contributions to the renewable energy (RE) pool.

Nomenclature			
kWh	Kilowatt hour (3.6×106 J)	MWh	Megawatt hour (1000 kWh)
kha	1000 ha	t	Metric ton (1000 kg)
txx%	Ton at xx % moisture content	td	Dry ton (0% moisture)
kt	1000 ton	MC	Moisture content
Energy conversion factor (*q*p,net,d):			
1 td straw=17.57 GJ4 =4.88MWh			

*Correspondence to: beh@skogoglandskap.no

Straw is a solid agricultural residue with applications including fodder mix for ruminants, animal bedding, building insulation, and energy feedstock. It has a net calorific value of ~4.7–4.6 MWh t^{-1} at a typical moisture content at utilization of 14–20%.[5] The International Energy Agency (IEA) first developed an overview of the potential of straw as a source of energy in participant countries in 1983.[6] This trig-gered R&D in the neighboring countries of Denmark and Sweden, where the annual use has steadily increased, and presently represents a robust intake of 900 kt_d in Denmark.[5] Central-southern Sweden has about 830 $kt_{18\%}$ available for energy purposes,[7] but this potential remains largely unexploited, with just 50–70 kt currently utilized in the most prolific county of Scania.[8] Germany by comparison has a resource of 6.9–11.4 Mt, depending on sustainability scenario[9] and Ukraine approximately 1.2 Mt.[10] Norway, a contributor to the initial IEA report,[6] has a land area of 304,250 km^2 of which 3.3% is agricultural land, 2.7% (~821.5 k ha) is arable[11] and 300 k ha is used for cultivating cereal and oil seeds.[12] The country has no industrial-level use of straw as an energy resource, but an increasing number of farmers have set up heating plants to provide heating of poultry barns, greenhouses, or other heat-demanding production facilities. The gross available amount of straw in Norway is estimated to be around 500–700 kt_d (2.5–3.5 TWh), of which about 100 kt_d is annually used for fodder and animal bedding.[13] Approximately 335–469 kt_d could be considered to be sustainably available, according to the EEA's recommended retention thresholds

Advances in Bioenergy: The Sustainability Challenge, First Edition. Edited by Peter D. Lund, John Byrne, Göran Berndes and Iacovos A. Vasalos.

Originally published in: *WIREs Energy Environ* 2014, 3: 411–418 doi: 10.1002/wene.107

of 33–37%.[2] However, the variable weather conditions in the fall set a narrow time frame for the straw harvesting operation, and the technically utilizable volume may lie somewhere below this.

The current national target for Norway is to become carbon neutral by 2050,[14] including an increase in the use of bioenergy by 14 TWh (50.4 PJ) by 2020 without compromising agricultural output. This means that the primary resources for bioenergy in Norway will be low-value wood grades and by-products from agriculture and food production. The space heating profile for Norway differs considerably from neighboring countries, with only around 2% of households being connected to district heating networks, and 8% with common or individual central heating. Around 95% of households use direct electric heating,[15] often supplemented by wood stoves, while air-to-air heat pumps are rapidly becoming the predominant alternative to radiators.

This article describes the current development of straw supply and the use of straw for energy in Norway. The article builds on a recent research project on straw for energy in Norway, where the focus was on productivity and costs of straw supply, workload involved in operating the heating plants and the overall economy of the straw heating system.

STRAW SUPPLY

Straw distinguishes itself from wood fuels (e.g., chips, briquettes, wood pellets) by a lower energy density (0.35–0.63 MWh m^{-3} baled), a higher content of ash, chlorine, and alkaline metals (forming aggressive salts in the flue), and lower ash melting temperature (e.g., 700–840°C for cereal straw ash vs 1000–1200°C for wood ash[16,17]). As for wood fuels, storage stability (mould formation, decay) and the net calorific value are correlated with water content. These properties influence logistical options, supply costs and conversion plant design and management.

Similarly, the straw supply chain consists of a number of consecutive operations, where each operation implies an appropriate change of the form, properties, or position of the straw. The supply chain starts at the cereal field and ends in the furnace, where the major supply chain design parameters include annual volume to be supplied, bearing capacity of the field soil, transport and vehicle legislation, and fuel quality requirements of the heating plant.[18] Baling is the most common way of collecting the straw and is done to promote material handling and storage, increasing straw density from 45–50 kg m^{-3} to approximately 110–150 kg m^{-3}.[19] Prior to baling, raking is an option to both enhance drying of the straw and to merge several windrows to one. Figure 26.1 shows how the unit costs accumulate through

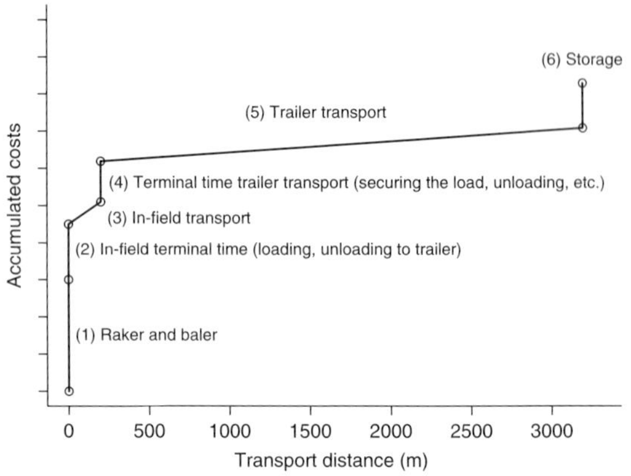

Figure 26.1 Illustration of how the unit costs accrue through the supply chain

a common supply chain. Raking, baling, and storage are stationary processes and cost drivers, whereas transport costs are distance dependent.

The capacity for a process (e.g., baling) is often given in tons $hour^{-1}$, while the capacity of a transport unit is given in ton-km h^{-1}. When analyzing entire supply chains or sequences of operations, these fractions are inverted to input (hours, currency or energy) per production unit (t, m^3, or MWh) making the calculations additive, as the total input per production unit can be summarized directly.

NORWEGIAN FARMERS USING STRAW FOR HEATING

The performance of seven local supply chains located in south-eastern and central Norway (N 59 17′–64 00′, E 10 40′–11 42′) were observed over a period of 3 years to estimate straw yields, supply costs, operating costs of the heating plants, and to collect experiences from supply solutions and energy conversion and distribution. Straw was collected from wheat and barley cultivations. Six farmers used round balers (Ø 150 cm, w 120 cm), and one used a square baler (Hesston or high density baler). The farmers reported crop yield (kg grain per hectare and MC) and straw yield from each field, together with a description of harvesting and transport equipment, transport distances, and time consumption for the different work tasks. The straw bales were stored either in a storehouse, under a tarpaulin cover, or wrapped in plastic film. Six heating plants were batch-fed, implying that one to three bales are burned at once. These plants normally have a higher rated capacity than the heat demand, and surplus heat is temporarily stored in an insulated buffer tank on top of the furnace.

One heating plant was equipped with a bale shredder and conveying system for continuous firing. All plants were equipped with an energy meter to monitor heat delivery from the plant.

Operational costs for the various work tasks (baling, transport, etc.) were estimated from reported time consumption and cost rates provided by Norwegian agricultural periodicals.[20,21] Annual capital costs were estimated for the heating plant and storehouse using the annuity method. This implies a fixed annual cost of capital and depreciation over the depreciation period, which was set to 30 years for buildings and 20 years for technical installations. The salvage value was set to 0, and the inflation-adjusted nominal rate to 5%.

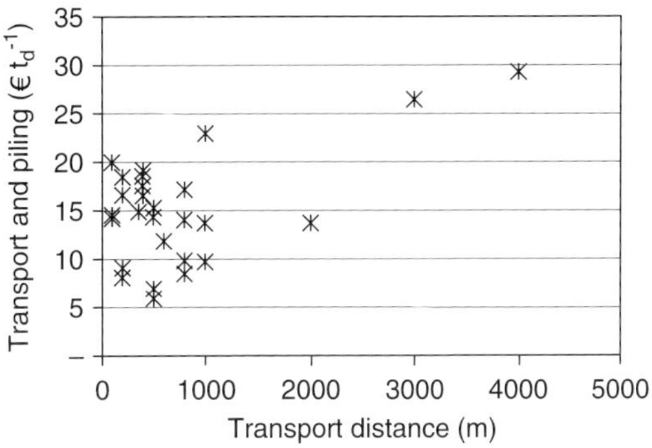

Figure 26.2 Cost of transport and piling of round bales.

Yield Levels, Baling Costs, and Bale Properties

The straw yield was in the range 1.5–3 t_d ha^{-1}, and only weakly correlated ($R^2 = 0.2$) to grain yield.[22] This yield level is lower than what is reported from test plots in Norway (2.0–3.9 t_d ha^{-1})[13] and Denmark (3.2–4.2 t_d ha^{-1})[23] and rules of thumb from neighboring countries (~3 t ha^{-1}),[24,25] but conforms with results from 249 farm assessments in the UK (2–3 t ha^{-1}).[26] Modern cereal varieties allocate more biomass in the seeds and less in the straw; in a crop test in 2001, crops originating from modern varieties had 20–25% less biomass in the straw compared with varieties originating from 1900.[13] Stubble height and raking have an even higher impact on the straw yield, as the straw is substantially more solid at the bottom than at the top,[13] and because raking tears the straw and blades so that more biomass is left in field. A higher stubble height and more raking is assumed in practical situations than in scientific yield trials.

Raking was done on 30% of the collection area. There may be several motives for raking; heavy rainfall makes the otherwise airy windrow collapse, negatively affecting drying and handling at baling. Secondly, merging several windrows may speed up the baling process and reduce soil compaction of the baler. The farmers in this study claimed that they would increase their use of raking, despite the additional costs, as this operation allows washing the straw (by rain) and improves the drying process. But they also underlined the importance of avoiding several consecutive raking operations on the same field, as this reduces the straw yield considerably.

The moisture content of the straw when baled and placed in seasonal storage was between 16 and 25% (wet basis), higher than the 14–20% typically experienced[5] . The density of the round bales was in the interval 84–139 kg_d m^{-3}, with an average of 108 kg_d m^{-3}. There was no traceable relationship between bale density and species origin. The density of the square bales was not measured. In earlier reports these bales had a density 10–30% higher than round bales.[5,25]

Transport

The transport costs in Figure 26.2 are based on the reported time consumption and the hourly rate for man, machine, and equipment set by Norwegian agricultural periodicals.[20,21] The majority of the farmers collect straw from fields surrounding their farm and heating plant, therefore the transport distances were relatively short.

The time consumption was 14 ± 4 min t_d^{-1} for round bale transport using ordinary agricultural trailers and front loaders, and 5 min t_d^{-1} for square bale transport using a self-loading and stacking square bale trailer. For short transports (<2 km), the transport distance had little impact on transport costs while the variation in costs were rather high. The underlying reasons for this variation was not determined, but might be tied to (1) lower focus on transport efficiency on short transport distances (implying suboptimal choice of transport equipment), (2) that the layout at the seasonal storage affects maneuverability and time consumption, this will give relatively more variation on shorter rather than longer transport distances. Time consumption and cost models for straw supply are presented in Table 26.1. The reason for the relative high hourly rate for the big baler and the self-loading big-bale trailer is the low annual use of these specialized machines.

Seasonal Storage

Because of the low energy density and the importance of maintaining low moisture content, a low-cost, well-functioning storage is crucial for economy and user satisfaction. The straw bales were stored either in a storehouse, under a tarpaulin cover, or wrapped in plastic film. The

Table 26.1 Time Consumption and Cost Models for the Different Work Tasks in the Supply Chain

		Specific Time Consumption, min odt^{-1}		
Task	Equipment	Round Bale[1]	Square Bale[2]	Hourly Rate, €
Raking	Rake	6.5	6.5	67
Baling	Round baler	21	—	75
	Big baler	—	6.2	180
Transport cost function	Tractor with loader (four bales)	7.1+14.1d[3]	6.5+12.9d[3]	63.75
	Tractor, 2.4m×4m gravel trailer	14.1+4.6d3	12.9+2.9d3	68.75
	Tractor, 2.4m×8.5m bale trailer	14.1+1.3d[3]	12.9+1.2d[3]	71.25
	Tractor, self-loading trailer	—	4.3+1.6d[3]	187.5

[1] 1Ø=1.5m, V=2.12m3, mass=212 kgd.
[2] 2h;w=1.3m; 1.2 m, V=3.74m3, mass=430 kgd.
[3] 3d is the transport distance (km) from the field to the seasonal storage.

main advantage of the two latter options is obviously the low capital cost. However, those using tarpaulin or plastic cover in Norway reported major challenges of retaining proper fuel quality during the winter. The bales got moist either from below by surface water or from above by condensation or both, even when stored on a well-drained area. Similar experiences are reported from Sweden.[25] During the project period the share of farmers having a storehouse for the straw bales doubled from 0.43 to 0.86. To estimate storage cost per ton (or MWh), one needs to include the annual capital costs and the total dry mass stored. The interactions between investment cost, interest rate, utilization, and bale density are illustrated in the nomogram in Figure 26.3. The establishment costs for the new storehouses were in the range 10–42 € m^{-3} space volume. The spatial utilization, i.e., bale volume compared with gross available storage volume, was in the range 63–74% for round bales and 90% for the square bales.

Heat Production and Plant Management

The farmers reported energy production and time consumption for each occasion stoker work was carried out at the plant. There was no clear relationship between stoker work and plant size, and a large variation in time consumption for the batch-fed plants (Figure 26.4), most of which seems to be related to storage facility design. Using a tarpaulin cover one has to open and seal the tarpaulin every time new bales are to be fed to the plant. This might also imply snow removal and rejecting bales that have become wet. The continuously fed plant had a far lower time consumption, which is natural as 12 big bales were placed on a belt conveyer each time, and sweeping and ash removal was done automatically.

The total heating costs for the Norwegian farmers are shown in Table 26.2. Within the supply chain, the cost of baling, transport, and storage are of equal magnitude. Square bale supply is 20% cheaper than round bale supply. The higher capital cost of the more advanced continuously fed heating plant is offset by the reduced time consumption for stoker-work and increased plant efficiency.

The total cost of straw firing was in the range 36–42 € MWh^{-1}, which is 40–46% of the current cost of electric heating (~90 € MWh^{-1}).

Box 26.1 EMERGING CONVERSION TECHNOLOGIES FOR STRAW FUEL

While this article deals only with combustion of straw for heat generation, considerable efforts have been made to develop more sophisticated conversion technologies for lingocellulosic feedstocks like straw in recent decades.[27] Gasification (gas from substoichiometric combustion) and digestion (gas from anaerobic digestion) are promising technologies, where the product (gas) can be used for heat, heat and power, or vehicle fuel. New pretreatment methods, enzymes, and yeasts also open the option of ethanol fermentation of lignocellulosic biomass.[28,29] Besides the more diverse energy carriers produced, these technologies also provide increased recovery of nutrients[30] and may alleviate the need for compensatory fertilizing on cereal fields. In some of these technologies (e.g., anaerobic digestion), the straw in fact can be used twice; first as animal bedding in deep straw beds, and then as an energy resource in the digestion chamber. And best of all; the bedding phase may improve the digestion phase, and the digestion phase improves the fertilizing properties of the straw (and manure) for subsequent use on the fields.[31]

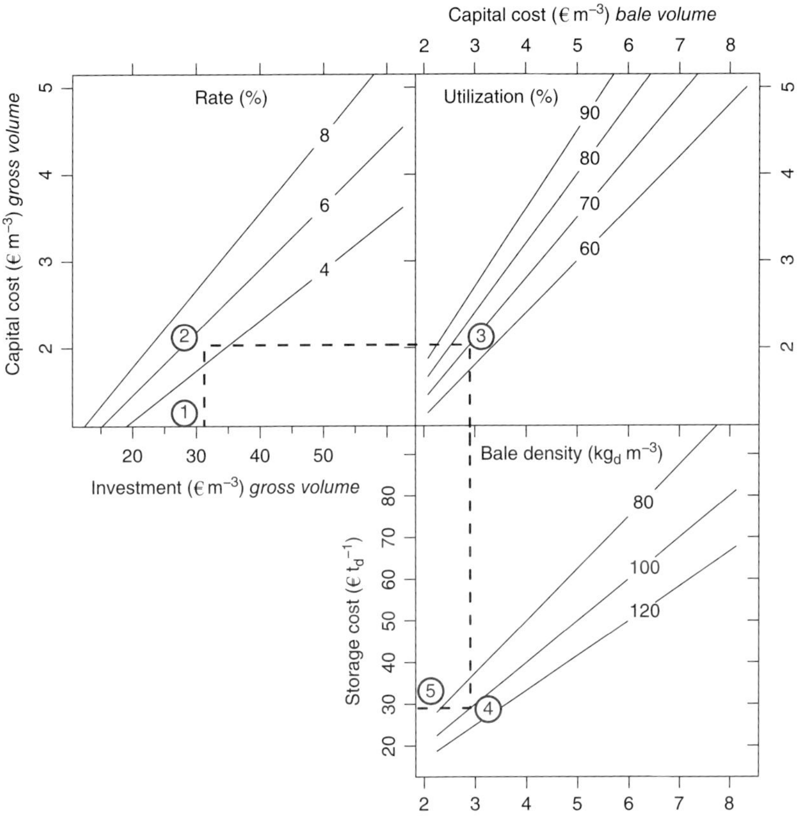

Figure 26.3 Nomogram illustrating how storage cost (per ton) is related to investment, rate of interest, packing density (utilization), and bale density. At a given investment (1) and rate (2) it provides the annual capital cost per gross storage volume. At a given space utilization level (3), i.e., net bale volume compared with gross storage volume, and bale density (4), the capital cost per ton dry material can be read (5).

DISCUSSION

The farmers in this study over a 3-year period always managed to collect straw of fair quality for their heating plants although MC in bales of 16–25% exceed the 14–20% common elsewhere.[5] Costs of supply to farm gate ranged from 13.30 to 16.50 € MWh^{-1}. Comparative costs in the UK were 9.31 € MWh^{-1} although much variation was noted,[32] whereas a supply cost of 7.7 € MWh^{-1} was assumed in a recent Swedish study[7] . At a smaller scale, the use of straw for heat production has been a sound strategy for Norwegian farmers to utilize own resources and detach themselves from cost fluctuations on other energy commodities—this notion is supported by their willingness to make further investments in storage facilities.

Large-scale utilization (CHP or district heating) would require significantly more organization and commitment through contractual delivery, while reducing feedstocks

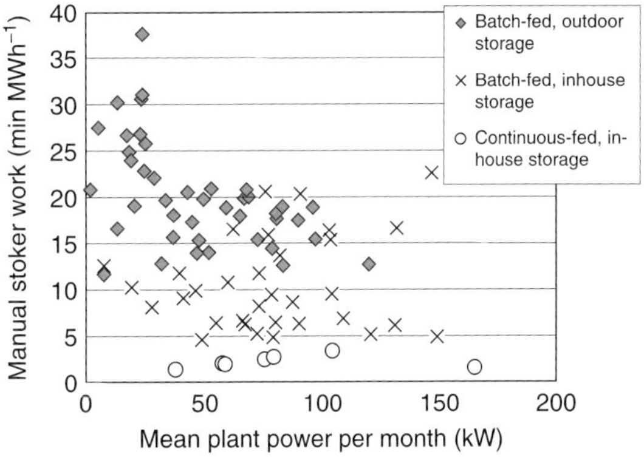

Figure 26.4 Time consumption for firing, sweeping, and ash removal per MWh, grouped by plant technology and straw storage conditions.

Table 26.2 Total Cost of Heat Production, € MWh−1

Activity	Square Baler, Cont. Firing	Round Baler, Batch Firing
Raking	1.54	1.54
Baling	3.95	5.58
Transport	3.56	3.88
Storage	4.25	5.52
Sum supply cost	13.30	16.53
Plant efficiency correction	*0.75*	*0.85*
Real supply cost	17.73	19.44
Stoker work	2.08	5.21
Plant capital cost	18.28	14.24
Sum cost, € MWh^{-1}	38.10	38.89

available for farm-level energy systems. Procurement costs for large-scale plants could be expected to increase at a higher rate with increasing required volume than in neighboring countries, because of the often narrow and elongated procurement areas characterized by the Norwegian topography. However, the transition from small scale to large scale requires a change in motivation from the purely economic to the legislative and political.[7]

The case of Denmark shows how crucial a constructive political framework, implementation of policy instruments, and cooperative supply organizations are for large-scale mobilization of the straw feedstock.[8] In Sweden, which has a highly developed district heating network, extensive water-borne heating in buildings, and shares many similarities in supply structure with Denmark, only roughly 10% of the 830 k t resource is currently utilized for energy. The lesson for Norway being that the development of effective supply chains alone is not likely to unlock the straw feedstock potential for larger-scale utilization.

Straw yields are an important driver in supply economics, but there is considerable variation in yields reported in research. Larsen et al.[23] clarify this in showing how yields differed between and within species, cultivars, and years (climatic conditions). This implies a considerable margin for error in making comparisons across regions, countries, or years. It also emphasizes the importance of adjusting forecasted yields during the growing season, in planning operational aspects of straw supply.

CONCLUSIONS

The total cost of straw firing was about 40–46% of the current cost of electric heating (~90 € MWh^{-1}). This confirms that there is sufficient justification for expansion of small-scale straw-based heating facilities for economic reasons alone, and especially so if electricity prices increase during cold periods.

Baling, storage, and stoking costs for round bales were considerably higher than for square bales. Assuming that most farmers would choose to minimize the frequency and duration of stoking the boiler, it can be concluded that for energy purposes, a square-bale supply and continuously fed combustion chain is preferable to one based on round bales.

The capital cost of storage is the single largest expense element not directly related to conversion. The development of conversion technologies less sensitive to the moisture content of the fuel would both contribute to reducing the temporal and qualitative constraints on procurement while simultaneously improving the economic feasibility of the straw-to-energy pathway.

ACKNOWLEDGMENTS

The publication was compiled as a contribution to IEA Task 43, *Biomass Feedstocks for Energy Markets*, and funded by the Norwegian Bioenergy Innovation Centre, CenBio. CenBio is jointly funded by the Research Council of Norway (193817/S60) and the participating research and industrial partners.

REFERENCES

1. EC. Directive 2009/28/EC of the European Parlia-ment and of the Council of 23 April 2009 on the promotion of the use of energy from renewable sources, 2009. Available at: http://eur-lex.europa.eu/LexUriServ/LexUriServ.do?uri=Oj:L:2009:140:0062:en:PDF. (Accessed August 16, 2013).
2. EEA. How much bioenergy can Europe produce without harming the environment? *EEA Report 2006*, vol 7, 2006, 72 p. Available at: http://www.eea.europa.eu/publications/eea_report_2006_7. (Accessed October 5, 2013).
3. EEA. EU bioenergy potential from a resource-efficiency perspective. *EEA Report 2013*, 2013, 60 p. Available at: http://www.eea.europa.eu/publications/eu-bioenergy-potential. (Accessed October 5, 2013).
4. CEN/TC 335—solid biofuels. EN 14961–1:2010 solid biofuels—fuel specifications and classes—part 1: general requirements. *Annex B.5—typical values for virgin straw materials*, 2010, 40 p. Available at: http://www.standard.no/en/. (Accessed September 15, 2013).
5. Nikolaisen L, Nielsen C, Larsen MG, Nielsen V, Zielke U, Kristensen JK, Holm-Christensen B. *Straw for Energy Production: Technology–Environment–Economy*. 2nd ed.; 1998, 53 p (Also published in Danish and German, 1998). Available at: http://www.videncenter.dk/. (Accessed September 8, 2013).

6. Kjus O. Utilization of straw as a source of energy. Preparatory IEA-project. Status and planning report. Report from Norway. *Stensiltrykk Ser A* 1983, 727:10.
7. Ekman A, Wallberg O, Joelsson E, Börjesson P. Possibilities for sustainable biorefineries based on agricultural residues—a case study of potential straw-based ethanol production in Sweden. *Appl Energy* 2013, 102:299–308.
8. Voytenko Y, Peck P. Organisational frameworks for straw-based energy systems in Sweden and Denmark. *Biomass Bioenergy* 2012, 38:34–48.
9. Weiser C, Zeller V, Reinicke F, Wagner B, Majer S, Vetter A, Thraen D. Integrated assessment of sustainable cereal straw potential and different straw-based energy applications in Germany. *Appl Energy 2014*, 114:749–762.
10. Voytenko Y, Peck P. Organization of straw-to-energy systems in Ukraine and Scandinavia. *Biofuels Bioprod Biorefining* 2011, 5:654–669.
11. World Bank. *Databank—Agriculture & Rural Development.* Available at: http://data.worldbank.org/topic/agriculture-and-rural-development. (Accessed August 10, 2013).
12. SSB (Statistics Norway). Grain and oilseed crops; table 04613 2013. Available at: https://www.ssb.no/en/jord-skog-jakt-og-fiskeri/statistikker/korn. (Accessed August 22, 2013).
13. Riley H, Åsven M, Todnem J, Eltun R. Halm som biobrensel—Tilgjengelige halmmengder, halmbehov til dyreforög strø/talle, samt konsekvenser av halmfjerning for jordas bæreevne og kvalitet. *Bioforsk Rapport* 2012, 58 p. Available at: http://www.bioforsk.no/ikbViewer/page/tjenester/publikasjoner/publikasjon?p_document_id=100889. (Accessed August 10, 2013).
14. Norwegian Climate Policy. Norwegian Ministry of the Environment 2007. Available at: http://www.regjeringen.no/nb/dep/md/dok/regpubl/stmeld/2006-2007/Stmeld-nr-34-2006-2007-.html?id=473411. (Accessed August 17, 2013).
15. SSB (Statistics Norway). Energy consumption in households, 2009. Available at: http://www.ssb.no/en/husenergi. (Accessed October 10, 2013).
16. Toscano G, Corinaldesi F. Ash fusibility characteristics of some biomass feedstocks and examination of the effects of inorganic additives. *J Agric Eng* 2010, 41: 13–19.
17. Olanders B, Steenari B-M. Characterization of ashes from wood and straw. *Biomass Bioenergy* 1995, 8:105–115.
18. Nilsson D, Hansson PA. Influence of various machinery combinations, fuel proportions and storage capacities on costs for co-handling of straw and reed canary grass to district heating plants. *Biomass Bioenergy* 2001, 20:247–260.
19. Nielsen V. Teknik til halmbjærgning siden 1950. *DJF Rapport* 2003, 28 p. Available at: http://web.agrsci.dk/djfpublikation/djfpdf/djfma95.pdf. (Accessed October 4, 2013).
20. Berger H. Leiekjøringspriser. *Norsk Landbruk*, vol. 07/10, 2010. Available at: http://www.norsklandbruk.no/media/1660529/leiekj%C3%B8ringspris%202010.pdf. (Accessed August 14, 2013).
21. Agjeld J. Maskinleie Prisliste. *Bedre gardsdrift*, vol. 8, 2010. Available at: http://www.gardsdrift.no/asset/25817/1/25817_1.pdf. (Accessed October 8, 2013).
22. Belbo H. Halm som biobrensel. *Rapport fra skog og landskap* 2011, 24 p. Available at: http://www.skogoglandskap.no/filearchive/rapport_22_11_halm_som_biobrensel.pdf. (Accessed August 12, 2013).
23. Larsen SU, Bruun S, Lindedam J. Straw yield and saccharification potential for ethanol in cereal species and wheat cultivars. *Biomass Bioenergy* 2012, 45:239–250.
24. Presning og håndtering av halm i storballer. *Videnblad* 1998. Available at: http://www.videncenter.dk/Videnblade-dok/VB120.pdf. (Accessed August 9, 2013).
25. Bernesson S, Nilsson D. Halm som energikalla¨. *Rapport—miljö, teknik ock landtbruk* 2005, 7: 88p. Available at: http://publikationer.slu.se/Filer/SLU_BT_R2005_07_Halm_Bernesson_Nilsson_tryckfil.pdf. (Accessed August 9, 2013).
26. Glithero NJ, Wilson P, Ramsden SJ. Straw use and availability for second generation biofuels in England. *Biomass Bioenergy* 2013, 55:311–321.
27. Faaij A. Modern biomass conversion technologies. *Mitig Adapt Strat Glob Chang* 2006, 11:335–367.
28. Biswas R, Uellendahl H, Ahring BK. Conversion of C6 and C5 sugars in undetoxified wet exploded bagasse hydrolysates using *Scheffersomyces (Pichia) stipitis* CBS6054. *AMB Express* 2013, 3p.
29. Vaaje-Kolstad G, Westereng B, Horn SJ, Zhanliang L, Xhai H, Sørlie M, Eijsink VGH. An oxidative enzyme boosting the enzymatic conversion of recalcitrant polysaccharides. *Science* 2010, 330:219–222.
30. Skøtt T. Straw to energy—status, technologies and Innovation in Denmark 2011. 2011. Available at: http://www.inbiom.dk/download/viden_biomasse/halmpjeceuk_2011.pdf. (Accessed August 9, 2013).
31. Jørgenssen PJ. Praktisk anvendelse av dybstrøelse som substrat paå biogasanlæg. 2013. Available at: http://www.ens.dk/sites/ens.dk/files/undergrund-forsyning/vedvarende-energi/bioenergi/biogas-taskforce/anvendelse_af_dybstroeelse_til_biogasproduktion.pdf. (Accessed August 9, 2013).
32. Littlewood J, Murphy RJ, Wang L. Importance of policy support and feedstock prices on economic feasibility of bioethanol production from wheat straw in the UK. *Renew Sustain Energy Rev* 2013, 17: 291–300.

27

Transport Sector in Ireland: Can 2020 National Policy Targets Drive Indigenous Biofuel Production to Success?

Egle Gusciute, Ger Devlin,* Fionnuala Murphy and Kevin McDonnell

Biosystems Engineering Department, University College Dublin, Belfield, Dublin, Ireland

INTRODUCTION

Biofuels are renewable fuels produced from biomass. They produce fewer emissions than the conventional fossil fuels and can be subdivided into three groups: biodiesel, bioethanol, and pure plant oil (PPO). Biofuels, a renewable source of fuel, have received much attention in the last decade due to the rise in oil prices, the increase in greenhouse gas (GHG) emissions from fossil fuels, and the need for security of energy supply produced in a sustainable and competitive manner. In Ireland in 2009, biofuels were mostly imported rather than produced indigenously.[1]

Transport Sector in Ireland

Transport is the fastest growing sector in Ireland, in terms of primary energy demand and carbon dioxide emissions (CO_2).[2] In 2007, transport accounted for 43% of final energy demand and 36% (17,014 kt CO_2) of Ireland's energy-related CO_2 emissions (higher than any other sector).

Figure 27.1 illustrates the primary energy demand and carbon dioxide emissions by sector from 1990 to 2010. Although almost all sectors have experienced a decline, the increase in the transport sector is significant. From 1990 to 2010, the average growth of energy use in transport sector is 4.3% per annum, which totals to 132% over the 20 years.[3] Overall the transport sector consumes slightly less than one third of all energy in Ireland.[3]

The transport sector is entirely depended on oil imports. In 2008, 89% of Ireland's energy was imported; in comparison, only 55% of energy was imported in the European Union (EU) as a whole.[1] The fact that Ireland's energy imports have remained at around 90% since 2001[1] has serious implications for the security of energy supply. In 2009, oil accounted for 60% of total energy imports representing the 5th highest in the EU in 2009.[1] Ireland therefore must look to indigenous alternative renewable fuels if it is to reduce its import energy dependency.

Figure 27.2 demonstrates Ireland's dependency on imports, especially diesel. Imports have decreased since 2008; nevertheless, this is mostly due to the decrease in the economic activity rather than a reduced dependency on imports.

Thus, policy implementations and mechanisms are necessary in the transport sector if Ireland is to reach its 2020 EU targets, and guarantee security of energy supply and a competitive and sustainable energy in the future.

Transport final energy demand for 2020 is 5375 ktoe.[5] Achieving the biofuels target would require 345 ktoe by 2020 (14.4 PJ).

*Correspondence to: ger.devlin@ucd.ie

Advances in Bioenergy: The Sustainability Challenge, First Edition. Edited by Peter D. Lund, John Byrne, Göran Berndes and Iacovos A. Vasalos.

Originally published in: *WIREs Energy Environ* 2014, 3: 419–430 doi: 10.1002/wene.84

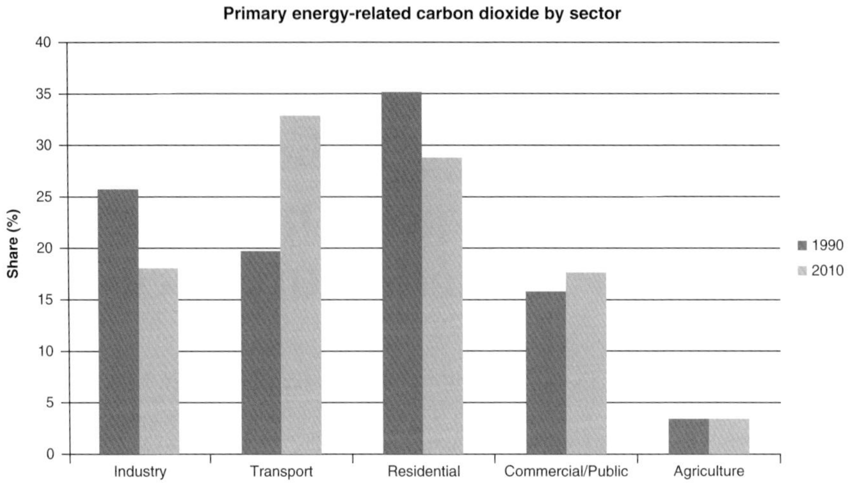

Figure 27.1 Primary energy-related carbon dioxide by sector. Created using data from Ref 3.

POLICY

EU Directives

In 2007, the EU agreed new climate and energy targets, 20–20–20 by 2020, 20% reduction in GHG emissions; 20% increase in energy efficiency; and 20% of the EU's energy consumption to be from renewable sources by 2020. The EU supports the use of biofuels through two main directives: The directive on the promotion of the use of biofuels and other renewable fuels for transport or Biofuels Directive (2003/30/EC)[6] and the Renewable Energy Directive (2009/28/EC).[7]

The Biofuels Directive

The Biofuels Directive 2003/30/EC entered into force in May 2003 and is primarily concerned with the promotion of the use of biofuels in the transport sector. Each member state was required to replace 5.75% of all transport fossil fuels with biofuels by 2010. The directive also set an intermediate target of 2% by December 2008. The Irish Government White Paper committed to achieving the 5.75% of road and rail transport energy from renewable sources by 2010[8] but this was later revised to 3%.[9] This was replaced with Directive 2009/28/EC detailing targets of 10% by 2020 for every member EU state.[3]

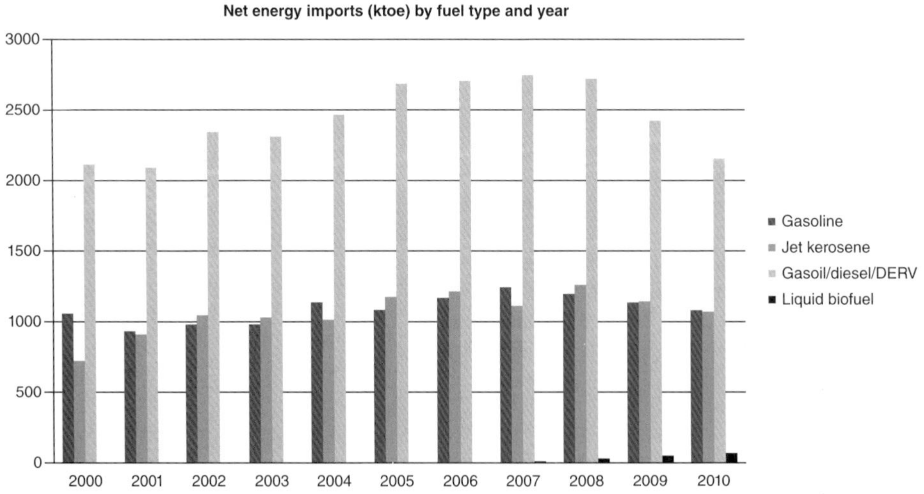

Figure 27.2 Fuel imports (2000–2010). Created using data from Ref 4.

The Renewable Energy Directive

The EU has committed to reduce GHG emissions under the Kyoto Protocol.[10] In order to reach the 8% binding target on 1990 levels between 2008 and 2012 and to develop a sustainable energy plan for Europe, a renewables directive came into force in 2009, the Fuel Quality Directive 2009/30/EC. This directive outlines targets to be achieved by 2020; 20% of total energy to come from renewable sources, a 20% reduction in GHG emissions, and a 20% increase in energy efficiency.[11] Each member state was assigned specific targets in order to achieve the overall target. Ireland's target is set at 16% by 2020.[7] A 10% target is set for all member states to be achieved in the transport sector from renewable sources.[7] Because of the concerns regarding food security and land use change, the European Commission has decided to limit the contribution of food-based biofuels to 5% of the overall transport target.[12] Any fuel above this target must not be based on food crops. The fuel quality directive also outlines sustainability criteria for biofuels that monitor how and where they are produced. All biofuels must achieve at least 35% GHG emissions reduction from 2013 in comparison with conventional fossil fuels. This figure rises to 50% in 2017, and further to 60% for biofuels produced in installations in which production started on or after January 2017.[7]

National Policies and Support Mechanisms

The government first outlined its commitment to reaching the 2020 targets in the Government White Paper on Energy in 2007. It set national targets, in line with the EU targets, committing to a 20% target to be achieved across the electricity, heat, and transport energy sectors by 2020. In the transport sector, renewables would account for 5.75% of all road transport by 2010 and 10% by 2020. The National Climate Change Strategy 2007 to 2012, detail further measures in which Ireland would meet its Kyoto commitments and enable Ireland to meet the 2020 targets.[13] A National Biofuels Obligation was set at 5% by 2010 for all fuel suppliers. However, in 2008 it was lowered to 4% by 2010 due to concerns with the impact of biofuels on food prices.[1] According to the National Oil Reserve Agency (NORA) statistics, only 2.2% target was met in 2010. The fuel blend reached 2.87% for biodiesel in motor fuel and 3.42% for bioethanol in petrol for 2012.[14]

The Irish government outlined its commitment to sustainable energy production in the publications mentioned above and it introduced various policy support schemes and mechanisms.

Mineral Oil Tax Relief Scheme

The Mineral Oil Tax Relief Scheme (MOTR) was a pilot scheme introduced in 2005 and granted motor tax relief to approved biofuel suppliers. It developed into a more substantial scheme in 2006 that ran until 2010. It was designed to incept a national biofuels industry by offering tax incentives whereby producers could sell the biofuels without excise duty, thus making it cheaper than the conventional fossil fuel alternative. This was mostly targeted toward captive transport fleets that used their own fuel tanks on site or in the truck or bus depots. The total excise derogation would stand to the cost the taxpayer »205 million. MOTR II was introduced in 2006 and ran until 2010. As a scheme, it failed to reach the desired results and outcome as; (1) only 16 companies were granted a place in the scheme; (2) many companies did not have facilities that were required to produce biofuels, thus although they were under the scheme, they did not produce anything; (3) changing market conditions and the availability of cheaper imported alternatives made it difficult to compete in the commercial market.[15] The uptake of MOTR II was slow, with less than 28% of the relief used by the end of 2009.[15] Although this scheme was always going to be temporary, and the sector would have had to survive on its own in the commercial market, without the exemption from the excise duty, producing biofuel at competitive market prices is proving difficult.

Biofuels Obligation Scheme 2010

In 2010, when MOTR II scheme ended, the Biofuels Obligation Scheme was introduced. A subsidy scheme was replaced by an obligation scheme, under which all road transport fuel suppliers are obliged to use biofuel in the fuel mix (4 L of biofuel in every 100 L of transport fuel) to ensure that a certain percentage is represented in the annual sales.[15,16] The scheme is administered by the NORA and began on July 1st 2010. The starting penetration rate is 4% per annum and will be increased over time.[15,16] Under the BOS scheme, a fuel supplier must apply to NORA (National Oil Reserves Agency) for a cert every time it puts 1 L of biofuel on the market. The certs are totaled at the end of the year and if the supplier has put enough biofuel on the market then the certs will testify this. If not, then the fuel supplier must either buy certs from a surplus fuel supplier or either pay a buy-out levy for approximately 45 cent/L. The share of transport energy from biofuels has increased from 1 ktoe in 2005 (0.03%) to 92 ktoe in 2010 (2.4% in energy terms).[3] Although this scheme has so far increased the use of biofuels, it has not necessarily increased the production of indigenous biofuels as preblended fuels are imported at more competitive prices. Overall the scheme has resulted in major fuel companies bypassing smaller indigenous producers. Furthermore, the value of the biofuel certs issued to biofuel producers cannot be determined until the end of the year.[17]

Vehicle Registration and Annual Motor Tax Change

Under the Kyoto Protocol, Ireland is legally bound to meet its set target of 13% reduction in GHG emissions above 1990 levels in the period 2008 to 2012.[3] Transport is the largest CO_2 emitting sector. In 2010, CO_2 emissions were 129% higher than in 1990 (4.2% average annual growth rate), falling for the first time in 2008 by 1.8%. In the 2008 budget, it was announced that the vehicle registration (VRT) and annual motor tax systems would base the tax rates on the specific CO_2 emissions (CO_2 grams per kilometer—g/km) rather than engine size. This incentive came into effect in July 2008 and is aimed at encouraging the consumer to purchase more fuel efficient vehicles with lower GHG emissions.[3] Of the 86,932 new private cars licensed in 2011, 23,246 (27%) were petrol based, whereas 61,730 (71%) were diesel. Before the change in this tax-based system in 2008, the trend for petrol cars was always higher than diesel. In 2008, 92,298 petrol cars and 50,560 diesel cars were newly licensed. This began to swap around from 2009 onwards when the trend was 22,802 petrol cars and 30,645 diesel in 2009. In 2010, the figures were 27,124 petrol cars and 53,998 diesel cars.[18]

The National Renewable Energy Action Plan 2010

The government published the National Renewable Energy Action Alan in 2010 in accordance with the requirements of the EU directive 2009/28/EC. It sets out the strategic approach measures that will be used to meet Ireland's 16% energy target of renewables by 2020. In reference to transport, the Government plans to reduce the dependency on oil imports by increasing the use of biofuels and the use of electric vehicles.[16]

Policy Constraints

EU Sustainability Criteria

The increased use of biomass for biofuel production has led to concerns regarding the sustainability of this practice. Concerns surround the methods of cultivating and producing biofuels, particularly in regard to actual GHG emissions reductions in comparison with fossil fuels, and in concerns with land use change due to increased demand for arable land for biomass production. In order to ensure the sustainability of biofuel used to achieve the targets in the EU, the European Commission proposed a set of sustainability criteria in the Directive 2009/28/EC on the promotion of the use of energy from renewable sources. The sustainability criteria consist of the following main points:

- The directive lays out certain GHG emissions reductions to be achieved from the use of biofuels. In the case of biofuels and produced by installations that were in operation on 23 January 2008, GHG emissions savings must be at least 35% from 2013. This figure rises to 50% in 2017, and further to 60% for biofuels produced in installations in which production started on or after January 2017.
- The raw materials sourced for biofuel production, from within the EU or from third countries, should not be obtained from land with high biodiversity value, land with a high carbon stock, or land that was peatland in 2008.

These criteria, though undoubtedly good for the sustainable production of biofuels, may restrict growth of the biofuel production industry in Ireland, as biofuels must meet certain minimum criteria.

Common Agricultural Policy

The common agricultural policy (CAP) has mixed effects on the growth of energy crops. On one hand, energy crop production is supported by the energy crops scheme that provides grant aid, and by nonfood set aside policy. On the other hand, the land available for energy crop production is restricted by the cross-compliance policy. Cross-compliance restricts the conversion of grassland to arable cropping. Cross-compliance regulations require that the ratio of the area of permanent pasture to the total agricultural area of each Member State must not decrease by 10% or more from the 2003 reference ratio (EC, 2004). Ireland is therefore under obligation not to allow any significant reduction in the total area of permanent pasture (91% of agricultural land area in Ireland),[19] which restricts the type of energy crops that can be grown.[20]

A reform of the CAP is due with new rules coming into effect in 2014, which may affect the production of biomass in the EU.

BIOFUELS

The two liquid biofuels that penetrate the transport sector in Ireland are bioethanol and biodiesel, both 'first generation' fuels due to the type of feedstock and technologies used to produce them.

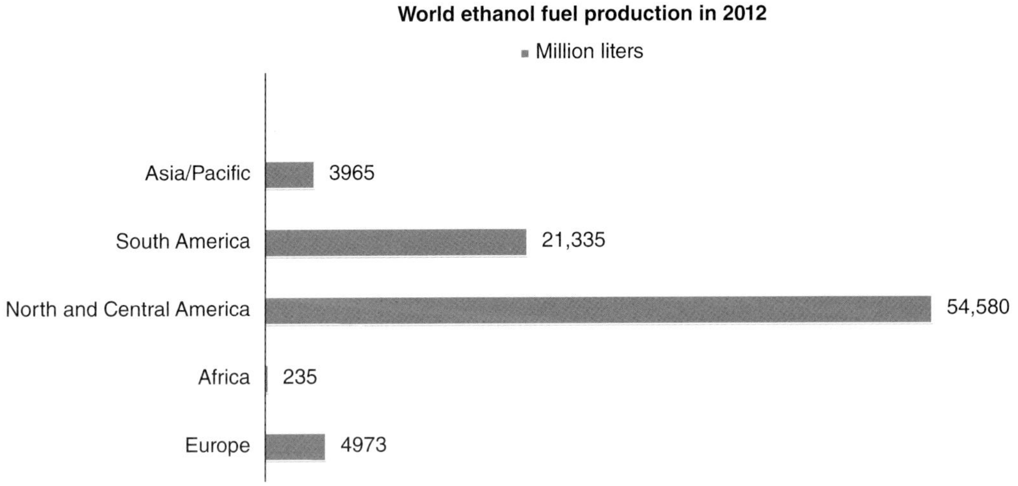

Figure 27.3 World ethanol production. Created using data from Ref 30.

Bioethanol

Bioethanol is a biofuel produced by agricultural feedstocks that are high in sugar or starch such as sugarcane, sugar beet, corn (maize), and wheat, as well as lignocellulosic crops.

Bioethanol suitable for EU blending requirements can be produced via a number of pathways. The general production pathway involves pretreatment of feedstocks, a saccharification process to release the fermentable sugars from polysaccharides, fermentation of released sugars, and finally a distillation step to separate ethanol from the liquid product.[21,22] The technology used to liberate the sugar molecules, and where the sugar is liberated from, are the two factors which generally differentiate between one production process and another. Ethanol production from sugar crops such as sugar beet and cane is the most straightforward process as the sugar molecules are relatively easy to liberate.[21] Bioethanol production from starch crops is more complex as it requires a pretreatment step. These two processes are 'first generation' processes. Ethanol production from lignocelluloisc biomass, such as agricultural residues, is a 'second generation' process, requiring a more complex production system involving either thermochemical or biochemical processes.[23]

Bioethanol is suitable for use in conventional petrol engines, without any modification, when blended up to a rate of 5% with fossil petrol fuel. It can be blended up to 85% (known as B85) for use in engines, which have been specifically modified for this purpose.[24] Several cities in Europe are using higher blends of biofuels in 'captive fleets' such as public transport systems and in employee vehicles.[6]

The largest world producer of bioethanol is the USA,[25] followed by Brazil,[26] as outlined in Figure 27.3. Brazil produces bioethanol from sugar cane, which accounts for 40% of overall fuel use in Brazil.[21] The USA is currently the largest producer of bioethanol, whereas at the same time, it is also the largest importer of Brazilian bioethanol.[26] Costs of producing bioethanol in Brazil are less than half of those in Europe, approximately \$0.20/L for new plants, in comparison with approximately \$0.55/L, including subsidies in Europe.[27] Production of bioethanol from sugar cane (or sugar beet in Ireland's case) could prove to be a low cost source of biofuel in the future,[28] with overall ethanol production expected to increase as the production costs will decrease faster than those of biodiesel. There are efforts now to revive sugar beet production in Ireland and to develop a bioethanol facility that requires 1 million tonnes of sugar beet and 56,000 tonnes of grain to produce 154,000 tonnes of sugar and 50 million L of ethanol at an estimated plant construction cost of »350–»400 million.[29]

Biodiesel

Biodiesel is a biofuel produced from oil-rich crops such as oil-seed rape, soybeans, and jatropha, or residues such as recovered waste vegetable oils (RVO) and animal fats (tallow). A number of different production pathways can be followed to produce boiodiesel. Firstly, the vegetable oil must be extracted from the feedstock using chemical or mechanical means. Following extraction, the oil-rich mixture is treated to remove impurities such as large particles (pieces of plant tissue, mucilages, resinous products), fine

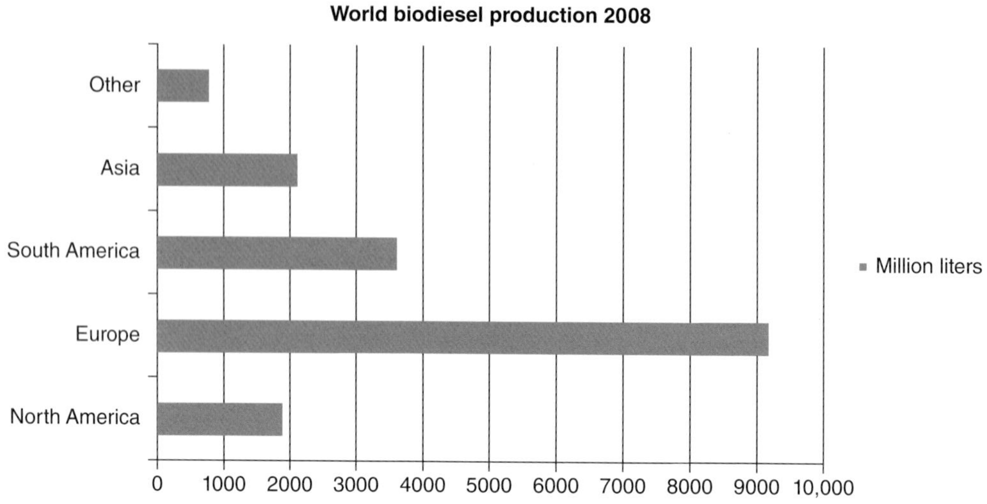

Figure 27.4 World biodiesel production. Created using data from Ref 34.

colloidal particles (invisible in transparent oil), free fatty acids produced by acylglycerols hydrolysis and colored and semi-volatile substances dissolved in the oil.[31] The oil from the feedstock is then processed by transesterification to produce methyl ester. Transesterification is a chemical process that requires the use of alkaline, acid or enzymatic catalyzers and ethanol or methanol, and produces fatty acid (biodiesel) as the main product and glycerine as a by-product.[21] A number of different pathways can be followed depending on type of catalyst, biphasic or monsphasic reaction systems, and pressure and temperature, depending on the feedstock used.[32]

Biodiesel can be used in conventional compression-ignition diesel engines, normally between a 5% and 7% blend to avoid any effect on engine performance.[24] Biodiesel can be used as a pure fuel with only minor modifications to the engine compared with pure bioethanol.[33]

The EU is currently the largest producer of biodiesel, accounting for approximately 52% of worldwide production (Figure 27.4).[34]

Indigenous Production

As outlined previously, there is very little production of indigenous biofuels in Ireland. The two main sources of feedstock for ethanol production in Ireland are sugar beet and wheat.[35] Despite the availability of these resources, there is no ethanol producing facility in the country except for Carbery based in Co Cork who produce and export ethanol from whey, which is a by-product of their cheese making plant. All of the ethanol contained in fuel blends is currently imported.

Ethanol production in Ireland faces a major obstacle; it requires government subsidies to be able to produce it at competitive prices when compared with traditional fossil fuels.[35] For example, the proposed ethanol production site at Mallow required economic support in order to produce ethanol at competitive prices and under the conditions of MOTR II it could have only survived if further incentives or excise relief had been available.

A further limitation on the development of the biofuels industry in Ireland is the amount of land required to produce it. There are several competing uses for agricultural land in Ireland; arable land for crop production (9%), grassland for beef and dairy production and set aside (91%).[19] Production of biofuels on land traditionally used for food production or for recreation could cause a significant public outcry over the use of this land for biofuels, as well as conflicting with the CAP as outlined above.[24] Recent discussions in Europe suggest that the target for crop based biofuels will be changed from a minimum of 10% to a maximum of 5%. This will undermine any potential investments for biofuel industries in Ireland and beyond. Murphy and Power[36] have estimated that to meet the Biofuels Directive of 5.75% using biodiesel produced from rapeseed oil, 6.3% of Irish agricultural land is required to produce the crop, this equates to 70% of arable land. If the target is met using ethanol produced from wheat, 3.9% of Irish agricultural land is required.

Green Biofuels Ireland (GBI), based in New Ross, County Wexford, operate a state-of-the-art 30,000 tonnes capacity plant that uses RVO and tallow to produce 34.5 million L of biodiesel annually. When compared with the GHG emissions from fossil diesel, this biodiesel makes an 83% saving, important for EU sustainability criteria.

The biodiesel is sold to major oil distributors in both the Republic and Northern Ireland.[17]

Tallow Availability

Tallow refers to the inedible animal fats produced as a by-product to the renderingindustry and produced by the rendering process. Singh et al.[19] takes into consideration the different existing markets for tallow, and predicts half of the tallow available for sale in 2020 (19,000 tonnes) will be converted to biodiesel. This results in a practical energy of 0.715 PJ per annum.

RVO Availability

RVO is a waste product of the food industry. Singh et al.[19] predicts that, with better collection networks and waste management strategies, two thirds of available RVO will be collected in 2020 and that 75% of this will be converted to biodiesel. This equates to 0.45 PJ of practical energy per annum.

Rapeseed

Oilseed rape, a crop grown for its high oil yield, is one of the primary feedstocks or biodiesel production in Ireland. It is grown in rotation, 1 year in every 4 or 5, with conventional agricultural crops such as wheat.[37] The PPO is extracted from the seed of the rape plant by crushing, it is then filtered and can be used in modified vehicle engines without further processing, or can be further process into biodiesel.[38] The process residue is compacted into rape cake, a high-protein animal feed, thus improving the sustainability of the system.[17]

Oilseed rape is becoming progressively more attractive at farm level, with prices increasing, resulting in the crop becoming a key profit generator.[39] As such, the area of oilseed rape planted has almost doubled from 6300 ha in 2009 to 12,400 in 2011, resulting in the production of 55,800 tonnes in 2011, which is a 97% increase on 2009 figures. The yields of the crop have also improved dramatically from 3.7 tonnes/ha in 2009 to 4.5 tonnes/ha in 2011. A rise of 22%.[18] Despite the increasing levels of oilseed rape production, the rapeseed biodiesel production is currently present with GBI Ltd. There has recently been a new biodiesel plant developed in Wicklow Port, Co Wicklow with potential capacity of 36,000 tonnes, which will see a market for tallow and used cooking oil. With the end of the MOTR scheme in 2010, plants which had been producing rape methyl ester were shut down as they could no longer survive without support.[17,39] With the appropriate supports, indigenous production of biodiesel from oilseed rape has the potential to thrive. Hamelinck et al.[40] have estimated that the realistic potential production of oilseed rape in the medium term is 10–15 kha, equating to 17.3 million L or 0.6 PJ per annum at the higher end. This equates to 24.6% of the total biodiesel blended in motor fuel in 2012. 70,292,126 L of biodiesel and 55,648,948 L of bioethanol was blended in Ireland in 2012.[14]

ALTERNATIVES

Compressed Natural Gas

Compressed natural gas (CNG) is a promising alternative fuel that offers benefits relative to liquid fossil fuels. CNG is a form of fossil energy so cannot be classified as a renewable source of energy, however it has environmental benefits when compared with gasoline and diesel and poses a lesser affect to contaminate land, air or water if there is a leak.[41] In addition, CNG powered vehicles emit 85% less NOX, 70% less reactive HC's, and 74% less CO than similar gasoline powered vehicles.[42] An advantage over current electric vehicles is that CNG can be used in heavy vehicles.[43]

CNG is replacing traditional fossil fuels in transport systems across the world. There are 15 million natural gas vehicles operating around the world with almost 20,000 natural gas fuelling stations in operation.[44] Ireland has considerable natural gas resources, 9.11 billion m^3,[45] thus, CNG represents an indigenous source of fuel that can reduce reliance on imported gasoline and diesel. Bus Eireann has announced the first trial of the country's first natural-gas-powered bus that will be running on the streets of Cork City. This trial is being carried out in partnership with Bord Gais Networks with the ultimate aim being the purchase of a fleet of natural gas vehicles.[46]

Instead of natural gas, biogas can also be used, thus making the fuel renewable.[43] As such, the production of biogas has received attention as a renewable source of energy for transport.[47,48]

Grass Biomethane

Biomethane has emerged as a potential source of biofuel amid concerns over the availability of land required to produce energy crops from biofuel production. Biomethane is produced by upgrading the biogas, itself produced by the anaerobic digestion of a range of feedstocks, to the same standard as natural gas. Biogas can be produced by the anaerobic digestion of biomass such as green waste, grass, food waste, and animal slurry.[20]

The feasibility of utilizing grass for biomethane production has been examined in a number of papers.[35,36,49–52]

Grass is a crop with considerable yields in the Irish climate. Smyth et al.[53] have shown that grass biomethane has a very good energy balance (comparable to palm oil biodiesel), does not require land use change, and does not affect farming practices. Furthermore, it is estimated that 10% of Irish grasslands could fuel over 55% of the Irish private car fleet.[54] In Ireland, grasslands account for 3.94 Mha (90% of the agricultural land). By 2020, 0.39 Mha of agricultural grassland may be surplus to animal feed requirements due to destocking of the national herd. This refers to a destocking of the cattle herd under the National Climate Change Strategy 2007[13] where the agricultural sector must reduce its GHG emissions by 2.4 Mt of CO2e—in this case, the CO2e coming from the methane emissions of cattle which are responsible for 86.6% of methane emissions from ruminant animals. If the cattle herd is reduced, mainly in the beef sector as opposed to dairy then 0.39 Mha of agricultural grassland may be surplus by 2020. According to Singh et al.,[19] 0.39 Mha of surplus grass has the potential to generate 47.58 PJ/a energy, and if by a conservative estimate, only 25% of this area (0.1 Mha) was used to produce biomethane, there would be the potential to generate 11.9 PJ energy.

Biofuels produced from grass may also earn double credits toward the EU renewable energy targets as grass is considered to be a ligno-cellulosic material.[55]

Algae

The cultivation of algal biomass for the production of third generation biofuels has received increasing attention in recent years as they can be produced in the marine environment and on nonarable lands, thus bypassing the food versus fuel debate. There are two types of algae; seaweeds (macroalgae) and phytoplanktons (microalgae).

Macroalgae are appropriate for biofuel production, however they do not generally contain lipids and are being considered for the natural sugars and other carbohydrates they contain, making them suitable for biogas and ethanol production rather than biodiesel. Macroalgae has a negative lower heating value and a high moisture content (ca 80–85%), as such the most appropriate method of processing for energy is fermentation by anaerobic digestion, to create biogas, or ethanol fermentation.[56]

Currently, the majority of seaweed collection in Ireland is for human consumption and for hydrocolloid production, with 29,000 tonnes harvested in 2006 for Arramara Teo, a state owned company.[56] Ireland has considerable macroalgal resources, with an estimation of 3 million tonnes of standing kelp stock, however this estimation is highly uncertain.[56] Barriers exist to the further exploitation of marine biomass, in particular any potential effects on marine biodiversity.[56]

Microalgae are particularly suited to biofuel production due to their high photosynthetic growth rate, high lipid content, low land usage and high carbon dioxide absorption.[57] Algae absorb freely available sunlight and can utilize waste streams to provide essential nutrients for cultivation. Algae can convert waste CO_2 from power plant exhaust gas to organic biomass which can then be converted to energy.[56] Moreover, municipal wastewater streams can be harvested to provide additional nutrients.[58]

Biorefinery-type processes may be the most economical method of producing algal biofuels as several commercial products can be obtained from the algal biomass. Lipids can be extracted for biodiesel production, whereas other products can be fermented to produce ethanol and biogas. It is also possible to produce protein-rich feed for both animal and human consumption.[56] Bulk markets for the coproducts are potentially available.[58]

Further research and development is required into the production of algal biofuels, as such any contribution to Irelands 2020 targets is likely to be minimal. Bruton et al.[56] have stated that up to 447 TJ of energy may be generated from macroalgae by 2020 (approximately 0.2% of current national road-fuel demands). Similarly, it has been estimated that about 79 TJ could come from microalgae resources by 2020.

Cynar

Cyn-diesel is a synthetic diesel fuel produced by the Cynar Process from the pyrolysis of a wide range of waste plastics including; unsorted and unwashed including commercial and industrial packaging as well as heavily contaminated agricultural wastes.[59] The system consists of stock in-feed system, pyrolysis gasification chamber, contactor, condensers, centrifuge, oil recovery line, off-gas cleaning, and adulterant removal. A Cynar plant can produce up to 9500 L of high-grade synthetic fuel from 10 tonnes of waste plastics (polyethylene, polypropylene, and polystyrene), with systems ranging from 10 to 20 tonnes per day. The Cyn-diesel is essentially equivalent to regular diesel. Blends of Cyn-diesel with fossil diesel have been tested in accordance with EU fuel standards specified in the EN 590 standard in order to determine its suitability for use in a diesel fuel blend, with blends up to 40% complying with the standards.[59]

Alternative fuels produced from nonbiomass waste are not incorporated in both European and Irish biofuel policies at present. However, they represent an ideal opportunity to divert waste from landfill while producing a clean fuel that can contribute to renewable energy targets. It is important that alternative fuels are considered in any policy updates to allow their contribution, to the European and national renewable energy targets.

Electric Vehicles

The Renewables Energy Directive requires 10% of transport to be generated from renewable sources. Considering the substantial resources of ocean and wind energy available in Ireland, the Government has set a target of 10% of all vehicles to be electric vehicles by 2020.[16] Foley et al.[60] have estimated that 284,383 electric vehicles will be required in the vehicle fleet by 2020 to meet this target.[60] New electric vehicles registered in 2010 numbered 23 in total. This figure rose to 48 in 2011 and stands at 16 for the first 6 months of 2012.[61] The rate of adoption of electric vehicle will need to increase significantly toward 2020 to meet the target. If this target is met, it can be expected that the electric vehicle will initially produce 1.5% of energy from renewable sources.[2] This not only would help Ireland to reach the 10% renewable targets by 2020 but also reduce CO_2 emissions, provided the electricity is produced in an environmentally friendly way. In 2009 the Government introduced grant support and tax incentives to encourage electric vehicle uptake; up to »5000 for the purchase of battery electric vehicles (BEVs) and up to »2500 for the purchase of plug-in hybrid electric vehicles (PHEVs), available from 2011 to 2012, with zero VRT on BEVs and VRT relief of up to »2500 for PHEVs.[1] This scheme will only be applicable to small and light vehicles.

In 2010, ESB installed the first on-street charge points in Dublin and has set the following targets to be met in the future; 2000 home charge points, 1500 public charge points, and 30 fast charge points. Two Irish companies, M2C and JTM Power are developing new charging infrastructures. The SEAI has also initiated the Aran Islands Project, which involves developing technologies and methods to use high amounts of wind or ocean energy available on the islands to power electric vehicles. There are eight Mega ECity electric cars currently operating on the islands on trial basis and it is expected that up to 24 will be in operation in the future.

GHG EMISSIONS REDUCTIONS

As discussed previously, the EU has introduced sustainability criteria regarding the GHG emissions associated with biofuel production, which may have a significant impact on the types of biofuels viable for use in the future. From 2013 to 2017, biofuels must achieve a minimum 35% reduction in GHG emissions versus fossil fuels. According to Table 27.1, which summarizes the GHG emissions from the production of indigenous biofuels in Ireland, only biodiesel

Table 27.1 GHG Emissions Associated with Indigenous Biofuel Production in Ireland

	Total GHG Emissions (kg CO_2-eq/GJ)	% Reduction in GHG Emissions Versus Fossil Fuels
Biodiesel		
Rapeseed[1]	62.16	29
Tallow[2]	40	54
Recovered vegetable oil[2]	27.11	69
Bioethanol		
Wheat, barley, sugar beet[3]	69	28
Biomethane		
Wheat, barley, sugar beet[3]	64	27
Grass[4]	69.4	21

[1] Ref 63.
[2] Ref 64.
[3] Ref 65.
[4] Ref 62.

produced from residues meets this minimum reduction. In 2017, the targeted reduction increases to 50%, which both biodiesels from residues can meet. However, beyond 2017 the target reaches 60%, which only biodiesel from RVO can meet. As it is, biodiesel from oil seed rape, bioethanol, and biomethane from wheat, barley and sugar beet, and biomethane from grass, all fail to meet even the minimum sustainability criteria. This shows the improvements necessary in current biomass production and processing methods required to increase the sustainability of these biofuels. For example, Korres et al.[62] found that by improving current biomethane production by using electricity from wind, improving digester configuration, and by using a vehicle optimized for gaseous fuel the GHG reduction for grass biomethane can be increased to 54%.

DISCUSSION AND CONCLUSION

- An analysis of the potentially available feedstocks in Ireland by 2020 is carried out in this paper. Table 27.2 shows that if all available resources considered are utilised, Ireland can achieve approximately 98.5% of the biofuel target by 2020.
- Any fuels contributing to the EU target will have to meet minimum GHG reductions. This will put further pressure indigenous biofuel production as the only fuel that meets the minimum 60% GHG emissions reduction, coming into force in 2017, is RVO biodiesel, which contributes only 4.2% to the overall biofuel target.
- Ireland only has a very small indigenous production of biodiesel, 34.5 million L per annum. A newly developed

Table 27.2 Potential Availability of Biofuels by 2020

Biofuel	Quantity (PJ)	% Contribution
Target	14.4	100
Current fuels		
Tallow	0.715	5
RVO	0.45	3
Rapeseed	0.6	4.2
Potential fuels		
Grass biomethane	11.9	82.6
Algae	0.526	3.7
Total	14.191	98.5

biodiesel plant, Irish Production Biofuels Ltd. has the capacity for 36,000 tonnes of biodiesel per annum to be sold into the Irish fuel market through RVO and tallow. Ireland currently has no indigenous ethanol industry but Carbery in Co Cork produce ethanol from whey, which is a by-product of their main cheese manufacturing facility, but this ethanol is currently exported after the excise relief was removed after MOTR II. As such, all policy efforts have so far failed to create an indigenous biofuels industry in Ireland, causing the country to remain heavily reliant on fuel imports.

- The MOTR schemes I and II effectively failed in the sense that they did not meet the targets of biofuel incorporated into the market in the 4 years that it existed. There were also issues around many of the successful applicants who could sell the biofuel without the excise never actually producing any fuel indigenously in the timeframe but simply imported the biofuel and pocketed the subsidy. When the subsidy was finished it effectively terminated any OSR operations also.
- The Biofuel Obligation Scheme is a slight improvement in that it is getting the biofuel into the market. The original target of 4.166% (volume) has been easily reached and as of January 2013 a new target of 6.383% has been introduced. But this is not necessarily driving the home-grown biofuel as fuel suppliers are mostly importing the fuel preblended. What potentially should happen is that the main fuel suppliers should liase with indigenous biofuel producers and blend the fuel themselves. This approach would help drive the success of the industry in Ireland and also develop a more secure supply of biofuel into the market.

ACKNOWLEDGEMENT

This study was funded under the Charles Parsons Energy Research Program (Grant Number Grant Number 6C/CP/E001) of Science Foundation Ireland (SFI).

REFERENCES

1. SEAI. Energy Security in Ireland: A Statistical Overview, 2011. Available at: http://www.seai.ie/Publications/Statistics_Publications/EPSSU_Publications/Energy_Security_in_Ireland/Energy_Security_in_Ireland_A_Statistical_Overview.pdf. (Accessed August 8, 2012).
2. Howley M, Dennehy E, Ó Gallachóir B. Energy In Transport 2009 Report, 2009. Available at: http://www.seai.ie/Publications/Statistics_Publications/EPSSU_Publications/Energy_in_Transport/Energy_In_Transport_2009_Report.pdf. (Accessed August 8, 2012).
3. Howley M, Dennehy E, Holland M, Ó Gallachóir B. Energy in Ireland 1990–2010: 2011 Report, 2011. Available at: http://www.seai.ie/Publications/Statistics_Publications/Energy_in_Ireland/Energy_in_Ireland_1990-2009.pdf. (Accessed August 8, 2012).
4. SEAI. Energy Statistics Database, 2012. Available at: http://www.cso.ie/px/sei/database/sei/sei.asp. (Accessed August 8, 2012).
5. Clancy M, Scheer J. Energy Forecasts for Ireland to 2020–2011 Report, 2011. Available at: http://www.seai.ie/Publications/Statistics_Publications/EPSSU_Publications/Energy_Forecasts_for_Ireland_for_2020_-2011_Report.pdf. (Accessed August 2012).
6. EC. Directive 2003/30/EC of the European Parliament and of the Council of 8 May 2003 on the Promotion of the Use of Biofuels or Other Renewable Fuels for Transport, 2003. Available at: http://eur-lex.europa.eu/LexUriServ/LexUriServ.do?uri=OJ:L:2003:123:0042:0046:EN:PDF. (Accessed August 9, 2012).
7. EC. Directive 2009/28/EC of the European Parliament and of the Council of 23 April 2009 on the Promotion of the Use of Energy from Renewable Sources and Amending and Subsequently Repealing Directives 2001/77/EC and 2003/30/EC, 2009. Available at: http://eur-lex.europa.eu/LexUriServ/LexUriServ.do?uri=Oj:L:2009:140:0016:0062:en:PDF. (Accessed August 9, 2012).
8. DCMNR. Delivering a Sustainable Energy Future for Ireland. The Energy Policy Framework 2007–2020. Government White Paper, 2007. Available at: http://www.dcenr.gov.ie/NR/rdonlyres/B0D3D168-3DCD-43E9-8D87-F4666239049F/0/EnergyWhitePaper12March2007.pdf. (Accessed August 8, 2012).
9. DCENR. Public Consultation on the Biofuels Obligation Scheme, 2008. Available at: http://www.dcenr.gov.ie/NR/rdonlyres/EC1AD24F-304F-4894-85C1-865715A6DE70/0/BiofuelsObligationScheme.pdf. (Accessed August 8, 2012).
10. UN. Kyoto Protocol to the United Nations Framework Convention on Climate Change, 1997. Available at: http://unfccc.int/resource/docs/convkp/kpeng.pdf. (Accessed August 10, 2012).
11. EC. Renewable Energy Road Map—Renewable Energies in the 21st Century: Building a More Sustainable Future, 2007. Available at: http://eur-lex.europa.eu/LexUriServ/LexUriServ.do?uri=COM:2006:0848:FIN:EN:PDF. (Accessed August 10, 2012).

12. Lewis B, Kambas M. EU Commission to Cap Food-Based Biofuels in Major Shift. Reuters, 2012. Available at: www.reuters.com/article/2012/09/17/us-eu-biofuel-idUSBRE88G0IL20120917. (Accessed August 10, 2012)
13. NCCS. Ireland National Climate Change Strategy 2007–2012, 2007. Available at: http://www.environ.ie/en/Publications/Environment/Atmosphere/FileDownLoad,1861,en.pdf. (Accessed August 9, 2012).
14. NORA. National Oil Reserves Agency, 2012. Available at: www.nora.ie. (Accessed August 9, 2012).
15. DCMNR. Regulatory Impact Assessment: Biofuel Obligation Scheme, 2010. Available at: http://www.nora.ie/biofuels-obligation-scheme.141.html. (Accessed August 9, 2012).
16. DCENR. National Renewable Energy Action Plan Ireland: Submitted under Article 4 of Directive 2009/2008/EC, 2010. (Accessed August 9, 2012).
17. Hearne J. Turning the Tapon Biofuels. Irish Examiner, 2012. Available at: http://www.irishexaminer.com/ireland/turning-the-tapon-biofuels-193845.html#T7q8nzis_Gk.email. (Accessed August 10, 2012).
18. CSO. © Government of Ireland 2012 Material Compiled, Edited and Presented by the Central Statistics Office. Reproduction is authorised, except for commercial purposes, provided the source is acknowledged. Print ISSN ISBN 978-1-4064-2664-9, 2012. (Accessed August 11, 2012).
19. Singh A, Smyth BM, Murphy JD. A biofuel strategy for Ireland with an emphasis on production of biomethane and minimization of land-take. *Renew Sustain Energy Rev* 2010, 14:277–288.
20. Smyth BM, Ó Gallachóir BP, Korres NE, Murphy JD. Can we meet targets for biofuels and renewable energy in transport given the constraints imposed by policy in agriculture and energy? *J Clean Product* 2010, 18:1671–1685.
21. Nigam PS, Singh A. Production of liquid biofuels from renewable resources. *Progress Energy Combust Sci* 2011, 37:52–68.
22. Mielenz JR. Ethanol production from biomass: technology and commercialization status. *Curr Opin Microbiol* 2001, 4:324–329.
23. Gnansounou E. Production and use of lignocellulosic bioethanol in Europe: current situation and perspectives. *Bioresour Technol* 2010, 101:4842–4850.
24. Hammond GP, Kallu S, McManus MC. Development of biofuels for the UK automotive market. *Appl Energy* 2008, 85:506–515.
25. RFA. Accelerating Industry Innovation: 2012 Ethanol Industry Outlook, 2012. Available at: http://www.ethanolrfa.org/pages/advanced-ethanol. (Accessed August 16, 2012).
26. Azadi H, de Jong S, Derudder B, De Maeyer P, Witlox F. Bitter sweet: how sustainable is bio-ethanol production in Brazil? *Renew Sustain Energy Rev* 2012, 16:3599–3603.
27. IEA. World Energy Outlook 2006: Fact Sheet—Biofuels, 2006. Available at: http://www.iea.org/publications/freepublications/publication/fs_biofuels.pdf. (Accessed August 16, 2012).
28. Demirbas A. Competitive liquid biofuels from biomass. *Appl Energy* 2011, 88:17–28.
29. Griffin D. Efforts under way to revive sugarbeet industry. Irish Times 2012. Available at: http://www.irishtimes.com/business/sectors/agribusiness-and-food/efforts-under-way-to-revive-sugar-industry-1.3484. (Accessed December 19, 2012)
30. RFA. Global Ethanol Production to Reach 85.2 Billion Litres in 2012. Available at: http://www.ethanolrfa.org/news/entry/global-ethanol-production-to-reach-85.2-billion-litres-in-2012/. (Accessed August 16, 2012).
31. Santori G, Di Nicola G, Moglie M, Polonara F. A review analyzing the industrial biodiesel production practice starting from vegetable oil refining. *Appl Energy* 2012, 92: 109–132.
32. Haas MJ, McAloon AJ, Yee WC, Foglia TA. A process model to estimate biodiesel production costs. *Bioresour Technol* 2006, 97:671–678.
33. Wiesenthal T, Leduc G, Christidis P, Schade B, Pelkmans L, Govaerts L, Georgopoulos P. Biofuel support policies in Europe: lessons learnt for the long way ahead. *Renew Sustain Energy Rev* 2009, 13:789–800.
34. OECD-FAO. Chapter 3 – Biofuels. In: Agricultural Outlook 2011–2020, 2011. Available at: http://www.oecd.org/site/oecd-faoagriculturaloutlook/. (Accessed August 17, 2012).
35. Travers J. Mallow Sugar Factory Ethanol Production Evaluation Study, 2006.
36. Murphy JD, Power NM. An argument for using biomethane generated from grass as a biofuel in Ireland. *Biomass Bioenergy* 2009, 33:504–512.
37. Teagasc. Winter Oilseed rape, 2009. Available at: http://www.teagasc.ie/crops/oilseed_rape/Fact_Sheet_WOSR_August_2009.pdf. (Accessed August 16, 2012).
38. Thamsiriroj T, Murphy JD. Can rape seed biodiesel meet the European Union sustainability criteria for biofuels? *Energy Fuels* 2010, 24:1720–1730.
39. Doyle A. Oilseed rape still key crop profit generator for 2012. *Irish Farmers J* 2012. Available at: http://www.farmersjournal.ie/site/farming-Oilseed-rape-still-key-crop-profit-generator-for-2012-13549.html. (Accessed August 17, 2012)
40. Hamelinck C, van den Broek R, Rice B, Gilbert A, Ragwitz M, Toro F. Liquid Biofuels Strategy for Ireland, 2004. Available at: http://www.sei.ie/uploadedfiles/InfoCentre/LiquidbiofuelFull.pdf. (Accessed August 17, 2012).
41. Semin RAB. A technical review of compressed natural gas as an alternative fuel for internal combustion engines. *Am JEng Appl Sci* 2008, 1:302–311.
42. Goyal P, Sidhartha. Present scenario of air quality in Delhi: a case study of CNG implementation. *Atmos Environ* 2003, 37:5423–5431.
43. Engerer H, Horn M. Natural gas vehicles: an option for Europe. *Energy Policy* 2010, 38:1017–1029.
44. NGVKB. 2011 Natural Gas Vehicle Statistics Published. Available at: http://www.iangv.org/2012/05/2011-natural-gas-vehicle-statistics-released/. (Accessed August 20, 2012).
45. CIA. The World Factbook: Ireland. Available at: https://www.cia.gov/library/publications/the-world-factbook/fields/2179.html. (Accessed August 20, 2012).

46. English E. Natural gas bus hits the streets in bid to cut fuel bill. *Irish Examiner* 2012. Available at: http://www.irishexaminer.com/ireland/natural-gas-bus-hits-the-streets-in-bid-to-cut-fuel-bill-201037.html. (Accessed August 20, 2012).
47. Börjesson P, Mattiasson B. Biogas as a resource-efficient vehicle fuel. *Trends Biotechnol* 2008, 26:7–13.
48. Murphy JD, McCarthy K. The optimal production of biogas for use as a transport fuel in Ireland. *Renew Energy* 2005, 30:2111–2127.
49. Asam Z-u-Z, Poulsen TG, Nizami A-S, Rafique R, Kiely G, Murphy JD. How can we improve biomethane production per unit of feedstock in biogas plants? *Appl Energy* 2011, 88:2013–2018.
50. Browne J, Nizami A-S, Thamsiriroj T, Murphy JD. Assessing the cost of biofuel production with increasing penetration of the transport fuel market: a case study of gaseous biomethane in Ireland. *Renew Sustain Energy Rev* 2011, 15:4537–4547.
51. Nizami AS, Orozco A, Groom E, Dieterich B, Murphy JD. How much gas can we get from grass? *Appl Energy* 2012, 92:783–790.
52. Thamsiriroj T, Smyth H, Murphy JD. A roadmap for the introduction of gaseous transport fuel: a case study for renewable natural gas in Ireland. *Renew Sustain Energy Rev* 2011, 15:4642–4651.
53. Smyth BM, Murphy JD, O'Brien CM. What is the energy balance of grass biomethane in Ireland and other temperate northern European climates? *Renew Sustain Energy Rev* 2009, 13:2349–2360.
54. Smyth BM, Murphy JD, O'Brien CM. What is the energy balance of grass biomethane in Ireland and other temperate northern European climates? *Renew Sustain Energy Rev* 2009, 13:2349–2360.
55. Singh A, Pant D, Korres NE, Nizami A-S, Prasad S, Murphy JD. Key issues in life cycle assessment of ethanol production from lignocellulosic biomass: challenges and perspectives. *Bioresour Technol* 2010, 101:5003–5012.
56. Bruton T, Lyons H, Lerat Y, Stanley M, Rasmussen MB. A Review of the Potential of Marine Algae as a Source of Biofuel in Ireland, 2009. Available at: http://www.seai.ie/Publications/Renewables_Publications_/Bioenergy/Algaereport.pdf. (Accessed August 22, 2012).
57. Singh A, Olsen SI. A critical review of biochemical conversion, sustainability and life cycle assessment of algal biofuels. *Appl Energy* 2011, 88:3548–3555.
58. Singh A, Nigam PS, Murphy JD. Mechanism and challenges in commercialisation of algal biofuels. *Bioresour Technol* 2011, 102:26–34.
59. Murphy F, McDonnell K, Butler E, Devlin G. The evaluation of viscosity and density of blends of Cyn-diesel pyrolysis fuel with conventional diesel fuel in relation to compliance with fuel specifications EN 590:2009. *Fuel* 2012:112–118.
60. Foley A, Gallachoir BO, Leahy P, McKeogh E. Electric vehicles and energy storage – a case study on Ireland. In: *Vehicle Power and Propulsion Conference, 2009. VPPC '09. IEEE*; 2009.
61. Office CS. Releases and Publications—Transport. Available at: http://www.cso.ie/en/releasesandpublications/transport/archive/. (Accessed July 20, 2012).
62. Korres NE, Singh A, Nizami A-S, Murphy JD. Is grass biomethane a sustainable transport biofuel? *Biofuels Bioprod Biorefin* 2010, 4:310–325.
63. Thamsiriroj T, Murphy JD. Is it better to import palm oil from Thailand to produce biodiesel in Ireland than to produce biodiesel from indigenous Irish rape seed? *Appl Energy* 2009, 86:595–604.
64. Thamsiriroj T, Murphy JD. The impact of the life cycle analysis methodology on whether biodiesel produced from residues can meet the EU sustainability criteria for biofuel facilities constructed after 2017. *Renew Energy* 2011, 36:50–63.
65. Power NM, Murphy JD. Which is the preferable transport fuel on a greenhouse gas basis; biomethane or ethanol? *Biomass Bioenergy* 2009, 33:1403–1412.

28

Prospects for Domestic Biofuels for Transport in Sweden 2030 Based on Current Production and Future Plans

Maria Grahn[1]* and Julia Hansson[2]

[1]Division of Physical Resource Theory, Department of Energy and Environment, Chalmers University of Technology, Göteborg, Sweden
[2]Climate & Sustainable Cities, IVL Swedish Environmental Research Institute, Stockholm, Sweden

RENEWABLE ENERGY IN THE SWEDISH TRANSPORT SECTOR

In 2012, the use of renewable energy in the Swedish transport sector amounted to 12.6% when calculated according to the so-called Renewable Energy Directive (RED) calculation rules (i.e., as share of the use of gasoline, diesel, biofuels used in road and rail transport, and electricity in land transport, and with the contribution from biofuels produced from wastes, residues, nonfood cellulosic material, and lig-nocellulosic material considered double the actual amount).[1,2] This was the largest share of renewable fuels for transport among the EU countries,[3] and unless the use of renewable fuels decreases and/or the energy demand in the transport sector increases substantially or the calculation rules in the RED change, Sweden will meet the target of 10% renewable energy in the transport sector for 2020 stipulated in the RED.

Ethanol, biodiesel options (FAME, fatty acid methyl ester; HVO, hydrotreated vegetable oil) and biogas are the main biofuels used for transport in Sweden; DME (dimethyl ether) and ETBE (an octane-increasing additive for gasoline that can be based on ethanol) are used in much lower volumes.

In 2012 the total use of biofuels for transport in Sweden (6.95 TWh, see Table 28.1) corresponded to about 8% of the total use of gasoline, diesel, natural gas, and biofuels for road transport.[4] About 47% of the ethanol was used for low blending, with the remaining share as E85 and ED95.[4] About 90% of the biodiesel in the transport sector was used for low blending, and 70% of this consisted of FAME.[4] Gas for vehicles is pure biogas, pure natural gas, or a mixture that varies depending on location and time of year. The share of biogas of the total gas use for vehicles amounted to about 60%.[4] More generally, the share of petrol has decreased and diesel increased in recent years.[4] In total, the use of energy for domestic transport in Sweden had decreased in 2012 by 3% compared with 2010.[4]

In addition to EU policies such as RED, Sweden has and has had several different national policies stimulating the production and use of renewable fuels for transport, as well as policies for associated vehicles, e.g., energy and carbon tax exemption for renewable fuels, vehicle tax exemption for green cars (replacing a green car premium), benefits value for certain green cars, obligation for fuel retail outlets to offer renewable fuels, investment aid for biogas and other renewable gases, carbon dioxide (CO_2) differentiated

*Correspondence to: maria.grahn@chalmers.se

Advances in Bioenergy: The Sustainability Challenge, First Edition. Edited by Peter D. Lund, John Byrne, Göran Berndes and Iacovos A. Vasalos.

Originally published in: *WIREs Energy Environ* 2014, 4: 431–446 doi: 10.1002/wene.138

Table 28.1 The Use of Biofuels for Transport in Sweden in 2012[4]

Biofuels for Transport	TWh (Share of Total in Parenthesis)
Ethanol	2.4 (34%)
of which low blending	1.13 (16%)
of which E85/ED95	1.27 (18%)
Biodiesel	3.73 (54%)
of which low blending	3.35 (48%) of which FAME: 2.35 and HVO: 1.0
Other	0.38 (6%)
Biogas (million Nm^3)	0.8 (12%)

annual vehicle tax, super green car premium.[5,6] A quota system for biofuels for transport is expected to become operative in 2014, and the energy and carbon tax exemption rules have changed (mainly because tax exemption is more costly for the government, see Ref 7).

The ambition of the Swedish Government is for Sweden in 2030 to have a vehicle fleet that is independent of fossil fuels, e.g., by a transition to sustainable renewable fuels, major developments in electrification, and measures that reduce the demand for transport.[8] The overall national vision for 2050 is a Sweden with no net emissions of greenhouse gases (GHG) to the atmosphere. A national roadmap for 2050 is being developed.[9] An official survey of different options for 'fossil fuel independence' has identified measures to reduce emissions and dependence on fossil fuels in the transport sector in line with the vision for 2050.[10]

In Sweden, in 2012, slightly more than 50% of the ethanol used was produced domestically, mainly based on domestic resources.[11] The remaining share originates primarily in other EU countries, and about 80% of the total ethanol use was based on cereals (primarily wheat and corn).[12] Of the biodiesel (FAME and HVO) used, 68% was produced domestically.[11] The FAME was primarily based on feedstocks (rapeseed) from other EU countries, and about 40% of the total use of HVO originates from Swedish raw tall oil.[12] The demand for biofuels and feedstock is expected to increase further in response to EU and global targets.

A review of different kinds of existing scenarios and visions, for the contribution of biofuels in the Swedish transport sector in 2030 (e.g., Refs 9 and 13–16), shows that the contribution of biofuels in the Swedish fuel mix differs considerably among the included studies.[6] In 2030, the amount of biofuels for transport in Sweden varies between 4 and 53 TWh, corresponding to 6–56% of the current energy demand for road transport in Sweden (the endpoints of the interval are represented by Refs 17 and 18). More information about the included studies can be found in Ref 6 Although the existing studies have different aims, delimitations, approach and assumptions (e.g., regarding import of biofuels) and are thus not completely comparable, the review indicates that it, as expected, is difficult to predict the possible contribution of biofuels in the future Swedish transport sector.

This study differs from existing scenarios and visions as it takes its starting point from current production and actual expansion plans made by existing and potential Swedish biofuel producers, mapped primarily through personal communication. The potential future contribution of domestically produced biofuels for transport in Sweden to 2030 is estimated from a combination of the mapped information of the prospects from the biofuel actors and scenario analysis, which takes into account uncertainties regarding the realization of the plans. Results should be of interest to policy makers and other actors evaluating the prospects for renewable fuels. In this paper the term *biofuels* is used to denote biofuels for transport.

BIOFUEL OPTIONS

Biomass can be categorized by chemical composition. Different elements are better suited for different processes that convert biomass into biofuels of use in the transportation sector. The main biofuels production options currently discussed in Sweden are illustrated in Figure 28.1.

MAPPING THE PROSPECTS FOR DOMESTIC PRODUCTION OF BIOFUELS FOR TRANSPORT

The prospects for domestic production of biofuels for transport are mapped based on contact with actual and potential Swedish biofuel producers and the relevant literature review, see also Ref 6. The mapping was performed during 2012–2013, with some updates until February 2014. We have not assessed the probability of realizing the various biofuel plans (or the credibility of the information). Historically, Swedish biofuel producers' plans have changed (e.g., regarding size and start year); an alternative scenario is constructed to meet these uncertainties.

We have chosen to describe all types of facilities, including canceled projects as well as research and pilot plants (even if producing none or insignificant amounts of biofuels) as these indicate interest in developing biofuel conversion technologies. However, these kinds of facilities are not included in the scenarios. Note that the large number of small-scale commercial plants, producing anaerobic digestion biogas (AD biogas), are not mapped individually.

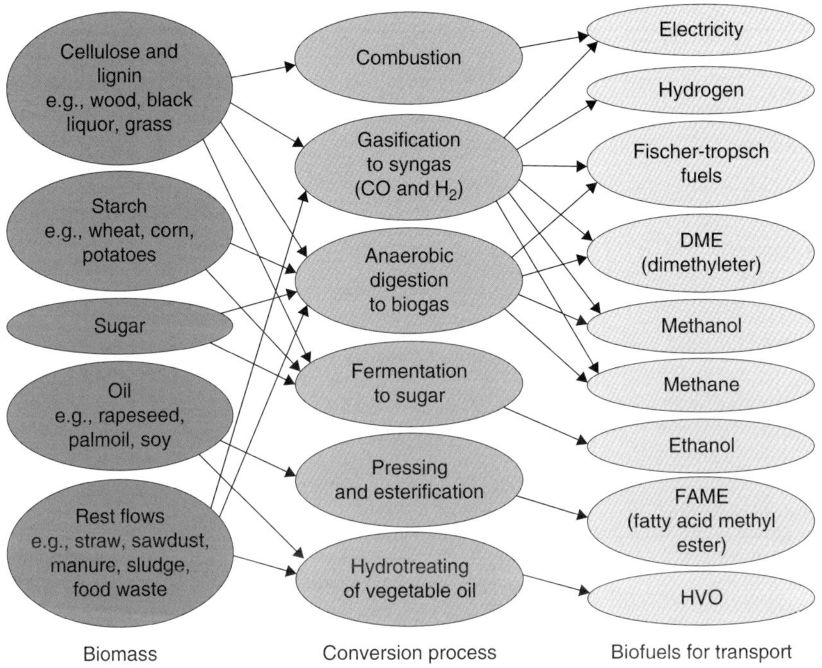

Figure 28.1 Biofuels production options illustrating the suitable conversion processes for various kinds of biomass. To avoid making the figure too difficult to read we have chosen to exclude arrows pointing out that all types of biomass can be combusted.

Summary of Existing and Planned Production Capacity for Biofuels for Transport in Sweden

A summary of the mapping of existing and planned production capacity for biofuels is presented in Tables 28.2–28.4. The individual facilities (among which canceled projects as well as research and pilot plants are included) are described in the following sections. In general, we have used the terms regarding status (e.g., demonstration, pilot, etc.) used by the actors.

Table 28.2 Current and Planned Ethanol Production Capacity in Sweden

Production Facilities	Feedstock	Estimated Capacity (GWhfuel/year)[1]	Actual or Planned Start
SEKAB, Örnsköldsvik	Cellulose	65	1940
Agroetanol 1, Norrköping	Grain	390	2001
Agroetanol 2, Norrköping	Grain	1130	2008
ST1, Gothenburg	Food waste	34[2]	End of 2014
Nordisk Etanol och Biogas AB, Karlshamn	Grain/cellulose	770/1540[3]	No sooner than 2015/2020
NBE Sweden, Sveg	Cellulose	445	No sooner than 2015
Current capacity, rounded to nearest 5 GWhfuel/year		1585 (65)[4]	
Further planned capacity, rounded to nearest 5 GWhfuel/year		1250–2020 (480–2020)[4]	
Total current and planned capacity		2835–3605 (545–2085)[4]	

Two test facilities: SP demo in Örnsköldsvik and NBE Sweden in Sveg are described in the text.
[1]Calculated assuming 21 MJ/L ethanol in lower heating value.[52]
[2]Calculated assuming 27 MJ/kg ethanol in lower heating value.[52]
[3]The expansion plan for 2020 has grain-based ethanol being replaced by cellulose-based ethanol.
[4]Figures in brackets point out the cellulosic part of the summaries.

Table 28.3 Current and Planned Production Capacity of Synthetic Fuels via, e.g., Gasification

Production Facilities	Feedstock/Product	(GWh_{fuel}/year)	Estimated Capacity Actual or Planned Start
LTU Green fuels Piteå (3 MW)	Black liquor/DME	6	2010
Göteborg Energi, GoBiGas1, Gothenburg (20 MW)	Solid biomass/methane	180	2013
Göteborg Energi, GoBiGas2, Gothenburg (100 MW)	Solid biomass/methane	1000	No sooner than 2016
Värmlandsmetanol, Hagfors (110 MW)	Forest residues/methanol	600	2016
E.ON, Bio2Gas (200 MW)	Solid biomass/methane	1600	2018
Vallvik biofuel, Rottneros AB (About 200MW)	Black liquor/methanol	750	No sooner than 2017
Rottneros biorefinery, Rottneros AB (About 200 MW)	Forest residues/methanol	750	No sooner than 2017
Biorefinery Norrtorp, SAKAB et al., Kumla (250 MW)	Forest residues/ methanol, methane	1300 methanol + 480 methane	No sooner than 2023
Current capacity		186	
Further planned capacity		6480	
Total current and planned capacity		6666	

One test facility, Chalmers gasifier (4 MW), and three canceled projects Värnamo demo (18 MW), Domsjö fabriker (200 MW), and BillerudKorsnäs (200 MW) are described in the text.

Cereal-Based Ethanol

The domestic production capacity for cereal-based ethanol amounted to 1230 GWh_{fuel} (210,000 m^3) per year in 2010, and the actual production was about 1130 GWh_{fuel}/year (194,000 m^3) of ethanol in 2010 and about 1100 GWh_{fuel}/year (188,000 m^3) ethanol in 2011 (Frida Källström, Lantmännen Energi 2012, personal communication). Currently the two Agroetanol facilities in Norrköping produce grain-based ethanol.

Since 2010, the technical production capacity has increased to 390 GWh_{fuel}/year (67,000 m^3) and 1,130 GWh_{fuel}/year (193,000 m^3), respectively, due to a range of improvements for a total of about 1,520 GWh_{fuel}/year (260,000 m^3).

The facility of Nordisk Etanol och Biogas AB in Karlshamn, with a planned initial production of 770 GWh_{fuel} (130,000 m^3) ethanol per year, plans to gradually increase to 1540 GWh_{fuel} (260,000 m^3) ethanol per year (Mikael Runesson, Nordisk Etanol och Biogas AB, 2013, personal communication). Within about 5 years from the production start date, the plan is to partly produce ethanol based on straw or other cellulose-based feedstock. The by-products and waste will be used for biogas production (supplemented with straw, it will yield about 600 and 1000 GWh_{fuel}/year, respectively) (Mikael Runesson, Nordisk Etanol och Biogas AB, 2013, personal communication). The environmental permit has been granted, agreements for, e.g., feedstock, energy, and digested residues have been reached, and production could start as early as 2015 (Mikael Runesson, Nordisk Etanol och Biogas AB, 2013, personal communication). The ongoing task is to find financing.

Table 28.4 Current and Planned Production Capacity of Biodiesel (FAME and HVO)

Production Facilities	Feedstock	Capacity (GWh_{fuel}/year)	Actual or Planned Start
Energigårdarna facility, Karlshamn (RME[1])	Rapeseed	500	2006
Perstorp, Stenungssund (RME[1])	Rapeseed	1700	2007
Preem, Gothenburg (HVO[2])	Tall oil	1000	2010
Preem, Gothenburg (HVO[2])	Unspecified renewable sources	Up to 1000	2016
Total current capacity		3200	
Further planned capacity		Up to 1000	
Total current and planned capacity		Up to 4200	

[1]RME stands for rapeseed methyl ester, which is one type of FAME fuel.
[2]HVO stands for hydrotreated vegetable oil.

Ethanol Based on Cellulose or Waste/Residues

Domsjö Fabriker in Örnsköldsvik has produced ethanol from cellulose in the form of sugar from the production of sulfite pulp since the 1940s. Production capacity is about 65 GWh_{fuel}/year (11,000 m^3). The ethanol is bought by SEKAB, but not all of it is used for transport (Lina Wännström, SEKAB, 2011, personal communication).

SEKAB E-technology AB has run a demonstration plant for cellulose-based ethanol (called Etanolpiloten as it initially was a pilot plant) since 2004, with a capacity of 1 GWh_{fuel}/year (100–150 m^3) ethanol per year (in part financed by the Swedish Energy Agency). The actual production has been a few dozen m^3 per year. Since 2011, SEKAB E-Technology AB has phased out operation engagement due to reduced need in the context of research activities. Technical development has been successful; the next step is to commercialize and bring the developed technology to full-scale production (Sune Wännström, SEKAB, 2012, personal communication) Minor activity took place at the plant during 2012, and in spring 2013 it was announced that the plant will operate for at least another year led by SP Technical Research Institute of Sweden and supported by the Swedish Energy Agency and the Swedish Governmental Agency for Innovation Systems (Vinnova). The plan was to convert the demonstration plant into a biorefinery, Biorefinery Demo Plant.[19]

The experimental facility by NBE Sweden in Sveg that ran between 2008 and 2010 produced small quantities of ethanol for research purposes but is no longer in use (Lars Fritz, NBE Sweden, 2012, personal communication) NBE Sweden plans to build a full-scale ethanol plant based on forest resources integrated with a combined heat and power plant in Sveg, with a production capacity of 445 GWh_{fuel}/year (75,000 m^3) ethanol, but since they have a permit for production of 595 GWh_{fuel}/year (80,000 tons), the annual production may increase over the years (Lars Fritz, NBE Sweden, 2012, personal communication). NBE Sweden estimates that their plant can operate in 2015 at the earliest, given that they receive enough funding rather promptly. It is possible that the plant will be built in steps, i.e., that the plant will not be full scale in 2015 (Lars Fritz, NBE Sweden, 2012, personal communication)

As already mentioned, the planned facility of Nordisk Etanol och Biogas AB in Karlshamn is expected to convert from cereal- to cellulose-based ethanol in 2020, if realized.

The energy company ST1 has announced plans for an ethanol plant in Gothenburg, based on by-products and waste from the food industry, to be ready by the end of 2014 and estimated to deliver 34 GWh_{fuel} (4500 tons) ethanol per year as well as animal feed in the form of pellets. ST1 has also indicated a potential for more ethanol production facilities in the future, but so far no figures are available (Stefan Samuelsson, ST1, 2013, personal communication).

Anaerobic Digestion Biogas

In 2011, a total of 1473 GWh_{fuel} of anaerobic digestion biogas (AD biogas) was produced by 233 facilities (50% of this was produced in Skåne, Stock-holm, and Västra Götaland).[20] About half of the total production was upgraded (mainly biogas from wastewater treatment plants and co-digestion facilities), and the majority of this was used as transport fuel. There is still an unused potential. Kågeson and Jonsson[21] estimate the amount of biogas from digestion of by-products and waste that can be upgraded to transport fuel to be up to 2–3 TWh_{fuel}. The total potential for AD biogas is estimated at about 11 TWh_{fuel} in Ref 22 and 15 TWh_{fuel} in Ref 23, which can be used for heat and electricity generation or upgraded and used as transport fuel.

Since AD biogas is mainly produced in numerous small facilities widely distributed in Sweden, it will generally be analyzed at the aggregate level, instead of analyzed at the level of individual facilities as the other larger biofuels production facilities. However, as already mentioned, the planned facility of Nordisk Etanol och Biogas AB in Karlshamn is expected to produce a large amount of AD biogas (600 GWh_{fuel}/year in 2015 and 1000 GWh_{fuel}/year in 2020, if realized), and the contribution from this facility will be taken into account separately.

Synthetic Biofuels

Several synthetic fuels can be produced from syngas from gasification of biomass, e.g., methanol, DME, biomethane, and Fischer-Tropsch fuels (synthetic gasoline, diesel, aviation fuels, etc.).

Gothenburg Biomass Gasification Project, GoBiGas is a project to produce biomethane by (indirect) gasification of biofuels and forestry waste, led by Göteborg Energi (and run in partnership with E.ON). An initial 20-MW facility (which is expected to deliver about 180 GWh_{fuel} biomethane per year when tuned and the availability has reached the expected level) has been constructed in Gothenburg and started operating in October 2013 (Lars Holmquist, Göteborg Energi, 2013, personal communication). Wood pellets are gasified currently, but other biofuels and forest residues will be used. Given that this first stage is successful, the plan is to build another 100 MW which would give nearly 1 TWh_{fuel} biomethane per year. The second stage is expected to start operation in 2016 at the earliest (Lars Holmquist, Göteborg Energi, 2013, personal communication). GoBiGas has had financial support from the Swedish Energy Agency (maximum granted aid SEK 222 million). The second stage of GoBiGas has been selected for financial support by the EU within the NER 300 program (which could provide assistance corresponding to 50% of the incremental cost for 5 years).

E.ON also has plans to build a biomass gasification plant (Bio2G) in Sweden at about 200 MW, which given full production corresponds to about 1600 GWh_{fuel}/year (Björn Fredriksson Möller, E.ON, 2012, personal communication). The schedule depends on how the financial and political conditions develop. The applications for permits have been prepared and options for sites assessed; therefore, the plant could be completed and start operations about 4 years after the decision is taken, i.e., in 2018 at the earliest (Björn Fredriksson Möller, E.ON, 2012, personal communication).

Chemrec has had a demonstration plant (about 3 MW) for the development of DME based on gasified black liquor in Piteå, operating since 2010. Since 2011, the plant has been producing on a more regular basis, currently at 150 days per year at a production capacity of 6 GWh_{fuel}/year [4 tons DME per day (Ingvar Landälv, Chemrec, 2012, personal communication)]. The plant has been supported by the Swedish Energy Agency (Ingvar Landälv, Chemrec, 2012, personal communication). The DME is sold to 10 DME-driven trucks that AB Volvo has developed and is testing. Since 2013, activities at the demonstration plant are led by Luleå Technical University under the name LTU Green fuels.[24]

For several years, until summer 2012, there was also a biofuels production project via gasification from black liquor at a full-scale (200 MW) facility in Örnsköldsvik in connection to Domsjö Fabriker, about 70 times the capacity in Piteå (based on the technology from Chemrec AB). The planned production capacity corresponded to approximately 1 TWh_{fuel}/year (100,000 metric tons per year of DME or 140,000 tons methanol per year) (Ingvar Landälv, Chemrec, 2012, personal communication). The project had a grant from the Swedish Energy Agency with a limit of SEK 500 million (about Euro 50 million) and expected cost of more than SEK 3 billion (Ingvar Landälv, Chemrec, 2012, personal communication). In 2011, the new owners of Domsjö Fabriker, Aditya Birla Group, decided not to invest in the project owing to the lack of long-term political decisions on policy.

Vallvik Biofuel is another 200 MW biofuels production project planning to produce methanol via gasification of lignocellulose in the form of black liquor. The planned production capacity is 750 GWh_{fuel} (140,000 tons) of methanol per year. Rottneros AB, supported by Chemrec AB, leads this project (indicating that it builds on the technology developed by Chemrec). With conducive long-term political decisions and the funding needed, Vallvik Biofuel is expected to be operating in a few years, at the earliest by 2017 (Ragnar Lundberg, Rottneros AB, 2012, personal communication). In addition, Rottenros AB has another 200 MW project with Tyréns and 2GenAB, Rottneros Biorefinery, where lignocellulose from forest biomass will be used to produce methanol by gasification. The plant is expected to be the same size as Vallvik Biofuel or somewhat larger (Ragnar Lundberg, Rottneros AB, 2012, personal communication) Provided the political climate is supportive and creates the prerequisites for funding, this plant could also be operating in a few years (Ragnar Lundberg, Rottneros AB, 2012, personal communication).

VärmlandsMetanol AB plans to build a methanol plant based on gasification of forest resources in the municipality of Hagfors (at about 110 MW). The plant is expected to produce 600 GWh_{fuel} of methanol per year (375,000 L of methanol per day corresponding to about 137,000 m^3 per year at full production); the plant will also provide district heating.[25] The expected start of operations is 2016; VärmlandsMetanol is searching for funding systematically.[25]

BillerudKorsnäs has analyzed the possibilities for large-scale production of pyrolysis oil from forest residues at the Skärblacka pulp and paper mill, resulting in a production of 750 GWh_{fuel} (160,000 tons) of pyrolysis liquid per year.[26] The pyrolysis process produces different types of hydrocarbons that can be upgraded to biofuels for transport. The project has been selected for financial support from the EU in the NER 300 program. However, in December 2013, BillerudKorsnäs announced the discontinuation of the project in its current form. Their analysis had shown that the technology is available but the present commercial conditions and in the near future are not sufficiently attractive.[27] The project has been canceled.

During 2012/2013 Sakab AB (which handles hazardous waste) and VärmlandsMetanol AB, E.ON Gasification Development AB, E.ON Värme Sverige AB, Peab, Structor AB, and the municipality of Kumla performed a pilot study of a 250-MW biorefinery coproducing methanol, biomethane, and district heating by gasification of forest resources as wood chips as well as by-products and waste (Biorefinery Norrtorp). There are three different output options proposed, either (1) 1700 GWh_{fuel} (300,000 tons) of methanol per year; (2) 1300 GWh_{fuel} methanol per year and 480 GWh_{fuel} of biomethane per year, or (3) 1800 GWh_{fuel} of biomethane per year (Raimo Huhtala, Sakab AB, 2012, personal communication). The plant is planned in connection to the industrial area of Sakab AB in Kumla, but the next step is an economic analysis. Given a decision to proceed with the project and the necessary funding, the plant could operate commercially in about 10 years, no sooner than 2023 [Raimo Huhtala, Sakab AB, 2012, personal communication; Erik Rensfelt, Växjö Värnamo Biomass Gasification Centre (VVBGC) 2012, personal communication].

Växjö Värnamo Biomass Gasification Centre, VVBGC, and its 18-MW biomass gasification plant in Värnamo are state-owned through a holding company of Linnaues University. After upgrading and restart during the EU-project CHRISGAS 2004–2010, the plan was to transform the plant into a flexible research, demonstration, and training unit (Margareta Thyselius, VärmlandsMetanol, 2012, personal communication). The Swedish Energy Agency had planned to finance part of the reconstruction and the first project (gasification of biomass to syngas), provided industry also invested. When this failed, the Swedish Energy Agency decided to stop financial support in 2011, and the facility is currently not in use.

At Chalmers University of Technology, an indirect fluidized bed gasifier, with a capacity of 2–4 MW, was connected to an already existing 12 MW circulating fluidized bed boiler in 2008, with the boiler acting as a heat source for the integrated gasifier. The purpose of the facility is to prove the reliability of the integration concept and test different types of solid fuels.[28] Produced biofuels are for research purposes only and not used as transportation fuels.

Plants not included in this overview are, e.g., the following two small-scale facilities: (1) the pilot plant for pressurized entrained flow biomass gasification (PEBG) of biomass particles from forest residues that has been assessed at ETC, Piteå[29] and (2) the indirect gasification technology Woodroll in Köping (at 150 kW developed by Cortus). The plan for a 5-MW Woodroll gasifier, with upscaling to 25 MW in a second step[26], is neither part of the overview owing to lack of information.

BIODIESEL: FAME

There are several producers of FAME in Sweden (e.g., Perstorps BioProducts AB, Energigårdarna Eslöv/Ecobränsle, Karaby bioenergi HB, Norups gård AB, and SoilOil i Häckeberga AB); however, only two are relatively large. The facility of Perstorp in Stenungssund has a yearly production capacity of about 1700 GWh_{fuel}/year (180,000 m^3) and the facility of Energigårdarnas/Ecobränsle in Karlshamn has a yearly production capacity of about 500 GWh_{fuel}/year (50,000 m^3), but neither one has produced at full capacity in recent years[5] (Christian Bundy, Ecobränsle, 2012, personal communication; Susanne Eckersten, Perstorp BioProducts, 2012, personal communication). In 2011, the production from Perstorp in Stenungsund amounted to about 1200 GWh_{fuel}/year (126,000 m^3) and the production from Energigårdarnas/Ecobränsle amounted to 65 GWh_{fuel}/year (7,000 m^3) (Christian Bundy, Ecobränsle, 2012, personal communication; Susanne Eckersten, Perstorp BioProducts, 2012, personal communication). Perstorp reports that one reason for not producing at full capacity is that the FAME fuel quality that is suitable for the Nordic climate (low content of monoglycerides) requires facilities to be run at a slower pace. The facility in Karlshamn reports that after a total stop in 2008, the annual production has increased each year, since the facility started to operate again in 2009, and production has been adjusted to the demand. The majority of the feedstock for the domestic FAME production is imported rapeseed.[12]

There seems to be no new plans for future biodiesel facilities in Sweden. Instead, some of the small producers stopped producing in 2012 and 2013 (Carl-Johan Andersson, Soiloil, 2012, personal communication). The future for FAME in Sweden will depend on the outcome of discussions on how to include indirect land use change in the EU sustainability criteria for biofuels; FAME might not meet the future requirements for reduced climate impact.

Biodiesel: HVO

HVO (hydrotreated vegetable oil) is a form of biodiesel that can be produced in a separate facility and by an integrated process in petroleum refineries resulting in a biomass-based share of the conventional fossil-based fuel. Biological oils, such as raw tall oil and slaughter waste, are used for HVO production, but given proper pretreatment many other feedstocks, e.g., straw, manure, algae, by-products, and waste from forest and pulp, and all kinds of biological oils can be used (Thomas Ögren and Helene Samuelsson, Preem, 2012, personal communication).

The Swedish HVO based on raw tall oil is produced in two steps. First the intermediate product raw tall diesel is separated from the raw tall oil, in a Sunpine facility (Johan Lundbäck, Sunpine, 2012, personal communication). Then the raw tall diesel is processed further to HVO diesel fuel by Preem.

The Swedish production of HVO from raw tall oil amounted to about 670 GWh_{fuel} (67,000 m^3), in 2011 (Thomas Ögren and Helene Samuelsson, Preem, 2012, personal communication). The production capacity, for HVO based on raw tall oil, at Preem is 1000 GWh_{fuel} (100,000 m^3) of HVO diesel per year (Thomas Ögren and Helene Samuelsson, Preem, 2012, personal communication). The total Swedish production of raw tall oil corresponds to about 2–2.5 TWh (Swedish Forest Industries Federation (Lina Palm, Skogsindustrierna, 2012, personal communication). This corresponds to about 1.3–1.75 TWh raw tall diesel (with negligible energy losses in the last conversion step this representing the theoretical maximum level of HVO from domestic tall oil). With a larger share of sustainable and acceptable feedstock for HVO, the Swedish production of HVO could increase; Preem recently announced that they will almost double their HVO production by 2016 using different types of renewable sources.[30]

Other Biofuels for Transport under Development

Biofuels based on algae and synthetic fuels produced from biomass-based CO_2 and water using electricity as the main energy source, so-called electrofuels, are examples of biofuels that have a great supply potential but currently need further research and development before being available at large scale. CO_2 released in the production of biofuels comes in a pure and concentrated form, which makes it advantageous to use in the production of electrofuels. The most discussed types of electrofuels are methane (a concept denoted 'power-to gas') or methanol (which can be converted into DME) and the most discussed algae-based biofuels are methane or FAME fuels.

The vehicle industry is interested in electrofuels (not only biomass-based). For example, Audi has invested in a power-to-gas facility in Werlte, Germany, that opened in 2013 with a production capacity of 10 GWh_{fuel} (1000 tons) of synthetic methane per year[31] and Volkswagen recently highlighted electrofuels in the form of methane as an important future complement to conventional natural gas and other biomass-based methane.[32] See Refs 33 and 34 for more on electrofuels. The interest in converting fast-growing algae into biofuels has increased, in, e.g., in the United States, where a large number of companies (e.g., Exxon Mobile, Shell, and BP) have declared themselves to be in the algae business.[35] See Ref 36 for more on algae-based biofuels.

The petrochemical sector is also studying other options for large-scale production of biofuels in the form of fossil fuels with substantial renewable content. One concept studied by, e.g., Preem involves lignin-derived bio-oils and pyrolysis oils.[37,38] Conversion of lignin precipitate to a bio-oil that in turn can be fed into the refinery is between the laboratory and pilot scales, implying that it will not be realized for more than a decade.[37]

As there are still large uncertainties about these biofuel pathways, they are not included in the analysis of this study.

SCENARIOS FOR BIOFUELS PRODUCTION IN SWEDEN UNTIL 2030

The mapping of existing and planned production capacity for biofuels for transport in Sweden shows that many different types of biofuels are produced currently or are planned to enter the market during the coming decade. There are plans for the production of cellulose-based ethanol, methanol, DME, methane, HVO as well as AD biogas. We use this overview as a basis for developing scenarios for potential biofuels production in Sweden until 2030.

The large numbers of small-scale commercial plants that make up the expansion of AD biogas are generally analyzed at the aggregate level, whereas the large-scale plants that make up the expansion of other biofuel systems are analyzed individually. We construct two scenarios to represent two different plausible outcomes, where we in Scenario 1 assume that all ongoing mapped plans for additional biofuels production are realized and that the expansion continues, whereas we in Scenario 2 illustrate a less ambitious alternative where producers' plans are delayed combined with a slower expansion pace. Detailed information for the scenarios is given below.

In both scenarios, already existing facilities (apart from noncommercial research facilities and terminated projects) are assumed to continue to deliver the current amounts of biofuels during the entire time period. However, we are aware that proposed changes in biofuel policies may cause difficulties for a few existing fuel options to meet future restrictions and may lead to some existing facilities being phased out. Currently, the EU is considering developing policies related to the GHG performance of a biofuel's entire life cycle (this would include factors for indirect land use change, iLUC, and a stricter level for GHG reductions as well as limiting the use of crop-based biofuels). Such policies may affect the eligibility of, first and foremost, FAME fuels, see for example the proposal for an amendment of the RED and the Fuel Quality Directive[39] from the EU Commission.[40] The causes behind land use change are multiple, complex, interlinked, and change over time (cf. Refs 41 and 42), leading to uncertain quantifications and making related policy development a complex task. The second scenario, though represent a precautious future where different factors, including policies, may change the prerequisites and delay investments.

In both scenarios, all biofuels produced are assumed to be used in the road-based transport sector. All planned facilities are assumed to produce the amount of biofuels per year reported by the actors. In reality, the annual biofuels production could be somewhat reduced if the production capacity is not used to the expected extent. Further, it is assumed that biomass supply potential is not a limiting factor for the levels of biomass used in these scenarios until 2030, but this will be discussed. It is important to understand that neither of the scenarios are predictions of what will happen; they illustrate how biofuels production could develop under different assumptions.

Table 28.5 Scenario 1 Assumptions for the Continued Expansion of Domestic Biofuels Production after Planned Investments, as well as the Results for 2030

	Start Year[1]	Growth Factor	Amount of Biofuels in Start Year (TWh_{fuel}/year)	Amount of biofuels 2030 (TWh_{fuel}/year)
Ethanol grain	2013	0	1.5	1.5
Ethanol cellulose + waste	2021	1.05	2.1	3.4
FAME	2013	0	2.2	2.2
HVO	2017	1.05	2.0	4.0
Biogas (anaerobic digestion)	2014	1.10	0.8	5.1[2]
Biomethane	2024	1.05	3.3	4.6
DME/methanol	2024	1.05	3.4	4.8

[1] The year after the last planned facility, indicating that no facilities will be built earlier if not currently planned.
[2] Apart from the growth at the aggregate level, individual contributions from cogenerated AD biogas in the facility of Nordisk Etanol och biogas are added in 2015 and 2020.

Scenario 1: Full Realization and Continued Expansion

In Scenario 1, all planned facilities are assumed to be realized and start to produce the stated amount of biofuels in the year reported by the actors, and existing biofuel facilities are assumed to remain in production at current levels until 2030 (production from non-commercial research facilities and terminated projects are not included in the scenarios). Current production along with the realization of the production plans results in a potential contribution of almost 18 TWh_{fuel} in 2023. However, in Scenario 1, besides the implementation of the planned investments, slightly different continued expansion paths for the different types of biofuels are assumed. In the mapping of production plans, no indications for investment plans for grain-based ethanol or crop-based FAME fuels were identified. Therefore, the growth rates for these two biofuel options are set to zero, and they will thus remain at current size during the entire time period, in both scenarios. As described above, AD biogas is treated differently than the other biofuels: the numerous small-scale commercial plants are not analyzed individually but at the aggregate level. For AD biogas upgraded to fuel quality for transport, an increase by 10% per year until 2030 is assumed. In addition to this growth assumption, the specific contribution from the facility of Nordisk Etanol och biogas is added. This will lead to an AD biogas use of approximately 5 TWh_{fuel} in the Swedish transport sector, in 2030, which is in line with the assumptions in Refs 22 and 23 that roughly half of the produced AD biogas will continue to be upgraded to transport fuel quality. For remaining biofuels production options, we assume 5% growth per year. An Scurve describes the technological maturity of a product, i.e., how fast new technologies enter the market, from the first laboratory phase, through different growth phases such as introduction, expansion, break through until fully commercialized. The choice of growth rates, 5 and 10%, comes from a rough estimation of where on the Scurve the different fuel options are.

As start year, for the continued expansion of each type of biofuels production, the year after the last planned facility is ready to operate is chosen, indicating that no new facilities will be built earlier than that, if not currently planned. The assumed start year and growth factor for each biofuel type is presented in Table 28.5, where the amount of biofuels in the start year as well as the result for 2030 can also be found. Figure 28.2 shows the results for the development of domestic biofuels production for the entire time period.

The decrease of grain-based ethanol, in 2020, is because the facility 'Nordisk Etanol och Biogas' is assumed to change from grain-based ethanol to cellulose-based ethanol in 2020 (applies also to Scenario 2).

In Scenario 1, the potential contribution from domestically produced biofuels amounts to approximately 16 TWh_{fuel} and 26 TWh_{fuel} in Sweden in 2020 and 2030, respectively (the contribution from crop-based fuels remains below 4 TWh_{fuel} during the entire time period). Assuming a conversion efficiency from biomass to biofuels of roughly 50% implies that approximately 30 and 50 TWh of biomass feedstocks are needed for the domestic production of biofuels for transport in Sweden in 2020 and 2030, respectively. For comparison, the total use of biofuels for transport in the EU in 2012 corresponds to about 170 TWh.[3]

Scenario 2: Delayed Full Realization and Slower Continued Expansion

From experience, we know that Swedish biofuel projects sometimes are delayed by several years.[6] In Scenario 2, we assumed that the start year for all ongoing planned facilities is delayed by 5 years and that investments in, and implementations of, subsequent additional biofuels production

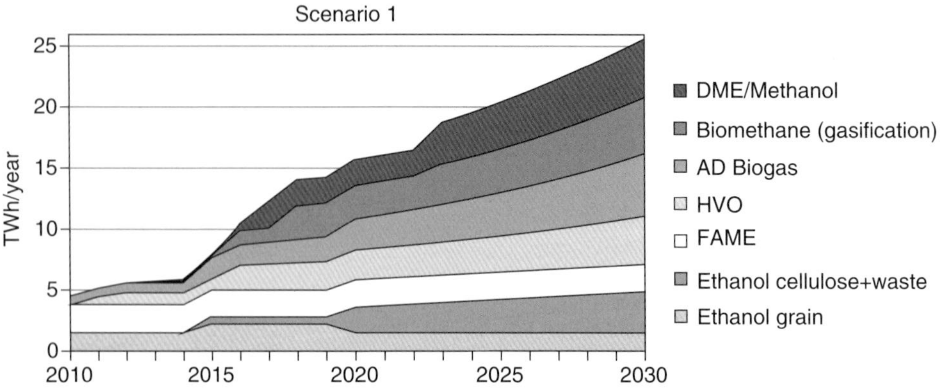

Figure 28.2 Domestically produced biofuels for transport in Sweden 2010–2030, in Scenario 1, based on current production, the realization of future plans, and with assumptions for the implementation of subsequent additional biofuels production.

after the planned facilities follow a slower pace until 2030. Delayed full realization and slower continued expansion could be the result of general instabilities in the economic system as well as uncertainties in how biofuel policies will be designed.

The assumptions for the continued expansion of biofuels production after the realization of the mapped production capacities in Scenario 2 are presented in Table 28.6. The results for 2030 are presented in Table 28.6, while results for the entire time period are shown in Figure 28.3.

In Scenario 2, the potential contribution of domestically produced biofuels amounts to approximately 8 TWh_{fuel} and 20 TWh_{fuel} biofuels in Sweden in 2020 and 2030, respectively (the contribution from crop-based fuels again remains below 4 TWh_{fuel} during the entire time period). Assuming a conversion efficiency from biomass to biofuels of 50% implies that approximately 15 and 40 TWh of biomass feedstocks are needed for the domestic production of biofuels for transport in Sweden in 2020 and 2030, respectively.

The assumption that the start year for all planned facilities is delayed by 5 years is rather arbitrary, and if the start year for all planned facilities instead were 10 years delayed, and the development thereafter follows the same pattern as in Scenario 2, the potential contribution of domestically produced biofuels would be 6 TWh_{fuel} and 15 TWh_{fuel} in 2020 and 2030, respectively.

Potential Biofuel Share of Total Energy Demand for Swedish Road Transport

In order to compare the estimated potential contribution of domestic biofuels with the total future energy demand

Table 28.6 Scenario 2 Assumptions for the Continued Expansion of Domestic Biofuels Production after Planned Investments, as well as the Results for 2030

	Start (year)[1]	Growth Factor	Amount of Biofuels Start Year (TWh_{fuel}/year)	Amount of Biofuels 2030 (TWh_{fuel}/year)
Ethanol grain	2013	0	1.5	1.5
Ethanol cellulose + waste	2026	1.025	2.1	2.7
FAME	2013	0	2.2	2.2
HVO	2022	1.025	2.0	3.1
Biogas (anaerobic digestion)	2019	1.050	0.8	3.6[2]
Biomethane	2029	1.025	3.3	3.6
DME/methanol	2029	1.025	3.4	3.8

[1] In this scenario it is assumed that the realization of the planned biofuels production capacity will be delayed by 5 years compared to reported plans (setting years 2014–2019 identical to 2013).

[2] Apart from the growth at the aggregate level, individual contributions from cogenerated AD biogas in the facility of Nordisk Etanol och biogas are added in 2020 and 2025.

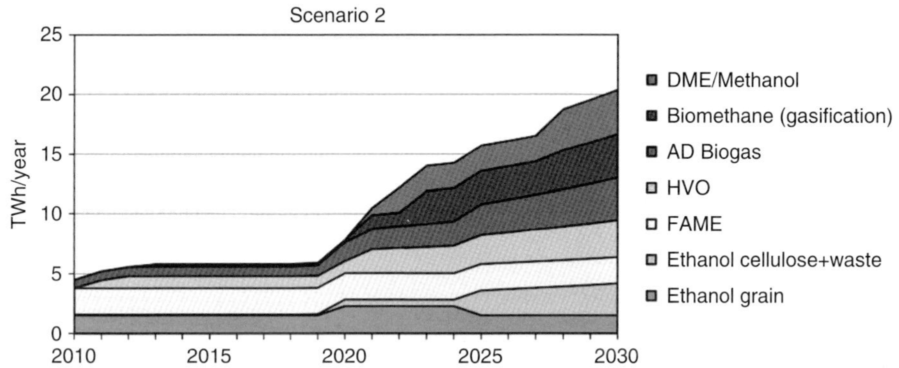

Figure 28.3 Domestically produced biofuels for transport in Sweden 2010–2030 based on current production, the realization of future plans and a slower pace for the implementation of subsequent additional biofuels production. All plans are assumed to be delayed by 5 years compared to what is currently reported.

for road transport in Sweden, an estimation of the latter is needed. The actual energy demand for road transport will depend on, e.g., the development of energy efficiency related to transport and behavioral change related to energy use. Recent official estimations of the energy demand for road transport in Sweden in 2020 and 2030, generated by Swedish authorities, but with different approaches and aims, can be found in Refs 14 and 43. Both these estimations will be used for the comparison.

Demand Scenario A, based on Ref 43, represents a long-term prognosis based on current vehicle trends as well as policies. Demand scenario B, based on Ref 14, represents a situation in which Sweden focuses on meeting the vision of no net emissions of GHG to the atmosphere by 2050. Energy demand scenario B represents a relatively low energy demand, which is motivated by the implementation of measures promoting energy efficiency, as well as reducing the demand for transport, and includes a substantial contribution from electric vehicles. Demand scenario A represents a higher energy demand and includes a limited amount of electric vehicles. The actual energy demand for road-based transport estimated in the two different demand scenarios for 2020 and 2030 are presented in Table 28.7. Table 28.7 also presents the shares of the energy demand for Swedish road transport to which the potential domestically produced biofuels estimated in Scenarios 1 and 2 correspond. Depending on the biofuel scenario and demand scenario, the potential biofuel share of the total energy demand for road transport is 10–30% in 2020 and 26–79% in 2030. The high ends of the intervals represent the estimated energy demand for road transport if Sweden is to meet the national vision of no net emissions of GHG to the atmosphere in 2050. The wide range within the intervals indicates that the possible future biofuel share, to a high degree, depends on the estimated total future energy demand, which is very uncertain.

DISCUSSION

Owing to its policy measures,[6] Sweden had the largest share (12.6%) of renewable fuels for transport among EU countries in 2012 (calculated according to the RED). In this paper, we have assessed the potential future contribution of domestically produced biofuels for transport in Sweden by 2030, based on our mapping of existing and planned domestic production capacity for biofuels and developing two potential biofuels production scenarios.

Mapping exercises always deliver a snapshot of reality, in this case how actual and potential Swedish biofuel producers, during 2012–2013, viewed the prerequisites for their future biofuels production (and how this information was shared and available to the authors). In reality, plans often tend to change with time, which is illustrated to a limited extent in our Scenario 2. We have not assessed the probability that a given biofuel plan will be realized (or the credibility of the information), and we have not analyzed the probabilities for the different scenarios to be realized, as it is difficult to predict which plans will be realized (and when) and what plans might be added. However, the information has been compared and checked with other references when available. The authors have for example been involved in the governmental investigation on fossil fuel independence in the Swedish transport sector in which Swedish biofuel actors have provided and checked the same information.[10] Changes in a biofuel producer's plans (e.g., regarding size, start year, and status) will influence the estimation of potential future contribution of domestically produced biofuels. Some of the included potential biofuels

Table 28.7 Estimated Potential Biofuels Production Capacity in the Two Biofuel Scenarios (1 and 2) Compared to Two Different Estimations of the Future Energy Demand (A and B) for Swedish Road Transport in 2020 and 2030

	Biofuel Scenario 1		Biofuel Scenario 2	
	2020	2030	2020	2030
Biofuels production capacity (TWh/year)	16	26	8	20
Demand scenario A (TWh/year)[1]	80	78	80	78
Demand scenario B (TWh/year)[2]	54	33	54	33
Comparison A. Biofuel share of energy demand for Swedish road transport, assuming demand A	0.20	0.33	0.10	0.26
Comparison B. Biofuel share of energy demand for Swedish road transport, assuming demand B	0.30	0.79	0.15	0.61

For Comparison, the energy demand for Swedish road transport was approximately 84 TWh in 2012.[4]
[1]Reference scenario for the long-term prognosis in Ref 43.
[2]Scenario 1 in Ref 14.

production may not be realized or be replaced by other projects. However, our study is of interest to policy makers and other actors evaluating the prospects for renewable fuels.

Compared to a similar assessment of the prospects for Swedish production of biofuels for transport made during 2009–2010,[44] there are some new projects, but most of the projects known in 2010, are still active, although the time tables for building the facilities generally have been somewhat delayed, for example, the facilities by NBE Sweden, Nordisk Etanol och Biogas AB, ST1, and VärmlandsMetanol AB. The actors mainly refer to global economic instabilities limiting investments as the reason for the delay. In general, according to the biofuel industry actors, in particular the lack of long-term political decisions has prevented a more progressive development of biofuels production in Sweden (long-term economic support is demanded), not the pace of technology development. Political signals (including international signals) influence the possibility to obtain funding. The lead time for developing biofuels production facilities (including building permits, environmental permits, etc.) is also of importance.[6]

Some existing biofuel technologies are not seen in the scenarios in this study, for example, Fischer-Tropsch (FT) fuels. No Swedish plans for FT were identified in the mapping exercise. However, for the time period beyond currently planned facilities, FT fuels (or other types of biofuels) could replace the types of biofuels currently shown in the scenarios. The scenarios also do not show a contribution from algae-based biofuels, biomass-based electrofuels, or other biomass-based fuel technologies that currently are in an early phase of development. If these technologies develop at a more rapid pace before 2030 than currently expected, the potential contribution for domestically produced biofuels in Sweden for 2030 may exceed the estimations we have presented.

Not all the domestically produced biofuels will necessarily be used in the Swedish road transport sector. The demand for biofuels from other transport sectors, such as sea-based transport and aviation, as well as demand from the chemical industry, could increase, see e.g., Refs 45–47. For road transport, it is possible to use vehicles driven by hydrogen or electricity as a complement to biofuels, when goals for reduced CO_2 emissions are to be achieved. But from a technical perspective, it is considered more challenging to use electricity and hydrogen for sea-based transport and aviation. Therefore, the domestically produced amounts of biofuels may be used in other end-use sectors in Sweden as well as in other countries.

Sweden currently imports considerable amounts of biofuels for transport per year (approximately 3 TWh/year in 2011).[48] However, the future contribution from import of biofuels for transport is uncertain, and Sweden is sometimes considered a potential export country for biofuels for transport.[14] For simplicity, we compared the total estimated biofuels production potential to the estimated future Swedish road transport demand. This exercise should not be interpreted as suggesting there will be no export of Swedish biofuels in the future. Market prices will likely determine where the biofuels end up. There could also be an increased demand for biomass from the heat and power sector, aiming to reduce the use of fossil fuels, limiting the amount of biomass that is available for the transport sector.

Are the domestic biomass resources approximately sufficient to meet the estimated potential biofuels production? The latest estimate of the Swedish additional biomass supply potential (available for energy) reports 50–60 TWh of additional biomass per year in the short term and at least 75–90 TWh in 30–50 years.[10,26] The estimated potential biofuels production corresponds to approximately 15–50 TWh per year of biomass, in the period 2020–2030. That scenario also represents a substantial increase compared with the present biomass use for transport in Sweden. In 2010, the Swedish use of biofuels for transport corresponded to approximately 10 TWh of biomass of which slightly more than 2 TWh was domestically produced.[5] However, it seems that the estimated domestic biomass supply could meet the estimated demand from domestic biofuels production, at least in theory.

Of the estimated 50–60 TWh of additional biomass per year (potentially available for energy) in Sweden, the majority could be in the form of lignocellulosic and waste feedstock (however, some amount of the underlying agricultural area could also be used to produce grain and other more traditional agricultural crops).[10,26] The underlying estimation assumes a mix of energy crops. The estimated potential biofuels production corresponds roughly to a biomass demand of approximately 32–43 TWh lignocellulosic and waste feedstock in 2030, while the production of grain-based ethanol and FAME does not increase in 2030 compared with that in 2010. For comparison, the overall environmentally compatible biomass supply potential in Sweden in 2030 is estimated to be about 157 TWh in Ref 49.

In reality, all land-using activities will impact the bioenergy supply potential. Apart from food, feed, and grazing, the competition for biomass and biofuels from other sectors, as well as the demand for biofuels in other countries, will influence how large share of the domestic biomass supply potential is used for domestic biofuels production in the future. The location of the biofuel facilities in relation to the location of the biomass supply is also of importance.

The actual contribution of biofuels for road transport in Sweden in the future will depend on many factors, such as the development of policies (national as well as at the EU level), global developments for fossil fuels and biofuels, the future energy demand, the production and distribution costs for biofuel facilities, the possibility for the Swedish vehicle fleet to use the produced biofuels, and the global competition for biomass and biofuels. The future for Swedish biofuels production will for example depend on if there is support for new domestic biofuels production as proposed in Ref 10.

The fuel infrastructure and the vehicle fleet may limit the amount of new types of fuels used. Some of the fuel types may be used as blends in conventional fuels, but it is challenging to blend in more than 10–20%, to meet biofuel shares higher than that, modified vehicles need to enter the market. The average lifespan for cars in Sweden is 17 years[50], meaning that the majority of cars sold today will still be around in 2030, unless incentives are introduced to phase out vehicles at a more rapid pace. In January 2012, there were in total 2885 fuel stations in Sweden.[51] Fuel infrastructure is well developed (1694 locations)[51] for the fuel blend E85 (85% ethanol and 15% gasoline) mainly as a result of a Swedish law requiring fuel stations of a certain size to supply at least one renewable fuel. The developed E85 infrastructure, in combination with a well-established low-blend market, indicates that the ethanol production seen in our scenarios will not be limited by infrastructure. In 2012, most HVO–diesel blends were a mix of 23% HVO and 7% FAME in conventional diesel, available at more than 700 locations.[6] There are no technological limitations for the share of HVO that a diesel engine can run on, which indicates that the HVO production seen in our scenarios will not be limited by existing infrastructure. For methane, DME, and pure FAME fuels, there is some current infrastructure (132, 4, and 22 locations respectively).[51] An expansion of methane and DME requires new infrastructure to be built. A drawback for methane is the relatively costly distribution system. In Sweden, no methanol is currently used in the road transport sector. However, it is possible to blend 3% methanol in gasoline, and adjusted vehicles can run on M85 or M100. To be able to use the methanol shown in the scenarios, new infrastructure needs to be built.

The contribution of the estimated potential biofuels production to the total Swedish energy demand for transport depends to a large extent on how energy demand develops, which differs considerably among studies. Achieving the estimated energy demand in Ref 14 requires a relatively large adjustment of the transport sector.

Finally, we have not assessed the investment costs associated with each of the mapped biofuel plans or the need for investors and other funding possibilities, but this is an interesting task for future analyses.

CONCLUSION

In addition to current biofuels production (mainly wheat-based ethanol, rapeseed-based FAME, AD biogas, and HVO based on tall oil), there are plans for cellulose-based ethanol, food waste-based ethanol, methane, methanol, and further expansion of HVO production in Sweden. Some of the projects have been delayed relative to the original plans. In particular, the lack of long-term political decisions, not limits in technology development, is what has prevented a more progressive development of biofuels production in Sweden, according to the Swedish biofuel industry actors.

The potential domestic Swedish biofuels for transport production is approximately 18 TWh_{fuel} in 2023, assuming continued domestic production of biofuels at current levels and the realization of all the ongoing mapped plans for additional biofuels production. The potential domestic biofuels production reaches about 26 TWh_{fuel} in 2030 with an assumed continued expansion of biofuels production.

For comparison, the current Swedish energy demand for road-based transport is approximately 84 TWh.[4]

The estimated potential domestic Swedish biofuels production corresponds to about 20–30% in 2020 and 33–79% in 2030 of the future Swedish energy demand for road transport, when comparing with two official scenarios representing distinctly different developments of the transport and related energy demand.

However, if the realization of the mapped biofuels plans is delayed, canceled and/or the continued implementation of additional biofuel capacity is slower than assumed above, the potential domestic biofuels production in Sweden is reduced substantially.

REFERENCES

1. Government Offices of Sweden. Sweden's second progress report on the development of renewable energy pursuant to Article 22 of Directive 2009/28/EC, 2013. Members States' progress reports, 2013 reports translated in English. Available at: http://ec.europa.eu/energy/renewables/reports/2013_en.htm. (Accessed May 3, 2014).
2. European Parliament and Council. Directive 2009/28/EC of the European Parliament and of the council of 23 April 2009 on the promotion of the use of energy from renewable sources and amending and subsequently repealing Directives 2001/77/EC and 2003/30/EC, 2009.
3. European Commission. Eurostat energy statistics, Shares 2012 results, 2013. Available at: http://epp.eurostat.ec.europa.eu/portal/page/portal/energy/other_documents. (Accessed March 12, 2014).
4. Swedish Energy Agency. Transportsektorns energianvändning, 2012 (The use of energy in the transport sector in 2012). ES 2013:02, 2013. Available at: www.energimyndigheten.se (In Swedish). (Accessed April 10, 2014).
5. Government Offices of Sweden. Sweden's first progress report on the development of renewable energy pursuant to Article 22 of Directive 2009/28/EC, 2011. Members States' progress reports, 2011 reports translated in English. Available at: http://ec.europa.eu/energy/renewables/reports/2011_en.htm. (Accessed January 10, 2014).
6. Hansson J, Grahn M. Utsikt för förnybara drivmedel i Sverige (Prospects for renewable fuels in Sweden) - Uppdatering och utvidgning av studien "Möjligheter för förnybara drivmedel i Sverige till år 2030 av Grahn och Hansson, 2010", IVL Report B2083, 2013. Available at: http://spbi.se/wp-content/uploads/2013/03/IVL_B2083_2013_final.pdf (In Swedish). (Accessed November 5, 2013).
7. Swedish National Audit Office. Biodrivmedel för bättre klimat – Hur används skattebefrielsen? (Biofuels for transport for a better climate – How is the tax exemption used?), *RiR* 2011:10. 2011. (In Swedish).
8. Government offices of Sweden. Regeringens proposition (Government bill) 2008/2009:163: En sammanhållen Klimat och Energipolitik – Energi (A coherent Climate and Energy policy - Energy), 2009. Available at: www.regeringen.se/sb/d/11033/a/122785 (In Swedish). (Accessed February 12, 2014).
9. Swedish Environmental Protection Agency. Underlag till en färdplan för ett Sverige utan klimatutsläpp 2050 (Basis for a roadmap for a Sweden without greenhouse gas emissions in 2050), Report 6537, 2012. Available at: http://www.naturvardsverket.se/Documents/publikationer6400/978-91-620-6537-9.pdf?pid=4965 (In Swedish). (Accessed December 18, 2013).
10. Swedish Government Official Reports. Fossilfrihet på väg (On its way to fossil fuel independence). SOU 2013:84, 2013. Available at: http://www.regeringen.se/sb/d/108/a/230739 (In Swedish). (Accessed January 20, 2014).
11. Swedish Energy Agency. Övervakningsrapport avseende skattebefrielse för biodrivmedel år 2012 (Supervision report of the tax exemption for biofuels in 2012). 2013-03-19, 2013. Available at: http://www.energimyndigheten.se/PageFiles/26233/2013-0026%20Rapport%20skattebefrielse%20f%C3%B6r%20biodrivmedel%20%C3%A5r%202012.pdf (In Swedish). (Accessed November 5, 2013).
12. Swedish Energy Agency. Hållbara biodrivmedel och flytande biobränslen under 2012 (Sustainable biofuels for transport and liquid biofuels during 2012). ET2013:06, 2013. Available at: www.energimyndigheten.se (In Swedish). (Accessed January 15, 2014).
13. Swedish Energy Agency. Långsiktsprognos 2010 (Long term prognosis 2010). ER 2011:03, 2011. Available at: www.energimyndigheten.se (In Swedish). (Accessed March 10, 2013).
14. Swedish Transport Administration. Delrapport transporter - Underlag till färdplan 2050 (Partial report – Basis for the roadmap for 2050). Publication no. 2012:224. ISBN: 978-91-7467-418-7, November 2012. Available at: http://publikationswebbutik.vv.se/upload/6930/2012_224_Delrapport_transporter_under lag_till_fardplan_2050.pdf (In Swedish).
15. Profu. Roadmap för ett fossilbränsleoberoende transportsystem år 2030 (Roadmap for a fossil independent transport system 2030). For Svensk Energi and Elforsk. Version 2012-11-14, 2012. Available at: www.elforsk.se (In Swedish). (Accessed February 5, 2013).
16. KNEG. Hinder och drivkrafter för minskad klimatpåverkan från godstransporter (Obstacles and driving forces for reduced climate impact from freight transport). KNEG resultatrapport 2012. Klimatneutrala godstransporter på väg (KNEG), Trafikverket, Chalmers, Göteborgs miljövetenskapliga centrum (GMV), 2012. (In Swedish).
17. Swedish Transport Administration. Målbild för ett transportsystem som uppfyller klimatmål och vägen dit (Objectives for a transport system that fulfills climate targets and the way there). Publication number: 2012:105, 2012. Available at: http://publikat ionswebbutik.vv.se/shopping/ShowItem____5664.aspx. (Accessed May 7, 2013).
18. Swedish Government Official Reports. Slutbetänkandet från utredningen om förnybara fordonsbränslen: Introduktion av förnybara fordonsbränslen (Final report from the official investigation of renewable transport fuels:

Introduction of renewable transport fuels), SOU 2004:133, 2004. Available at: http://www. regeringen.se/content/1/c6/03/67/01/bb423867.pdf (In Swedish). (Accessed February 19, 2013).
19. SP. Biorefinery Demo Plant in Örnsköldsvik, 2013. www.sp.se/sv/index/services/biorefinerydemoplant/Sidor/default.aspx. (Accessed April 15, 2014).
20. Swedish Energy Agency. Produktion och användning av biogas år 2011 (Production and use of biogas in 2011). ES 2012:08, 2012. Available at: www.energimyndigheten.se (In Swedish). (Accessed January 12, 2013).
21. Kågeson P, Jonsson L. Var inom transportsektorn får biogasen störst klimatnytta? (Where in the transport sector will the largest climate benefit of biogas be achieved?). Report from Centre for transport studies, CTS Working Paper 2012:18, Stockholm, 2012.
22. Nordberg U. Biogas – Nuläge och framtida potential. *Värmeforsk*, ISSN 1653-1248. Stockholm, 2006.
23. Linné, M., Ekstrandh, A., Englesson, R., Persson, E., Björnsson, L., Lantz, M. Den svenska biogaspotentialen från inhemska restprodukter. Avfall Sverige, Svenska Biogasföreningen, Svenska Gasföreningen, Svenskt Vatten. Lund, 2008.
24. Luleå Tekniska Universitet, LTU Green Fuels, 2013. Available at: www.ltu.se/org/tvm/Avdelningar/LTU-Green-Fuels. (Accessed April 15, 2014).
25. Börjesson P, Lundgren J, Ahlgren S, Nyström I. Dagens och framtidens hållbara biodrivmedel - Under-lagsrapport från f3 till utredningen om FossilFri Fordonstrafik (Current and future sustainable biofuels for transport – Report from f3 to the governmental investigation on Fossil fuel free vehicle transport). f3 Report 2013:13, 2013. Available at: http://www.f3centre.se/sites/default/files/f3_borjesson_et_al_dagens_och_framtidens_hallbara_biodrivmedel__slutversion_rev_130620.pdf. (Accessed November 14, 2013).
26. BillerudKorsnäs. 2013. Available at: www.billerud.se/Media/Pressmeddelanden/2013/BillerudKorsnas-avbryter-forstudie-rorande-produktion-av-gron-olja-starker-fokus-mot-smartare-forpackningar/(InSwedish). (Accessed April 2, 2014).
27. Fredriksson Möller B, Gillberg B, Huhtala R, Reinholdtzon A, Westlind G. 2013. Idestudie Bioraffinaderi Norrtorp 2013-09-17. (In Swedish).
28. Business Region Göteborg, Chalmers Gasifier, 2010. Available at www.businessregiongoteborg.com/download/18.5cdb5cc912845f4210b8000716/Chalmers _(Accessed April 3, 2014).
29. Marklund M, Weiland F, Hedman H, Westerlund S, Öhman M, Lundgren J. Renewable motor fuels from the forest industry via Pressurized Entrained flow Biomass Gasification. Final Report to the Swedish Energy Agency, 2012. Can be ordered from the Swedish Energy Agency (PO Box 310, SE-631 04 Eskilstuna, Sweden or registrator@energimyndigheten.se).
30. Preem. Pressrelease "Preem investerar 300 miljoner i ny produktionskapacitet för biodrivmedel", 2014. Available at: www.preem.se/templates/page____12120.aspx?newspageid=12143. (Accessed April 4, 2014).
31. Audi e-gas. Energy turnaround in the tank, 2013. Available at: www.audi.com/content/com/brand/en/vorsprung_durch_technik/content/2013/10/energy-turnaround-in-the-tank.html. (Accessed April 9, 2014).
32. VW. Volkswagen Group Strategies. 0% Emission, 100% emotions, The road to Electromobility. 2012. Available at: www.volkswagenag.com/content/vwcorp/info_center/en/publications/2013/01/0_Emissions_100_Emotions.bin.html/binarystorageitem/file/Final_VW_EMob_20120514_komplett_EN.pdf. (Accessed September 12, 2013).
33. Mohseni F. Power to gas: bridging renewable electricity to the transport sector. Licentiate thesis, KTH Royal Institute of Technology, Stockholm, 2012.
34. Nikoleris, A, Nilsson L. Elektrobränslen: en kunskapsöversikt. Report No 85. ISBN. 978-91-86961-11-4, Lunds University, Department of Technology and Society, Environmental and Energy Systems Studies, 2013. Available at: http://lup.lub.lu.se/luur/download?func=downloadFile&recordOId=3738215&fileOId=3738230 (In Swedish).
35. Ny Teknik. Charlotta von Schultz. Världens hajp runt algbaserat bränsle (The algae fuel hype), 2009. Available at: www.nyteknik.se/nyheter/innovation/forskning_utveckling/article674108.ece (Accessed April 13, 2014). (In Swedish).
36. Slade R, Bauen A. Microalgae cultivation for biofuels: Cost, energy balance, environmental impacts and future prospects. *Biomass Bioenergy* 2013, 53:29–38.
37. Eriksson S. Towards a fossil free refinery. Presentation at the LU Biofuels Day, 21 May, 2013. Lund, Sweden.
38. Holland P. Preems strategi för att utveckla klimatvänliga biodrivmedel från restprodukter (Strategy of Preem to develop climate friendly biofuels from by-products and waste). Presentation by Petter Holland, president and CEO of Preem at Årets Bioenergidagar (organised by Svebio), Göteborg, Sweden, November 5, 2013.
39. European Commission. Proposal for a Directive of the European Parliament and of the Council amending Directive 98/70/EC relating to the quality of petrol and diesel fuels and amending Directive 2009/28/EC on the promotion of the use of energy from renewable sources, COM(2012) 595 final, Brussels 17.10.2012, 2012.
40. Berndes G, Ahlgren S, Börjesson P, Cowie AL. Bioenergy and land use change: state of the art. *WIREs: Energy Environ* 2013, 2:282–303. doi: 10.1002/wene.41.
41. Di Lucia L, Ahlgren S, Ericsson K. The dilemma of indirect land-use changes in EU biofuel policy - an empirical study of policy-making in the context of scientific uncertainty. *Environ Sci Policy* 2012, 16:9–19.
42. Swedish Energy Agency. Långsiktsprognos 2012 (Long term prognosis 2012). ER 2013:03, 2012. Available at: www.energimyndigheten.se (In Swedish). (Accessed October 15, 2013).
43. Grahn M, Hansson J. Möjligheter för förnybara drivmedel i Sverige till år 2030. (Possibilities for renewable fuels for transport in Sweden to 2030). Report for the Swedish Petroleum Institute, 2010. Available at: http://spbi.se/word

press/wp-content/uploads/2010/09/Chalmersrapport-SPI.pdf (In Swedish). (Accessed January 8, 2013).
44. European Parliament and Council. Directive 2009/30/EC of the European Parliament and of the Council of 23 April 2009 amending Directive 98/70/EC as regards the specification of petrol, diesel and gas-oil and introducing a mechanism to monitor and reduce greenhouse gas emissions and amending Council Directive 1999/32/EC as regards the specification of fuel used by inland waterway vessels and repealing Directive 93/12/EEC, 2009.
45. Ringbeck J, Koch V. Aviation biofuels: a roadmap towards more carbon-neutral skies. *Biofuels* 2010, 1:519–521.
46. Lichtenthaler FW, Peters S. Carbohydrates as green raw materials for the chemical industry. *C R Chim* 2004, 7:65–90.
47. Bengtsson S, Fridell E, Andersson K. Environmental assessment of two pathways towards the use of biofuels in shipping. *Energy Policy* 2012, 44:451–463.
48. Swedish Energy Agency. Analys av marknaderna för biodrivmedel - Aktuella marknadsfrågor för första och andra generationens biodrivmedel (Anaysis of the market for biofuels – Current market issues for first and second generation biofuels). ER 2012:29, 2012. Available at: www.energimyndigheten.se (In Swedish). (Accessed February 9, 2013).
49. European Environment Agency (EEA). How much bioenergy can Europe produce without harming the environment? EEA Report No 7/2006, ISBN 92–9167 –849-X, ISSN 1725–9177, 2006. Available at: http://www.eea.europa.eu/publications/eea_report_2006_7.
50. BIL Sweden. Trade organization for manufacturers and importers of cars, trucks and buses in Sweden, 2013. Available at: www.bilsweden.se (Accessed March 1, 2013).
51. Svenska Petroleum och Biodrivmedel Institutet. Statistics fuel stations, 2013. Available at: http://spbi.se/statistik/forsaljningsstallen/forsaljningsstallen-med-fornybara-drivmedel (Accessed March 1, 2013). (In Swedish).
52. European Commission. Background facts to directive 2009/28/EC of the European parliament and of the Council of 23 April 2009, on the promotion of the use of energy from renewable sources and amending and subsequently repealing Directives 2001/77/EC and 2003/30/EC, 2014. Available at: www.notisum.se/rnp/eu/lag/309L0028.htm (Accessed June 10, 2014).

29

Land and the Food–Fuel Competition: Insights from Modeling

Sylvia Prieler,* Günther Fischer and Harrij van Velthuizen

International Institute for Applied Systems Analysis, Food and Water Research, Laxenburg, Austria

INTRODUCTION

A shift toward low-carbon economies is inevitable when climate change mitigation is a prime concern. Transport fuels account for about one-fifth of CO_2 emissions from fossil fuel combustion.[1] Until 2035, transport energy demand is projected to increase by 40–50% when current or announced policies are implemented.[2] Using non-fossil fuels combined with substantial improvements in fuel use efficiency will be fundamental elements for achieving at the same time ambitious greenhouse gas (GHG) emission reduction targets. Biofuels, alternatives to gasoline and diesel that are derived from biomass, have the potential to substantially reduce GHG emissions of the transport sector. In addition, biofuels offer improved energy security through diversification of the fuel mix and may foster economic development in rural regions.

The transition to a low-carbon economy intensifies the energy–agriculture linkage and adds a new dimension to the agricultural system, heightens resource competition in the food sector, and also provides new opportunities for rural communities. The recent increases in fossil energy prices, combined with political targets and mandates for biofuels, have turned a growing number of agricultural products as competitive feedstocks into the energy market. In the future, food, feed, and energy feedstock markets are expected to integrate more closely.[3–5]

*Correspondence to: prieler@iiasa.ac.at

The additional demand from the energy sector has contributed to reversing the process of declining agricultural prices observed in the decades before 2000.[6–11] In the longer term, expanded demand and increasing prices for agricultural commodities may represent opportunities for agricultural and rural development. However, higher food prices pose an immediate threat to the food security of poor net food buyers. Many of the world's poor spend more than half of their household incomes on food, and even in rural areas the majority of the poor are net purchasers of food.[6]

Soaring prices in international commodity markets, growing international investments for land, notably in Sub-Saharan Africa, and potential impacts on food security and the environment have raised concerns and drawn the attention of scientists, policymakers, and the general public. According to press reports, foreign investors expressed interest in around 56 Mha of land globally in less than a year. Of these, 29 Mha were in Sub-Saharan Africa.[12]

The magnitude of the lifecycle GHG emission reduction achieved with biofuels compared to using fossil fuels depends on volume of biofuels produced, biofuel conversion pathways, and technology as well as on type and mode of biomass feedstock production. The majority of lifecycle analysis studies conclude that, when excluding land use change, producing biofuels from current feedstocks via efficient first-generation conversion processes results in emission reductions in the range of 20–60% relative to fossil fuels.[13–15] A notable exception is Brazilian ethanol from sugarcane with 80% or more saving due to energy

Advances in Bioenergy: The Sustainability Challenge, First Edition. Edited by Peter D. Lund, John Byrne, Göran Berndes and Iacovos A. Vasalos.

Originally published in: *WIREs Energy Environ* 2013, 2: 447–464 doi: 10.1002/wene.55

use including cogeneration of sugarcane residues and by-products.[6,16] In addition, the emissions due to land use changes directly or indirectly associated with expanding feedstock production can considerably reduce the potential GHG savings of bioenergy.[15,17] Foregone biofuels GHG benefits due to associated land use changes have been central in the debate of the effectiveness and sustainability of biofuels.[18–20] Conversion of land and changed land management practices for the production of biofuel feedstocks (direct land use change) as well as displacement of agricultural activities to other areas induced by biofuel initiatives and causing land use change somewhere else [indirect land use changes (iLUCs)], may lead to substantial losses of carbon stocks in soils and vegetation. An example of iLUC is soybean production in Brazil, which has been identified as an important driver of deforestation as it induces land settlers (e.g., cattle ranchers and small holders) to advance further into the Amazon forest to open up new lands.[21–23] Though up to now mainly driven by livestock grazing land demand, this tendency is likely to increase with rising demand for soybean oil to produce biodiesel.

The central importance of direct and iLUC effects for an accurate accounting of the impacts of increased biofuel production on GHG emissions is generally acknowledged.[15] However, approaches and calculation methods are debated. From a climate perspective, the question is not whether the GHG rating of a biofuel is above or below that of petroleum fuel, but whether net climate forcing increases or decreases as a result of producing more biofuels.[24]

Such concerns have contributed to the development of sustainability criteria in the biofuel policies of the European Union[14] and the United States.[25,26] A key element is compliance with defined minimum lifecycle GHG emission reduction compared to fossil fuel use. The role of iLUCs in the GHG balance of biofuels is widely debated[24] and polices may adapt to forthcoming analyses.

This paper intends to contribute to better understanding the energy–food security–environment nexus. It starts from an overview of the current trends in biofuel consumption and adds to the discussion around the food–fuel debate by applying IIASA's (International Institute for Applied Systems Analysis) Ecological–Economic Modeling Framework to study impacts of historic biofuel production on food and feed markets and environment, and to assess scenarios for future biofuel expansion.

BIOFUEL CONSUMPTION—CURRENT STATUS, TRENDS, AND OUTLOOK

Biofuel Production Development

Globally biofuel production has increased rapidly during the last decade, from 10 Mtoe (0.4 EJ) in 2000 to nearly 60 Mtoe (2.5 EJ) in 2010. Biofuel production has been dominated by three major producers: the United States of America (maize-based ethanol), Brazil (sugarcane ethanol), and the European Union (mostly biodiesel production, primarily using rapeseed) (Figure 29.1a). In recent years the number of countries engaged in biofuel production and consumption, as well as trade, has been increasing (Figure 29.1b).

For example, in response to the government's B7 blending mandate (7% biodiesel in total diesel consumption) and increasing demand from the world market, biodiesel

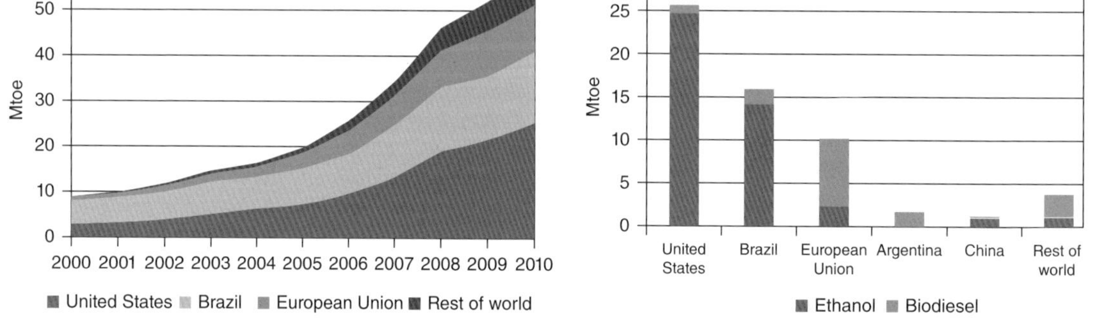

Figure 29.1 Development and state of global biofuel production. (a) Expansion of biofuel production 2000–2010. (b) Ethanol and biodiesel production in 2010. [Figure 1a is based on data, with permission, from Ref 27. Copyright 2011, BP. Figure 1b is based on data, with permission, from Ref 28. Copyright 2011, REN21 (original data were converted from liters to Mtoe).]

production in Argentina, using soybeans, developed in only four years to a production of 1.7 Mtoe (71 PJ) in 2010. Today, Argentina is the largest biodiesel exporter with more than half of its total biodiesel production shipped primarily to Europe.[29]

Second-Generation Biofuels

First-generation biofuels produced from conventional agricultural crops are generally considered as a first step only toward the development of renewable liquid energy sources for transportation.[30] Advanced or second-generation biofuels include fuels produced from lignocellulosic biomass. These nonedible feedstocks avoid direct competition with food uses. Residues and by-products of agricultural and forestry production constitute a low-cost source of biomass feedstocks without competing for cropland. However, such feedstock sources are limited and ambitious bioenergy production targets require expansion of the range of feedstocks including lignocellulosic woody and herbaceous energy feedstocks. These plantation crops are not necessarily competing with food and feed production but may nevertheless compete for agricultural land. Biofuel production from algae could provide another option for biofuel production without competing for land.[31,32]

Despite substantial research and development activities pilot and demonstration plants, which have been set up in North America and Europe, the near- and mid-term potential for commercial-scale application of second-generation biofuel conversion technologies remain rather uncertain.

Biofuel Forecasts and Scenarios

The OECD-FAO Agricultural Outlook 2011–2020[33] projects that the use of agricultural output as feedstock for biofuels will continue its strong growth, which is largely driven by national and regional biofuel mandates and support policies. By 2020, an estimated 13% of the global coarse grain harvest, 15% of vegetable oil production, and 30% of the global sugarcane production are projected to be used for biofuel production. Higher fossil oil prices may induce even faster growth of biofuel production, and, at sufficiently high oil prices, biofuel production in many countries may become competitive even without subsidies and policy support.

The International Energy Agency (IEA) annually compiles and publishes the World Energy Outlook (WEO). It describes regional transport energy futures including biofuel consumption for three scenarios, which differ according to government policies on energy and climate change. The WEO-2011[1] central 'New Policies Scenario' assumes that recent government commitments are implemented in a cautious manner. The '450 Scenario' traces an energy path consistent with meeting the goal of limiting the temperature rise to 2°C and the 'Current Policy Scenario' is based on existing policies.

By 2035, biofuel consumption in the '450 Scenario', 'New Policies Scenario', and 'Current Policies Scenario' amounts to, respectively, 370 Mtoe/y (15.5 EJ/y), 202 Mtoe/y (8.4 EJ/y), and 159 Mtoe/y (6.7 EJ/y). Despite the differences in overall biofuel usage, demand for conventional crop-based biofuels approaches similar levels in the three scenarios toward 2035 due to different assumptions on the role of advanced biofuels. For instance, in the '450 Scenario' by 2035 advanced biofuels, derived from waste, lignocellulosic or other nonfood crops, make up nearly 70% of the total biofuel usage.

The IEA Energy Technology Perspectives presents the 'BLUE Map' scenario[34,35] as an option to demonstrate how to create a more secure and sustainable energy future. In this scenario, the use of biofuels reaches 745 Mtoe/y (31 EJ/y) in 2050 (or 27% of total transport fuel demand); and after 2020, biofuel growth is mostly from advanced biofuels. Except for sugarcane ethanol, all conventional crop-based biofuels will phase out after 2040.

In summary, the IEA scenarios of the 2011 World Energy Outlook assume that the use of first-generation biofuels will stabilize at around 150 Mtoe/y (6.3 EJ/y) after 2030. Thereafter, second-generation biofuels are assumed to provide a significant share of biofuels. Two key questions emerge.

First, what is the impact on food markets and the environment of the envisioned increase of first-generation biofuels to around 150 Mtoe/y (6.3 EJ) by 2035 from a current 60 Mtoe/y (2.5 EJ)?

Second, the growing importance of second-generation biofuels will result in significant demand for lignocellulosic biomass feedstocks including dedicated energy crops grown on 'surplus' land. How much land is potentially available for energy crop production when assuming that for food security reasons the current cropland will be devoted to food and feed crops, and forest areas will be preserved to protect biodiversity and terrestrial carbon stocks?

MODELING FRAMEWORK AND SCENARIO APPROACH

For the analysis of the global agricultural system a state-of-the-art ecological–economic modeling framework is applied comprising of two major components, the FAO/IIASA Global Agro-ecological Zone (AEZ) model[36,37] and the IIASA World Food System (WFS), an applied general equilibrium model. The WFS has been calibrated and validated over past time windows and successfully reproduces regional consumption, production, and trade of major agricultural commodities.[38–41]

This modeling framework, adapted and expanded for resource use and by-product generation of biofuel production, forms the basis for scenario evaluation of the impacts of alternative biofuel deployment pathways on food and agriculture at the national, regional, and global levels. The simulations differentiate between different sources of feedstocks for transport biofuel production, based respectively on biochemical conversion of sugar or starch crops for bioethanol or vegetable oil from oil crops for biodiesel production.

The use of feedstocks depends on the type of biofuel as well as the country or region where feedstocks are environmentally adapted. A rule-based downscaling methodology is applied to allocate the results of the WFS simulations to the spatial grid of the resource database for the analysis and quantification of environmental implications including land use change.

The WFS simulations capture both direct and iLUCs, by modeling responses of consumers and producers to agricultural commodity price changes caused by the additional demand for crops due to biofuel feedstock use. Our approach accounts for expansion of cropland, where relevant, it considers production intensification on existing agricultural land as well as biofuel co-product utilization for animal feed.

Details of the methodology are described elsewhere[10,42,43] and in the Supporting Information.

Scenario Overview and Assumptions

To analyze the impact of future biofuel expansion on various world food system indicators and on land use, simulations were carried out on a yearly basis from 1990 to 2035 for the following three scenarios.

REF: An initial baseline assessment serves as 'neutral' point of departure to which alternative biofuel scenarios are compared for assessing their impacts. The reference scenario REF uses the historical biofuel development until 2010 and thereafter keeps biofuel feedstock use constant at the 2010 level.

WEO-2011: The regional transport energy futures of the WEO-2011[2] 'New Policies Scenarios' were used as representing a likely future pathway for energy and transport futures. Biofuel consumption increases more than threefold compared to current levels reaching just over 200 Mtoe (8.4 EJ) and meeting 8% of road transport fuel demand by 2035 (up from about 3% at present). Second-generation biofuels are assumed to enter the market by 2020 and provide 20% of all biofuels by 2035 leaving an equivalent of 162 Mtoe (6.8 EJ) to be produced from conventional crop-based feedstocks.

WEO-2011-hP (Higher agricultural productivity): The land use implications of a large and rapid increase in crop-based biofuel production will depend on future yield growth and land availability. Simulated agricultural productivity growth results from (endogenously modeled) increased input use and multicropping intensity and an exogenously determined regional and crop-specific technology factor derived from FAO projections and selected country studies. Due to the lack of widely accepted empirical estimates of induced technological growth of different crops, the WFS simulations implemented an alternative scenario of technological progress where it is assumed that the additional demand for biofuel feedstocks would enhance yield growth beyond levels assumed in Scenario WEO-2011.

In addition, the results are based on the following assumptions used in the model simulations.

Socioeconomics: To assess agricultural development over the next decades requires some coherent assumptions about how key socioeconomic drivers might evolve over that period. Food demand is determined by population numbers, projected incomes and associated dietary changes. Population development and economic growth are based on UN projections calibrated by country or regional group to match basic assumptions of the FAO perspective study 'Agriculture Towards 2015/30'.[44]

Land use restrictions: Demand for food, feed, and biofuel feedstocks combined with assumed improvements in agricultural technology determine agricultural land requirements. Assumptions on the magnitude and sources of agricultural land conversion are important for the dynamics of long-term agricultural prices as well as for environmental impacts. The WFS simulations do not permit expansion of agricultural land into protected areas as defined in the AEZ database.[36] No other restrictions were assumed, such as regulations for preventing conversion of forests to cropland.

The model calculations assume that lignocellulosic feedstocks are obtained either from various agricultural and forestry wastes or from dedicated energy crops grown on suitable and unprotected grassland/woodland. Lignocellulosic feedstocks, in this way, do not compete for cropland. Grassland and woodland availability for lignocellulosic feedstocks to produce second-generation biofuels are also dealt with in this paper.

BIOFUELS AND CURRENT CROPLAND USE

Current Cropland Utilization

Agricultural land has always been utilized for a variety of products including food, feed, fiber, and crops for industrial products (Figure 29.2). More than half of the global

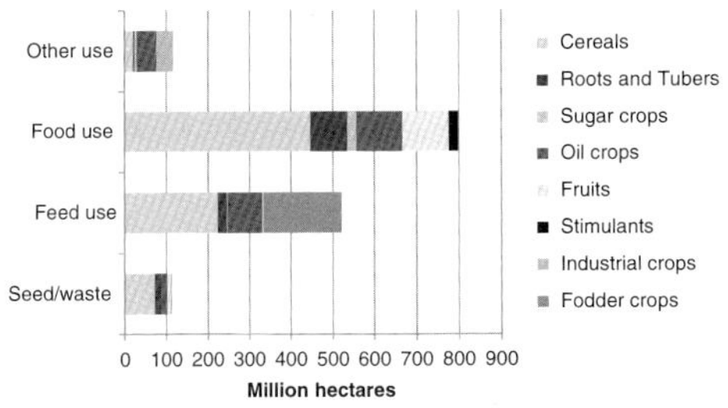

Figure 29.2 Global utilization of cropland, by commodity group, 2008. 'Seed/Waste': Land used for seed production requirements and land equivalents for losses due to on farm waste; 'Other use' includes industrial crops (e.g., cotton, tobacco, natural rubber), and oil crops, cereals, and sugar crops for industrial products (e.g., soap, cosmetics, biofuel). (Created from author's calculations based on Ref 45.)

1.5 billion hectares cropland in 2008 was used directly for human consumption and about one third for feed and fodder production. More than 110 million hectares (Mha) (8%) of cropland were used for industrial purposes including for example vegetable oils for the oleo-chemical industry (e.g., soap, detergents), starch as adhesive in papermaking, or sugar, starch and vegetable oil for biofuel production. Some 7% of cropland is associated with seed production or losses due to on farm waste.

Available statistics do not consistently report production and utilization of cropland for biofuel production. On the basis of the recorded amounts and types of biofuels and achieved national yields, we estimate current cropland in use for biofuel feedstock production at about 25 Mha globally.

Biofuel By-Products for Livestock Feed

By-products generated during biofuel conversion are in part suitable as livestock feed. These by-products can free up land that would otherwise be required for grazing or fodder production and has to be accounted for in the balance of cropland needed for biofuels.

For instance, maize-based ethanol production results in protein-rich co-products, such as distillers' dried grains with solubles (DDGS). In the United States DDGS production has surged along with the expanding biofuel production. On the basis of figures from the U.S. Department of Agriculture,[46] we estimate current land replacement for DDGS coproduct utilization in the United States at about 3 Mha. For comparison US cropland area amounts to 173 Mha, of which around 40 Mha are used for maize production.

Oilseed crushing produces protein-rich cakes and meals for livestock feed in addition to vegetable oil required for biodiesel production. Currently, about 40% of the annual vegetable oil production is traded. The oilseed markets are still dominated by food and other non-food uses, although the growing biodiesel industry had a significant impact. In recent years, more than 90% of the demand growth of vegetable oil in the EU was caused by biofuel demand. The increasing absorption of domestically produced rapeseed oil for biodiesel in the EU, and subsequent shortage of vegetable oil supplies for food products, has shifted the EU from being a small net-exporter of rapeseed oil to a major net-importer (as of 2005).

The WFS model simulates the impacts of biofuel co-products on feed use, prices and land use. Oil crops traditionally have been cultivated for producing both vegetable oil and protein livestock feed. DDGS is a relatively new commodity and current utilization rates for livestock feed are not widely reported. For every ton of ethanol produced about one ton of DDGS is produced. In the biofuel scenarios a growing share of DDGS enters protein feed commodity markets starting from 30% in 2000 up to 65% in 2035.

Increasing Use of Cropland for Biofuel Feedstock Production (Backcasting Scenario Analysis)

Backcasting scenario analysis has been used to quantify the price and resource use impacts of demand growth for biofuel feedstocks in recent years. The outcomes of scenarios with historical biofuel production levels were compared to a simulation for 2000–2008 without considering the historical expansion of biofuels. The difference found was interpreted as an estimate of the market impacts of historical biofuel development and policies.

Results indicate that biofuel production expansion has indeed contributed to widening the demand-supply gap in 2008, which explains a part of the observed historical price increases. Comparing the additional use of cropland in 2008 to the respective additional production of transport biofuels (increment since 2000) gives an indication of the associated resource use per additional unit of biofuels produced. According to these simulations, by 2008 an additional use of cropland land of more than 8 Mha are attributable to biofuel feedstock demand. In comparison, over the same period 2000–2008 the increase in global cropland was 17 Mha.

The comparison with total additional production of transport biofuel of 37 Mtoe (1.5 EJ) since 2000 suggests that an average 0.21 Mha were converted for each additional Mtoe of biofuel produced in 2008. Note that this figure is representing a relatively short simulation period with a fast expansion of biofuel production, especially after 2005. It has contributed significantly to increases of agricultural prices in 2007/08.

REFERENCE PROJECTION UNTIL 2035 WITHOUT BIOFUEL EXPANSION

Results of the reference scenario simulations, without biofuel expansion after 2010, are summarized below. These were used to evaluate the relative impacts of increasing biofuel consumption in the biofuel scenarios described later.

Agro-Economic Developments

With still considerable population growth to reaching more than 8 billion in 2035 and economic growth rates of about 2% and 5% annually in the developed and developing world respectively, cereal demand continues to grow as well, from 2.5 billion tons in 2010 to 3.1 billion tons in 2030, with somewhat growing net imports by developing countries.

During the period from the late 1970s to the early 1990s, real prices of agricultural commodities declined by a factor of more than two and then stagnated until about 2002 by when food prices started to rise. The REF scenario is characterized by fairly modest increases of agricultural world market prices during the 2000 to 2035 period. In part, these price increases of international prices are also a consequence of an assumed further reduction of agricultural support and protection measures.

The global value added of crop and livestock production went up by 34% between 2010 and 2035, which corresponds to an average annual increase of 1%. However, as growth in the non-agricultural sectors is higher than growth in agriculture, the share of agriculture in total GDP decreases on a global level from 4% in 2010 to 2.8% in 2035. The REF scenario projects also a strong decline of agriculture contribution to total GDP in the developing world namely from 9.9% in 2010 to 4.4% in 2035. This is mainly due to a strong anticipated growth of the non-agricultural sectors in the larger Asian economies.

Cropland Use and Harvested Area

In the model simulations some 1.6 billion ha of land is used in 2010 for crop production of which nearly 1 billion ha are located in the developing countries. Yield growth and production intensification on existing agricultural land are insufficient in the REF scenario to meet the increased demand for food and feed and global cropland expands by 86 Mha between 2010 and 2035. While aggregate cropland use in developed countries remains fairly stable, all of the net increases occur in developing countries. Sub-Saharan Africa and South America together account for 85% of the cropland expansion.

Cropland represents the amount of physical land used for crop production. In practice, part of the land is left idle or fallow, and part of the cropland is used to produce multiple crops within one year. The implied cropping intensity (defined as the ratio of harvested over cropland) increases from about 86% in 2010 to 91% in 2035 albeit with large regional differences (Table 29.1).

Land Use Changes

The full integration of the land resource information (AEZ) and the agricultural expansion requirements resulting from changing demand, supply and productivity (WFFs Model) permits computation of land balances. Besides cropland increases, land will also be required for expanding built-up areas and infrastructural uses for a growing population. Both cropland and built-up areas increase at the expense of forest, grassland/woodland and other land.

Between 2010 and 2035 cropland and urban land in the REF scenario expand by 86 Mha and 43 Mha respectively. Assuming no specific land use restrictions (beyond preserving protected areas), an estimated 48 Mha comes from forest conversion and another 81 Mha grass/woodland areas are converted to meet land demand for cropland and built-up areas in the REF scenario. Most land use

Table 29.1 Cropland and Harvested Area, Scenario REF

	Million Hectares (Mha)						Harvested/Cropland		
	Cropland			Harvested Area			Cropping Intensity		
Scenario REF	2010	2020	2035	2010	2020	2035	2010	2020	2035
Developed	607	605	608	434	446	461	0.72	0.74	0.76
Developing	968	1010	1057	921	980	1047	0.95	0.97	0.99
Rest of World	41	40	37	35	35	36	0.85	0.88	0.92
World	1616	1655	1702	1391	1461	1543	0.86	0.88	0.91
Expansion		39	86		70	152			

[Source: IIASA World Food System reference scenario (REF) simulations, January 2012. Reproduced with permission from the International Institute for Applied Systems Analysis (IIASA).]

conversions occur in developing countries in particular in Latin America and Sub-Saharan Africa.

In our simulations average annual deforestation rates due to agricultural land expansion and urbanization amount to 2.2 Mha per year between 2010 and 2020 and 1.7 Mha per year in the period 2020 to 2035. In comparison, the historical rate of total annual net deforestation in the 1990s was estimated by FAO at 8.3 Mha and for the first decade of this century 5.2 Mha.[47]

IMPACTS OF EXPANDING BIOFUEL PRODUCTION

Biofuel Scenario Formulation

In Scenario WEO-2011 global biofuel consumption in 2035 global biofuel use is 203 Mtoe (8.5 EJ), of which 162 Mtoe (6.8 EJ) are produced with first-generation technologies and feedstocks (Figure 29.3). Second-generation technologies are assumed to enter the market after 2020 and reach 20% of global biofuel supply by 2035. In OECD countries higher shares of second-generation technologies in total biofuels are assumed compared to the developing world.

The expansion of biofuel use in Scenario WEO-2011 requires by 2035 production of additional 102 Mtoe (4.3 EJ) first-generation biofuels compared to 2010. Three fourth are for bioethanol based on cereals and sugar crops and one fourth for biodiesel produced from vegetable oil. Table 29.2 shows the quantities of agricultural commodities used as feedstocks for transport biofuel production in respectively 2010, 2020, and 2035.

The increased agricultural productivity of the biofuel scenario variant WEO-2011-hP was implemented by adjusting the technological progress in agriculture. For this purpose, the countries were grouped based on available yield gap estimates by comparing actual agricultural production of the year 2000 with yield potentials derived from the agro-ecological zones calculations.[36] The country grouping and technical factor modifications in relation to REF (and WEO-2011) are shown in Table 29.3. Yields start becoming higher compared to their levels in REF as of 2000. It should be noted that for all countries the resulting agricultural productivity is still well below the regions biophysical potentials.

Impacts on Agricultural Markets and People at Risk of Hunger

When simulating scenarios with increased demand for food staples due to the production of first-generation biofuels, resulting market imbalances push commodity prices upward (Table 29.4) with the exception of the commodity 'protein feed' where increased biofuel production may result in lower prices due to large amounts of by-products generated, i.e., livestock feeds from starch-based ethanol

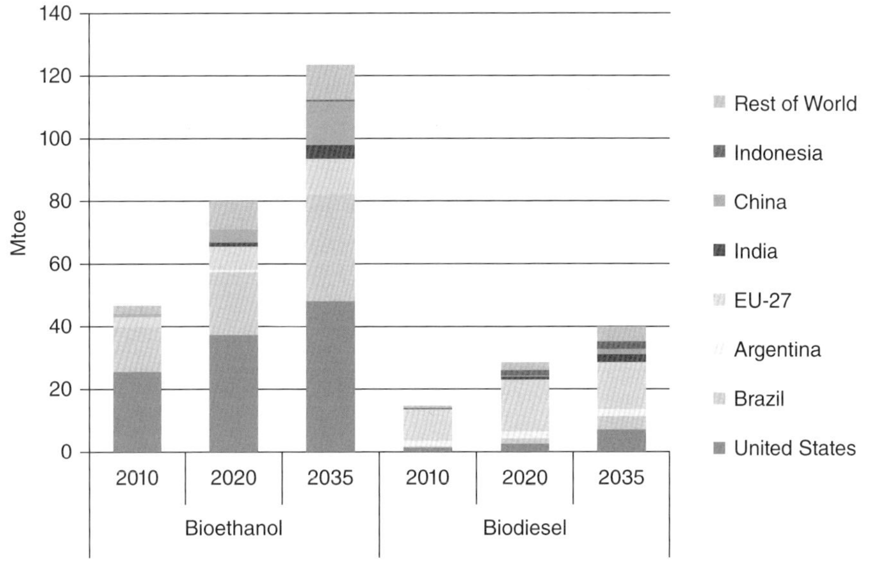

Figure 29.3 Regional distribution of crop-based biofuel use in the biofuel scenarios WEO-2011 and WEO-2011-hP Source: IIASA World Food System biofuel scenario based on World Energy Outlook 2012[2]

Table 29.2 Global Use of Agricultural Commodities for Biofuel Production in the Biofuel Scenarios

	2010	2020	2035	Increase 2010–2035
Cereals (million tons)	109	181	270	161
Other food[1] (million tons sugarcane equiv.)	588	1091	1716	1128
Vegetable oil (million tons)	17	34	47	30

[1]In WFS sugarcane and cassava (for bio-ethanol) are included in the category 'Other food'.
[Source: IIASA World Food System simulations, January 2012. Reproduced with permission from the International Institute for Applied Systems Analysis (IIASA)]

Table 29.3 Additional Productivity Growth Assumptions in Biofuel Scenario Variant WEO-2011-hP

	Additional Increases in Technical Factor of Aggregate Crop Yield Growth Relative to REF		
Country Group	2025	2035	2050
Group 1: High additional productivity growth Sub-Saharan Africa	+7.5%	+12. 5%	+20%
Group 2: Medium additional productivity growth India, Pakistan, Indonesia, Thailand, Mexico, Argentina, Central and South America, North Africa, Far East Low Income, Middle East Low Income	+4%	+6.4%	+10%
Group 3: No additional productivity growth all countries not mentioned in Groups 1 and 2 (includes all developed countries and China and Brazil)	No Increase		

Table 29.4 Impacts of Biofuel Scenarios on Agricultural Prices

	Change of Price Index Relative to Reference Scenario REF			
	WEO-2011		WEO-2011-hP	
Percentage Scenario	2020	2035	2020	2035
Crops	12	15	7	4
Cereals	13	17	8	7
Other crops	11	14	6	2
Livestock	3	2	2	− 1
Wheat	13	16	9	9
Rice	5	8	− 2	− 7
Coarse grains	19	24	14	13
Bovine and ovine	4	2	2	0
Dairy	5	4	3	− 1
Other meat	2	1	1	− 1
Protein feed	− 7	− 17	− 9	− 20
Other food	13	16	7	4
Nonfood	2	2	− 7	− 11

[Source: IIASA World Food System simulations, January 2012. Reproduced with permission from the International Institute for Applied Systems Analysis (IIASA).]

production (i.e., DDGS) and protein meals and cakes from crushing of oilseeds for biodiesel production. Having access to cheap feed sources also results in only reduced increases of livestock product prices.

Improvements in agricultural productivity, as assumed in scenario WEO-2011-hP, substantially moderate the price increases due to biofuel production. Overall, the crop price index increases for scenario WEO-2011 and WEO-2011-hP by 12% and 7% in 2020, and by respectively 15% and 4% in 2035. Price increases vary by commodity, and mainly occur for coarse grains (mostly maize), wheat and 'other crops' (oil crops and sugar).

Table 29.5 Annual Aggregate Yield Growth between 2010 and 2035

	Baseline	Biofuel Scenarios	
Percentage	REF	WEO-2011	WEO-2011-hP
North America	0.76	0.95	0.92
Europe and Russia	0.61	0.71	0.62
Pacific OECD	0.66	0.75	0.68
Sub-Saharan Africa	1.69	1.75	2.05
Latin America	0.89	0.95	1.00
Middle East and North Africa	1.56	1.66	1.70
Asia, East	0.83	0.83	0.81
South/Southeast Asia	1.26	1.28	1.46
Rest of world	1.04	1.07	1.08
Developed	0.66	0.79	0.73
Developing	1.00	1.04	1.15
World	0.93	0.99	1.01

[Source: IIASA World Food System simulations, January 2012. Reproduced with permission from the International Institute for Applied Systems Analysis (IIASA).]

The assumed additional productivity growth in the developing countries increases the region's competitive position and stimulates higher production overall. As a consequence, the developed countries loose in market shares.

Between 2010 and 2035 the number of people at risk of hunger decreases in the REF scenario by 30% or about 300 million people. Although major improvements in food production and food security occur also in the two biofuel scenarios, in comparison with the REF scenario, the WEO-2011 and WEO-2011hP scenarios are less favorable for food security, respectively 61 and 4 million more people are estimated to be at risk of hunger.

Agricultural Productivity

Overall agricultural productivity growth can be captured by an indicator of a country's produced crop volume (using international price weights for aggregation) per unit of cropland. In the REF scenario aggregate yields are projected to increase between 1990 and 2035 respectively by 73% and 42% in developing and developed countries. Between 2010 and 2035 annual growth rates of aggregate yield in REF range from 1.7% in Africa to 0.6% in Europe (Table 29.5).

Price increases resulting from additional crop demand in the biofuel scenario WEO-2011 provide an incentive for farmers to invest in increasing crop production. This is reflected in increases in aggregate yields compared to the REF scenario. The assumed higher technological progress achieved in the developing countries in WEO-2011-hP improves the developing world's competitive position, but provides less economic incentives for yield increases in developed countries and China and Brazil.

Land Use Impacts

Land conversion is explicitly modeled to maintain full consistency between the spatial AEZs approach used for appraising land productivity on the one hand, and the expansion of cropland as determined in the WFS simulations on the other.

Cropland Expansion

In the reference projection REF cropland expansion (for food and feed) amounts to 86 Mha between 2010 and 2035. Figure 29.4 shows the extents and regional distribution of changes in the use of cropland due to additional biofuel production. For the biofuel scenario WEO-2011, by 2035, an additional 19 Mha (compared to cropland use in the REF scenario) is put into cultivation for producing feedstock for the first-generation biofuels. The additional cropland used per additional biofuel energy produced with first generation feedstocks and conversion technologies in WEO-2011 is 0.21 Mha/Mtoe up to 2020, and decreases for the longer term 2010–2035 to 0.19 Mha/Mtoe.

The additional crop productivity growth assumptions in Scenario WEO-2011-hP decrease cropland expansion to 6 Mha in the shorter term (by 2020). In the longer term, by 2035, the effect of the assumed increased productivity in the developing countries, significantly moderates the cropland expansion. The additional 8 Mha that are converted to arable land compared to REF by 2035 imply an additional cropland requirement of only 0.08 Mha/Mtoe for first-generation biofuel use because feedstock production can expand mostly on existing cropland. This shows that land use impacts of biofuel use may very well depend on

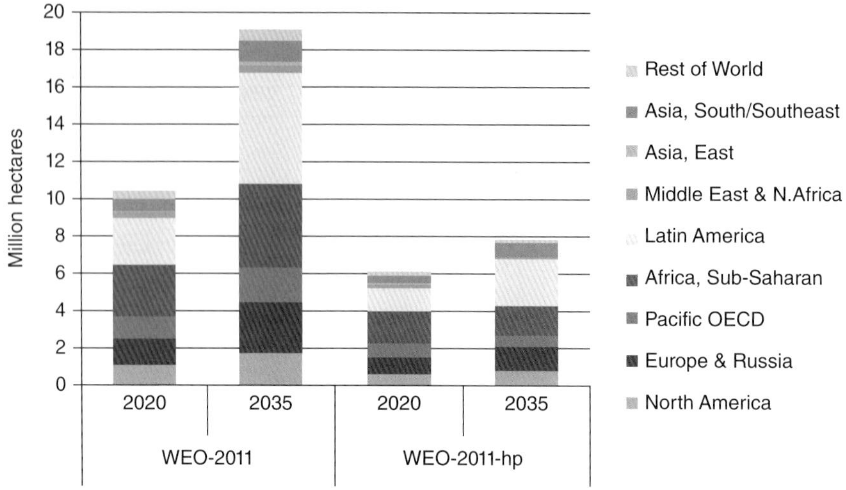

Figure 29.4 Additional cropland use in the biofuel scenarios by 2020 and 2035, relative to REF. [Source: IIASA World Food System simulations, January 2012. Reproduced with permission from the International Institute for Applied Systems Analysis (IIASA).]

whether such additional crop demand will trigger lasting overall yield improvements in agriculture.

Deforestation

Additional cropland expansion induces some additional deforestation caused directly or indirectly by land demand due to first-generation biofuel feedstock production. Estimates of scenario WEO-2011 indicate that by 2035 additional biofuel feedstock production may be responsible for up to 6 Mha of forest loss. Most of additional deforestation would occur in Africa, Latin America, and Southeast Asia. Under the higher agricultural productivity growth assumptions in WEO-2011-hP additional deforestation due to biofuels is reduced to less than 3 Mha.

It should be noted that production of lignocellulosic feedstocks for the second-generation production chains is assumed to occur in available grassland and woodland areas, without causing additional deforestation or using cropland.

Land Required for Second-Generation Biofuels

Low-cost crop and forest residues, wood processing wastes, and the organic fraction of municipal solid waste can all be used as lignocellulosic feedstocks for second-generation biofuel conversion routes. Using residues and wastes for the production of biofuels would require well-designed logistical systems but no additional land would be needed.

In regions, with limited availability of residues and where large and increasing amounts of feedstocks are demanded, additional land will be needed for establishing plantations of perennial energy grasses or short rotation tree systems. By 2035, the biofuel scenarios project 20% of biofuels or 41 Mtoe (1.7 EJ) to come from second-generation biofuels.

Indicative biofuel yields for the second-generation conversion routes are between 6 to 18 dry tons biomass required for each ton oil equivalent (toe) biofuel.[48] An average typical yield of 10 dry tons per hectare is reasonable and achievable for lignocellulosic feedstock production in 2035. Assuming that by 2035 for each toe of second-generation biofuels an average of 10 dry tons biomass is needed and lignocellulosic feedstock yields are 10 t/ha/y, the respective land requirements would be 41 Mha, when all feedstocks for second-generation biofuels were from dedicated energy plantations. In the section 'Grassland availability' we discuss availability of grassland and woodland for potential biofuel feedstock production.

Greenhouse Gas Emission Savings

As climate benefits and GHG emission savings are stated in the political and scientific debate as a prime goal of biofuel use, a substantial net reduction of GHGs over the whole lifecycle of biofuel production is considered important to meet climate mitigation objectives. Biofuel use avoids carbon emissions by substituting for fossil fuel use. Depending on feedstock and biofuel supply chain, some carbon is

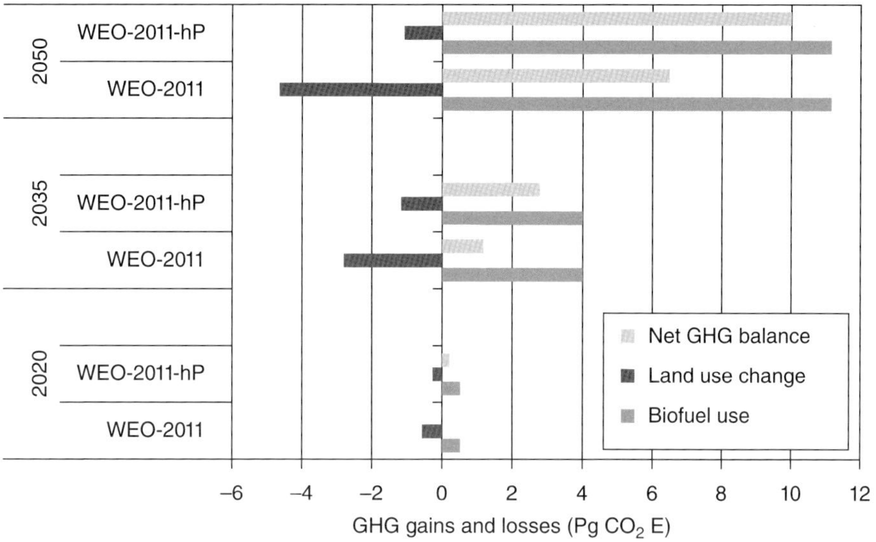

Figure 29.5 Cumulative net greenhouse gas savings of biofuel scenarios for 2020, 2035 and 2050. (Source: IIASA World Food System simulations, January 2012.)

emitted during the production process (e.g., transport, energy use in production process, fertilizer use). In addition, direct and iLUCs can lead to substantial losses of terrestrial carbon stocks.

Figure 29.5 shows the net accumulated GHG savings of the two biofuel scenarios due to expansion of first- and second-generation biofuels after 2010 for respectively the periods 2010–2020, 2010–2035 and 2010–2050. As the years beyond 2035 are not included in the IEA's World Energy Outlook 2011 further assumptions on biofuel use development were adopted for the simulations. Between 2035 and 2050 global biofuel use is assumed to increase by almost 40% to 286 Mtoe (12 EJ) and more than 60% of this increase is from second-generation biofuels.

The net GHG balance of a biofuel scenario (shown in green, bar "Net GHG balance") is determined by the GHG savings (using factors established by lifecycle analysis of several representative biofuel production chains) achieved with biofuels replacing gasoline and diesel (blue bar "Biofuel use") minus the GHG emissions caused by direct and iLUCs (red bar "Land use change").

Lifecycle analysis determines the amount of GHG saved by replacing one MJ of biofuel with one MJ of fossil fuel. We use 50% GHG saving for all feedstocks, except for sugarcane where a 75% saving is applied.[13,14,16,49] These commonly used factors are subject to uncertainty and depend on especially chosen technology and energy source used in the biofuel production process. GHG savings due to "Biofuel use" are calculated as avoided emissions from substituted fossil fuel use minus emissions generated during biofuel production.

The study methodology projects spatially explicit agricultural land uses. A carbon accounting method, based on IPCC Tier 1 approaches,[50] was used to quantify for each scenario changes of vegetation and soil carbon pools. While this method is consistent with the recommended approach for GHG inventories, it should be noted that large uncertainties are involved in estimating regional carbon pools.

Carbon losses from vegetation and soils due to land use changes (deforestation and grassland conversion) occur mainly at the time of land conversion. In contrast, GHG savings resulting from the replacement of fossil fuels with biofuels accumulate only gradually with time. Hence, the anticipated time frame and period of biofuel production is of crucial importance for the assessment of net biofuel GHG savings.

In the short and medium term until 2020 and 2035, the projected biofuel consumption in WEO-2011 results in no or little GHG savings as the length of the time period is hardly sufficient to compensate for carbon lost due to additional land use conversion. The assumed higher agricultural productivity in developing countries in WEO-2011-hP can provide biofuel feedstocks with less land conversion resulting in higher achievable GHG savings. The cumulative net GHG emission savings due to biofuel use between 2010

Table 29.6 Additional Land Use and Associated Carbon Emissions Resulting from Biofuel Use [Source: IIASA World Food System simulations, January 2012. Reproduced with permission from the International Institute for Applied Systems Analysis (IIASA).]

	Additional Biofuel Use in Target Year			Additional Land Use Compared to REF			Land Use Change Carbon Emissions per Additional Biofuel Use (20 Year Discounting)	
	First Generation	Second Generation	Total	First Generation	Second Generation	Total	First Generation Only	First and Second Generations
Time Period Scenario	Mtoe			Mha			g CO_2 equ / MJ	
2010 to 2020								
WEO-2011	47.2	1.5	48.7	10.4	1.5	11.9	52	51
WEO-2011-hP	47.2	1.5	48.7	6.1	1.5	7.6	28	28
2010 to 2035								
WEO-2011	101.9	41.5	143.4	19.0	41	60.0	44	35
WEO-2011-hP	101.9	41.5	143.4	7.8	41	48.8	12	12

and 2035 amount to 2.8 Pg CO_2 equivalent in WEO-2011-hP, i.e significantly higher than the 1.1 Pg CO_2 equivalent in WEO-2011. For comparison, in 2009 alone, the road transport sector was estimated to produce 4.9 Pg CO_2 emissions globally.[51]

By 2050 the amount of second-generation biofuels increases cumulative net GHG savings via biofuels use while at the same time only little additional land use conversion is required. For this longer time horizon we estimate the accumulated GHG savings of 6 to almost 10 Pg CO_2 equivalent depending on assumed agricultural productivity increases in developing countries.

Table 29.6 summarizes the additional biofuel use and associated land use requirements for the biofuel scenarios and presents land use change carbon emissions per additional biofuel use. Carbon emissions from additional land use change are annualized over 20 years following the EU Renewable Energy Directive.[14]

In the short-term period, 2010 to 2020, we calculate land use related carbon emissions of 51 to 28 g CO_2 equivalent per additional MJ of biofuel use in the target year for WEO-2011 and WEO-2011-hP, respectively. The increasing importance of second-generation biofuels and increasing agricultural yields over time reduce these coefficients to 35 and 12 g CO_2 equivalent calculated for the longer time horizon 2010 to 2035. One MJ of fossil fuel used in transport releases 90.3 g CO_2. Thus the land use effects are of major significance for the expected gains of shifting from fossil fuels to renewable biofuels.

For comparison a recent study, which assessed the impacts of the EU biofuels renewable transport energy targets,[52] calculated land use emissions, computed over a 20 year period, of 38–40 g CO_2 eq/MJ. It should be noted that the choice of 20 years for the amortization of land use change related emissions is arbitrary but consistent with the EU's Renewable Energy Directive framework.

GRASSLAND AVAILABILITY FOR BIOFUEL FEEDSTOCK PRODUCTION

The large volumes of biofuel production envisaged in various transport energy scenarios will considerably increase the demand for biomass. Making feedstocks available in a sustainable way, without compromising food security and biodiversity, requires a sound policy framework and land use planning.

Current cropland is concentrated on the world's most fertile areas and for reasons of food security these should be preserved primarily for future food production. Forest preservation is key for maintaining terrestrial carbon stocks and securing biodiversity. In this situation, underutilized grassland/woodland areas may provide an acceptable and sustainable land use option for the cultivation of lignocellulosic energy crops.

Satellite derived land cover interpretations have been used together with statistical data of the year 2000 on cropland and forest land from the FAO to compile a consistent spatial characterization of each land grid-cell (at 5 by 5 minute longitude/latitude) for seven major land use categories.[36] The class termed 'grassland and woodland' (GRWL) (Figure 29.6) comprises all areas that are not one of (1) cropland (separate for rain-fed and irrigated); (2) forest land; (3) barren or very sparsely vegetated; (4) urban land and land used for housing and infrastructure; or (5) water. Such GRWL areas amount to a global total of 4.6 billion hectares. This compares well with similar land use data sets.[53] For comparison, the FAO statistics report

Figure 29.6 Occurrence (%) of grass- and woodland. The map shows the share of grass/woodland in 5 min latitude/longitude grid cells (see text). [Reproduced with permission from Ref 36 Copyright 2012, International Institute for Applied Systems Analysis (IIASA).]

'permanent meadows and pastures', a subset of GRWL, amounting to a global total of 3.4 billion hectares.[45]

Mobilizing potential global biomass supplies depends on overcoming various challenges related to land ownership and current utilization, investments and infrastructure development, social barriers and environmental sustainability. Larger and more concentrated areas dedicated to energy crop production increase the utilization of equipment, reduce transportation and harvesting costs and are seen as essential for scaling-up second generation biofuel production to industrial levels.

Figure 29.7 presents an overview of the suitability and spatial concentration of the global GRWL. For the latter, the share of GRWL in a grid-cell is used as a proxy. It also indicates protected GRWL areas of more than 0.5 billion hectares or 11% of the total. Of the remaining areas 2.7 billion hectares occurr in grid-cells where GRWL share exceeds 50%. These areas are characterized by widely varying productivty ranging from marginally suitable (mS) with average yields of 5 ton dry matter or less per hectare to very suitable (VS) areas potentially producing 20 tons per hectare or more (Figure 29.8). Suitability and biomass productivity of rain-fed lignocellulosic feedstock production in each grid-cell was derived from the GAEZ databases.[36]

Another 1.3 billion hectares of unprotected GRWL is distributed in more heterogenous landscapes with GRWL covering 30–50% (in green) or less than 30% of the grid-cells (in red). Less than half of unprotected GRWL or 1.8 billion hectares is assessed as agronomically suitable (Land suitability classes VS+S+MS+mS) for rain-fed production of lignocellulosic feedstocks (middle bar in Figure 29.7).

Farm economics and logistics of biomass feedstock collection will favor production of lignocellulosic energy crops in areas of high GRWL concentration and where rain-fed yields of 10 dry ton per hectare or more are possible. Most such areas are concentrated in South America and Sub-Saharan Africa. In addition, such areas can also be found in Central America, the United States, Europe and northern Australia.

We estimate that some 1.1 billion hectares of unprotected GRWL is very suitable, suitable or moderately suitable for lignocellulosic feedstock production (lower bar in Figure 29.7) with rain-fed production potentials of 10 dry tons per hectare or more. About two thirds of those (or 672 Mha) are in GRWL areas with share >50% (blue hatched area). These areas could provide excellent resources for industrial lignocellulosic energy crop production. However, current accessibility and livestock use may restrict development of feedstocks for second-generation biofuel production chains.

Table 29.7 provides a further characterization of unprotected GRWL in terms of broad classes of accessibility and ruminant livestock density. Accessibility is defined as the travel time to the nearest city of 50,000 or more people by land (road/off road) or water (navigable river, lake, ocean) in the year 2000.[54] An important current use of grasslands is livestock grazing. Therefore, grid-cells were tabulated by classes of ruminant livestock densities based on FAOs gridded livestock of the world.[55]

Industrial scale lignocellulosic feedstock production may favor areas of low livestock densities in close proximity to markets. Table 29.7 indicates a total of 246 Mha (sum of extents shaded in green) of GRWL with such favorable characteristics, i.e., with a high rain-fed production potential (over 10 ton/ha), a high concentration of GRWL (>50% of grid-cell), close proximity to markets (<6 hours travel time) and relatively

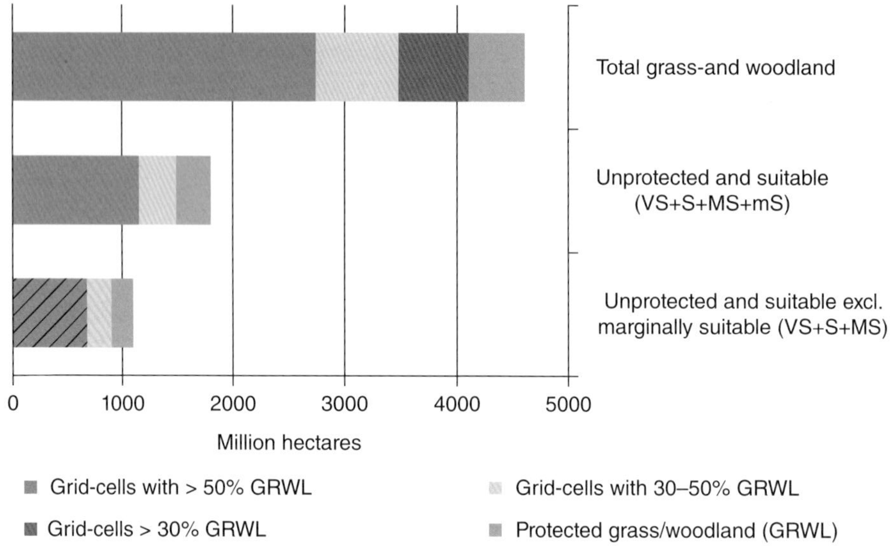

Figure 29.7 Suitability of global grass/woodland areas for lignocellulosic feedstock production, by grass/woodland concentration. Suitability refers to achievable rain-fed yields. VS: Very Suitable; S: Suitable; MS: Moderately suitable; mS: Marginally suitable; The share of grass- and woodland (GRWL) in 5 min latitude/longitude grid cells is used as proxy for GRWL concentration. Table 7 shows a further characterization of the hatched blue area.

low livestock population (<25 Reference Livestock Unit (RLU) / km2).

High demand for lignocellulosic feedstocks may warrant investment in improved road infrastructure and shorten longer travel times to closest markets. A common stocking density (e.g., in Brazil) is 100 RLU per km^2. However other land uses are likely to occurr next to GRWL use. This suggests that more than 50 RLU may be classified as high and 25–50 as medium stocking density, which might limit the large-scale industrial energy crop produciton. Changes in live-stock feed management such as improved grassland and addition of feed crops may permit large-scale lignocellulosic feedstock production along with livestock production. Another 229 Mha (shown in yellow) are

Figure 29.8 Potential rain-fed yield of lignocellulosic feedstocks in current areas with dominantly grassland and woodland. Dominantly includes 5 min longitude/latitude grid-cells with over 50% grassland and woodland. [Reproduced with permission from Ref 36. Copyright 2012, International Institute for Applied Systems Analysis (IIASA).]

Table 29.7 Accessibility and Livestock Density in Unprotected Grassland/Woodland Suitable[1] for Production of Lignocellulosic Feedstocks in Grid Cells with GRWL Share Exceeding 50%

Million Hectares	Travel Time to Closest Market[2]				
Ruminant Livestock Density[3]	<3 h	3–6 h	6–12 h	>12 h	Total
Very low: < 10 RLU/km^2	45	71	66	62	244
Low: 10–25 RLU/km^2	59	71	38	13	181
Medium: 25–50 RLU/km^2	54	47	24	7	132
Very high: > 50 RLU/km^2	60	38	14	3	115
Total	218	227	141	86	672

[1]Suitable here includes the suitability classes Very Suitable (80–100% of maximum attainable yield), Suitable (60–80%), and Moderately Suitable (40–60%);
[2]Defined as travel time to nearest city of 50,000 or more people in year 2000[54].
[3]Reference Livestock Unit (RLU).

agronomically suitable with sizeable ruminant livestock currently present or lacking good access (between 6 and 12 h). These areas could be developed for production of lignocellulosic feedstocks, but would require appropriate changes in livestock feed management and improved transport infrastructure.

CONCLUSIONS

Utilization of crops as feedstocks for first-generation biofuels has increased the integration of energy and agricultural markets, with significant effects on land utilization. Globally cropland comprises some 1500 Mha and is concentrated on the world's most fertile areas. Between 2000 and 2008 about half of global cropland expansion or 8 Mha can be attributed to land demand for biofuel expansion.

In the central 'New Policies Scenario' of the World Energy Outlook 2011 (WEO-2011) biofuel consumption increases from a current 60 Mtoe (2.5 EJ) to 200 Mtoe (8.4 EJ) until 2035. Second-generation non-crop based biofuels are assumed to enter the market in 2020 and supply 20% of total biofuels by 2035. In our simulations of the world food system, the additional crop demand from biofuels use increases crop prices by about 10 to 15% compared to a world where biofuel production would stabilize at the level of 2010. Even then, price increases in the livestock sector are fairly modest due to cheap feed sources resulting from large amounts of by-products generated during biofuel production.

By 2035, an additional 19 Mha of cropland is put into cultivation due to first-generation biofuel use. Land use emissions (from deforestation and grassland conversion) are only 30% lower than the emissions saved by avoiding fossil fuel use. As a consequence in this scenario, only a modest net GHG saving can be achieved by replacing biofuels with fossil fuel use.

However, land conversion and food security implications of the simulated biofuel increases strongly depend on assumed agricultural productivity growth rates.

As demonstrated in our scenario simulation, additional people at risk of hunger due to higher commodity prices caused by first-generation feedstock demand for biofuel production can be avoided with higher agricultural productivity growth in developing countries, which would increase the region's competitive position, stimulate higher production and gain market shares compared to the developed world.

The cumulative net GHG balance of biofuel consumption also improves when higher productivity growth is assumed for the developing world. In this case, the cumulative net GHG emission savings due to biofuel use between 2010 and 2035 amounted to 2.8 Pg CO_2 equivalent. Hence during this 25-year period first-generation biofuels can only make a modest contribution to lowering emissions from the transport sector.

For biofuels' GHG benefits to materialize, it is important to achieve yield gap reduction in developing countries, to carefully monitor the speed of biofuel expansion and to implement enforceable land use restrictions, especially avoiding additional deforestation.

Ambitious bioenergy production targets as described in the IEA 'Blue Map' Scenario (745 Mtoe or 31 EJ biofuels use by 2050) will result in significant demand for biomass feedstocks including dedicated lignocellulosic energy crops grown on surplus land. We estimate between 246 and 475 Mha of grassland and woodland, to be agronomically suitable for lignocellulosic feedstock production, with achievable rain-fed yields of at least 10 tons dry matter per hectare, to be in areas where grasslands dominate, with good accessibility and relatively low ruminant livestock density. These areas, on initial analysis, might be well suited for developing industrial scale second-generation biofuels. Regional and local studies are required to gain a more detailed understanding of the potential and limitations of grassland and woodland areas for lignocellulosic energy crop production.

The biofuel energy potential of available grassland and woodland is determined by technical progress on biofuel yields of second-generation conversion technologies (liters biofuel per ton biomass) and agronomic biomass yields (biomass per hectare). An indicative biofuel yield between 0.9 Mtoe/Mha (assuming 110 l biofuel per ton biomass and 12 ton biomass per hectare) and 2.7 Mtoe/Mha (i.e., 300 l biofuel per ton biomass and 18 ton biomass per hectare) suggests for the 246 Mha grassland and woodland well suitable for industrial-scale lignocellulosic energy crop production a bioenergy yield of 222 to 667 Mtoe (9–28 EJ). This compares to a current transport fuel use of 2283 Mtoe (95 EJ), which is projected to increase to 3257 Mtoe (136 EJ) by 2035. In other words, with these assumptions, we estimate by 2035 some 7–20% of liquid transport fuels could be based on second-generation biofuels with biomass sourced from grassland and woodland areas. However, in the future, grassland and woodland areas may also be used for biomass production required for renewable heat and electricity energy produced for the stationary sector.

REFERENCES

1. International Transport Forum (ITF). *Reducing Transport Greenhouse Gas Emissions. Trends & Data 2010.* Paris: OECD/ITF; 2010, pp 94 http://www.internationaltransportforum.org/Pub/pdf/10GHGTrends.pdf.
2. IEA (International Energy Agency). *World Energy Outlook 2011* Paris: OECD/IEA; 2011, pp 645.
3. Schmidhuber J. *Biofuels: An Emerging Threat to Europe's Food Security? Impact of an Increased Biomass Use on Agricultural Markets, Prices and Food Security: A Longer-Term Perspective.* Paris: Notre Europe; 2007.
4. Tyner WE, Taheripour F. Policy options for integrated energy and agricultural markets. In: Paper Presented at the Transition to a Bio-Economy: Integration of Agricultural and Energy Systems conference on February 12–13; 2008, at Westin Atlanta Airport.
5. Ewing M, Msangi S. Biofuels production in developing countries: assessing tradeoffs in welfare and food security. *Environ Sci Policy* 2009, 12:520–228.
6. FAO (Food and Agriculture Organization of the United Nations). *The State of Food and Agriculture. Biofuels: Prospects, Risks and Opportunities.* Rome: FAO; 2008, pp 138.
7. Rosegrant MW, Zhu T, Msangi S, Sulser T. Global scenarios for biofuels: impacts and implications. *Rev Agri Econ* 2008, 30/3:495–505.
8. Msangi S, Sulser T, Rosegrant MW, Valmonte-Santos R, Ringler C. Global scenarios for biofuels: impacts and implications. Global Bioenergy Partnership. Available at: http://www.globalbioenergy.org/uploads/media/07_Global_Scenarios_for_Biofuels_Impacts_and_Implications_01.pdf. (Accessed December 12, 2011).
9. Banse M, van Meijl H, Tabeau A, Woltjer G, Hellmann F, Verburg PH. Impact of EU biofuel policies on world agricultural production and land use. *Biomass Bioenergy* 2011, 35:2385–2390.
10. Fischer G, Hizsnyik E, Prieler S, Shah M, van Velthuizen H. *Biofuels and Food Security.* Vienna, Austria: The OPEC Fund for International Development (OFID) and International Institute of Applied Systems Analysis (IIASA); 2009, 228 pp. Available at: http://www.iiasa.ac.at/Research/LUC/Homepage-News-Highlights/Biofuels Report Final.pdf
11. Timilsina GR, Beghin JC, van der Mensbrugghe D, Mevel S. The impacts of biofuels targets on land-use change and food supply: a global CGE assessment. *Agri Econ* 2012, 43/3:315–332.
12. Deininger K, Byerlee D, Lindsay J, Norton A, Selod H, Stickler M. Rising global interest in farmland. can it yield sustainable and equitable benefits? The World Bank; 2011. Available at: http://siteresources.worldbank.org/INTARD/Resources/ESW_Sept7_final_final.pdf. (Accessed January 10, 2012).
13. Fritsche UR, Wiegmann K. Ökobilanzierung der Umweltauswirkungen von Bioenergie-Konversionspfaden. Expertise für das WBGU-Hauptgutachten Berlin: Welt im Wandel—Bioenergie und nachhaltige Landnutzung; 2008 Available at: http://www.wbgu.de/wbgu_jp2008_ex04.pdf.
14. European Commission. Directive 2009/28/EC of the European Parliament and of the Council of 23 April 2009 on the promotion of the use of energy from renewable sources and amending and subsequently repealing Directives 2001/77/EC and 2003/30/EC. Available at: http://eur-lex.europa.eu/LexUriServ/LexUriServ.do?uri = Oj:L:2009:140:0016:0062:en:PDF; 2009. (Accessed January 11, 2012).
15. IPCC. In: O. Edenhofer, R. Pichs-Madruga, Y. Sokona, K. Seyboth, P. Matschoss, S. Kadner, T. Zwickel, P. Eickemeier, G. Hansen, S. Schlömer, C. von Stechow (eds). IPCC SR-REN: Full Report. Renewable energy Sources and Climate Change Mitigation. Special Report of the International Panel on Climate Change (IPCC). Prepared by Working Group III of the Intergovernmental Panel on Climate Change. Cambridge, UK: Cambridge University Press; 2011, pp 1075.
16. Macedo IC, Seabra JEA, Silva JEAR. Green house gases emissions in the production and use of ethanol from sugarcane in Brazil: the 2005/2006 averages and a prediction for 2020. *Biomass Bioenergy* 2008 32:582–595.
17. Berndes G, Bird N, Cowie A. *Bioenergy, Land Use Change and Climate Change Mitigation.* Background Technical Report. Paris: IEA Bioenergy; 2011, pp 62.
18. Searchinger T, Heimlich R, Houghton RA, Dong F, Elobeid A, Fabiosa J, Toggoz S, Hayes D, Yu T-H. Use of US croplands for biofuels increases greenhouse gases through emissions from land use changes. *Science* 2008.319:5867,1238–1240.

19. Fargione J, Hill J, Tilman D, Polsaky S, Hawthorne P. Land clearing and the biofuel carbon debt. *Science* 2008;319:1235–7.
20. Kim S, Bruce ED. Indirect land use change for biofuels: Testing predictions and improving analytical methodologies. *Biomass Bioenergy* 2011, 35:3235–3240.
21. FBOMS. Relation between expansion of soy plantations and deforestation—understanding the dynamics. Executive Summary. Report prepared by the Forest Work Group of the Brazilian Forum of NGOs and Social Movements for Environment and Development (FBOMS). Friends of the Earth, Sao Paulo; 2005.
22. Naughton-Treves L. Deforestation and carbon emissions at tropical frontiers: a case study from the Peruvian Amazon. *World Dev* 2004, 32:173–190.
23. Fearnside PM. Soybean cultivation as a threat to the environment in Brazil. *Environ Conserv* 2001, 28:23–38.
24. Creutzig F, Popp A, Plevin R, Luderer G, Minx J, Edenhofer O. Reconciling top-down and bottom-up modelling on future bioenergy deployment. Nat Climate Change 2012, 2/5:320–327.
25. United States Environment Protection Agency (U.S. EPA). *Energy Independence and Security Act* (EISA) *of 2007*. Washington, DC: EPA; 2007.
26. U.S. EPA. *Renewable Fuel Standard Program. U.S.* Washington, DC: EPA, 2009. March 12, 2009. Available at: http://www.epa.gov/otaq/fuels/renewablefuels/index.htm.
27. BP Statistical Review of World Energy June 2011. Available at: http://htpp://www.bp.com/statisticalreview (Accessed December 5, 2011).
28. REN21 Renewables 2011. *Global Status Report*. Available at: http://www.ren21.net/Portals/97/documents/GSR/REN21_GSR2011.pdf. (Accessed January 10, 2012).
29. USDA Foreign Agricultural Service. *Argentina—Biofuels Annual. GAIN Report*; 2011. Available at: http://gain.fas.usda.gov/Recent%20GAIN%20Publications/Biofuels%20Annual_Buenos%20Aires_Argentina_7–6–2010.pdf. (Accessed January 11, 2012).
30. Eisentraut A. *Sustainable Production of Second-generation Biofuels—Potential and Perspectives in Major Economies and Developing Countries*. Paris: OECD/IEA; 2010, pp 217.
31. Stokstad E. Engineered superbugs boost hopes of turning seaweed into fuel. *Science* 2012 335: 273. Available at: http://www.sciencemag.org/content/335/6066/273.full.pdf.
32. Wargacki AJ, Leonard E, Win MN, Regitsky DD, Santos CNS, Kim PB, Cooper SR, Raisner RM, Herman A, Sivitz AB, et al. An engineered microbial platform for direct biofuel production from brown macroalgae. *Science* 2012, 335:308–313.
33. OECD/FAO. *OECD-FAO Agricultural Outlook 2011–2020*. OECD Publishing and FAO; 2011. doi: 10.1787/agr_outlook-2011-en.
34. IEA 2010. *Energy Technology Perspectives 2010 – Scenarios & Strategies to 2050*. Paris: IEA; 2010, pp 650.
35. IEA. *Technology Roadmap: Biofuels for Transport*. Paris: OECD/IEA; 2011.
36. Fischer G, Nachtergaele F, Prieler S, Teixeira E, Tóth G, van Velthuizen H, Verelst L, Wiberg D. *Global Agro-Ecological Zones* (GAEZ *v3.0) – Model Documentation*. International Institute for Applied systems Analysis (IIASA), Laxenburg. Rome, Italy: Austria and the Food and Agriculture Organization of the United Nations (FAO); 2012. Available at: http://typo3.fao.org/fileadmin/user_upload/gaez/docs/GAEZ_Model_Documentation.pdf.
37. Fischer G, van Velthuizen H, Shah M, Nachtergaele FO. *Global Agro-ecological Assessment for Agriculture in the 21st Century. Methodology and Results*. Laxenburg, Austria: IIASA; 2002. Available at: http://www.iiasa.ac.at/Admin/PUB/Documents/RR-02–002.pdf.
38. Fischer G, Frohberg K, Parry ML, Rosenzweig C. Climate change and world food supply, demand and trade: who benefits, who loses? *Global Environ Change* 1994, 4:7–23.
39. Fischer G, Frohberg K, Keyzer MA, Parikh KS. *Linked National Models: A Tool for International Policy Analysis*. Kluwer Academic Publishers; 1988, pp 214.
40. Parikh KS, Fischer G, Frohberg K, Gulbrandsen O. *Towards Free Trade in Agriculture*. International Institute for Applied Systems Analysis. Dordrecht: Martinus Nijhoff Publishers; 1988.
41. Fischer G, Shah M, Tubiello FN, van Velhuizen H. Socio-economic and climate change impacts on agriculture: an integrated assessment of agriculture, 1990–2080. Phil Trans R Soc B 2005, 360:2067–2083; doi:10.1098/rstb.2005.1744.
42. Fischer G, Prieler S. Impacts of biofuel expansion on world food systems and the environment. A global agricultural ecological-economic modelling framework for scenario analysis. ELOBIO Deliverable 5.2 and 5.3. Intelligent Energy Europe 2010. 66 pp. Available at: http://www.elobio.eu/fileadmin/elobio/user/docs/D5.2_5.3.pdf
43. Fischer G, Shah M, van Velthuizen H. *Climate Change and Agricultural Vulnerability*. Laxenburg, Austria: IIASA; 2002. Available at: http://www.iiasa.ac.at/Admin/PUB/Documents/XO-02–001.pdf.
44. Bruinsma, J. (ed). World agriculture: towards 2015/2030. An FAO Perspective. Easrthscan, London / FAO, Rome; 2003.
45. Food and Agricultural Organization. *Statistics FAOSTAT*. 2012 Available at: http://faostat.fao.org/ (Accessed November 25, 2011).
46. Linwood AH, Baker A. *Estimating the Substitution of Distillers' Grains for Corn and Soybean Meal in the U.S. Feed Complex*. United States Department of Agriculture. FDS-11-I-01; 2011. Available at: http://www.ers.usda.gov/Publications/FDS/2011/09Sep/FDS11I01/FDS11I01.pdf. (Accessed January 11, 2012).
47. FAO. *Global Forest Resources Assessment 2010*. Rome: FAO; 2010.
48. IEA. (2008). *From first to second-generation biofuel technologies. An overview of current industry and R&D activities*. Paris: OECD/IEA; 2008.
49. Edwards R, Larive J-F, Beziat J-C. Life cycle GHG. Well-to-wheels analysis of future automotive fuels and powertrains in the European context. Version 3c, July 2011. CONCOWE/ECAR/JRC Report EUR 24952 EN – 2011. Available

at: http://iet.jrc.ec.europa.eu/about-jec/downloads. (Accessed December 12, 2011).

50. Penman J, Gytarsky M, Hiraishi T, Krug T, Kruger D, Pipatti R, Buendia L, Miwa K, Ngara T, Tanabe K, Wagner F. Good practice guidance for land use, land-use change and forestry. IPCC National Greenhouse Gas Inventories Programme and Institute for Global Environmental Strategies. Kanagawa, Japan; 2003. Available at: http://www.ipcc-nggip.iges.or.jp/public/gpglulucf/gpglulucf.html. (Accessed December 12, 2011).
51. IEA. CO_2 emissions from fuel combustion. Highlights. 2011 Edition. Paris: OECD/IEA; 2011.
52. Laborde D. Assessing the land use change consequences of european biofuel policies. Final Report. *ATLASS* Consortium. 2011. Available at: http://trade.ec.europa.eu/doclib/docs/2011/october/tradoc_148289.pdf.
53. Erb KH, Gaube V, Krausmann F, Plutzar C, Bondeau A, Haberl H. A comprehensive global 5 min resolution land-use data set fort he year 2000 consistent with national census data. *J Land Use Sci* 2007, 2:191–224
54. Nelson A. Estimated travel time to the nearest city of 50,000 or more people in year 2000. Ispra Italy: Global Environment Monitoring Unit – Joint Research Centre of the European Commission; 2008. Available at: http://bioval.jrc.ec.europa.eu/products/gam.
55. Wint GRW, Robinson TP. *Gridded Livestock of the World 2007*. Rome: FAO; 2007, pp 131.

FURTHER READING/RESOURCES

International Energy Agency. *Bioenergy, Land Use Change and Climate Change Mitigation. Background Technical Report*. Paris: IEA Bioenergy; 2011.

United Nations Environment Program (UNEP). *Towards Sustainable Production and Use of Resources: Assessing Biofuels*. Paris: UNEP; 2009, pp 118.

30

The Impact of Biofuel Demand on Agricultural Commodity Prices: A Systematic Review

U. Martin Persson*

Physical Resource Theory, Department of Energy & Environment, Chalmers University of Technology, Göteborg, Sweden

INTRODUCTION

Against a backdrop of nearly three decades of declining or stable food prices, world market prices of basic agricultural commodities spiked in the years of 2007–2008, with prices for wheat and corn increasing by about 60%, rice by 124%, fats and oils by 91%, and beef by 26%[1] (see Figure 30.1). With the majority of the world's low- and middle-income countries (and the majority of households in them[2]) being net buyers of food, higher prices had a significant impact on welfare; it is estimated that around 100 million people who would otherwise have been lifted out of poverty and hunger due to economic growth remained poor and undernourished due to the rise in the 2007–2008 food price.[3,4]

Taking the world by surprise, the 2007–2008 food price crisis sparked both public upheavals across the world and an intense—and sometime heated—debate on the causes and consequences of the price increases. In the center of the spotlight was the diversion of agricultural produce from food consumption to use as feedstock for biofuel production in developed countries, mainly the European Union (EU) and the United States. International organizations such as the World Bank and the International Monetary Fund (IMF) claimed that the lion's share of the price increases was due to increased biofuel demand,[5] with the United Nation's special rapporteur on the right to food, Jean Ziegler, going as far as describing the diversion of arable land to the production of biofuel feedstock as a 'crime against humanity' and calling for a 5-year ban on biofuel production. The OECD and FAO, in their annual agricultural outlook,[6] took a more moderate stance, while the United States and the EU tried to downplay the role of biofuels, with the European Commission arguing that its modest use of cereal for the production of ethanol was a 'drop in the ocean' and 'not something to shake the markets'.[5]

The impact on agricultural commodity prices is central also to another debate surrounding biofuel policies and targets, that of indirect land-use changes (iLUCs) caused by increased biofuel demand and leading to emissions that offset or possibly negate the climate benefits of biofuels. These land-use changes are mediated through changes in market prices for biofuel feedstocks.[7] Therefore, understanding the effect of biofuel demand on agricultural commodity prices is important not only to assess poverty and malnutrition implications of increased biofuel use, but also for appraising its environmental benefits.

While several studies have reviewed analyses of how bioenergy and biofuel demand affect food prices,[8–13] these studies have not been systematic, resulting in only a partial coverage of the literature on this topic. More importantly, the reviews all find that results on the sensitivity of food

*Correspondence to: martin.persson@chalmers.se

Advances in Bioenergy: The Sustainability Challenge, First Edition. Edited by Peter D. Lund, John Byrne, Göran Berndes and Iacovos A. Vasalos.

Originally published in: *WIREs Energy Environ* 2014, 4: 465–482 doi: 10.1002/wene.155

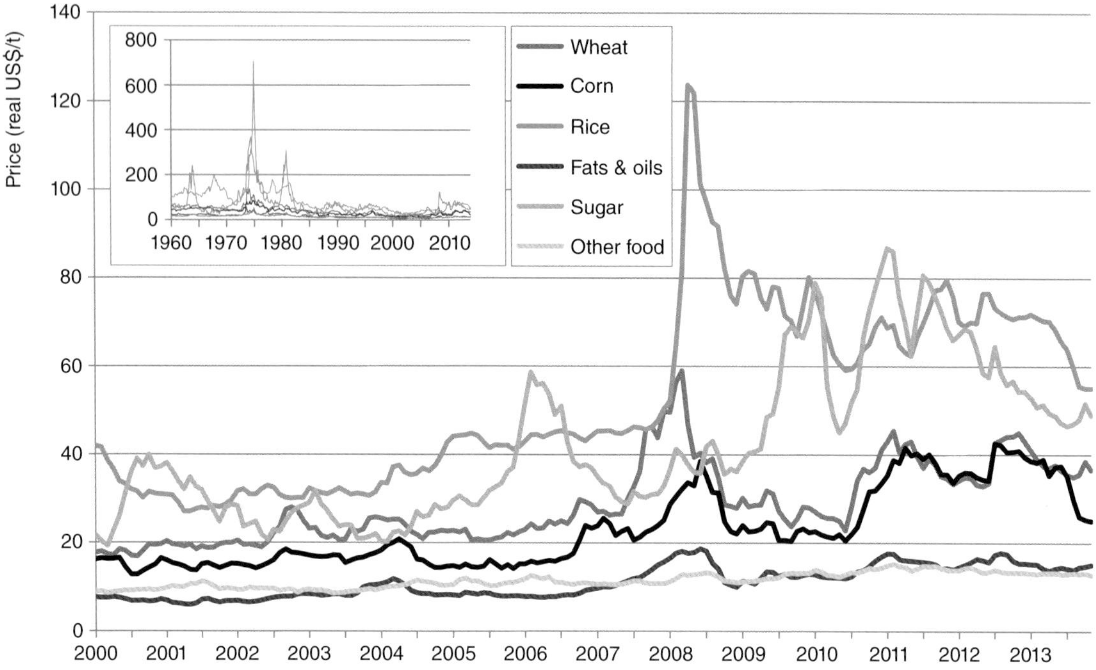

Figure 30.1 Real (inflation adjusted) world market prices for main crops and other food items from January 1960 to November 2013. Source: Data from the World Bank.[h]

prices to bioen-ergy demand differs substantially between the covered studies but do not provide a comprehensive analysis of what drives these differences in results. The aim of this study is to systematically review studies that have tried to assess quantitatively the impact of increased bioen-ergy demand on agricultural commodity prices and to use a theoretical framework to explore the reasons for differing results. The review covers 121 studies using different methods (primarily economic equilibrium models) for exploring the relationship between bioenergy demand and agricultural commodity prices, and includes studies with global, regional, and local scope.

MATERIALS AND METHOD

A Systematic Review Approach

As systematic review is 'a review of a clearly formulated question that uses systematic and explicit methods to identify, select, and critically appraise relevant research, and to collect and analyze data from the studies that are included in the review.'[14] This review seeks to shed light on the following question: How does the price of agricultural commodities—at global or regional level—respond to changes in demand for bioenergy or biofuels?

To identify studies that try to answer this question, a search in the multidisciplinary bibliographic databases Web of Knowledge (www.webof knowledge.com) and Scopus (www.scopus.com) for the keywords 'bioenergy'/'biofuel'/'ethanol'/ 'biodiesel', 'food'/'agricultural', and 'price', was carried out in June 2012. The search covered all types of publications in the two databases, was restricted to the period 2000–2012, and yielded 2068 unique records. After screening the title and abstract for references to the effect of biofuel or bioenergy demand on agricultural commodity markets, 509 records remained. Full-text resources in English (where these could be obtained; 402 records) were then screened again and records kept were those that: (30.1) contained *original* research that (30.2) presented *quantitative estimates* of how an increased demand for bioenergy affects agricultural commodity prices at the global, regional, or local level. After this screening 83 studies remained. An additional 38 studies were identified, primarily from citations in the screened full-text resources that fulfilled the two stated criteria for inclusion. The full review protocol, including the search queries used and a flow diagram describing the selection process, can be found in Appendix S1 (Supporting Information), and the full EndNote library including all the identified studies from the original search can be obtained from the author.

Table 30.1 Summary of the Studies Included in this Review in Terms of Methodology Applied and the Year Published

Method	No. Studies
Modeling studies	112
Of which[1]: partial equilibrium (PE) models	83
General equilibrium (CGE) models	30
Statistical studies	9
Publication year	
Before 2000	3
2000–2006	13
2007–2012	105

[1]Total of PE and CGE models sums up to more than total number of modeling studies as one study included both a PE and a CGE model.

Tables 30.1 and 30.2 summarize main characteristics of the 121 studies included in the review sample, in terms of methodology applied, year published, and geographical scope. Details on each individual study included in the review can be found in Appendix S1 of this article. With the exception of a limited number of statistical studies, all studies use economic (market-equilibrium) simulation models. This is not surprising given that an estimate of the impact of biofuels demand on agricultural commodity prices in practice requires studies to present prices not just under a scenario with a given amount of biofuels demand, but also a 'counterfactual' scenario with zero—or less—biofuel demand, in order to quantify the price impacts associated with changes in biofuel demand. While this could also be achieved in through econometric analyses, including biofuel demand as an explanatory variable, the time span over which biofuel production has reached levels where it affects agricultural commodity markets at a large scale is in most cases too short to allow for robust statistical inferences.

As expected, the bulk (87%) of studies have been published in 2007 or after, following the increasing interest in and debate about global biofuel policies. There is a strong geographical bias in terms of biofuel demand and price changes assessed, with a focus on increased biofuel demand in the United States and the EU. Close to half of the studies (n = 58) solely analyze the impact of US biofuel policies on US commodity prices. The focus on US and EU biofuel demand can be justified based on the fact that these regions accounted for much of the increase in biofuel demand in the latest decade. The lack of studies reporting impacts on agricultural commodity prices in developing regions could be perceived as more problematic, although, given that it is there that the strongest negative impacts of rising prices are expected, both in terms of poverty and malnutrition and in terms of increased incentives for land-use changes.

Theoretical Framework and Data Gathered from the Studies

The basic mechanism through which increased demand for biofuels affects food prices is straightforward: by diverting arable land away from the production of agricultural commodities for food and feed markets, the supply to these decreases, causing prices in agricultural commodity markets to rise. In the biofuels debate, it is sometimes argued that the problem is that we use food (e.g., corn or wheat) to produce biofuels, and not nonfood feedstocks such as jatropha or cellulose. However, as long as the production of feedstock requires agricultural land it incurs competition between food and fuel. Higher demand for biofuels will therefore tend to drive up food prices, whether the actual feedstock can be eaten or not (although, as will be

Table 30.2 Geographical Scope of the Studies Included in the Review in Terms of in Which Region the Modeled Biofuel Demand Shock Occurs and in Which Region Price Effects are Analyzed

			Prices in							
		Tot. no.	World	USA	EU	Brazil	China	India	ROW	Local
			44	70	14	9	6	3	10	3
Biofuel demand shock in	Global	16	15	1	1	1	1	1	2	0
	USA	84	19	68	6	5	2	1	3	3
	EU	26	16	9	13	5	2	1	3	0
	Brazil	10	6	3	2	5	1	0	1	0
	China	4	1	0	0	0	3	0	0	0
	India	2	1	0	0	0	0	1	0	0
	ROW	7	2	0	0	0	0	0	6	0

'Tot. no.' refers to the number of studies assessing biofuel demand shocks/prices in a given region. Because a single study may assess the impact of biofuel demand changes in more than one region on prices in more than one region, these numbers are generally lower than the row/column sums. For the same reason, the total number of studies in this table exceeds the number of studies in the review.

shown below, if the yield of second generation feedstocks are higher, this will lessen the competition for land and hence the effect on food prices).

As demand for arable land for biofuel production increases, effectively reducing the supply of agricultural commodities previously grown on that land and thereby shifting the supply curve for food and feed markets inward, prices increase, leading to three potential adjustments in agricultural markets: (30.1) consumers (both intermediate and final) respond to higher prices by lowering demand for agricultural commodities, and farmers respond to price increases by (30.2) taking new agricultural land into production and (30.3) by intensifying production to increase yields on existing lands. The larger these responses are, the lower the equilibrium price change. Formally, following Hertel,[15] we can express the relative price response (Δp, in percent) due to a biofuel-induced supply shock (ΔS_b, percentage increase in total feedstock supply) as

$$\Delta p = \Delta S_b / (\eta_D + \eta_A + \eta_Y) \tag{30.1}$$

where η_D is the price elasticity of demand and η_A and η_y are the area and yield price elasticities (often referred to as the extensive and intensive margins of supply), respectively.[a] These unit-free elasticities express the percentage change in supply or demand resulting from a 1% change in price. From Eq. (30.1), it is clear that the price response due to increased biofuel demand will be determined by the size of the biofuel demand shock and the sum of the demand and supply elasticities.

That the size of the biofuel demand shock should influence the price response appears obvious. However, the fact that previous reviews have not normalized estimated price effects to the size of the demand shock is one reason why they have had difficulties comparing results from different studies in a comprehensible way. In order to make results comparable, a 'biofuel multiplier' is constructed here by dividing the price change for a given agricultural commodity by the (absolute) size of the biofuel demand shock in energy terms. That is, a multiplier for the effect of corn ethanol on corn prices is defined as the percentage increase in corn prices divided by the concomitant shock in annual corn ethanol production in exjoules (EJ/year). The latter is adjusted for the assumed biofuel yield (in liters of biofuel per ton of feedstock) to reflect that a higher yield will imply that a smaller amount of corn or rapeseed is withdrawn from the food market for a given demand of ethanol or biodiesel, lowering the price response.

To calculate the biofuel multiplier, data are collected from each study on the changes in prices of agricultural commodities due to changes in biofuel demand, the magnitude of the biofuel demand change, and the assumed biofuel yield. Note that many studies run more than one scenario, with differing levels of biofuel demand, and for each (given availability of data) a biofuel multiplier is calculated.

Many studies simulate demand shocks for a number of biofuels simultaneously, which will introduce an error in the estimated multipliers if the demand for different feedstocks competes for the same agricultural land. For instance, there is a clear competition for land between corn and soy in the United States, implying that the price for soy may be affected more by the demand for corn for ethanol than that of soy for biodiesel, as the former in most scenarios is much larger. Calculating a biofuel multiplier soy biodiesel would in such a case grossly overstate the effect of biodiesel demand on soy prices. As it is not possible to adjust multipliers to remove this effects, for each scenario we calculated a biofuel multiplier either for corn ethanol or soy biodiesel (depending on which was larger), but not for both. The analyses of the effect of EU biofuel policies on agricultural markets, however, nearly all simulated shocks in biodiesel and (wheat) ethanol demand of similar magnitude, and because of a much smaller number of studies on EU biofuel demand, it was decided to calculate multipliers for both biofuels, acknowledging the potential over-estimation due to cross-price effects. Furthermore, in scenarios with increasing demand for different biofu-els in different regions (e.g., corn ethanol in the United States, biodiesel in the EU and sugarcane ethanol in Brazil), interaction effects were deemed small enough to allow a calculation of multipliers for more than one biofuel. Note, however, that in over half of the scenarios a single biofuel accounted for more than two-thirds of the total demand shock, and in 44% of the scenarios one biofuel accounted for more than 90% of the total demand shock, implying that cross-price effect in most cases is expected to be small.

Following Eq. (30.1), we should expect the biofuel multiplier to be determined by the elasticity of supply and demand in the economic models. These elasticities may be explicit parameters in some models [e.g., simple partial equilibrium (PE) models], while in others they are the implicit result of the model structure and range of model parameters [e.g., large computable general equilibrium (CGE) models].[b] By collecting data on the relative changes in nonbiofuel (food and feed) demand for agricultural commodities, as well as relative changes in total production and harvested area of biofuel feedstocks due the biofuel supply shock, demand, area, and yield elasticities in the models were estimated. These implicit elasticities are easily calculated (given the availability of data) and should better reflect how price sensitivity differs across models, compared to comparing solely the direct parameters values for different model elasticities.[18]

Table 30.3 Data Collected from the Studies Included in the Review

1	Change in agricultural commodity prices [%]
2	Size of the biofuel demand shock [EJ]
3	Baseline biofuel demand [EJ]
4	Assumed biofuel yield [l/t feedstock]
5	Change in nonbiofuel (food and feed) demand for agricultural commodities [%]
6	Change in total production of the biofuel feedstock crop [%]
7	Change in harvested area of the biofuel feedstock crop [%]
8	Utilization of biofuel byproducts accounted for [Y/N]
9	Bilateral trade specification [Integrated world market/Armington]

Previous studies have also illustrated the importance of accounting for byproducts from biofuel production (e.g., distillers grains from corn ethanol production or soybean and rapeseed meal from biodiesel production) when estimating price impacts, as byproduct utilization reduces the effective amount of feedstock being withdrawn from the market due to biofuel production.[19] Information on whether studies included this mechanism or not was therefore also collected. Similarly, studies of iLUC have illustrated that differences in how bilateral trade is modeled affects where and how much land expansion occurs in response to biofuel demand. We therefore also identified whether studies assumed a single world price for agricultural commodities, or whether they modeled agricultural commodities from different countries as heterogeneous goods (the so-called Armington assumption), implying that there is not a single world market price for each commodity.

Table 30.3 summarizes the data this review aimed to collect from each of the studies it covers. In the case that some of this information was not presented in the full-text resource, the authors of the study were contacted via e-mail and asked to provide the missing information. In all, biofuel multiplier is calculated in 433 different biofuel scenarios. In 257 of those, data presented in the studies or provided by the authors allowed the estimation of demand elasticities, whereas total supply, yield, and area elasticities could be estimated in 335, 182, and 185 scenarios, respectively.

RESULTS—BIOFUEL MULTIPLIERS

Figure 30.2 displays the estimated commodity price multipliers for the main biofuels (corn, wheat, sugar, and cellolusic ethanol and biodiesel) and commodity markets (US, EU, and world market). Starting with the effect of US corn ethanol demand on domestic corn price, most studies (80%) find that corn prices increase by 10–70% for an annual increase in corn ethanol demand by 1 EJ. There are a few studies that exhibit stronger price effects. However, these are primarily studies that analyze the effects of changing ethanol demand in a setting where corn production is purely stochastic and does not respond to price changes[20–22] (i.e., planting decisions have been made and only weather will determine yields) and are therefore more indicative of the impact of changes in ethanol demand on price volatility, rather than long-term equilibrium prices.

Excluding the latter studies (and other outliers) from the sample gives an average corn price multiplier of 36%/EJ/year, with a standard deviation (σ) of 18%/EJ/year ($n = 218$). If averaged over studies instead of individual scenarios (to reduce the risk of studies presenting many scenarios skewing results), the mean multiplier is reduced somewhat, to 32%/EJ/year ($n = 44$; $\sigma = 18$%/EJ/year). To put these numbers in perspective, between 2000 and 2008 annual US corn ethanol production increased by about 0.68 EJ (or 29 million liters) and the domestic corn price increased by about 85%.[c] According to the results here, the increased demand for corn ethanol alone (i.e., not including interaction effects with other factors affecting corn prices) was responsible for 11–43% of the price increase (results are similar also during the 2000–2012 period).

Figure 30.2 also displays the 'ripple effects'[23] increased demand for corn for ethanol production has on prices for other agricultural commodities, resulting mainly from competition for land on the supply side and substitution on the demand side. For instance, farmers switching from planting soy to corn in response to higher corn prices reduce overall soy supply and raise soy prices; higher corn prices induce substitution of corn for wheat in livestock feed markets, increasing wheat prices; increases in corn, soy, and wheat prices raise the price for livestock feed and, therefore, lead to higher meat and dairy prices. On average across studies, the increase in soy and wheat prices is roughly half those estimated for corn, whereas meat price increases are about one fifth of those for corn.

Compared to the impacts of US corn ethanol, there are much fewer studies from which the effect of biodiesel demand on oilseeds and vegetable oil markets can be estimated. On average, models estimate that EU prices for rape and other oilseeds increase by 68%/EJ/year of increased biodiesel demand ($n = 26$; $\sigma = 219$%/EJ/year), although this is substantially reduced to 25%/EJ/year ($\sigma = 25$%/EJ/year) if one outlier is removed. Similarly, the average biodiesel price multiplier for vegetable oil is 528%/EJ/year ($n = 4$; $\sigma = 718$%/EJ/year), but reduces to 171%/EJ/year ($\sigma = 96$%/EJ/year) if the same outlier is removed. The latter estimate would imply that the increased

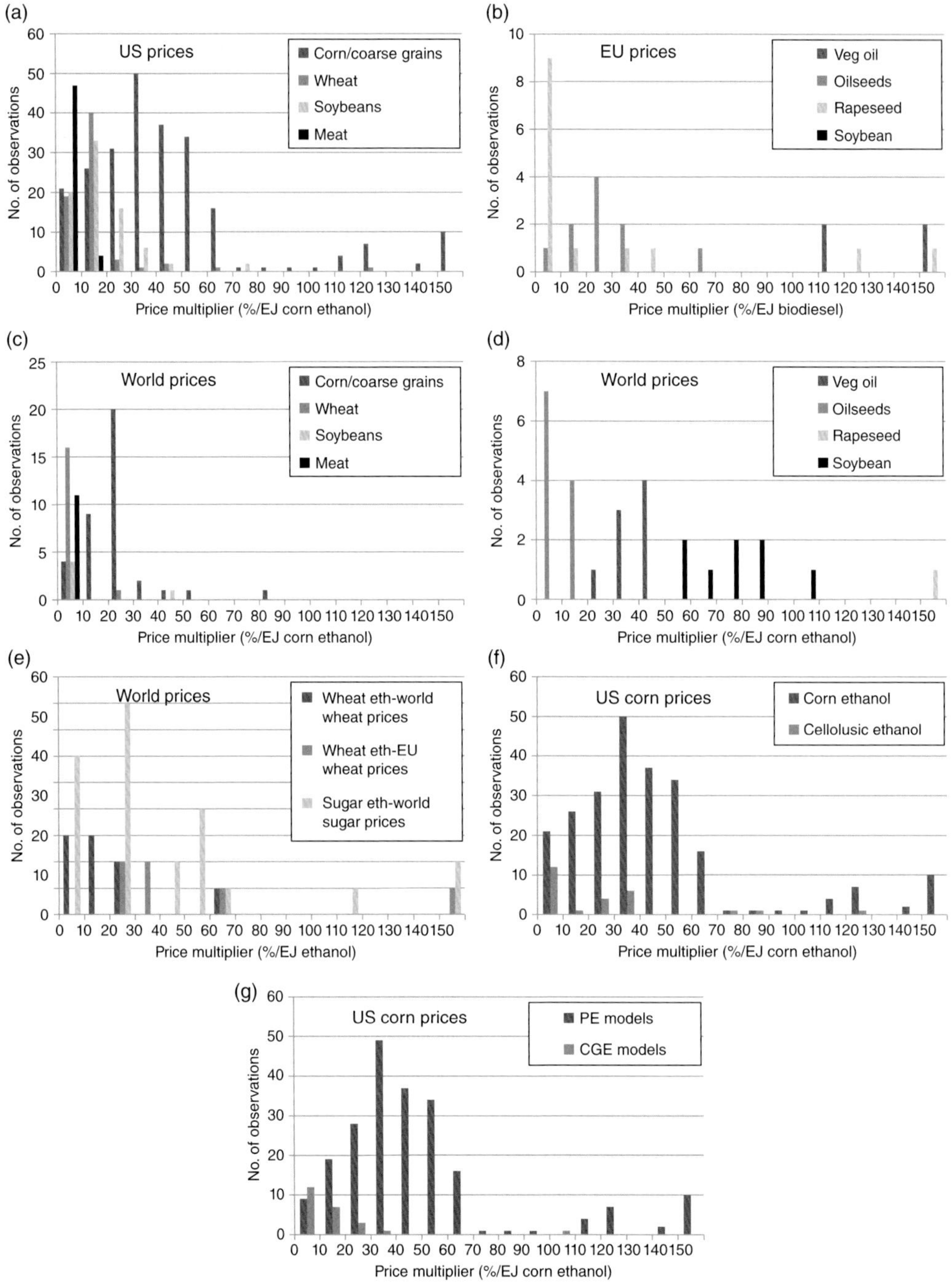

Figure 30.2 Histograms of estimated biofuel multipliers (percentage change in agricultural commodity price divided by size of biofuel demand shock), as calculated from the model studies included in the review. Panels (a)–(e) display multipliers for the main biofuels (corn, sugar and wheat ethanol, and biodiesel from vegetable oils) for different market prices. Panel (f) reports the difference in estimated corn price impacts for corn ethanol versus second generation, cellulosic ethanol in the United States, whereas panel (g) displays the differences in estimated US corn ethanol multipliers from partial versus general equilibrium models. PE, partial equilibrium; CGE, computable general equilibrium; EJ, exjoules.

EU biodiesel production of 0.26 EJ/year in the 2002–2008 period alone accounted for about a third of the increase in vegetable oil prices during that time span.[d]

At the level of global markets, estimated impacts are substantially smaller. The average ethanol multiplier for corn or coarse grain prices is 23%/EJ/year (n = 38; σ = 15%/EJ/year) and the average biodiesel multipliers for oilseeds and vegetable oil are 8%/EJ/year (n = 11; σ = 7%/EJ/year) and 38%/EJ/year (n = 8; σ = 9%/EJ/year), respectively.

The results also show that the impact of second-generation ethanol based on cellolusic feedstocks (primarily switchgrass) on agricultural commodity prices are smaller than those from first-generation (corn) ethanol, with an average biofuel multiplier of 25%/EJ/year (n = 26; σ = 31%/EJ/year). These result reflects the fact that all but one of the studies model a fully competitive land market, so that although cellulosic crops are allowed (but no restricted) to expand into marginal agricultural land (i.e., grazing or idle land), cellulosic crops compete with food crops pushing up their prices. The degree to which cellulosic crops are grown on marginal land ranges between studies, from nearly all to less than half. Thus, while the higher yield of second-generation biofuels alleviates the commodity price impacts due to increased land competition, it does not eliminate it. If the United States was to supply 16 billion gallons of cellulosic ethanol (1.4 EJ/year) by 2022, as mandated by the current Renewable Fuel Standard (RFS) legislation, the studies reviewed here suggest that this would increase corn prices by approximately 35%.

Finally, Figure 30.2 shows that PE models generally estimate a larger sensitivity of commodity prices to biofuel demand than CGE models; the difference for the corn/coarse grain ethanol multiplier is large, with an average multiplier from PE models of 39%/EJ/year (n = 195; σ = 17%/EJ/year) compared to 12%/EJ/year (n = 23; σ = 9%/EJ/year) for CGE models (excluding the above-mentioned outliers). The difference in results between PE and CGE models has also been recognized in previous reviews on this topic[24–26] and has primarily been attributed to the fact that CGE models account for interaction between different sectors of the economy, allowing a shock in one sector to be absorbed through adjustment in the entire economy, leading to smaller price impacts. Below we scrutinize this, and other, explanations for the large difference in estimated biofuel impacts on prices in more detail.

EXPLORING AND EXPLAINING DIFFERENCES IN RESULTS BETWEEN STUDIES

The most striking feature of the results presented above is the large differences between studies in terms of how sensitive agricultural commodity prices are to increased demand for biofuels. Here the reasons for these differences are explored, based on the determinants of price impacts identified above: supply and demand elasticities, PE versus CGE modeling approach, treatment of byproducts, and bilateral trade specification.

Accounting for Biofuel Byproduct Use

Both the production of ethanol from corn and biodiesel from oilseeds generate byproducts that can be utilized as feed in the livestock industries. This does not only contribute positively to the profitability of biofuels, but also substantially affects the impact biofuel demand has on agricultural commodity markets.[19,27] Each ton of corn used to produce ethanol production yields approximately 320 kg of dried distillers grains with solubles (DDGS) and as a result of increased ethanol production in the United States, livestock producers substituted about 16 million tons of corn for DDGS over the 2001–2008 period (DDGS increasing from 1.3 to 10.3% of US feedstuffs).[27] Not accounting for DDGS feed utilization will thus overstate the amount corn withdrawn from the feed market due to ethanol demand and can lead to overestimating the resulting corn price impacts by as much as 50%.[19]

Oilseed processing generates two products, oil and meal (e.g., soybean or rapeseed meal), where the former is used to produce the biodiesel and the latter constitutes high-protein feed ingredient. As a result of increased rapeseed oil production for biofuel production in the EU, rapeseed meal supply nearly doubled in the 2001–2008 period, causing livestock producers to switch away from soy meal as the relative price of rapeseed meal dropped.[27] This effect explains why the estimated price multipliers for vegetable oil are much higher than those for oilseeds (Figure 30.2) and also why these studies find that biodiesel demand has a negligible, or even negative, impact on meat prices.

From the above examples, it is obvious that a correct estimation of the biofuel demand impact on food prices requires accounting for byproducts. However, 95% of the estimated corn ethanol multipliers and all of the estimated biodiesel multipliers come from models that do account for the utilization of byproducts, so this cannot explain the large differences in price effects between models found above.

Supply and Demand Elasticities

The theoretical framework predicts that the divergent estimates of biofuel multipliers can be explained by differences in supply and demand elasticities in the equilibrium models. Figure 30.3 provides a first test of this

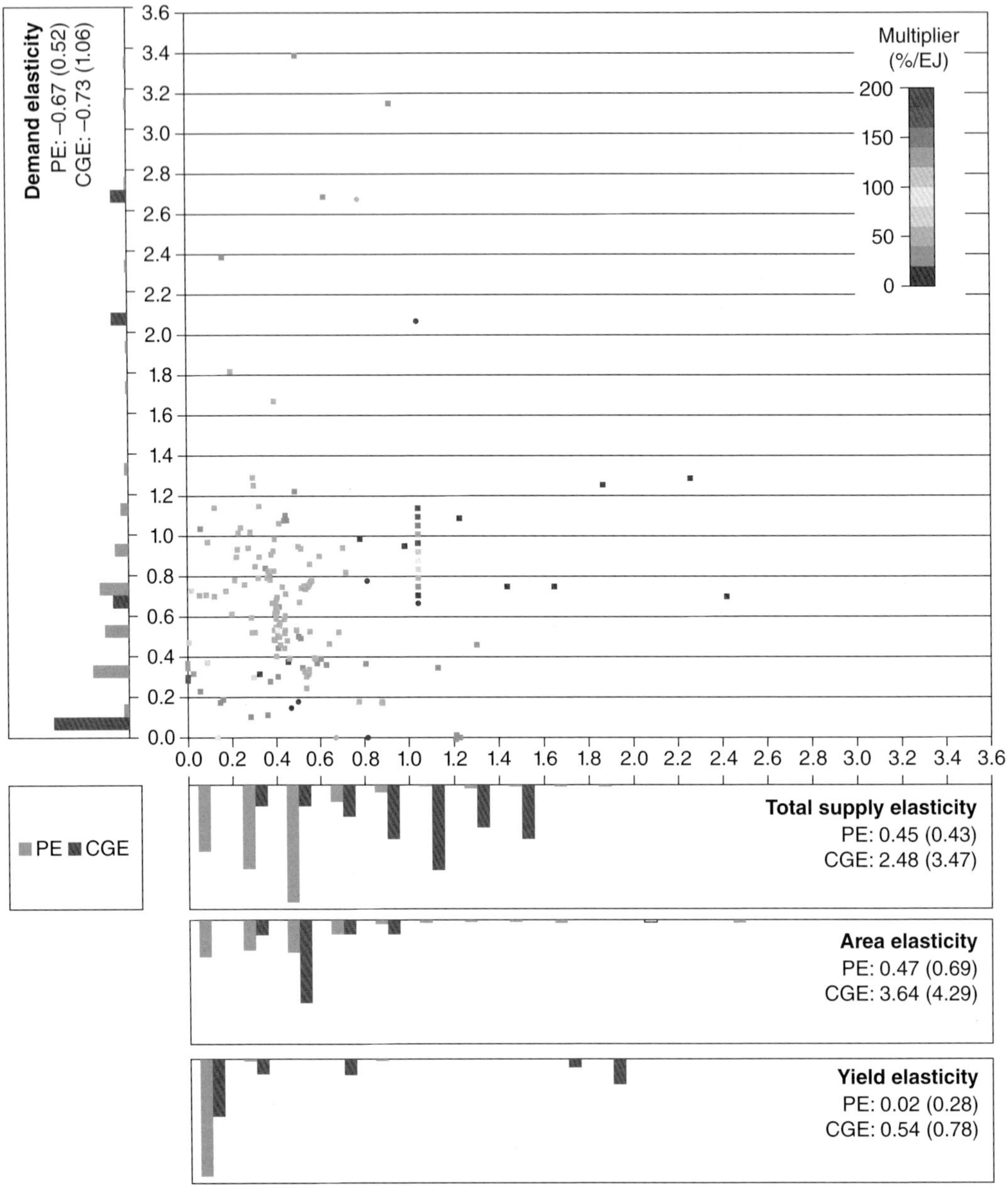

Figure 30.3 Estimated corn ethanol multipliers for corn or coarse grain prices as a function of the models demand and supply elasticities. The histograms on the left and bottom show the distribution of demand and supply elasticities from the models reviewed, differentiating between partial (PE) and general (CGE) equilibrium models, as well as the average elasticities for each of the two model categories (standard deviations in parenthesis). EJ, exjoules.

hypothesis; it displays the estimated corn ethanol multipliers for corn or coarse grain prices as a function of the estimated implicit supply and demand elasticities in each model. As can be seen, lower multipliers are generally associated with higher supply and/or demand elasticities, in line with the theoretical prediction.

More extreme estimates of corn ethanol multipliers are also easily explained by the models implicit elasticity

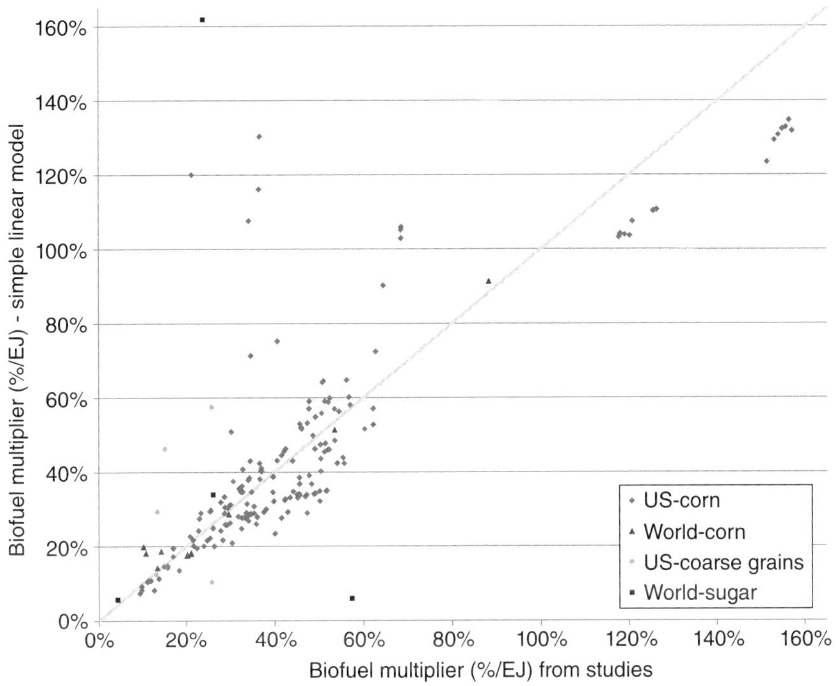

Figure 30.4 The figure plots the estimated biofuel multipliers from the models reviewed (horizontal axis) against the same measure estimated using an extremely simple model—one linear supply curve and one linear demand curve—calibrated using supply and demand elasticities from the larger model. The closer the data points are to the diagonal line, the more accurately the simple model reproduces the result from the full models. EJ, exjoules.

assumptions, both on the high and low end of the spectrum. As noted above, price effects exceeding 100%/EJ/year are nearly all the result of models assuming a zero supply elasticity. Similarly, analyses predicting very low price impacts from increased biofuel demand (i.e., corn ethanol multipliers in the order of a few percent per EJ/year)[28,29] stem from models exhibiting unrealistically strong supply responses (elasticities in the order of 7.7-11.6).

Figure 30.4 provides even stronger evidence that models' supply and demand elasticities is the main determinant of the sensitivity of commodity prices to biofuel demand. Here the estimated biofuel multipliers from the reviewed PE and CGE models are compared with the same multiplier estimated from the simplest PE model of an agricultural commodity market one can conceive: a single linear supply curve and a single linear demand curve (see Appendix S1 for derivation). For each observation for which both the biofuel multiplier and implicit supply and (non-biofuel) demand elasticities could be calculated, the simple model was calibrated to historical market data (price and total production for the year nearest to that modeled) using the implicit supply and demand elasticities estimated from the larger model. The price change in the simple model due to a reduction in biofuel demand (of the same size in relative terms as that applied in the larger model) was then calculated, as well as the resulting biofuel multiplier.

In Figure 30.4, the biofuel multipliers from the simple model are plotted against those from the reviewed studies. As can be seen, with the exception of the sugar and coarse grain ethanol multipliers (for which there are few sample points), the simple model predicts the estimated biofuel multipliers from the more complex model exercises well. That is, the difference in estimated sensitivity of agricultural commodity prices to biofuel demand can to a large extent be explained by differences in demand and supply elasticities across models.[e]

Put differently, a very simple model can give a good estimate of the impact that biofuel demand has on agricultural commodity prices, given that we know the relevant supply and demand elasticities. Consequently, it is interesting to compare the implicit supply and demand elasticities from the models reviewed (see Figure 30.3 and Table 30.4) with empirical estimates of these parameters from the literature. First note, however, that apart from US corn elasticities

Table 30.4 Estimated Average Supply and Demand Elasticities from the Model Studies in the Review

Crop/Elasticity	World	USA	EU	Brazil
Corn				
Total demand	−0.21 ($n = 11$; $\sigma = 0.13$)	−0.69 ($n = 183$; $\sigma = 0.51$)		
Total supply	0.85 ($n = 17$; $\sigma = 0.94$)	0.49 ($n = 191$; $\sigma = 0.47$)		
Yield	0.17 ($n = 13$; $\sigma = 0.21$)	−0.01 ($n = 110$; $\sigma = 0.27$)		
Area	0.30 ($n = 13$; $\sigma = 0.12$)	0.52 ($n = 104$; $\sigma = 0.74$)		
Vegetable oil (demand)/oilseeds (supply)[1]				
Total demand	−0.40 ($n = 5$; $\sigma = 0.23$)	−0.48 ($n = 14$; $\sigma = 0.24$)		
Total supply	2.18 ($n = 12$; $\sigma = 2.25$)	0.55 ($n = 8$; $\sigma = 0.53$)	2.03 ($n = 13$; $\sigma = 0.85$)	
Yield	0.37 ($n = 7$; $\sigma = 0.75$)	0.02 ($n = 5$; $\sigma = 0.03$)	0.29 ($n = 3$; $\sigma = 0.46$)	
Area	0.67 ($n = 4$; $\sigma = 0.18$)	0.33 ($n = 5$; $\sigma = 0.33$)	0.87 ($n = 3$; $\sigma = 0.75$)	
Sugar crops				
Total demand	−0.37 ($n = 15$; $\sigma = 1.16$)			−1.08n = 6; $\sigma = 1.94$)
Total supply	0.67 ($n = 20$; $\sigma = 1.50$)			1.65 ($n = 18$; $\sigma = 0.45$)
Yield	0.03 ($n = 23$; $\sigma = 0.14$)			0.05 ($n = 3$; $\sigma = 0.08$)
Area	0.64 ($n = 19$; $\sigma = 0.1.51$)			1.51 ($n = 3$; $\sigma = 0.29$)

No. observations and standard deviations in parentheses.
[1]For USA, demand and supply elasticities refer to soybeans.

the basis for comparing estimated elasticities across models with empirical estimates is relatively thin (less than 20 observations; see Table 30.4) and even for US corn elasticities, which has been the focal point of much of the iLUC debate, empirical estimates from the recent past are rare.[30,31]

Starting with the demand side, one of the most comprehensive assessments of demand elasticities for agricultural commodities, utilizing a dataset covering 114 countries, find that demand generally is more inelastic the higher the income level; e.g., for cereals, demand elasticities range from −0.37 in low-income countries, to −0.18 in middle-income countries and −0.01 in high-income countries.[32] The elasticities for fats and oils are −0.39, −0.21, and −0.07, for low-, middle-, and high income countries respectively. However, note that these elasticity estimates cannot be directly compared to the own price elasticities for individual crops obtained from the models here because they only covers final consumption (e.g., not feed demand) and because there most likely are considerable cross-price elasticities (e.g., consumers shifting from corn to other cereals as the price of corn increases).

Reference 30 estimates price flexibility of demand (the inverse of demand elasticity) for corn and soybeans in the United States in the period 1981–2010, arriving at elasticities of −0.74 (95% confidence interval −0.57 to −1.04) for corn and −0.97 (95% confidence interval −0.68 to −1.72) for soybeans. Although the average US corn demand elasticity across the model studies reviewed here (−0.69; see Table 30.4) is within the empirical range, close to 60% of the estimated implicit elasticities lie outside of this range; almost half of the studies display a more inelastic demand (i.e., an elasticity larger than −0.57). Similarly, models tend to underestimate the demand elasticity for soybeans in the United States, with close to 80% of studies displaying more inelastic demand than the range given by the empirical analysis (i.e., elasticity larger than −0.68), although the sample here is small.

The total supply response to higher prices can be divided up into a change in yield (intensive margin) and a change in harvested area (extensive margin). Starting with the former, whether yields respond to prices has been a hotly debated topic in relation to estimates of iLUCs from increased biofuel demand, as the share of increased supply that will come from area expansion will depend on the ratio between the area and yield elasticities.

As can be seen in Figure 30.3, most PE models assume that the yield elasticity is nil, whereas many CGE models (e.g., GTAP) assume that price increases do lead to net increases in yields.[f] As can be seen in Table 30.4, on average across models, yield elasticities are assumed to be relatively large for corn and oilseeds globally, and for oilseeds in the EU. The empirical evidence is mixed, but most studies do not support a statistically significant price impact on agricultural yields.[31] Reference 33 analyzes the yield elasticity of corn, soybeans, wheat, and rice in main producer countries but is only able to find a positive yield elasticity for soybean in Brazil, for all other crops and countries the elasticity is insignificant or negative. Reference 34, however, does find statistically significant yield elasticities for corn, soybeans, and wheat in the United States, with point estimates of 0.15, 0.07, and 0.43, respectively.

Contrary to yield elasticities, empirical analyses consistently find that as prices for crops rise, the harvested area increases. However, estimated elasticities still differ in magnitude; Reference 34 finds elasticities for harvested area of corn and soybeans in the United States both to be around 0.5 whereas Ref 33 estimates an US acreage elasticity for corn in the range 0.25–0.30 and for soybeans in the range 0.3–0.35. Combining these figures with the estimated yield elasticities results in total US supply elasticities in the range 0.25–0.65 for corn and 0.30–0.57 for soybeans. Close to 60% of the implicit US corn supply elasticities estimated for the models reviewed here falls within the empirical range, and the average elasticity (0.49; see Table 30.4) lies in the middle of this range.

To summarize, it has been showed here that differences in elasticities between models to a large extent can explain the differences in estimated biofuel multipliers across studies and that compared to empirical estimates the models reviewed here tend to underestimate the demand response to higher corn and soy prices in the United States, whereas supply responses are mostly in line with the best available empirical evidence. It is, however, very important to make the distinction between short-run and long-run elasticities.

Both yields and area are likely to be more responsive to price changes over the long-term—implying that the long-run price effect of biofuel demand is likely to be smaller than the short-run effect—but most empirical studies focus on short-run (e.g., annual) elasticities.[31]

As noted also by several others studies, recent and robust analyses on demand and supply elasticities for agricultural commodities are scarce, even for the United States, and hence the empirical evidence cannot help much in reducing the uncertainty regarding the commodity price impacts due to biofuel demand. For instance, if one uses the simple, two-equation, corn market model introduced above and the supply and demand elasticity ranges from the literature to estimate the impact of increased corn ethanol demand in 2000–2008 on 2008 US corn prices, the result is a price increase of 16–40% (corresponding to a corn ethanol multiplier of 22–54%/EJ/year).[d] Put differently, the simple model predicts that 30–63% of the increase in US corn prices in 2000–2008 can be attributed to increased biofuel demand. This range is not narrower (although somewhat higher) than the 14–43% range reported above from the reviewed model studies.

Modeling Partial Versus General Equilibrium

As noted, PE models on average predict stronger price responses to increased biofuel demand compared to general equilibrium models. Primarily this appears to stem from supply being more price elastic in CGE models than in PE models, as demand elasticities does not differ much between the two types of models (see Figure 30.3). Does this difference in supply elasticities stem from general equilibrium effects that PE models are unable to capture or simply from different assumption regarding the magnitude of these elasticities?

The route through which general equilibrium effects would impact supply is through changes in agricultural input prices, as other sectors of the economy adjusts to higher food prices and increased supply of biofuels. Most importantly, the increased biofuel supply will lower the demand for gasoline and diesel, while increasing the demand for land, causing the supply curve for agricultural commodities to shift as the cost for these inputs change.

The first of these effects is relatively minor; across all the models included in the review that includes a link between biofuels and oil markets and report impacts, the average reduction in petrol prices is 4%/EJ/year of increased biofuel supply ($n = 25$; $\sigma = 14$%/EJ/year). In a study[35] that specifically analyzes the impact of biofuel demand on fuel prices, estimated impacts are even smaller (1.3-2.1%/EJ/year), especially if accounting for the noncompetitive (cartel) nature of international oil markets. If one assumes that roughly one fourth of the total cost of US corn production comes from fuel expenditures (including indirect fuel use through fertilizer and pesticide inputs),[36] accounting for the fuel price reduction due to increased biofuel supply should reduce the corn ethanol impact on corn prices in the United States by roughly 0.3-1%/EJ/year. This effect is far too small to be able to explain the difference in estimated US corn ethanol multiplier between CGE and PE models (12%/EJ/year vs 39%/EJ/year on average, after removing outliers).

Many PE models do not explicitly model land markets,[37] thereby missing the key component of the bioenergy-food price nexus: competition over limited land resources. Some land competition may still be captured by PE models' cross-price elasticities for the area harvested for different crops (although these dynamics are calibrated to relatively small historical perturbations in prices and quantities, with limited applicability to large biofuel demand shocks), and over two-thirds of the included PE models also model the expansion of total cropland in some way (e.g., through inclusion of other land-use sectors or land-supply curves). Across the sample analyzed here, PE models exhibit a lower average area elasticity than CGE models (after removal of outliers, 0.32 ± 0.25 for PE models vs 0.58 ± 0.20 for CGE models; see also Figure 30.3), although it is unclear to which extent this stems from differences in elasticity assumptions and the structural model differences.

Finally, equilibrium effects on other factor prices (e.g., labor and capital) in CGE models is likely to have a minor effect on the supply response as the links to these factor markets are more indirect and price impacts therefore can be expected to be smaller (see, e.g., results in Ref 24). In conclusion, the differences in supply elasticities between CGE and PE models are mainly a result of different assumptions, although the structural differences in terms of treatment of land markets may contribute to the higher supply elasticities found in CGE models.

International Trade Specification

The extent to which price increases in the region experiencing a biofuel supply shock is transmitted to agricultural commodity markets in other regions is crucial for assessing poverty and malnutrition impacts as well as the implications in terms of iLUC (as it primarily is higher commodity prices in tropical countries with carbon-rich forests that risk creating stronger incentives for deforestation). However, most studies report price impacts solely for the main developing regions (EU and USA) or for world market prices. This largely stems from many models covering only the United States and that in the global (mainly PE) models the main approach to modeling price transmission in global markets is assuming a perfectly integrated world market (IWM) with single clearing price for a given commodity at the global level.

However, CGE models typically adopt another approach that asserts that products originating from different places are imperfect substitutes (the so-called Armington assumption), implying that there is not a single world market price for any given agricultural commodity. In the Armington model, more of the market adjustment due to a supply or demand shock is forced to be absorbed in the region where the shock emanates and regions more closely linked to it through trade (i.e., major importers and competing exporters), leading to a 'stickiness' in trade patterns.[16,38]

In contrast to the IWM assumption, in which the price impact of a biofuel shock will be the same in all regions, regardless of the region where the supply/demand shock occurs, the Armington assumption will generally imply higher price changes in the region where the shock occurs, with the price transmission to other regions depending on the strength of previous trade ties. For instance, Refererence 27 analyzing the impact of US corn ethanol and EU biodiesel mandates on world markets finds that coarse grain prices increase by 12.6% in the United States, followed by regions being major competitors with the United States in global markets (e.g., 6.4% in Brazil, 4.6% in Oceania/Australia, and 3.6% in Canada), and only between 1 and 2% in most other regions of the world (and higher for major importers, such as Japan). Similarly, oilseeds prices increase most in the EU (19.1%), followed by the palm oil producers Indonesia and Malaysia (18.6%) and the soybeanpro-ducers USA (11.2%) and Brazil (10.8%). Results from other studies[39–41] that use the Armington assumption show a similar discrepancy between price increases in the region of the biofuel shock and global markets (export weighted average world price).

On the one hand, there is some empirical supporting the Armington representation of bilateral trade in agricultural commodities; Reference 38 uses data on coarse grain trade in 35 major exporters and importers in the period 1975–2002 to show that an Armington specification predicts changes in trade more accurately than the IWM assumption. On the other hand, studies have shown that increasingly there is a high level of price transmission from international to national markets in most developing country regions, especially for the widely trade cereals (wheat, corn, and rice); the main exceptions being West and Southern Africa.[42–44]

WHAT CAN CURRENT STUDIES SAY ABOUT THE PRICE IMPACTS OF A LARGE-SCALE GLOBAL EXPANSION OF BIOFUEL DEMAND?

If biofuels is to make a substantial contribution to global climate mitigation it must be scaled up considerably from today's relatively modest levels. In 2011–2012, global biofuel production reached about 2.5 EJ/year,[45] but long-term climate stabilization scenarios put global bioenergy demand in the order of 50–300 EJ/year, demanding some 0.5-1 global hectares (Gha) of land.[46,47] To which extent can the studies reviewed here help inform a discussion about the impacts of such an increase in global land demand for biofuel or bioenergy production on agricultural commodity markets?

First note that apart from a few exceptions discussed below, all biofuel scenarios analyzed here incur relatively small biofuel demand shocks compared to long-term scenarios for bioenergy demand, see Figure 30.5; in three-quarters of the scenarios, the total biofuel demand is even less than current global production levels (partly stemming from the fact that most of the studies assess the impacts of biofuel demand in a single region, mainly the United States). If demand and supply curves were linear one could simply extrapolate the price multipliers estimated from these limited biofuel shocks to ever increasing biofuel demand. However, basic intuition holds that both demand and supply should be nonlinear for large shocks; as more and more feedstocks are diverted from food and feed markets, demand is going to become increasingly inelastic, and on the supply side one will eventually start to approach biophysical limits (i.e., limited availability of arable land to expand into and theoretical potentials limiting the yield increases that can be achieved).

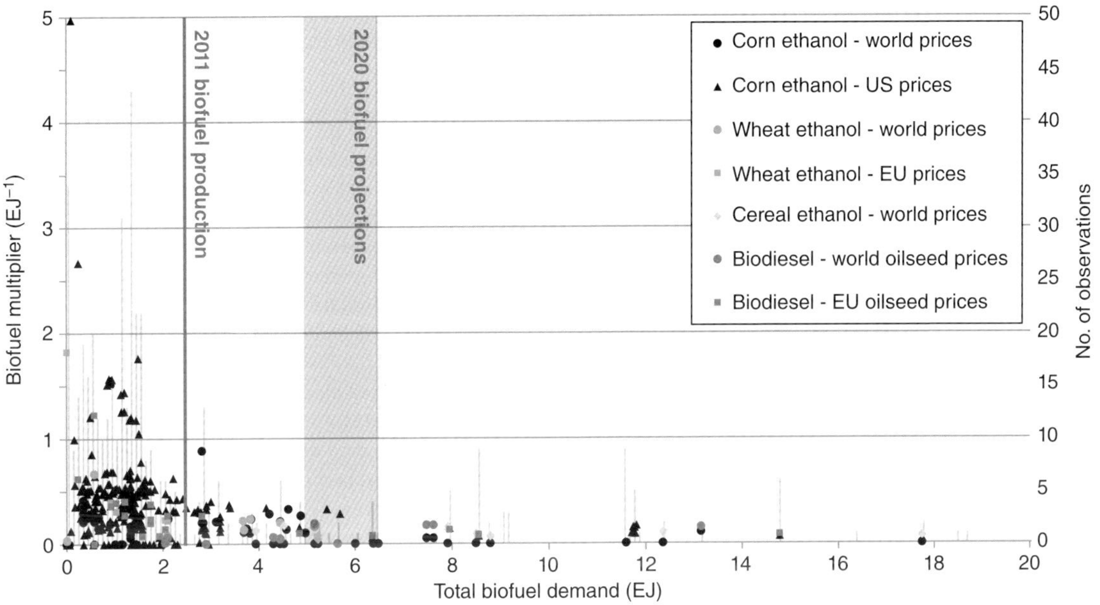

Figure 30.5 Estimated biofuel multipliers for the main biofuels as a function of the total biofuel demand in the corresponding model scenarios. The green line shows the current (2011) annual global biofuel production[45] and the green-shaded area near-term (2020) projections for global biofuel demand, primarily based on existing and proposed national biofuel mandates and support policies.[48,49] EJ, exjoules.

Consequently, we should expect marginal biofuel multipliers to increase with the size of the biofuel demand shock. This effect is evident for instance in the results from Ref 50, analyzing the impact of second-generation ethanol on US commodity markets, where the corn price multiplier increases from 3%/EJ/year for a biofuel demand of 11.6 EJ/year in 2020, to 5%/EJ/year as the biofuel demand increases to 18.6 EJ/year in 2100. Similar results are found in a few other studies.[51,52]

The results from all studies reviewed here (see Figure 30.5), however, appear to suggest that biofuel multipliers decrease with the size of the biofuel demand shock. This is also the case for most individual studies that vary the size of the biofuel demand. The latter can be explained by: (1) differences in the size of the biofuel shock is in most cases small, (2) models are either linear by construction or close to linear for small shocks,[18] and (3) in dynamic models where biofuel demand increases over time, so do usually the implicit supply and demand elasticities in the models due to system inertia that is relaxed over time, and thus biofuel multipliers appear to decrease with larger shocks, whereas in reality the results show how impacts are dampened over time.

There are a few studies in the sample that do model the impact of much larger bioenergy demand and also find very modest impacts on food prices: Reference 53 estimates that global coarse grain prices increase by about 50% for a bioenergy demand shock of 150 EJ/year, corresponding to a biofuel multiplier of less than 1%/EJ/year. References 54 and 55 report even smaller impacts, the former finds that food prices increase by less than 10% in most world regions for a bioenergy demand shock of around 300 EJ/year and the latter that global crop prices increase by around 5% for a bioenergy demand shock of 200 EJ/year.[g]

These large discrepancies in estimated sensitivity of food prices to bioenergy demand between studies analyzing short-term, small biofuel demand shocks and those assessing long-term, large-scale shocks are again most likely a result of differing assumptions regarding supply elasticities between models. Reference 54 assumes that on top of exogenous technological change in agricultural production (average yields increasing by 80% over the coming century), yields are very responsive to price increases; in the bioenergy demand scenario yields increase by an additional 20% for a food price increase in the order of 10%, implying a long-run implicit yield elasticity of around 2.

In Ref 55, the large increase in land devoted to bioenergy production—0.8 Gha globally, corresponding to roughly half the current cropland area—occurs through expansion into natural forest land and pastures, in combination with greatly intensified production on the latter. Similarly, Ref 53 sees roughly 2 GHa land in bioenergy production in year 2100, most of which comes from expansion into currently unman-aged land. These large increases in land devoted to crop production, both for food and bioenergy, with relatively small corresponding increases in crop prices, suggest very elastic supply of agricultural land in these models.

These results point to the importance of how land markets are modeled (or not) when assessing long-term implications of a large-scale increase in bioenergy production, as land competition is at the very heart of the biofuel-food price nexus. Several models illustrate the importance of explicitly accounting for land markets when assessing the impacts of biofuel demand on food prices. In Ref 27 disregarding the land competition between cropland and pastures (effectively assuming perfectly elastic demand for pastureland) reduces the price increase of coarse grains due to biofuel demand by half, and that of oilseeds by 30%. Reference 54 shows that not allowing cropland to expand into current forestland reduces the supply of potential cropland by three-quarters, and dramatically increases the food price impacts of increased biofuel demand. Reference 56 similarly shows that forcing biofuel expansion in the model to occur on 'marginal' (currently uncultivated but not forested) land largely prevents biofuel demand from affecting food prices.

As noted in the discussion on PE versus CGE models, most PE models do not explicitly model arable land markets.[37] As this modeling approach puts no limitation on total availability of arable land or allow for modeling land competition arising from new crops (i.e., second-generation biofuel feedstock crops) it is not suited for analyzing the impact of long-term, large-scale increases in bioenergy demand. However, even in models that do include land markets, this is a weak point. For instance, many of the CGE models most extensively used to assess the price and land-use impacts of biofuel demand—e.g., GTAP-BIO[16] and ENVISAGE[57]—solely model the land competition among cropland, grazing land, and production forests, not allowing the conversion of currently unmanaged land. Studies that do allow for expansion of crop and bioenergy production onto currently unmanaged land either do so by simplified assumptions regarding land-transformation elasticities based on historic data[25,55] or on top-down assessments of availability of potential cropland based on, e.g., climate, soil, and terrain data.[58,59] A recent, bottom-up assessment[60] has questioned the validity of the latter estimates of potentially available cropland, arriving at much lower numbers when accounting for various constraints and tradeoffs associated with land conversion.

DISCUSSION AND MAIN FINDINGS

Despite the recent interest in assessing the impact biofuel demand has had, and will have, on agricultural commodity markets, this review has shown that there is still considerable uncertainty around how prices are affected by biofuel demand. Three areas have been identified where effort needs to be directed to improve models and data in order to better predict both current and future impacts of biofuel demand.

First, results have clearly shown that estimated price impacts of increased biofuel demand crucially depend on the elasticities of demand and supply but that there is considerable uncertainty around the value of these parameters. Of particular concern is the limited empirical evidence for the price elasticities of supply outside of the developed regions[7] and in the long run. For example, the discussion on the responsiveness of yields to prices has focused almost solely on empirical evidence from the US corn producers (in the short run)[17,31] and, at least in the GTAP model, the resulting yield elasticity has then been applied to all crops and all regions.[16]

Given that the largest potential to increase agricultural supply—both in terms of expanding harvested area and closing yield gaps[61]—lies in developing regions, a stronger and broader empirical foundation for basing the elasticity of supply in models upon is sorely needed. To predict accurately the impacts of future, more large-scale increases in biofuel demand, we also need to understand how elasticity of demand and supply evolve over time. On the demand side, this will mainly be shaped by the level of economic growth in developing regions, as demand elasticities have been shown to decrease strongly with rising incomes.[62] For both demand and supply, policies and institutions are likely to play an important role. On the supply side, historical evidence shows a complementarity between price incentives and provision of public goods (e.g., investments in agricultural research, infrastructure, etc.), implying that if price increases take place concurrent with increases in public spending on agriculture, the supply response can be expected to be much larger.[63]

Second, the issue of price transmission in international agricultural markets appears to warrant closer attention—especially if we are to gauge accurately the poverty and land-use change impacts of future biofuel-induced increases in agricultural prices—as the distribution of price impacts across regions differs widely between the two major approaches taken in the literature today (IWM and Armington). The one study[38] that tests these competing hypotheses empirically—finding support for the Armington assumption—assesses the short-term (annual) changes in bilateral trade flows over a long-time period (1975-2002). It is likely that the price differentials predicted by the Armington model will decrease over time, especially given the rapidly increasing integration of world agricultural markets as a result of advances in information and communication technology in conjunctions with reductions in governmental distortions to agricultural production, consumption, and trade.[64] It is therefore questionable if the support for the Armington approach also holds when assessing long-term equilibrium impacts of increased biofuel demand, especially for scenarios extending decades into the future.

Third, the modeling of land markets is one of the weak points in current modeling of the biofuel-food price nexus. If we are to gauge accurately the impacts of large-scale bioenergy demand on agricultural commodity markets it is crucial to capture land competition and especially the forces (including policies[65]) that shape the future expansion of cropland into currently uncultivated land—mainly forests and so-called 'marginal' cropland. Doing so, however, requires both better data and better theories for incorporating land-transformation processes in economic models.

In order to improve our understanding of how biofuel demand impact agricultural commodity markets, what is also needed—in addition to better empirical underpinnings of models—is joint efforts to pry open the black boxes that many of the large-scale models included in this review often are treated as. Far too many studies, simply focus on the quantitative outputs (e.g., price changes) of single model runs without attempting to understand or explain the model dynamics that give rise to those results, compare outputs with empirical data, or conduct parametric and structural sensitivity analyses (there are of course exceptions; for an excellent example, see Ref 16).

Learning more about what explains differences in results also requires better communication of model results, allowing readers to appraise the validity of the model outputs and compare outcomes across studies. Many of the studies included in this review do not present key model inputs (magnitude of the biofuel demand shock) or outputs (changes in demand, cropland areas, and total production that allows for estimating implicit demand and supply elacticities).

Finally, it is worth emphasizing that while increases in biofuel demand will inevitably affect agricultural commodity markets, this review has also illustrated that policy also plays an important role in shaping those impacts. The key determinants of price impacts, the responsiveness of supply, and demand to price changes can be effected by policies, e.g., through public sector investments in agriculture that increases agricultural yields (i.e., particularly crucial where yield gaps are largest today, among, credit-constrained smallholders in developing regions lacking access infrastructure for marketing their produce[66]) or by affecting the possibilities for expanding agriculture into currently uncultivated areas. The latter, however, also stresses that we may be facing difficult tradeoffs between the environmental and socioeconomic impacts of increased biofuel demand, as restrictions on land-use change will imply larger price changes, especially for large-scale increases in global biofuel production.

CONCLUSION

This review has shown that although there is unequivocal evidence—supported by basic economic theory—that increased demand for biofuels leads to higher prices in agricultural commodity markets, there is still considerable uncertainty regarding the exact magnitude of those impacts, both in the short and long run. The share of recent increases in US corn prices that is due to increased corn ethanol demand is estimated to be in the range of 14–43%, whereas increased EU biofuel demand is estimated to have accounted for roughly a third of the vegetable oil price hike in the 2000–2008 period, although the uncertainty span for the latter is very large due to a small number of observations.

What primarily is needed in order to narrow down the uncertainties in these estimates is better empirical foundations of economic models, especially regarding the key determinants of price impacts: the responsiveness of demand and supply to changes in agricultural commodity prices. We also need better representation of land markets in economic models, especially the representation of land-use change and agricultural expansion into currently uncultivated areas, as well as a better understanding about price transmission in international markets for agricultural commodities.

ACKNOWLEDGMENTS

The preparation of this study has been supported by the Swedish Energy Agency (STEM) and by the Norden Top-level Research Initiative subprogram 'Effect Studies and Adaptation to Climate Change' through the Nordic Centre of Excellence for Strategic Adaptation Research (NORD-STAR). I am most grateful to all the authors who dug through the archives and old hard drives to provide me with data from their studies, without them this review would not have been possible. Comments from Christian Azar and two anonymous reviewers are also gratefully acknowledged.

NOTES

[a]A full derivation of this expression can be found in the technical appendix to the long version of Thomas Hertel's paper, at http://www.agecon.purdue.edu/pdf/Hertel_AAEA_Presidential_Address_LongVersion.pdf.

[b]Note, although, that modelers often choose the parameter settings in computable general equilibrium models to calibrate the implicit elasticities to empirical data (see, e.g., Refs 16 and 17).

[c]Prices taken from the US Department of Agriculture (http://www.ers.usda.gov/data-products/feed-grains-database/feed-grains-yearbook-tables.aspx) and adjusted for inflation. Data on corn ethanol production from the Renewable Fuels Association (http://www.ethanolrfa.org/pages/statistics).

[d]Price source IMF, downloaded from Index Mundi (http://www.indexmundi.com/commodities/) and adjusted for inflation. Data on biodiesel production from the European Biodiesel Board (http://www.ebb-eu.org/prev_stats_production.php).

[e]Performing a regression with the biofuel multipliers from the studies as dependent variable and the estimates from the simple model as explanatory variables gives a very good fit ($r^2 = 0.73$), with a statistical significance of the model (f-test) in the order of 10^{-54}.

[f]Note that increasing prices generally will affect *average* yields through two channels: (1) yields on existing land might increase, as a response to farmers increasing non-land inputs, and (2) production expands on new land (i.e., on native vegetation or replacing other crops) that may exhibit different yields than that on existing farmland (i.e., in well-developed agricultural areas expansion will generally occur on land less suitable for production for a given crop, displaying lower yields). The yield elasticities as estimated from the reviewed studies, as well as in the empirical estimates discussed, are the net of these two effects.

[g]Note that the aggregate price changes for all crops or all food commodities, as reported by these studies, cannot be directly compared to the changes in prices for individual commodities reported above; e.g., between 2000 and 2008, when real-world market prices for rice, wheat, and corn rose by 90–145%, aggregate food prices rose by 86%.

[h]There is also a difference between demand for bioenergy and biofu-els, as the energy conversion efficiency from biomass feedstock to biofuel may be around 50% for ethanol (for biodiesel from vegetable oils, it is, however, much higher) and the biofuel multipliers calculated above are per EJ/year of biofuel, not EJ/year of feedstock.

REFERENCES

1. World Bank. Prices (Pink Sheet): Monthly world prices of commodities and indices. 2014. Available at: http://siteresources.worldbank.org/INTPROSPECTS/Resources/334934-1304428586133/PINK_DATA.xlsx. (Accessed January 7, 2014).
2. Bryngelsson DK, Åhlen A, Azar C, Persson UM. The effect of food-price movements on African households. *Int J Agr Resour Govern Ecol* 2012, 9:121–146.
3. Ivanic M, Martin W. Implications of higher global food prices for poverty in low-income countries. *Agr Econ* 2008, 39:405–416.
4. Headey D. *Was the Global Food Crisis Really a Crisis? Simulations Versus Self-Reporting*. Washington, D.C.: International Food Policy Research Institute; 2011.
5. Ciaian P, Kancs DA. Interdependencies in the energy-bioenergy-food price systems: a cointegration analysis. *Resour Energy Econ* 2011, 33:326–348.
6. Organisation for Economic Co-operation and Development (OECD), United Nation's Food and Agriculture Organization (FAO). OECD-FAO Agricultural Outlook 2008–2017. 2008. Available at: http://www.fao.org/es/esc/common/ecg/550/en/AgOut2017E.pdf. (Accessed June 27, 2012).
7. Hertel TW, Tyner WE. Market-mediated environmental impacts of biofuels. *Global Food Security* 2013, 2:131–137.
8. Chakravorty U, Hubert M-H, N0stbakken L. Fuel versus food. *Annu Rev Resour Econ* 2009, 1:645–663.
9. Gerber N, Eckert Mv, Breuer T. The impacts of biofuel production on food prices: a review. In: *ZEF-Discussion Papers on Development Policy*, 2008.
10. Kretschmer B, Bowyer C, Buckwell A. *EU Biofuel Use and Agricultural Commodity Prices: A Review of the Evidence Base*. London & Brussels: Institute for European Environmental Policy (IEEP); 2012, 75.
11. Pfuderer S, Davies G, Mitchell I. Annex 5: The role of demand for biofuel in the agricultural commodity price spikes of 2007/08. In: *Agricultural Price Spikes 2007/2008: Causes and Policy Implications*. London: Department for Environment Food and Rural Affairs (DEFRA); 2010, 72.
12. Zilberman D, Hochman G, Rajagopal D, Sexton S, Timilsina G. The impact of biofuels on commodity food prices: assessment of findings. *Am J Agr Econ*. 2012. doi:10.1093/ajae/aas037.
13. Timilsina GR, Shrestha A. How much hope should we have for biofuels? *Energy* 2011, 36:2055–2069.
14. Moher D, Liberati A, Tetzlaff J, Altman DG, The PG. Preferred reporting items for systematic reviews and meta-analyses: the PRISMA statement. *PLoS Med* 2009, 6:e1000097.
15. Hertel TW. The G supply and demand for agricultural land in 2050: a perfect storm in the making? *Am J Agr Econ* 2011, 93:259–275.
16. Golub AA, Hertel TW. Modeling land use chang impacts of biofuels in the GTAP-BIO framework. *Clim Change Econ* 2012, 3:1250015.

17. Keeney R, Hertel TW. The indirect land use impacts of United States biofuel policies: the importance of acreage, yield, and bilateral trade responses. *Am J Agr Econ* 2009, 91:895–909.
18. Edwards R, Mulligan D, Marelli L. *Indirect Land Use Change from Increased Biofuels Demand: Comparison of Models and Results for Marginal Biofuels Production from Different Feedstocks*. Ispra: European Commission Joint Research Centre; 2010.
19. Taheripour F, Hertel TW, Tyner WE, Beckman JF, Birur DK. Biofuels and their by-products: global economic and environmental implications. *Biomass Bioenerg* 2010, 34:278–289.
20. Babcock BA. *The Impact of US Biofuel Policies on Agricultural Price Levels and Volatility. Issue Paper No. 35, International Centre for Trade and Sustainable Development (ICTSD)*, 2011, 38.
21. Babcock BA, Barr K, Carriquiry M. Costs and benefits to taxpayers, consumers, and producers from U.S. ethanol policies. *CARD Staff Report Series* 2010.
22. McPhail LL, Babcock BA. Short-run price and welfare impacts of federal ethanol policies. 2008, 40. Available at: http://www2.econ.iastate.edu/research/webpapers/paper_12943.pdf. (Accessed October 22, 2013).
23. Naylor RL, Liska AJ, Burke MB, Falcon WP, Gaskell JC, Rozelle SD, Cassman KG. The ripple effect: biofuels, food security, and the environment. *Environment* 2008, 49:30–43.
24. Wiggins S, Fioretti E, Keane J, Khwaja Y, McDonald S, Levy S, Srinivasan CS. Review of the indirect effects of biofuels: economic benefits and food insecurity. 2008. Available at: https://files.pbworks.com/download/S4hhX13P15/np-net/12639257/ODI%20%282008%29%20Review%20indirect%20effects%20biofuels-%20Economic%20benefits_and_food_insecurity.pdf?ld=1. (Accessed May 15, 2012).
25. Al-Raffai P, Dimaranan B, Laborde D. European Union and United States biofuel mandates: impacts on world markets. Technical Notes 2010, 77.
26. Timilsina GR, Shrestha A. Biofuels: markets, targets and impacts. *Policy Research Working Paper Series* 2010,49. Available at: http://ssrn.com/paper=1645735. (Accessed March 30, 2012).
27. Taheripour F, Hertel TW, Tyner WE. Implications of biofuels mandates for the global livestock industry: a computable general equilibrium analysis. *Agr Econ* 2011,42:325–342.
28. Campiche JL, Bryant HL, Richardson JW. Long-run effects of falling cellulosic ethanol production costs on the US agricultural economy. *Environ Res Lett* 2010, 5:014018.
29. Dicks MR, Campiche J. Ugarte DdlT, Hellwinckel C, Bryant HL, Richardson JW. Land use implications of expanding biofuel demand. *J Agr Appl Econ* 2009, 41:435–453.
30. Adjemian MK, Smith A. Using USDA forecasts to estimate the price flexibility of demand for agricultural commodities. *Am J Agr Econ* 2012, 94:978–995.
31. Berry ST. Biofuels policy and the empirical inputs to GTAP models. 2011. Available at: http://www.arb.ca.gov/fuels/lcfs/workgroups/ewg/010511-berry-rpt.pdf. (Accessed December 2, 2012).
32. Muhammad A, Seale Jr JL, Meade B, Regmi A. International evidence on food consumption patterns. *US Department of Agriculture, Economic Research Service, Technical Bulletin.* Bulletin number: 1929. 2011.
33. Berry ST, Schlenker W. Technical report for the ICCT: empirical evidence on crop yield elasticities. 2011, 25.
34. Huang H, Khanna M. An econometric analysis of U.S. crop yield and cropland acreage: implications for the impact of climate change. In: *Agricultural & Applied Economics Association (AAEA), CAES, and WAEA Joint Annual Meeting 2010*, 2010, 34. (Accessed December 02, 2013).
35. Hochman G, Rajagopal D, Zilberman D. The effect of biofuels on the international oil market. *Appl Econ Perspect Pol* 2011, 33:402–427.
36. Schnepf, R. Energy use in agriculture: background and issues. 2004, 44. Available at: http://www.nationalaglawcenter.org/assets/crs/RL32677.pdf. (Accessed November 27, 2012).
37. Kretschmer B, Peterson S. Integrating bioenergy into computable general equilibrium models—a survey. *Energ Econ* 2010, 32:673–686.
38. Villoria NB, Hertel TW. Geography matters: international trade patterns and the indirect land use effects of biofuels. *Am J Agric Econ* 2011, 93:919–935.
39. Hertel TW, Golub AA, Jones AD, O'Hare M, Plevin RJ, Kammen DM. Effects of US maize ethanol on global land use and greenhouse gas emissions: estimating market-mediated responses. *BioScience* 2010, 60:223–231.
40. Kretschmer B, Narita D, Peterson S. The economic effects of the EU biofuel target. *Energ Econ* 2009, 31:S285–S294.
41. Tabeau A, Banse M, Woltjer G, Meijl HV. Impact of the EU biofuels directive on the EU food supply chain. *J Food Prod Market* 2011, 17:373–385.
42. Brown ME, Tondel F, Essam T, Thorne JA, Mann BF, Leonard K, Stabler B, Eilerts G. Country and regional staple food price indices for improved identification of food insecurity. *Global Environ Chang* 2012, 22:784–794.
43. Conforti P. *Price Transmission in Selected Agricultural Markets*. Rome: Food and Agricultural Organization of the United Nations (FAO); 2004, 91.
44. Minot N. *Transmission of World Food Price Changes to Markets in Sub-Saharan Africa*. Washington, D.C.: International Food Policy Research Institute (IFPRI); 2011, 44.
45. BP. Statistical review of world energy June 2013. 2013, 48. Available at: http://www.bp.com/statisticalreview. (Accessed May 18, 2012).
46. Berndes G, Hoogwijk M, van den Broek R. The contribution of biomass in the future global energy supply: a review of 17 studies. *Biomass Bioenerg* 2003, 25:1–28.
47. Dornburg V, van Vuuren D, van de Ven G, Langeveld H, Meeusen M, Banse M, van Oorschot M, Ros J, van den Born GJ, Aiking H. Bioenergy revisited: key factors in global potentials of bioenergy. *Energ Environ Sci* 2010, 3:258–267.
48. E4Tech. Biofuels review: scenario development for the renewable fuels agency. 2008, 13. Available at: http://webarchive.nationalarchives.gov.uk/20110407094507/http://renewablefuelsagency.gov.uk/sites/renewablefuelsagency.gov.uk/files/_documents/E4tech_Scenarios_report.pdf. (Accessed January 7, 2014).

49. OECD/FAO. OECD-FAO agricultural outlook 20112020. 2011, 196. Available at: 10.1787/agr_outlook-2011-en. (Accessed January 7, 2012).
50. Johansson DJA, Azar C. A scenario based analysis of land competition between food and bioenergy production in the US. *Clim Change* 2007, 82:267–291.
51. Rosegrant MW, Zhu T, Msangi S, Sulser T. Global scenarios for biofuels: impacts and implications. *Rev Agr Econ* 2008, 30:495–505.
52. Yang J, Huang J, Qiu H, Rozelle S, Sombilla MA. Biofuels and the Greater Mekong Subregion: assessing the impact on prices, production and trade. *Appl Energ* 2009, 86:S37–S46.
53. Gillingham KT, Smith SJ, Sands RD. Impact of bioenergy crops in a carbon dioxide constrained world: an application of the MiniCAM energy-agriculture and land use model. *Mitig Adapt Strat Glob Chang* 2008, 13:675–701.
54. Popp A, Dietrich JP, Lotze-Campen H, Klein D, Bauer N, Krause M, Beringer T, Gerten D, Edenhofer O. The economic potential of bioenergy for climate change mitigation with special attention given to implications for the land system. *Environ Res Lett* 2011, 6:034017.
55. Gurgel A, Reilly JM, Paltsev S. Potential land use implications of a global biofuels industry. *J Agr Food Ind Org* 2007, 5:9.
56. Havlík P, Schneider UA, Schmid E, Böttcher H, Fritz S, Skalský R, Aoki K, Cara SD, Kindermann G, Kraxner F, et al. Global land-use implications of first and second generation biofuel targets. *Energy Policy* 2011, 39:5690–5702.
57. Timilsina GR, Beghin JC, van der Mensbrugghe D, Mevel S. The impacts of biofuels targets on land-use change and food supply: a global CGE assessment. *Agr Econ* 2012, 43:315–332.
58. Prins AG, Eickhout B, Banse M, van Meijl H, Rienks W, Woltjer G. Global impacts of european agricultural and biofuel policies. *Ecol Soc* 2011, 16:49.
59. Havlík P, Schneider UA, Schmid E, Böttcher H, Fritz S, Skalský R, Aoki K, Cara SD, Kindermann G, Kraxner F, et al. Global land-use implications of first and second generation biofuel targets. *Energy Policy* 2011, 39:5690–5702.
60. Lambin EF, Gibbs HK, Ferreira L, Grau R, Mayaux P, Meyfroidt P, Morton DC, Rudel TK, Gasparri I, Munger J. Estimating the world's potentially available cropland using a bottom-up approach. *Global Environ Chang* 2013,23:892–901.
61. Johnston M, Licker R, Foley J, Holloway T, Mueller ND, Barford C, Kucharik C. Closing the gap: global potential for increasing biofuel production through agricultural intensification. *Environ Res Lett* 2011, 6:034028.
62. Seale J, Regmi A, Bernstein J. *International Evidence on Food Consumption Patterns*. Washington, D.C.: United States Department of Agriculture (USDA); 2003, 70.
63. Schiff M, Montenegro CE. Aggregate agricultural supply response in developing countries: A survey of selected issues. *Econ Dev Cult Change* 1997, 45:393–410.
64. Anderson K. Globalization's effects on world agricultural trade, 1960–2050. *Philos T Roy Soc B* 2010, 365: 3007–3021.
65. Persson UM. Conserve or convert? Pan-tropical modeling of REDD-bioenergy competition. *Biol Conserv* 2012,146:81–88.
66. Jayne TS, Mather D, Mghenyi E. Principal challenges confronting smallholder agriculture in Sub-Saharan Africa. *World Dev* 2010, 38:1384–1398.

FURTHER READING

Azar C. Biomass for energy: a dream come true ... or a nightmare? *Wiley Interdiscip Rev Clim Change* 2011, 2:309–323.

Persson UM. Socio-economic consequences of increased biomass demand. In: Sandén B, Pettersson K, eds. System *Perspectives on Biorefineries.* Göteborg: Chalmers University of Technology, Version 2.1; 2013. Available at: http://www.chalmers.se/en/areas-of-advance/energy/cei/Pages/Systems-Perspectives.aspx. (Accessed January 15, 2014).

Smith P, Haberl H, Popp A, K-h E, Lauk C, Harper R, Tubiello FN, de Siqueira PA, Jafari M, Sohi S, et al. How much land-based greenhouse gas mitigation can be achieved without compromising food security and environmental goals? *Global Change Biol* 2013, 19:2285–2302.

31

How do Sustainability Standards Consider Biodiversity?

Oskar Englund* and Göran Berndes

Division of Physical Resource Theory, Department of Energy and Environment, Chalmers University of Technology, Gothenburg, Sweden

INTRODUCTION

Meeting the increasing demand for biomass-based products without creating unacceptable socioeconomic and environmental consequences is a great challenge.[1,2] Human societies already use roughly half the planet's land surface, producing biomass with a total energy content equivalent to about 20% of the total global net primary productivity.[3] Human land use has caused extensive land degradation and biodiversity loss, as well as emissions to air and water that contribute to, e.g., eutrophication, acidifcation, stratospheric ozone depletion, and climate change.[4–6] A growing awareness of the possible size and land use consequences of the rapidly growing bioenergy sector has prompted demands that resources and feedstocks be put to best use and that environmental and social effects of changing production systems (positive and negative) be understood as production grows.[7] Promotion of bioenergy offers considerable opportunities for the agricultural and forestry sectors, which can fnd new markets for their products and also make economic use of biomass previously considered mere waste. Further, many options exist for developing synergies among different land uses,[8,9] as well as for shaping land use systems with less impacts on land, water and biodiversity.[10–14] However, a given objective often conficts with others. Thus, the development of land use often takes place in the context of trade-offs among more or less incompatible objectives, and promoting sustainable land use realistically requires balancing objectives in terms of environmental and socioeconomic factors.[15,16]

Governance is an essential component of a sustainable energy system. Legislation and regulation as well as sustainability standards and certifcation schemes are required to guide deployment of bioenergy production systems. Bioenergy feedstock producers in the private sector, as well as governmental and nongovernmental organizations, have taken initiatives to develop criteria/indicators for sustainable biomass production. These can for instance be applied in legislation, such as the EU Renewable Energy Directive (EU-RED), and in sustainability certifcation schemes. These governance systems can help mitigate negative impacts and promote best management practices, and their use also contributes to shaping the way land is used to produce food and biomaterials.

A variety of generic sustainability certifcation schemes exist for agriculture and forestry, but they can also be crop-specifc or relate to a specifc end use of biomass, e.g., bioenergy. In addition, a number of non-operational sustainability standards exist, which are developed to guide or infuence other actors involved in developing operational standards, such as certification schemes or policymakers. Such guidelines have been developed by, e.g., International Tropical Timber Organization (ITTO), for sustainable management of tropical forests; International Federation of

*Correspondence to: oskar.englund@chalmers.se

Advances in Bioenergy: The Sustainability Challenge, First Edition. Edited by Peter D. Lund, John Byrne, Göran Berndes and Iacovos A. Vasalos.

Originally published in: *WIREs Energy Environ* 2015, 4: 483–506 doi: 10.1002/wene.118

Organic Agriculture Movements (IFOAM), for organic agriculture; and the Global Bioenergy Partnership (GBEP), for sustainable bioenergy feedstock production. Many sustainability standards exist, both mandatory and voluntary, and with varying scope. They also differ in how they prioritize different aspects of sustainability. For example, some may be very focused on the environmental performance of a production system, while others focus more on social aspects.

Studies show that there are many challenges associated with the current status of sustainability certifcation and standards, including the heterogeneity of systems.[17–21] According to noncertifed producers, main barriers include high administrative complexity, high costs, and small market advantages.[22,23] In addition, stakeholders along bioenergy supply chains may need to comply with different standards to maintain market access and to comply with legislative mandates. Consumers who try to make environmentally conscious purchasing decisions and regulatory agencies and governments involved in enforcing sustainability standards may fnd it diffcult to manage a wide range of systems that use different criteria/indicators. Thus, the proliferation of schemes and standards has lead to confusion among actors involved, market distortion and trade barriers, an increase in commodity costs, and questions about the adequacy of systems in place and how to develop systems that are effective and cost effcient.[22,24–26] A recent study undertaken to monitor the actual implementation process of sustainability certifcation of bioenergy found that there is no global/common defnition of how the sustainability concept should be translated into practice, i.e., how to measure sustainability and which criteria/indicators to use.[22] The study called for a globally harmonized approach and establishment of a common language, including terminology, to describe sustainability and how it should be verifed/documented.

Biodiversity presents a challenge for sustainability certifcation. While there is wide support for the objective to conserve biodiversity (e.g., the Convention on Biological Diversity has 193 parties and 168 signatures)[27] operationalization into guiding principles, criteria/indicators, and legislation is complicated. For example, in 2009, the EU-RED established that raw materials used for the production of biofuels and bioliquids may not be produced on land that had the status of highly biodiverse grassland in or after January 2008. At the time of writing (June 2013), the European Commission is still in the process of operationalizing elements of the biofuel sustainability criteria, including clarifying some of the requirements that need to be met with respect to the biodiversity criteria, e.g., in relation to highly biodiverse grasslands.

This article presents an assessment of how biodiversity is considered in different types of sustainability standards. First, biodiversity is defned and strategies for biodiversity conservation are discussed. Then, standards for sustainable production of biomass in agriculture and forestry are evaluated on how they consider biodiversity, i.e., how they attempt to prevent actions that can threaten biodiversity and support actions that can conserve it. It is also assessed how sustainability standards address the conversion of certain ecosystem types. A broad set of standards is included—relating to either (1) sustainable agricultural management; (2) sustainable forest management; or (3) sustainable production of biofuel feedstocks. Similarities and differences within and between the three categories are identifed. Finally, key barriers to, and challenges for, certifcation schemes are discussed and recommendations are made for further development of sustainability standards.

BACKGROUND

Biodiversity has been defned in many different ways. Often, it is considered equal to *species diversity/richness*. However, this is an oversimplifcation.[28] The Convention of Biological Diversity (CBD) and the Millennium Ecosystem Assessment (MEA) highlight the complexity of biodiversity by defning it as: 'the variability among living organisms from all sources including, *inter alia*, terrestrial, marine, and other aquatic ecosystems and the ecological complexes of which they are part; this includes diversity within species, between species and of ecosystems'.[29,30] CBD describes biodiversity by using three principal levels: ecosystems, species, and genes; the variation within these levels is described with reference to specifc concepts, e.g., at the species level, richness, abundance and function.[29] Here, biodiversity is defned using four components: *species diversity*, *ecosystem diversity*, *genetic diversity*, and *functional diversity* (Table 31.1).

Species diversity is what most people associate with the term biodiversity and also the aspect that is most commonly referred to when preservation issues are discussed.[28] *Ecosystem diversity* is, along with functional diversity, the most important factor when assessing the stability of an ecosystem at the landscape level.[30] *Genetic diversity* refers to the variation at the level of polymorphism in a population. A large genetic diversity gives a better chance to adapt to changing conditions, such as climate change. Some of the individuals in the population are in that case more likely to have allelic variants that are suited for the new conditions and will be able to pass those on to their offspring.[31] Species that contribute to biogeochemical cycles (e.g. water-, matter-, carbon-, and nitrogen cycling) in similar ways, i.e., that have similar functional traits,

Table 31.1 Definition of Biodiversity, Using Four Components

	Species Diversity	Ecosystem Diversity	Genetic Diversity	Functional Diversity
Definition	The variety of species in an ecosystem, and their relative abundance, Refs 28–32	The variety of terrestrial and aquatic ecosystems in a region, Refs 28–31, 33	The variety of genetic material within species, Refs 28–31, 34	The variety of functional traits in an ecosystem (the variety of ways an ecosystem can respond to changing conditions), Refs 28, 30, 35

are called functionally equivalent species and form functional groups.[36] An ecosystem with few functional groups and/or with few functionally equivalent species in each group will be more vulnerable to environmental change. Therefore, both the number of functional groups (i.e. the number of functional traits) and the number of functionally equivalent species in each group contribute to *functional diversity*. Fewer species in general (a decline in species diversity) may result in fewer functionally equivalent species and therefore a decline in functional diversity and higher vulnerability to environmental changes.[28,30,35]

Preservation of unmanaged areas is a common strategy for conserving biodiversity in, e.g., legislation, international agreements, and certifcation standards. However, what is considered 'natural biodiversity' is in many cases a consequence of centuries of human infuence. Cropland and pastures constitute fundamental elements in many landscapes, and the composition of species and ecosystems in such landscapes has successively changed from that of earlier stages, when there was less infuence from human activities.[29] Therefore, it is important to examine how agricultural and forest management can be shaped so as to support biodiversity conservation in *production landscapes*. Ecoagriculture research is one approach, attempting to clarify how agricultural landscapes can support a higher level of biodiversity, with neutral or even positive effects on agricultural production and livelihoods.[37] Landscape ecology,[38] countryside biogeography,[39] and GIS-based landscape modelling,[40] are other examples of disciplines and tools that can advance our understanding of interactions between managed and unmanaged areas, and the effects of different production systems on biodiversity.[37]

ASSESSING BIODIVERSITY CONSIDERATIONS IN SUSTAINABILITY STANDARDS

In order to build on the collective scientifc knowledge of sustainability standards in relation to biodiversity and ecosystem services, a review was made of 205 scientifc articles (published 2006–2012) that were identifed based on database searches.[a–d] A collection of 35 articles was considered relevant (Table 31.2). Most of these papers focused on biodiversity; no paper was found with a primary focus on ecosystem services, although in four papers it was jointly or secondarily addressed. Given that biodiversity can be defned in different ways,[28] it is notable that 16 of the 35 papers did not defne or describe biodiversity. Of the remaining papers, four discussed species diversity, one discussed genetic diversity, and the rest of the papers discussed more than one aspect of biodiversity. Furthermore, most of the relevant studies concerned forest management (23/35), primarily in boreal and temperate forests (20/35). Two studies concerned agricultural production systems, and one single study investigated both sectors. The remaining studies primarily concerned cocoa or coffee cultivation. Two papers were particularly relevant: Marjokorpi and Salo[60] benchmarked eight standards and guidelines for sustainable forest plantation management, applicable in tropical and subtropical regions, against a reference standard based on a categorization of key concepts of biodiversity. Hennenberg et al.[70] benchmarked forest and agricultural management standards, relevant for bioenergy, against each other.

On the basis of the fndings of the review, scope and methodology were developed with the ambition to produce results and insights beyond what was found in the reviewed articles. The assessment includes benchmarking a selection of sustainability standards against a benchmark standard, developed to represent a case with very high ambitions concerning biodiversity conservation, covering main agricultural and forestry related activities that infuence biodiversity. This allows for comparisons within a sector-category of standards, as well as between standards categories, i.e., between different sectors.

Selection of Sustainability Standards

Four different categories of standards were considered: (1) standards for certifcation of sustainable forest

Table 31.2 Studies Investigating Sustainability Standards in Relation to Biodiversity and/or Ecosystem Services

Source	Type of Land Management	Primary Focus (biodiversity or ecosystem services)	Secondary Focus (biodiversity or ecosystem services)	Biodiversity Aspects	Biome
Ref 41	Forest	Biodiversity	—	Unspecified	Boreal
Ref 42	Forest	Biodiversity	—	Unspecified	Global
Ref 43	Forest	—	Biodiversity	Unspecified	Tropical
Ref 44[1]					
Ref 45	Forest	Biodiversity	—	Unspecified	Boreal
Ref 46	Forest	Biodiversity	—	Unspecified	Tropical
Ref 47	Cocoa forest gardens	Biodiversity	Ecosystem services	Species diversity	Tropical
Ref 48	Forest	Biodiversity	—	Species diversity	Tropical
Ref 49	Forest	Biodiversity	—	Unspecified	Boreal
Ref 50	Coffee	—	Biodiversity	Genetic diversity	Tropical
Ref 51	Forest	—	Biodiversity	Unspecified	Boreal
Ref 52	Forest	Both	—	Unspecified	Temperate
Ref 20	Forest	—	Both	Unspecified	Global
Ref 53	Forest	—	Biodiversity	Several	Global
Ref 54	Coffee	Biodiversity	—	Several	Tropical
Ref 55	Coffee	Biodiversity	—	Several	Tropical
Ref 56	Forest	Biodiversity	—	Unspecified	Global
Ref 57	Forest	Biodiversity	—	Several	Boreal
Ref 58	Forest	Biodiversity	?[2]	?[2]	Temperate
Ref 59	Forest	?[2]	?[2]	?[2]	Temperate
Ref 60	Forest	Biodiversity	—	Several	Tropical
Ref 61	Forest	Biodiversity	—	Several	Temperate
Ref 62	Forest	Biodiversity	—	Unspecified	Global
Ref 63	Coffee	Biodiversity	—	Species diversity	Tropical
Ref 64	Forest	Biodiversity	—	Species diversity	Boreal
Ref 65	Forest	Biodiversity	—	Several	Boreal
Ref 66[1]					
Ref 67	Forest	—	Biodiversity	Several	Tropical
Ref 68	Forest	Both	—	Several	Boreal
Ref 69	Short rotation coppice	—	Biodiversity	Unspecified	Subtropical and temperate
Ref 70	Bioenergy feedstocks (forest and agriculture)	Biodiversity	Ecosystem services	Unspecified	Global
Ref 71	Agriculture (sugarcane)	—	Biodiversity	Unspecified	Subtropical
Ref 72	Agriculture	—	Biodiversity	Unspecified	Temperate
Ref 73	Coffee	Biodiversity	—	Unspecified	Tropical
Ref 74	Forest	Biodiversity	—	Several	Boreal

[1]Paper written in a language of which the authors lack translation capacity.
[2]Unknown. Full text unavailable at the time of assessment

management; (2) standards for certifcation of sustainable agricultural management; (3) standards for certifcation of sustainable production of specifc crops commonly used as biofuel feedstock; (4) standards for sustainable production of unspecifed biofuel feedstock. In addition, guidelines for development or implementation of standards that can be sorted under (1–4) were also considered. A total of 26 standards were selected for the assessment, including 11 forest management standards, 9 agriculture management standards, and 6 biofuel-related standards (Table 31.3). All selected standards include a set of principles and criteria/indicators, or the equivalent (standards often differ in their terminology), indicating each standard's requirements for production to be considered sustainable or responsible.

Table 31.3 Overview of Standards Included in the Study

Scheme/Organization	Assessed Standard	Abbreviation Used	Type[1]
Forest management			
Forest Stewardship Council	FSC Principles and Criteria For Forest Stewardship (FSC-STD-01-001 v.4-0 EN), Ref 75	FSC	1
Sustainable Forestry Initiative	Requirements For The SFI 2010–2014 Program, Ref 76[5]	SFI	1
Finnish Forest Certification System	FFCS 1002–3:2003, Criteria for Certification of Holdings of Individual Forest Owners (FFCS 2003), Ref 77[5]	FFCS	1
Malaysian Timber Certification System	MC&I(2002) and MC&I(Forest Plantations),[5] Refs 78, 79[4]	MTCS	1
Canadian Standards Association	Z809-08 Sustainable Forest Management, Ref 80[5]	CSA-SFM	1
Green Gold Label	GGLS5 – Forest Management Criteria, Ref 81	GGLS5	1
Naturland	Naturland Standards for Organic Forest Management, Ref 82	Naturland Forest	1
International Tropical Timber Organization	Revised ITTO criteria and indicators for the sustainable management of tropical forests, Ref 83	ITTO	5
African Timber Organization/ITTO	ATO/ITTO principles, criteria, and indicators for the sustainable management of African natural tropical forests, Ref 84	ATO/ITTO	5
ITTO/International Union for Conservation of Nature	ITTO/IUCN guidelines for the conservation and sustainable use of biodiversity in tropical timber production forests, Ref 85	ITTO/IUCN	5
Ministerial Conference on the Protection of Forests in Europe	(1) ANNEX 1 OF THE RESOLUTION L2: Pan-European Criteria and Indicators for Sustainable Forest Management; (2) Improved Pan-European Indicators for Sustainable Forest Management, Refs 86, 87	PEOLG	5
Agricultural management			
Global Partnership for Good Agricultural Practices	Control Points and Compliance Criteria (1) All Farms Base, (2) Crops Base, (3) Combinable Crops, Refs 88–90[4]	GLOBALGAP	2
KRAV – Swedish Organic Agriculture	Regler för KRAV-certifierad produktion januari 2011, Ref 91[4]	KRAV	2
European Union	(1) Council Regulation (EC) No 834/2007 of 28 June 2007; (2) Commission Regulation (EC) No 1235/2008 of 8 December 2008; (3) Commission Regulation (EC) No 889/2008 of 5 September 2008, Refs 92–94	EU Organic	2
United States Department of Agriculture	National Organic Program, Ref 95	USDA-NOP	2
Green Gold Label Agricultural Source	GGLS2: Agricultural Source Criteria, Ref 96	GGLS2	2
Fairtrade	Fairtrade Standard for small producer organizations, Ref 97	Fairtrade	2
Naturland	Naturland standards on Production (2011), Ref 98	Naturland production	2
International Federation of Organic Agriculture Movements	The IFOAM Norms for Organic Production and Processing, version 2005, Ref 99[2]	IFOAM	5
Sustainable Agriculture Network/Rainforest Alliance	Sustainable Agriculture Standard, version 2, Ref 100	SAN/RA	2

(Continued)

Table 31.3 Overview of Standards Included in the Study

Scheme/Organization	Assessed Standard	Abbreviation Used	Type[1]
Biofuel related			
Roundtable on Sustainable Palm Oil	RSPO Principles and Criteria for Sustainable Palm Oil Production, 2007, Ref 101[3]	RSPO	3
Roundtable on Responsible Soy	(1) RTRS Standard for Responsible Soy Production Version 1.0; (2) RTRS EU-RED Compliance Requirements for Producers, version 3.0, Refs 102, 103[3]	RTRS	3
Bonsucro	Bonsucro Production Standard Including Bonsucro EU Bonsucro Production Standard, 2011, Ref 104[3,4]	Bonsucro	3
Roundtable on Sustainable Biofuels	(1) Consolidated RSB EU-RED Principles & Criteria for Sustainable Biofuel Production; (2) Indicators of Compliance For the RSB EU-RED Principles & Criteria, Refs 105, 106[3]	RSB	4
International Sustainability & Carbon Certification	ISCC 11-03-15 V 2.3: EU Sustainability Requirements for the Production of Biomass, Ref 107[3]	ISCC	4
Greenergy	Greenergy Brazilian Bioethanol verification programme, Ref 108[3]	Greenergy	3

[1]Type 1: standards for certifcation of sustainable forest management; Type 2: standards for certifcation of sustainable agricultural management; Type 3: standards for certifcation of sustainable production of specifc crops commonly used as biofuel feedstock; Type 4: standards for sustainable production of unspecifed biofuel feedstock; Type 5: guidelines for development or implementation of standards that can be sorted under (1-4).
[2]A new standard is under development intended to be a Type 2 standard.
[3]EU Renewable Energy Directive (EU-RED) approved.
[4]Standard updated since assessment.
[5]Endorsed by PEFC (Programme for the Endorsement of Forest Certifcation, http://www.pefc.org).

Development of the Benchmark Standard

Sustainability standards should address actions that may cause biodiversity impacts as well as strategies for preventing biodiversity impacts. Agriculture and forestry systems differ concerning the extent to which associated management activities affect biodiversity, but many management practices are common to both (e.g., land clearing and preparation, planting, and harvesting). Specifc conservation strategies may also be relevant in both agriculture and forestry. Therefore, actions that negatively affect biodiversity and strategies to protect biodiversity can be formulated so as to be relevant for both agriculture and forest management. Consequently, one general biodiversity-focused benchmark standard was developed for the purpose of facilitating an assessment of the extent to which both agricultural and forestry standards consider the key elements of biodiversity conservation.

Actions that *threaten* biodiversity include (1) habitat destruction and fragmentation; (2) habitat degradation and modification; (3) overexploitation (including over-hunting, and overconsumption of resources necessary for wild species to survive); (4) introduction of nonnative (invasive) species; (5) pollution (including human induced changes in climate); and (6) human population growth (or rather, the corresponding potential increase in resource consumption).[31,109,110] It should be noted that forest fragmentation is a type of degradation process. But since the reference standard includes criteria covering *actions* rather than *effects* of actions, and since it needs to be relevant for both forestry and agriculture, the division of actions as above was judged most suitable.

Conservation biology provides principles and tools for preserving and restoring biodiversity. The general principles developed by the scientifc organization Society for Conservation Biology include nine *strategies* that need to be combined in order to conserve nature. These are: (1) protect species at risk of extinction; (2) designate ecological reserves; (3) lessen the human impact on natural systems; (4) restore ecosystems that have been degraded; (5) augment populations with individuals raised in cultivation or captivity; (6) control the number of individuals harvested in nature; (7) prevent establishment of nonnative species, and eliminate nonnative species that have become established; (8) understand and participate in the policy-making process; and (9) educate others about the importance of conservation.[109,111]

The *benchmark standard* was developed using seven *principles* based on the above described threats and strategies, under which 26 *criteria* were defned and sorted (Table 31.4). The criteria were intended to translate the broadly formulated principles into concrete actions

Table 31.4 Reference Standard Used for Benchmarking of Selected Standards (benchmark standard)

Principles	Criteria	Consideration levels: Considered	Consideration levels: Partly Considered
1. Endangered species: Threatened and endangered species shall be preserved, within and around the production area.	1.1. Threatened and endangered species within the production area shall be identified and Protected 1.2. T Threatened and endangered species around the production area shall be considered	1.1 and 1.2	1.1 only
2. Habitat destruction and fragmentation: All parts of the production chain shall be managed in such a way that the destruction and fragmentation of natural habitats is minimized.	2.1. Habitat destruction is avoided 2.2. Specific strategies for avoiding habitatfragmentation need to be applied 2.3. HCV areas need to be identified and preserved	2.1-2.3	At least one
3. Habitat degradation and modification: All chemicals and fertilizers as well as the production techniques in the entire production chain, shall be chosen in such ways that they do not contribute to the degradation and modification of habitats around the production area or the alteration of functional diversity in and around the production area.	3.1. The use of chemicals for pest management (pesticides) is restricted, to avoid substances that can be harmful for untargeted species. A list of prohibited chemicals should be provided. 3.2. Guidance for pesticide application is provided, to avoid contamination of surrounding terrestrial and aquatic ecosystems 3.3. Guidance for fertilization is provided, to minimize nutrient leaching to surrounding terrestrial and aquatic ecosystems 3.4. Buffer zones are required to protect watercourses 3.5. Water resources protected 3.6. Soil erosion prevented 3.7. Soil quality maintained 3.8. Waste management applied 3.9. Recycling applied	At least six of 3.1-3.7	At least four of 3.1–3.7
4. Overexploitation: The production system shall be managed sustainably over time. Overharvesting of production species shall be avoided and soil fertility maintained.	4.1. Long-term sustainability considered 4.2. Sustainable harvest rates identified and applied 4.3. Crop-rotation applied	All relevant[1]	At least one relevant[1]
5. Invasive species and GMOs: Exotic species or GMOs shall not be used, in order to avoid disturbing natural ecosystems and the genetic diversity in populations of native species.	5.1. Native species preferred over exotic 5.2. Measures taken to prevent introduction of invasive species 5.3. GMOs prohibited	5.1-3	At least one
6. Energy use and GHG emissions: Net GHG emissions from biomass production shall be neutral or negative. Therefore, energy use needs to be minimized, fossil fuels avoided and the natural carbon stock maintained or enhanced.	6.1. Energy use minimized 6.2. Fossil energy avoided[2] 6.3. Carbon stock maintained or enhanced	6.1-3	At least one
7. Research, awareness, and education: Improvements in biodiversity conservation are dependent on continuous research and increased environmental awareness among consumers and people living in or near sensitive ecosystems. Proper education for workers is key for successful implementation of certification criteria.	7.1. Research supported7.2. Awareness spread 7.3. Education to workers provided	7.1-3	At least one

GHG, Greenhouse gas; GMO, Genetically modifed organism; HCV, high conservation value.
[1] '4.2 Sustainable harvest rates' relevant for forest management; '4.3 crop rotation' relevant for agricultural management.
[2] Should not be interpreted as fossil fuels are prohibited—but that producers are encouraged to avoid them.

applicable to both agriculture and forest management. The development of the benchmark criteria was supported by the literature review and a pre-assessment of the standards selected for assessment, in order to ensure a compatible format. This pre-assessment also ensured that all biodiversity related principles, criteria/indicators, and potential guidance for compliance in the assessed standards were covered by suitable benchmark criteria.

The selected standards were individually compared with the benchmark standard and for each benchmark criterion it was determined whether a specifc standard was *compliant* or not. In addition, based on the specifcs of compliance with the criteria sorted under a given principle, each standard was given a numerical value for that principle, refecting the degree to which the standard *considers the principle* (0: principle disregarded; 1: principle considered in part; 2: principle fully considered, cf. Table 31.4). The overall biodiversity *stringency* of a given standard was fnally obtained by summing the compliance values given for all seven principles, for that standard. Standards reaching a sum of 10 or more were classifed as *Stringent*, and standards reaching a sum of six or less were classifed as *Unstringent*.

Assessment of Ecosystem Conversion

Given that land conversion may induce adverse effects on biodiversity, it was investigated how the standards addressed conversion of certain types of ecosystems, namely: (1) tropical and subtropical forests; (2) temperate forests; (3) boreal forests; (4) wetlands; (5) grass-, shrub- and woodlands; and (6) degraded land. A pre-assessment was performed for the purpose of identifying how the standards address conversion of natural ecosystems, in general terms. It showed that the assessed standards address land conversion in four principal ways: *protecting* (no production allowed); *restricting* (production is allowed, provided that properties of ecosystems are not altered); *encouraging* (production on certain lands is encouraged); and *requiring a high conservation value* (HCV) *assessment* (*restricting* production on, or fully *protecting*, HCV land). All standards were then assessed to clarify how they address conversion of the ecosystem types given above (i.e., by either protecting, restricting, encouraging, or requiring an HCV assessment).

ASSESSMENT OUTCOME

Below, the degree of compliance with criteria, the overall *consideration* of principles, and the *stringency* of the assessed standards are presented. The results for the three categories of standards are then compared. Finally, the results of the ecosystem conversion assessment are presented.

Standards for Sustainable Forest Management

On average, the 11 assessed forestry standards comply with 60% of the relevant benchmark criteria, complying with 8 (Green Gold Label S5; GGLS5) to 19 (Naturland) of the 25 criteria (Table 31.5).

Benchmark criteria that were *well considered* overall by these standards (i.e., criteria where at most one of the assessed standards failed to reach compliance) include those related to (1) endangered species within the production area; (2) habitat destruction; (3) HCV areas; (4) water resources; (5) erosion; (6) soil quality; and (7) long-term sustainability.

Fewer than half of the assessed forestry standards reached compliance with criteria related to: (1) pesticide application; (2) fertilization (3) waste management; (4) recycling; (5) invasive species; (6) GMOs; (7) energy use; (8) fossil energy; (9) research; and (10) awareness.

Four principles are *poorly considered* overall by the assessed forestry standards. Only one standard fully complies with *Energy use and GHG emissions*; only two standards fully comply with *Habitat degradation and modification* and *Invasive species and GMOs*; and only three standards fully comply with *Research, awareness, and education*. Overall, the standards comply well with the principles *Endangered species*, *Habitat destruction and fragmentation*, and *Overexploitation* (Table 31.6).

The results indicate that Forest Stewardship Council (FSC), Sustainable Forestry Initiative (SFI), Malaysian Timber Certifcation System (MTCS), Naturland, and International Tropical Timber Organization and International Union for Conservation of Nature (ITTO/IUCN) are *Stringent*, from a biodiversity perspective. GGLS5 and PEOLG are *Unstringent* (Tables 31.5 and 31.6).

Standards for Sustainable Agriculture

On average, the nine assessed agricultural standards comply with 61% of the benchmark criteria, complying with 7 (Green Gold Label S2; GGLS2) to 23 (Fairtrade) of the 25 criteria (Table 31.7).

Benchmark criteria that were *well considered* overall by these standards (i.e., criteria where at most one of the assessed standards failed to reach compliance) include those related to (1) habitat destruction; (2) pesticide use; (3) fertilization; (4) water resources; (5) erosion; (6) soil quality; (7) waste management; (8) long-term sustainability; and (9) crop rotation.

Fewer than half of the assessed standards reached compliance with criteria related to: (1) endangered species within the production area; (2) endangered species around the production area; (3) habitat fragmentation; (4) native

Table 31.5 Forestry Standards: Compliance with Benchmark Criteria

Principles	Criteria	FSC	SFI	FFCS	MTCS	CSA-SFM	GGLSB	Naturland	ITTO	ATO/ITTO	ITTO/IUCN	PEOLG
1. Endangered species	Endangered species within the production area protected	•	•	•	•	•	•	•	•	•	•	•
	Endangered species around the production area considered		•	•	•			•	•		•	
2. Habitat destruction and fragmentation	Habitat destruction avoided	•	•	•	•	•	•	•	•	•	•	•
	Habitat fragmentation avoided	•		•	•	•		•	•	•	•	
	HCV areas identified and protected	•	•	•	•	•	•	•	•	•	•	•
3. Habitat degradation and modification	Pesticide use restricted	•	•	•	•			•		•	•	
	Guidance for pesticide application provided		•	•					•			
	Guidance for fertilization provided to avoid nutrient leaching			•				•				
	Buffer zones required	•	•	•	•			•	•	•	•	
	Water resources protected ·	•	•	•	•	•	•	•	•	•	•	•
	Soil erosion prevented	•	•		•	•	•	•	•	•	•	•
	Soil quality maintained	•	•		•	•	•	•	•	•	•	•
	Waste management required	•	•		•					•		•
	Recycling required			•	•							
4. Overexploitation	Long-term sustainability considered	•	•	•	•	•	•	•	•	•	•	•
	Sustainable harvest rates identified and applied	•		•	•	•	•	•	•	•	•	
	Crop-rotation applied N/R	N/R	N/R	N/R	N/R	N/R	N/R	N/R	N/R	N/R	N/R	
5. Invasive species and GMOs	Native species preferred over exotic	•	•	•	•	•		•		•	•	
	Measures taken to prevent introduction of invasive species	•	•		•	•		•			•	
	GMOs prohibited	•				•		•		•		
6. Energy use and GHG	Energy use minimized							•				
	Fossil energy avoided							•				
	Carbon stock maintained or enhanced	•	•		•	•		•	•	•		•
7. Research, awareness, and education	Research supported		•								•	•
	Awareness spread		•	•							•	•
	Education to workers provided	•	•		•	•				•	•	•

'•', compliance; ATO/ITTO, African Timber Organization and International Tropical Timber Organization; CSA-SFM, Canadian Standards Association and Sustainable Forest Management; FSC, Forest Stewardship Council; FFCS, Finnish Forest Certification System; GGLS5, Green Gold Label S5; GMO, xxx; HCV, high conservation value; ITTO, International Tropical Timber Organization; MTCS, Malaysian Timber Certification System; PEOLG, Pan European Operational Level Guidelines; SFI, Sustainable Forestry Initiative.

Table 31.6 Forestry Standards: Compliance with Benchmark Principles

Principles	FSC	SFI	FFCS	MTCS	CSA-SFM	GGLS5	Naturland	ITTO	ATO/ITTO	ITTO/IUCN	PEOLG
1. Endangered species	+/–	+	+	+	+/–	+/–	+	+	+/–	+	+/–
2. Habitat destruction and fragmentation	+	+/–	+	+	+	+/–	+	+	+	+	+/–
3. Habitat degradation and modification	+/–	+	+/–	+/–	–	–	+	+/–	+/–	+/–	–
4. Overexploitation	+	+/–	+	+	+	+	+	+	+	+	+/–
5. Invasive species and GMOs	+	+/–	+/–	+/–	+	–	+	–	+/–	+/–	–
6. Energy use and GHG	+/–	+/–	–	+/–	+/–	+/–	+	+/–	+/–	–	+/–
7. Research, awareness, and education	+/–	+	+/–	+/–	+/–	–	–	–	+/–	+	+

ATO/ITTO, African Timber Organization and International Tropical Timber Organization; CSA-SFM, Canadian Standards Association and Sustainable Forest Management; FSC, Forest Stewardship Council; FFCS, Finnish Forest Certifcation System; GGLS5, Green Gold Label; GHG, xxx; GMO, xxx; ITTO, International Tropical Timber Organization; MTCS, Malaysian Timber Certifcation System; PEOLG, xxx; SFI, Sustainable Forestry Initiative. Green color (+) indicates considered; yellow color (+/-) indicates partly considered; orange color (-) indicates disregarded.

species; (5) invasive species; (6) energy use; (7) fossil energy; (8) carbon stock; (9) research; and (10) education.

Four principles are *poorly considered* overall by the assessed agricultural standards. No standard fully complies with *Invasive species and GMOs*, and only two standards comply fully with *Endangered species*, *Energy use*, and *GHG emissions* and *Research, awareness, and education*, respectively. Overall, the standards comply well with *Overexploitation* and *Habitat degradation and modification*, and the same can be seen for *Habitat destruction and fragmentation*, although to a lesser extent (Table 31.8).

The results indicate that Fairtrade, Sustainable Agriculture Network/Rainforest Alliance (SAN/RA), Naturland, and KRAV are *Stringent*, from a biodiversity perspective. Global Partnership for Good Agricultural Practices (GLOBALGAP), European Union Organic (EU Organic), National Organic Program (NOP), and GGLS2 are *Unstringent* (Tables 31.7 and 31.8).

Biofuel-Related Sustainability Standards

On average, the assessed biofuel-related standards comply with 72% of the benchmark criteria, complying with 13 (International Sustainability & Carbon Certifcation; ISCC) to 21 (Roundtable on Responsible Soy; RTRS) of the 25 criteria (Table 31.9).

Benchmark criteria that were *well considered* overall by these standards (i.e., criteria where at most one of the assessed standards failed to reach compliance) include those related to (1) endangered species within the production area; (2) habitat destruction; (3) HCV areas; (4) pesticide use; (5) pesticide application; (6) buffer zones; (7) water resources; (8) erosion; (9) soil quality; waste management; (10) long-term sustainability; (11) invasive species; and (12) carbon stock.

Fewer than half of the assessed standards reached compliance with criteria related to: (1) native species; (2) GMOs; (3) fossil energy; (4) research; and (5) education.

Two principles are *poorly considered* overall by the assessed biofuel-related standards. No standard fully complies with *Invasive species and GMOs* or with *Research, awareness, and* education. Overall, the standards comply well with *Endangered species*, *Habitat destruction, and fragmentation*, *Habitat degradation and modification* and *Overexploitation* (Table 31.10).

The results indicate that Roundtable on Sustainable Palm Oil (RSPO), RTRS, Bonsucro, and Greenergy are *Stringent*, from a biodiversity perspective. ISCC is *Unstringent* (Table 31.9 and 31.10).

Comparison of Outcomes for the Three Categories of Standards

The assessed biofuel-related standards reached the highest level of compliance, complying on average with 72% of the benchmark criteria, compared to 61% for the agricultural standards and 60% for the forestry standards. There are large variations between the standards categories regarding compliance with some benchmark criteria (Table 31.11, Figure 31.1). In some cases (e.g., guidance for fertilization), this can be explained by the differences

Table 31.7 Agricultural Standards: Compliance with Benchmark Criteria

Principles	Criteria	GLOBALGAP	KRAV	EU Organic	NOP	GGLS2	Fairtrade	Naturland	IFOAM	SAN/RA
1. Endangered species	Endangered species within the production area protected		•				•	•		•
	Endangered species around the production area considered		•				•			
2. Habitat destruction and fragmentation	Habitat destruction avoided	•	•	•	•		•	•	•	•
	Habitat fragmentation avoided						•	•	•	•
	HCV areas identified and protected	•	•				•	•	•	•
3. Habitat degradation and modification	Pesticide use restricted	•	•	•	•	•	•	•	•	•
	Guidance for pesticide application provided		•	•	•		•	•	•	
	Guidance for fertilization provided to avoid nutrient leaching	•	•	•	•		•	•	•	•
	Buffer zones required		•				•	•	•	•
	Water resources protected	•	•		•	•	•	•	•	•
	Soil erosion prevented	•	•	•	•	•	•	•	•	•
	Soil quality maintained	•	•	•	•	•	•	•	•	•
	Waste management required	•	•	•	•	•	•	•	•	•
	Recycling required	•	•				•	•		•
4. Overexploitation	Long-term sustainability considered	•	•	•	•	•	•	•	•	•
	Sustainable harvest rates identified and applied	N/R	N/R	N/R	N/R	N/R	N/R	N/R	N/R	N/R
	Crop-rotation applied	•	•	•	•		•	•	•	•
5. Invasive species and GMOs	Native species preferred over exotic							•		
	Measures taken to prevent introduction of invasive species						•			
	GMOs prohibited		•	•			•	•	•	•
6. Energy use and GHG	Energy use minimized	•	•				•			•
	Fossil energy avoided	•	•				•			•
	Carbon stock maintained or enhanced		•					•	•	•
7. Research, awareness and education	Research supported						•			•
	Awareness spread	•				•	•	•		•
	Education to workers provided						•			•

'•', compliance; EU Organic, European Union Organic; GLOBALGAP, Global Partnership for Good Agricultural Practices; GGLS2, Green Gold Label S2; GHG, xxx; GMO, xxx; HCV, high conservation value; IFOAM, International Federation of Organic Agriculture Movements; KRAV, xxx; NOP, National Organic Program; SAN/RA, Sustainable Agriculture Network/Rainforest Alliance.

Table 31.8 Agricultural Standards: Compliance with Benchmark Principles

Principles	GLOBALGAP	KRAV	EU Organic	NOP	GGLS2	Fairtrade	Naturland	IFOAM	SAN/RA
1. Endangered species	–	+	–	–	–	+	+/–	–	+/–
2. Habitat destruction and fragmentation	+/–	+/–	+/–	+/–	–	+	+	+	+
3. Habitat degradation and modification	+/–	+	+/–		+	+	+	+	+
4. Overexploitation	+	+	+		+/–	+	+	+	+
5. Invasive species and GMOs	–	+/–	+/–	–	–	+/–	+/–	+/–	+/–
6. Energy use and GHG	+/–	+	–	–	–	+/–	+/–	+/–	+
7. Research, awareness, and education	+/–	–	–	–	+/–	+	+/–	–	+

EU Organic, European Union Organic; GLOBALGAP, Global Partnership for Good Agricultural Practices; GGLS2, Green Gold Label S2; GHG, xxx; GMO, xxx; IFOAM, International Federation of Organic Agriculture Movements; KRAV, xxx; NOP, National Organic Program; SAN/RA, Sustainable Agriculture Network/Rainforest Alliance. Green color (+) indicates considered; yellow color (+/–) indicates partly considered; orange color (–) indicates disregarded.

between forest management and agricultural management. In other cases (e.g., preferred use of native species), the reason is less clear. Criteria that were *well considered* across all three standards categories include those related to (1) habitat destruction; (2) water resources; (3) soil erosion; (4) soil quality; and (5) long-term sustainability. Overall *poorly considered* criteria include those related to (1) fossil energy and (2) support for research.

Table 31.9 Biofuel-related Standards: Compliance with Benchmark Criteria

Principles	Criteria	RSPO	RTRS	Bonsucro	RSB	ISCC	Greenergy
1. Endangered species	Endangered species within the production area protected	•	•	•	•		•
	Endangered species around the production area considered	•	•	•			•
2. Habitat destruction and fragmentation	Habitat destruction avoided	•	•	•	•	•	•
	Habitat fragmentation avoided	•	•	•	•		
	HCV areas identified and protected	•	•	•	•	•	•
3. Habitat degradation and modification	Pesticide use restricted	•	•	•	•	•	•
	Guidance for pesticide application provided	•	•	•	•		•
	Guidance for fertilization provided to avoid nutrient leaching		•	•		•	
	Buffer zones required	•	•	•	•	•	•
	Water resources protected	•	•	•	•	•	•
	Soil erosion prevented	•	•	•	•	•	•
	Soil quality maintained	•	•	•	•	•	•
	Waste management required	•	•	•	•	•	•
	Recycling required	•	•	•		•	
4. Overexploitation	Long-term sustainability considered	•	•	•	•	•	•
	Sustainable harvest rates identified and applied	N/R	N/R	N/R	•		N/R
	Crop-rotation applied	N/R	•		•		•
5. Invasive species and GMOs	Native species preferred over exotic	•			•		
	Measures taken to prevent introduction of invasive species GMOs prohibited	•	•	•	•		•
6. Energy use and GHG	Energy use minimized	•	•	•	•		
	Fossil energy avoided	•	•				
	Carbon stock maintained or enhanced	•	•	•	•	•	•
7. Research, awareness, and education	Research supported			•			
	Awareness spread	•	•	•			
	Education to workers provided				•	•	

'•', compliance; GHG, xxx; GMO, xxx; HCV, high conservation value; ISCC, International Sustainability & Carbon Certifcation; RSPO, Roundtable on Sustainable Palm Oil; RTRS, Roundtable on Responsible Soy; RSB, Roundtable on Sustainable Biofuels.

Table 31.10 Biofuel-Related Standards: Compliance with Benchmark Principles

Principles	RSPO	RTRS	Bonsucro	RSB	ISCC	Greenergy
1. Endangered species	+	+	+	+/–	–	+
2. Habitat destruction and fragmentation	+	+	+	+	+/–	+/–
3. Habitat degradation and modification	+	+	+	+	+	+
4. Overexploitation	+	+	+/–	+	+/–	+
5. Invasive species and GMOs	+/–	+/–	+/–	+/–	–	+/–
6. Energy use and GHG	+	+	+/–	+/–	+/–	+/–
7. Research, awareness, and education	+/–	+/–	+/–	–	+/–	+/–

GHG, xxx; GMO, xxx; ISCC, International Sustainability & Carbon Certifcation; RSPO, Roundtable on Sustainable Palm Oil; RTRS, Roundtable on Responsible Soy; RSB, Roundtable on Sustainable Biofuels. Green color (+) indicates considered; yellow color (+/–) indicates partly considered; orange color (–) indicates disregarded.

Further, the forestry and biofuel-related standards are generally more stringent than the agricultural standards regarding endangered species, both within and around the production area. The same result was found for habitat fragmentation and requirements for HCV assessments. This is reasonable for forestry standards, since forests subject to some degree of management typically host natural ecosystems to a larger degree than agriculture. It was less expected that biofuel-related standards would differ from agricultural standards, since most of the assessed biofuel-related

Table 31.11 Commonly Well Considered Criteria and Commonly Nonconsidered Criteria, for the Three Standard Types, Respectively

Principles	Criteria	Forest	Agriculture	Biofuels
1. Endangered species	Endangered species within the production area protected	+	–	+
	Endangered species around the production area considered		–	
2. Habitat destruction and fragmentation	Habitat destruction avoided	+	+	+
	Habitat fragmentation avoided		–	
	HCV areas identified and protected	+		
3. Habitat degradation and modification	Pesticide use restricted		+	+
	Guidance for pesticide application provided	–		+
	Guidance for fertilization provided to avoid nutrient leaching	–	+	
	Buffer zones required			+
	Water resources protected	+	+	+
	Soil erosion prevented	+	+	+
	Soil quality maintained	+	+	+
	Waste management required	–	+	+
	Recycling required	–		
4. Overexploitation	Long-term sustainability considered	+	+	+
	Sustainable harvest rates identified and applied		N/R	
	Crop-rotation applied	N/R	+	
5. Invasive species and GMOs	Native species preferred over exotic		–	–
	Measures taken to prevent introduction of invasive species		–	+
	GMOs prohibited	–		–
6. Energy use and GHG	Energy use minimized	–	–	
	Fossil energy avoided	–	–	–
	Carbon stock maintained or enhanced		–	+
7. Research, awareness, and education	Research supported	–	–	–
	Awareness spread	–		
	Education to workers provided		–	–

'+', benchmark criteria complied with by all the assessed standards, with one exception allowed; '–', benchmark criteria complied with by less than 50% of the assessed standards; GHG, xxx; GMO, xxx.

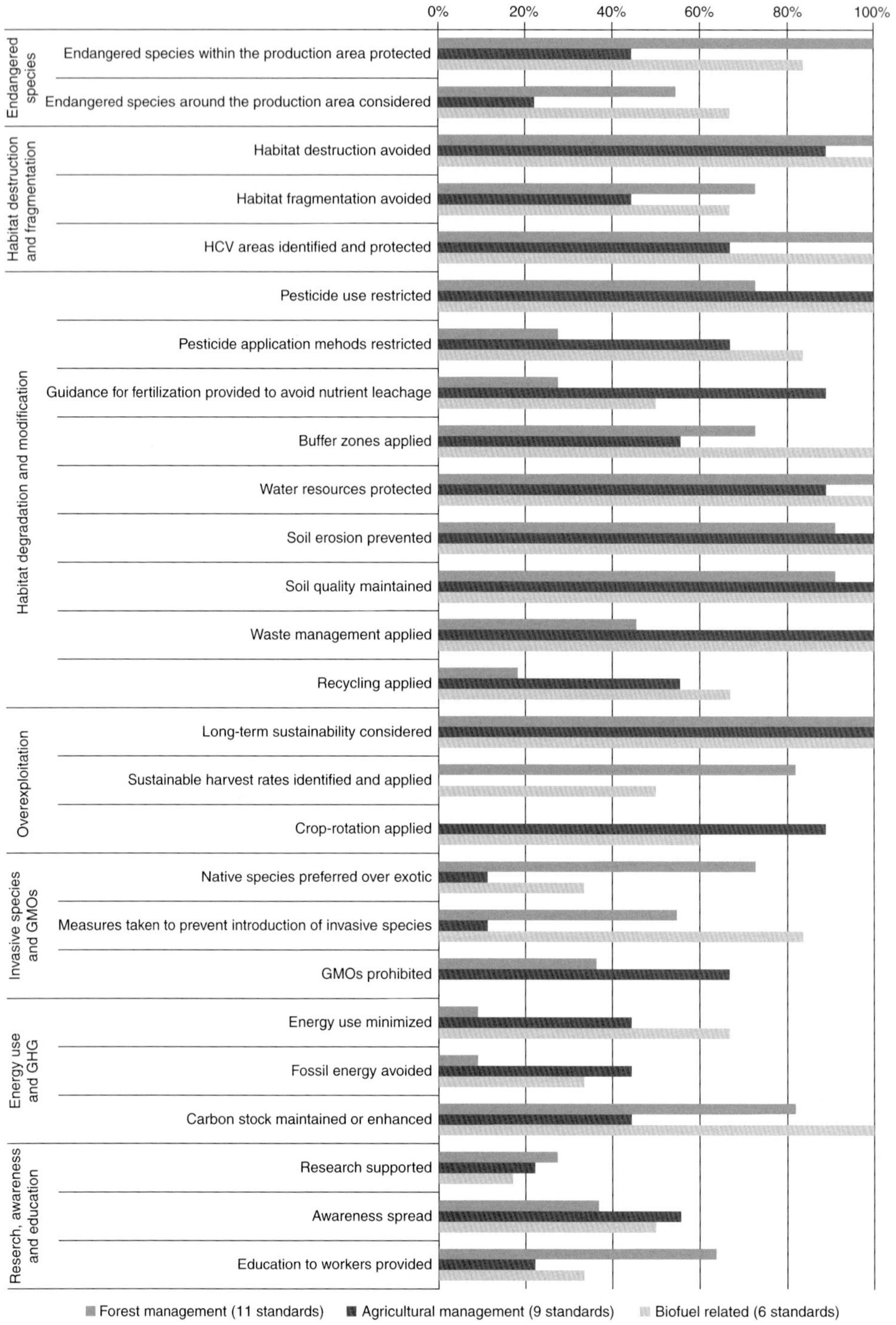

Figure 31.1 Percentage of benchmark criteria complied with by the three standard types, respectively.

standards refer to agriculture to a higher degree than to forestry.

The agricultural and biofuel-related standards are somewhat more stringent than the forestry standards concerning the use of pesticides. This is reasonable since pesticides are more commonly used in agriculture than in forestry. In addition, since agricultural products are used for human consumption, pesticides in agriculture may be of higher concern than in forestry. Guidance for pesticide application is more common in the biofuel-related standards than in the agricultural standards, even though several of the biofuel-related standards are not explicitly aimed at agricultural production.

All biofuel-related standards require buffer zones, while the forestry and agricultural standards are less stringent in that regard. Requirements of waste management and recycling are more common in the agricultural and biofuel-related standards than in the forestry standards. This is reasonable since agriculture is more intensive than forestry and thereby uses more material and inputs, and thus causes more waste. All but one of the biofuel-related standards, about half of the forestry standards, and only one of the agricultural standards, require measures to prevent the introduction of invasive species.

Energy use and biospheric carbon stocks are considered to a higher degree in the biofuel-related standards than in the forestry and agricultural standards. Given that GHG emissions reduction should be an important objective in land use in general, this can be considered a weakness in forestry and agricultural standards. Perhaps surprisingly, the biofuel-related standards typically do not promote renewable energy. Spreading awareness is considered to a higher degree in the agricultural and biofuel-related standards than in the forestry standards, while the opposite is the case for education and training of workers.

Habitat destruction and fragmentation and *Overexploitation* are principles *well considered* by the assessed forestry standards. *Poorly considered* principles include *Energy use and GHG* and *Research, awareness, and education* (Figure 31.2). *Habitat degradation and modification* and *Overexploitation* are principles *well considered* by the assessed agricultural standards. *Poorly considered* principles include *Endangered species*, *Invasive species and GMOs*, *Energy use and GHG* and *Research, awareness, and education* (Figure 31.2). *Habitat destruction and fragmentation* and *Habitat degradation and modification* are principles *well considered* by the assessed biofuel-related standards. *Poorly considered* principles include *Invasive species and GMOs* and *Research, awareness, and education* (Figure 31.2).

Ecosystem Conversion Assessment

The ways the assessed standards address conversion of seven main ecosystem types is presented in Table 31.12. All of the eleven forestry standards address ecosystem conversion. For forested land, conversion is typically regulated by requiring that *HCV* areas be identifed and protected. Exceptions include Naturland (which *restricts* conversion of forested land), and GGLS5 (which *protects* forested land from conversion into plantation forest). Among the assessed forestry standards, only the ITTO/IUCN standard covers a nonforest ecosystem (wetlands) in its required HCV assessment—the other standards cover forest ecosystems only. In addition, few of the standards address conversion of nonforested land by other means than requiring HCV assessments: Wetlands are *protected* by SFI and Naturland, and conversion is *restricted* by FFCS and PEOLG. Conversion of grass, shrub-, and woodlands is *encouraged* by PEOLG, while Naturland *protects* such ecosystem types. The remaining forestry standards do not consider nonforest ecosystems. Thus, in some cases, it may be possible to convert highly biodiverse grasslands or wetlands into certifed plantation forests.

Six of the nine agricultural standards address ecosystem conversion (exceptions: EU Organic, NOP, and GGLS2). Contrary to the forestry standards, the agricultural standards that address ecosystem conversion typically do not rely solely on HCV assessments for identifying no-go areas (exception: Fairtrade). Instead, they require that all natural vegetation remain unmanaged. Thus, in some cases (i.e., EU Organic, NOP, and GGLS2), farmers may be able to convert any ecosystem type into certifed cropland, while in other cases (i.e., all the other assessed agricultural standards), farmers are basically not able to convert any type of unmanaged land into certifed cropland.

All six biofuel-related standards consider ecosystem conversion. While the forestry standards rely on HCV assessments and focus primarily on forested lands, and agricultural standards either protect everything or nothing, the biofuel-related standards consider all the ecosystem types, either by requiring protection or a combination of protection and HCV assessments. RSPO, e.g., protects forested land and wetlands, requires an HCV assessment for grasslands, and encourages conversion of degraded lands. This is likely a direct effect of the sustainability requirements in EU-RED, which state that forested lands, wetlands, and highly biodiverse grasslands are no-go areas for biofuel feedstock production, if the products are to be sold on the EU-RED market. Since all six biofuel-related standards are RED-approved, areas that are no-go according to RED are also no-go areas in these standards.

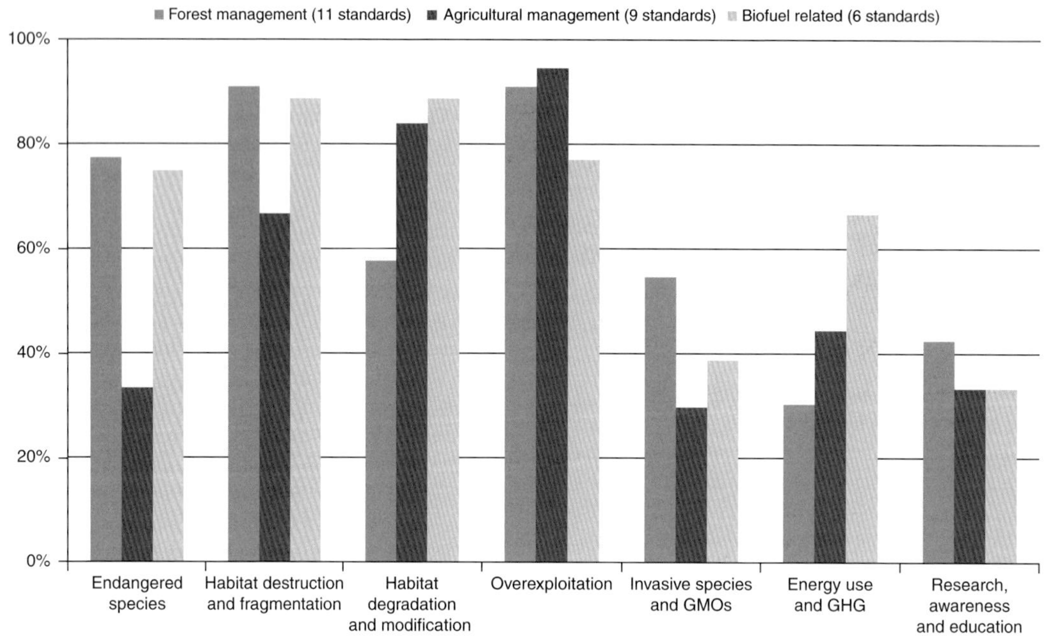

Figure 31.2 Percentage of benchmark criteria complied with on a principle level for the three standard types, respectively. Well considered principles are those with >80% criteria complied with in total; Poorly considered principles are those with <50% criteria complied with in total.

Overall, fve standards encourage restoration of degraded land; three forestry standards (ATO/ITTO, ITTO/IUCN and PEOLG) and two biofuel-related standards (RSPO and RTRS). Four standards protect degraded lands in different ways; one forest standard (Naturland) and three agricultural standards (GLOB-ALGAP, IFOAM, and SAN/RA). Only one of the fve assessed RED-approved standards (RTRS) encourages restoration of degraded land, even though the latter measure is included in RED as an option for earning GHG emissions savings bonuses. This may be explained by the fact that it is not yet possible to earn such bonuses, due to the delay in developing a defnition of 'degraded land'.

SUMMARY AND DISCUSSION OF ASSESSMENT OUTCOME

In summary, the assessed biofuel-related standards had the highest level of compliance with the benchmark standard, complying on average with 72% of the benchmark criteria, compared to 61% for the agricultural standards and 60% for the forestry standards. Fairtrade and SAN/RA (agriculture), and RSPO and RTRS (biofuel) were the most stringent, while GGLS5 and PEOLG (forest), GLOBALGAP, EU Organic, NOP, and GGLS2 (agriculture), and ISCC (biofuel) were the least stringent.

In general, the assessed standards consider *habitat destruction*, *-fragmentation*, *-degradation*, *-modification,* and *overexploitation* well, while *invasive species and GMOs, research, awareness, and education*, and *Energy use and GHG* are poorly considered.

There are notably large differences in stringency between some standards having a similar scope. For example, IFOAM, which sets the 'norms' for organic agriculture, is signifcantly more stringent than either EU Organic or NOP. In addition, KRAV endorses EU Organic, even though KRAV classifes as *Stringent* and EU Organic as *Unstringent*. Further, the SFI standard, which is a forest industry initiative, shows similar stringency as the FSC standard, which is often regarded as more thorough in its coverage of ecological issues.[51] Furthermore, the high stringency in the Fairtrade standard, and to some extent also SAN/RA, was unexpected, as these are perceived to primarily focus on social aspects.

Some of the differences in stringency can be explained by differences concerning biophysical properties or legal contexts in countries or regions in which the standards originate or for which they are intended. For example, standards targeting production in countries that have less

Table 31.12 Restrictions in Conversion of Six Ecosystem Types, in the Assessed Standards

	Tropical and Subtropical Forests	Temperate Forests	Boreal Forests	Grass-, Shrub- and Woodlands	Wetlands	Degraded Land
Forest management						
FSC	HCV	HCV	HCV			
SFI	HCV	HCV	HCV		P	
FFCS	N/R	HCV	HCV		R	
MTCS	HCV	N/R	N/R			
CSA-SFM	N/R	HCV	HCV			
GGLS5	P[1]	P[1]	P[1]			
Naturland	R	R	R	P	P	R
ITTO	HCV	N/R	N/R			
ATO/ITTO	HCV	N/R	N/R			E
ITTO/IUCN	HCV	N/R	N/R		HCV	E
PEOLG	N/R	HCV	HCV	E	R	E
Agricultural management						
GLOBALGAP				P[2]	P[2]	P[2]
KRAV	P	P	P	P	P	
EU Organic						
NOP						
GGLS2						
Fairtrade	HCV	N/R	N/R	HCV	HCV	
Naturland	P	P	P	P	P	
IFOAM	P	P	P	P	P	P
SAN/RA	P	P	P	P	P	P
Biofuel related						
RSPO	P	N/R	N/R	HCV	P	E
RTRS	P	P	N/R	HCV P[3]	P	E[3]
Bonsucro	HCV P[4]	N/R	N/R	HCV P[4]	HCV P[4]	
RSB	P	P	P	P	P	
ISCC	P	P	P	P	P	
Greenergy	P	N/R	N/R	P	P	

ATO/ITTO, African Timber Organization and International Tropical Timber Organization; CSA-SFM, Canadian Standards Association and Sustainable Forest Management; EU Organic, European Union Organic; FSC, Forest Stewardship Council; FFCS, Finnish Forest Certifcation System; GGLS5, Green Gold Label S5; GGLS2, Green Gold Label S2; GLOBALGAP, Global Partnership for Good Agricultural Practices; HCV, high conservation value; IITO/IUCN, International Tropical Timber Organization and International Union for Conservation of Nature; ISCC, International Sustainability & Carbon Certifcation; KRAV, xxx; MTCS, Malaysian Timber Certifcation System; PEOLG, xxx; RSPO, Roundtable on Sustainable Palm Oil; RTRS, Roundtable on Responsible Soy; RSB, Roundtable on Sustainable Biofuels; SAN/RA, Sustainable Agriculture Network/Rainforest Alliance; SFI, Sustainable Forestry Initiative. Besides for 'degraded land', all ecosystem types refer to unmanaged areas. 'P' indicates 'protected'; 'R' indicates 'restricted'; 'E' indicates ''encouraged; 'HCV' indicates that conversion can be allowed, provided that the area is not regarded as HCV; 'N/R' indicates 'not relevant', i.e., the ecosystem type is not affected by the assessed standard.

[1]For establishments of plantations only.

[2]Preservation of 'unproductive' nonforested land is recommended.

[3]To comply with the 'RTRS EU-RED Compliance Requirements for Producers'.

[4]To comply with the 'Bonsucro EU Production Standard'.

stringent environmental legislation and/or limited enforcement capacity—and where agriculture or forestry sectors are still expanding onto natural vegetation—likely need to be more stringent than standards developed for countries where the existing protection of remaining natural ecosystems is effective and where the land conversion pressure is low. Nevertheless, the fnd-ings demonstrate that standards that are seemingly similar can be rather heterogeneous, and also that standards perceived as mainly associated with certain aspects can be highly stringent in regard to other aspects.

Regarding ecosystem conversion, forestry standards typically only protect areas that are considered HCV. They also tend to limit the HCV assessment requirements to include forested land only, i.e., they do not prevent conversion of highly biodiverse grasslands or wetlands into certifed plantation forests.[e] Agricultural standards cover more ecosystem types and typically do not provide for much fexibility:

specifc ecosystem types are either no-go areas or there are no conversion restrictions at all. The infexibility that several of the agricultural standards apply may result in areas that could be benefcially converted into sustainable cultivation, such as some degraded grasslands, not being available. The biofuel-related standards, as already noted, are infuenced by EU-RED and cover ecosystem conversion comprehensively, using a combination of HCV requirements and strict protection measures. Finally, some standards (EU Organic, NOP, and GGLS2) do not restrict land conversion at all. This may not be a large problem in countries with stringent legislation and suffcient enforcement capacity, but in countries where this is lacking, natural vegetation may be converted into certifed agriculture, impacting biodiversity.

The fairness of the comparison between forestry and agricultural standards may be questioned, since certain benchmark criteria may be more connected to one type of land use than the other. This was however considered when interpreting the results.

It should be noted that the assessment outcome cannot be used as a basis for understanding the *effectiveness* of the standards. *First*, the assessment outcome clarifes whether a particular standard complies with a benchmark criterion, but it does not provide information about how this benchmark criterion is addressed. For instance, one standard may only require that a particular issue be *considered*, while another standard addresses the same issue by requiring *measurements*, *assessments*, *monitoring*, etc.—still, both these standards would be assessed as complying with the corresponding benchmark criteria. However, we observed that standards having a high level of compliance with the benchmark criteria are typically also more concrete in their way of addressing them. *Second*, in order to fully understand the effectiveness of sustainability standards to conserve biodiversity, feld studies are necessary to measure actual results. *Third*, stringent sustainability standards may appear advantageous from a biodiversity perspective, but the effectiveness in conserving biodiversity depends on the relative importance of production under such standards compared to other production. Naturally, stringent certifcation schemes have to be widely applied in order to have a real effect on biodiversity conservation.

Given that certifcation involves additional costs (e.g., increased production cost, additional administration and fees to the certifcation scheme, certifcation body and auditors), biomass production meeting sustainability standards may require support (e.g., establishment of markets for biofuels or other products that are based on such biomass) in order to become established and grow to signifcant scale.

However, where such 'premium markets' become established, production complying with more stringent standards may still not become signifcant in scale if production complying with less stringent standards also meets the requirements associated with the premium market and can support a market price that is too low from the perspective of the more stringent standards. Biofuels and other biomass products may also be produced based on many different feedstocks and some alternatives that comply with stringent standards may outcompete other more costly alternatives complying with the same standard. For example, sugarcane ethanol meeting the EU-RED requirements may outcompete more expensive European cereal ethanol meeting the same requirements. If this happens, EU-RED helps little in protecting biodiversity affected by European cereal production, which would then primarily be associated with food markets.

The biodiversity impact of sustainability standards may be impacted by unintended effects. For example, biomass production complying with sustainability standards may reach lower yields than conventional production (or production under less stringent standards). *Ceteris paribus*, lower yields mean that more land has to be brought into production to meet the demand and associated land use change may result in biodiversity impacts.[43]

The extent to which sustainability standards protect biodiversity also depends on whether they stimulate innovation in land use aiming at protecting or enhancing biodiversity in the production landscape. Studies indicate that structuring and managing agricultural landscapes to host wild biodiversity can have neutral and even positive effects on agricultural output and livelihoods.[37] There are many examples of how perennial herbaceous and woody crops can be integrated into existing agricultural landscapes so as to enhance biodiversity and reduce impacts on water and soils.[14]

CONCLUSIONS AND RECOMMENDATIONS

Standards are not permanent, but they are continuously developed to refect the state of understanding including availability of data and tools. Some standards are subject to frequent incremental updates while others undergo less frequent but more substantial changes. All the assessed standards can, to a varying degree, be improved to better consider biodiversity. The benchmark standard presented in this paper could be used as one basis in the development of more concrete criteria/indicators that ft into the scope of individual standards. The further development of sustainability standards should aim for increased harmonization and reduction of heterogeneity of systems, while staying relevant for their intended production system. A balance needs to be found between stringency and comprehensiveness on the one hand and feasibility from a biomass-producer perspective on the other hand. It is important to avoid unnecessary requirements that increase administrative burden and cost without improving conservation outcome. Requirements that are too restricting/demanding may slow implementation and even prevent biomass production under the sustainability standard from reaching meaningful scale.

The use of generic standards, which have to be comprehensive in order to cover relevant aspects for a broad set of production systems, could be facilitated by (1) specifying for which production systems a criterion is valid, (2) defning different quantitative requirements for different production systems, or (3) developing substandards for different production systems. Heterogeneity of systems can also be reduced by considering multiple end uses of biomass in individual standards, when appropriate. SAN/RA, e.g., was initially applicable for food products only, but is now endorsed by RSB for production of biofuel feedstock. Consequently, producers can now sell their certifed products at a premium on both the food market and the biofuel market, and traders can buy SAN/RA certifed products and sell them on the EU-RED market.

Effectively realizing the objective to promote sustainable bioenergy requires fexibility in implementation to refect local circumstances. In some countries, national legislation may already cover some of the requirements in a standard. If it is recognized that existing national legislation is suffciently comprehensive and strict—and widely complied with—producers in such a country may be exempted from their reporting obligations concerning these requirements. Transitional rules might be used in situations where immediate full compliance with sustainability requirements is judged unattainable; less stringent initial requirements combined with a plan for development towards full compliance may be preferred where the alternative is more disruptive production for noncertifed markets. A concrete example can be seen in the Fairtrade standard, which requires compliance with a basic set of criteria in the frst year and then gradually extends compliance requirements to include an increasing number of criteria over time. This approach makes the frst step into certifcation surmountable but, as the entire standard is highly comprehensive, constant improvement inevitable.

Similarly, allowing for some degree of fexibility may also help in addressing certain unwanted effects, such as yield losses and nitrogen leaching associated with a transition to certifed production and exclusion of suitable lands irrespective of conservation outcome. The possibility to produce for certifed markets might be an incentive for developing land development plans balancing protection and production objectives in areas where much of the land is covered by ecosystems that are not allowed to be used under a sustainability standard. Such land development plans—where a certain share of these ecosystems are used for biomass production—might be preferred where the land conversion pressure is high.

The further development of certifcation standards could also consider approaches to stimulate innovation in land use that contributes positively to biodiversity conservation. Achieving this likely requires involvement of innovative practitioners, scientists, and indigenous land managers that are adapting, designing, and managing diverse types of landscapes to generate positive co-benefts for production, biodiversity, and local people. This work may increasingly need to consider adaptation needs arising as a consequence of climate change. A new role for actors involved with sustainability certifcation could be to facilitate processes where specifc tools are used to engage stakeholders in defning and prioritizing sustainability objectives—including biodiversity conservation—in a specifc local/regional context. Initiatives that show how this could take place include, e.g., ToSIA,[112] and attempts to identify and certify so-called 'low-iLUC' production.

As previously discussed, land management is characterized by trade-offs, and standards therefore need to consider the goals and objectives of different stakeholders in order to be effective. Therefore, it is necessary for standard developers to involve a wide range of stakeholders in the development process. By doing so, a standard can be developed to reach acceptance among all stakeholders, be it NGOs, landowners, biofuel producers, or traders. This process can also contribute to the development of shared views among diverse stakeholders involved in the public debate about bioenergy sustainability in general.

ENDNOTES

[a]Scopus database. KEY((certif* OR standard*) AND (*diversity OR richness) AND (agricultur* OR forest* OR produc* OR biomass)) AND DOC-TYPE(ar OR re) AND PUBYEAR>2005 AND (LIMIT-TO(SUBJAREA,"AGRI"))

[b]Scopus database. KEY((certif* OR standard*) AND (ecosystem service* OR ecosystem function*)) AND DOCTYPE(ar OR re) AND PUBYEAR>2005 AND (LIMIT-TO(SUBJAREA, "AGRI"))

[c]Web of Science database. Topic = (((certif*) AND (biodiversity OR diversity OR richness) AND (agri-cultur* OR forest* OR biomass))) Refned by: Web of Science Categories = (ECOLOGY OR ENVIRONMENTAL SCIENCES OR FORESTRY OR BIODIVERSITY CONSERVATION OR AGRICULTURE MULTIDISCIPLINARY) Timespan = 2006-2012. Databases = SCI-EXPANDED. Lemmatization = On

[d]Web of Science database. Topic = ((certif* AND (ecosystem service* OR ecosystem function*) AND (agricultur* OR forest* OR biomass))) Refned by: Web of Science Categories = (ECOLOGY OR ENVIRONMENTAL STUDIES OR AGRONOMY OR FORESTRY OR ENVIRONMENTAL SCIENCES OR BIODIVERSITY CONSERVATION OR AGRICULTURAL ENGINEERING OR MULTIDISCI-PLINARY SCIENCES)Databases = SCI-EXPANDED Timespan = 2006-2012 Lemmatization = On

[e]The FSC system is presently under revision to extend the HCV assessment to also cover ecosystem types other than forests.

ACKNOWLEDGMENTS

The authors are grateful for the very helpful comments from three anonymous reviewers. Financial support from Vattenfall, E.ON and the Swedish Energy Agency are gratefully acknowledged.

REFERENCES

1. Smith P, Haberl H, Popp A, Erb KH, Lauk C, Harper R, Tubiello FN, de Siqueira PA, Jafari M, Sohi S, et al. How much land-based greenhouse gas mitigation can be achieved without compromising food security and environmental goals? *Glob Chang Biol* 2013, 19:2285–302.
2. Lambin EF, Meyfroidt P. Global land use change, economic globalization, and the looming land scarcity. *Proc Natl Acad Sci USA* 2011, 108:3465–72.
3. Haberl H, Erb KH, Krausmann F, Gaube V, Bondeau A, Plutzar C, Gingrich S, Lucht W, Fischer-Kowalski M. Quantifying and mapping the human appropriation of net primary production in earth's terrestrial ecosystems. *Proc Natl Acad Sci USA* 2007, 104:12942–7.
4. Ellis EC. Anthropogenic transformation of the terrestrial biosphere. *Philos Trans Roy Soc A Math Phys Eng Sci* 2011, 369:1010–35.
5. Steffen W, Crutzen PJ, McNeill JR. The anthropocene: are humans now overwhelming the great forces of nature. *Ambio* 2007, 36:614–21.
6. Zalasiewicz J, Williams M, Haywood A, Ellis M. The anthropocene: a new epoch of geological time? *Philos Trans Roy Soc A Math Phys Eng Sci* 2011, 369:835–41.
7. Chum H, Faaij A, Moreira J, Berndes G, Dhamija P, Dong H, Gabrielle B, Goss Eng A, Lucht W, Mapako M, Masera Cerutti O, McIntyre T, Minowa T, Pin-goud K, Bioenergy. In: Edenhofer O, Pichs-Madruga R, Sokona Y, Seyboth K, Matschoss P, Kadner S, Zwickel T, Eickemeier P, Hansen G, Schlömer S, von Stechow C eds. *IPCC Special Report on Renewable Energy Sources and Climate Change Mitigation*. Cambridge / New York: Cambridge University Press; 2011.
8. Egeskog A, Berndes G, Freitas F, Gustafsson S, Sparovek G. Integrating bioenergy and food production – a case study of combined ethanol and dairy production in Pontal, Brazil. *Energy Sustain Dev* 2011, 15:8–16.
9. Dale BE, Bals BD, Kim S, Eranki P. Biofuels done right: land effcient animal feeds enable large environmental and energy benefts. *Environ Sci Technol* 2010, 44:8385–9.
10. Parish ES, Hilliard MR, Baskaran LM, Dale VH, Griffths NA, Mulholland PJ, Sorokine A, Thomas NA, Downing ME. Multimetric spatial optimization of switchgrass plantings across a watershed. *Biofuels Bioprod Biorefn* 2012, 6:58–72.
11. Gopalakrishnan G, Negri MC, Salas W. Modeling biogeochemical impacts of bioenergy buffers with perennial grasses for a row-crop feld in Illinois. *Glob Change Biol Bioenergy* 2012, 4:739–50.
12. Berndes G, Börjesson P, Ostwald M. Multifunctional biomass production systems –an overview with presentation of specifc applications in India and Sweden. *Biofuels Bioprod Biorefn* 2008, 2:16–25.
13. Garg KK, Karlberg L, Wani SP, Berndes G. Jatropha production on wastelands in India: opportunities and trade-offs for soil and water management at the watershed scale. *Biofuels Bioprod Biorefn* 2011, 5:410–30.
14. Dimitriou I, Baum C, Baum S, Busch G, Schulz U, Köhn J, Lamersdorf N, Leinweber P, Aronsson P, Weih M, Berndes G, Bolte A. Quantifying environmental effects of short rotation coppice (SRC) on biodiversity, soil and water. IEA Bioenergy Task 43; 2011. Report No.: 2011:01. Available at: http://ieabioenergytask43.org/wp-content/uploads/2013/09/IEA_Bioenergy_Task43_TR2011-01.pdf (Accessed March 6, 2014).
15. Busch G. GIS-based Tools for Regional Assess ments and Planning Processes Regarding Potential Environmental Effects of Poplar SRC. *Bioenergy Res* 2012, 5:584–605.
16. Gopalakrishnan G, Negri CM, Snyder S. Redesigning agricultural landscapes for sustainability using bioenergy crops: quantifying the tradeoffs between agriculture, energy and the environment. *Aspect Appl Biol* 2011, 112:139–46.
17. Englund O, Berndes G, Fredrikson F, Dimitriou J. Meeting sustainability requirements for src bioenergy: usefulness of existing tools, responsibilities of involved stakeholders, and recommendations for further developments. *Bioenergy Res* 2012, 5:606–620.
18. Junginger M, van Dam J, Zarrilli S, Ali Mohamed F, Marchal D, Faaij A. Opportunities and barriers for international bioenergy trade. *Energy Policy* 2011, 39:2028–2042.
19. O'Connell D, Braid A, Raison J, Handberg K, Cowie A, Rodriguez L, George B. Sustainable production of bioenergy – a review of global bioenergy sustainability frameworks and assessment systems. Australia: Rural Industries Research and Development Corporation, 2009. Available at: https://rirdc.infoservices.com.au/downloads/09-167. (Accessed September 16, 2013).
20. Stupak I, Lattimore B, Titus BD, Smith CT. Criteria and indicators for sustainable forest fuel production and harvesting: a review of current standards for sustainable forest management. *Biomass Bioenergy* 2011, 35:3287–308.

21. van Dam J, Junginger M, Faaij A. From the global efforts on certifcation of bioenergy towards an integrated approach based on sustainable land use planning. *Renew Sustain Energy Rev* 2010, 14:2445–72.
22. Pelkmans L, Goovaerts L, Smith CT, Joudrey J, Stupak I, Englund O, Junginger M, Goh CS, Chum H, Cowie A. Monitoring sustainability certifcation of bioenergy, task 4: recommendations for improvement of sustainability certifed markets. EIA Bioenergy; 2013. Available at: http://www.bioenergytrade.org/downloads/iea-sust-cert-task-4-fnal2013.pdf. (Accessed September 16, 2013).
23. Goovaerts L, Pelkmans L, Goh CS, Junginger M, Joudrey J, Chum H, Smith CT, Stupak I, Cowie A, Dahlman L, Englund O, Goss Eng A. Monitoring sustainability certifcation of bioenergy, task 1: examining sustainability certifcation of bioenergy. IEA Bioenergy; 2013. Available at: http://www.bioenergytrade.org/downloads/iea-sust-cert-task-2-fnal2013.pdf. (Accessed September 16, 2013).
24. Buytaert V, Muys B, Devriendt N, Pelkmans L, Kretzschmar JG, Samson R. Towards integrated sustainability assessment for energetic use of biomass: a state of the art evaluation of assessment tools. *Renew Sustain Energy Rev* 2011, 15:3918–33.
25. Magar SB, Pelkonen P, Tahvanainen L, Toivonen R, Toppinen A. Growing trade of bioenergy in the EU: Public acceptability, policy harmonization, European standards and certifcation needs. *Biomass Bioenergy* 2011, 35:3318–27.
26. van Dam J, Junginger M. Striving to further harmonization of sustainability criteria for bioenergy in Europe: recommendations from a stakeholder questionnaire. *Energy Policy* 2011, 39:4051–66.
27. Convention on Biological Diversity. List of Parties; 2013. Available at: http://www.cbd.int/convention/parties/list/#tab=0. (Accessed September 16, 2013).
28. Hamilton AJ. Species diversity or biodiversity? *J Environ Manage* 2005, 75:89–92.
29. Secretariat of the Convention on Biological Diversity. *Handbook of the Convention on Biological Diversity Including Its Cartagena Protocol on Biosafety*. 3rd ed. Montreal: Secretariat of Convention; 2005.
30. Millennium Ecosystem Assessment. Ecosystems and human well-being. In: *Biodiversity Synthesis*. Washington, DC: World Resources Institute; 2005. Available at: http://www.maweb.org/documents/document.354.aspx.pdf. (Accessed September 16, 2013).
31. Miller G. *Living in the Environment: Principles, Connections, and Solutions*. 15th ed. Belmont CA: Thomson Brooks/Cole; 2007.
32. Bisby F. Characterization of biodiversity. In: Heywood VH, Watson RT, eds. *Global Biodiversity Assessment*. Cambridge: The United Nations Environment Programme and Cambridge University Press; 1995, 21–106.
33. Heywood VH, Baste I. Introduction. In: Heywood VH, Watson RT, eds. *Global Biodiversity Assessment*. Cambridge: The United Nations Environment Programme and Cambridge University Press; 1995, 1–19.
34. Namkoong G, Boyle T, Gregorius H, Joly H. Testing criteria and indicators for assessing the sustainability of forest management: genetic criteria and indicators. Working paper No. 10. Bogor, Indonesia: CIFOR; 1996.
35. Cadotte MW, Carscadden K, Mirotchnick N. Beyond species: functional diversity and the maintenance of ecological processes and services. *J Appl Ecol* 2011, 48:1079–1087.
36. Leibold MA, McPeek MA. Coexistence of the niche and neutral perspectives in community ecology. *Ecology* 2006, 87:1399–1410.
37. Scherr SJ, McNeely JA. Biodiversity conservation and agricultural sustainability: towards a new paradigm of "ecoagriculture" landscapes. *Philos Trans R Soc Lond B Biol Sci* 2008, 363:477–94.
38. Wu J. Landscape ecology, cross-disciplinarily, and sustainability science. *Landsc Ecol* 2006, 21:1–4.
39. Daily G, Ehrlich P, Sánchez-Azofeifa GA. Countryside biogeography: use of human-dominated habitats by the avifauna of southern Costa Rica. *Ecol Appl* 2001, 11:1–13.
40. Dushku A, Brown S, Pearson T, Shoch D, Howley B. Remote sensing. In: Scherr SJ, McNeely JA, eds. *Farming with Nature: The Science and Practice of Ecoagriculture*. Washington DC: Island Press; 2007.
41. Sverdrup-Thygeson A, Borg P, Bergsaker E. A comparison of biodiversity values in boreal forest regeneration areas before and after forest certifcation. *Scand J Forest Res* 2008, 23:236–43.
42. Wintle BA, Lindenmayer DB. Adaptive risk management for certifably sustainable forestry. *For Ecol Manage* 2008, 256:1311–9.
43. Karsenty A, Gourlet-Fleury S. Assessing sustainability of logging practices in the Congo Basin's managed forests: the issue of commercial species recovery. *Ecol Soc* 2006, 11(1):26. Available at: http://www.ecologyandsociety.org/vol11/iss1/art26/. (Accessed September 16, 2013).
44. Zapf R, Schultheiss U, Doluschitz R. Assessment of sustainability – common requirements and comparative evaluation of the systems RISE, KSNL and the "DLG certifcation system for sustainable agriculture". *Ber Landwirtsch* 2009, 87:402–427.
45. Koskela E, Ollikainen M, Pukkala T. Biodiversity conservation in commercial boreal forestry: the optimal rotation age and retention tree volume. *Forest Sci* 2007, 53:443–452.
46. Dennis RA, Meijaard E, Nasi R, Gustafsson L. Biodiversity conservation in Southeast Asian timber concessions: a critical evaluation of policy mechanisms and guidelines. *Ecol Soc* 2009, 13:25. Available at: http://www.ecologyandsociety.org/vol13/iss1/art25/. (Accessed September 16, 2013).
47. Bisseleua DHB, Missoup AD, Vidal S. Biodiversity conservation, ecosystem functioning, and economic incentives under cocoa agroforestry intensifcation. *Conserv Biol* 2009, 23:1176–84.
48. Felton A, Wood J, Felton AM, Hennessey B, Lindenmayer DB. Bird community responses to reduced-impact logging in a certifed forestry concession in lowland Bolivia. *Biol Conserv* 2008, 141: 545–555.

49. Johansson J, Lidestav G. Can voluntary standards regulate forestry?—assessing the environmental impacts of forest certifcation in Sweden. *Forest Policy Econ* 2011, 13:191–198.
50. Wiersum KF, Gole TW, Gatzweiler F, Volkmann J, Bognetteau E, Wirtu O. Certifcation of wild coffee in Ethiopia: experiences and challenges. *Forests Trees Livelihoods* 2008, 18:9–21.
51. Clark M, Kozar J. Comparing sustainable forest management certifcations standards: a meta-analysis. *Ecol Soc* 2011, 16:3. Available at: http://www.ecologyandsociety.org/vol16/iss1/art3/. (Accessed September 16, 2013).
52. Ioras F, Abrudan IV, Dautbasic M, Avdibegovic M, Gurean D, Ratnasingam J. Conservation gains through HCVF assessments in Bosnia-Herzegovina and Romania. *Biodivers Conserv* 2009, 18:3395–3406.
53. Lattimore B, Smith CT, Titus BD, Stupak I, Egnell G. Environmental factors in woodfuel production: opportunities, risks, and criteria and indicators for sustainable practices. *Biomass Bioenergy* 2009, 33:1321–1342.
54. Gove AD, Hylander K, Nemomisa S, Shimelis A. Ethiopian coffee cultivation—implications for bird conservation and environmental certifcation. *Conserv Lett* 2008, 1:208–216.
55. Philpott SM, Bichier P, Rice R, Greenberg R. Field-testing ecological and economic benefts of coffee certifcation programs. *Conserv Biol* 2007, 21:975–85.
56. Garrelts H, Flitner M. Governance issues in the ecosystem approach: what lessons from the Forest Stewardship Council? *Eur J Forest Res* 2011, 130:395–405.
57. Elbakidze M, Angelstam P, Andersson K, Nordberg M, Pautov Y. How does forest certifcation contribute to boreal biodiversity conservation? Standards and outcomes in Sweden and NW Russia. *For Ecol Manage* 2011, 262:1983–95.
58. Miller DA, Wigley TB, Miller KV. Managed forests and conservation of terrestrial biodiversity in the southern United States. *J Forest* 2009, 107:197–203.
59. Hickey GM, Nitschke CR. Monitoring sustainable forest management in the Pacifc Rim Region. *J Sustain Forest* 2007, 24:245–78.
60. Marjokorpi A, Salo J. Operational standards and guidelines for biodiversity management in tropical and subtropical forest plantations – how widely do they cover an ecological framework? *Silva Fenn* 2007, 41:281–297.
61. Suzuki N, Olson DH. Options for biodiversity conservation in managed forest landscapes of multiple ownerships in Oregon and Washington, USA. *Biodivers Conserv* 2008, 17:1017–39.
62. Visseren-Hamakers IJ, Glasbergen P. Partnerships in forest governance. *Glob Environ Chang* 2007, 17: 408–19.
63. Anand MO, Krishnaswamy J, Das A. Proximity to forests drives bird conservation value of coffee plantations: implications for certifcation. *Ecol Appl* 2008, 18:1754–63.
64. McGeoch MA, Schroeder M, Ekbom B, Larsson S. Saproxylic beetle diversity in a managed boreal forest: importance of stand characteristics and forestry conservation measures. *Divers Distrib* 2007, 13:418–29.
65. Lõhmus A, Kraut A. Stand structure of hemiboreal old-growth forests: characteristic features, variation among site types, and a comparison with FSC-certifed mature stands in Estonia. *For Ecol Manage* 2010, 260:155–165.
66. Marchai D, Stappen F, Schenkel Y. Sustainable production criteria and indicators for solid biofuels. *Biotechnol Agron Soc Environ* 2009, 13:165–176.
67. Lagan P, Mannan S, Matsubayashi H. Sustainable use of tropical forests by reduced-impact logging in Deramakot Forest Reserve, Sabah, Malaysia. *Ecol Res* 2007, 22:414–21.
68. Eriksson S, Hammer M. The challenge of combining timber production and biodiversity conservation for long-term ecosystem functioning—a case study of Swedish boreal forestry. *For Ecol Manage* 2006, 237:208–217.
69. Smeets EM, Faaij AP. The impact of sustainability criteria on the costs and potentials of bioenergy production–applied for case studies in Brazil and Ukraine. *Biomass Bioenergy* 2010, 34:319–33.
70. Hennenberg KJ, Dragisic C, Haye S, Hewson J, Semroc B, Savy C, Wiegmann K, Fehrenbach H, Fritsche UR. The power of bioenergy-related standards to protect biodiversity. *Conserv Biol* 2010, 24:412–23.
71. Smeets E, Junginger M, Faaij A, Walter A, Dolzan P, Turkenburg W. The sustainability of Brazilian ethanol—an assessment of the possibilities of certifed production. *Biomass Bioenergy* 2008, 32:781–813.
72. Zedler PH, Anchor T, Knuteson D, Gratton C, Barzen J. Using an ecolabel to promote on-farm conservation: the Wisconsin Healthy Grown experience. *Int J Agric Sustain* 2009, 7:61–74.
73. Tejeda-Cruz C, Silva-Rivera E, Barton JR, Sutherland WJ. Why shade coffee does not guarantee biodiversity conservation. *Ecol Soc* 2013, 15:13. Available at: http://www.ecologyandsociety.org/vol15/iss1/art13/. (Accessed September 16, 2013).
74. Timonen J, Siitonen J, Gustafsson L, Kotiaho JS, Stokland JN, Sverdrup-Thygeson A, Mönkkönen M. Woodland key habitats in northern Europe: concepts, inventory and protection. *Scand J Forest Res* 2010, 25:309–24.
75. FSC. FSC principles and criteria for Forest Stewardship: FSC-STD-01-001 (version 4-0). Bonn, Germany: Forest Stewardship Council; 2002. Available at: https://ca.fsc.org/download.principles-criteria-v4.7.pdf. (Accessed September 16, 2013).
76. SFI. Requirements for the SFI 2010–2014 Program. Standards, Rules for Label Use, Procedures and Guidance. Washington DC: Sustainable Forestry Initiative; 2010. Available at: http://www.sfprogram.org/fles/pdf/sfrequirements2010-2014pdf/. (Accessed September 16, 2013).
77. FFCS. FFCS 1002–3:2003, criteria for certifcation of holdings of individual forest owners. Unoffcial Translation (3 February 2004) of the Original Finnish Standard FFCS 1002–3:2003 text. Helsingfors: Finnish Forest Certifcation System; 2004. Available at: http://www.pefc.f/media/Standardit/FFCS_1002_3_2003ENG.pdf. (Accessed September 16, 2013).

78. MTCC. Malaysian Criteria and Indicators for Forest Management Certifcation [MC\&I(2002)]. Kuala Lumpur: Malaysian Timber Certifcation Council; 2002.
79. MTCC. Malaysian Criteria and Indicators for Forest Management Certifcation (Forest Plantations) [MC\&I(Forest Plantations)]. Kuala Lumpur: Malaysian Timber Certifcation Council; 2008. Available at: http://www.mtcc.com.my/wp-content/uploads/2013/01/21.pdf. (Accessed September 16, 2013).
80. CSA. Sustainable Forest Management Z809-08. Mississauga, Canada: Canadian Standards Association; 2010. Available at: http://www.csagroup.org/documents/codes-and-standards/publications/2419617.pdf. (Accessed September 16, 2013).
81. GGL. GGLS5 – Forest Management Criteria. Zwolle, The Netherlands: Green Gold Label; 2010. Available at: http://www.greengoldcertifed.org/data/docs/ggl5.pdf. (Accessed September 16, 2013).
82. Naturland. Naturland Standards for Organic Forest Management. Gräfelfng, Germany: Naturland; 1998. Available at: http://www.naturland.de/fleadmin/MDB/documents/Richt linien_englisch/Naturland-Standards-Organic-forest mana gement_1998-11.pdf. (Accessed September 16, 2013).
83. ITTO. Revised ITTO criteria and indicators for the sustainable management of tropical forests, including reporting format. ITTO Policy Development Series No 15. Yokohama: International Tropical Timber Organization; 2005. Available at: http://www.itto.int/direct/topics/topics_pdf_download/topics_id=9630000&no=1. (Accessed September 16, 2013).
84. ATO/ITTO. ATO/ITTO principles, criteria and indicators for the sustainable management of African natural tropical forests. ITTO Policy Development Series No 14. Yokohama/Libreville: African Timber Organization and International Tropical Timber Organization; 2003. Available at: http://www.itto.int/direct/topics/topics_pdf_download/topics_id=1550000&no=1. (Accessed September 16, 2013).
85. ITTO/IUCN. ITTO/IUCN Guidelines for the conservation and sustainable use of biodiversity in tropical timber production forests. ITTO Policy Development Series No 17. Yokohama/Gland: International Tropical Timber Organization and International Union for Conservation of Nature; 2009. Available at: http://www.itto.int/direct/topics/topics_pdf_download/topics_id=1918&no=0. (Accessed September 16, 2013).
86. MCPFE. Annex 1 of the Resolution L2: Pan-European criteria and indicators for sustainable forest management. In: *Ministerial Conference on the Protection of Forests in Europe*; 1998. Available at: http://www.foresteurope.org/docs/MC/MC_lisbon_resolution_annex1.pdf. (Accessed September 16, 2013).
87. MCPFE. Improved Pan-European Indicators for Sustainable Forest Management. In: *Ministerial Conference on the Protection of Forests in Europe*; 2002. Available at: http://www.foresteurope.org/docs/reporting/Vienna_Improved_Indicators.pdf. (Accessed September 16, 2013).
88. GlobalGAP. Integrated farm assurance – all farm base. Control Points and Compliance Criteria. Version 4.0, Ed. 4.0-1_FEB2012. Cologne, Germany: GlobalGAP; 2011.
89. GlobalGAP. Integrated farm assurance – crops base. Control Points and Compliance Criteria. Version 4.0, Ed. 4.0-1_FEB2012. Cologne, Germany: GlobalGAP.; 2011.
90. GlobalGAP. Integrated farm assurance – combinable crops. Control Points and Compliance Criteria. Version 4.0, Ed. 4.0-1_FEB2012. Cologne, Germany: GlobalGAP; 2011.
91. KRAV. Regler för KRAV-certiferad produktion januari 2011. Uppsala, Sweden: KRAV; 2011.
92. European Council. Council Regulation (EC) No 834/2007 of 28 June 2007 on organic production and labelling of organic products and repealing Regulation (EEC) No 2092/91. Brussels: European Council; 2007. Available at: http://eur-lex.europa.eu/LexUriServ/LexUriServ.do?uri=OJ:L:2007:189:0001:0023:EN:PDF. (Accessed September 16, 2013).
93. European Commission. Commission Regulation (EC) No 1235/2008 of 8 December 2008 laying down detailed rules for implementation of Council Regulation (EC) No 834/2007 as regards the arrangements for imports of organic products from third countries. Brussels: European Commission; 2008. Available at: http://eur-lex.europa.eu/LexUriServ/LexUriServ.do?uri=OJ:L:2008:334:0025:0052:EN:pdf. (Accessed September 16, 2013).
94. European Commission. Commission Regulation (EC) No 889/2008 of 5 September 2008 laying down detailed rules for the implementation of Council Regulation (EC) No 834/2007 on organic production and labelling of organic products with regard to organic production, labelling and control. Brussels: European Commission; 2008. Available at: http://eur-lex.europa.eu/LexUriServ/LexUriServ.do?uri=OJ:L:2008:250:0001:0084:EN:PDF. (Accessed September 16, 2013).
95. USDA. 7 C.F.R. Part 205: National Organic Program Regulations. Washington DC: United States Department of Agriculture; 2000. Available at: http://www.ecfr.gov/cgi-bin/text-idx?c=ecfr&sid=3f34f4c22f9aa8e6d9864cc2683cea02&tpl=/ecfrbrowse/Title07/7cfr205_main_02.tpl. (Accessed September 16, 2013).
96. GGL. GGLS2 – Agricultural Source Criteria. Zwolle, The Netherlands: Green Gold Label; 2010. Available at: http://www.greengoldcertifed.org/data/docs/ggl2.pdf. (Accessed September 16, 2013).
97. FLO International. Fairtrade Standard for Small Producer Organizations. Version: 01.05.2011. Bonn, Germany: FLO International; 2011. Available at: http://www.fairtrade.net/fleadmin/user_upload/content/2009/standards/documents/2012-07-11_SPO_EN.pdf. (Accessed September 16, 2013).
98. Naturland. Naturland Standards on Production. Gräfelfng, Germany: Naturland; 2011. Available at: http://www.naturland.de/fleadmin/MDB/documents/Richtlinien_engli sch/Naturland-Standards-Organicforest-management _1998-11.pdf. (Accessed September 16, 2013).
99. IFOAM. The IFOAM Norms for Organic Production and Processing, Version 2005. Bonn, Germany:

International Federation of Organic Agriculture Movements; 2005.

100. SAN. Sustainable Agriculture Standard. July 2010 (version 2). San José: Sustainable Agriculture Network/Rainforest Alliance; 2010. Available at: http://sanstandards.org/userfles/fle/SAN%20Sustainable%20Agriculture%20Standard%20July%202010.pdf. (Accessed June, 2012).
101. RSPO. RSPO Principles and Criteria for Sustainable Palm Oil Production, Including Indicators and Guidance. October 2007. Kuala Lumpur: Roundtable on Sustainable Palm Oil; 2007. Available at: http://www.rspo.org/fles/resource_centre/RSPO%20Principles%20&%20Criteria%20Document.pdf. (Accessed September 16, 2013).
102. RTRS. RTRS Standard for Responsible Soy Production Version 1.0. Buenos Aires: Roundtable on Responsible Soy; 2010. Available at: http://www.responsiblesoy.org/index.php?option=com_docman&task=doc_download&gid=290&Itemid=19&lang=en. (Accessed September 16, 2013).
103. RTRS. RTRS EU RED Compliance Requirements for Producers. Version 3.0_EN. Buenos Aires: Roundtable on Responsible Soy; 2011. Available at: http://www.responsiblesoy.org/index.php?option=com_docman&task=doc_download&gid=654&Itemid=63&lang=en. (Accessed September 16, 2013).
104. Bonsucro. Bonsucro Production Standard, Including Bonsucro EU Bonsucro Production Standard. London: Bonsucro; 2011. Available at: http://www.bonsucro.com/assets/Bonsucro_Production_Standard_March_2011_3.pdf. (Accessed September 16, 2013).
105. RSB. Consolidated RSB EU RED Principles \& Criteria for Sustainable Biofuel Production. Geneva, Switzerland: Roundtable on Sustainable Biofuels; 2011. Available at: http://rsb.org/pdfs/standards/RSB-EU-RED-Standards/11-05-10-RSB-STD-11-001-01-001-vers-2-0-Consolidated-RSB-EU-RED-PCs.pdf. (Accessed September 16, 2013).
106. RSB. Indicators of Compliance for the RSB EU RED Principles \& Criteria. Geneva: Roundtable on Sustainable Biofuels; 2010. Available at: http://rsb.org/pdfs/standards/11-03-08-RSB-PCs-Version-2.pdf. (Accessed September 16, 2013).
107. ISCC. Sustainability requirements for the production of biomass: ISCC 11-03-15 V 2.3-EU. Cologne, Germany: International Sustainability & Carbon Certifcation; 2011. Available at: https://www.cbh.com.au/media/300335/iscc%20202%20sustainability%20requirements.pdf. (Accessed September 16, 2013).
108. Greenergy. Greenergy Brazilian Bioethanol verifcation programme: Greenergy bioethanol sustainability criteria for Brazilian sugarcane STD DOC – 002 v.5 31 January 2011. London: Greenergy; 2011. Available at: http://ec.europa.eu/energy/renewables/biofuels/sustainability_schemes_en.htm. (Accessed September 16, 2013).
109. Trombulak SC, Omland KS, Robinson JA, Lusk JJ, Fleischner TL, Brown G, Domroese M. Principles of conservation biology: recommended guidelines for conservation literacy from the education committee of the society for conservation biology. *Conserv Biol* 2004, 18: 1180–90.
110. Millenium Ecosystem Assessment. *Ecosystems and Human Well-being: Current State and Trends, Volume*. Washington, DC: Island Press; 2013. Available at: http://www.maweb.org/documents/document.273.aspx.pdf. (Accessed September 16, 2013).
111. Soule M. What is conservation biology. *BioScience* 1985, 35:727–34.
112. TMUG - ToSIA Management and User Group. Tool for Sustainability Impact Assessment. Joensuu, Finland: European Forest Institute; 2013. Available at: http://www.northerntosia.org/fles/attachments/northerntosia/fnal_conference/fnal_copy_n_tosia_general_fier.pdf (Accessed March 6, 2014).

32

A Global Survey of Stakeholder Views and Experiences for Systems Needed to Effectively and Efficiently Govern Sustainability of Bioenergy

Inge Stupak,[1*] Jamie Joudrey,[2] C. Tattersall Smith,[3] Luc Pelkmans,[4] Helena Chum,[5] Annette Cowie,[6] Oskar Englund,[7] Chun Sheng Goh[8] and Martin Junginger[8]

[1]Department of Geosciences and Natural Resource Management, Faculty of Science, University of Copenhagen, Copenhagen, Denmark
[2]Faculty of Forestry, University of Toronto, Toronto, ON, Canada
[3]Department of Geography, University of Toronto, Toronto, ON, Canada
[4]Unit Separation and Conversion Processes, VITO NV, Mol, Belgium
[5]National Bioenergy Center, National Renewable Energy Laboratory, Golden, CO, USA
[6]NSW Department of Primary Industries, University of New England, Armidale, Australia
[7]Department of Energy and Environment, Division of Physical Resource Theory, Chalmers University of Technology, Göteborg, Sweden
[8]Energy & Resources, Copernicus Institute of Sustainable Development, Utrecht University, Utrecht, Netherlands

DEVELOPMENTS IN GOVERNANCE OF BIOMASS AND BIOENERGY SUSTAINABILITY

The use of biomass in modern energy production has increased dramatically over the last decade, owing to several important initiatives to combat climate change, such as the Kyoto Protocol, and an increased desire of many countries to seek energy security and independence. While the Kyoto Protocol does not prescribe specific instruments for achieving greenhouse gas (GHG) emission reductions, jurisdictional legislation launched in the early 2000s (Table 32.1) sought to increase the use of transportation biofuels and other renewables in energy supplies.[1–3] Subsequently, production and global biofuel trade increased rapidly, with the EU, USA, Brazil, and Argentina leading for liquid transportation biofuels,[1] and Europe and North America for solid biofuels.[2]

The increased international trade led to growing concerns over sustainability of biofuels and a variety of governance systems has emerged to regulate the bioenergy sector for maximization of the benefits and minimization of the possible negative impacts.[4] The general concept of governance is used in different ways,[5] but we use it in this paper in the broad sense of governance processes undertaken by governments, market actors, voluntary organizations or networks, that aim at steering formal or informal organizations, or territories, though laws, private certification, standards, or norms to achieve specified outcomes.[6] This concept of governance recognizes the interdependence of the

*Correspondence to: ism@ign.ku.dk

Advances in Bioenergy: The Sustainability Challenge, First Edition. Edited by Peter D. Lund, John Byrne, Göran Berndes and Iacovos A. Vasalos.

Originally published in: *WIREs Energy Environ* 2015, 507–534 doi: 10.1002/wene.166

Table 32.1 A Timeline of Sustainability Governance with Selected Key International Initiatives, Agreements, and Voluntary Certification Schemes[1]

Year	International and National Initiatives		Voluntary Certification	
	Initiative	Main Scope	Scheme	Main Scope
1973			California Certified Organic Farmers (CCOF)	Organic agriculture
1987	Brundtland Report on sustainable development released	Sustainability		
1990	Ministerial Conference on the Protection of Forests in Europe (MCPFE, now Forest Europe)	Forestry		
1992	Earth Summit at Rio de Janeiro	Sustainability		
1993	Convention on Biological Diversity (CBD)	Biodiversity		
1994	Montreal Process	Forestry	Forest Stewardship Council (FSC)	SFM
	United Nations Framework Convention on Climate Change (UNFCCC)	Climate change	Sustainable Forestry Initiative (SFI)	SFM
1996			Canadian Standards Association's SFM standards (CSA)	SFM
1997	Kyoto Protocol (KP) adopted	Climate change	GlobalGAP, Sustainable Agricultural Network	Good Agricultural Practices
			Sustainable Agriculture Network (SAN)	Good Agricultural Practices
1999			Programme for the Endorsement of Forest Certification Schemes (PEFC)	SFM
2002			CertFor, Chile	SFM
			CerFlor Brazil	SFM
			Green Gold Label (GGL)	Biomass supply chains
2004			Roundtable for Sustainable Palm Oil (RSPO)	Palm oil supply chains
2005	Kyoto Protocol (KP) enforced	Climate change		
2007	US Renewable Fuel Standard 2 (RFS2)	Transportation biofuels		
	California's Low-Carbon Fuel Standard (LFCS)	Transportation biofuels		
2008	Committee of Standardization (CEN) TC 383 for sustainability criteria initiated	Biofuels		
2009	EU-Renewable Energy Directive 2009/28/EC (EU-RED)	Renewable Energy including transportation biofuels		
	ISO TC 248 initiated	Biofuels		
2010	Report COM(2010)11—recommended sustainability requirements for solid biofuels	Biofuels for heat, power and cooling	Bonsucro	Sugar and ethanol supply chains
			International Sustainability & Carbon Certification (ISCC)	Biomass and bioenergy supply chains

Table 32.1 A Timeline of Sustainability Governance with Selected Key International Initiatives, Agreements, and Voluntary Certification Schemes[1] (*Continued*)

Year	International and National Initiatives		Voluntary Certification	
	Initiative	Main Scope	Scheme	Main Scope
2011	Global Bioenergy Partnership releases sustainability indicators for bioenergy (GBEP)	Biofuels	2BSvs	Grain biofuel supply chains
			Roundtable for Sustainable Biomaterials (RSB)	Biofuels and biomaterials supply chains
			RoundTable for Responsible Soy (RTRS)	Soy biofuel supply chains
2012			Initiative Wood Pellet Buyers (IWPB)	Supply chains for solid woody biomass in power generation

SFM: Sustainable Forest Management.
[1]Compiled from GIZ[58], Stupak et al.[40], and homepages of the various initiatives (see also Table 32.3 for URLs).

public, market-based and voluntary governing processes, and the relationships that may exist between them.[5]

For example, such governance emerged in response to the conversion of forests to produce and export Malaysian and Indonesian palm oil for biodiesel production, when the Roundtable on Sustainable Palm Oil (RSPO) was formed in the early 2000s.[7] Similarly, the Green Gold Label (GGL) was developed to address sustainability of solid biomass. Prominent examples of legislative sustainability requirements are those of the USA and the EU for transportation biofuels. The USA introduced minimum threshold values for GHG emission savings in its revised Renewable Fuels Standard (RFS2) in 2007,[8] and this regulation also only accepts certain types of biomass as eligible.[9] In 2009, the EU adopted the Renewable Energy Directive (EU-RED),[10] which together with the 2010 European Commission recommendations for solid fuels[11] include sustainability criteria for GHG emission savings, protection of land with high levels of biodiversity or carbon stocks, and criteria for environmentally sound agricultural practices as defined in the EU Common Agricultural Policy (EU-CAP), while the need and feasibility to include air, soil and water, social sustainability, and affordable food as mandatory criteria is being explored.[11,12]

During the late 2000s, a proliferation of public and private governance initiatives developed (Table 32.1), partly because private certification became an option for demonstrating compliance with the EU-RED sustainability criteria for liquid biofuels.[13] Several of the new private certification schemes[14] were aligned with the EU-RED criteria, but some also went beyond these criteria. A lack of consensus around new EU criteria for solid biomass has furthermore been a driver for emerging national initiatives (e.g., in the UK[15]). These newer bioenergy schemes and initiatives often have overlapping scopes with older certification schemes inaugurated since the 1990s, including those for sustainable forest management (SFM) or sustainable agricultural practices (Table 32.1). Certification with these schemes is accepted as part of the evidence required by some of the bioenergy schemes,[16] but scheme holders are also considering expansions and improvements to better address biomass and bioenergy sustainability.[17,18]

Gamborg et al.[4] discuss value disagreements and complexity of regulatory measures as two factors giving rise to friction in bioenergy governance, and recent reviews of the ability of certification systems to safeguard the sustainability of bioenergy also conclude that the increasing amount of overlapping public and private regulation has led to concerns about confusion among producers and consumers, redundant costs, barriers to trade and subsequent failure in achieving sustainability goals.[14,19] A 2006–2007 survey identified the need for increased harmonization of EU bioenergy policies, and concluded that certification is necessary to promote sound, sustainable development for bioenergy.[20] Another survey from 2009 concluded that a harmonized European bioenergy certification system is needed.[21] Other studies have reached similar conclusions,[22–24] but unresolved questions still remain. Specifically, which combination of governance mechanisms will be most efficient, fair and transparent; how harmonization can be achieved; and, how large a role can certification systems play? Although a large part of the global trade with wood pellets is currently certified[25] only 10% of global forests[26] and 14% of global palm oil production are certified.[7]

The role of certification in the future will, among other things, depend on its 'input' and 'output' legitimacy. Input legitimacy may be defined in a number of ways, but depends on the extent to which regulations are perceived as justified and related to their inclusiveness.[27] Risks of rent-seeking behavior and abuse in private regulation have been highlighted,[28] and also well-recognized private schemes are occasionally criticized for not being adequately inclusive[29] or effectively safeguarding sustainability.[30,31] It is important that a scheme's standards are broadly accepted, that there is equal representation and participation of relevant stakeholders, neutrality in how they are addressed, and procedural regularity and transparency of decisions.[27,32–34]

Output legitimacy may be defined by the extent to which regulation effectively solves the issues and concerns it was designed to address.[27] This relates to the comprehensiveness of the standards, and the suitability of the selected indicators for demonstrating the standards' sustainability goals.[27,35] At the practical level, robust and transparent tools should be available for measuring and assessing performance.[36–38] Due to constantly changing conditions and generation of new knowledge and experiences, certification schemes must continuously develop and improve in order to maintain legitimacy and competitive edge in the market.

The first-hand experiences possessed by involved public and private actors are especially critical to identifying benefits and challenges and suggesting improvements for governance of bioenergy sustainability. Therefore, a survey was designed with the objective of analyzing stakeholders' views, experiences, and ideas in relation to the governance challenges identified:

1. What is the usefulness of international, national, public and private regulatory governance mechanisms; and which are preferred and considered most effective for meeting sustainability goals for biofuels?
2. Did recent developments in certification systems and other forms of sustainability governance affect the sector's production, business and trade; and did certification costs and benefits impact decisions to get certified or not?
3. Which developments are most needed for voluntary certification schemes to meet future challenges in bioenergy-related markets and enable schemes to maintain or increase their legitimacy?

While previous survey studies focused on groups of downstream economic actors in the bioenergy supply chain, this survey included economic actors along entire biofuel supply chains, in addition to actors in the public sector, private certification schemes, standards development organizations, NGOs, professional associations and individuals with scientific expertise.

METHODOLOGY

Survey Design

A survey with three themes and a total of 59 questions was designed to address the objectives of the study, and determine if answers depended on the respondents' backgrounds [Table 32.2, see Appendix S1 (Supporting Information) for the full questionnaire]. Throughout the paper, Q2, Q3, and so on, refer to question numbers 2, 3, and so forth of the survey.

Questions typically required answers as 'yes', 'no', 'I don't know', or ratings such as 'not important', 'moderately important', 'important', 'very important', and 'not relevant'. Open-ended questions requested additional explanation to better understand selection of predefined answers.

To test differences in views and experiences among subgroups, we collected information on respondents' capacity (biomass, biofuel or energy producer, certification scheme, certification body, etc.; see Appendix S1 for the 18 defined capacities), (Q2), location (online system registration of country, with a manual check of accuracy), geographic experience (Q5), certification scheme familiarity (Q7), broader certification scheme experience (Q3), and duration of their bioenergy and certification experience (Q4). We aimed at having a balanced representation of respondents for the main capacities, from all continents, and certification schemes for different sectors (forestry, agriculture, biomass, and bioenergy).

The survey instrument was pretested, and in May 2012, the survey was sent to >1000 potential respondents from organizations worldwide. When the survey was closed in September 2012, a total of 196 responses had been recorded, with 192 determined as suitable for data analysis.

Respondent Backgrounds

The originally defined subgroups of background variables were aggregated into one-dimensional variables, with aggregation to a level that ensured an adequate number of degrees of freedom for statistical analyses.

Figures 32.1 and 32.2 show the different subgroups within each of the seven categorical background variables.

All capacities were represented among respondents, even if some were better represented than others (Figure 32.1). Most of the 192 respondents were located in

Table 32.2 Overview of Structure and Topics of the IEA Bioenergy Survey (cf. Appendix S1 for the Full Set of Questions, with Q2, Q3, etc. Referring to Question Numbers 2, 3, etc.)

No	Topic	Questions	Number of Questions
0	Your background	Q2–5 and Q7	5
Theme1: The mix of governance mechanisms			
1	Governance of the sustainability of bioenergy systems	Q8–17	10
2	Noncertified land	Q18–19	2
3	Harmonization and compliance	Q20–27	8
Theme 2: Impacts on markets, trade and costs			
4	Trade and markets	Q28–34	7
5	Cost and adaptation	Q35–42	8
Theme 3: Credibility and efficacy of voluntary certification			
6	Standards and scheme development and revision	Q43–49	7
7	On-the-ground effectiveness of sustainability certification	Q50–55	6
8	Verification and auditing challenges	Q56–61	6

Europe or North America (Figure 32.2(a)), but many had experiences from more than one region, including developing countries (Figure 32.2(b), Q5).

Many respondents selected RSB, ISCC, FSC, or a PEFC endorsed scheme as most familiar to them (Figure 32.2(c), Q7). Several respondents also selected other schemes approved under EU-RED, or schemes designed for pellets (IWPB, Drax, GGL, LBE, and ENplus). The remaining group included, e.g., eight respondents to CSBP. Several respondents indicated that they also had broader certification experiences (Figure 32.2(d), Q3), often with RSB, ISCC, Bonsucro, REDCert, RTRS, 2BSvs, CSBP, FSC, or PEFC endorsed schemes (especially SFI, CSA, ATFS, and AFS). A list of acronyms is available in Table 32.3, and we refer to existing reviews for further description of the individual systems.[14,19,39,40]

Most respondents had >5 years' experience with bioenergy (Figure 32.2(e), Q4), and >1 years' experience with

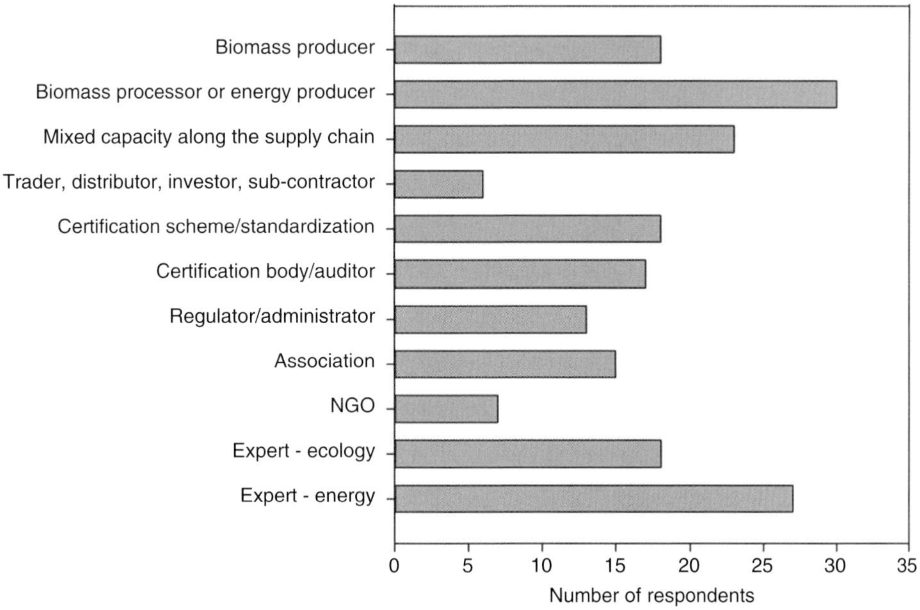

Figure 32.1 The distribution of the 192 respondents by their capacities, with groups of capacities redefined based on Q2.

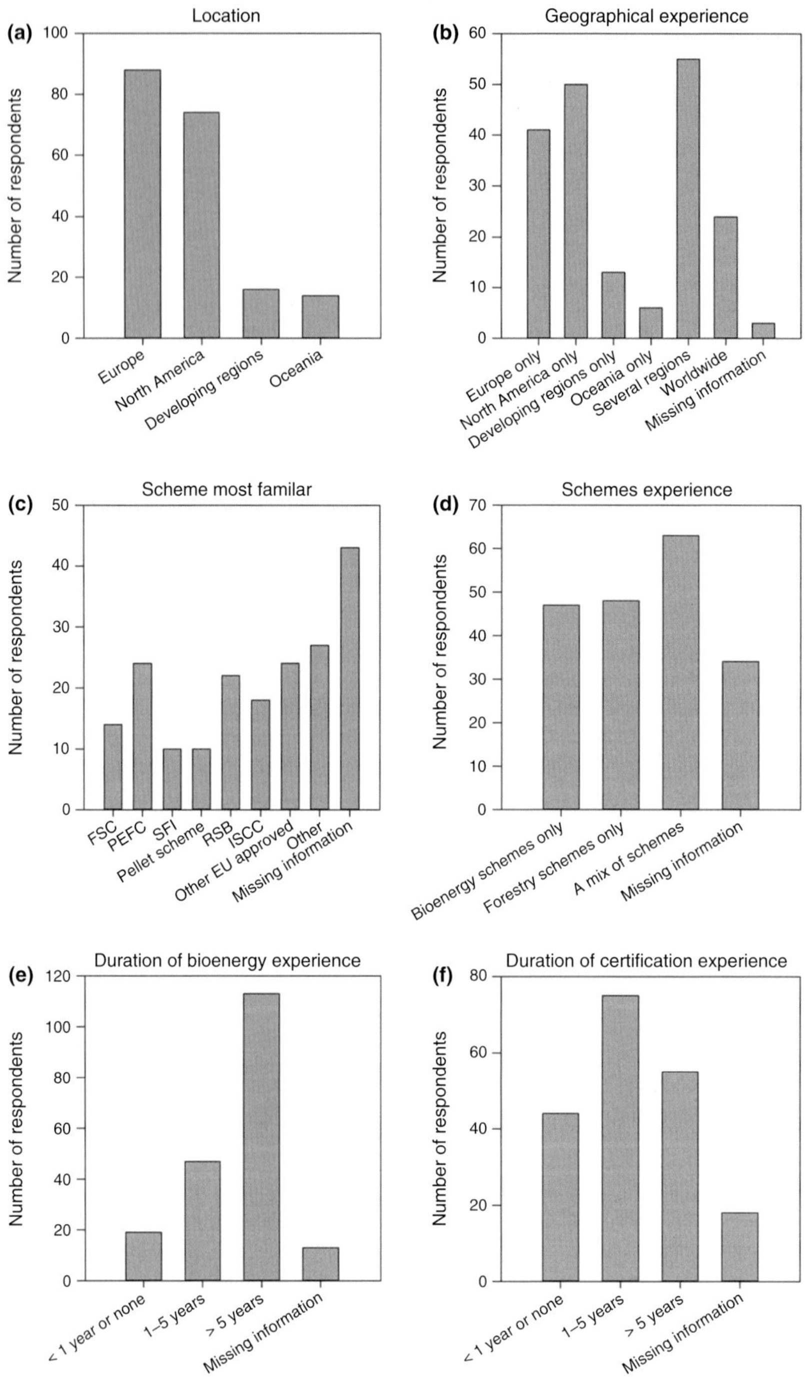

Figure 32.2 The distribution of the 192 respondents by (a) the geographical region in which they are located, (b) the geographical regions from which they have experience, (c) the single voluntary certification they selected as the one most familiar to them, (d) the types of voluntary certification schemes with which they more broadly have experience, (e) duration of their bioenergy experience, and (f) duration of their certification experience.

Table 32.3 Acronyms, Full Names, and URL of the International and National Initiatives Mentioned in the Text

Acronym	Full Name	URL
International conventions		
Cartegena	The Cartagena Protocol on Biosafety to the Convention on Biological Diversity	http://bch.cbd.int/protocol
CBD	Convention on Biological Diversity	http://www.cbd.int/
CCD	United Nations Convention to Combat Desertification	http://www.unccd.int/
CITES	The Convention on International Trade in Endangered Species of Wild Fauna and Flora	http://www.cites.org/
EU-ELC	The European Landscape Convention	http://www.coe.int/t/dg4/cultureheritage/heritage/Landscape/default_en.asp
ILO	International Labour Organization Conventions	http://www.ilo.org/global/standards/introduction-to-international-labour-standards/conventions-and-recommendations/lang–en/index.htm
IPCC	Intergovernmental Panel on Climate Change	http://www.ipcc.ch/
KP	The Kyoto Protocol	http://unfccc.int/kyoto_protocol/items/2830.php
Ramsar	Ramsar Convention	http://www.ramsar.org/
UNFCCC	United Nations Framework Convention on Climate Change	http://unfccc.int/
Legislation		
BAEZ	Brazilian Agroecological Zoning	http://www.planalto.gov.br/ccivil_03/projetos/PL/2009/msg764-090917.htm
ESD	The European Union Effort Sharing decision	http://ec.europa.eu/clima/policies/effort/
EU-CAP	The European Union Common Agricultural Policy	http://ec.europa.eu/agriculture/
EU-RED	The European Union Renewable Energy Directive	http://ec.europa.eu/energy/renewables/biofuels/biofuels_en.htm
EU-TR	The European Union Timber Regulation	http://ec.europa.eu/environment/eutr2013/
FLEGT	The European Union Forest Law Enforcement, Governance and Trade	http://www.euflegt.efi.int/
FQD	The European Union Fuel Quality Directive	http://ec.europa.eu/environment/air/transport/fuel.htm
Lacey Act	The US Lacey Act	http://eia-global.org/lacey/
LFCS	California Low Carbon Fuel Standard	http://www.energy.ca.gov/low_carbon_fuel_standard/
Natura 2000	Natura 2000 network	http://ec.europa.eu/environment/nature/natura2000/index_en.htm
RFS/RFS2	The US Renewable Fuel Standard	http://www.epa.gov/otaq/fuels/renewablefuels/
UK-RTFO	The UK Renewable Transport Fuels Obligation	https://www.gov.uk/renewable-transport-fuels-obligation
International processes		
African Dry Zone	The Dry-Zone Africa Process on Criteria and Indicators for Sustainable Forest Management	http://www.fao.org/forestry/ci/16609/en/, http://www.fao.org/docrep/004/ac135e/ac135e00.htm#Contents
ATO	African Timber Organization	http://www.fao.org/forestry/ci/16609/en/, http://www.comifac.org/
Dry Forest Asia	Regional Initiative for the Development and Implementation of National Level Criteria and Indicators for the Sustainable Management of Dry Forests in Asia	http://www.fao.org/forestry/ci/16609/en/, http://www.fao.org/docrep/004/ac135e/ac135e00.htm#Contents
Forest Europe	Forest Europe	http://www.foresteurope.org/
GBEP	Global Bioenergy Partnership	http://www.globalbioenergy.org/
ITTO	The International Tropical Timber Organization	http://www.itto.int/
Lepaterique	Lepaterique Process of Central America on Criteria and Indicators for Sustainable Forest Management	http://www.fao.org/forestry/ci/16609/en/, http://www.fao.org/docrep/004/ac135e/ac135e00.htm#Contents
MCPFE	Ministerial Conferences on Protection of Forests in Europe (now Forest Europe)	http://www.foresteurope.org/

Table 32.3 Acronyms, Full Names, and URL of the International and National Initiatives Mentioned in the Text (*Continued*)

Acronym	Full Name	URL
Montreal	The Montréal Process	www.montrealprocess.org/
Near East	The Near East Process	http://www.fao.org/forestry/ci/16609/en/, http://www.fao.org/docrep/004/ac135e/ac135e00.htm#Contents
Tarapoto	The Tarapoto Proposal of Criteria and Indicators for Sustainability of the Amazon Forest	http://www.fao.org/forestry/ci/16609/en/, http://www.fao.org/docrep/004/ac135e/ac135e00.htm#Contents, http://otca.info/portal/
Certification and standardization		
2BSvs	Biomass Biofuel, Sustainability voluntary scheme	http://en.2bsvs.org/
AFS	Australian Forestry Standard	http://www.forestrystandard.org.au/
ATFS	American Tree Farm System	https://www.treefarmsystem.org/
BonSucro	BonSucro	http://www.bonsucro.com/
CCOF	California Certified Organic Farmers	http://www.ccof.org/
CEN TC383	CEN TC383/EN 16214 Sustainably produced biomass for energy applications by the European Committee for Standardization	http://www.cen.eu/
CertFor	CertFor, Chile	http://www.certfor.org/
CerFlor	CerFlor, Brazil	http://www.inmetro.gov.br/qualidade/cerflor.asp
CSA	Canadian Standards Association	http://www.csasfmforests.ca/
CSBP	Council on Sustainable Biomass Production	http://www.csbp.org/
Drax	Drax Sustainability Policy for Biomass	http://www.drax.com/biomass/sustainability_policy/
ENplus	ENplus by European Pellet Council	http://www.enplus-pellets.eu/
Florida Master Logger	Florida Master Logger	http://floridaforest.org/programs/master-logger/
FSC	Forest Stewardship Council	https://ic.fsc.org/
Georgia Master Harvester	Georgia's Timber Master Harvester Program	http://ga-mth.forestry.uga.edu/
GGL	Green Gold Label	http://www.greengoldcertified.org/
ISCC	International Sustainability and Carbon Certification	http://www.iscc-system.org/
ISO TC 248	ISO TC248/ ISO 13065 Sustainability Criteria for Bioenergy by the International Organization for Standardization	http://www.iso.org/iso/standards_development/technical_committees/other_bodies/iso_technical_committee.htm?commid=598379
IWPB	Initiative Wood Pellet Buyers (now Sustainable Biomass Partnership, SBP)	http://www.laborelec.be/ENG/services/biomass-analysis/initiative-wood-pellet-buyers-iwpb/, http://www.sustainablebiomasspartnership.org/
LBE	Laborelec Biomass Verification Procedure	http://www.laborelec.be/ENG/biomass-verification-procedure/
NTA8080	NTA 8080 Sustainably Produced Biomass	http://www.sustainable-biomass.org/
PEFC	Programme for the Endorsement of Forest Certification Schemes	http://www.pefc.org/
RAL-GGWL	RAL-Gütegemeinschaft Wald- und Landschaftspflege e.V.	http://www.ral-ggwl.de/
REDcert	REDcert Gesellschaft zur Zertifizierung nachhaltig erzeugter Biomasse mbH	http://www.redcert.org/
RSB	Roundtable on Sustainable Biomaterials	http://rsb.org/
RSPO	Roundtable on Sustainable Palm Oil	http://www.rspo.org/
RTRS	Round Table on Responsible Soy	http://www.responsiblesoy.org/
SAN	Sustainable Agriculture Network	http://san.ag/web/
SFI	Sustainable Forestry Initiative	http://www.sfiprogram.org/
SmartLogging	SmartLogging by the Rainforest Alliance	http://www.rainforest-alliance.org/forestry/certification/smart-logging
TREES	TREES (TRaining, Extension, Enterprises, Sourcing) by the Rainforest Alliance	http://www.rainforest-alliance.org/work/community-forestry

Table 32.4 Relationships between Background Variables, as Indicated by the *P*-Values of Chi-Square Tests for Equal Proportions[1]

Variable 1 → Variable 2 ↓	Capacity	Region	Geographical Experience	Scheme Most Familiar	Scheme Experience
Capacity	—	0.3025	0.1413	0.0187	0.0012
Region	0.3025	—	<0.0001	0.0002	0.0412
Geographical experience	0.1413	<0.0001	—	<0.0001	0.0062
Scheme most familiar	0.0187	0.0002	<0.0001	—	<0.0001
Scheme experience	0.0012	0.0412	0.0062	<0.0001	—

[1] H_0: Respondents are equally distributed with regard to variable 1 among all groups separated by variable 2.

certification (Figure 32.2(f), Q4). As many as 45 respondents had >5 years' experience with both bioenergy and certification, 42 had >5 years' bioenergy experience and 1–5 years' certification experience, and 29 respondents had 1–5 years' of both bioenergy and certification experiences.

To support the interpretation of differences among groups of respondents, we analyzed the background variables to see if proportions were equal, e.g., if respondents from a certain region tend more often to select a certain certification scheme as their most familiar. We applied a generalized logit model, using the SAS procedure for logistic regression models (LOGISTIC). The SAS Statistical Package version 9.3[41] was used for these and subsequent statistical analyses.

Respondents' capacity and geography were related to their certification scheme experiences (Table 32.4). For example, forest biomass producers and processors of solid biomass were more often knowledgeable of forestry schemes, and industries based on agricultural biomass were more often familiar with bioenergy schemes. More Europeans had experiences with bioenergy schemes, compared to North Americans.

Statistical Data Analyses

Responses to questions with predefined answers were analyzed against all background variables separately, except answers to questions of the third theme (Table 32.2), which were only analyzed against the selected certification scheme (Q7). However, results from tests of the effect of duration of bioenergy or certification experiences (Q4) were left out, as they were judged to be less interesting and sensible.

A varying share of the 192 respondents answered to the different questions. Some of the questions were only relevant to certain subgroups of respondents (themes 2 and 3), and otherwise testing of selected questions did not indicate other than values being missing completely at random. The reliability and validity of the questions and responses was also considered by evaluating the consistency of the open text comments with predefined answers, and through extensive discussions among co-authors regarding consistency of the responses with our understanding, the literature and the general scientific discourse.

Based on these observations and considerations, respondents providing no answer, answering 'I don't know', 'not relevant', or 'none', depending on the question, were excluded from the dataset prior to analyses. A binary logistic model was applied to binomial data and a proportional odds model to ordinal multinomial data,[42] using a SAS procedure for generalized models (GENMOD) and a procedure to create one-way to n-way frequency and contingency tables and chi-square statistics (FREQ). For analyses comparing the rating of different alternatives, the REPEATED statement of the GENMOD procedure was applied to take account of different alternatives being rated by the same person. When testing multiple pairwise differences in the same analysis, the Bonferroni adjustment for multiple comparisons was applied to adjust *P*-values. The pairwise relationships between all background variables were analyzed using a chi-square test for equal proportions (FREQ).

Throughout the paper, $P < 0.05$ is referred to as a significant difference, and $0.05 < P < 0.15$ sometimes as a tendency. For questions requiring rating of alternatives, we used an indicator to illustrate the differences between alternatives. It was obtained by converting the rating values to numeric values and then calculating the weighted average, cf. notes to tables. When many free text comments were given, these were organized thematically and summarized.

Further details on pretesting, missing values, and consideration of reliability and validity can be found in Appendix S2.

SURVEY RESPONSES

The Mix of Governance Mechanisms

Sustainability Governance of Bioenergy Systems

The EU-RED was rated as significantly more important than 28 other initiatives for sustainable land management

Table 32.5 Respondents' Rating of the Importance of Various National and International Initiatives, Including Legislation, Conventions, Standards, and International SFM Processes (Q8). See Table 32.3 for an Explanation of Abbreviations

		Very Important	Important	Moderately Important	Not Important		
Initiative	N[1]	%				Indicator[2]	Grouping[3]
EU-RED	141	63	26	7	4	2.5	a
RFS2	114	54	24	16	6	2.3	ab
ISO	119	32	45	18	6	2.0	bc
EU-CAP	106	36	35	24	6	2.0	bc
UNFCCC	138	35	37	19	9	2.0	bc
BAEZ	75	36	33	20	11	2.0	bcd
LCFS	105	34	30	25	11	1.9	bcde
CBD	112	29	36	26	9	1.9	bcde
Montreal	75	27	41	23	9	1.9	bcde
EUTR	85	26	44	15	15	1.8	cde
CEN	100	22	43	27	8	1.8	cdef
KP	148	26	34	28	11	1.8	cdef
ITTO	65	18	45	26	11	1.7	cdefg
Natura 2000	80	23	39	25	14	1.7	cdefg
ILO	105	26	33	26	15	1.7	cdefg
GBEP	92	18	39	29	13	1.6	cdefg
MCPFE	63	19	37	32	13	1.6	cdefg
Ramsar	90	18	34	30	18	1.5	defgh
CITES	95	14	33	35	19	1.4	defghi
NTA8080	83	11	35	35	19	1.4	efghi
CCD	74	15	24	42	19	1.4	efghi
EU-ELC	77	9	35	38	18	1.4	efghi
ATO	39	10	41	18	31	1.3	efghi
Cartagena	73	7	38	30	25	1.3	efghi
Dry Forest Asia	35	6	43	20	31	1.2	efghi
African Dry Zone	35	6	34	26	34	1.1	fghi
Tarapoto	34	3	32	32	32	1.1	ghi
Lepaterique	30	3	23	37	37	0.9	hi
Near East	30	3	17	43	37	0.9	i

[1] *N* is the number of respondents that rated the importance of the relevant initiative, leaving out missing observations and respondents answering 'I don't know'.
[2] Rating indicator calculated as a weighted average value when 'not important' = 0, 'moderately important' = 1, 'important' = 2, and 'very important' = 3.
[3] 3Initiatives that have the same letter were not rated significantly different ($P < 0.05$).

or bioenergy (Table 32.5, Q8). Respondents familiar with only bioenergy schemes rated the importance of EU-RED significantly higher than those familiar with only forestry schemes, as did Europeans compared to North Americans. The RFS2 was rated almost as important as the EU-RED. A significant majority of respondents agreed that the listed initiatives do not adequately ensure the sustainability of bioenergy (68% of 116 respondents, Q9).

Respondents said that jurisdictional legislation significantly influences their business or organization (Q10), mentioning energy-related legislation such as EU-RED, LCFS, green electricity legislation (Austria), carbon taxes, and energy funds (Australia). Legislation to combat illegal logging (Lacey Act, EUTR, and FLEGT) was also mentioned, as was legislation for forestry and protection of environment, nature, and biodiversity.

International initiatives and jurisdictional regulation together were not considered sufficient to ensure sustainability by 26% of 142 respondents, while 51% said it is partly enough, and 23% that it is adequate (Q11). It was, e.g., commented that national approaches to bioenergy sustainability are lacking, best management practices (BMPs) should be updated to include biomass harvesting, or that challenges remain in developing countries. Most respondents thought additional requirements should be implemented using a mix of voluntary initiatives and regulatory requirements (65% of 147 respondents, Q12). Biomass producers were significantly more often in favor of voluntary approaches, while energy experts and those involved in regulation, administration, and standardization were significantly more often in favor of regulatory requirements.

Several respondents also identified subcontractor certification systems that apply to their business (Q13), for example RAL-GGWL (Germany), SmartLogging, Georgia and Florida master logger programs, and farmer outgrower schemes. Of 38 respondents, 84% said that such schemes help achieve sustainability of bioenergy (Q14). Comments included that subcontractor schemes or training programs are crucial for on-the-ground implementation of BMPs, while others said they are not ideal because quantitative criteria are not always included. Many respondents indicated that EU-approved (ISCC, RSPO) and forestry schemes (PEFC, FSC) include subcontractor responsibilities.

A significant majority of respondents believed that voluntary certification schemes involving third-party audits are essential to ensure bioenergy sustainability (74% of 124 respondents, Q15). Some commented that voluntary certification sets higher standards than legislation, is more adaptable to regional differences and new knowledge, and is better at engaging stakeholders. Some also felt that voluntary certification can adapt faster than laws, and that the experiences gained may provide a basis for future legislation. Some said, however, that in regions with strong governance, certification may add cost without any real benefits, that jurisdictional regulation has stronger enforcement, can have impact at larger scales and may be more cost efficient.

A significant majority of respondents considered certification to be important worldwide, rather than in any specific region (Table 32.6, Q16). Respondents with experiences from only bioenergy schemes significantly more often thought worldwide certification is important compared to those with experiences from only forestry schemes.

Table 32.6 Number of Respondents Indicating that Voluntary Certification is Important in a Specific Region of the World (Q16)

Region	N^1	Grouping[2]
Worldwide	73	a
South America	32	b
South-East Asia	30	bc
Sub-Saharan Africa	27	bcd
South Asia	26	bcde
North America	21	bcde
Russia and former USSR	21	bcde
Central America	20	bcde
East Asia	19	bcde
Central and East Europe	18	bcde
North Africa	17	bcde
North Europe	12	bcde
Oceania	11	bcde
West Europe	10	cde
Middle East	9	de
South Europe	7	e

[1] *N* is the number of respondents answering that certification is important in the relevant region, leaving out missing observations and respondents answering 'not relevant' or 'I don't know'.
[2] The number of respondents indicating the importance of a region were not significantly different for regions which have the same letter ($P < 0.05$).

Noncertified Land

Significantly more respondents said that noncertified land is a problem for both agriculture (72% of 74 respondents, Q18) and forestry (78% of 92 respondents, Q18) than those saying it is not. European participants more often thought it a problem for agriculture than North Americans, and similarly for respondents with worldwide experiences compared to other respondents. A respondent commented that management plans are already required for most farms in the USA to participate in government programs[a].

Significantly fewer economic operators (biomass, biofuel, and energy producers) said that noncertified land is a problem for forestry compared to other respondents. Some said there is little difference on-the-ground where adequate regulations exist (USA), and that certification in such cases first of all serves to qualify for certified markets. One respondent mentioned that certification cannot create market differentiation if all forests are certified.

Group or regional certification, national verification methods and fiber sourcing programs were rated most effective for addressing noncertified land (Table 32.7, Q19). Economic operators were significantly less convinced that group certification is effective compared to other respondents. Respondents involved in certification rated national verification methods significantly less effective compared to respondents from other capacities. Europeans rated certification of entrepreneurs and introduction of international criteria in national legislation as more effective than did North Americans. One respondent suggested few harmonized criteria with cross-compliance, as in the EU-CAP, or income tax as control systems.

Redundancy, Harmonization, and Compliance

Of 109 respondents, 87% said that overlap and redundancy of schemes is a problem, while 13% said that all schemes have added value (Q20). However, 54% of 91 respondents (Q23) still found clear advantages to keeping the diversity of schemes, with 42 respondents indicating both redundancy and advantages to keeping the diversity. Explanations included needs to harmonize common features,

Table 32.7 Respondents' Rating of Different Governance Tools as Effective Alternatives or Support Tools to Voluntary Certification (Q19)

Alternative/ Support Tool	N^1	High	Moderate %	Low	Indicator[2]	Grouping[3]
Group or regional certification	82	66	26	9	2.6	a
National verification methods	81	56	37	7	2.5	ab
Fiber/biomass sourcing programs	74	53	36	11	2.4	ab
Certification of entrepreneurs	79	56	28	16	2.4	ab
Introduction of international criteria in national legislation	74	53	23	24	2.3	ab
Stepwise certification service	75	47	32	21	2.3	ab
Bilateral or multilateral agreements	79	38	43	19	2.2	ab
Multilateral initiatives as GEBP	67	27	49	24	2.0	b
Noncertification Systems	73	25	36	40	1.9	b

[1] N is the number of respondents rating the effectiveness of the relevant alternative/support tool, leaving out missing observations and respondents answering 'not relevant' or 'I don't know'.
[2] Rating indicator calculated as a weighted average value when 'low' = 1, 'moderate' = 2, and 'high' = 3.
[3] Alternatives/support tools that have the same letter were not rated significantly different ($P < 0.05$).

while allowing regional-specific criteria, or allow different approaches to evolve, while letting competition between schemes ensure that only the best endure.

There was concern that redundancy will lead to serious problems in the market (Table 32.8, Q21), for example that only few can understand differences in criteria and procedural rules among schemes. This enables shopping for cheaper and less rigorous systems, thereby decreasing credibility, uptake and impact. Costs are furthermore added with multiple training programs for certifiers, or when a company must certify against overlapping schemes. Schemes with specific approaches to feedstocks, regions, or end-uses were mentioned as inherently leading to redundancy.

Table 32.8 Distribution between Respondents Answering 'Yes' or 'No' that Redundancy in Certification Schemes Results in a Challenge (Q21)

Challenge	N^1	Yes %	No %	Grouping[2]
Confusion about sustainability definitions	92	95	5	a
Lack of consistency	88	90	10	a
Trade barriers	82	82	18	ab
Unwarranted discrimination in the use of raw materials	69	78	22	ab
Lack of transparency	87	70	30	b

[1] N is the number of respondents answering to the relevant challenge, leaving out missing observations and respondents answering '?'.
[2] Challenges that have the same letter were not assessed significantly different ($P < 0.05$).

There was no significant difference in how respondents overall rated five suggested solutions for redundancy (Table 32.9, Q22). However, respondents experienced with only bioenergy schemes rated mutual recognition as more relevant than did other respondents. Respondents

Table 32.9 Respondents' Rating of the Best Solutions to Challenges Caused by the Redundancy of Certification Schemes, cf. Table 32.8 (Q22)

Solution	N^1	High	Moderate %	Low	Indicator[2]	Grouping[3]
Mutual recognition	98	48	29	21	2.3	a
Introduce meta-standards	92	40	25	27	2.1	a
Reduce the number of voluntary schemes	94	37	30	27	2.1	a
Include international criteria in national legislation	92	34	30	28	2.1	a
Use future ISO or CEN standards	89	34	19	36	2.0	a

[1] N is the number of respondents rating a certain predefined solution, leaving out missing observations and respondents answering 'I don't know' or 'not relevant'.
[2] Rating indicator calculated as a weighted average value when 'low' = 1, 'moderate' = 2, and 'high' = 3.
[3] Solutions that have the same letter were not rated significantly different ($P < 0.05$).

selecting ISCC and RSB tended to rate inclusion of international criteria in national legislation as more relevant than especially respondents to SFI. Suggested solutions included letting the market solve the problems once minimum requirements have been set, and that the EU-RED recognition mechanism introduce requirements for standard-setting processes and compliance systems (e.g., ISEAL membership). Significantly more respondents said there are conflicting values between sectors which hampers harmonization (87% of 78 respondents, Q24). Respondents selecting a PEFC scheme tended to see less conflicting values compared to respondents selecting other schemes.

Respondents more often thought it important that certification schemes comply with EU-RED than with other international or jurisdictional initiatives (Table 32.10, Q25), especially European respondents. One respondent suggested that forest certification be complementary to broader initiatives, but still align with international processes and protocols if practical; it reduces companies' work to show compliance with broader initiatives.

Of 57 respondents, 60% said they had not found difficulties in satisfying the requirements of international processes or protocols (Q26). Respondents experienced with only bioenergy schemes significantly more often said they had found difficulties than respondents with experiences from forestry or mixed schemes. Some difficulties in relation to EU-RED included lack of farmer education, exclusion of primary forests, and the EU Effort Sharing Decision considering bioenergy as climate neutral, while EU-RED and FQD do not. A respondent also suspected that sections addressing municipal solid waste as a feedstock in RFS are vague and difficult to apply.

Table 32.10 Number of Respondents Recommending that Certification Schemes Comply with a Certain Initiative (Q25)[1]. See Table 32.3 for an Explanation of Abbreviations

	Yes	Not indicated
Initiative	%	
EU-RED	61	39
ILO	35	65
RFS2	32	68
EU-CAP	29	71
UNCCCF	27	73
CITES	26	74
ISO	26	74
CBD	24	76
Other	23	77
EUTR	21	79
KP	21	79
LFCS	19	81
Ramsar	18	82
Natura 2000	15	85
CCD	14	86
GBEP	14	86
CEN	13	87
BAEZ	12	88
ITTO	12	88
Montreal	12	88
MCPFE	7	93
Cartagena	6	94
EU-ELC	5	95
ATO	4	96
African Dry Zone	4	96
NTA8080	4	96
Tarapoto	4	96
Dry Forest Asia	2	98
Near East	2	98
Lepaterique	1	99

[1]There was a total of 84 respondents answering to this question, hereof 22 indicating 'I don't know'. No grouping showing significantly different responses among initiatives were included, as not all P-values for pairwise comparisons were available. However, all P-values for pairwise comparison with EU-RED were available and with $P < 0.05$.

Impacts on Markets, Trade, and Costs

Trade and Markets

A significant majority of respondents said they had observed changes in production or trade for their business, or the ones they audit, from 2008 to 2012 (70% of 50 respondents, Q28). This figure was 60% of 29 economic operators, and 94% of 17 respondents of other capacities (e.g., certification businesses), excluding associations.

Positive changes were experienced by 43% of 37 respondents, 11% experienced negative changes, and 46% changes that were neither positive nor negative (Q30). Positive experiences included increased biofuel trade from Brazil to the USA, increased market size, or distribution and production capacity due to RFS and LCFS. Some experienced improved market access after certification, or better registration and communication in the market, and less use of pesticides forbidden in destination markets. Some certification companies invested in additional business units and auditor education. An energy company started up a biogas certification scheme. Negative experiences included failure in meeting sustainability standards (biofuels from Argentina, biomass from native forests, and weed waste in Australia[b]), volatile demands for certified products, and increased administration costs.

Cost and Adaptation

There was no significant difference in ranking of reasons for not being certified (Table 32.11, Q35). However,

Table 32.11 Respondents' Rating of Reasons Why their Feedstock is Not Certified (Q35)

Reason	N[1]	High Importance	Medium Importance	Minor Importance	Indicator[2]	Grouping[3]
			%			
No significant market advantages	27	59	30	11	2.5	a
Administrative complexity	22	50	36	14	2.4	a
Too costly to become certified	23	52	30	17	2.4	a
No suitable certification scheme	21	48	24	29	2.2	a

[1] *N* is the number of respondents to the relevant predefined reason for not being certified, leaving out missing observations and respondents answering 'not relevant'.
[2] Rating indicator calculated as a weighted average value when 'minor importance' = 1, 'medium importance' = 2, and 'high importance' = 3.
[3] Issues that have the same letter were not rated significantly different ($P < 0.05$).

almost all respondents selecting FSC or RSB said that high administrative complexity is a very important reason. Other reasons given were lack of necessity, requirements already included in existing legislation or through other measures (Canada and USA), or that markets are too new and quantities too low.

Many respondents said that regulatory sustainability requirements affect their decision to enter a market (62% of 39 respondents, Q36). Comments indicated that businesses currently adjust their practices as necessary, even if several respondents commented that entering a market will ultimately depend on the balance between costs and benefits. Other respondents found regulations helpful for guaranteeing the sustainability of their production and the stability of their business. However, one respondent indicated that it is only worthwhile entering a market if the methods for assessing whether or not a product meets requirements are consistent and clear, as vague requirements may change unexpectedly and prevent future planning.

Legislative requirements and building a green business profile were rated highest among suggested motivating factors for seeking certification, although respondents were divided (Table 32.12, Q37). Additional motivations included clients' demands, risks of losing reputation, economic profits, or concern for environment and wildlife.

No cost type was rated significantly higher or lower, and respondents were divided as to whether a certain cost is considered high, medium or low (Table 32.13, Q39). Among 38 respondents, 63% said, however, that the costs of certification are significant compared to the value gains or profit margins of their commodity (Q40). Significantly more respondents from developing countries and North America said costs are significant compared to Europeans. When respondents indicated how their organization deals with the costs imposed by certification, there was a tendency for respondents to prefer to increase their commodity prices rather than accept lower profits (Table 32.14, Q41).

Table 32.12 Respondents' Rating of Reasons Why they Would Get Sustainability Certified (Q37)

Reason	N[1]	High Importance	Medium Importance	Minor Importance	Indicator[2]	Grouping[3]
			%			
Legislative requirements	34	68	6	26	2.4	a
Green business profile	37	59	19	22	2.4	a
General market access	36	47	47	6	2.4	ab
Maintain current market shares	31	45	32	23	2.2	ab
Access to niche markets	35	43	34	23	2.2	ab
Increased market shares	34	41	29	29	2.1	ab
Price premium for certified products	37	43	22	35	2.1	ab
Improve management systems	32	19	38	44	1.8	b

[1] *N* is the number of respondents to the relevant reason, leaving out missing observations and respondents answering 'not relevant'.
[2] Rating indicator calculated as a weighted average value when 'minor importance' = 1, 'medium importance' = 2, and 'high importance' = 3.
[3] Issues that have the same letter did not differ significantly ($P < 0.05$).

Table 32.13 The Respondents' Rating of Costs Items Associated with Certification (Q39)

Source of Costs	N^1	High	Medium	Minor	Indicator[2]	Grouping[3]
		%				
Services of the certifier and auditing company	29	55	28	17	2.4	a
Costs related to the extra administration	29	48	31	21	2.3	a
Fees to the certification body	31	45	35	19	2.3	a
Fees to the certification scheme	33	39	42	18	2.2	a
Increased feedstock prices due to purchasing certified materials	25	36	28	36	2.0	a
Costs related to changed management practices	31	23	45	32	1.9	a

[1]*N* is the number of respondents to the relevant source of costs, leaving out missing observations and respondents answering 'not relevant'.
[2]Rating indicator calculated as a weighted average value when 'minor' = 1, 'medium' = 2, and 'high' = 3.
[3]Cost sources that have the same letter were not rated significantly different ($P < 0.05$).

Credibility and Efficacy of Voluntary Certification

Scheme Revision and Scope

Of 64 respondents, 67% indicated that the scheme they selected has been revised within the last 5 years for bioenergy concerns (Q43). Respondents selecting a PEFC scheme more often said it had not, except for SFI (Table 32.15, Q43). A respondent mentioned that SFI added new bioenergy-relevant language to its standards. Bioenergy schemes had undergone revisions to adapt to national conditions (RSPO, Colombia), EU-RED (ISCC, RTRS, and NTA8080) or a multiple of feedstocks (RSB).

Respondents selecting a bioenergy scheme more often saw a need for future revisions compared to respondents selecting a forestry scheme. They especially saw a need to include periodic revisions and future legislative requirements, but were more satisfied with current GHG methodology (Table 32.16, Q44). Most respondents rejecting a need to include regular standard revisions said it is already included (92% of 26 respondents from both types of schemes). Among those that thought revision of GHG methodology is not needed, 80% of 25 respondents selecting a bioenergy scheme said it is already included, while 92% of 12 respondents selecting a forestry scheme said it is not needed. Respondents to forestry schemes commented that GHG is beyond the scope of forest certification and should be left for specialized schemes or national regulators.

Respondents selecting a bioenergy scheme more often saw a need for revisions to address land management sustainability criteria (biodiversity, soil, water, etc.) compared to respondents selecting a forestry scheme (Table 32.16). Two respondents to ISCC commented that requirements will not go beyond those of EU-RED, but questioned whether water issues will be included in EU-RED in the future. One respondent mentioned that PEFC will undergo revisions in 2015 to decide whether land management criteria should be improved, and another that ATFS certification relies on development of BMPs at the local (state) level, which all have periodic revisions. Among those not seeing a need to improve sustainability criteria for land management, most said it is already included, while some

Table 32.14 Respondents' Ranking of Options to Describe How their Company Deals with Increasing Costs Due to Certification (Q41); 5 is Most Important and 1 is Least Important

Option	N^1	5	4	3	2	1	Indicator[2]	Grouping[3]
		%						
		%						
Increase commodity prices	14	50	14	7	29	–	3.9	a
Seek efficiency improvements	10	20	40	20	20	–	3.6	a
Negotiate less expensive feedstock prices	8	13	38	38	13	–	3.5	a
Change to less expensive feedstock supplier	9	44	11	11	11	22	3.4	a
Accept lower profits	11	9	18	27	–	45	2.5	a

[1]*N* is the number of respondents to a the relevant option, leaving out missing observations.
[2]Rating indicator calculated as a weighted average of values 1–5.
[3]Issues that have the same letter did not differ significantly ($P < 0.05$), with $P = 0.074$ for a pairwise test if 'increase commodity prices' is significantly different from 'accept lower profits'.

Table 32.15 Distribution between Respondents Answering 'Yes' or 'No' When Asked If their Certification Scheme Had Been Revised within the Last 5 Years to Increase its Relevant to Bioenergy Production Systems (Q43)

Scheme/Groups of Schemes	N^1	Yes (%)	No (%)	Grouping[2]
PEFC scheme	8	13	88	a
FSC	8	63	38	a
Other EU approved	10	70	30	a
Other	7	71	29	a
RSB	11	73	27	a
Pellet scheme	5	80	20	a
ISCC	11	82	18	a
SFI	4	100	-	b

[1] N is the number of respondents selecting the relevant scheme or group of schemes, leaving out respondents answering 'I don't know / not relevant'.
[2] Schemes/groups of schemes which have the same letter are not significantly different ($P < 0.05$).

commented that such revisions are in process (RTRS, RSB, IWPB, and LBE).

Few respondents suggested revisions to address indirect land-use change (iLUC), but some noted that criteria for iLUC are needed, as some feedstocks are iLUC drivers, and that it should be included, if included by EU-RED. Respondents rejecting revisions to include iLUC most often said it is not needed or that iLUC should be handled at the policy and global market level.

A respondent to RTRS suggested it is important to include social criteria, if expansion of soy production is allowed. Some said that various chain-of-custody initiatives have been undertaken to improve traceability, and that online traceability systems are being considered. Respondents furthermore mentioned that ISCC may expand its scope to include solid biomass for heat, power, and other energy or nonenergy uses.

Stakeholder Involvement

Most respondents felt adequately involved in the standard development and revision process of the scheme they selected (73% of 73 respondents, Q45). One respondent commented that not all have access to politicians that create regulations (for which the scheme ensure compliance). A producer indicated it is too time-consuming to be involved, and that being informed is adequate.

Of 67 respondents, 57% said there is a proper balance among stakeholders involved in standard development and revisions (Q46). Some commented that NGOs and small-scale producers in developing countries should be more involved (e.g., ISCC, RSB, IWPB, and RSPO), or that small-scale producers are under-represented in SFI in south-eastern USA. Others called for more involvement of people with hands-on industry experience (ENplus, TREES, 2BSvs, RTRS, PEFC Sweden).

A significant majority of respondents saw a need to engage with other schemes and organizations to coordinate, supplement, or harmonize the efforts of their own scheme (78% of 64 respondents, Q47). Ensuring consistent verification and auditing was rated as the most urgent issue to be addressed through such collaboration (Table 32.17, Q48).

Table 32.16 Distribution between Respondents Answering 'Yes' or 'No', Overall and for Respondents Selecting a Bioenergy Scheme or a Forestry Scheme, Respectively, When Asked If their Scheme Needs Revision Regarding a Certain Issue (Q44)

	Overall				Bioenergy[3]			Forestry[3]			
		Yes	No			Yes	No		Yes	No	
Issue	N^1	%		Grouping[2]	N^1	%		N^1	%		Test[3]
Periodic revisions should be conducted	70	63	37	a	39	82	18	22	23	77	***
Future legislative requirements	57	61	39	ab	33	79	21	17	18	82	***
Handling of conflicting criteria	39	56	44	abc	22	64	36	11	36	64	n.s.
Future management system requirements	54	46	54	abc	34	50	50	13	23	77	n.s.
Sustainability criteria for land management	69	45	55	abc	40	53	47	20	15	85	**
Expanded supply chain scope	58	41	59	abc	35	49	51	15	13	87	*
iLUC concerns	56	41	59	abc	34	50	50	14	0	100	***
Sustainability criteria for the Chain of Custody	58	40	60	bc	35	46	54	17	12	88	*
GHG calculations	64	38	63	c	39	36	64	16	25	75	n.s.

[1] N is the number of respondents answering to the relevant issue, leaving out missing observations and respondents answering 'I don't know'.
[2] Issues that have the same letters are not significantly different ($P < 0.05$).
[3] Test result of differences in the distribution of answers between respondents selecting a bioenergy and a forestry scheme, respectively, where n.s.: non-significant difference, $^{*}0.01 < P < 0.05$, $^{**}0.001 < P < 0.01$, $^{***}P < 0.001$.

Table 32.17 Distribution between Respondents Rating If a Certain Issue is Important to Address in Consultations with External Stakeholders (Q48)

Issue	*N*[1]	High	Medium	Minor	Indicator[2]	Grouping[3]
			%			
Ensure consistent verification and auditing	71	63	27	10	2.5	a
Jointly meeting sustainability criteria and goals along the supply chain	69	43	41	16	2.3	ab
Improve the Chain of Custody	66	24	52	24	2.0	abc
More clarity in responsibilities along the supply chain (sustainability aspects)	64	25	50	25	2.0	abc
Meeting external administrative requirements	66	27	46	27	2.0	bc
Improved consideration of external stakeholders views	66	26	41	33	1.9	c

[1]*N* is the number of respondents answering the relevant issue, leaving out missing observations.
[2]Rating indicator calculated as a weighted average value when 'minor' = 1, 'medium' = 2, and 'high' = 3.
[3]Issues that have the same letters are not significantly different ($P < 0.05$).

Respondents selecting a bioenergy scheme rated 'jointly meeting sustainability criteria and goals along the supply chain' as more important than those selecting a forestry scheme. Other suggestions for collaboration included creation of a global platform for trading certified products, benchmarking and mutual recognition. It was also commented that schemes are competitors and that violation of anti-trust laws must be observed. Others commented that systems currently need to consolidate themselves, that it gets increasingly complicated as more actors get involved, or that informal collaboration is more helpful.

On-the-Ground Effectiveness

There were significant differences in how respondents rated the effectiveness of their scheme in achieving different criteria (Table 32.18, Q50). Respondents selecting a bioenergy scheme significantly more often said that their scheme is effective in achieving criteria for GHG emissions reductions, energy balances, and ecosystem carbon stocks compared to those selecting a forestry scheme. Some respondents still found major weaknesses in the GHG balance calculations of their scheme (55% of 62 respondents, Q52). Respondents selecting ISCC were most convinced there are no weaknesses compared to respondents selecting RSB.

Concerns included, e.g., calculations not adequately taking account of iLUC, lack of information about real behavior of agricultural systems, use of default values leading to under-estimation of GHG emissions or reducing the incentive to seek further emission reductions, cf. also 3.4.1.

Significantly more respondents selecting a forestry scheme said that their scheme is effective in achieving criteria for biodiversity, soil, water and social issues (Table 32.18, Q50). Comments indicated that these issues are sometimes only included indirectly in bioenergy schemes via criteria for direct land-use changes (RSB), promotion of good agricultural practices (ISCC), or they rely on existing regulation (RSB). However, it was also commented that criteria for direct land use change do not capture the biodiversity benefits of existing agricultural land.

Relatively fewer respondents said that their schemes are effective in achieving criteria for air quality and competition with raw material use. Explanations given were for example that air quality is already more effectively dealt with by legislation; that it is difficult to address by certification schemes; or that such criteria have emerged in standards, but they are still new and experiences are few. Comments also included that it is difficult to address; that competition with other raw material is not a problem for waste; or that it is only a problem if policy distorts the markets.

Most respondents answered that they had not encountered attempts to falsify sustainability claims (81% of 69 respondents, Q54). However, some new cases of mistakes in mass balance calculations due to lack of training, certified raw material claims among operators without a chain-of-custody, logo use violations, or continued sales despite certificate withdrawals. A respondent informed that FSC is seeking to address the latter through an Online Claims Platform.

Many respondents thought that certification had caused unintended consequences (68% of 63 respondents, Q51), which were often positive. However, one respondent mentioned exclusion of certain feedstocks in Australia due to sustainability criteria. Similarly RFS2 does not accept material from areas planted after December 2007, which may discourage reforestation in areas where it is needed for conservation purposes.

Table 32.18 Distribution between Respondents, Overall and for Respondents Selecting a Bioenergy Scheme or a Forestry Scheme, Respectively, When Asked Whether their Certification Scheme is Effective in Achieving Different Sustainability Criteria or Related Issues (Q50)

	Overall				Bioenergy[3]			Forestry[3]			
		Yes	No			Yes	No		Yes	No	
Criterion	N^1	%		Grouping[2]	N^1	%		N^1	%		Test[3]
Traceability	69	88	12	a	36	86	14	26	96	4	n.s.
Biodiversity	71	83	17	ab	36	72	28	27	93	7	*
Soil	70	83	17	ab	34	74	26	27	93	7	*
Water	71	79	21	abc	35	69	31	27	85	15	n.s.
Social aspects	72	75	25	abc	37	62	38	27	90	11	*
Ecosystem carbon stocks	65	70	30	abcd	34	79	21	23	48	52	*
GHG emissions	71	65	35	bcd	41	85	15	20	10	80	***
Air quality	65	65	35	cd	34	65	35	23	57	43	n.s.
Energy balance	65	55	45	cd	36	77	28	22	14	86	***
Competition with other raw materials	61	46	54	d	31	55	45	22	32	68	n.s.

[1] N is the number of respondents answering to the relevant criterion.
[2] Criteria that have the same letters are not significantly different overall ($P < 0.05$).
[3] Test result of the differences in the distribution of answers between respondents selecting a bioenergy and a forestry scheme, respectively, where n.s.: non-significant difference, $^{*}0.01 < P < 0.05$, $^{**}0.001 < P < 0.01$, $^{***}P < 0.001$.

Verification and Auditing Challenges

Among 43–45 respondents, 64% rated their scheme as highly effective with regard to 'auditor competency', while the corresponding figures were 55, 48, and 47% for 'detecting and correcting noncompli-ances', 'auditing frequency' and 'sampling intensity', respectively (Q56). Between 30 and 40% of the respondents said the effectiveness is medium with regard to these four issues. There were no significant differences in how the four elements were rated. Comments indicated that new auditors sometimes lack important multi-disciplinary competences, and that noncompliances are possible as operators are not always obliged to reveal comprehensive documentation to the auditor. Another commented that sufficient smallholder sampling can be a challenge, and audit frequency based on risk levels was suggested as a solution.

Auditing difficulties did not differ significantly among sustainability issues (Table 32.19, Q57), but 20 respondents selecting RSB, other EU-approved schemes or FSC

Table 32.19 Respondents' Rating of Difficulties in Auditing for Different Sustainability Criteria and Related Issues (Q57)

		High	Medium	Minor		
Criterion	N^1		%		Indicator[2]	Grouping[3]
Ecosystem carbon stocks	35	34	57	9	2.3	a
Competition with other raw material uses	33	30	58	12	2.2	a
Biodiversity	48	38	42	21	2.2	a
GHG emissions	40	35	45	20	2.2	a
Social aspects	47	36	38	26	2.1	a
Energy balance	36	31	47	22	2.1	a
Supply chain control	49	27	51	22	2.0	a
Soil & water	45	24	47	29	2.0	a
Traceability	47	21	51	28	1.9	a
Air quality	34	21	47	32	1.9	a
Internal Management systems[4]	49	22	39	39	1.8	a

[1] N is the number of respondents answering to the relevant criterion, leaving out missing observations and respondents answering 'not relevant'.
[2] Rating indicator calculated as a weighted average value when 'minor' = 1, 'medium' = 2, and 'high' = 3.
[3] Sustainability criteria that have the same letters are not significantly different ($P < 0.05$).
[4] Documentation, training, etc.

Table 32.20 Distribution between Respondents Rating the Magnitude of the Costs for Providing Information for Verification for Different Sustainability Criteria and Related Issues (Q58)

		High	Medium	Minor				
Criterion	N^1		%		Indicator[2]	Grouping[3]	N_D[1]	Test[5]
Ecosystem carbon stocks	26	54	31	15	2.4	a	21	B<F*
Biodiversity	33	42	45	12	2.3	ab	28	B>F**
GHG emission reductions	26	38	35	27	2.1	abc	21	n.s.
Supply chain control systems	34	29	44	26	2.0	abc	29	n.s.
Soil and water	36	31	42	28	2.0	abc	31	B>F*
Internal management systems[4]	36	33	33	33	2.0	abc	31	F>B*
Traceability / place of origin	37	24	49	27	2.0	abc	32	n.s.
Competition with other raw materials uses	25	20	52	28	1.9	abc	20	n.s.
Energy balance	25	32	28	40	1.9	abc	20	n.s.
Air quality	31	26	35	39	1.9	bc	26	n.s.
Social aspects	34	18	41	41	1.8	c	29	n.s.

[1]*N* is the number of respondents answering the relevant criterion, leaving out missing observations and respondents answering 'Not relevant'. Similarly, *N*D is the number of respondents answering to a certain issue excluding those in the group 'other schemes'.
[2]Rating indicator calculated as a weighted average value when 'minor'=1, 'medium'=2, and 'high'=3.
[3]Sustainability criteria that have the same letters are not significantly different ($P<0.05$).
[4]Documentation, training, etc.
[5]Test result of the difference in rating of cost magnitudes between respondents selecting a bioenergy (B) or a forestry (F) scheme, respectively. B>F: respondents selecting a bioenergy scheme rated the costs to be higher than did those selecting a forestry scheme; F>B, respondents selecting a forestry scheme rated the costs to be higher than did those selecting a bioenergy scheme, where n.s.: non-significant, $^*0.01 < P < 0.05$, $^{**}0.001 < P < 0.01$, $^{***}P < 0.001$.

generally rated difficulties in conducting audits higher than 24 respondents selecting ISCC, PEFC, or SFI. Respondents commented that criteria for ecosystem carbon (SFI) and social issues (RSB), and raw material competition are poorly covered and not often audited, including a comment recommending that the latter should be handled by the market and not by governance.

Costs for verifying criteria for ecosystem carbon stocks and biodiversity were rated highest by respondents (Table 32.20, Q58). Overall, respondents selecting different schemes evaluated the auditing costs differently in the following order (from high to low costs): other schemes > RSB > other EU-approved > ISCC > FSC > PEFC > SFI > pellet schemes. Respondents selecting a bioenergy scheme more often considered auditing costs of ecosystem carbon stocks, biodiversity, soil and water as high as compared to respondents selecting a forestry scheme. The opposite was found for internal management systems.

The need for new or improved verification tools was not evaluated significantly different among sustainability criteria. Respondents were divided on whether improvements are needed or not (Table 32.21, Q59). However, across criteria respondents selecting different schemes tended to see the need for improving verification tools differently in the following order (from a high to low 'yes'/'no' ratio): FSC > ISCC > Other > Other EU-approved > PEFC > pellet schemes > RSB > SFI. Introduction of simple

Table 32.21 Distribution between Respondents Voting If there is a Need to Develop New Tools for Verification of Sustainability Criteria (Q59)

		Yes	No		
Criterion	N^1		%	Grouping[2]	Test[4]
Ecosystem carbon	36	42	58	a	n.s.
GHG emissions	37	46	54	a	n.s.
Energy balance	34	47	53	a	n.s.
Biodiversity	40	53	48	a	n.s.
Soil and water	38	53	47	a	n.s.
Competition with other raw materials	34	53	47	a	n.s.
Air quality	35	60	40	a	n.s.
Traceability	40	60	40	a	n.s.
Social aspects	41	61	39	a	n.s.
Supply chain control systems	37	62	38	a	n.s.
Internal management systems[3]	38	71	29	a	*

[1]*N* is the number of respondents answering to a certain issue, leaving out missing observations and respondents answering 'Not relevant'.
[2]Sustainability criteria that have the same letters are not significantly different ($P < 0.05$).
[3]Documentation, training, etc.
[4]Test results of differences among those saying 'yes' and 'no', where n.s.: non-significant, $^*0.01 < P < 0.05$, $^{**}0.001 < P < 0.01$, $^{***}P < 0.001$.

internationally recognized forest inventory methodologies, such as the IPCC tier 1 approach, were suggested for improving ecosystem carbon monitoring (FSC). Elaboration of lists of high conservation value areas were suggested for dealing with biodiversity protection in southeastern USA, as it is a major concern for nonindustrial private forest landowners in this region (PEFC). RSB respondents said existing tools for soil and water are complex and regionally specific, but made no specific suggestions on how to solve this challenge. It was mentioned that customers often ask for more traceability information than can be supplied (SFI). A suggestion for improvements in this regard included online tracking systems (FSC).

Of 35 respondents, 43% said they found consistency among the data, tools, models, and guidelines used for verification along the whole supply chain (Q60). Respondents selecting a bioenergy scheme tended more often to reject this than those selecting a forestry scheme. Comments indicated inconsistencies in auditing practices for different criteria, in risk evaluation, and among public and private geospatial data. One person suggested that auditors explicitly reference the tools they use for verification, while another said that although there are no written auditing guidelines, many assessors apply the same methodology to different schemes and sectors.

DISCUSSION OF SURVEY RESPONSES

Despite efforts to request feedback from stakeholders globally, several major production regions, such as South America and Asia, were not adequately represented to draw solid conclusions on their views and experiences. More respondents from RSPO would also have been desirable, while difficulties in getting a balanced representation among, e.g., different capacities may simply reflect that some groups are smaller or that fewer persons with a certain group possess the required expert knowledge. Keeping these limitations in mind, this section discusses the achievements, challenges, and solutions in relation to the development of sustainability governance for the bioenergy sector as identified by this survey. We also identify knowledge gaps and research questions, which may in the future lead to tests of specific questions for rigorously defined target populations.

The Mix of Governance Mechanisms

Co-Regulation and the Role of Certification

Co-regulation, in the form of legislation which uses market-based certification schemes to demonstrate compliance, was widely supported by respondents to this survey. Such forms of governance may conserve public resources and allow authorities to make use of a private agent's expertise, with a possible drawback being the vested interests of the private agents.[43] Studies have shown that private certification systems and public policies may interact synergistically,[44,45] but studies on the outcomes of such interactions, when actors are linked in a principal-agent relationships are lacking, cf. 4.2 and 4.3.

Like this study, the EU consultation on sustainability criteria for solid and gaseous biomass showed overall support for new mandatory criteria, with the exception of public authorities in some forest-rich Member States.[46] Biomass producers within our survey were also less in favor of new mandatory sustainability requirements than other respondents. A reason can be that the biomass producing sector sees a duplication of the existing legislation and other work that has been undertaken, as is the case for forestry in the Canadian provinces.[47,48] In the USA, widely implemented BMPs are also sometimes considered adequate for forestry.[49] A respondent's comment, cf. 3.2.2., suggested that existing requirements for agricultural management plans are adequate in the United States, even if such plans are only required for areas participating in various environmental programs.[38] Interestingly, the process to enhance biomass sustainability criteria in the UK considers a system that can approve whole regions.[50] This system should evaluate case by case whether existing evidence is adequate and sufficiently rigorous. This may allow smaller producers to stay within more informal, trust-based governance systems, which are predominant in private forestry, especially in developed countries.[51] Combining such considerations with the market's demand for prescriptive regulation has been identified as a major challenge for SFM,[45] and it applies to sustainable bioenergy production, too.

Noncertified Land

Optimism regarding the role of certification in biomass sustainability governance in this study may be due to the success of forest certification in some developed countries such as Canada[52] and Finland,[53] or belief that new linking of jurisdictional regulation with private international certification may change the situation, cf. also 4.2.1.

Group certification, which was rated as promising for increasing small holders' accessibility to certification, is already a component of several certification schemes.[39,54] It can significantly reduce costs, but successful application depends on the existence of organizations that can provide outreach, secure funding, understand the rules, and handle documentation.[55] It is therefore not surprising that it tends to be more helpful for groups in Northern than Southern regions.[29] Schepers[29] nevertheless also recommends continued developments of group certification as a means to approach forests in the South (cf. also 4.1.1.). Several other solutions were suggested as useful, but responses were often split. This may indicate that the best solution depends on the situation or it may reflect conflicting values between sectors. Future research should clarify which solutions are appropriate for differing circumstances.

Redundancy and Harmonization

While we found broad consensus that the redundancy of certification schemes cause significant challenges, there was less consensus on how to solve these challenges, with conflicting values between sectors again being one likely explanation. The bioenergy sector tended to recommend legislative criteria more often than the forest sector, and the situation might be comparable to experiences from international agreements, where strict rules and strong enforcement often leads to a small group of signatories.[56] A narrow focus increases the chance of success (cf. regulation for timber legality), but sustainability standards are generally broad in scope. Meta-standards are alternatives to one global agreement, with the advantage that they to some extent allow the industry to make use of the systems they already apply. Furthermore, they reduce the likelihood of conflict with WTO trade rules.[43,57] Non-governmental meta-standards have also proven useful for international harmonization of national certification scheme standards, as in the case of PEFC.[43]

As respondents to bioenergy schemes in this study, respondents to a survey on EU-approved schemes also recommended mutual recognition among schemes, although they were concerned that inferior schemes may allow companies to green wash or that it will remove possibilities to demonstrate commitment to higher sustainability standards.[23,58] Mutual recognition has greater potential in fragmented markets, as for PEFC national schemes that do not compete internationally, or when schemes are complementary in scope. Gan and Cashore[59] suggest formal partnerships of complementary schemes, where forestry schemes verify biomass production and harvesting, and bioenergy schemes focus on conversion, distribution, and consumption.

As in this survey, others have argued that with time, less competitive schemes are likely to disappear.[56] Competition may push for improvements in scheme inclusiveness, effectiveness, and cost efficiency, and for a better balance among these factors.

Impacts on Markets, Trade, and Costs

Trade and Markets

We qualitatively confirmed expectations expressed in a previous study,[21] that European respondents will more often see benefits to their trade and business compared to actors from other regions. A study by Goh et al.[25] conducted concurrently with this survey reported examples of Russian wood pellets and liquid biofuels from Argentina and Southeast Asia being rejected because criteria for reduction of GHG emissions were not met, and mentioned that future EU criteria for solid biomass may restrict pellet imports from Russia (GHG emission reductions) and Canada (rejection of material from primary forest). However, no general short-term changes were anticipated as procurement strategies for wood pellets involve long-term contracts.

The literature also suggests that possible effects of sustainability certification on liquid biofuel trade also appear to have been outweighed by other factors.[25] Decreasing participation in the EU markets among developing countries is rather due to EU import tariffs, lack of noticeable price premiums, a high domestic demand for ethanol in traditional exporting countries, such as Brazil,[60] support for agriculture in importing countries, and, for exporting countries in Africa, lack of domestic markets to compensate volatile international demand.[61]

We did not as intended (Q31–33) obtain good quantitative data on certified trade, and even if Goh et al.[25] were successful in collecting some information on certified trade through interviews, they came to the same conclusion. Online platforms for certified trade, such as FSC's Online Claims Platforms, could be developed to include such information.

Benefits and Costs

The main motivations for getting certified for our (mostly) European and North American respondents were legislative requirements and green business profile strategies. A study of FSC chain-of-custody certificate holder perceptions similarly found that legislative requirements of EUTR and the US Lacey Act are important drivers of forest certification in Vietnam, while support for sustainable forestry, environmental protection and Corporate Social Responsibility (CSR) goals are important in China.[62] This suggests the existence of two different types of actors: some that merely wish to fulfil minimum requirements of the legislation to access new markets, and some that care about responsible production and are willing to exceed minimum legislative requirements.

Price premiums were previously thought to be a main motivation to get certified, but it was rated among the least important benefits of certification in this study. Similar findings exist for forest certification in Europe,[63] the USA,[64] and Asia.[62] Price-premiums also appear small for certified biodiesel and almost nonexistent for certified bioethanol.[60]

While these results were for manufacturers or traders (chain-of-custody), another study addressing wood producers (SFM) showed significant price premiums for high quality tropical logs, but not for low quality logs.[65]

Certification costs were, however, considered as significant in this study, with no significant differences among respondents in different capacities. However, Pacini et al.[60] and van Dam et al.[39] find that the required number of auditing days and costs for showing compliance are highest for biomass producers, often located in developing countries. For the wood pellet trade between Canada and Europe the certification cost distribution may vary, sometimes being significant especially for the biomass producers, while at other times buyers pay all the additional costs.[66] The study by Goh et al.[25] suggests that for liquid biofuels, the additional cost of certification appears relatively less significant compared to price of the feedstock, e.g., vegetable oils, maybe because the relevant schemes are often limited to the EU-RED requirements. Although market actors hesitate to get certified, more are likely to participate in the future. As costs are considered significant, it will be critical for the legitimacy of certification systems to examine if the benefits and costs are distributed in a fair manner along the whole supply chain, and which are the tradeoffs between effectiveness and cost efficiency.

Legitimacy, Credibility, and Efficacy of Voluntary Certification

Scheme Revision and Scope

Respondents selecting a bioenergy scheme generally saw a greater need for revisions of these schemes compared to respondents selecting a forestry scheme, probably reflecting that bioenergy schemes are newer in development. Forestry schemes already have a well-developed identity and well-proven systems, and most respondents did not see a need for revisions. We note that forestry schemes undertake regular (e.g., every 5 years) reviews and revisions of standards as a part of normal practice. However, a few respondents recommended including for example GHG and other sustainability criteria at the supply chain level for these schemes. Social criteria and EUTR requirements have already been added to the PEFC chain-of-custody.[67] Altogether this suggests that forestry chain-of-custody standards may develop from being mere tracking systems to being supply chain sustainability standards.

Respondents to this survey were skeptical whether iLUC should be addressed by certification schemes, and similar reservations exist for the closely related food security criterion in another study.[68] One problem is that responsibilities for iLUC effects are rarely solely driven by a particular organization and are likely to be due to cross-sectoral interactions.[69,70] This survey and other studies suggest that iLUC needs to be addressed at larger scales through government mechanisms.[24,71,72] Alternative means to include iLUC in certification are also being explored, e.g., by ISO[73] and RSB.[74] However, these approaches are likely not adequate to deal with all concerns related to iLUC.

Legitimacy and Stakeholder Involvement

This survey indicated that most certification schemes allow for involvement of all types of stakeholders, although unequal access to legislators was mentioned as a barrier. Lima et al.[75] find that this is especially a problem when voluntary certification schemes are used transnationally to demonstrate compliance with national or regional legislation. This conflicts with emerging international agreements on good governance,[34] as countries may be affected without having influence on the policies causing the impacts.

As in this survey, others also find that local stakeholders are often involved mainly in implementation of certification.[76] Using RSB as an example, Lima[75] concludes that in private biofuel regulatory systems, representatives from developing countries are often large energy and agribusinesses, not smaller stakeholders. Similar observations have been made for forestry schemes,[45] RTRS,[77] and RSPO,[78] with one explanation being that introduction of regulatory structures is often driven by large-scale interests in commercial production. In-depth comparative analyses of the balance and quality of the stakeholder participation might help determine the factors leading to successful, equitable involvement.

On-the-Ground Effectiveness

Respondents to forestry schemes were generally confident that their schemes effectively achieve criteria for biodiversity, soil, water and social aspects. Studies confirm that forest certification improves forest management practices,[44] and habitat suitability and diversity.[79,80] However, Elbakidze et al.[80] still find variations between regions in Sweden and Russia under FSC certification, due to different levels of ambition rather than differences in local conditions (e.g., level of forest fragmentation or land-use history, cf. 4.1.1). International generic indicators are under development to ensure consistent application of the FSC standards among countries,[81] but avoiding standards with different levels of ambition, while allowing for local adaptation, is still an important challenge to schemes' legitimacy.

Respondents to bioenergy schemes were less convinced that their schemes are effective in achieving land management and social criteria and saw more needs for revision. Hennenberg et al.[82] also observed that bioenergy standards fall short in promoting agricultural practices with low negative impacts on bio-diversity. Others find more rigor in bioenergy than forestry standards, e.g., in preventing conversion of high biodiversity value grass- and wetlands.[23] However, inventories of on-the-ground impacts of bioenergy schemes likely have not been conducted yet, and they are still scarce for forestry schemes.

Respondents to forestry schemes more often said their schemes are not effective for the ecosystem carbon criterion, although this has long been included in the Forest Europe and Montreal intergovernmental SFM processes, on which also the international PEFC international standard relies.[67] However, carbon issues are only emerging in forest certification standards,[40] while it is already partly included under EU-approved schemes' GHG calculations,[10] see also 4.3.4.

It appeared from comments that some respondents interpreted effectiveness of certification as 'comprehensiveness of the standards' or 'quality of the auditing'. However, these do not per se ensure that sustainability goals are being reached on-the-ground. It will also depend on the effectiveness of the selected indicators. This likely demonstrates how difficult it is to obtain knowledge about on-the-ground effectiveness, even from deeply involved people. Angel et al.[83] also conclude that the largest gap in the literature on global environmental standards is research on the impacts of those standards, and Meidinger et al.[84] point out that in forest certification and environmental law, 'we lack anything remotely approximating evaluation research'. Expensive approaches such as monitoring and rigorously designed scientific studies appear to be the only way forward if impact should be documented under a variety of conditions.

Verification and Auditing Challenges

Certification schemes often require that auditors attend specific training workshops,[39,67,85] which might explain the relatively high confidence in auditor competences seen in this study. However, some studies show that the specific outcomes of FSC forest certification depend on the individual auditors' knowledge, skills and attitudes,[86] or that, e.g., SFI auditors are more rigorous than FSC auditors in their expectations for continuous improvement.[87] Perhaps this links to the larger difficulties in auditing observed among respondents to FSC and RSB compared to ISCC, PEFC, and SFI in this study. It may also link to differences in the ease of criteria interpretation. A field test of forestry standards showed some degree of difficulty in the interpretation of >40% of the criteria of the FSC Maritimes standard, 22% in FSC Draft standard for Finland and 10% in SFI standard, while for PEFC Finland and UKWAS, the criteria presented no difficulties in interpretation.[88] It is a question if criteria of for example FSC and RSB are less well designed, or if they are more complex and ambitious, thus requiring a higher level of knowledge and experience to make assessments under local conditions.

For ecosystem carbon stocks, GHG balances, and biodiversity, the auditing difficulties, costs and the need to develop verification tools appeared inter-linked by respondents' slightly higher rating compared to other sustainability issues. Other studies assert that improved tools are needed for assessing bio-diversity conservation,[80] ecosystem carbon,[89] GHG performance,[90] and agricultural practices.[38] Increased harmonization of the protocols, tools, models, and guidelines used for on-the-ground verification and public availability of these tools could pave the road for more consistency, transparency, and credibility of verification procedures.

The effectiveness of certification schemes in finding and correcting noncompliances was evaluated as low-moderate by 45% of respondents. We find this rating surprisingly low, but it may relate to opinions that sampling frequency and intensity could also be improved. Another reason may be auditors only pushing for continual progress, while the standard requirements are performance-based,[91] or noncompliances being tolerated because the requirements are generally perceived as not fair or practical.[92] Some studies use the number of detected noncompliances and their correction as a measure of the effectiveness of forest certification,[91,93] but this study suggests that it is also relevant to address the amount of nondetected non-compliances during regular audits.

It appears to be an important observation that when metastandards, such as EU-RED, tend to harmonize schemes' substantive requirements, increased competition on procedural rules may arise.[56] These are more complicated to understand than sub-stantive rules, and it becomes harder to identify which are the more rigorous schemes. Certification systems generally use third party auditing, which is a strong compliance system,[16,94] while use of cost-reducing self-declaration by farmers is relatively common as part of EU-RED approved schemes' compliance systems.[39] According to the Steering Committee of the State-of-Knowledge Assessment of Standards and Certification,[54] standards systems are currently exploring how self-assessment can be combined with second- and third-party audits to streamline the certification process. Such analyses will be useful for designing future compliance systems, especially if also differences in on-the-ground effectiveness are addressed.

CONCLUSION

This survey revealed a broad support for existing and new co-regulation among stakeholders, but also that low share of certified land is seen as a challenge for both forestry and agriculture. Group certification and verification methods for the national or regional scale were among the most popular suggestions to meet this challenge, but co-regulation in itself may also be an incentive for increasing certified land areas and volumes in the bioenergy markets in the future. Meta-standards were also found useful, and by endorsing different schemes and allowing case-by-case evaluations of existing evidence as proof of compliance with sustainability requirements, they may bridge the gaps to respondents being less convinced that more regulation is needed, and observed disagreement on how to deal with redundancy and overlap among schemes. Diverging views appeared partly rooted in conflicting values among sectors and geographical regions.

Most respondents said that the consequences of introducing new sustainability governance had resulted in positive or no changes to their production and trade, although this is uncertain for developing countries. Understanding more precisely how bioenergy trade streams are affected by sustainability regulations at the macro-scale will, however, require new systems that can gather aggregated data on certified trade globally. Online claims platforms could be developed to include such data. Legislative requirements and promoting green business profiles especially motivated market players to seek certification, and costs of certification were generally seen as a barrier. Actors motivated by legislative requirements are likely to select less costly schemes that include only minimum requirements, while those seeking a green business profile are likely seeking more ambitious and costly schemes.

Respondents generally thought that certification schemes allow for involvement of different types of actors, even if underrepresentation of small holders or people with industrial hand-on experience were noted in some cases. Also, power inequities arise from unequal access to policy decision making, especially for affected developing countries, or actors with less lobbying capacity.

Respondents representing a bioenergy scheme generally thought their schemes need to develop in several aspects, for example in introducing regular standard revisions and handling conflicting criteria. Respondents representing a forestry scheme were more satisfied with the quality and scope of their scheme, although there was some interest in addressing ecosystem carbon stocks and sustainability criteria in chain-of-custody. Respondents to both types of schemes saw difficulties in dealing with iLUC, suggesting that it sit at another scale, but some respondents to bioenergy schemes still thought it must be addressed. Respondents to bioenergy schemes were confident that their scheme is effective especially in achieving the criteria for GHG and energy balances, while respondents to forestry schemes were confident that their schemes achieve criteria for biodiversity, soil, water and social aspects. There were differences among schemes in how difficulties and costs of auditing and the need for new verification tools were evaluated. Such differences might be related to the level of ambitions, complexity and flexibility of the indicators, and the extent to which the scheme dynamically revise and improve their substantive and procedural standards.

It was thus apparent that there are a number of tradeoffs to be made between a scheme's cost level, inclusiveness, quality of substantive and procedural rules, and the subsequent effectiveness on-the-ground. With the increasing complexity of regulatory systems, involving both public and private actors, knowledge about these tradeoffs is a precondition to improve transparency and effectiveness and thus ensure the legitimacy of governance complex regulating sustainability of bioenergy. The findings and experiences from this survey may be helpful for formulating a number of research questions that can form the basis for indepth interviews, or can be tested with improved survey instruments applied to more rigorously defined populations.

NOTES

[a] Management plans are required for farms receiving government subsidies under the Conservation Stewardship Program (CSP), the Environmental Quality Incentives Program (EQIP), the Wildlife Habitat Incentives Program (WHIP), and the Conservation Reserve Program,[38] but we are not aware of how large proportion of the agricultural area in the US that is participating under these programs.

[b] It is still pending whether biomass from native forests in Australia will be eligible under the Renewable Energy Target (RET) scheme in the future. The government announced a review of the RET scheme in February 2014 and the use of biomass from native forests is one of the questions addressed. An outcome of the review is the review panel's support for 'the Government's election commitment to reinstate the eligibility of native forest wood waste as a renewable energy source'.[95]

ACKNOWLEDGMENTS

The authors gratefully acknowledge the IEA Bioenergy Executive Committee for providing funding to make the project possible, and the Nordic Energy Research

(ENERWOODS) and the EUDP program under the Danish Energy Agency for supporting the work with supplementary funding. We sincerely thank the key organizations that generously prioritized giving us very helpful feedback for improving the survey design and questions. FSC, PEFC, and the INBIOM network kindly published the invitation to participate in their newsletters, and colleagues provided us with relevant contacts from their networks. Torben Martinussen, University of Copenhagen, was indispensable in providing statistical advice and feedback and suggestions from two anonymous reviewers were greatly appreciated and helped improve the paper. Finally, we would especially like to thank the respondents of the survey, for providing thoughtful and engaging answers, and helping to move the discussion beyond the identification of the challenges, toward solutions.

REFERENCES

1. Lamers P, Hamelinck C, Junginger M, Faaij A. International bioenergy trade—a review of past developments in the liquid biofuel market. *Renew Sustain Energy Rev* 2011, 15:2655–2676.
2. Lamers P, Junginger M, Hamelinck C, Faaij A. Developments in international solid biofuel trade—an analysis of volumes, policies, and market factors. *Renew Sustain Energy Rev* 2012, 16:3176–3199.
3. Ziolkowska J, Meyers WH, Meyer S, Binfield J. Targets and mandates: lessons learned from EU and US biofuels policy mechanisms. *AgBioForum* 2010, 13:398–412.
4. Gamborg C, Anker HT, Sandøe P. Ethical and legal challenges in bioenergy governance: coping with value disagreement and regulatory complexity. *Energy Policy* 2014, 69:326–333.
5. Stoker G. Governance as theory: five propositions. *Int Soc Sci J* 1998, 50:17–28.
6. Bevir M. *Theory of Governance*. Berkeley and Los Angeles, CA: University of California Press; 2013, 263.
7. Roundtable on Sustainable Palm Oil (RSPO). Trans-forming the market to make sustainable palm oil the norm, 2013. Available at: http://www.rspo.org/file/IG-1%20(Low%20Res).pdf. (Accessed October 1, 2014).
8. United States Environmental Protection Agency (US-EPA). The Renewable Fuels Standard (RFS), 2014. Available at: http://www.epa.gov/otaq/fuels/renewablefuels/. (Accessed October 1, 2014).
9. Schnepf R, Yacobbuci BD. Renewable Fuel Standard (RFS): overview and Issues. Congressional Research Service, 7–5700, www.crs.gov, 2013, Report No. R40155. Available at: http://fas.org/sgp/crs/misc/R40155.pdf. (Accessed October 1, 2014).
10. European Union. Directive 2009/28/EC of the European Parliament and of the Council of 23 April 2009 on the promotion of the use of energy from renewable sources and amending and subsequently repealing Directives 2001/77/EC and 2003/30/EC. Rep. L 140/16. Official Journal of the European Union, L 140/16, June 5, 2009. Available at: http://eur-lex.europa.eu/. (Accessed October 1, 2014).
11. European Commission. Report from the Commission to the Council and the European Parliament on sustainability requirements for the use of solid and gaseous biomass sources in electricity, heating and cooling. Com(2010)11 final, 2010. Available at: http://ec.europa.eu/energy/renewables/bioenergy/sustainability_criteria_en.htm. (Accessed October 1, 2014).
12. Diaz-Chavez R, Kunen E, Walden D, Fingerman K, Arya L, Chalmers J, Kretschmer B, Poláková J, Farmer A, Bowyer C, et al. *Mandatory Requirements in Relation to Air, Soil, or Water Protection: Analysis of Need and Feasibility*. Windrock International, Institute for European Environmental Policy (IEEP), EcoFys, Utrecht, The Netherlands BV; 2013. Available at: http://ec.europa.eu/energy/renewables/studies/doc/2013_tasks3and4_requirements_soil_air_water.pdf. (Accessed October 1, 2014).
13. European Commission. Renewable Energy: Biofuels—Sustainability schemes. Available at: http://ec.europa.eu/energy/renewables/biofuels/sustainability_schemes_en.htm. (Accessed October 1, 2014).
14. van Dam J, Junginger M, Faaij APC. From the global efforts on certification of bioenergy towards an integrated approach based on sustainable land use planning. *Renew Sustain Energ Rev* 2010, 14:2445–2472.
15. Department of Energy & Climate Change (DECC). Sustainability standards for electricity generation from biomass, 2013. Available at: https://www.gov.uk/sustainability-standards-for-electricity-generation-from-biomass. (Accessed October 1, 2014).
16. Goovaerts L, Pelkmans L, Goh CS, Junginger M, Joudry J, Chum H, Smith CT, Stupak I, Cowie A, Dahlman L et al. Task 1: Examining sustainability certification of bioenergy—strategic inter-task study: monitoring sustainability certification of bioenergy. IEA Bioenergy, 2013. Available at: http://www.bioenergytrade.org/publications.html#Monsum. (Accessed January 27, 2014).
17. Programme for Endorsement of Forest Certification (PEFC). Forest certification is key to ensuring sustainability of biomass, 2012. Available at: http://pefc.org/news-a-media/general-sfm-news/1083-. (Accessed October 1, 2014).
18. Sustainable Forestry Initiative (SFI). The SFI Program and Bioenergy from Wood. SFI-00001, 2013. Available at: http://www.sfiprogram.org/files/pdf/bioenergyfeb2011lowresversionapdf/. (Accessed January 27, 2014).
19. van Dam J, Junginger M, Faaij A, Jürgens I, Best G, Fritsche U. Overview of recent developments in sustainable biomass certification. *Biomass Bioenergy* 2008, 32:749–780.
20. Magar SB, Pelkonen P, Tahvanainen L, Toivonen R, Toppinen A. Growing trade of bioenergy in the EU: public acceptability, policy harmonization, European standards and certification needs. *Biomass Bioenergy* 2011, 35:3318–3327.
21. van Dam J, Junginger M. Striving to further harmonization of sustainability criteria for bioenergy in Europe: recommendations from a stakeholder questionnaire. *Energy Policy* 2011, 39:4051–4066.

22. Alakangas E, Junginger M, van Dam J, Hinge J, Kerännen J, Olsson O, Porsö C, Martikainen A, Rathbauer J, Sulzbacher L, et al. EUBIONET III - Solutions to biomass trade and market barriers. *Renew Sust Energ Rev* 2012, 16:4277–4290.
23. Englund O, Berndes G. How do Sustainability Standards Consider Biodiversity? *WIREs Energy Environ* 2015, 4:26–50, doi:10.1002/wene.118.
24. Scarlat N, Dallemand JF. Recent developments of biofuels/bioenergy sustainability certification: a global overview. *Energy Policy* 2011, 39:1630–1646.
25. Goh CS, Junginger M, Joudry J, Chum H, Pelkmans L, Smith CT, Stupak I, Cowie A, Dahlman L, Englund O et al. Task 3: Impacts of sustainability certification on bioenergy markets and trade—strategic inter-task study: monitoring sustainability certification of bioenergy. IEA Bioenergy, 2013. Available at: http://www.bioenergytrade.org/publications.html#Monsum. (Accessed January 27, 2014).
26. Fernholz K, Kraxner F. Certified forest products markets, 2011–2012. In: *UNECE/FAO Forest Products Annual Market Review, 2011–2012*. New York and Geneva: United Nations; 2012, 107–116.
27. Mena S, Palazzo G. Input and output legitimacy of multi-stakeholder initiatives. *Bus Ethics Q* 2012, 22:527–556.
28. Cafaggi F, Renda A. Public and private regulation: mapping the Labyrinth. CEPS Working Documents, 2012. Available at: http://www.ceps.eu/book/public-and-private-regulation-mapping-labyrinth. (Accessed October 1, 2014).
29. Schepers D. Challenges to legitimacy at the Forest Stewardship Council. *J Bus Ethics* 2010, 92:279–290.
30. Enviva. Enviva LP Plans Second Wood Pellet Facility in North Carolina, 2011. Available at: http://www.envivabiomass.com/press-releases/enviva-lp-plans-second-wood-pellet-facility-in-north-carolina/. (Accessed October 1, 2014).
31. Natural Resources Defense Council (NRDC). Enviva's wood pellet mill in Ahoskie, North Carolina threatens endangered ecosystems and wildlife, 2013. Available at: http://www.nrdc.org/energy/forestnotfuel/files/enviva-wood-pellets-FS.pdf. (Accessed October 1, 2014).
32. Koppell JGS. Global governance organizations: legitimacy and authority in conflict. *J Public Adm Res Theory* 2008, 18:177–203.
33. Schouten G, Glasbergen P. Creating legitimacy in global private governance: the case of the Roundtable on Sustainable Palm Oil. *Ecol Econ* 2011, 70:1891–1899.
34. United Nations Economic and Social Commission for Asian and the Pacific (UNESCAP). What is good governance? 2009. Available at: http://www.unescap.org/pdd/prs/ProjectActivities/Ongoing/gg/governance.asp. (Accessed October 1, 2014).
35. Partzsch L. The legitimacy of biofuel certification. *Agr Hum Values* 2011, 28:413–425.
36. Englund O, Berndes G, Fredrikson F, Dimitriou I. Meeting sustainability requirements for SRC bioenergy: use-fulness of existing tools, responsibilities of involved stakeholders, and recommendations for further developments. *Bioenerg Res* 2012, 5:606–620.
37. Buytaert V, Muys B, Devriendt N, Pelkmans L, Kret-zschmar JG, Samson R. Towards integrated sustainability assessment for energetic use of biomass: a state of the art evaluation of assessment tools. *Renew Sust Energy Rev* 2011, 15:3918–3933.
38. Endres JM. Agriculture at a crossroads: energy biomass standards and a new sustainability paradigm? *Univ Ill Law Rev* 2011, 2:503–547.
39. van Dam J, Ugarte S, Iersel SV. *Selecting a Biomass Certification System—A Benchmark on Level of Assurance, Costs and Benefits*. NL Agency, Utrecht, The Netherlands; 2012.
40. Stupak I, Lattimore B, Titus BD, Smith CT. Criteria and indicators for sustainable forest fuel production and harvesting: a review of current standards for sustainable forest management. *Biomass Bioenergy* 2011, 35:3287–3308.
41. The SAS Institute (SAS). *SAS Statistical Package Version 9.3, 2002–2010*. Cary, NC: SAS; 2010.
42. Jensen A, Sørensen H. *Lecture Notes for Applied Statistics*. Copenhagen: Department of Natural Sciences, Life Science Faculty, University of Copenhagen, Frederiksberg, Denmark; 2007.
43. Lin J. Governing biofuels: a principal-agent analysis of the European Union biofuels certification regime and the clean development mechanism. *J Environ Law* 2011, 24:43–73.
44. Forrer J, Mo K. From certification to supply chain strategy: an analytical framework for enhancing tropical forest governance. *Organ Environ* 2013, 26: 260–280.
45. McDermott C, Cashore B, Kanowski P. *Global Environmental Forest Policies—An International Comparison*. London; Washington, DC: The Earthscan Forest Library, Earthscan; 2010.
46. European Commission. Results of the public consultation on additional sustainability measures at EU level for solid and gaseous biomass used in electricity, heating and cooling. European Commission, Directorate-General for Energy, ENER C.1, Regulatory Policy & Promotion of Renewable Energy, 2011. Available at: http://ec.europa.eu/energy/renewables/consultations/doc/20110329_biomass_consultation_report.pdf. (Accessed October 1, 2014).
47. Berch SM, Bulmer C, Curran M, Finvers M, Titus B. Provincial government standards, criteria, and indicators for sustainable harvest of forest biomass in British Columbia: soil and biodiversity. *Int J For Eng* 2012, 23:33–37.
48. Waito B, Johnson L. A national scan of regulations & practices relevant to biomass harvesting. World Wildlife Fund (WWF)—Canada, The Forest Products Association of Canada (FPAC), 2010. Available at: http://assets.wwf.ca/downloads/biomass_report_2010.pdf. (Accessed October 1, 2014).
49. Kittler B, McDow W, Larson B. *Pathways to Sustainability*. Washington, DC; Milford, CT: Environmental Defense Fund and Pinchot Institute; 2012 Available at: http://www.pinchot.org/gp/Pathways_to_Sustainability. (Accessed October 1, 2014).
50. Department of Energy and Climate Change (DECC). *Government Response to the Consultation on Proposals to Enhance the Sustainability Criteria for the Use of Biomass*

Feedstocks under the Renewables Obligation (RO). London: DECC; 2013. Available at: https://www.gov.uk/government/uploads/system/uploads/attachment_data/file/231102/RO_Biomass_Sustainability_consultation_-_Government_Response_22_August_2013.pdf. (Accessed October 1, 2014).
51. Pappila M. Forest certification and trust—different roles in different environments. *For Policy Econ* 2013, 31: 37–43.
52. IEA Bioenergy. The Science-Policy Interface on the Environmental Sustainability of Forest Bioenergy— A Strategic Discussion Paper, IEA Bioenergy, Report ExCo:2013:03, 2013. Available at: http://www.bioenergytrade.org/publications.html. (Accessed October 1, 2014).
53. Programme for Endorsement of Forest Certification (PEFC). PEFC Council Information Register, 2014. Available at: http://www.pefcregs.info/statistics.asp, (Accessed October 1, 2014).
54. Steering Committee of the State-of-Knowledge Assessment of Standards and Certification. *Toward sustainability: The Roles and Limitations of Certification*. Washington, DC: RESOLVE, Inc.; 2012. Available at: http://www.resolv.org/site-assessment/files/2012/06/Report-Only.pdf. (Accessed October 1, 2014).
55. Crow S, Danks C. Why certify? Motivations, outcomes and the importance of facilitating organizations in certification of community-based forestry initiatives. *Small-scale Forestry* 2010, 9:195–211.
56. Heyvaert V. Regulatory competition—accounting for the transnational dimension of environmental regulation. *J Environ Law* 2012, 35:1–31.
57. Endres JM. Clearing the air: the meta-standard approach to ensuring biofuels environmental and social sustainability. *Va Envtl Law J* 2010, 28:73–120.
58. Deutsche Gesellschaft für die Internationale Zusammenarbeit (GIZ). *Recognition of Private Certification Schemes for Public Regulation: Lessons Learned from the Renewable Energy Directive*. GIZ GmbH Bonn, Eschborn, Germany; 2013. Available at: http://www.wwf.de/fileadmin/fm-wwf/Publikationen-PDF/BiofuelsWorkshop_GIZ_RecognitionOf_PrivateCertificationSchemes.pdf. (Accessed October 1, 2014).
59. Gan J, Cashore B. Opportunities and challenges for integrating bioenergy into sustainable forest management certification programs. *J Forestry* 2013, 111:11–16.
60. Pacini H, Assunção L, van Dam J, Toneto J. The price for biofuels sustainability. *Energy Policy* 2013, 59:898–903.
61. Walter A, Segerstedt A. International trade of biofuels: current trends and the potential role of Africa. In: Janssen R, Rutz D, eds. *Bioenergy for Sustainable Development in Africa*. Netherlands: Springer; 2012, 147–162.
62. Bowers T, Eastin I, Ganguly I, Cao J, Seol M. Forest certification in Asia: the changing marketplace for value-added wood product manufacturers in China and Vietnam. *Forestry Chron* 2012, 88:578–584.
63. Owari T, Juslin H, Rummukainen A, Yoshimura T. Strategies, functions and benefits of forest certification in wood products marketing: perspectives of Finnish suppliers. *For Policy Econ* 2006, 9:380–391.
64. Espinoza O, Buehlmann U, Smith B. Forest certification and green building standards: overview and use in the U.S. hardwood industry. *J Clean Prod* 2012, 33: 30–41.
65. Kollert W, Lagan P. Do certified tropical logs fetch a market premium? A comparative price analysis from Sabah, Malaysia. *For Policy Econ* 2007, 9:862–868.
66. Goh CS, Junginger M. SolidStandards, enhancing the implementation of quality and sustainability standards and certification schemes for solid biofuels (EIE/11/218). D5.2a. Case studies of sustainably certified solid biomass supply chains—Green Gold Label. Utrecht University, 2013. Available at: http://www.solidstandards.eu/images/Sustainability/D5.2a.p. (Accessed January 27, 2014).
67. Programme for Endorsement of Forest Certification (PEFC). PEFC International Standards, 2014. Available at: http://www.pefc.org/standards/technical-documentation/pefc-international-standards-2010. (Accessed October 1, 2014).
68. Buchholz T, Luzadis VA, Volk TA. Sustainability criteria for bioenergy systems: results from an expert survey. *J Clean Prod* 2009, 17 (suppl 1):S86–S98.
69. Geist HJ, Lambin EF. Proximate causes and underlying driving forces of tropical deforestation. *BioScience* 2002, 52:143–150.
70. Lambin EF, Meyfroidt P. Global land use change, economic globalization, and the looming land scarcity. *Proc Natl Acad Sci U S A* 2011, 108:3465–3472.
71. Auld G, Gulbrandsen LH, McDermott CL. Certification schemes and the impacts on forests and forestry. *Annu Rev Environ Res* 2008, 33:187–211.
72. Diaz-Chavez RA. Assessing biofuels: aiming for sustainable development or complying with the market? *Energy Policy* 2011, 39:5763–5769.
73. ISO. ISO/PC 248 Sustainability criteria for bioenergy. Available at: http://www.iso.org/iso/standards_development/technical_committees/other_bodies/iso_technical_committee.htm?commid=598379. (Accessed October 1, 2014).
74. van de Staaij J, Peters D, Dehue B, Mayer S, Schueler V, Toop G, Junquera V, László M. Low Indirect Impact Biofuel (LIIB) Methodology, Version 0, Eco-Fys, RSB Secretariat at EPFL. WWF International, Gland, Switzerland, 2012. Available at: http://www.ecofys.com/files/files/12-09-03-liib-methodology-version-0-july-2012.pdf>;. (Accessed October 1, 2014).
75. Lima MGB. Biofuel governance and international legal principles: is it equitable and sustainable? *Melbourne J Int Law* 2009, 10:470–492.
76. Roberge A, Bouthillier L, Mercier J. The gap between theory and reality of governance: the case of forest certification in Quebec (Canada). *Soc Nat Res* 2011, 24:656–671.
77. Elgert L. Certified discourse? The politics of developing soy certification standards. *Geoforum* 2012, 43:295–304.
78. von Geibler J. Market-based governance for sustainability in value chains: conditions for successful standard setting in the palm oil sector. *J Clean Prod* 2013, 56:39–53.

79. Azevedo JC, Wu XB, Messina M, Williams J, Fisher R. The role of the sustainable forestry initiative in forest landscape changes in Texas, USA. In: Lafortezza R, Sanesi G, Chen J, Crow T, eds. *Patterns and Processes in Forest Landscapes*. Netherlands: Springer; 2008, 273–295.
80. Elbakidze M, Angelstam P, Andersson K, Nordberg M, Pautov Y. How does forest certification contribute to boreal biodiversity conservation? Standards and outcomes in Sweden and NW Russia. *For Ecol Manag* 2011, 262:1983–1995.
81. FSC. IGI International Generic Indicators. Available at: http://igi.fsc.org/a-short-history.33.htm. (Accessed October 1, 2014).
82. Hennenberg KJ, Dragisic C, Haye S, Hewson J, Semroc B, Savy C, Wiegmann K, Fehrenbach H, Fritsche UR. The power of bioenergy-related standards to protect biodiversity. *Conserv Biol* 2010, 24:412–423.
83. Angel DP, Hamilton T, Huber MT. Global environmental standards for industry. *Annu Rev Environ Res* 2007, 32:295–316.
84. Meidinger EE. Forest certification as environmental law making by global civil society. In: Meidinger E, Elliott C, Oesten G, eds. *Social and Political Dimensions of Forest Certification*. Remagen: Oberwinter; 2003, 293–329. Available at: www.forstbuch.de.
85. Forest Stewardship Council (FSC). *FSC International Standard: General Requirements for FSC Accredited Certification Bodies—Application of ISO/IEC Guide 65:1996 (E), FSC-STD-20-001 (Version 3–0) EN*. FSC, Bonn, Germany; 2009.
86. Maletz O, Tysiachniouk M. The effect of expertise on the quality of forest standards implementation: the case of FSC forest certification in Russia. *For Policy Econ* 2009, 11:422–428.
87. Sample VA, Price W, Mater CM. Certification on public and university lands: evaluations of FSC and SFI by the forest managers. *J Forestry* 2003, 101:21–25.
88. UPM. *Parallel Field Testing of Forest Certification Standards: A Project to Promote a Global Increase in the Use of Certified Wood*. UPM, Forestry and Wood Sourcing Environmental Forestry Affairs Helsinki, Finland. Available at: http://standards.nsf.org/apps/group_public/download.php?document_id=10103. (Accessed October 1, 2014); 2005.
89. Ecofys. Accounting for Carbon. Mapping the Land-scape of Climate Assurance Systems: A Report Based on the Study "Mapping Climate Assurance Systems: Final Report" produced by Ecofys and Commissioned by GTZ. GTZ, Federal Ministry of Economic Cooperation and Development, Bonn, Germany, 2010. Available at: http://www.isealalliance.org/sites/default/files/Accounting%20for%20Carbon%20-%20Draft%20Final%20Report%202.pdf. (Accessed October 1, 2014).
90. Hennecke AM, Faist M, Reinhardt J, Junquera V, Neeft J, Fehrenbach H. Biofuel greenhouse gas calculations under the European Renewable Energy Directive—a comparison of the BioGrace tool vs the tool of the Roundtable on Sustainable Biofuels. *Appl Energy* 2013, 102:55–62.
91. Teitelbaum S, Wyatt S. Is forest certification delivering on First Nation issues? The effectiveness of the FSC standard in advancing First Nations' rights in the boreal forests of Ontario and Quebec, Canada. *For Policy Econ* 2013, 27:23–33.
92. Ramcilovic-Suominen S, Hansen CP. Why some forest rules are obeyed and others violated by farmers in Ghana: instrumental and normative perspective of forest law compliance. *For Policy Econ* 2012, 23: 46–54.
93. Mrosek T, Balsillie D, Schleifenbaum P. Field testing of a criteria and indicators system for sustainable forest management at the local level: case study results concerning the sustainability of the private forest Haliburton Forest and Wild Life Reserve in Ontario, Canada. *For Policy Econ* 2006, 8:593–609.
94. Lloyd R, Calvo V, Laybourn C. Ensuring Credibility and Effectiveness: Designing Compliance Systems in CSO Self-Regulation. One World Trust, London, UK, 2010. Available at: http://www.oneworldtrust.org/csoproject/images/documents/SRI_Compliance_Mechanisms.pdf. (Accessed October 1, 2014).
95. Warburton D, Fisher B, Veld S, Zema M. Renewable Energy Target Scheme, Report of the Expert Panel. Commonwealth of Australia, 2014. Available at: https://retreview.dpmc.gov.au/ret-review-report-0. (Accessed October 1, 2014).

Index